1.15 miles	$\cong$ 1 minute (1′) of latitude $\cong$ 1 nautical mile
1.6449	= coefficient for 90% standard deviation
3.141 592 654	= π
*6 miles	= length, width, of normal township
*10 sq. chains (ch²)	= (Gunter's) 1 acre
*15° longitude	= width of one time zone = 360°/24 hr
15°F	changes length of 100-ft steel tape by 0.01 ft
*16½ ft	= 1 rod = 1 pole = 1 perch = ¼ ch (Gunter's)
*20°C	= standard temperature (Celsius) in taping = 68°F
23°26½′	= maximum declination of sun at solstices
23ʰ56ᵐ04.091ˢ	= length of sidereal day in mean solar time, and 3ᵐ55.909ˢ solar time short of mean solar day; also 3ᵐ56.555ˢ sidereal time short of mean solar day
*24 hr	= 360° of longitude
*25.4 mm	= 1 in. (U.S. standard foot of 1959)
*36	= number of sections in normal township
*50	= Beaman arc reading for 0° vertical angle (30 in old arcs)
57°17′44.8″	= 1 radian (rad) = 57.295 779 51°
*66 ft	= length of Gunter's chain = 100 links (lk)
69.1 miles	$\cong$ 1° latitude
*80 ch	= (Gunter's) 1 mile
100	= usual stadia ratio
101 ft	$\cong$ 1 second (1″) of latitude
300	$\cong$ stadia ratio in some precise levels
333⅓	$\cong$ stadia ratio in some precise levels, for yard rods
*400 grads	= 360°
*480 ch	= width and length of normal township
490 lb/ft³	= density of steel for tape computations
*640 acres	= one normal section of 1 mile²
6076.10 ft	= 1 nautical mile
*3600/3937	= ratio U.S. yd/m for old legal (1866) and surveyor's foot
4,046.9 m²	= 1 acre
*6400 mils	= 360°
5,729.577 951 ft	= radius of 1° curve, arc definition
5,729.650 686 ft	= radius of 1° curve, chord definition
10,000 km	= distance from equator to pole (basis for length of meter)
*43,560 ft²	= 1 acre
206,264.806 25 sec	= 1 radian = cot 1 sec = 180°/π in sec
299,792.5 km/sec	= speed of light, and other electromagnetic waves, in vacuum
1,650,763.73	= wave lengths of krypton gas in vacuum, 1960 meter length
6,356,583.8 m	= earth's polar semi-axis (Clarke ellipsoid 1866)
6,378,206.4 m	= earth's equatorial semi-axis (Clarke ellipsoid 1866)
20,906,000 ft	= mean radius of earth = 3960 miles
29,000,000 lb/in.²	= Young's modulus of elasticity for steel

* Denotes exact value. All others correct to figures shown.

THE SURVEYING HANDBOOK

THE SURVEYING HANDBOOK

SECOND EDITION

edited by Russell C. Brinker and Roy Minnick

CHAPMAN & HALL

I(T)P   An International Thomson Publishing Company

New York • Albany • Bonn • Boston • Cincinnati • Detroit • London • Madrid • Melbourne • Mexico City
Pacific Grove • Paris • San Francisco • Singapore • Tokyo • Toronto • Washington

Cover design: Trudi Gershenov

Copyright © 1995
By Chapman & Hall
A division of International Thomson Publishing Inc.
I(T)P The ITP logo is a trademark under license

Printed in the United States of America

For more information, contact:

Chapman & Hall
One Penn Plaza
New York, NY 10119

International Thomson Publishing
Berkshire House 168-173
London WC1V 7AA
England

International Thomson Editores
Campos Eliseos 385, Piso 7
Col. Polanco
11560 Mexico D.F. Mexico

Thomas Nelson Australia
102 Dodds Street
South Melbourne, 3205
Victoria, Australia

International Thomson Publishing Gmbh
Königwinterer Strasse 418
53228 Bonn
Germany

Nelson Canada
1120 Birchmount Road
Scarborough, Ontario
Canada M1K 5G4

International Thomson Publishing Asia
221 Henderson Road #05-10
Henderson Building
Singapore 0315

International Thomson Publishing - Japan
Hirakawacho Kyowa Building, 3F
1-2-1 Hirakawacho-cho
Chiyoda-ku, 102 Tokyo
Japan

1 2 3 4 5 6 7 8 9 10 XXX 01 00 99 98 97 96 95

Library of Congress Cataloging-in-Publication Data
   The surveying handbook/editors, Russell C. Brinker and Roy Minnick.
      —2nd ed.
      p.   cm.
   Includes bibliographical references and index.
   ISBN 0-412-98511-X
   1. Surveying, I. Brinker, Russell C. (Russell Charles).
II. Minnick, Roy.
TA555.887 1994   91-5241
526.9--dc20   CIP

British Library Cataloguing in Publication Data available

Please send your order for this or any other Chapman & Hall book to
Chapman & Hall, 29 West 35th Street, New York, NY 10001, Attn: Customer Service Department.
You many also call our Order Department at 1-212-244-3336 or fax you purchase order to 1-800-248-4724.

For a complete listing of Chapman & Hall's titles, send your request to
Chapman & Hall, Dept. BC, One Penn Plaza, New York, NY 10119.

Contents

3 Measurement Errors   20
Bro. B. Austin Barry, FSC

4 Linear Measurements   42
Kenneth S. Curtis

7   Leveling

Robert J. Schultz

8   Instrument Adjustments   *140*
Gerald W. Mahun

9   Traversing   *156*
Jack B. Evett

10 Survey Drafting   *180*
Edward G. Zimmerman

Preface

The first edition of *The Surveying Handbook*, although a ground breaker, was widely accepted. However, surveying is a dynamic profession with each new development just one step ahead of the next, and updating became critical. In addition, the editors received constructive criticism about the first edition that needed to be addressed. So, while the objective of *The Handbook* remains intact, the logical evolution of the profession, along with the need to recognize constructive criticism, led to the second edition.

New chapters have been added on water boundaries, boundary law, and geodetic positioning satellites. The chapter on land data systems was rewritten to provide a dramatic updating of information, thus broadening the coverage of *The Handbook*. The same may be said for the state plane coordinate chapter. The material on public lands and construction surveying was reorganized as well. Appendices were added to tabulate some information that was buried in the earlier edition in several places. Numerous other changes were incorporated to help the handbook retain its profession-wide scope, one step beyond the scope of an upper-division college textbook. Along with the most sophisticated techniques and equipment, the reader can find information on techniques once popular and still important.

Four new authors are welcomed to the list of contributors: Grenville Barnes, R. B. Buckner, Donald A. Wilson, and Charles D. Ghilani.

The editors and publisher feel confident that a second edition of *The Surveying Handbook* meets the objectives of broad, thorough coverage and current information, while recognizing the valuable advice and suggestions of first edition users.

RUSSELL BRINKER
ROY MINNICK
February 25, 1993

Preface to the First Edition

THE SURVEYING HANDBOOK has been written to fill the need for a comprehensive volume on professional surveying. In the past, similar books have been filled primarily with tables more readily obtained from other sources, while several of the more recent versions concentrate on a single area of the profession and are published by the American Congress on Surveying and Mapping, American Society of Photogrammetry, and American Society of Civil Engineers. The 36 chapters in this volume were prepared by 35 contributors, generally based on their special fields.

Obviously, even the largest handbook could not cover every phase of surveying in complete depth. But sufficient material is given herein to provide surveyors and others with suitable information outside their specialty field. It can then be determined whether a full-sized special book on a subject area is needed. Some surveying equipment sales and repair shops stock a small number of textbooks. Customers have asked ''Why can't I get just one volume to refresh and guide me instead of having to buy half a dozen books?'' It is hoped this volume will eliminate that problem.

Based on advance publication interest, surveyors, civil, agricultural and other engineers, foresters, architects, archeologists, geologists, small home builders, realtors, title companies, and lawyers will find useful information in THE SURVEYING HANDBOOK.

Abundant figures and tables are included in this volume. References to textbooks, technical journals, and magazines will help readers find additional sources of specific information desired. Profuse footnotes have been used only in Chapter 31, The Role of the Surveyor in Land Litigation: Pretrial. At most chapter ends, superscript numbers refer to the list of References and Notes, thereby retaining a cleaner appearance and reducing awkward typesetting.

THE SURVEYING HANDBOOK is written in an easy-to-read style that avoids word repetition and other excess verbiage. A handbook is supposed to be practical and that has been the goal of Contributors and Editors. Many Contributors have written their own textbooks or parts thereof, and nearly all are frequent authors of technical papers.

Contacting Contributors residing in 19 different states by telephone and letters has been an unexpected challenge for the Editors who have been so heavily dependent upon the contributors' efforts and cooperation. Their expenditure of time and funds for the extremely small stipend paid handbook contributors by publishers is greatly appreciated.

In addition to the typical textbook chapters, special ones on Survey Drafting; Mining Surveys; Optical Tooling (Industrial Applications); Land Descriptions; Pre-Trial Preparation; Courtroom Techniques; Survey Business Management; Surveying Charges, Contracts, Liability; Land Information Systems; and Surveying Profession, Registration, Associations, are included.

This handbook is the result of the labors over the last $5\frac{1}{2}$ years. Although basic principles of survey measurement remain the same, technology and sources of information may change. Recently, NGS has published portions of NAD-83, several states have adopted new state plane coordinate systems, and data storage and retrieval methods at primary survey information sources have been modernized. The surveying profession is not static, but is constantly changing in response to modern technology.

RUSSELL BRINKER
ROY MINNICK

Contributors

GRENVILLE BARNES
Assistant Professor
Surveying and Mapping
    Program
Civil Engineering Department
University of Florida
Gainesville, FL

BRO. B. AUSTIN BARRY, F.S.C.
Professor of Civil Engineering
Manhattan College, Bronx, NY

RUSSELL C. BRINKER, P.E.
Adjunct Professor of Civil
    Engineering (Retired)
New Mexico State University
Las Cruces, NM

JOHN BRISCOE
Attorney at Law
San Francisco, CA

R. B. BUCKNER, Ph.D., L.S.
Surveying Program Chair
East Tennessee State University
Johnson City, TN

EARL F. BURKHOLDER, P.L.S., P.E.
Consulting Geodetic Engineer
Klamath Falls, OR

BOYD L. CARDON, R.L.S.
Professor of Mathematics
Ricks College, Rexburg, ID

FRANK T. CAREY, L.S., R.L.S.
Boundary Officer
California State Lands Commission
Instructor, Sacramento City Colege

KENNETH S. CURTIS, L.S.
Professor Emeritus of Surveying
    and Mapping
Purdue University
West Lafayette, IN

RICHARD L. ELGIN, Ph.D., L.S., P.E.
Elgin, Knowles & Senne, Inc.,
Elgin Surveying & Engineering, Inc.,
Rolla, MO

JACK B. EVETT, Ph.D., P.E., L.S.
Professor of Civil Engineering
The University of North Carolina at
    Charlotte
Charlotte, NC

ROBERT J. FISH, R.L.S.
Phoenix, AZ

CHARLES D. GHILANI, Ph.D.
Surveying Program Chair
Pennsylvania State University

DAVID W. GIBSON, L.S.
Associate Professor
Surveying and Mapping
   Program
Civil Engineering Department
University of Florida
Gainsville, FL

E. FRANKLIN HART, L.S., P.E.
Professor of Civil Engineering
   Technology
Bluefield State College
Bluefield, WV

LARRY D. HOTHEM
Manager
GPS Research and Applications
Geometronics Standards Section
USGS National Mapping Division
Reston, VA

ANDREW KELLIE, Ph.D., L.S.
Associate Professor of Engineering
   Technology
Murray State University
Murray, KY

DAVID R. KNOWLES, Ph.D., L.S., P.E.
Elgin, Knowles & Senne, Inc.
Professor of Civil Engineering
University of Arkansas
Fayetteville, AR

GERALD W. MAHUN
Assistant Professor of Civil
   Engineering
University of Wisconsin, Platteville
Platteville, WI

PORTER W. MCDONNELL, P.E., L.S.
Professor of Surveying
Metropolitan State College
Denver, CO

DAVID F. MEZERA, Ph.D., P.L.S., P.E.
Associate Professor of Civil and
   Environmental Engineering
University of Wisconsin, Madison
Madison, WI

ROY MINNICK, L.S., R.L.S.
Tidelands and Waterways
First American Title Company
Santa Ana, CA

DENNIS MOULAND, P.L.S.
Cadastral Consultants, Inc.
Glenwood, CO

CARLOS NAJERA, L.S., R.L.S.
Boundary Officer
California State Lands Commission
Instructor, Sacramento City College

CAPTAIN DONALD E. NORTRUP
National Oceanic and Atmospheric
   Administration
Norfolk, VA

JOHN S. PARRISH, L.S.
Chief
Branch of Cadastral Survey
Bureau of Land Management
Carson City, NV

JAMES P. REILLY, Ph.D.
Academic Department Head
Surveying Department
New Mexico State University
Las Cruces, NM

WALTER G. ROBILLARD, L.S., R.L.S.
Attorney at Law
Atlanta, GA

ROBERT J. SCHULTZ, P.E., P.L.S.
Professor of Civil Engineering
Oregon State University
Corvallis, OR

JOSEPH H. SENNE, Ph.D., P.E.
Elgin, Knowles & Senne, Inc.
Professor Emeritus of Civil
 Engineering
University of Missouri, Rolla
Rolla, MO

M. LOUIS SHAFER, L.S., R.L.S.
Chief of Surveys, District III
California Department of
 Transportation
Marysville, CA

F. HENRY SIPE, L.L.S. #1
Consulting Land Surveyor
Elkins, WV

BRYANT N. STURGESS, L.S., R.C.E.
Boundary Officer
California State Lands Commission

WAYNE VALENTINE, P.E., L.S.
Geometronics Group Leader
U.S. Forest Service
Missoula, MT

ELLIS R. VEATCH II, P.S.
Senior Trainer
Ashtech, Inc.
Sunnyvale, CA

DONALD A. WILSON, R.L.S., R.P.F.
Land Boundary Consultant
Newfields, NH

PAUL R. WOLF, Ph.D.
Department of Civil and
 Environmental Engineering
University of Wisconsin, Madison
Madison, WI

EDWARD G. ZIMMERMAN, L.S., R.L.S.
Senior Land Surveyor
Supervisor of Survey Training and
 Professional Development
California Department of Transportation
Sacramento, CA

1

Surveying Profession, Registration, and Associations

Walter G. Robillard

1-1. INTRODUCTION

This chapter provides information about the professional organizations and their role in professional surveying. Addresses of key organizations are included in that direct contact can be made to obtain further information.

1-2. OVERVIEW

Prior to formation of the American Congress of Surveying and Mapping (ACSM) in the 1930s, surveying was an important part of civil engineering and had an appropriate number of courses in college civil engineering curricula. The American Society of Civil Engineers (ASCE) was the primary sponsor of surveying technical papers and continues to include them in the monthly civil engineering magazine and periodically in a journal of surveying engineering. Recently, an engineering surveying manual prepared by the Committee on Engineering Surveying of the Surveying Engineering Division has been published and is available for purchase from the ASCE.

The ACSM, through its quarterly journal and bimonthly bulletins, provides excellent articles on all pertinent items that along with its semiannual national meetings make member-

ship essential. Other worthy publications include the *Point of Beginning* (*P.O.B.*) *Magazine* and *Professional Surveyor*.

When civil engineering professional registration was first legislated in the early part of the 20th century, the civil engineering license included surveying privileges. Gradually, separate licensing of surveyors became the law in most parts of the United States.

1-3. THE FUTURE

Challenges of the future—e.g., space exploration, oceanographic research, urban and land planning and development, ecology and the use and search for natural resources—are dependent on and interrelated to the fields of mapping, charting, geodesy, and surveying.

1-4. BACKGROUND OF SURVEYING AND MAPPING

Records of land surveys date back to the Babylonian era, 3000 or more years ago. Boundary stones were used during those times to mark property in the valleys of the Tigris and Euphrates Rivers.

Geological relics from about 3000 B.C. are still preserved, depicting certain physical features of ancient Babylonia. Town plans of Babylon survive that date back to 2000 B.C.

In ancient Egypt, the valley of the Nile River was flooded frequently, and boundary stones were often shifted or washed away. The Egyptians developed a system of surveying through which they were able to perpetuate the boundary and property lines of that rich area.

Some surveys of ancient times relate to those of today. During the construction of the Aswan Dam on the Nile River, surveyors established precise points for use as guides in cutting, moving, and reassembling the statues at Abu Simbel. This was necessary in order to preserve the beauty and harmony of the original design and construction, which in turn depended significantly on measurements made by the surveying techniques of ancient times.

In the second century of the Christian era, Ptolemy introduced and named the system of latitude and longitude. The Vinland map, which is thought to have been made about A.D. 1440, delineated Iceland, Greenland, and a land mass called Vinland that represented the North American mainland. In 1594, Mercator devised geometrically accurate map-projection systems.

In the United States, the public-land system of townships, ranges, and sections was developed in 1784. In 1803, Lewis and Clark explored and surveyed the country along the Missouri River and west to the Pacific Ocean. Hassler and Blunt led the way in coastal charting in the 1850s; the Powell, Fremont, Hayden, King, and Wheeler surveys of the 1860s opened the development of the American West. Significant developments in aerial photogrammetry, as applied to surveying and mapping, occurred in the 1920s and are still going on. The 1960s brought the beginning of manned space exploration, climaxed by the landing on the moon. The Surveyor I through VII series of satellites contributed much valuable data leading up to that highly successful moon landing. In the 1970s, exploration of space by the United States continued with orbiting surveys of Mars, and the spectacular landing of the space vehicle on that planet, followed by transmission of both photos and detailed data concerning the surface. Surveyors always have been closely identified with exploration and the growth in complexity and sophistication of the cultural development that follows exploration.

1-5. THE SURVEYING PROFESSION

Surveyors are licensed by each state, usually under the authority of a board of registration. The addresses of the boards are listed in Appendix 1.

Laws governing the practice of surveying are enacted at state level. With the exception of public-land surveys, there are almost no federal laws regulating survey practice and no federal license or registration.

Qualifications for surveyors vary from state to state, but generally a pattern of six years of prescribed experience and a 16-hour written examination are the requirements for registration. Many states use portions of examinations prepared by the National Council of Engineering Examiners supplemented by a portion prepared by the state to test on specific state laws.

1-6. SURVEYING LITERATURE

A substantial body of literature about surveying exists. Booksellers specializing in surveying are listed in Appendix 3.

1-7. SURVEYING EDUCATION

Surveying degrees are offered by only a few colleges in the United States. More common is the two-year program offered by community colleges.

2

Surveying Field Notes, Data Collectors

Russell C. Brinker

2-1. INTRODUCTION

Surveying is defined in the 1978 *ASCE Manual No. 34: Definitions of Surveying and Associated Terms* prepared by a joint committee of the ASCE and ACSM as

"(1) The science and art of making all essential measurements in space to determine the relative positions and points and/or physical and cultural details above, on, or beneath the earth's surface and to depict them in usable form, or to establish the position of points and/or details. Also, the actual making of a survey and recording and/or delineation of dimensions and details for subsequent use. (2) The acquiring and/or accumulation or qualitative information and quantitative data by observing, counting, classifying, and recording according to need."

Examples are traffic surveying and soil surveying.

Manually or electronically made field notes are necessary to document surveying results. In this chapter, basic principles of good notekeeping will be discussed, detailed suggestions listed, and simple examples given. Many special noteforms have been designed to fit the specific requirements of various federal, state, city, and county agencies, large companies, property surveyors, and other organizations. Some of these specialized noteforms are included in later chapters. No single style is universally accepted and termed the "standard," even for a job as common as differential leveling. Diverse field conditions, equipment and personnel, and special needs cannot be served by rigid arrangements—e.g., property surveys often require recorders to improvise different noteforms. Tables of some surveying terms, abbreviations, and symbols used in noteforms are presented at the end of this chapter. A short list of surveying textbooks and other references is also provided.

2-2. IMPORTANCE OF FIELD NOTES

Field notes are the only truly permanent and original records of work done on a project. Monuments and corners set or found may be moved or destroyed, and maps prepared from notes sometimes show incorrect distances, angles, and locations of details. Obviously, one notekeeping error can ruin the accuracy and credibility of the succeeding steps: computing and mapping. Written documents (deeds) can

3

jumble numbers and directions, and computer operators have been known to introduce their own mistakes. Original field notes are, therefore, the court of last resort.

A notekeeper's job is often the key assignment in a surveying field party; hence the party chief, who presumably is the most experienced and competent member, often assumes that responsibility. Numerical data must be recorded, sketches drawn, descriptions prepared, and mental calculations quickly made (as in first-order three-wire precise leveling), while one or more people shout things as they move around. On property surveys, the party chief, while keeping notes, may roam along boundary lines to get information. In a two-member differential-leveling unit, the notekeeper is also the instrument operator.

Property survey notes introduced as key-exhibit evidence in court cases can be a critical factor in decisions affecting land transfers by future generations. Land values continue to increase, so accuracy and completeness of surveys and notes are vital. The cross-referenced notes in a land surveyor's files become the saleable "good will" of the business.

The investment worth of surveying notes depends on the time and cost to reproduce any field work, plus the loss caused by their unavailability if immediately needed. The name, address, and telephone number of the person who prepared the notes and company that owns the field book must be lettered in India ink on the outside and inside cover. If a reward will be paid for return of the book if lost, it should be stated.

Because of possible omissions and copying errors, only original notes may be admitted in court cases, since they are the "best" evidence. Copies must always be clearly identified as such. Measurements not recorded at the time they were made or entered later from memory—which is even worse than copying from a scratch bit of paper—are definitely unreliable.

Since the time and date of erasures are always questionable and possible cause for rejection of the notes, erasures must not be made *on recorded measurements*. Also, the originally recorded material may later be found useful and correct. A pencil line should be run through a wrong number without destroying its legibility and the correct value placed above or below the deleted number. Part or all of a page to be canceled should be voided by drawing diagonal lines across it, but without making any part illegible, and prominently marked VOID. Erasing a nonmeasured line for a topographic sketch while in the field may be justified.

2-3. ESSENTIALS OF SUPERIOR NOTES

Five primary features are considered in evaluating field notes:

1. *Accuracy.* This is the most important factor in all surveying procedures, including note-keeping.
2. *Composition.* Noteforms suitable for each project, with column headings arranged in order of readings and sufficient space provided for sketches and descriptions without crowding, promote accuracy, completeness, and legibility.
3. *Completeness.* A single omitted measurement or detail can nullify an entire set of notes and delay computing or plotting. On projects far from the office, time and money are wasted when returning to the field for missing data. Before leaving the survey site, notes must be carefully reviewed for closure checks and possible overlooked items.
4. *Clarity.* Planning logical field procedures before leaving the office enables a notekeeper to record measurements, descriptions, and sketches without crowding. Mistakes and omissions become more obvious, which helps to eliminate costly office errors in computing and drafting.
5. *Legibility.* Notes must be decipherable and understandable by all users, including those who have not visited the survey area. Neat, efficient-appearing notes are more likely to represent professional-quality measurements and inspire confidence in the field data.

2-4.　FIELD BOOKS

Field books used in professional work contain valuable information acquired at considerable cost; they must survive rough usage and difficult weather conditions and last indefinitely. Various types are available, but bound books —the longtime standards with sewed binding, hard stiff covers of leatherette, polyethylene, or covered cardboard, and 80 leaves—are generally selected.

Stapled, sewed, and spiral-bound books are not suitable for most professional work. Duplicating field books may be convenient for jobs requiring progressive transfer of notes from field to office. The original sheet can be detached while a copy is retained in the field book. The loose-leaf original pages are filed in special binders.

Loose-leaf books have both advantages and disadvantages. The advantages include (1) a flat working surface; (2) the capacity to separately file individual project notes, thereby facilitating indexing and referencing, instead of wasting a partly filled book; (3) removable pages for shuttling between field and office; (4) easy insertion of preprinted noteforms, tables, diagrams, formulas, and other useful material; (5) the ability to carry different rulings in the same book; and (6) lower overall cost because the cover can be reused. Disadvantages are possible loss of some loose sheets and having the project data divided between field and office.

2-5.　TYPES AND STYLES OF NOTES

Surveying field notes can be divided into four basic types: tabulations, sketches, descriptions, and combinations. The combination method is most common because it fits so many overall needs.

One axiom applies to all four types: If doubtful about the need for certain data, include them and make a sketch. A supplementary proverb, "One picture [sketch] is worth 10,000 words," might well have been written to guide surveying notekeepers. Preprinted noteforms for particular groups of surveys often use arrangements comparable to those illustrated in this text.

Left- and right-hand field book pages are generally paired and share the same number. The left page is commonly ruled in six columns for tabulations, with notes and sketches on the right-hand page. Column headings proceed from left to right in the order readings are taken and minor calculations made. Figures 2-1 and 2-2 are basic notes presented for illustrative purposes *only* to show two different tabulation arrangements on the same page.[2]

In Figure 2-1, distances between hubs are recorded *between* the hub letters, names, or numbers. Measurements *to* a hub, in stations, are placed *opposite* the hub. A sketch on the right-hand page may help but not be necessary, so the notes could be tabulations only. For a simple example of traverse distances, angles, and bearings, everything could be put on a sketch along with other information if only single angles are measured.

Figure 2-2, a combination of type, demonstrates that it is easier to follow the "open" style differential-leveling notes having a $(+)$ sight and height of instrument (HI) on *one* line, followed by the $(-)$ sight and elevation on the *next* one, rather than the "closed" type, which puts all four values on the *same* line. This is especially helpful when a less experienced person uses a noteform to check something.

The project title can run across the tops of both pages or be confined to the left one. The upper right corner of the right-hand page, away from descriptions and sketches, is a good place for these standard items: date, weather, party, and equipment type with serial number.

1. *Date, time of day* (AM *or* PM), both *starting* and *finishing times* are necessary for record purposes. The number of hours spent in the field on a job may help to assess the precision attained, work delays, and other factors.

2. *Weather conditions* ranging from extremely high to ultralow temperatures, sunshine, fog,

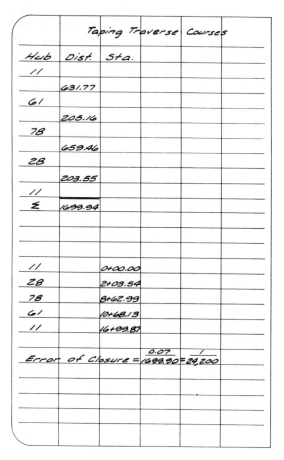
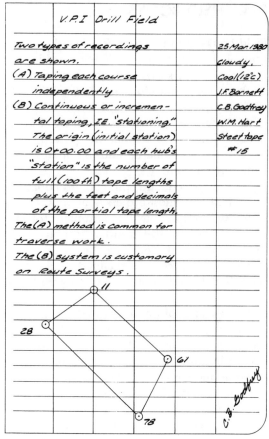

Figure 2-1.   Taping traverse courses.

wind, dust, or rain and snowstorms can adversely affect surveying results. Equipment is designed to operate under deplorable conditions; humans are not, and precision suffers. Weather information is one source to consider when investigating necessary corrections and erratic closures.

3. *Party names* with initials and assignments are essential for present and future reference. Court cases and questions regarding survey details may require contacting the field personnel many years after a survey has been completed. Common job symbols are $\barA$ for instrument operator, Ø for rodpersons, and N for notekeeper.

4. *Equipment make*, *model*, *type*, and *serial number*, along with adjustment condition, govern survey accuracy when properly handled by an experienced operator. Knowing which

instrument was employed may assist in isolating and correcting certain errors. For example, built-in errors of some electromagnetic distance-measuring instruments (EDMIs) and automatic level compensators are currently being investigated and corrections applied to various old surveys. However, unrecorded repairs and later adjustments, along with the lack of instrument identification on some projects, can prevent application of proper modifications.

Bench-mark descriptions should begin with the general location starting with state, county, town, or a familiar area if not otherwise covered by the job description or title. They then proceed through a recognizable feature, such as a street, bridge, building, etc., to a descrip-

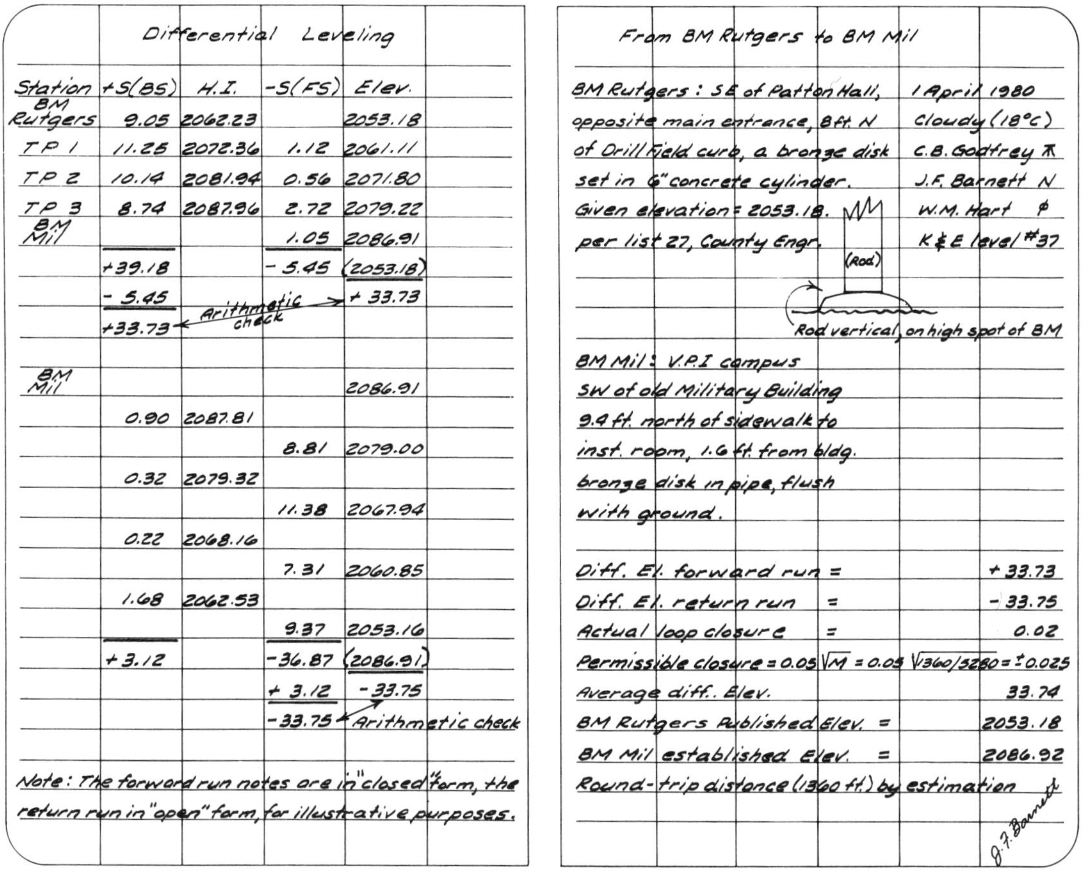

	+S(BS)	H.I.	-S(FS)	Elev.
Differential Leveling				
Station	**+S(BS)**	**H.I.**	**-S(FS)**	**Elev.**
BM Rutgers	9.05	2062.23		2053.18
T.P 1	11.25	2072.36	1.12	2061.11
T.P 2	10.14	2081.94	0.56	2071.80
T.P 3	8.74	2087.96	2.72	2079.22
BM Mil			1.05	2086.91
	+39.18		-5.45	(2053.18)
	-5.45			+33.73
	+33.73	Arithmetic check	+33.73	
BM Mil				2086.91
	0.90	2087.81		
			8.81	2079.00
	0.32	2079.32		
			11.38	2067.94
	0.22	2068.16		
			7.31	2060.85
	1.68	2062.53		
			9.37	2053.16
	+3.12		-36.87	(2086.91)
			+3.12	-33.75
			-33.75	Arithmetic check

Note: The forward run notes are in "closed" form, the return run in "open" form, for illustrative purposes.

From BM Rutgers to BM Mil

BM Rutgers : SE of Patton Hall,   1 April 1980
opposite main entrance, 8 ft. N   Cloudy (18°C)
of Drill Field curb, a bronze disk   C.B. Godfrey ⊼
set in 6" concrete cylinder.   J.F. Barnett N
Given elevation = 2053.18.   W.M. Hart φ
per list 27, County Engr.   K&E level #37

(Rod)

Rod vertical, on high spot of BM

BM Mil: V.P.I campus
SW of old Military Building
9.4 ft. north of sidewalk to
inst. room, 1.6 ft. from bldg.
bronze disk in pipe, flush
with ground.

Diff. El. forward run =	+33.73
Diff. El. return run =	-33.75
Actual loop closure =	0.02
Permissible closure = 0.05 $\sqrt{M}$ = 0.05 $\sqrt{360/5280}$ = ±0.025	
Average diff. Elev.	33.74
BM Rutgers Published Elev. =	2053.18
BM Mil established Elev. =	2086.92
Round-trip distance (1360 ft.) by estimation	

Figure 2-2. Differential leveling (open and closed styles).

tion of the object or mark itself. When used later in the field book, it need only be referred to by page number and not described again.

Bench-mark names, such as Rock, Bridge, Hydrant, etc., provide a clue to the location and may reduce the number of ties required to find the point. On long lines, numbers in sequence are often preferable, but they can be subject to mistakes and do not give a recognition key for the mark.

Symbols and abbreviations save notekeeping time and space. There are standards for common items, but if unusual ones are required, they must be identified the first time employed.

A useful instrument for notekeepers is a moderately priced, small, dependable camera. Photographs of found and set monuments, bench-mark locations, fence lines, and other items methodically described and referenced at the time they are taken, can eliminate or simplify lengthy lettered notes. The camera's position and aiming direction should be indicated by a symbol in the field book. All prints must be numbered, signed, and dated, then mounted in an album to be filed with the project field notes.

2-6. AUTOMATIC RECORDING

New *data collector* models are now available for use with *electronic distance-measuring instruments*, theodolites, and total-station equipment to automatically record distance and angle measurements electronically. Data collectors display and record measurements, in some cases by merely pushing a button. Reading and transcribing errors are eliminated in both field and office. Stored data are transferred from the collector to a field or office computer, then on to a printer that makes working plots and convenient page-wide printouts. Figure 2-3 shows the Lietz SDR2 electronic field book; Figure 2-4 defines the Lietz SDR2 flowchart from a recorder to the computer and other units in the assembly.

The K & E Vectron and Auto Ranger II provide a readout display for visual checking and transfer the panel measurements to a field computer, without manual input, for calculation and storage. The field computer can be interfaced with an office computer for printouts and permanent records.

The Wild TC1 recording attachment uses a magnetic tape cassette to store a complete block of measured data with built-in checks by touching a key. A cassette reader transfers the data to desk, mini, and large computers, and the information can then be transmitted from field to office by telephone.

The ABACUS' SDC71 ⓉⓂ Survey data collector provides all the power needed for survey applications, from easy, quick sequences to more complex projects. It is not tied to a single-brand total station, works well with most automated field equipment, and can be used independently as an electronic field book. The SDC71 allows the user to reduce source data to field coordinates and upload data to a variety of CAD systems.

Kern's Alphacord recording unit registers and stores automatically measured data that

Figure 2-3. Lietz SDR2 electronic field book. (Courtesy of the Lietz Company.)

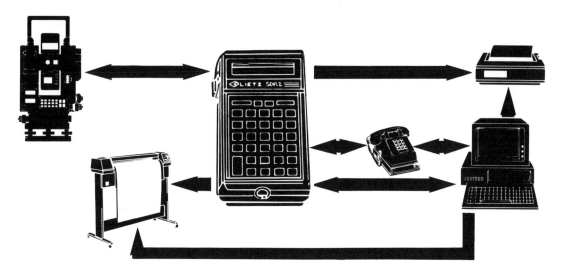

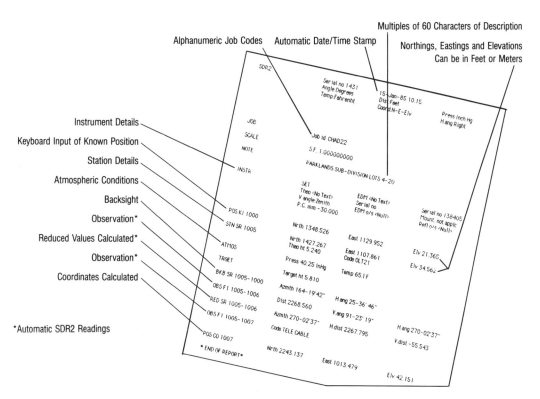

Figure 2-4. SDR2 flowchart. (Courtesy of the Lietz Company.)

along with keyed information is fed directly to a computer. While still in the field, the stored items can be relayed to a cassette tape recorder for later processing, thus making the unit's storage capacity available for repeated use.

The Magnavox 502 Georeceiver satellite surveyor automatically assures that valid data are properly recorded on magnetic cassette tape. Visual display permits manual recording of site position for the survey field notes.

All the data collectors briefly here have storage capacity for a normal day's operation. Data collectors are best suited for use with total-station instruments on projects providing many data to be passed on to a computer and other accessories. Among items to be considered are size, weight, power consumption, range of suitability for possible use with equipment from other manufacturers, storage media and capacity, manual and automatic data entry, clarity of display, ready access to repair facilities, initial and maintenance costs, and future upgrading feasibility.

Numerous changes and improvements have been made since a 1981 article, "Evaluating Data Collectors,"[3] was published in *P.O.B. Magazine*. The author tabulated six devices by name, manufacturer, physical characteristics, power supply, whether or not a stand-alone unit, communication, compatible equipment, data display, storage medium, memory size, availability, and price. The points discussed and evaluation checklist (see Table 2-1) are still pertinent.

A recent article, "Measuring the Productivity of Data Collectors and Total Stations" by Tom Donahue, evaluates four data-collection systems indoors at the Minneapolis Metro Dome. A Citation CI 450 top-mounted slope distance meter, representative of the equipment owned by a majority of surveying firms, with a Lietz TM 1A optical theodolite and standard field book were used for a base comparison.

Each system was timed and its performance measured over a typical traverse/topo project. Evaluations were also conducted to measure ease of use, flexibility, versatility, and overall performance. Six traverse points on the Metro Dome's field level and approximately 135 topo points on the field and lower deck areas were chosen. To more closely mirror field conditions, topo was shot from three different stations. The crew consisted of one equipment operator, two rod people and one person to set up stations.

Did the total stations and data collectors significantly increase productivity? The answer is

Table 2-1.   Data collectors evaluation checklist

Evaluation Checklist

1. Physical Characteristics
  (a) Dimensions of the unit
  (b) Total weight
  (c) Number of components in the system

2. Power Supply
  (a) Number and type of batteries required
  (b) Battery life per charge
  (c) Backup power supply

3. Interfacing the Data Collector to More Than One Type of Total Station
4. Method of Transmitting the Data to a Computer
5. Capacity of Data Collector

6. Type of Storage Medium
  (a) Solid-state
  (b) Tape

7. Features of Data Collector
  (a) Editing data
  (b) Computation
  (c) Checking data sequences
  (d) Type of data that can be entered

8. Type of Display
  (a) Light-emitting diode (LED)
  (b) Liquid crystal display (LCD)

9. Maintenance of Data Collector
10. Upgradability of the System

Source: Courtesy of *P.O.B. Magazine.*

yes and no. For radial surveys, with lots of data to be collected and where many shots can be taken from each traverse station, this type of equipment will more than pay back its original cost. However, where traversing only, or traversing with a small number of sideshots/station, an EDMI and field book would be almost as productive. The difference being that radial surveying is shot intensive, while traversing is almost set up and move. Consider the fact that it took the EDMI system approximately 40 to 50 seconds per shot to read and record topographic information versus 2.5 to 13 seconds for the total-station systems. All total-station/data collector systems finished the entire project in less than 1 hour and 40 minutes, while the EDMI/field book system took 3 hours and 45 minutes. That's 2.3 times slower than the slowest total station! Had the project consisted of the traverse stations only, the difference would have been minimal.

Other major findings of these evaluations were:

1. The weak link in the data collection chain is the transfer to a software package. This often involves coordination between different manufacturers and may change the field capabilities of a data collector. It is, therefore, imperative to evaluate this and other links in the system before purchasing.

2. For extensive radial surveying, a total station is two to three times faster than a mount-on EDM. This is mainly due to pointing time and one-button measurement.

3. For extensive radial surveying, a data collector is 35 to 50 percent faster than a field book. This assumes all the measured data can be electronically transferred to the collector. If some, or all, of the data need to be manually keyed in, then productivity is reduced to that of a field book.

4. The fastest total station had a coarse measurement mode, high-speed electronics, and wide beam width that made it ideal for radial stakeout and topographic surveys.

5. The best data collector was the only one which had built-in computation software and was also capable of calculating the traverse precision ratio in the field. Other built-in computation capabilities included resection, field stakeout, coordinate computation, and much more. In addition, readable field notes could be transferred directly to a printer without a computer.[4]

The advent of electronic recording has not diminished the need for highly competent notekeepers and field books. Since sketches, nonnumerical information, and descriptions must still be hand-prepared, the rapidly made measurements may increase a notekeeper's burdens on topographic and property surveys. But this responsibility is lessened somewhat by merely pointing, then just pushing buttons, instead of reading and recording. Cost of a data collector is an important factor for small surveying firms when considering how to enhance field and office equipment owned or contemplated for purchase.

Standard field books are readily accepted everywhere in court cases. Conversely, questions must be answered about magnetic tapes since they can be altered, erased by power mishaps or human error, lost, suffer deterioration, or make identification of the original versus a copy difficult. Also, electronic data collectors may present a possible hazard in underground surveys (see Chapter 29).

2-7. NOTEKEEPING POINTERS

Basic points, some previously mentioned, are listed as practical guides for notekeepers:

1. Letter the name, address, and telephone number of the field book's owner in India ink on the front and inside cover. State whether a reward will be paid for the return of a lost book.

2. Number all pages before first use of a field book. Left- and right-hand pages are paired and share the same number.

3. Employ the Reinhardt system of lettering for clarity, speed, and simplicity. Do not mix upper- and lowercase letters. Larger-size and uppercase lettering should be reserved for more important features.

4. Use a 3-H or harder pencil; keep it sharp, bear down.

5. Start a new day's work on a fresh page. For some projects in which large complicated sketches must be expanded, other considerations may be overriding.

6. Always record measurements immediately in the field book—not on scrap paper for later copying to improve appearance (a costly and dangerous act). "Rite in the Rain" field books now have waterproof paper specifically designed to accept field notes in wet or humid weather, even during rainstorms!

7. Carry a straightedge for ruling lines, a small protractor, and scales.

8. Make sketches to general proportion rather than exactly to scale or without advance planning. Keep in mind that preliminary estimates of the space required are often too small.

9. When in doubt, use sketches instead of tabulations.

10. Avoid crowding: It is one of the most common mistakes. If helpful, use several right-hand pages for descriptions and sketches to match a single left-hand page of tabulations, and vice versa.

11. If clarity is thereby improved, exaggerate details on sketches.

12. When possible, line up descriptions and sketches with corresponding data. For example, the beginning of a bench-mark description should be on the same line as its elevation.

13. Keep tabulated figures inside and off the column rulings with decimal points in line vertically.

14. Place a zero before the decimal point for numbers smaller than one—i.e., record 0.67 instead of just .67.

15. Record notes in an order that will facilitate office computations and mapping. For example, in stadia topography, number and read detail points in a clockwise rotation.

16. Letter measurements parallel with or perpendicular to sketch lines so they cannot be misunderstood. Machine-drawing type dimension lines are rarely used in surveying sketches.

17. Show the precision of measurements by recording significant zeros. Enter 2.60 instead of 2.6 if the reading was actually determined to hundredths.

18. Do not try to change a recording error by writing one number over another to transform a 3 to a 5, or a 7 to a 9.

19. Record what is read. Never "fudge" observations or closures. Surveying is an art and science. Art can stand retouching, science cannot.

20. Record aloud numbers given for recording. For a distance of 172.58, call out "one, seven, two, point five, eight" for verification.

21. To eliminate gross errors, make a mental estimate of all measurements before receiving and recording them.

22. Show essential computations made in the field so they can be checked later.

23. For compactness, employ conventional symbols and abbreviations. If not standard, identify them when first used or in a special table.

24. Use explanatory notes when they are pertinent.

25. Do not erase measured data—lines of a sketch can be deleted in the field.

26. Run a single line through an erroneous number and record the correct value above or below it.

27. To void a page, draw diagonal lines from opposite corners and letter VOID prominently without obscuring a number or any part of a sketch.

28. Letter COPY in large letters diagonally on copied notes but keep the lettering off a sketch or any numbers.

29. Run notes down the page except on route surveys where they progress upward to agree with sketches made while looking in the forward direction.

30. Review the notes, make all possible arithmetic checks, compute closures and error ratios, and record them before leaving the field. On large projects employing several parties, satisfactory closures indicate completed work and facilitate assignments for the next day.

31. Place a north arrow at and pointing to the top or left side of every page if possible since notes and drawings are read from the bottom or right side. A meridian arrow must be shown.

32. Title and index each project. Cross-reference every new job or continuation of a previous one by the client's organization, property owner, and description.

33. On all original notes, sign surname and initials in the lower right corner of the right-hand page. This is equivalent to signing a check and accepting responsibility for it.

2-8. ADDITIONAL BASIC NOTEFORMS

Additional noteforms covering basic and more advanced surveying operations are illustrated in Figures 2-5 through 2-16 (pp. 13–18) and in later chapters. They can serve as examples on

		Differential Leveling (3-wire) with Foot Rod			
BS(+)	Half-Interval	Mean(+)	FS(-)	Half-Interval	Mean(-)
(1) BM 29			TP 1		
8.266	161		3.491	171	
8.105	165	8.1037	3.320	168	3.3210
7.940	326		3.152	339	
(2) TP 1			TP 2		
6.574	216		4.623	204	
6.358	217	6.3577	4.419	200	4.4203
6.141	759	14.4614	4.219	743	7.7413
(3) TP 2			BM 30		
6.203	182		2.819	188	
6.021	178	6.0223	2.631	185	2.6320
5.843	1119	+20.4837	2.446	1116	-10.3733
		-10.3733			
		+10.1104 ft.(D.E.)			

The notekeeper must each time inform the instrument-man that the two half-intervals match acceptably within 3 (maybe 4) 1000ths before the rod or the instrument is moved. If there is a bad matching, the notekeeper demands a complete re-reading. Cumulative summation of the half-interval columns enables cross-comparison of each set-up to maintain a cumulative balance of the sight distances, fore and aft, for the run.

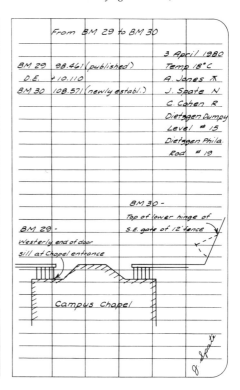

From BM 29 to BM 30

		3 April 1980
BM 29	98.461 (published)	Temp 18°C
D.E.	+10.110	A. Jones ⊼
BM 30	108.571 (newly establ.)	J. Spate N
		C. Cohen R
		Dietagen Dumpy
		Level # 15
		Dietagen Phila.
		Rod # 19

BM 30 –
Top of lower hinge of S.E. gate of 12' fence

BM 29 –
Westerly end of door sill at Chapel entrance

Campus Chapel

Figure 2-5. Differential leveling, three-wire.

	Reciprocal Leveling				
Sight	B.S.	F.S.	Sight	B.S.	F.S.
BM Stocker	1.467		BM Stocker		1.766
BM Monument		9.875			1.738
		9.897			1.753
		9.886			1.760
		9.892			1.758
		9.901			1.740
		9.888			1.748
		9.887			1.752
		9.885		Mean =	1.752
	Mean =	9.889	BM Monument	10.221	
D.E.		-8.422		+8.469	
Mean Diff. Elev.		-8.445			
El. BM Stocker		396.178	(Known from records)		
El. BM Monument		387.733	(Established)		

COPY

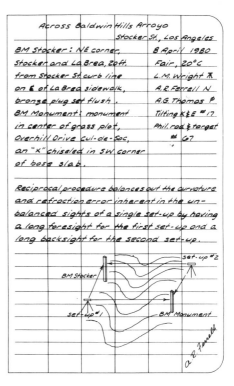

Across Baldwin Hills Arroyo
Stocker St., Los Angeles

BM Stocker: NE corner, Stocker and La Brea, 2 ft. from Stocker St. curb line on E of La Brea sidewalk, bronze plug set flush.
BM Monument: monument in center of grass plot, Overhill Drive Cul-de-Sac, an "x" chiseled in SW corner of base slab.

8 April 1980
Fair, 20°C
L.M. Wright ⊼
A.R. Ferrell N
A.G. Thomas P
Tilting K&E #17
Phil. rod & target #67

Reciprocal procedure balances out the curvature and refraction error inherent in the unbalanced sights of a single set-up by having a long foresight for the first set-up and a long backsight for the second set-up.

set-up #2
BM Stocker
set-up #1
BM Monument

Figure 2-6. Reciprocal leveling.

	Profile		Across		
Station	B.S.(+)	H.I.	F.S. on TP(-)	F.S.(-)	Elev.
BM Rutgers	3.93	2057.11			2053.18
0+00.0				3.09	2054.02
+06.0				3.23	2053.88
+06.5				3.72	2053.39
+26.5				4.23	2052.88
+46.5				4.89	2052.22
+46.8				4.35	2052.76
0+50				4.6	2052.5
+52.5				4.84	2052.27
1+00				8.4	2048.7
+50				10.3	2046.8
2+00				12.4	2044.7
TP 1	1.34	2046.37	12.08		2045.03
2+50				3.3	2043.1
3+00				4.4	2042.0
+50				4.6	2041.8
4+00				5.2	2041.2
+50				5.7	2040.7
5+00				5.9	2040.5
+50				5.9	2040.5
6+00				5.5	2040.9
+50				5.7	2040.7
TP 2	6.88	2047.95	5.30		2041.07
7+00				7.1	2040.9

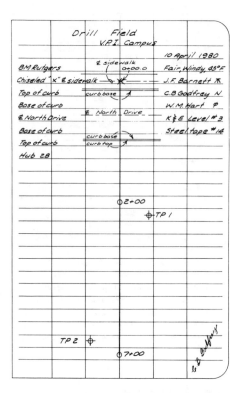

Figure 2-7. Profile leveling across drill field.

Borrow-Pit Leveling for Madison Factory Site.

Point	+S(BS)	H.I.	-S(FS)	Elev.	Cut	N*
BM Bridge	7.72	829.88	—	(822.16)	—	—
A0			5.0	824.9	9.9	1
A1			5.9	824.0	9.0	2
A2			6.8	823.1	8.1	2
A3			7.6	822.3	7.3	1
B0			4.2	825.7	10.7	2
B1			5.1	824.8	9.8	4
B2			7.6	822.3	7.3	4
B3			8.5	821.4	6.4	2
C0			3.0	826.9	11.9	1
C1			3.9	826.0	11.0	3
C2			5.2	824.7	9.7	4
C3			9.8	820.1	5.1	2
D1			2.1	827.8	12.8	2
D2			4.4	825.5	10.5	4
D3			10.1	819.8	4.8	2
E1			0.3	829.6	14.6	1
E2			3.8	826.1	11.1	2
E3			11.8	818.1	3.1	1
BM Bridge	7.71	(check)				

* N = Number of rectangles the point touches.

$$\text{Vol.(cu. yds.)} = \frac{\text{area of rect.}}{27}\left(\Sigma h_1 + 2\Sigma h_2 + 3\Sigma h_3 + 4\Sigma h_4\right)$$

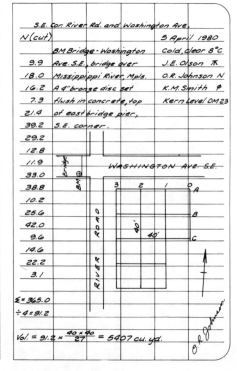

Figure 2-8. Borrow-pit leveling.

Closing The Horizon; Direction Instrument Method				
Point Occ.	Point Obs.	Direction Reading	Angle	Adjusted Angle
61	78	8°24'		
61	11	104°22'	95°58'	95°58.7'
61	11	105°57'		
61	ALT	221°25'	115°28'	115°28.7'
61	ALT	223°57'		
61	78	12°29'	148°32'	148°32.6'
		Σ =	359°58'	360°00'
		Discrepancy =	0°02'	

Point Occ.	Obs.	Tel.	Direction Reading	Mean	
61	78	D	109°58'54.0"		
	78	R	289°58'53.5"	53.8"	0°00'00.0"
61	11	D	205°57'47.6"		
	11	R	25°57'48.0"	47.8"	95°58'54.0"
61	ALT	D	321°26'24.6"		
	ALT	R	141°26'23.7"	24.2"	211°27'30.4"

Angles:
78 to 11	95°58'54.0"
11 to ALT	115°28'36.4"
ALT to 78	148°32'29.6"

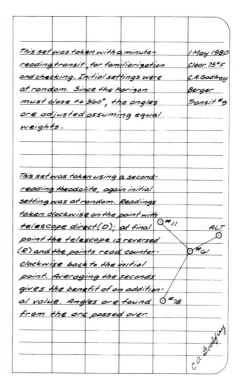

This set was taken with a minute-reading transit, for familiarization and checking. Initial settings were at random. Since the horizon must close to 360°, the angles are adjusted assuming equal weights.

1 May 1980
Clear 75°F
C.A.Godfrey
Berger
Transit #9

This set was taken using a second-reading theodolite, again initial setting was at random. Readings taken clockwise on the point with telescope direct (D); at final point the telescope is reversed (R) and the points read counterclockwise back to the initial point. Averaging the seconds gives the benefit of an additional value. Angles are found from the arc passed over.

C.A. Godfrey

Figure 2-9. Closing the horizon; direction instrument method.

Double Direct Angles: Traverse #7				
Hub	Single ∢	Double ∢	Mean ∢	Adjusted Angles
11	91°53'	183°45'	91°52.5'	91°52.0'
28	88°12'	176°23'	88°11.5'	88°11.0'
78	84°00'	167°59'	83°59.5'	83°59.0'
61	95°59'	191°57'	95°58.5'	95°58.0'
		Σ =	360°02.0'	360°00'

Σ interior angles = $(n-2)180° = 360°$
Permissible closure (specified) = $±02'\sqrt{n} = ±04'$ so the angles are suitable for adjustment. Specifications for the job sets the permissible closure, type of instrument to be used, etc, as required by the accuracy needed for the traverse. In this case a one-minute-reading transit and polygon closure of $±02'\sqrt{n}$ were specified.

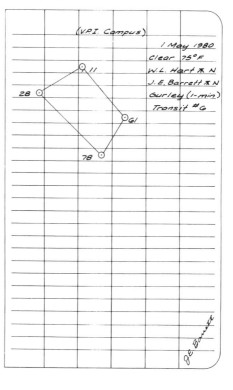

(V.P.I. Campus)

1 May 1980
Clear 75°F
W.L. Hart ∏ N
J.E. Barrett ∏ N
Gurley (1-min)
Transit #6

J.E. Barrett

Figure 2-10. Double direct angles.

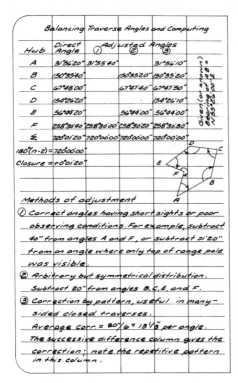

Balancing Traverse Angles and Computing						Bearings (Azimuths)					

Hub	Direct Angle	Adjusted Angles ①	②	③	
A	31°36'20"	31°55'40"		31°56'10"	
B	150°35'40"		150°35'20"	150°35'20"	
C	67°48'00"		67°47'40"	67°47'50"	
D	154°26'20"			154°26'10"	
E	56°44'20"		56°44'00"	56°44'00"	
F	258°30'40"	258°30'00"	258°30'20"	258°30'30"	
Σ	720°01'20"	720°00'00"	720°00'00"	720°00'00"	

180°(n-2) = 720°00'00"
Closure = +0°01'20"

Methods of adjustment

① Correct angles having short sights or poor observing conditions. For example, subtract 40" from angles A and F, or subtract 01'20" from an angle where only top of range pole was visible.

② Arbitrary but symmetrical distribution. Subtract 20" from angles B, C, E, and F.

③ Correction by pattern, useful in many-sided closed traverses.
Average corr = 80"/6 = 13⅓" per angle. The successive difference column gives the correction; note the repetitive pattern in this column.

Method Hub ③	Aver. Corr.	Rounded to 10"	Successive diff.	
A	13⅓"	10"	10"	
B	26⅔"	30"	20"	
C	40"	40"	10"	
D	53⅓"	50"	10"	
E	66⅔"	70"	20"	
F	80"	80"	10"	

1 May 1980
C.H. Dunn

Bearing Calculation	Azimuth Calculation
AB N 55°26'00" E +(SW)	AB 55°26'00"
B 150°35'20" +'	BA 235°26'00"
S 206°01'20" +	+B + 150°35'20"
	26°
BC N 26°01'20" E +(SW)	BC 386°01'20"
C 67°47'40" +'	CB 206°01'20"
S 93°49'00" +	+C + 67°47'40"
CD N 86°11'00" W -(SE)	CD 273°49'00"
D 154°26'20" +	DC 93°49'00"
DE S 68°15'20" W +(NE)	+D +154°26'20"
E 56°44'00" +	DE 248°15'20"
N 124°59'20" +	ED 68°15'20"
EF S 55°00'40" E -(NW)	+E + 56°44'00"
F 258°30'20" +	EF 124°59'20"
N 203°29'40" +	FE 304°59'20"
FA S 23°29'40" W +(NE)	+F 258°30'20"
A 31°56'20" +	FA 203°
	563°29'40"
AB N 55°26'00" E +	AF 23°29'40"
(check)	+A + 31°56'20"
	AB 55°26'00" (check)

Given (or known)
Bearing AB =
N 55°26'00" E

Figure 2-11.   Balancing traverse angles and computing bearings (azimuths).

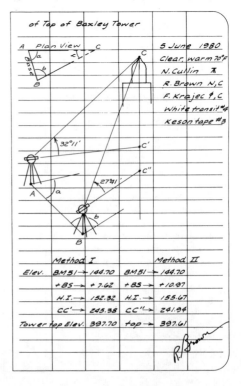

Trigonometric Leveling to Find Elevation						of Top of Baxley Tower	

Inst. at	Object	Horiz. Circle	Vert. Angle	Rod Reading	Elev.
A	C	0°00.0'	+32°11'		
	B	88°10.2'			
	BM 51	———	0°00'	7.62	(144.70)
B	A	0°00.0'			
	C	57°40.8'	+27°41'		
	BM 51	———	0°00'	10.97	(144.70)

Length AB = 259.01 ft.

Computation

Angle	Value	Method I	
a	88°10.2'	AC = AB $\frac{\sin b}{\sin C}$ =	389.91
b	57°40.8'	CC' = AC tan 32°11' =	245.38
Σ	145°51.0'		
	180°00.0'	Method II	
C	34°09.0'	BC = AB $\frac{\sin a}{\sin C}$ =	461.16
		CC" = BC tan 27°41' =	241.94

Plan View

5 June 1980
Clear, warm 70°F
N. Cullin 🔭
R. Brown N,C
F. Krajec ⚐,C
White transit #4
Keson tape #3

32°11'
27°41'

	Method I	Method II
Elev.	BM 51 → 144.70	BM 51 → 144.70
	+BS → + 7.62	+BS → +10.97
	H.I. → 152.32	H.I. → 155.67
	CC' → 245.38	CC" → 241.94
Tower top Elev.	397.70	top → 397.61

R. Brown

Figure 2-12.   Trigonometric leveling.

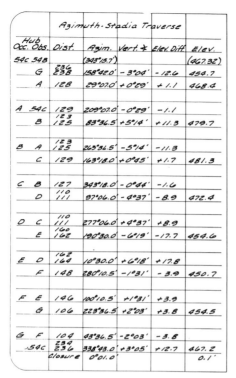

Azimuth-Stadia Traverse

Hub Occ.	Obs.	Dist.	Azim.	Vert. ∠	Elev. Diff	Elev.
54c	54B		(345°15.7')			(467.32)
	G	236 / 238	158°42.0'	-3°04'	-12.6	454.7
	A	128	29°07.0'	+0°29'	+1.1	468.4
A	54c	129 / 123	209°07.0'	-0°29'	-1.1	
	B	125	83°36.5	+5°14'	+11.3	479.7
B	A	123 / 125	263°36.5'	-5°14'	-11.3	
	C	129	163°18.0'	+0°45'	+1.7	481.3
C	B	127	343°18.0'	-0°44'	-1.6	
	D	110 / 111	97°06.0'	-4°37'	-8.9	472.4
D	C	110 / 111	277°06.0'	+4°37'	+8.9	
	E	160 / 162	190°30.0'	-6°19'	-17.7	454.6
E	D	162 / 164	10°30.0'	+6°18'	+17.8	
	F	148	280°10.5'	-1°31'	-3.9	450.7
F	E	146	100°10.5'	+1°31'	+3.9	
	G	106	223°36.5	+2°03'	+3.8	454.5
G	F	104	43°36.5'	-2°03'	-3.8	
	54c	234 / 236	338°43.0'	+3°05'	+12.7	467.2
	closure		0°01.0'			0.1'

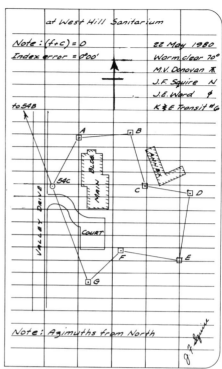

Figure 2-13. Azimuth-stadia survey.

Topographic Details by Stadia

Point	Dist	Azim.	Vert. ∠	Diff in Elev.	Elev.
☉ @ 54c	(El. 467.32)			h.i. = 4.8	
54B	—	(343°15.7')			
CB 1	100	45°52'	+0°59'	+1.7	469.0
CB 2	71	91°10'	+0°58'	+1.2	468.5
CB 3	95 / 94	123°09'	-2°25'	-4.0	463.3
E Dr. 4	39	196°37'	-3°15'	-2.2	465.1
ER 5	49	347°41'	-1°01'	-0.9	466.4
ER 6	102	354°07'	5.1 at 0°00'	-0.3	467.0
E Cr. 7	106 / 105	137°51'	-4°03'	-7.8	459.5
E Cr. 8	115 / 114	149°27'	-3°59'	-8.0	459.3
CP 9	108 / 107	341°15'	-3°16'	-6.1	461.2
A	128	29°07.0'	+0°29'	+1.1	468.4
54B	—	(343°15.7')			
☉ @ A	(El. 468.4)			h.i. = 4.7	
54c	129	(209°07.0')	-0°29'	-1.1	467.3
CB 12	47	129°15'	3.8 at 0°00'	+0.9	469.3
CP 13	39	150°45'	+1°28'	+1.0	469.4
CP 14	60 / 58	58°07'	+9°53' 9.7	+6.2	474.6
FL 15	89 / 88	32°00'	+6°57'	+10.7	479.1
FL 16	76	0°14'	+3°36'	+4.8	473.2
FC 17	83	337°02'	+1°11'	+1.7	470.1
CP 18	54	331°45'	4.4 at 0°00'	+0.3	468.7
CP 19	41	305°10'	-3°12'	-2.3	466.1
54c	—	(209°07.0')			

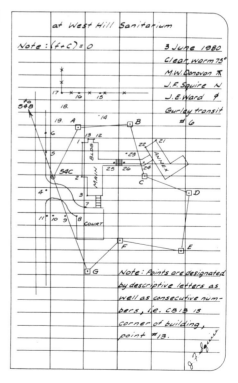

Figure 2-14. Topographic details by stadia.

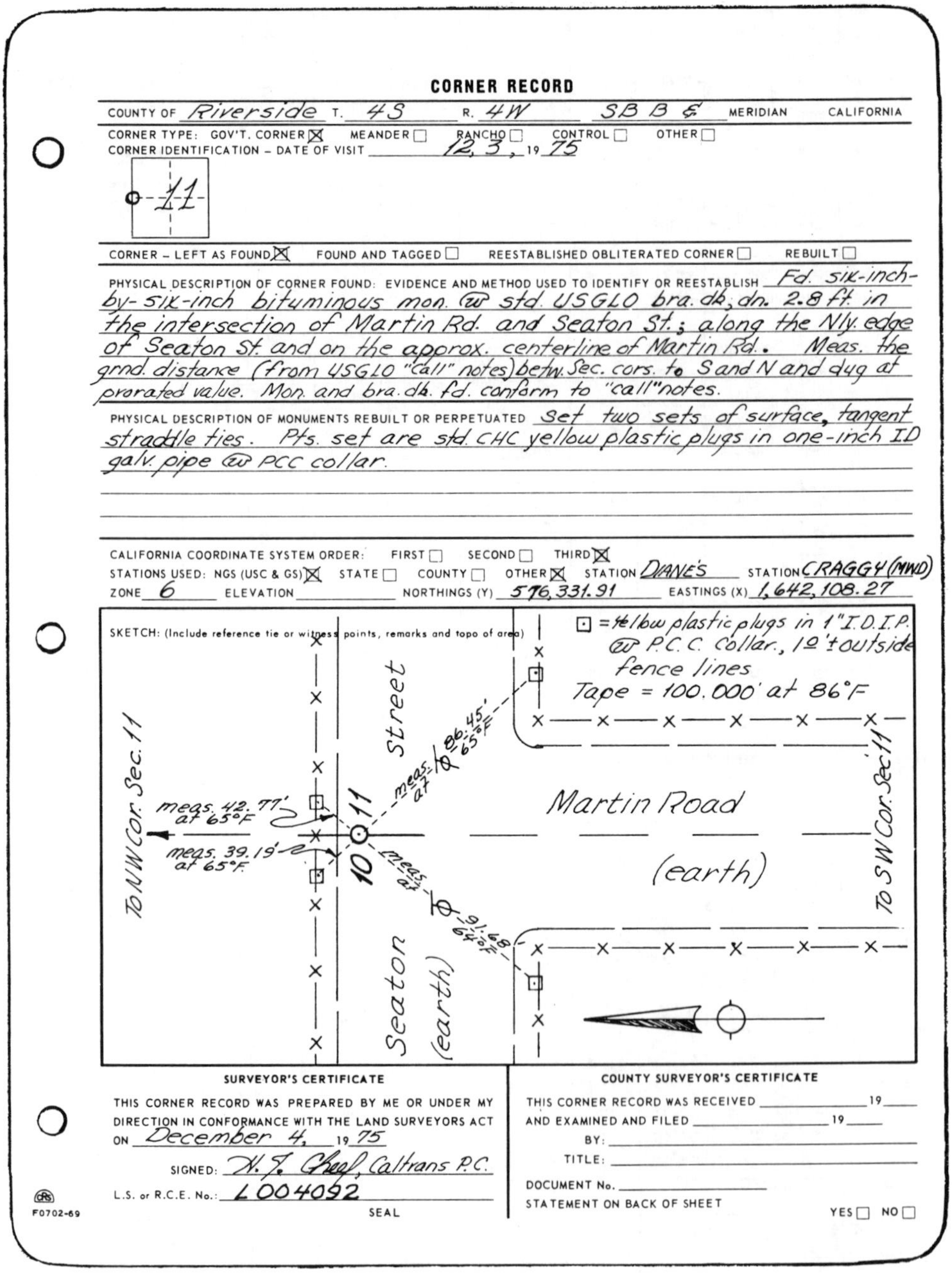

CORNER RECORD

COUNTY OF *Riverside* T. *4S* R. *4W* *SB B & M* MERIDIAN CALIFORNIA

CORNER TYPE: GOV'T. CORNER ☒ MEANDER ☐ RANCHO ☐ CONTROL ☐ OTHER ☐

CORNER IDENTIFICATION – DATE OF VISIT *12, 3*, 19 *75*

11

CORNER – LEFT AS FOUND ☒ FOUND AND TAGGED ☐ REESTABLISHED OBLITERATED CORNER ☐ REBUILT ☐

PHYSICAL DESCRIPTION OF CORNER FOUND: EVIDENCE AND METHOD USED TO IDENTIFY OR REESTABLISH *Fd. six-inch-by-six-inch bituminous mon. @ std. USGLO bra. dk; dn. 2.8 ft. in the intersection of Martin Rd. and Seaton St.; along the Nly. edge of Seaton St. and on the approx. centerline of Martin Rd. Meas. the grnd. distance (from USGLO "call" notes) betw. Sec. cors. to S and N and dug at prorated value. Mon. and bra. dk. fd. conform to "call" notes.*

PHYSICAL DESCRIPTION OF MONUMENTS REBUILT OR PERPETUATED *Set two sets of surface, tangent straddle ties. Pts. set are std. CHC yellow plastic plugs in one-inch ID galv. pipe @ PCC collar.*

CALIFORNIA COORDINATE SYSTEM ORDER: FIRST ☐ SECOND ☐ THIRD ☒

STATIONS USED: NGS (USC & GS) ☒ STATE ☐ COUNTY ☐ OTHER ☒ STATION *DIANE'S* STATION *CRAGGY (MWD)*

ZONE *6* ELEVATION ______ NORTHINGS (Y) *576,331.91* EASTINGS (X) *1,642,108.27*

SKETCH: (Include reference tie or witness points, remarks and topo of area)

SURVEYOR'S CERTIFICATE

THIS CORNER RECORD WAS PREPARED BY ME OR UNDER MY DIRECTION IN CONFORMANCE WITH THE LAND SURVEYORS ACT ON *December 4*, 19 *75*

SIGNED: *H. F. Cheel, Caltrans P.C.*

L.S. or R.C.E. No.: *L 004092*

SEAL

COUNTY SURVEYOR'S CERTIFICATE

THIS CORNER RECORD WAS RECEIVED ______ 19___

AND EXAMINED AND FILED ______ 19___

BY:

TITLE:

DOCUMENT No.

STATEMENT ON BACK OF SHEET YES ☐ NO ☐

F0702-69

Figure 2-15. Corner record.

which more complicated and special survey forms are built, with individual preferences exercised. Preprinted and more advanced types used by various agencies and organizations are included in pertinent chapters. All figures in this chapter except 2-3 and 2-4 are excerpted from *Noteforms for Surveying Measurements*.

NOTES

1. *ASCE Manual No. 34: Definitions of Surveying and Associated Terms*. NY, NY 1984. ACSM and ASCE.

2. R. C. Brinker, B. A. Barry, and R. Minnick. 1981. *Noteforms for Surveying Measurements*, 2nd ed. Rancho Cordova: Landmark Enterprises, CA, pp. 12, 14.

3. T. Donahue. 1986. Measuring the productivity of data collectors and total stations. *P.O.B. Magazine 11*(2), 85.

REFERENCES

McLean, J. E., Jr. 1981 Choosing an EDM data collector. *P.O.B. Magazine 7*(1), 36.

Kavanagh, S. J., and G. Bird. 1992 *Surveying Principles and Applications*, 3rd ed. Englewood Cliffs, NJ: Prentice Hall.

Wolf, P. R. and R. C. Brinker. 1994 *Elementary Surveying*, 9th ed. New York: Harper Collins.

3

Measurement Errors

Bro. B. Austin Barry, FSC

3-1. INTRODUCTION

Surveying is the art and science of making measurements. The notion of "exact" measurement, of "perfect" result, of "accurate" work is quickly dispelled when trying to duplicate an angle or distance measurement or difference of elevation. It is also evident when different people make the same measurement.

3-2. READINGS

Generally, when reading any graduated scale, the final digit is estimated—an appraisal of the distance between fine-scale graduations, such as 6.27 in Figure 3-1. This can be the end of a 50-ft tape, with graduation in tenths and half-tenths also marked, or a rod reading taken for elevation of a point.

Note that 6.27 would be the estimate of most observers, not 6.26 or 6.28, though these figures would almost surely be estimated by some others. Obviously, if extra care is warranted, a scale with finer graduations—say, to thousandths—might be used and the readings made with a magnifier, probably to tenthousandths of a foot. Such readings, if repeatedly made by an observer or observers,

might vary more widely in the last digit (estimated). Such readings made to ten-thousandths instead of hundredths are more accurate than those of Figure 3-1, and the apparently wider fluctuation in the last place is not nearly as serious.

3-3. REPEATED READINGS

Assume a series of observed readings using the fine graduations and magnifier that permits readings to ten-thousandths of a foot (see Table 3-1). If the readings had been taken to thousandths only, all would have been listed as 6.276; if to hundredths, 6.28; if to tenths only, 6.3 units. In this case, the best value attainable is the arithmetic mean or average of the set. It would be recorded as 6.27603 or perhaps 6.2760 units.

It is never possible to obtain absolutely correct fourth- or fifth-decimal-place figures in the example simply because the method of measuring is not sufficiently refined. An exact value does exist but cannot be identified. The objective is to get what may be termed a best available result by refined measurements and techniques of successive readings.

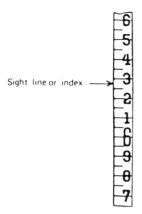

Sight line or index →

Figure 3-1. Interpolation.

3-4. BEST VALUE

Because an average is the "best available value," it can be used, although without assurance that it is correct or incorrect. Since in a set of measurements the true answer is unknown, it must be concluded that the mean value—and, in fact, any of the 10 measured values in Table 3-1—contains an error. In this context, "error" means the difference between a measured and true (or correct) value. Note that this does not apply to a "count" of bolts, cans, cartons, etc.

Results differ, perhaps only slightly, but it means no measurement in a set can be selected as the correct figure or exact result. By examining the range of measured values, the worst ones can be eliminated and those retained that cluster close together. It is the

Table 3-1. Repeated readings

1	6.2763
2	6.2757
3	6.2761
4	6.2760
5	6.2761
6	6.2758
7	6.2760
8	6.2764
9	6.2759
10	6.2760
	Mean = $\overline{6.27603}$

mean or average of these more reliable bunched values that is logically accepted as a "best" value.

3-5. ACCURACY AND PRECISION

Accuracy is descriptive of exactness or trueness of a measurement, its correctness. *Precision*, on the other hand, describes the closeness to one another of several measurements for the quantity. It speaks of the measurer's care and acumen and the instrument quality. Precision is revealed only by repeating measurements and then observing discrepancies among the results and variations of each from the set's mean.

3-6. ERRORS IN MEASUREMENT

In surveying, many measurements of quantities are made. Each contains errors: *systematic* (*cumulative*) and *accidental* (random). It is never possible to find a correct or true value for the quantity being measured, as opposed to *counts* of chaining pins, plumb bobs, level rods, etc. However, a reliable value is obtained if systematic errors are corrected and accidental ones studied for sign and size.

3-7. SYSTEMATIC ERRORS

Systematic errors in a measurement are proportionate to some influencing cause. When evaluated for size and sign, they can be corrected and eliminated.

Example 3-1. A 100-ft steel tape standard at 68°F will be shorter when used at 28°F by an amount

$$E_t = kL \, \Delta t = (0.00000645)(100.000)(68 - 28)$$

$$= 0.0258 \text{ ft}$$

When used to lay out a 560-ft length on construction, the required distance is

$$560.000 + (5.60)(0.0258) = 560.000 + 0.1445$$
$$= 560.144 \text{ ft.}$$

For the same temperature conditions, a taped measurement between two fixed hubs on the job that is reported as 346.842 ft must be corrected by

$$E_t = (0.00000645)(346.842)$$
$$\times (68 - 28) = 0.0895 \text{ ft}$$

The corrected field length is 346.842 − 0.0895 = 346.753 ft.

Example 3-2.   A reading on a distant level rod is affected by earth curvature and atmospheric refraction $E_{CR} = 0.574$ ft (M^2), where M is miles. If the reading seen from 325 ft is 6.354, the corrected rod result equals $6.354 - 0.574(325/5280)^2 = 6.352$ ft.

Example 3-3.   If a velocity meter or pressure gauge is calibrated and found to consistently read 10% high, all readings can be corrected by 10%. In a similar way, the vertical circle index error of a transit can be applied to each vertical angle reading as a correction.

Systematic errors also can be corrected by compensation procedures or devices.

Example 3-4.   In extending line *AB* on the ground to a point *C* by setting up a transit on *B*, backsighting to *A*, and plunging the telescope to set *C*, a maladjustment of the transit might place *C* to the left of its proper location. By repeating the procedure, starting with inverted telescope, point *C* will fall to the line's right. Correct placement of point *C* is midway between the two.

Example 3-5.   In differential leveling, an instrument whose line of collimation is not parallel to the bubble-tube axis will give correct results if the foresight and backsight distances are kept equal, thus compensating in each pair of sightings for instrument error, as well as curvature, and refraction.

Example 3-6.   In using electronic distance-measuring instruments (EDMIs), the velocity of light is affected by air temperature, atmospheric pressure, and vapor pressure, so distances must be corrected by a calculated sum. To correct directly for length errors, it is possible to modulate the EDMI circuitry for meteorological and environmental conditions.

For any measurement to give a true value, all systematic errors must be identified, analyzed, and corrected. Every source should be examined, since systematic errors can be natural, personal, or instrumental. Until all are isolated and corrected, accidental-error theory has no application. All that follows assumes systematic errors have been eliminated by proper corrections. However, note that evaluating and applying a correction still leave room for accidental errors.

3-8.   ACCIDENTAL ERRORS

Accidental errors in measurements are random in nature, probably small rather than large, and equally liable to be plus as minus. They do not accumulate, but are partly compensating in nature. Logically, by repeating a measurement and calculating the mean (average) of several measurements, a safer and better value for the quantity is secured.

3-9.   ERRORS VERSUS VARIATIONS

Because accidental errors are random and unpredictable, they cannot be evaluated or quantified; thus, corrections to counteract them are indeterminate. Making successive measurements, however, and comparing the results disclose differences in values. Studying the mean of a set of several values and their variations v from the mean indicates the reliability of each value. It is logical to assume unseen and unknowable errors x behave like visible and understandable variations v, so the mean of many measurements should be close to the quantity's correct value. Reliance is placed on variations to judge the mean value's nearness to truth.

3-10. DISTRIBUTION OF ACCIDENTAL ERRORS

A large set of measurements of a quantity can be represented in a bar graph called a *histogram* (Figure 3-2). Connecting the bar tops by a faired curve, a frequency distribution curve, permits visual representation of the measurements and their variation from the average or mean. Observation shows that

1. Small variations from the mean value occur more frequently than large ones.
2. Positive and negative variations of the same size are about equal in frequency, rendering their distribution symmetrical about a mean value.
3. Very large variations seldom occur.

These three characteristics can be seen in Figure 3-2, where variations are plotted. The three

are also characteristics of accidental errors, for the very existence of variations is explainable only by the presence of accidental errors in measurements. Therefore, it is not only convenient but also permissible to speak almost interchangeably of variations and errors.

The *normal* or *Gaussian* distribution is the most important of many possible distributions, since it has a wide range of practical applications. It is sometimes called the bell-shaped distribution, which typifies measurement distributions in practice. The histogram and frequency curve of Figure 3-2 are symmetric and shaped like a bell, thus that indicating the set of measurements is a normal distribution. The following mathematical model adequately describes such a distribution:

$$y = \frac{1}{\sqrt{(2\pi)}\sigma} \cdot e^{-(x-u)^2/2\sigma^2} \qquad (3\text{-}1)$$

The plot of this is called the *normal distribution curve;* if the height of the curve is standardized so that the area underneath it is equal to unity, then the graph is called a probability curve (Figure 3-3).

Focusing on the three obvious variation traits, Figure 3-4 depicts a set of several measurements of a quantity—say, the distance taped between two monuments. The true but unknowable length is indicated as "true value," with an error x_i, and the measurement's mean is shown with its variation v_i. The average error, while unknown, is shown as

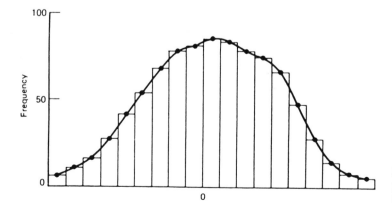

Figure 3-2. Histogram with superimposed frequency distribution curve.

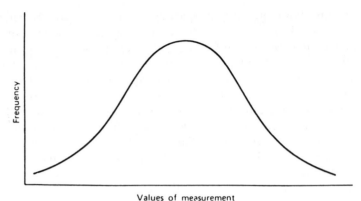

Figure 3-3. Normal distribution curve.

$\Sigma x/n$. Although surveyors and engineers can never equate a variation from the mean v_i with an error x_i, the pattern of occurrence of errors (invisible, unknown) is reasonably assumed to follow that of variations (visible, knowable). It is seen that

$$\underset{\text{(arithmetic mean)}}{\overline{X}} = \underset{\text{(true value)}}{\overline{X}_0} \underset{\text{(mean error)}}{-\frac{\Sigma x}{n}} \quad (3\text{-}2)$$

and also

$$\underset{\text{(any variation)}}{v_i} = \underset{\text{(corresponding error)}}{x_i} \underset{\text{(mean error)}}{-\frac{\Sigma x}{n}} \quad (3\text{-}3)$$

When n is larger, it signifies that the mean measured value $\overline{X}$ is closer to the true value (mean of the population). It is also apparent that variations then become more nearly equal

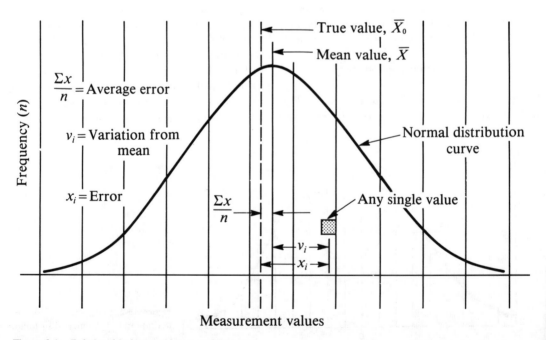

Figure 3-4. Relationship between error and variation.

to errors. Though the true magnitude of a measured quantity is never determinable, it can be ascertained as closely as required by taking enough measurements. If the range of variations narrows to a small value, the range of errors also constricts, rendering the mean value $\overline{X}$ predictably close to the true one $\overline{X}_0$. This hypothesis enables observers to speak confidently of true value and true error.

3-11. STANDARD DEVIATION

A practical indicator used to describe the reliability or worth of a set of repeated measurements is the *standard deviation*, defined as

$$\sigma_s = \pm\sqrt{\frac{\Sigma v^2}{(n-1)}} \qquad (3\text{-}4)$$

Thus, if n measurements of a quantity are obtained, each made in the same manner, the mean value of the set can be employed and given a degree of acceptance by citing the set's standard deviation. Table 3-2 (see p. 25) illustrates the computation of precision for a set of

Table 3-2. Measure of precision for set A

Measured Value	Variation v	v^2
165.861	−0.003	0.000009
165.866	+0.002	0.000004
165.860	−0.004	0.000016
165.864	0.000	0.000000
165.863	−0.001	0.000001
165.865	+0.001	0.000001
165.864	0.000	0.000000
165.863	−0.001	0.000001
165.863	−0.001	0.000001
165.866	+0.002	0.000004
165.864		
Mean		$\Sigma v^2 = 0.000037$

$$\sigma_s = \pm\sqrt{\frac{0.000037}{(10-1)}} = \pm 0.002$$

Best value (mean) = 165.864
Measure of precision $\sigma_s = \pm 0.002$

10 measured values. Today, good hand calculators perform this task easily through the keys marked $\Sigma +$, $\overline{X}$, and σ.

Statistical theory, borne out by extended measurement observations, enables helpful

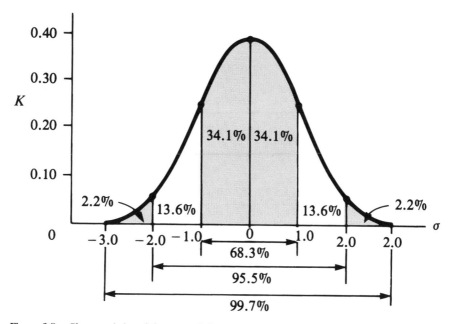

Figure 3-5. Characteristics of the normal distribution curve.

conclusions about a set of measurements. The normal distribution curve (see Figure 3-5) has the following characteristics:

1. The area beneath the entire curve is unity or 100% probability that all measurements will fall somewhere in the curve's range.
2. The area beneath the curve bounded by $\pm \sigma_s$ is 68.3%, and $\pm \sigma_s$ is the 68.3% "error."
3. The area beneath the curve bounded by a $\pm 2\sigma_s$ is 95.5%, i.e., $\pm 2\sigma_s$ represents the 95.5% "error."
4. The area beneath the curve bounded by $\pm 3\sigma_s$ is 99.7%, so $\pm 3\sigma_s$ defines the 99.7% "error."

Stated differently, another measurement should fall with a 68.3% probability within the $\pm \sigma$ bounds; a 95.5% probability between $\pm 2\sigma$; and a 99.7% probability inside $\pm 3\sigma$. These limits are referred to as a one-sigma or 68.3% confidence level, a two-sigma or 95.5% confidence level, and a three-sigma or 99.7% confidence level. Values used for other confidence levels are given in Table 3-3.

3-12. USES OF STANDARD DEVIATION

There is a practical use for standard deviation in comparing sets of measurements of a quantity.

Example 3-7. Comparing set *A* (Table 3-2) with set *B*, we obtain

$$\text{Set } A: \ 165.864 \pm 0.002$$
$$\text{Set } B: \ 165.867 \pm 0.006$$

The standard deviation for set *A* suggests it is a better one, although both have validity. Now assume the following two additional sets:

$$\text{Set } C: \ 165.862 \pm 0.007$$
$$\text{Set } D: \ 165.864 \pm 0.004$$

To find the weighted mean of all four sets, a weight is accorded each set proportional to the inverse square of its standard deviation. Of the four, it is clear the greatest confidence can be placed in set *A*, the least in set *C*.

Set	X	σ_s	Weight Factor	Weight	$Wt(X - 165.860)$
A	165.864	± 0.002	$(1/0.002)^2 = 250,000$	12.25	0.049
B	165.867	± 0.006	$(1/0.006)^2 = 27,778$	1.36	0.010
C	165.862	± 0.007	$(1/0.007)^2 = 20,408$	1.00	0.002
D	165.864	± 0.004	$(1/0.004)^2 = 62,500$	3.06	0.012
				17.67	0.073

Weighted mean = $165.860 + 0.073/17.67 = 165.860 + 0.004 = 165.864$

Table 3-3. Size of error in a single measurement of a set

Name of Error	Symbol	Value	Certainty (%)	Probability of Larger Error
Probable	E_p	$0.6745\sigma_s$	50	1 in 2
Standard deviation	σ_s	$1.0\sigma_2$	68.3	1 in 3
90% error	E_{90}	$1.6449\sigma_s$	90	1 in 10
Two-sigma or 95.5% error	$2\sigma_s$	$2\sigma_s$ or $3E_p$	95.5	1 in 20
Three-sigma or 99.7% error	$3\sigma_s$	$3\sigma_s$	99.7	1 in 370
Maximum*	E_{max}	$3.29\sigma_s$	99.9 +	1 in 1000

*Some authorities regard the 95.5% error as the "maximum error." Neither view is absolutely correct, since the theoretical maximum error is $\pm\infty$, which does not occur in practice. It is, then, a good practical decision to use the 95.5 or 99.9$^+$% error as the "practical" maximum that is tolerable.

Another practical use for standard deviation is to determine whether one set of measurements is significantly different from another set. If the discrepancy between the means of the two sets is not more than twice a value called σ_{DIFF}, they can be accepted as measurements of the same quantity.

$$\sigma_{\text{DIFF}} = \sqrt{(\sigma_s)^2_A + (\sigma_s)^2_B} \qquad (3\text{-}5)$$

Example 3-8. Suppose two sets are compared as follows:

Set E: 165.848 ± 0.006,

$$\sigma_{\text{DIFF}} = \sqrt{(0.006)^2 + (0.010)^2}$$

Set F: 165.867 ± 0.010, $\quad 2\,\sigma_{\text{DIFF}} = 0.024$

$\overline{0.019}$ (difference)

Conclusion: Set E is not significantly different from set F, since 0.019 is not more than twice σ_{DIFF} (= 0.024). Therefore, sets E and F can be regarded as valid measurements of the same quantity, and inspection indicates they can be combined in the weighted mean procedure.

Set	X	σ_s	Weight Factor	Weight	$W_i(X - 165.860)$
A	165.864	±0.002	250,000	25.0	0.100
B	165.867	±0.006	27,778	2.8	0.020
C	165.862	±0.007	20,408	2.0	0.004
D	165.864	±0.004	62,500	6.2	0.025
E	165.848	±0.006	27,778	2.8	−0.034
F	165.867	±0.010	10,000	1.0	0.007
				$\overline{39.8}$	$\overline{0.122}$

Weighted mean = 165.860 + 0.122/39.8 = 165.860 + 0.003 = 165.863

3-13. VARIANCE AND STANDARD DEVIATION

Variance is another measure of scatter among measured values in a set of measurements. Preferred by some users, it is simply the square of the standard deviation; thus,

$$\text{Variance} = \sigma_s^2 = \frac{\Sigma v^2}{(n - 1)} \qquad (3\text{-}6)$$

Comparing the four sets in Section 3-12, for instance, would show the following:

Set	Variance V	
A	0.000004 or	1/250,000
B	0.000036 or	1/27,800
C	0.000049 or	1/20,400
D	0.000016 or	1/62,500

Other measures exist, but are not covered here:

1. Standard error of the standard deviation = $\pm\sigma_s/\sqrt{2n}$.
2. Standard error of the variance = $\pm\sigma_s^2\sqrt{2n}$.
3. Standard error of coefficient of variation = $V/\sqrt{2n}$.
4. Standard error of the median = $1.25\sigma_s/\sqrt{n}$.

3-14. USE OF STANDARD SPECIFICATIONS FOR PROCEDURE

In measurements of any kind, reliance is placed on an established procedure that has been used repeatedly, many hundreds of times, to establish the validity of results. Thus, when standard specifications for a task are followed, only a limited number of measurements is needed to be certain the standard deviation from a shortened set is acceptable. Experience shows a larger number of measurements does not give a greatly different result for the mean or average value of the set.

Confidence inspired by following a fixed procedure or specifications also applies to instruments used in making measurements. A specific EDMI is advertised to give distances accurately to $\pm(5$ ppm $+ 4$ cm) because many thousands of test observations were made by the manufacturer. Thus, any measurement with this instrument by experienced personnel can be accorded the same accuracy. For instance, a length measured as 1543.02 m has an error value of $\pm(0.0077 + 0.04) = \pm 0.048$ m. This is properly regarded as the standard deviation of the measurement. The measurement precision can be stated as $\pm 0.048/1543$ or 1 in 32,100.

Example 3-9. The Kern DM502 accuracy has improved from $\pm(5$ mm $+ 5 \times 10^{-6}$ D) to $\pm(3$ mm $+ 5 \times 10^{-6}$ D) by recent technical advances. These limits are the one-sigma (68.3%) values obtained by analysis of many measurements. Using this instrument in the prescribed manner assures it is part of the large family or population of measurements already made and thus able to share in that established reliability.

3-15. DISTRIBUTION OF ACCIDENTAL ERRORS

Virtually all surveying measurement errors conform to a pattern called normal distribution. The theoretically perfect normal distribution (normal probability curve) shown in Figure 3-3 is the plot of equation

$$y = (h/\sqrt{\pi})e^{-h^2 x^2} \tag{3-7}$$

the familiar bell-shaped curve. It is symmetrical, with flatness or peakedness dependent on error sizes (variations from true value).

Natural phenomena and surveying measurements follow the same law of normal

Gaussian distribution for heights of 17-year-olds in a school system; weights of apples gathered from a single tree; weights of babies at birth; and repeated distance, angle, or level measurements. Other distribution patterns result from imposed influences and are mostly Poisson distributions. Examples are the arrival of ships or trains, traffic grouping on a street, incidence of storms, telephone demand, and road accidents. Plotting Poisson distributions or predicting results is not possible by the methods used here, which depend on normal (natural, uninfluenced) distribution.

To study a large set of surveying measurements, plot them as a histogram (bar graph) and superimpose a curve connecting the tops of the bars (see Figure 3-2). If this curve looks bell-shaped, its normally distributed results establish confidence that the rules of probability are fulfilled.

3-16. PLOTTING THE NORMAL PROBABILITY CURVE

To facilitate comparison, the *normal probability curve* can be plotted at a scale consistent with the histogram. The normal distribution equation is rendered in the form

$$y = \frac{KnI}{\sigma_s} \tag{3-8}$$

where I is the class interval, and the following values of K are used:

x	K
$\bar{x}$ (mean)	0.39894
$\bar{x} \pm 0.5\sigma_s$	0.35206
$\bar{x} \pm 1.0\sigma_s$	0.24197
$\bar{x} \pm 1.5\sigma_s$	0.12953
$\bar{x} \pm 2.0\sigma_s$	0.05399
$\bar{x} \pm 2.5\sigma_s$	0.01753
$\bar{x} \pm 3.0\sigma_s$	0.00443
$\bar{x} \pm \infty$	0.00000

(If individual values are used, $I = 1$; if grouped by 2s or 5s, etc., $I = 2$ or 5, etc.)

Example 3-10. In testing an automatic level instrument, the marker recorded a set of

439 rod readings, with the mean $\overline{X} = 6.5782$ and $\sigma_s = \pm 0.00304$ ft. The ordinates for the superimposed normal distribution curve are

x	y
$\bar{x}$ (mean)	57.5
$\bar{x} \pm 0.5\sigma_s$	50.8
$\bar{x} \pm 1.0\sigma_s$	34.9
$\bar{x} \pm 1.5\sigma_s$	18.7
$\bar{x} \pm 2.0\sigma_s$	7.8
$\bar{x} \pm 2.5\sigma_s$	2.5
$\bar{x} \pm 3.0\sigma_s$	0.6
$\bar{x} \pm \infty$	0.0

The normal curve superimposed on the histogram (Figure 3-6) shows a quite good fit.

3-17. MEANING OF STANDARD DEVIATION

The σ_s indicates that any next measurement introduced in this set should, with 68.3% certainty, fall within the $\pm \sigma_s$ range of the mean

$\overline{X}$. It pertains to this sample, not the population as a whole. A sample, or set, representing the whole population distribution will have a mean value $\overline{X}$, but not one equal to the true mean of the whole population. Nor will another sample, or another, etc. Every set likewise has its own standard deviation, not that of the entire population, and unlike other sets (Figure 3-7). It is clear also that an average of the means of several sets will get closer to the population mean. The logic of this is that working with the whole population progressively by sets, the average of all means must ultimately equal the population mean. It follows, therefore, that if the sample is larger, the mean of the sample will more likely approach the true value of the population mean.

3-18. MEANING OF STANDARD ERROR

The standard deviation of any set or sample of n items yields $\sigma_m = \pm \sigma_s / \sqrt{n}$, which is the

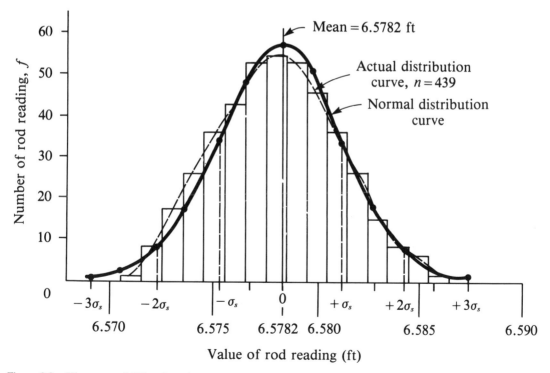

Figure 3-6. Histogram of 439 rod readings with actual distribution curve and normal distribution curve plotted.

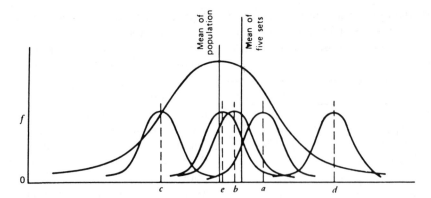

Figure 3-7. Relationship of samples to population: *a* through *e* are the means of sets of individual measurements.

standard error of the mean. This is a measure of the sample validity. It signifies that with 68.3% confidence the set's mean lies within $\pm \sigma_m$ of the true value of the population mean, or with 95.5% confidence it lies within $\pm 2\sigma_m$, etc.

If several sample sets of measurements are taken from the same population, the sample means vary somewhat from sample to sample and will themselves form a sampling distribution (of $\overline{X}$). The standard deviation of the group means is less than the σ_s of any sample. To find the standard deviation of the population of measurements of a particular quantity, work from one set, assuming that the population would, if entirely covered by samples, be found to have a mean value near that of the one sample, and a standard deviation of the means of all the similar samples repeated enough to cover the entire population. Recalling that the variance of the set is σ_s^2, then the variance of the means of sets is

$$\text{Variance } (\overline{X}) = \text{variance } (\Sigma x_1/n)$$

$$= 1/n^2 \text{ variance } (\Sigma x_i)$$

But variance $(x_i) = \sigma^2$ for all i

$$\text{Variance } (\overline{X}) = (1/n^2)(\sigma^2 + \sigma^2 + \cdots + \sigma^2)$$

$$= n\sigma^2/n^2 = \sigma^2/n$$

Thus,

$$\sigma_m = \frac{\sigma_s}{\sqrt{n}}$$

Although nature's distribution of apple sizes or men's weights may yield a mean value and standard deviation showing the whole population's scatter, surveying measurements are different. Surveying measurements do not exist until they are made; they cannot include all possible measurements of an angle, length, or difference of elevation. Therefore, from one sample, use the mean and standard error as an indication of correct value—e.g., the mean of the population of measurements if all possible measurements were made. The σ_m is an indicator of the standard deviation of the population, all based on a reasonable number of measurements in a single set.

Example 3-11. Set A of Table 3-2 would yield $\sigma_m = \pm(0.002/\sqrt{10}) = 0.0006$ ft. With 68.3% assurance, the true value is within ± 0.0006 of 165.864 ($= \overline{X}$), or between 165.863 and 165.865 ft.

Example 3-12. In the 439 measurements of Example 3-10, the mean $\overline{X}$ is 6.5782 and $\sigma_s = \pm 0.00304$. Then

$$\sigma_m = \pm 0.00304/\sqrt{439}$$

the "maximum" error (Table 3-3) is $3.29\sigma_m$ or ± 0.000477, which gives a 99% confidence that the true value lies between 6.5777 and 6.5787 ft. Such a large number of measurements would be unlikely except in testing a procedure or developing a new instrument.

Example 3-13. From records of some 30 surveys of similar precision, a single mean value of difference of elevation is sought, along with a measure of its precision (accuracy). The following values are grouped randomly into six sets, size five, for study:

	1	2	3	4	5	6	The 30-Set
	3.717	3.622	3.651	3.775	3.611	3.697	
	3.621	3.594	3.632	3.583	3.622	3.564	
	3.753	3.609	3.661	3.656	3.527	3.595	
	3.558	3.695	3.524	3.633	3.648	3.616	
	3.675	3.659	3.623	3.577	3.706	3.639	
$\overline{X}$	3.665	3.636	3.618	3.645	3.623	3.622	3.635
σ_s	0.077	0.041	0.055	0.080	0.065	0.050	0.060
σ_m	0.034	0.018	0.025	0.036	0.029	0.022	0.011

Range bracketing the population mean with a 68.3% confidence:

	1	2	3	4	5	6	The 30-Set
	3.631	3.618	3.593	3.609	3.594	3.600	3.624
	to	to	to	to	to	to	to
	3.699	3.654	3.643	3.681	3.652	3.644	3.646

Any one of the six X and σ_m values is representative of the mean and standard deviation of the whole population. For further comparison, the $\overline{X}$ and σ_m values are shown for the 30 measurements regarded as a single sample set; it is seen that each of the six smaller sets has a range that brackets the $\overline{X}_{30}$ ($= 3.635$). Further, each set's mean is a valid contender for the population mean, which can, of course, never be known for sure. This example also demonstrates that a small sample can and sometimes must be used, but a larger sample gives a more refined result.

3-19. PLOTTING THE NORMAL DISTRIBUTION IN OTHER FORMS

The shape of a normal curve (bell-shaped) depends on the standard deviation σ_s, which spreads out the curve when it is larger. Whatever the mean and standard deviation, however, one in three observations will lie beyond one standard deviation from the mean, one in 20 beyond two, etc. Drawing the cumulative distribution curve in another form makes some things clearer. This is done by plotting "percentage-smaller-than" against values of the measurement.

Example 3-14. A set of 16 angle measurements, tallied in ascending order, (Table 3-4), is indicated by percentages calculated on the basis of $(n + 1)$ for reasons to be explained later.

$n = 16,$ calculated mean $= 134°37'19.21''$
$n + 1 = 17,$ $\sigma_s = \pm 03.48''$
Mean minus $\sigma_s = 134°37'15.73''$
Mean plus $\sigma_s = 134°37'22.69''$

This small array of numbers is grouped in ascending order with class width of $02''$ and a histogram plotted. Connecting the tops to form a frequency distribution curve (Figure 3-8) shows it is not bell-shaped or satisfactory.

The 16 results are better portrayed on a cumulative *frequency distribution curve* (the S-curve, Figure 3-9). This curve can be held to virtually a straight line between the 15.8% and 84.2% values (the $\pm \sigma_s$ limits) and made to pass through the plotted points in a "best-fit"

Table 3-4. Set of 16 angle measurements

X	f	f	%	Class Width	Class Mean	f
134°37'13.8″	1	1	5.9			
14.8″	1	2	11.8	13.5–15.5″	14.5″	2
15.8″	1	3	17.6	15.5–17.5″	16.5″	4
16.1″	1	4	23.5	17.5–19.5″	18.5″	3
16.8″	1	5	29.5	19.5–21.5″	20.5″	4
17.4″	1	6	35.3	21.5–23.5″	22.5″	1
17.5″	1	7	41.2	23.5–25.5″	24.5″	1
19.1″	1	8	47.1	25.5–27.5″	26.5″	1
19.4″	1	9	52.9			
20.3″	1	10	58.8			
20.8″	1	11	64.7			
20.9″	1	12	70.6			
21.4″	1	13	76.5			
23.4″	1	14	82.4			
24.0″	1	15	88.2			
26.0″	1	16	94.1			

manner. Then the points outside this range, usually spoken of as the 15 to 85% range, will tail off to form an S-curve. Observing the curve at 50% shows the mean value, which should verify that previously calculated. The 15 and 85% points are values marking the 68.3% limits of certainty.

A still better way to plot a set of values is on arithmetic probability paper designed to plot any normal or *Gaussian* distribution as a straight line. There are two ways to use such paper for a meaningful plot and examine the scatter:

1. Plot the individual points, percentage-smaller-than versus the actual values; draw a best-fit straight line, and read $\bar{X}$ (mean) at 50% and the limits of σ_s at the 15.8 and 84.2% lines.

2. Calculate the mean and σ_s values, draw a straight line through them, and finally plot individual values to see how well they fit.

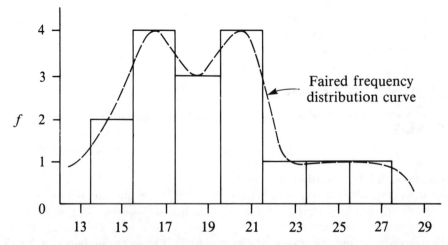

Figure 3-8. Histogram and faired frequency distribution curve for set of 16 angle measurements.

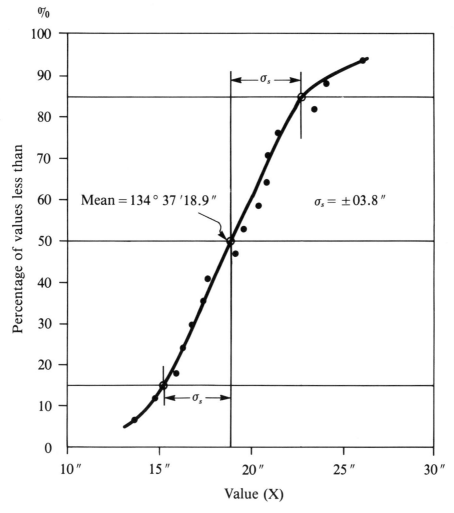

%

Figure 3-9. Cumulative frequency distribution curve for set of 16 angle measurements.

Example 3-15. The 16 angle values are plotted (Figure 3-10) by method a and the curve drawn as a best-fit line. Percentages are calculated using $(n + 1)$ in the denominator to plot the end point(s) of the curve. Comparison of the angle measurements with those of Example 3-14 shows these results. From the plot on arithmetic probability paper,

$$\text{Mean} = 134°37'19.2'', \qquad \sigma_s = 04.0''$$

From calculation,

$$\text{Mean} = 134°37'19.21'', \qquad \sigma_s = 03.5''$$

The plotted points are seen to fit a faired straight line decently well, so they can be visually adjudged to conform with normal distribution and be accepted for any further statistical treatment. The scaled outlines of Figure 3-10 can be photocopied and used for arithmetic probability paper.

3-20. PROPAGATION OF ACCIDENTAL ERRORS

Basic to the combined effect of accidental errors is their tendency to cancel themselves

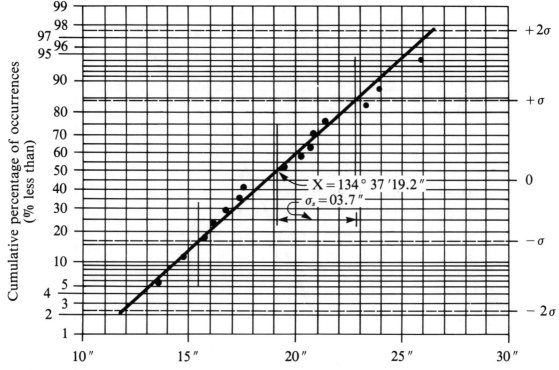

Figure 3-10. Plot of 16 angle measurements to find mean $\overline{X}$ and standard deviation σ_s of the set.

out (as likely to be $+$ as $-$) and their propensity to cluster about the mean—more frequently small, seldom large. This gives rise to the law of compensation

$$E_{\text{total}} = \pm\sqrt{E_1^2 + E_2^2 + E_3^2 + E_4^2 + \cdots + E_n^2} \quad (3\text{-}9)$$

Example 2-16. If a 100-ft steel tape is corrected for temperature to $\pm5°F$, a random error of ±0.0032 ft results. Comparison with a standard tape may expose an error of ±0.002 ft. When applying tension within 2 lb, there is a possible error of ±0.0013 ft. Therefore, total effective error for each tapelength from these three sources is

$$E_{\text{total}} = \pm\sqrt{0.0032^2 + 0.002^2 + 0.0013^2}$$
$$= \pm0.0040 \text{ ft}$$

3-20-1. Addition

When adding measured quantities having known errors, the error of the sum is

$$E_s = \pm\sqrt{e_1^2 + e_2^2 + e_3^2 + \cdots + e_n^2} \quad (3\text{-}10)$$

Example 3-17. If the steel tape of the previous example is known to have a ±0.0040-ft error each time it is used, the error in taping 1600 ft would be

$$E_s = \pm\sqrt{16(0.0040)^2} = \pm4(0.0040)$$
$$= \pm0.016 \text{ ft}$$

Example 3-18. The error in the sum of two measured angles is computed similarly.

Angle AOB = 15°32'18.9" $\pm$ 05"

Angle BOC = 67°17'45.0" $\pm$ 15"

Angle AOC = 82°50'03.9" $\pm$ 15.8"

$$(= \pm\sqrt{05^2 + 15^2})$$

3-20-2. Subtraction

The same rule applies as in addition, readily understood by considering subtraction as negative addition: similar to Equation (3-10).

Example 3-19. If a known line AB is 1867.857 $\pm$ 0.018 ft long and a segment AP on the line is taped to be 195.009 $\pm$ 0.010 ft, then the remaining segment PB is

$$1672.848 \pm (\sqrt{0.018^2 + 0.010^2} = \pm 0.021) \text{ ft}$$

3-20-3. Multiplication

When multiplying two or more measured quantities having accidental errors, this general equation for relative errors applies

$$E_{product} = \pm A \cdot B \sqrt{\left(\frac{E_A}{A}\right)^2 + \left(\frac{E_B}{B}\right)^2} \quad (3\text{-}11)$$

Example 3-20. A field measured as 160.881 ± 0.026 ft long by 75.007 ± 0.001 ft wide has an error in its area $(12,067.20 \text{ ft}^2)$ of

$$\pm 12,067.20 \sqrt{\left(\frac{0.026}{160.881}\right)^2 + \left(\frac{0.011}{75.007}\right)^2}$$

$$= \pm 2.63 \text{ ft}^2$$

3-20-4. Division

The error in a quotient of two measured quantities is

$$E_{quotient} = \pm \frac{A}{B} \sqrt{\left(\frac{E_A}{A}\right)^2 + \left(\frac{E_B}{B}\right)^2} \quad (3\text{-}12)$$

Example 3-21. If the area of the rectangular plot is somehow known to be $49,650 \pm 10$ ft^2 and the width dimension measured several times is found to be 175.62 ± 0.46 ft, the calculated length dimension is 282.72 ± 0.74 ft, since the error is

$$\pm 282.72 \sqrt{\left(\frac{10}{49,650}\right)^2 + \left(\frac{0.46}{175.62}\right)^2}$$

$$= \pm 0.74 \text{ ft}$$

3-20-5. Other Operations

The volume of a rectangular tank or bin whose three dimensions are measured has an error

$$E_{volume} = \pm L \cdot W \cdot H$$

$$\times \sqrt{\left(\frac{E_L}{L}\right)^2 + \left(\frac{E_W}{W}\right)^2 + \left(\frac{E_H}{H}\right)^2} \quad (3\text{-}13)$$

A cube, if all three sides are measured, will follow the same pattern. However, if only one

side A is measured and the other dimensions are regarded as equal, the volume error is

$$E_{volume} = \pm A^3 \sqrt{\left(\frac{E_A}{A}\right)^2 \times 3}$$

$$= \pm A^2 E_A \sqrt{3} \quad (3\text{-}14)$$

The general form for error in raising to a power any quantity containing an error is

$$E_{power} = \pm A^n \sqrt{n \left(\frac{E_A}{A}\right)^2}$$

$$= \pm E_A \cdot A^{(n-1)} \sqrt{n} \quad (3\text{-}15)$$

The volume of a measured sphere, because only one dimension is measured, follows the power rule just given.

$$\text{Volume of sphere} = \tfrac{4}{3}\pi r^3$$

$$E_{sphere} = \pm (\text{vol}) \sqrt{n \left(\frac{E_r}{r}\right)^2}$$

$$= \pm \frac{4\sqrt{3}}{3} \pi r^2 E_r \quad (3\text{-}16)$$

Example 3-22. If a sphere's radius is measured as 10.00 ± 0.08 ft, the calculated volume is 4188.8 ft^3 and the error will be

$$E_{vol} = \pm \frac{4\pi}{3} 10^3 \sqrt{3\left(\frac{0.08}{10}\right)^2} = \pm 58.0 \text{ ft}^3$$

The error in volume of a cylindrical tank can be found by first analyzing the area of the circular end, then working with it, the length, and the errors in both.

Example 3-23. A cylindrical tank of diameter 10.00 ± 0.02 ft is 30.00 ± 0.04 ft long. End area is $\tfrac{1}{4}\pi D^2 = 78.540$ ft^2 and volume is $78.540 \times 30.00 = 2356.2$ ft^3.

$$E_{area} = \pm \tfrac{1}{4}\pi 10^2 \sqrt{2\left(\frac{0.02}{10}\right)^2} = \pm 0.222 \text{ ft}^2$$

$$E_{vol} = \pm (30.00)(78.540)$$

$$\times \sqrt{\left(\frac{0.222}{78.540}\right)^2 + \left(\frac{0.04}{30.00}\right)^2}$$

$$= \pm 7.36 \text{ ft}^3$$

The area of a right triangle whose altitude and leg are both measured has an area error computed similarly.

Example 3-24.   The sides of a right triangle are measured as 100.000 ± 0.021 and 35.000 ± 0.012 ft. Area is $\frac{1}{2}ab = 1750.00$ ft^2.

$$E_{\text{area}} = \pm\frac{1}{2}ab\sqrt{2\left(\frac{E_a}{a}\right)^2 + \left(\frac{E_b}{b}\right)^2}$$

$$= \pm 1750\sqrt{\left(\frac{0.021}{100}\right)^2 + \left(\frac{0.012}{35}\right)^2}$$

$$= \pm 0.70 \text{ ft}^2 \tag{3-17}$$

3-21.   AREA OF A TRAVERSE

The error in area of a closed traverse can be found similarly, using the *relative accuracy* concept. If the traversing procedure prescribed for, say, a 1/5000 accuracy is followed, the area is obtained by multiplying total latitude by meridian distances. If we assume that a traverse area already obtained is 1,062,323 ft^2 and the sum of the latitudes equals 1867.812 ft, the meridian distances total

$$1,062,323/1867.812 = 568.753 \text{ ft}$$

The error ascribable to each factor is

$$E_{\text{latitude}} = \pm 1,967,812/5,000 = \pm 0.374 \text{ ft}$$

$$E_{\text{meridian}} = \pm 568.763/5,000 = \pm 0.114 \text{ ft}$$

$$E_{\text{area}} = \pm 1,062,323\sqrt{\left(\frac{0.374}{1,868}\right)^2 + \left(\frac{0.114}{569}\right)^2}$$

$$= \pm 300.9 \text{ ft}^2$$

It may be noted that this is the same as

$$E_{\text{area}} = \pm(\text{area})\sqrt{\left(\frac{1}{5000}\right)^2 + \left(\frac{1}{5000}\right)^2}$$

$$= \pm 300.5 \text{ ft}^2$$

or, more simply,

$$E_{\text{area}} = \pm(\text{area})(1/5000)\sqrt{2} = \pm 300.5 \text{ ft}^2$$

Had the traverse been run to 1/10,000 accuracy,

$$E_{\text{area}} = \pm(\text{area})(1/10,000)\sqrt{2} = \pm 150.2 \text{ ft}^2$$

Common traverse relationships between linear and area accuracies are given in Table 3-5.

3-22.   ERRORS AND WEIGHTS

Results obtained from different measurements of the same quality can be combined to find a *weighted mean* by assigning proportionately greater weights to measurements, or measurement sets, that have smaller standard deviations σ_s or smaller standard errors σ_m. Weights should be inversely proportional to the square of the sigma quantities.

Example 3-25.   Several sets of linear measurement are listed. It is desired to find the combined weighted mean.

Table 3-5.   Summary table of traverse area error

Order/Class	Linear Accuracy	Area Accuracy
First	1/100,000	$(1/100,000)\sqrt{2} = 1/70,700$
Second/I	1/50,000	$(1/50,000)\sqrt{2} = 1/35,300$
Second/II	1/20,000	$(1/20,000)\sqrt{2} = 1/14,100$
Third/I	1/10,000	$(1/10,000)\sqrt{2} = 1/7070$
Third/II	1/5000	$(1/5000)\sqrt{2} = 1/3530$

Set	X	σ_s	Weight Ratio	Weight w	$w \cdot (X - 387)$
A	387.071	± 0.025	$(1/0.025)^2 = 1,600$	5.38	0.382
B	387.126	± 0.042	$(1/0.042)^2 = 567$	1.91	0.240
C	387.080	± 0.019	$(1/0.019)^2 = 2,770$	9.32	0.746
D	387.112	± 0.058	$(1/0.058)^2 = 297$	1.00	0.112
				17.61	1.480

Weighted mean = $387.000 + 1.480/17.61 = 387.084$ ft

Example 3-26. A weighted mean value of elevation difference between two bench marks can be calculated if a set of level runs by one party is combined with that of another party, each set being the result of several runs having a calculated standard deviation.

Set	X	σ_s	Weight Ratio	w	$w(X - 166)$
A	167.212	± 0.182	$(1/0.182)^2 = 30.190$	1.000	1.212
B	166.978	± 0.071	$(1/0.071)^2 = 198.373$	6.571	6.380
				7.571	7.592

Weighted mean = $166.000 + 7.592/7.571 = 167.003$ ft.
(The unweighted mean would be 167.095 ft.)

Example 3-27g. In differential leveling, one party made a run using second-order/class I methods, another followed the same route using third-order/class II methods. The weighted mean of the two results is found through the published relative accuracy values.

Set	Difference of Elevation (m)	Order/ Class	Relative Accuracy	Weight Ratio	w	$w(X - 41)$
Q	41.0962	2nd/I	± 1.0 mm$\sqrt{K}$	$\left(\dfrac{1}{1.0K}\right)^2 = \dfrac{1}{K}$	1.69	0.1626
R	41.1076	3rd/II	± 1.3 mm$\sqrt{K}$	$\left(\dfrac{1}{1.3K}\right)^2 = \dfrac{0.59}{K}$	1.00	0.1076
					2.69	0.2702

Weighted mean = $41.0000 + 0.2707/2.69 = 41.1004$ m. (The unweighted mean would be 41.1019 m.)

Example 3-28. A bench mark is to be established using three different level runs following varied procedures and contrasting specifications. A weighted mean is required, with weights fixed inversely proportional to the square of the published relative errors. Distances in kilometers are shown.

Run	Elevation m	k	Order/Class	Relative Error e	Weight Ratio $1/e^2$
A	46.1672	1.5	2nd/I	± 1.0 mm$\sqrt{1.5}$ = ± 1.225	0.667
B	46.2107	2.0	3rd	± 2.0 mm$\sqrt{2.0}$ = ± 2.828	0.125
C	46.1810	1.5	2nd/II	± 1.3 mm$\sqrt{1.5}$ = ± 1.592	0.395

Run	Weight	Weight (Elevation − 46)
A	5.336	0.8922
B	1.000	0.2107
C	3.160	0.5720
	9.496	1.6759

Weighted mean = 46.0000 + 1.6759/9.496 = 46.1765 m

3-23. CORRECTIONS

Corrections for measured quantities should be inversely proportional to their weights, directly proportional to the squares of accidental errors or standard deviations or standard errors.

Example 3-29. Angles of a triangle were measured with different instruments. Lacking a better guide, errors assigned to each point are the nominal capabilities of different instruments used at these points.

Point		Instrument Type	E	E^2	Correction Index	Correction	Adjusted X
A	70°13′50″	10-sec	$\pm 10''$	100	1	1/41(40) = 01″	70°13′49″
B	58°45′20″	20-sec	$\pm 20''$	400	4	4/41(40) = 04″	58°45′16″
C	51°01′30″	1-min	$\pm 60''$	3600	36	36/41(40) = 35″	51°00′55″
	180°00′40″				41		180°00′00″

Example 3-30. A line *AB* is measured by first-order methods, and then again in two segments *AC* and *CB* by third-order methods. Adjusted values are needed for all three measured distances. (The value of *AB* must obviously equal the sum of *AC* plus *CB*, finally.)

Line	X	Precision Fraction	e^*	$1/e^2$	w	$w(X - 2104)$
AC	1416.912					
CB	687.901					
	2104.813	1/5000	0.421	5.643	1	0.813
AB	2104.697	1/20,000	0.105	90.299	16	11.152
Diff =	0.116				17	11.965

*$(1/5000)(2104.813) = 0.421$
$(1/20,000)(2104.697) = 0.105$
Weighted mean $= 2104 + 11.965/17 = 2104.704$

Since $(AC + BC)$ must be adjusted downward by $(2104.813 - 2104.704 = 0.109)$ to equal this weighted mean, the two segments (proportionately) are

$$AC = 1416.912 - (1417/2105)(0.109)$$

$$= 1416.839$$

$$CB = 687.901 - (688/2105)(0.109)$$

$$= 687.865$$

Sum $= 2104.704$ (check)

The weight ratios could be obtained directly from the expressions of precision; thus,

Constant $\times e^2$	Weight Ratio ($= 1/e^2$)
Inverse of $(1/5000)^2 = 40.0 \times 10^{-9}$	25×10^6 or 1
Inverse of $(1/20,000)^2 = 2.5 \times 10^{-9}$	400×10^6 or 16

Another way to arrive at these new values is as follows, noting that adjustments are made in inverse ratio to the weights:

Segment	e	$1/e^2$	w	Adjustment Ratio	Adjustment*
AC + CB	0.421	5.643	1	$1/1 = 1.0000$	0.109
AB	0.105	90.70	16	$1/16 = 1.0625$	0.007
				1.0625	0.116

*Adjustment calculations:
   $(1/1.0625)(0.116) = 0.109$
$(0.0625/1.0625)(0.116) = 0.007$
              Sum $= 0.116$

3-24. SIGNIFICANT FIGURES IN MEASUREMENTS

3-24-1. Exact Versus Doubtful Figures

When a measurement is made, all digits in the result are exact if they are obtained by counting or finding a point that lies between two markers. Digits are doubtful when they result from estimating. *Significant figures* include all exact digits, plus a doubtful one. Several rules and conventions will be given.

3-24-2. Use of Zero

A zero is not significant when it serves merely to place the decimal.

1. 0.00584 contains three significant figures, seen more clearly if written as 5.84×10^{-3}.

2. 34,000 mi has two significant figures, unless it is clearly intended that the value is exact. Writing it 3.4×10^4 is not common, but is better usage (scientific form), or as 34×10^3 (engineering form).

When used otherwise, a zero is significant.

3. 4.6007, 9,1030, or 4100.0 each has five significant figures.

4. 0.076130 also has five significant figures and could well be written 7.6130×10^{-2} or 76.130×10^{-3}.

3-24-3. Rules of Thumb for Significant Figures

The following are reasonable rules and conventions:

1. Unless some precision indicator is affixed—e.g., standard deviation—the usual interpretation for the last (doubtful) digit is plus or minus one-half a unit in the last column. Thus a measured length of 81.713 means the range of uncertainty extends from 81.7125 to 81.7135 units.

2. Although use of only one doubtful figure in the final result is anticipated, it is desirable to use two doubtful digits throughout the calculation and round off only at the end.

3. Adding and subtracting several measured values limit the result to show no more than the least valid item. The following examples are obvious:

6.27	56.17	
4.3	11.036	367.796
13.876	79.3015	−28.7
24.4	146.51	339.1

4. In multiplying or dividing, the result must not be credited with more significant digits than appear in the term with the smallest number of significant figures, as shown here:

$$6.7153 \times 4.67 = 31.4 \quad (\text{not } 31.360451)$$

$$(86.85 \times 10^4)^2$$
$$= 754.3 \times 10^9 \quad (\text{not } 754.29225 \times 10^9)$$
$$850.436/4.56 = 186 \quad (\text{not } 186.499123)$$

However, 8 and 9 are almost two-digit numbers, and occasionally an extra digit in the product is warranted; thus,

$$9.703 \times 4.07062 = 39.497$$

$$(\text{instead of just } 39.50)$$

5. Calculators sometimes convey a false sense of precision, so care must be taken to cut back and properly round off the final result, like this

$$\tfrac{1}{2}(87.645 \times 8.6305)$$

$$= 756.42 \text{ or } 756.420 \quad (\text{not } 756.4201725)$$

3-24-4. Rounding Off

When dropping excess digits, raise the last one and retain it if the discarded quantity is greater than one-half, or leave it unchanged if the discarded quantity is smaller than one-half; thus,

4.796 becomes 4.80 or 4.8 or 5

8.512 becomes 8.51 or 8.5 or 9

If the quantity to be discarded is exactly 5, round off the preceding digit to the nearest even value; thus,

> 10.675 becomes 10.68 or 10.7
> 10.685 becomes 10.68 or 10.7
> 10.695 becomes 10.70 or 10.7
> 10.705 becomes 10.70 or 10.7
> 10.6749 becomes 10.67, and
> 10.6751 becomes 10.68

3-24-5. Using Exact Values

The procedures described apply to quantities resulting from measurements. If exact values are implied in a statement—e.g., a 2000-ft radius curve—the number of significant digits is not limited. This and others are stated and discussed here.

A field of 89,102.6 ft^2 is properly converted to acres using the following exact conversion factor:

89,102.6/43,560

> = 2.04551 acres   (but not 2.04551423)

If a 2000-ft radius is specified for a circular curve, this sets the degree of curve D at

$$\frac{5729.577951}{2000 \text{ (exact)}} = 2.864788976°$$

> or   2.8647890 to 8 digits

This comes from a ratio in the circle

$$\frac{D \text{ (degree of curve)}}{360°} = \frac{100 \text{ ft}}{2\pi R \text{ ft}}$$

> or $R = 5729.577951/D$

With a good calculator, the value of π is given to eight, nine, or 10 valid digits, and the 2000 ft is implied to be equipped with an endless row of exact digits (zeros). Preserving these digits is necessary to find the length of a circular curve. Assuming an intersection angle of $43°47'34'' = 43.79278°$ and a 2000-ft radius, we obtain

$$L_c = 43.79278/2.864789$$

$$= 15 + 28.656 \quad \text{(stationing designation)}$$

This requires seven digits, minimally, and inexactness can readily occur if the two angular values are carelessly truncated early on.

REFERENCES

BARRY, B. A. 1978 *Errors in Practical Measurement in Science, Engineering, and Technology*. Rancho Cordova, CA: Landmark Enterprises.

BUCKNER, R. B. 1983 *Surveying Measurements and Their Analysis*. Rancho Cordova, CA: Landmark Enterprises.

CRANDALL, K. C., and R. W. SEABLOOM. 1970 *Engineering Fundamentals in Measurements, Probability, Statistics, and Dimensions*. New York: McGraw-Hill.

4

Linear Measurements

Kenneth S. Curtis

4-1. INTRODUCTION

Surveyors are fundamentally concerned with the measurement of horizontal and vertical distances and angles and, more recently, with direct positioning. These, then, are used in various combinations in traversing, triangulation, trilateration, mixed-mode operations, mapping, layout staking, leveling, etc.

Linear distance measurement can be achieved by (1) *direct comparison measurement* with a tape, either fully supported on the ground or suspended in catenary; (2) *optical distance-measurement methods* by remote angular observation on a variable- or fixed-base length held horizontal or vertical, such as in tacheometry, stadia, or subtensing; and (3) *electromagnetic distance instruments* utilizing the travel time of radio or light waves converted to distance. This chapter covers direct tape measurement and optical distance measurement. Chapter 5 discusses electromagnetic distance-measuring instruments (EDMIs).

To a surveyor, the word distance usually refers to the horizontal length between two points projected onto a horizontal plane. Many measuring devices yield slope distances, which must be converted to horizontal. Maps and land areas are based on horizontal measurements or dimensions. Whether to measure a distance by pacing, taping, stadia, or with a

highly accurate EDMI depends on the accuracy required, and this, in turn, depends on the purpose of the measurement. Only a person thoroughly familiar with all types of measuring techniques can choose the optimum and most cost-effective procedure.

4-2. UNITS OF LINEAR MEASUREMENT

Several methods are used to measure distances. They range from rather inaccurate estimates to very precise instrumental procedures. Most early measurement units were derived from physical dimensions associated with parts of the human body. For example, the cubit, digit, palm, hand, span, foot, yard, pace, and fathom can be traced to human anatomy. Many others, such as the rod, pole, perch, chain, furlong, mile, and league, are extensions of these basic units (Table 4-1). Three barleycorns laid lengthwise equaled one inch, 12 of them equaled one foot. Many old units have been discarded in favor of the basic ones, foot and meter. Much of the world has now converted to the meter-decimal system (SI units) as illustrated in Table 4-2. Numerous English-related countries, such as the United States, remain slow to completely convert from the foot (English) system.

Table 4-1. Units of length

Unit	Inches (in.)	Feet (ft)	Yards (yd)	Rods (rd)	Chains (ch)	Meters (m)
1 inch	1	0.08333	0.02778	0.00505	0.00126	0.02540
1 foot	12	1	0.3333	0.0606	0.01515	0.3048
1 yard	36	3	1	0.1818	0.04545	0.9144
1 rod	198	16.5	5.5	1	0.25	5.0292
1 chain	792	66	22	4	1	20.1158
1 mile	63,360	5280	1760	320	80	1609.35
1 meter	39.37	3.281	1.094	0.199	0.04971	1

The primary standard of length in the United States, the National Prototype Meter 27, a 90% platinum and 10% iridium bar, is housed at the National Bureau of Standards (NBS) in Gaithersburg, MD. It is identical in form and material with the International Prototype Meter deposited at the International Bureau of Weights and Measures at Sevres, France, and also with other national prototype meters distributed in 1889, in accordance with a treaty known as the Convention of the Meter, dated May 20, 1875. The meter was originally conceived as 1/10,000,000 part of a meridional quadrant of the earth.

In 1960, the official definition of the meter was redefined as a length equal to 1,650,763.73 wavelengths in a vacuum of the radiation of the orange-red light of a krypton-86 atom. The International Prototype bar was abrogated in favor of a natural and indestructible standard thought to have an accuracy adequate for metrology's modern needs. However, in 1983, the meter was redefined again as the distance traveled by light in a vacuum during 1/229,792,458 sec. It is claimed that this new definition allows the meter to be defined 10 times more accurately and achieves the goal of using time, the most accurate basic measurement, to define length.

In the United States, since 1893, the yard has been defined in terms of the meter by the following relations: 1 yd = 3600/3937 m or 1 m equals 39.37 in. exactly. This legal ratio is used to define the "U.S. Survey Foot." In 1959, after several years of discussion, the United Kingdom and United States agreed to establish a new uniform relationship between the yard and meter as

1 yard = 0.9144 meter exactly

or 1 foot = 0.3048 meter   (international foot)

or 1 inch = 25.4 millimeter

Since the new value of the yard is smaller by two parts per million than the 1893 yard, only in large-scale geodetic survey data is the difference important.

The land-surveying profession is uniquely sensitive to metric system usage because many problems can arise in converting recorded

Table 4-2. Metric units of length

Unit	Micrometers (μm)	Millimeters (mm)	Centimeters (cm)	Decimeters (dm)	Meters (m)	Kilometers (km)
1 micrometer	1	0.001	0.0001			
1 millimeter	1000	1	0.1	0.01	0.001	
1 centimeter	10,000	10	1	0.1	0.01	0.00001
1 decimeter	100,000	100	10	1	0.1	0.0001
1 meter	1,000,000	1000	100	10	1	0.001
1 kilometer			100,000	10,000	1000	1

measurements from nonmetric units to SI units. It may appear quite simple: Just use a 30-m tape instead of a 100-ft tape. However, over several hundred years, land descriptions have been recorded in the English system of units or the Spanish vara (Table 4-3), and future generations may never completely get away from these historical units.

For a fee, the metrology division of the NBS provides calibration of line standards of length and measuring tapes. Using Invar base-line tapes and highly accurate electronic distance-measuring instruments, the National Geodetic Survey (NGS) has established nearly 200 calibration base lines (CBLs) across the United States, providing a means to detect constant and scale errors in measuring instruments. A typical CBL layout consists of four monuments located at 0, 150, 430, and 1400 m, relatively positioned with accuracies approaching one part per million.

4-3. DIRECT COMPARISON DISTANCE-MEASUREMENT METHODS

4-3-1. Pacing

Rough estimates of distances can be made by eye, based on experience in observing commonly used dimensions, such as 100 ft, 100 yd, or a city block. Distances can also be scaled from using a variety of map measures, maps, or aerial photographs.

A better approximation is obtained by walking the distance with a natural or artificial *pace*. The length of a human pace varies and few can develop a 3-ft artificial place to mea-

Table 4-3. Units used in land measurements

1 Gunter's chain (100 links) = 66 ft = 4 rods or poles or perches
80 Gunter's chains = 1 mile
1 vara = 32.993 in. in Mexico, 33 in. in California, and $33\frac{1}{2}$ in. in Texas

sure distances, without creating fatigue. Therefore, it is best to determine your natural pace by walking over a course of known length —300 to 900 ft—several times, to standardize the pace.

Many factors can affect pace length, including slope and roughness of the terrain, shoe weight and clothing type, pacing speed, fatigue, and age. In addition, it is difficult to keep count of the steps. Sometimes, strides (two paces) are counted using a notched stick or mechanical tally register. A *pedometer* strapped to the pacer will automatically record the distance covered in miles after it has been adjusted to the wearer's pace. A similar instrument, called a *passometer*, automatically records the number of paces.

With a little practice, a good pacer can attain results within 1% of the true distance (1/100). No special equipment is required for its many practical applications, one of which is detection of blunders that can occur in taping or in other more accurate distance-measuring procedures.

4-3-2. Odometer or Measuring Wheel

The *odometer* is a device similar to the distance recorder in an automobile speedometer. It is attached to a wheel of known circumference and rolled over the distance to be measured. Results obtained depend on the topography and surface irregularities, but on level smooth ground may yield 1/200 accuracy. Measuring wheels serve as rough checks on more accurate measurements and can be useful in reconnaissance and preliminary surveys. Wheels of 2, 4, or 6 ft in circumference are most popular. They include precision totalizers or counters, which can be reset; one model has a battery-powered electronic totalizer and an LCD counter (Figure 4-1). Some measuring wheels can be attached to a vehicle with a rear-mounted hitch, allowing longer distances to be measured while moving at speeds of up to 8 mph.

Figure 4-1. Measuring wheel. (Courtesy of Rolatape Corporation.)

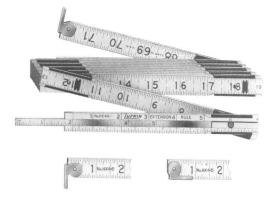

Figure 4-2. Folding rule and short power hardware tape. (Courtesy of Lufkin and the Cooper Group.)

4-3-3. Folding Rules and Hardware Tapes

Folding rules and short power tapes, sometimes referred to as hardware tapes, are variously made, differently graduated (Figure 4-2), and used on all types of building construction sites. A rule, being stiff, can be held in any position desired whereas a tape is flexible and generally needs two people for measuring. Some tapes are graduated in feet, inches, and eighth-inch (or sixteenth-inch), or feet, inches, and decimals. Except on building construction, surveyors generally ignore inches and work in feet and decimals (tenths and hundredths). Since some tapes are also graduated in metric units, surveyors need to carefully check the units before using any tape.

4-3-4. Woven or Fiberglass Tapes

The flexibility of *woven* or *fiberglass tapes* makes them extremely effective under many conditions where steel tapes are impractical.

However, due to moisture and temperature, all woven tapes are liable to shrink or stretch and frequently should be compared with steel tapes to determine their accuracy and actual measuring length. Woven tapes, usually 50 to 150 ft long, are a combination of dacron fibers and coatings that have the stability of fiberglass and flexible strength of polyester. They feature high dielectric strength for safety on construction sites near high-tension circuits. One maker of woven nonmetallic tapes reinforces the first 9 in. of line with green plastic—green indicating nonconductivity. Short tapes are normally enclosed in a case. Some cloth tapes have fine metal strands of wire woven lengthwise into their fabric and are truly metallic tapes. They should not be used around electrical units.

A relatively new fiberglass tape, made of thousands of strands of glass fibers coated with polyvinyl chloride, is flexible, strong, nonconductive, and will not need a temperature correction (Figure 4-3). Under normal use, with tension lower than 5 lb, a correction to compensate for elasticity is seldom required. When greater tension is applied, some small corrections are needed; e.g., 0.02 in. per 3 ft at 11 lb and 0.04 in. per 3 ft at 22 lb. These tapes are available in lengths of 50 to 300 ft in a metal case or an on open-type reel. They are practical for locating details in mapping or checking reference distances.

4-3-5. Steel and Invar Tapes

The most common tapes used in surveying practice are steel ribbons (*band chains*) of constant cross section, usually varying in width from $\frac{1}{4}$ to $\frac{3}{8}$ in. and in thickness from 0.008 to 0.025 in. Normal lengths are 100, 200, or 300 ft. Metric tapes of the same thickness and width usually are 30, 50, 60, or 100 m long (Figure 4-4).

Graduations and identifying numbers are stamped either on soft (babbitt) metal previously embossed at the tape divisions or etched in the tape metal. Riveted, heavy-plated brass end-clips or rings provide a place to attach leather thongs, tension handles, or hooks to allow one person to make measurements unassisted (Figure 4-3). Most steel tapes come on reels and are stored on them. If a reel proves awkward, the tape can be removed from it and, when not in use, wound up into 5 ft-loops

to form a figure 8, and then "thrown" into a circle about 8 in. in diameter. The common 100-ft steel tape weighs from $1\frac{1}{2}$ to $2\frac{1}{2}$ lb, depending on thickness and width. If a steel tape gets wet, it should be wiped dry with a cloth and again with an oily cloth. Steel tapes are quite rugged, but if tightened with kinks in them, they break rather easily.

Tapes are marked in many ways to satisfy user desires. For example, some tapes have the last foot of each end divided into decimals, but others have an extra subdivided foot added to the zero end. Tapes with an extra foot are called *add tapes;* those without an extra foot are termed *cut tapes.* The latter type is becoming extinct because the subtraction required for each measurement is a possible source of error. Several variations are available such as divisions subdivided through their entire length. Others have zero points about $\frac{1}{2}$ ft from the end, or both end points at the outer edges of the end loops instead of being on a line itself. Before using them, surveyors must be completely familiar with the divisions and markings of all tapes.

Steel tapes expand or contract due to changes in temperature. Nickel-steel alloy tapes, known as *Invar* (which has a coefficient of thermal expansion about $\frac{1}{30}$ that of steel), *Lovar*, and *Minvar*, are used in high-precision surveying on geodetic base lines and as a standard of comparison for other working

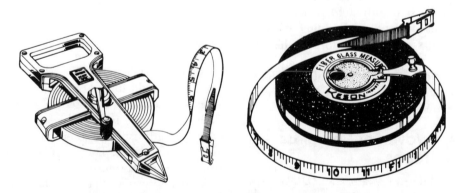

Figure 4-3. Fiberglass measuring tapes (in an open reel case and a metal case). (Courtesy of Keson Industries, Inc.)

Figure 4-4. Steel tape on open reel and Invar tape in wooden case. (Courtesy of Lufkin and the Cooper Group.)

tapes. They are almost always wound on an oak plywood reel (Figure 4-4). The nickel-steel alloy tapes are relatively insensitive to temperature, but the metal is soft, somewhat unstable, easily broken, and their cost is perhaps 10 times that of ordinary tapes.

Some tape equipment companies offer the option of a graduated thermometer scale, which corresponds to the tape contraction and expansion, as a variable terminal mark of the tape. The distance measured then depends on the prevailing temperature. Also, there are separate 6-in. wooden rules graduated with temperature corrections for 50- and 100-ft steel tapes.

Another handy device is the *topographic trailer tape* (Figure 4-5), used in conjunction with an *Abney hand level* to obtain horizontal distances by measuring along slopes. The tape, approximately $2\frac{1}{2}$ ch long, is basically 2 ch plus a distance on the trailer equal to the number of graduations indicated by the topographic arc reading. The total length thus measured on a slope equals a horizontal dis-

tance of 2 ch (132 ft). If the same procedure is carried out by reading the tape's reverse side, the distance is 1 ch (66 ft). It is a perfect tape for surveyors, foresters, and mappers to get slope corrections.

A device of historical importance in the United States is the *Gunter's chain* (Figure 4-6), which had extensive use in land surveying and the public land surveys during the 1700s and 1800s. This basic chain was 66 ft long and divided into 100 parts or links. Each link was equal to 0.66 ft or 7.92 in. and made of heavy

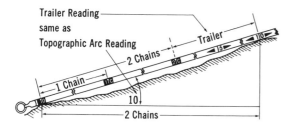

Figure 4-5. Topographic trailer tape. (Courtesy of Keufel & Esser Co.)

Figure 4-6. Gunter's chain. (Courtesy of Keuffel & Esser Co.)

wires connected by loops and three connecting rings with end handles. Intermediate tags of various design identified every tenth link. The handles had a length adjustment feature to compensate for chain lengthening due to wear on the 600 to 800 connecting wearing surfaces. Distances were recorded in chains and links or chains and decimals; e.g., 20 ch 12.4 lk or 20.124 ch. The Gunter's chain, 66 ft long, therefore, was $\frac{1}{80}$ of a mile (4 rods). Ten chains square is equal to 43,560 ft^2 or 1 acre —a very useful system.

Subsequently, an engineer's chain of 100 ft with 100 1-ft links was developed. Chains were replaced by development of the 100-ft steel tape. However, the chain unit remains a fundamental part of land-surveying practice, and a steel tape graduated in chain units is available. The term *chaining* continues to be used interchangeably with *taping*, even though a tape is used in the measurement.

4-3-6. Taping Accessories

To measure distances accurately with a tape, a number of so-called accessories are necessary or desirable. Although wooden stakes and tacks probably provide greater accuracy, *chaining pins* or *taping arrows* generally are used to mark the tape ends or intermediate points on the ground. They are also helpful as tallies to count the number of tape lengths in a given line. The pins, made of heavy steel wire, are usually 14 in. long, pointed at one end with a round loop at the other, and brightly painted with alternate red and white bands. A standard set consists of 11 pins on a steel ring or in a leather *quiver*, which can be attached to a surveyor's belt. After 1000 ft, the rear tapeperson is holding 10 pins if standard procedures are followed.

Since steel tapes are calibrated to measure correctly when under a definite tension, for precise measurements a *tension handle* or *spring balance* is attached to one end of the tape and a desired tension or pull applied. They are also used to counteract the effect of *sag* when measuring without a fully supported tape. The usual spring balance reads up to 30 lb in $\frac{1}{2}$-lb increments or 15 kg in $\frac{1}{4}$-kg calibrations. Without a tension handle, tapepersons have to estimate the proper pull.

Since tension must always be applied to a tape, especially at intermediate points, wrapping it around one hand is not recommended. Instead, a tape *clamping handle* should be used, permitting tension to be applied by a scissors-type grip, which does not slip or damage the tape. Without using a clamp, the tape could be slightly bent and kinked. Kinks, once introduced, cannot be entirely straightened out and create weak spots where future fracture will likely occur.

Plumb bobs are employed to place the tape directly over a point when the tape must be suspended above it. The commonly employed plumb bob is a fine quality, accurately centered brass bob that comes in various sizes, varying in weight from 6 to 18 oz, with a fine hardened steel point. All bobs provide for attaching a cord to the top; some carry replacement steel points in the bob. For convenience, plumb bobs are normally carried in a sewn leather sheath and sometimes fastened to a *gammon reel*, which provides instant rewind of the plumb-bob string, up and down adjustment, and an accurate sighting target (Figure 4-7).

Range poles are used to mark ground point locations and the direction of a line on which taping must proceed. They normally are 1 to $1\frac{1}{4}$ in. in diameter, either round, octagonal, or with deep corrugations to diffuse surface glare. Usually, they are 6 to 8 ft long or in multiple sections totaling perhaps 12 ft and equipped with a steel pointed shoe and shank. Made of wood, metal, or fiberglass, they are alternatively painted red and white in 1-ft, or 50-cm, sections and can be used for rough measurements. They are not javelins and should not be used to loosen rocks or stakes. Some equipment companies offer a short tripod to support range poles.

When striving for high accuracy, *pocket thermometers* are used to obtain air temperature and, it is hoped, provide an adequate estimate

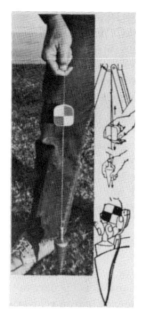

Figure 4-7. Plumb bob fastened to a gammon reel. (Courtesy of Lietz Co.)

of the tape temperature during measurement. A common type is 5 in. long with a scale reading to 2°F (from −30°F to +120°F), carried in a protective metal case with pocket clip. There are also *tape thermometers* available that can be easily fastened directly to the tape and should ensure a more accurate tape temperature. Several years ago, a company offered a *tape temperature corrector*, a thermometer mounted on the standards of a transit. It had a scale in decimals of a foot for temperature corrections to be applied to a 100-ft steel tape. After reading the thermometer, the tapeperson can measure long or short as indicated.

Tape-repair kits contain sleeve splices, a combination hand punch and splicing tool, and eyelet rivets to facilitate field repairs of broken tapes. A simple and rapid method for emergency repairs is also available, consisting of sheet-metal sleeves coated with solder and flux to be fitted over the broken ends and hammered down tightly. Then, using heat from a match, the tape is securely fastened together. Repaired tapes should be used only on rough work.

To keep the tape ends at equal elevations when measuring over rough or sloping terrain, a simple *hand level* is used. It consists of a bubble mounted on a metal sighting tube and reflected by a 45° mirror or prism into the tube so the bubble can be observed at the same time as the terrain. If the slope angle or percentage of grade is desired, an Abney-type hand level or clinometer with a graduated vertical arc attached to its side is available (Figure 4-8).

Other taping accessories include detachable tape-end hooks with serrated face grips; tape-end leather thongs, tape rings or tape handles; and various size reels sometimes sold separately from the tapes.

4-3-7. Taping Procedures

Taping techniques and procedures vary because of differences that exist in kinds of tapes available, terrain traversed, project requirements, long-established practices, and personal preference. Regardless of the methods used, surveyors must be masters of their operations and fully understand the consequences of different techniques. Tapes are employed for two fundamental measurements: (1) to measure the distance between two existing physical points; or (2) to lay out and mark a distance called for in plans, specifications, land descriptions, or plats.

As pointed out earlier, a horizontal distance is the desired result. There are three basic methods of taping: (1) *horizontal taping*, (2) *slope taping*, and (3) *dynamic taping* (Figure 4-9). In horizontal taping, the tape is held

Figure 4-8. Abney-type hand level and clinometer. (Courtesy of Keuffel & Esser Co.)

horizontally and the end or intermediate points transferred to the ground or other surface. In slope taping, the tape length, or a portion thereof, is marked on the inclined surface supporting it, the slope determined, and the corresponding horizontal distance computed. In dynamic taping, a sloping taped distance from a transit or theodolite spindle is measured along with the vertical angle, and the horizontal distance subsequently calculated.

In all operations, careful attention must be given to proper tape support, proper alignment, use of correct tension, skill in handling plumb bobs and placement of pins, accounting for temperature variations, and any other factors that might affect accuracy of the result.

4-3-8. Taping over Level Ground

If we assume a level line void of tall grass and underbrush, the 100-ft tape can be laid on the ground, thus fully supported throughout. Under these conditions, the proper tension, generally applied, is about 10 lb. If the ground is irregular or contains obstructions, the tape may have to be suspended in catenary, plumb bobs used, and more tension applied.

To ensure a measurement is kept on line, range poles are placed at the terminal points and sometimes at intermediate ones, depending on terrain and visibility. The rear tapeperson keeps the head tapeperson on line by eye using hand signals or voice communication. If more accuracy is desired, a transit or theodolite is used to maintain alignment.

Taping can be accomplished by two people: a rear tapeperson and head tapeperson. The head tapeperson carries the zero end of the tape forward, aligns it, and after a 10-lb pull is applied, marks the distance by sticking a taping pin at right angles to the tape but at approximately a 45° angle with the ground. This permits, if necessary, more exact plumbing over the point with a transit or taping bob. As the measurement progresses, the taping pins are collected by the rear tapeperson and used as tallies to keep track of the full tape lengths. Careful attention should also be given to marking the zero and terminal points on a tape, which may be graduated in different ways. This is particularly important for fractional tape lengths.

When the end of a line is reached, the head tapeperson stops and the rear tapeperson moves up to the last pin set. The tape is moved until a full foot mark is opposite the pin and

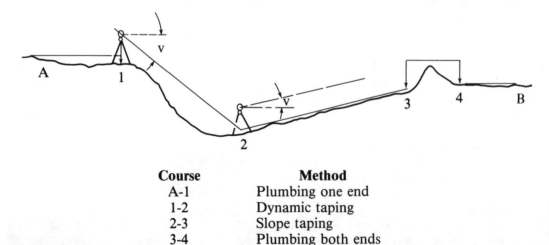

Course	Method
A-1	Plumbing one end
1-2	Dynamic taping
2-3	Slope taping
3-4	Plumbing both ends
4-B	Level tape on flat ground

Figure 4-9.   Horizontal, slope, and dynamic taping. (Adapted from photo, courtesy of Lufkin and the Cooper group.)

the terminal point falls within the end-foot length, usually subdivided into tenths and perhaps hundredths. The rear tapeperson notes the foot mark number held, the head tapeperson's reading is subtracted, if using a tape having the first foot subdivided (a *subtract* or *cut tape*), or added if employing a tape with an additional subdivided foot (an *add tape*). Most taping errors or mistakes are made in measuring the fractional tape lengths, or in keeping a correct tally of full tape lengths (Figure 4-10).

4-3-9. Horizontal Taping on Sloping or Uneven Ground

In measuring on sloping or uneven ground, it is standard practice to hold the tape horizontal and use a plumb bob at one or both ends. More tension, usually 20 to 25 lb, must be applied in order to obtain a 100-ft horizontal distance. Plumbing the tape above 5 ft is difficult, and wind can make accurate work impossible. Bracing both forearms tightly against the body reduces swaying and jerking the tape.

When a full 100-ft measurement becomes impossible, it is divided into subsections of shorter lengths totaling 100 ft. Referred to as *breaking tape*, this is most important on steeper sloping ground. A hand level removes the guesswork from estimating whether the tape ends are at the same elevation. *Taping downhill is preferable*, since it is easier to set a forward pin with the plumb bob than to keep the rear end plumbed over a set pin with tension on the tape.

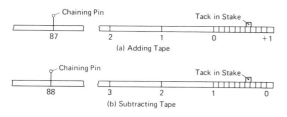

Figure 4-10. Reading partial tape lengths. (Courtesy of Brinker/Wolf.)

4-3-10. Slope Taping

Instead of breaking tape every few feet on a steep but uniform slope, it may be desirable to tape along the slope, determine the slope angle with an Abney level or transit or the elevation difference, and then compute the horizontal distance (see Section 4-3-12). Considerable practice is required for field personnel to do accurate taping in hilly or rolling terrain.

4-3-11. Dynamic Taping

Dynamic taping is similar to slope taping but is done from the transit horizontal axis (Figure 4-10) and is sometimes referred to as the *transit-and-floating-tape* technique. It is best accomplished with a fully graduated tape, sometimes on distances up to 200 or 300 ft. A transit is set up over the beginning mark and the first taping point established on line ahead. The head and rear tapepersons combine to measure and record the slope distance from the new taping point to the transit axis. A vertical angle is read from the transit and the horizontal distance calculated. The instrument is moved ahead to the first taping point, a second one set forward on line, and the procedure repeated. A person holding at midtape should use a tape clamp to avoid kinking or bending the tape. This method is surprisingly fast, accurate, and permits measurements to be made across typical obstacles.

By exercising strict attention to some basic concepts, surveyors can achieve a high degree of precision in taping. Several of the techniques noted may be necessary to accurately measure a distance of only several hundred feet. Another good practice is to measure each distance twice—forward and back—perhaps with a different tape.

4-3-12. Systematic Errors in Taping

The total error present in a measurement made in the field is equal to the algebraic sum of all random (accidental) errors and all systematic errors contained therein. Every attempt must be made to identify any systematic errors and apply corrections to nullify them. Systematic errors in taping are caused by the following conditions that may exist during measurement:

1. Tape not its nominal length.
2. Temperature of tape during measurement is not that at which it was standardized.
3. Tension (pull) applied to the tape is not the same as that when standardized.
4. Measurements were made along slopes instead of in a horizontal plane.
5. Tape was not fully supported throughout but in catenary (suspended with sag).

To achieve a prescribed relative accuracy in taping measurements, the raw observed data have to be corrected to get the true or best length, since measurements are seldom made under ideal conditions; adjustments must be made for pull, temperature, and mode of support.

Incorrect Length of Tape

Although tapes are precisely manufactured, they become worn, kinked, stretched, or improperly spliced after breaks and should be checked periodically against a standard. This can be accomplished in several ways. Most surveying offices either have a special tape to be used only for checking or standardizing other tapes, or maintain permanent marks 100.00 ft apart to check working tapes.

For higher-precision work or to maintain a standard tape, the NBS will, for a fee, issue a certificate for a submitted tape, giving its length to the nearest 0.001 ft at 68°F (20°C) for any specific tension and support conditions. Also, various other governmental agencies maintain the capability to standardize tapes as a service to the public. The NGS has established a number of base lines around the country where tapes and EDMIs can be calibrated by surveyors.

Applying corrections caused by incorrect length of tape is a simple matter but should be

carefully considered. Assume that the actual length of a 100-ft tape is 99.98 ft and a distance between fixed points measured with this tape was recorded as 1322.78 ft. Since each full tapelength was short by 0.02 ft, the correct length is

$$(100.00 - 0.02) \times 13.2278 = 1322.78 - 0.26$$
$$= 1322.52 \text{ ft}$$

However, if a certain distance is to be established, such as in staking out, with a tape known to be too short, the reverse is true, so add the 0.26 ft—i.e., lay out a length of 1323.04 ft. In approaching this problem of incorrect length of tape, Table 4-4 is useful in making the corrections.

Correction for Temperature

The coefficient of thermal expansion of the steel used in common tapes is 0.00000645 per unit length per 1°F. If the length of a tape is known at some standardized temperature, T_0 (NBS uses 68°F or 20°C), the correction can be obtained from

$$C_t = 0.00000645 \, (T_f - T_o) \, L \qquad (4\text{-}1)$$

As an example, if a steel tape known to be 100.00 ft long at a standardized temperature of 68°F is to be used in the field at 43°F (T_f), the actual length of the tape is 0.0097 ft short. This causes a change in length of about 0.01 ft for each change in temperature of 15°F, which is frequently used in approximate calculations. An astounding example is the change in temperature of 75°F between measurements in the summer and winter in some areas. The length of a 100-ft steel tape changes 0.05 ft due to this temperature difference, equivalent to a discrepancy of 2.6 ft in a mile. Errors in taping due to temperature are frequently overlooked

Table 4-4.   Corrections for incorrect tape length

When Tape Is	To Lay Out a Distance	To Measure a Line Between Fixed Points
Too Long	Subtract	Add
Too Short	Add	Subtract

by inexperienced surveyors, but obviously they cannot be disregarded. In SI units the coefficient of thermal expansion of steel is 0.0000116 per unit length per degree C. The corrections can be plus or minus in sign.

Tape temperatures are difficult to measure, particularly on partly cloudy summer days. *Tape thermometers* clamped on near the ends do not contribute to their sag, and so are most reliable. Some tapes have a terminal graduation that varies with temperature.

Corrections for Tension or Pull

Steel tapes are standardized at some specified tension and, being elastic, change length due to variations in the tension applied. An ordinary 100-ft steel tape stretches only about 0.01 ft for an increase in tension of 15 lb. If *spring balances* are used to maintain the prescribed pull, errors caused by tension variations are negligible. Without a spring balance, the tension applied usually varies either above or below the standard and can be considered an accidental error and disregarded in all but precise measurements. *Inexperienced tapepersons are likely to apply tension lower than the standardized figure.*

The formula for tension correction is derived using the modulus of elasticity E, which is the ratio of unit stress to unit strain. The total correction for elongation C_P of a tape length L is calculated from the expression

$$C_P = \frac{(P_f - P_o)L}{AE} \qquad (4\text{-}2)$$

in which P_f is the applied field pull in pounds, P_o the standard tension, A the tape cross-sectional area, and E the modulus of elasticity of the tape—taken as 29,000,000 lb per square in. for steel. The cross-sectional area can be calculated from the tape length, its weight, and the specific weight of steel.

Corrections for Slope

In reducing slope measurements to their horizontal lengths—required in slope and dynamic taping—a correction must be applied to the measured distance equal to the difference between the hypotenuse s and side d of a right triangle having its vertical side equal to h.

When the difference in elevation of the two points h is measured, the correction for slope C_h can be obtained from the formula

$$C_h = \frac{h^2}{2s} + \frac{h^4}{8s^3} \qquad (4\text{-}3)$$

Usually, the first term in the equation is sufficient. If the vertical angle α is measured along with the slope distance, the following formula applies:

$$C_h = s(1 - \cos \alpha) - s \text{ vers } \alpha \qquad (4\text{-}4)$$

This correction has a negative sign since the hypotenuse is always longer than the other side.

Corrections for Sag

A steel tape suspended and supported only at the end points takes the form of a *catenary curve*. Obviously, the horizontal distance between its ends will be shorter than for a tape fully supported throughout its entire length. The difference between the curve and chord lengths is the sag correction and always has a negative sign. Sag is related to weight per unit length and the applied tension. For a tape supported at its midpoint, the total effect of sag in the two spans is considerably smaller. More intermediate supports further reduce the sag to zero when fully supported.

The following formula is used to compute the sag correction C_s:

$$C_s = \frac{w^2 L^3}{24 P^2} = \frac{W^2 L}{24 P^2} \qquad (4\text{-}5)$$

where w is weight of tape in pounds per foot, L the unsupported length between supports, $W = wL$ the total weight of tape between supports, and P the total applied tension in pounds.

These formulas show that greater tension is required for a tape not supported throughout. In fact, there is a theoretical normal tension for each tape that increases its length exactly by the shortening due to sag. This pull can be determined practically using an actual mock-up or by theoretically employing the following formula in a trial-and-error method:

$$P_f = \frac{0.204W\sqrt{AE}}{\sqrt{P_f - P_o}} \qquad (4\text{-}6)$$

Sources of Error and Mistakes in Taping

In addition to the major sources of systematic error in taping, three other conditions must also be carefully monitored: (1) faulty alignment, which has an adverse effect on accuracy—thus, a tape held 1.5 ft off line in 100 ft causes an error of 0.01 ft in distance; (2) taping pins must be set in proper position; and (3) sag is difficult to evaluate when a strong wind blows on an unsupported tape.

Several possible common mistakes made in taping and recording cannot be tolerated and include (1) faulty tallying, (2) misreading the tape graduations, (3) improper plumbing, (4) reversing or misunderstanding the calls in recording numbers, or dropping or adding one foot, and (5) mistaking the end mark.

Taping continues to be a basic operation in many aspects of surveying because it is fast and easy for measuring relatively short distances.

4-4.   OPTICAL DISTANCE-MEASUREMENT METHODS

4-4-1.   General Introduction

The alternative to using ground or catenary taping or electromagnetic methods of distance measurement is to use optical methods. Although some surveyors only relate optical methods to conventional stadia procedure, there is a large family of instruments, methods, and procedures generally classified as optical distance measurement.

Several of these methods were popular in the past. Now, with the advent of electromag-netic distance technology and its versatility, most indirect optical methods—with the possible exception of stadia tacheometry—have been relegated to the status of little-used substitutes. They have, nevertheless, played an important role in attempts to develop distance-measuring equipment that would be rapid although not as accurate. Some had only limited practical application, others were very costly, a few quite complicated, and a number had great potential but surveyors did not recognize their capabilities.

Electronic distance instruments are generally rather expensive and not useful over the range of distances used in optical measurement. Therefore, tapes, EDMIs, and optical instruments complement rather than displace one another.

The term *tacheometry* or *tachymetry* means "rapid" measurement. Actually, any measurement made rapidly could be considered tacheometric but general practice is to include only optical measurements by stadia, subtense bars, etc. Thus, exceedingly fast EDMIs, covered in Chapter 5, are not included in this term.

Only general coverage is attempted in this text; however, two British books published in 1970 are completely dedicated, in great detail, to optical distance measurement. Smith groups various optical devices into the following categories[1]:

1. Instruments on the rangefinder principle
   (a) Fixed-base
   (b) Fixed-angle
2. Theodolites
   (a) Conventional stadia tacheometry
   (b) Tangent tacheometry
   (c) Wedge attachments to theodolites
3. Self-reducing tacheometers
   (a) Using vertical staves-diagram tacheometers
   (b) Using horizontal staves-double-image tacheometers
4. Subtense bar
5. Planetable alidades
6. Miscellaneous items

As will be apparent later in a discussion of principles, optical distance measurement really only combines a fixed quantity with a variable one. All optical instruments involve an angle and a base, which are either fixed or variable; also, the base may be horizontal or vertical. Smith summarizes the possibilities in a convenient form (Table 4-5).

4-4-2. Stadia Tacheometry

Stadia tacheometry is the commonly known procedure that utilizes two supplementary horizontal (stadia) lines placed at equal distances above and below the central horizontal line in an instrument's telescope. Usually, they are short lines to differentiate them from the longer main horizontal one. A graduated vertical rod is sighted and the intercept between the stadia lines read. Modern internal-focusing instruments have a fixed distance between the stadia lines so when the telescope is horizontal and the rod vertical, distance D from the instrument center to the rod equals 100 times the stadia intercept S as in Figure 4-11.

$$D = 100S \qquad (4\text{-}7)$$

In some instruments, the stadia multiplier k is 333.

It is not always possible to keep the telescope horizontal; more commonly, the line of sight is inclined. The stadia intercept S then must be multiplied by the cosine of the vertical angle α to make AB perpendicular to D (see Figure 4-12).

$$D = 100AB \qquad (4\text{-}8)$$
$$D = 100S \cos \alpha \qquad (4\text{-}9)$$

and

$$H = D \cos \alpha = S\,100\,(\cos^2\alpha) \qquad (4\text{-}10)$$
$$V = D \sin \alpha = S\,100\,(\sin \alpha \cos \alpha)$$
$$= S\,100\,(\tfrac{1}{2} \sin 2\alpha) \qquad (4\text{-}11)$$

These formulas are referred to as the *inclined-sight stadia formulas* and form the basis for several stadia reduction (computing) devices.

It should be noted that some older instruments have external-focusing telescopes and the distance X (see Figure 4-13) is proportional to the stadia intercepts—from the external principal focus of the lens. To obtain D, the distances $f + c$ must be added. Their sum is usually about 1 ft and must be applied when using the older external-focusing instruments. It is almost zero and is ignored for the newer internal-focusing type.

4-4-3. Stadia Reduction Devices

The inclined-sight stadia Equations (4-10) and (4-11) must be solved many times when the *stadia method* is used to locate features in transit-stadia and planetable mapping. The desired vertical and horizontal distances V and

Table 4-5. Optical distance-measuring devices

Section	Form	Base	Angle	Base	Resulting Distance
1	Fixed-base rangefinder	Fixed	Variable	Horiz.	Slope
2	Fixed-angle rangefinder	Variable	Fixed	Horiz.	Slope/Horiz.
3	Stadia tacheometry	Variable	Fixed	Vertical	Slope
4	Tangent tacheometry	Fixed	Variable	Vertical	Slope
5	Wedge attachment	Variable	Fixed	Horiz.	Slope/Horiz.
6	Diagram tacheometers	Variable	Fixed	Vertical	Horiz.
7	Double-image tacheometers	Variable	Fixed	Horiz.	Horiz.
8	Subtense bar	Fixed	Variable	Horiz.	Horiz.
9	Planetable alidades	Variable	Fixed	Vertical	Horiz.

Source: Courtesy of J. R. Smith. "Optical Distance Measurement," Granada Publishing Ltd., England, 1970.

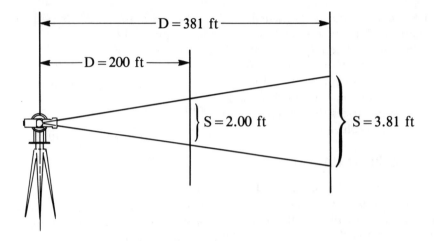

Figure 4-11. Stadia with level sight. (Adapted from photo, courtesy of Keuffel & Esser Co.)

H come from the equations. Although not true with modern hand-held programmable calculators, using the formulas in direct application was slow and tedious. Therefore, many reduction devices were designed to facilitate the computing. These include (1) stadia tables (Tables 4-6 and 4-7), (2) diagrams, (3) slide rules, (4) *stadia circles*, (5) *Beaman arcs*, (6) self-reducing curved stadia lines, and (7) cam-operated movable reticle lines. Instruments using curved stadia lines and cams will be discussed later under diagram tacheometers.

Stadia tables were included in surveying textbooks for years. They assume a stadia in-

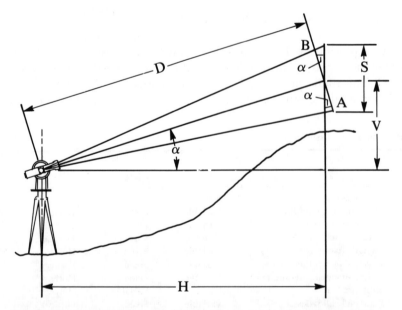

Figure 4-12. Inclined-sight stadia. (Adapted from photo, courtesy of Keuffel & Esser Co.)

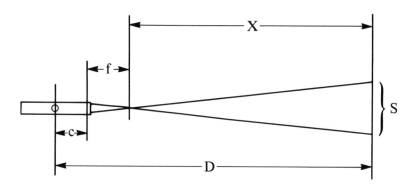

Figure 4-13. Geometry of external-focusing telescope. (Adapted from photo, courtesy of Keuffel & Esser Co.)

Table 4-6. Horizontal corrections for stadia intercept 1.00 ft

Vert. Angle	Horiz. Cor. for 1.00 ft	Vert. Angle	Horiz. Cor. for 1.00 ft	Vert. Angle	Horiz. Cor. for 1.00 ft
0°00′		5°36′		8°02′	
	0.0 ft		1.0 ft		2.0 ft
1°17′		5°53′		8°14′	
	0.1 ft		1.1 ft		2.1 ft
2°13′		6°09′		8°26′	
	0.2 ft		1.2 ft		2.2 ft
2°52′		6°25′		8°38′	
	0.3 ft		1.3 ft		2.3 ft
3°23′		6°40′		8°49′	
	0.4 ft		1.4 ft		2.4 ft
3°51′		6°55′		9°00′	
	0.5 ft		1.5 ft		2.5 ft
4°15′		7°09′		9°11′	
	0.6 ft		1.6 ft		2.6 ft
4°37′		7°23′		9°22′	
	0.7 ft		1.7 ft		2.7 ft
4°58′		7°36′		9°33′	
	0.8 ft		1.8 ft		2.8 ft
5°17′		7°49′		9°43′	
	0.9 ft		1.9 ft		2.9 ft
5°36′		8°02′		9°53′	
					3.0 ft
				10°03′	

Results are correct to the nearest foot at 100 ft and to the nearest 1/10 ft at 100 ft, etc.

With a slide rule, multiply the stadia intercept by the tabular value and subtract the product from the horizontal distance.

Example: vertical angle, 4°22′; stadia intercept, 3.58 ft.

$$\text{Corrected horiz. dist.} = 358 - (3.58 \times 0.6) = 356 \text{ ft}$$

Table 4-7 gives the vertical heights for a stadia intercept of 1.00 ft. With a slide rule, multiply the stadia intercept by the tabular value.

Example: vertical angle, 4°22′; stadia intercept, 3.58 ft.

$$\text{Vertical height} = 3.58 \times 7.59 = 27.2 \text{ ft}$$

Table 4-7. Vertical heights for stadia intercept 1.00 ft

Min.	0°	1°	2°	3°	4°	5°	6°	7°	8°	9°
0	0.00	1.74	3.49	5.23	6.96	8.68	10.40	12.10	13.78	15.45
2	0.06	1.80	3.55	5.28	7.02	8.74	10.45	12.15	13.84	15.51
4	0.12	1.86	3.60	5.34	7.07	8.80	10.51	12.21	13.89	15.56
6	0.17	1.92	3.66	5.40	7.13	8.85	10.57	12.27	13.95	15.62
8	0.23	1.98	3.72	5.46	7.19	8.91	10.62	12.32	14.01	15.67
10	0.29	2.04	3.78	5.52	7.25	8.97	10.68	12.38	14.06	15.73
12	0.35	2.09	3.84	5.57	7.30	9.03	10.74	12.43	14.12	15.78
14	0.41	2.15	3.89	5.63	7.36	9.08	10.79	12.49	14.17	15.84
16	0.47	2.21	3.95	5.69	7.42	9.14	10.85	12.55	14.23	15.89
18	0.52	2.27	4.01	5.75	7.48	9.20	10.91	12.60	14.28	15.95
20	0.58	2.33	4.07	5.80	7.53	9.25	10.96	12.66	14.34	16.00
22	0.64	2.38	4.13	5.86	7.59	9.31	11.02	12.72	14.40	16.06
24	0.70	2.44	4.18	5.92	7.65	9.37	11.08	12.77	14.45	16.11
26	0.76	2.50	4.24	5.98	7.71	9.43	11.13	12.83	14.51	16.17
28	0.81	2.56	4.30	6.04	7.76	9.48	11.19	12.88	14.56	16.22
30	0.87	2.62	4.36	6.09	7.82	9.54	11.25	12.94	14.62	16.28
32	0.93	2.67	4.42	6.15	7.88	9.60	11.30	13.00	14.67	16.33
34	0.99	2.73	4.47	6.21	7.94	9.65	11.36	13.05	14.73	16.39
36	1.05	2.79	4.53	6.27	7.99	9.71	11.42	13.11	14.79	16.44
38	1.11	2.85	4.59	6.32	8.05	9.77	11.47	13.17	14.84	16.50
40	1.16	2.91	4.65	6.38	8.11	9.83	11.53	13.22	14.90	16.55
42	1.22	2.97	4.71	6.44	8.17	9.88	11.59	13.28	14.95	16.61
44	1.28	3.02	4.76	6.50	8.22	9.94	11.64	13.33	15.01	16.66
46	1.34	3.08	4.82	6.56	8.28	10.00	11.70	13.39	15.06	16.72
48	1.40	3.14	4.88	6.61	8.34	10.05	11.76	13.45	15.12	16.77
50	1.45	3.20	4.94	6.67	8.40	10.11	11.81	13.50	15.17	16.83
52	1.51	3.26	4.99	6.73	8.45	10.17	11.87	13.56	15.23	16.88
54	1.57	3.31	5.05	6.79	8.51	10.22	11.93	13.61	15.28	16.94
56	1.63	3.37	5.11	6.84	8.57	10.28	11.98	13.67	15.34	16.99
58	1.69	3.43	5.17	6.90	8.63	10.34	12.04	13.73	15.40	17.05
60	1.74	3.49	5.23	6.96	8.68	10.40	12.10	13.78	15.45	17.10

tercept of 1 ft and list values in parentheses in Equations (4-10) and (4-11) for various values of vertical angle α. To compute H and V, appropriate tabular figures are obtained for α and multiplied by S. Some tables list corrections that are multiplied by S and subtracted from $100S$ to get values of H.

The sole purpose of the 10-in. Kissam slide rule or Cox (circular) stadia rule is to obtain H, V, and/or horizontal corrections by using appropriate S and α values.

A stadia circle consists of two special H and V scales attached to planetable alidades and some transits, on request, which permits the observer to read the same values obtainable from stadia tables instead of the vertical angle. These numbers must then be multiplied by S. Since spacing of the graduations is irregular, a vernier is not needed. A stadia arc is essential in planetable mapping.

To avoid minus readings, the V-index reading on level sights is 50 on most arcs, so 50 must be subtracted from each reading to obtain the true multiplier. The H multiplier normally is near 100. The Beaman arc is the same as a stadia circle, except the horizontal scale H yields a percentage correction that must be subtracted from $100S$. The vertical scale V is the same as on a stadia circle. All the reduction devices were developed to ease the burden and make computing the great volume of stadia side-shots simpler. Self-reducing tacheometers having curved stadia lines or cams are also based on the inclined-sight stadia formulas.

4-4-4. Accuracy of Stadia Measurements

The ordinary level rod graduated to hundredths of a foot can be used in stadia work for maximum sight lengths of 300 to 400 ft. For longer sights, special stadia rods with graduations to 0.1 ft, yd, or m are easier to read and more satisfactory. Observational technique is an individually developed procedure to ensure efficient data measurement, recording, and rod movement.

For most large-scale mapping and contour intervals, horizontal distances to the nearest foot and elevations nearest 0.1 ft suffice, so stadia easily meets this requirement. Stadia-read distances are normally in the 1/300 to 1/500 range, although by carefully calibrating the stadia interval factor F/I, using a target attachment, and making repeated measurements, accuracies as high as 1/1000 to 1/2000 have been recorded. For measuring longer distances, where the full stadia interval does

not fall on a typical rod, a half- or quarter-interval can be read and used to obtain distance. This is particularly useful for small-scale mapping with a planetable.

4-4-5. Diagram Tacheometers

The diagram tacheometer (or self-reducing tacheometer) utilizes a diagram of *reduction curves*, projected into the field of view, and allows horizontal distance H and height difference V to be read directly as intercepts on a graduated vertical rod (Figures 4-14 and 4-15). These reduction curves, based on the inclined-sight stadia formulas, on a diagram plate rotate when the telescope is tilted by turning a series of planet gears. The horizontal multiplication constant is always 100, but there are four height-reading curves, depending on the vertical angle involved. Multiplication factors of 10, 20, 50, and 100 are used.

Rod intercepts are read separately from a base zero-curve line. The instrument is a standard theodolite with this added facility for making stadia reductions. A standard leveling rod can be used, but a special tacheometric staff with a double wedge-shaped mark at 1 m (or 4 ft) above the base allows the zero curve to be set accurately. A staff with an extendable

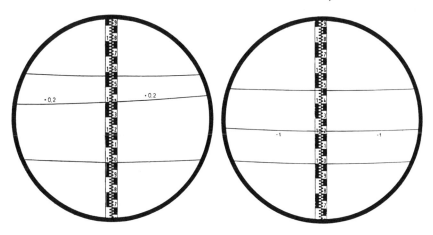

Distance: 57.2 m
Diff. of elevation: $+0.2 \cdot 40.1 = +8.02$ m

Distance: 48.5 m
Diff. of elevation: $-1 \cdot 21.7 = -21.7$ m

Figure 4-14. Reduction tacheometer with vertical staff. (Courtesy of Wild Heerbrugg Instruments, Inc.)

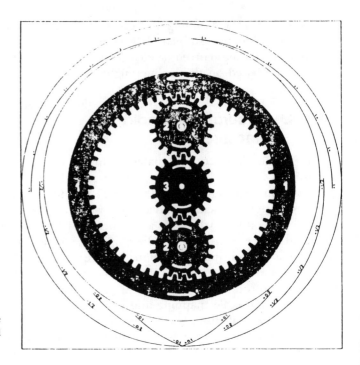

Figure 4-15. Planet gears and diagram plate with reduction curves. (Courtesy of Wild Heerbrugg Instruments, Inc.)

leg is available for sighting at a definite height of instrument.

Another self-reducing tacheometer similar to the diagram instrument has a fixed and movable reticle along with a cam (Figure 4-16). In this instrument, the lines are straight but the cam is based on the inclined-sight stadia formulas. This mechanical reduction system has a fixed reticle containing a vertical and horizontal line. A second horizontal line is on a movable reticle. As the telescope inclination is changed, a cam geared to the horizontal axis raises or lowers the movable reticle, so a rod intercept between two horizontal lines yields the horizontal distance. After switching with a knurled ring (to change the cam configuration), the rod intercept supplies a vertical distance. Again, the instrument is a standard theodolite with an added reduction mechanism.

4-4-6. Planetable Alidades

Planetable alidades invariably contain some type of stadia reduction system. As examples, the Keuffel & Esser standard alidade uses a stadia circle with H and V scales; the newer self-indexing model has a separate optical scale-reading eyepiece to view the three scales, including the typical elevation angle (Figure 4-17). Berger and Gurley alidades use the closely related Beaman arc. Modern Swiss-made Wild and Kern alidades employ the diagram tacheometric self-reducing method with curved-reduction lines. A horizontal multiplier of 100 and three multipliers for difference in elevation (20, 50, and 100) are appropriately marked on the reticle and read directly.

4-4-7. Optical Wedge Attachments

A distance-measuring wedge is a simple theodolite accessory. When attached to the telescope objective housing, a slope distance reading is taken on a horizontal staff set up at the target point. A counterweight screwed tightly to the eyepiece end holds the telescope in balance. Besides the fixed-angle wedge, a parallel-plate micrometer functions similarly to that used in precise leveling, except it measures in a horizontal rather than vertical plane.

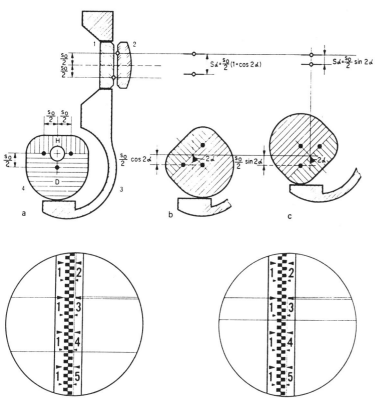

Horizontal distance reading 15.6 m
The switching ring is set at D

Height difference reading 6.4 m
The switching ring is set at $\varDelta$ H
Sighting height on the rod 1.30 m

Figure 4-16. Mechanical reduction system employing a cam-operated movable reticule with straight lines. (Courtesy of Kern Instruments, Inc.)

Of the two optical elements, the wedge supplies a fixed-angle δ (normally $34'22.6''$) and the parallel-plate micrometer aids in resolution of the fine-reading part of the variable base b (Figure 4-18). Therefore, a slope distance and the required horizontal distance are obtained by using a measured vertical angle for reduction.

In optical distance measurement, a distance D is derived from the parallactic angle intercepting a staff length d. The glass wedge covers only the middle section of the objective lens. Therefore, it deflects only those rays that actually pass through the wedge, while the others go straight through. When sighting a horizontal staff, two images are seen displaced

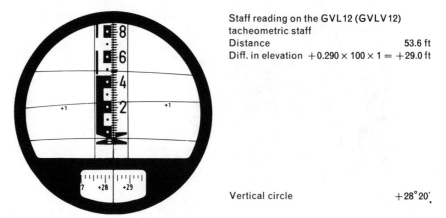

Staff reading on the GVL 12 (GVLV 12)
tacheometric staff

Distance   53.6 ft

Diff. in elevation $+0.290 \times 100 \times 1 = +29.0$ ft

Vertical circle $+28°20'$

Figure 4-17. Modern planetable alidade with reduction curves. (Courtesy of Wild Heerbrugg Instruments, Inc.)

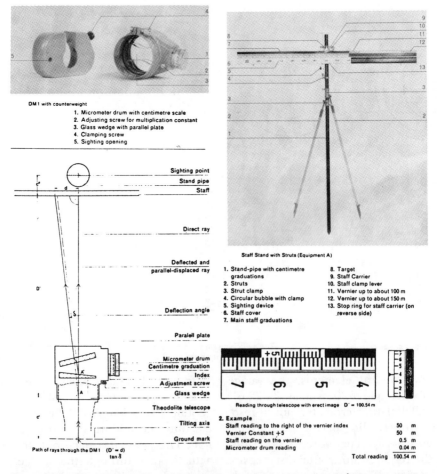

DM 1 with counterweight

1. Micrometer drum with centimetre scale
2. Adjusting screw for multiplication constant
3. Glass wedge with parallel plate
4. Clamping screw
5. Sighting opening

Sighting point
Stand pipe
Staff

Direct ray

Deflected and
parallel-displaced ray

Deflection angle

Paralell plate

Micrometer drum
Centimetre graduation
Index
Adjustment screw
Glass wedge

Theodolite telescope

Tilting axis

Ground mark

Path of rays through the DM 1   (D' = d)
tan δ

Staff Stand with Struts (Equipment A)

1. Stand-pipe with centimetre   8. Target
   graduations   9. Staff Carrier
2. Struts   10. Staff clamp lever
3. Strut clamp   11. Vernier up to about 100 m
4. Circular bubble with clamp   12. Vernier up to about 150 m
5. Sighting device   13. Stop ring for staff carrier (on
6. Staff cover   reverse side)
7. Main staff graduations

Reading through telescope with erect image   D' = 100.54 m

2. Example
Staff reading to the right of the vernier index   50   m
Vernier Constant +5   50   m
Staff reading on the vernier   0.5   m
Micrometer drum reading   0.04   m

Total reading 100.54 m

Figure 4-18. Distance-measuring wedge attachment with horizontal staff. (Courtesy of Wild Heerbrugg Instruments, Inc.)

in relation to each other by a similar *d*. This variable spread represents 1/100 of the slope distance D (multiplication constant = 100).

The lower half of a horizontal staff consists of the main graduations (lines with 1-cm intervals); the upper half has two verniers with graduation intervals of 0.9 cm. The inner vernier is used to measure distances from 10 to 100 m and the outer one (with the distinguishing sign of +5) for distances from 60 to 150 m. The +5 indicates an added constant of 50 m. The accuracy obtainable with the wedge attachment falls in the range of 1/5000 to 1/10,000—much better than with stadia tacheometry. A wedge attachment is employed mainly in measuring traverse sides. The time involved setting up the horizontal staff rules out its use in multiple side-shot detailing.

4-4-8. Double-Image Tacheometers

These are similar in operation to wedges except the effect of a vertical angle is automatically eliminated. A "fixed" angle δ is obtained by an optical device consisting of two rotating wedges, which together with a micrometer system (using a rhombic prism) reduces slope distance to the horizontal ($D \cos \alpha$). The micrometer is used only to resolve the fine reading part of the distance. The same horizontal staff is employed with the double-image tacheometer.

Some double-image tacheometers are equipped with a device to set the rotating wedges so their effect is eliminated when the telescope is horizontal, and the deflection increases proportionally to $\sin \alpha$ for telescope inclinations. Then the staff reading gives $D \sin \alpha$ corresponding to the difference in elevation between the instrument and staff. Maximum range is about 150 m and accuracies are comparable to the wedge attachments. Their principal advantage is the automatic reduction of readings.

4-4-9. Subtense Tacheometry

The subtense method essentially consists of accurately measuring the variable angle subtended by a fixed horizontal distance or base (Figure 4-19). Generally, the standard subtense bar is 2 m long and consists of a metal tube with a target at each end controlled by an Invar wire, so it is unaffected by temperature changes. The bar is mounted horizontally on a tripod and its small triangular targets can be internally illuminated for night operation. A small optical sight ensures that the bar is perpendicular to the sight line.

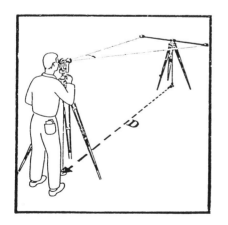
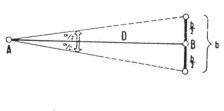

Figure 4-19. Measuring with a subtense bar. (Courtesy of Wild Heerbrugg Instruments, Inc.)

The distance D is derived from the following formula:

$$D = b/2 \cot \alpha/2 \qquad (4\text{-}12)$$

Since base length b is exactly 2 m, distance D totally depends on the horizontal angle measurement α. To obtain accuracies of 1/5000, sights must be limited to 400 to 500 ft and the angle measured to $+1''$ of arc with a precise 1-sec theodolite. Several sets of angles should be read to ensure this $1''$ accuracy. Tables furnished by subtense-bar makers reduce or eliminate computations.

The method's principal advantage over taping lies in its ability to measure over rough terrain, across gullies and wide streams. Furthermore, no slope correction is necessary, since the horizontal angle subtended by the bar is independent of the inclination of the sight line; therefore, the horizontal distance is obtained directly. Since introduction of electronic distance-measuring instruments, the subtense method is no longer able to compete.

4-4-10. Tangential Tacheometry

A variation on stadia tacheometry is called tangent tacheometry, sometimes described as vertical subtense. Two targets are set a known distance apart on a vertical rod. Then by measuring the subtended angle *and* vertical angle, the required horizontal distance and difference in height can be computed.

4-4-11. Rangefinders

Rangefinders are fixed-base or fixed-angle types (Figure 4-20). From a fixed-base b and fixed angle at A, the system's optics are manipulated to vary angle at $B(\alpha)$ until the partial images of Y seen through A and B are coincident in the instrument's field of view. Distance $AY(L)$ is then a direct function of variable α and constant b with angle A normally arranged as a right angle.

In fixed-angle rangefinders, angle B of α is fixed and distance b is varied by sliding a prism unit at B along the bar until the two partial images of Y seen through A and B are coincident. Both fixed-base and fixed-angle rangefinders provide the slope distance, but the vertical angle must be used to get a horizontal equivalent.

Wild makes a small hand-held rangefinder, TMO, with a fixed 25-cm base, and a larger tripod-mounted coincidence rangefinder, TM2, with a long 80-cm base. Depending upon range, measuring accuracies range from 1/10 to 1/150. The TM2 has a measuring range from 300 to 5000 m. Their distinct advantage over other methods is that inaccessible dis-

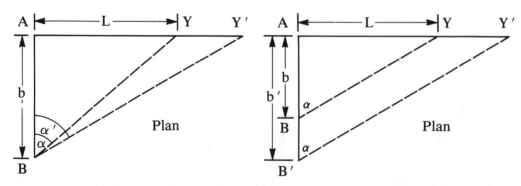

Figure 4-20. Geometry of fixed-base and fixed-angle rangefinders. (Courtesy of R. J. Smith, "Optical Distance Measurement," Granada Publications Ltd., England, 1970.)

tances can be measured and a rodperson is not required.

REFERENCES

ALLAN, A. L. 1977 An alidade for use with electro-optical distance measurement. *Survey Review* (July).

American Congress on Surveying and Mapping 1978. *Metric practice guide for surveying and mapping*, 12 pages, A.C.S.M., 210 Little Falls Street, Falls Church, Virginia 22046.

ASPLUND, E. C. 1971 A review of self reducing, double image tacheometers. *Survey Review* (Jan.).

BAUER, S. A. 1949 The use of geodetic control in surveying practice. *Surveying and Mapping 9*(3), 187–191. Falls Church, VA.

BERRY, D. W. 1954 Tacheometry with the Ewing stadialtimeter. *Surveying and Mapping 14*, 479–483. Falls Church, VA.

BIESHEUVEL, H. 1955 The double image telemeter. *Empire Survey Review 13*(98), 160–164.

BOLTER, W. H. 1956 The self-indexing alidade. *Surveying and Mapping 16*, 157–163. Falls Church, VA.

BRINKER, D. M. 1971 Modern taping practice versus electronic distance measuring. *ACSM Fall Convention Papers*, pp. 354–356. Falls Church, VA.

COLCORD, J. E. 1971 Tacheometry in survey engineering, *Journal of the Surveying and Mapping Division*, A.S.C.E., *97*(SU1), 39–52. New York, NY.

COLCORD, J. E., and CHICK, F. H. 1968 Slope taping. *ASCE Journal of the Surveying and Mapping Division 94*(SU2), 137–148. New York, NY.

CURTIS, K. S. 1976 *Optical distance measurement.* Indiana Society of Professional Land Surveyors (ISPLS) Surveying Publication Series No. 7. Indianapolis, IN.

DEUMLICH, F. 1982 *Surveying instruments.* (transl.). Berlin/New York: de Gruyter.

DRACUP, J. F., FRONCZEK, C. J., and TOMLINSON, R. W. 1977 (Revised by Spofford, P. R. 1982). Establishment of calibration base lines. *NOAA Technical Memorandum NOS NGS-8*, NOAA, (August). Rockville, MD.

EMERY, S. A. 1954 Nomography applied to land surveying. *Surveying and Mapping 14*(1), 59–63. Falls Church, VA.

FRONCZEK, C. J. 1977 (Reprinted with corrections 1980). Use of calibration base lines. *NOAA Technical Memorandum*, NOS NGS-10, NOAA, (Dec.). Rockville, MD.

GOLLEY, B. J., and SNEDDON, J. 1974 Investigation of dynamic taping. *Journal of the Surveying and Mapping Division, ASCE 100*(SU2), 115–122.

HARRINGTON, E. L. 1955 The stepping method of stadia. *Surveying and Mapping 15*(4), 460. Falls Church, VA.

HIDNERT, P., and KIRBY, R. K. 1953 New method for determining linear thermal expansion of invar geodetic surveying tapes. *J. Research NBS 50* 179, RP2407.

HODGES, D. J., and GREENWOOD, J. B. 1971 *Optical distance measurement.* Butterworth and Co., England.

Instruction Manual, *Self-reducing engineer's tacheometer*, KI-RA, Kern & Co., Ltd., Aaru, Switzerland.

JUDSON, L. V. 1956 *Calibration of line standards of length and measuring tapes at the national bureau of standards;* National Bureau of Standards Circular No. 572.

KENNGOTT, R. 1980 EDM calibration baseline at Westchester Community College, Valhalla, New York, *ACSM Paper Proceedings*, Fall Technical Meeting, Niagara Falls, New York, (Oct.), pp. LS-2-F-1 to 8.

KISSAM, P. 1961 Stadia will cut your costs. *Engineering Graphics 1*(5), 6.

KISSAM, P., et al. 1940 *Horizontal control surveys to supplement the fundamental net;* A.S.C.E. Manual No. 20. New York, NY.

LOVE, J., and SPENCER, J. 1981 Explanation and availability of calibration base line data. *ACSM Paper Proceedings*, Fall Technical Meeting, San Francisco, California, (Sept.), pp. 209–217.

Lufkin Rule Company. 1972. *Taping techniques for engineers and surveyors*, Apex, North Carolina.

MALTBY, C. S. 1954 Alidade development in the geological survey. *Surveying and Mapping 14*, 173–184.

MELTON, G. D. 1969 Devices for distance measurement—The Wild RDS. *Proceedings of the Second Annual Land Surveyors' Conference*, University of Kentucky, pp. 54–60.

MILLER, C. H., and ODUM, J. K. 1983 Calculator program for reducing alidade or transit stadia traverse data. *Surveying and Mapping 43*(4), 393–398. Falls Church, VA.

MOFFITT, F. H. 1975 Calibration of EDM's for precision measurement. *Surveying and Mapping 35*(2), 147–154. Falls Church, VA.

MOFFITT, F. H. 1958 The tape comparator at the University of California. *Surveying and Mapping 18*(4), 441–444. Falls Church, VA.

MORSE, E. D. 1956 The practical application of close tolerance in survey work. *Report of 5th Annual Texas Surveyors Association Short Course*, pp. 59–69.

MORSE, E. D. 1951 Concerning reliable survey measurements. *Surveying and Mapping 11*(1), 14–21. Falls Church, VA.

MUSSETTER, W. 1953 Stadia characteristics of the internal focusing telescope. *Surveying and Mapping 13*(1), 15–19. Falls Church, VA.

MUSSETTER, W. 1956 Tacheometric surveying. *Surveying and Mapping 16*(Part I), 137–156; (Part II), 473–487. Falls Church, VA.

O'QUINN, C. A. 1976 Florida's electronic distance measuring equipment base lines. *ACSM Paper Proceedings*, 36th Annual Meeting, Washington, DC, (Feb.), pp. 118–127.

POLING, A. C. 1965 A taped base line and automatic meteorological reading instruments for the calibration of electronic distance measuring instruments. *International Hydrographic Review 42*(2), 173–184. Monaco.

RICK, J. O. 1975 ACSM student chapter offers tape comparison service to surveyors. *Surveying and Mapping 25*(3), 239–243. Falls Church, VA.

SAASTAMOINEN, J. 1959 Tacheometers and their use in surveying. *Canadian Surveyor 14*(9), 444–453.

SHEA, H. J. 1946 Second-order taping through use of taping bucks. *Surveying and Mapping 6*(2), 103–109. Falls Church, VA.

SHEPARD, C. D. 1932 Nomographic chart for steel tapes. *Civil Engineering Magazine 2*(7), 440–442. ASCE, New York, NY.

SMIRNOFF, M. V. 1952 The use of the subtense bar. *Surveying and Mapping 12*(4), 390–392. Falls Church, VA.

SMITH, J. R. 1970 *Optical distance measurement*. Granada Publishing Ltd., England.

STANLEY, D. R. 1952 The microptic alidade. *Surveying and Mapping 12*, 25–26. Falls Church, VA.

TURPIN, R. D. 1954 A study of the use of the Wild telemeter DM1 for a closed traverse. *Surveying and Mapping 14*, 471–477. Falls Church, VA.

WAGNER-SMITH, R. W. 1961 Errors in measuring distances by offsets. *Surveying and Mapping 21*(1), 73–77. Falls Church, VA.

WOLF, P. R., WILDER, B., and MAHUN, G. 1978 An evaluation of accuracies and applications of tacheometry. *Surveying and Mapping 38*(3), 231–244. Falls Church, VA.

5

Linear Measurements: EDM Instruments

Porter W. McDonnell

5-1. INTRODUCTION

Electronic measurement is a modern method of precise and rapid determination of slope (line-of-sight) distances. Some electronic distance-measuring instruments (EDMIs) are designed to reduce the slope distances to horizontal distances.

EDMIs can be used over water and from one high point to another—e.g., over buildings, etc. They have maximum ranges varying from 500 to 64,000 m, or 0.3 to 40 mi. The introduction of electronic distance measuring has simplified traversing work: The selection of traverse stations can be made without concern for the feasibility of taping the sides. The tedious, slow process of clearing the entire traverse line and proceeding 100 ft at a time for the measurement is eliminated. In some cases, electronic traversing now takes the place of triangulation for horizontal control over large areas. Construction layout work can be done from a single central station (radial line stakeout), and short-range EDMIs may be incorporated into theodolite designs to create all-purpose instruments usually known as "total stations."

The many instruments available differ in detail, but nearly all depend on the precision of a quartz crystal oscillator and determination of distance by measuring a "phase shift." (Two recent models, not covered here,[1] make use of a "transit time technique.") An instrument is set up at one end of the distance to be measured. It transmits a beam of infrared light *or* microwave, which serves as a carrier for the waves used for measurement. The beam is received at the other end of the distance by a reflector, when using an infrared beam, or by another electronic instrument if employing a microwave beam. In either case, the beam is returned to the master instrument. A *reflector* consists of one or more "corner prisms" (Figure 5-1) or, for short distances, just a molded scotch light or bicycle reflector. The reflectors are designed to return the light beam, even if they are only approximately pointed toward the source. For the microwave beam, the remote instrument is similar to the master instrument and thus more than a passive reflector. Light sources other than infrared (visible lasers, incandescent light) have also been used.

Several infrared instruments are shown in Figures 5-2, 5-3, and 5-4, and a microwave instrument is seen in Figure 5-5. An EDMI "modulates" the light or microwave beam to pulsate at each of several different frequencies and an increased amplitude. These *pattern* frequencies permit the comparison of several

Figure 5-1. Single- and multiple-prism reflectors used to receive and return light beams in electronic distance measurements. They are shown in tilting mounts. (Courtesy of the Lietz Co.)

phase shifts of returning beams. An on-board computer deduces distance from these data—i.e., by a phase comparison of the returning beam with an internal branch of the outgoing beam.

5-2. PRINCIPLES

Sections 5-3 through 5-6 present a discussion of the principles of electronic distance-measuring instruments, based on a paper by Gort.[2] It refers to the HP 3800 instrument for

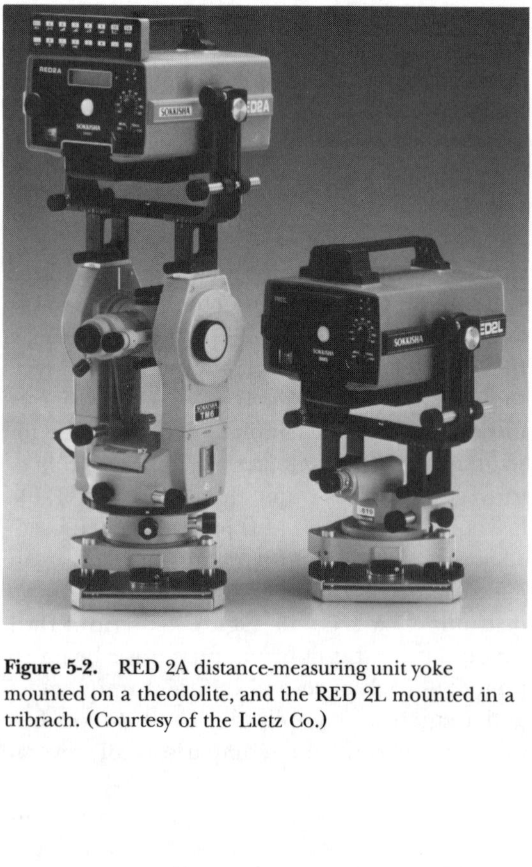

Figure 5-2. RED 2A distance-measuring unit yoke mounted on a theodolite, and the RED 2L mounted in a tribrach. (Courtesy of the Lietz Co.)

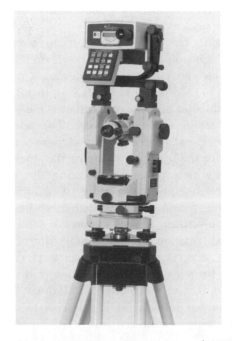

Figure 5-3. Stinger infrared EDM instrument. (Courtesy of IR Industries, Inc.)

Figure 5-4. Kern DM 503 mounted on an electronic theodolite. The unit may be used on conventional theodolites as well. It features transmitting optics *above* the telescope and receiving optics *below* it. (Courtesy of Kern Instruments, Inc.)

which Gort had the design responsibility. Hewlett-Packard no longer makes EDMIs, but the following explanation illustrates the function of current models as well as those of the recent past.

5-3. THE APPLICATION OF MODULATION

In general, the measurement of any distance is accomplished by comparing it to a multiple of a calibrated distance, e.g., by using a 100-ft tape. In electronic distance meters, the same comparison principle is used: The calibrated distance is the wavelength of the modulation on a carrier (light or microwave). In the HP 3800, the effective wavelength is a precise 20 ft, which is related to the modulation frequency by

$$\lambda = v/f \qquad (5\text{-}1)$$

where

λ = modulation wavelength

v = velocity of light

f = modulation frequency

Figure 5-5. Microwave-type EDM instrument, the Tellumat CMW20. (Courtesy of Teludist, Inc.)

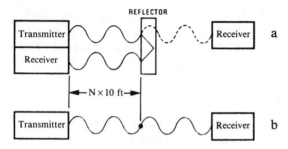

Figure 5-6. Distance being measured is a multiple of 10 ft; the wavelength is 20 ft. (Courtesy of Hewlett-Packard Co.)

Suppose the distance to be measured is an exact multiple n of 10 ft as shown in Figure 5-6a. The total optical path, however, will be $2n \times 10$ ft, which is shown by folding out the reflector-to-receiver path (Figure 5-6b). As the total path is $2n \times 10$ ft, the total phase delay will be $n \times 360°$. (Each 20-ft wavelength represents a full 360° phase delay.) The phase difference between a transmitted beam and received beam is also $n \times 360°$, which cannot be distinguished from a 0° phase difference. Figure 5-6b shows the sine wave reaching the receiver at the same point in a cycle as when it was transmitted.

In general, the distance to be measured may be expressed as $n \times 10 + d$ ft. Figure 5-7 shows the total optical path by folding out the returned beam for clarity. The total phase delay ϕ between transmitted and received signal becomes

$$\phi = n \times 360° + \Delta\phi$$

in which $\Delta\phi$ equals the phase delay due to the distance d. As $n \times 360°$ is equivalent to 0° for

a phase meter, the angle can be measured and will represent d according to the relation

$$d = \left(\frac{\Delta\phi}{360} \times 10 \right) \text{ ft}$$

In order to find the number n of 10-ft multiples, a 200-ft modulation wavelength is used next. This results in another ambiguity, of multiples of 100 ft. Of course, the procedure can be repeated with a 2000-ft wavelength to resolve this ambiguity, and so on.

5-4. THE INDEX OF REFRACTION OF AIR

The accuracy of distance measurements depends, among other things, on the calibration of the measurement unit. Since the modulation wavelength is used as a measurement unit in electronic distance meters, this wavelength has to be accurately established. Equation (5-1) shows that the accuracy in λ depends on v and f. The velocity of light in air may be expressed as

$$v = \frac{c}{n} \tag{5-2}$$

where c = vacuum velocity of light and n = index of refraction of air.

As c is a universal constant, only n has to be determined in order to find v. The index n is a function of the wavelength used, and the

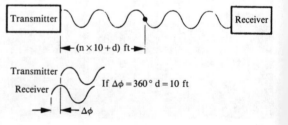

Figure 5-7. Distance being measured is *not* a multiple of 10 ft; the returning signal is out of phase with the transmitted beam. (Courtesy of Hewlett-Packard Co.)

density and composition of the air. Thus, it is affected by the atmospheric pressure and temperature at the time of measuring. As the accuracy of a distance measurement depends on determination of this index, the question arises: How accurately must the pressure and temperature be known? Figure 5-8a shows the relationship between the error in a distance measurement and inaccuracy of the temperature. Figure 5-8b displays the error as a function of pressure measurement inaccuracies. These graphs, used together, demonstrate that a 10° high estimate of temperature combined with a 1-in. low-pressure estimate causes a distance error of 15 ppm, or 1 part in 67,000—sufficiently accurate for most survey work.

In the case of microwave instruments, the humidity also is a factor (see Section 5-7).

5-5. DECADE MODULATION TECHNIQUE

As mentioned previously, a 360° phase delay equals 10 ft in the phase meter at the highest

modulation frequency (24.5 MHz). The HP 3800, however, uses four different modulation frequencies in order to measure a full 10,000 ft without ambiguity. The four modulation frequencies are related in decade steps and yield phase-meter constants, which are 360° = 10 ft at one end of the scale and 360° = 10,000 ft at the other end. The fraction of 10 ft is determined first, using the 24.5 MHz modulation; the fraction of 100 ft is then determined by the 2.45 MHz modulation; the fraction of 1000 ft by applying 245 kHz modulation; and the fraction of 10,000 ft by using 24.5 kHz modulation. Table 5-1 shows this in compact form and indicates how the ambiguity is resolved for a distance of 6258.31 ft.

With respect to multiples of 10,000 ft, a readout can be ambiguous. The range of the HP 3800 is specified as 10,000 ft for good viewing conditions in the daytime, but it is possible to measure longer distances under favorable atmospheric conditions. The instrument would not distinguish between 6258.31 and 16,258.31; in fact, it is limited to a six-digit display.

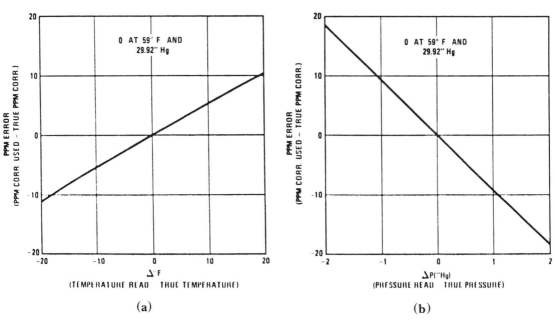

Figure 5-8(a) and (b). Distance errors in ppm caused by using incorrect temperature and atmospheric pressure. (Courtesy of Hewlett-Packard Co.)

Table 5-1.   Decade modulation for HP 3800

Modulation Frequency	Phase-Meter Constant	Distance Incl. Ambiguity	HP 3800 Readout
24.5 MHz	$360° = 10$ ft	$n_1 \times 10 + 8.31$ ft	xxx8.31
2.45 MHz	$360° = 100$ ft	$n_2 \times 100 + 58.31$ ft	xx58.31
245 kHz	$360° = 1000$ ft	$n_3 \times 1000 + 258.31$ ft	x258.31
24.5 kHz	$360° = 10,000$ ft	$n_4 \times 10,000 + 6258.31$ ft	6258.31

5-6. ENVIRONMENTAL CORRECTION

It was evident from the description of the decade modulation technique and readout system that the effective modulation wavelength must be kept accurately at 20 ft or its decade multiples. If a fixed modulation frequency is used, the wavelength would vary in accordance with changes in the index of refraction of air. The HP 3800 modulation frequencies can be slightly varied to compensate for index changes and thus keep the effective wavelength an accurate 20 ft. The master oscillator in the frequency generator is set manually by a control on the power unit. Figure 5-9 gives a correction to be dialed in as a function of temperature and pressure. This should be

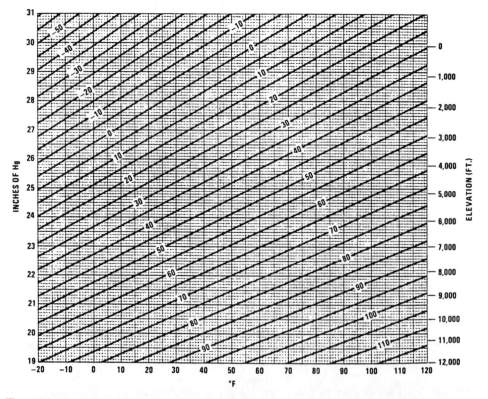

Figure 5-9.   A ppm correction chart for atmospheric conditions. (Courtesy of Hewlett-Packard Co.)

done before the distance measurement, as it calibrates, in effect, the instrument's distance unit. Newer instruments permit the simple keying in of temperature and pressure, eliminating the need for a diagram or graph.

5-7.   LIGHT BEAM VERSUS MICROWAVE

In general, the use of microwave beams offers the advantages of penetration through fog or rain and, usually, a longer range. Two instruments—transmitting and receiving—are required, as opposed to a single instrument and reflecting prism. The two units also provide a speech link. Microwave instruments are considerably more sensitive to humidity than light-beam models. At normal temperature, an error of 1 part in 100,000 will be introduced by a 2.7°F (or 1.5°C) error in the difference between wet- and dry-bulb thermometer readings. This effect increases on warmer than normal days. However, corrections for meteorological conditions (temperature, pressure, and humidity) are easily made. Microwave beams are wider than light beams and can present difficulties in underground surveys, indoor work, or measurements made close to a water surface. Narrower beams have been used in recent years. A microwave instrument appears in Figure 5-5.

Light-wave instruments are much more common. In a 1984 article by McDonnel in *P.O.B. Magazine*, all EDMIs were tabulated listing 47 models.[3] Only two models use microwaves. Of the remaining 45 instruments, two employ helium neon lasers as a light source and 43 use infrared. In addition, all total-station instruments are of the infrared type. For short distances, such as a half-mile, atmospheric conditions are not as important and approximate values of temperature and pressure are often sufficient (see Sections 5-4 and 5-6). Humidity is not a factor. Most light-wave models are limited to shorter ranges than mi-

crowave devices, but the HeNe laser instruments will measure up to 25,000 m using nine prisms (the K & E Ranger V-A) or 64,000 m to a bank of 30 prisms (K & E Rangemaster III).

5-8.   INSTRUMENTAL AND REFLECTOR ERRORS

The National Geodetic Survey (NGS) recommends that electronic distance-measuring instruments be checked occasionally. For this purpose, they have assisted in establishing a number of calibration base lines in each state. Their specifications for these lines state that each one should be about a mile long and have intervisible monuments at 500 ft, 1150 ft, 2600 ft, and the ends.

Such a base line can be used to detect an instrument's frequency shift. An improperly tuned instrument may have a frequency error, making it comparable to a tape of incorrect length. In addition, it is possible to make a frequency check by employing a frequency counter, available from the manufacturer or other electronics supplier. Such a check should be made at regular intervals if high-order work or surveys with long lines are being performed.

The corner cube prisms used with light-wave instruments (see Figure 5-1) have a so-called "effective center." The location of this center is not geometrically obvious because it involves the fact that light travels more slowly through glass than air. The effective center will be behind the prism itself and generally not directly over the station to which the measurement is desired. Thus, there is a "reflector constant," sometimes amounting to 3 or 4 cm, to be subtracted from the measurement. The amount is accurately determined by the manufacturer. It may be partially or totally offset by advancing the electrical center of the transmitting unit during manufacture. In the case of a particular instrument, e.g., there is a correction of $+320$ mm at the transmitting unit. If it is used with a reflector that has a constant of 40 mm, a combined correction of 280 mm is required. This correction may be dialed into the unit and automatically added by an internal computer before a distance is displayed. Field determination of the constants is possible but not usually necessary. A discussion of this topic and errors caused by untilted or misaligned prisms is provided in a technical report by Kivioja and Oren.[4]

5-9. SLOPE CORRECTION

EDM instruments all measure slope (line-of-sight) distances only. Reduction of inclined measurements to horizontal distance is accomplished in several ways, sometimes by a built-in microprocessor. The *vertical* projection of the slope distance (a difference in elevation) may also be determined. The latter process is a modern variation of trigonometric leveling.

A simple right-triangle reduction may be applied if the distance is short or the precision needed is modest. The effects of earth curvature and atmospheric refraction become important for longer distances. For short ones, the horizontal component is

$$S \cos \alpha \quad \text{or} \quad S \sin Z \qquad (5\text{-}3)$$

where S is the slope distance, α the vertical angle, and Z the zenith angle. If the elevations of both ends of the measurement are known, the elevation difference h will serve as a basis for slope reduction. The horizontal distance is

$$S - \frac{h^2}{2S} - \frac{h^4}{8S^3} \qquad (5\text{-}4)$$

in which the last term is usually negligible.

Consideration must be given to the mounting position of an EDM unit and the design of the prisms-and-target combination. For example, an EDMI may be telescope-mounted, causing its electrical center to rotate forward

or backward about the horizontal axis of the telescope for an inclined sight. A special "telescope-mount target" (Figure 5-10) may be used to allow the prism reflector to be positioned at the right distance above the painted target and tilted at a similar angle. The vertical or zenith angle is measured to the painted target and the slope distance read to the prism. An optical sight (*collimator*) is shown above the prism housing in the figure. The horizontal distance is calculated by Equation (5-3). The target in Figure 5-11 provides a vertical pole with threads on top for mounting a prism and an adjustable painted target directly below it. It is intended for use with a yoke-mounted EDM unit, such as those in Figures 5-2 and 5-3, which have their own tilting axes (vertically above that of the theodolite). Slope reduction again involves the simple Equation (5-3) in the preceding paragraph.

The target intended for use with a yoke-mounted EDMI (Figure 5-11) is much less expensive than the telescope-mount model (Fig-

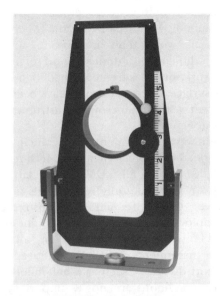

Figure 5-10. Prism holder and target designed for use with an EDM instrument mounted on the telescope of a theodolite. The prism holder is set above the target at the same offset as exists between EDMI and telescope. (Courtesy of SECO Manufacturing Co., Inc.)

Figure 5-11. Target pole threaded on top for a prism holder, designed for use with a yoke-mounted EDM instrument. The target is set below the prism at the same vertical offset as exists between the axes of the yoke and the telescope. (Courtesy of SECO Manufacturing Co., Inc.)

ure 5-10). It is easy to program a pocket calculator to make a slope reduction in which a telescope-mounted EDMI is used with the less expensive, nontilting target. The formula is

$$S \cos \alpha - d \sin \alpha \qquad (5\text{-}5)$$

where d is the distance from the theodolite axis to the top-mounted EDMI axis and α the vertical angle with its proper sign. It is sufficiently accurate to place the painted target below the prism at the same distance d. Theoretically, it should be at distance $d \sec \alpha$.

There is practically no distance error if a telescope-mounted EDMI reads an inclined distance to a prism, and then the theodolite measures the slope *to the prism* rather than a painted target below it. If $S = 100.00$ ft, $\alpha = 5°17'$, and $d = 0.50$ ft, the error in horizontal distance will be 0.001 ft (too small to measure), and it decreases for longer measurements. In Equation (5-5), vertical angle α will be differ-

ent and the last term, $-d \sin \alpha$, will not be needed. Thus, Equation (5-3) may be used.

Instruments such as the Kern DM503 in Figure 5-4 transmit the beam *above* the telescope barrel and receive the returned beam *below* it. They require a tall rectangular prism. The telescope is pointed to the middle of the prism.

Curvature of the earth and refraction of the atmosphere must be considered when measuring long lines. In 1977, Gort described an example involving a slope measurement of 1085.276 m (about two-thirds of a mile).[5] The zenith angle was $78°11'42''$. Ignoring the curvature and refraction correction would have caused a distance error of 1 part in 30,000. The effect on the computed difference in elevation would have been about 7.5 cm (or 0.25 ft). Gort was reporting on the development of the Hewlett-Packard 3820A Total Station instrument (no longer in production), which contained a microprocessor for computing the corrections. The equations used by the on-board computer were

$$\text{Horizontal distance} = S(\sin Z - E_1 \cos Z) \quad (5\text{-}6)$$

$$\text{Vertical distance} = S(\cos Z + E \sin Z) \quad (5\text{-}7)$$

in which

$$E_1 = \frac{0.929S \sin Z}{6,372,000}$$

and

$$E = \frac{0.429S \sin Z}{6,372,000}$$

In the latter terms, 6,372,000 is the earth's radius. For the case discussed, the horizontal distance is 1062.287 m, the vertical distance 222.103 m. For trigonometric leveling over short distances, disregarding curvature and refraction, the vertical distance would, of course, be $S \cos Z$.

If we assume the same zenith angle, but a slope distance of only 305.000 m (about 1000 ft), the horizontal distance is 298.549 m without a curvature and refraction correction, and 298.546 m with it. The vertical distance is 62.397 m by right-triangle trigonometry and 62.403 with the correction.

Equation (5-6) uses the instrument elevation as the datum. Thus, if the sight is long and steeply inclined, as in the first example, a reciprocal observation (from the other end) will give a different answer. The two answers should agree if each is reduced to a common datum. The distance found in the example, 1062.287, would be 1062.324 at the datum of the higher station. Each can be reduced to sea level by the following factor:

$$\frac{6,372,000}{6,372,000 + H}$$

where H is the station elevation in meters. Equations (5-6) and (5-7) use an average coefficient of refraction (0.071).

5-10.   ACCURACY SPECIFICATIONS

Equipment manufacturers usually list accuracy as a standard deviation or mean square error (nearly equivalent concepts). The specification given is a two-part quantity: a constant uncertainty (independent of distance) and a parts-per-million term (proportionate to distance). Figure 5-12 depicts this graphically for a typical instrument and also shows the claimed ranges to single and triple prisms. The stated accuracy is $\pm(5 \text{ mm} + 5 \text{ ppm})$ for the Topcon GTS-2. The 90% error, equal to 1.64 times the standard deviation, is also plotted, indicating an "allowable error" for which the instrument could be considered suitable.

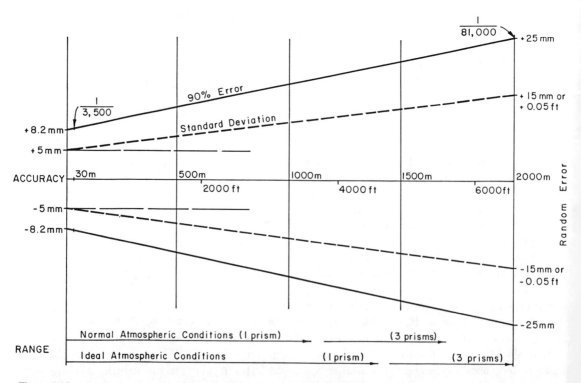

Figure 5-12.   Graph of accuracy and range specifications for Topcon GTS-2.

For short measurements, the constant part of the random error is significant and exceeds the errors of ordinary taping. The figure shows that at 30 m nearly all the error is the constant type and the 90% error is 1 : 3,500. At 2000 m (a maximum range), two-thirds of the error is the proportionate kind, and the total error is 1 : 81,000. Thus, as in angle measurements, short traverse sides should be avoided.

Figure 5-12 may be said to present a "worst case." The various sources of error are complex and do not necessarily add together in the manner shown. Periodicity and calibration errors are parts of the first term (5 mm) and are systematic for a given instrument measuring a particular distance, but will vary with other lengths. Principal factors in the second term (5 ppm) are *noise* due to heat shimmer, which is random, and crystal oscillator error that is systematic. Some references show the first and second terms in the accuracy statement to propagate as two random errors. At 2000 m, the maximum range in Figure 5-12, the standard deviation would be

$$\sigma = \sqrt{5^2 + \left[\frac{5(2,000,000 \text{ mm})}{1,000,000}\right]^2} = \pm 11.2 \text{ mm}$$

instead of ±15 mm.

The random errors discussed here are in addition to those caused by inaccurate estimates of temperature and atmospheric pressure, and incorrect centering of instrument and target. Systematic errors can be caused by improper calibration, inexact pointing on the reflector (for a peak return signal), and assuming an incorrect prism constant.

5-11. TOTAL-STATION INSTRUMENTS

Hewlett-Parkard invented the name Total Station to promote its Model 3810A, which sensed vertical angles electronically, as well as distances, and automatically applied sines and cosines to generate elevations and horizontal distances. The name caught on, either because such an exciting new type of instrument seemed to deserve a special new name, or surveyors forgot that the old term *tacheometer* or *tachymeter* would be appropriate (as would electronic tacheometer). Gradually, the name Total Station was applied to instruments that combined an electronic distance-measuring instrument and a theodolite, but had no electronic angle sensing or automatic data reduction. (Pentax simply called this an "EDM theodolite.") With its various meanings, the term Total Station became as common as dumpy level and began appearing without capital letters.

In 1983, *P.O.B. Magazine* recognized the popularity of the name total station but saw a need to define and name three separate types, as follows:

Manual total station. Both distance measuring and angle measuring make use of the same telescope optics (coaxial). Slope reduction of distances is done by optically reading the vertical angle (or zenith angle) and keying it into an on-board calculator or any pocket calculator (see Figure 5-13).

Semi-automatic total station. Contains a vertical angle sensor for automatic slope reduction of distances (without keying in the slope angle). Horizontal angles are read optically.

Automatic total station. Both horizontal and vertical angles are read electronically for use with slope distances in a data collector or internal computer.[6]

Note that a theodolite with a mount-on EDM unit was *not* classified as a total station, except in the case of an electronic (digitized) theodolite, in which a modular total station was the design commitment, as in Figure 5-4.

Manual total stations are theodolites with built-in EDM units. Typically, these neat compact instruments weigh less than traditional

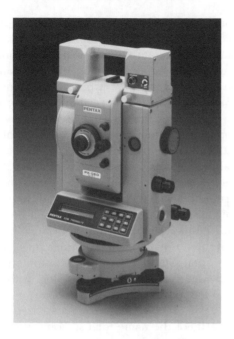

Figure 5-13. Manual total station, combining an EDM instrument, theodolite, and preprogrammed calculator in a single instrument. After zenith angle is keyed in, display shows slope distance, horizontal distance, and vertical distance in turn. Instrument offers tracking mode, stakeout mode, and distance-averaging mode. Optional cordless control unit is available. (Courtesy of Pentax Corp.)

American vernier transits. The model in Figure 5-13, and some of its competitors, have a detachable carrying handle containing the battery.

Vertical angle sensors (as on semiautomatic models) may function only in the telescope-direct position and over a limited range, such as ±40°. Thus, it may be necessary to override the sensor at times.

While the instruments are "total" in many respects, *it is important to note that a solar observation will damage the EDM unit unless an objective filter or Roelofs prism protects the optics.*

The manual and semiautomatic total stations represent an amazing advance in convenience and portability in the short history of the popular short-range electronic distance-measuring instruments. (The HP 3800 was introduced in 1971.) Even more dramatic and

revolutionary, however, are the automatic total stations. The Zeiss Elta 2, e.g., permits "free stationing." The instrument is set up on a new station (position unknown). A random combination of direction measurements or combined direction and distance measurements is used for up to five targets. The target coordinates are read from memory and the station coordinates are calculated, using a least-squares adjustment if there is redundant data. Also displayed is the standard error of the coordinates and adjusted scale factor of the system. The Elta 2 can perform a similar operation to obtain the height of instrument (HI) above datum using a distance and slope angle to one or up to five bench marks. The measurements, of course, are sensed electronically and need not be keyed in manually.

The Kern E1 (Figure 5-4) and E2 contain tilt sensors and thus are able to apply a correction to measured horizontal angles, if the instrument is not level. This is also true on the Geodimeters 140, 420, and 440, some Jena models, and the HP 3820. The Kern E2 and HP 3820 will even display the amount of inclination, so a surveyor can level the instrument exactly, if preferred, and they also correct for the effects of an inclined horizontal axis.

The Kern models, Geodimeter 140, the MK-III, and Omni I are designed to make full use of the powerful HP 41CX calculator. Values generated by the total station can be transferred directly to the calculator and stored for computation. The Kern remote receiver (on the sight rod) is also depending on this calculator to determine required orthogonal offsets in stakeout work.

Automatic total stations are frequently used with data collectors. Generally, data collectors are electronic supplements to the conventional field book, permitting a convenient interface with a computer and plotter, and remote transmission of data by telephone, using an acoustic modem (see Figure 5-14).

Because the horizontal circle is read electronically, the instrument can display the angle in degrees or grads, or subtract the

Figure 5-14. Data collector or electronic field book. Geodat 126 uses HP-41CX calculator as "administrative system" (display, keyboard, calculating capacity, and memory module) and adds HP Interface Loop, RS-232C interface, and additional memory for storing measurement data, file handling, and special programs. (Courtesy of Geodimeter.)

angle from 360° to display a counterclockwise angle. It is important to note that the smallest angular unit displayed (1″, e.g.) is not a manufacturer's claim of accuracy. One manufacturer will, at extra cost, modify its instrument for greater accuracy and yet not change the smallest displayed unit.

The combination instruments (manual, semiautomatic, and automatic total stations) are revolutionizing field procedures for all kinds of surveying.

NOTES

1. P. W. McDonnell. 1985. Total station survey. *P.O.B. Magazine 10* (6) (Aug.–Sept.), 20.

2. P. W. McDonnell. 1984. EDM instruments survey. *P.O.B. Magazine 9* (3) (Feb.–March), 17.

3. L. A. Kivioja, and W. A. Oren. 1981. A new correction to EDM slope distances in applications of untilted corner reflectors, effects of misaligned reflectors and determination of reflector constants. Final Technical Report CE-G-81-1, Purdue University, School of Civil Engineering, West Lafayette, IN.

4. A. F. Gort. 1977. The Hewlett-Packard 3820A electronic total station. Proceedings of the Fall Convention, ACSM. Washington DC.

5. P. W. McDonnell. 1983. Total station survey. *P.O.B. Magazine 8* (6) (Aug.–Sept.), 16.

REFERENCE

Bird, R. G. 1989. *EDM Traverses, Measurement, Computation, Adjustment.* New York: John Wiley & Sons.

6

Angle Measurement: Transits and Theodolites

Edward G. Zimmerman

6-1. INTRODUCTION

Three dimensions or combinations thereof must be measured to locate an object with reference to a known position: (1) horizontal length, (2) difference in height (elevation), and (3) angular direction. This chapter discusses the design and uses of surveyors' transits and theodolites to measure horizontal and vertical angles.

6-2. ANGULAR DEFINITION

An angle is defined as the difference in direction between two convergent lines. A *horizontal* angle is formed by the directions to two objects in a horizontal plane, or by lines of intersection in the horizontal plane with the vertical plane containing the objects (Figure 6-1). In surveying, one of the directions that forms a *vertical* angle is usually either (1) the direction of the vertical (*zenith*) (hence, the angle is termed the *zenith distance*) or (2) the line of the vertical plane in which the angle lies with the horizontal plane, therefore called an *angle of elevation* (+), or *angle of depression* (−) (Figure 6-2).

6-3. UNITS OF ANGULAR MEASUREMENT

The *sexagesimal system* uses angular notation in increments of 60 by dividing the circle into 360 deg; degrees into 60 min; and minutes into 60 sec. Therefore, a complete circle contains 360°, 21,600′, and 1,296,000″. This angular system is employed almost exclusively by surveyors, engineers, and navigators in the United States, as well as extensively in other parts of the world.

The *centesimal system* of angular measurement is based on a circle of 400 increments or *grads* (400^g); 100 centesimal minutes (100^c) per grad; and each centesimal minute split into 100 centesimal records (100^{cc}). The ease of addition and subtraction expressed in centesimal form leads to a decimal notation. Thus, 210^{g}71^{c}84^{cc} is noted as 210.7184^g. This method is used widely throughout Europe.

The *mil system* divides a circle into 6400 increments or *mils*. An angle of 1 mil subtends an arc of approximately 0.98 unit on a circle of 1000-unit radius.

A *radian* is the angle subtended at the center of a circle by an arc equal in length to the circle's radius. It is equal to $360°/2\pi$ or approximately 57°17′44.8″. Table 6-1 lists conversions between the four systems described.

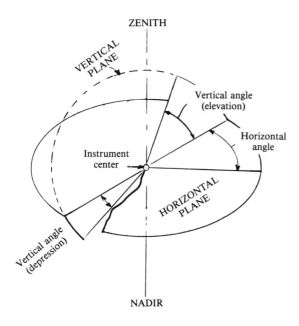

ZENITH

VERTICAL PLANE

Vertical angle (elevation)

Horizontal angle

Instrument center

HORIZONTAL PLANE

Vertical angle (depression)

NADIR

Figure 6-1. Horizontal and vertical angles and respective reference planes.

6-4. BEARINGS AND AZIMUTHS

Directions of lines being surveyed must be determined and tied to a fixed line of known direction commonly defined as a reference meridian. A meridian is either a real survey line or an imaginary reference line to which all courses of a survey are angularly related.

Four basic classifications of meridians commonly used by surveyors are as follows:

1. *Astronomic meridian.* A line on the earth's surface having the same astronomic longitude at every point (see Figure 6-3).
2. *Magnetic meridian.* The vertical plane in which a freely suspended magnetized needle, under no transient artificial magnetic disturbance, will come to rest.
3. *Grid meridian.* A line through a point parallel to the central meridian or *y*-axis of a rectangular coordinate system.
4. *Assumed meridian.* An arbitrarily chosen line with a directional value assigned by the observer.

The direction of a line is the horizontal angle from a reference meridian and can be expressed as an azimuth or bearing.

6-4-1. Bearings

The bearing angle of a line is measured from the north or south terminus of a reference meridian to the east or west, giving a reading always smaller than 90°. Bearings can be astronomic, geodetic, grid, assumed, computed, forward, backward, and in property surveys, record, or deed bearings. In Figure 6-4, the bearing of a line *OA* in the northeast quadrant is measured clockwise from the meridian. Thus, its bearing angle noted is N 45° E. Likewise, the bearings are measured counterclockwise, so for line *OB* it is S 37°43′ W. Line *OD* in the northwest quadrant has a bearing angle of N 47°25′ W.

6-4-2. Azimuths

Azimuths are angles measured clockwise from any reference meridian and range from 0 to 360° (Figure 6-4). They do not require letters to identify their quadrant. For example, the azimuth of line *OA* is 45°; line *OB*, 123°17′, line *OC*, 217°43′; and line *OD*, 312°35′. Figure 6-4 also shows that azimuths can be calculated readily from bearings, and vice versa. In plane

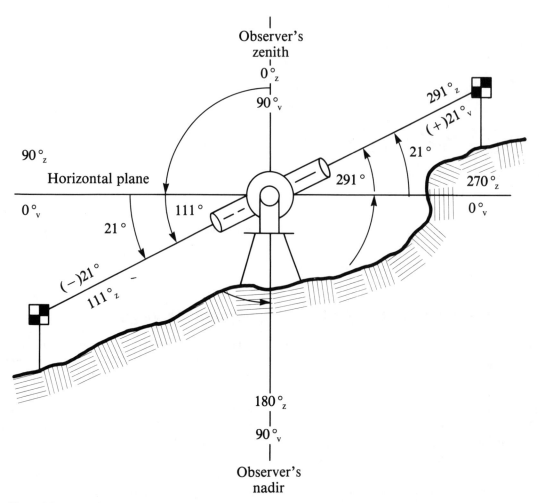

Figure 6-2. Vertical and zenith angles.

Table 6-1. Comparison of various angular definitions.

	Degrees	Grads	Mils	Radians
1 deg =	1	1.11111	17.77778	0.017453
1 grad =	0.9	1	16.0	0.015708
1 mil =	0.05625	0.0625	1	0.0009875
1 rad =	57.29578	63.66198	1018.59164	1

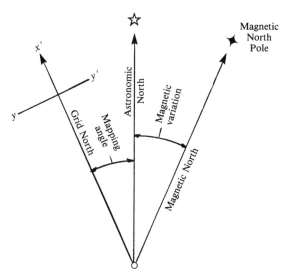

Figure 6-3. Comparison of meridians.

surveying and navigation, azimuths usually are measured from north, but in geodesy and the military services they are referenced to south.

Every line has two directions that differ by 180°. In Figure 6-5, the forward bearing from point K to point L is N 26° E, and the back bearing is S 26° W; likewise, the forward azimuth of line KL is 26° AZ_N, and the back azimuth is 205° AZ_N.

6-5. OPERATIONAL HORIZONTAL ANGLES

Three types of horizontal angles, shown in Figure 6-6, are defined as follows:

1. *Interior angles* are measured clockwise or counterclockwise between two adjacent lines of a closed polygonal figure.
2. *Deflection angles*, right or left, are measured from an extension of the preceding course and the "ahead" line. It must be noted whether the deflection angle is right (R) or left (L).
3. *Angles to the right* are turned from the back line in a clockwise or right-hand direction to the "ahead" line.

Angles are normally measured with a surveyor's transit or theodolite, but they can also be obtained with a sextant, compass, alidade, planetable, or tape. This chapter considers only angular measurements by transit or theodolite.

6-6. HISTORY AND DEFINITION OF SURVEYOR'S TRANSIT AND THEODOLITE

A transit is the surveying instrument having a horizontal circle divided into degrees, minutes, seconds, or other units of circular measurement, and has an alidade that can be reversed in its support without being removed. It is equipped with a vertical circle or arc. Transits are used to measure horizontal and vertical angles, differences in elevation, and horizontal distances. Modern transits may vary in appearance or construction from earlier counterparts, but their principles of operation and use are comparable.

The surveyor's transit probably originated in England during the 16th century. Reference was made in an early engineering essay to a "Topographical Instrument" that appears to be an ancestor of the modern transit. The first American-made transit was most likely constructed by William Young of Philadelphia, in 1831. However, it has been reported that Edmund Draper also constructed a transit in Philadelphia in the same year.

Theodolites are precision surveying instruments consisting of an alidade with a telescope and an accurately graduated circle, and equipped with the necessary levels and optical-reading circles. The glass horizontal and vertical circles, optical-reading system, and all mechanical parts are enclosed in an alidade section along with three leveling screws contained in a detachable base or tribrach. The convenience and inherent precision of theodolites have greatly expanded their use,

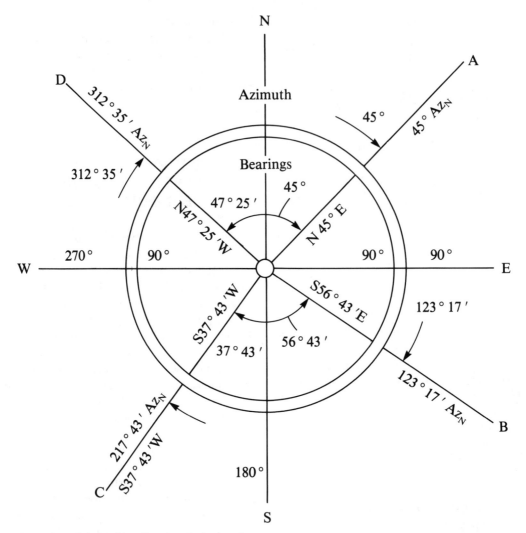

Figure 6-4. Relationship of bearings and azimuths.

and on most types of surveys, they now largely replace transits.

The first optically read theodolite was produced in the early 1900s. Dr. Heinrich Wild designed the then-revolutionary instrument to replace the cumbersome, awkwardly read transit-type theodolites. This new instrument allowed an operator to simultaneously observe both sides of a horizontal circle by means of an optical micrometer, and to determine the arithmetical mean to within a few seconds.

Wild's first optically read theodolite, known as the Wild TH-1, became available in 1923

and was the ancestor of the modern Wild T-2 theodolite. In 1935, Wild also designed an instrument manufactured by the Kern Company. This theodolite was christened the Kern DK-2 and became the forerunner for the Kern Company's present line of optical theodolites.

The pioneer theodolites, as well as their modern counterparts, share the following basic features: (1) They are compact, lightweight, and easy to operate; (2) are shock-, weather-, and dustproof; (3) have high pointing and reading accuracy; and (4) use glass circles and precise graduations that permit small instru-

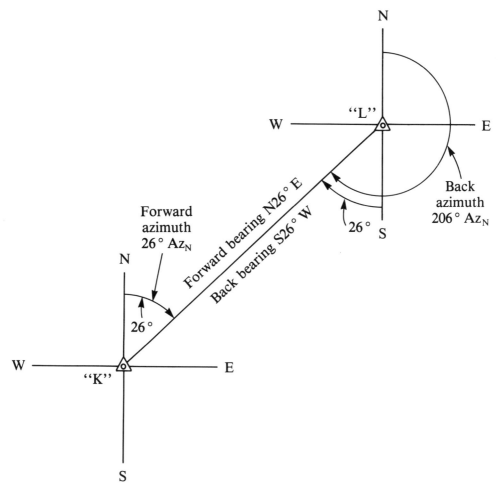

Figure 6-5.   Back or forward bearings and azimuths.

ments to be used in triangulation work. A wide variety of theodolites is available for modern use, most of them based on Wild's original design.

6-7. ANGLE-DISTANCE RELATIONSHIP

Surveyors must know several relationships between an angular value and its corresponding subtended distance. Memorizing two simple trigonometric functions, $\sin 1' = \tan 1' = 0.00029$ (approx.) and $\sin 1° = \tan 1° = 0.0175$

(approx.), permits a quick manual or hand calculator check on angle-distance relationships.

1″ of arc

> = 1 ft at 40 mi or 0.5 m at 100 km (approx.)

1′ of arc = 1 in. at 340 ft (approx.)

1′ of arc

> = 0.03 ft at 100 ft, or 3 cm at 100 m (approx.)

Surveyors must strive to maintain a balance in precision for angular and linear measurements. In Table 6-2, if distances in a survey are

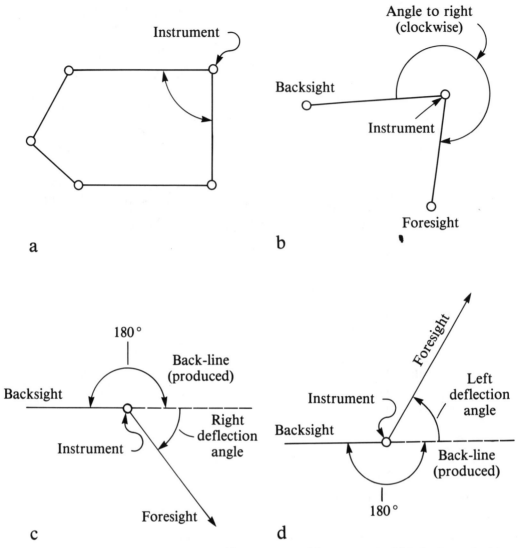

Figure 6-6. Different types of horizontal angles. (a) Interior angle. (b) Angle to right. (c) Deflection angle right. (d) Deflection angle left.

Table 6-2. Comparison of angular and linear errors.

Standard Error of Angular Measurement	Linear Error in		Accuracy Ratio
	1000 ft	300 m	
05′	1.454	0.436	1/688
01′	0.291	0.087	1/3440
30″	0.145	0.044	1/6880
20″	0.097	0.029	1/10,300
10″	0.048	0.015	1/20,600
05″	0.024	0.007	1/41,200
01″	0.005	0.001	1/206,000

to be measured with a relative precision of 1 part in 20,000, the angular error should be limited to 10″ or smaller.

6-8. TRANSIT

6-8-1. Surveyor's Transit

A transit can be called "the universal surveying instrument" because of its many uses. Surveyors and engineers employ transits to (1) measure or lay out horizontal angles and directions; (2) determine vertical angles, differences in elevation (leveling), and distances indirectly (tacheometry); and (3) prolong straight lines. The horizontal and vertical angles read have their vertex at the "instrument center," a point located at the intersection of the instrument's vertical and horizontal axes, as shown in Figure 6-7.

When properly setup and leveled over a survey point, the instrument's axis is identical to a vertical line between the point and its zenith. The horizontal axis is, of course, perpendicular to its vertical axis. Thus, the line of sight (pointing of the telescope) can be rotated simultaneously in both horizontal and vertical planes while maintaining a correct geometric relationship. The line of sight is used to turn a vertical or horizontal angle and measure both with one set of pointings.

6-8-2. Telescope

A transit telescope closely resembles that of a level, as described in Chapter 7. Most level and transit telescopes are equipped for an erecting image, but to obtain superior optical qualities, some high-precision instruments have an inverting eyepiece with fewer lenses. Modern transit telescopes are the internal-focusing

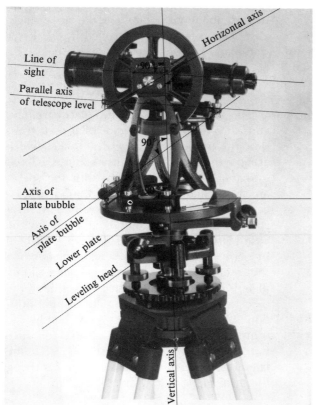

Figure 6-7. Geometry of the transit. (Courtesy of Teledyne-Gurley Co.)

type, generally have a 20- to 30-diameter magnifying power, and are equipped with stadia cross wires (see Figure 6-8 for typical cross-wire patterns). Stadia measurements with transits are discussed in Chapter 22.

A telescope is mounted on the transit's horizontal axis by axles resting in bearings on top of standards that are integral with the upper plate (see Figure 6-9). Adjustable bearings ensure a truly horizontal axis, so the telescope can be rotated 360° in an accurate vertical plane. The telescope is locked in any vertical position by a telescope clamp screw and fine pointing made by turning the telescope tangent screw. A vertical circle connected to the horizontal axis is read by means of the vertical vernier attached to one of the standards. A telescope level tube with a sensitive vial is fastened to the telescope's underside allowing the transit to also be used as a level.

6-8-3. Upper Plate

The alidade of a transit consists of a horizontal circular plate attached to a vertical spindle that allows the upper plate to rotate about the instrument's vertical axis. Two adjustable level vials, one parallel to the telescope and the other at right angles to it, are connected to the plates along with two verniers designated as A and B, set 180° apart. The level vials are also perpendicular to the vertical axis, bringing it to vertical when the vials are leveled.

6-8-4. Lower Plate

A lower plate (Figure 6-7) mounted on a hollow (outer) *spindle* has a graduated horizontal circle on its upper face. The upper plate or inner spindle perfectly fits the outer spindle of the lower plate, which rests in the tapered bore of the leveling head. With the exception of two observation windows, in which verniers and graduated horizontal circle meet and are viewed, the lower plate is completely covered by the upper plate. A clamp on the lower plate locks the inner and outer spindles so the upper and lower plates can be turned about the outer spindle as a single unit. This rotation is controlled by the lower clamp and tangent screw.

6-8-5. Leveling Head

The leveling head is a two-piece structure consisting of a base plate and collar that screws onto the tripod head. A vertical socket, which accepts the lower-plate outer spindle, is built into the base plate. Integral with the socket are four "spider" arms located 90° apart, accepting the leveling screws. The vertical socket-leveling screw unit is attached to the base plate by a half ball-and-socket joint held in place by a sliding plate beneath it. When the leveling screws are tightened, the sliding plate is pulled up against the underside of the base to hold the transit in a horizontal position. If the screws are loosened slightly, the transit head can be shifted a small distance horizontally on the base plate. Thus, a fine lateral adjustment is achieved when centering the instrument over a survey point.

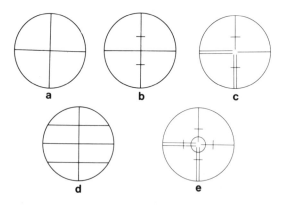

Figure 6-8.   Cross-wire patterns. (a) Standard pattern on less precise transits. (b) Upper and lower wires for stadia; shorter to avoid confusion. (c) Double wires allow centering rather than covering sighted object at distance. (d) Extended stadia lines. (e) For direct solar observation; diameter of circle is 15′45″.

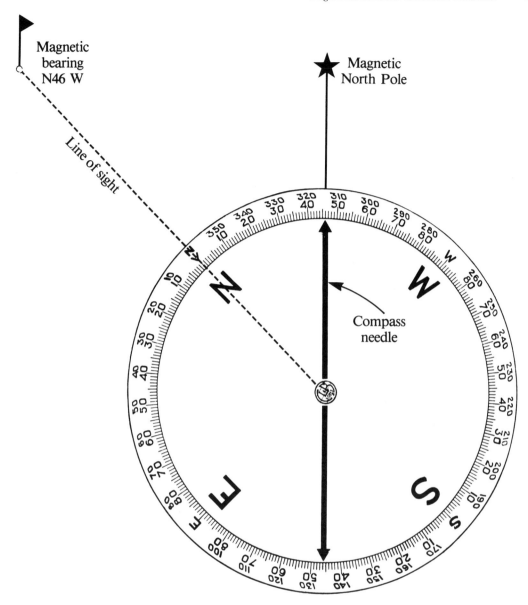

Figure 6-9. Transit compass rose.

Leveling the instrument by lengthening one set of screws and shortening the other rotates the instrument's vertical axis about the ball-and-socket joint. A short chain with a hook for the plumb-bob string is attached to the lower end of the spindle.

As the transit is leveled, the spindle axis becomes a continuation of the plumb line and places the instrument center directly over the survey point. Some transits are equipped with an optical plummet to pass a line of sight through a prism downward along the vertical axis of the instrument. When the transit is leveled, the plummet's line of sight is vertical, so the instrument can be centered over a survey point without using a plumb bob.

Summary

A brief review of the functions of the transit's clamps and tangent screws follows:

1. Vertical rotation of the telescope is controlled by the vertical motion clamp and vertical tangent screw, generally mounted on the right-hand standard of the transit.

2. The upper plate clamp locks the upper and lower circles together; the upper tangent screw permits a small differential rotation between the two plates.

3. A lower plate clamp locks it to the leveling head; the lower tangent screw rotates the lower plate in small increments relative to the leveling head.

4. If the upper plate clamp is locked and the lower one unlocked, the upper and lower plates rotate as a unit, thereby enabling the sight line to be pointed at an object with a preselected angular value set on the plates.

5. With the lower clamp locked and the upper clamp loose, the upper plate can be rotated about the lower one to set a desired angular value. By locking the upper clamp, an exact reading or setting is attained by turning the upper tangent screw.

6-8-6. The Compass

The compass needle is made of magnetized steel and equipped with a cup-type jeweled center bearing to support the needle on a sharply pointed steel pivot post, thus allowing the needle to rotate freely in response to the earth's magnetic field. The needle assembly is contained within a glass-covered compass box attached to the upper plate. The steel mounting post is identical to the vertical axis of the instrument.

A lifter raises the needle off the pivot when not in use. If allowed to rest on the pivot while the instrument is being transported, the compass needle bearing will be quickly worn or damaged. When released, the needle rotates until it is aligned with the magnetic north (or south) pole.

The compass circle or "rose" directly beneath the needle is graduated, as shown in Figure 6-9. Note that east and west are reversed in the compass circle. The needle does not move during an observation; it remains stationary, pointing at the magnetic pole, while the circle (and line of sight) rotate beneath the needle. Thus, it provides the sight-line direction without need to switch the E and W letters when recording a magnetic bearing.

On transits with a compass circle fixed to the upper plate, cardinal N and S points of the circle are in the same vertical plane as the sight line; directions observed are therefore magnetic. Other transits are equipped with a movable compass circle, so a known or assumed declination can be set by rotating the circle. Directions are then referred to true north (or true south in the southern hemisphere).

At best, the expected accuracy of a compass bearing is no better than about 1°. For additional techniques and precautions in using a compass, see Chapter 21.

6-8-7. Tripod

Most surveying instruments are mounted on a three-legged stand known as a tripod. It consists of two parts: (1) the upper component or tripod head and (2) a set of three legs.

The tripod head has a male thread, usually $3\frac{1}{2}$ in. in diameter with eight threads per inch, on which a transit is secured. Smaller or older instruments may have a 3-in.-diameter thread size. Most theodolites employ a different mounting system that has a special $\frac{5}{8}$-in.-diameter bolt with 11 threads per inch. Transit and theodolite tripods are not interchangeable; however, the ensuing discussion applies to transits and theodolites, as well as tripods for levels, sighting targets, and other surveying instruments.

Tripod legs are attached by adjustable tension hinges, permitting the legs to swing on a line radial to the tripod head's center. Legs may be fixed in length or adjustable, solid construction or split. A nonfixed-leg tripod is preferable because it is simpler to level the plates when setting up the instrument over a fixed point. Tripods are generally $4\frac{1}{2}$ to 5 ft tall. Special extension legs are available that raise the telescope height to 8 or 10 ft for sights over brush or other obstructions. All tripods have pointed metal feet, which are pushed into soft ground by stepping on a metal spur fixed at the lower end of each foot.

To assure the best results from a surveying instrument, a stable tripod setup is required. Always plant tripod feet firmly in the ground to form an angle of approximately 60° between the leg and ground, with the telescope at eye-level height, if possible. When setting up on a sloping surface, one tripod leg should be pointed uphill to provide a stable instrument. On a hard surface, tripod feet should be set in a crack or depression, or anchored to each other by a chain loop or wooden framework. This prevents the legs from sliding outward and collapsing the tripod.

Attention must be given to proper tripod maintenance. Bolts attaching the metal feet to legs must be tightened periodically to maintain rigidity. Leg hinges should be tight enough to almost hold an unsupported leg in a horizontal position. If too tight, tripod hinges may bind then unexpectedly release, ruining an accurate setup.

6-8-8. Graduated Circles

Transits are equipped with *horizontal* and *vertical circles*, generally constructed of brass, bronze, or aluminum. During manufacture, graduation lines and numbers are engraved on the circles by machine and filled with black or white paint for visibility.

Vertical circles are commonly divided into quadrants of 90° each, with the number of degrees increasing to 90° in both directions from a 0° mark. The 0° mark of each set of quadrants coincides with the instrument's horizontal plane. The circle is divided into half-degree increments with every 10° division line numbered (Figure 6-10).

Horizontal circles are graduated either in half-, one-third, or one-quarter-degree increments, with every 10° division line numbered. Different circles available are graduated from (1) 0 to 360° in a clockwise direction; (2) 0 to 360° in both directions; or (3) 0 to 360° in a clockwise direction with four 0 to 90° quadrants superimposed, as shown in Figure 6-11. Most surveyors prefer the 0 to 360° in both directions style.

6-8-9. Verniers

Verniers are additional scales used to more accurately read transit circles. One vernier attached to the left-hand transit standard is used in conjunction with the vertical circle. Two other verniers are integrated with the upper plate, 180° apart, for reading the horizontal circle. A vernier enables readings of the circle much closer than the smallest circle division allows. Vernier principles are thoroughly discussed in Chapter 7 and apply to a transit vernier.

Before using a vernier, determine the value of the smallest interval on the adjacent graduated horizontal or vertical circle. Verniers are designed to subdivide the scale unit into an equal number of vernier divisions. The angular value of the smallest vernier division is its *least count*. For example, in Figure 6-12b, the smallest vernier division is 30″; hence, it is classified as a 30-sec least count vernier. Most verniers are 1′, 30″, or 20″ least count. Examples of various verniers are given in Figure

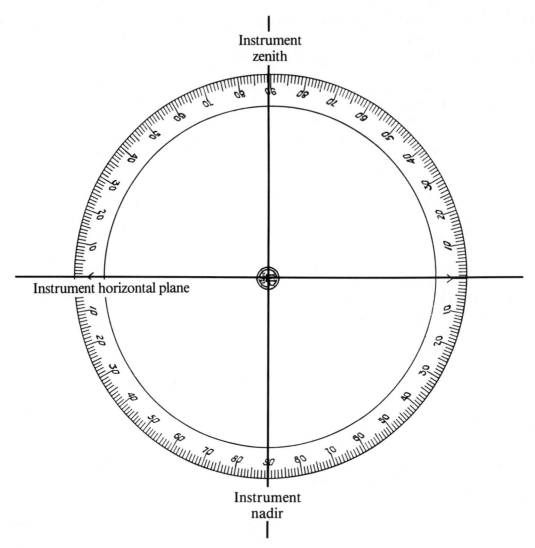

Figure 6-10.   Vertical circle for transit.

6-12; note that all verniers can be read in either direction. This type of vernier is a *double-direct* vernier.

To determine (measure) a horizontal angle, read the horizontal circle as close as its least count permits with the index (zero) mark of the vernier. Next, read the value of the vernier line that is coincident with any division line on the circle; only one vernier division line will exactly align with a circle division line. Com-

bine the two values read to determine the final value. For example, in Figure 6-12b, when we read left to right, the vernier's index mark falls between 130°00′ and 130°20′ on the circle, to provide a "rough" reading of 130°10′. Coincidence occurs at 09′30″ on the vernier. Combining 130° and 09′30″ equals 130°09′30″, the final reading.

The vernier must always be read in the direction that the angle is being turned. Proper

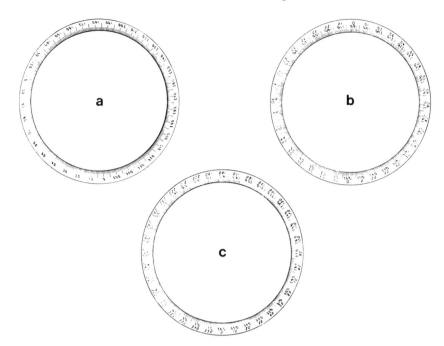

Figure 6-11. Horizontal circles for transits. (a) Graduated 0 to 360°, clockwise. (b) Graduated 0 to 360°, both directions. (c) Graduated 0 to 360°, clockwise, quadrants superimposed.

reading direction is assured by matching slopes of the numbers on the circle and vernier.

A magnifying glass must always be used when reading a vernier. An experienced operator using a glass can consistently estimate to one-half the vernier's least count. To avoid parallax error, observers must always position their eyes directly above the matching lines. Looking at division lines on either side of and immediately adjacent to the index mark is helpful to verify that a selected line is the correct choice. If two sets of vernier and scale lines seem to almost coincide, so a symmetrical pattern exists, interpolating a middle number is reasonable.

6-8-10. Transit Operation

To remove a transit from its carrying case, hold either the leveling head or both standards. *Do not lift* by grasping the telescope.

While holding the transit in both hands (with all motion clamps loose), attach the instrument to a securely positioned tripod. Tighten the threads enough to prevent slack, but do not bind the leveling head on the tripod by overtightening. The transit/tripod unit can be carried by folding together the legs and placing it on your shoulder. Keep the tripod feet forward and the weight of the unit comfortably balanced. Indoors or in bushy terrain, the tripod should be carried under your arms in a horizontal position with the transit forward, giving you more control in avoiding obstacles.

6-8-11. Instrument Setup

On reaching a setup location, extend the legs and center the tripod/transit by eye over the survey point. Attach the plumb-bob string to a hook under the leveling head, using a

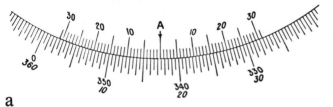

GRADUATED 30 MINUTES READING TO ONE MINUTE
DOUBLE DIRECT VERNIER

Reading L to R = 17° 25 '
R to L = 342° 35 '

a

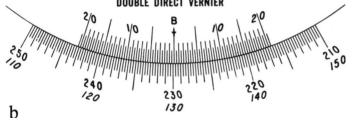

GRADUATED 20 MINUTES READING TO 30 SECONDS
DOUBLE DIRECT VERNIER

Reading L to R = 130° 09 ' 30 "
R to L = 229° 50 ' 30 "

b

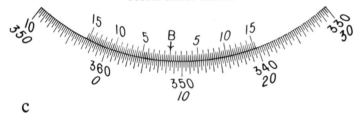

GRADUATED 15 MINUTES READING TO 20 SECONDS
DOUBLE DIRECT VERNIER

Reading L to R = 08° 24 ' 20 "
R to L = 351° 35 ' 40 "

c

Figure 6-12.   Transit verniers.

slipknot to create a loop of string. Slide the slipknot up or down to adjust the line length so the plumb bob is approximately $\frac{3}{8}$ in. above the survey point. Bring the plumb bob into rough alignment over the point by moving one tripod leg at a time in a radial direction. The leveling plate is now brought to roughly horizontal by shifting an appropriate tripod leg in a circumferential direction. If an adjustable leg tripod is being used, combine lengthening or shortening the legs, with movement, to rough level and center.

Align the plate levels over the opposite leveling screws and turn each screw of a pair in opposite directions (lengthening one, shortening the other) with the thumb and forefinger of each hand. *Do not overtighten;* keep both screws in light contact with the leveling plate. A bubble will travel in the same direction as rotation of the left thumb. After both bubbles have been centered, slightly loosen a screw of each set to ease pressure on the leveling plate and shift the head laterally to bring the plumb bob exactly over the survey station. Retighten the screws, check plate bubbles for level, and relevel if necessary. Rotate the upper plate 180° and recheck for level. If the bubbles move off center by more than one division, an adjustment is necessary. See Chapter 8 for checking and adjustment procedures.

A transit equipped with an optical plummet is set up in a similar fashion. The instrument is roughly centered with a plumb bob and leveled. The plumb bob is removed and final centering achieved with the special cross-slide tripod head on optically plumbed transits. After optically centering the instrument, it is necessary to relevel. The plummet's line of sight is also the vertical axis of a transit, so the instrument must be recentered after releveling. An optical-plummet transit is faster to set up because the time needed to adjust and damp the plumb bob's swing is eliminated. It has a more accurate means of centering, particularly on a windy day.

When transit operations have been completed, center the leveling head, equalize lengths of the leveling screws, and lightly tighten all clamps prior to "boxing" or putting away the instrument. At this time, check the transit to make sure it is clean and relatively dust-free. Be certain a cap is placed over the objective lens and the instrument fits securely in the case. If undue pressure is required to close the carrier, check to see whether the transit is correctly positioned. Any moisture on the instrument should be dried prior to putting it away.

6-8-12. Measuring a Horizontal Angle with a Transit

A horizontal angle is measured by first setting the plates to read 0°00′, then backsighting along the reference line from which the angle is to be measured. After the plates are "zeroed" using the upper motion clamp, upper tangent screw, and vernier, the line of sight is brought onto the backsight point by turning the lower clamp and lower tangent screw. The upper clamp is loosened and the telescope rotated independently of the circle until the line of sight is on the foresight target (see Figure 6-13). The angular reading is obtained from the circle and vernier and recorded.

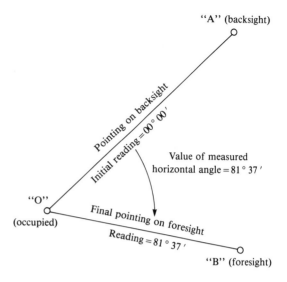

Figure 6-13. Single measurement of a horizontal angle.

A step-by-step procedure follows:

1. Loosen both clamps and bring the 0° circle mark roughly opposite the vernier index mark. This can be accomplished quickly by rotating the lower plate with fingertips pressed against its bottom surface.

2. Tighten upper clamp and bring the zero (0°00′) mark of the circle into precise alignment with the vernier index, using the upper tangent screw. When upper clamp is tight, circle and vernier (upper plate) are locked together as one rotating unit.

3. With lower clamp still loose, point telescope (rotating upper and lower plate unit by hand) to backsight.

4. Tighten lower clamp and use lower tangent screw to align the vertical cross wire with backsight.

5. Loosen upper clamp and rotate upper plate until the telescope is roughly aligned with foresight.

6. Tighten upper clamp and finish pointing by aligning vertical cross wire on the foresight by using upper tangent screw.

7. Using a magnifying glass, read and record angular value indicated by vernier.

6-8-13. Laying out a Horizontal Angle with the Transit

In Figure 6-14, line *IC* will be laid out using angular reference to line *IA* employing the following procedure:

1. Set up transit at *I* and zero plates.
2. Point telescope along line *IA* and sight backsight *A*. Achieve fine pointing with lower tangent screw.
3. Loosen upper clamp, rotate upper circle by hand, and find the preselected angle.
4. Tighten upper clamp and perfect reading with upper tangent screw.

The line of sight is now in the required direction and point *C* can be established on line *IC*.

6-8-14. Measuring Horizontal Angles by Repetition

Repeated measurements of an angle increase accuracy over that obtained from a single measurement. To measure a horizontal angle by repetition, obtain an initial reading with a transit or repeating theodolite, as discussed in Section 6-8-12. Then, continue measuring as follows:

1. Read and record value for the initial angle, as noted in Section 6-8-12.

2. Loosen lower clamp, plunge (transit) the telescope, rotate upper/lower plate unit, and point to backsight.
3. Tighten lower clamp and perfect backsight, pointing with the lower tangent screw. The telescope is now inverted and aligned on backsight, with the initial angle reading remaining set on the horizontal circle.
4. Loosen upper clamp, rotate upper plate, and point at foresight.
5. Tighten upper clamp and complete foresight pointing, using the upper tangent screw.
6. A second angular measurement is accumulated on the plates and read as the sum of the first and second angle. Divide the sum by 2 (or the number of repetitions) to determine average value of the angle.

Repetitions are continued until the required measurement accuracy is met. Repeated sightings are fashioned in even-numbered sets with the telescope plunged on alternate observations. The initial and final readings of a set are made with the same precision. The mean of a set (final reading/number of repetitions) has a precision exceeding that afforded by the vernier least count and scale graduations. Assume an angle 63°21'21" is measured with a 30" transit. A single observation can be read correctly to within 30" or 63°21'30" (possible error limit ± 15"). Measured twice, the observed reading

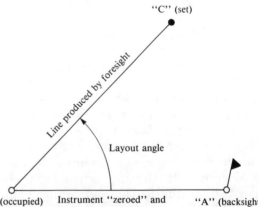

Figure 6-14. Layout horizontal angle by single sighting.

(to $\pm 15''$) is $126°42'30''$. Divided by 2, the average is $63°21'15''$, correct to one-half the vernier's least count and has an error limit of $\pm 07''$. Measured four times, the $253°25'24''$ reading averaging $63°21'22''$ is correct to within one-quarter of the least count, with a possible error of $\pm 03''$.

As computed, the expected accuracy of a measurement is in direct proportion to the number of observations. However, factors including possible eccentricity in instrument centers, errors in plate graduations, instrument wear, and random errors associated with reading, setting, and pointing an instrument fix a practical limit on the number of repetitions at about six or eight. Beyond this number, there is little or no appreciable increase in accuracy.

Systematic instrument errors result because, in actual practice, geometrical relationships of a transit or theodolite (Section 6-8-1) cannot be exactly maintained. When operations with repeating instruments include direct and inverted pointings, systematic errors of the same magnitude occur in opposite directions, thus largely canceling them out. Also, repeating an angular measurement provides a check for, and exposes, reading blunders.

In summary, measuring angles by repetition (1) improves accuracy, (2) compensates for systematic errors, and (3) eliminates blunders.

6-8-15. Laying Out Angles by Repetition

Establishing an angle with accuracy greater than can be expected from a single pointing is accomplished by adaptations of methods detailed in the preceding section. In Figure 6-15, *IA* is an existing reference line, *IB* the line to be established.

An instrument is set up at *I*, plates zeroed, and the sight line directed to a backsight on *A*. Turn specified angle *AIB* as accurately as the least count permits, fixing *IB'*. Temporary point *B'* is set at the required horizontal distance *IB*. Angle *AIB'* is then measured by repetition enough times to attain a desired accuracy (Section 6-8-14). Compare the value obtained for angle *AIB'* with that of angle *AIB*. Any difference is angle *BIB'*, the angular correction needed to locate *B* within specified limits. Usually, the correction angle is too small to be laid off with a transit, so a direct offset is measured along an arc of a circle formed by radius *IB*. A convenient method of accomplishing the offset computation is to multiply length *IB* by angle *BIB'* converted to radian measure. For example (see Figure 6-15), to set *B* 725 ft from *I* on a line $47°28'$ right of line *AI* to an angular accuracy of $\pm 05''$ using a 1-min transit:

1. Set up instrument at *I*, level, zero plates, and backsight on *A*.

2. Turn off $47°28'$ right and establish *B'* at 725 ft.

3. Measure angle *AIB* six times by repetition and read total measurement of $284°46'$.

4. The mean of total angle ($284°46'/6$) is $47°27'40''$, accurate to $\pm 05''$.

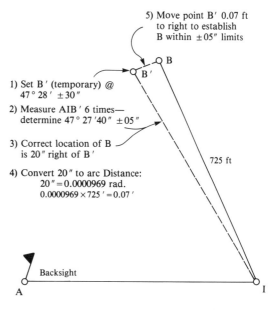

5) Move point B' 0.07 ft to right to establish B within $\pm 05''$ limits

1) Set B' (temporary) @ $47°28' \pm 30''$

2) Measure AIB' 6 times— determine $47°27'40'' \pm 05''$

3) Correct location of B is 20'' right of B'

4) Convert 20'' to arc Distance:
   $20'' = 0.0000969$ rad.
   $0.0000969 \times 725' = 0.07'$

725 ft

Backsight

A

I

B

B'

Figure 6-15. Layout horizontal angle by repetition.

5. Comparing angle *AIB'* with angle *AIB* discloses a difference of 20″ to the left.

6. Compute the correction arc: 20″ (= 0.00009696 rad) × 725 ft = 0.07 ft.

6-8-16. Extending a Straight Line

Often, it is required to prolong a straight line forward from an existing point. Set up a transit or theodolite at the end of the line and locate a new point on the line ahead. As shown in Figure 6-16, straight line *HI* must be extended to *J*. However, *J* is well beyond the range of visibility, but line *IJ* can be established from a setup on *I*. The transit is backsighted to *H*, the upper and lower plates clamped, telescope plunged, a new point *I* sighted, and a point *I* set on the extended line. This procedure is repeated until the line reaches *J*. It is more accurate to plunge the telescope than turn 180° with the horizontal circle.

If the telescope is misaligned or horizontal axis not level, it will be necessary to use the "double-centering" method of prolonging a line as shown in Figure 6-17. To establish *C*, the instrument is set at *B*, the upper and lower plates locked, and a backsight taken to *A*. The telescope is plunged and a temporary mark made at *C*. Then, with the telescope remaining inverted, the alidade is rotated 180°, again backsighted on *A*, the telescope plunged, and a second mark *C'* established. The two foresights will have equal and opposite errors if the instrument is not in perfect adjustment. Therefore, the correct location of *C* will lie midway between C_1 and C_2.

6-8-17. Establishing Line Beyond an Obstruction

Figure 6-18a depicts a simple method to extend a line around an obstruction. This procedure should be used only on low-accuracy surveys. For greater accuracy, the method in Figure 6-18b is employed. A 90° angle is turned at *A* to establish *A'* at a suitable offset from the survey line. *B'* is set at 90° and the same offset distance as *AA'* creating line *A'B'* parallel with *AB*. Line *A'B'* is then extended by setting *C'* and *D'*, which are now occupied to establish *C* and *D* at 90° and offset distance *BB'*. The survey distance is obtained from the total length of line segments *AB*, *B'C'*, and *CD*. To ensure reliability, 90° angles at *A*, *B*, *C'*, *D'* must be accurately turned. It is also important to make lines *AB* and *C'D'*, and offset distance *AA'* as long as practical.

Another method of bypassing obstacles is shown in Figure 6-19c. The transit is set up at *B* and deflection angle *α*, no larger than needed to clear the obstruction, is turned to locate *C*. Point *C* is occupied and an angle 2 *α* deflected in the opposite direction from the first angle. *D* is established on the resulting prolongation of *AB* at the same distance as *BC*. Occupying *D* and turning a deflection angle *x* in the same direction as angle *B* produce line *ABD* ahead to *E*. It is necessary

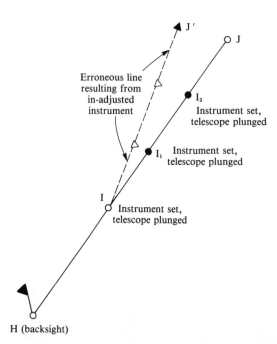

Figure 6-16.   Extend straight line, single plunge.

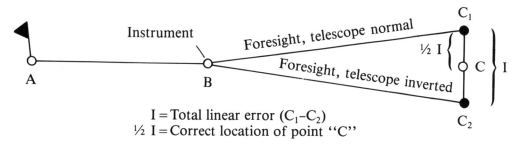

$I = $ Total linear error $(C_1 - C_2)$
½ $I = $ Correct location of point "C"

Figure 6-17.   Double centering.

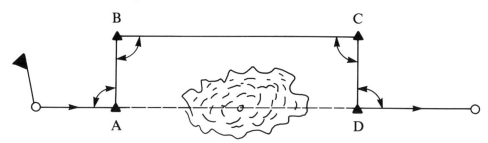

Turn 90° angles @ A, B, C, D
Make distance AB and CD equal

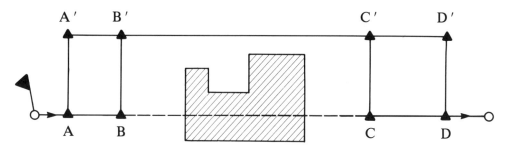

Turn 90° angles @ A, B, C ', D '

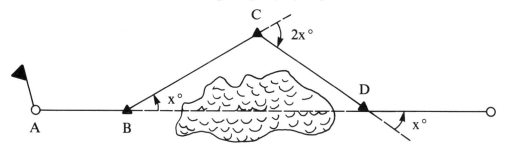

Turn x° deflection (left) @ B; turn 2 x° deflection (right)
@ C; establish D @ a distance equal to B–C; turn x° deflection
(left) @ D.

Figure 6-18.   Extending straight lines beyond obstructions.

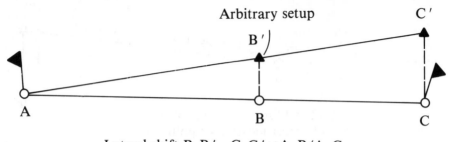

Lateral shift B–B' = C–C' × A–B/A–C

Figure 6-19.   Wiggling in.

to calculate distance *BD*. Lengths *AB* and *CD* are equal so a convenient equation to use is

$$BD = 2 \times BX \times \cos \alpha \qquad (6\text{-}1)$$

6-8-18.   Wiggling In

It may be necessary to set up on a line between two points not intervisible, but both discernible, from an intervening setup. Estimate the alignment by eye and set up the transit. Backsight to the far point and plunge the telescope to check on the nearer one (Figure 6-19). Measure the resulting length *CC'* and use it to estimate the lateral shift required to bring the instrument on line. This operation may have to be repeated a few times until the final shift is within the instrument's sliding head limit. When the correct position is attained, check the alignment by double centering the transit between terminal points of the line.

6-8-19.   Vertical Angles

Before measuring a vertical angle, check for index error by carefully leveling the instrument and centering bubble in telescope vial with the vertical clamp and tangent screw. Check proper adjustment by rotating the upper plate 180°. If the bubble remains centered, the instrument is in correct adjustment. Read vertical circle with telescope bubble centered; if its reading is not 0°00', an index error exists, so all measured vertical angles must be cor-

rected. For example, if a +00°03' index error is found, it has to be subtracted from all altitude (+) angles or added to depression (−) angles.

To begin readings, place the horizontal cross wire exactly on the point being observed by using the vertical clamp and tangent screw. The angle is read on the vertical circle and vernier, then any index error is applied.

For a transit equipped with a 360° vertical circle, vertical angles are observed with the telescope in both direct and inverted positions. The mean of angles read is correct, free of index error.

Some modern transits have a movable vertical vernier, which is controlled by a separate tangent screw and level bubble. The vernier is referenced to horizontal when the bubble is centered and, therefore, minimizes any index error. With transits of this type, the vernier bubble is leveled prior to reading vertical angles.

6-8-20.   Sources of Error

Detailed definitions of errors affecting measurements made with transits and theodolites are discussed in Chapter 3. Classifications investigated here are (1) instrumental errors, (2) personal errors, and (3) natural errors.

Instrument Errors

Proper functioning of transits and theodolites depends on a precise geometrical relationship between instrument components. Although accurately manufactured and assembled, instruments will not achieve and maintain this relationship without provisions for adjustment. If an instrument is in proper adjustment, geometry is maintained within acceptable limits. However, when maladjustment is present, unacceptable errors may result unless certain operational procedures are followed. Sources of instrumental errors are as follows:

1. *Eccentricity of verniers.* If the difference between A and B vernier readings is not exactly 180° and constant around the entire horizontal circle, the verniers are offset. In this case, use only vernier A or the mean of both verniers. If the difference varies for circle positions, eccentricity of instrument centers is indicated. To compensate for this condition, observe angles at several other positions on the circle, mean the A and B vernier readings, and determine averages of observed values.

2. *Imperfect graduations.* Only high-precision measurements are affected by this condition. If angular measurements are distributed around the circle and verniers meaned, this error can be reduced to a minimum.

3. *Plate bubble adjustments.* The vertical axis of an instrument will not be truly vertical if maladjusted plate bubbles are used to level an instrument. Any inclination from vertical creates a variable error in both horizontal and vertical measurements that cannot be equalized by direct and inverted observations. A possible compensation is to center plate bubbles and rotate the upper plate 180°. Observe the length of bubble run-out and bring each bubble back one-half that distance toward the center by manipulating the leveling screws. This will bring back the vertical axis to near vertical. Do not move leveling screws until all angular measurements of a set have been completed.

4. *Line of sight not perpendicular to horizontal axis.* When an instrument's telescope having this error is plunged, the sight line describes a cone of error, whose center is the intersec-

tion of sight line and horizontal axis. The error is most apparent in measuring a horizontal angle on a backsight at a markedly different angle of elevation than for the foresight, or when plunging the telescope to prolong a line or turn deflection angles. All these errors can be minimized by meaning equal numbers of readings with the telescope direct and reversed.

5. *Horizontal axis not perpendicular to vertical axis.* If this error is present, the line of sight will describe an inclined plane when the telescope is plunged. A horizontal angle measured with the backsight and foresight at different elevations will be erroneous. This error is similar to point 4 but it can be controlled by direct and plunged telescope pointings.

6. *Line of sight not parallel with axis of telescope level vial.* This condition creates an error when an instrument is used for spirit leveling. It is eliminated by balancing lengths for fore- and backsights. An error is also introduced when observing vertical angles, but compensated for by meaning equal numbers of direct and inverted sightings.

Summary

A brief review follows:

(a) Instrument errors from inadjustment of plate bubbles or the horizontal axis affect vertical angles and increase in magnitude as vertical angles get larger.

(b) Instrument error misaligning the sight line is greatest when plunging the telescope. No error results if the telescope is not plunged when measuring a horizontal angle between back- and foresight points at the same elevation.

(c) All instrument errors are systematic but can be decreased to an acceptable level by meaning equal numbers of readings with the telescope in normal and inverted positions. Half the readings are too large, half too small. Averaging the sum gives the correct angle.

Systematic errors are kept to a minimum by keeping instruments in correct adjustment.

veyors can perform certain adjustments in the field, as outlined in Chapter 8. However, if an instrument is damaged by being dropped or from an accidental blow, it should be sent to a professional repair facility.

Personal Errors

These errors result from limitations of human eyesight and judgment and are considered to be accidental. Examples are described in the following list:

1. *Not centering the instrument over an occupied point*. This affects all horizontal angles measured there. Error magnitude varies inversely with the lengths of courses observed. A transit set 0.04 ft off center to measure an angle between sides 200 ft long causes an error of approximately 1 min. For 2000-ft sides, the error is about 6 sec. Thus, although reasonable care should be taken in instrument centering over a station, spending extra time in positioning the transit perfectly is unnecessary, particularly on long sights.

2. *Not sighting directly on a point*. This error has the same effect mentioned in the previous point. A pointing 0.10 ft off a target 350 ft away produces an angular error of approximately 1 min. The same pointing error for a mark 2000 ft distant is roughly 10 sec. Greater care must be exercised on shorter sight distances, and a narrower object (plumb-bob string or pencil point) used.

3. *Misreading vernier*. An accidental error occurs when the observer does not use a magnifying glass, or reads the scale and vernier graduations in a nonradial direction. An experienced instrumentperson can correctly estimate to one-half the vernier's least count.

4. *Improper focusing (parallax)*. Care must be taken to sharply focus the cross wire and objective-lens images. Horizontal and vertical angles suffer in accuracy when improper focus causes parallax.

5. *Level bubbles not centered*. Plate bubbles should be checked frequently during operations, but not releveled during a measurement set (as they are in differential, profile, and other leveling projects). If an instrument is accidentally bumped, relevel and begin the interrupted operation again.

6. *Displacement of tripod*. Survey personnel must exercise care when walking around an instrument. If set up on soft ground, it can easily be displaced by one step near a tripod foot. The instrument can also be disturbed if contacted by loose clothing or a carelessly carried tool. If moved, reset it and repeat the work in progress when the disaster occurred.

To summarize, personal errors are accidental and cannot be entirely eliminated, only reduced in number and magnitude through proper techniques. They constitute the major factor in angular measurement inaccuracy. The prime sources of personal errors are caused by failing to exactly read and set vernier and micrometer scales and in not making perfect pointings on targets.

Natural Errors

Natural errors are defined as those created by the following:

1. Poor visibility resulting from rain, snow, fog, or blowing dust.

2. Sudden temperature changes, causing uneven expansion or contraction of instruments or tripods.

3. Unequal refraction deforming the line of sight or inducing a shimmering effect (heat waves) that make it difficult to accurately observe targets.

4. Settlement of tripod feet on hot pavement, or soft and soggy ground.

5. Gusty or high-velocity winds that vibrate or displace an instrument, move plumb-bob strings, and make sighting procedures difficult.

In summary, natural errors generally are not enough to affect work of ordinary precision. To lessen their effect in higher-order, certain steps can be taken, such as reducing temperature changes and refraction problems by shielding the transit from the sun with an umbrella, or performing work at night. When surveying in soft or swampy areas, support tripod legs on long wooden stakes driven into any unstable ground. Always discontinue work when weather conditions become unreasonably severe.

6-9.   OPTICAL THEODOLITES

6-9-1.   Descriptions

Two types of theodolites are available: double-center and directional. Both share certain features, but each has unique operating principles. Differences in them will be discussed.

Compared with transits, theodolites are compact and generally weigh only about 10 lb. Their vertical axis is cylindrical and rotates on precision ball bearings. The horizontal and vertical circles are made of glass and have precisely etched graduation lines and numerals. Optional models have graduations in degrees, grads, or mils. The circles, optical-reading system, and mechanical parts are totally enclosed within a weather- and dust-proof housing. All circle readings and bubble position checks can be made from the eyepiece end of the telescope, thereby eliminating unnecessary movement around the instrument.

Telescopes are usually short, fully transitable, and equipped with a large objective lens; they contain a glass reticle having an exactly engraved set of cross lines, and have internal focusing to provide sharp views even at relatively short ranges. The alidade can be detached from its mounting or tribrach.

Three screws supporting the tribrach are used to level the instrument in concert with a circular (*bull's-eye*) bubble on the tribrach, and a single level is mounted on the alidade. The vertical circle is equipped with either an indexing level bubble or automatic compensator to establish a horizontal reference plane minimizing "index error."

Angles are read through an optical system consisting of a microscope and series of prisms. An adjustable mirror on the outside of the instrument housing reflects light into the reading system; a battery-powered light provides illumination for night work.

6-9-2.   Repeating Theodolites

This type of theodolite is constructed with a double vertical axis similar to a transit, cylindrical rather than conical as in a transit. Repeating theodolites are equipped with upper and lower circle clamps and tangent screws and can be used to measure angles by repetition.

Generally, horizontal and vertical circles are graduated in 20-sec or 1-min increments, both circles having the same least count. Some theodolites read directly to 6, 10, or 30 sec. Most horizontal circles are divided from 0 to 360° in a clockwise direction. Vertical circles are also graduated from 0 to 360°, 0° corresponding to the instrument's zenith. With the telescope level, in normal position, a zenith angle of 90° is read; in an inverted position, the zenith angle is 270°. Figures 6-20 and 6-21 show typical repeating theodolites. Each of these models is equipped with automatic compensators to correctly orient the vertical circle; both have a telescope magnification of 30 × and 30″/2 mm plate-bubble sensitivity. Both types use optical plummets.

The Wild T-1 (Figure 6-20) can be read directly to 6″ with estimation to 3″ for both circles. Figure 6-20 shows the field of view observed through the T-1's reading microscope. Zenith angles appear in the upper window (V); horizontal angles are seen in the lower window (Hz); and the micrometer readings viewed in the middle opening. To read an angle, rotate the micrometer knob (located on the right standard) and center a circle graduation line between the double index marks. Direct numerical reading of the micrometer setting then is found in the right-side window.

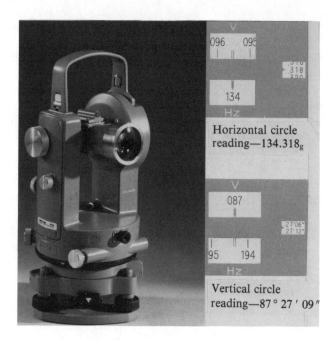

Horizontal circle
reading—134.318$_g$

Vertical circle
reading—87° 27′ 09″

Figure 6-20. T-1 repeating theodolite. (Courtesy of Wild Heerbrugg Instruments, Inc.)

The angle depicted in Figure 6-20 is

Upper window (zenith angle)	87°
+ Right window (micrometer setting)	27′09″
Final reading (to nearest 03″) =	87°27′09″

Both circles appear simultaneously in the reading microscope, so the Hz circle is yellow, the V circle white.

The Kern K1-S (Figure 6-21) is identical in basic operation to the Wild T-1 except for its circle-reading method. Figure 6-21 presents

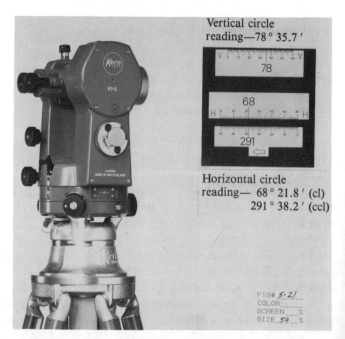

Vertical circle
reading—78° 35.7′

Horizontal circle
reading— 68° 21.8′ (cl)
291° 38.2′ (ccl)

Figure 6-21. Kern K1-S repeating theodolite. (Courtesy of Kern Instruments, Inc.)

the K-1's reading microscope field of view. Upper window (V) displays a zenith angle, middle opening (H) the clockwise horizontal angle, and bottom window (arrow to left) exhibits the counterclockwise horizontal angle. Observed angles are direct scale observations, read to 0.5 min with estimation to 0.1 min. Thus, the readings in Figure 6-22 are

Zenith	78°35.7′
Horiz. (clockwise)	68°21.8′
Horiz. (counterclockwise)	291°38.2″

Basic angular measurement operations with a repeating theodolite are identical to those for a transit. Sections 6-8-14 through 6-8-16 are also applicable to repeating theodolites. On older theodolites, a spirit level attached to the vertical circle must be centered prior to reading a zenith angle. Modern theodolites are equipped with an automatic compensator to minimize inclination of the vertical axis in zenith angle measurement.

6-9-3. Directional Theodolites

A directional theodolite is not equipped with a lower motion. It is constructed with a single vertical axis and cannot accumulate angles. It does, however, have a horizontal circle positioning drive to coarsely orient the horizontal circle in any desired position. A directional-type theodolite with plate-bubble sensitivity of, generally, 20″/2 mm division is more precise than repeating theodolites.

Directions, rather than angles, are read. After sighting on a point, the line direction from instrument to object is noted; when a pointing is taken on the next mark, the difference in directions between them is the included angle. Optical-reading systems of direction instruments permit an observer to simultaneously view the circle at diametrically opposite positions, thus compensating for any circle eccentricities.

Theodolites shown in Figures 6-22 and 6-23 are typical of directional theodolites. Both are

Figure 6-22. Wild T-2 directional theodolite. (Courtesy of Wild Heerbrugg Instruments, Inc.)

equipped with a micrometer scale that provides horizontal and vertical circle readings directly to 1″ (with estimation to 0.1″), automatic compensators for vertical circle orientation, and optical plummets.

Figure 6-22 shows the field of view through the reading microscope of a Wild T-2 theodolite. The upper window exhibits vertical lines above and below a thin horizontal line, representing simultaneous readings on opposite sides of the horizontal or vertical circle. Rotation of the micrometer knob brings a set of lines into coincidence and moves a pointer in the middle window to display circle readings directly to 10 min. Single minutes and seconds estimated to the nearest 0.1 sec are obtained through the bottom window. The reading in Figure 6-22 is

Middle window (after coincidence)	94°10′
+Lower window (after coincidence)	02′44″
Final reading	94°12′44″

Choice of a vertical circle (white) or horizontal circle (yellow) is determined by the position of a selector knob, located on the instrument's right standard.

A different format is used for reading the Kern DKM-2AE. Figure 6-23 illustrates the view seen in its reading microscope. The upper window displays a vertical circle reading in degrees (large number) and nearest 10 min (framed by cursor). The vertical (V) and horizontal (H) coincidence scales are located directly below the top window. Each contains a pair of double index lines superimposed over a pair of single lines regulated by the micrometer control knob. To read an angle, obtain coincidence of the selected circle by rotating the micrometer knob and centering the single line within the double index marks. Degrees and tens of minutes are read in the upper window: the large upper window for vertical angles, the large lower window for horizontal angles. The direction in Figure 6-23 is read as follows:

Coincidence achieved	
Top window (vertical)	85°30′
+Bottom window (sec/min)	35′14″
Zenith angle	85°35′14″

When observing horizontal angles, use the third window from the top (H) to achieve coincidence and read the horizontal angle immediately below.

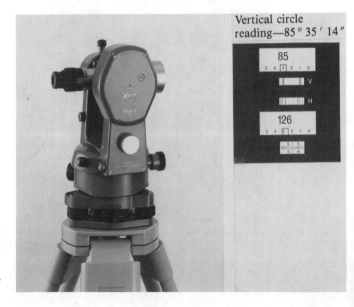

Figure 6-23. Kern DKM 2-A directional theodolite. (Courtesy of Kern Instruments, Inc.)

6-9-4. Electronic Theodolites

Recent developments in electronics have been incorporated into surveying instrument manufacturing, and several digital theodolites are now on the market. They are designed to automatically read, record, and display horizontal and vertical angles. Their basic design and operation is similar to those previously discussed; however, these electronic models can display results digitally or store data directly in an electronic data recorder for later retrieval, computing and plotting by an office computing system.

Figure 6-24 shows the Wild T2000, a newly developed electronic theodolite. When equipped with a Wild Distomat EDMI, the T2000 becomes a total station, capable of

measuring and displaying horizontal and vertical angles, horizontal distance, and elevation difference. An on-board computer reduces slope distances and corrects horizontal distances for earth curvature and refraction. Coordinates for a currently occupied station are calculated if starting values are entered for the initial traverse point.

T2000 uses a dynamic angle-measuring system, allowing a full scan of the rotating circle during each measurement. Graduations around the entire circle are considered during each observation, eliminating the effect of graduation errors and any residual eccentricity of the circle and axis. Internal precision of the measuring system is about $\pm 0.2''$. However, the manufacturer states that when atmospheric and pointing errors are considered,

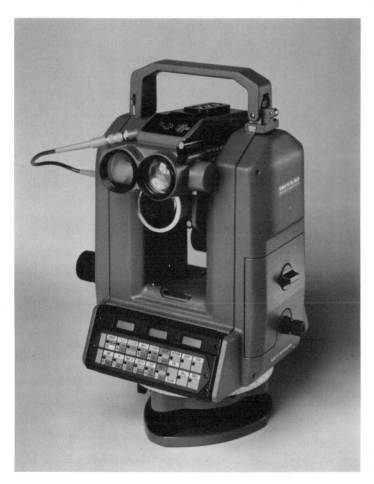

Figure 6-24. Wild T-2000 electric theodolite equipped with an EDMI. (Courtesy of Wild Heerbrugg Instruments, Inc.)

the standard deviation for a mean normal and inverted pointing is $\pm 0.5''$.

6-9-5. Theodolite Setup

Roughly level and center the tripod over the point to be occupied. Lift the theodolite from its case by the standards or, if equipped, the lifting handle. Place the instrument on the tripod and attach it by firmly tightening the tribrach attachment bolt. The instrument can be centered like a transit (horizontal location by plumb bob, fine centering by optical plummet), but the following method using an adjustable-leg tripod is faster.

After selecting a setup point, place one tripod leg about 2 ft beyond it. Next, grasp the other two legs and, while looking through the optical plumb, position legs so the point occupied is visible in the plummet eyepiece. Push legs firmly into the ground and, while looking through the plummet, bring the cross wires over the setup point by adjusting the level screws. Next, roughly center the bull's-eye level by changing the tripod leg lengths to approximately center and level the instrument over a point. The setup is completed by carefully leveling the theodolite and then precisely positioning it over the mark by lateral shifting on the tripod head. During final centering of the instrument, do not rotate it; if the tripod head is not level, rotation will take a theodolite out of level.

Theodolites are equipped with three leveling screws, a single plate level, and a circular bull's-eye level used mainly for coarse setup purposes. Final leveling is done with the screws and plate bubble in the following four steps:

1. Rotate instrument and align the plate-bubble axis with two leveling screws.
2. Center plate bubble by adjusting the screws, rotate the instrument 90°, and use the third screw to recenter bubble.
3. Repeat those procedures, then reverse to make a final leveling check.
4. If the alidade bubble moves more than two divisions upon reversal, it must be readjusted.

6-9-6. Horizontal Angles with Directional Instruments

As noted in Section 6-9-3, a directional theodolite reads directions or "positions" on its horizontal circle. The difference in directions to two points is the angle included between them.

The following methods are used with the Wild T-2 theodolite, but can also be employed with other directional instruments. Figure 6-25 is a diagram of the measurement procedure for a theodolite set up at Q. With the horizontal clamp loose, make a rough pointing to K, tighten clamp, and perfect pointing with the horizontal tangent screw. Coincidence is achieved with the micrometer knob and line direction QK observed as $12°31'16''$. Next, loosen the horizontal clamp, rotate instrument, and point roughly on L. Tighten horizontal clamp, and using the horizontal tangent screw perfect pointing and observe line direction QL as $76°11'39''$. Subtracting QL from QK gives an angle $63°40'23''$.

Note that no attempt was made to set the horizontal circle on zero, although the theodolite has a control to move it to a predetermined approximate position. This control is very coarse and does not permit fine settings. Trying to make exact settings of the "seconds" portion of any position is not recommended.

Measuring Horizontal Angles by Repetition

Repetitive direction measurement requires each line to be observed with the telescope in direct and reversed positions. The directions are meaned and the results used to calculate an angle. A complete set of direct and reverse observations to a point is termed a *position*. An example set of field notes for directions to points H, I, and J, with a Wild T-2 theodolite setup at K is shown in Figure 6-26.

The following measurement procedure refers to those noted:

1. Loosen horizontal clamp and point at H, the left-most station of the set, designated

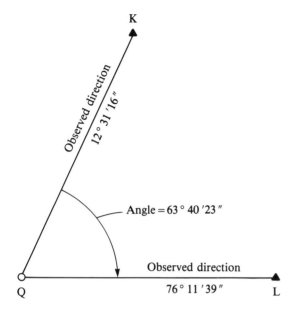

K

Observed direction

12°31′16″

Angle = 63°40′23″

Observed direction
76°11′39″

Q   L

Figure 6-25. Horizontal angle with direction
instrument. Single measurement.

the initial point. Initializing on *H* permits directions to be read and recorded in a clockwise sequence.

2. Lock the horizontal clamp and use the circle-drive knob to set approximately 00°00′10″ on the circle.

3. Perfect pointing and read the directions to *H* (00°00′34″). Loosen horizontal clamp and point on *I*; tighten clamp and finish pointing.

4. Read and record the direction to *I* (202°21′53″).

5. Use the same routine to observe and record the direction to *J* in completing the position's first step.

6. Next, loosen the horizontal clamp, rotate alidade 180°, reverse the telescope, and point again to *J*.

7. Read and record the direction, which will differ by approximately 180° from the first reading.

8. With scope inverted, sight to *I* and *J* (counterclockwise), completing the first position or set of angles.

In this example, it was required to complete two positions. Distribute the readings uniformly around the circle by making a second initial position pointing at *H*, with approxi-

mately 90°20′20″ set on the circle. Perfect pointing, read and record the direction to *H* (90°22′29″). Complete a second position by measuring the remaining directions in the set with the telescope direct and reversed.

Always reduce notes prior to leaving all occupied stations by meaning direct and inverted observations for every pointing in the position, then determine the mean direction for each position. Compute all the meaned directions by subtracting the value of the meaned initial direction (*KH*) to get final directions. If any position varies from the mean of all positions by more than ±5″, reject it and reobserve that particular position.

6-9-7. Zenith Angles with a Theodolite

Unlike transits, theodolites are not equipped with a telescope level. One of two different mechanical arrangements is used to orient a theodolite's vertical circle to its zenith.

The first method, used on older instruments and more precise theodolites, is to attach a spirit level on the vertical circle. A separate tangent screw rotates the circle about its horizontal axis to center the bubble and

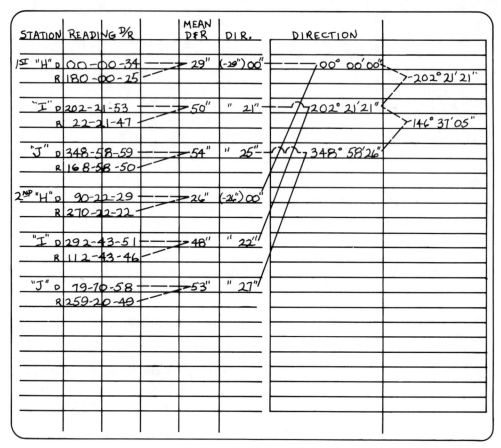

Figure 6-26. Example field notes for horizontal angle by repetition with direction instrument.

minimize index error. The general procedure in using this instrument type is to point the horizontal cross wire on an observed object, then center the bubble orienting the circle. Neglecting to center the bubble introduces an accidental error of unknown size into a measured angle.

The second indexing method utilizes an *automatic compensator* responding to the influence of gravity. With the theodolite properly leveled, the compensator is free to bring the vertical-circle index to its true position in much the same way an automatic level compensator functions. Automatic compensators are generally of two types: (1) *mechanical*, whereby a suspended pendulum controls prisms directing light rays of the optical reading system or

(2) an *optical* system, in which the optical path is reflected from the level surface of a liquid.

Zenith observations generally follow this routine:

1. Point instrument on object to be observed.
2. With the telescope in direct mode and vertical clamp loose, set horizontal cross wire on object, tighten vertical clamp, and perfect pointing with vertical tangent screw.
3. If the instrument is not equipped with an automatic compensator, index the vertical circle by centering bubble, then read and record the vertical angle.
4. Rotate alidade 180°, loosen vertical clamp, repoint on target with telescope inverted, and read and record the observed zenith angle.

5. A mean zenith angle is obtained by first adding the direct and reverse readings to obtain the algebraic difference between their sum and 360°, then dividing this difference by 2, and algebraically adding the result to the first (direct) series measurement.

6. The result is the zenith angle corrected for any residual index error.

If greater reliability is required, repeat the steps and determine mean values. The result of an individual set should agree with those of at least two sets within the following limits:

$$1'' \text{ theodolite } \pm 5''$$
$$20'' \text{ theodolite } \pm 12''$$
$$1' \text{ theodolite } \pm 12''$$

If these limits are not met, read additional sets until the rejection limit is satisfied.

Zenith (Z) angles can readily be converted to vertical (V) angles. For example, 79°51'14″ Z is resolved:

$$(90°00'00'' - 79°51'14'') = +10°08'46'' \text{ V}$$

A 101°31'06″ Z equals

$$(90°00'00'' - 101°31'06'' V) \text{ or } -11°31'06'' V$$

Zenith angles from the inverted sightings are 273°16'47″ Z equals

$$(273°16'47'' - 270°00'00'') = +03°16'47'' \text{ V}$$

and 264°21'32″ Z equals

$$(264''21'32'' - 270°00'00'') = -05°38'28'' \text{ V}$$

Figure 6-2 illustrates the relationships.

6-9-8. Horizontal Angle Layout by Directional Theodolites

Layout of horizontal angles can be done in the following manner:

1. Set up, level, and center the instrument over a selected point; observe and record the direction to reference backsight.

2. Add the required angle to the reference direction to determine the direction of line being established.

3. Put this answer in the instrument, using the micrometer knob to set the desired single minutes and seconds.

4. Loosen the horizontal clamp and rotate alidade until degrees and tens of minutes are roughly located in the microscope.

5. Lock the horizontal clamp and finish setting, employing the horizontal tangent screw to bring appropriate division lines into coincidence.

6. Sight through the telescope to mark a point on the established line.

7. Loosen the horizontal clamp, rotate alidade 180°, invert telescope, and sight to newly set point.

8. Read and record the direction, loosen horizontal clamp, point on backsight, determine direction, and calculate the angle.

9. If the second angle differs by more than 10 sec from a first (layout) angle, repeat the entire procedure from a new start.

10. If a wide variance continues on additional repetitions, check the theodolite for maladjustment.

6-9-9. Forced Centering

Most modern theodolites are mounted in a detachable tribrach that permits the instrument to be quickly interchanged with an EDM reflector or sight pole without disturbing integrity of the tripod/tribrach setup. To take full advantage of the interchangeability and "forced-centering" operation, a survey crew should be equipped with three or more tripod/tribrach sets and the necessary adapter hardware.

A list of steps in the forced-centering procedure follows:

1. On completing observations at a station, the theodolite is detached from its tribrach, leaving a tripod/tribrach unit centered and leveled over the station.

2. The theodolite is carried ahead to the next station and attached to a tripod/tribrach

from which the foresight target has been removed.

3. The rearmost or former backsight unit is picked up and carried forward to a new station to be observed, set up and a foresight target fastened.

Advantages of forced centering are obvious: Instead of three separate setups at every station (foresight, theodolite occupation, and backsight), only a single placement of the tripod/tribrach unit is necessary. Two opportunities for accidental setup errors have been eliminated.

6-9-10. Expected Accuracy of Theodolites

Results derived from testing, manufacturer's technical specifications, and conservative assumptions indicate that the accuracy of measurements made by experienced personnel, under favorable conditions and using instruments in good adjustment, are reasonably expected to be within the following limits.

For a 1-sec theodolite, most angles measured should have a probable error not more than

(a)	One position (1 direct, 1 reverse)	$\pm 4''$
(b)	Two positions	$\pm 3''$
(c)	Four positions	$\pm 2''$

For a 1-min theodolite, the maximum error in most angles measured by repetition is

(a)	Turned twice	$\pm 7''$
(b)	Four repetitions	$\pm 4''$
(c)	Six repetitions	$\pm 3''$
(d)	Twelve repetitions	$\pm 2''$

7

Leveling

Robert J. Schultz

7-1. INTRODUCTION

Leveling is a process to determine the vertical position of different points below, on, or above the ground. In surveying operations, vertical elevations and vertical control are generally derived independently of horizontal control. Some modern positioning devices, termed total stations, allow simultaneous determination of spatial coordinates. *Elevation* is the vertical distance above a well known datum or arbitrary reference surface. Elevations are helpful for the placement of a water drain line to provide free gravity flow, construction of a sports playing field, and among other applications, the vertical layout of a roadbed to allow a smooth flow of trucks and trains, which must ascend or descend sloping terrain.

Usually, elevation measurements are made above a specific reference surface, such as *mean sea level*. This surface may be defined as the position of the ocean if all currents and tides cease to exist. It is then projected under the land surface. It is a surface on which gravity measurements would all be the same value, and hence it may be called an equipotential gravity surface and, more specifically the *geoid*. The earth's gravity field decreases with distance above mean sea level. Scientific studies have located this surface by such varying techniques as continuous *tide gage* readings and calculations employing artificial variations in satellite orbital elements.

7-2. DEFINITIONS

Figure 7-1 demonstrates a basic vocabulary of words used in leveling literature. All surfaces shown are referenced to the physical plumb-bob line at a point, such as A. This line is the direction that the string of a free-hanging plumb bob takes in a still atmosphere. Conventional leveling equipment is constructed to place the telescope line of sight in a plane perpendicular to the plumb-bob line.

A *vertical line* follows the direction of gravity (plumb-bob line) through any point, such as B. If points A and B are several miles apart, curvature of the earth causes plumb-bob lines through A and B to converge. Because of curvature, mass density changes and hidden masses below the earth's surface, all vertical lines are not parallel, even at close spacing. Generally, however, gravity variations are small, and these lines can be considered parallel in most applications. Surveys performed under

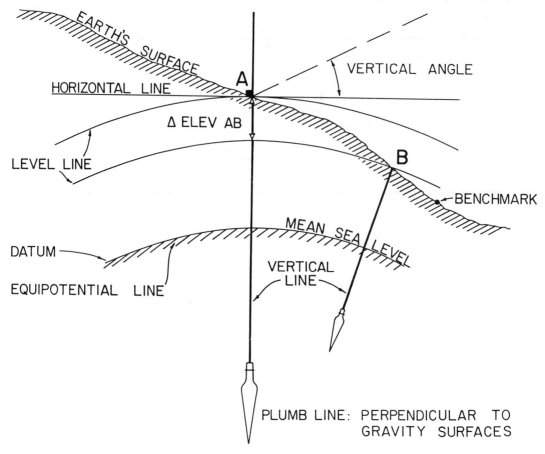

Figure 7-1. Leveling nomenclature.

this assumption are considered to be *plane* surveys.

A *horizontal line* is perpendicular to the vertical line at a point under consideration. Spirit or pendulum-type levels make the line of sight horizontal, hence, the sight line can be rotated in the horizontal plane when an instrument is properly set up.

A *level line* is a line in a level surface. A level surface has all points perpendicular to the direction of gravity and hence is curved. To many people, horizontal and level surfaces are synonymous, but surveyors draw a clear distinction between them and attempt to measure vertical distances between level surfaces to obtain elevations.

The *difference in elevation* between two points *A* and *B* is expressed as vertical distance

through point *A* along a vertical line to the intersection of a level line or surface through point *B*. The elevation of point *B* in Figure 7-1 is found by subtracting the difference in elevation *AB* from *A* to give *B* a lower elevation. Points below mean sea level are considered negative quantities.

A *bench mark* (BM) is a permanent point of known elevation. It is located by arbitrary assignment of a fixed elevation or extension of vertical control. High-order field work based on a beginning reference surface, such as mean sea level, usually provides the basis for the network. The BM should be permanent, stable, and a recoverable object such as a brass cap, steel pipe, or man-made object listed with a description in some public agency's table of vertical-control data. The most commonly used reference datum in the United States is the *National Geoditic Vertical Datum of 1929,* formerly known as the *Sea Level Datum of 1929.*

7-3.   DIRECT LEVELING

Elevations of points are determined by direct and indirect means. Most vertical control for engineering construction is accomplished by *direct leveling.* In this method, elevation differences between a continuous short series of horizontal lines is determined by direct observations on graduated rods, using an instrument equipped with a sensitive spirit level, or a pendulum-type "automatic" level. Figure 7-2 illustrates this procedure.

First- and second-order vertical-control stations are established by the National Geodetic Survey (NGS) along lines at approximately 1-km spacings to form grids 50 to 100 km square. Other federal, state, city, and county organizations provide control of lower order. The U.S. Geological Survey prepares Control Survey Data maps at a scale of 1:250,000, showing major federal lines for use by local communities and surveyors. Those desiring information on first- and second-order NGS lines can write or call: The Director, NGS Information Center, NOAA, Rockville, MD 20852, to order data sheets listing adjusted mean sea-level bench marks included in the national network.

Many state highway offices also provide additional information about local vertical control. It is common to use these given data and run a line of levels to a construction job site, thus providing a temporary bench mark to control all other site elevations. Farmers wishing to drain their fields may assign an arbitrary elevation to a fixed point or rock outcrop in the area and refer all elevations to it.

7-4.   PROCEDURES IN DIFFERENTIAL LEVELING

From its written description, a beginning bench mark is located; then 10 to 100 m from the bench mark, the instrument is set up and leveled. A backsight (BS) to the bench mark is observed on a plumbed graduated rod, and the height of instrument (HI) determined by the formula HI = elev BM + BS. The BS may be negative if the rod is held in a mine survey against the tunnel roof. A foresight (FS) reading is taken on the rod held at any suitable point, thus creating a *temporary bench mark* (TBM) or *turning point* (TP). To eliminate systematic instrumental errors, the backsight and foresight should be approximately equal in length. Then elev TP (or TBM) = HI − FS. A series of such setups is taken until the final permanent bench mark is reached. The instrument can be set up and operated this way, by a skilled operator, under most topographic

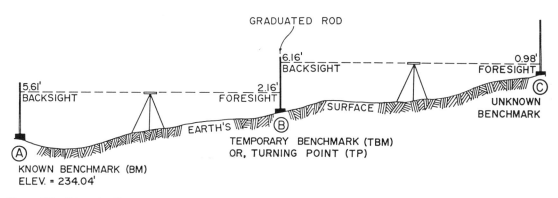

Figure 7-2.   Direct leveling.

conditions. Briefly stated, differential leveling consists of taking one backsight and one foresight of approximately equal length to achieve high-precision leveling.

In Figure 7-3, the height of a building is $B''B = AB' \tan \alpha + AB' \tan \beta = AB'(\tan \alpha + \tan \beta)$. Vertical angles above the horizon are positive $(+)$, those below it are negative $(-)$. Thus, elev B = elev A + HI + $AB' \tan \beta$, where β is a negative angle and elev B'' = elev A + HI + $AB' \tan \alpha$, where α is a positive angle.

7-5. TRIGONOMETRIC LEVELING

An indirect technique for measuring elevation differences is to read the vertical angle and distance to a point using a clinometer, transit, or theodolite. For short distances, plane surveying principles are applied.

In Figure 7-3, the height of a building is $B''B = AB' \tan \alpha + AB' \tan \beta = AB'(\tan \alpha + \tan \beta)$. Vertical angles above the horizon are positive $(+)$, those below it are negative $(-)$. Thus, elev B = elev A + HI + $AB' \tan \beta$, where β is a negative angle and elev B'' = elev A + HI + $AB' \tan \alpha$, where α is a positive angle.

In this procedure, precision of vertical angles α and β and linear measurement AB'

must be compatible to yield an accurate answer. Table 7-1 shows the relations between linear and angular errors, and Table 7-2 the precision of computed values.

In the right triangle $AB'B''$ (Figure 7-3), if α is 5° exactly and the horizontal distance AB' is 1000 ft exactly, the vertical distance $B'B''$ equals 87.49 ft. An error of 1 sec in the vertical angle measurement would have little effect on the height $B'B''$. However, an error of 1 min would cause errors of 0.15, 0.29, and 0.44 ft at 500, 1000, and 1500 ft, respectively. If the horizontal distance measurement AB' is also in error, the accuracy of the height determination $B'B''$ would decrease owing to a combination of the angular- and distance-measurement errors. In general, these errors are kept about equal to one another, thus forming a circle of error and creating a small plus or minus allowable tolerance in the actual height determination.

7-6. EARTH SHAPE CONSIDERATIONS

Over long distances, the effects of *earth's curvature* and *atmosphere refraction* must be considered in leveling, as shown in Figure 7-4. A properly set up and leveled instrument, in adjustment, has its line of sight perpendicular

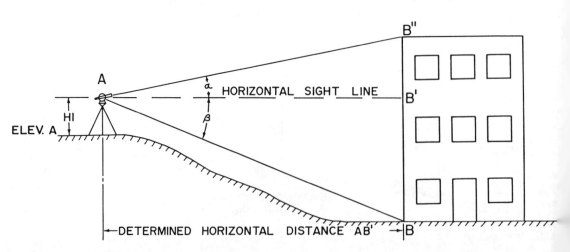

Figure 7-3. Elevation by vertical angles.

Table 7-1.   Relation between linear and angular errors

Allowable Angular Error for Given Linear Precision			Allowable Linear Error for given Angular Precision				
Precision of Linear Measurements	Allowable Angular Error	Least Reading in Angular Measurements	Allowable Linear Error in				Ratio
			100′	500′	1000′	5000′	
$\frac{1}{500}$	6′53″	5′	0.145	0.727	1.454	7.272	$\frac{1}{688}$
$\frac{1}{1000}$	3′26″	1′	0.029	0.145	0.291	1.454	$\frac{1}{3440}$
$\frac{1}{5000}$	0′41″	30″	0.015	0.073	0.145	0.727	$\frac{1}{6880}$
$\frac{1}{10,000}$	0′21″	20″	0.010	0.049	0.097	0.485	$\frac{1}{10,300}$
$\frac{1}{50,000}$	0′04″	10″	0.005	0.024	0.049	0.242	$\frac{1}{20,600}$
$\frac{1}{100,000}$	0′02″	5″	0.002	0.012	0.024	0.121	$\frac{1}{41,200}$
$\frac{1}{1,000,000}$	0′00.2″	2″	0.001	0.005	0.010	0.048	$\frac{1}{103,100}$
		1″		0.002	0.005	0.024	$\frac{1}{206,300}$

to the plumb line and, except for the atmospheric refraction, the line of sight would lie in a horizontal plane. The earth's curved level surface departs from the horizontal by a distance c.

The normal ellipsoidal earth model has doubled curvature with independent radii in the *meridian* and *prime vertical*, which is at 90° to the meridian. If a spherical earth is assumed and low precision satisfactory, a single radius of approximately 20.9×10^6 ft can be assumed. A more precise radius depends on the observer's *latitude* and the sight-line *azimuth* of the observation.

A practical expression for curvature is $c = 0.667M^2$, where c is the earth's curvature in feet, and M the distance in miles. The coefficient 0.667 contains appropriate factors for geometry, unit conversion, and the earth's radius.

Due to a difference in density, an optical sight line passing through the atmosphere refracts or bends back toward the earth. This refraction effect r is usually taken as one-seventh the effect of curvature and helps compensate for that factor. The refraction correction requires knowledge of temperature, pressure, and relative humidity, which are difficult to evaluate over long distances, so simplifying assumptions are generally used.

The combined effect of curvature and refraction is given by

$$(c + r) = h = 0.574M^2$$

The following shows that $(c + r)$ increases rapidly with distance:

	$(c + r)$ Effect				
Distance	200 ft	500 ft	1000 ft	1 mi	2 mi
h ft	0.001	0.005	0.021	0.574	2.296

Table 7-2. Precision of computed values

Size of Angle and Function		Angular Error				
		1′	30″	20″	10″	5″
		Precision of Computed Value Using Sine or Cosine				
Sin 5° or cos 85°		$\dfrac{1}{300}$	$\dfrac{1}{600}$	$\dfrac{1}{900}$	$\dfrac{1}{1800}$	$\dfrac{1}{3600}$
10	80	$\dfrac{1}{610}$	$\dfrac{1}{1210}$	$\dfrac{1}{1820}$	$\dfrac{1}{3640}$	$\dfrac{1}{7280}$
20	70	$\dfrac{1}{1250}$	$\dfrac{1}{2500}$	$\dfrac{1}{3750}$	$\dfrac{1}{7500}$	$\dfrac{1}{15,000}$
30	60	$\dfrac{1}{1990}$	$\dfrac{1}{3970}$	$\dfrac{1}{5960}$	$\dfrac{1}{11,970}$	$\dfrac{1}{23,940}$
40	50	$\dfrac{1}{2890}$	$\dfrac{1}{5770}$	$\dfrac{1}{8660}$	$\dfrac{1}{17,310}$	$\dfrac{1}{34,620}$
50	40	$\dfrac{1}{4100}$	$\dfrac{1}{8190}$	$\dfrac{1}{12,290}$	$\dfrac{1}{24,580}$	$\dfrac{1}{49,160}$
60	30	$\dfrac{1}{5950}$	$\dfrac{1}{11,900}$	$\dfrac{1}{17,860}$	$\dfrac{1}{35,720}$	$\dfrac{1}{71,440}$
70	20	$\dfrac{1}{9450}$	$\dfrac{1}{18,900}$	$\dfrac{1}{28,330}$	$\dfrac{1}{56,670}$	$\dfrac{1}{113,340}$
80	10	$\dfrac{1}{19,500}$	$\dfrac{1}{39,000}$	$\dfrac{1}{58,500}$	$\dfrac{1}{117,000}$	$\dfrac{1}{234,000}$
		Precision of Computed Value Using Tan or Cot				
Tan or cot 5°		$\dfrac{1}{300}$	$\dfrac{1}{600}$	$\dfrac{1}{900}$	$\dfrac{1}{1790}$	$\dfrac{1}{3580}$
10		$\dfrac{1}{590}$	$\dfrac{1}{1180}$	$\dfrac{1}{1760}$	$\dfrac{1}{3530}$	$\dfrac{1}{7050}$
20		$\dfrac{1}{1100}$	$\dfrac{1}{2210}$	$\dfrac{1}{3310}$	$\dfrac{1}{6620}$	$\dfrac{1}{13,250}$
30		$\dfrac{1}{1490}$	$\dfrac{1}{2980}$	$\dfrac{1}{4470}$	$\dfrac{1}{8930}$	$\dfrac{1}{17,870}$
40		$\dfrac{1}{1690}$	$\dfrac{1}{3390}$	$\dfrac{1}{5080}$	$\dfrac{1}{10,160}$	$\dfrac{1}{20,320}$
45		$\dfrac{1}{1720}$	$\dfrac{1}{3440}$	$\dfrac{1}{5160}$	$\dfrac{1}{10,310}$	$\dfrac{1}{20,630}$
50		$\dfrac{1}{1690}$	$\dfrac{1}{3390}$	$\dfrac{1}{5080}$	$\dfrac{1}{10,160}$	$\dfrac{1}{20,320}$
60		$\dfrac{1}{1490}$	$\dfrac{1}{2980}$	$\dfrac{1}{4470}$	$\dfrac{1}{8930}$	$\dfrac{1}{17,870}$
70		$\dfrac{1}{1100}$	$\dfrac{1}{2210}$	$\dfrac{1}{3310}$	$\dfrac{1}{6620}$	$\dfrac{1}{13,250}$
80		$\dfrac{1}{590}$	$\dfrac{1}{1180}$	$\dfrac{1}{1760}$	$\dfrac{1}{3530}$	$\dfrac{1}{7050}$
85		$\dfrac{1}{300}$	$\dfrac{1}{600}$	$\dfrac{1}{900}$	$\dfrac{1}{1790}$	$\dfrac{1}{3580}$

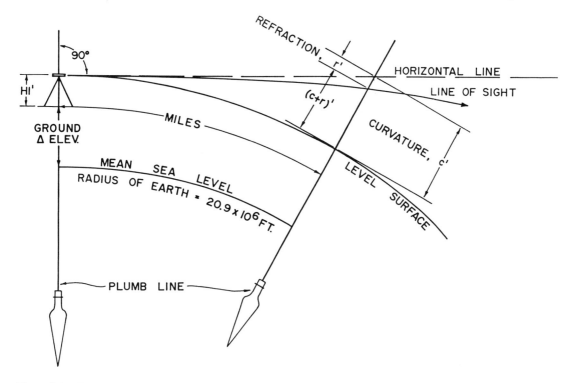

Figure 7-4. Curvature and refraction.

This effect is generally neglected in construction surveys. In precise control surveys performed by the direct technique, backsights and foresights are closely balanced to reduce or cancel this systematic error, as shown in Figure 7-5, where $(c + r)_1$ cancels $(c + r)_2$ when $d_1 = d_2$.

Another way to compensate for this error is *reciprocal leveling*, where sightings must be taken across a gorge, canyon, or river. An instrument is set up on both sides of the obstacle, and the level rods are read simultaneously to cancel or reduce the errors. This is accomplished by meaning the results. An assumption

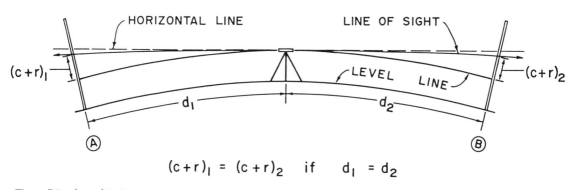

Figure 7-5. $(c + r)$ Balancing effect.

made in reciprocal leveling is that refraction at both ends of the line is the same, although this is seldom the case.

7-7.   INDIRECT LEVELING

In *trigometric leveling*, a similar curvature and refraction condition exists, as illustrated in Figure 7-6. The elevation of point C above A' is $BD + DE - BC$. For long sights, triangle $A'BD$ is still considered to be a right triangle and horizontal distance $A'D$ is taken as the level surface length AE. In current EDM practice, it is usually convenient to measure the slope distance $A'B$. Vertical distance BD is calculated as $A'D \tan \alpha$ and DE is the equivalent curvature and refraction correction $(c + r)$.

A theoretical way to overcome this $(c + r)$ correction is to simultaneously observe with two theodolites at A and C, thereby cancelling the systematic errors. Rarely are refraction conditions identical at both ends of a line, but averaging helps to distribute the error and is the generally accepted measuring technique for long-distance trigometric leveling.

7-8.   OTHER LEVELING PROCEDURES

In addition to standard field surveying instruments, specialty items such as *barometers, lasers,* and the *global positioning system* (GPS) (Chapter 15) with receivers and antennas are available for unique leveling tasks. Air pressure changes with height, but precision surveying *altimeters* have been developed that use changes in barometric pressure to determine elevation differences from a base station with a precision of 2 ft. For small area surveys, a single altimeter may suffice for local photogrammetric control or rough spot-elevations, where easy access exists for vehicles. The procedure sets the altimeter to a known base elevation, and then readings are taken at desired locations. Finally, on returning to the base station, any difference in this reading from the original is distributed linearily around the loop.

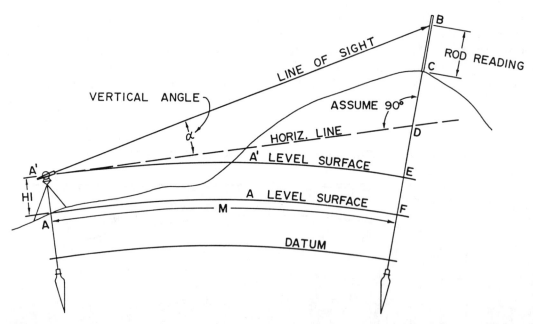

Figure 7-6.   Trigonometric leveling.

An improved technique uses two altimeters. One is placed at the base bench mark and its changes noted, so time-dependent corrections can be applied to the roving instrument. Additional altimeters permit a leapfrogging technique to produce higher accuracy.

Construction site lasers have been developed that hold line for horizontal control but also provide vertical elevations. Alignment lasers can be placed in sewers, water lines, and construction trenches, and positioned on a slope. A target attached to the pipe end shows the low-power beam and permits pipe ends to be positioned on the proper design slope.

Another laser device can be leveled to rotate the beam over the area in a horizontal plane. Elevations to the nearest one-tenth of a foot on the building site are determined. A sliding light-sensing device attached to the side of the leveling rod can be moved up or down to receive the rotated laser light and show how far the rod's end is below the laser plane. Some systems allow an audio as well as a visual determination of the foresight. Systems of this type can be run by an equipment operator and do not require a two-person surveying crew.

The GPS consists of earth *satellites*, which transmit ultrastable signals with timing using cesium and rubidium atomic clocks, ground-based equipment to monitor the satellites, and a receiver to convert the signals into positions in a given reference frame. The ground equipment needed to use the system consists of an antenna, receiver, and computer that passively receives the signals from four or more satellites and computes a three-dimensional position.

7-9. INSTRUMENTATION

Leveling instruments are designed to have the sight line in or near the level surface; this is accomplished in a variety of ways by both simple and complex devices. For example, building inspectors may wish to check a con-

crete flat slab to determine if high and low points are within $\pm 1/8$ or $\pm 1/4$ in. of the mean surface elevation stated in job specifications. By flooding the slab with water, it is a simple task to measure the puddle depth with a carpenter's rule and sight across the water to estimate ridge heights. Early Roman builders used water leveling troughs with sight vanes to level aquaducts and tunnels—crude engineering, but projects were constructed and civil works accomplished with such elementary devices. Modern instrumentation improves the precision of layout and construction.

7-10. HAND LEVEL

A basic instrument used in construction today is the *hand level*. It consists of a spirit bubble and sighting horizontal wire in a telescope having zero or 2 × magnification. Figure 7-7 shows a spirit bubble vial.

Low-cost hand levels generally have a sighting chamber with no optics and only a horizontal cross wire and mirror to show the bubble image on the wire. Sights are considered level when the bubble is centered on the wire. Low-powered optics are often introduced into the hand level for work up to 50 or 75 ft. Unaided sights are generally used up to 30 ft for elevations to 0.1 ft, the normal specification figure for grading around building construction sites. Demands for more accuracy are met by typical engineers' levels, which include *dumpy* and *automatic levels*.

7-11. LEVELING VIALS

Level vials are rated by their *sensitivity*, the degree of tilt that moves the bubble through one 2-mm (or 0.01 ft or 0.1 in. in older instruments) division etched on the vial, or by the radius of curvature ground on the inside by their manufacturer. A short radius causes the bubble to be insensitive, but suitable. A hand-level bubble might have only one center grad-

GLASS VIAL
with circular curvature

PAINTED GRADUATIONS
at 2 mm spacings

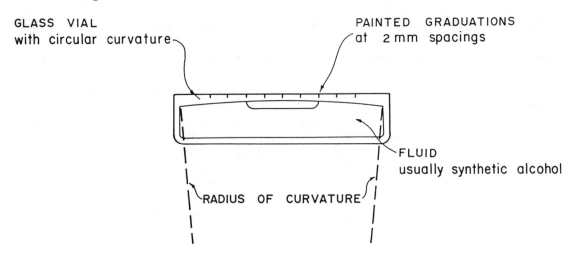

RADIUS OF CURVATURE

FLUID
usually synthetic alcohol

Figure 7-7. Bubble vial.

uation, so the observer estimates when the bubble is centered. A more precise instrument, the dumpy level, has a 20″ of arc/2-mm divisions or 68-ft radius bubble sensitivity. For first-order instruments, the values can be 2″/2 mm at a 680-ft radius. Special construction techniques are employed to grind the curvature into the glass vial at these high sensitivity values.

Two techniques are available to test bubble sensitivity. The laboratory procedure is to set the bubble into a level bubble trier, which is a slotted piece of wood hinged at one end and mounted on a precise screw. The vial is placed

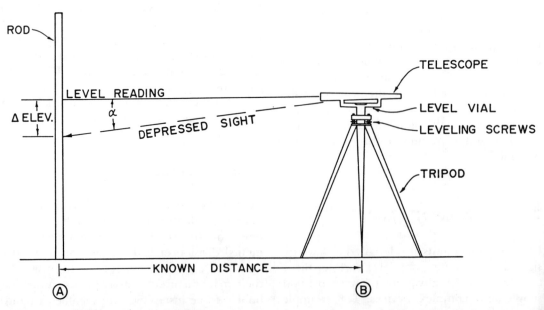

Figure 7-8. Testing bubble vial sensitivity.

in the apparatus slot and raised or lowered by the screw. When the screw thread pitch is known, the number of turns of the screw to move the bubble through a known number of graduations tells the bubble sensitivity.

A field technique is shown in Figure 7-8. Here an engineer's level is set up some known distance from a leveling rod and a reading is taken. The level is then tilted through a known number of divisions on the bubble, and a new reading is taken on the rod. The sensitivity of the instrument s is given by

$$s = \tan^{-1} \alpha \left[\frac{\Delta\text{elev}}{(AB)(\# \text{ of div})} \right]$$

where Δelev and AB are in the same units, and s is usually expressed in minutes or seconds of arc/2-mm division.

The fluid inside bubble vials is usually synthetic alcohol, which will not freeze at low temperatures. With the extremely precise instruments used for first-order work, such as a universal theodolite or precise level, the quantity of fluid in the bubble tube can be regu-

lated by a special chamber attached to the vial. This ability to change the length of the bubble allows the instrument operator to compensate for temperature changes and use the most sensitive portion of the vial.

7-12. ENGINEERS' LEVELS

A series of engineers' levels with sensitive level bubbles, high telescope magnification, and fine-pitch leveling screws have evolved for construction and precise surveying work. Modifications include prisms instead of level bubbles, spherical ball-seat leveling surfaces rather than leveling screws, and tilting telescopes instead of fixed ones. Figure 7-8 shows a tripod, the leveling screws to tilt the instrument head so the leveling bubble or leveling device can be centered, and a telescope permitting long sights to be made.

A modern level is shown in Figure 7-9. Light rays enter the telescope *objective lens*, which has a special light-gathering coating on

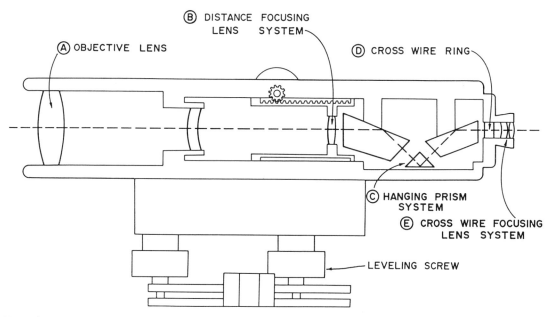

Figure 7-9. Automatic leveling telescope.

its outer surface, and pass through a distance-focusing movable lens that allows the image of a distance object to be in focus on the cross-wire ring. The pendulum system shown has two fixed prisms and a freely hanging third prism that passes the horizontal rays onto the cross-wire ring. To allow the pendulum system to hang free, a low-sensitivity spirit bubble is mounted on the instrument, and rough leveling of the instrument produces an "automatic leveling" of the sight line.

Different manufacturers have designed various systems to accomplish automatic leveling; instruments so constructed are called *automatic levels*. At the telescope's eyepiece end, a magnifying lens system exists enabling the instrument operator to simultaneously view the cross-wire ring and distant image. If focusing is not performed exactly, a condition known as *parallax* can exist and cause small reading errors by the observer.

In older telescopes, the cross wires were spider webs cemented to a ring. Newer instruments have lines etched on a glass reticle that sits in a metal holder. The holder can be repositioned to correct for systematic errors in the hanging prism system.

7-13. THE DUMPY LEVEL

The engineer's dumpy level is an older instrument that contains leveling screws, a spirit bubble, and a telescope (Figure 7-10). The four leveling screws are worked in opposite pairs to position the bubble at the vial center. The bubble follows the directional movement of the left thumb. Automatic levels generally have three leveling screws and the telescope elevation is changed slightly by raising or lowering one or two leveling screws. Four-screw instruments rotate around a spherical ball seat, and once leveled, they can be releveled to the original position if the bubble wanders away from the center position.

A telescope *line of sight* is the straight line through the center of the objective lens and intersection of the horizontal and vertical cross wires. Instrument design requires the line of sight to be made parallel to the axis of the level bubble vial—i.e., the line tangent to the radius of curvature of the vial at the marked center location when the instrument is exactly leveled. Because it is difficult to construct an instrument with a fixed telescope and bubble, each part can be shifted slightly so adjustments

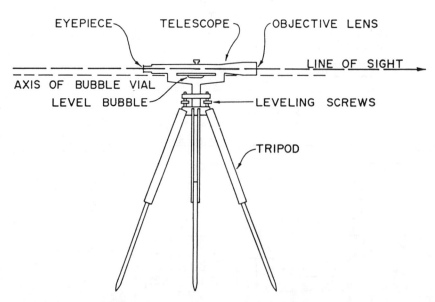

Figure 7-10. Dumpy level.

to remove systematic instrumental error may be made in the shop or field. On construction instruments such as the old *wye level*, the dumpy and the automatics, their telescopes have approximately 32 × magnification, and sensitivity of the bubble vials is about 20″/2-mm divisions.

added to the rod reading. This system can be very exact and repeated for a statistical determination of the difference to be added to the rod reading. Automatic instruments used in construction can be fitted with a removable optical micrometer and approach the accuracy of special, extremely precise, automatic levels used for first-order leveling work.

7-14. FIRST-ORDER LEVELS

For first-order work, a spirit-bubble-type instrument with greater magnification and bubble sensitivity has been constructed with a horizontal pin around which the telescope can rotate when driven by an extremely fine-pitch screw. This tilting level also has a reticle containing three horizontal wires, so the horizontal distance to the leveling rod can be read by stadia (see Chapters 6 and 22). To assist in obtaining an exact middle-wire reading, an *optical micrometer* is built onto the end of the telescope; this arrangement is shown in Figure 7-11. A *planoparallel lens* serves as the *objective lens* and can be tilted through a range equal to the 5- or 10-mm graduations on the leveling rod. An observer moves the micrometer to make the line of sight fall exactly on a rod graduation. The micrometer reading is then

7-15. LEVELING RODS

To determine the height of instrument above a bench mark, a graduated rod is held vertically on the point and a reading taken. For precise work, rod graduations are in the SI system (meters), whereas on construction and other work in the United States, markings are in decimal feet. When a hand level is used, a carpenter's rule graduated in feet and inches might be employed. Rods come in a variety of sizes and shapes, but the *Philadelphia* type has an advantage, since the rodperson can independently check instrument operator readings. These rods generally come in two pieces, which allow readings up to 12 or 13 ft to be taken and checked by setting a target on the graduations, which are shown to 1/100 of a foot (see Figure 7-12).

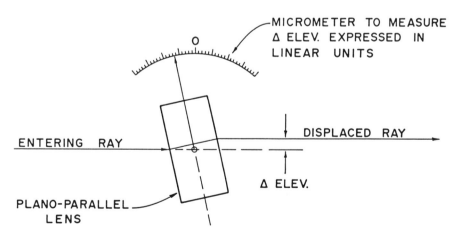

Figure 7-11. Optical micrometer.

The foot lengths are indicated by red numbers, and the 0.01-ft wide graduations are painted black on a white background. The intersection of the black and white portions of the graduations defines the value. As shown in the red 3-ft portion of the rod, the horizontal cross-wire reading is 3.64 ft. A vernier target can be slipped over the rod and readings taken to 0.001 ft with the direct-reading vernier for instrument sight distances of 200 ft or shorter. This target can be used on the lower portion of the rod, and a technique called *high rod*, employing a downward-reading scale on the back of the rod, permits the rodperson to verify high-rod readings with a properly set target and vernier on the rod's back side.

The metric-faced construction rods are usually color-coded orange for meter readings with centimeter spacings. The instrument operator reads the meters, decimeters, and centimeters directly from the rod. Then, if a fine reading is desired, the number of millimeters is estimated. A target with scale provides a millimeter check estimate.

On precise work, all readings are taken in the SI system. The rods are of special one-piece construction approximately 3 m long, which presents special shipping and carrying problems. The rod face is made of Invar steel and held to the face under tension. A thermometer is usually provided so a temperature correction can be applied. The smallest graduations on the rod face are in centimeters or 5-mm values. The optical micrometers on first-order instruments allow repeatable readings to 0.1 mm, the equivalent of 0.0003 ft or 0.00003 in. Because of these small values, plumbing the rod is usually done with an attached bull's-eye level or special holder and special care given to placing the rod on a hard permanent surface. *Portable turning points* are carried into the field and used to provide a stable platform for the rods.

7-16. NOTEFORMS

Different noteforms are used for various standard direct-leveling procedures. When running a third-order control project, traverse (when suitable), around a construction site, a closed noteform is generally used (see Table 7-3 and Figure 2-2).

This noteform contains one *backsight* (+ S) and one *foresight* (− S) per setup and illustrates differential leveling. The *height of instru-*

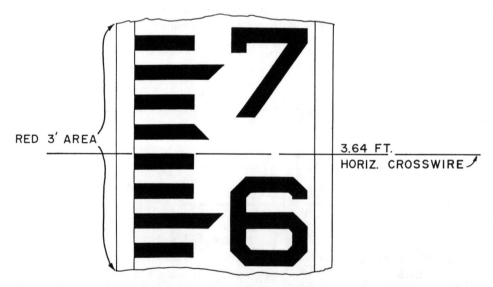

Figure 7-12. Philadelphia rod face detail.

Table 7-3. Establishing height of green pin (closed noteform)

STA	+ S	HI	− S	ELEV (ft)
BM 235	3.47	239.23	—	235.76
TP-1	9.10	246.62	1.71	237.52
TP-2	7.91	253.46	1.07	245.55
Green Pin	3.07	252.57	3.96	249.50
TP-3	0.13	245.59	7.11	245.46
TP-4	1.90	240.16	7.33	238.26
BM 235			4.41	235.75
	+ 25.58		− 25.59	− 235.76
			+ 25.58	− 0.01✔
			− 0.01✔	

Table 7-4. Left-hand page, partial set of three-wire notes (in mm)

STA	Thread Backsight	Mean	Thread Interval	Sum of Intervals
209	1216			
	1108	1108.3	108	215
	1001		107	
	3325			
	1326			
TP-1	1237	1237.0	89	178
	1148	2345.3	89	393
	7036			

ment (HI) is determined for each setup but not done in precise leveling where only the difference in elevation between two bench marks is desired. As shown, the Σ backsights − Σ foresights = final elevation − initial elevation. Accidental errors in leveling do not usually allow a circuit that starts and stops at the same point to generally close with the same value. This misclosure would have to be adjusted out by some technique consistent with the caliber of the work.

In first- or second-order control work, the notes include readings on a *precise level's* upper, middle, and lower wire, thereby allowing several checks and providing a stadia distance. One page of the field book is devoted to backsights, the other to foresights. The *stadia-interval factor* for the instrument must be known, and acceptable precomputed values for the maximum length of sight, maximum differences in lengths observed per setup, and maximum differences in cumulative distances for a section are computed and recorded by the notekeeper. A partial set of notes is shown in Table 7-4.

7-17. PROFILE LEVELING

When topographic conditions will not allow differential leveling or low precision is suffi-

cient to accomplish the job, *profile leveling* is used in combination with differential leveling. Side shots with no checks can be taken in determining the profile of a road and/or the elevation of a manhole. An example of profile notes is shown in Table 7-5.

A separate column of *intermediate foresights* (IFS) was listed, with *side shots* taken to stations 0 + 00, 1 + 00, 2 + 00 and 2 + 52 on Highland Street. A station in highway work is 100 ft on the ground and Sta 0 + 00 is the beginning point. Sta 1 + 00 is a point 100 ft away and usually on the street centerline. Manhole (MH) number 17 might be a grade shot taken to the top or to an invert in the manhole. A note or sketch on the right-hand page of the field book would clarify the situation. The nonchecked profile shots are recorded to only 0.1 ft, which is consistent with street subgrading work.

7-18. PRECISE LEVELING

A precise level without an optical micrometer was used for the notes in Table 7-4. The rod reading was estimated to the nearest millimeter at the three-wire positions and intervals between the upper-middle and middle-lower wires recorded in column four. A check is performed here, since these intervals should be the same, but they are usually allowed to

Table 7-5.   Highland Street profile

STA	+S	HI	−S	−IFS	ELEV (ft)
BM 235	3.16	238.92	—	—	235.76
TP-1	7.71	244.87	1.76		237.16
0 + 00				1.7	243.2
1 + 00				3.5	241.4
2 + 00				7.1	237.8
2 + 52				6.7	238.2
MH #17				10.6	234.3
TP-2	7.15	250.51	1.51		243.36
Green Pin			1.00		249.51
	+18.02		−4.27		−235.76
	−4.27				+13.75✔
	+13.75✔				

differ by up to 3 mm. Since the upper interval is greater by one unit than the lower one, the mean of 1108.3 is entered in column three. A check of this mean is obtained by adding the three thread backsights to 3325 and dividing by 3. The sum of the intervals is recorded for use with the right-hand page to ensure that the proper sight lengths have been maintained and balance properly. The final product from the note page is the *mean sum* of the backsights, 2345.3 mm, as shown. The difference in elevation is obtained by adding algebraically the mean sum of the foresights from the field book's right-hand page.

Many variations exist in noteforms and instrumentation. Optical micrometers require special notes, as do rods that have offset graduations on the left and right side to help improve the work by providing a statistical value and check against gross mistakes. Large governmental agencies and private firms generally use standard in-house noteforms written for computer reduction, if not already recorded in that format in an electronic field-data collector.

7-19. PROFILES IN HIGHWAY DESIGN

As noted earlier, the purpose of leveling is to locate objects in the vertical direction. Some standard leveling applications include (1) road profiles, (2) sewer and drainage design, (3) *borrow pits*, and (4) simple mapping.

A profile is a vertical section through the surface of the ground along any fixed line. In highway design work, it is important to know elevations along the proposed route centerline and plot them both along the centerline and at right angles to it. Profile levels are run over the proposed centerline. They are plotted in a form similar to that shown in Figure 7-13, and the road slope (grade or gradient) is selected by comparing centerline cuts and fills on trial grade lines.

The horizontal and vertical scales of a profile are generally different. The abscissa (*x*-axis) is usually noted in stations and the ordinate (*y*-axis) usually in feet above a datum. If the data are put into a computer graphics system and viewed on the computer screen, optimum road grades can be determined without a need for profile paper and design drafting boards.

After the proposed centerline is chosen, a designer will normally want to calculate the volumes of cut-and-fill materials required for the project. This information can be obtained in the field with a hand level or engineer's level and tape through a process known as *slope staking*. It can also be done photogrammetrically in a computer system that contains a *digital terrain model* at a fixed-grid spacing. In either case, stakes must be positioned in the field for construction control. See Chapter 24 for more details on noteforms.

7-20. LEVELING FOR WATER AND SEWER DESIGN

A similar but different application is in the design of a sewer or gravity-feed water system. The basic principle used is that water flows downhill, and a leveling process establishes the relative elevations of critical ground elevations. For a sewer line, a set of profile levels is taken along the proposed sewer location and a ground profile similar to Figure 7-13 drawn. Next, the survey engineer picks the critical depth and slope locations for manholes and prepares a set of profile drawings for contractor inspection and bidding. In constructing a sewer line, leveling is again used to locate the depth of trench cut and to place the pipe at its design elevations. In locating the manhole's pipe elevations, *inverts* are first established as

these occur at breaks in grade. Then batter boards are erected over the trench to contain horizontal and vertical control. To set the inside bottom of each piece of pipe at its correct elevation, workers *stringline* the pipe and measure vertically from the string. On larger jobs, a surveying laser is set up in the trench or pipe and the beam directed down the centerline of the proposed pipe location to targets attached to the pipe. Construction surveying consists of many applications where leveling plays a major role in the building process (see Chapter 24).

7-21. EARTHWORK QUANTITIES

An application of leveling to earthwork quantity calculations is in borrow-pit volumes (see Figures 5-8 and 7-14).

After it has been determined that suitable materials are to be removed from the ground for placement at another site, a surveyor lays out a horizontal grid of equal spacing, and

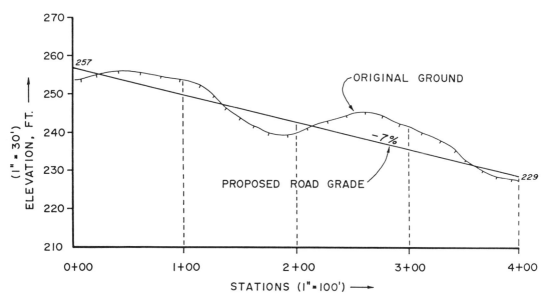

Figure 7-13. Highland Street profile.

numbers and letters the grid lines to reference the intersection points as 2-B, 3-C, etc. Level sights are taken on the grid corners and other points at breaks in the ground slopes, from a single instrument setup if possible, or by running a level circuit. Ground elevations are calculated. During and after excavation, whenever earthwork quantities are desired (perhaps for payment purposes), the intersection points are releveled and the volume of material removed from the regular prisms is determined by the following equation:

$$V = \ell \times w \, \frac{(\Sigma h_1 + 2\Sigma h_2 + 3\Sigma h_3 + 4\Sigma h_4)}{4 \times 27}$$

Here ℓ and w represent the individual length and width grid dimensions and the h's represent the difference in elevations at the number of prism corners to be counted in the calculations. In Figure 7-14, 2-B is the change in elevation of one corner of prism a, 2-C would be a corner common to prisms a and b, and 3-C would be a corner common to the four prisms a, b, c, and d. When the edge of excavation does not fall along the grid line, a wedge of earth might exist that should be included in the calculations. If the excavation is rough, or a better determination of irregular sides and bottom of the pits is needed, grid spacing should be reduced and volumes determined on smaller size prisms of earth.

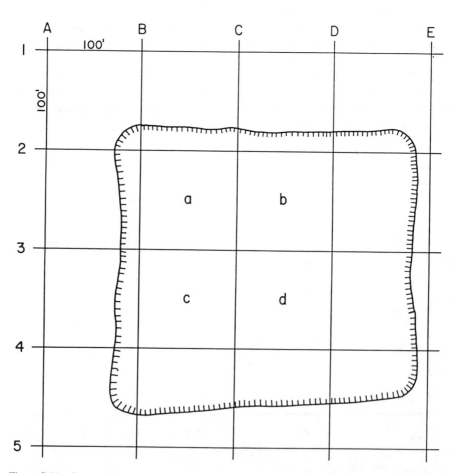

Figure 7-14. Borrow pit.

7-22. MAPPING

In laying out a subdivision or a house lot, the survey engineer should map the topographic surface to best decide how the terrain can be used to enhance the house setting. This can be done by the grid technique for areas that are flat or small in size (see Figure 7-15). In this example, the ground elevations on the grid points have been established by leveling and a map drawn showing a series of contour lines that represent points of similar elevations. These contours present ground configurations and can be envisioned by the designer and owner to assist in house location.

7-23. ERRORS

As in all surveying work, when instruments are used, it is possible to make *blunders, systematic errors,* and *accidental errors* (see also Sections 3-5 to 3-10). Proper notekeeping and appropriate field procedures should eliminate the first two,

while multiple readings can reduce the third to a minimum. Examples of blunders in leveling include (1) using the wrong point for a bench mark, (2) reading the rod incorrectly, (3) reading on the stadia cross wire instead of the middle wire, and (4) transposing numbers in field books. Because of the repetitive nature of direct leveling, it is important that the instrument operator and notekeeper not slip into bad habits or shortcuts to speed up the work. When looking for a control bench mark from which to begin a survey, the descriptions are sometimes vague and the wrong point can be selected as the beginning monument. This can be checked by locating two known bench marks and leveling between them. In addition to checking an error in monument identification, this procedure will also disclose any differential settlement of one or both points.

Leveling rods that are not self-checking can cause reading errors when the horizontal cross wire is close to a foot or meter mark. The reading 1.92 ft can easily be cited as 2.92 ft because the instrument operator sees the large red 2 numeral by the 0.92 reading and may

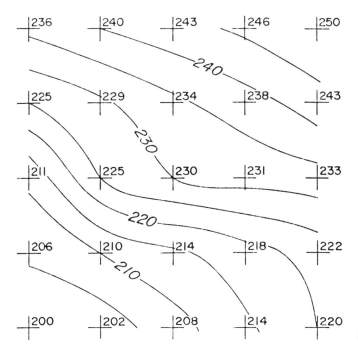

Figure 7-15. Contour lines.

fail to note that the reading is in the 1-ft interval of the rod. The Philadelphia leveling rod, when used with target, will catch this type of blunder.

Leveling instruments that contain stadia cross wires can also result in the wrong wire being read. Even setting the target will not eliminate this error unless the reader turns his or her head aside and conscientiously resights the reticle with this error in mind. Manufacturers have shortened the stadia wires in an attempt to help distinguish them from the horizontal cross wire, which runs over the entire field of view.

7-24. SYSTEMATIC ERRORS AND ADJUSTMENTS

All surveying instruments have systematic errors that result from improperly assembling components to meet certain geometric conditions, such as the line of sight being parallel to the axis of the level bubble vial. Manufacturers have built adjustable components into the

equipment, so field or laboratory adjustments can be performed to place the equipment in near-perfect alignment. Systematic errors are reduced or eliminated by instrument adjustments and proper field procedures.

Modern leveling equipment has two geometric conditions that must be corrected or serious systematic errors will accumulate on a long level line. The first requirement is to make the axis of the bull's-eye level on an automatic level, or the axis of the spirit level on a dumpy level perpendicular to the vertical axis (see Figure 7-16). As shown, the vertical axis of the level is tilted and the bubble centered in the tube. The level bubble axis *AB* is now rotated 180° so end *A* goes to *A'*, and *B* to *B'*. The bubble will not stay in the tube center and the distance that the bubble is off center is twice the error present. The bubble should be brought halfway back with an adjusting pin and then placed in the center of the vial using the leveling screws. This adjustment will make the vertical axis truly vertical.

Only after the first adjustment is completed may the second condition be adjusted—i.e., to

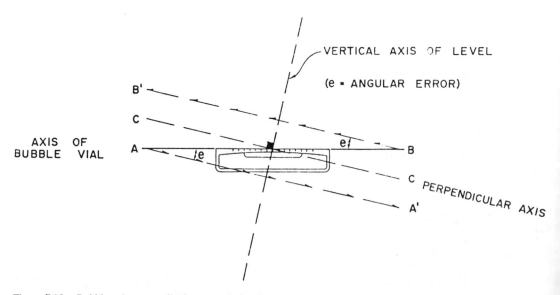

Figure 7-16. Bubble axis perpendicular to vertical axis.

make the telescope's line of sight parallel to the axis of the level bubble vial (see Figure 7-17).

If the line of sight is not parallel to the bubble-vial axis, it will sight up or down and give an erroneous rod reading. In differential leveling, this error can be canceled out by keeping the backsight and foresight distances equal. In profile leveling, the error increases with distance from the instrument. *Different-leveling* procedures should therefore be used and the bubble always placed in the tube center for each sighting. This means the leveling screws on older four-screw dumpy levels can be readjusted for each sight, and the sight distances balanced by some technique—e.g., pacing, stadia, counting rail lengths or concrete highway expansion joints.

Before and during a run of profile levels, it is best to level the instrument and perform the level-bubble-axis test by rotating the telescope 180° and bringing the bubble *one-half* way back to the center with the leveling screws. Then, take the profile readings and the values should be satisfactory except for atmospheric and refraction corrections.

A check on this second geometric condition is called the *peg test* and can be performed in a variety of ways. One precise variation of the technique is to set up the level midway between two stakes 200 ft apart and determine the true difference in elevation between the stakes. The instrument is then placed at some convenient distance, say, 20 ft, behind one of the stakes and a reading taken on the first one. If we assume that the instrument is in adjustment, the rod reading on the second stake is calculated, and for a precise determination, a correction is made for curvature and refraction. The rod is then placed on the distant stake and read. It will equal the calculated value for an instrument in adjustment or be greater or smaller if the line of sight is up or down. The telescope reticle can be adjusted to a correct position based on the measured value. For more details on instrument adjustment, see Chapter 8.

7-25. SPECIFICATIONS

In order to keep systematic errors at a minimum, agencies have developed specifications for various orders of work. Table 7-6 has been used by the NGS, NOAA, and generally accepted by persons wishing to fulfill first- or second-order vertical-control surveys. These specifications show three orders of work: first-, second-, and third-order, with classes I and II listed under the first two. First-order, first-class pertains to the basic framework of the national network and is the most precise work performed. Third-order work is used on small engineering projects.

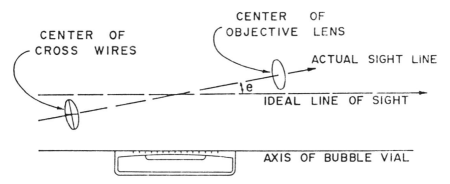

Figure 7-17. Line-of-sight condition.

Table 7-6. Classification, standards of accuracy, and general specifications for vertical control (1974)

Order Class	First I	First II	Second I	Second II	Third
Classification	Class I, Class II	Class I	Class II		
Minimal observation method	Micrometer	Micrometer	Micrometer or 3-wire	3-wire	Center wire
Section running	SRDS* or DR or SP	SRDS or DR or SP	SRDS or DR† or SP	SRDS or DR‡	SRDS or DR§
Difference of forward and backward sight lengths never to exceed					
Per setup (m)	2	5	5	10	10
Per section (m)	4	10	10	10	
Maximum sight length (m)	50	60	60	70	90
Minimum ground clearance of line of sight (m)	0.5	0.5	0.5	0.5	0.5
Even number of setups when not using leveling rods with detailed calibration	Yes	Yes	Yes	Yes	—
Determine temperature gradient for the vertical range of the line of sight at each setup	Yes	Yes	Yes	—	—
Maximum section misclosure (mm)	$3\sqrt{D}$	$4\sqrt{D}$	$6\sqrt{D}$	$8\sqrt{D}$	$12\sqrt{D}$
Maximum loop misclosure (mm)	$4\sqrt{E}$	$5\sqrt{E}$	$6\sqrt{E}$	$8\sqrt{E}$	$12\sqrt{E}$

Single-run methods				
Reverse direction of single runs every half-day	Yes	Yes	Yes	—	—
Nonreversible compensator leveling instruments				
Off-level/relevel instrument between observing the high and low rod scales	Yes	Yes	Yes	—	—
3-wire method				
Reading check (difference between top and bottom intervals) for one setup not to exceed (tenths of rod units)	—	—	2	2	3
Read rod 1 first in alternate setup method	—	—	Yes	Yes	Yes
Double scale rods				
Low-high scale elevation difference for one setup not to exceed (mm)					
With reversible compensator	0.40	1.00	1.00	2.00	2.00
Other instrument types					
Half-centimeter rods	0.25	0.30	0.60	0.70	1.30
Full-centimeter rods	0.30	0.30	0.60	0.70	1.30

* SRDS, single-run, double simultaneous procedure; DR, double-run; SP, spur, less than 25 km, double-run; D, shortest length of section (one way) in km; E, perimeter of loop in km.
† Must double-run when using 3-wire method.
‡ May single-run if line length between network control points is less than 25 km.
§ May single-run if line length between network control points is less than 10 km.

Instruments acceptable include automatic, tilting, or geodetic types. The rods comprise those containing Invar scales and steel-face rods. The lines should be double-run forward and backward, and monuments spaced at a maximum of 1- to 2-km sections in larger loops. The limiting length of sight is 50 to 90 m.

To minimize possible tilt in the line of sight, the maximum difference in length between a forward and backward sight is restricted, as well as the cumulative difference in lengths. A factor based on inclination of the line of sight is applied to the final unbalanced length to correct fieldwork for any tilted sight error.

When a survey starts at a point and loops back to the same point, the accidental errors in reading, sighting, and atmospheric conditions are proportional to the number of setups and/or distances between bench marks. The combined error can be estimated based on the maximum-length-of-sight restriction, a closure calculated using a coefficient for the order of work, and the square root of the distance around the loop or length of a section.

In large networks, the area loops intersect at certain common station points. Work done in each loop will contain the separate loop errors, and elevations at the common points must be adjusted to provide consistent data. These adjustments are usually performed by the method of least squares suited to surveying data that follow normal error-theory distribution (see Chapters 3 and 6).

7-26. PRECISE THEORIES

When performing precise first-order surveys, special note should be made of the fact that the earth is not spherical but lumpy and shaped like an oblate spheroid. Because the earth is spinning, gravity values vary in the north-south direction from 978 gal at the equator to 983 gal at the poles. Named for Galileo, a *gal* is a gravity unit equal to 1 cm/sec^2. The earth has a molten core and hard crust that bulges at the equator, making distance from the center of the earth to the equator longer than that from the center to the pole (see Figure 7-18 on p. xxx). A review of the figure shows that gravity values decrease as you go farther from the center of the earth, and the gravity value at the pole is higher than at the equator. Because of the difference in the *a* semimajor and *b* semiminor lengths, the

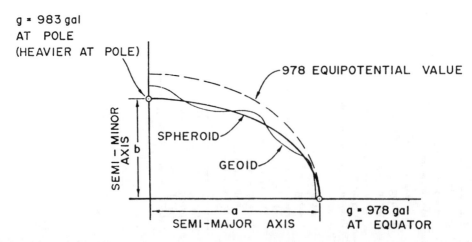

Figure 7-18.   Earth's bulge.

rate of change of the gravity field streamlines in the north-south direction of the pole. A problem results because leveling instruments are set up on the earth's surface, and the gravity field changes at a different rate, with elevation in going from the equator to the pole.

7-27. GEOIDAL HEIGHT

Leveling data should be referred to the *geoid*. The difference between the topographic surface and geoid is known as a *geoidal height* (see Figure 7-19). For horizontal surveying computations, positions are referred to the pure mathematical spheroidal surface. A geoidal height and spheroidal height differ by undulations in the geoid. A geoidal height expressed in linear units is called an *orthometric* height. For first-order work run in the north-south direction, a special *orthometric correction* is required. It can be taken from nomograms, which require using a section's mean elevation and mean latitude to get orthometric correction per minute of change in latitude.

A second and more precise method of determining the size of correction is to measure the gravity values at the bench marks on the level line. This can lead to a new system of measurements that better explains the physical earth situation (see Figure 7-20). It assumes a north-south direction and shows a change in latitude on the earth. No corrections are applied to measurements taken in the east-west direction.

Points *a* and *b* on the topographic surface are vertically above points *A* and *B* on the geoid. The *equipotential gravity surfaces* converge going northerly. This means that plumb lines *curve* because they are perpendicular to the gravity field and hence lines *aA* and *bB* are shown slightly curved in the figure. It also means that length *aA* does not equal *bB* because the gravity surfaces streamline.

Leveling from *B* to *A* on the geoid will show no change in elevation. However, leveling from *B* to *b*, then along the level surface *ba*, which is in the north-south direction, and finally leveling from *a* to *A* will result in an elevation change equal to the distance (*bB* − *aA*). The second result will be different from direct leveling of *BA*. Thus, by using normal recording techniques, two different route-depended elevations will be obtained by leveling from *B* to *A*.

To overcome this recording problem, a different measuring system has been derived that accounts for the linear change in height and change in the gravity field. The result is a

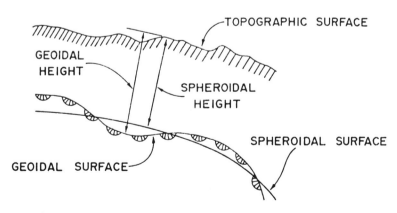

Figure 7-19. Reference heights.

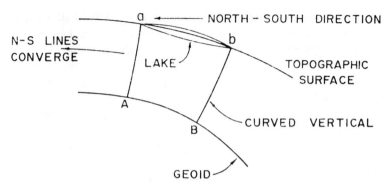

Figure 7-20. Orthometric problem.

Table 7-7. Adjusted surface gravity, leveling elevations, geopotential numbers

Station	Surface Gravity (milligal)	Leveling Elevations (m)	Geopotential Number (geopotential unit)
Corvallis, OSU-PC	980 573.14*	77.142	75.643
τβπ	980 575.49	71.337*	69.951
U54	980 573.21	76.787†	75.295
Corvallis, OSU-KL	980 573.81	73.336	71.911
College	980 573.31	72.219	70.816
RM2 College	980 573.28	71.880	70.484
RM1 College	980 573.28	72.757	71.344

* Fixed from published data before adjustment.
† Fixed through leveling observations.

geopotential number

$$\text{GPU} = \int_A^a g\,dh$$

where g is a variable acceleration due to gravity and dh the linear elevation change. This record system requires knowledge of the gravity field by direct measurement or interpolation.

An example GPU calculation follows (see Table 7-7). The leveling elevation of station τβπ is given as 71.337 m and the adjusted surface gravity value, in millegal, is 980,575.49. The product of these numbers yields a GPU value for station τβπ of 69.951. This number better expresses the leveling condition at τβπ than the given elevation, but a great deal of work must be expended to determine an ad-

justed surface gravity value. Hence, this type of recording is generally used only for first-order leveling.

REFERENCES

ANDERSON, E. G. 1979 Are primary leveling networks useless? 39th Annual ACSM Meeting. Washington, D.C.

BALAZS, E. I. 1981 The 1978 Houston-Galveston and Texas Gulf Coast vertical control surveys. *Surveying and Mapping* 41(4), 401.

BALAZS, E. I., and C. T. WHALEN. 1977 Test results of first-order class III leveling. *Surveying and Mapping* 37(1), 45.

BERRY, R. M. 1976 History of geodetic leveling in the United States. *Surveying and Mapping* 36(2), 137.

_____ 1977 Observational techniques for use with compensator leveling instruments for first-order levels. *Surveying and Mapping 37*(1), 17.

BOAL, J. D., F. W. YOUNG, and R. MAZAACHI. 1984 Geometric aspects of vertical datums. 44th Annual ACSM Meeting. Washington, D.C.

CADDESS, H. N., and G. M. COLE. 1983 Precision leveling river crossing technique using reciprocal vertical angles. 43rd Annual ACSM Meeting. Washington, D.C.

COOK, K. L., and R. J. SCHULTZ. 1971 Understanding geopotential numbers. 37th Annual ACSM Meeting. Washington, D.C.

HOLDAHL, S. R. 1983 Correction for leveling refraction and its impact on definition of the North American datum. *Surveying and Mapping 43*(2), 23.

——— 1984 Aspects of a new height system for North America. 44th Annual ACSM Meeting. Washington, D.C.

HUETHER, G. 1983 Employment and experiences with NI 002 from Jenoptik Jena for the motorized leveling. 43rd Annual ACSM Meeting. Washington, D.C.

HUSSAIN, M., and R. D. HEMMAN. 1985 Accuracy evaluation of laser levels. 45th Annual ACSM Meeting. Washington, D.C.

KIVIOJA, L. A. 1985 Hydrostatic leveling. 45th Annual ACSM Meeting. Washington, D.C.

LIPPOLD, H. R., JR. 1980 Readjustment of the national geodetic vertical datum. *Surveying and Mapping 40*(2), 155.

QUINN, F. H. 1976 Pressure effects on Great Lakes vertical control. *ASCE Journal of the Surveying and Mapping Division 102*(SU1), 31–38. New York.

REMONDI, B. W. 1986 Performing centimeter-level surveys in seconds with GPS carrier phase: Initial results. 46th Annual ACSM Meeting. Washington, D.C.

8

Instrument Adjustments

Gerald W. Mahun

8-1. INTRODUCTION

Surveying instruments are very durable, but delicate and precise pieces of equipment. No matter how well an instrument has been adjusted, rough handling, temperature variations, humidity, and a host of other factors can quickly affect its precision. The safest rule a surveyor can follow is to keep an instrument adjusted, but then use it as if it is not adjusted.

A surveyor should not adjust equipment unless it is needed, since minor corrections may not be necessary or possible. The rules in this chapter outline procedures whereby surveyors first test equipment to determine if an adjustment is necessary, and the corrections are then carried out only if needed.

Nearly all adjustments of levels, transits, and theodolites are based on the *principle of reversion*. Reversing the instrument position by rotation in a horizontal or vertical plane doubles any error present, enabling a surveyor to directly determine how much correction is needed. If the instrument is badly out of adjustment, it may be necessary to repeat all steps to reduce the size of error each time.

As a rule, older levels and transits are easy to work with because of their simple, open construction. Automatic levels and theodolites are more difficult, owing to their use of *compensators*, prisms, and glass circles. Many of their adjustments must only be done by qualified specialists; however, this chapter will outline those procedures surveyors can, with practice, perform.

A log book should be maintained for each instrument, stored in the instrument case or office, and an entry made noting the date and type of adjustment each time one is performed. This serves two purposes: (1) It reminds surveyors to periodically check an instrument's adjustment and (2) If one particular adjustment is consistently required, it indicates repair is necessary. A good instrument, properly adjusted and handled, should last a lifetime and spend a minimum number of hours in a repair facility.

8-2. CONDITIONS FOR ADJUSTMENT

Before making an adjustment, it is wise to ensure that any instrumental error tested for and found is a result of the equipment's condition and not the test's deficiency. To properly test and adjust equipment, the following rules should be followed:

1. Perform adjustments on a cloudy windless day, free of heat, if possible. Avoid situations where the sight line passes alternately through sun and shadow. Allow up to 30 min for the instrument temperature to stabilize, if there is a significant difference between the temperature at the storage and adjustment locations. The instrument should be shaded from any direct sun rays.

2. Make sure that all tripod hardware is snug, so the tripod will not shift under the instrument weight. Spread the tripod feet well apart and press the shoes firmly into the ground. Do not set up on a hard surface, as there is a chance a leg could either slide or get kicked out.

3. Choose a relatively flat area that provides flat sights for at least 200 ft in opposite directions.

4. Locate all adjusting nuts and screws and clean any threads that might be dirty. Most tools needed for older instruments consist of adjusting pins of various sizes. Test-fit the pins to see which adjusting nuts they are for. Do not use undersize pins, as they will ream out the holes in the adjusting nuts. Screwdrivers and wrenches, if needed, should be test-fit also. In any case, do not use a pair of pliers to grip a nut or screw. Adjusting pins can be readily fashioned from flush-cut nails that are carefully filed down to size. If any adjusting nut hole has been reamed, it can be carefully drilled out with a twist drill bit. The opposite end of the bit can then be used as an adjusting pin. When adjusting, do not overtighten the screws or nuts.

5. Perform adjustments in the proper sequence, as most are dependent on previous ones.

6. During and after the adjustments, handle the instrument carefully. Rough handling may negate any adjustments performed.

7. Refer to the instrument manual for any special adjustments. This is especially true for theodolites and automatic levels.

8-3. BREAKDOWN OF ADJUSTMENTS

Two types of adjustments are made on most surveying instruments: (1) preliminary and (2) principal. Preliminary adjustments are those performed each time an instrument is used and should habitually be checked each time the instrument is set up.

Principal adjustments are more detailed and are made only when a test indicates a need for them. They should be checked periodically to determine any possible instrument errors.

Sections of this chapter explain the different types of surveying equipment and their preliminary and principal adjustments. Surveyors should be capable of successfully adjusting most equipment by following the procedures explained.

8-4. GENERAL DEFINITIONS

Most instrument adjustments are partially dependent on the position of an air bubble or the intersection of a set of cross hairs under a given condition. To interpret these positions and relate them to an adjustment procedure, it is important to understand some of the mechanical aspects and common terms associated with the various surveying instruments.

The bubble tube or level vial is a sealed glass tube nearly filled with a nonfreezing, fast-moving, quite stable liquid—commonly, purified synthetic alcohol. The upper inner surface of the bubble tube is circular in the tube length's direction. The tube top is etched with graduations used to center the bubble and determine how far the bubble moves off center when the bubble tube is reversed. Generally, one end of the tube is fixed in position when mounted on the instrument; the other end can be raised or lowered by adjusting nuts or screws. The bubble-tube axis is an imaginary longitudinal line tangent to the midpoint of the upper inner curved surface of the bubble tube (Figure 8-1).

Figure 8-1. Bubble tube.

The cross hairs consist of very fine filaments of etched lines on a flat glass plate. They are placed on a reticle ring inside the telescope, forward of the eyepiece. The reticle ring is held in position by four capstan-headed screws (Figure 8-2) that pass through elongated holes in the telescope tube, so if one or more screws is loosened, the reticle ring can be rotated through a small angle. The reticle ring is moved vertically by loosening the top (or bottom) screw and then tightening the bottom (or top) screw, and horizontally by loosening the left (or right) screw and tightening the right (or left) screw.

An instrument's line of sight is defined as a line passing through the cross hairs' intersection and the optical center of the telescope's objective lens. The optical center is a fixed point, but the line of sight can be moved by shifting or rotating the cross hairs.

The vertical axis of an instrument is defined as the line about which the instrument rotates in a horizontal plane. It coincides with the spindle axis and a freely suspended plumb line attached to the instrument.

Transits and theodolites also have a horizontal axis. It is a line about which the telescope rotates in a vertical plane. This axis coincides with that of the horizontal cross arm supporting the telescope. The correct axes relationships for a properly adjusted level, transit, and theodolite are shown in Figures 8-3 and 8-4.

8-5. ADJUSTMENT OF LEVELS

The two preliminary adjustments required of all levels are to (1) eliminate parallax and (2) properly position the cross hairs.

Parallax

When working with any instrument telescope, an observer simultaneously views two images. One is the object focused on by the

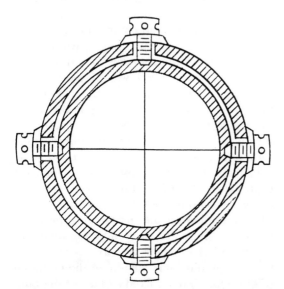

Figure 8-2. Reticle ring.

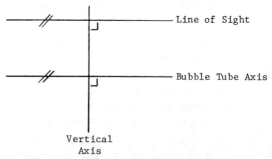

Figure 8-3. Axes relationship of a level.

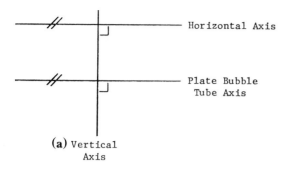

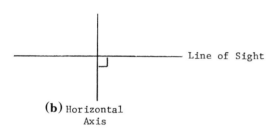

Figure 8-4. Axes relationship for a transit/theodolite. (a) Front view. (b) top view.

telescope, and the second is the cross hairs' image. Both must come to focus on a single plane—i.e., at the back of the observer's eye in order to be seen clearly. If this condition is not met, parallax exists.

To test for parallax, the telescope should be focused on some distant well-defined object. While viewing the object through the telescope, the observer's eye is shifted slightly horizontally and vertically to check for any movement of the cross hairs relative to the object. If the cross hairs do not so move over the object, an adjustment is not necessary. If the cross hairs do appear to move on the object, parallax exists and must be corrected.

To eliminate parallax, the telescope's focusing knob is rotated so everything is out of focus except the cross hairs. Then, using the eyepiece focusing ring located at the telescope's rear, adjust the focus of the cross hairs until they are sharp and well defined. Refocus the telescope on a distinct object and again

check for parallax. Repeat the procedure as necessary until parallax is eliminated, after which the instrument will not need to be adjusted again. However, if another surveyor uses the equipment, he or she should also test for and clear any parallax, since vision varies from person to person.

Cross Hairs

Equipment manufacturers attempt to make the vertical and horizontal cross hairs truly perpendicular to each other, but for older instruments this condition is less likely to have been met. It is important that the horizontal cross hair of a level be truly horizontal when the instrument is leveled so, if necessary, any part of this cross hair can be used to obtain true rod readings. If under these conditions the vertical cross hair is not exactly vertical, surveyors should understand that this situation will not affect the performance of the level in determining elevations.

To test the horizontal cross hair, first level the instrument, then check for and eliminate any parallax. Using one end of the horizontal cross hair, take a reading on a level rod. With the horizontal slow motion of the instrument, rotate the level so the horizontal cross hair is sighted to its other end on the level rod. Check to see if there has been any change in the vertical position of the cross hair with respect to the initial rod reading. If there is no change, the cross hair is truly horizontal. If the cross hair has moved above or below the initial rod reading, then it is not truly horizontal and must be adjusted.

To adjust, note the distance the cross hair has moved above or below the initial rod reading. Then, slightly loosen two adjacent reticle adjusting screws. While sighting through the telescope, rotate the reticle until the end of the cross hair is moved back half the length it was off the initial rod reading. Tighten the two reticle adjusting screws.

Check the adjustment by repeating the test. It is important to note that the horizontal cross hair was rotated about its center so a new initial rod reading must be taken. Repeat as necessary until the rod reading at both ends of the horizontal cross hair is the same.

If the vertical and horizontal cross hairs are truly perpendicular, the vertical cross hair must now be truly vertical. To check this, after the horizontal cross hair has been adjusted, a sight is made on a freely suspended plumb line with the vertical cross hair. If the vertical cross hair does not coincide exactly with the plumb line, a note should be made in the instrument's log book.

8-5-1. Dumpy Level: General Information

The dumpy level consists of a telescope, bubble tube, and leveling head containing the spindle (Figure 8-5).

8-5-2. Dumpy Level: Principal Adjustments

Due to their simple construction, dumpy levels have only two principal adjustments: (1) bubble tube and (2) line of sight.

Bubble Tube

The purpose of this adjustment is to make the bubble-tube axis perpendicular to the instrument's vertical axis. To test this condition, set up the instrument so the bubble tube is directly over two opposite leveling screws and carefully center the bubble. Rotate the instrument 90° to place the bubble tube over the remaining pair of leveling screws and again center the bubble. Rotate the instrument 180° to reverse the tube's position. If the bubble runs off center, an adjustment is necessary.

The distance the bubble moves represents twice the error present. To correct, bring the bubble back halfway by turning the adjusting nuts at one end of the bubble tube. Recenter the bubble using the two leveling screws in line with the tube. Rotate the instrument 90° and center the bubble using the other pair of leveling screws. Provided the adjustment was done correctly, the bubble will remain centered as the instrument is rotated. If the bubble runs again, repeat the adjustment until it stays centered.

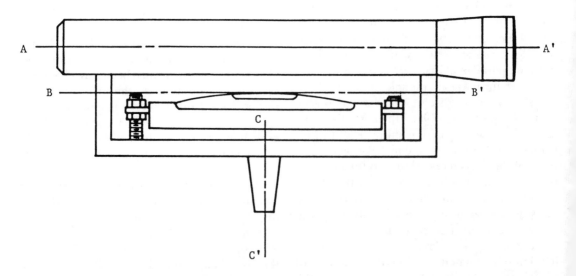

```
A-A'   Line of Sight
B-B'   Bubble Tube Axis
C-C'   Vertical Axis
```

Figure 8-5. Dumpy level.

Line of Sight

The purpose of this adjustment is to make the line of sight perpendicular to the instrument's vertical axis. The method used to test this condition is called the two-peg test or, simply, the *peg test*.

Level the instrument at a point C midway between two stakes A and B, which should be at least 200 ft apart (Figure 8-6). Assume an elevation for A and take a backsight (BS) on a level rod held there. Rotate the level and read a foresight (FS) on a rod held on B. Because the instrument is halfway between A and B, any error caused by an inclined or depressed line of sight is the same in both rod readings. The true elevation of B with respect to the assumed elevation at A is obtained by adding the BS to A's elevation, to get the height of the instrument (HI), then subtracting the FS. The error is both added and subtracted, thereby canceling itself out. The only elevation in error is the HI.

The instrument is then moved to a point D on the opposite side of B from A with the eyepiece end of the telescope within a few inches of a rod held on B. After leveling the instrument, a short BS is taken on B, looking backward through the telescope objective lens. The cross hairs will not be visible, but a pencil point held against the rod can be centered in the field of view to get a reading. After rotating the instrument, a normal FS is taken on A. The HI is the elevation of B plus the BS and since the distance is very short, the HI is essentially without error. The computed FS for A is then simply the HI minus A's elevation. If the computed and observed FS are not the same, an adjustment is necessary.

To correct the error, loosen the top (or bottom) and tighten the bottom (or top) reticle screws to move the cross hairs vertically until the observed FS matches the computed one. Always loosen one screw first and then tighten the second to prevent reticle ring warping. After the cross hairs have been moved to the correct position, test to make sure the horizontal cross hair is still horizontal. To check the adjustment, repeat the test, setting up the instrument behind A.

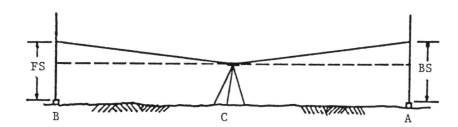

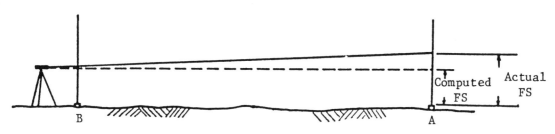

Figure 8-6. Peg test.

8-7-3.   Wye Level: General Information

The wye level differs from the dumpy in that the telescope, with the bubble tube, is removable from the two wyes in which it is mounted. The wyes are circular clamps mounted on a support bar that is attached to the vertical axis of the instrument (Figure 8-7). One or both wyes has two adjusting nuts, which allow it to be moved up or down with respect to the support bar. When the clamps are opened, the telescope is free to roll in the wyes. An imaginary line connecting the center of each wye defines their axis, i.e., the line about which the telescope rolls.

8-5-4.   Wye Level: Principal Adjustments

Since the wye level has more components than the dumpy, it has additional sources of error. The conditions to test are the line of sight, lateral adjustment, bubble tube, and wyes.

Line of Sight

The purpose of this adjustment is to make the line of sight coincide with the axis of the wyes. To test this condition, open the wyes, level the instrument, and sight a distinct point with the cross hairs. While sighting, roll the

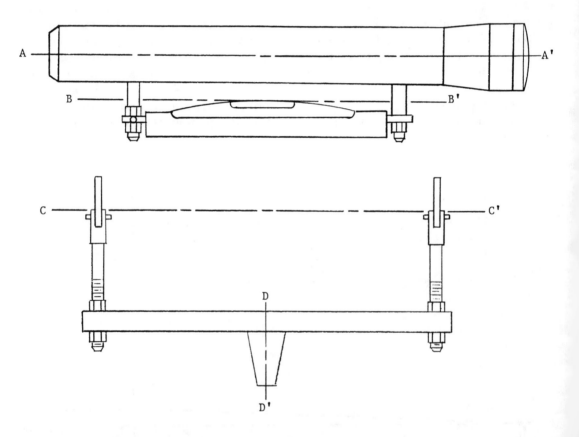

```
A-A'   Line of Sight
B-B'   Bubble Tube Axis
C-C'   Axis of the Wyes
D-D'   Vertical Axis
```

Figure 8-7. Wye level.

telescope 180° in the wyes so the bubble tube is above the telescope. An adjustment is necessary if the cross hairs have moved off the sighted point.

The length the cross hairs move above or below the point represents twice the error. Bring the cross hairs back half the distance by loosening the top (or bottom) and tightening the bottom (or top) reticle screws. Check by resighting a distinct point and repeating the test. Repeat as necessary until the adjustment is complete. Roll the telescope back to its correct position, clamp the wyes, and check to ensure the horizontal cross hair is still horizontal.

Lateral Adjustment

This adjustment makes the axis of the wyes, line of sight, and bubble-tube axis lie in the same vertical plane when the instrument is leveled.

At one end of the bubble tube is a set of adjusting nuts (as on a dumpy) and a set of capstan-headed screws perpendicular to them. These screws shift one end of the tube horizontally when the telescope is in its normal position. To check this adjustment, open the wyes and level the instrument. Roll the telescope approximately 30° in the wyes so the bubble tube, viewed from the rear, is in the five o'clock position. Any length the bubble runs off center represents the full error. To adjust, loosen one and tighten the other capstan-headed screw until the bubble is brought back to center. Check by rolling the telescope so the bubble tube is in the seven o'clock position. Repeat the adjustment if the bubble runs.

Bubble Tube

The purpose of this adjustment is to make the bubble-tube axis parallel to both the line of sight and axis of the wyes. To test, open the wyes, rotate the instrument placing the bubble tube directly over two opposite leveling screws, and center the bubble. Rotate the instrument 90°, placing the tube over the remaining pair of leveling screws, and again center the bub-

ble. Carefully remove the telescope from the wyes, turn it end for end, and replace it. An adjustment is necessary if the bubble runs off center.

The length the bubble runs represents twice the error. To correct, bring it back halfway using the bubble-tube adjusting nuts. Check by releveling the instrument and repeating the test. If the bubble runs, repeat the adjustment until it remains centered.

Wyes

This adjustment makes the axis of the wyes perpendicular to the instrument's vertical axis. If the preceding adjustments have been carried out correctly, this will also make the line of sight and bubble-tube axis perpendicular to the vertical axis.

Center the bubble first over one pair of opposite leveling screws and then over the remaining pair. Rotate the instrument 180° and check the bubble run. The length of movement represents twice the error present. Correct by bringing the bubble back halfway with the wye adjusting nuts. Readjust as necessary until the bubble stays centered in all positions.

8-5-5. Automatic Level: General Information

Automatic levels differ from dumpy and wye levels in having a compensating device that maintains a horizontal line of sight when the instrument is approximately leveled. Automatic levels also have three leveling screws, instead of four, and a circular bubble whose upper inner surface is spherical and has etched a bull's-eye on it. This bull's-eye generally defines the limits within which the compensator will maintain a horizontal line of sight.

At first glance, automatic levels appear to be complicated devices that a surveyor should not attempt to adjust. However, except for the compensator, the instrument is relatively simple in design and a few adjustments can be easily performed with satisfactory results.

8-5-6. Automatic Level: Preliminary Adjustments

Preliminary adjustments for automatic levels are the same as those for the dumpy and wye: (1) parallax and (2) cross hair. On dumpy or wye levels, the reticle adjusting screws are easy to find; on automatic levels, they tend to be elusive, but generally are located under a cover just forward of the telescope eyepiece. Some automatic levels will have only one or two reticle screws. If there is only one, the horizontal cross hair has been preset at the factory and should not be rotated. If there are two, a surveyor may be able to rotate the reticle. Generally speaking, since these screws are well-shielded, the cross hairs will stay in adjustment but should be periodically checked. If this condition is not met, then the cross-hair intersection should be used in taking rod readings or the instrument should be sent to a repair facility.

An additional preliminary check for automatic levels concerns the compensator. If dust or humidity enters the compensator or the instrument is excessively jarred, the compensating mechanism may stick and give erroneous rod readings. To test for this, carefully level the instrument and take a rod reading on a solid point. While sighting, tap a leg of the tripod. This will cause the compensator to swing, moving the cross hairs off the reading and then back to it. If the cross hairs return to the original reading, the compensator is working properly. If they do not, the compensator is sticking. Tap again to check.

In the event the compensator sticks, the surveyor should not attempt to fix it. After a cover is removed, the problem will worsen more as dust or moisture find their way into the compensator, and the level must be sent to a repair facility for proper adjustment.

8-5-7. Automatic Level: Principal Adjustments

Due to their simple design, automatic levels have only two principal adjustments: (1) circular bubble and (2) line of sight.

Circular Bubble

Unlike the ordinary bubble-tube axis, a circular bubble has a plane tangent to the midpoint of its upper inner surface. For proper adjustment, this plane must be perpendicular to the instrument's vertical axis. To test this condition, center the bubble in the bull's-eye using the leveling screws, then rotate the instrument 180°. It requires adjustment if the bubble moves out of the bull's-eye.

In order to correct this error, the bubble must be brought back half the distance it ran. The circular bubble housing should have a set of three or four adjusting screws located on its top or bottom. By turning one or more of these screws, bring the bubble back halfway. Relevel the instrument and repeat the test.

Line of Sight

The purpose of this adjustment is to make the line of sight perpendicular to the instrument's vertical axis. The test and adjustment procedure are the same as those used for the dumpy level line-of-sight adjustment (peg test). To move the cross hairs vertically on instruments having only one or two reticle adjusting screws, the screw at the six or twelve o'clock position is turned. The reticle is spring-loaded at the opposite side, so it is forced to move when the screw is turned. This adjustment can only be performed correctly if it has been determined that the compensator is functioning properly.

8-5-8. Tilting Level: General Information

Tilting levels are three-screw instruments consisting of a telescope, circular bubble, sensitive bubble tube, and leveling head. The telescope is mounted so that it can be tilted by rotating a drum located beneath the eyepiece. This feature allows the instrument to be precisely leveled each time a reading is taken.

The preliminary adjustments are the same as those for the dumpy level, except that there may be an auxiliary telescope for observing the bubble tube—in which case, it too must be checked for parallax.

8-5-9. Tilting Level: Principal Adjustments

Because of its simple design, there are only two principal adjustments for the tilting level: (1) circular bubble and (2) precise bubble tube.

Circular Bubble

The purpose of this adjustment is to make the plane of the circular bubble perpendicular to the instrument's vertical axis.

The test and adjustment of the circular bubble are dependent on how it is mounted on the instrument. If it rotates with the telescope about the vertical axis, use the same procedure as that for an automatic level. If it does not rotate with the telescope, an adjustment is not really necessary, since the precise bubble tube is used to obtain a horizontal line of sight.

Precise Bubble Tube

This adjustment makes the precise bubble-tube axis parallel to the instrument's line of sight. To test for this condition, use the dumpy level peg-test procedure. When the correct FS to give a horizontal line of sight is computed, the cross hairs are brought to that reading by rotating the telescope's tilting drum. The precise bubble is then centered using the bubble-tube adjusting nuts. If the bubble is the coincident type, the adjustment makes the bubble's two ends coincide.

8-6. ADJUSTMENTS OF TRANSITS AND THEODOLITES

The two preliminary adjustments for transits and theodolites are to (1) eliminate parallax and (2) properly position the cross hairs.

Parallax

This adjustment is the same as for levels.

Cross Hairs

In leveling, it is important to have the horizontal cross hair truly horizontal. Since transits and theodolites are used primarily for angle measurement, it is more important to have the vertical cross hair truly vertical in order to use any part of it for sighting.

To test the vertical cross hair, first level the instrument, then check for and eliminate any parallax. Sight on a freely suspended non-swinging plumb line. If the vertical cross hair does not coincide with the plumb line, an adjustment must be made by rotating the reticle. Loosen two adjacent reticle screws and rotate the cross hairs until the vertical one coincides with the plumb line. Retighten the screws and check the adjustment by reversing the position of the telescope and repeating the test.

After the adjustment has been made, perpendicularity of the horizontal and vertical cross hairs should be checked by using the horizontal cross-hair test described for levels. The result should be recorded in the instrument's log book.

8-6-1. Transit: General Information

The primary function of transits is to measure horizontal and vertical angles. Figure 8-8 shows the instrument axes to be adjusted.

Traditional transits are of an open design with all adjusting screws and nuts exposed. Basic transit design has been modified on newer instruments to incorporate additional features. Some use an optical plummet in lieu of a plumb line, whereas others have been modified to the point where they resemble a theodolite more than a transit. These instruments have some advantages of both, and since they are a cross between the two designs, their adjustments are a cross between those for transits and theodolites.

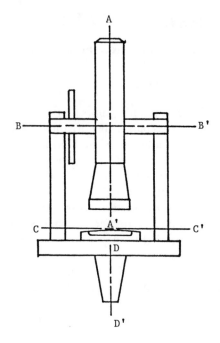

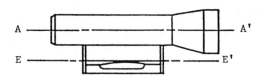

A–A'   Light of Sight
B–B'   Horizontal Axis
C–C'   Plate Bubble Tube Axis
D–D'   Vertical Axis
E–E'   Telescope Bubble Tube Axis

Figure 8-8.   Transit/theodolite.

8-6-2.   Transit: Principal Adjustments

Transits have more axes than do levels; therefore, there are more adjustments to be made. The principal adjustments for transits are (1) plate bubble tubes, (2) line of sight, (3) horizontal axis, (4) telescope bubble tube, (5) vertical vernier, and (6) horizontal vernier test.

Plate Bubble Tubes

Most traditional transits have two-plate bubble tubes mounted at right angles to each other. Each tube has its own axis, which must be adjusted to make it perpendicular to the instrument's vertical axis. Newer instruments may have only a single-plate bubble tube.

To test the plate bubble tubes, set up the instrument so each bubble tube is in line with two diagonally opposite leveling screws. Center each bubble separately using the corresponding pair of leveling screws, then rotate the transit 180°, and check the bubbles' runs. If one or both bubbles run, an adjustment is necessary—the amount of movement representing twice the error present.

To correct, bring each bubble back halfway, using the bubble-tube adjusting nuts or screws. Relevel the instrument and repeat until the bubbles remain in place as the transit is rotated.

Line of Sight

The purpose of this adjustment is to make the line of sight perpendicular to the horizontal axis. This will allow true straight-line extension when transiting the telescope.

The method used to test this adjustment is to extend a straight line on relatively flat terrain by *double centering* (Figure 8-9). Set up the instrument and select or set a distinct point A at a distance of at least 100 ft. Point A and the instrument point define a straight line that is to be extended. Backsight on A locking both horizontal motions; then reverse the telescope and set a point B at a distance of at least 200 ft. If the transit is in adjustment, point B will be on the extension of the straight line. Rotate the transit about its vertical axis and backsight on A with the telescope now reversed. Lock both horizontal motions; then reverse the telescope and set a point C at a distance $AC = AB$. If the transit is in adjustment, points B and C will coincide exactly.

If B and C do not coincide, the true extension of the straight line will pass through a point D halfway between B and C. Distance CB represents four times the error present.

To adjust the transit, while still sighting on C, move the cross hairs horizontally one-fourth

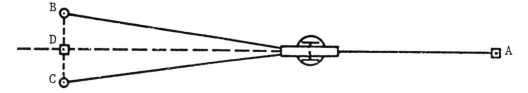

Figure 8-9. Double centering.

distance *CB* in the direction of *B*. To move the cross hairs, slightly loosen the top reticle adjusting screw, then alternately loosen and tighten the side reticle screws; check by repeating the test. After the adjustment is completed, recheck the vertical cross hair and adjust if necessary.

Horizontal Axis

This adjustment makes the horizontal axis perpendicular to the vertical axis of the transit. To test this condition, set up and level the instrument approximately 20 ft from a tall vertical wall. Raise the telescope to a vertical angle of approximately 30° and sight some distinct point *A* on the wall. Plunge the telescope to horizontal—i.e., to a vertical angle of approximately 0°—and mark a point *B* on the wall. Rotate the instrument 180°, reverse the telescope, and resight *A*. Plunge the telescope to horizontal and mark a point *C* on the wall (Figure 8-10). If *B* and *C* do not coincide, the horizontal axis needs to be adjusted.

One end of the horizontal axis must be raised or lowered by means of an adjusting screw at the end of the horizontal cross arm. This moves a saddle in which the cross arm is seated and held in place by a clamp. Loosening the clamp and turning the adjusting screw shifts the horizontal axis.

To determine the required length of movement, mark a point *D* halfway between *B* and *C*. This places *D* vertically beneath *A*. Sight *D*, then raise the telescope to *A* where, because of the instrument error, the cross hairs' intersection will miss point *A*. Using the horizontal-axis adjusting screw, raise or lower the end of the cross arm until the cross hairs are brought to *A*. Tighten the clamp and

check the adjustment by repeating the test. After completion, recheck the vertical cross hair and adjust as necessary.

A word of caution: Do not overtighten the clamp, since this can apply too much pressure on the cross arm, preventing the telescope from rotating freely.

Telescope Bubble Tube

The purpose of this adjustment is to make the axis of the telescope bubble tube parallel to the line of sight.

To perform this adjustment, follow the peg-test procedure described for dumpy levels. The telescope bubble must be centered, using

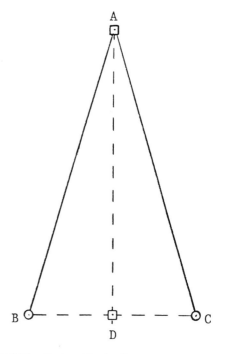

Figure 8-10. Horizontal-axis adjustment.

the vertical slow motion, for each backsight and foresight reading. When the correct FS to give a horizontal line of sight is computed, the cross hairs are tilted up or down to that reading using the vertical slow motion. The telescope bubble is then brought to center using the bubble-tube adjusting nuts or screws. Check by repeating the peg test.

Vertical Vernier

The purpose of this adjustment is to ensure that the vertical vernier is reading exactly 0°, when the line of sight is horizontal. This adjustment is performed in one of two ways, depending on the type of vertical vernier arrangement used on the transit.

METHOD 1.   This is used if there is no vertical vernier bubble tube. Set up and level the instrument; then using the vertical lock and slow motion, center the telescope bubble. If the vertical vernier does not read exactly 0°, carefully loosen the vernier mounting screws, move it to a reading of exactly 0°, and retighten.

This adjustment is much more difficult than it seems because when the mounting screws are loosened, the vernier "falls away" from the vertical circle. When retightening, it is important that the vernier not rub against the circle and a gap is not left between the two.

To avoid shifting the vernier, reading a vertical angle direct and reverse and using the average will cancel out any errors. In cases where it is not possible or practical to read angles direct and reverse, as in stadia work, an *index error* should be applied. An index error is the angle on the vertical vernier, when the telescope bubble is centered. It should be recorded, with its correct mathematical sign and telescope orientation, in the instrument's log book and case. On newer transits where the vertical vernier is not readily accessible, an index error should be used or the instrument sent to a repair facility.

METHOD 2.   This is used if the vertical vernier has a bubble tube. Set up and level the instrument; then using the vertical lock and slow motion, center the telescope bubble. Set the vernier to a reading of exactly 0° using the vernier slow-motion screw and center the vernier bubble using the bubble-tube adjusting screws.

Horizontal Verniers Test

The purpose of this test is to determine if the two horizontal verniers on a transit are truly 180° apart. The transit is designed to provide the option of reading one or the other of the A and B horizontal verniers when measuring angles. An error in horizontal-angle measurement is introduced if the two verniers are read alternately, and they are not 180° apart. To test for this condition, lock the A vernier at exactly 0° and read the B vernier. Record any error. Repeat the procedure for at least three more readings spread evenly around the horizontal circle.

If the error is consistent, within reading ability, one of the verniers is off. An adjustment is not easily made, so the transit should be sent to a repair facility, or only one vernier should be used consistently when measuring angles.

If the error is not consistent, this may indicate that the spindles are worn or the plates are warped. If that is the case, the transit must be sent to a repair facility.

8-6-3.   Theodolite: General Information

A theodolite's main function is the same as that of a transit: measuring horizontal and vertical angles. The instrument differs from the transit in having an optical plummet, optical-reading system, a circular bubble, only a one-plate bubble tube, and three leveling screws. Most adjusting screws and nuts are located under protective covers.

Figure 8-8 shows the basic axes to be adjusted. Some of these adjustments also apply to the newer-style transits, which physically resemble theodolites. In addition to the preliminary transit adjustments for parallax and cross hairs' position, a theodolite must also have the parallax cleared in its optical plummet and angle-reading telescope.

8-6-4. Theodolite: Principal Adjustments

The principal adjustments of theodolites are similar to those of transits: (1) plate bubble tube, (2) circular bubble, (3) line of sight, (4) horizontal axis, (5) telescope bubble tube, (6) vertical circle, and (7) optical plummet.

Plate Bubble Tube

The purpose of this adjustment is to make the axis of the plate bubble tube perpendicular to the vertical axis of the theodolite. To test, set up the theodolite and roughly level it using the circular bubble. Rotate the instrument so the plate bubble-tube axis is parallel to a line through two leveling screws, and carefully center the bubble using them. Rotate the instrument 90° and center the bubble using only the remaining leveling screw. Rotate the theodolite 180° and check for bubble run. An adjustment must be made if the bubble runs for a distance representing twice the error present.

To correct, bring the bubble back halfway using the bubble-tube adjusting nuts or screws. Repeat until the bubble remains stationary as the theodolite is rotated.

Circular Bubble

The circular bubble is used to roughly level a theodolite and allow the use of the optical plummet. This adjustment makes the plane of the circular bubble perpendicular to the instrument's vertical axis.

To test the circular bubble, level the theodolite using the plate bubble tube. If the circular bubble is not centered in the bull's-eye, it needs to be adjusted. Carefully center the circular bubble in the bull's-eye using the circular bubble adjusting nuts or screws.

Line of Sight

The purpose of this adjustment is to make the line of sight perpendicular to the horizon-tal axis. The procedure used is the same as that described for transit line-of-sight adjustment.

Horizontal Axis

This adjustment makes the horizontal axis perpendicular to the instrument's vertical axis. The procedure used is the same as that described for a transit horizontal-axis adjustment.

Telescope Bubble Tube

This adjustment makes the axis of the telescope bubble tube parallel to the line of sight. Generally, theodolites do not have telescope bubble tubes. For those theodolites and newer-style transits that do, use the procedure described for a transit telescope bubble-tube adjustment.

Vertical Circle

The purpose of this adjustment is to ensure that the vertical circle of a theodolite is correctly oriented, with respect to gravity, when vertical angles are read. This is accomplished either by (1) an automatic compensator or (2) a vertical-circle bubble tube.

To test a theodolite, level it and if it has a vertical-circle bubble tube, carefully center it using the bubble centering screw. Read a direct and reverse vertical angle to a selected point A. A vertical angle, for the purposes of this test, is defined as measured with respect to the horizon in a vertical plane. Angles of inclination are considered positive, angles of depression negative. The instrument is in need of adjustment if the direct and reverse vertical angles are not equal. Averaging the two readings gives the correct vertical angle.

If the theodolite has a compensator for circle orientation, it should be properly adjusted at a repair facility. An index error can be computed, recorded in the log book, and applied to each single vertical angle.

If the theodolite has a vertical-circle bubble tube, resight on point A and, using the bubble centering screw, set the correct vertical angle on the reading system. The effect of this is to slightly rotate the vertical circle while leaving the cross hairs set on *A*. This will also cause the vertical-circle bubble to run. Recenter the bubble using the bubble-tube adjusting nuts or screws. Check by reading direct and vertical angles to *A* again, repeating the adjustment as necessary.

Optical Plummet

The optical-plummet sight will only be truly vertical if the instrument is level, and the line of sight of the plummet is coincident with the instrument's vertical axis. An optical plummet is either built into the tribrach or upper instrument assembly. In the first case, the plummet remains fixed in position as the instrument is rotated; in the second, it rotates with the theodolite.

Both types can be tested using a plumb bob. Level the instrument and hang a plumb bob below it. Carefully mark a point on the ground directly beneath the plumb bob and remove it. Sight through the optical plummet and check the plummet reference mark with respect to the ground mark. An adjustment is necessary if the marks do not coincide. Four plummet adjusting screws are located just forward of the eyepiece and may be under some sort of cover. Turn the appropriate adjusting screws—first loosening, then tightening—to bring the plummet reference mark to the ground mark.

If an optical plummet is mounted in the upper assembly, it can also be tested by leveling the theodolite over a ground point using the optical plummet, then rotating 180°. If the plummet mark moves off the ground point, use the adjusting screws to bring it back halfway. Check by repeating the test.

8-7. OTHER SURVEYING EQUIPMENT

8-7-1. Tribrach

Tribrachs are the most versatile of surveying instruments and should be periodically tested and adjusted. Tribrachs use a circular bubble for leveling and may or may not have a built-in optical plummet.

The tribrach should be attached to a compatible theodolite, if possible. To test and adjust the circular bubble and optical plummet, follow the procedures explained under principal adjustments of theodolites.

If a compatible theodolite is not available, a tribrach can still be tested and adjusted if an extra circular bubble or striding level is available. The tribrach is leveled using one of these, and its circular bubble is brought to center using the adjusting screws. The optical plummet can be adjusted using a plumb line, as previously explained under principal adjustments of theodolites.

8-7-2. Rod Level

To test a *rod level*, hang a plumb bob from a firm overhead support and mark a point on the ground directly beneath it. Raise the plumb bob high enough to just clear a short section of range pole. Attach the rod level by screwing or taping to the pole and then place the range pole tip on the ground point, centering its top beneath the plumb bob. Use the bubble adjusting screws to center the bubble if necessary.

8-7-3. Striding Level

Striding levels should be tested and adjusted on the transits or theodolites with which they were designed to be used. The transit or theodolite should first have its horizontal axis adjusted. Carefully level the instrument and place the striding level on its cross arm. If the bubble runs toward one end or the other, center it using the adjusting nuts at one end of the bubble tube.

8-7-4. Tripod

The tripod is an often-overlooked piece of surveying equipment. It serves as a platform for various instruments. For example, a theodolite, no matter how well-adjusted, cannot be expected to give good results if the tripod supporting it is unstable.

Shoes must be rigidly attached to the tripod's legs to prevent shifting under the weight of an instrument. Secure fastening clamps are necessary to avoid leg slippage on extension-leg tripods. Bolts connecting legs to the tripod head should be tightened firmly but not tight enough to disallow easy folding of the legs. Metal tripods should be checked for dents that could affect sliding of the extension legs. Wooden tripods must be inspected for cracks and flat spots under the clamps, worn areas on the wood's protective coating refinished, and to prevent swelling, any moisture wiped off immediately.

8-8. CLEANING EQUIPMENT

Surveying equipment is frequently used in relatively hostile environments; dirt and water are its worst enemies. Proper maintenance includes not only periodic adjustments, but also regular cleaning. Instruments can be sent to a repair facility for a thorough cleaning and lubrication, but a surveyor *can* do a few things to keep equipment in good condition.

As soon as possible, dirt and water must be removed from external instrument parts with a mild general household cleaner, cotton swabs, and pipe cleaners. Pay particular attention to clamp screws, leveling screws, and exposed metal joints. If water gets inside a telescope, resist the temptation to go in after it; opening a telescope allows dust to get in, and on older instruments the cross hairs are fragile and easily destroyed.

A soft camel's hair brush works best for cleaning lenses. If lenses are streaked, a lint-free cloth and some optical-quality glass cleaner are necessary. Newer optics are coated and can be damaged by excessive rubbing or using a household glass cleaner.

Surveyors should avoid the temptation to oil or grease equipment. Lubricants attract dust like a magnet, accelerating wear. Thorough lubrication and internal cleaning should be done by a repair facility.

8-9. SHIPPING EQUIPMENT

When shipping equipment to a repair facility, it is important to pack it properly. Instrument cases alone are not designed for shipping purposes and, therefore, should be put in a sturdy container with a generous amount of packing material. If the container is dropped, the packing material rather than the instrument will absorb the shock.

The shipping and return addresses should appear in at least two different places on the exterior of the container and be included inside in the event the external addresses are destroyed or obliterated. Labels identifying the contents as *fragile precision equipment* should also appear in multiple locations on the exterior. A letter explaining in detail the problems with the equipment should be inside the container.

Equipment being shipped must be insured. Surveying equipment represents a large investment, and all possible measures should be taken to protect it.

REFERENCES

BOUCHARD, H., and F. H. MOFFITT. 1992. *Surveying*, 9th ed. New York: Harper Collins.

LAPCZYNSKI, D. 1980. Keep it clean. *P.O.B. Magazine* 6(1), 34.

LAPCZYNSKI, D. 1981. Sending instruments for service. *P.O.B. Magazine* 6(2), 25.

LOMMEL, G. E., H. RUBEY, and M. W. TODD. 1958. *Engineering Surveys: Elementary and Applied*, 2nd ed. New York: Macmillan.

SMITH, F. R. 1982. How to check the adjustment of a rod level. *P.O.B. Magazine* 7(5), 30.

WOLF, P. R., and R. C. BRINKER. 1994. *Elementary Surveying*, 8th ed. New York: Harper Collins.

9

Traversing

Jack B. Evett

9-1. INTRODUCTION

A traverse is a series of consecutive straight lines along the path of a survey, the lengths and directions of which are or have been determined by field measurements. The surveying performed to evaluate such field measurements is known as *traversing*. Although often used in land and route surveying, it is also employed in other types of surveying.

The end points of traverse lines, known as traverse stations or "hubs," are commonly marked in the field by wooden stakes with tacks in the top, steel rods, or pipes driven into the ground. On blacktop or concrete pavement, traverse stations can be located by driving a nail into the blacktop or by chiseling or painting an "X" or other mark on the concrete. On a map or plat, traverse stations may be marked with a small circle. A small triangle denotes a control station.

There are two basic types of traverses: (1) *open* and (2) *closed*. Both originate at a point of known location. An open traverse terminates at a point of unknown position; a closed traverse finishes at a point of fixed location. Figure 9-1 illustrates an open traverse that might represent a proposed highway or pipeline location. Figure 9-2 shows two closed

traverses. In Figure 9-2a, *ABCDE* represents a proposed highway route, but the actual traverse begins at known location 1 and ends at fixed location 2. This type of closed traverse is known as "geometrically open, mathematically closed." In Figure 9-2b, *ABCDEA* represents a parcel of land for which the actual traverse begins and ends at known point *A*. This type of closed traverse is "geometrically and mathematically closed." (Subsequent citations to closed traverses in this chapter refer to geometrically and mathematically closed ones.)

Although open traverses sometimes are used on route surveys, such as highway or pipeline locations, they should be avoided because an independent check for errors and mistakes is not available. The only means of verifying an open traverse is to repeat all measurements and computations (not an independent check). For closed traverses, independent mathematical means of checking both measured angles and distances are available (see Sections 9-3 and 9-4) and should be utilized to verify survey accuracy. Whenever open traverses of the type shown in Figure 9-1 are encountered, if possible they should be transformed to either (1) geometrically open, mathematically closed ones (Figure 9-2a) by extending the traverse to beginning and ending points of known loca-

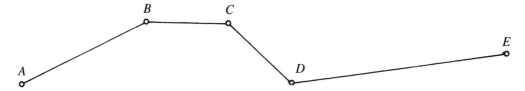

Figure 9-1. Open traverse.

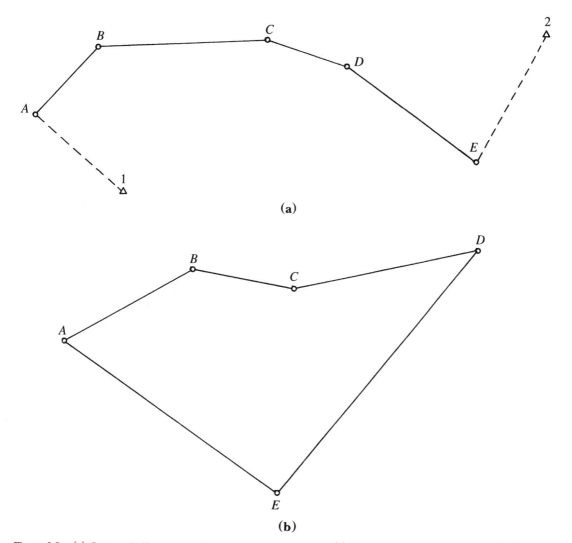

(a)

(b)

Figure 9-2. (a) Geometrically open, mathematically closed traverse. (b) Geometrically and mathematically closed traverse.

tion, or (2) geometrically and mathematically closed ones (Figure 9-2b), by continuing the traverse so it ends on its beginning point.

9-2. FIELD MEASUREMENTS IN TRAVERSING

As stated in the previous section, traversing involves measuring both the lengths and directions of lines. Details regarding these procedures in the field have been presented in earlier chapters; however, some considerations applicable to traversing are presented in this section.

Lengths of traverse lines can be determined by any convenient method, but most measurements are made with electronic devices or by taping. In closed traverses, the lengths of lines are measured, recorded, and shown on a map or plat. On open traverses, it is common practice to locate stations by their total distances from the starting point. Distances are then noted in "stations" and "pluses" (a full station is 100 ft). In Figure 9-1, if *A* is station 0 + 00 and the distance from *A* to *B* is 569.8 ft, station *B* becomes 5 + 69.8. For length *BC* equal to 744.5 ft, station *C* is 13 + 14.3.

Directions of traverse lines can be determined relative to a reference direction (such as north) by reading bearings or azimuths, or measuring interior angles, deflection angles, or angles to the right (preferred) or to the left. Bearings and azimuths are obtained by sighting the transit's telescope along a line and noting the compass reading. A deflection angle is formed at a traverse station by an extension of the previous line and the succeeding one. The numerical value of a deflection angle must always be followed by R or L to indicate whether it was turned right or left from the previous traverse line extended. An angle to the right is read at a traverse station by backsighting along the previous line and measuring the clockwise angle to the next point. In Figure 9-3, the deflection angle is 33°33′ R; the angle to the right is 213°33′.

Closed traverse—e.g., land boundary surveys—are usually run by measuring and recording interior angles, such as *ABC*, *BCD*, etc., in Figure 9-2b. Open traverses—e.g., route surveys—are more commonly run using either deflection angles or angles to the right.

Since bearings read in the field are not highly accurate, traversing is generally done by measuring and recording interior angles, deflection angles, or angles to the right, all of which are determinable to the nearest minute or smaller relatively quickly with an ordinary transit or theodolite. However, bearings are generally used in computing latitudes and departures as well as closure (see Section 9-4)

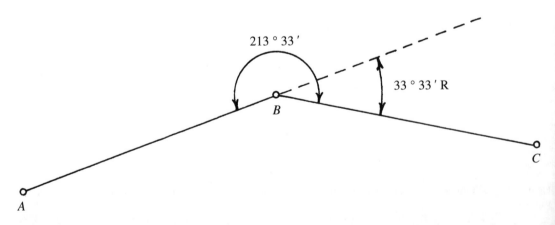

213 ° 33 ′

33 ° 33 ′ R

B

C

A

Figure 9-3. Deflection angle and an angle to the right.

and directions are frequently indicated on maps and plats by bearings. Therefore, although bearing (compass) readings may be made and recorded in the field for checking purposes only, actual bearings used in making closure calculations are often computed from appropriate angle readings—e.g., interior angles; those given on maps and plats are generally secured from the closure calculations.

9-3. ANGLE MISCLOSURE AND BALANCING

For closed traverses, an excellent verification of angular measurements is available, as the sum of the interior angles of a closed polygon Σ is

$$\Sigma = (n - 2)180° \qquad (9\text{-}1)$$

where n is the number of sides (or angles) in the polygon. Hence, if the sum of the measured interior angles of a closed traverse is equal to Σ as computed using Equation (9-1), the accuracy of each measured angle is assured with reasonable certainty. (It is always possible that compensating errors or mistakes were made.)

Because of imperfections in equipment and errors made by surveyors, it is not unusual for the sum of the measured angles to differ from Σ. The numerical difference between the computed sum and Σ is known as the *angle misclosure*. An angle misclosure of 1 or 2 min might ordinarily be considered tolerable, but larger values are not. Permissible misclosure c can be computed using the formula

$$c = K\sqrt{n} \qquad (9\text{-}2)$$

where n is the number of sides (or angles) and K a fraction of the least division of a transit vernier or the smallest graduation on a theodolite scale. A commonly used value of K is 1 min. If this value is reasonable, permissible misclosure for a nonagon is $1'\sqrt{9}$, or $3'$, and

an angle sum that falls within the range 1259°57' to 1260°03' would be acceptable.

If the angle misclosure for a closed traverse is greater than the permissible figure as determined from Equation (9-2), a surveyor should remeasure each angle in order to achieve acceptable misclosure. If the angle misclosure is within the permissible range, the angles should be balanced or adjusted so their sum is equal to the correct geometric total—i.e., the number determined by Equation (9-1). Angle balancing can be done utilizing arbitrary adjustments, average adjustments, or adjustments based on measuring conditions. Details of these methods follow.

An *arbitrary* adjustment of traverse angles is commonly used for most ordinary traverses. Thus, if the misclosure is $1'$, that figure is put in a suspect angle (if there is one), otherwise in any angle. For a $2'$ misclosure, the entire correction might be inserted in one angle, or $1'$ each in two angles.

The *average* adjustment method divides misclosure by the number of angles and applies the result to all angles. When following this system, care must be taken not to give a false impression of angle precisions. For example, if the misclosure of a nonagon is $3'$, the average adjustment would be $3'/9 = 20''$. For original measurements made to the nearest minute, it is inappropriate to change each angle by $20''$. Instead, a correction of $1'$ can be applied to every third angle, thereby avoiding more serious distortion of the traverse. If, however, original measurements were made to the nearest $20''$ —by "repetition" or using a better instrument—it is reasonable to apply corrections of $20''$ to every angle.

When warranted, adjustments can be made based on known measuring conditions. If the sight line along one traverse side is partially obstructed, thereby making accurate sighting difficult, the angle misclosure can be divided into two equal parts and applied to each angle having this line as a common side. Or, if two lines forming an angle are both much shorter than all other traverse sides, a larger error is more likely to occur there. Hence, that angle deserves the total adjustment.

An important factor in most surveys is to maintain comparable precision in angle and distance measurements (see Table 6-2). Note that errors in angle measurements are not related to their size, whereas errors in distance measurements increase as lines lengthen.

9-4. TRAVERSE MISCLOSURE AND BALANCING

Once a closed traverse survey has been completed, its accuracy must be checked. If required, the survey should be balanced or adjusted to effect perfect closure—i.e., geometric consistency among angles and lengths. The first step in this process is to determine angle misclosure and balance the angles (see Section 9-3). This step ensures the correct total for angular measurements, but additional computations are needed to assess the effects on traverse accuracy by including distance measurements and probably balancing the survey for them. This step is normally done by computing "latitudes" and "departures" for use in various computations.

9-4-1. Latitudes and Departures

The *latitude* of a line is its orthographic projection on the north-south axis of the survey. In terms of an ordinary rectangular coordinate system, latitude is the y-coordinate of a line secured by multiplying its length by the bearing angle cosine. North latitudes are considered positive, south ones negative.

The *departure* of a line is its orthographic projection on the east-west axis of the survey —i.e., the x-coordinate on an ordinary rectangular system found by multiplying its length by the bearing angle sine. East departures are considered positive, west ones negative.

The basis for using latitudes and departures to check and adjust a traverse survey is that the algebraic sums of both latitudes and departures must equal zero for a closed traverse.

If both algebraic sums are zero, the survey is balanced, and its overall accuracy accepted. As with angle misclosure, it is not unusual for latitude and departure algebraic sums to differ from zero, so the discrepancy is the *latitude misclosure* and, for departures, the *departure misclosure*. Their combined result, known as *linear misclosure*, is determined by computing the square root of the sum of the squares of latitude and departure misclosure. A final parameter used in analyzing traverse surveys is their *precision*, determined by dividing linear misclosure by traverse perimeter and expressing the quotient in reciprocal form. Typically, the denominator of the precision is rounded to the nearest 100, or the nearest 10 if the denominator is relatively small.

Precision is used to judge whether or not the linear misclosure of a traverse is permissible. For a given survey, permissible precision may be prescribed by state or local law. For example, North Carolina requires a minimum of 1/10,000 for "urban land surveys" and 1/5000 for "rural and farmland surveys." In some cases, permissible precision may be specified in the contract under which a surveying project is being performed.

Example 9-1. Table 9-1 gives the lengths and bearings, as computed from measured interior angles, of a five-sided closed traverse survey. Determine the latitudes and departures, linear misclosure, and precision of the survey.

The computations for solving this problem are shown in Table 9-1. The latitude of AB was determined by multiplying its length by the cosine of its bearing angle—i.e., $647.25 \times \cos 56°25' = 358.03$. Its sign is negative because the latitude is south. The departure of AB was determined similarly using the sine function. Computations for the other four lines were made in the same manner. Note from Table 9-1 that the linear misclosure and precision were determined to be 0.85 ft and 1/2400, respectively.

Table 9-1. Computation of latitudes and departures

Station	Bearing	Length (ft)	Latitude (ft)	Departure (ft)
A				
	S 56°25′ W	647.25	− 358.03	− 539.21
B				
	N 32°00′ E	300.95	255.22	159.48
C				
	N 28°52′ W	318.18	278.64	− 153.61
D				
	N 82°02′ E	555.02	76.92	549.66
E				
	S 3°49′ W	252.61	− 252.05	− 16.81
A				
		2074.01	+ 0.70	− 0.49

Linear misclosure = $\sqrt{(0.70)^2 + (-0.49)^2}$ = 0.85 ft.
Precision = 0.85/2074.01 or 1/2400.

9-4-2. Traverse Balancing

If linear misclosure for a closed traverse survey is greater than the permissible limit prescribed by law, contract, or the like, lengths and, if necessary, angles of the traverse must be remeasured in order to get more accurate information and a permissible misclosure. If linear misclosure is within the permissible amount, the survey should be balanced, or adjusted, by distributing linear misclosure throughout the traverse to close the figure. Methods for balancing traverses include (1) arbitrary method, (2) Crandall method, (3) least-squares method, (4) transit rule, and (5) compass rule.

When an arbitrary method is used, latitudes and departures are adjusted based on a surveyor's judgment. If there is justification to believe the measurement of one traverse line is less reliable than all others, it would be reasonable to adjust only the latitude and departure of that line, forcing the latitude and departure algebraic sums to zero. The Crandall and least-squares methods follow prescribed computations based on probability theory. Both the transit and compass rules apply proportional adjustments.

With the transit rule, adjustments are applied to respective latitudes in proportion to their lengths; thus the longer a latitude, the greater is its adjustment, and vice versa. Similarly, adjustments are applied to respective departures in proportion to their lengths. Adjustments can be computed using the following formulas:

$$\frac{\text{Adjustment in latitude } AB}{\text{Latitude misclosure}} = \frac{\text{latitude of } AB}{\text{absolute sum of latitudes}} \quad (9\text{-}3)$$

$$\frac{\text{Adjustment in departure } AB}{\text{Departure misclosure}} = \frac{\text{Departure of } AB}{\text{Absolute sum of departures}} \quad (9\text{-}4)$$

For simplicity in computations, the formulas can be rearranged to the form

$$\text{Adjustment in latitude } AB = \text{latitude of } AB \times \frac{\text{latitude misclosure}}{\text{absolute sum of latitudes}}$$

since the misclosure/absolute sum ratio is a constant for all latitudes in a particular traverse.

Similarly, adjustments by the compass rule are applied to both latitudes and departures in proportion to the lengths of the lines. In other words, the longer a line, the greater are its latitude and departure adjustments, and vice versa, as shown in the following formulas:

$$\frac{\text{Adjustment in latitude } AB}{\text{Latitude misclosure}}$$

$$= \frac{\text{length of } AB}{\text{perimeter of traverse}} \quad (9\text{-}5)$$

$$\frac{\text{Adjustment in departure } AB}{\text{Departure misclosure}}$$

$$= \frac{\text{length of } AB}{\text{perimeter of traverse}} \quad (9\text{-}6)$$

The compass rule, relatively simple to apply, is the most often employed method for balancing traverses.

Example 9-2. Balance the traverse of Example 9-1 by the compass rule.

The computations for solving this problem are shown in Table 9-2. The adjustment for latitude *AB* was determined according to Equation (9-5) by multiplying the latitude misclosure by the length of *AB* and dividing by the perimeter of the traverse—i.e., $0.70 \times 647.25/2074.01 = 0.22$. Its sign, as well as those for adjustments of all other latitudes, is negative because the latitude misclosure is positive; therefore, each individual latitude must be made algebraically smaller. The adjustment for departure *AB* was determined similarly; its sign is positive because the departure misclosure is negative.

After adjusted latitudes and departures have been determined, revised lengths and bearings for the various traverse lines can be computed trigonometrically. The adjusted length of a line may be determined by finding the square root of the sum of the squares of the adjusted latitude and departure of that line. The adjusted bearing angle may be computed as the arctangent of the quotient of departure divided by latitude. The quadrant in which the bearing falls can be determined by observing the signs of the latitude and departure.

Example 9-3. Determine the adjusted lengths and bearings of the traverse lines for which adjusted latitudes and departures were computed in Example 9-2.

The computations for solving this problem are shown in Table 9-3. The adjusted length of *AB* was computed by taking the square root of the sum of the squares of its latitude and departure, i.e.,

$$\sqrt{(-358.25)^2 + (-539.06)^2} = 647.25 \text{ ft}$$

Table 9-2. Adjusted latitudes and departures by the compass rule

Station	Computed		Adjustment		Adjusted	
	Latitude	Departure	Latitude	Departure	Latitude	Departure
A						
	−358.03	−539.21	−0.22	+0.15	−358.25	−539.06
B						
	255.22	159.48	−0.10	+0.07	255.12	159.55
C						
	278.64	−153.61	−0.11	+0.08	278.53	−153.53
D						
	76.92	549.66	−0.19	+0.13	76.73	549.79
E						
	−252.05	−16.81	−0.08	+0.06	−252.13	−16.75
A						
	+0.70	−0.49			0.00	0.00

All values are in ft.

Table 9-3. Adjusted lengths and bearings

Station	Adjusted Latitude (ft)	Departure (ft)	Adjusted Length (ft)	Bearing
A				
	−358.25	−539.06	647.25	S 56°24′ W
B				
	255.12	159.55	300.90	N 32°01′ E
C				
	278.53	−153.53	318.04	N 28°52′ W
D				
	76.73	549.79	555.12	N 82°03′ E
E				
	−252.13	−16.75	252.69	S 3°48′ W

Its bearing angle was determined by finding the arctangent of the quotient of departure divided by latitude—i.e.,

$$\text{arctan} (539.06/358.25) = 56°24'.$$

Since both latitude and departure are negative, this bearing falls in the southwest quadrant, S 56°24′ W. Computations for the remaining lines were made in the same manner.

9-5. RECTANGULAR COORDINATES

In map plotting, area computing, as well as in other applications, it is sometimes convenient to locate a line by giving rectangular coordinates for its end points with respect to a reference coordinate system. Rectangular coordinates for a point with respect to a common *x-y* coordinate system have two numbers separated by a comma and enclosed in parentheses. The first number indicates distance to the point measured from the *y*-axis parallel to the *x*-axis; the second gives distance to the point from the *x*-axis parallel to the *y*-axis. In surveying, coordinates may be referred to north-south and east-west axes (meridians) with the north coordinate given first.

Coordinates of each corner of a traverse can be determined readily if (adjusted) latitudes and departures are known. In comput-

ing coordinates, it is necessary to have a starting point—i.e., one corner having known coordinates. The starting point may be referenced to a known coordinate system (such as the state plane coordinate system), or assumed coordinates may be used.

It should be clear from Figure 9-4 that, given the coordinates of one point, say, *A*, the *x*-coordinate of *B* is equal to the *x*-coordinate of *A* plus (or minus) the departure of *AB*. Similarly, the *y*-coordinate of *B* can be found by adding (or subtracting) the latitude of *AB* to the *y*-coordinate of *A*. Coordinates of all corners of a closed traverse can be calculated in the manner just described by beginning at a point *A* with known (or assumed) coordinates and proceeding around the traverse to point *A*. If latitudes and departures were "balanced," the original and calculated coordinates of *A* should be the same, thereby affording a good, but not perfect, check on the computations.

Example 9-4. If coordinates of traverse point *A* in Example 9-3 are N 2000.00, E 1200.00 ft, determine the coordinates of the other traverse corners.

The computations for solving this problem are shown in Table 9-4. The N-coordinate of *B* was determined by adding the latitude of *AB* to the N-coordinate of *A*—i.e., −358.25 + 2000.00 = 1641.75. The E-coordinate of *B* was

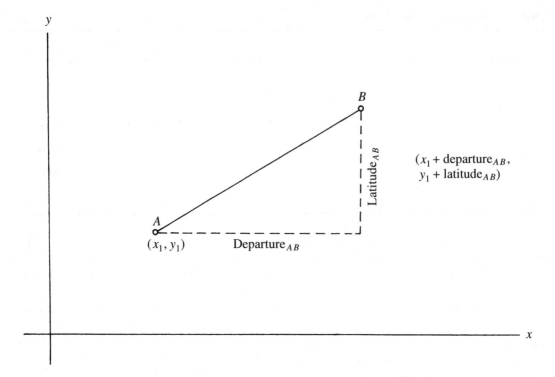

Figure 9-4.   Illustration of computation of coordinates.

computed by adding the departure of *AB* to the E-coordinate of *A*—i.e., $-539.06 + 1200.00 = 660.94$. Coordinates of remaining points were found in the same manner. Starting and calculated coordinates of point *A* are the same, indicating that the computed coordinates are probably correct.

9-6.   MISSING DATA

Preceding sections demonstrated how traverse misclosure can be determined if lengths and bearings of all lines have been measured. The premise for computing misclosure is that the algebraic sums of both latitudes and depar-

Table 9-4.   Computation of station coordinates

Station	Latitude	Departure	N-coordinate	E-coordinate
A			2000.00	1200.00
	− 358.25	− 539.06		
B			1641.75	660.94
	255.12	159.55		
C			1896.87	820.49
	278.53	− 153.53		
D			2175.40	666.96
	76.73	549.79		
E			2252.13	1216.75
	− 252.13	− 16.75		
A			2000.00	1200.00

All values are in ft.

tures are zero for perfect closure. This premise can be used to calculate a maximum of two "missing data" for a closed traverse—lengths of two lines, bearings of two lines, length and bearing of the same line, or length of one line and bearing of another—if *all other* bearings and lengths are known. The algebraic sums of both latitudes and departures, some of which will be unknown or in terms of unknowns, must equal zero. Then two simultaneous equations with two unknowns can be solved to find the unknown values.

Probably the most common application of this procedure is calculating the bearing and length of a single traverse line, when all others have been measured. This problem is easy to solve since the line's latitude and departure must force the traverse total latitudes and departures to equal zero. After the latitude and departure of the missing line have been computed, its length and bearing are readily obtained by the methods described in Section

9-4-2 and illustrated in Example 9-3. Example 9-5 illustrates this type of procedure.

If lengths of two different lines of a traverse are unknown, the method described yields two simultaneous equations with two unknowns, which can generally be solved directly. In some cases, however, two equations with two unknowns do not provide a unique solution. An example occurs when the bearings of two adjacent traverse lines are unknown. As illustrated in Figure 9-5, two sets of bearings for *BC* and *CD* will close the figure for the same values of all line lengths and other bearings. Both sets of bearings can be obtained by solving the simultaneous equations.

The procedure described here to solve for missing data should not ordinarily be used for surveying traverses because it negates any check on their accuracy. Surveyors might measure the lengths and bearings of all traverse lines except one, then compute them for the remaining line. But any errors in the field

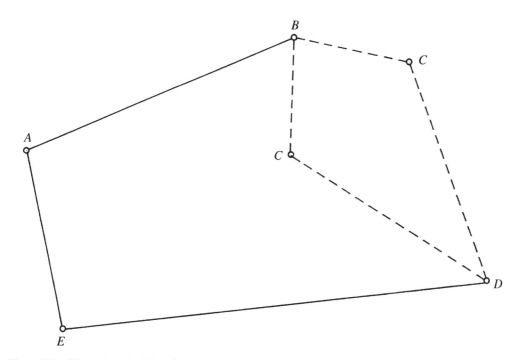

Figure 9-5.   Illustration of missing data.

work are thrown into the computed bearing and length of the "unknown" line, so surveyors might be unaware of large mistakes made in the measurements. Unfortunately, this practice is sometimes followed with catastrophic results, if large measurement mistakes were made.

Solving for missing data may be warranted in some cases, however. For example, suppose it is necessary to establish a field line between two points *A* and *B* through heavily wooded terrain, as in Figure 9-6, and neither the length nor direction from *A* to *B* is known. One possible solution would be to run a random line from *A* toward *B*. If it does not pass precisely through *B* (a likely prospect), compute a corrected direction and rerun the line. This could be difficult if obstructed by trees and/or brush. An alternative might be to run a random traverse from *A* to *B* along a rela-

tively clear path, then compute a length and direction for line *AB* to lay it out in the field. Such a random traverse is shown in Figure 9-6 (*A* to *D* to *C* to *B*).

Another example of effective use when computing missing data is partitioning land into separate tracts. For example, suppose *ABCDEFGHA* in Figure 9-7 has been surveyed, and balanced latitudes and departures are known for each line. It is desired to divide this tract into three smaller tracts by cutoff lines *H* to *C* and *G* to *D*. Lengths and directions of these cutoff lines—to stake them in the field —can be ascertained by the methods presented in this section.

Example 9-5. Suppose the scenario described previously to define a direction and length for line *AB* in the field is followed by running a random traverse along the path *BCDA* shown in Figure 9-6. Bearings computed from deflection angles and the measured lengths of lines *BC*, *CD*, and *DA* are given in Table 9-5. Find the length and bearing of line *AB*. All lengths are in feet.

Computed latitudes and departures are shown in Table 9-5. Equate latitude and departure sums to zero.

$$x + 522.75 + 735.18 + 232.21 = 0$$
$$y - 352.82 - 60.87 + 190.29 = 0$$
$$x = \text{latitude of } AB = -1520.14$$
$$y = \text{departure of } AB = 233.40$$

$$\text{Length of } AB = \sqrt{(-1520.14)^2 + (233.40)^2}$$
$$= 1536.47 \text{ ft}$$
$$\text{Bearing angle of } AB = \arctan(223.40/1520.14)$$
$$= 8°22'$$

Since latitude is negative and departure is positive, the bearing of *AB* is S 8°22′ E.

9-7. AREA COMPUTATIONS

One of the reasons for running and computing closed traverses is to define areas. Land is

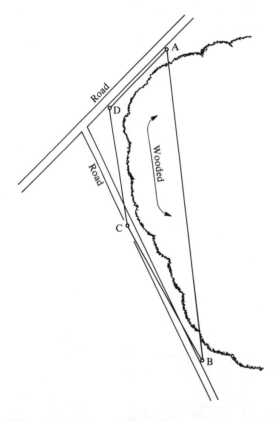

Figure 9-6.   Illustration of missing data.

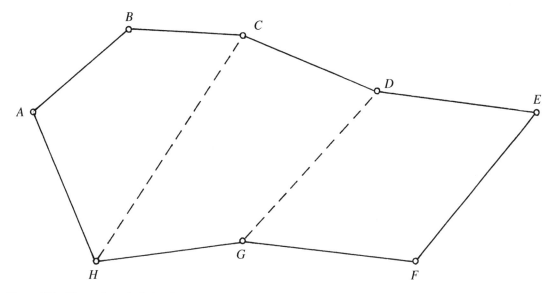

Figure 9-7. Illustration of missing data.

ordinarily bought and sold on a basis of cost per unit area. For this reason as well as many others, an acurate determination of a tract area is often necessary.

If distances are measured in feet, area is generally computed in square feet. For large areas, particularly those related to land in the United States, area is commonly expressed in *acres:* 1 acre = 43,560 ft^2. When distance is measured in meters, area is computed in square meters or *hectares:* 1 hectare = 10,000 m^2. The relationship between acres and hectares is 1 hectare = 2.471 acres or 1 acre = 0.4047 hectare.

Three means of computing traverse area are presented in this section: (1) the *double meridian distance* (DMD) method, (2) coordinate method, and (3) the use of a *planimeter*.

9-7-1. DMD Method

The DMD method requires that latitudes and departures of traverse boundary lines be known, as they are after a traverse has been checked for misclosure and balanced.

The meridian distance of a line is the perpendicular distance from the line's midpoint to a reference meridian (north-south line). In Figure 9-8, *FD* is

Table 9-5. Data for finding length and bearing of line *AB*

Station	Bearing	Length (ft)	Latitude (ft)	Departure (ft)
A				
			x	*y*
B				
	N 32°33′ W	655.75	552.75	− 352.82
C				
	N 4°44′ W	737.70	735.18	− 60.87
D				
	N 39°20′ E	300.22	232.21	190.29
A				

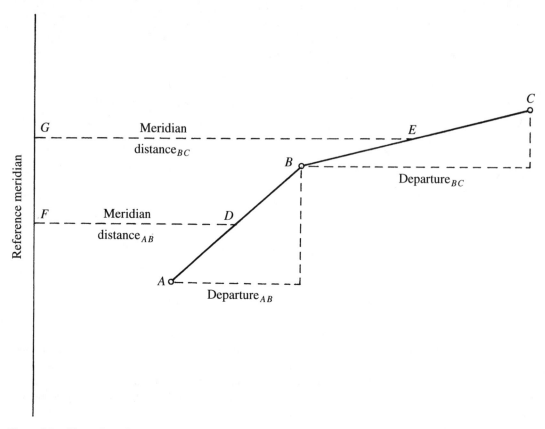

Figure 9-8.   Illustration of computation of meridian distance.

the meridian distance of line *AB*, and *GE* the meridian distance of line *BC*. Mathematically, the meridian distance of *BC* is equal to the meridian distance of *AB*, plus half the departure of *AB* plus half the departure of *BC*.

In order to avoid working with half-departures, surveyors use the double meridian distance—i.e., twice the meridian distance—in making computations. Obviously, the DMD of *BC* is equal to the DMD of *AB* plus the departure of *AB* plus the departure of *BC*. This can be generalized to say that *the DMD of a traverse line is equal to the DMD of the previous line plus the departure of the previous line plus the departure of the line itself*. If the reference meridian is moved to pass through *A* in Figure 9-8, the DMD of *AB* is equal to its departure.

Summarizing the preceding discussion gives the following rules for computing DMDs for a closed traverse:

1. The DMD of the first line is equal to the departure of the first line. (If the "first line" is chosen as the one that begins at the westernmost corner, negative DMDs can be avoided.)

2. The DMD of each succeeding line is equal to the DMD of the previous line plus the departure of the previous line plus the departure of the line itself.

As a means of providing a check on DMD computations, if departures have been balanced, the last line's DMD should be equal in magnitude but opposite in sign to its departure.

Once DMDs have been determined, traverse area can be computed by multiplying the DMD of each line by its latitude, summing the products, and taking half the absolute value of the sum. This computation gives traverse area, but the proof is not demonstrated here.

Example 9-6. Find the area of the closed traverse of Example 9-3 by the DMD method.

Computed DMDs and DMD × latitude products are shown in Table 9-6. The DMD of *AB* was set equal to the departure of *AB* (-539.06), and the DMD of *BC* determined by adding the DMD of *AB*, departure of *AB*, and departure of *BC*,

$$-539.06 - 539.06 + 159.55 = -918.57$$

DMDs of remaining lines were calculated in the same manner. Note that the DMD of the last line *EA* (16.75) is equal in magnitude but opposite in sign to the departure of *EA* (-16.75). Values in the last column were obtained by multiplying the latitude of each line by its DMD—i.e., for line *AB* (-358.25) × (-539.06) = 193,118.

The traverse area is 339,239/2 = 169,620 ft^2, or 3.894 acres. The minus sign indicates only that the DMDs were calculated by a sequence around the traverse in a clockwise direction (instead of counterclockwise). Carrying out acreage beyond three decimal places is probably the limit, since 0.0001 acre represents 4.36 ft^2. Note that in many deeds, the acreage stated is qualified by "more or less" to cover *small* errors only.

A check on the calculated area can be made by employing *double parallel distances* (DPDs). The DPD for any traverse course is equal to *the DPD of the previous line plus the latitude of the previous line plus the latitude of the line itself.* (The DPD of the first course may be set equal to the latitude of the first course.) The traverse area can be computed by multiplying the DPD of each line by its departure, summing the products, and taking half the absolute value of the total.

Standard tabular forms are available for computing latitudes and departures, adjusted latitudes and departures, DMDs and areas on a single sheet.

9-7-2. Coordinate Method

The area of a closed traverse can be determined by this method if the coordinates of each corner are known.

The computational procedure in applying the coordinate method is to multiply the *x*-coordinate of each corner by the difference between adjacent *y*-coordinates, add the resulting products, and take half the absolute value of the sum; *y*-coordinates must be taken in the same order when obtaining the differ-

Table 9-6. Computation of area by DMDs

Station	Latitude (ft)	Departure (ft)	DMD (ft)	DMD × Latitude (ft^2)
A				
	-358.25	-539.06	-539.06	193,118
B				
	255.12	159.55	-918.57	$-234,346$
C				
	278.53	-153.53	-912.55	$-254,173$
D				
	76.73	549.79	-516.29	$-39,615$
E				
	-252.13	-16.75	16.75	$-4,223$
A				
				$-339,239$

ence between adjacent y-coordinates. This process can be expressed in equation form as

$$A = [x_1(y_2 - y_n) + x_2(y_3 - y_1)$$
$$+ \cdots + x_n(y_1 - y_{n-1})]/2 \qquad (9\text{-}7)$$

The coordinate method may be applied by substituting coordinates into Equation (9-7), but the method is expedited by listing them in the following form and securing sums of the products of all adjacent diagonal terms taken: (1) down to the right, i.e., x_1y_2, x_2y_3, etc., and (2) up to the right, i.e., y_1x_2, y_2x_3, etc.

$$\frac{x_1 \; x_2 \; x_3}{y_1 \; y_2 \; y_3} \cdots \frac{x_n \; x_1}{y_n \; y_1}$$

The traverse area is equal to half the absolute value of the difference between these two sums. In applying this procedure, note that the first coordinate listed must be repeated at the end of the list.

Example 9-7.   Find the area of the closed traverse of Example 9-3 (and Example 9-4) by the coordinate method.

First, list the coordinates determined in Example 9-4 in the format indicated previously.

$$\frac{2000.00 \; 1641.75 \; 1896.87}{1200.00 \; 660.94 \; 820.49}$$

$$\times \frac{2175.40 \; 2252.13 \; 2000.00}{666.96 \; 1216.75 \; 1200.00}$$

The sum of the products of adjacent diagonal terms taken down to the right

[i.e.,   (2000.00)(660.94)

$$+ (1641.75)(820.49) + \ldots]$$

is   9,283,530, and that of products taken up to the right

[i.e.,   (1200.00)(1641.75)

$$+ (660.94)(1896.87) + \ldots] = 8{,}944{,}292$$

The difference between sums is 9,283,530 − 8,944,292, or 339,238. Thus, the traverse area is 339,238/2 = 169,619 ft^2, or 3.894 acres.

9-7-3.   Area by Planimeter

Applying the DMD method requires knowing latitudes and departures, where the coordinate method needs coordinates for each corner. Both methods are limited to use with areas bounded by straight lines, but many parcels have some curved boundaries. For example, one or more boundaries of a land tract may follow a meandering roadway or creek. Such curved boundaries can be converted to a number of small straight-line segments suitable for the DMD or coordinate method.

An alternative means of finding land area utilizes a *planimeter*. Its operation does not require latitudes and departures, coordinates, or straight-line boundaries. Unlike the other methods, however, a scale drawing of the tract for which area is to be determined must be available.

A planimeter (see Figure 9-9) is a mechanical device that integrates area and records the answer as an operator traces the boundary of a figure with the pointer. An ordinary planimeter consists of two arms. One, the anchor arm, has a weight with a sharp point at its free end. The other, a scale bar, has a pointer at its free end. Near where the two arms join are a graduated drum, disk, and vernier.

To measure the area of a tract on a scale drawing, the anchor point is secured at some convenient location on the drawing, preferably outside the area to eliminate applying a polar constant, and the pointer set over a specific traverse boundary corner. An initial four-digit reading is taken; the first digit is read from the disk, the next two from the drum, and the last one on the vernier. The operator then carefully moves the pointer around the traverse boundary until the starting point is reached. A straightedge may be used to guide the pointer around the traverse, but ordinarily it is moved meticulously around the boundary freehand. At this time, another

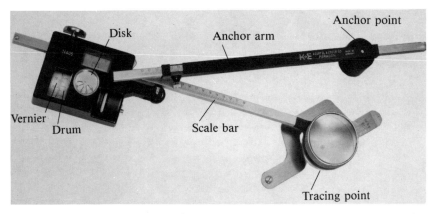

Disk Anchor arm Anchor point

Vernier Drum Scale bar Tracing point

Figure 9-9. Mechanical planimeter. (Courtesy of Cubic Precision, K & E Electro-Optical Products.)

reading is taken. The difference between initial and final readings, scaled if necessary, gives the traverse area. The boundary is then traced in the opposite direction back to the starting point, where the reading should be within a few digits in the fourth place as a check.

Although the procedure for finding area by planimetering sounds simple, caution must be exercised if accurate results are to be obtained. Since the area obtained by a planimeter is not necessarily an exact value (the same area measured twice will often yield slightly different results), it is good practice to trace a figure several times and take an average of the results thus obtained. It is also desirable to trace the figure one or more times in the opposite direction and average these values also. Unless absolutely sure of the planimeter

scale constant, its value should be verified prior to determining a desired area. This can be accomplished easily by tracing a figure of known area, such as a 5-in. square drawn to scale by the user and the diagonals measured to assure exactly a 25-in.^{2} area. One final admonition: As noted previously, the anchor point is preferably positioned outside the traverse. If positioned inside, a polar constant (usually provided by the manufacturer) must be added.

A *mechanical planimeter* is shown in Figure 9-9. *Electronic planimeters*, similar in operation to mechanical planimeters, present results in digital form on a display console with the ability to give answers directly in units of acres or hectares. Figure 9-10 displays an electronic planimeter.

Figure 9-10. Planix electronic planimeter. (Courtesy of the Lietz Co.)

```
C     THIS PROGRAM BALANCES A CLOSED-TRAVERSE SURVEY BY THE COMPASS RULE.
C     IT CAN BE USED FOR DISTANCES MEASURED IN EITHER FEET OR METERS.
C
C     INPUT DATA MUST BE SET AS FOLLOWS.
C
C     DATA LINE 1    COLUMN 1      ENTER 0 (ZERO) OR BLANK IF DISTANCES
C                                  ARE IN FEET. ENTER 1 (ONE) IF DISTANCES
C                                  ARE IN METERS.
C                    COLUMNS 2-9   ENTER DATE, IF DESIRED.
C                    COLUMNS 10-80 ENTER TITLE, IF DESIRED.
C
C     DATA LINE 2    COLUMNS 1-2   ENTER DESIGNATOR FOR BEGINNING POINT OF
C                                  FIRST TRAVERSE LINE.
C                    COLUMNS 4-5   ENTER DESIGNATOR FOR ENDING POINT OF
C                                  FIRST TRAVERSE LINE.
C                    COLUMNS 6-20  ENTER NUMBER INCLUDING DECIMAL GIVING
C                                  LENGTH OF FIRST TRAVERSE LINE.
C                    COLUMN 25     ENTER N IF BEARING OF FIRST TRAVERSE
C                                  LINE IS NORTH; ENTER S IF IT IS SOUTH.
C                    COLUMNS 27-28 ENTER NUMBER (RIGHT-ADJUSTED WITHOUT
C                                  DECIMAL) GIVING NUMBER OF DEGREES IN
C                                  BEARING ANGLE OF FIRST TRAVERSE LINE.
C                    COLUMNS 30-31 ENTER NUMBER (RIGHT-ADJUSTED WITHOUT
C                                  DECIMAL) GIVING NUMBER OF MINUTES IN
C                                  BEARING ANGLE OF FIRST TRAVERSE LINE.
C                    COLUMN 33     ENTER E IF BEARING OF FIRST TRAVERSE
C                                  LINE IS EAST; ENTER W IF IT IS WEST.
C                    COLUMNS 35-44 ENTER NUMBER INCLUDING DECIMAL GIVING
C                                  X-COORDINATE OF BEGINNING POINT.
C                    COLUMNS 45-54 ENTER NUMBER INCLUDING DECIMAL GIVING
C                                  Y-COORDINATE OF BEGINNING POINT.
C
C     DATA LINE 3    ENTER DATA FOR SECOND TRAVERSE LINE (OMIT COORDINATES
C                                  OF BEGINNING POINT) IN SAME FORMAT AS
C                                  THAT FOR FIRST TRAVERSE LINE (SEE DATA
C                                  LINE 2 ABOVE).
C
C     SIMILARLY, ENTER DATA FOR SUCCEEDING TRAVERSE LINES IN SAME FORMAT.
C
      INTEGER UNITS
      DIMENSION DATE(2),TITLE(12),NL(100),NR(100),D(100),BL(100),
     $NDEG(100),NMIN(100),BR(100),ALAT(100),ADEP(100),XCOOR(100),
     $YCOOR(100)
   99 FORMAT(I1,2A4,11A6,A5)
   98 FORMAT(A2,1X,A2,F15,2(4X,A1,1X,I2,1X,I2,1X,A1,1X,2F10.2)
   97 FORMAT(*1*,11A6,A5,2A4,///,11X,*GIVEN DATA*,//,11X,*=====*,//
     $*  LINE        LENGTH, FT        BEARING*,//,*
     $*-----*)
```

Figure 9-11. Program for analyzing traverses. (Adapted from J. B. Evett, 1979, *Surveying*, New York: John Wiley & Sons, p. 147.)

```
96 FORMAT(1X,A2,'-',A2,F9.2,4X,A1,1X,A1)
95 FORMAT(////,20X,'SURVEY ADJUSTED BY COMPASS RULE',/,20X,'==============
  $===============',///,' POINT LINE   LENGTH   BEARING
  $ LATITUDE   DEPARTURE   X-COORDINATE  Y-COORDINATE',/,
  $ ----- ---- ------ ------- -------- --------- ------------ ------------)
94 FORMAT(2X,A3,58X,F10.2,5X,F10.2,/,
  $2I3,1X,A1,2X,F10.2,2X,F10.2)
93 FORMAT(///,' LINEAR MISCLOSURE = ',F6.2,' FT',//,' PRECISION = 1/'
  $,I5,//,' AREA = ',F11.0,' SQ FT  OR',F11.3,' ACRES')
87 FORMAT('1',11A6,A5,2A4,///,11X,'GIVEN DATA',/,11X,'==========',//
  $' LINE      LENGTH   M   BEARING',/,' ---- ')
83 FORMAT(///,' AREA = ',F11.0,' SQ M   OR',F11.3,' M',//,' PRECISION = 1/',
  $' LINEAR MISCLOSURE = ',F6.2,' M',//,' PRECISION = 1/',
  $,F11.3,' HECTARES')
  DSUM=0.
  SUMD=0.
  SUML=0.
  AREA=0.
  PI=3.14159265
  READ(5,99)UNITS,DATE,TITLE
  DO 100 J=1,100
  N=J-1
  IF(J.EQ.1)READ(5,98)NL(J),NR(J),D(J),BL(J),NDEG(J),NMIN(J),BR(J),
  $XBEG,YBEG
  IF(J.NE.1)READ(5,98,END=101)NL(J),NR(J),D(J),BL(J),NDEG(J),NMIN(J),
  $BR(J)
  DEG=NDEG(J)
  AMIN=NMIN(J)
  BRAD=(DEG+AMIN/60.)*PI/180.
  ALAT(J)=D(J)*COS(BRAD)
  IF(BL(J).EQ.'S')ALAT(J)=-ALAT(J)
  SUML=SUML+ALAT(J)
  ADEP(J)=D(J)*SIN(BRAD)
  IF(BR(J).EQ.'W')ADEP(J)=-ADEP(J)
  SUMD=SUMD+ADEP(J)
  DSUM=DSUM+D(J)
100 CONTINUE
101 IF(UNITS.EQ.0)WRITE(6,92)TITLE,DATE
  IF(UNITS.EQ.1)WRITE(6,87)TITLE,DATE
  WRITE(6,96)(NL(J),NR(J),D(J),BL(J),NDEG(J),NMIN(J),BR(J),J=1,N)
  WRITE(6,95)
  EC=SQRT(SUML*SUML+SUMD*SUMD)
  PC=DSUM/EC
  JPC=PC+.5
  DO 102 J=1,N
  ALAT(J)=ALAT(J)-SUML*D(J)/DSUM
  ADEP(J)=ADEP(J)-SUMD*D(J)/DSUM
  D(J)=SQRT(ALAT(J)**2+ADEP(J)**2)
```

Figure 9-11. (*Continued*)

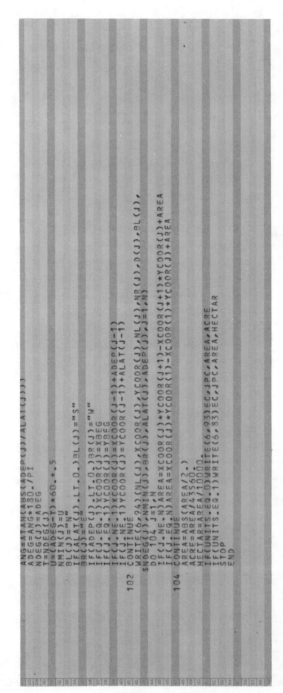

```
      ANG=ATAN(ABS(ADEP(J)/ALAT(J)))
      ADEG=ANG*180./PI
      NDEG(J)=ADEG
      T=NDEG(J)
      U=(ADEG-T)*60.+.5
      NMIN(J)=U
      BL(J)="N"
      IF(ALAT(J).LT.0.)BL(J)="S"
      BR(J)="E"
      IF(ADEP(J).LT.0.)BR(J)="W"
      IF(J.EQ.1)XCOOR(J)=XBEG
      IF(J.EQ.1)YCOOR(J)=YBEG
      IF(J.NE.1)XCOOR(J)=XCOOR(J-1)+ADEP(J-1)
      IF(J.NE.1)YCOOR(J)=YCOOR(J-1)+ALAT(J-1)
  102 CONTINUE
      WRITE(6,94)(NL(J),XCOOR(J),YCOOR(J),NL(J),NR(J),D(J),BL(J),
     $NDEG(J),NMIN(J),BR(J),ALAT(J),ADEP(J),J=1,N)
      DO 104 J=1,N
      IF(J.NE.N)AREA=XCOOR(J)*YCOOR(J+1)-XCOOR(J+1)*YCOOR(J)+AREA
      IF(J.EQ.N)AREA=XCOOR(J)*YCOOR(1)-XCOOR(1)*YCOOR(J)+AREA
  104 CONTINUE
      AREA=ABS(AREA/2.)
      ACRE=AREA/43560.
      HECTAR=AREA/10000.
      IF(UNITS.EQ.0)WRITE(6,83)EC,JPC,AREA,ACRE
      IF(UNITS.EQ.1)WRITE(6,83)EC,JPC,AREA,HECTAR
      STOP
      END
```

Figure 9-11. (*Continued*)

174

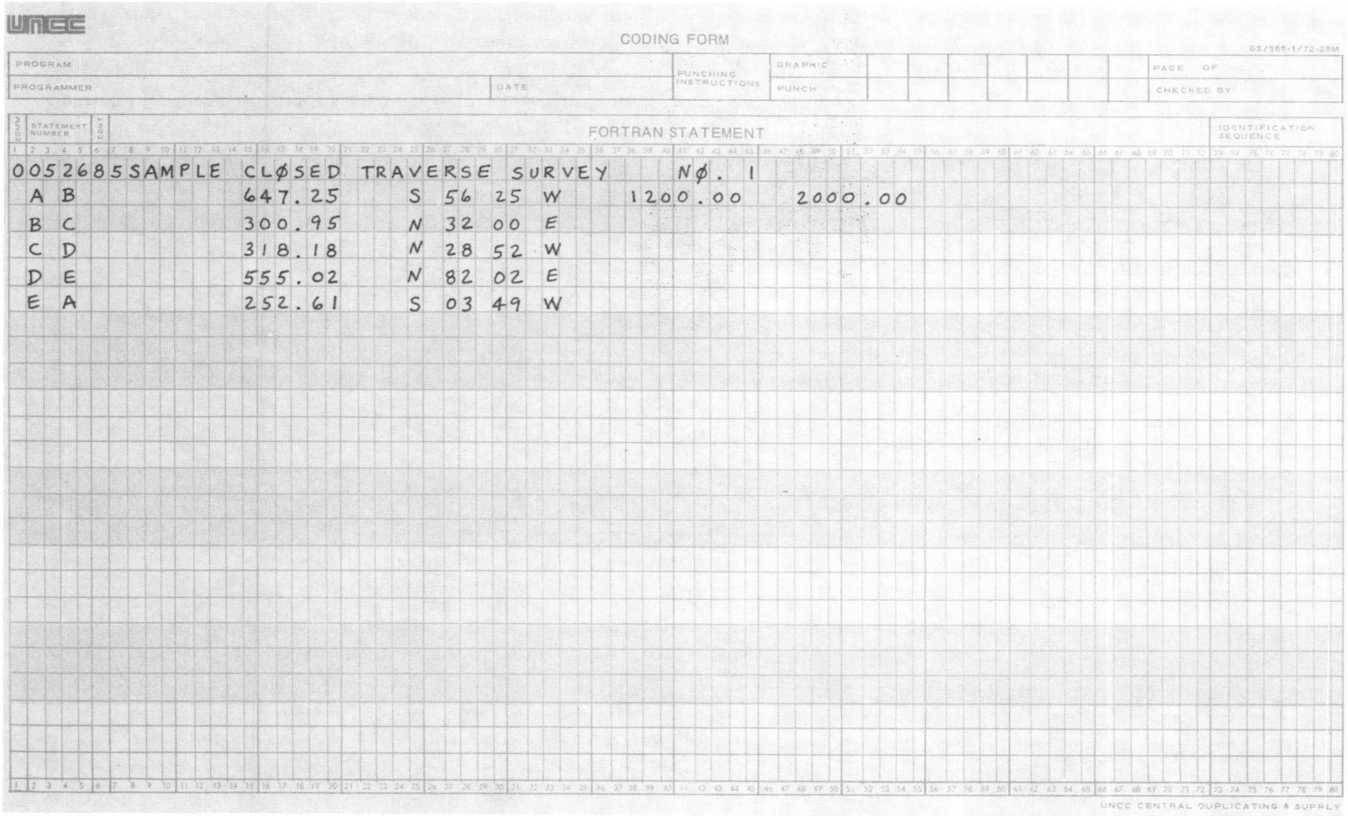

CODING FORM

PROGRAM

PROGRAMMER

DATE

PUNCHING INSTRUCTIONS

GRAPHIC

PUNCH

PAGE OF

CHECKED BY

FORTRAN STATEMENT

STATEMENT NUMBER					
005 2685	SAMPLE CLOSED TRAVERSE SURVEY	NO. 1	1200.00	2000.00	
A B	647.25	S 56 25 W			
B C	300.95	N 32 00 E			
C D	318.18	N 28 52 W			
D E	555.02	N 82 02 E			
E A	252.61	S 03 49 W			

IDENTIFICATION SEQUENCE

UNCC CENTRAL DUPLICATING & SUPPLY

Figure 9-12. Input data for sample closed traverse survey no. 1.

175

SAMPLE CLOSED TRAVERSE SURVEY NUMBER 1 05-26-85

GIVEN DATA

LINE	LENGTH, FT	BEARING
A-B	647.25	S 56 25 W
B-C	300.95	N 32 0 E
C-D	318.18	N 28 52 W
D-E	555.02	N 82 2 E
E-A	252.61	S 3 49 W

SURVEY ADJUSTED BY COMPASS RULE

POINT	LINE	LENGTH	BEARING	LATITUDE	DEPARTURE	X-COORDINATE	Y-COORDINATE
A						1200.00	2000.00
	A-B	647.24	S 56 24 W	-358.25	-539.06		
B						660.94	1641.75
	B-C	300.90	N 32 1 E	255.12	159.55		
C						820.49	1896.87
	C-D	318.05	N 28 52 W	278.54	-153.53		
D						666.96	2175.40
	D-E	555.12	N 82 3 E	76.73	549.80		
E						1216.75	2252.14
	E-A	252.69	S 3 48 W	-252.14	-16.75		

LINEAR MISCLOSURE = 0.87 FT

PRECISION = 1/ 2392

AREA = 169621. SQ FT OR 3.894 ACRES

Figure 9-13. Output data for sample closed traverse survey no. 1.

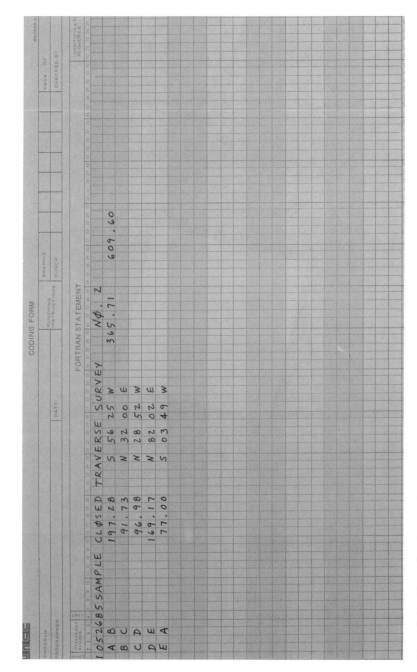

Figure 9-14. Input data for sample closed traverse survey no. 2.

```
SAMPLE CLOSED TRAVERSE SURVEY NUMBER 2                              05-26-85

        GIVEN DATA
        ==========
LINE    LENGTH, M   BEARING
----    ---------   -------
A-B       197.28    S 56 25 W
B-C        91.73    N 32 20 E
C-D        96.93    N 28 52 W
D-E       169.17    N 82 52 E
E-A        77.00    S  3 49 W

                   SURVEY ADJUSTED BY COMPASS RULE
                   ===============================
POINT  LINE  LENGTH    BEARING    LATITUDE  DEPARTURE  X-COORDINATE  Y-COORDINATE
-----  ----  ------    -------    --------  ---------  ------------  ------------
A                                                         365.76        609.60
       A-B   197.28  S 56 24 W    -109.19    -164.30
B                                                         201.46        500.41
       B-C    91.72  N 32  1 E      77.76      48.63
C                                                         250.09        578.17
       C-D    96.94  N 28 52 W      84.90     -46.80
D                                                         203.29        663.07
       D-E   169.20  N 82  3 E      23.39     167.58
E                                                         370.87        686.46
       E-A    77.02  S  3 48 W     -76.86      -5.11

LINEAR MISCLOSURE = 0.26 M

PRECISION = 1/ 2434

AREA =  15759. SQ M      OR      1.576 HECTARES
```

Figure 9-15. Output data for sample closed traverse survey no. 2.

9-8. PROGRAMMED TRAVERSE COMPUTATIONS

The various traverse computations presented in this chapter are among the most common, extensive, and important ones made in surveying. Practicing surveyors prepare these calculations on a daily basis.

Many years ago, surveyors had to make such computations manually, using slide rules or trigonometric tables and logarithms. Subsequently, large mechanical calculators capable of performing addition, subtraction, multiplication, and division became available, but trigonometric tables were still required. In the 1960s and 1970s, high-speed digital computers and hand-held calculators, some programmable, greatly increased the surveyor's computational capability. In addition to making rapid computations, both computers and calculators had built-in trigonometric functions, so trigonometric tables were no longer needed. In terms of routine computations, subsequent advances in computer hardware and software continue to make life easier for surveyors.

There are numerous computer programs available for surveyors to use in analyzing traverses. For example, the program in Figure 9-11 is written in FORTRAN and designed for card input, but it could easily be modified to another language for other kinds of input or for use on microcomputers. For input, the program receives the length and bearing of each line and gives as output adjusted lengths, bearings, latitudes, departures, and coordinates, as well as linear misclosure, precision, and traverse area. The program can be used for distances measured in either feet or meters; comments at the program's beginning tell how input data must be arranged to utilize the program.

For demonstration purposes, the computer program was run using input data from Example 9-7 (lengths and bearings of a closed traverse survey). Input data prepared for use in the program are shown in Figure 9-12. Output from the program, which includes answers found previously in Examples 9-1, 9-2, 9-3, 9-4, and 9-7 (the program computes the area by the coordinate method), is given in Figure 9-13. Examination of this output reveals that the answers closely verify those obtained by calculation in the examples, with some slight variations resulting from round-off errors in the manual computations.

As a final demonstration, lengths of the traverse lines in Example 9-1 (used as input to the computer program) were converted from feet to meters and the program run again for the same traverse with lengths in meters, along with original bearings. Input data for this application of the program are shown in Figure 9-14 and output in Figure 9-15.

10

Survey Drafting

Edward G. Zimmerman

10-1. INTRODUCTION

A sketch, map, or graphic display, is often the only visible product of a surveyor's work. Therefore, the importance of presenting the client with a nice-appearing, professionally done graphic product cannot be overemphasized. Attractiveness, accuracy of plot, legibility, and clearly imparted information are vital in creating a survey drafting product worthy of professional respect.

Survey drafters differ from their engineering and architectural counterparts in being not only familiar with applying principles of drafting and graphics, but also able to comprehend survey instrumentation and methods of measurement. Field notes must be reduced and interpreted and information sources researched so a drafter can construct a comprehensive graphic representation of actual facts and conditions.

A survey drafter is a multitalented technician. He or she should have training in the basics of mechanical drawing and mathematical ability extending through trigonometry. Since much of the work prepared will originate from either a set of field notes or rough sketch with some penciled-in instructions, the drafter needs to be familiar with fundamental principles of survey operations. Actual field experience is important in developing an awareness of the methods of recognizing, collecting, and recording field data and will assist the draftsperson in translating the data into a finished drawing.

Drafters do not spend all their time drawing or tracing maps. Depending on the type and priority of workloads, survey drafters may also calculate and check traverse surveys, reduce and plot cross-section and topography field notes, calculate earthwork quantities, and prepare material estimates for construction projects. Often, with a smaller firm, the draftsperson may be assigned to field duties as workloads respond to seasonal fluctuations.

10-2. SURVEY DRAWINGS

Survey drawings fall into three general categories: (1) property and control maps reflecting surveys made to establish or reestablish ownership lines or survey control networks; (2) topographic maps showing elevations, natural and artificial features, and form of the earth's surface; and (3) construction maps made to provide and control horizontal and vertical location, alignment, and configuration of construction work.

An accurate plot of survey information plus legibility and attractiveness determine a map's usefulness. Most maps show few dimensions, and a person using them must rely on a scale, protractor, or drafting machine to determine intermediate dimensions. Unlike mechanical or civil-engineering drawings, survey maps are irregular and not readily drawn by traditional "T square and triangle" methods.

All survey drawings are made to be copied or reproduced. Therefore, drafters should be aware of the infinite variety of reproduction possibilities and applications available through graphic processes, photographic enlargement and reduction, and exhibit preparation. The quality of a finished drawing, type of reproduction equipment, quality of the printed product, and economic requirements are all factors influencing the choice of reproduction method.

Computer-assisted drafting (CAD) is changing mapmaking much as electronic instrumentation is revolutionizing field surveying. Interactive drafting systems, including a cathode-ray display tube and an automatic plotter, are combined with a computer for processing data recorded in the field on an interfaceable data collector. In larger agencies and companies, automated drafting units are fast becoming standard equipment. Operation of a CAD system, such as the one shown in Figure 10-1, requires yet another level of training and ability in a survey drafter but will not be addressed in this chapter.

10-3. MAP SCALE

Map scale is the term used to define the ratio of distances represented on a map to actual ground distances. When a drawing is made to a chosen scale, all dimensions—distance, direction, and difference in elevation—will be in correct relationship and accurately represent the actual figure.

The scale of a map should be indicated by both numerical and graphic means. Numerical scales may be either representative, in which one unit on the map represents a certain number of the same units on the ground—e.g., 1/400, 1 : 400—or equivalent, in which a statement indicates that 1 in. on a map equals a whole number of feet on the ground—e.g., 1 in. = 4000 ft.

Figure 10-2 shows an example of graphic scales. Since drawing paper may change dimensions over time or be distorted by reproduction processes, a graphic scale should be placed on maps to provide a constant check of the exact scale.

The ranges of scales are defined as (1) large scale, 1 in. = 100 ft or larger; (2) medium scale, 1 in. = 100 to 1000 ft; and (3) small scale, 1 in. = 1000 ft or smaller. Table 10-1 provides a guide in choosing the scale for a particular map.

10-4. MAP DRAFTING

Most maps fall into two general classifications: (1) those maps that show land ownership and become part of the public record and (2)

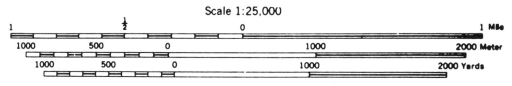

Figure 10-1. Typical computer-aided drafting system. (Courtesy of Hewlett-Packard Inc.)

Figure 10-2. Typical graphic or bar scale.

those that show land form and are used as the basis for design and construction of structural facilities, both private and public.

Preparation of a preliminary map or manuscript is the first two phases of map drafting. First, the manuscript is carefully laid out, plotting all control lines and points with the utmost accuracy. This map is drawn with a hard-grade pencil on a high-quality drafting film that produces sharp and precisely located line work. Features comprising the final map should be plotted on the manuscript in the following sequence: (1) Lay out control points and lines; (2) plot details; (3) compile topography and other detail work; and (4) finish the map, complete with lettering and all notes.

Lettering and symbolization need not be of finished quality, just accurately located.

Second, following checking and revisions, the manuscript can be placed on a "light table" and the final map traced in ink on a stable-base mylar drafting film. The preferred sequence is to (1) complete all the lettering, notes, and title and (2) finish the line work or topography.

Scribing, another method of producing a final map, is gaining in popularity. In this process, lines from the manuscript are photographically transferred onto a sheet of drafting film coated with an opaque surface. Using specially designed scribing tools, the drafter scrapes and cuts the coating to reproduce all

Table 10-1. Selection of map scales

Type of Map and Use	Equivalent (ft per in.)	Representative (ratio, 1/...)
Design		
Civil improvements	10–50	120–600
General construction	40–200	480–2000
Property/boundary (dependent on figure size)	50–500	600–6000
Topographic/planimetric		
Small site	10–50	120–600
Large site	40–200	480–2400
Urban	200–1000	2400–12,000
Regional	500–2000	6000–24,000

the original manuscript's features and line work.

10-5. DATUMS FOR MAPPING

All measurements made by surveyors to determine and depict elevations and horizontal positions should relate to a datum of reference. In the 48 contiguous United States and Alaska, the *American Datum of 1983* is used. These reference figures are made available through the state plane coordinate system of a particular state. Most states have a plane coordinate system based on either a Lambert conformal-conic projection, or transverse Mercator projector system. Adopted in the 1930s, the systems use the U.S. survey foot (1 ft = 1200/3937 m) as the standard of measurement unit.

Elevations in the United States are referred to a vertical datum or reference surface based on mean sea level—i.e., the *North American Vertical Datum of 1988* (NAVD 88). This datum is determined from the average elevations of 26 sea-level tidal stations in the United States and Canada. Coordinates of horizontal control stations and elevations of benchmarks throughout the United States are available in published form from the National Ocean Survey (NOS). Instructions and specifications for the use of both datums are also available from the NOS.

10-6. TOPOGRAPHIC MAPS

A topographic map is a graphic representation of a portion of the earth's surface as it existed on a certain date. It is drawn from field survey data or aerial photographs and shows, by notation or symbol, all natural or artificial land features, including boundaries, cities, roads, railroads, pipelines, electric lines, buildings, and vegetation. Land forms are depicted by contour lines. A topographic map without contour lines is defined as a planimetric map. See Figure 10-3 for a typical map.

10-7. TOPOGRAPHIC MAP CONSTRUCTION

A topographic map should be drawn in three phases: (1) Develop horizontal control, producing a framework for plotting details; (2) plot all points of known elevation and locations of artificial or natural features; and (3) construct contour lines from plotted points of elevation, drawing all features and symbols.

Discussions in the following sections address the work involved in conventional line drawings. The availability of photogrammetrically based plotting systems makes production of larger topographic maps by automated equipment more economical. Most large surveying and engineering firms and agencies use this method. Automated drafting will be covered in later sections.

10-7-1. Plotting Control

Control points and lines for topographic as well as other survey maps can be plotted by one of several methods. The selection of which method to use is guided by the field survey format and form in which field information is forwarded to the drafter.

A traverse or control survey can be plotted by laying out a series of angles and scaled lines. Angles are plotted by a drafting machine, protractor, or coordinates, or constructed by methods described in Section 10-7-5.

10-7-2. Drafting Machine

A drafting machine combines all the functions of a straightedge, triangle, scale, and protractor in one convenient unit (Figure 10-4). It is constructed so that any movement of the machine head is in a parallel motion, but the horizontal and vertical scales retain an initial base-line orientation, ready for use at every position on a drawing. The protractor and vernier scale allow the ruling scales to be rotated to any desired angular value, usually

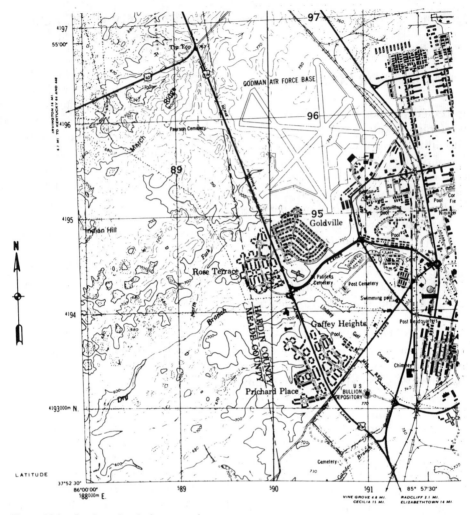

Figure 10-3.   Portion of typical topography map.

within ±1 min for either angle construction or instrument orientation.

10-7-3. Protractor

Protractors are a tool vital to mapping and topographical plotting work, and are available in several configurations (Figure 10-5). Most are graduated in half-degree increments and vary in size from 4- to 10-in. diameters. To use, simply align the protractor base and center mark with a base line and angle point, then scale the required angle on the protractor's periphery. Although quick and convenient to use, protractors are not accurate enough for plotting precise control networks.

10-7-4. Coordinates

Plotting by coordinates is a simple and accurate method, although it requires that all information be coordinated and an accurately gridded base map used for plotting. An accurate grid pattern is laid out at an appropriate scale with squares having an even dimension, say, 100, 500, or 1000 ft. Label each *x* and *y* line with its grid value to avoid plotting blunders.

Figure 10-4. Drafting machine and table. (Courtesy of Alvin & Co., Inc.)

By scaling from the appropriate *x* and *y* line, each point can be plotted to an accuracy of 0.01 to 0.02 in. Each point is marked by a small circle and connected with straight lines to outline the figure. After plotting, measure the bearing and distance between plotted points to check plotting accuracy and detect any blunders. Every point is plotted independently of the others so compounding plotting errors cannot occur.

10-7-5. Alternate Methods of Plotting Angles

Tangent Method

This method consists of extending the back tangent a scaled number of units beyond a point. The length of extension times the natural tangent of the angle to be constructed yields an offset distance, which is then scaled along a perpendicular erected at the end of the extension (Figure 10-6).

Chord Method

This method is similar to the tangent procedure except that a tangent arc is struck,

rather than a perpendicular. The chord distance for the angle to be plotted is calculated and scaled along the arc (Figure 10-7).

10-7-6. Plotting Details

Although details need not be plotted with the same accuracy as control points, objects must be placed on the map within allowable error standards. The choice of plotting methods for details depends on field procedures for gathering the information. For example, if topography was obtained by a radiation survey, then details should be plotted with a drafting machine or protractor and scale. If a survey was made by the *checkerboard* method, a grid constructed on the map sheet is used to plot details. Surveys from a total station can be plotted by radiation or coordinates.

10-7-7. Characteristics of Contours

For proper delineation and interpretation of a topographic map, it is important for a drafter to be familiar with the nature of contour lines, as shown in Figure 10-8. Their

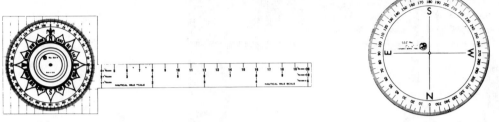

Grid Compass Nautical Protractor   **Circular Compass Protractor**

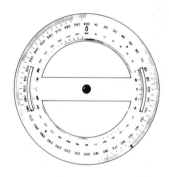

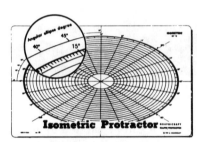

Combination Circular Protractor   **Isometric Protractor**

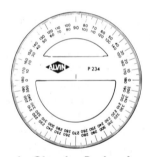

Academic Circular Protractor   **Academic Protractors**

Figure 10-5.   Surveying engineering protractors. (Courtesy of Alvin & Co.. Inc.)

principal characteristics are as follows:

1. Contour lines spaced closed together represent steeper slopes; lines farther apart indicate a gentler slope.
2. Uniform spacing of lines represents a uniform, even slope.
3. Mounds or depressions are portrayed by closed contour lines. Depression contours

will have inward-facing, radial tick marks to avoid confusion.

4. Contour lines always close.
5. Contour lines never cross or merge into each other, except, e.g., in the case of overhanging rock ledges.
6. Contour lines are perpendicular to the direction of the represented slope.

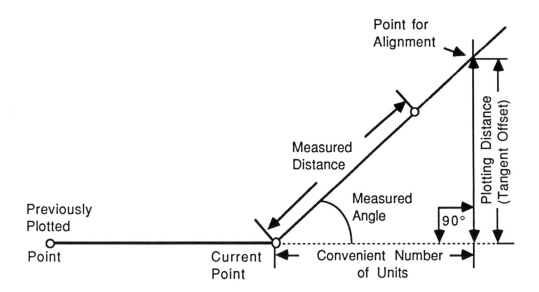

TAN Measured Angle X Number of Units = Plotting Offset.

Figure 10-6. Construct angle: Tangent method.

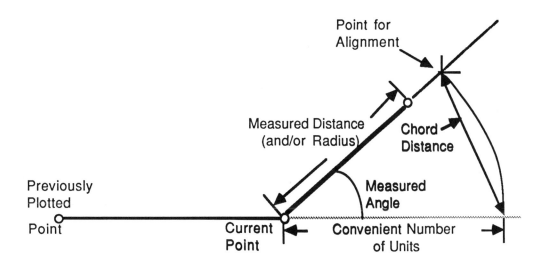

Measured Angle = Central Angle of Arc.
Chord = 2R X SIN 1/2 Central Angle.

Figure 10-7. Construct angle: Chord method.

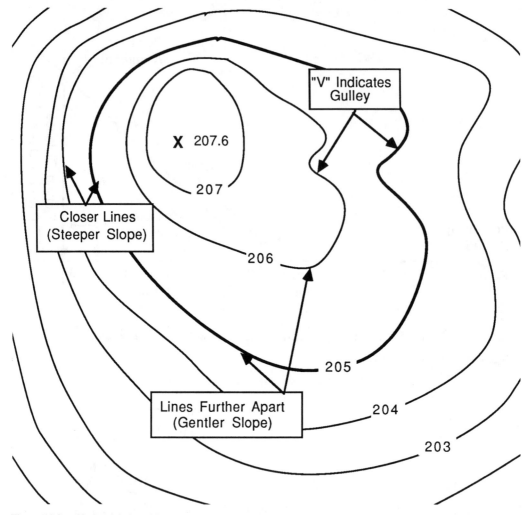

Figure 10-8.   Characteristics of contours.

10-7-8. Plotting Contours

When detail plotting is completed, the map sheet will show planimetric detail and elevations of pertinent points. The drafter's next operation is to draw contour lines, guided by previously plotted elevations. A contour line represents an even unit of elevation, generally in multiples of 1, 2, 5, 10, 20, 50, or 100 ft. Selection of a contour interval depends on map scale, details to be presented, and severity of the terrain being mapped. Table 10-2 shows suggested contour intervals. Contour lines are drawn only for those elevations divisible by the contour interval.

Each line is sketched freehand, using smooth flowing curves at direction changes to more nearly represent a natural formation. Every fifth line should be drawn heavier and numbered with the elevation. When a fairly level area is depicted, it is advisable to number all contour lines.

10-7-9. Drawing Contours by Interpolation

Interpolation is a procedure to locate contour lines in their correct proportional position between adjacent points of elevation. This is done by assuming the terrain's slope be-

Table 10-2.   Selection of contour interval

Use of Topographic Map	Equivalent Scale (ft per in.)	Contour Interval (in ft)
Smaller sites	10–50	1–5
Larger sites	40–200	1–10
Urban	200–1000	2–20
Regional	500–2000	5–100

tween any two elevations is constant, and therefore, intervals in elevation will translate to horizontal distances. As illustrated in Figure 10-9, a drafter can accomplish this in one of several ways:

1. *Estimation.* An experienced topographer can make mental calculations as well as estimate positions of the lines.

2. *Direct calculation.* Measure the intervening horizontal distance and proportion the correct position for each line.

3. *Mechanical interpolation.* A rubberband marked with a uniform series of marks can be stretched to find the correct interval for each line. Also, spacing dividers can be used to proportion contour lines. This drafting instrument is constructed with 11 legs arranged to subdivide a distance spanned into equal parts.

4. *Graphic.* A transparent piece of drafting film with converging lines, as shown in Figure 10-10, can be pivoted between points of elevation to find correctly proportioned positions.

10-7-10.   Topographic License

Engineering is tempered with art to create a contour map convening the natural appearance of land forms to the map viewer. Mapping art is defined as using discretion and judgment in expressive placement of lines between control points, which reflect locations determined by engineering methodology. Topographic license is a drafting talent that can be attained by only a proper combination of field experience, drawing practice, and training.

10-8.   CONSTRUCTION MAPS

Maps of this type can vary from a simple plot plan for a residence to a major engineering project, such as a dam or freeway, and maps assisting civil engineers in project design. There are as many different layouts for construction maps as there are for other types of maps; however, all types have a title, typical scale, lettering and symbols, or other recognized guidelines.

10-8-1.   Earthwork Cross Sections

Cross sections are generally used for design purposes and to prepare estimates for earthwork projects. They are easily plotted on preprinted paper having 1-in. grids divided into 0.1-in. increments (see Figure 10-11). Cross sections can be plotted from field books or scaled from contour maps.

When working with field notes, plot sections in a vertical column, using the same stationing sequence shown in the notes. Also, follow the indicated right/left placement. Vertical plotting scale should be at least four times as large as horizontal scale to exaggerate and achieve useful vertical separation within each section. When cross sections are scaled from a topographic map, draw the center or base line, mark full stations on it, and lay out cross-section lines at each station. Scale along every cross-section line, noting distances where these lines cross a contour, then plot the distance out and elevation as though they came from a field book.

Cross sections with lines plotted on them representing finished grade become closed

MATHEMATICAL RELATIONSHIP:

$$\frac{\text{Horiz. Dist.}}{\text{Vert. Diff.}} = \frac{42\text{-ft}}{6.5\text{-ft}} = \frac{6.5\text{-ft}}{1\text{-ft}} = \frac{0.65\text{-ft}}{0.1\text{-ft}}$$

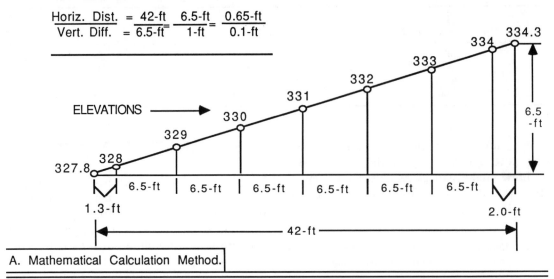

A. Mathematical Calculation Method.

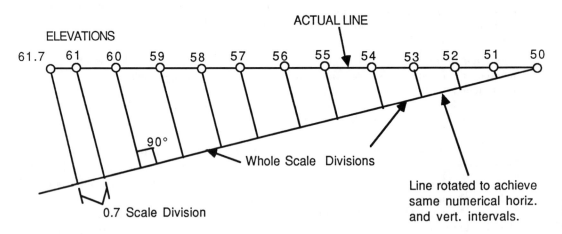

B. Proportionate Triangles Method.

Figure 10-9. Interpolation of contour intervals.

figures, so their areas can be determined by computation, planimetering, and other methods. Additional computations by the *average-end-area* method produce reasonably accurate volumes of a series of adjoining sections. A plot of this type should also include field-book references, referrals to the surveyor's vertical and horizontal control data, and scale.

10-8-2. Plan and Profile Maps

This type of map—drawn to depict details of linear-type construction projects, such as streets and highways, railroads, and pipelines —is used by builders to guide them in construction. As shown in Figure 10-12, the *plan* portion of the drawing is the horizontal layout

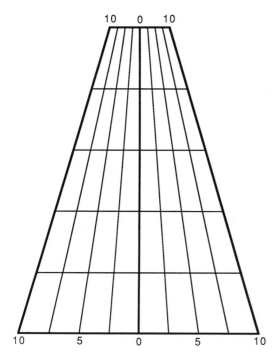

10 0 10

10 5 0 5 10

Figure 10-10. Clear plastic interpolator (may be rotated to fit the distance between two controlling points of elevation).

of the project, the *profile* view shows the vertical plane or *grade* information. Either preprinted paper or mylar sheets can be used. Generally, the sheet's top half is blank for a plan view, and the bottom half ruled for a profile view. Preruled lines facilitate rapid and accurate plotting of vertical features, such as flow lines of pipes, grade lines of highways, and natural ground profiles. The following items should be included on a plan and profile map:

1. Basis of stations (usually on centerline)

2. Vertical datum and available bench marks

3. Centerline alignment and control information

4. Lengths and widths of all features

5. Underground structural details

6. Construction notes pertaining to required materials and methods

The plan view portion is drawn by tracing previously plotted control maps or plotting survey data from field books or maps. Features of the project are clearly outlined, along with survey control and dimensional information sufficient for construction layout. Topographic features and contour lines are sometimes included to give further information for estimating and layout purposes.

Profile views include a natural ground line drawn along the project centerline and also grade lines with slope percentages for project features. Grades for aboveground construction are shown for the tops of finished surfaces, but underground grades control the structure's invert.

10-8-3. Site or Grading Plans

In contrast to plan and profile maps depicting long narrow strips of development, a site plan represents multisided figures showing parking lots, buildings, shopping centers, etc. This variety of map is drawn as a plan view but must have sufficient elevation information to enable the designer, estimator, and builder to complete their work. Site plans are developed from boundary and topographic surveys and must reflect proper design criteria conforming to local agency standards covering drainage, ingress and egress, etc. (see Figure 10-13).

Generally, a site or grading plan is prepared in two stages: (1) a preliminary map showing topography and boundary information for use by the engineer to create and finalize his or her design and (2) the final map reflecting completed design information and employed to actually construct the project.

10-8-4. As-Built Maps

At the conclusion of construction projects, *as-built* field surveys are made to locate all features of the project. Rarely is a project completed without making "field adjustments" or modifications to the original design. The survey map can usually be made by highlight-

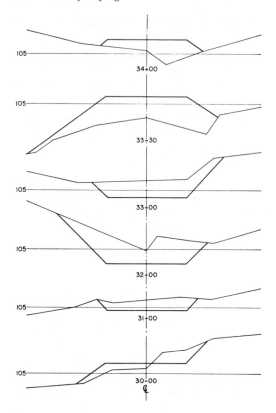

Figure 10-11. Cross-section plot. Grid lines have been omitted for clarity.

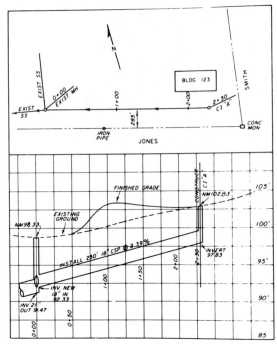

Figure 10-12. Portion of a plan and profile map.

ing or superimposing the changes on a copy of the original construction plans.

The map should show any changes or modifications to the original project design, including notations, locations, and dimensions, and referencing to the primary survey control system. To avoid confusion, as-built maps should be clearly designated as such. In future engineering relative to the completed project, they will become a permanent and useful information source.

10-9. BOUNDARY MAPS

A boundary map is drawn to depict ownership facts disclosed by research from public records

and/or field surveys. They usually delineate currently owned parcels and parcels being created for future sales. When relatively large portions of the earth are shown, boundary maps are sometimes called cadastral maps.

10-9-1. Subdivision Maps

Subdivision maps portray a tract of land that has been divided into several smaller parcels. The effect of the map, when made part of the public record, creates a basis of ownership for each individual parcel of lot as it is sold. A typical subdivision map is shown in Figure 10-14; however, maps of this type may vary in conformance to local agency requirements.

Subdivision maps are best traced in ink on mylar from a preliminary map that has been plotted by coordinates. Information shown on these maps should include (1) the scale, (2) survey data references, (3) ownership and title

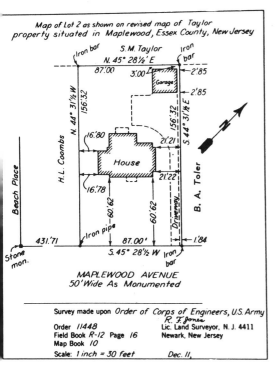

Figure 10-13. Typical site plan.

data, and (4) zoning and other regulatory statements.

10-9-2. Parcel Maps

This map is similar in format and purpose to a subdivision map, but is intended for either a combined or single parcel of land. It must also conform to local regulations and is drawn in the same format as a subdivision map (see Figure 10-15).

10-9-3. Record of Survey

A record of survey map generally does not show a transfer of property. It is intended to show a survey performed and then made part of the public record. This map, although similar to those described in the preceding sections, does not have to conform to the same regulatory requirements (see Figure 10-16).

10-10. INFORMATION SHOWN ON MAPS

Information provided on survey maps obviously varies, depending on their purpose. However, some important items are required on all surveying and engineering maps and are discussed in this section.

10-10-1. Scales

Map scale should be indicated both as a statement—e.g., 1 in. = 400 ft—and shown graphically. It should be placed on the map for quick location by the viewer, preferably close to the title block. See the example in Figure 10-17.

10-10-2. Meridian Arrows

If at all possible, the top edge of a map should represent north, but configuration or size may require a different orientation for the drawing. Regardless of the figure's position, the basis of the map's reference meridian is indicated by a stylized arrow or arrows such as shown in Figure 10-18. The true meridian is defined by a full arrow and other meridians —grid, magnetic, etc.—can be identified with a half arrow and letter designation. It is advisable to show angular difference (rotation) between two or more meridians.

10-10-3. Lettering

Survey drawings can be lettered in two different styles, the choice generally determined by their prospective use. For professional appearance, the lettering style and methods used on a map should be consistent throughout the entire drawing. Single-stroke Reinhardt-style machine-guided lettering is used for drawings intended to remain in the office or be used strictly by other surveyors or engineers. Maps produced for or utilized by the general public can be lettered freehand, using either Rein-

PLAT OF

CARNATION VILLA

LOTS 13 & 14, CAMELLIA ACRES (15 BM 21)
CITY OF SACRAMENTO, CALIFORNIA
JANUARY, 1986   SCALE: 1"=40'
TIMOTHY S. TRAIN, LAND SURVEYOR
SHEET 2 OF 2

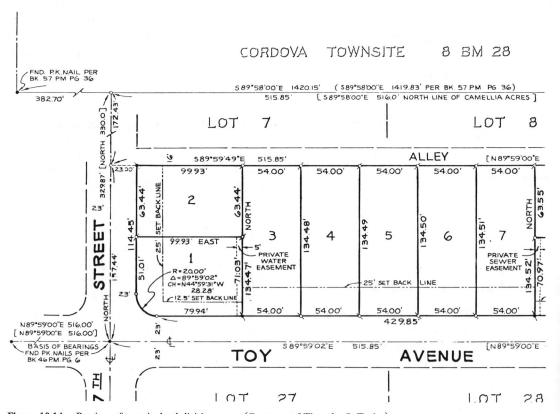

Figure 10-14.   Portion of a typical subdivision map. (Courtesy of Timothy S. Train.)

hardt letters or any attractive style that can be produced speedily and consistently (see Figure 10-19 for examples).

Adhesive-backed film transfer media can be typed on, cut out, and pasted on a drawing. This process is particularly time-saving for tables or lengthy statements on a map. A machine that prints a wide variety of fonts and sizes on opaque or transparent adhesive-backed tape is shown in Figure 10-20.

10-10-4. Titles

The purpose of a title is to identify the drawing and thus provide information for indexing. It may be neatly lettered freehand,

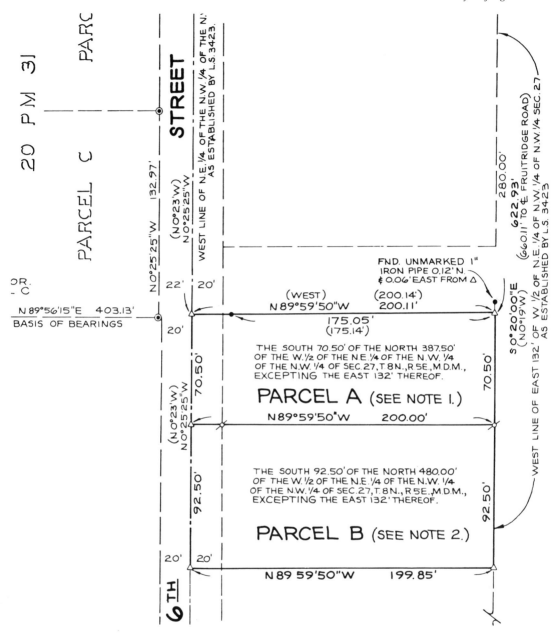

Figure 10-15. Portion of a typical parcel map. (Courtesy of Timothy S. Train.)

mechanically lettered, or preprinted commercially on standard size drawing sheets.

To be esthetically pleasing, a title block may be placed anywhere on the map sheet, but is ordinarily located in the lower right-hand corner. Care should be taken to keep the title

block size symmetrical and in proper proportion to the map size. Use conventional lettering with variations in weight and size to emphasize important parts of the title.

Elements of a title block include (1) name of the firm or agency, (2) type of map, (3)

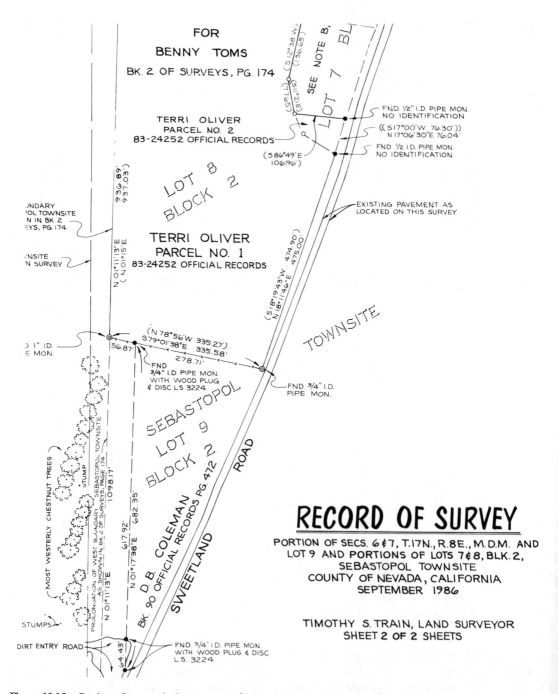

Figure 10-16. Portion of a record of survey map. (Courtesy of Timothy S. Train.)

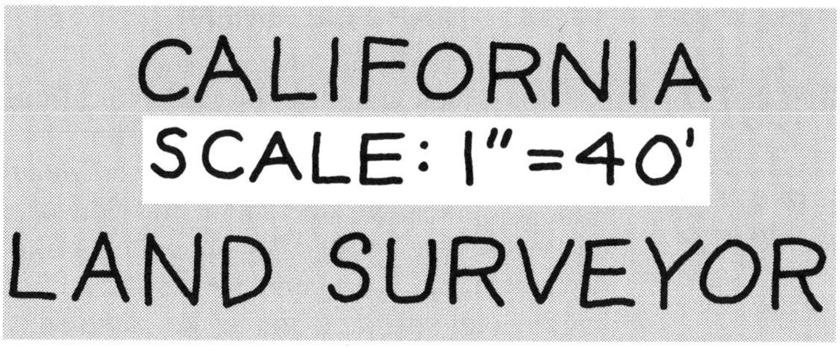

CALIFORNIA
SCALE: 1"=40'
LAND SURVEYOR

Figure 10-17. Statement of scale placed on map. (Courtesy of Timothy S. Train.)

indexing information, (4) name and location of the project, (5) scale and contour interval, (6) names of key personnel responsible for the map, and (7) map completion date (see Figure 10-21). It is common practice for most firms and agencies to use preprinted map sheets with frames and title blocks in a standard format and location. All the drafter need do is fill in the blank spaces.

10-10-5. Notes and Legends

Notes are placed on a map to provide explanations for special features unique to it. They should be brief yet impart adequate information to preclude misinterpretation. Information contained in notes should refer to items such as (1) project basis of bearings, (2) datums of both horizontal and vertical con-

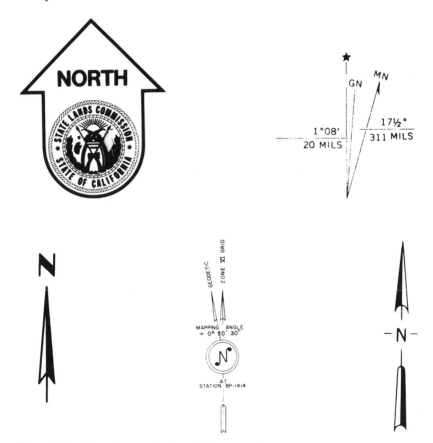

Figure 10-18. Several examples of meridian arrows.

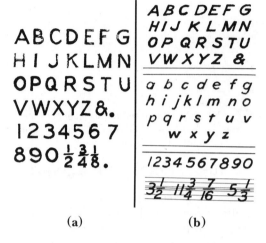

ABCDEFG
HIJKLMN
OPQRSTU
VWXYZ&.
1234567
890½⅜⅛.

ABCDEFG
HIJKLMN
OPQRSTU
VWXYZ &

a b c d e f g
h i j k l m n o
p q r s t u v
w x y z

1234567890

3½ 11¾ 7⁄16 5⅓

(a)   (b)

Figure 10-19. General lettering examples. (a) Vertical single-stroke Gothic capitals, and numerals. (b) Inclined single-stroke Gothic.

trol, and (3) references to old maps or data referred to in compiling the map.

A legend provides the key to symbols representing topographic or other features that otherwise would require lengthy or repetitive explanatory notes. Notes and legends should be located in close proximity to the title to aid the viewer in quickly locating them (see Figure 10-22 for an example).

10-10-6. Symbols

Because it is not possible to show all features on a small map, many can be represented by symbols. Several hundred standard symbols have been developed to represent topographic and other features unique to surveying. A few of these are shown in Figure 10-23 and inside the back cover. Employing symbols permits plotting many features in their correct locations without hopelessly crowding the map. Accurate positions of the features are plotted and a representative symbol drawn or conveniently cut from a preprinted sheet of adhesive-backed material and simply pasted on the map.

10-11. MAP PLACEMENT

A map's pleasing appearance helps to inspire confidence and dependability in the user's mind. Nothing detracts more from an otherwise acceptable product than omission of a heavy line border, and/or an out-of-balance placement of a drawing. Generally, a 1-in. margin is used on all four sides, although a 2-in. space may be necessary on the left edge if a number of sheets are to be bound together or hung on a rack. The title block is usually integrated with the lower right-hand corner border lines.

To properly place and fit a traverse and topographic details on a map sheet, first compute the maximum total of N-S latitudes and E-W departures. For a 24 × 30-in. sheet with 1-in. margins on three sides, and 2 in. on the left edge, the available mapping space is 22 × 27 in.

Assume that the total of the largest N-S latitudes is 1280 ft and for the E-W departures is 1810 ft. The largest scale possible is 1280/22 or 1 in. = 58 ft in the N-S direction and 1810/27 or 1 in. = 67 ft in the E-W space. Rounding off to a usable multiple of 10, 1 in. = 80 ft fits a standard scale.

Having the most westerly, northern, and southern stations (or topographic details) at the same distance from the borders produces an orderly appearance. The title box in the lower right-hand corner and any special notes placed directly over it will counterbalance the "weight" of the traverse and details. Special symbols, if any, are located above the notes. This arrangement enables a map user to quickly select the desired map by title and then check for special information before moving on to the drawing.

Placing the north arrow near the top right-hand side of the sheet also helps to balance it. Maps are read from the bottom or right-hand side, so the meridian is generally parallel with the right-hand border. On maps that are not required to match others on one or more sides, the meridian can be rotated to accom-

KROY 190
Helvetica Medium
ABCDEFGHIJKLMNOPQRSTUV
WXYZ abcdefghijklmnopqrstuv
wxyz 0123456789 .,?!&“”()’-:/#
%$£+−×÷=±□;@°*•'"

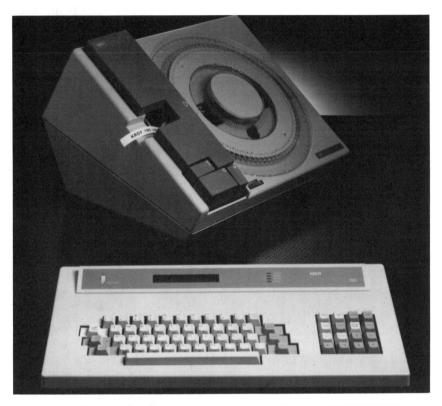

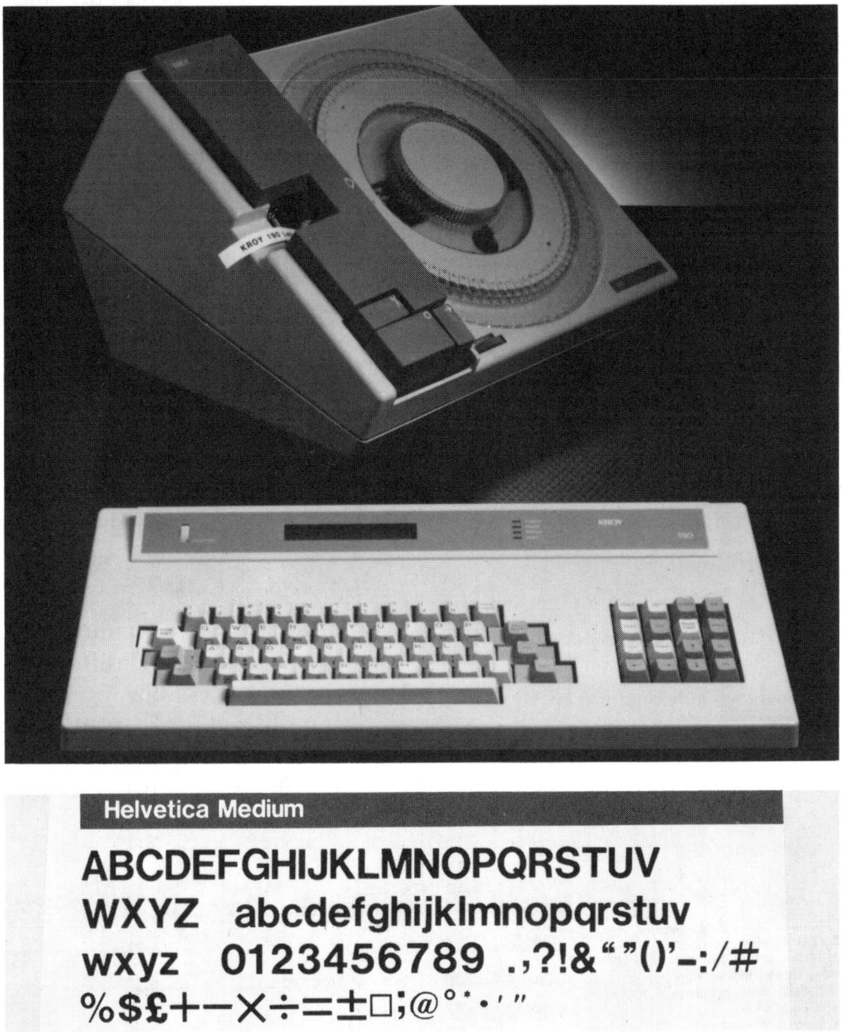

Figure 10-20. Kroy 190 (TM) Lettering System and typical text. (Courtesy of KROY Inc.)

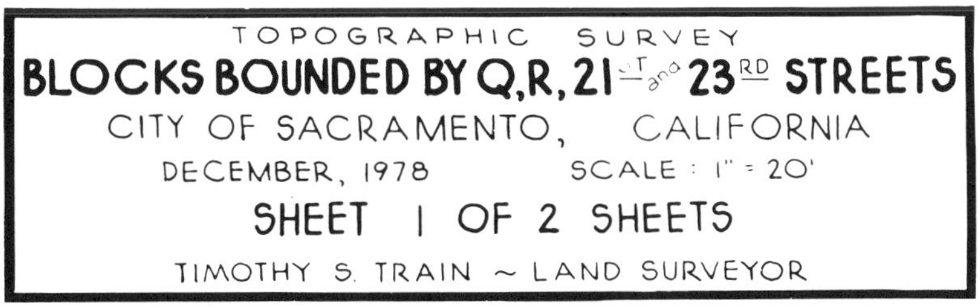

TOPOGRAPHIC SURVEY
BLOCKS BOUNDED BY Q,R, 21ST and 23RD STREETS
CITY OF SACRAMENTO, CALIFORNIA
DECEMBER, 1978 SCALE : 1" = 20'
SHEET 1 OF 2 SHEETS
TIMOTHY S. TRAIN ~ LAND SURVEYOR

Figure 10-21. Typical title block for map. (Courtesy of Timothy S. Train.)

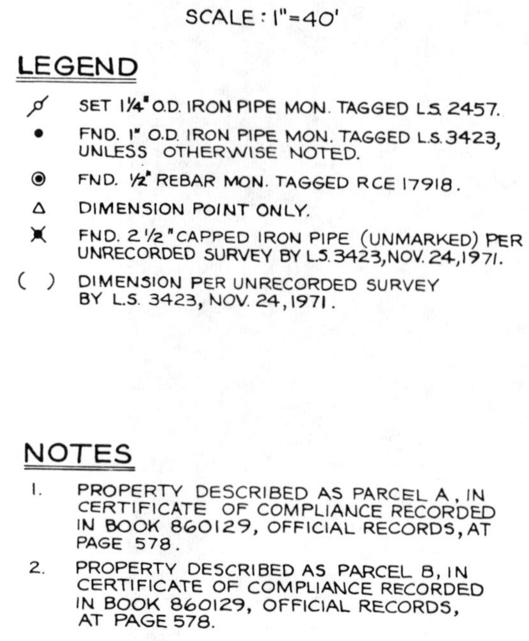

SCALE : 1"=40'

LEGEND

⚲	SET 1¼" O.D. IRON PIPE MON. TAGGED L.S. 2457.
●	FND. 1" O.D. IRON PIPE MON. TAGGED L.S. 3423, UNLESS OTHERWISE NOTED.
◉	FND. ½" REBAR MON. TAGGED R.C.E. 17918.
△	DIMENSION POINT ONLY.
✖	FND. 2½" CAPPED IRON PIPE (UNMARKED) PER UNRECORDED SURVEY BY L.S. 3423, NOV. 24, 1971.
( )	DIMENSION PER UNRECORDED SURVEY BY L.S. 3423, NOV. 24, 1971.

NOTES

1. PROPERTY DESCRIBED AS PARCEL A, IN CERTIFICATE OF COMPLIANCE RECORDED IN BOOK 860129, OFFICIAL RECORDS, AT PAGE 578.

2. PROPERTY DESCRIBED AS PARCEL B, IN CERTIFICATE OF COMPLIANCE RECORDED IN BOOK 860129, OFFICIAL RECORDS, AT PAGE 578.

Figure 10-22. Typical map notes and legend. (Courtesy of Timothy S. Train.)

modate an odd-shaped figure and/or to have property, street, or traverse lines parallel with the border.

Preparing a scaled sketch map on tracing paper to outline the most distant features is often desirable. The sketch can then be shifted and rotated on the final map sheet, assuring good positioning.

10-12. DRAFTING MATERIALS

Three choices of drafting media are available for today's survey drafter: (1) paper, (2) cloth, and (3) film. Sufficient differences exist between and within each medium to allow the draftsperson a wide choice of material to fit the requirements of any drafting application.

10-12-1. Paper

Drafting papers are manufactured in a wide selection. They are classified by strength, longevity, and erasability, and are categorized as opaque (drawing) or transparent (tracing) paper. For sketching single-use exhibits, or for a high-quality map, opaque paper is virtually

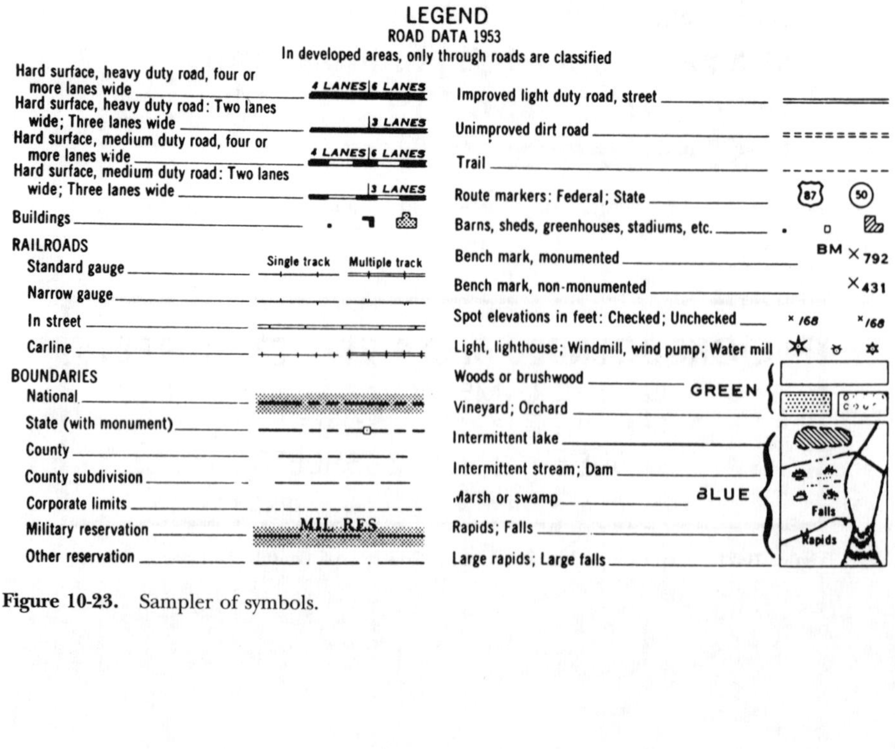

LEGEND
ROAD DATA 1953
In developed areas, only through roads are classified

Hard surface, heavy duty road, four or more lanes wide	4 LANES|6 LANES
Hard surface, heavy duty road: Two lanes wide; Three lanes wide	3 LANES
Hard surface, medium duty road, four or more lanes wide	4 LANES|6 LANES
Hard surface, medium duty road: Two lanes wide; Three lanes wide	3 LANES
Buildings	▪ ◣ ⌂

RAILROADS

	Single track	Multiple track
Standard gauge		
Narrow gauge		
In street		
Carline		

BOUNDARIES

National	
State (with monument)	
County	
County subdivision	
Corporate limits	
Military reservation	MIL RES
Other reservation	

Improved light duty road, street	
Unimproved dirt road	
Trail	
Route markers: Federal; State	87 50
Barns, sheds, greenhouses, stadiums, etc.	▪ □ ▨
Bench mark, monumented	BM ×792
Bench mark, non-monumented	×431
Spot elevations in feet: Checked; Unchecked	×168 ×168
Light, lighthouse; Windmill, wind pump; Water mill	☀ ♂ ✿
Woods or brushwood	GREEN
Vineyard; Orchard	
Intermittent lake	
Intermittent stream; Dam	
Marsh or swamp	BLUE
Rapids; Falls	
Large rapids; Large falls	

Figure 10-23. Sampler of symbols.

ignored. Transparent papers have wide applications as tracing paper on which a master drawing is to be produced; after checking and revision, it is used in reproduction processes to provide any number of copies. In addition to good actinic transparency, a high-quality tracing paper must withstand repeated handling and have a high degree of permanence.

10-12-2. Cloth

Cloth or "linen" combines features of transparency, surface quality, strength, and permanence. This medium resists repeated erasures, with little or no loss in surface quality. Linen provides a highly receptive surface for pencil, ink, and typing. Ink erasures can be made with a vinyl eraser or a gentle abrasive, such as Bon-Ami, and then wiped clean with a moist cloth. Although the advantages of cloth outweigh those of paper, it is generally not used for economic reasons except to meet a particular project specification or application.

10-12-3. Film

The advantages of a polyester drafting medium include high transparency, dimensional stability, and resistance to tearing, heat, and aging. In addition, it is waterproof, highly receptive to pencil, ink, typewriting, and paste-up processes, easily erased, and can be coated with an opaque material for use in scribing.

Recognizing the time and expenses accumulated in any survey project, we see that it is advisable to invest in the slightly more costly best-quality medium to guarantee excellent appearance and the highest permanence of survey drawings.

10-13. REPRODUCTION OF MAPS AND DRAWINGS

Two distinct processes are used for reproduction purposes: (1) copying or (2) duplicating. Copy machines are suitable for either line or pictorial work, but they operate at relatively slow speeds; hence, cost per copy is accordingly high.

By comparison, duplicating machines are designed to produce large numbers of copies at high speed, resulting in a lower cost per copy. Offset presses and stencil machines are examples of duplicating processes. Although in some instances these machines can be employed in combination with a copy process, reproduction requirements dictated by survey drawings are best met by one of the various copying procedures. Therefore, duplication processes will not be addressed in this text.

10-13-1. Copying Processes

Diazo, photographic, and electrostatic are the most popular types of copy processes. At one time, "blueprinting" was the most popular method, but it has now been largely replaced by the diazo process.

A complete line of reproduction and copy services is available from companies that specialize in this work. If volume warrants, a survey firm might consider the convenience afforded by using an in-house reproduction unit. Many small diazo units are available, as are limited-format copy cameras that do not require darkroom development processes.

10-13-2. Diazo

In this process, paper coated with diazonium salt is exposed to light that has passed through a transparent original. The exposed sheet is then developed by exposure to an alkaline-based solution such as ammonia. Where light has passed through the tracing, the diazonium salt breaks down, leaving a blank area on the print; on remaining areas not exposed to light, the salts and ammonia combine to produce an opaque dye, leaving an image of the original tracing lines. This image is not absolutely permanent, will fade when exposed to sunlight, and the exposed paper does not have very good drafting qualities. A positive original produces a positive copy and likewise a negative tracing will yield a negative copy. A diazo copy is a high-contrast image with fine detail and ideal for document reproduction.

10-13-3. Photographic

This process consists of using a precise camera to photograph an original document or drawing. The exposed copy film is developed into a negative from which a variety of prints can be made. A copy camera is usually mounted on a track along with a vacuum-frame copy board to hold the document being copied. The camera has a large format and is equipped with a high-quality lens. The copy board is movable relative to the camera, allowing a wide latitude of reduction of enlargement possibilities.

Film and paper prints, both positive or negative, clear or matte film positives, and a photographic negative are products available through this process. Film prints are provided on a stable-based material that is also a high-quality drafting surface. This process yields the most accurate and versatile product, but it is also the most expensive; a large, complex camera and darkroom facilities are required.

10-13-4. Electrostatic

This process, also known as *xerography*, depends on an electrostatically charged aluminum drum to deposit powder onto copy paper. The drum is given a positive charge that is partially dissipated by light reflected from the document to be copied. The charge portion remaining attracts negatively charged powder, creating an intermediate image on the drum. Copy paper with a negative charge is brought into contact with a drum that transfers the image onto the paper, which is then fixed or permanently fused into the paper by heat.

Electrostatic copy machines require no chemical processing, so a dry finished copy is obtained in very few seconds. The more expensive machines have lens systems capable of reducing and enlarging originals. Copies can be made on virtually any type of paper, including plain paper, offset masters, and transparent paper or film. Recently developed machines are capable of producing full-color copies.

Proper selection of a copy procedure should be based on intended use, availability, and permanence. Consideration should also be given to the cost and time required to produce a copy.

10-14. AUTOMATED DRAFTING

CAD is transforming mapmaking much as electronic instrumentation is revolutionizing field surveying. Interactive drafting systems include a cathode-ray display tube and automatic plotter, and are driven by a computer to process field data. Digital information is recorded in the field on an electronic data collector that interfaces with the CAD system. This system is fast becoming standard equipment in larger agencies and companies.

Advantages offered by various automated machines now available are many: Human error is all but eliminated, production is greatly increased by the obvious speed advantages, and a consistent and accurate map product results. Once a map has been complied from information input on type or magnetic disk, some or all of the data may be used on future projects. For example, initial data from a topographic/boundary survey can be stored and later retrieved to make earthwork estimates, develop a site/grading plan, produce a boundary map, and compile an as-built map.

State-of-the-art survey instrumentation collects huge quantities of information on tape or other data-storage devices. This information in the x, y, and z dimensions can be channeled into a computer to produce a digitized terrain model. The model, in turn, is fed to an automated drafter, which automatically turns out a map in preselected format and specifications by the computer operator.

10-15. SOURCES OF MAPPING ERRORS

Mapping errors can generally be traced to the following sources:

1. Inaccurate linework from use of a blunt or too-soft pencil
2. Inaccurate angular plotting with a protractor
3. Inaccurate linear plotting with the scale
4. Selection of mapping scale or contour interval unsuitable for map requirements
5. Drafting media affected by moisture or climatological change

10-16. MISTAKES AND BLUNDERS

Blunders differ from errors and result from carelessness and poor judgment. The following are a few examples of mistakes:

1. Poor linework or lettering, creating ambiguities
2. Using the wrong scale
3. Setting incorrect angles on a drafting machine or protractor
4. Misinterpretation of field notes
5. Misorientation in all or portion of a plot
6. Inappropriate choice of drafting media

10-17. MAPPING STANDARDS AND SPECIFICATIONS

Federal mapping agencies have adopted standards to control expected map accuracy by specifying the maximum error permitted in horizontal positions and elevations shown on maps. A map conforming to these specifica-tions can use the statement, "This map complies with national map accuracy standards."

The mapping standards state that the following maximum errors are permitted on maps:

Horizontal Accuracy

For maps at scales larger than $1:20,000$ (1 in. = 1667 ft), not more than 10% of the well-defined points tested shall have a plotting error in excess of 1/30 in. For maps of smaller scales, the error factor is 1/50 in. Well-defined points are characterized as easily located in the field and capable of being plotted to within 0.01 in.

Vertical Accuracy

No more than 10% of elevations tested shall be in error more than one-half the contour interval.

REFERENCES

ASCE. 1972 Selection of maps for engineering and planning. Task committee for preparation of a manual on selection of map types, scales, and accuracies for engineering and planning. *Journal of the Surveying and Mapping Division* (SUl).

_____. 1983 Map uses: scales and accuracies for engineering and associated purposes. Report of the ASCE Surveying and Mapping Division Committee on Cartographic Surveying, New York.

MOFFIT, F. H., and H. BOUCHARD. 1992 *Surveying*, 9th ed. New York: Harper Collins.

DAVIS, R. E., F. S. FOOTE, J. M. ANDERSON, and E. M. MIKHAIL. 1981 *Surveying—Theory and Practice*, 6th ed. New York: McGraw-Hill.

SLOANE, R. C., and J. M. MONTZ. 1943 *Elements of Topographic Drawing*. New York: McGraw-Hill.

WATTLES, G. W. 1981 *Survey Drafting*. Orange, CA. Wattles Publications.

WOLF, P. R., and R. C. BRINKER. 1994 *Elementary Surveying*, 9th ed. New York: Harper Collins.

11

Triangulation

M. Louis Shafer

11-1. INTRODUCTION

Triangulation is the surveying technique in which unknown distances between stations may be determined by trigonometric applications of a triangle or triangles. In triangulation, one side called the baseline and at least two interior angles of the triangle must be measured. When all three interior angles are measured, accuracy of the calculated distances is increased and a check provided against any measurement error.

The most basic use of triangulation can be found in surveys of the public domain. Although the use of electronic measuring instruments has eliminated most requirements for this type of triangulation, the 1973 *Manual of Surveying Instructions* made the following statement:

> Triangulation may be used in measuring distances across water or over precipitous slopes. The measured base should be laid out so as to adopt the best possible geometric proportions of the sides and angles of the triangle. If it is necessary to determine the value of an angle with a precision of less than the least reading of the vernier, the method of repetition should be employed.

A complete record of the measurement of the base, the determination of the angles, the location and direction of the sides, and other essential details is entered in the field tables, together with a small diagram to represent the triangulation. In the longer and more important triangulations, all of the stations should be occupied, if possible, and the angles should be repeated and checked to a satisfactory closure; the latter may be kept within 0'20" by careful use of the one-minute transit.

In line practice the chainmen are frequently sent through for taped measurement over extremely difficult terrain, but with the length of the interval verified by triangulation. This is done to ensure the most exact determination of the length of the line while also noting the intervening topographic data.[1]

The use of triangulation or *trigonometry* has been addressed by various public-land survey instructions since "Instructions for Deputy Surveyors, E. Tiffin, Surveyor General, United States, 1815" for ascertaining distances across "insuperable obstacles" such as rivers and canyons. If it is necessary to retrace an original survey across such obstacles, the original field notes are essential to determine how the distance was measured. The following table is an example of public-land survey field notes for triangulation across a lake:

Field Record	Chains	Final Field Notes
At A $\dfrac{54°29'}{3} = 18°09'40''(-02'')$	27.80	To the south shore of Grand Lake, bears N 62° E and S 48° W. Set an iron post, 3 ft long, 1 in. in diam., 28 in. in the ground, for meander cor. of frac. secs. 13 and 18, with brass cap marked. To make a triangulation across the lake, I designate the above meander cor. point A and set a flag B at point for meander cor. on north shore of lake, also a flag C on the north shore that from point A bears N 18°09'38'' E; the base BC bears S 81°44'11'' E, 16.427 chs. dist., the mean by two sets of chainmen, by 1st set = 16.425 ch, by 2nd set = 16.429 ch, longer base impracticable; the angle subtended at point C = 80°06'11''; all angles by three repetitions with error of 0'20'' balanced to 180°. Distance across lake = 51.92 ch.
At B $\dfrac{245°13'}{3} = 81°44'20''(-09'')$		
At C $\dfrac{240°19'}{3} = \dfrac{80°06'20''(-09'')}{180°00'20''(-20'')}$		

$$\text{Dist.} = 16.427\,\frac{\sin 80°06'11''}{\sin 18°09'38''}$$

$$
\begin{array}{ll}
\log 16.427 & = 1.215558 \\
'' \sin 80°06'11'' & = 9.993488 \\
\hline
 & 1.209046 \\
\end{array}
$$

$$
\begin{array}{ll}
'' \sin 18°09'38'' & = 9.493710 \\
'' \qquad\; 51.92 & = 1.715336 \\
\quad +27.80 \\
\hline
\quad\;\; 79.72 \\
\end{array}
$$

79.72   The north shore of lake, bears S 82° E and N 75° W.

11-2. GEODETIC TRIANGULATION

A wider spread and intricate application of triangulation are used for the horizontal control required over a vast area, when traversing does not provide the high uniform accuracy desired. A geodetic triangulation survey, in which stations are miles apart, must consider the earth's size and shape. It is performed primarily by the National Geodetic Survey (NGS) (formerly the U.S. Coast and Geodetic Survey, a branch of the U.S. Department of Commerce). Over the past two centuries, a net of triangulation stations, related to each other, has been developed over the entire continen-tal United States. The system can be used as starting control for any triangulation survey that may be undertaken by a private surveyor, engineer, or public entity.

Triangulation is also employed for control in large metropolitan areas and on major construction projects. This chapter describes the procedures and instructions necessary to develop a triangulation network and retrieve information on stations already established. All procedures and instructions shown in this chapter follow NGS standards, and any triangulation conforming to these procedures and standards can be indexed into its system.

This chapter is written with the knowledge that modern technology is on the verge of

PERMISSIBLE FIGURES

(A) Simple quadrilateral.—The simple quadrilateral is the best figure, and it should be employed wherever possible. It combines maximum strength and progress with a minimum of essential geometrical conditions when approximately equilateral or square and therefore the square quadrilateral is the perfect figure. It has a strength factor,

$$\frac{D - C}{D} \text{ of } 0.6.$$

(B) Four-sided central-point figure with one diagonal.—When one diagonal of the quadrilateral is obstructed, a central point, which is visible from the four corners can be inserted. This figure requires the solution of two side equations and five angle equations, and hence adds to the labor of adjusting. Its strength factor is 0.56.

(C) Four-sided central-point figure without diagonal.—At times, neither diagonal can be made visible and the figure becomes a simple four-sided central-point quadrilateral with a strength factor of 0.64. The central point in this case should be carefully located to maintain the strength of the R_1 chain of triangles. An excellent location is near one side line and about midway along it. If too near the side line, however, refraction errors may be almost the same for the closely adjacent lines, and furthermore, the R_2 value will be so large as to be of little value as a check on lengths computed through the R_1 triangles.

(D) Three-sided central-point figure.—This is a simple and usually very strong figure. It is often used to compensate for a great variation in length of the side lines of adjacent quadrilaterals, and to quickly change the direction of the scheme. Its strength factor is 0.60 and the equations required for its adjustment are the same as for a regular quadrilateral.

(E) Five-sided figure with four diagonals.—This figure may be considered as a four-sided central-point figure with one diagonal, in which the central point falls outside the figure. It is used to afford a check when either a diagonal or a side line is obstructed. It has the same strength factor, 0.56, as the above four-sided central-point figure with one diagonal, (B), and requires the same adjustment equations and precautions against making any of the angles too small. This figure can often be used by the observing party when a side line of a quadrilateral is found to be obstructed.

(F) Five-sided figure with three diagonals.—This figure is similar to the four-sided central-point figure, (C), except that the central point falls outside the figure. The strength factor is 0.64.

(G) Five-sided central-point figure with two diagonals.—This figure is an overlap of a central-point quadrilateral and a simple quadrilateral, and is the most complicated figure employed. It has been used to carry the scheme over difficult or convex areas. This figure can generally be made very strong. Its strength factor is 0.55.

(H) Five- and six-sided central-point figures without diagonals.—Any polygon with a central point, having separate chains of triangles on either side of the central point, will give a double determination of length, since it is permissible to carry the two lengths through the same triangle provided different combinations of distance angles are employed. However, the five- and six-sided central-point polygons are the only ones that should receive consideration, and they are inferior to the simpler quadrilaterals. The factors of strength are 0.67 for five sides and 0.68 for six.

Figure 11-1. Types of figures used in triangulation and strength factors. (*Manual of Reconnaissance for Triangulation.* SP225, Government Printing Office: Washington, D.C., 1938.)

making conventional triangulation obsolete in most instances. The use of the global positioning system (GPS) is becoming commonplace in the establishment of control networks. GPS involves the use of satellites.

GPS not only develops higher accuracy, but also does not require lines of sight between stations. It does, however, require a clear overhead horizontal view. The use of GPS is discussed in Chapter 15.

Almost all geodetic triangulation involves a series of triangles called a triangulation system or triangulation network to complete the control of a selected area. Control is carried from one known base line through several triangles before another base line must be established or checked into. The number depends on the "strength of figures," which will be discussed later. Tighter control is obtained by using a series of quadrilaterals, requiring three other stations to be visible from each station instead of the two necessary when using triangles.

Figure 11-1 is an excerpt from the *Manual of Reconnaissance for Triangulation*, special Pub-

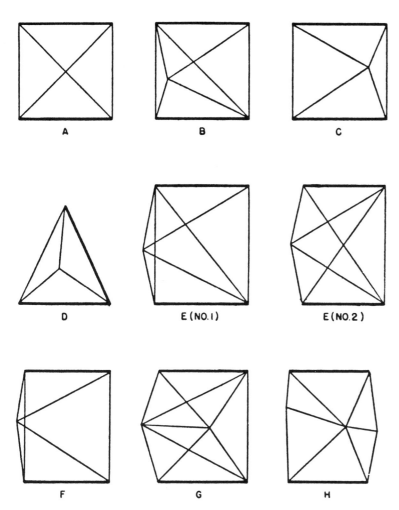

A   B   C

D   E (NO.1)   E (NO.2)

F   G   H

Figure 11-1. (*Continued*).

lication No. 225.[3] It shows various permissible figures and their strength factors for determining the strength of figures and how often base lines or check-in distances are needed. This is covered later in the strength-of-figures section.

11-3. CLASSIFICATION AND SPECIFICATIONS OF TRIANGULATION

In 1984, the Federal Geodetic Control Committee (FGCC), with approval of the Office of Management and Budget (OMB), published *Classification, Standards of Accuracy, and General Specifications of Geodetic Control Surveys.*[2] This publication outlines permissible tolerances for triangulation surveys acceptable by order and class. Three orders of accuracy are listed, with the second and third orders subdivided into class I and class II. The purpose and accuracy required dictate the tolerances allowed.

First-order classification (Table 11-1) demands the highest accuracy and is recommended for primary national networks, special surveys to study movements in the earth's crust, and metropolitan area surveys. Minimum rela-

Table 11-1. Classification of triangulations

Classification	First-Order	Second-Order Class I	Second-Order Class II	Third-Order Class I	Third-Order Class II
Recommended spacing of principal stations	Network stations seldom less than 15 km. Metropolitan surveys 3 to 8 km and others as required.	Principal stations seldom less than 10 km. Other surveys 1 to 3 km or as required.	Principal stations seldom less than 5 km or as required.	As required.	As required.
Strength of figure R_1 between bases					
Desirable limit	20	60	80	100	125
Maximum limit	25	80	120	130	175
Single figure Desirable limit					
R_1	5	10	15	25	25
R_2	10	30	70	80	120
Maximum limit					
R_1	4 10	25	25	40	50
R_2	15	60	100	120	170
Base measurement Standard error	1 part in 1,000,000	1 part in 900,000	1 part in 800,000	1 part in 500,000	1 part in 250,000
Horizontal direction					
Instrument	0″.2	0″.2	0″.2} { 1″.0	1″.0	1″.0
Number of positions	16	16	8 } or {12	4	2
Rejection limit from mean	4″	4″	5″ } { 5″	5″	5″
Triangle closure					
Average not to exceed	1″.0	1″.2	2″.0	3″	5″.0
Maximum seldom to exceed	3″.0	3″.0	5″.0	5″.0	10″.0
Side checks In side equation test, average correction to direction not to exceed	0″.3	0″.4	0″.6	0″.8	2″
Astro azimuths					
Spacing-figures	6–8	6–10	8–10	10–12	12–15
No. of obs./night	16	16	16	8	4
No. of nights	2	2	1	1	1
Standard error	0″.45	0″.45	0″.6	0″.8	3″.0
Vertical angle observations Number of and spread between observations	3 D/R–10″	3 D/R–10″	2 D/R–10″	2 D/R–10″	2 D/R–20″
Number of figures between known elevations	4–6	6–8	8–10	10–15	15–20
Closure in length (also position when applicable) after angle and side conditions have been satisfied, should not exceed	1 part in 100,000	1 part in 50,000	1 part in 20,000	1 part in 10,000	1 part in 5000

Source: FGCC. 1974. ''Classifications, Standards of Accuracy, and General Specifications of Geodetic Control Surveys.''

tive accuracy between directly connected points for first-order is 1 part in 1000,000. Spacing between first-order stations should be more than 15 km for primary network stations, 3 to 8 km in a metropolitan survey.

Second-order, class I standards are recommended for control surveys established between tracts bounded by the primary national control network. They strengthen the network and provide more stations for local applications. These standards should also be used in metropolitan areas where a closer net is required than allowable for first-order. The minimum relative accuracy between adjacent stations for second-order, class I is 1 part in 50,000. Stations strengthening the primary network should be not closer than 10 km apart, whereas those in metropolitan areas should be 1 to 3 km apart.

Second-order, class II standards are used to establish control along coasts and inland waterways, interstate highway systems, and large land subdivisions and construction projects. Second-order, class II standards can also be used for the further breakdown of second-order, class I nets in metropolitan areas. Second-order, class II triangulation contributes to, but is supplemental to, the primary national network. The minimum relative accuracy between adjacent points is 1 part in 20,000. Supplemental stations to the primary net should be at least 5 km apart while other second-order, class II stations are set up as required.

Third-order, class I and II standards establish control for local area projects, such as small engineering jobs on local improvements and developments and small-state topographic mapping. Third-order triangulation extends higher-order control and can be adjusted. National network stations set for third-order work may be spaced as required to satisfy project needs. The minimum relative accuracy between adjacent stations for third-order, class I is 1 part in 10,000; for third-order class II, 1 part in 5000.

In order to obtain a certain classification, definite specifications must also be met. For triangulation, the following four specifications are required:

1. Specifications for Instruments

Classification	Type of Instrument
First-order	Optical-reading theodolite micrometer readings, smaller than 1 sec, Examples: Wild T-3 and Kern DKM-3.
Second-order class I and II	Optical-reading theodolite micrometer reading of 1 sec. Examples: Wild T-2, Kern DKM-2, and Zeiss Th-2.
Third-order	Good-quality transit or repeating theodolite. Same quality theodolite required for second-order is recommended to eliminate the extra effort needed to obtain the specified accuracies for repeating instruments.

2. Specifications for Number of Observations Required (One observation includes direct and reverse reading.)

Classification	Number of Observations
First-order	16*
Second-order, class I	16*
Second-order, class II	8 with $0'.2''$ instrument* 12 with $1''.0$ instrument*
Third-order, class I[†]	4 with $1''.0$ instrument*
Third-order, class II[†]	2 with $1''.0$ instrument*

*To minimize collimation circle errors and micrometer irregularities, initial settings utilizing the entire circle are necessary. (See Table 11-2 for initial settings.)

†When we use a transit or repeating theodolite, one to eight sets of six observations are required, depending on instrument type (Table 11-3).

3. Rejection Limit Specifications

Classification	Rejection Limit from Mean
First-order	$\pm 4''$
Second-order, class I	$\pm 4''$
Second-order, class II	$\pm 5''$
Third-order, class I	$\pm 5''$
Third-order, class II	$\pm 5''$

(Continued)

Table 11-2. Circle settings: directional theodolites

Two positions of circle

	10-min Micrometer Drum		
1	0°	00′	10″
2	90	05	40

Four positions of circle

	5-min Micrometer Drum			10-min Micrometer Drum			Wild T-3 Circle	Micrometer Readings	(units)
1	0°	00′	40″	0°	00′	10″	0°	00′	15
2	45	01	50	45	02	40	45	00	45
3	90	03	10	90	05	10	90	02	15
4	135	04	20	135	07	40	135	20	45

Six positions of circle

1	0	00	10	0	00	10	0	00	15
2	30	01	50	30	01	50	30	00	35
3	60	03	30	60	03	30	60	00	50
4	90	00	10	90	05	10	90	00	15
5	120	01	50	120	06	50	120	00	35
6	150	03	30	150	08	30	150	00	50

Eight positions of circle

1	0	00	40	0	00	10	0	00	10
2	22	01	50	22	01	25	22	00	25
3	45	03	10	45	02	40	45	00	35
4	67	04	20	67	03	55	67	00	50
5	90	00	40	90	05	10	90	00	10
6	112	01	50	112	06	25	112	00	25
7	135	03	10	135	07	40	135	00	35
8	157	04	20	157	08	55	157	00	50
1	0	00	40	0	00	10	0	00	10
2	15	01	50	15	01	50	15	00	25
3	30	03	10	30	03	30	30	00	35
4	45	04	20	45	05	10	45	00	50
5	60	00	40	60	06	50	60	00	10
6	75	01	50	75	08	30	75	00	25
7	90	03	10	90	00	10	90	00	35
8	105	04	20	105	01	50	105	00	50
9	120	00	40	120	03	30	120	00	10
10	135	01	50	135	05	10	135	00	25
11	150	03	10	150	06	50	150	00	35
12	165	04	20	165	08	30	165	00	50

Sixteen positions of circle

1	0	00	40	0	00	10	0	00	10
2	11	01	50	11	01	25	11	00	25
3	22	03	10	22	02	40	22	00	35
4	33	04	20	33	03	55	33	00	50
5	45	00	40	45	05	10	45	00	10
6	56	01	50	56	06	25	56	00	25
7	67	03	10	67	07	40	67	00	35
8	78	04	20	78	08	55	78	00	50
9	90	00	40	90	00	10	90	00	10
10	101	01	50	101	01	25	101	00	25
11	112	03	10	112	02	40	112	00	35
12	123	04	20	123	03	55	123	00	50
13	135	00	40	135	05	10	135	00	10
14	146	01	50	146	06	25	146	00	25
15	157	03	10	157	07	40	157	00	35
16	168	04	20	168	08	55	168	00	50

Table 11-3. Number of observations using a transit and circle settings

Accuracy Class	Transit	Number of Observations	Number of Sets	Spread between D & R and Sets Not to Exceed
Third-order, class I triangulation	10″	6 D & R	2–3	4″
	20″	6 D & R	4–5	5″
	30″	6 D & R	6–8	6″
Third-order, class II triangulation	10″	6 D & R	1–2	5″
	20″	6 D & R	2–3	6″
	30″	6 D & R	3–4	7″

Transit and repeating type instruments

The circle settings

Sets	Instrument 10″ Setting			Instrument 20″ Setting			Instrument 30″ Setting		
	0°	00′	00″	0°	00′	00″	0°	00′	00″
1	0°	00′	00″	0°	00′	00″	0°	00′	00″
2	90	05	30	90	10	20	90	10	30
1	0	00	00	0	00	00	0	00	00
2	60	03	30	60	06	20	60	06	30
3	120	07	00	120	13	00	120	13	00
1				0	00	00	0	00	00
2				45	05	20	45	05	30
3				90	10	00	90	10	00
4				135	15	20	135	15	30
1				0	00	00	0	00	00
2				36	04	20	36	04	30
3				72	08	00	72	08	00
4				108	12	20	108	12	30
5				144	16	00	144	16	00
1							0	00	00
2							30	03	30
3							60	07	00
4							90	10	30
5							120	14	00
6							150	17	30
1							0	00	00
2							25	02	30
3							51	05	30
4							76	08	00
5							102	10	30
6							128	14	30
7							153	17	00
1							0	00	00
2							22	02	30
3							45	05	00
4							67	07	30
5							90	10	00
6							112	12	30
7							135	15	00
8							157	17	30

4. Triangle Closure Specification for Triangulation Net*

Classification	Average Not to Exceed	Maximum Seldom to Exceed
First-order	1".0	3".0
Second-order, class I	1".2	3".0
Second-order, class II	2".0	5".0
Third-order, class I	3".0	5".0
Third-order, class II	5".0	10".0

*This is a simple field check to indicate the accuracies of triangulation observations.

11-4. PLANNING

After the triangulation project limits are determined, it is necessary to select the station sites. The first step is to collect all pertinent data, including various scaled maps and information on any existing triangulation stations in or near the area.

One of the handiest tools for planning the project and field reconnaissance is a set of topographic maps. U.S. Geological Survey Quadrangle (quad) maps are excellent and easily obtained from local surveying or map supply firms. They show contours, roads, trails, improvements, and some primary triangulation stations in the area. Quad sheets also have geodetic and rectangular coordinate control and are available for the entire United States.

In most cases, any triangulation project considered today has enough existing primary triangulation stations in the general vicinity for beginning control and data for necessary checks. These data can be obtained from the Director, National Geodetic Information Center, NOS, NOAA, Rockville, MD. Included diagrams will show locations of geodetic control stations, lines of sight between stations, and horizontal control data sheets giving directions to monuments, other visible stations, and grid and geodetic stations' coordinates. With these data, existing triangulation stations not given on a topographic map can be plotted to show their location and availability for a proposed project. If existing control is not available, base lines must be established, as discussed later in the chapter.

After all existing horizontal control on topographic maps is plotted, the general location of all triangulation stations needed for the project will be known before setting foot in the field. Contours determine the probable visibility between proposed stations, so profiles can be drawn to assure that the required clearance is available, and exact heights of towers required to provide acceptable clearance considering obstructions and the earth's curvature. Obstructions can become problems in all types of terrain, but curvature of the earth is noted mainly in flat lands and over long distances. The earth's curvature and refraction are discussed in detail in Chapter 7. For planning a triangulation project, the effect of curvature and refraction, which have an approximate relation to each other, can be determined by the formula

$$h \text{ (ft)} = 0.574M^2 \text{ (mi)} \qquad (11\text{-}1)$$

where h is the height in feet that a line, horizontal at the point of observation, will be above a level surface at a distance of M statute miles. Table 11-4 lists the corrections of curvature and refraction for 1 to 60 mi.

If the line of sight between two stations extends across flat terrain and towers of equal heights can be constructed at the stations, the heights necessary to compensate for curvature and refraction can be determined by the formula

$$h = 0.574\left(\frac{M}{2}\right)^2 \qquad (11\text{-}2)$$

in which h is the height above ground at both stations (in ft), and M the distance between stations (in mi).

The formula for working in the metric system is

$$h \text{ (m)} = 0.0675\left(\frac{k}{2}\right)^2 \text{ (kilometers)} \quad (11\text{-}3)$$

By constructing towers at both stations, the height needed is only one-fourth that required if a tower is constructed at only one station.

Table 11-4. Correction for earth's curvature and refraction

Distance	Correction	Distance	Correction	Distance	Correction	Distance	Correction
Miles	Feet	Miles	Feet	Miles	Feet	Miles	Feet
1	0.6	16	146.9	31	551.4	46	1214.2
2	2.3	17	165.8	32	587.6	47	1267.7
3	5.2	18	185.9	33	624.9	48	1322.1
4	9.2	19	207.2	34	663.3	49	1377.7
5	14.4	20	229.5	35	703.0	50	1434.6
6	20.6	21	253.1	36	743.7	51	1492.5
7	28.1	22	277.7	37	785.6	52	1551.6
8	36.7	23	303.6	38	828.6	53	1611.9
9	46.4	24	330.5	39	872.8	54	1673.3
10	57.4	25	358.6	40	918.1	55	1735.8
11	69.4	26	388.0	41	964.7	56	1799.6
12	82.7	27	418.3	42	1012.2	57	1864.4
13	97.0	28	449.9	43	1061.0	58	1930.4
14	112.5	29	482.6	44	1111.0	59	1997.5
15	129.1	30	516.4	45	1162.0	60	2065.8

The height calculated by this formula corrects only for curvature and refraction. The total heights necessary to have a satisfactory line of sight need to be increased enough to reduce the horizontal refraction caused by unequal air currents along the terrain.

Whether a line of sight will clear an obstruction can be determined by the formula

$$e = e_1 + (e_2 - e_1)\frac{d_1}{d_1 + d_2} - 0.574 d_1 d_2 \quad (11\text{-}4)$$

where

e = elevation of the line at obstruction (in ft)

e_1 = elevation of the lower station (in ft)

e_2 = elevation of the higher station (in ft)

d_1 = distance from the lower station to obstruction (in mi)

d_2 = distance from obstruction to the higher station (in mi)

$0.574 d_1 d_2$ = correction for curvature and refraction

Example 11-1. From a topographic map, it is planned to locate stations A and B 10 mi apart. The elevation of A is 20 ft and the elevation of B 90 ft. On line, 3 mi from A, is a ridge with an elevation of 40 ft. To determine the elevation of the line of ob-

struction, use the following formula:

$$e = 20 + (90 - 20)\frac{3}{(3 + 7)}$$

$$- 0.574(3)(7)$$

where e = 28.1 ft for the line of obstruction and the elevation of obstruction is 40 ft; therefore, stations A and B are not intervisible from the ground.

To determine the height above stations A and B necessary to see over the obstruction, subtract the elevation of the line of obstruction from the obstruction's elevation. In this case, $40 - 28.1 = 11.9$ ft. Thus, a tower of 11.9 ft, plus a required clearance necessary over the obstruction to reduce horizontal refraction, should be constructed over each station. If it is feasible to construct towers over these stations, the locations are satisfactory. Otherwise, new sites must be selected.

11-5. STRENGTH OF FIGURES

The accuracy of a triangulation net depends on not only the methods and precision used in making observations, but also the shapes of figures in the net. The system to measure the accuracy of shapes is known as *strength of figures*. Distance angles are those opposite the known and required sides. The accuracy or relative strength of a triangle is expressed as a number —i.e., the smaller the number, the greater the relative strength. To qualify as a certain classification, the sum of all numerical values of relative strength through a series of triangles between base lines cannot exceed the set standard. These standards are listed in Table 11-1 under strength of figures.

R is the standard symbol for strength of figures. When figures of a net are other than triangles (Figure 11-1), more than one scheme can be used to calculate the required side. R_1 and R_2 indicate that the strength of figure through the best-shaped triangles and second-best-shaped triangles, respectively. The summation $(\Sigma)^1$ of the R_1 and R_2 values determines when a base line is necessary to comply with the classification specifications.

The formula for strength of each figure (known side to required side) is

$$R = \frac{D - C}{D}(A^2 + AB + B^2) \quad (11\text{-}5)$$

where

D = number of directions observed in each figure

C = number of conditions to be satisfied in each figure

A and B = logarithmic differences for 1 sec of distance angles A and B in units of the sixth decimal place

In determining D, directions along the known side of the figure are not included. C can be determined by the following formula:

$$C = (n' - s' + 1) + (n - 2s + 3) \quad (11\text{-}6)$$

where

n = number of lines observed in both directions (including known side)

s' = number of stations occupied

n = total number of lines (including known side)

s = total number of stations

In triangle ABC (Figure 11-2) where AB is known, all stations are occupied, and $D = 4$

$$C = (3 - 3 + 1) + (3 - 2(3) + 3) = 1$$
$$\frac{D - C}{C} = \frac{4 - 1}{4} = 0.75$$

In the strength-of-figure formula, $D - C/D$ is referred to as the strength factor and given for the various figures listed in Table 11-1. The values for $\delta_A^2 + \delta_A \delta_B + \delta_B^2$ have been tabulated and listed in Table 11-5 To use this table, find the factor by locating the smaller distance angle across the top of the table and large distance angle down the left side.

Example 11-2. Determination of strength of figures (Figure 11-3). If side MN and all interior angles are known, find R_1, R_2 to required side OP.

Possible schemes are (1) (MNP, NOP), (2) (MNO, MOP), (3) (MNP, MOP), and (4) (MNO, NOP).

$$D = 10 \text{ and } C = (6 - 4 + 1)$$
$$+ (6 - 2(4) + 3) = 4$$

Strength factor $= D - C/D = 10 - 4/10 = 0.60$

For scheme $MNP, NOP : MNP$ distance angles are $75°, 64°$ and NOP distance angles are $83°, 53°$.

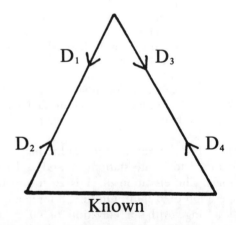

Figure 11-2. Determination of strength factors.

The strength of figure

°	10°	12°	14°	16°	18°	20°	22°	24°	26°	28°	30°	35°	40°	45°	50°	55°	60°	65°	70°	75°	80°	85°	90°
10	428	359																					
12	359	295	253																				
14	315	253	214	187																			
16	284	225	187	162	143																		
18	262	204	168	143	126	113																	
20	245	189	153	130	113	100	91																
22	232	177	142	119	103	91	81	74															
24	221	167	134	111	95	83	74	67	61														
26	213	160	126	104	89	77	68	61	56	51													
28	206	153	120	99	83	72	63	57	51	47	43												
30	199	148	115	94	79	68	59	53	48	43	40	33											
35	188	137	106	85	71	60	52	46	41	37	33	27	23										
40	179	129	99	79	65	54	47	41	36	32	29	23	19	16									
45	172	124	93	74	60	50	43	37	32	28	25	20	16	13	11								
50	167	119	89	70	57	47	39	34	29	26	23	18	14	11	9	8							
55	162	115	86	67	54	44	37	32	27	24	21	16	12	10	8	7	5						
60	159	112	83	64	51	42	35	30	25	22	19	14	11	9	7	5	4	4					
65	155	109	80	62	49	40	33	28	24	21	18	13	10	7	6	5	4	3					
70	152	106	78	60	48	38	32	27	23	19	17	12	9	7	5	4	3	2	2	1	1		
75	150	104	76	58	46	37	30	25	21	18	16	11	8	6	4	3	2	2	2	1	0	0	
80	147	102	74	57	45	36	29	24	20	17	15	10	7	5	4	3	2	1	1	1	0	0	0
85	145	100	73	55	43	34	28	23	19	16	14	10	7	5	3	2	2	1	1	0	0	0	0
90	143	98	71	54	42	33	27	22	19	16	13	9	6	4	3	2	1	1	1	0	0	0	
95	140	96	70	53	41	32	26	22	18	15	13	9	6	4	3	2	1	1	0	0	0		
100	138	95	68	51	40	31	25	21	17	14	12	8	6	4	3	2	1	1	0	0			
105	136	93	67	50	39	30	25	20	17	14	12	8	6	4	3	2	1	1	0	0			
110	134	91	65	49	38	30	24	19	16	13	11	7	5	3	2	2	1	1	1				
115	132	89	64	48	37	29	23	19	15	13	11	7	5	3	2	2	1	1					
120	129	88	62	46	36	28	22	18	15	12	10	7	5	3	2	2	1						
125	127	86	61	45	35	27	22	18	14	12	10	7	5	4	3	2							
130	125	84	59	44	34	26	21	17	14	12	10	7	5	4	3								
135	122	82	58	43	33	26	21	17	14	12	10	7	5	4									
140	119	80	56	42	32	25	20	17	14	12	10	8	6										
145	116	77	55	41	32	25	21	17	15	13	11	8											
150	112	75	54	40	32	26	21	18	16	15	13	9											
152	111	75	53	40	32	26	22	19	17	16													
154	110	74	53	41	33	27	23	21	19														
156	108	74	54	42	34	28	25	22															
158	107	74	54	43	35	30	27																
160	107	74	56	45	38	33																	
162	107	76	59	48	42																		
164	109	79	63	54																			
166	113	86	71																				
168	122	98																					
170	143																						

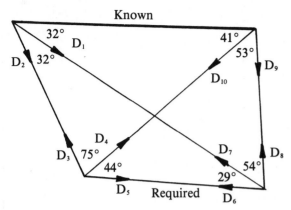

Figure 11-3. Determination of strength of figures.

In Table 11-5, locate distance angles 75° on the left side, 64° across the top, and by intersection of these two angles and interpolation, the factor is found to be 2. Next, locate 83° on the left side, 53° across the top; the factor is listed as 3. Add $2 + 3 = \delta_A^2 + \delta_A\delta_B + \delta_B^2 = 5$.

$$R(MNP, NOP) = D - C/D(\delta_A^2 + \delta_A\delta_B + \delta_B^2)$$

$$= 0.60 \times 5 = 3.0$$

If we complete the same process for the remaining three schemes, the following tabulation can be made:

Scheme	Common Side	Distance Angles	$\delta_A^2 + \delta_A\delta_B + \delta_B^2$	$\dfrac{D-C}{D}$	R
MNP	NP	75°, 64°	2		
NOP		83°, 53°	3		
			$\Sigma = \overline{5}$	×0.60	$3 = R_1$
MNO	MO	54°, 94°	2		
MOP		119°, 32°	9		
			$\Sigma = \overline{11}$	×0.60	$7 = R_2$
MNP	MP	75°, 41°	8		
MOP		29°, 32°	39		
			$\Sigma = \overline{47}$	×0.60	28
MNO	NO	54°, 32°	19		
NOP		44°, 53°	11		
			$\Sigma = \overline{30}$	×0.60	18

Since the more accurate scheme has a smaller R value, scheme MNP, NOP becomes R_1 and scheme MNO, MOP is R_2. Referring to the standards of accuracy shown in Table 11-1, we see that this figure qualifies for first-order triangulation. If a network was developed with all figures having these same R_1 and R_2 values, then a base line is necessary after two figures to qualify as first-order, or a base line is required after eight figures to meet the standards of second-order, class I.

11-6. BASE LINES

Measured lines that control a triangulation net are referred to as base lines. The frequency with which base lines are needed depends on strength of the geometric figures, and since lengths of all required lines are derived from the base lines, great care must be taken to ensure their accuracy. A base line may be between two existing control stations or independently measured within the triangulation net.

Plotting triangulation schemes on the reconnaissance maps can determine when base lines are necessary, and whether existing stations are adequate or new ones must be established.

With the acceptance of electronic distance-measuring instruments (EDMIs) for measuring base lines, the procedure has been greatly simplified. Pre-EDMI base-line measurements required a fairly level leg between stations and use of special Invar tapes and supports. The favorite location was along a railroad tangent, thereby restricting their selection. With EDMIs, a base line can be placed wherever there is a clear line of sight. Care must be taken that the EDMI used meets all requirements and specifications for the class order of triangulation desired (see Chapter 5 on EDMIs).

11-7. FIELD RECONNAISSANCE

Field reconnaissance crews are responsible for contacting property owners, finding all existing control, and selecting the location of the proposed stations.

It is important to use proper methods in contacting owners of lands to be entered. Generally, an owner will consent to the use of his or her property if asked beforehand but will resent the intrusion without prior consent. If visited prior to any field reconnaissance, an owner will probably have no objection to future use as long as all conditions agreed on are met.

A property owner should be fully informed on the procedure to be followed in completing the triangulation survey, including (1) monuments to be set, (2) type of stands and towers required, (3) probability of night observations, (4) number of times the station will be used, and (5) settlement for any damage that may occur to his or her property as a result of monumenting the station. A good motivation for allowing placement of a station on private

land is to name it after the owner. Having a set of monuments with one's name gives a sense of importance and pride.

The owner should be contacted each time before land is entered unless a blanket consent has been given, especially if there are livestock, buildings, or equipment on the land. It is also advisable to talk with the local law enforcement agency, as this could prevent later embarrassment.

After descriptions of existing control stations have been acquired, field recovery can begin. All stations within the National Geodetic Survey control networks have a *station description* that includes distances and directions of reference azimuth marks from a main station mark, geodetic and state coordinate position of the mark, and a "to-reach" description tied to a nearby permanent landmark such as the local post office (Figure 11-4). Since these descriptions may have been written and not updated for more than 50 years, finding stations can require a diligent search, or even include looking for accessory marks as well as the main station.

When a station is found, a "Report on Condition of Survey Mark" form should be filled out and sent to the NOS (Figure 11-5). If the station sought is not found, the thoroughness of the search should be reported; if recovered, the marks condition should be reported along with any differences discovered in the to-reach description. When approved by the NOS, changes are incorporated into the station description. For a station recovered exactly as previously described, with the monuments and marks in good condition, and measurements between marks in agreement, a statement to this effect will suffice. Any changes concerning station marks or to-reach descriptions must be addressed on the "Report of Condition" note.

Proposed triangulation station sites can be researched after existing stations have been recovered and proposed schemes laid out on the map. Initially, these sites are visited to see if they are acceptable as station locations. Fac-

Department of Commerce
U.S. Coast and Geodetic Survey
Form 825
Rev. Aug. 1964
Name of Station: BIG BAR
Chief of Party: Walter R. Helm

Description of Triangulation Station

State: California   County: Butte
Year: 1949   Described by: L. A. Critchlow

Note	Height of Telescope above Station Mark 15.67 Meters	Height of Light above Station Mark 18.56 Meters
4	Surface-station mark, Underground-station mark	DISTANCES AND DIRECTIONS TO AZIMUTH MARK, REFERENCE MARKS AND PROMINENT OBJECTS WHICH CAN BE SEEN FROM THE GROUND AT THE STATION

	OBJECT	BEARING	DISTANCE feet	DISTANCE meters	DIRECTION °	′	″
	BUCK (USGS)				00	00	00.0
12c	Azimuth Mark	E	about 1.0 mile		40	52	19.2
12c	R.M.# 2	W	57.093	17.402	222	47	01
	Big Bar Lookout tower	W	128.001	(39.015)	231	59	09.0
12c	R.M.# 1	NE	65.493	19.962	355	32	44

Detailed description:

The station is located on a high, timbered ridge on the east side of the Feather River Canyon, about 20 miles airline north-northeast of Oroville, about 10 miles airline east of Paradise, about 2 miles airline southeast of Pulga and on the same high point on which the *Big Bar Lookout* is located.

To reach the station from the post office in Oroville, go northwest on Oak Street for 2 blocks to a stop sign. Turn right on State Highway # 24 and go 28.0 miles to a bridge and a sign "NORTH FORK FEATHER RIVER BRIDGE 12-38." Keep straight ahead on State Highway # 24, crossing the bridge, and go 0.4 mile to a side road on the right. Turn right as per sign "BUCKS LAKE 23" and go 6.5 miles to a 5-way intersection of roads. Turn sharp right as per sign "BIG BAR L.O." and go 4.1 miles to a triangle blazed tree and the azimuth mark on the right. Keep straight ahead and go 1.2 miles to a fork. Turn right as per sign "BIG BAR LOOKOUT 1" and go 0.8 mile to the summit and the station.

The station mark is a standard disk stamped *"big bar 1949"* set in a drill hole in a boulder that projects about 4 inches above the ground.

Reference mark number 1 is a standard disk stamped "BIG BAR NO 1 1949" set in a drill hole in a boulder that projects about 3 inches above the ground. It is about 2 feet lower than the station.

Reference mark number 2 is a standard disk stamped "BIG BAR NO 2 1949" set in a drill hole in a boulder that projects about 6 inches above the ground. It is 58.6 feet east of the southeast leg of the lookout tower and about 2 feet higher than the station.

The azimuth mark is a standard disk stamped "BIG BAR 1949" set in a drill hole in a boulder that projects about 16 inches above the ground. It is about 21 feet west of the center of the road and 23.2 feet north of a 24-inch pine tree with a triangle blaze.

Big Bar Lookout Tower is located on the highest point of the hill and is built on ground that is about 2 feet higher than the station. The tower is in the early stages of construction. The point measured to and cut in is a punch mark in the center and top of the middle base I-beam.

Figure 11-4. Station description (horizontal control data from USC & GS).

tors to be checked include (1) visibility to the other station sites, (2) accessibility, (3) probable permanence of the station marks, and (4) acceptance and future plans of the property owners.

The most important item to verify is visibility. Numerous factors affecting visibility may not be evident on the reconnaissance maps, but do show up on a field check of the site. If we consider the height of towers available for

REPORT ON CONDITION OF SURVEY MARK

Form Approved: OMB No. 41—R1923
Approval Expires: April 1978

Name or Designation: _____ Year Established: _____

State: _____ County: _____ Organization Established by: _____

Distance and direction from nearest town: _____

Description published in: *(Line, book, or quadrangle number)* _____

Mark searched for or recovered by: Name - _____

Organization - _____

Date of report _____ Address - _____

Condition of marks: List letters and numbers found stamped in (not cast in) each mark.

Mark stamped:	Condition:

Marks accessible? ☐ Yes ☐ No   Property owner contacted? ☐ Yes ☐ No

Please report on the thoroughness of the search in case a mark was not recovered, suggested changes in description, need for repairing or moving the mark, or other pertinent facts:

Witness Post? Yes ___ No ___

Witness Post set ___ feet ___ of ___ mark.

Witness Post set ___ feet ___ of ___ mark.

If additional forms are needed, indicate number required.

NQAA FORM 76-91 (3-77)   U.S. DEPARTMENT OF COMMERCE · NATIONAL OCEANIC AND ATMOSPHERIC ADMINISTRATION · NATIONAL OCEAN SURVEY

Figure 11-5. Report on condition of survey mark.

use, new construction, tall trees, or other intervening obstructions can make sight lines to proposed or existing stations impossible. Visibility must be checked thoroughly, since any error can cause time-consuming delays when observations begin, possibly disrupting the entire triangulation scheme.

The ability to get materials and equipment to a station location for setting monuments and occupying stations is another important factor. Many triangulation stations require some walking to reach the mark, but this should be kept to a minimum. A station requiring excessive hiking or climbing not only adds an element of danger for surveyors, but also makes the station useless to all but those few willing to assume the extra effort and risk. It is not always possible to avoid inaccessible stations, but the main idea is to make them usable.

The property owner is a principal factor in determining accessibility. If contacted beforehand, the owner will probably permit the vehicular entry of his or her property and explain or show the easiest way to a specific location, in addition to providing keys or combinations to any locked gates.

Permanence of marks is never assured, but many factors should be considered to make their location more reliable. Things to look for include (1) soil condition, (2) possibility of construction or development, (3) farming activities, and (4) changing ownership of private or public land.

Ground condition where a station is placed should be stable. Sand hills, large gravel deposits, and unstable rock outcroppings should be avoided, as well as areas subject to continuous frost action or erosion. Reliability of monuments set in these conditions is greatly diminished owing to the high probability of movement.

Any area subject to proposed development or construction is an unfavorable site for a station, and although it is impossible to guarantee this will not happen, the odds can be greatly increased with a little research. By checking master plans and zoning from the planning agency with jurisdiction over the area, the likelihood of future development can be determined. The topography of an area will also indicate possible future development. In addition, always contact property owners to find out their future plans for the land.

In farming areas, station locations are favorable along a fence line between property owners, near farm buildings, and along groves maintained for shade or wind breaks. Cultivated fields are dangerous since crops may be destroyed every time the station is used, and a surface monument cannot be maintained because of repeated cultivation.

Lands owned by public entities are generally ideal station locations. Locating a monument along the right-of-way fence of established roadways or railroads and within a public park usually ensures its permanence. It is always necessary for property owners to be satisfied with station sites; otherwise, their permanence and availability may be in jeopardy. This policy holds true with a public entity for stations on public land, as well as for private owners.

11-8. SETTING STATION MARKS

After the triangulation net has been established and approved, station monuments are set. Each station should consist of a station mark, at least two reference marks, and one azimuth mark. The azimuth mark may be eliminated if another station is readily visible from the ground.

Station marks, the main monuments from which observations are made, have disks placed both underground and on the surface whenever possible. Both of the bronze disks have the station name and year established stamped on them before being set. The station mark can be set in a concrete monument, boulder, or rock outcropping, or brazed to the end of a pipe positioned in the ground.

The concrete monument is poured in place and consists of a subsurface and surface monument. Figure 11-6 shows the dimensions and configuration of a concrete monument set in the ground. The surface monument is positioned over the subsurface one by use of a plumbing bench. Care must be taken while pouring the concrete not to disturb this bench. When using a rock outcropping to set a station mark, it should be a hard, solid part of the main ledge, not a detached fragment. A hole is drilled deep enough to accommodate the disk stem, and the rock surface chipped in a diameter large enough to countersink the disk surface. The disk is attached with cement or epoxy.

At least two surface reference marks should be set for each station: bronze disks marked with an arrow showing direction to the station mark and labeled REFERENCE MARK. The disks are stamped with the station name and date and numbered serially clockwise from north—i.e., RM No. 1 would be the first reference mark looking angularly clockwise from the north direction. Lines from reference marks should intersect the station mark as close to 90° as possible, ensuring a good intersecting angle to recover or relocate the station mark. The lines from reference marks to a station monument should be clear and kept shorter than 30 m to make taping easy, but long enough to ensure direct visibility from an observing tower. Since the function of reference marks is to check or relocate the station mark, they are placed in locations least susceptible to being disturbed, e.g., set in rock outcrops, solid permanent concrete structures such as retaining walls, or concrete monuments at least 30 in. long and 12 in. in diameter.

An azimuth mark is set for each station, primarily for usage by local surveyors. It should be in a location visible from the station mark, using an ordinary ground tripod setup and approximately $\frac{1}{4}$ to $\frac{1}{2}$ mi from the station mark. Azimuth marks are usually set along a right-of-way leading to the station mark. Construction of the monument is the same as for reference marks, except the disk is labeled AZIMUTH MARK. It has an arrow, like the reference-mark disks, set pointing to the station mark. Although an accurate distance may be measured between the azimuth mark and station mark, it is not mandatory and its line is basically to establish direction.

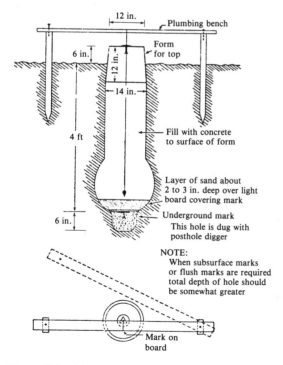

Figure 11-6. Diagram of concrete triangulation station monument.

11-9. SIGNAL BUILDING

Before actual observations can begin, observing stands or towers (signals) may have to be constructed over the station mark. Usually, these signals are only constructed at stations involved with immediate observations and removed as the observing crew moves on. Types of signals vary from a basic surveying tripod to a 100-ft-plus steel or Bilby tower.

Steel or Bilby towers are specially designed for triangulation. They consist of an inner and a structural steel tripod that are independent of each other. The inner tripod supports the theodolite; the outer tripod is for the observer's platform and signal lamp, which is placed 10 ft above the base plate of the inner tripod. An observer's tent is also placed around the outer tower. Both towers are built simultaneously from the ground up. These towers are classified by the heights of the inner tower; they come in heights of 37, 50, 64, 77, 90, 103, and 116 ft.

Bilby towers were a necessity to the NGS for extending triangulation over heavily wooded parts of the country, but for most localized projects wooden towers should be sufficient. The NGS has separate building crews that construct towers ahead of the observing crews. Smaller public surveying departments and private surveying companies will probably rely on the same personnel to both observe and build towers. Depending on the size and purpose of a project, stands and towers may be constructed over every station before any observations are made or constructed in conjunction with observations. A time schedule should be worked out in either case to ensure the even flow of the project and determine the personnel necessary. Care should be taken when constructing a tower to be certain that a vertical leg does not obstruct the lines of sight to other stations.

11-10.   OBSERVATIONS

The NOS has a standard format for notekeeping on a triangulation survey, including *Observations of Horizontal Directions* (NOAA 76-52), *Observations of Double Zenith Distances* (NOAA 76-156), *Abstracts of Directions* (NOAA 76-86), *Abstracts of Zenith Distances* (NOAA 76-135), and *List of Directions* (NOAA 76-72).[4-8]

11-10-1.   Description of Triangulation Station

Figure 11-4 describes a station mark and how to find it, and lists bearings and distances to the azimuth mark and reference monuments.

11-10-2.   Observations of Horizontal Directions

Notes are entered in this book as stations are occupied. All data are completed on top of each page along with other pertinent remarks as notes are taken. A complete set of notes is entered for each station before notes on another one are begun. A complete set of notes in recorded order consists of (1) to-reach description and schematic of the station monuments (Figure 11-7), (2) description of and measurements between monuments (Figure 11-8), (3) observation of the marks and intersecting stations (Figure 11-9), and (4) observations of the main and supplemental stations (Figure 11-10).

Figure 11-8 describes each mark at the station, including measurements to topographic features and between the station mark and reference marks. The identification stamped on a monument must be stated exactly—i.e., do not note "stamped WYNECOOP RM NO 2, 1974." if the disk shows "WYNECOOP NO 2, 1974," The notes "1a, 7a" referred to are standard numbered notes used by the NGS and its predecessor, the U.S. Coast and Geodetic Survey, in its publication *Manual of Geodetic Triangulation*[9] (Table 11-6).

Figure 11-7 lists observations on marks and intersection stations taken at station Wynecoop. Pertinent remarks to be recorded include (1) names and duties of the observing party members, (2) weather conditions, and (3) height and type of signal. The remarks column is used to record conditions affecting observations. Positions that need to be reobserved are noted at the end of regular observations. The position number is recorded in the first column and refers to the initial circle settings listed in Table 11-2.

Complete names and descriptions of station marks and intersecting stations observed are entered in the second column for the first position; abbreviated names may be used on subsequent positions, but numbers or letters cannot be used in place of abbreviations. Intersecting stations—e.g., Williams Water Tank

(*Text continues on p. 229*)

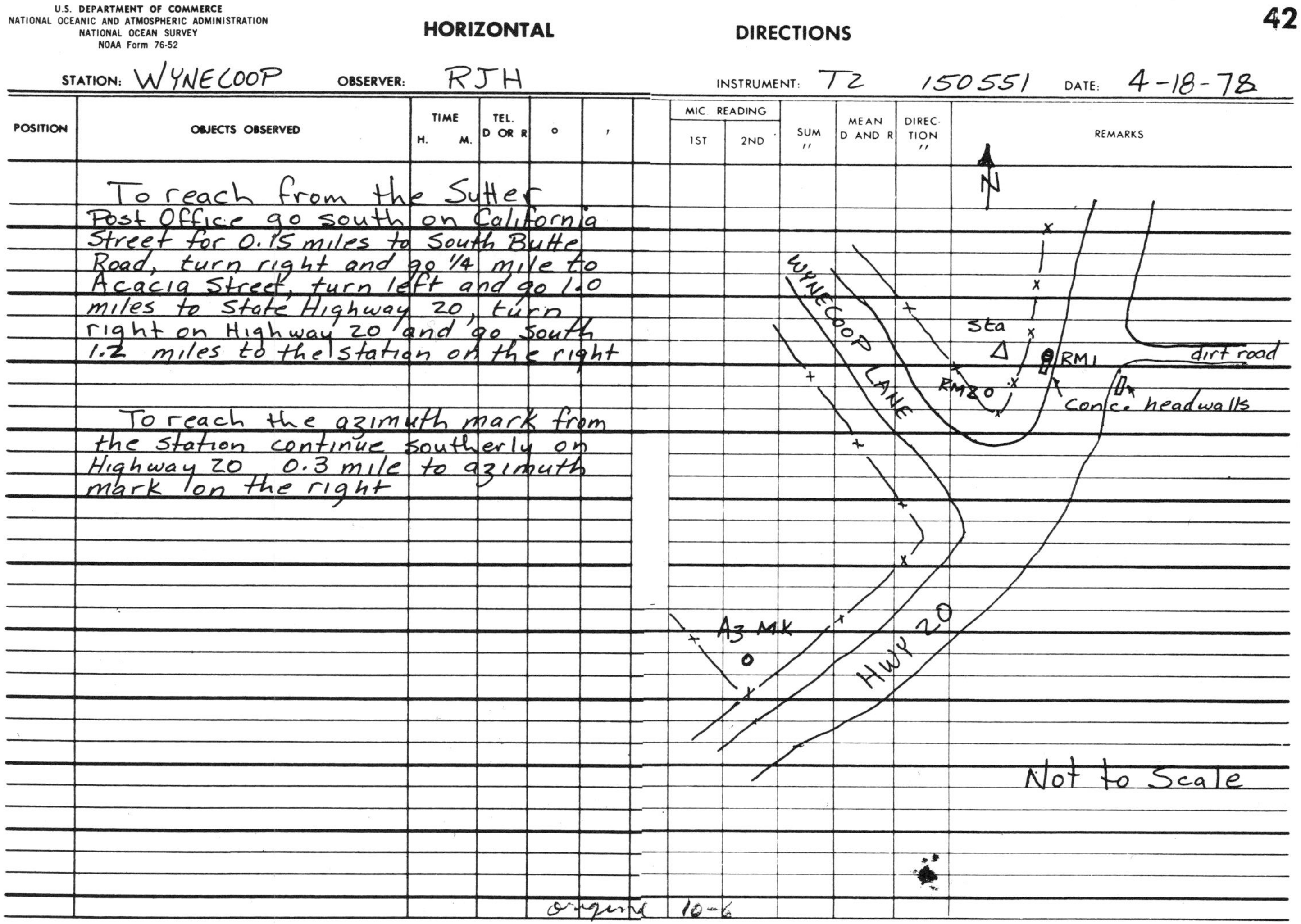

Figure 11-7. To-reach description and sketch.

U.S. DEPARTMENT OF COMMERCE
NATIONAL OCEANIC AND ATMOSPHERIC ADMINISTRATION
NATIONAL OCEAN SURVEY
NOAA Form 76-52

HORIZONTAL   DIRECTIONS

STATION: WYNECOOP   OBSERVER: RJH   INSTRUMENT: TZ 150551   DATE: 4/18/78

POSITION	OBJECTS OBSERVED	TIME H. M.	TEL. D OR R	°	'	MIC. READING 1ST	2ND	SUM ''	MEAN D AND R	DIREC-TION ''	REMARKS
	STATION										
	Stamped: WYNECOOP	1972									Stamped: WYNECOOP NO 1 1972
	Note: 1b, 7a										Set flush in N end Concrete Headwall
	Projects: 0.2 feet										Projects: 0.2 ft above Edge pavement
	48.5 ft W of Center Hwy 20										15.2 ft E of center HWY 20
	18.3 ft W of wire fence										8.5 ft W wire fence
	20 ft S center dirt road on E side Hwy 20										
	89.8 ft NE Power Pole						Held	8.200 m			26.82 ft
								.025			X.3048
								8.175 m			.21456
											.07728
											8.04760
											8.774736 m
	AZIMUTH MARK										
	Stamped: WYNECOOP	1972									
	Note: 17a										
	Projects: 0.8 ft above O.G.										
	34.2 ft NW Center HWY 20										
	4.0 ft NW wire fence										
	4.0 ft NE wire fence										
	BM 1 to BM 2										
	Held 27.900 m	91.41 ft									
	Cut .039	X.3048									
	27.861	73128									
		36564									
		274230									
	27.861	27.006176 8 m									
							RM 2				
							Stamped: WYNECOOP NO 2 1972				
							Note: 11b				
							Projects: 0.2 ft				60.2 ft W of Center HWY 20
											3.0 ft NE of wire fence
											5.6 ft E of power pole
							Held	26.600 m			82.27 ft
							Cut	.001			X.3048
								26.599 m			69816
											34908
											261.8100
											26.599896 m

Figure 11-8. Mark descriptions.

U.S. DEPARTMENT OF COMMERCE
NATIONAL OCEANIC AND ATMOSPHERIC ADMINISTRATION
NATIONAL OCEAN SURVEY
NOAA Form 76-52

HORIZONTAL   DIRECTIONS

STATION: WYNECOOP   OBSERVER: RJH

INSTRUMENT: T 2 150551   DATE: 4/18/78

POSITION	OBJECTS OBSERVED	TIME H. M.	TEL D OR R	°	'	MIC. READING 1ST	2ND	SUM "	MEAN D AND R "	DIREC-TION "	REMARKS
1	BALDY (1939)	1640	D	00	00	09	10	19	11.2		Rec: C Nelson
			R	180	00	13	13	26			
	Williams Water Tank (Gray) 3.0 mi NE		D	92	12	46	46	92	45.5	34.3	92° 12'
			R	272	12	49	49	98			
	WYNECOOP RM No1 (E) (1972)		D	119	56	21	22	43	22.0	10.8	119° 56'
			R	299	56	23	22	45			
	WYNECOOP RM No2 (S) (1972)		D	210	10	55	56	111	55.8	44.6	210° 10'
			R	30	10	56	56	112			
	WYNECOOP AZ MK (SW) (1972)		D	250	12	03	03	06	03.8	52.6	250° 11'
			R	70	12	04	05	09			
2	BALDY		D	45	02	36	37	73	37.8		
			R	225	02	39	39	78			
	Williams Water Tank		D	137	15	16	16	32	16.0	38.2	92° 12'
			R	317	15	16	16	32			
	RM No1		D	164	58	44	45	89	45.2	07.4	119° 56'
			R	344	58	46	46	92			
	RM No2		D	255	13	20	20	40	20.5	42.7	210° 10'
			R	75	13	21	21	42			
	AZ MK		D	295	14	33	32	65	33.5	55.7	250° 11'
			R	115	14	35	34	69			√RJH

Figure 11-9. Observation of marks and intersection stations.

U.S. DEPARTMENT OF COMMERCE
NATIONAL OCEANIC AND ATMOSPHERIC ADMINISTRATION
NATIONAL OCEAN SURVEY
NOAA Form 76-52

HORIZONTAL **DIRECTIONS**

STATION: WYNECOOP   OBSERVER: RJH   INSTRUMENT: T-2 150551   DATE: 4/18/78

POSITION	OBJECTS OBSERVED	TIME H. M.	TEL. D OR R	°	'	MIC. READING 1ST	2ND	SUM	MEAN D AND R	DIREC-TION	REMARKS
											Weather: Clear & Cool
											Obs: RJ Hand
											Rec: C Nelson
1	BALDY 1939	1810	D	00	00	10	12	22			Ht Stand 3.61 m
		1817	R	180	00	14 14	28		12.5		Ht Inst 3.85 m
	GAS 1972		D	21	14	46	47	93			
			R	201	14	47	48	95	47.0	34.5	21° 14'
	SALT 1972		D	68	22	17	17	34			
			R	238	22	20	21	41	18.8	063	58° 22'
	CACHE 1939		D	247	57	03	03	06			
			R	67	57	04	06	10	04.0	51.5	247° 56'
2	BALDY	1829	D	22	02	39	39	78			
		1821	R	202	02	41	41	82	40.0		
	GAS		D	43	17	12	12	24			
			R	223	17	13	15	28	13.0	33.0	21° 14'
	SALT		D	80	24	45	45	90			
			R	260	24	49	49	97	46.8	06.8	58° 22'
	CACHE		D	269	59	32	33	65			
			R	89	59	33	33	66	32.8	52.8	247° 56'
											JRJH

Figure 11-10. Observation of main scheme stations.

Table 11-6. Standard numbered notes for description of marks.

The following notes have been used for many years in published descriptions and other publications of the CGS and later by NGS.

Surface marks

Note 1. A standard triangulation-station disk set in the top of (a) a square block or post of concrete, (b) a concrete cylinder, (c) an irregular mass of concrete.

Note 2. A standard triangulation-station disk cemented in a drill hole in outcropping bedrock, (a) and surrounded by a triangle chiseled in the rock, (b) and surrounded by a circle chiseled in the rock, (c) at the intersection of two lines chiseled in the rock.

Note 3. A standard triangulation-station disk set in concrete in a depression in overcropping bedrock.

Note 4. A standard triangulation-station disk cemented in a drill hole in a boulder.

Note 5. A standard triangulation-station disk set in concrete in a depression in a boulder.

Note 6. A standard triangulation-station disk set in concrete at the center of the top of a tile (a) that is embedded in the ground, (b) that is surrounded by a mass of concrete, (c) that is fastened by means of concrete to the upper end of a long wooden pile driven into the marsh, (d) that is set in a block of concrete and projects from 12 to 20 in. above the block.

Underground marks

Note 7. A block of concrete about 3 ft below the ground containing at the center of its upper surface (a) a standard triangulation-station disk, (b) a copper bolt projecting slightly above the concrete, (c) an iron nail with the point projecting above the concrete, (d) a glass bottle with the neck projecting a little above the concrete, (e) an earthenware jug with the mouth projecting a little above the concrete.

Note 8. In bedrock, (a) a standard triangulation-station disk cemented in a drill hole, (b) a standard triangulation-station disk set in concrete in a depression, (c) a copper bolt set in cement in a drill hole or depression, (d) an iron spike set point up in cement in a drill hole or depression.

Note 9. In a boulder about 3 ft below the ground (a) a standard triangulation-station disk cemented in a drill hole, (b) a standard triangulation-station disk set in concrete in a depression, (c) a copper bolt set with cement in a drill hole or depression, (d) an iron spike set with cement in a drill hole or depression.

Note 10. Embedded in earth about 3 ft below the surface of the ground (a) a bottle in an upright position, (b) an earthenware jug in an upright position, (c) a brick in a horizontal position with a drill hole in its upper surface.

Reference marks

Note 11. A standard reference-mark disk, with the arrow pointing toward the station, set at the center of the top of (a) a square block or post of concrete, (b) a concrete cylinder, (c) an irregular mass of concrete.

Note 12. A standard reference-mark disk, with the arrow pointing toward the station, (a) cemented in a drill hole in outcropping bedrock, (b) set in concrete in a depression in outcropping bedrock, (c) cemented in a drill hole in a boulder, (d) set in concrete in a depression in a boulder.

Note 13. A standard reference-mark disk, with the arrow pointing toward the station, set in concrete at the center of the top of a tile (a) that is embedded in the ground, (b) that is surrounded by a mass of concrete, (c) that is fastened by means of concrete to the upper end of a long wooden pile driven into the marsh, (d) that is set in a block of concrete and projects from 12 to 20 in. above the block.

Previously used notes 14 and 15 referred to seldom used types of witness marks and are purposely omitted.

Azimuth marks

Azimuth-mark notes are almost identical to reference-mark notes 11 through 13, which have been previously used for azimuth marks. The following numbers 16, 17, and 18 refer specifically to azimuth-mark disks.

Note 16. A standard azimuth-mark disk, with the arrow pointing toward the station, set at the center of the top of (a) a square block or post of concrete, (b) a concrete cylinder, (c) an irregular mass of concrete.

Note 17. A standard azimuth-mark disk, with the arrow pointing toward the station, (a) cemented in a drill hole in outcropping bedrock, (b) set in concrete in a depression in outcropping bedrock, (c) cemented in a drill hole in a boulder, (d) set in concrete in a depression in a boulder.

Note 18. A standard azimuth-mark disk, with the arrow pointing toward the station, set in concrete at the center of the top of a tile (a) that is embedded in the ground, (b) that is surrounded by a mass of concrete, (c) that is fastened by means of concrete to the upper end of a long wooden pile driven into the marsh, (d) that is set in a block of concrete and projects from 12 to 20 in. above the block.

—should be described completely, including the structure's local name and description of the part observed, approximate distance from the occupied station, and bearing (E, NE, etc.) from the occupied station. Beginning and ending times for each position are entered under the time head in the third column, using military time from 0000 (midnight) to 2359.

The fourth column describes the telescope position—D for direct or R for reverse—followed by the direction observed with two micrometer readings taken for each position (see Section 11-11-3). The sum of the micrometer readings is entered under the sum heading while the mathematical mean, to the nearest 0.1 sec, is noted in the following column, under mean. When using the Wild T-3 theodolite, the sum of the two direct micrometer readings and that of the two reverse micrometer readings are meaned. On the Wild T-2 and other comparable directional theodolites, the entry is the mean of all four micrometer readings. Seconds of direction observed from the initial station to each object are entered under "direction" (subtract the mean of initial from the mean of observed).

All pertinent information discussed earlier is entered under remarks. Degrees and minutes of direction observed from the initial station are also listed under this column for each direction observed.

After observations of marks and intersecting stations are completed, observations of the main and supplemental scheme begin (Figure 11-9). Recording main and supplementary observations is done in the same manner as for mark and intersecting station observations. Reobservations are entered after the original set. When the theodolite is checked for level, it should be recorded under remarks and any adjustment noted.

After each page or set of observations is recorded, it must be completely reviewed, all items checkmarked, and each page initialed, and corrections also checked and initialed. Entries must be in ink, and no erasures are

permitted. Errors are deleted by a single line through the incorrect entry and the correction placed above the deletion. When filled, an alphabetized index is recorded in the front of the book.

11-10-3. Observations of Double Zenith Distances

Typical recordings for vertical angles or double zenith distances observed as given in Figure 11-11 are self-explanatory.

11-10-4. Abstracts

After each set of horizontal distances and zenith distances has been observed and the field book checked, an abstract is recorded to make certain all observations are within required tolerances. Abstracts should be completed in ink, without erasures.

The "abstract of directions" (Figure 11-12) lists all observations taken directly from the "horizontal directions" book, except those taken on a wrong object or any resulting from a kicked tripod, which are rejected immediately in the book. Any observation so far from the apparent mean as to be considered a blunder may be rejected before calculating the initial mean. Once all observations have been entered on the abstract, the columns are summed and a mean angle is derived. Any observation outside the required tolerance is rejected. That position is then reobserved and the new observation entered on the abstract. The mean of the observations is recalculated using the new observation. If, when we use the new mean, the rejected observation and new observation both fall within the required tolerance, they are meaned and used as the observed seconds for that position. A rejected observation still outside required tolerance is enclosed in parentheses and an R written behind, meaning that it is rejected and not used to calculate the mean direction.

U.S. DEPARTMENT OF COMMERCE
NATIONAL OCEANIC AND ATMOSPHERIC ADMINISTRATION
NOAA FORM 76-156
(REV. 11-77)

DOUBLE ZENITH DISTANCES

Station W/WECOOP State California Instrument T2 150551

Observer R JH County Sutter Date 4/18/78

Object Observed	Time	Level O.	Level E.	Circle Right or Left	Circle Reading	Verniers A	Verniers B	Verniers Mean	Zenith Distance °	Zenith Distance '	Zenith Distance "	Remarks
BALDY (1939) (Light) 1.33m	1205			L	89° 58'	58	58	58.0				Rec C Nelson cool
				R	270 00	56	56	56.0				Ht of Stand 3.61m
				DZD	179 58			02.0	89	59	01.0	Ht of Inst 3.85m
				L	89 59	03	03	03.0				
				R	270 00	51	51	51.0				
				DZD	179 58			12.0	89	59	06.0	
GAS (Light) 1.27m	1220			L	89 07	19	19	19.0				
				R	270 52	36	36	36.0				
				DZD	178 14			43.0	89	07	21.5	
				L	89 07	05	05	05.0				
				R	270 52	45	45	45.0				
				DZD	178 14			20.0	89	07	10.0	(R)
				L	89 07	21	21	21.0				
				R	270 52	29	29	29.0				
				DZD	178 14			52.0	89	07	26.0	

Do not write in this margin

JDRG

Figure 11-11. Recording double zenith distances.

			STATIONS OBSERVED					
WYNECOOP				ABSTRACT OF DIRECTIONS				
STATE California		COMPUTED BY CN		DATE 4/18/78		VOLUME NO.		
OBSERVER RJH		CHECKED BY DRG		INSTRUMENT NO. TZ 150551		SHEET____ OF____		

POSITION NO.	BALDY (1939)	GAS	SALT	CACHE (1939)	Williams Water Tank	WYNECOOP RM No 1	WYNECOOP RM No 2	WYNECOOP AZ MK
(INITIAL)	0° 00′	21 14	58 22	247 56	92 12	119 56	210 10	250 11
	″	″	″	″	″	″	″	″
1	0.00	34.5	06.3	51.5	34.3	10.8	44.6	52.6
2	0.00	33.0	06.8	52.8	38.2	07.4	42.7	55.7
3	0.00	28.8	10.2	47.2	39.2	12.2	48.6	51.0
4	0.00	31.0	07.5	50.8	37.0			56.6
5	0.00	30.2	11.2	52.0				
6	0.00	33.5	09.5	48.0				
7	0.00	29.1 (25.2)R	08.0	49.8				
8	0.00	32.8	08.2	53.2				
9	0.00							
10	0.00							
11	0.00							
12	0.00							
13	0.00							
14	0.00							
15	0.00							
16	0.0							
SUM,		12.9	67.7	5.3	28.7	30.4	15.9	15.9
MEAN,		31.61	08.46	50.66	37.2	10	45	54.0
COR. FOR ECC.,								
DIRECTION,								

Figure 11-12.   Abstract of directions.

The following is a summary of the rejection limits:

1. First-order: 4 sec
2. Second-order, third-order, and azimuth marks: 5 sec
3. Reference marks: 20 sec
4. Intersection station: 5 to 10 sec, depending on the sharpness and nearness of an object

The "abstract of zenith distances" (Figure 11-13) lists all the vertical angles observed from a station. Besides the headings, the data in columns 1, 2, 3, 5, and 8 are entered by the observing party from information recorded in the "observations of double zenith distances" field book. Heights above the station mark, dates, and stations the lights are shown to are entered at the bottom of the abstract, as the information becomes available. The rest of the information is entered or calculated later.

11-11. FIELD OBSERVATIONS

An observing party should plan to arrive at the station early enough to (1) write the station description; (2) measure the distances between the reference and station marks; (3) check for stability and collimate the stand or tower; (4) set up the instrument; (5) observe vertical angles; and (6) observe horizontal directions to the azimuth mark, reference marks, and intersection stations before dark.

11-11-1. Station Setup

While driving to the station, a final check should be made of the to-reach description, including the approximate distance from azimuth mark to a station mark. On reaching the station mark, the to-reach description and schematic are recorded in the "observation of horizontal directions" book.

Next, distances between reference marks and the station mark are measured independently, in hundredths of a foot and thousandths of a meter. A 30-m tape, marked in meters on one side and feet on the other, is used to make the measurements. They are made horizontally and recorded immediately in observation of horizontal directions. Measurements are repeated until a check of 0.003 m is obtained between the meter and feet readings (Figure 11-12).

11-11-2. Vertical Angles

Normally, the first set of observations is for the vertical angles, measured from the zenith and referred to and recorded as double zenith distances in "observations of double zenith distances." Reciprocal observations should be made at all occupied stations, simultaneously if possible. If reciprocal observations are made on different days, it is best to try to make them under the same climactic condition and during the same hour of the day. This will lessen the effect caused by varying refraction. The best time to observe vertical angles and the hours of least refraction are between noon and 4:00 PM.

First-order and second-order, class I triangulation require three individual sets of observations of the double zenith distance. All other classifications require two. Observations are made with the middle of the horizontal cross hair sighted on the object. A full direct and reverse set of observations must be completed before another set is started to the same station. Two or more consecutive direct or reverse pointings to the same station are not made. This is to ensure that each pointing is separate and distinct. Sets may be completed on one object at a time, or completed with the telescope in one position on several objects in succession and then reversed.

11-11-3. Horizontal Directions to Reference Marks, Azimuth Marks, and Intersection Stations

Part of the daylight activities at a station is observing horizontal directions to the reference and azimuth marks and any intersection stations. Whenever possible, the initial station used for these marks should be the same as that used on the main scheme observations.

ABSTRACT OF ZENITH DISTANCES

Station __WYNECOOP__   State __CALIFORNIA__

Observer __BJH__   Instr. __T-2 150551__   Height of Stand __3.61 M__

DATE	HOUR	OBJECT OBSERVED	OBJECT ABOVE STATION =o	TELESCOPE ABOVE STATION =t	DIFF. OF HEIGHTS t−o	REDUCTION TO LINE JOINING STATIONS	OBSERVED ZENITH DISTANCE	CORRECTED ZENITH DISTANCE
			Meters	Meters	Meters	"	° ' "	° ' "
4/18/78	1205	BALDY (1939)	1.33	3.85	+2.52		89 59 01.0	
							06.0	
							89 59 03.5	
4/18/78	1220	GAS	1.27	3.85	+2.58		89 07 21.5	
							10.0 (R)	
							26.0	
							89 07 23.8	
4/18/78	1242	SALT	3.77	3.85	+0.08		90 01 15.0	
							22.0	
							90 01 18.5	
4/18/78	1255	CACHE	7.63	3.85	−3.78		89 56 47.0	
							42.0	
							89 56 44.5	

DATE	LIGHT SHOWN TO STATION	HEIGHT OF LIGHT* ABOVE STATION	DATE	LIGHT SHOWN TO STATION	HEIGHT OF LIGHT* ABOVE STATION
		Meters			Meters
4/16/78	BALDY (1939)	3.78	4/23/72	CACHE	3.78
4/19/78	GAS	3.78			
4/18/78	SALT	2.95			✓ DRG

DO NOT WRITE IN THIS MARGIN.

Figure 11-13.   Abstract of zenith distances.

Three positions are observed on reference marks; four positions are observed on azimuth marks and intersection stations. Any positions on the reference marks in excess of 20 sec from the mean are rejected and reobserved, whereas any positions of the azimuth mark or intersection stations in excess of 5 sec from the mean are rejected and reobserved.

Any necessary reobservations are made after the primary four positions have been completed and checked. A circle setting for the reobserved position will be the same as for the rejected position.

11-11-4. Horizontal Directions to Main and Supplemental Stations

The best time to start observing on main and supplemental stations is at dusk, as soon as all sighting lights become visible. Care must be taken to make sure that lights are pointed as accurately as possible. By this time, all other duties of the observing party should be completed. If vertical angles could not be observed earlier, they should be measured after the main scheme directions are taken. Stations used for the initial observation should have an easily discernible and reliable light for uninterrupted use. The distance to it should be long enough to ensure no effect due to wind, local refraction, or a slightly mispointed light, but not so far away as to cause an interruption because the light is faint or hazy. If possible, pick an initial that is convenient for observing directions in a clockwise rotation.

Observation is completed in the same manner as that used for reference marks and intersection stations, using the required number of positions for the classification sought. Pointing should be quick but not hurried. Deliberate pointing should be avoided as this tends to reduce accuracy and greatly increases the time required to complete a position.

Lightkeepers should keep their own set of notes, listing what sights were used on which stations. They should also list the sight height above the station mark. They are responsible

for making sure lights are constant and pointed as accurately as possible, and they should be in constant touch with the observing crew.

11-12. CONCLUSION

Many factors of triangulation have not been considered in this chapter, such as eccentricity of stations, reduction to center, and the use of striding levels for observation with a vertical angle over 2°. When we consider a triangulation project, the *Manual of Geodetic Triangulation*, Publication No. 247, is available from the U.S. Government Printing Office, or any successor to this publication will be helpful. Publication No. 248 is the guide for NGS triangulation crews and covers the complete procedures required in the field and not reduction process.

NOTES

1. U.S. Department of Interior Bureau of Land Management. 1973. *Manual of Surveying Instructions*. *U.S.* Government Printing Office, Washington, D.C.

2. FGCC. 1984. Classification, standards of accuracy, and general specifications of geodetic control surveys. Rockville, MD.

3. *Manual of Reconnaissance for Triangulation.* SP225, U.S. Government Printing Office: Washington, D.C., 1938.

4. NOS. *Observations of horizontal directions.* 1976. NOAA 76-52, Rockville, MD.

5. NOS. *Observations of double zenith distances.* 1976. NOAA 76-52, Rockville, MD.

6. NOS. *Abstracts of directions.* 1976. NOAA 76-86, Rockville, MD.

7. NOS. *Abstracts of zenith distances.* 1976. NOAA 76-135, Rockville, MD.

8. NOS. *List of directions.* 1976. NOAA 76-72, Rockville, MD.

9. *Manual of Geodetic Triangulation.* Publication No. 247, U.S. Government Printing Office, 1971. Washington, D.C.

10. *Manual of Geodetic Triangulation.* 1971. Publication No. 247, U.S. Government Printing Office, Washington, D.C.

12

Trilateration

Bryant N. Sturgess and Frank T. Carey

12-1. INTRODUCTION

Trilateration is a method of control extension, control breakdown, and control densification that employs electronic distance-measuring instruments (EDMIs) to measure the lengths of triangles sides rather than horizontal angles, as in triangulation. The triangle angles are then calculated based upon measured distances by the familiar law of cosines. Trilateration consists of a system of joined and/or overlapping triangles usually forming quadrilaterals or polygons, with supplemental horizontal angle observations to provide azimuth control or check angles. Zenith angles are required when elevations have not been established or differential leveling is not contemplated, in order to reduce slope distances to a common reference datum.

With the development of EDMIs, trilateration has become a very practical highly accurate, and precise means of establishing and/or expanding horizontal control.

12-2. USE OF TRILATERATION

Trilateration is commonly employed to study gradual and secular movements in the earth's crust in areas subject to seismic or tectonic activity, to test and construct defense and sci-

entific facilities, and on high-precision engineering projects. It is also used in control expansion or densification for future metropolitan growth; coastline control; inland waterways; control extension; densification for land subdivisions and construction; and deformation surveys of dams, geothermal areas, structures, regional/local tectonics, and landslides. Trilateration can be used for a simple low-order topographic survey covering a small area, or on large projects for the design and/or construction of highways, bridges, dams, or even to extend topographic mapping control from small local tracts to regional areas. It can be a simple process with single-line measurements using ordinary off-the-shelf EDMIs and support equipment. Or, it can be a complex process employing highly refined EDMIs, with special measures for determining the refractive index correction and eccentric measurements to an eccentric or offset bar, at either the reflector or instrument station with instrument occupations at both ends of the line.

12-3. ADVANTAGES OF TRILATERATION

Trilateration is a practical and highly accurate means of rapid control extension. When properly executed, it is superior to both triangulation and traverse for special-purpose precise surveys and often is the preferred method, because of its advantageous cost-benefit ratio and potential.

Basic trilateration is less expensive than classical triangulation and, under most conditions, more accurate. Trilateration permits controlling large and small geographical areas with a minimum number of personnel. It is not required to measure lines with all sights simultaneously in position, as with triangulation, unless the procedure of line pairs is being employed. Trilateration also provides necessary scale control lacking in triangulation.

12-4. DISADVANTAGES OF TRILATERATION

Trilateration has a smaller number of internal checks compared with classical triangulation, in which each quadrilateral contains two diagonals called *braces* (hence, braced quadrilateral) and has four triangle closures, three of which are independent. Additionally, there are other checks consisting of agreements between common sides of the triangles (*side-equation* tests). The number of checks or redundancies in a braced quadrilateral in triangulation is four. In trilateration, there is only one.

Trilateration can be reinforced by modified observational techniques to provide the same number of redundancies as triangulation by employing the group or ratio method of length measurement. In order to employ this technique, the length of a line must be measured from both ends. Cost then becomes the limiting factor. Geometric restrictions, which limit the selection of locations where stations can be established, may cause difficulty in fully utilizing trilateration networks to place control at specially needed sites. Pure trilateration cannot be performed when precise angle measurements must be taken to intersect reference objects, azimuth marks, and reference monuments. In such cases, a combination of trilateration and triangulation has to be employed.

Higher-order trilateration (first- and second-order, class I) requires sampling of meteorological conditions to be commensurate in precision with the distance measurements. It is paramount that appropriate techniques be employed to guarantee reasonable sampling. First-order and super first-order trilateration necessitate implementing meteorological sampling techniques, perhaps by flying the sight line with a light aircraft to sample temperature, humidity, and barometric pressure, or utilizing an EDMI with a two-color laser system to determine the refractive index correction.

Trilateration is inherently more expensive than traverse. In addition, on large geodetic control projects the inventory of reflectors for measuring long lines must be sufficient in numbers to guarantee maximum signal strength to the EDMI. Reflector arrays have to be manned and are significantly more expensive than *show lights* required for triangulation.

12-5. COMPARISON OF TRIANGULATION AND TRILATERATION

Triangulation and trilateration are both forms of control extension, control expansion, or control breakdown (densification). Triangulation is a method of surveying in which the stations are points on the ground, forming vertices of triangles comprising chains of quadrilaterals or polygons. Within these triangles the angles are observed by theodolite, and the lengths of sides determined by successive computations through the chain of triangles. Scale is provided by at least two stations having known positions in the first quadrilateral of the chain, or by a base line connecting two stations in the first quad. The positions of all remaining stations in the first quad, and any in a chain of quads, are computed in terms of measured angles and known positions at the chain's beginning and end.

In triangulation, distances are computed from angle observations; in trilateration, angles are calculated from distance observations. Triangulation gives impressive redundancy when compared with trilateration. For example, in a simple triangle with all angles measured, one redundant measurement yields one condition equation or one degree of freedom. Given the same triangle in trilateration, unless three distances are measured from each end, there are no redundant measurements.

Similarly, a braced quadrilateral with all angles measured provides four redundancies; the same figure in pure trilateration yields only one redundancy. Table 12-1 tabulates the redundancies possible with various figures. All points given in the example are presumed to be intervisible.

It can be seen that the number of redundancies in triangulation is significantly higher than in trilateration. This does not necessarily mean triangulation provides a more accurate or precise positional solution than trilateration. On the contrary, when trilateration networks are given an adequate design configuration by avoiding angles smaller than the specified minimum, by using proper field techniques and properly matched calibrated equipment, and by adding check angles when appropriate, the final results can be significantly superior to triangulation, with lower cost.

12-6. CONSTRUCTION OF BRACED QUADRILATERALS

Well-shaped geometrical figures are required for both arc and area networks. For arcs, quadrilaterals must approximate a square with both diagonals measured. When only a single diagonal can be observed, a center point must be visible from the four vertices of the quad. In area systems, well-shaped triangles containing angles seldom smaller than 15° are mandatory for first- and second-order surveys. If these conditions cannot be met, then one or more of the large angles in the quad must be observed by theodolite. On occasion, geographical constraints will not allow the "ideal" quadrilateral configuration having angles larger than 25°.

Figures can contain angles of 5 to 7° smaller than specified while maintaining line accuracies. However, this is permissible only on second-order, class II or third-order, class I and II surveys. It is desirable to include horizontal-angle readings to increase redundancy and improve accuracy in cases where conditions of deficient angle sizes are present. First-order and second-order, class I surveys should not have deficient angle sizes (within the same range of 18 to 20°) unless compensated for by angle measurements.

Since few engineers and surveyors are engaged in projects that establish new primary positions on arc or transcontinental networks,

Table 12-1.   Geometrical redundancies

	Quadrilateral	Pentagon	Hexagon
Number of lines	6	10	15
Number of triangles	4	10	20
Number of triangles used in the computations	3	6	10
Number of check triangles	1	4	10
Number of geometrical conditions in trilateration	1	3	6
Number of geometrical conditions in triangulation	4	9	16

Source: FGCC 1979. "Classification, Standards of Accuracy, and General Specifications of Geodetic Control Survey." Silver Spring MD.

only area networks and small-area-type projects will be addressed,

Figures 12-1, 12-2, and 12-3 show typical polygons that can be used in trilateration. The center-point polygon and center-point quadrilateral are more expensive to design, but may be required owing to topographical constraints. In a high-order control network, the familiar braced quadrilateral, reinforced with angular measurements at selected stations, provides an economical means of increasing redundancies in trilateration.

A trilatered center-point polygon composed of single triangles with no diagonals observed contains only a single condition or degree of freedom, regardless of the number of triangles contained in the figure. For each additional line measured, one more condition is added.

The following are methods of increasing redundancy by one condition or degree of freedom, thus improving reliability by adding:

1. An azimuth measurement other than the one required to orient the system.

2. An independent distance measurement accomplished through triangulation, e.g.

3. Two degrees of freedom by including a station of known position, other than the station required to index and originate the computations.

4. An angle observation; for each one, an additional condition or degree of freedom is introduced.

The more conditions designed into a figure network, the greater reliability obtained for

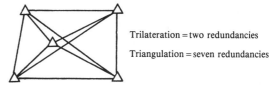

Figure 12-2.   Center-point quadrilateral.

the final position of each point within the system.

A good policy to follow is a semimarriage of trilateration and triangulation by measuring all distances and incorporating angle observations generally at the station with the largest angle. Sufficient redundancy is gained by one such occupation to make the expenditure cost-effective. For instance, one occupation of this type on a conventional braced quadrilateral increases the redundancy from one to two. However, it must be remembered that when certain EDMIs are properly tuned and calibrated, they will surpass the performance of any theodolite (short distances excluded) in determining angles through distance observation. Consequently, when check angles are specified, the proper angle observation weight must be determined accurately to avoid biasing the figure. Therefore, unless the check angle is placed in the proper perspective, it will yield an extra redundancy but can degrade the mathematical fit of the figure unless properly weighted.

Another means of increasing redundancies in a conventional braced quadrilateral for higher-order trilateration surveys, such as first

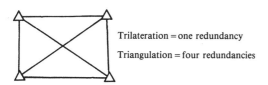

Figure 12-1.   Braced quadrilateral.

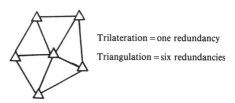

Figure 12-3.   Center-point polygon.

and second, class I, is to incorporate distance measurements taken alternately from line ends with measurements to EDMI reflectors mounted on an eccentric bar. Observing distances from each end provides a field check on a prior line measurement, without lengthy computations. Ideally, lengths should be measured from each end on different days. However, if production or scheduling will not permit this, measurements can be taken on the same day, but at a spaced interval such as morning and afternoon, or day and night. The advantage of this method is that different meteorological conditions exist at various times, so when meaned, they should yield a result closer to the true distance.

A full marriage of trilateration and triangulation, in which all distances are measured from each end and every station is occupied by theodolite, will provide the strongest and theoretically most precise means of establishing horizontal control. This method predictably is the most costly; however, for those projects demanding the ultimate in accuracy and precision, there is no better method with the possible exception of the global positioning system (GPS).

12-7. NETWORK DESIGN

The geometric figures used in trilateration are usually braced quadrilaterals and center-point polygons. However, because of a lack of internal checks, established specifications and standards must be adhered to rigorously. The minimum specified angle sizes should not be reduced unless check angles are observed at those locations.

Theoretically, the basic quadrilateral should be square in configuration, and a central polygon should have equal angles. Field conditions may make achieving this desirable arrangement difficult or impossible. The sample trilateration networks shown in Figures 12-4 and 12-5 are typical of new networks.

12-8. RECONNAISSANCE

Field reconnaissance for trilateration is similar to traversing and triangulation reconnaissance. It is equally essential in trilateration and triangulation that all existing and proposed new stations be plotted on a base map, so the network configuration can be verified and checked for conformance with specifications. In areas where the triangles fall below specifications, owing to geographical constraints, occupation for angle observations can be planned to add reinforcement to the figures and redundancies in the mathematical solution. For a detailed description of reconnaissance, refer to Section 11-7, "Field Reconnaissance."

General guidelines to follow in layout of a trilateration net are:

1. Avoid situations where the sight line passes in close proximity to a ridge, saddle, tree top, or other obstruction between the EDMI and reflector station. Unless adequate clearance is maintained, the light beam will be reflected and weakened by heat waves, thus reducing the range and diminishing the reliability of distance measurements.

2. Line selections that go over large bodies of water must clear them by at least 50 ft; otherwise, range and accuracy suffer.

3. Attempting EDMI measurements near high-voltage transmission lines or microwave relays (even two-way radio communication with the EDMI station) can cause erroneous measurements. Avoid EDMI setups close to such energy sources. In most cases, a distance of 200 ft is a reasonable minimum.

In station-to-station communication between survey party members, avoid any transmissions from the EDMI station during line measurements. The precautions mentioned in item 3 do not necessarily apply to all EDMIs. It is prudent to first determine if the equipment selected for the project is susceptible to these electromagnetic and static forces, before choosing a questionable site.

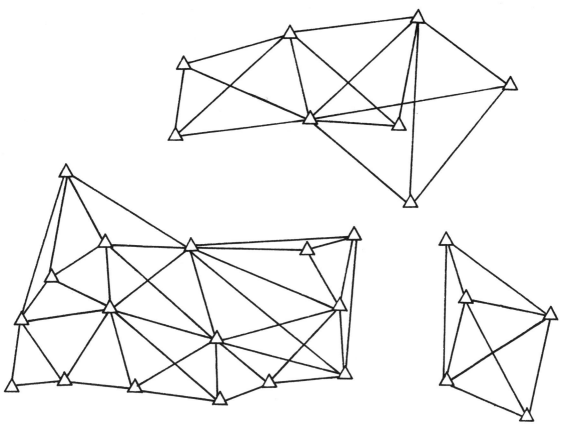

Figure 12-4.   Existing trilateration networks.

12-9.   SPECIFICATIONS AND TRILATERATION

In the United States, standards of accuracy for geodetic control surveys are prepared by the Federal Geodetic Control Committee (FGCC) with subsequent review by the American Congress on Surveying and Mapping (ACSM), the American Society of Civil Engineers (ASCE), and the American Geophysical Union (AGU).

Table 12-1, which is taken from the FGCC publication entitled *Classification, Standards of Accuracy, and General Specifications of Geodetic Control Surveys*, provides a summary of various standards of accuracy for triangulation; Table 12-2 shows standards for trilateration control. They are available from the National Ocean

Survey (NOS) in Rockville, MD Table 12-3 lists trilateration standards for field operations covering various orders of surveys.

It is paramount that reference be made to another publication, *Specifications to Support Classification, Standards of Accuracy, and General Specifications of Geodetic Control Surveys*, by the FGCC, also published by the NOS.[2] Both publications should be consulted before designing a trilateration control network. The standards provided in these publications are for three orders of accuracy: first, second, and third. Second- and third-order are subdivided into two classes: I and II.

First-order trilateration is used mainly in (1) arc networks, (2) transcontinental primary control, (3) network densification in large metropolitan areas, (4) highly accurate surveys

Figure 12-5. Plat of geysers geothermal field (descriptions of stations have been omitted).

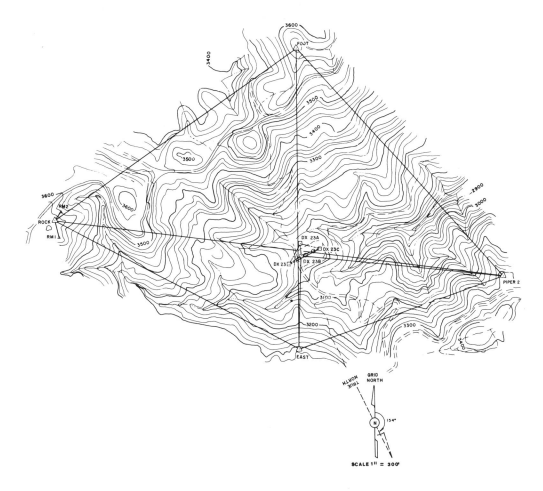

Figure 12-5. (*Continued*).

NOTE: The source of contours is drawing number 3250 entitled "Unit 17 Project Area, Upper Squaw Creek" by Union Oil Co. of California, Geothermal Division.

for defense, (5) sophisticated engineering control projects, and (6) surveys for monitoring earth deformation. First-order is the *primary* horizontal control and provides the basic framework for a national control net.

Second-order, class I horizontal control connects first-order arcs and is the principal framework for control densification. This order is also employed for large engineering projects as well as control expansion for metropolitan areas and, in some cases, deformation surveys.

Second-order, class II surveys are used as the basic framework for big photogrammetric projects, large-scale subdivisions, and spacious construction projects. This order is occasionally utilized to control interstate highway systems.

Third-order, class I and II trilateration surveys are a supplemental-type control. This lowest order is employed primarily to control small engineering projects, construction, and photogrammetric, hydrographic, and topolographic projects.

Table 12-2. Classification, standards of accuracy, and general specifications for horizontal control

TRILATERATION

Classification	First-Order	Second-Order		Third-Order	
		Class I	Class II	Class I	Class II
Recommended spacing of principal stations	Network stations seldom less than 10 km. Other surveys seldom less than 3 km.	Principal stations seldom less than 10 km. Other surveys seldom less than 1 km.	Principal stations seldom less than 5 km. For some surveys, a spacing of 0.5 km between stations may be satisfactory.	Principal stations seldom less than 0.5 km.	Principal stations seldom less than 0.25 km.
Geometric configuration§ Minimum angle contained within, not less than	25°	25°	20°	20°	15°
Length measurement Standard error*	1 part in 1,000,000	1 part in 750,000	1 part in 450,000	1 part in 250,000	1 part in 150,000
Vertical angle observations† Number of and spread between observations	3 D/R—5"	3 D/R—5"	2 D/R—5"	2 D/R—5"	2 D/R—5"
Number of figures between known elevations	4–6	6–8	8–10	10–15	15–20

Astro azimuths**					
Spacing-figures	6–8	6–10	8–10	10–12	12–15
No. of obs./night	16	16	16	8	4
No. of nights	2	2	1	1	1
Standard error	0″.45	0″.45	0″.6	0″.8	3″.0
Closure in position‡					
After geometric conditions have been satisfied should not exceed	1 part in 100,000	1 part in 50,000	1 part in 20,000	1 part in 10,000	1 part in 5000

*The standard error is to be estimated by

$$\sigma_m = \sqrt{\frac{\Sigma y^2}{n(n-1)}}$$

where σ_m is the standard error of the mean, y a residual (i.e., the difference between a measured length and the mean of all measured lengths of a line), and n the number of measurements. The term "standard error" used here is computed under the assumption that all errors are strictly random in nature. The true or actual error is a quantity that cannot be obtained exactly. It is the difference between the true value and measured value. By correcting each measurement for every known source of systematic error, however, one may approach the true error. It is mandatory for any practitioner using these tables to reduce to a minimum the effect of all systematic and constant errors so that real accuracy may be obtained. (See page 267 of Coast and Geodetic Survey, Special Publication No. 247, *Manual of Geodetic Triangulation* Revised edition, 1959, for definition of "actual error.")

**The standard error for astronomic azimuths is computed with all observations considered equal in weight (with 75% of the total number of observations required on a single night) after application of a 5-sec rejection limit from the mean for first- and second-order observations.

†See FGCC *Detailed Specifications on Elevation of Horizontal Control Points* for further details. These elevations are intended to suffice for computations, adjustments, and broad mapping and control projects, not necessarily for vertical network elevations.

‡Unless the survey is in the form of a loop closing on itself, the position closures would depend largely on the constraints or established control in the adjustment. The extent of constraints and actual relationship of the surveys can be obtained through either a review of the computations, or a minimally constrained adjustment of all work involved. The proportional accuracy or closure (i.e., 1/100,000) can be obtained by computing the difference between the computed value and fixed value, and dividing this quantity by the length of the loop connecting the two points.

§See FGCC *Detailed Specifications on Trilateration* for further details. *Source: FGCC Specifications to Support Classification, Standards of Accuracy, and General Specifications of Geodetic Control Surveys*, Silver Spring, MD.

Table 12-3. Trilateration standards for field operations

	First-Order	Second-Order		Third-Order	
		Class I	Class II	Class I	Class II
Length measurement standard error	1 part in 1,000,000	1 part in 750,000	1 part in 450,000	1 part in 250,000	1 part in 150,000
PPM (total error budget)	1 ppm	1.3 ppm	2.2 ppm	4 ppm	6.7 ppm
"Atmospheric sampling accuracy requirements"					
Temperature (T)	±0.2°F/±0.1°C (+0.1 ppm)	±0.9°F ±0.5°C (±0.5 ppm)	±1.8°F ±1.0°C (±1.0 ppm)	±1.8°F ±1.0°C (±1.0 ppm)	±3.6°F ±2.0°C (±2.0 ppm)
Barometric pressure (P)	±0.01" HG, 10' alt. ±0.25 mm HG (±0.1 ppm)	±0.05" HG, 50' alt. ±1.3 mm HG (±0.5 ppm)	±0.10" HG 100' alt. ±2.5 mm HG (±1 ppm)	±0.10" HG, 100' alt. ±2.5 mm HG (±1 ppm)	±0.20" HG, 200' alt. ±5.0 mm HG (±2 ppm)
Humidity (E)	Suf. sampling to yield an accuracy of 0.1 ppm.	Suf. sampling to yield an accuracy of 0.1 ppm.	Sampling is not nec. Use avg. +0.4 ppm.	Sampling is not nec. Use avg. +0.4 ppm.	Sampling is not nec. Use avg. +0.4 ppm.
EDMI					
Centering	±0.5 ppm ±1/2 mm	±1 ppm ±1/2 mm	±2 ppm ±1 mm	±3 ppm ±1 mm	±5 ppm ±1 mm
Recommended atmospheric sampling procedure	5	5	5	6	6
Distance measurement procedure					
Eccentric bar 4 positions measurements (0, +, −, 0) 10 REPS each position	Yes	Yes	Yes	Yes	No (10 repetitions on center)
Measure from both ends of the line	Yes	Yes	Yes	No	No
Measure on different days	Yes	Yes	No	NA	NA
Different time of day (morning/afternoon) or (day/night)	Yes	Yes	Yes	NA	NA
Use of conventional tripods	No	No	Yes	Yes	Yes
Use of 4-ft stands	Yes	Yes	preferable	No	No
Minimum distance**	10,000 m†	5000 m†	5000 m	2500 m	1250 m

*Represents the lowest acceptable method.
**EDMI specification of ±(S mm + 1 ppm).
†With frequency monitoring. *Source:* FGCC *Specifications to Support Classification, Standards of Accuracy, and General Specifications of Geodetic Control Surveys,* 1980) Silver Spring, MD.

244

12-10. CHECK ANGLES

It is good practice to include economically justified angle observation in a trilateration network to increase the number of redundancies within the scheme. For first- and second-order, class I networks, only first-order universal theodolites such as the Wild T-3, Wild T-2000S, or Kern DKM-3 should be employed. The level of precision and accuracy required for this order of work precludes the use of any lower-order instruments. First-order instruments on second-order, class II surveys can be useful and productive if suitable equipment and thoroughly trained personnel are available. The Wild 2000, Wild T-2E, Kern DKM-2AE, Aus Jena 010A, or Lietz TM-1A are representative second-order universal theodolites.

Table 12-3 recommends that 4-ft wooden stands (see Figure 12-4b) be utilized for all first- and second-order trilaterations. Table 12-1 lists the number of positions and least count required in various classes of triangulation to be compatible with the trilateration specifications shown in Table 12-2. Targets used in conjunction with check-angle observations should be good quality, with no phase. Ideally, directional lights or 360° lights should be used exclusively (see Figure 12-13).

On lines where the inclination (vertical angle) of the observed line is over 5° from the horizontal, striding-level readings should be made and corrections applied to the observations for this error. Because of their course divisions, most newer second-order theodolites do not have a sufficiently sensitive plate level to allow meaningful corrections.

Older second-order instruments permit a striding-level vial to sit atop (astride) the theodolite vertical axis. For the most part, modern second-order instruments with automatic verticle-circle indexing do not provide a striding level, but can be precisely leveled using special procedures noted in their respective manuals. These methods employ an automatic vertical-circle indexing provision of the theodolite. First-order geodetic theodolites, such as the Wild T-3 and Kern DKM-3, have plate levels sufficiently sensitive to determine corrections commonly called the "correction for inclination of the standing axis." It is a function of the following:

1. Value of one division on the plate (or striding level) is usually in seconds of arc.
2. Number of graduations the standing axis is out of plumb.
3. Tangent of sight-line inclination.

The correction formula for C is

$$C = d(W - E)\tan h \qquad (12\text{-}1)$$

where

C = correction in arc seconds

d = bubble value in seconds of arc per level-vial graduation

W = arithmetic difference of the direct and reverse readings of the west (or left) end of the bubble

E = arithmetic difference of the direct and reverse readings of the east (or right) end of the bubble

h = vertical angle of the sight line to the target, positive if the station is above the horizon, negative if below

Example 12-1. Given: Arc seconds per vial division $d = 6.4''$, inclination of the sight line $h = 10°20'$, and the following bubble readings, compute W, E, and C.

	Circle Left (Direct)	Circle Right (Reverse)
	+	−
Direct	3.5	30.0
Reverse	28.0	1.0

$$W: 28.0 - 3.5 = 24.5$$
$$E: 30.0 - 1.0 = 29.0$$
$$W - E = 24.5 - 29.0 = -4.5$$
$$C = d/4[(W) - (E)]\tan h$$
$$C = 6.4''/4(-4.5)(0.182)$$
$$C = -1.31''$$

The correction applied to the line (direction) on which the inclination of the standing axis is read. If the correction sign is minus, the correction is subtracted; if plus, it is added.

When bubble readings are taken for inclination correction, it is important that the instrument be kept as nearly level as possible, caused by possible inconsistencies in uniformity of the level vial. For the example given, the dislevelment angle is extreme, so the instrument should be releveled before the next round begins. As noted, divisions should be estimated to 0.1 of its smallest graduation.

The recorder can perform a helpful check to guard against errors in reading the plate and striding levels by noting the difference between the left- and right-hand bubble readings. This indicates the relative *length* of the bubble, which should be consistent for both direct and reverse orientations and not disagree by more than 0.3 divisions. If it does not check, then either an observing or recording error has occurred, and the source should be identified by reobserving the bubble if the error is detected during the occupation.

It is possible for the bubble length to vary during a night's work because of changes in temperature and barometric pressures; however, these changes will be slow and practically unnoticeable. Bubble lengths should remain relatively static, enabling an observer to get a feel for the general trend, so errors in readings are spotted instinctively.

12-11. ZENITH ANGLE OBSERVATIONS

Determining elevation differences by employing zenith angle measurements and slope distance is accurate and rapid, provided that (1) the timing of observations is carefully considered, (2) atmospheric conditions are favorable, (3) good sights and observing techniques are employed, and (4) the second-order (or better) universal theodolite is in good repair and ad-

justment. The times of day when refraction is at its worst should be avoided.

Minimum refraction occurs from noon to 3 PM and the maximum from 9 PM to midnight. During the period from noon to 3 PM, refraction tends to be relatively constant. The poorest times for observing are between 8 and 9 AM, and between 6 and 7 PM, because during these periods refraction is changing most rapidly. When simultaneous reciprocal observations are used, the paired observations at each end should be completed within a total elapsed time of 15 min or less to avoid changing refraction conditions. This technique practically eliminates the effects of curvature and refraction.

The following list should be followed in selecting techniques and parameters for observations. It gives the order of preferred methods with the best listed first:

1. Simultaneous reciprocal observations, noon to 3 PM. (Two instruments, combined ET for all observations is $Z = 15$ min.)

2. Reciprocal observations, noon to 3 PM, different days.

3. Simultaneous reciprocal observations, 9 PM to midnight. (Two instruments, combined ET for all observations is $Z = 15$ min.)

4. Reciprocal observations, 9 PM to midnight, different days.

5. Simultaneous reciprocal observations at any other time. (Two instruments, combined ET for all observations is $Z = 15$ min.)

6. Reciprocal observations, with no coordination of time.

7. Nonreciprocal observations.
   - (a) Noon to 3 PM.
   - (b) 9 PM to midnight.
   - (c) Any other time.

Table 12-2 lists the recommended number of repetitions and rejection limits for each class of trilateration. However, users are advised to use 3 D/R or more and a rejection limit of ± 5 arc seconds or less in all classes.

12-12. TRIGONOMETRIC LEVELING

In order to reduce observed slope distances to a common datum, elevations must be known at a line's terminal points. Since trilateration establishes the position of new stations, elevations are seldom known. The most common method of determining elevations is, of course, differential leveling. It is the most accurate but unfortunately the most expensive. An acceptable alternative to differential leveling is trig leveling. Both methods are very successful when sound specifications and operational procedures are followed (see Section 7 for a discussion of trig leveling and the computation of differences in elevations based on zenith angle and slope distance observations).

Table 12-3 lists the number of observations and allowable spread between zenith angle or vertical angle sets, together with the allowable variation from known elevations.

12-13. ELECTRONIC DISTANCE MEASURING INSTRUMENTS

An EDMI is the heart of any trilateration network. It is the workhorse of any modern surveying project and ultimately controls the raw data secured. EDMIs employed in the United States are generally the light-wave and microwave types. Visible-light-wave EDMIs include the Rangemaster and Ranger series marketed by Keuffel & Esser Company and the geodimeter produced and marketed by the AGA Corporation of Stockholm, Sweden. Nonvisible infrared light-wave types are produced by nearly all EDMI manufacturers.

Microwave is a second type of carrier wave. However, the effects of meteorological conditions on this type of electromagnetic wave are very extensive, compared with visible and non-visible light waves, so microwave does not have widespread usage in geodetic control surveying. For example, humidity contributes, as a worst possible case, about $\frac{1}{2}$-ppm error to light-wave-type EDMIs, but a comparable level of humidity can cause 70 ppm or more in microwave instruments.

Most short-range (< 3 km) EDMIs on the U.S. market today employ the gallium-arsenide electroluminescent diode (GA-AS diode) as an infrared light source for the projected beam. No separate light modulation is necessary because the desired intensity or pattern of modulation is obtained directly from the diode by RF (reference frequency) controlled voltage. Some medium-range instruments employ a modified GA-AS diode as a semiconductor lasing infrared light source.

12-14. EDMI INTERNAL ERRORS

EDMI manufacturers list an error statement typically in the form $\pm(A + B)$, where A is expressed in millimeters or decimal feet and B in representative parts per million (ppm). These specified errors are usually ± 1 standard deviation (the standard error or one-sigma, σ) that is the error to be normally expected for a single observation when the EDMI is in good working order and properly calibrated. The expression is an informative message from manufacturer to consumer, indicating the relative reliability to be expected from a single observation. The true value of A as determined by base-line testing should generally be equal to or smaller than the stated standard error.

The range of all errors should never exceed 3σ. If this limit is surpassed, it could indicate that the EDMI is in need of repair or maintenance. The unique error statement for an individual EDMI is not static and will change with time due to the aging of electronic components, faulty optical alignment of the instrument caused by an external shock or blow, or a frequency change resulting from mechanical and/or electronic sources. It is imperative that sufficient base-line calibration checks be made and regular shop maintenance scheduled at a factory-authorized repair facility. Normally, a base-line check should be done after any shop work is completed and at the completion of a project.

In considering the two groupings of internal errors inherent in EDMIs, the first one *A* consists of a family of errors present for all measurements within the operating range of the instrument that is not a function of distance. Thus, the error is the same whether the EDMI is measuring 1 m or 6 km. These internal errors are listed as follows:

1. Instrument offset.
2. Cyclic errors (nonlinear).
3. Instrument resolution.
4. Instrument repeatability.
5. Pointing error.
6. Reflector, instrument calibration.

The second group of internal errors *B* results from a drift of the crystal frequency standard used to control the modulation frequency. Several of the more common causes of this shift are due to (1) loosening of an adjustment screw, (2) electronic aging, (3) extreme temperature conditions in the case of a crystal not provided with an oven or a temperature compensating system, and most commonly (4) inadequate warm-up time to allow the crystal and other electronic components to stabilize. For example, laboratory frequency monitoring has shown that inadequate warm-up time for various infrared and lasing instruments can produce frequency errors of up to 4 ppm or more. Usually, a warm-up period of 20 mm is sufficient for frequency stabilization.

Naturally, good judgement must prevail when determining which lines need a full warm-up period and which do not. A line of 10,000 m, for instance, should get full warm-up since an arbitrary error of 4 ppm would represent $10,000 \times 4 = 0.040$ m. A line of 100 m, on the other hand, would produce only 0.0004 m, which is considerably outside the sensitivity range of most EDMIs.

Consider first the *A* group of internal errors. Items 1 and 6 are systematic in nature and should be treated together. Comparison measurements on an NGS-established base line will give an indication of the sign and magnitude of this error, provided that items 2 through 5 are not excessive. Cyclic error is systematic and normally measured before an EDMI leaves the factory. Data are available from some manufacturers on request. Cyclic error is not constant throughout an EDMI's lifespan and will change with age. A repair facility should be available that can accomplish this calibration, as well as determine instrument resolution, repeatability, and pointing errors by applying appropriate tests. The reduction or elimination of these error sources is a project for a factory or repair facility.

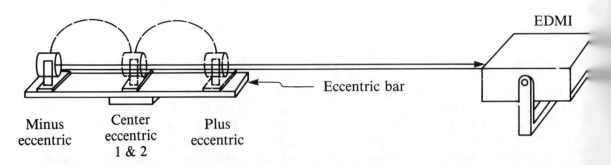

Figure 12-6.   Eccentric measurements.

Instrument resolution can be tested by using an eccentric bar (Figure 12-6). The EDMI and reflector array are set up at a convenient distance of approximately 300 m and multiple measurements taken to the prism on center, repointing for each shot, while very carefully noting maximum signal response. The prism is moved to a minus eccentric position and a similar series of shots taken, then turned to a plus eccentric position, and finally back to center. An average of the minus eccentric measurements reduced by the offset distance should equal the centers' average. The plus eccentric measurements' average added to the offset distance must also equal the average of the centers. Any difference between the measured center—those centers determined by applying the offset correction to the minus and plus eccentrics—should agree to within a millimeter or smaller. (See Figure 11-14a for a detail of the eccentric bar.)

The measurement procedure and check of conditions in using the eccentric bar are as follows:

1. Order of measurement is CTR1, $-$, $+$, CTR2.

2. Each eccentric position is read 10 times.

3. Repoint the EDMI after each set of five shots.

4. Center ECC_1 must equal center ECC_2 within $\pm A$ group error (see Section 12-15).

5. Minus eccentric $-$ the offset distance (0.150 m) must equal center ECC_1 within $\pm A$ group error.

6. Plus ecentric $+$ the offset distance must equal center ECC_1 within $\pm A$ group error.

7. When all three checks are made, all four observations are averaged.

8. If A group errors are shown to be ± 5 mm, then the corrected plus and minus eccentric shots should be equal to or smaller than center ECC_1 ± 5 mm.

Instrument repeatability and pointing error can be checked during the same setup for instrument resolution. The repeatability test is performed by pointing the instrument for maximum signal response. Then activate the EDMI and record a series of multiple measurements without disturbing the instrument pointing. There would be no change of distance on the display greater than $\pm$ one least-count unit. Statistically, 95% of the measurements should remain unchanged, with 5% having no greater difference than $\pm$ one least-count number on the display.

Pointing error is an apparent change in the distance display caused by the bundle of light rays returning to the EDMI not focusing at the proper location on the diode. The technique for determining the magnitude of pointing error is shown in Figure 12-7.

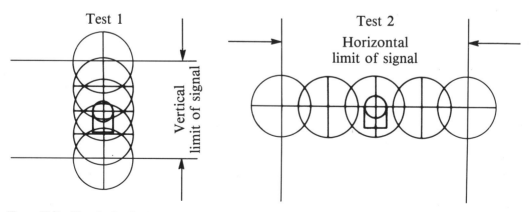

Figure 12-7.   Magnitude of pointing error.

Test 1 begins by maximizing the signal horizontally with the horizontal cross hair below the prism. Using the vertical motion, drop the cross hair until the signal fades out. Then raise the vertical wire until the signal is strong enough to cycle the EDMI and record the distance and vertical angle (or vote the vertical position of the cross hair with respect to the prism). Continue to test with approximately five distance measurements until the signal is lost on the upper side of the prism. Test 2 is identical to test 1, except the signal is maximized vertically, and the EDMI turned in azimuth left to right with the horizontal motion. Distance measurements are recorded, and the vertical cross-hair position is noted with respect to the prism.

If there is a change in the distance readout within ± one prism width or height, special pointing techniques should be standardized for the EDMI to normalize all distance measurements. It is very important that the pointing technique used in a repair facility for calibration be identical to the procedure employed on the base line and in the trilateration net. The technique chosen should be mutually decided on by the repair technician and operator in order to capitalize on the unique design characteristics of an individual EDMI and the way the instrument will be utilized in the field.

A technique that proved very successful on the HP 3800A and HP 3810B, and should work on all EDMIs of similar design for systematic pointing, is as follows: (1) Using the vertical wire, slit the prism (or center the vertical cross hair on the maximum signal, using the horizontal motion) and drop the cross hair vertically until the signal stops. (2) Elevate the cross hair very slowly with the vertical motion only; watch the signal return and stop at the point of the *first* maximum signal return.

If test 1 and 2 do not show a change in distance with ± one prism width or height, the standard pointing technique recommended by the manufacturer should be followed.

It is not necessarily possible to totally eliminate the first group of internal errors (group A), but they can be minimized. Items 1, 2, and 6 are systematic and may approach zero, but 3, 4, and 5 can be reduced by a savvy repair technician. If item 5 is a problem, it can be eliminated with better pointing techniques.

The second element to consider is the *B* group of internal errors—those caused by a drift in the frequency standard of the crystal used to control the modulation frequency. If the instrument is thoroughly warmed-up and the crystal frequency is part of a temperature compensating system, the frequency error is minimal; however, all frequency errors are systematic. The resulting distance error can be corrected if the magnitude of frequency error is known. On higher-order projects, it might be advisable to monitor the frequency periodically throughout the job, with an appropriately sensitive frequency counter. On other projects, checking the frequency at its beginning and end is sufficient.

A good policy to follow immediately prior to initiating a trilateration project is to send the EDMI to a reputable shop for maintenance and calibration, with special emphasis given to the items just discussed. After shop work, the next step in the calibration sequence is a trip to an NGS-established base line for further testing and calibration. It is imperative that the same techniques of distance measurement are utilized on the base line as on the trilateration network, preferably with the same prism or prism type and the same observer.

A series of measurements is taken (usually 10) with repointing after each group of five in all four eccentric positions of the eccentric bar (center, minus, plus, then center). After the first line is measured, the prism is advanced to the next baseline monument. When every line has been measured and all refractive index corrections and systematic errors are applied, each measured distance is compared with that established by the NGS and the difference computed for every line. The resulting error is tabulated and a standard deviation computed.

If a systematic error remains, it will show up by analyzing the variances. The scatter should be smaller than ± 5 mm, and roughly an equal number of lines will measure short as long. If the variances are predominately plus or predominately minus, the error plot should be adjusted up or down to establish a line centered on zero error. The distance a graph moved up or down is then treated as another total offset correction for measured distances.

The error plot should not have a sloping shape—i.e., an error with respect to distance showing a rate of change, since this suggests the presence of a frequency error.

Selecting an EDMI for trilateration is not a simple task: There are many variables to consider, such as the following:

1. The range of distances within which the EDMI will be utilized.
2. The manufacturer's stated error for the EDMI.
3. Reliability and reputation of the manufacturer.
4. Durability and accuracy of the EDMI, as reported by users and repair facilities.
5. Temperature range within which the EDMI will maintain its design error specification — i.e., is its crystal temperature compensated to maintain frequency standard through a wide range of temperatures?
6. Price.

A common error specification claimed by various manufacturers for medium-range infrared EDMIs is $\pm (5$ mm $+ 5$ ppm$)$. The EDMI specifications are not adequate for third-order, class II trilateration as demonstrated in Example 12-2.

Example 12-2. Can an EDMI with a specification of $\pm (5$ mm $+ 5$ ppm$)$ be utilized on a third-order, class I trilateration project in which the minimum distance is 3000 m?

The answer is no! Referring to Table 12-3 under ppm (total error budget), we can see

that third-order, class I has a total error budget of ± 4 ppm. The B group error statement of ± 5 ppm as claimed by the manufacturer already exceeds the allowable error of ± 4 ppm for third-order, class I, without even considering the A group errors, centering errors, or refractive index correction (RIC) errors.

Example 12-3. Can the same EDMI be used on a third-order, class II project with the minimum distance being 3000 m?

If we assume that the maximum EDMI error does not exceed the manufacturer's claimed error statement, and further, that the ppm's listed for T, P, e^1, and centering, etc., per Table 12-3 are held constant, then

STEP 1. Convert the A group error to ppm and combine with B group error.

$$\text{ppm}_{\text{EDMI}_E} = \pm (5 \text{ mm} + 5 \text{ ppm})$$

$$A \text{ group error}_{\text{ppm}} = \frac{0.005 \text{ m} \times 10^6}{3000 \text{ m}}$$

$$= 1.67 \text{ ppm}$$

A group error$_{\text{ppm}}$ + B group error$_{\text{ppm}}$

$$= \pm (1.67 \text{ ppm} + 5 \text{ ppm})$$

$$= \pm 6.67 \text{ ppm}$$

Therefore, $\text{ppm}_{\text{EDMI}_E}$ at 3000 m $= \pm 6.67$ ppm.

STEP 2.

$$\Sigma \text{ppm}_E$$

$$= \sqrt{\text{ppm}_{\text{EDMI}}^2 + \text{ppm}_T^2 + \text{ppm}_P^2 + \text{ppm}_e^2 1 + \text{ppm}_{CE}^2}$$

where Σppm_E is the total error budget in ppm, ppm_{EDMI} the error statement for the

Table 12-4. Parts per million changes in length for pressure and temperature

	Pressure error or change of
	±0.10″ HG
	±100′
1 ppm change in length	±2.5 mm HG
	Temperature error or change of
	±1.8°F
	±1.0°C

EDMI as determined by baseline calibration and laboratory measurement of frequency, ppm_T the estimate of temperature measurement uncertainities in ppm (see Table 12-4), ppm_P the estimate of barometric pressure uncertainities in ppm (see Table 12-4), $ppm_e 1$ the estimate of vapor pressure uncertainities (humidity) in ppm (see Table 12-4), and ppm_{CE} the centering error in ppm (see Tables 12-3 and 12-5).

$$\left(\frac{10^6 \sqrt{0.001^2 \, m + 0.001^2} \; m}{3000 \; m} \right.$$

$$= \text{ppm for} \pm 1 \text{ mm centering error} \left. \vphantom{\frac{10^6}{3000}} \right)$$

$$\Sigma ppm_E = \sqrt{6.67^2 + 2^2 + 2^2 + 0.4^2 + 0.5^2}$$

$$\Sigma ppm_E = \pm 7.3 \text{ ppm}$$

The EDMI specifications are not adequate for third-order, class II trilateration. An in-spection of the data shows that the element contributing most to the high error of ±7.3 ppm is the *B* group specification concerning frequency. If it is possible to lower this source of error by reading the actual frequency and calculating corrections, or adjusting the frequency to near perfection, this EDMI would be capable of both third-order, class II and class I measurements at a minimum distance of 3000 m.

If the error specification can be adjusted to ±(5 mm + 1 ppm), the following improvement could be realized:

$$ppm = 2.67^2 + 2^2 + 2^2 + 0.4^2 + 0.5^2$$

$$pp = \pm 3.94 \text{ ppm}$$

The new ppm resulting from improving the error specification of the EDMI now has shown dramatic results, improved its predicted per-

Table 12-5. Distance versus centering error

	Centering Error of ±1 mm		Centering of ±1/2 mm	
Distance (m)	$ppm^{\pm}_{CE}$	Error Ratio 1 ± Part in ...	$ppm^{\pm}_{CE}$	Error ratio 1 ± Part in ...
10,000	0.1	7,071,068	0.1	14,142,136
5000	0.3	3,535,534	0.1	7,071,068
2500	0.6	1,767,767	0.3	3,535,534
1000	1.4	707,107	0.7	1,414,214
500	2.8	353,553	1.4	707,106
300	4.7	212,132	2.4	424,264
200	7.1	141,421	3.5	282,842
100	14.1	70,711	7.1	141,422

formance in line measurement, and conforms to third-order, class I specifications.

12-15. EDMI EXTERNAL ERRORS

A number of factors causing errors in EDMI measurement will be discussed.

12-15-1. Refractive Index Correction

Table 12-4 lists the magnitude of errors in pressure and temperature sampling that cause a ± 1-ppm change in the distance. Chapters 4 and 6 deal with the equations for determining the refraction index correction resulting from changes of temperature, atmospheric pressure, and water vapor pressure, from the standard. A discussion of these equations will not be repeated. However, a discussion on sampling techniques will be addressed.

The accuracy and precision of an EDMI measurement are a direct function of how carefully the atmospheric conditions are determined along the beam path. It is typical procedure to simply measure temperature at the EDMI height and pressure with an inexpensive barometer, probably uncalibrated, and ignore the effect of water-vapor pressure. This method will not be effective for even the lowest order of trilateration unless distances are very short.

12-15-2. Temperature Measurement

Measuring temperatures with a thermometer gives figures only for the instrument and not necessarily the air temperature. Unless shielded from sunlight, a thermometer is heated by radiation and shows an erroneous temperature. A thermometer carelessly hung on a low bush can pick up thermal radiation from the ground, as will dangling it from the survey truck's door handle or exterior mirror.

Nearby heated objects affect temperature readings. If we assume that the thermometer is shielded from radiation, standardized, and the sight line from the EDMI to the reflector is equally distant above the ground throughout the beam length, sampling tripod-height temperatures at both ends of a line should be adequate. Shorter distances will ensure smaller temperature errors. Sight lines are seldom a constant height above the ground. Within the first 30 ft above the ground, temperatures can vary as much as 9°C (16.2°F) or more. This corresponds to a 9-ppm error if a ground temperature is erroneously used.

Isotherms are imaginary lines connecting points of equal temperature. A characteristic of an isotherm is that it generally tends to follow the ground profile. Typically, an isotherm for a given temperature is likely to be at a lower height above the ground on a hilltop than on the valley floor. Figure 12-8 is a graphic representation of this phenomenon.

In most situations, the temperatures will be more accurate if taken on each end of the line at a height of 25 to 30 ft above the EDMI and reflector. The most representative temperature samplings are obtained on overcast days, with light to moderate winds mixing the atmosphere.

The primary source of error in determining the *index of refraction correction* is the difference that exists between the observed temperature and actual average temperature over the line's length. A typical sight line passes from one high point to another, or from a high point to a low point at variable heights above the earth's surface. Since these variations in height above the ground occasionally have large differences and least since in atmospheric temperature varies with altitude, temperature measurements at the end points may not create an accurately representative model. Hence, scheduling field operations during periods when meteorological conditions are more conducive to accurate temperature sampling will provide improved representative temperature models.

Figure 12-9 charts temperature versus height above the ground surface and time of day in

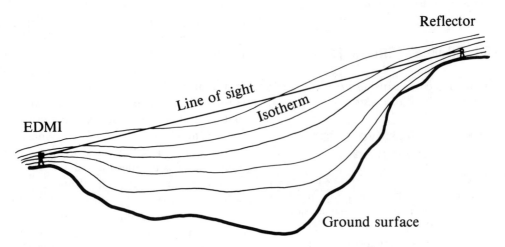

Figure 12-8.   Isotherms.

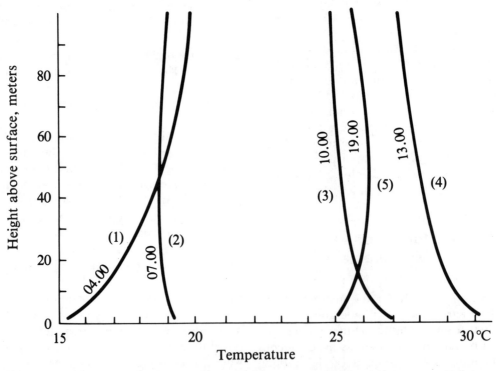

Figure 12-9.   Daily cycle of temperation variation. (Adapted from P. S. Carnes, 1961). *Temperature Variations in the First 200 Feet of the Atmosphere in an Arid Region*, Missile Meteorology Division, U.S. Army Signal Missile Support Agency, New Mexico.

lowest 100 m at White Sands Missile Range, New Mexico. At an air sampling height of 10 m—e.g., an average line height above the ground of 40 m—the largest error in temperature during the daytime is 1°C (1 ppm) around 1300 hr, and the smallest error occurring in the daytime is $\frac{1}{4}$°C ($\frac{1}{4}$ ppm) around 0700 hr.

Given overcast conditions with sufficient wind to stir and mix the air, the graph approaches 0700, as shown in Figure 12-9, where near-ground temperatures in the 10 m range would closely approximate those temperatures much higher up (to 80 m $\pm$). If elevation differences are extreme and weather conditions not conducive to good sampling at 10 m, other tactics must be employed to determine the refractive index corrections, such as a two-color laser EDMI, multiwavelength distance-measuring instrument (MWDMI); the ratio method; or aircraft monitoring, etc. Using temperature-sampling procedures at the 10-m height should suffice for second-order, class II and lower.

The following conditions should be considered before writing specifications for field operations:

1. Night temperatures will tend to be cooler near the ground than above it. Consequently, if the average height of the sight line above ground is higher than the sampling of height, temperature readings taken at night under clear skies will be cooler than the true average sight-line temperature, resulting in EDMI distances that are shorter than the true lengths.

2. Daytime temperatures tend to be warmer near the ground than higher up. If the average height of the sight line above ground is higher than the sampling height, temperature readings taken during the daytime under sunny skies will be warmer than the true average temperature along the sight line. Therefore, the resulting EDMI distance is longer than the true distance.

3. Overcast days or nights, with moderate winds, provide the best conditions for accurate temperature sampling.

4. Temperature and pressure samplings are most representative when taken along the line of sight.

The following list notes the preferred methods of sampling the atmosphere to determine the refractive index correction. Items tallied are shown in the order of accuracy from best to worst. The methods correlate with Table 12-3.

1. MWDMI.

2. Ratio method.

3. Aircraft monitoring of temperatures and pressure along the line of sight.

4. Balloon-suspended temperature thermistors.

5. Thermistors and aspirators erected on a mast at the EDMI and reflector.

6. Temperature and pressure samplings at the height of the EDMI and reflector.

7. Temperature and pressure sampling at the EDMI only.

Temperature readings can be taken with thermometers, but such use is not recommended except in measuring relative humidity. Portable digital-reading battery-operated temperature indicators manufactured by Weathertronics Qualimetrics, Inc. are relatively inexpensive. Such temperature sensors are accurate to ± 0.1°F or 0.16°C and have ventilated sensors and cables that can reach short distances, up to 50 ft. Also available are similar units that measure both temperature and relative humidity.

Some fabrication is necessary to properly utilize the temperature-sensing equipment. First, an appropriate aspirator must be designed that has the following features: (1) circulates air past the temperature sensor, (2) shields the sensor from solar radiation, (3) is lightweight, and (4) is mountable on an extending mast. Second, an appropriate extending mast must have the following features: (1)

is lightweight, (2) is extendable to 30 ft or higher, and (3) is mountable on a truck bumper or pickup bed.

12-15-3. Aspirator

A design was developed incorporating the use of a 5-in. PVC elbow with a reducer to 4 in. on one end and 12-in. section of 4-in. PVC glued into the reducer. A 12-V DC motor with fan was mounted inside the elbow and a piece of $1\frac{1}{2}$-in. PVC was glued to the 12-in section of the 4-in. PVC to act as a receiver for the mast. Then, all were covered with insulating foam to guard against solar radiation. A hole was bored to receive the temperature-indicator sensor and placed to position it in the aspirator's airstream. The motor was powered via a long extension attaching to the truck battery with alligator clips; it is designed to exhaust from the elbow when the sensor was placed upwind from the motor (see Figure 12-11).

Figure 12-10.   12-V aspirator; extended mast; truck-mounting assembly; meteorological box.

12-15-4. Mast

The mast is an aluminum extension pole originally designed as a handle for a tree pole saw. It consists of five or more sections of aluminum tubing, 5 to 6 ft in length, with different diameters to permit the unit to "telescope." Each section is held aloft by a friction collar. (See Figure 12-10.) Use caution when extending the mast. *Do not extend the mast when in the vicinity of power lines.*

12-15-5. Truck Mounting

A piece of $1\frac{1}{2}$-in. pipe, 36 in. long with a $1\frac{1}{2}$-in. floor flange bolted to a truck bumper, front or rear, or bolted to the bed of a pickup, serves as a holder for the mast and aspirator. The aspirator, mast, and truck mount can be assembled using locally available materials. The temperature assembly described has been used, under ideal conditions, with good results in performing first-order trilateration; however,

the project was a minigeodetic network consisting of lines no longer than 4 km and noncritical changes in elevation. The described system is adequate for second-order, class II and lower-order surveys. Under favorable atmospheric conditions—overcast, light-to-moderate winds—the system can exceed second-order, class II specifications. (See Figure 12-11.)

12-15-6. Barometric Pressure Measurement

Barometric pressure must be measured at each end of the line with a sensitive, high-quality temperature-compensated barometer. Also, the barometer should reliably differentiate the least division shown in Table 12-3 under atmospheric sampling accuracy requirements for barometric pressure. The barometers should be periodically calibrated against a mercury column, preferably at the beginning and end of the project—more often if the

Figure 12-11. Another example of instruments shown in Figure 12-10.

project lasts longer than several weeks or the barometers have been subjected to hard usage. (See Figure 12-12.)

During the project, a daily check should be made on all barometers used for line measure-ment operations to determine if they are read-ing the same ($\pm$ one least count). For exam-ple, if one of the instruments does not agree with the others, it can be safely reindexed in the field to match them. If more than one barometer reads differently, all should be re-calibrated before proceeding with the project. Surveying barometers are delicate precision instruments and should be shaded from the sun and treated with the same care due any scientific equipment.

12-15-7. Humidity

The effect of humidity on visible light and infrared EDMIs is admittedly small and does not exceed 1×10^6 under the worst condi-tions. However, second-order, class I and first-order trilateration must have atmospheric observations made at each end of the line to determine the size of correction for water-vapor pressure e^1. The equipment is inexpen-sive and consists basically of two thermome-ters, one a regular unmodified thermometer and the other adapted to include an appara-tus, usually a sleeve with an absorbent material that is moistened with water.

Both thermometers are usually aspirated by a small battery-powered fan to evaporate the water and cool the thermometers to an appar-

Figure 12-12. Meteorological box contain-ing surveying barometer, thermometer, and electronic temperature-sensing equipment with 40-ft cable.

ent lower temperature when a difference between the thermometers is noted. A correction is then computed by the equation given in Section 12-4. A choice is available between a conventional sling psychrometer or newer state-of-the-art portable battery-powered types with digital readout and electronic sensor.

12-15-8. Electronic Interference

Care must be taken to avoid potential EDMI errors when operating in the proximity of transmission lines, microwave, or radio communication. Radio transmission also can severely affect electronic atmospheric-sensing equipment, so all such transmission should be shunned as well during temperature measurements.

12-15-9. Reflector Correction

Reflector correction for individual differences between units due to slight differences in their offsets has been discussed previously in Section 12-14. The reflector correction is both a systematic and an external error, relative to the EDMI, with a sign that can be either plus or minus.

12-16. REFLECTORS

Any reflecting surface can be used from a common shaving mirror to precision-ground planoparallel optical surfaces, highway reflectors, reflective tape, mercury surfaces, and, of course, the conventional retroreflecting prism, provided that the following conditions are met: (1) The exact reflector constant/offset is known, (2) the reflected light source is sufficiently strong to enable an EDMI to function within the expected accuracy and precision standards, and (3) the reflected beam is free of ambiguous stray light from a surface other than the intended interior ones of the reflector.

Typically, surfaces such as reflective tape, plane mirrors, and highway reflectors are offset no more than 1 mm. On the other hand, retroreflective prisms vary from -40 mm to -70 mm to -30 mm, depending on the manufacturer. It must be emphasized that these offsets are *nominal*. The true offset can be in disagreement by as much as ± 1 mm to ± 3 mm by actual measurement. If the project standards—i.e., error budget—can tolerate this uncertainty, then by all means disregard the true offset and employ the nominal. However, it is prudent surveying practice to know the exact magnitude of an individual reflector offset. This can be measured in combination with calibration tests given an EDMI while on the test base line, but only if the EDMI has sufficient resolution to measure a least count of ± 0.001 ft, and can reliably detect relative displacements of the same size.

The HP 3800 has been utilized for this purpose, with good results. This test is performed at relatively short distances of about 300 ft.

The same reflector submitted with an EDMI for laboratory calibration, zeroing (making the total offset correction equal zero), and later taken to a base line for validation measurements is selected for standardizing all other reflectors which need calibration. After approximately 20 measurements are made to the calibrated reflector, all others are inserted successively and the measurements repeated. The difference between the mean of the first measurements using a calibrated reflector is subtracted from those taken with each individual unit. The result is a reflector correction *RC* added to or subtracted from the slope distance to correct for the offset difference from the standardized prism.

The next step is to (1) run the trilateration EDMI and calibrated reflector through an NGS base line to validate the total offset correction *TOC* established during laboratory calibration earlier or (2) determine the magnitude and sign of the *TOC* for the EDMI and calibrated reflector as a unit in the event there has been a change.

Multiple shots are taken, with the EDMI being repointed after each distance measurement with a sufficient number of repetitions read to arrive at a realistic mean value. Then the difference is computed between the mean of the measured distance minus the NGS-established distance published for the base line. This results in one *TOC* determination for each base-line distance. Several different distances on the base line must be measured to ensure that the best value of the EDMI/reflector combination *TOC* has been determined. Once the *TOC* and reflector correction for the calibrated reflector are known, the corrected slope distance for a line is computed by the following relationship:

Corrected slope distance = measured slope distance
$$\pm\ TOC \pm RC$$

Each reflector should be individually numbered for identification. The corresponding value of *RC* established for the specific reflector is then recorded for future use. Usually, it will be observed that all reflectors from a manufacturer tend to have *RC*s approximately the same size. Differences of as much as 7 mm have been found between reflectors of different manufacturers, although each claims to have the same offset. Caution, therefore, should be exercised when mixing reflectors, unless the *RC* value is known.

If maximum-rated distances are to be measured, the reflectors must be premium quality, clean, reasonably accurately pointed, and in good repair. For lines inclined more than 15°, the use of tilting prisms is recommended.

Centering an EDMI and reflectors over the marks is extremely important and requires the tribrachs to be in perfect adjustment; otherwise, precision calibration will be meaningless. Centering the reflector over a sufficient number of base-line monuments ensures getting the best possible value for the instrument/reflector combination *TOC*.

12-17. CENTERING

The Wild-type tribrachs with attached optical plummet, manufactured by Wild, Geotech, Lietz, Topcorn, and others, are very strong, well-designed, and durable. However, unless reasonable care is taken in transporting and day-to-day handling during field usage, the bull's-eye bubble and mechanical adjustment system of the optical plummet can be jarred out of calibration. Extreme variations in temperature contribute to shifts in the tribrach zero, owing to extremely unbalanced tension on the adjusting screws. Read the manufacturer's adjustment instructions carefully and then follow them precisely. The tolerable centering error ± 1 mm. (0.003') more or less, depending on the class of survey and line length (see Table 12-5 on page 252).

Since centering errors are only a part of the total error allowed in distance measurements, it naturally follows that the influence of centering errors on the total must be held to a minimum. Table 12-5 serves as a guideline for determining when distances become critical in considering the effect of centering and establishes centering standards for project control specifications. For instance, Tables 12-2 and 12-4 show that trilateration standards for a second-order, class II distance of 500 to 1000 m *might* be attained if the centering is done to $\pm 1/2$ mm or lower. As a practical note, it is very difficult to adjust a Wild-type tribrach finer than 1/2 mm because the cross hairs on some optical plummets appear nearly 1/2 mm wide at nominal tripod heights.

If extremely high precision is required on short lines, necessitating a centering error of 0.5 mm or smaller, operators should incorporate a centering-rod-type system instead of the familiar tribrach style with a fixed optical plummet. Centering-rod types are currently manufactured by both Wild and Kern. Other tribrach designs allow the optical plummet to be rotated in azimuth while viewing the cross hair, and they can be centered to within 1/2 mm without much difficulty.

The following is a list of general centering guidelines:

1. Check the bull's-eye level bubble optical plummet twice each week or more often as conditions dictate.

2. Tribrachs should be stored and transported —even from station to station—in a suitably protective container that safeguards the tribrach from moisture, physical shocks, and vibration.

3. Do not press the surface of the bull's-eye bubble or touch the vial unless the bubble is being adjusted.

4. When leveling for final centering, look squarely down on the bull's-eye to avoid any parallax and make absolutely sure the bubble is concentric with the reference circle on the glass.

5. If a tribrach is dropped or given any form of physical shock, it must be checked before being used for a distance or angle measurement.

6. Follow the manufacturer's instructions for centering adjustment. Pay particular attention to the tightening and loosening sequence for capstan screws and locks.

7. In some cases where high precision is important, the use of a level vial, such as one in a tribrach carrier or theodolite, can be utilized for leveling and centering the tribrach. This is always superior to using the bull's-eye on the tribrach.

Example 12-4. Given a centering error of ± 1 mm and line of 5000 m in a length, what is the probable effect of centering errors, expressed in parts per million (ppm) and error ratio (*ER*)?

To find a solution, two setups are required for each line: one at the EDMI station and another at the reflector station. If we employ the standard statistical method of error propagation, which states that the final error is equal to the square root of sums of the squares of the individual errors, the following expressions apply:

$$\text{ppm}_{CE} = \frac{10^6\sqrt{CE_1^2 + CE_2^2}}{D} \qquad (12\text{-}2)$$

$$ER = \frac{D}{\sqrt{CE_1^2 + CE_2^2}} \qquad (12\text{-}3)$$

where *ER* is the error ratio expressed in the familiar form " 1 part in ... " resulting from the assigned centering error, ppm_{CE} the parts per million resulting from the assigned centering error, *D* the measured distance, CE_1 the assigned centering error of the EDMI, and CE_2 the assigned centering error of the reflector. Then for $D = 5000$ m,

$$\text{ppm}_{CE} = \frac{10^6\sqrt{0.001^2 \text{ m} + 0.001^2 \text{ m}_c}}{5000 \text{ m}}$$

$$= \pm 0.28 \text{ ppm}$$

$$ER = \frac{5000 \text{ m}}{\sqrt{0.001^2 \text{ m} + 0.001^2 \text{ m}}}$$

$$= 3,535,534 \text{ or } 1 \text{ part in } 3,500,000 \text{ parts}$$

Table 12-5 tabulates a centering error of ± 1 mm and $\pm 1/2$ mm, respectively, utilizing Equations (12-2) and (12-3) with distance as the argument.

12-18. INSTRUMENT SUPPORTS

NGS-type instrument supports should be employed for first-order and second-order, class I control surveys. These so-called 4-ft stands are economical and easy to fabricate and provide a superbly stable instrument and target base for measuring both distances and angles. (See Figure 12-14b for construction details.) According to Table 12-3, standard tripods are advisable for only third-order, class I, class II and perhaps second-order, class II projects. If a theodolite setup is required for check-angle measurements on second-order, class II, or third-order, class I and II, it is advisable to construct a 4-ft stand for theodolite occupations.

It is possible to forego the use of stands in pure trilateration, if available conventional tripods are (1) in good repair; (2) in good adjustment; and (3) substantial and sturdy enough to support the EDMI, reflectors, etc., without displacement from weight and/or wind.

If any angle observations are contemplated, 4-ft stands should be built. The increased accuracy and precision, plus ease of setting up, will offset the time and cost of their construction. An additional advantage of employing the stand is that its height above every mark remains constant throughout the project. Thus, an instrument or reflector height always equals the height of the stand above the mark plus the incremental height of the instrument or reflector above the stand. After the stand height has been measured and checked, it need not be remeasured throughout the project, as would be required each time a new setup is made with a conventional tripod.

It is always good practice to measure all heights in feet and meters, independently, to provide a necessary cross-check. It is not difficult to obtain figures that agree within ± 1 mm (0.003 ft). High-order, small-scale precise trilateration networks (minigeodetic networks) of 2 mi or shorter require precision in measurement of the instrument and reflector heights above the marks. In trig leveling, the accuracy of HIs and HSs are especially critical on short lines. The heights of instruments and sights should be taken to ± 1 mm. On lines longer than 2 mi, the requirement can be relaxed to ± 1 cm or more, depending on precision needs.

12-19. ACCESSORIES

If 4-ft stands are used for instrument supports, several items are needed to attach the angle- and distance-measuring equipment to the stand. The suggested designs shown on Figures 12-13 and 12-14a have evolved through many years of trial and error and are patterned from equipment manufactured by the NGS.

Tribrach Plate

The tribrach plate is attached by wood screws directly to the wooden cap of the stand. Before fastening it, the plate is centered exactly over the monument with a vertical colli-

mator or the optical plummet of a tribrach, and then screwed down and checked. The plate is constructed of $\frac{3}{8}$-in. aluminum stock and cut triangular in shape, 14 in. on each side, with a $\frac{5}{8}$-in. hole drilled slightly oversized in the center of the plate. Additionally, six holes are drilled around the perimeter and countersunk to allow loose passage of no. 8 wood screws, while permitting firm attachment of the tribrach plate to the stand's wooden cap.

Attachment Bolt

The attachment bolt is a $\frac{5}{8} \times 11 \times 1$-in. hex-head bolt, which serves as a means of attaching all prisms, lights, and tribrach of the eccentric bar to a tribrach plate. The only machining necessary is a $\frac{5}{16}$-in. hole bored through the bolt from end to end (Figure 12-9). This hole allows the optical plummet sight line on the tribrach to pass through the bolt and permit viewing the mark below.

Eccentric Bar

The eccentric bar is constructed of $\frac{5}{8} \times 2\frac{3}{4} \times 15$-in. aluminum stock (Figures 12-10 and 12-11). Three $\frac{5}{8}$-in.-diameter holes are bored on the centerline at a spacing of 0.492 ft. (0.150 m). The spacing of 0.150 m was chosen to keep the bar a manageable size and weight. The purpose of the eccentric bar is to (1) provide redundant measurements, (2) provide an internal check on the EDMI, (3) detect blunders, and (4) check EDMI resolution.

Stud Bolt

The purpose of a stud bolt is to secure the prism in the eccentric bar. It is a $\frac{5}{8} \times 11$-in. threaded bolt $1\frac{1}{16}$-in. long, turned to a proper diameter to allow a slip-fit into each hole in the eccentric bar. The slip-fit must not permit any wobble or lateral movement of the prism. The bolt is designed with one threaded end to screw into a prism case, and one unthreaded end to allow the prism to be lifted vertically and changed from one eccentric position to another.

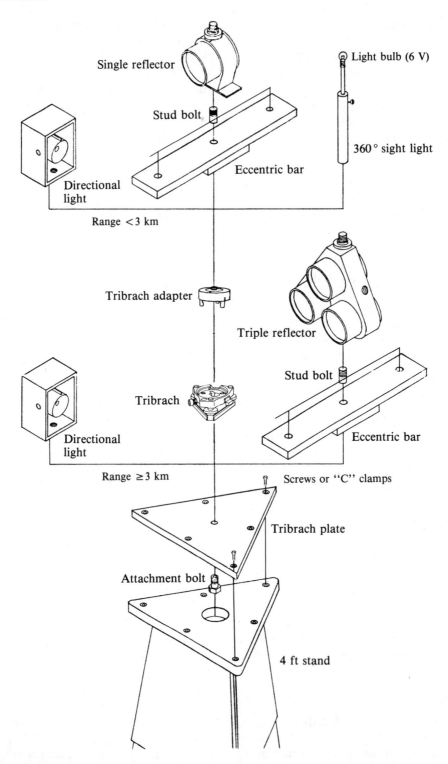

Single reflector

Light bulb (6 V)

Stud bolt

360° sight light

Directional light

Eccentric bar

Range <3 km

Tribrach adapter

Triple reflector

Stud bolt

Tribrach

Directional light

Eccentric bar

Range ≥3 km

Screws or "C" clamps

Tribrach plate

Attachment bolt

4 ft stand

Figure 12-13. Survey target system.

TARGET SYSTEM DETAIL

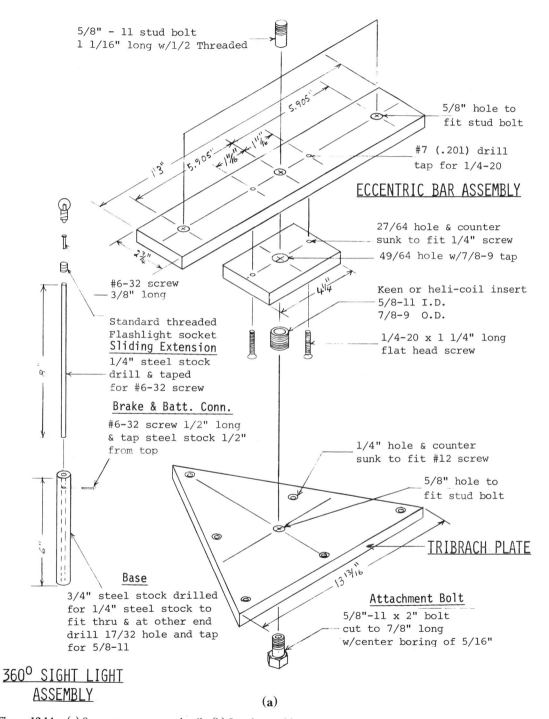

5/8" - 11 stud bolt
1 1/16" long w/1/2 Threaded

5.905"

5.905"

1⁄3"

1 1⁄16"

1 1⁄16"

5/8" hole to
fit stud bolt

#7 (.201) drill
tap for 1/4-20

ECCENTRIC BAR ASSEMBLY

27/64 hole & counter
sunk to fit 1/4" screw

49/64 hole w/7/8-9 tap

Keen or heli-coil insert
5/8-11 I.D.
7/8-9 O.D.

1/4-20 x 1 1/4" long
flat head screw

2¾"

4¼"

#6-32 screw
3/8" long

Standard threaded
Flashlight socket
Sliding Extension
1/4" steel stock
drill & taped
for #6-32 screw

8"

Brake & Batt. Conn.
#6-32 screw 1/2" long
& tap steel stock 1/2"
from top

1/4" hole & counter
sunk to fit #12 screw

5/8" hole to
fit stud bolt

TRIBRACH PLATE

6"

13 13⁄16"

Base
3/4" steel stock drilled
for 1/4" steel stock to
fit thru & at other end
drill 17/32 hole and tap
for 5/8-11

Attachment Bolt
5/8"-11 x 2" bolt
cut to 7/8" long
w/center boring of 5/16"

360° SIGHT LIGHT
ASSEMBLY

(a)

Figure 12-14.   (a) Survey target system details. (b) Stand assembly.

4 FOOT STAND

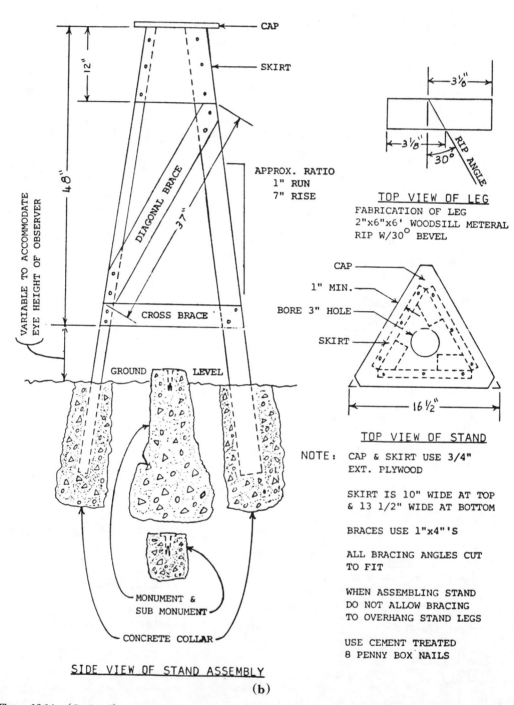

APPROX. RATIO
1" RUN
7" RISE

TOP VIEW OF LEG
FABRICATION OF LEG
2"x6"x6' WOODSILL METERAL
RIP W/30° BEVEL

TOP VIEW OF STAND

NOTE: CAP & SKIRT USE 3/4"
EXT. PLYWOOD

SKIRT IS 10" WIDE AT TOP
& 13 1/2" WIDE AT BOTTOM

BRACES USE 1"x4"'S

ALL BRACING ANGLES CUT
TO FIT

WHEN ASSEMBLING STAND
DO NOT ALLOW BRACING
TO OVERHANG STAND LEGS

USE CEMENT TREATED
8 PENNY BOX NAILS

SIDE VIEW OF STAND ASSEMBLY

(b)

Figure 12-14. (*Continued*).

360° Sight Light

The 360° sight light is designed to be used as a triangulation target. The light is powered by a 6-V battery source with a rheostat-controlled power lead. It provides a very superior target for short- to medium-length lines. The maximum distance for daytime use is approximately 1 km; for nighttime work, approximately 6 km or more with full power.

The sight light consists of two basic parts: (1) the base and (2) a sliding extension. The base, a $\frac{3}{4}$-in.-diameter steel stock, bored $\frac{1}{4}$-in.-diameter longitudinally and on one end bored and threaded to $\frac{5}{8} \times 11$ in. for attaching to a standard tribrach adaptor, or for connecting directly to the tribrach plate by the attachment bolt. Additionally, the base is drilled and tapped for a no. 6×32 screw, which is used for a brake on the extension, and as an electrical pole for one side of the battery. The second principal part is a sliding extension, turned for a smooth slip-fit into the base, with one end drilled and tapped to receive the 6×32 screw attaching the threaded flashlight socket to the extension.

On long lines, where centering can be more lax, the show lights or reflectors may be bolted directly to the tribrach plate if the plate is sufficiently level to prevent eccentricities detrimental to the desired classification standard being followed. Typically, the 4-ft stand cap should be set level when the stand is installed, regardless of the project accuracy requirement.

These accessories are inexpensive to fabricate and require only simple machining any home craftsman can do. The aluminum stock is available in sheet form and can be sheared or cut to the proper size for a nominal charge at most sheet-metal facilities.

12-20. DISTANCE REDUCTION AND TRILATERATION ADJUSTMENTS

The two adjustments commonly used in trilateration are (1) condition equations involving differences in angles or areas and (2) the variation of coordinates method. For low-order networks 1 : 15,000 or lower, adjustments similar to the compass and transit rules can be employed but are not recommended.

Condition equations for adjusting trilateration were developed by the late Earl S. Belote, a geodesist with the U.S. Coast and Geodetic Survey. Condition equations use the differences of the angles calculated from the sides and evolved by differentiating the basic equations to compute those angles. Changes in the lengths can be expressed as follows:

$$dA''$$
$$= \frac{(ada - a \cos Cdb - a \cos BDc)}{2}(\text{area})\sin 1''$$
$$db''$$
$$= \frac{(-b \cos Cda + bdb - b \cos Adc)}{2}(\text{area})\sin 1''$$
$$dc''$$
$$= \frac{(-c \cos Bda - c \cos Adb - Cda)}{2}(\text{area})\sin 1''$$

In a trilateration network, the number of equations is equal to $N - 2S$, N being the number of lines measured and S the number of new stations. If ties to additional stations are contemplated or required, additional equations will be necessary. Therefore, it is suggested that if connecting ties are to be utilized, the variation of coordinates method should be used. In Chapter 16 the theory and application of the variation of coordinates method of adjusting a quadrilateral are described in detail.

12-21. FIELD NOTES

Figure 12-15 shows the field notes for a meteorological data observation, Figure 12-16 covers zenith angle field data, and Figure 12-17 is a set of EDMI field measurements for multireadings/zenith angles.

STATE LANDS DIVISION

DISTANCE MEASUREMENT DATA
METEOROLOGICAL DATA OBSERVATION

DATE: 4 APRIL 1984

STA. ROCK (INSTRUMENT) TO STA. ENUF (REFLECTOR)

INITIAL DATA		FINAL DATA	
1735	TIME	1750	
45.8 °F	TEMP	45.1 °F	
26.65	PRESSURE INCHES / HG	SAME	
3240 FT	PRESSURE ALTITUDE	SAME	
4.960 FEET	HS	1.512	METERS

REFLECTORS

TYPE WILD TILTING SINGLE

NO. OF REFLECTORS 1

REFLECTOR NO. WILD #2

EXTENSION USED: YES (NO)

COMMENTS: GROUND TEMP 46°F
ASP. HEIGHT = 20 FT.

STAND HEIGHT	4312	1314
REFL. HEIGHT	0.648	198
HS =	4.960	1512

Figure 12-15. Distance-measurement data-meteorological data.

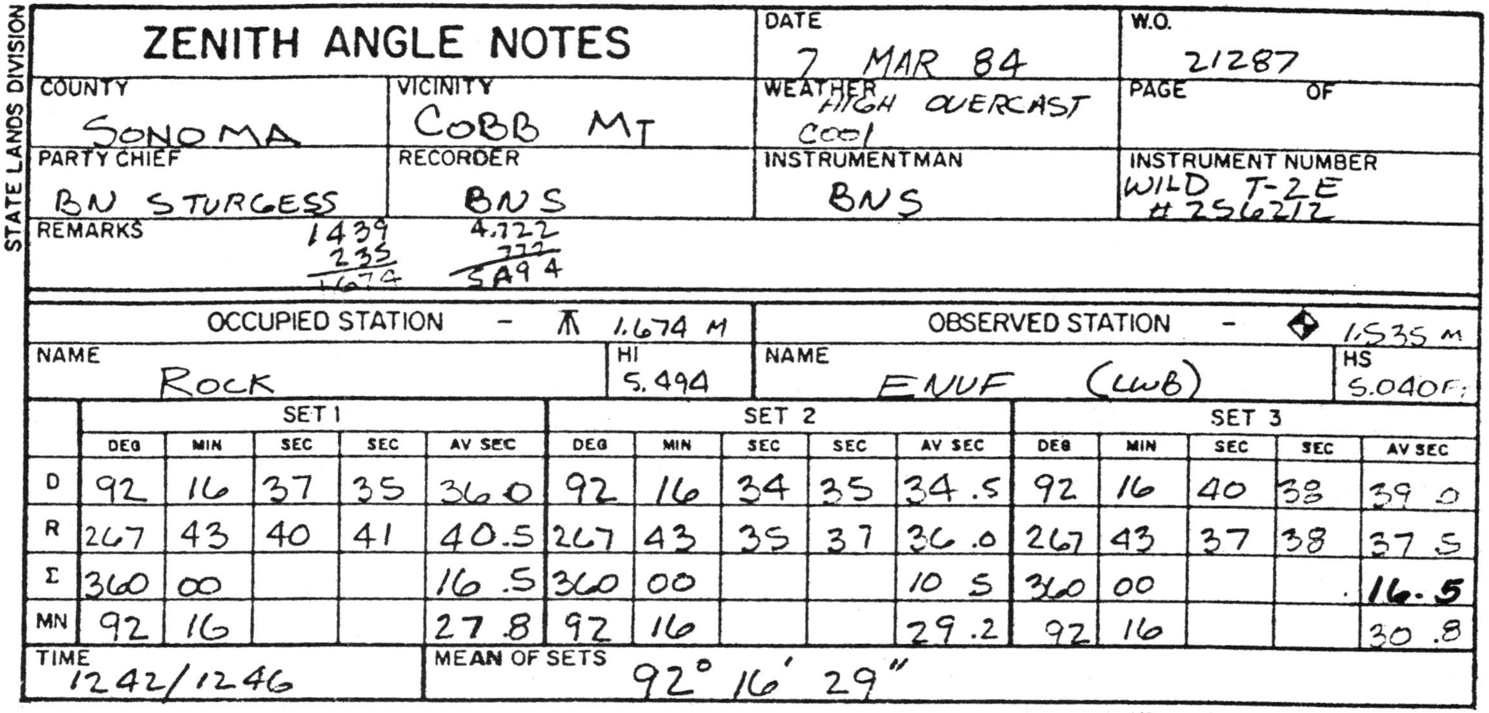

ZENITH ANGLE NOTES

COUNTY	SONOMA	VICINITY	COBB MT
DATE	7 MAR 84	W.O.	21287
PARTY CHIEF	BW STURGESS	RECORDER	BWS
WEATHER	HIGH OVERCAST COOL	PAGE	OF
INSTRUMENTMAN	BWS	INSTRUMENT NUMBER	WILD T-2E #256217

REMARKS:

```
 1439        4722
  235         772
 1674        5494
```

OCCUPIED STATION – ROCK   T 1.674 M   HI 5.494

OBSERVED STATION – ENUF (WB)   1535 m   HS 5.040 F

NAME: ROCK   NAME: ENUF (WB)

	SET 1 DEG	MIN	SEC	SEC	AV SEC	SET 2 DEG	MIN	SEC	SEC	AV SEC	SET 3 DEG	MIN	SEC	SEC	AV SEC
D	92	16	37	35	360	92	16	34	35	34.5	92	16	40	38	39.0
R	267	43	40	41	40.5	267	43	35	37	36.0	267	43	37	38	37.5
Σ	360	00			16.5	360	00			10 5	360	00			16.5
MN	92	16			27.8	92	16			29.2	92	16			30.8

MEAN OF SETS   92° 16' 29"

TIME 1242 / 1246

Figure 12-16.   Zenith angle notes.

EDM FIELD MEASUREMENT
FOR MULTI READINGS/ZENITH ANGLES

STATE LANDS COMMISSION

DATE 20 MAR 84	W.O. 21287
CHECKED	DATE

COUNTY LAKE	VICINITY COBB MTN.	WEATHER CLEAR COOL W'LY O/5	PAGE OF
CHIEF OF PARTY B.N. STURGESS	RECORDER BNS	OBSERVER BNS	SIGHT SETTER C. WISHMAN
OCCUPIED STATION ENUF CSLC 1976	HI 5.157 FT / 1.571 M ELEV 4000.000	EDM MODEL & NO. HP 3810 B	THEO. MODEL & NO.
OBSERVED STATION Rock CSLC 1980	HS 5.382 FT / 1.640 M ELEV 4423.405	PRISM TYPE & NO. WILD SINGLE #1 NO EXTENSION	PRISM CONSTANT

ECC. POSI.	σ	CENTER	σ	− ECC	σ	+ ECC	σ	CENTER
1		10740.657		10741.145		10740.171		10740.653
2		.653		.145		.168		.653
3		.650		.139		.155		.650
4		.650		.152		.148		.650
5		10740.653		10741.139		10740.145		10740.650
6		10740.653		10741.145		10740.158		10740.647
7		.650		.155		.151		.647
8		.647		.145		.145		.653
9		.657		.159		.148		.653
10		10740.650		10741.155		10740.151		10740.653
MEAN		10740.652		10741.148		10740.154		10740.651
ECC. COR.		0		−.492		+.492		0
DISTANCE		10740.652		10740.656		10740.646		10740.651

EDM READINGS

Figure 12-17. EDM field measurement.

STATE OF CALIFORNIA

	START		FINISH		MEAN (4 SETS)	
	EDM	PRISM	EDM	PRISM		10,740.651
Time	0827		0842		PPM CORR.	− .001
Temp. Air	60.8°F	59.7°F	61.5°F	59.5°F	PRISM CORR.	− .011
Temp. Grd.	62°	59°	61°	62°	FINAL SLOPE DIST.	10,740.639
Press.	26.78"HG	26.36"HG	NC	NC		
Press. Alt.	3100	3530	NC	NC		
PPM Calc.	+31.88		+31.94			
PPM Mean						
PPM Dialed	+32					

NOTES: SIGNAL =65 NO ATTEN. TEMP @ +30'

HEIGHT OF STAND   4.312 FT.   1.314 M

HEIGHT OF EDM ABOVE TOP OF STAND   .845FT.   .257M

HEIGHT OF EDM ABOVE MARK (HI)   5.157FT.   1.571M

ZENITH ANGLE		START	FINISH	MEAN		
HI	TIME					
HS	MEAN OF SET	°	'	"		

	SET 1					SET 2					SET 3				
	DEG.	MIN.	SEC₁	SEC₂	AV. SEC.	DEG.	MIN.	SEC₁	SEC₂	AV. SEC.	DEG.	MIN.	SEC₁	SEC₂	AV. SEC.
D															
R															
Σ															
MN															

Figure 12-17. *(Continued)*.

269

NOTES

1. FGCC 1979. "Classification Standards of Accuracy, and General Specifications of Geodetic Control Survey." Silver Spring MD.
2. FGCC 1975. *Specifications to Support Classification, Standards of Accuracy, and General Specifications of Geodetic Control Surveys*. [1975] 1980 (revised). Federal Geodetic Control Committee Washington DC.

REFERENCES

BOMFORD, G. [1971] 1977 *Geodesy*, 3rd ed. New York: Oxford University Press.

BRYAN, D. G. 1980 Scale variation in trilateration adjustment applied to deformation surveys. M. S. Thesis, Virginia Polytechnic Institute and State University, Blacksburg, VA.

BURKE, K. F. 1971 Why compare triangulation and trilateration. 31st Annual ACSM Meeting, March 7–12. Washington D.C.

CARNES, P. S. 1961 Temperature variations in the first two hundred feet of the atmosphere in an arid region. Missle Meteorology Division, U.S. Army, Signal Missile Support Agency, N.Y. Alamogordo.

CARTER, W. E., and J. E. PETTEY 1981 Report of survey for McDonald Observatory, Harvard Radin Astronomy Station and vicinity. NOAA Technical Memorandum NOS NGS 32. Rockville MD.

DAVIS, R. E., F. S. FOOTE, J. M. ANDERSON, and E. M. MIKHAIL. 1981 *Surveying Theory and Practice*. New York: McGraw-Hill.

DRACUP, J. F. [1969] 1976 Suggested specifications for local horizontal control surveys (revised).

———— 1976 Tests for evaluating trilateration surveys. Proceedings of the ASCM Fall Convention, Seattle, WA, Sept. 28–Oct. 1.

———— 1980 Horizontal control. NOAA Technical Report NOS 88 NCS 19. Rockville MD.

DRACUP, J. F., and C. F. KELLEY. [1973] 1981 Horizontal control as applied to local surveying needs. ASCM Publication (reprint).

DRACUP, J. F., C. F. KELLEY, G. B. LESLEY, and R. W. TOMLINSON. 1979 *Surveying Instrumentation and Coordinate Computation Workshop Lecture Notes*, 3rd ed. ACSM. Falls Church VA.

FRONCZEK, C. J. 1977 Use of calibration base lines. NOAA Technical Memorandum NGS-10. Rockville MD.

GOSSETT, CAPTAIN F. R. 1959 Manual of geodetic triangulation. Special Publication No. 247. USGPO Wash. D.C.

GREENE, J. R. 1977 Accuracy evaluation in electro-optic distance measuring instruments. *Surveying and Mapping*. Qtrly, September Falls Church VA. Vol. 37, No. 3 P. 247.

INGHAM, A. E. 1975 *Sea Surveying*. New York: John Wiley & Sons.

KELLY, M. L. 1979 Field calibration of electronic distance measuring devices. Proceedings of the ACSM, 39th Annual Meeting, March 18–March 24. Washington D.C.

LAURILA, S. H. 1976 *Electronic Surveying and Navigation*, New York: John Wiley & Sons.

MEADE, B. K. 1969 Corrections for refractive index as applied to electro-optical distance measurements. U.S. Department of Commerce, Environmental Science Services Administration, Coast and Geodetic Survey. Rockville MD.

———— 1972 Precision in electronic distance measuring. *Surveying and Mapping*. Vol. 32, No. 1 p. 69.

MOFFITT, F. H., and H. BOUCHARD. 1975 *Surveying*, 6th ed. New York: Harper Collins.

ROBERTSON, K. D. 1975 A method for reducing the index of refraction errors in length measurement. *Surveying and Mapping Journal*.

———— 1979 *The Use and Calibration of Distance Measuring Equipment for Precise Measurement of dams*. Fort Belvoir, VA: U.S. Army Corps of Engineers, E.T.L. (revised).

13

Geodesy

Earl F. Burkholder

13-1. INTRODUCTION

Literally, the word geodesy means "dividing the earth"; however, by usage its meaning now includes both science and art. The science of geodesy is devoted to determining the earth's size, shape, and gravity field. The art of geodesy utilizes scientific data in a practical way to (1) obtain latitude, longitude, and elevation of points; (2) compute lengths and directions of lines on the earth's surface; and (3) describe the trajectory of missiles, satellites, or other spacecraft. It is not intended here to designate a given activity as being either science or art, but to recognize a legitimate difference in emphasis that may exist in various areas of geodesy.

13-2. DEFINITIONS

1. *Geometrical geodesy*. Concerned with the size and shape of the earth's mean-sea-level surface.
2. *Physical geodesy*. Relates the earth's geophysical internal constitution to its corresponding external gravity field.
3. *Satellite geodesy*. Deals with satellite orbits, tracking existing satellites, and predicting the trajectory of a given missile, satellite, or spacecraft.

4. *Geodetic astronomy*. Chronicles the changing position of the stars and other celestial objects. Although listed separately, it overlaps geometrical and satellite geodesy and is not discussed further here. Additional information can be found in Chapter 17 in the section on field astronomy and in other texts on geodetic astronomy.

13-3. GOALS OF GEODESY

It is not practical to mathematically describe the earth's entire topographical surface. However, one goal of geodesy is to obtain a mathematical model that best approximates the earth's mean-sea-level surface. The model most commonly used is an *ellipsoid*, formed by rotating an ellipse about its minor axis. (In the past, such a figure has been referred to as a *spheroid*. For the purposes of this handbook, the two terms can be used interchangeably.) The ellipse major axis is in the equatorial plane; the minor axis coincides with the earth's spin axis. Often an ellipsoid is defined by the length of its semimajor axis a and the semiminor axis b, Figure 13-1a. However, in Section 13-6, the ellipsoid is also defined in other ways.

The earth's mean-sea-level surface is called the *geoid*, shown in Figure 13-1b. The geoid extends under the land masses and is the

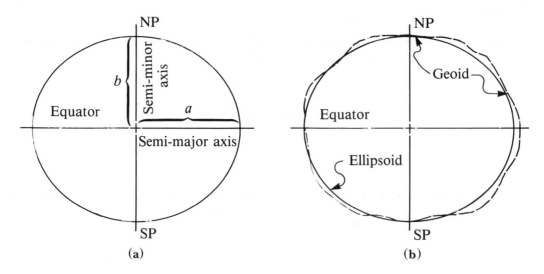

Figure 13-1. (a) Mathematical ellipsoid. (b) Physical geoid.

mean equilibrium level to which water would rise in a transcontinental canal. The geoid does not follow the ellipsoid exactly, but undulates from it by as many as 100 m. For this reason, the earth's mean-sea-level shape has been referred to as a lumpy potato. Additionally, there is an identifiable bulge in the geoid of 10 to 15 m in the southern hemisphere, giving rise to the earth being described as pear-shaped. On the other hand, despite mountains and ocean trenches, the earth is nearly spherical and by comparison is smoother than an orange. If its diameter at the equator was 10 m, the distance South Pole to North Pole would be shorter by only 0.034 m.

A second goal of geodesy is to describe the location of points on the earth's surface relative to the equator (latitude), an arbitrary meridian (longitude), and mean sea level (elevation). Thus, the rotational ellipsoid is indispensable in providing a framework for geodetic control networks. These networks, such as that in Figure 13-2, define the latitude and longitude of control points throughout the world. Previously, geodetic surveying operations were confined to continental land masses, and precise intercontinental ties were

impossible. Now with space-age technology, geodetic surveying activities are conducted on a global scale, so any two points on the earth can be tied together.

A third goal of geodesy is to determine the earth's external gravity field. Isaac Newton, Christian Huygens, and others in the middle 1600s recognized that the earth's shape is influenced by gravity. Since then, much scientific research has been devoted to the earth's geophysical attributes. This aspect of geodesy is important to surveyors because the geoid—the mean sea level to which elevations are referenced—is actually defined by an equipotential surface. Since the distance between equipotential surfaces is defined in terms of the work required to move a unit mass from one to the other, the perpendicular distance between two level surfaces is not constant but varies from the equator to the pole. Precise differential-leveling computations must accommodate that difference.

It is not possible or practical to cover all aspects of geodesy in one chapter of this surveying handbook. Therefore, the remainder of this chapter will be devoted to (1) a brief history of geodesy, (2) artful applications of

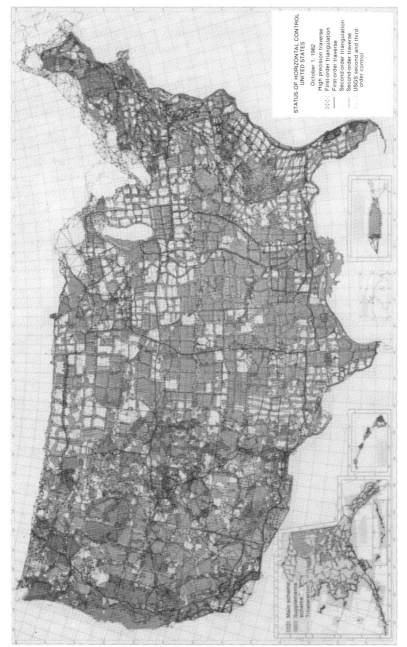

Figure 13-2. Status of horizontal control in the United States (Courtesy of National Geodetic Survey, NOAA/National Ocean Survey.)

geometrical geodesy, and (3) one brief section each on physical and satellite geodesy.

13-4. HISTORY OF GEODESY

Who first pondered the extent of the earth beyond the horizon? Who first realized inferences about out planet could be drawn from star observations? Although answers to these questions can only be conjectured, it is known that Pythagoras (b. 582 B.C.) declared the earth to be a globe, and Aristotle (384–322 B.C.) concluded that the earth must be spherical. However. an Alexandrine scientist named Eratosthenes (276–195 .C) is given credit for first determining the earth's size; admittedly, his measurements were crude by today's standard, but the method correct for this assumption of a spherical earth. The length be obtained for the earth's circumference was only about 16% too large.

Little was recorded about geodesy from the time of Eratosthenes until after the Middle Ages. However, a new epoch of geodesy began in the early 1600s with the invention of telescopes, publication of 14-place logarithms, and applications of triangulation to arc measurement. Later developments include the theory of gravity, differential and integral calculus, standardization of lengths and techniques of least-squares adjustment.

In 1615, a Dutchman, Willebrord Snellius, measured an arc over 80 mi long with a series of 33 triangles. The distance he obtained for the earth's radius was too small by about 3.4%. Next a Frenchman, Jean Picard, measured an arc on the meridian through Paris in 1669–70 and obtained a length for the earth's radius too large by only 0.7%.

Later, Picard's work was extended north to Dunkirk and south to Collioure by the Cassini brothers. The total latitude difference from Dunkirk to Collioure is 8°20′, but the arc was completed in two segments: the parts north and south of Paris. The length of one degree of latitude—and subsequently the earth's radius—for the northern part was found to be shorter than for the southern one. Hence, based on the triangulation arc through Paris, the Cassini brothers concluded, and even insisted, that the earth is not a sphere but elongated at the poles.

In 1687, Issac Newton published his law of gravitation, in which he stated that the earth is flattened at the poles; Figure 13-3 illustrates his logic. The force of gravity experienced by a plumb bob near the earth's surface is the vector sum of gravitational attraction and centrifugal force due to the earth's rotation. A level surface (sea level) is always perpendicular to the direction of the gravity vector, and a plumb bob points toward the earth's center only if the observer is standing at the equator

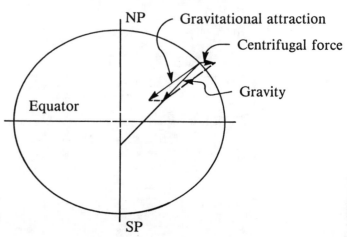

Figure 13-3. Gravity, the vector sum of gravitational attraction and centrifugal force.

or poles. Realizing this, Newton concluded that the earth must be a rotational ellipsoid flattened at the poles.

An ensuing dispute between the British (followers of Newton) and French (Cassini followers) regarding the earth's true shape was settled by two geodetic surveying expeditions sponsored by the French Academy of Science. In 1735, the first party went to the equatorial region of Peru (present-day Ecuador) to make arc measurements. In 1736, the second party went to the northern latitude of Lapland (present-day Finland). Results of the expeditions showed quite conclusively that the earth is flattened at the poles as stated by Newton (see Figure 13-4).

The lack of a universal length standard was a problem that plagued early geodesists and still affects modern interpretation of early efforts. In the late 1700s, two Frenchmen, Delambre and Mechain, were charged with determining the meridian arc distance, equator to pole, as accurately as possible. That distance was then set as *10 million meters*, the length standard now accepted worldwide. Table 13-1 shows the values for early measurements of the earth's size and shape, ending with results obtained by Delambre and Mechain.

Since the early 1800s, there have been numerous determinations of the earth's size and shape, and some are still in use (Table 13-2). If the earth was truly a homogeneous rotating fluid as postulated by Newton, one would expect the numbers in Table 13-2 to agree better than they do. However, since the internal density of the earth is not uniformly distributed, a "best-fitting" ellipsoid for any area of the earth will not necessarily be best fitting elsewhere. Consequently, practical applications of geodesy to survey control networks have been based on different ellipsoids, depending on the part of the world (or continent) in question. The Clarke Spheroid of 1866 was used as the reference ellipsoid for geodetic datums in the United States from 1879 to 1983 (Table 13-3).

With the advent of satellite triangulation and Doppler point-positioning, it has become possible to obtain parameters of an earth-

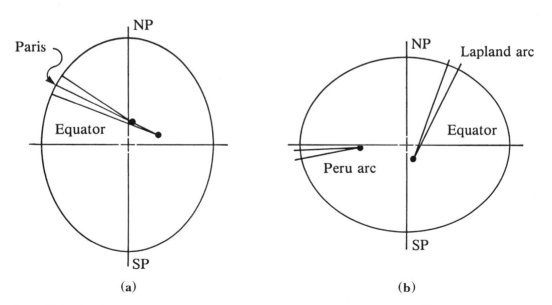

Figure 13-4. Comparison of Cassini arcs in France and the arcs in Lapland and Peru. (a) Cassini's prolate spheroid. (b) Newton's oblate spheroid.

Table 13-1. Early determinations of the earth's size and shape

Investigator	Approximate Date	Meridian Quadrant Arc Length (m)	Flattening
Eratosthenes	200BC	11,562,500	—
Willebrord Snellius	AD 1615	9,660,000	—
Jean Picard	1670	10,009,081	—
Cassini Brothers	1700	10,042,652	−1 : 66
French Academy of Science	1750	10,000,157	1 : 310.3
Delambre and Méchain	1800	10,000,000	1 : 334

Source: I. I. Mueller and K. H. Ramsayer, 1979, *Introduction to Surveying*, Frederick Ungar Publishing Co., New York, p. 148.

Table 13-2. Ellipsoids and area where used

Name	Semimajor Axis (m)	$1/f$	Used in
Everest 1830	6 377 276.345	300.801 7	India
Bessel 1841	6 377 397.155	299.152 8	China, Japan, Germany
Clarke 1866	6 378 206.4	294.978 7	North and Central America
Modified Clarke 1880	6 378 249.145	293.465	Africa
International 1924	6 378 388	297	Europe
Krasovskiy 1942	6 378 245	298.3	Former Soviet Republic and adjacent countries
Australian National 1965	6 378 160	298.25	Australia
South American 1969	6 378 160	298.25	South America

Source: I. I. Mueller and K. H. Ramsayer, 1979, *Introduction to Surveying*, Frederick Ungar Publishing Co., New York, p. 148.

Table 13-3. Geodetic datums used in the United States

Datum Name	Year Adopted	Reference Ellipsoid	Remarks
New England Datum	1879	Clarke 1866 $a = 6,378.206.4$ m $b = 6,356,583.5$ m	First official U.S. Datum, datum origin: station Principo in Maryland
U.S. Standard Datum	1901	Clarke 1866	Datum origin moved to Meades Ranch in Kansas
North American Datum	1913	Clarke 1866	A name change only to reflect adoption by Canada and Mexico
North American Datum of 1927	1927	Clarke 1866	A general readjustment holding location of station Meades Ranch
North American Datum of 1983	1983	Geodetic Reference System of 1980 $a = 6378137.000$ m $1/f = 298.257222101$	An extensive readjustment on a new reference ellipsoid having its origin at the earth's center of mass

centered ellipsoid based on a global best fit. The *North American Datum of 1983* is a comprehensive readjustment of the North American continent geodetic horizontal-control networks. The various national systems are tied to a worldwide geometric satellite network computed on the new ellipsoid, the Geodetic Reference System of 1980, adopted by the 17th General Assembly of the International Union of Geodesy and Geophysics Meeting in Canberra, Australia, December 1979. The adoption and use of an earth-centered ellipsoid make accurate global mapping possible. Additionally, the geodetic position of any point is fixed, working ''from the whole to the part'' on a global scale.

13-5. GEOMETRICAL GEODESY

13-5-1. Geometry of the Ellipsoid

As stated in Section 13-3, the ellipsoid is obtained by rotating an ellipse about its minor axis. The minor axis coincides with the earth's spin axis; the major axis sweeps out an equatorial plane as the ellipse is rotated. Any cross section of the ellipsoid containing both poles is a *meridian section* and shows the form of the original ellipse as illustrated in Figure 13-5a. The meridian section through Greenwich, England, Figure 13-6, is taken as the reference meridian. All other meridians are tied to this *prime meridian* by their *longitude*, the angular difference between meridian sections Figure 13-5b. Longitude starts with 0° at the prime meridian and increases eastward to 360° for a complete revolution. However, it is common practice in the Western Hemisphere to use longitude increasing *westward* from Greenwich to 180°W at the international date line. A word of caution: If west longitude is employed, it should be so noted to avoid confusion with the higher practice of east longitude 0 to 360°.

A position on a meridian is defined by its *geodetic latitude*, the angular distance north or south of the equator. The geodetic latitude goes from 90°S ($-90°$) at the South Pole to 90°N ($+90°$) at the North Pole. As shown in Figure 13-5a, the *normal* is perpendicular to the ellipse tangent and goes from a point of tangency to the spin axis. The angle ϕ that a

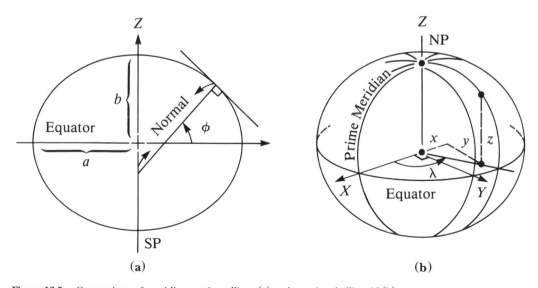

Figure 13-5.   Comparison of meridian section ellipse (a) and rotational ellipsoid (b).

Figure 13-6. Astraddle prime meridian in Greenwich, England, (Courtesy of Engineering Surveys Ltd., West Byfleet, England.)

normal makes with the equatorial plane is the point's geodetic latitude.

There is also an orthogonal three-dimensional coordinate system associated with the ellipsoid (Figure 13-5b). The origin is at the intersection of the spin axis and equatorial plane. A two-dimensional *XY*-plane lies in the equatorial plane with the positive *X*-axis pointing toward the prime meridian. The *Y*-axis is at longitude 90°E and the positive *Z*-axis points toward the North Pole. For this reason, the following discussion of a two-dimensional ellipse will be in terms of the *XZ* prime-meridian plane instead of an *XY*-plane commonly used for two-dimensional coordinates.

13-5-2. The Two-Dimensional Ellipse

A two-dimensional ellipse is a conic section described as the path of a point moving in a plane, so the total distance to two points, called foci, remains constant. The equation of an ellipse in the *XZ*-plane is given by

$$\frac{X^2}{a^2} + \frac{Z^2}{b^2} = 1 \qquad (13\text{-}1)$$

where a is the ellipse semimajor axis and b the semiminor axis.

Flattening of the ellipse and its eccentricity are defined in terms of a and b as

$$\text{flattening } f = \frac{(a-b)}{a} = 1 - \frac{b}{a} \quad (13\text{-}2)$$

$$\text{eccentricity } e = \frac{\sqrt{a^2 - b^2}}{a} \quad (13\text{-}3)$$

$$\text{second eccentricity } e' = \frac{\sqrt{a^2 - b^2}}{b} \quad (13\text{-}4)$$

Two parameters are required to define an ellipse. Previously, the semimajor axis a and semiminor axis b have been used; however, an ellipse can be defined equally well by a and e, e', or f. Current practice defines the ellipsoid size and shape with the semimajor axis a and reciprocal flattening $1/f$. Given these two parameters, the eccentricity and semiminor axis are

$$e^2 = 2f - f^2, \quad e = \sqrt{e^2} \quad (13\text{-}5)$$

$$b = a(1 - f) = a\sqrt{1 - e^2} \quad (13\text{-}6)$$

Equations (13-5) and (13-6) are obtained by substitution and the algebraic manipulation of (13-2) and (13-3).

Construction of an Ellipse

There are three ways to construct an ellipse of any size and shape. The first method is a mechanical one using a piece of string and a pencil. Since the sum of the distances from each focus to a point on the ellipse is constant, a curve can be drawn by anchoring the string ends at the foci, taking up the slack with a pencil, and tracing one-half of the ellipse while keeping the string taut. The second half is drawn by taking up the slack in the opposite direction, rather than wrapping the string around one focus. The following items are illustrated in Figure 13-7:

1. Total string length is twice the ellipse semimajor axis is shown by the pencil being positioned at point P_1.

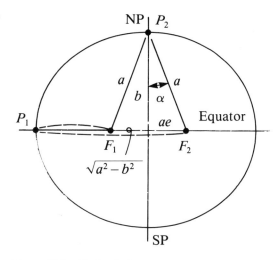

Figure 13-7. Mechanical construction of ellipse: String method.

2. Separation of the foci from the origin determines the minor axis length. If both foci are at the origin, the semiminor and semimajor axes are identical, so the ellipse becomes a circle ($e = 0$). If the two foci are separated so the string becomes taut, the semiminor axis length goes to zero and the ellipse degenerates to a straight line ($e = 1$).

3. Stopping the pencil at point P_2 on the minor axis produces a symmetrical figure, and the distance from each focus to P_2 is a, one-half the total string length. The resulting right triangle is solved for the distance focus to origin as $\sqrt{a^2 - b^2}$.

$$\text{Sin } \alpha = \frac{\sqrt{a^2 - b^2}}{a}$$

from Figure 13-7 is the same as e, the eccentricity (α is the *angular eccentricity*). Note: The distance between each focus and the origin is given by the product ae.

The second method of constructing an ellipse is a graphical one. First, two circles are drawn. The radius of the outer circle is a and the inner one b. Next, any number of radial lines are drawn as shown in Figure 13-8. Finally, lines are drawn parallel to the X-axis from the intersection of the radial line and

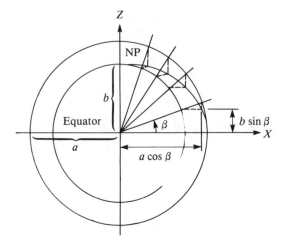

Figure 13-8. Graphical construction of ellipse showing parametric (reduced) latitude.

inner circle and parallel to the Z-axis from the intersection of the radial line with the outer circle. The intersection of these two lines from the same radial line falls on an ellipse, which is formed by plotting a sufficient number of points and "connecting the dots." Two ellipse properties illustrated in Figure 13-8 are as follows:

1. The angle between the equatorial plane (X-axis) and radial line is the *parametric latitude* β. In some geodetic literature it is called "reduced latitude."

2. The X- and Z-coordinates of a point on an ellipse are given, respectively, by

$$X = a \cos \beta \quad \text{and} \quad Z = b \sin \beta \quad (13\text{-}7)$$

The third and perhaps the most efficient method of constructing an ellipse is to compute X- and Z-coordinates for a sufficient number of points and plot them directly. Coordinates are computed using a, the semimajor ellipse axis; e^2, the eccentricity squared; and ϕ, the geodetic latitude of a point. The

following are equations for X and Z:

$$X = \frac{a \cos \phi}{(1 - e^2 \sin \phi)^{1/2}} \quad (13\text{-}8)$$

$$Z = \frac{a(1 - e^2) \sin \phi}{(1 - e^2 \sin^2 \phi)^{1/2}} \quad (13\text{-}9)$$

Important ellipse properties in Figure 13-9 are as follows:

1. It is doubly symmetrical. Coordinates need to be computed for one quadrant only.

2. N, the normal, goes from the ellipse to the spin axis.

3. The normal length is $X/\cos \phi$. Using Equation (13-8), we can write it as

$$N = \frac{a}{(1 - e^2 \sin^2 \phi)^{1/2}} \quad (13\text{-}10)$$

Three Types of Latitude

Three types of latitude are routinely encountered in geometrical geodesy. Geodetic latitude and parametric latitude have already been discussed. The third one is *geocentric latitude*. ψ, the angle between the equatorial plane and a line from the ellipse center to a surface point (Figure 13-10a). Tan ψ is obtained directly as Z/X. The geocentric latitude, geodetic latitude, and parametric latitude are related by substituting values of X and Z as contained in Equations (13-7), (13-8), and (13-9).

$$\text{Tan } \psi = \frac{Z}{X} = \frac{b \sin \beta}{a \cos \beta}$$

$$= \frac{b}{a} \tan \beta = (1 - e^2)^{1/2} \tan \beta \quad (13\text{-}11)$$

$$\text{Tan } \psi = \frac{Z}{X}$$

$$= \frac{a(1 - e^2)\sin \phi / (1 - e^2 \sin^2 \phi)^{1/2}}{a \cos \phi / (1 - e^2 \sin^2 \phi)^{1/2}}$$

$$= (1 - e^2)\tan \phi \quad (13\text{-}12)$$

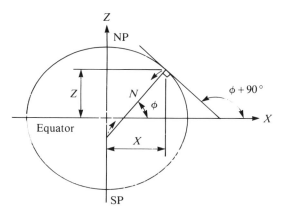

Figure 13-9. Mathematical construction of ellipse.

Equations (13-11) and (13-12) can be summarized as

$$\text{Tan } \psi = (1 - e^2)^{1/2} \tan \beta$$

$$= (1 - e^2)\tan \phi \qquad (11\text{-}13)$$

Comparison of Equation (13-13) with Figure 13-10b shows that the three latitudes are identical at the equator and poles. Between the equator and poles, the geocentric latitude is smaller than the parametric and geodetic latitudes, whereas the geodetic latitude is larger than either the geocentric or parametric latitudes. The maximum difference between geodetic and geocentric latitude occurs when the former is more than 45°, while the latter is smaller than 45°.

Radius of Curvature in Meridian Section

If the meridian section was spherical—i.e., if *e* equals zero—the ellipsoid would be a sphere and the radius of curvature the same at any point. Since the meridian section is an ellipse, its radius of curvature is not constant, but changes with increasing latitude. The *radius of curvature M* at any point in the meridian section is obtained by taking the first and second derivatives of Equation (13-1) and substituting those expressions in the general equation for radius of curvature given in Equation (13-14).

$$M = \left[1 + \left(\frac{dZ}{dX} \right)^2 \right]^{3/2} \bigg/ \frac{d^2Z}{dX^2} \qquad (13\text{-}14)$$

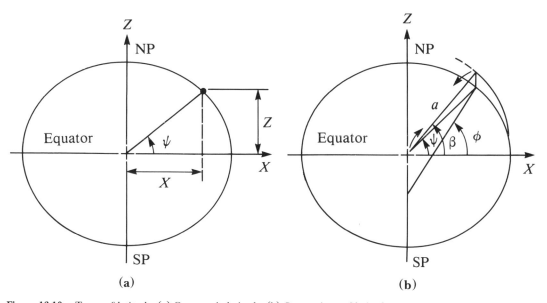

Figure 13-10. Types of latitude. (a) Geocentric latitude. (b) Comparison of latitudes.

The result is Equation (13-15), which gives the meridian radius of curvature for a specific ellipse at any geodetic latitude.

$$M = \frac{a(1 - e^2)}{(1 - e^2 \sin^2 \phi)^{3/2}} \qquad (13\text{-}15)$$

Length of Meridian Arc

Early determinations of the earth's size were made by measuring a portion of a meridian arc and comparing that length to the angle subtended at the earth's center. (The angle was usually found by astronomical observations.) Different values for the length of one degree of arc at various latitudes implied that a meridian section of the earth was ellipsoidal. Having selected an ellipsoid as a model for the earth, we compute the arc length by integrating the differential geometry elements in Figure 13-11, where the arc length of differential elements dS equals the instantaneous radius of curvature M times the differential change in geodetic latitude $d\phi$ ($d\phi$ in rad).

$$dS = M d\phi \qquad (13\text{-}16)$$

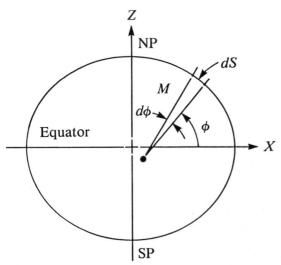

Figure 13-11.   Differential elements of meridian arc length.

Meridian arc length from one latitude to another is obtained by integrating Equation (13-17) between selected limits.

$$S_{\phi_1 \to \phi_2} = \int_{\phi_1}^{\phi_2} M d\phi \qquad (13\text{-}17)$$

The value of M from Equation (13-15) is substituted in Equation (13-17) and the constant portion moved outside the integral to get

$$S_{\phi_1 \to \phi_2}$$

$$= a(1 - e^2) \int_{\phi_1}^{\phi_2} (1 - e^2 \sin^2 \phi)^{-3/2} d\phi \quad (13\text{-}18)$$

Equation (13-18) is an elliptical integral and cannot be integrated in closed form—i.e., the expression inside the integral must be expressed in a series expansion containing ever-smaller terms that can then be integrated individually. A solution is obtained by including all those terms that make a difference in the answer to the accuracy desired. Terms beyond those are dropped.

Evaluation of the series expansion of an elliptical integral gives the meridian arc length of a specified ellipsoid as

$$S_{\phi_1 \to \phi_2} = a(1 - e^2) \Bigg[A(\phi_2 - \phi_1)$$

$$- \frac{B}{2}(\sin 2\phi_2 - \sin 2\phi_1)$$

$$+ \frac{C}{4}(\sin 4\phi_2 - \sin 4\phi_1)$$

$$- \frac{D}{6}(\sin 6\phi_2 - \sin 6\phi_1)$$

$$+ \frac{E}{8}(\sin 8\phi_2 - \sin 8\phi_1)$$

$$- \frac{F}{10}(\sin 10\phi_2 - \sin 10\phi_1) \Bigg] \quad (13\text{-}19)$$

where

$$A = 1 + \frac{3}{4}e^2 + \frac{45}{64}e^4 + \frac{175}{256}e^6 + \frac{11,025}{16,384}e^8 + \frac{43,659}{65,536}e^{10} + \cdots$$

$$B = \frac{3}{4}e^2 + \frac{15}{16}e^4 + \frac{525}{512}e^6 + \frac{2205}{2048}e^8 + \frac{72,765}{65,536}e^{10} + \cdots$$

$$C = \frac{15}{64}e^4 + \frac{105}{256}e^6 + \frac{2205}{4096}e^8 + \frac{10,395}{16,384}e^{10} + \cdots$$

$$D = \frac{35}{512}e^6 + \frac{315}{2048}e^8 + \frac{31,185}{131,072}e^{10} + \cdots$$

$$E = \frac{315}{16,384}e^8 + \frac{3465}{65,536}e^{10} + \cdots$$

$$F = \frac{693}{131,072}e^{10} + \cdots$$

Note: In Equation (13-19), the latitude difference in the A coefficient term is in radian units. Also, if limits 0 and 90° are chosen for ϕ_1 and ϕ_2, the length of a meridian quadrant is

$$S_{0° \rightarrow 90°} = a(1 - e^2)\left[A\frac{\pi}{2}\right] \qquad (13\text{-}20)$$

Other methods for computing meridian arc length expand the elliptical integral in terms of e'^2 instead of e^2 as listed here.

13-5-3. The Three-Dimensional Ellipsoid

Elements of a two-dimensional ellipse have been discussed, but the earth is three-dimensional. Point position on an ellipsoid surface is defined by three-dimensional coordinates or by latitude and longitude (Figure 12-5b). Given a point on an ellipsoid surface, there are several three-dimensional elements to be considered.

The Normal Section

A *normal section* is created by intersecting a plane containing the normal at a point and the ellipsoid. The plane, which can be oriented in any azimuth, is sometimes illustrated as the vertical plane that rotates about a theodolite's standing axis. (This is correct if the deflection of the vertical is zero. The standing axis of a theodolite is perpendicular to the geoid, but the normal is perpendicular to the ellipsoid. The difference is the *deflection of the vertical.*) A normal section having an azimuth of 90° (or 270°) at a point defines the *prime vertical* plane through it. The "normal" computed by Equation (13-10) is the prime vertical instantaneous radius of curvature. The normal section radius of curvature at a point on the ellipsoid in any azimuth is given by Euler's theorem, as follows:

$$R_\alpha = \frac{MN}{M \sin^2 \alpha + N \cos^2 \alpha}$$

$$\text{or} \quad \frac{1}{R_\alpha} = \frac{\cos^2 \alpha}{M} + \frac{\sin^2 \alpha}{N} \qquad (13\text{-}21)$$

where M is the radius of curvature in the meridian section, N the radius of curvature in the prime vertical, and α the normal section azimuth.

Radius of curvature properties for normal sections in various azimuths include:

1. $R_\alpha = M$ for a normal section in azimuth 0°.
2. $R_\alpha = N$ for a normal section in azimuth 90°.

3. $R_{30°} = R_{150°} = R_{210°} = R_{330°}$ due to symmetry. (Values of R_α repeat mirror-fashion with respect to both axes throughout all four quadrants.)

4. Values of R_α will always be greater than M and smaller than N.

As the geodetic latitude changes, the lengths of M and N also change. Note what happens to values of M and N at the equator and poles. For the equator, substitute $0°$ in Equations (13-10) and (13-15).

$$N_{0°} = \frac{a}{(1 - e^2 \sin^2 0°)^{1/2}} = a$$

$$M_{0°} = \frac{a(1 - e^2)}{(1 - e^2 \sin^2 0°)^{3/2}} = a(1 - e^2)$$

At a pole, substitute $\pm 90°$ in Equations (13-10) and (13-15).

$$N_{90°} = \frac{a}{(1 - e^2 \sin^2 90°)^{1/2}}$$

$$= \frac{a}{(1 - e^2)^{1/2}} = \frac{a^2}{b}$$

$$M_{90°} = \frac{a(1 - e^2)}{(1 - e^2 \sin^2 90°)^{3/2}}$$

$$= \frac{a}{(1 - e^2)^{1/2}} = \frac{a^2}{b}$$

It is obvious that M and N are equal at the poles, because the prime vertical of a given meridian is itself a meridian section. The ellipsoid radius of curvature at the poles is the same in all azimuths.

$$c = \frac{a^2}{b} = \text{the polar radius of curvature} \quad (13-22)$$

Length of a Parallel

A *parallel* of constant latitude on the ellipsoid describes a small cicle, as opposed to a great circle, whose plane is parallel to the equatorial plane. A parallel crosses all meridians at a 90° angle (Figure 13-12a) and is a circle whose radius equals $N \cos \phi$ (Figure 13-12b). Since a parallel is a circle, its length is simply $2\pi r$. Partial length of parallel L_p can be computed as a proportionate part of the total circumference or calculated directly as a product of the radius ($r = N \cos \phi$) times the subtended angle in radians. The subtended angle is the longitude difference between meridian sections.

$$L_p = (\lambda_2 - \lambda_1)N \cos \phi = \Delta\lambda N \cos \phi \quad (13-23)$$

Ellipsoid Surface Area

Ellipsoid surface area is computed by integrating the differential-area elements in figure 13-13. The differential area dA is the product of the differential meridian length times the differential parallel length. Area is obtained by performing a double integration of Equation (13-24).

$$dA = (M d \phi)(N \cos \phi d\lambda) \quad \text{or}$$

$$\text{Area} = \int_{\lambda_1}^{\lambda_2} \int_{\phi_1}^{\phi_2} MN \cos \phi \, d\phi \, d\lambda \quad (13-24)$$

Previous expressions for M and N, Equations (13-15) and (13-10), are substituted into Equation (13-24) and a double integration performed to obtain

$$\text{Area} = \frac{(\lambda_2 - \lambda_1)a^2(1 - e^2)}{2}$$

$$\times \left[\frac{\sin \phi}{(1 - e^2 \sin^2 \phi)} \right.$$

$$\left. + \frac{1}{2e} \ln \left(\frac{1 + e \sin \phi}{1 - e \sin \phi} \right) \right]_{\phi_1}^{\phi_2} \quad (13-25)$$

Equation (13-25) will give the ellipsoid surface area for any block defined by latitude and longitude limits. The entire ellipsoid surface area is computed by choosing limits of longitude from 0 to 2π radians and latitude limits

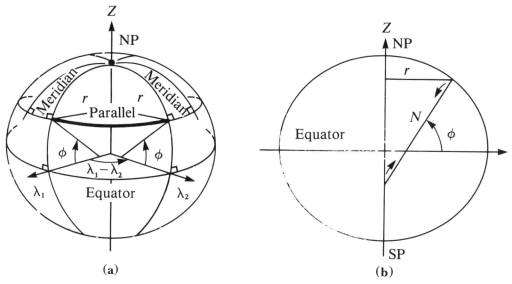

Figure 13-12.   Length of a parallel. (a) Parallels and meridians. (b) Radius of a parallel.

from $-90°$ at the South Pole to $+90°$ at the North Pole

Total area $= 2\pi a^2(1 - e^2)$

$$\times \left[\frac{1}{1 - e^2} + \frac{1}{2e} \ln\left(\frac{1 + e}{1 - e}\right) \right]$$

$$(13\text{-}26)$$

The Geodetic Line

The shortest distance between any two surface points on an ellipsoid is the *geodetic line*, also known as the *geodesic*. The geodetic line on the ellipsoid surface is analogous to a great circle arc on a sphere. When a geodetic line is drawn on a rectangular graticule of meridians and parallels, it appears as a curved line similar to a great circle.

Starting on the equator and traversing a geodetic line to the *antipole* (the point 180° from the beginning point), the route would go across either the North or South Pole and follow a meridian exactly. If the terminal point is several kilometers east or west of the an-

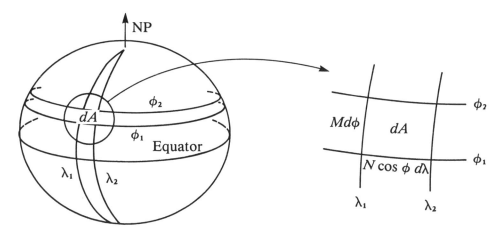

Figure 13-13.   Computation of ellipsoid surface area.

tipole, the geodetic line will not pass over either pole, but will reach some maximum (or minimum) latitude, where it will cross the meridian at 90°. Note also that if the geodetic line does not follow a meridian, it will cross the equator at some azimuth other than 0°. As the azimuth at the equator increases, the maximum latitude reached decreases (Figure 13-14), until the geodetic line stays on the equator.

Given a point on the equator, the shortest distance to the antipole is a geodetic line over a pole. The shortest distance to the *lift-off point* is along the equator (Figure 13-14). Any point between the antipole and lift-off point is reached following a geodetic line crossing the equator between 0 and 90°C; for the earth, the lift-off point is approximately 34 km from the antipole. Except for the geodetic line following a meridian or equator, the azimuth changes continuously with respect to the meridian it crosses. The azimuth of a geodetic line can be determined at any point of known latitude by using *Clairaut's constant*, defined as

$$N \cos \phi \sin \alpha = K \quad \text{(Clairaut's constant)} \quad \text{(13-27)}$$

Given the latitude of a point and geodetic line azimuth at the point, Clairaut's constant is

computed. That result is then used at other latitudes to solve for the geodetic line azimuth there. For example, in Figure 13-15, the latitude of point A is 42°15′28″.17621, point B is 42°20′16″.96171, and the geodetic line azimuth at point A is 55°16′28″.12. With the GRS 1980 ellipsoid (a = 6378137.0 m, e^2 = 0.006694380023), and Equation (13-10) for N, Clairaut's constant is computed as

$$\frac{a}{(1 - e^2 \sin \phi_A)^{1/2}} \cos \phi_A \sin \alpha_A$$

$$= \frac{3{,}879{,}837.711 \text{ m}}{0.9984852096} = 3{,}885{,}723.768 \text{ m}$$

The geodetic line azimuth at point B is obtained by rewriting Equation (13-27) as

$$\sin \alpha_B = \frac{\text{Clairaut's constant}}{N_B \cos \phi}$$

$$= \frac{3{,}885{,}723.768 \text{ m}}{4{,}721{,}791{,}697 \text{ m}} = 0.8229341778$$

The azimuth at point B is 55°22′46″.57. The geodetic line azimuth difference between points A and B is due to convergence of the meridians. Therefore, Clairaut's constant and

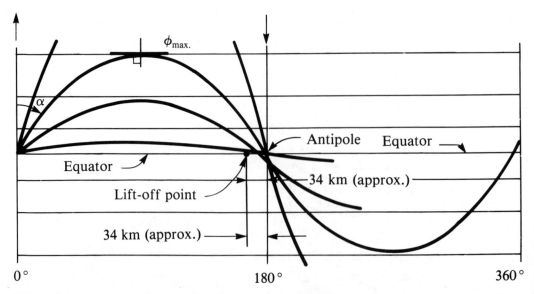

Figure 13-14. Geodetic lines around the earth.

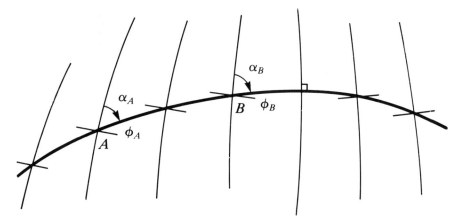

Figure 13-15. Geodetic line crossing meridians.

the geodetic line can be used to compute convergence between points.

$$\text{Covergence } (A \rightarrow B)$$

$$= \alpha_B - \alpha_A = 000°06'18''.45$$

Equations for convergence typically involve an approximation using the midlatitude of the line between points A and B. Clairaut's constant and a geodetic line provide a closed-formula method for determining convergence between points.

The maximum latitude reached by a geodetic line occurs where it crosses a meridian at 90°. Since we know the value of Clairaut's constant for the line and that $\sin 90° = 1.0$, it is possible to solve for the maximum latitude reached by a given geodetic line through writing Equation (13-27) as follows:

$$N_{\max} \cos \phi_{\max}(1.0)$$

$$= K \quad \text{(Clairaut's constant)}$$

from which—with considerable manipulation —the following can be written:

$$\cos \phi_{\max} = \frac{K(1 - e^2)^{1/2}}{(a^2 - K^2 e^2)^{1/2}}$$

$$= \frac{K}{(c^2 - K^2 e'^2)^{1/2}} \quad (13\text{-}28)$$

Clairaut's constant is not a unique property of a geodetic line. The constant remains unchanged along a parallel of latitude, although a parallel is not the shortest distance between two ellipsoid points.

Comparison of Geodetic Line and Normal Section

Due to a difference in direction of the normals at points A and B in Figure 13-16a, the normal section trace on the ellipsoid from A to B is different from the trace from B to A. The geodetic line between points A and B is not the trace of either normal section between the points. The geodetic line shown in Figure 14 when reverses its curvature only when it crosses the equator. However, when comparing the geodetic line with normal section traces between the points, it is impossible to show the geodetic line without giving it a double curvature (Figure 12-16c). The difference between a geodetic line azimuth A_g and the normal section azimuth A_n is given to a close approximation by

$$A_n - A_g = \frac{e^2}{12} \frac{S^2}{N_A^2} \cos^2 \phi_m \sin 2A_n \quad (13\text{-}29)$$

where e is the eccentricity of the ellipsoid, S the distance from point A to B, ϕ_m the mean latitude of line, and N_A the normal at point A.

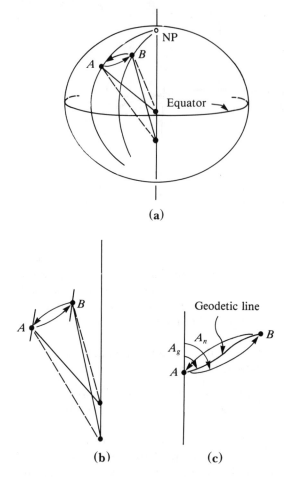

(a)

(b)   **(c)**

Figure 13-16. Comparison of normal plane curves and geodetic line. (a) Normal plane curves on the reference ellipsoid. (b) Normal plane curves: from A to B and B to A. (c) The geodetic line between A and B.

13-5-4. Geodetic Position Computation

A geodetic line can be used to compute the latitude and longitude of a point, given the distance and direction from a known point. The procedure is a *geodetic direct* computation. *Geodetic inverse* is employed to compute the distance and direction between points when the latitude and longitude of both are given. Details for performing both the "direct" and "inverse" geodetic position computations are presented by Jank and Kivioja in the Septem-

ber 1980 issue of *Surveying and Mapping*.[1] Their method utilizes the numerical integration of differential geometry elements shown in Figure 13-17. Clairaut's constant is used to determine the correct azimuth of each geodetic line element.

Although numerical integration may not be as quick as other methods for short lines, it is superior because any desired level of accuracy can be obtained regardless of line length. This is achieved by choosing a sufficiently small differential length element and programming a calculator or computer to do the repetitive calculations. According to Jank and Kivioja, centimeter accuracy can be expected for length elements up to 2 km long. If length elements are kept smaller than 200 m, millimeter accuracy can be attained.

The following Puissant Coast and Geodetic Survey formulas, used for geodetic direct and inverse computations, are quite accurate for lines up to 60 mi long.

Geodetic Direct

Given the latitude and longitude (ϕ_1, ϕ_2) of point 1, a geodetic line azimuth from north through point 1, α_1 and the distance S in meters along the geodetic line to point 2, find the latitude, longitude, and azimuth of the geodetic line at point 2.

$$\phi_2 = \phi_1 + \Delta\phi$$

$$\Delta\phi'' = SB\cos\alpha_1 - S^2C\sin^2\alpha_1$$
$$- D(\Delta\phi'')^2 - hS^2E\sin^2\alpha_1 \quad (13\text{-}30)$$

where

$$B = \rho/M_1 \quad (\text{sec per m})$$

$$h = SB\cos\alpha_1 \quad (\text{sec})$$

$$C = \frac{\rho\tan\phi_1}{2M_1N_1} \quad (\text{sec per m}^2)$$

$$D = \frac{3e^2\sin\phi_1\cos\phi_1}{2\rho(1 - e^2\sin^2\phi_1)} \quad (\text{per sec})$$

$$E = \frac{(1 + 3\tan^2\phi_1)(1 - e^2\sin^2\phi_1)}{6a^2} \quad (\text{per m}^2)$$

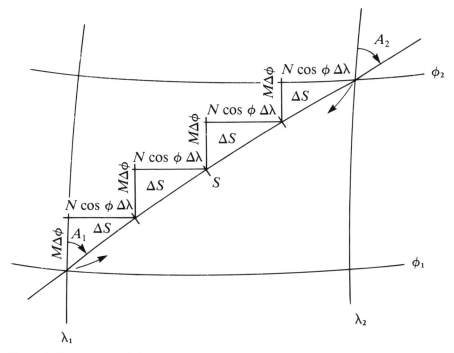

Figure 13-17. Differential elements of geodetic line computation.

The constants h, B, C, D, and E are computed using $\rho = 206264.8062470964$ sec per rad, M [Equation (13-15)] is the radius of curvature in the meridian, N [Equation (13-10)] the normal, and e^2 the ellipsoid eccentricity squared. Note that $\Delta\phi''$ appears on both sides of Equation (13-30), requiring an iterative solution. Use $\Delta\phi'' = $ zero for the first iteration.

With the latitude of point 2 known, the longitude is calculated.

$$\lambda_2 = \lambda_1 + \Delta\lambda \quad \text{(east longitude)}$$

$$\Delta\lambda'' = \frac{S\rho \sin \alpha_1}{N_2 \cos \phi_2} \qquad (13\text{-}31)$$

The azimuth from point 2 back to point 1 can be computed using Clairaut's constant; however, the Puissant formulas use the following:

$$\alpha_2 = \alpha_1 + \Delta\alpha + 180°$$

$$\Delta\alpha'' = \frac{\Delta\lambda'' \sin \phi_m}{\cos\left(\dfrac{\Delta\phi}{2}\right)}$$

$$+ (\Delta\lambda'')^3 \frac{\sin \phi_m \cos^2 \phi_m}{\rho} \qquad (13\text{-}32)$$

where

$$\phi_m = \left(\phi_1 + \frac{\phi_2}{2}\right)$$

Geodetic Inverse

Given the latitude and longitude of two points, it is required to find the geodetic line azimuth at each point and the distance from one to the other.

$$\Delta\phi = \phi_2 - \phi_1 \quad \text{and} \quad \Delta\lambda$$

$$= \lambda_2 - \lambda_1 \quad \text{(east longitude)}$$

$$x = \frac{\Delta\lambda'' N_2 \cos^2\phi_2}{\rho} = S\sin\alpha_1$$

$$y = \frac{1}{B}[\Delta\phi'' + Cx^2 + D(\Delta\phi'')^2$$

$$+ E(\Delta\phi'')x^2] = S\cos\alpha_1$$

$$\tan\alpha_1 = \frac{S\sin\alpha_1}{S\cos\alpha_1} = \frac{x}{y} \quad \text{(from north)} \quad (13.33)$$

$$S = \sqrt{x^2 + y^2} \quad \text{(in meters)} \quad (13.34)$$

13-6. GEODETIC DATA TRANSFORMATIONS

13-6-1. Use of a Model

This section discusses how geodetic directions and distances on the ellipsoid are obtained from field measurements. Since the ellipsoid is an abstract mathematical model, the data must be transformed from the actual measurement configuration to its equivalent representation on a model. The distance transformations shown in Figure 13-18 is an example. The measured slope distance must be transformed to an equivalent distance on the reference ellipsoid (the model) before it is used in a geodetic position computation.

13-6-2. Target Height Correction

When a target is sighted through a theodolite, the direction to it is the normal section azimuth from instrument to the target. If the target is not on the ellipsoid, there will be a difference in directions to it and to the station on the ellipsoid. The difference occurs because the normals at the target and theodolite are not parallel (Figure 13-19). The situation is analogous to sighting the top of a range pole

that is not held plumb over a point. A correction to the observed direction can be computed from Equation (13-35), as follows

$$\alpha = \alpha_h + \Delta\alpha$$

$$\Delta\alpha'' = \frac{\rho h e^2 \cos^2\phi_1}{2N_1(1 - e^2)}\left(\sin 2\alpha_h - \frac{S}{N_1}\sin\alpha_h \tan\phi_1\right)$$

$$(13\text{-}35)$$

where $\rho = 206264.8062470964$ sec of arc per rad, h is the target height above the ellipsoid, e^2 the ellipsoid eccentricity squared, ϕ the geodetic latitude of instrument station, α the normal section direction from point 1 to point 2, and α_h the normal section direction from point 1 to the target elevated above point 2.

Note the following items regarding the use of Equation (13-35):

1. The sign of the correction $\Delta\alpha''$ is determined by $\sin 2\alpha_h$. It is positive if the target is in the NE or SW quadrants and negative for the SE and NW quadrants.

2. Theodolite elevation is immaterial because the standing axis contains the vertical, and the instrument measures the dihedral angle.

3. The correction for target height is quite small ($\pm 0''.5$) for elevations under 4000 m, but could be significant on precise surveys if targets are located on high mountaintops.

13-6-3. Deflection-of-the-Vertical Correction

As shown in Figure 13-20b, a normal is perpendicular to an ellipsoid, but a vertical is perpendicular to the geoid. Deflection of the vertical (also called deviation of the vertical) is the difference in directions of the normal and vertical. Deflection of the vertical at a point can be scaled from an accurate geoid map showing contours of the geoid for a given area. Figure 13-20a shows part of the *Geoid Contour Map of the North American Datum of 1927*, printed by the U.S. Army Map Service.[2] The geoid is below the ellipsoid and slopes upward to the northeast, across central Ore-

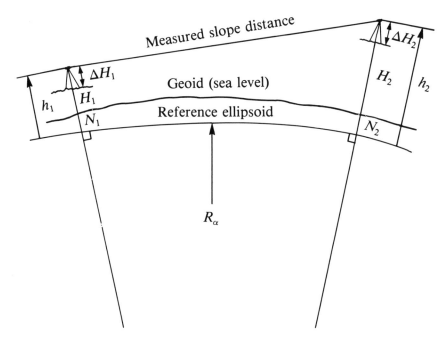

Figure 13-18. Elements of distance transformation.

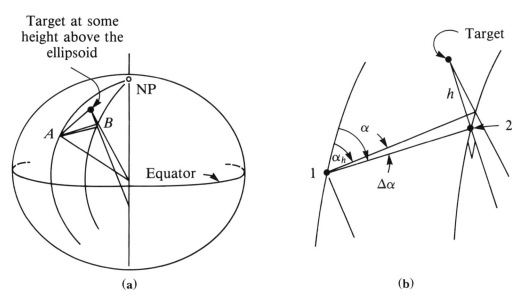

(a)   **(b)**

Figure 13-19. Difference in direction due to height of target. (a) Target height above ellipsoid. (b) Difference in observed azimuth due to height of the target.

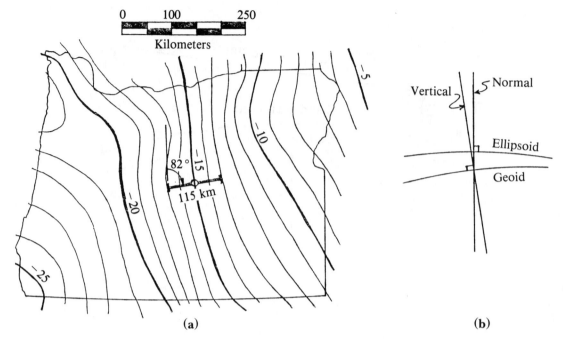

Figure 13-20. (a) Slope of geoid across the state of Oregon. (b) Deflection of the vertical. (Geoid Contours in North America; 1967 Washington D.C.; U.S. Army Map Service.)

gon. It rises 4 m in a scaled distance of 115 km. The azimuth of a line perpendicular to the geoid contours—the direction water would flow—scales 262°. From this data, approximately three significant figures, the total deflection of the vertical is 7.17 sec, and the NS and EW components are determined.

$$\text{NS: } xi = \xi = 7''.17 \cos(262°) = -1''.00$$

$$\text{EW: eta} = \eta = 7''.17 \sin(262°) = -7''.10$$

By convention, if the vertical above a theodolite is deflected into the northeast quadrant, both components are positive. Stated differently, if the geoid slopes upward to the north and east as in Figure 12-20b, both components are negative.

Deflection of the vertical is actually defined and determined using gravity measurements and physical geodesy techniques. The geoid map is compiled from aggregate deflection-of-the-vertical data and gravity measurements over the entire geodetic control network.

Geodetic scientists compile geoid maps and geodetic surveyors use geoid maps to relate geographic positions determined from astronomical observations to the point's geodetic position.

Equations (13-36), (13-37), and (13-38) give the relationships between astronomical and geodetic latitude, longitude, and azimuth, as follows:

$$\phi = \Phi - \xi \tag{13-36}$$

$$\lambda = \Lambda - \frac{\eta}{\cos \phi} \tag{13-37}$$

$$\alpha = A - \eta \tan \phi \tag{13-38}$$

where ϕ is the geodetic latitude, λ the geodetic longitude, α the geodetic azimuth, Φ the astronomical latitude, Λ the astronomical longitude, A the astronomical azimuth, ξ the NS component of deflection of the vertical, and η the EW component. Thus, if astronomical observations are made for the geographic position of a point, the geodetic latitude and longitude can be determined using equations (13-36) and (13-37). Additionally, the geodetic azimuth of a normal section from theodolite to target can be obtained from the line's astronomical azimuth using equation (13-38).

The steps required to transform an observed astronomical azimuth of a normal section to the corresponding geodetic line azimuth are the following:

1. Convert from the astronomical azimuth to the geodetic azimuth of the normal section, Equation (13-38).

2. Correct the normal section azimuth for the height of target above the ellipsoid, Equation (13-35).

3. Compute the geodetic line azimuth from the normal section azimuth, Equation (13-29).

13-6-4. EDMI Distance Transformation to the Ellipsoid

As shown in Figure 13-21, the slope distance measured at some elevation must be transformed to its equivalent ellipsoid distance before being used in geodetic position computations. The method and formulas shown for distance transformation are adapted from Appendix I of Fronczek's "Use of Calibration Base Lines."[3]

The ray path of an electronic distance-measuring instrument (EDMI) is not exactly a straight line. Due to electromagnetic wave refraction by the atmosphere, the distance measured by an EDMI must be corrected for the ray-path curvature to obtain a straight-line chord distance before it is transformed to the ellipsoid. Symbols used in the distance transformation are as follows:

a = semi major axis of reference ellipsoid
e^2 = eccentricity squared of reference ellipsoid
α = mean azimuth of line, referenced to N or S
ϕ = mean latitude of line
H_i = elevation of station 1 or 2 above the geoid
ΔH_i = theodolite or reflector height above the station mark

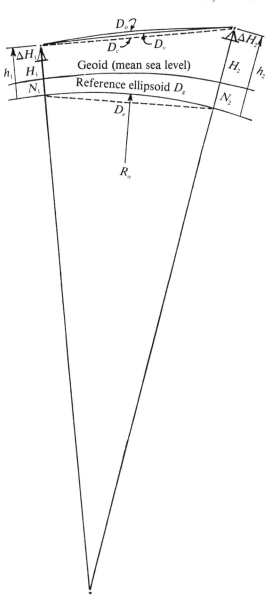

Figure 13-21. Transformation of measured distance to reference ellipsoid.

N_i = geoid height at station 1 or 2 (*Note:* In geodetic literature there is an unfortunate duplication in using the capital letter N to denote both the geoid height and radius of curvature in the prime vertical.)
h_i = instrument or reflector height above the ellipsoid
M = radius of curvature in the meridian

N = radius of curvature in the prime vertical
R_α = ellipsoid radius of curvature in azimuth α
k = index of refraction (for light-wave instruments $k = 0.18$, for microwave instruments $k = 0.25$)
D_o = observed slope distance corrected for temperature, pressure, eccentricity, reflector constant, and electrical center of EDMI
$D_v = D_o + I$ — second velocity correction
D_c = chord distance at instrument elevations
D_e = chord distance at ellipsoid surface
D_g = geodetic arc distance on the ellipsoid

The algorithm to transform an EDMI distance to geodetic arc distance is

$$M = \frac{a(1 - e^2)}{(1 - e^2 \sin^2 \phi)^{3/2}} \tag{13-15}$$

$$N = \frac{a}{(1 - e^2 \sin^2 \phi)^{1/2}}$$

$$(N = \text{radius of curvature}) \tag{13-10}$$

$$R_\alpha = \frac{MN}{M \sin^2 \alpha + N \cos^2 \alpha} \tag{13-21}$$

$$D_v = \frac{D_o - (k - k^2)D_o^3}{12 R_\alpha^2} \tag{13-39}$$

$$D_c = 2 \frac{R_\alpha}{k} \sin\left[\frac{D_1 k}{2 R_\alpha}\left(\frac{180}{\pi}\right)\right] \tag{13-40}$$

$$h_1 = H_1 + \Delta H_1 + N_1$$

$$(N = \text{geoid height}) \tag{13-41}$$

$$h_2 = H_2 + \Delta H_2 + N_2$$

$$(N = \text{geoid height}) \tag{13-42}$$

$$D_e = \left[\frac{\left(D_c^2 - (h_2 - h_1)^2\right)}{\left(1 + \frac{h_1}{R_\alpha}\right)\left(1 + \frac{h_2}{R_\alpha}\right)}\right]^{1/2} \tag{13-43}$$

$$D_g = 2 R_\alpha \left[\sin^{-1}\left(\frac{D_e}{2 R_\alpha}\right)\right]\left(\frac{\pi}{180}\right) \tag{13-44}$$

13-7. GEODETIC DATUMS

13-7-1. Regional Geodetic Datums

A geodetic datum is a mathematical model of the earth's figure on which geodetic computations are based. A *regional geodetic datum* is one that "fits" and is intended to be used in a specific area. Mitchell[4] defines a geodetic datum using geometrical geodesy concepts and the following five elements:

1. a = semimajor axis of the ellipsoid
2. b = semiminor axis of the ellipsoid
3. ϕ = latitude of the initial point
4. λ = longitude of the initial point
5. α = azimuth from initial point to another point

Unstated in the definition are assumptions that the ellipsoid and geoid are coincident at the initial point, and the earth's spin axis is parallel to the ellipsoid minor axis. The *North American Datum of 1927*, with its initial point at Meades Ranch in Kansas and based on the Clarke Spheroid of 1866, is an example of such a datum.

Further development and sophistication are reflected in Ewing and Mitchell[5] in which a regional geodetic datum is defined by:

1. a = semimajor axis of the reference ellipsoid
2. f = flattening of the reference ellipsoid
3. ξ_o = deflection of the vertical in the meridian at the datum origin, $\xi_o = \Phi_o - \phi_o$
4. η_o = deflection of the vertical in the prime vertical at the datum origin,

$$\eta_o = (\Lambda_o - \lambda_o)\cos \Phi$$

5. α_o = geodetic azimuth from the origin along an initial line of the network, $\alpha_o = A_o - \eta_o \tan \Phi$
6. N_o = geoid height at the datum origin—i.e., the distance between the reference and geoid
7.   The condition that the ellipsoid minor axis be parallel to the earth's spin axis

The *North American Datum of 1927* also fits this definition of a regional geodetic datum, with one exception—i.e., the initial azimuth from station Meades Ranch to station Waldo is published as 75°28′09″.64, but orientation throughout the network is controlled by geodetic azimuth determinations utilizing astronomical azimuths and deflection of the vertical at numerous stations (such a station is *Laplace station*).

The geoid height at station Meades Ranch was assumed to be zero in 1927, but subsequent observations and refinements in the network have shown residual components of deflection of the vertical exist there. Hence, the ellipsoid and geoid coincide at station Meades Ranch ($N_o = 0$), but the two surfaces are not tangent there.

13-7-2. Global Geodetic Datums

A regional geodetic datum has an initial point on the ellipsoid surface and is chosen for its approximation to the earth's shape for a particular area or continental land mass, but a *global geodetic datum* has its datum point at the earth's center of mass and a reference ellipsoid chosen on the basis of a global "best fit." Thus, points on any continent can be accurately related to any other points throughout the world that are tied to the same global datum.

Global and regional datums are not defined the same way. They both use a reference ellipsoid, but since the earth's shape is actually determined by forces of gravitational attraction and centrifugal acceleration, those physical geodesy elements, along with others, are employed to define a global geodetic datum. Moritz[6] gives the following as defining parameters of a global geodetic datum:

1. The datum origin is located at the earth's center of mass
2. The Z-axis is the direction of the Conventional International Origin (CIO) defining a mean North Pole
3. The X-axis is parallel to the zero meridian adopted by the Bureau International De L'Heure (BIH) and known as the Greenwich mean astronomical meridian
4. A reference ellipsoid is defined by
   (a) a = the semimajor axis
   (b) GM = the geometric gravitational constant (the Newtonian constant G times the mass M of the earth, including the atmosphere
   (c) J_2 = zonal spherical harmonic coefficient of second degree
   (d) ω = earth's angular velocity

Given the four physical geodesy parameters of a reference ellipsoid, the ellipsoid eccentricity is computed using Equation (13-45), as follows:

$$e^2 = 3J_2 + \frac{4}{15} \frac{\omega^2 a^3}{GM} \frac{e^3}{2q_o} \tag{13-45}$$

where

$$2q_o = \left(1 + \frac{3}{e'^2}\right)\arctan e' - \frac{3}{e'}; \; e'^2 = \frac{e^2}{(1 - e^2)}$$

Equation (13-45) has e, the eccentricity, on both sides of the "equals" signs, which means it must be solved iteratively even though it is in chosen form.

13-7-3. Parameters of Selected Regional Geodetic Datums

1. *North American Datum of 1927: Clarke Spheroid of 1866*

$a = 6,378,206.4$ m	$b = 6,356,538.8$ m
$\phi_o = 39°13′26″.686$ N	Origin: Meades Ranch, Kansas, USA
$\lambda_o = 98°32′30.″506$ W	
$\alpha_o = 65″28′09″.64$ from origin to station Waldo	

2. *European Datum: International Ellipsoid of 1924*

$$a = 6,378,388.0 \text{ m} \qquad 1/f = 297.00$$
$$\phi_o = 52°22'51''.45 \text{ N} \qquad \text{Origin: Helmert Tower, Potsdam, Germany}$$
$$\lambda_o = 13°03'58''74 \text{ E}$$

3. *Pulkovo Datum: Krassovski Ellipsoid of 1942*

$$a = 6,378,245 \text{ m} \qquad 1/f = 298.3$$
$$\phi_o = 59°46'18''.55 \text{ N} \qquad \text{Origin: Pulkovo Observatory, Leningrad, former Soviet republic}$$
$$\lambda_o = 30°19'42''.09 \text{ E}$$

4. *Tokyo Datum: Bessel Elipsoid of 1841*

$$a = 6,377,397.155 \text{ m} \qquad 1/f = 299.1528$$
$$\phi_o = 35°39'17''.51 \text{ N} \qquad \text{Origin: Tokyo Observatory, Japan}$$
$$\lambda_o = 193°44'40''.50 \text{ E}$$

13-7-4. Parameters of Selected Global Geodetic Datums

1. *Geodetic Reference System of 1967*

$$a = 6,378,135 \text{ m}$$
$$GM = 3.98603*10^{14} \text{ m}^3/\text{sec}^2$$
$$J_2 = 0.0010827$$
$$\omega = 7.2921151467*10^{-5} \text{ rad/sec}$$
$$1/f = 298.247 \quad \text{(computed and rounded)}$$

exact

2. *World Geodetic System of 1972*

$$a = 6,378,135 \text{ m}$$
$$GM = 3.986005*10^{14} \text{ m}^3/\text{sec}^2$$
$$J_2 = 0.001082616$$
$$\omega = 7.2921151467*10^{-5} \text{ rad/sec}$$
$$1/f = 298.26 \quad \text{(computed and rounded)}$$

exact

3. *Geodetic Reference System of 1980*

$$a = 6,378,137 \text{ m}$$
$$GM = 3.986005*10^{14} \text{ m}^3/\text{sec}^2$$
$$J_2 = 0.00108263$$
$$\omega = 7.29115*10^{-5} \text{ rad/sec}$$

exact

$$e^2 = 0.006694380022903416$$
$$1/f = 298.2572221008827$$
$$e'^2 = 0.006739496775481622$$
$$f = 0.003352810681183637$$
$$b = 6,356,752.314140347$$
$$c = 6,399,593.625864032$$

(computed to 16 significant figures)

13-8. PHYSICAL GEODESY

13-8-1. Gravity and Leveling

The study of physical geodesy concerns the earth's gravity field and spacing of equipotential surfaces. Due to the earth's eccentricity and gravity-field irregularities, vertical spacing between equipotential surfaces varies from point to point (Figure 13-22a).

The equipotential surface most commonly known and understood is mean sea level, which serves as a reference datum for the elevation of points on or near the earth's surface. The vertical distance from mean sea level to each bench mark is its *orthometric height*, determined by applying an "orthometric height correction" to the differences in elevation observed along a given line of precise levels. It has been said that differential levels is the simplest surveying concept to teach or understand, but the most difficult when considering long lines, high precision, and large differences in elevation. The orthometric height correction is one subtle concept that makes the statement true.

It is easy to accept orthometric height as a vertical distance from mean sea level to the equipotential surface, until one considers that the orthometric height of the water surface at the south end of Lake Huron is 5 cm higher than the same equipotential surface at the north end. How can the same water surface have two heights, when a loop of precise levels around the lake shows no difference in elevation? Additional clarification is required.

Another way to visualize the apparent discrepancy is to observe, in Figure 13-22b, the difference in elevation between points 1 and 2 by following low route A that is greater than along high route B. A loop from point 1 to point 2, along route A and back along route B, will fail to close because equipotential surfaces are not parallel.

For the Lake Huron example, an orthometric-height correction must be computed and applied to obtain a geopotential number that is the same for the entire lake surface. For a precise level line, the orthometric height cor-

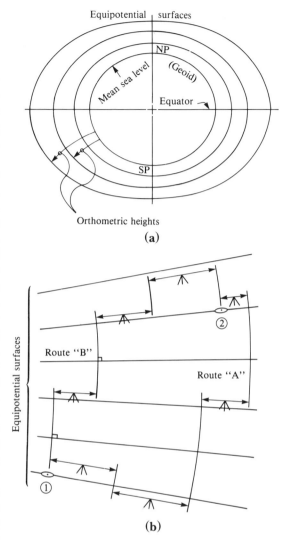

Figure 13-22. Orthometric heights related to precise leveling. (a) Equipotential surfaces. (b) Route-dependent leveling.

rection must be computed and applied to the observed differences in elevation to close the loop. This means that the correct difference in orthometric height between points 1 and 2 can be determined, irrespective of the route taken between the two points.

Orthometric height correction is a function of the force of gravity that, in turn, is related to the altitude, latitude, and longitude of a point. Hence, the further study of gravity and physical geodesy is vitally important to control surveyors and geodesists concerned with precise elevations over large areas. Adequate treatment of the topic is beyond this chapter's scope.

13-9. SATELLITE GEODESY

13-9-1. Geodetic Positioning

Although satellite geodesy is primarily concerned with orbits of satellites and other spacecraft, the use of satellite signals and space-age technology for geodetic positioning has revolutionized geodetic surveying practice. In the past, the line of sight between points was required to make triangulation measurements. However, with the launching of the Echo I satellite, intervisibility ceased to be critical because the satellite was photographed against a star background, simultaneously from two stations. Photographic images of the satellite were then analyzed to obtain a geometri-

cal tie between the two stations. A worldwide geometrical satellite triangulation net was completed using the Wild BC-4 ballistic camera system and Pageos satellites, but the program was discontinued in favor of better all-weather positioning systems.

Doppler Positioning with the Transit System

The next satellite positioning system to enjoy worldwide application was the Doppler positioning system, utilizing signals from a group of five satellites in polar orbits having approximately 107-min periods. The satellites broadcast two very stable frequencies (150 and 400 MHz), with orbital parameters and timing data phase-modulated on the signal. A stationary receiver on the ground (one type is shown in (Figure 13-23) receives a higher or lower frequency, depending on whether the satellite is moving toward or away from the receiver. This observed change in frequency (the *Doppler shift*) from numerous satellite passes is analyzed to determine a control point's geodetic

Figure 13-23. Magnavox MX 1502 geoceiver/satellite surveyor. Portable, battery-operated precise point-positioning and translocation system (Courtesy of Magnavox Advanced Products and Systems Co.)

position. If two Doppler receivers are used in pairs, with one receiver positioned over a known control point, the relative positions of unknown points can be determined to submeter accuracy, with data from approximately 25 acceptable satellite passes.

The Doppler positioning (transit) system was developed by the U.S. Navy for global navigation, and it became operational in January 1964. Since released for civilian use in July 1967, the transit system has proved to be very reliable for worldwide geodetic point positioning. Line of sight is no longer required between geodetic control points, and unlike satellite triangulation it can be used day or night, rain or shine. Additionally, Doppler receivers are quite portable, making it possible to establish geodetic positions in remote locations with minimum logistical support.

The GPS (NAVSTAR) Positioning System

Despite the functional success and heavy use of the transit system, another satellite positioning system has been developed. The Geodetic Positioning System (GPS) will involve a group of 18 satellites with 12-hr periods and provide 24-hr global coverage. Chapter 15 is devoted entirely to this subject.

NOTES

1. W. Jank, and L. A. Kivioja. 1980. Solution of the direct and inverse problems of reference ellipsoids by point-by-point integration using programmable pocket calculators. *Surveying and Mapping Journal* 40 (3).
2. Geoid Contour Map of North American Datum of 1927; Washington D.C., U.S. Army Map Service.
3. C. J. Fronczek. 1977. Use of calibration baselines. NOAA Technical Memorandum NOS NGS-10
4. H. C. Mitchell. 1948. Definition of terms used in geodetic and other surveys. USC & GS Special Publication No. 242, Washington, D.C..
5. E. Ewing, and M. M. Mitchell. 1970. *Introduction to Geodesy*. New York: Elsevier North-Holland.
6. H. Moritz. 1978. The definition of a geodetic datum. 2nd International Symposium on Redefinition of North American Geodetic Networks. Arlington, VA, April 24–28.

REFERENCES

BOMFORD, G. 1971. *Geodesy*, 3rd ed. London: Oxford University Press.

COLLINS, J. 1982–1983. The global positioning for surveying—today. *P.O.B Magazine* 8 (2).

FRONCEZK, C. J. 1967. Geoid contours in North America—from astrogeodetic deflections, 1927 North American datum. U.S. Army Map Service, Washington, D.C.

HEISKANEN, W. A., and F. A. VENNING MEINESZ. 1958. *The Earth and Its Gravity Field*. New York: Mc-Graw-Hill.

HEISKANEN, W. A., and H. MORITZ. 1967. *Physical Geodesy*. San Francisco and London: W. H. Freeman.

HOAR, G. J. 1982. *Satellite Surveying—Theory, Geodesy, Map Projections*. Torrance, CA: Magnavox Advanced Products and Systems.

JORDAN, W., and EGGERT, O. 1962. *Jordan's Handbook of Geodesy*. (Translated into English by M. W. Carta). Washington D.C.: U.S. Army Map Service.

MORITZ, H. 1980. Geodetic reference system 1980. *Bulletin Geodesique* (*Paris*) 54 (3).

MUELLER, I. 1964. *Introduction to Satellite Geodesy*. New York: Frederick Ungar Publishing Company.

SEPPELIN, T. O. 1974. The Department of Defense world geodetic system 1972. Proceedings of the International Symposium on Problems Related to the Redefinition of North American Networks. Frediction, New Brunswick, Canada, May.

SMITH, J. R. 1988. *Basic Geodesy—An introduction to the History and Concepts of Modern Geodesy Without Mathematics*. Rancho Cordora, CA: Landmark Enterprises.

STANSELL, T. A. 1978. The TRANSIT navigation satellite system—status, theory, performance, applications. Magnavox Government and Industrial Electronics Company, Report 5933, Torrance, CA, Oct.

TORGE, W. 1991. *Geodesy*. Berlin–New York: Walter DeGruyter.

VANIECK, P., and E. KRAKIWSKY, 1982. *Geodesy: The Concepts*. Amsterdam–New York–Oxford: North-Holland Publishing Company.

14

Inertial and Satellite Positioning Surveys

David F. Mezera and Larry D. Hothem

14-1. INTRODUCTION

Godetic control networks have classically been divided into two distinct categories: (1) horizontal and (2) vertical. Each network has its own respective set of monumented points—i.e., horizontal control (or "triangulation") stations and bench marks in horizontal and vertical networks, respectively. Similarly, classical control surveying methods are divided into two nearly independent categories: (1) horizontal methods that include traversing (Chapter 9), triangulation (Chapter 11), trilateration (Chapter 12), and combinations of the three; and (2) vertical methods that consist of differential and trigonometric leveling (Chapter 6).

More recently, methods have developed that overlap the two classical categories, enabling direct determinations of three-dimensional geodetic positions or position differences. The computer revolution, space-age spin-offs, and related technology have fostered the development of nonconventional geodetic surveying instruments, systems and techniques such as *electronic tachymeters* or *"total station" instruments*; *photogrammetric geodesy*; *laser ranging*; *very long base-line interferometry* (VLBI); and *inertial and satellite positioning*. Some of these are most appropriate for scientific applications—e.g., earth crustal-movement studies. However, geodetic positioning (*geopositioning*) surveys by inertial and satellite methods have proved feasible for many applications previously restricted by default to classical geodetic control surveying methods. Future developments will probably expand the number of potential applications, so reliance in these two nonconventional control surveying methods will increase rapidly.

14-1-1. Inertial Positioning

Inertial surveying involves determining position changes from acceleration and time measurements and sensing the earth's rotation and local vertical direction. Adapted and modified to meet surveying practice requirements, *inertial positioning systems* are relatively new evolutions. The first commercial unit became available for nonmilitary use in 1975. Despite the short period of time since its introduction, the inertial surveying system (ISS) has been quite widely accepted. Applicability to surveying, mapping, geodesy, and engineering projects has been proven by numerous successful applications. The U.S. Bureau of Land Management's Cadastral Survey Division employs several ISS units for extensive original subdivisions of public lands in Alaska. They have been used on many federal mapping control projects in Canada. Inertial methods have also provided control for various geophysical prospecting projects.

An ISS features a number of attractive advantages when compared with more conventional surveying instrumentation. The equipment is typically mounted in a vehicle such as a van or helicopter, so the measurement work proceeds rapidly, with only brief stops required at survey stations of interest and a few selected intermediate points. Clear sight lines between adjacent stations are not required as in conventional traverse and triangulation surveys. Therefore, an ISS is especially well-adapted to surveys that involve numerous points, long distances between them, or areas where sight-line lengths are restricted by structures, vegetation, or rugged terrain. Considerable savings in time and labor costs, key features of an ISS, can be realized for such projects.

14-1-2. Satellite Positioning

Geodetic surveying techniques dependent on artificial satellites began in the early 1960s. Soon after the first Sputnik satellite was launched in 1957, results of satellite tracking and orbit-determination activities led to investigations of possible satellite-aided navigation systems. These, in turn, quickly demonstrated the potential for accurate satellite-based systems for geopositioning surveys. Since that time, numerous different schemes have been proposed, and several systems have reached operational status. *Satellite triangulation*, *SECOR*, and *Doppler satellite positioning* were three of the earliest systems to be investigated and developed. Satellite triangulation involved the use of precise metric cameras set up at widely spaced stations to simultaneously photograph illuminated satellites against their respective star backgrounds. Geodetic positions were determined by methods similar to photogrammetric aerotriangulation. Sequential

collation of ranges (SECOR) was essentially a trilateration system, with distances (ranges) from ground stations to satellites measured electronically using radio signals. Distances from three known satellite positions to an unknown ground point were sufficient to determine its geodetic position.

Doppler satellite positioning methods superseded both satellite triangulation and SECOR by the early 1970s, when portable geodetic Doppler receivers became available. Improved accuracy was also possible using the Doppler method. The design and development of a new-generation satellite global positioning system (GPS) began in the 1970s and continued in the 1980s.

Like inertial systems, satellite positioning does not require optical line of sight between adjacent survey points. However, a satellite receiver cannot operate if nearby obstructions block the satellite's signals from reaching the antenna. Both Doppler and GPS receivers can be operated in a point (absolute) or relative positioning mode. Demonstrations based on developing GPS technology have indicated a potential for the relative positioning of points to subcentimeter accuracy, with just a few minutes of observation.

14-2. DEVELOPMENT OF INERTIAL SURVEYING SYSTEMS

Navigation systems based on inertial technology continue to be used extensively in ships, aircraft, and space vehicles. The marine gyrocompass was invented in 1906, providing a nonmagnetic navigation aid for ships. The horizon and direction gyro was introduced during World War I, enabling pilots to orient aircraft with respect to the earth. The capability to derive reliable travel distances from measured accelerations and time intervals was developed during World War II; combined with advanced gyroscope technology, it provided the basis for today's modern navigation systems. Inertial surveying systems have evolved from these.

14-2-1. Military Initiatives

In the early 1960s, the U.S. Army Engineer Topographic Laboratories (ETL) contracted with General Electric to study the development of an artillery surveying system based on inertial navigation concepts. Between 1965 and 1972, ETL contracts with Litton Systems, Inc. resulted in a successful working model called the position and azimuth determining system (PADS). Test accuracies of ±10 m in latitude, longitude, and height over 210-km open traverses strongly indicated that the PADS might be modified for geodetic surveying applications. By 1975, successive refinements by Litton for the U.S. Defense Mapping Agency (DMA) had produced, first, the inertial positioning system and, later, the rapid geodetic surveying system. The latter model was capable of position and elevation accuracies of ±1 m when used in a closed traverse mode. This conclusively demonstrated the feasibility of inertial accuracies sufficient for many surveying purposes. The capability of inertial systems for determination of gravity and deflections of the vertical was also verified by the Litton and DMA tests.

14-2-2. Commercial Development

By early 1975, Litton had developed, concurrently with its work for the DMA, the Auto-Surveyor, a commercial version of the inertial positioning system. Since then, ISS units have been used for production work by several government agencies and numerous private enterprises in the United States, Canada, and elsewhere. In addition to the Auto-Surveyor, systems have been developed and are currently marketed by Ferranti Ltd. inertial land surveyor (FILS), and Honeywell, Inc. (GEO-SPIN).

The purchase of an ISS requires a large investment, with current costs in excess of half a million dollars for complete systems. Despite this, they have proven to be cost-effective for organizations with work volumes large enough to efficiently utilize their capabilities. Inertial surveying services are now provided on a contract basis by several private consulting companies. Because of high purchase costs, most inertial survey users will subcontract to one of these organizations for services or lease of equipment.

Reliabilities of early ISS versions were somewhat low, a common shortcoming of any complex, new hardware development. As newer versions were introduced, reliabilities have gone up. There have been reported average ISS "downtimes" as low as 5%.

14-3. INERTIAL POSITIONING THEORY

An in-depth knowledge of the operational theory and detailed inner workings of an ISS are not needed to successfully use inertial technology for surveying. However, it is helpful to know the basic theory and general function of its principal components rather than simply regarding the device as a "black box." Such knowledge will improve understanding of the capabilities and limitations of an ISS, leading to a well-informed choice between available surveying methods. If an inertial survey is selected, this background information will help to optimize survey design and organization. This section, therefore, introduces the basic concepts of inertial positioning.

Accelerations and times are the fundamental quantities measured by an ISS. Acceleration is a vector quantity with magnitude and direction, both of which typically vary continuously as the system moves from one point to another along the survey. At each precisely measured time interval, acceleration components are measured in directions parallel to three mutually orthogonal axes.

Acceleration is the rate of velocity change with respect to time. If a constant acceleration is maintained for a short time, the incremental velocity during that interval equals the product of acceleration and time. From the example in Figure 14-1, an acceleration of 2 m/sec^2 occurring over the 1-sec interval $a - b$ produces a velocity increase of $2 \times 1 = 2$ m/sec. Repetition of such a numerical integration process yields the terminal velocity for each time interval. Acting over the 1-sec interval $d - e$ in Figure 14-1, a constant velocity of 5 m/sec produces an incremental travel distance of $5 \times 1 = 5$ m.

In an ISS, the three acceleration components are measured at very small time intervals —on the order of 60 times per second—so they can be numerically integrated to yield accurate travel-distance components. When the ISS stops at a point of interest, each of three orthogonal components of distance traveled to that point equals the sum of incremental components in that particular direction since the previous stop. The ISS is a relative positioning system—i.e., distance components are measured from an initial known reference position, and new points are located relative to that control point.

14-4. INERTIAL SYSTEM CONFIGURATION

An ISS is a complex instrument consisting of precisely made mechanical parts, electronic circuitry, control devices, and computer hardware and software. However, to assist in the general understanding of the system's operation, a brief description of its principal components will be given, with emphasis on each component's function rather than its physical construction.

14-4-1. Accelerometers

The devices that measure accelerations are called accelerometers. Each may be visualized as a small pendulum equipped with a feedback system (Figure 14-2). When the accelerometer is at rest or moving at a uniform velocity, the pendulum hangs at its rest position. The pendulum's inertia resists changes in velocity, so it will try to swing away from its rest position when acceleration occurs. A detector senses the pendulum motion's beginning and sends an amplified electrical signal to a forcing (or torquing) system that, in turn, directs application of a force (or torque) just sufficient to

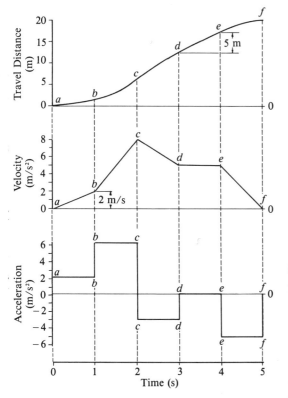

Figure 14-1. Acceleration, velocity, and travel distance as a function of time.

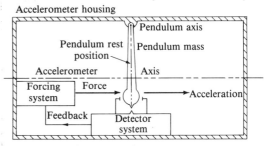

Figure 14-2. Simple accelerometer.

keep the pendulum from swinging. The feedback signal to the torquing system is a direct measure of applied force and, thus, acceleration also, since the two quantities are proportional according to Newton's first law, which states that force equals mass times acceleration. The electrical signal is scaled and digitized, then multiplied by the corresponding time interval to obtain the incremental velocity component.

Each accelerometer is fully sensitive only along its axis, the direction normal to both the pendulum's swing axis and its rest position axis (Figure 14-2). If acceleration is perpendicular to this axis, the instrument will detect nothing. If it occurs in any other direction, only that component parallel to the axis will be detected. Three orthogonally mounted accelerometers are thus required to measure the three constituent components of the total acceleration vector (Figure 14-3).

Without control or restraint or isolation from vehicle motions, the orthogonal accelerometer triad's orientation will vary with respect to inertial space as the ISS moves from point to point during a survey. To correctly combine incremental distance components between points for ISS position-change determinations, these orientation variations must be controlled and monitored.

14-4-2. Gyroscopes

Gryoscopic control is employed to stabilize the accelerometer triad in inertial space and provide a reference for detecting its attitude variations. A gyroscope, or gyro, has a symmetrical rotor—a wheel or ball of dense metal made to spin rapidly about its axis on bearings located in a gimbal housing. The rotor develops a large angular momentum and resists any forces that would tend to change its spatial orientation. Once a gyro has aligned itself in inertial space, a stable reference is provided that permits translations but resists rotations (Figure 14-4a).

Each gimbal also has a rotation axis at right angles to the rotor axis. It permits the spin-

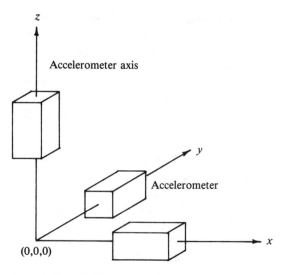

Figure 14-3.   Orthogonal accelerometer triad.

ning rotor to retain its original angular orientation, despite gimbal rotations about either the rotor or gimbal axis. This gyro is said to have a "single degree of freedom" since gimbal rotations about a third axis, mutually perpendicular to the other two, will change the rotor orientation. For a gyro with "two degrees of freedom," the gimbal axis itself rotates in bearings set in an outer gimbal, with a rotation axis perpendicular to both inner gimbal and rotor axes. This permits the rotor to retain its angular orientation during outer gimbal rotations in any direction (Figure 14-4b).

The gimbal rotation axes have detector-feedback-torquer devices, similar in function to those on the accelerometers, that serve to counteract unwanted rotations. The torquers may also be used to drive the gyros to a desired orientation.

14-4-3. Inertial Measuring Unit

The accelerometer triad and associated gyroscopes are mounted on an *inertial platform*, the heart of the inertial measuring unit (IMU). If single-degree-of-freedom gyros are used, there must be one corresponding to each ac-

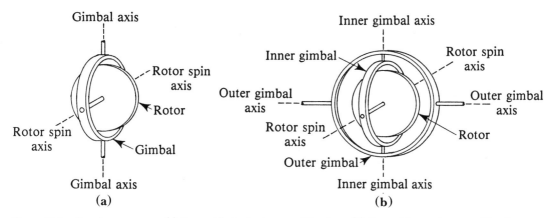

Figure 14-4. Simple gyroscopes. (a) Gyro with single degree of freedom. (b) Gyro with two degrees of freedom.

celerometer axis (Figure 14-5). Two gyros are sufficient if they each have two degrees of freedom. The accelerometer triad axes define the inertial platform axes, denoted either *x-y-z* or east-north-vertical. The whole inertial platform, of course, must also be gimbal-mounted with two degrees of freedom to provide for maintenance of its angular orientation in space, as the survey vehicle moves from one point to another. These gimbal rotation axes are also equipped with detector-torquer devices to monitor platform orientation varia-

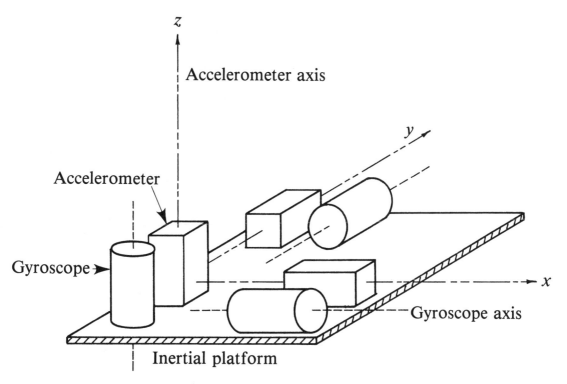

Figure 14-5. Inertial platform schematic.

tions and drive the platform to desired attitudes. An actual inertial platform, disconnected from the other IMU hardware, is shown in Figure 14-6. The IMU and associated parts are housed in a closed case that provides a thermally stabilized environment as well as protection from damage (Figure 14-7).

14-4-4. Control and Data-Handling Components

An *on-board computer* is an integral component of an ISS. The final computation and adjustment of the data are usually done on a larger off-line computer, but the on-board unit may perform the initial position calculation in "real time" as the survey proceeds. The computer controls and monitors system operation, and it may check and filter the raw measurement data before they are recorded for later

processing. It may compute misclosures when an inertial traverse closes on a control point, and it also directs the calibration and alignment of the platform at the initial point of a survey.

A *control and display unit*—typically mounted on or near the instrument panel of a survey vehicle—controls operations of the IMU and data-recording device. It allows communication with the on-board computer, initialization of calibration and measurement sequences, input of externally collected data, and visual readout of preliminary coordinates, closure errors, and other operational data (Figure 14-8).

A *data-recording unit*—typically a magnetic tape cassette recorder— is used to capture the raw or filtered measurement data for later processing and analysis. It must be a well-designed unit to ensure reliable operation in the survey vehicle's variable environment. Simi-

Figure 14-6. Inertial platform for Ferranti ISS. (Courtesy of Shell Canada Resources, Ltd.)

Figure 14-7. Complete Ferranti FILS II ISS. (Courtesy of Shell Canada Resources, Ltd.)

Figure 14-8. Control and display unit for Litton Auto-Surveyor.

larly, high-quality data tapes must be used for recording and preserving measurement information.

14-4-5. Survey Vehicle and Power Supply

An ISS may be transported in a variety of different vehicles. For ground operations, a van-type light-duty truck is suitable (Figure 14-9), but helicopters are also frequently used, especially in remote areas where ground travel may be difficult or impossible (Figure 14-10). For a ground vehicle, the main requirement is installation space for the necessary equipment. In addition to seats for the driver and operator, there must be room inside the vehicle for the inertial measuring unit, computer, recorder, batteries, and the control and display unit, plus space in the engine compartment for the power supply alternator. Four-wheel-drive capability or even a tracked vehicle may be required if off-road travel is necessary. A helicopter must be of sufficient size and power to safely accommodate both the equipment and personnel. For some applica-

tions, the driver/pilot may also serve as the operator.

A power supply system must be provided to support operation of the IMU and its thermally stabilized environment, the on-board computer, control and display unit, and data recorder. For a truck-type survey vehicle, this is typically a 24-V battery system with recharge provided by a special heavy-duty alternator powered by the truck's engine. For helicopter use, the ISS may be operated from the aircraft's 28V power system.

15-5. INERTIAL SURVEYING OPERATIONS

There are two primary system types used for inertial surveying: (1) space-stabilized and (2) local level. If an ISS is "space-stabilized," its inertial platform orientation is held fixed with respect to inertial space for the survey duration. With a local-level (or local-vertical, local-north) system, two platform axes are aligned parallel to local-vertical and north directions, respectively, so the third axis points eastward.

Figure 14-9.   Truck-mounted Litton DASH II ISS. (Courtesy of International Technology, Ltd.)

Figure 14-10. Ferranti FILS II inertial system mounted in Bell Jet Ranger helicopter. (Courtesy of Shell Canada Resources, Ltd.)

This local orientation is maintained and updated at each new point throughout the course of a survey. Much of the following description of ISS operation refers primarily to a local-level system.

14-5-1. System Calibration

Before beginning each day's operation, or each time the ISS is turned on anew, the system must be warmed up and calibrated. This calibration or alignment is mainly an automated sequence of self-tests initiated by the operator, but thereafter directed by the on-board computer under system software control. The calibration procedure may require an hour or more to complete. During this time, the IMU components are monitored for operational stability, and the gyro and accelerometer parameters tested to see if they fall within acceptable ranges. These parameters are recorded for later use during filtering and adjustment. The calibration sequence is repeated, perhaps several times, if the parameters do not meet specifications. Continued failure to successfully calibrate may necessitate system adjustment or repair.

In addition to daily calibrations, it is desirable to conduct periodic dynamic calibrations. Ideally, this would consist of inertial traversing over an L-shaped course, with traverse legs along cardinal directions and precise coordinates at the angle and terminal points. Analysis of the results will yield accelerometer and gyro parameter information.

14-5-2. Initial Orientation of the Inertial Platform

If the maximum sensitivity direction of a stationary accelerometer is perpendicular to the direction of local gravity, it will indicate zero acceleration. At the initial point of a survey with a local-level system, the platform leveling is accomplished by orienting the X- and Y-accelerometers to produce zero accelerations and aligning the Z-axis of the system along the local plumb line. If deflection of the vertical is known, platform orientation with respect to the geodetic reference is also defined.

A gyrocompassing procedure is used to align the sensitive direction of the Y-accelerometer toward local north. The earth's rotation causes a torque on the corresponding gyroscope axis, and the resulting precession concludes with it aligned in the plane of the earth's rotation axis (Figure 14-11a).

14-5-3. Compensation for Platform Movements

Without continuous correction of the gyroscope orientations, the local-level alignment of a platform is lost immediately because of the earth's rotation, even though the ISS remains stationary at the initial point. Compensation to maintain alignment with the geodetic reference is accomplished by the gyroscope torquers, as directed by the on-board computer (Figure 14-11b). The angular orientation correction is based on the earth's mean rotation rate and elapsed time since the initial orientation or previous correction.

When the platform is moved to another point, a similar loss of alignment occurs if the computer does not signal for orientation correction. But the required compensation is now composed of two parts: (1) allowance for earth rotation during the elapsed travel time and (2) compensation for the distance traveled (Figure 14-11c). With a space-stabilized ISS, compensations for both components of platform movement are computed and applied during the data processing, rather than torquing the gyros to reorient the platform during a field observation period.

Periodically during an inertial survey, it is necessary to stop the vehicle and initiate a control instruction sequence to inform the computer that the ISS is stationary, and thus all three current velocity values should be zero. This procedure is called a *zero velocity update* or ZUPT. Each accelerometer and gyro output is monitored and recorded as an indication of that component's drift since its initial calibration state. A ZUPT requires a stop duration of a minute or shorter.

14-5-4. Position Measurements

When the ISS reaches a point where a geodetic position is desired, a *station mark* is initiated. This position measurement procedure is similar to the ZUPT, except that in addition to the accelerometer and gyro data, preliminary coordinates are also recorded.

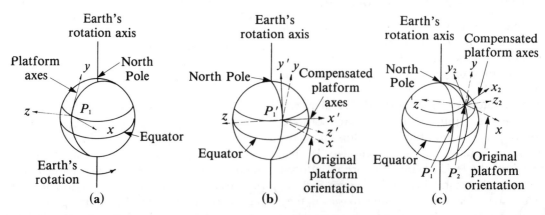

Figure 14-11. Compensation for inertial platform movements. (a) Local-level platform orientation at point P_1. (b) Compensation for rotation of point P_2 to P_1. (c) Compensation for rotation to point P_1 and platform travel to point P_2.

Since the basic point of reference for an ISS measurement is at the inertial platform's center and thus inaccessible, an offset reference mark is set on the exterior of the IMU by the instrument manufacturer. The ISS computer software is designed to account for components of this offset, in all position computations.

During field operations, offset measurements must be made since the reference mark cannot be set directly on the survey points. However, in most cases, the offsets are not measured directly to the reference mark on the IMU but for convenience and speed to a secondary reference point instead. This *measuring mark* may consist of a simple sighting

(a)

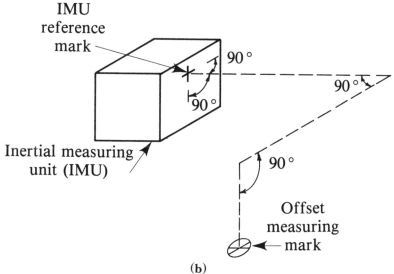

(b)

Figure 14-12. ISS offset measuring mark. (a) Offset measuring-mark sighting device beneath driver's door. (b) Measurement of offset components from IMU to measuring mark.

device mounted on the vehicle exterior—truck door or helicopter skid—for convenient use by the driver (see Figure 13-12a). For points that are not so accessible, offset direction, distance, and elevation difference may be measured from a protractor mounted on the front or rear of the vehicle (see Figure 14-13). For greater accuracy, especially for larger offsets, these components may be measured with an electronic tachymeter or total station instrument from a measuring mark located inside the vehicle (see Figure 14-14).

The offset components to the measuring mark may be determined using conventional surveying instruments (see Figure 14-12b). These values are input to the computer as constants to be applied to each position determined. Any additional offset from the measuring mark to a particular survey point also must be input, so the total offset from an inertial platform to the point will be included in the position computation. Regardless of the technique employed, care must be used for all offset measurements to ensure that desired positional accuracies are maintained.

For some surveys with less stringent accuracy requirements, both ZUPTs and station marks have been accomplished with a nearly stationary ISS is in airborne helicopter hovering over the point. The U.S. BLM has investigated the use of a similar technique for meandering water boundaries in Alaska.

14-6. INERTIAL DATA PROCESSING AND ADJUSTMENTS

Like all surveying measurements, those made with an ISS are subject to errors from instrumental, natural, and personal sources. Many procedures included in the calibration, measurement, and data-reduction/adjustment phases are designed to limit error magnitudes and reduce their effects on final survey results. The measurements are also subject to mistakes or blunders due to confusion or carelessness, but these should be discovered and eliminated through the use of repeated measurements and independent checks.

14-6-1. Error Sources and Characteristics

Nonorthogonality of the three accelerometer axes may cause the measured acceleration components to be incorrect. The daily ISS calibration prior to field observations can virtually eliminate this error if it remains con-

Figure 14-13.   Measuring offset from vehicle measuring mark to survey point. (Courtesy of Shell Canada Resources, Ltd.)

Figure 14-14. Offset measurement to survey point using electronic distance-measuring instrument. (Courtesy of Shell Canada Resources, Ltd.)

stant during the day's survey. The electronic signals are scaled and digitized, so if scaling is incorrect, double integrations of resultant erroneous accelerations result in incorrect travel-distance components. Errors in elapsed time measurements also result in incorrect travel distances, but time-measurement accuracy is very high, so these errors are very small.

Gyroscopes slowly drift away from their original alignments, thus allowing the inertial platform to drift. Most of this drift is a linear function of time that can be modeled and, therefore, eliminated during data reduction, but the small irregular component is more difficult to eliminate, and random position errors may result. Accelerometers also have generally linear drifts for which corrections can be made, but small, irregular drifts may contaminate accelerations, thus affecting computed distance components. Vehicle vibrations caused by the engine or strong winds may also cause accelerometer errors during a stop for a ZUPT or position determination.

Setting the measuring mark over a survey point or measuring the mark-to-point offset are processes subject to all the observation errors inherent in conventional surveying methods. The same is true of the procedure for finding the measuring-mark offset from the IMU reference mark. The ISS control unit operations, including data input, always provide opportunities for human mistakes or blunders, as do the data-reduction procedures. The computer software may be designed to automatically detect some large discrepancies, but as in all surveys, good practice will include careful and thorough checking of all input, output, and computed values.

14-6-2. Data Filtering

The on-board computer of an ISS may operate software that statistically filters raw measurement data as it is collected. The best estimates are obtained by a *Kalman filter*, an algorithm that weights each new data value based on the characteristics of previous measurements and prior knowledge of the nature and usual magnitudes of errors. Initial estimates of gyro and accelerometer error parameters, such as drift rates and biases, are supplied to the Kalman filter after the daily calibration procedure. As the survey proceeds, these parameters are updated at each stop for a ZUPT or station mark. This statistical filtering process significantly retards error growth in preliminary coordinates calculated by the on-board computer, so these real-time position estimates may be accurate enough for some purposes without further adjustment. The position error accumulates so rapidly in an ISS without a Kalman filter that rough field coordinates have little value. In such cases, it is necessary to wait for off-line postprocessing of the raw data.

14-6-3.   Horizontal and Vertical Positions

For each inertial traverse, coordinates of the beginning control station are input to the ISS. Following recommended practice, a survey concludes at a second control point with known coordinates, forming a closed traverse. The reliability of an open-ended traverse is always uncertain and should be avoided regardless of the accuracy sought.

At each desired survey point along a traverse, the ISS accumulates incremental travel-distance components from the previous point to obtain three-dimensional geodetic coordinates of the current point, relative to the initial control station. Coordinates are computed in terms of latitudes, longitudes, and elevations—although for display purposes and later use, latitudes and longitudes may be transformed into a local plane coordinate system such as state plane coordinates (i.e., output in terms of X, Y, elevation).

The vertical accelerometer measures accelerations due to the combined effects of vehicle motion and the local gravity vector. These two components must be separated to determine elevation differences between surveyed points. A gravity-effect estimate can be re-

moved from the vertical acceleration at each integration step. This estimate is derived from a mathematical model of the earth's gravity field or measured gravity value at the initial station. Elevation differences are computed using residual vertical accelerations.

For a space-stabilized ISS, each of the three accelerometers will detect a component of vertical acceleration, since platform orientation is not held fixed with respect to local vertical. In this case, the computer software must enable determination of the net vertical acceleration and separation of the gravity-field component, during either on-line processing by the on-board computer or off-line postprocessing.

14-6-4.   Data Adjustments

In following good practice, an inertial traverse begins at a known control station and ends at another. Closing a traverse on a known point provides a reliability check of intermediate survey points, since preliminary closing point coordinates—inertially derived—can be compared with known values. Any coordinate misclosures should be within both the tolerances expected for the ISS utilized and ranges specified for the particular survey. If either test fails, the traverse must be resurveyed.

If misclosures are within allowable limits, preliminary coordinates should be adjusted (or *smoothed*) by distributing accumulated errors of closure in proportion to elapsed travel time, travel distance (analogous to the compass rule traverse adjustment), or a time-distance combination. Smoothed coordinates may be determined by the on-board computer immediately on conclusion of the traverse run and recorded on the data tape. Alternatively, smoothed coordinates can be obtained during off-line postprocessing of the data.

If desired, detailed mathematical algorithms may be employed to rigorously model ISS error parameters and thus produce more accurate distributions of accumulated closure errors. Despite various attempts to improve modeling of the inertial measurement process, experience has shown that smoothed coordinate values typically exhibit sizable systematic error effects. To help reduce such systematic errors and provide desirable checks, each traverse should be run in both forward and reverse directions—e.g., from control point *A* to control point *B*, then from *B* back to *A*. Hereafter, a traverse run in this fashion will be referred to as a "single inertial traverse." This method is analogous to the common procedure of measuring with a theodolite in both direct and inverted telescope positions to cancel systematic instrumental errors from the mean angle value. Adjusted survey point coordinates can be taken as the weighted mean of smoothed values from forward and reverse traverse runs.

Uncertainties of adjusted survey point coordinates can be reduced somewhat by additional repetitions in the forward and back directions, just as any observed value may be statistically improved by repeated measurements. In any case, numbers of forward and reverse runs should be equal along any traverse line.

14-7. INERTIAL SURVEY DESIGN AND TYPICAL RESULTS

Because of the unique operational characteristics of an ISS, careful preparation and detailed organization are extremely important, both before and during an inertial survey.

14-7-1. Logistics

Since inertial system hardware is usually mounted semipermanently in a survey vehicle, accessibility of points to be positioned is a primary consideration during initial planning and design. When an ordinary ground vehicle is used, survey points must be either on or near traversable roadways or trails, depending on the offset measurement method utilized and terrain difficulty. Availability of a four-wheel-drive or all-terrain survey vehicle to transport the ISS allows more freedom in selecting survey point locations. Helicopters provide even greater flexibility, especially in remote areas where ground vehicle travel is difficult or impossible. However, the relatively large space required for helicopter landing may generate additional logistical problems, particularly in heavily forested areas.

Selecting and marking desired survey points are tasks best completed before ISS measurements commence. Since errors tend to build up as elapsed time increases, unnecessary delays should be avoided during inertial traversing. Because the number of points positioned by an inertial survey is usually large, work involved with reconnaissance, selection, marking, referencing and describing desired station locations comprises a significant portion of the total project effort. If many of the points must be permanently monumented survey stations, the labor, materials, and transportation required can make this project phase very costly in terms of both time and money.

Whether points are temporary or permanent, they must be clearly marked or flagged, so an ISS can be transported directly to each one without delay. In some cases, it may be necessary to run a "scout vehicle" in advance of the ISS unit to guide it to consecutive survey station locations. If approximate coordinates of a station are known, the navigational capability of an ISS may be used to locate the point marker. The display unit can be directed to indicate direction and distance to a desired point as the ISS approaches it. In congested areas, traffic-control measures may be necessary to ensure safety for the ISS vehicle and its occupants. Inertial surveys are sometimes conducted during late night and early morning hours to avoid peak traffic conditions.

14-7-2. Other Design Considerations

In addition to logistics, other factors must be considered in the design of an inertial survey. An ISS is a relative-positioning device; thus, two known geodetic control stations must be available to serve as inertial traverse beginning and ending points, respectively. Open-ended traverses and closed-loop traverses with only one control station should be avoided, since no satisfactory check or systematic error-correction technique exists for either of these survey types. In fact, if the purpose is to densify an area's existing geodetic control network, the U.S. Federal Geodetic Control Committee (FGCC) recommends that each inertial traverse tie into a minimum of four existing control points rather than just two. Additionally, each traverse should be run in both forward and reverse directions as described in Section 14-6-4. Accuracies of inertially derived positions cannot exceed those of control stations utilized for the survey and generally will be somewhat lower. Unchecked or suspect geodetic control should be avoided since erroneous control values incorporated into an inertial survey will degrade the accuracy of ISS results.

The direct relationship between elapsed time and ISS positional error accumulation constrains allowable traverse length. The desired quality of results limits allowable elapsed time intervals between both ISS stops at terminal control stations and consecutive ZUPT or station mark stop-points. Typical ZUPT intervals vary between about 1 and 8 min, depending on the accuracy level desired. Total elapsed time for a single inertial traverse is commonly limited to 1 or 2 hr or shorter for most applications. But, depending on the ISS vehicle type, inertial traverse lengths can easily reach 10 to 100 km or more.

An ISS produces the best results when vehicle travel approximates a straight-line path connecting the terminal control stations, desired intermediate points, and ZUPT stop-points. Point spacings and vehicle travel rate should each be nearly uniform and sharp horizontal direction changes ("doglegs") avoided. Normal BLM guidelines restrict any desired point along an inertial traverse to be within a corridor extending no more than 0.8 km either side of a straight line connecting the two terminal control stations. The FGCC recommends a maximum angular deviation from a straight-line path of 20 to 35°, depending on the desired accuracy.

For greatest reliability, each important survey point is linked to control stations by more than a single inertial traverse. Such a point should be connected to additional control stations by a crossing inertial traverse (see Figure 14-15a). Ideally, for a control densification or similar area-wide survey, known stations are

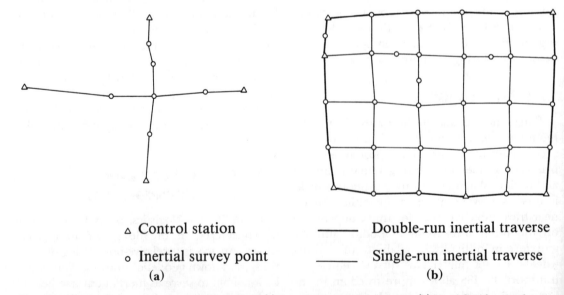

△ Control station   ———— Double-run inertial traverse

○ Inertial survey point ——— Single-run inertial traverse

(a)   (b)

Figure 14-15. Inertial traverse survey configurations. (a) Intersecting inertial traverses. (b) Inertial grid network.

located around the area perimeter, and desired survey point coordinates established using a series of crossing and interconnected inertial traverses surveyed in a uniform grid pattern (see Figure 14-15b). In practice, it is very difficult to design such perfect survey networks, especially for projects employing ground vehicles traveling on established roadways. However, to achieve maximum ISS accuracy, the network design process must pursue this ideal configuration, and inertial traversing must be done with great care, with a least-squares network adjustment used to obtain the most probable coordinate values. As with any type of survey, it is desirable to verify the inertial results on a sample of project points, making independent checks using another method.

14-7-3. Typical Inertial Survey Results

In a discussion of obtainable ISS accuracies, it is important to describe the survey conditions under which the results were produced. The possibility exists that uncorrected systematic errors remain even after a rigorous network survey and adjustment; it most especially exists for a single traverse with no cross-connecting inertial runs, where accuracy may be affected by such factors as ZUPT intervals, quality of existing control stations, and traverse shape, length, and duration.

For a Litton Auto-Surveyor ISS, a single, generally straight-line traverse of 2-hr duration can be expected to produce positional standard errors for points along the traverse path of about 0.40 m horizontally and 0.30 m vertically. For a similar but L-shaped traverse, the corresponding values are about 1.00 and 0.30 m, respectively. These results assume reasonable care in survey conduct, uniform ZUPT intervals of 3 to 5 min, and use of manufacturer's software. For a grid network of crisscrossing traverses (see Figure 14-15b), the corresponding positional standard errors are both expected to be about 0.15 m for a Litton ISS. This assumes adequate area perimeter control,

traverse durations each shorter than $1\frac{1}{2}$ hr, and a rigorous least-squares network adjustment. Some users have reported relative positional errors below the 0.10 m level under near-ideal or strictly controlled conditions.

When using a Ferranti FILS ISS in a helicopter with ZUPT intervals of 3 to 5 min was used, mean horizontal and vertical errors (relative to convetional survey positions) of about 0.25 and 0.15 m, respectively, have been obtained for generally straight-line traverse of no more than $1\frac{1}{2}$-hr duration and 30-km length. But with a ground vehicle operating in undulating or rugged terrain that necessitated circuitous and L-shaped traverses, the Ferranti ISS produced corresponding mean errors of 0.75 and 0.25 m, respectively—values quite similar to the expected errors for a Litton ISS. Likewise, a Ferranti FILS system should produce grid network accuracies comparable to those for an Auto-Surveyor, or about 0.15 m both horizontally and vertically.

For a Honeywell GEO-SPIN ISS in a ground vehicle, expected accuracies are about 0.15 m horizontally and 0.30 m vertically over a generally straight-line 51-km traverse. These numbers degrade to about 0.50 and 0.35 m, respectively, for a 39-km L-shaped traverse. For two intersecting traverses (as in Figure 14-15a) with lengths of 51 and 26 km, corresponding accuracies of 0.10 and 0.20 m can be expected. In one test, several nearly straight-line traverses were run with ZUPT intervals of 1 min and the results compared with positions obtained by conventional control surveying methods. For six traverses, each 4 km in length with survey mark points at about 0.4-km intervals, root-mean-square (RMS) errors were below 0.06 m, both horizontally and vertically. On two double-run 2-km traverses, corresponding RMS errors were 0.04 m or smaller. Thus, the potential of a GEO-SPIN ISS for establishing closely spaced control stations in a grid network (Figure 14-15b) is excellent. If traverses of such high quality are combined with a grid pattern and subjected to a rigorous adjustment, comparable accuracies should be attainable over much longer traverse lengths.

An ISS also has the potential to yield information about the earth's gravity field. Although not ordinarily needed for most survey purposes, gravity anomalies and deflections of the vertical may be derived from the observed data, provided that appropriate software configurations are available for both the on-board computer and postprocessing hardware. This gravity information may be a very important component of certain geodetic, geophysical, and geological surveys. Under good conditions, accuracies of 1.0 mGal and 0.5″ have been obtained for gravity anomalies and deflections of the vertical, respectively.

14-8. INERTIAL SURVEYING APPLICATIONS

Many different types of surveys have been successfully executed using inertial technology. In many instances, ISS usage has proved to be an economically viable alternative to conventional surveying methods, particularly for larger projects requiring geopositioning of many points over large areas.

14-8-1. Application Examples and Projections

As previously implied, most ISS projects could be characterized as control surveys. In such cases, an inertial survey provides a spatial framework of geodetically positioned points to support another activity, such as a more detailed survey utilizing a different technique.

An example of such a situation is a topographic mapping project utilizing photogrammetry (see Chapter 20). Such a project might be necessary for mapping a utility transmission-line corridor, a highway or railway right-of-way, or a new dam and reservoir site. An ISS can provide the required photo ground control points, positioned with respect to the area's existing geodetic control network. ISS accuracies would normally be adequate to support photogrammetric mapping. In remote areas,

the helicopter mode probably would be most efficient, but a ground vehicle could be used if a good road network exists.

The navigation capability of an ISS can be relied on to guide the survey vehicle to a preselected control point location. For example, as an aerial photography flight is planned, desirable photo control locations may be identified on a map and coordinates scaled. These values can be input to the on-board computer and used to direct the ISS to a desired point's immediate vicinity. A suitable point location can then be selected after a first-hand inspection of the area. While accurate position is observed with the ISS, this point is staked and flagged for a separate crew that will place the photo targets.

Cadastral surveying is another major application area for which an ISS is well-suited, particularly for large land parcels in remote areas. The BLM Cadastral Survey Division utilizes helicopter-borne Litton Auto-Surveyors to establish original U.S. public land survey system (PLSS) boundaries in Alaska. Inertial surveys have also been used to determine coordinate positions of remonumented PLSS section and quarter-section corners in several other states, such as Illinois and Wisconsin, so the geodetic and legal (real property) networks could be integrated.

Other extensive applications of inertial methods have included engineering, geophysical, and construction surveys. Seismic and gravity surveys for geophysical prospecting have utilized ISS geopositioning, including some surveys in offshore areas and positioning for drilling platforms. Inertial systems have also been used in helicopters in conjunction with laser profiling of ground terrain.

Research and development have been conducted using an ISS to determine the position and orientation of an aircraft and camera during an aerial photography flight; this system, when perfected, will greatly reduce photogrammetry ground control requirements. Some ISS use has developed in conjunction with mapping boundaries of natural resources, such as wetlands and soil types; similar inertial surveys could be employed on floodplain and land use/land cover mapping.

When the special characteristics of an ISS are considered, it seems likely there are numerous additional applications yet to be exploited. This is a self-contained system that has no line-of-sight restrictions, can move rapidly from point to point in a ground vehicle or helicopter, and produces three-dimensional geodetic positions with an accuracy adequate for many purposes. Better understanding of its capabilities should lead to increased future use.

14-8-2. Economic Considerations

Because of its high cost, the purchase of an ISS is only viable for an organization that has an applications volume sufficient to utilize its very high productivity. Measured in terms of output point positions, it may be as much as 5 to 20 (or even more) times that of conventional survey methods. However, alternatives to purchase exist, including inertial surveying services via contract and inertial equipment leasing.

Several factors may influence the cost of an inertial project. A significant part of the total expenditure may be the mobilization cost—the expense to ship equipment and transport personnel to the project site, install the ISS in a survey vehicle, and other overhead factors that are relatively fixed. The project area's geographical location with respect to the source of the inertial equipment or services can greatly influence these mobilization costs. Of course, the number and accuracy of survey points desired and project area size are important factors: The costs per point generally decrease rapidly as point numbers increase. Helicopter transportation, on a daily basis, will be much more expensive than use of a ground vehicle; however, increased productivity could make helicopter use much more economical. Advance preparation and planning by the client, as well as active participation during the actual inertial survey, will help keep down

the project cost. Because of the many expense variables involved, it is difficult to make generalized statements about inertial survey outlays. However, there have been many inertial projects reported to have a cost per point one-half to one-quarter (or even below) that of an equivalent conventional survey. Corresponding survey duration comparisons appear to be even more dramatic: There have been reports of as much as a 20-to-1 reduction compared with time estimates for corresponding conventional surveys.

14-9. DEVELOPMENT OF SATELLITE POSITIONING SYSTEMS

Shortly after the former Soviet republic launched its first Sputnik satellite in October 1957, its orbit was accurately determined by measuring and analyzing the Doppler frequency curve from a single satellite pass. Shortly thereafter, the U.S. Navy navigation satellite system (NNSS) was conceived. Early research and development was conducted at the Johns Hopkins University Applied Physics Laboratory. The U.S. Navy supported development of the NNSS, commonly referred to as the transit system, to provide accurate, all-weather, passive, worldwide navigation capability for its Polaris submarine fleet.

The first United States satellite to be used extensively for geodetic purposes, the Vanguard, was launched in 1958. The original navigation satellite and another designed to broadcast precise information on its own orbital position were orbited in 1960 and 1961, respectively. Since 1964, there has been at least one transit-equipped satellite continuously operational to provide Polaris navigation.

In addition to the transit system, several other satellite positioning systems developed concurrently during the 1960s. Photogrammetric satellite triangulation was used to establish a North American reference network of 21 widely spaced stations and a 45-station worldwide net that interconnected several previously unrelated areas. At several stations, calibrated metric cameras were synchronized to simultaneously photograph illuminated satellites against their star backgrounds. Some satellites carried flashing lights, others utilized sun illumination. Photographic coordinate measurements of satellite and star images were used to compute directions to the satellites, and geodetic positions of camera stations then found by classical astronomy techniques. Required field equipment was bulky and transportable only by large vans; data-reduction procedures long and complex; and positional accuracy limited to about 5 m, so satellite triangulation was replaced by other methods in the early 1970s.

The ranging method (satellite trilateration) was also utilized in the 1960s, and a phase-shift measurement technique, similar to that employed in EDM surveying instruments (see Chapter 5), used with the SECOR system. A ground station transmitted a phase-modulated electromagnetic signal to a satellite equipped with a SECOR transponder. The signal was received and retransmitted to the ground station, where its returned signal phase could be measured relative to the transmitted signal phase. The observed phase shift is a function of signal travel distance. If one unknown station and three known stations simultaneously measure the ranges to at least three satellite positions, the unknown station coordinates can be determined. Repeatability (precision) using SECOR was about 6 m, but for widely separated stations, relative accuracies as high as 1 : 100,000 were obtained by comparison with conventionally surveyed positions. Other methods also superseded SECOR because of its accuracy limitations and low mobility due to complex bulky ground station antenna and receiver equipment.

Another trilateration method developed during the 1960s, satellite laser ranging (SLR), became operational in the early 1970s, and continues in use today. Distances to satellites equipped with retroreflectors can be measured with 1-cm accuracies. Despite these ex-

cellent results, SLR use is limited primarily to crustal motion studies owing to the complexity of instrumentation and data-processing requirements.

The Doppler NNSS became available for civilian use in 1967, when details of receiver equipment and computation requirements were made public. Until then, usage had been restricted primarily to the navigation of U.S. Navy vessels. Although the control surveying potential of the transit system had been demonstrated, applicable Doppler receiver equipment was bulky with complex field operation and data-reduction procedures. But portable geodetic receivers were developed and available for testing by the early 1970s (see Figure 14-16 and 14-17). Initial results indi-

Figure 14-16. JMR-1 Doppler survey set in field operation. (Courtesy of JMR Instruments, Inc.)

cated that accuracies better than 1.5 m were possible, thus making many geodetic applications feasible. Improvements in hardware, software, and operational procedures have resulted in achievable accuracies of 0.5 to 1.0 m for point positioning and 15 to 30 cm for relative positioning. Today, the NNSS is used routinely for both navigation and geodetic control surveying purposes in the private sector.

In 1973, the success and future promise of transit led the U.S. Department of Defense to initiate design and development work on the navigation satellite timing and ranging (NAVSTAR). The first NAVSTAR satellites were orbited in 1978 to support testing and development of both the system's space and ground control segments and receivers for navigation and geodetic purposes.

14-10. SATELLITE ORBITS AND COORDINATE SYSTEMS

All satellite-based positioning systems require that several satellite coordinate positions either be known a priori or observed as an integral part of the measurement scheme. In either case, it is necessary to know the parameters used to define a satellite's orbit and elements of the coordinate system in which its positions will be expressed.

14-10-1. Satellite Orbits

The path around the earth of an orbiting artificial satellite is approximately elliptical (see Figure 14-18a). An ellipse is the locus of points with a constant sum of distances from the two fixed focal points. For a satellite orbit, one

Figure 14-17. Magnavox ANPRR-14 geoceiver. (Courtesy of Magnavox Advanced Products and Systems Co.)

focal point is at the geocenter (the earth's center of gravity). Ellipse size and shape are fixed by two parameters such as semimajor axis and eccentricity. The perigree point marks the satellite's closest approach to earth while the apogee is farthest away. If extended, a line connecting the perigee and apogee, the *line of apsides*, also passes through the ellipse's foci and coincides with the x_s-axis of an orbital coordinate system. This is a right-handed, three-dimensional Cartesian system with its y_s- and z_s-axes in and perpendicular to, respectively, the elliptical plane.

The orbital ellipse orientation, with respect to an astronomical or celestial coordinate system (see Chapter 17), may be defined by four

parameters: (1) the true anomaly, (2) argument of perigee, (3) inclination, and (4) the right ascension of the ascending node (see Figure 14-18b). These four parameters, plus the semimajor axis and eccentricity, define a smooth nonvarying elliptical path and are sometimes referred to as the *Keplerian orbital elements*. However, an orbit would follow such a path exactly only if the earth was a uniform-density sphere, and no disturbing forces acted on a satellite other than gravitational attraction toward the geocenter.

Of course, these conditions are not satisfied, and disturbing factors such as the following must be taken into account: (1) the earth's irregular gravity field; (2) attractions of the

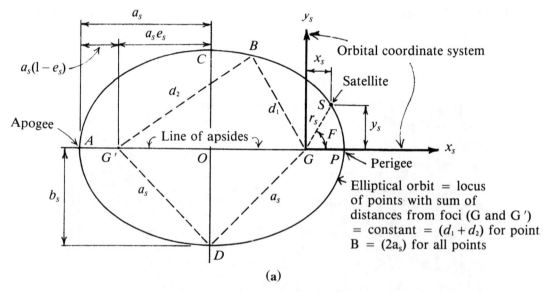

(a)

Figure 14-18. Satellite orbit definition and orientation.
(a) Elliptical satellite orbit and orbital coordinate system.
(b) Astronomical orientation of elliptical satellite orbit.

O = center of ellipse
G, G' = foci of ellipse
G = center of gravity of earth (geocenter) and origin of orbital coordinate system

a_s = semi-major axis = $\overline{OA} = \overline{OP}$

b = semi-minor axis = $\overline{OC} = \overline{OD}$

e_s = eccentricity = $\{(a_s^2 - b_s^2)^{1/2}/a_s\}$

F = true anomaly
r_s = distance from geocenter (G) to satellite
r_s = vector from geocenter (G) to satellite

$$r_s = \begin{bmatrix} x_s \\ y_s \\ z_s \end{bmatrix} = \begin{bmatrix} r_s \cos F \\ r_s \sin F \\ 0 \end{bmatrix}$$

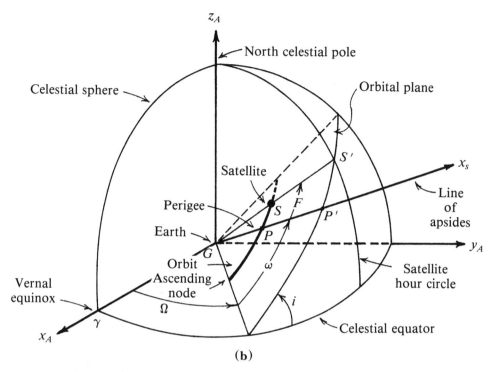

(b)

Figure 14-18. (*Continued*).

F = true anomaly

ω = argument of perigee

i = inclination of orbital plane

Ω = right ascension of ascending node

S' = satellite position projected on celestial sphere

sun, moon, and planets; (3) atmospheric drag; and (4) the sun's radiation pressure. The effects of these factors must be included in any mathematical model utilized to accurately compute or predict an orbit. The actual satellite locations, disturbed from an ideal elliptical path, may be expressed as a closely spaced time series of three-dimensional coordinate positions. Alternatively, locations may be defined by parameters for a mean elliptical orbit plus a time-variant set of corrections to be applied to obtain actual orbit positions.

14-10-2. Coordinate Systems

A satellite's orbital position must eventually be expressed in the same coordinate system used for a ground receiver station. To accomplish this, positions in the satellite orbital coordinate system (see Figure 14-18a) are transformed to an astronomical coordinate system (see Figure 14-18b) by rotating through three angles: (1) the argument of the perigee, (2) inclination, and (3) the right ascension of the ascending node. One additional rotation is required to transform satellite coordinates to a mean terrestrial system (see Figure 14-19). The rotation angle is Greenwich apparent sidereal time, a function of time and the earth's rotation rate; it is the angle between the vernal equinox and mean Greenwich meridian.

The mean terrestrial coordinate system has its origin at the earth's center of gravity and is

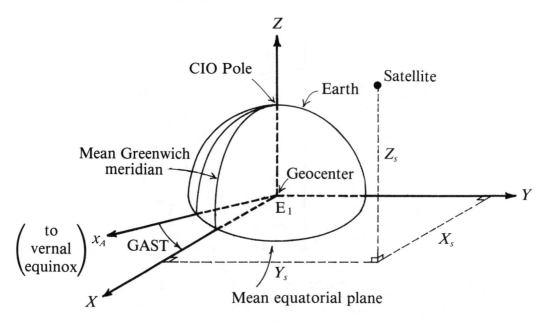

Figure 14-19.   Mean terrestrial coordinate system.

thus a geocentric system. The Z-axis passes through the Conventional International Origin (CIO) pole, a reference north pole that differs from the instantaneous pole because of "polar motion" (see Chapter 13). The X-axis is oriented in the meridian that passes through the mean Greenwich Observatory in England.

Geocentric Cartesian coordinates in the mean terrestrial system may be converted to geodetic latitudes, longitudes, and heights if a reference ellipsoid is selected. In most instances, however, the geodetic coordinates desired are those defined relative to the local geodetic datum, a nongeocentric system (see Chapter 13). For the general case, a three-dimensional coordinate transformation is required to convert receiver station coordinates from the mean terrestrial system to a geodetic datum.

14-11.   NAVY NAVIGATION SATELLITE SYSTEM

As of 1986, six operational satellites were in near-circular polar orbits, with small eccentric-

ity values and inclinations near 90°. The satellites orbit at approximately 1100-km altitudes and have orbital periods of about 107 min. For such a satellite, the earth rotates "beneath it" about 27° of longitude per orbit. Thus, each satellite passes within the line of sight of a particular ground station at least four times per 24 hrs, two each as that satellite passes from north-to-south and south-to-north. The orbits are usually spaced so that the interval between available satellites varies from about 35 to 100 min as a function of receiver latitude, with the longest interval at the equator.

Each satellite continuously broadcasts two coherent carrier frequencies, nominally 400 and 150 MHz, with both signals derived from the same ultrastable oscillator. Actual transmitted frequencies are about 80 ppm below nominal values, or 32 and 12 KHz below 400 and 150 MHz, respectively. The transmitter frequency must be stable, since the basic measurement made by a ground station receiver is a function of small frequency shifts in the received signal caused by the Doppler effect.

A satellite position is required for each observation, so orbital parameters must be known and observations accurately timed. A navigation message that defines the satellite's position is phase-modulated onto the 400-MHz signal, beginning and ending every even minute. This *broadcast ephemeris* message includes so-called fixed and variable parameters. Fixed parameters, included in each 2-min message, define a smooth, precessing elliptical orbit for a 12-hr interval. Variable parameters, updated every 2 min, are used to correct the elliptical path and find actual orbital coordinates at those times. Therefore, a transit satellite's signal consists of two stable carrier frequencies, precise timing marks, and a broadcast ephemeris message that defines its orbital coordinates as a function of time.

The basic NNSS ground support system consists of four tracking stations (Maine, Minnesota, California, and Hawaii) plus two computing and injection stations (Minnesota and California). Each tracking station observes Doppler data for signals transmitted from all satellites. The computing centers use the data for a particular satellite, plus historical tracking data, to compute parameters of the observed orbit and predict its orbit for the next 12 hr. Predicted orbit information is then transmitted from an injection station to update that satellite's memory.

14-12. GEODETIC DOPPLER RECEIVERS

Portable Doppler receivers have been developed and marketed by several firms, including Magnavox Corporation, JMR Instruments, Inc., Canadian Marconi Company, Decca Navigation Company, and Motorola, Inc. Doppler receiver system components include a receiver, antenna, data recorder, and power supply unit. The weight of a typical receiver system is in the 25- to 40-kg range, making it feasible to hand-carry or backpack the unit to locations inaccessible to vehicles (see Figures 14-20a and 14-21). A stable receiver oscillator generates a reference signal from which 400-

and 150-MHz frequencies are derived; thus, these receiver references are approximately 80 ppm higher than the respective satellite frequencies. This ocsillator also drives a clock that can be synchronized to time marks in a satellite ephemeris message.

Antenna design is important because the received signal strength is low. It must also be lightweight and transportable (see Figure 14-20b). After an antenna is set up and leveled, offsets—both horizontal and vertical—are carefully measured from the station mark to the antenna's effective center (see Figure 14-22).

Some early receiver models recorded observed data on punched paper tape (see Figure 14-17). Data recording and storage for most current models are accomplished by the use of magnetic tape cassettes (see Figure 14-23). At least one receiver utilizes a bubble memory cartridge for data collection. Receivers are designed for all-weather operation and may be left unattended in an automatic data-acquisition mode. Current systems operate on 12-V battery power and include a microprocessor that controls receiver operations, tests for equipment malfunctions, verifies recorded data, and even provides on-site position computations (see Figures 14-24 and 14-25). Modular construction provides for easy field repairs.

14-13. DOPPLER SATELLITE POSITIONING THEORY

Due to relative motion of the receiver antenna and satellite transmitter, the satellite's constant transmitted frequency—as a consequence of *Doppler shifts*—is no longer constant when received at a ground station antenna (see Figure 14-26). As a satellite approaches a receiver station, received Doppler frequency is greater than satellite-transmitted frequency. As the satellite passes and the receiver-satellite distance, or satellite range, begins to increase, the received frequency drops below the transmitted value. This Doppler shift is a function of signal propagation velocity and the time rate of change of satellite range, whereas the received Doppler-frequency-curve slope is a measure of satellite range.

Corresponding received Doppler and receiver reference frequencies are differenced to produce lower-frequency signals. The *beat frequency* for the 400-MHz signal is nominally 32 KHz, with a maximum variation of ± 9 KHz due to the Doppler effect. The receiver includes a counter to observe the number of beat-frequency cycles that occur during successive time intervals. Selected time intervals may be 2 min or shorter; commonly, nominal

30-sec intervals are used. During a typical satellite pass, 20 or more "30-sec Doppler counts" may be observed, the number depending on how long the satellite remains above the horizon. The maximum number of counts, about 36, would result from horizon-to-horizon tracking of a satellite that passed directly over the receiver. The beat-frequency cycle counts constitute the basic observables of a Doppler receiver.

Each Doppler count is a measure of the satellite-range change during a corresponding time interval—e.g., range difference $(r_2 - r_1)$ that occurs during time interval $(t_2 - t_1)$ in Figure 14-26a. Each range difference can be

(a)

Figure 14-20. (a) JMR-2000 global surveyor carried in standard backpack frame. (b) JMR-2000 global surveyor showing antenna storage. (Courtesy of JMR Instruments, Inc.)

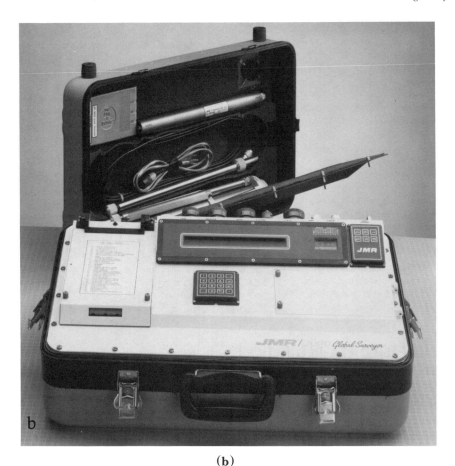

(b)

Figure 14-20. (*Continued*)

expressed mathematically as a function of electromagnetic wave propagation velocity, receiver reference frequency, Doppler count, Doppler-frequency shift, and time interval (see Equation 1 in Figure 14-26c).

A range difference is also a function of satellite coordinates and receiver antenna coordinates, since an individual range can be expressed geometrically in terms of end-point coordinates. The antenna position is the desired unknown, so estimated coordinates are substituted initially. Satellite coordinates are computed from the orbital parameters defined by broadcast ephemeris data. If 30-sec Doppler counts are used, required satellite positions corresponding to those intermediate times are interpolated from positions defined at 2-min time marks. By combining these two range-difference expressions (Equations 1 and 2 in Figure 14-26c), a nonlinear observation equation is formed (Equation 3 in Figure 14-26c). Each Doppler count corresponds to a similar equation, with antenna coordinates as only unknowns. These equations are linearized and solved by the least-squares method (see Chapter 16) to find corrections to the estimated coordinates. The solution is iterated until a three-dimensional set of antenna coordinates is found that best fits observed Doppler counts.

Figure 14-21. Motorola Mini-Ranger satellite survey system in carrying cases. (Courtesy of Motorola, Government Electronics Group.)

14-14. DOPPLER DATA-REDUCTION METHODS

Doppler surveys may be performed in two different modes: (1) point positioning or (2) relative positioning. Single point positioning can be done with one receiver any place on earth. A point's three-dimensional geodetic position in the satellite coordinate system can be computed with data from as few as two satellite passes. Relative positioning requires simultaneous operation of a minimum of two

Figure 14-22. Magnavox MX 1502 geoceiver satellite surveyor field site setup. (Courtesy of Magnavox Advanced Products and Systems Co.)

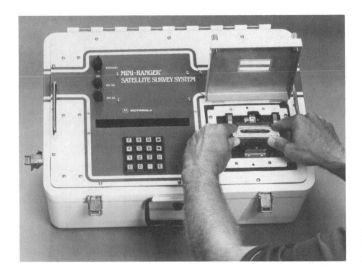

Figure 14-23. Data tape cassette for Motorola Mini-Ranger satellite survey system. (Courtesy of Motorola, Government Electronics Group.)

receivers: One occupies a station with known geodetic coordinates, the other is placed at an unknown point. The data-reduction process involves solving for the coordinate differences between the two stations and then applying them to the known station coordinates to find those of the unknown point. A relative-positioning Doppler survey may utilize more than two receivers if desired, thereby gaining additional reliability from the network configuration produced.

A two-dimensional "horizontal" position can be determined by using data collected during a single satellite pass, but a position so obtained may have up to a 30-m uncertainty. This figure can be greatly reduced by collecting and reducing the Doppler data for multiple satellite passes. Three-dimensional repeatability has been reported to be about 9 m for 10-pass solutions, 5 m for 50-pass solutions, and 2 m for 100-pass solutions. These results were obtained using orbital parameters from

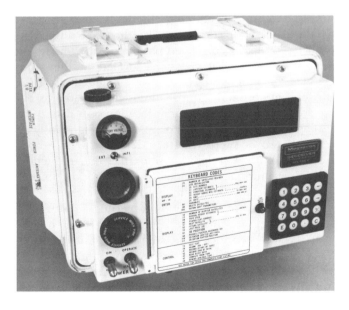

Figure 14-24. Receiver control panel for Magnavox MX 1502 geoceiver satellite surveyor. (Courtesy of Magnavox Advanced Products and Systems Co.)

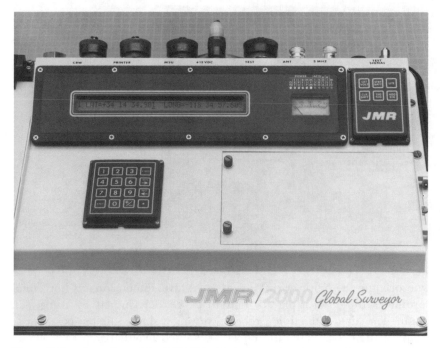

Figure 14-25.   Control panel for JMR-2000 global surveyor in operation. (Courtesy of JMR Instruments, Inc.)

the *broadcast ephemeris* and predicted values used on tracking data that is a minimum of 6 hr old at the time of the satellite pass. Errors in satellite orbital positions have a large effect on computed receiver station coordinates.

A postcomputed orbit, based on tracking data taken at about the same time as the observed satellite pass, will yield better results. Computed and distributed by the U.S. Department of Defense, this *precise ephemeris* is based on two days of tracking data obtained at over 20 stations distributed around the world. Point-positioning repeatability obtained using the precise ephemeris has been reported to be about 1 m or smaller for 10- to 33-pass solutions, whereas accuracies better than 1 m have been reported for multiple-pass solutions compared to external references such as the high-precision geodimeter traverse. The precise ephemeris has several disadvantages: (1) It is not generally available for public use; (2) it is only available for one or two of the satellites; and (3) receipt of ephemeris data, for autho-

rized users, is delayed as much as a week or more, so position computation cannot be done in "real time."

Doppler satellite relative positioning, in its simplest form, is referred to as *translocation*. Relative-positioning methods provide improved accuracy because they take advantage of highly correlated position errors for two or more receiver stations located in the same general area. These errors included effects of atmospheric propagation delays as well as orbit inaccuracies. Translocation uses data from each satellite pass observed simultaneously by two or more receivers. The *short-arc* method, another relative-positioning data-reduction technique, is similar to translocation, except that additional unknown parameters are included in the solution to permit adjusting the satellite orbit. The *semi short-arc* method likewise provides orbit adjustment, but it utilizes fewer additional parameters and yields a slightly less rigorous solution. Accuracies of 50 cm or better have been reported for multiple-

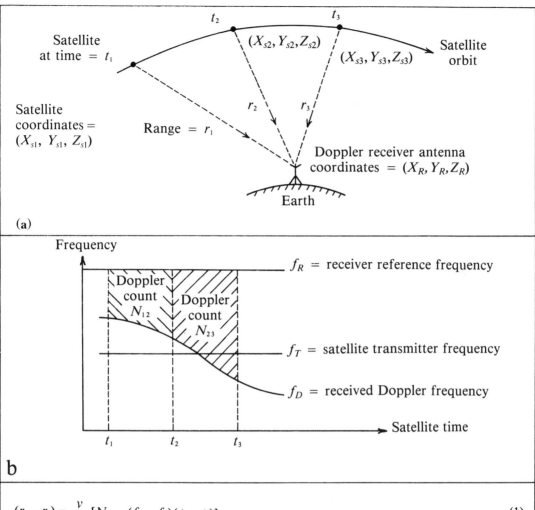

$$(r_2 - r_1) = \frac{v}{f_R} [N_{12} - (f_R - f_T)(t_2 - t_1)] \tag{1}$$

where: v = velocity of electromagnetic wave propagation;

$$(r_2 - r_1) = [(X_{s2} - X_R)^2 + (Y_{s2} - Y_R)^2 + (Z_{s2} - Z_R)^2]^{1/2}$$

$$- [(X_{s1} - X_R)^2 + (Y_{s1} - Y_R)^2 + (Z_{s1} - Z_R)^2]^{1/2} \tag{2}$$

$$\frac{v}{f_R} [N_{12} - (f_R - f_T)(t_2 - t_1)] = [\Delta X_2^2 + \Delta Y_2^2 + \Delta Z_2^2]^{1/2} - [\Delta X_1^2 + \Delta Y_1^2 + \Delta Z_1^2]^{1/2} \tag{3}$$

where: $\Delta X_i = (X_{si} - X_R)$; $\Delta Y_i = (Y_{si} - Y_R)$; $\Delta Z_i = (Z_{si} - Z_R)$.

(c)

Figure 14-26. Doppler satellite positioning theory. (a) Satellite-transmitter/receiver-antenna geometry. (b) Receiver Doppler count measurements. (c) Doppler positioning mathematical model.

pass solutions using translocation, and short-arc multiple-pass computations have yielded reported accuracies within 10 cm.

Although nominally a known value, the difference between ground reference and satellite-transmitted frequencies $(f_R - f_t)$ is treated as an unknown in a position solution. Small frequency drifts of each oscillator may be accounted for in this manner.

14-15.   GLOBAL POSITIONING SYSTEM

The GPS is a worldwide system of navigation satellites implemented by the U.S. Department of Defense (see Chapter 15).

14-16.   GPS EXPERIENCE AT THE NGS

The NGS's first experience with the GPS was in January 1983, as a participant of the FGCC's test of the Macrometer V-1000 survey system. In March 1983, the NGS acquired two Macrometer V-1000 GPS receivers. Test surveys were conducted from April through July; in September, the first of several operational control survey projects was begun, and in October 1983, a third Macrometer V-1000 was acquired.

In cooperation with the U.S. Defense Mapping Agency, the NGS conducted its first GPS satellite survey with Texas Instruments TI-4100 geodetic receivers in June 1984. In May 1985, two of seven TI-4100 GPS satellite receivers acquired by NGS were delivered; tests began that month, and the first operational survey started in June. Two more receivers were delivered by December 1985 and the remaining three units scheduled or use by the summer of 1986.

The NGS experience in GPS satellite surveying has been extensive, with projects carried out in various regions of the conterminous United States and Alaska. Over 50 GPS survey projects have been carried out since early 1983, involving the occupation of more than 500 stations. These GPS survey projects were executed to meet a wide range of control requirements or special survey investigations. The extensive experience gained from these surveys has clearly demonstrated that observations of GPS satellite signals yield very accurate three-dimensional relative-position data. Depending on spacing between stations, centimeter- to decimeter-level relative-position accuracies have been achieved. Because uncertainity levels in the order of centimeters can be obtained at a significant reduction in costs compared with conventional methods, the NGS has adopted GPS satellite surveying technology as the primary way to establish geodetic control. Consequently, this acceptance of GPS as a primary tool is affecting everyone involved in control, land, and engineering survey practices. Chapter 15 provides current practical survey techniques using GPS.

14-17.   CONCLUSIONS

Inertial and satellite geopositioning technologies are revolutionary developments that have already had a significant impact on surveying practices. Inertial surveying methods are particularly appropriate for large projects that require rapid repositioning of widely scattered points or a high density of points in a smaller area. A truck- or helicopter-mounted ISS has the potential for high productivity in moderate-to-high precision applications. Because of its absolute point positioning capability, the Doppler satellite method is well suited to the establishment of control networks in remote areas. Doppler surveys are also appropriate for relative positioning of widely separated control stations.

The impact of GPS surveying technology is being felt and will continue to advance rapidly. Surveyors need to be aware of its potential and learn as much as possible about the principles of GPS surveying practice. Although by no means a panacea, GPS satellite surveying technology is proving to be a very useful tool in satisfying general and special control surveying needs. In particular, it has been or is potentially applicable in the following:

1. Strengthening the existing national geodetic reference system (NGRS)
2. Establishing new horizontal and vertical control connected to the NGRS
3. Providing data to improve estimates of geoidal undulations
4. Determining reliable estimates for geoid height differences and useful values for orthometric heights at points not connected by differential leveling to the national vertical control network
5. Greater efficiency in establishing network control
6. Complementing and thereby enhancing the usefulness of other precise survey systems, such as the inertial survey and total station methods
7. Making high-precision measurements at the cm- to few-cm-level for purposes such as measuring horizontal motion, monitoring land subsidence and uplift and special engineering surveys for deformation studies, etc.

The GPS surveying technology is a revolutionary tool expected to dominate the state of the art for control and geodetic surveying through the end of this century.

REFERENCES

ASCE. 1983. Special issue on advanced surveying hardware. *Journal of Surveying Engineering 109 (2)*.

HOAR, G. J. 1982. *Satellite Surveying*. Torrance, CA: Magnavox Advanced Products and Systems.

The Institute of Navigation. 1980, 1984, 1987. *Global Positioning System*. Washington, D.C., Vols. I–III.

KING, R. W., E. G. MASTERS, C. RIZOS, A. STOLZ, AND J. COLLINS. 1985. *Surveying with GPS*. Kensington, Australia: School of Surveying, University of N.S.W.

Positioning with GPS-1985. Proceedings of the 1st International Symposium on Precise Positioning with the Global Positioning System, Vols. I and II. 1985. Rockville, MD.

Proceedings of the International Geodetic Symposium on Satellite Positioning, Vols. 1 and 2. 1976. Las Cruces, NM.

Proceedings of the 1st International Symposium on Inertial Technology for Surveying and Geodesy. 1977. Canadian Institute of Surveying, Ottawa, Canada.

Proceedings of the 2nd International Geodetic Symposium on Satellite Doppler Positioning, Vols. 1 and 2. 1979. Austin, TX.

Proceedings of the 2nd International Symposium on Inertial Technology for Surveying and Geodesy. 1981. Banff, Canada.

Proceedings of the 3rd International Geodetic Symposium on Satellite Doppler Positioning, Vols. 1 and 2. 1982. Las Cruces, NM.

Proceedings of the 3rd International Symposium on Inertial Technology for Surveying and Geodesy, Vols. 1 and 2. 1985. Banff, Canada.

Proceedings of the 4th International Geodetic Symposium on Satellite Positioning, Vols. 1 and 2. 1986. Austin, TX.

SCHERRER, R. 1985. *The WM GPS Primer*. Norcross, GA: WM Satellite Survey, Wild Heerbrugg Instruments.

STANSELL, T. A. 1978. *The Transit Navigation Satellite System*. Torrance, CA: Magnavox Government and Industrial Electronics.

Surveying Engineering Division of the ASCE 1985. Advanced positioning systems. *Engineering Surveying Manual*. New York, Chap. 20.

WELLS, D., ET AL. 1989. *Guide to GPS Positioning*. Fredericton, Canada: Canadian GPS Associates.

15

Global Positioning System Surveying (GPS)

Bryant N. Sturgess and Ellis R. Veatch II

15-1. INTRODUCTION

This chapter is not intended to be a text on global positioning surveying. The bibliography contains many other sources that will greatly expand one's knowledge of the topic to any depth or breadth required. The objective of the chapter is to provide information to eliminate problems commonly experienced by first time users of GPS technology. Some related topics are triangulation (Chapter 11), trilateration (Chapter 12), geodesy (Chapter 13), and inertial and satellite surveys (Chapter 14).

The navigation satellite time and ranging (NAVSTAR) was developed by the Department of Defense (DOD) as a three-dimensional, satellite-based, 24-h, global, all-weather navigational system.

The use of differential GPS as a surveying tool was not considered in its initial development and early satellite launches that began in February 1978. Surveying and other nonnavigational applications are added bonuses developed primarily by the private sector. GPS has eclipsed and will ultimately transcend the use of classical and conventional surveying instrumentation and techniques such as triangulation, trilateration, and traversing, while providing a highly accurate, precise, and cost-effective technology for control extension, control densification, airborne and terrestrial

photo control, deformation monitoring, hydrographic positioning, and a host of other applications. GPS has pushed many applications of conventional instrumentation into obsolescence. With internal system precision of better than 1 ppm, GPS is often several orders superior to the basic primary control on which the subsequent constrained adjustment of GPS observations is based. The National Geodetic Survey (NGS) now uses GPS exclusively for horizontal control work.

The planned constellation of 21 satellites and three spares in 12-h orbits at an altitude of 20,180 km will provide visibility of five or more satellites, 24 h a day to users over the globe. The satellites are arranged four to an orbital plain, with six circular planes inclines at 55° to the equator and skewed 60° between each other. The mandate is for mission control to maintain 21 operational satellites at all times, providing three-dimensional positioning around the clock as shown in Figure 15-1. The system will have three working spares to help ensure the continuation of this level of coverage.

This constellation array has been delayed, but by the end of 1991, the system contained 16 operating satellites with five prototype block I's and 11 production model block II's providing virtual 24-h satellite visibility sufficient for two-dimensional position, and 18-h three-di-

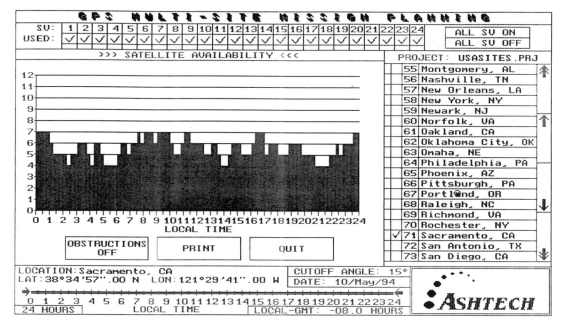

Figure 15-1. Projected satellite availability as of May 10, 1994, 15° cutoff.

mensional position in latitude, longitude, and height with a 10° cutoff angle above the horizon. With a 15° cutoff angle, the available time is reduced by $2\frac{1}{2}$ hr to a $15\frac{1}{2}$-hr three-dimensional observation period as in Figure 15-2.

Figures 15-1 and 15-2 are taken from typical mission planning software that shows satellite availability. Figure 15-1 demonstrates satellite availability for May 10, 1992 at Sacramento, CA when three-dimensional GPS measurements were possible using four or more satellites. Also shown are periods when only two-dimensional (x, y) measurements were possible with three satellites, and times shown with black shading when the constellation was not sufficient for surveying purposes.

15-2. GPS BASICS

Position computations with GPS are done by ranging to the satellites. The satellite positions are known—i.e., they are computed from the orbit parameters in the broadcast ephemeris.

By intersecting the ranges (analogous to distances) from multiple satellites, the position of the receiver antenna can be calculated, in effect, performing a classic resection in which the known stations are the satellites.

At least four satellites are required to compute a three-dimensional position. The ranges from three satellites would provide a good position if the timing were perfect. However, two clocks are being used: the satellite clock and receiver clock. If the receiver clock is different when compared to the satellite clock, and it always is, the computed position will be wrong by the amount the receiver clock is offset from the satellite clock. A fourth range is required to solve for the receiver clock offset to compute a position.

The GPS system is based on time. Precise atomic clocks control the frequency of the carrier signals and timing of code and message modulations. The code itself does not convey information; it is simply a set of unique patterns that are used to identify the satellites and pick the signal from the background noise.

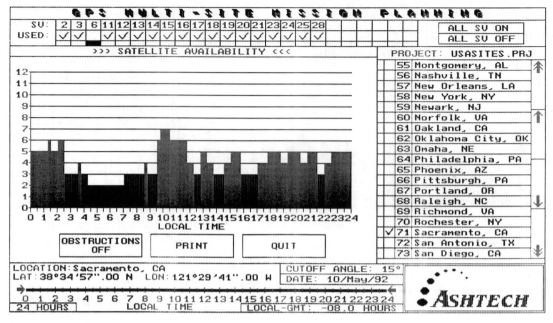

Figure 15-2. Satellite availability as of May 10, 1992, 15° cutoff.

Each satellite has its own code and signature. The range to a satellite is measured by correlating the code generated by the satellite with the same code generated by the receiver to determine the time shift between the two codes, or the transmission time of the satellite code, as illustrated in Figure 15-3.

The transmission time multiplied by the speed of light provides the distance, or range, to the satellite. This measurement contains several sources of error and is therefore termed a pseudorange, meaning a false range or range with error. These errors are primarily clock errors, but also include orbit error and signal delays caused by the ionosphere and

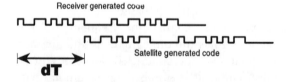

Figure 15-3. Code correlation. Time shift (dT) between the receiver-generated code and satellite-generated code.

troposphere. These errors limit the accuracy of GPS in computing point positions.

By using two receivers, many of the errors in the system can be cancelled. In differential navigation or *code-phase ranging*, the differencing is done by computing range corrections derived by comparing the observed ranges with the expected ranges based on the known position of the base receiver. The range corrections are then broadcast and applied by the remote receiver in its position computations. The resulting positions are accurate to 1 to 10 m as opposed to the 10- to 100-m accuracy of a standalone receiver. Although providing an improvement in position accuracy, pseudorange corrections are still quite inadequate for surveying use. There is, however, a second kind of ranging available.

Carrier-phase ranging, using the wavelength of the underlying carrier frequency, is employed for surveying measurements. Whereas code-phase ranging is straightforward (time of transmission multiplied by speed of light equals the distance to the satellite), carrier-phase ranging introduces an ambiguity into the observations. The receiver can measure the phase of a carrier wavelength very accurately, but it cannot directly measure the whole number of wavelengths associated with the first measurement of phase. This unknown number of cycles is referred to as the phase ambiguity or integer bias. The combination of the phase measurement and integer bias equals the initial range to satellites. The elements of a carrier base range are illustrated in Figure 15-4.

This situation is somewhat analogous to measuring a long distance with a surveyor's canyon chain when the head chainperson is out of voice contact with the rear chainperson. The hundreths are recorded precisely at the head end of the chain, but the exact number of feet are not known until the head and rear chaining notes are later combined. By knowing the orbits of the satellites and keeping track of the change in ranges throughout the observation session, the integer biases can be computed, thereby determining which foot the hypothetical rear chainperson was actually reading at each satellite.

By refining the ruler and differencing the carrier-phase measurement made with two receivers, not only are common errors removed, but accuracy is substantially increased. Millimeter measurements are made possible using a system originally designed for several-meter accuracy.

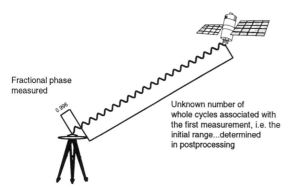

Fractional phase measured

0.906

Unknown number of whole cycles associated with the first measurement, i.e. the initial range...determined in postprocessing

Figure 15-4.   The carrier-phase range.

Three-dimensional base lines, or vectors, are the result of these efforts. Any GPS survey project is simply a collection of three-dimensional (spatial) vectors.

15-3.   BASIC NETWORK DESIGN

U.S. Geological Survey quadrangle maps at a scale of $1:24,000$ ($7\frac{1}{2}$ min) remain a most useful tool for preplanning. However, other types of maps and resources, such as commercial road maps, county atlases, state highway maps, county assessors' maps, and aerial photography, are all valuable for providing additional information in those areas where the quadrangles are either not topical, or outdated. In any event, select the most current map and plot all available (existing) horizontal and vertical control of the highest order available in the area and obtain network diagrams, station descriptions, and to-reaches from the appropriate government agencies such as the NGS, county surveyor, or local highway department. Plot the location of the horizontal and vertical control on the quad and then add the locations for which GPS positions are desired. So far, this is no different than the preplanning formerly required for conventional instrumentation for triangulation (Chapter 11) or trilateration (Chapter 12).

The location and relative distribution of the existing control are not as rigorous with GPS as with earlier conventional terrestrial technology. GPS is very sensitive to satellite geometry and less sensitive to network geometry. An example of good satellite geometry would be, say, four or more satellites spread out over the sky in different quadrants, as compared to the same satellites bunched together near the zenith.

GPS does not require station intervisibility, except in those cases where a check azimuth is required on an adjacent control net, or for purposes of establishing an eccentric (ECC) to an existing monument that is not possible to occupy with GPS due to local obstructions screening the satellites. In those instances where an ECC is necessary, two intervisible GPS stations are required in order to provide for the necessary backsight for conventional measurements. The minimum acceptable distance between the GPS stations for subsequent conventional terrestrial work is dependent on the accuracy required for the resulting azimuth. Table 15-1 provides a handy reference for determining station spacing.

There are factors that must be considered in designing the network. For example, the following conditions need to be given consideration before any network design can begin:

1. Standards and specifications for project: What minimum accuracy is required to accomplish the needs of the project while providing for necessary redundancy.

2. Distribution of existing horizontal and vertical control.

3. Minimum and maximum existing network control.

4. Minimum and maximum secondary station spacing.

It is a good idea to include as many existing network horizontal control points in the GPS schedule as economically possible. The number is independent on the relative size of the GPS project and funding envelope. Commonly a minimum of three existing NAD 83 horizontal control stations of 1 : 100,000 are included.

If NAD 83 stations are not available, then select the highest order in the area. On larger projects such as shown in Figure 15-5, existing horizontal control stations were used to control the network, providing maximum geometrical redundancy with good success. It is best to err on the side of generosity in the selection of horizontal and vertical project control. Place as much as can be afforded.

The heights produced from GPS surveys are referenced to the ellipsoid, not the geoid. Constantly, it is necessary to include bench marks referenced to NGVD in the GPS observation scheme as well. The selection of at least three bench marks spaced somewhat evenly in the area might be adequate on smaller projects of approximately 15 to 20 km. However, in projects where the separation between the ellipsoid and geoid are irregular, as could be expected in foothills or mountainous areas, more benches will be necessary to allow for accurate modeling should GPS be used to generate orthometric elevations (NGVD) for photo control or other engineering projects. It is important that the selected bench marks used to control the GPS survey have not been disturbed, and the elevation of the BM is consistent with that of adjacent BMs in terms of epoch and loop adjustment. It may be necessary to validate the elevation with a conventional level run if either stability or elevation is in doubt.

Network design can be a very individualized process. Probably the most common method is to simply link stations by sessions, leap frog-

Table 15-1.   Station Spacing

| Station Spacing (m) | Azimuth Accuracy Arc Seconds (one sigma) | | | | |
| | 1 | 2 | 4 | 6 | 10 |
	+ −	mm		Confidence	
100	—	—	—	3	5
200	—	2	4	6	10
300	—	3	6	9	14
400	2	4	8	12	19

ging from session to session until all stations have been connected. This tends to be a random method and often results in networks with one or more unnecessary sessions. The following design process is an attempt to be more orderly. It is based on, but does not rigorously follow, NGS guidelines. These guidelines can be found in *Geometric Geodetic Accuracy Standards and Specifications for Using GPS Relative Positioning Techniques*.[1]

The first step in designing an efficient network is to connect the stations into loops of nontrivial base lines as shown in Figure 15-6. This provides the framework of the network and allows loop closures according to NGS specification. In Figure 15-6, the network has been connected into three loops. The loops should contain no more than 10 base lines and not exceed 100 km in perimeter.

When creating the loops, try to keep them as "boxy" as possible. Avoiding a series of parallel loops will allow more flexibility in the second step.

The second step connects the nontrivial lines together in sessions. Each session should contain $N - 1$ (where N is the number of receivers being used) of the nontrivial loop base lines. In Figure 15-7, four receivers are being used and each session contains three loop base lines. The network is complete in seven sessions with no extra work; therefore, there are 21 nontrivial base lines in our network. Conversely, if you know the number of nontrivial base lines in your network, you know exactly how many sessions you need to complete the observations.

Try several approaches to each network. Networks do not always work out as cleanly as the example above. The addition of one other station in the network above would create two more nontrivial base lines and necessitate another session.

A good way to practice is to put a group of dots on a sheet of paper and make several copies. Try different loops and session groups. The example in Figure 15-7 is based on using four receivers; try laying out the same network

for three receivers (two nontrivial lines per session instead of three). Practice leads to efficiency.

The term nontrivial as used above to describe the base lines in the loop framework is somewhat misleading and has been misused. As used here, it denotes the minimum number $(N - 1)$ of base lines necessary to connect all stations in a session. Whether or not there are any trivial base lines is dependent on the processing method. The trivial base lines in Figure 15-7 are shown as dashed lines. These lines are only trivial if the processing software considers the data from all receivers at the same time, i.e., multi-base-line processing, resulting in true zero loop closures. If the base lines were processed independently, as is usually the case, the resulting lines $[(N*(N - 1))/2]$ are correlated, but they are not trivial. A recommended method is to process all base lines independently and use them in the network. In redundancy is strength. A fundamental rule in GPS surveying is, "Think network, not traverse."

A summary of recommendations for controlling a GPS network follows:

1. *Horizontal.* Technically two stations will provide an azimuth and a scale. However, there would be no redundancy for either azimuth or scale. It is recommended that the project be bracketed with NAD 83 horizontal control, three stations minimum, whereas four or more horizontal stations are recommended for larger control projects.

2. *Vertical.* Three stations are sufficient to solve for the two tilt biases: east-west and north-south. A fourth vertical provides a necessary redundancy. The tilt defines the slope of the geoid in relation to the ellipsoid and may be sufficient to provide reliable elevations in small areas where the geoid is fairly consistent. In areas where gravity anomalies are suspected or in projects over large areas, using a geoid model will be necessary to provide good elevations. The geoid model may be improved by comparing elevation differences and ellipsoid height differences

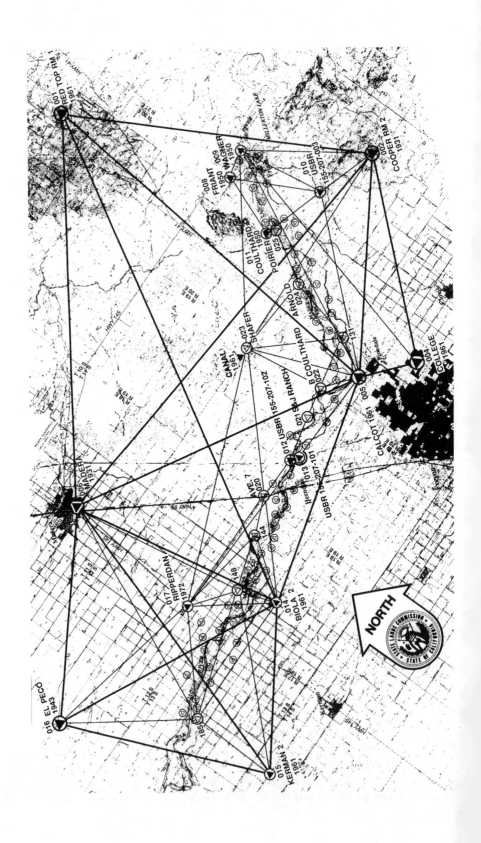

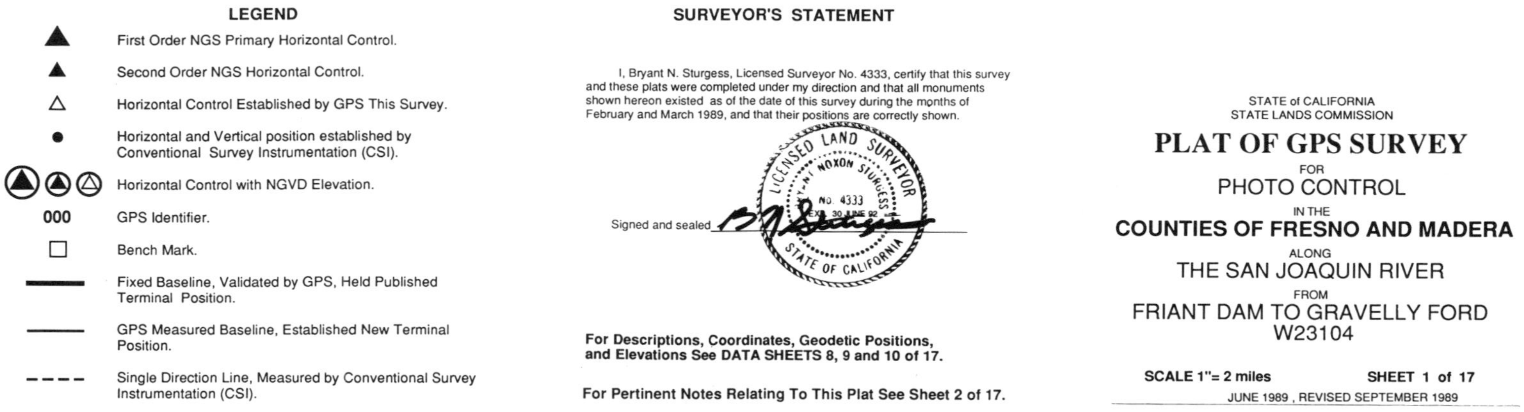

LEGEND

▲ First Order NGS Primary Horizontal Control.

▲ Second Order NGS Horizontal Control.

△ Horizontal Control Established by GPS This Survey.

● Horizontal and Vertical position established by Conventional Survey Instrumentation (CSI).

◉ Horizontal Control with NGVD Elevation.

000 GPS Identifier.

□ Bench Mark.

▬ Fixed Baseline, Validated by GPS, Held Published Terminal Position.

▬ GPS Measured Baseline, Established New Terminal Position.

- - - - Single Direction Line, Measured by Conventional Survey Instrumentation (CSI).

Figure 15-5. Sample control project.

SURVEYOR'S STATEMENT

I, Bryant N. Sturgess, Licensed Surveyor No. 4333, certify that this survey and these plats were completed under my direction and that all monuments shown hereon existed as of the date of this survey during the months of February and March 1989, and that their positions are correctly shown.

Signed and sealed

For Descriptions, Coordinates, Geodetic Positions, and Elevations See DATA SHEETS 8, 9 and 10 of 17.

For Pertinent Notes Relating To This Plat See Sheet 2 of 17.

STATE of CALIFORNIA
STATE LANDS COMMISSION

PLAT OF GPS SURVEY

FOR

PHOTO CONTROL

IN THE

COUNTIES OF FRESNO AND MADERA

ALONG

THE SAN JOAQUIN RIVER

FROM

FRIANT DAM TO GRAVELLY FORD
W23104

SCALE 1"= 2 miles   **SHEET 1 of 17**

JUNE 1989 . REVISED SEPTEMBER 1989

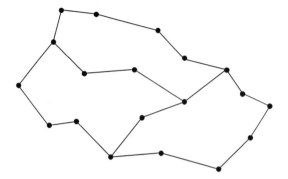

Figure 15-6. Nontrivial base line loops.

between known bench marks. Terrain changes can provide clues to the existence of gravity anomalies. For example, a gravity anomaly might be indicated in an area marked by the transition of valley floor to foothills, or by foothills to mountains. Use caution in mountainous regions by providing a generous number of known elevations. NGS specifications outline the frequency of multiple occupations required for a typical network. For example, 10% of the stations require triple occupation, whereas 5% of the base lines require double measurement loops. Thirty% of the new stations, 100% of the vertical control, and 25% of the horizontal control, any azimuth pairs, all require double station occupation. Designing a network to meet NGS specifications often results in more sessions and is, consequently,

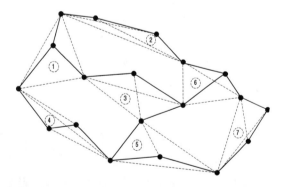

Figure 15-7. Trivial base lines (dashed) and nontrivial base lines (solid).

more expensive. Current NGS specifications may be obtained by directly contacting the agency. See Appendix 2 for its address.

15-4. RECONNAISSANCE

GPS reconnaissance in many respects is similar to conventional reconnaissance with one major exception: Usually, GPS does not require stations to be intervisible as in conventional terrestrial surveying. There are times, however, when intervisibility is a requirement. One such occasion would be when an independent validation of either the GPS system or primary control net is included for purposes of furnishing an independent check on the GPS measurements and/or adjustment. Until GPS technology is a universally accepted and recognized technique by the courts, this might serve as a method of convincing either the judge or jury that the results are what they are claimed to be. As time goes on and the use of GPS becomes a surveying technique used by more surveying organizations, this idea of validation will become of lesser and lesser importance. Another possible situation where intervisibility is a requirement occurs when a backsight is necessary in a GPS network to provide an azimuth for use by conventional terrestrial survey measurements to position a point that is not possible to occupy with a GPS antenna due to obstructions restricting line of sight to the satellites.

It is essential that all primary control be validated at some time during the GPS campaign to ensure no disturbance of the station. Usually, the checking of published measurements from the reference monuments (RMs) to the station will satisfy at least a part of the validation relating to site stability or possible disturbance. There is another validation that should take place in any primary network before published primary control positions are accepted in the final constrained adjustment of GPS base lines. That is the direct measurement between primary network control stations using GPS receivers. This may seem to be an unnecessary luxury, or perhaps overkill. However, the direct measurement between primary control stations serves as the best evidence of regional stability and fit with the control possible.

Pre–NAD 83 (NAD 27, e.g.), it was always a good idea to determine if the primary control being used for the campaign was entirely within the same NGS network adjustment. Post–NAD 83, this practice is no longer as rigid. Most NAD 83 positions have been validated to better than 1 part in 100,000. Nevertheless, sufficient occupations should be planned to verify the adjustment before the control is used. Usually, the GPS campaign will result in so-called free (unconstrained) adjustments that are superior to the constrained (fixed) ones. This is a situation rarely, if ever, experienced with conventional terrestrial surveys except those of the highest order.

15-4-1. Public Contact Strategy

More often than not, existing primary control necessary for referencing the GPS campaign, as well as monumentation that must be established during the campaign, is located on private (or government) property belonging to someone other than the client. In these cases, proper permission or notification for entry should be obtained. Securing permission of the property owner to enter private property, is not only good business, but it is polite, and in some locales, it will keep the surveyor out of legal complications. A diligent attempt must be exercised to secure proper permission. Failing this, the owner could be sent a letter of intent to enter the property to a survey with some explanation of the survey's purpose.

15-4-2. Site Selection

GPS sites must be selected free of significant obstructions and multipath conditions. Multipath is caused by an object or a media

that causes signal reflection. Multipath conditions can result from metallic objects located above the antenna plane such as buildings, signs, semitrailers, tankers, chain-link fences, all of these can be a common source of multipath. In short, avoid sites where signal-reflecting material is above the plane of the antenna. Figure 15-8 is an actual NGS primary control station site where a metal-sided fire lookout tower was a source of not only severe multipath, but also potential satellite signal blockage covering nearly 40° in azimuth and nearly 45° in altitude (see Figure 15-9).

Objects interfering with a direct signal from the satellite during the planned observation periods cause cycle slips. When a satellite signal is obstructed in any way, tracking is interrupted. However, when lock to the satellite is regained, the fractional part of the measured phase is restored. The lost integer count will be repaired during postprocessing. The signals propagate from the satellite and are transmitted to the receiver antenna along the line of sight. These signals cannot penetrate water, soils, walls, trees, buildings, or other obstacles. Standing between the receiver antenna and satellite can interrupt signals. Consequently, while the data are being collected, the operator should stay away from the antenna to avoid excessive blockages of signals resulting in cycle slips. Cycle slips are a loss of count and could be one cycle or a billion cycles (1 billion cycles represents 1 sec of signal transmission). Occasional cycle slips are not a problem and will be automatically repaired by the postprocessing software. Excessive cycle slips or blockages should be avoided. Figure 15-9 shows a polar plot of the terrain depicted in Figure 15-8, with the track of the satellites superimposed.

In this example, it can be seen that at approximately 0030 hours, satellite no. 11 will disappear behind the fire lookout structure. Accordingly, should GPS observations be planned between 2320 and 0020 hours, satellite no. 11 will be visible. However, multipath will still remain as a significant problem due to the metal slab-sided structure and positions of

Figure 15-8.   NGS station "Red top" showing a metal-sided building, the source of severe multipath conditions.

the balance of satellites from azimuth 90 to 320°. An eccentric station was necessary at this location to eliminate potential multipath conditions.

The adage "monuments are where you find them" remains true today. Established monuments are either clear for GPS measurements or they are not. Often, the latter is the common case, as in the example shown in Figure 15-8. In the event the sky is obstructed, a location must be selected for the setting of an eccentric station that has a clear view of the sky from the mask limit of, say, 15 to 20° to the zenith.

Site selection for new monumentation can be determined after the consideration of a few basic elements. It is not a complex operation and in many ways has similarities to conventional field methods. The site must be con-

ducive to the installation of the type of monumentation specified for the campaign. Future access to the site must be considered as well as monument survivability. If the location forms a part of the overall GPS network, then consideration must be given to its long-term preservation for subsequent use by others. It is only fair that the property owner be advised that other surveyors might require the use of this new station in the future with the possible result that the initial setting of the station might establish a precedence for continued usage.

If it is likely that the new station will be needed in the future for other possible activities, anticipate the effect tree growth will have on satellite visibility, or the relative safety of the site from construction activities, vandalism, or road widening, e.g. It is not possible to plan

California State Lands Commission
Survey Unit

GPS VISIBILITY CHART

Station: RED TOP   Latiitude: 37° 07' 41"
Date: 18 JAN 90   Longitude: 119° 46' 36"
By: B.N.S.   Elevation: 1,840'

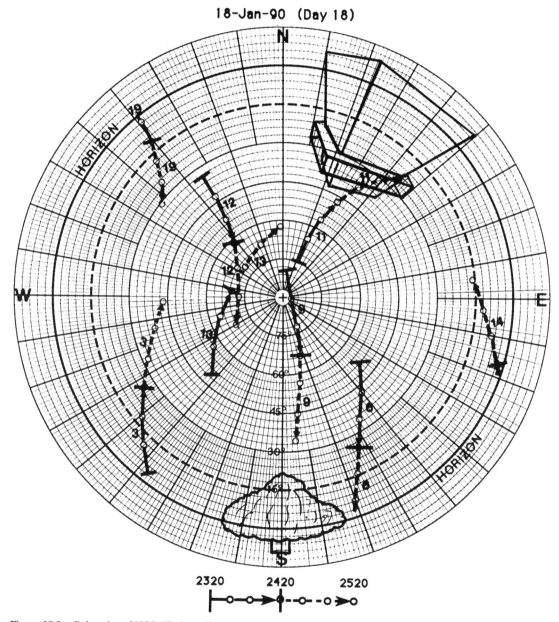

18-Jan-90 (Day 18)

Figure 15-9. Polar plot of NGS "Red top."

for all contingencies, but a greater percentage of new monumentation will survive the perils of development, agricultural activities, etc. and be usable for the next surveying job if some though is given to site selection.

Since GPS allows a high rate of productivity, easy station accessibility for the receiver operators is an essential element, no matter which technique for GPS measurement is selected, be it rapid-static, static, pseudokinematic, or kinematic. It is necessary for the receiver operators to get to the station in the least amount of time possible in order for the project to proceed with the smallest operational cost and highest productivity. The selection of GPS sites must be maximized for easy setups, while avoiding unnecessary walking to the station when a slight change of location would result in a drive-to-station. Avoid locations that require the excessive unlocking and locking of gates if possible. If access by key is a requirement, it would be advisable to have a duplicate set made for *each* receiver operator. By taking this precaution, no access complications will occur since all operators can access any locked area. This is one less problem to address and allows full flexibility in station assignments during the GPS measurement phase of the campaign.

Give consideration to the time of year when the actual observations will be scheduled. Actually, the reconnaissance should be followed as closely as possible by the GPS schedule with a minimum of delay because conditions rapidly change in the field. Reconnaissance done in the spring of the year might indicate a particular station and site with its resulting polar plot to be clear of obstruction, but its GPS occupation 2 or 3 wk later could fail due to the growth of arboreal foliage, or perhaps construction activities.

If the campaign is scheduled during the rainy season, determine if the site might be subject to ponding by water, or made inaccessible due to mud, and that vehicular access will be possible along the planned access route under these conditions. As late as 1990, due to limited satellites and inadequate geometry, it

was a given that at least part of the year GPS operations would be conducted during nighttime, or part day and part night depending on the time of the year. In 1992, yearround daytime use was possible in the midlatitudes due to the increased numbers of satellites made available during 1990 and 1991 providing 24-hr two-dimensional coveraged 18-hr three-dimensional position in latitude, longitude, and height with a 10° cutoff angle. With a 15° cutoff, the available time is reduced by $2\frac{1}{2}$ hr to a $15\frac{1}{2}$-hr observational period. What this means is that at the present and for the future, nighttime GPS will be necessary only for certain applications. One such application where such a schedule would be advantageous is in an area of high daytime traffic, or a hazardous location that if occupied at night would make the site less perilous for the receiver operator. Another application of nighttime operations would be the case of maximizing productivity by incorporating both daytime and nighttime missions when time constraints are a factor and maximum productivity is paramount.

With proper site selection and planning, nighttime observations can approach productivity levels associated with daylight. If proper consideration has not been given to site selection, access, safety, or logistical considerations, nighttime operations, as well as daytime GPS operations, will not be productive.

15-4-3. Monumentation

The criteria used for the selection of sites and setting of existing control monuments during the era when control extension utilized conventional instrumentation were different than presently is the case for GPS. Earlier networks required line-of-sight visibility between adjacent network stations, sometimes only obtainable from the erection of tall inner and outer observation towers placed over the stations. Generally, clear visibility from 10° above the horizon to the zenith was not a requirement except in cases where a Laplace observation or an astronomic azimuth was a requirement. GPS, on the other hand, requires no station intervisibility, but line of sight to the satellites.

In some ways, GPS is no different than conventional control densification or extension in the selection of the type of monumentation for installation. Depending on the intended use of the new station, the monumentation may be of temporary design for short-term use such as ECC to a property corner or photo control point that cannot be occupied due to satellite signal interference caused by the proximity of buildings, trees, or other obstructions. The selection of monumentation in this example could be typically a hub, nail and tin, pipe, or rebar. For those stations requiring some degree of permanence, consideration should be given to one or a combination of commercial ultrastable three-dimensional types, as in Figure 15-10, constructed of aluminum or stainless steel rods driven to refusal, then capped, collared, and covered for identification and protection. In some cases, due to local site requirements, it might be necessary to bury the monument to ensure survivability. Generally, the use of poured-in-place monuments has fallen into disfavor as permanent GPS stations or even bench marks, for that matter, due to possible vertical instability, so careful consideration should be given to the actual planned use of the monument. A per-

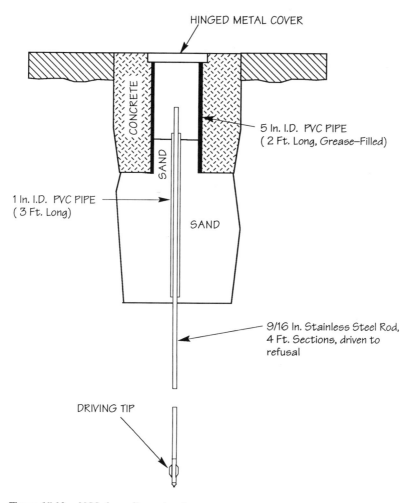

HINGED METAL COVER

CONCRETE

5 In. I.D. PVC PIPE
(2 Ft. Long, Grease-Filled)

SAND

1 In. I.D. PVC PIPE
(3 Ft. Long)

SAND

9/16 In. Stainless Steel Rod,
4 Ft. Sections, driven to
refusal

DRIVING TIP

Figure 15-10.   NGS three-dimensional monument.

manent GPS control station must be stable in three dimensions. Epoxying or grouting a monument in bedrock or a rock outcrop is a desirable and stable installation. The difference in actual cost between a temporary-type monument and one of more long-lasting design is negligible. Consider the future possible use of the station and remember that GPS is a precise three-dimensional measurement technology that must have vertical as well as horizontal stability.

Contrary view is that with the ease and accuracy of establishing high-precision points with GPS, it is not cost-effective to spend the time and resources in establishing superpermanent marks.

14-4-4. Recovery Notes

Recovery descriptions remain an invaluable resource to the GPS campaign. They had their beginning in the earliest days of triangulation and, as is the case with any good idea, they will be with us into the future. As long as there is a need to search for and recover control stations, there will be a need for some form of document describing access to the monument, its location and description. The recovery note (Figure 15-11) is an essential tool to the mission planner and receiver operator who must get to the station on time for the scheduled sessions to be a success. The recovery description is a narrative describing the station and gives particulars about its site. It contains specific instructions guiding the operator to the station, usually from a common starting point with other such stations on the project. It should contain specifics such as street names, landmarks, and cardinal directions and distances associated with the instructions for left, right, or straight-ahead movements. Additionally, it ought to include hazard warnings such as aggressive dogs, livestock, gate combinations, instructions for the closing gates, dust on crops, possible driving hazards, or anything that will allow the receiver operator to reach the station site in the least amount of time with as little risk as possible, while observing

safety and any special requests by the property owner. If property owers or caretakers were contacted en route, these references with phone numbers, addresses, or other appropriate comments should be noted.

If the survey work is to be conducted during darkness, the recovery description should reflect instructions that give consideration to the special difficulties of nighttime operations. There ought to be references to land falls, other factors while en route that will confirm the observer is on the right track. An aid for nighttime operations is adding accessories such as highway reflectors, reflective tape, reflective paint, etc. on fence posts, trees, gates, etc. to guide the operator to the station; these again should be noted in the recovery description. If nighttime operators are necessary, driving times should be buffered if the reconnaissance and to-reach preparation were done during the daytime. The GPS receiver operator probably will be unfamiliar with the site and station. The accuracy and completeness of the to-reach will determine the success of the GPS receiver operator in locating the assigned site and station.

15-4-5. Station Notes

The station noteform, as in Figure 15-12, used to document all recoveries; to-reaches; contacts with the public; names, addresses, and phone numbers; and all contacts resulting from recovery efforts. This information (along with the to-reach description) is essential and is entered into the database for later use during mission planning for GPS. This field noteform can also be used for all station recoveries and descriptions. The station notes and recovery descriptions are used by the mission planner in his or her tour of the project for purposes of familiarization with the primary control and new GPS stations. The receiver operator also uses these same documents to find the station under severe time constraints during the GPS measurement phase. It is important that the to-reach be clearly written and the station recovery notes be sufficiently

SURVEY NOTES

PARTY CHIEF	RECORDER	RODMAN	CHAINMEN
BN STURGESS	BNS		HEAD
			REAR

REMARKS	INSTRUMENTS	WEATHER
		OVERCAST, COLD

REMARKS

RECOVERY NOTES " BIOLA-2, 19(6)"

STATION " TO REACH "

TO REACH THE STATION FROM THE NEW
POST OFFICE IN BIOLA GO SOUTH ON BIOLA
STREET FOR .05 MILES TO THE INTERSECTION
OF "G" STREET. CROSS THE INTERSECTION
PASSING THE FIRE STATION, THE OLD RR.
RIGHT-OF-WAY AND GO 0.15 MILE TO THE
CURVE TO THE LEFT ("I" STREET) AND
THE STATION.

VISIBILITY TO THE NORTH IS TOTALLY
OBSCURED BY BUILDINGS. THE WATER TANK
IS NOT VISIBLE. IN ORDER FOR BIOLA-2 TO
BE UTILIZED BY CONVENTIONAL GEODETIC
MEASUREMENTS AT LEAST 50 FEET OF BIOLA
IS NEEDED. BIOLA-2 AZIMUTH MARK COULD
BE USED AS AN ECC. 10 TO 15 FEET AT THE
AZIMUTH WOULD HAVE MOST OF MAIN SCHEME
VISIBLE. HOWEVER 10 FEET WOULD STILL
BE NEEDED AT BIOLA-2.

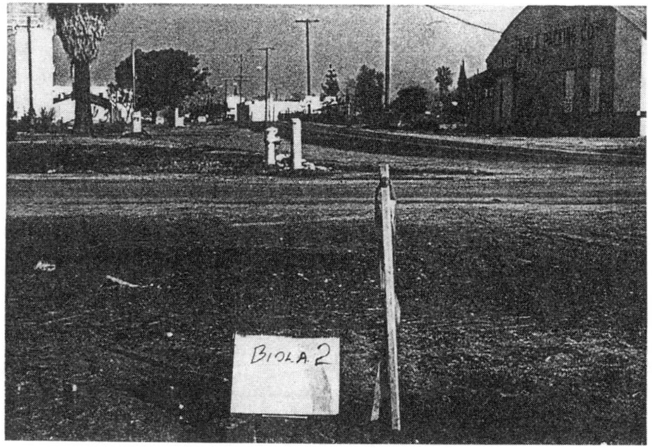

BIOLA 2

W.O.	COUNTY	VICINITY	DATE	CHECKED	DATE	PAGE	OF
21005	FRESNO	BIOLA	15 DEC 87			3	3

FORM 40.20 (4-75)

Figure 15-11. A Recovery Description.

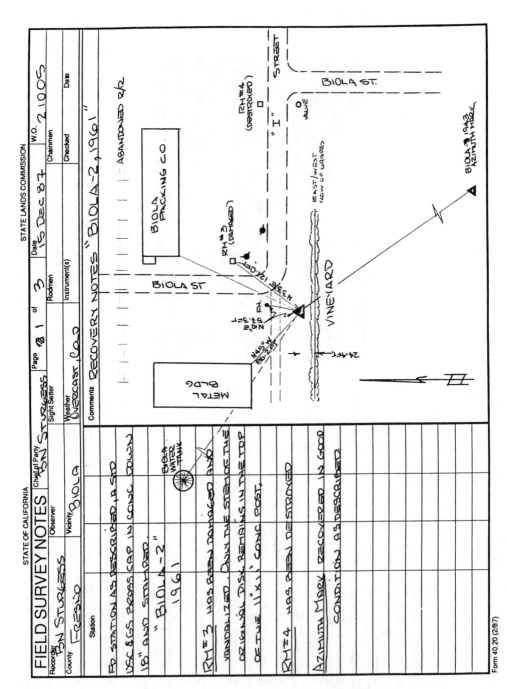

Figure 15-12. Station notes.

detailed to represent the site of the monument so as to be recognizable to a person unfamiliar with the location. At this time, station notes have a benefit. As a practical consideration, station notes often are combined with the recovery description on the same sheet or page to provide a concise document for recovery uses.

During recovery, the station site should be marked in a distinctive manner in order to be instantly recognizable by the receiver operator. Plastic flagging if eaten by livestock can be hazardous; consequently, its use should be considered carefully.

15-4-6. Polar Plot

The polar plot (see Figure 15-9) shows the sky line around the prospective GPS station in graphic form. The plot details the height of obstructions in degrees above the horizon and the azimuth of obstructions between the station and sky. It is essential that a plot be produced when possible obstructions are unavoidable as part of the reconnaissance agenda in order that mission planning proceed.

At new GPS stations being established, it is desirable to select sites free of obstructions that could interrupt satellite signals. It is not always possible to select a GPS site absolutely clear of obstructions from the horizon to the zenith clear around the compass. It follows also that few existing monuments for GPS positioning are completely free from possible obstructions to the satellites. If the sites are clear, there is not need whatsoever for a polar plot.

If the reconnaissance person knows when the GPS observations are planned to be made, the predicted locations of the satellites can be noted on the polar plot in the field and a site selected that allows those sectors of the sky to be clear even though the balance of the sky is occluded. The use of a conventional transit equipped with a plate compass is the preferred instrument to make the necessary measurements. Other options are acceptable, such as a Brunton compass, or perhaps an older-style

vane compass on a Jacob staff or a reconnaissance compass used in conjunction with an Abney hand level. Do not forget to allow for declination because the compass measurements must be a "true" bearing. The measurements need not be any more precise than + or − 1°. Remember that even obstructions such as leaves on a tree can cause a loss of lock, so be sure to outline the entire outside edge of the tree when taking measurements for the polar plot. The plot need not go below 10° above the horizon. Make certain to measure the azimuth to both sides of power poles and signs as well as the tops. The instrument should be set as high as possible above 5 ft to more closely represent the height of the receiver antenna during the data collection phase of the campaign. The use of a contrasting color is recommended for plotting the obstructions. Typical polar plot field notes should contain provisions for the following information:

1. Station name
2. GPS ID number (this is an ID number assigned for the individual campaign)
3. Observer
4. HI of instrument
5. Declination set into the instrument

In the event polar data will be later entered into a mission planning software program, a time-saving alternative to the preparation of the polar plot diagram is a set of field notes detailing the same written information as would be represented on the diagram. The field notes should have provisions for entering the following information:

1. Bearing to object
2. Vertical angle to the object
3. Description of the object or portion of the object being observed
4. GPS ID number or name
5. Name of observer
6. HI of instrument
7. Declination set into the instrument

This method is quicker for those times when the observer is the only person available at the site and is a time-saver when it comes to the task of data entry later in the office. Unless this type of summary is prepared, the data entry will involve individually inspecting and scaling each point detailed on the polar plot for the bearing and vertical angle and subsequent computer entry.

The required precision of all horizontal and vertical angle measurements is + or −1°; field notes recording measured values are not required. Any feature that may obstruct satellite radio transmissions must be shown. This may include such factors as trees, power poles, transmission lines or towers, structures, terrain (hills), etc. In the case of power poles, be sure to include a measurement to the top of the pole. Do not go overboard in the number of observations to outline a feature: This wastes time. For economy, use the minimum number to outline an object.

The information provided by the polar plot or field notes can either be used directly by the mission planner as in the past, or it can be entered into a database that is part of the mission planning software provided by the manufacturer. Once entered, an image can be assembled on the screen that duplicates the mask measured in the field, while the satellite ephemeris provides a template of the available satellites on the julian date and time selected. This allows the mission planner to select times when satellites will not be obscured by features such as trees, power poles, signs, structures, etc.

15-4-7. Photos and Rubbings

During the reconnaissance phase, it is desirable for documentation purposes to photograph the station site, the monument either found or set, and any other feature at the location that may help to recover the station in the future. Station notes as shown in Figure 15-13 are a diagram prepared that documents the site and shows a planimetric view of the location together with any reference points

(RPs) set or reference measurements to aid in either the resetting of the station should it be destroyed or disturbed, or recovery at a future date. The photos and station notes form a very powerful team for the preservation and perpetuation of monumentation. However, as time goes on, the future value of such precautions will become questionable as the ease of reestablishing positions by GPS evolves into the realm of real time.

The technique of taking rubbings consists of laying a page of blank field notes over a monument such as a brass cap and rubbing the surface of the disk with a soft lead pencil or lumber crayon, thus producing an exact image of the monument. The use of this technique can occur at the time of the preparation of the station notes, but is primarily useful when the station is visited on mission day by the receiver operator. This rubbing provides proof that the receiver was on the station designated for the session and not a reference mark. Even the most careful of surveying assistants manage, on occasion, to set up on a reference monument instead of the designated station. Of course, this method is not practical, e.g., when the monument is a galvanized pipe or rebar without any characteristic that serves to make this monument distinctively different from others. It is recommended that rubbings be used at least on the primary control stations.

15-5. GPS FIELD OPERATIONS

All GPS surveying is dependent on the measurement of the carrier phase and the solution of the ambiguities utilizing the basic carrier that is a 19-cm sine wave. A phase measurement is a portion of this basic carrier. A fractional phase can be directly determined, but the number of complete cycles (integer ambiguity) from the satellite is not directly measured. This is determined over a long observation period in static measurements, or in pseudokinematics by occupying a station twice dur-

STATE OF CALIFORNIA — STATE LANDS COMMISSION — STATE LANDS DIVISION

SURVEY NOTES

PARTY CHIEF	NAKAMOTO B.N. STURGESS
RECORDER	YOUNG
RODMAN	NAKAMOTO
CHAINMEN	HEAD NAKAMOTO; REAR YOUNG
REMARKS	SAN JOAQUIN RIVER * 13170803
INSTRUMENTS	K&E TRANSIT 208241; 2055 [illegible] 2 37543; K&E 100' CHAIN MES 1400; TENSION @ 15 LBS.
WEATHER	TEMP 104°; CLEAR, HOT

STATION * 13170803

TO REACH STATION from junction Rd. 145 & Ave. 7, go south on Rd. 145 to Ave. 6, turn right (west) onto Ave. 6 for 5.3 mi. to Rd. 22, turn left (South) onto Rd 22 for 1 mi. to intersection of Ave. 5 & Rd. 22. Continue South on Rd. 22 for ½ mi. to a dirt road, turn left (east) onto dirt road bordered by grape vines for .7 mi. to a curve to the right, continue south for .15 mi, via right (west) .10 mi, via left (south) .05 mi., via right (west) .2 mi to station.

Station is located on the northerly side of dirt road between grape vines and dirt road. Irrigation well is approx. 50' west of station.

Set * 6 rebar 4" down RP1 nail & tin on north face of pole, RP2 - paint mark on concrete pipe, RP 3 nail & tin on south face of power pole. STATION is 2" alum cap stamp CSLC PHOTO CONTROL * 13170803.

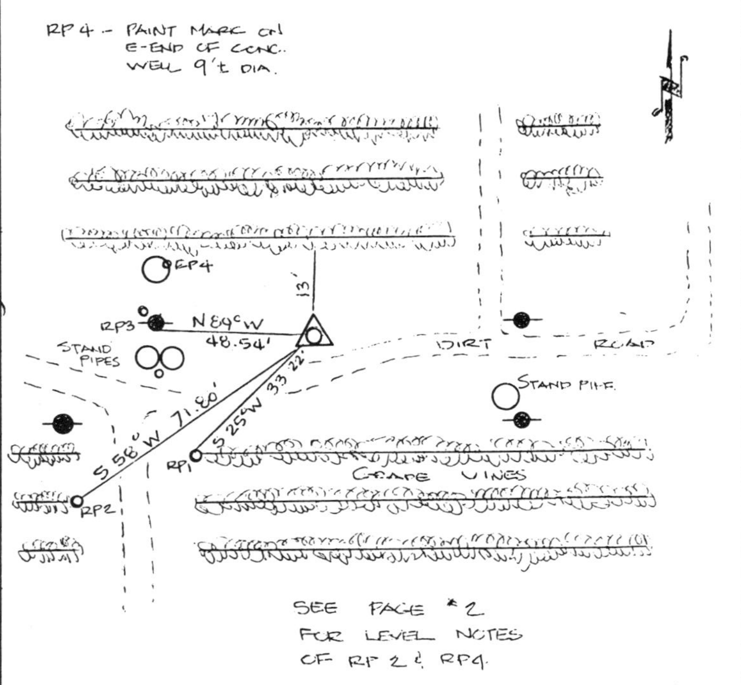

ALL BEARINGS TRUE DECLINATION = 18°30'

W.O.	COUNTY	VICINITY	DATE	CHECKED	DATE	PAGE	OF
W 23104	MADERA	MADERA	9-6-88			1	2

FORM 40.20 (4-75)

Figure 15-13. Sample GPS field notes, long-form.

ing the same session spaced by 1 hr for 30 to 60 epochs (about 5 to 10 min). In kinematics, the ambiguities are solved by initializing on a known base line that may have been created by an antenna swap. This allows for the instantaneous solution of the rover position.

15-5-1. Statistics

The term *static* was adopted because this technique requires receivers to remain stationary on a monument for a session length approaching an observing time of about 45 min (135 epochs) to 90 min (270 epochs) or more, as required depending on the standards and specifications for the campaign. The overall length of the static occupation is further dependent on the charging geometry of the position of the satellites with respect to the receivers. During times of rapidly changing *positional dilution of precision* (PDOP), shorter observation times can yield the same accuracy as longer occupation times with low, relatively unchanging levels of PDOP.

At the culmination of a session, the usual process is for one receiver to remain in place while the remaining receivers all move to a new location. The "fixed" receiver provides the hinge or pivot point for extending the control into the new session.

Typical uses of the static GPS technique might be in the areas of surveys of the highest order for such purposes as deformation, crustal motion studies, or high-order statewide GPS reference control nets, where the precision of 1 part in 1,000,000 or better is common. The static technique might also be employed in those cases where the spacing of the stations is so geographically distant that multiple sessions are not a practical consideration.

The number of sessions possible in a mission day is governed by the number of four (or more) satellite windows with acceptable PDOP, the length of satellite windows with acceptable PDOP, and travel time between the stations. The productivity of static procedures is also dependent on other variables such as weather, traffic, and luck.

The static technique is a common choice for the first-time user of GPS. It is the easiest in terms of training staff, or mission planning, but is the lowest in productivity of all possible techniques. For those networks requiring long occupation times because of high-precision specifications, or stations spaced far apart making a second session impossible during the same satellite window because of lengthy drive times, this is the only choice available. Statics can be accomplished with either dual- or single-frequency receivers.

The standards and specifications control productivity as well, by impacting the length of time necessary for an occupation. The higher the order, the more epochs are required. The same condition exists for long base lines in which extended time on station is a requirement.

The sample GPS field notes shown in Figure 15-13 are typical for static observations. This noteform would not be practical for other GPS field techniques in which high productivity would be impaired by the 2 or 3 min necessary to complete this form. However, for static measurements the observer is on station for at least 45 min so the time liability is not a problem in this case.

15-5-2. Rapid-static

Rapid-static is not a technique in itself. Rapid-static is simply an improvement in the software algorithms that currently use the additional information available with P-code to resolve the phase ambiguities faster. Simply stated, rapid-static is static observations for a shorter span of time. Occupations of 5 to 10 min take the place of occupations for 45 to 90 min. Current implementations utilize the P-code, but it is possible to perform rapid-static measurements with codeless dual-frequency data. Future software improvements may even make it possible to perform short sessions with single-frequency receivers. Rapid-static is certainly more productive than static or simultaneous rover pseudokinematic and twice as productive as pseudokinematic, because it requires only a single observation and half the total observation time. Rapid-static rivals kinematics in productivity and does not require that lock be maintained. The current implementations of rapid-static rely on the P-code, and one must realize that stated DOD policy is to employ antispoofing (AS) when the GPS system is fully operational. Antispoofing may or may not be turned on at all times. However, selective availability (SA) is turned on all the time, and it is reasonable to assume that the DOD will turn on antispoofing all the time. When AS is turned on, the P-code will essentially be useless to civilians without the information to decrypt the signal. Therefore, it is reasonable to assume that purchasing dual-frequency P-code receivers at this time should only be done if the GPS project has sufficient numbers of stations to amortize the cost differential prior to the implementation of AS.

15-5-3. Pseudokinematics

Pseudokinematic (PK), also known as false kinematic, pseudostatic, broken-static etc., can be the equal in accuracy and precision to 1-hr static GPS. PK is only a field concept and actually is a modified static method. There is no difference in the processing when compared to statics. The mechanics of PK is closely related to static in not only the postprocessing of the data, but also the common reliance on the changing of satellite geometry over a span of time. Static employs an unbroken period of observations of an hour or more to solve for the integer ambiguities. Pseudokinematics, on the other hand, uses short intervals of data separated by changes in the geometry of the satellites over the span of approximately 1 hr. In effect, PK is simply a static session in which the middle has been removed. PK does not require that lock be maintained between GPS setups, is not as productive as K or rapid-static, but can be used in locations where kinematics cannot. PK does not require dual frequency, as does rapid-static.

One variant of the technique employs a receiver on a known station (fixed station) and

one or more rover receiver operators visiting a series of unknown stations twice. The use of a base station guarantees common time measurements whenever the rover is observing on a point. One occupation is done initially on each station and then each station is revisited in the same original order about an hour after the first occupation. The technique is based on the changing geometry of the satellites during an interval of 1 h (or more) that solves for integer ambiguities. It is not necessary to maintain lock on the satellites when moving between stations. However, it is recommended that the receiver remain on to avoid unnecessary warm-up time. The receiver on the base station remains in place during the complete session, never being shut down and collecting data during the entire period. The rover receivers occupy different unknown points twice for 30 to 60 epochs at 10-sec epoch intervals (about 5 to 10 min) at each station, with a return revolution through the same stations about 1 h later during the same session, collecting data for another 30 to 60 epochs as before. During the second tour of stations, each site should have a minimum of four satellites common with the first revolution at that same station. Each individual receiver antenna height must be the same on the second tour as it was on the first. This requires the use of fixed-height antenna poles!

The recommended epoch interval of 10 sec with 6 observations per minute for 5 min is subject to some variation, depending on data quality and postprocessing results. In some cases, it might be advisable to increase the total time on station during each visit by several minutes.

The sample GPS field notes shown in Figure 15-14 are typical for high-production-type observations such as rapid-statics, pseudokinematics, kinematics, and other methods. This noteform contains considerably less data than that shown in Figure 15-13, which could pose a problem during postcampaign analysis in attempting to isolate an error.

Another variation of pseudokinematics called simultaneous rover pseudokinematics

FIELD DATA SHEET
GLOBAL POSITIONING SYSTEM OBSERVATION
California State Lands Commission

Survey: SAN JOAQUIN RIVER	W.O.: 24073		page _1_ of _1_
Station Name: USBR 155-207-102 TRSQ: —	Local Julian Day: 167		Start Date: 16 JUNE 91
GPS Station Number: 0012	Session #: A		Observers: BOB LEA
Latitude: /	Longitude: /		Elevation: /
Receiver #:: 003 Serial #: ASHTECH	Micro Antenna #: 003	Cable Length: 10 meters	Epoch Cycle: SET 10 sec
Power Battery: X Other:	Weather: CLEAR, COOL, CALM		
Leave Time: 0 FIELD OFFICE @ 0000	Refpos: /		Vehicle: /

Mission-Session	SV's	Time PST PDST	Wet Bulb Temp.	Dry Bulb Temp.	Humidity%	Pressure	Slant Ft.- meters	H I
BEGIN A91.167	19,02,14,11, 6,16,18	0845	59°F	72°F	—	Ft. 270' Hg" 29.71	5'10⁵/₁₆"	1.786m
END						Ft. Hg"		
						Ft. Hg"	CHECK END SESSION	
A91.167	19,02,11,06, 15,16,18	1120	88°F	101°F	—	Ft. 270' Hg" 29.71	5'10⁵/₁₆"	1.786m

Description of Monument and Remarks:
FD. CSLC BRASS CAP IN CONCRETE STAMPED
"USBR 155-207-102, 1961"
RESET 1988 LS4333

* LOST LOCK ON SV 14 & 06 DURING TRAIN PASSAGE — BOTH SATS WERE
LOW ON HORIZON @ 4° & 9° RESPECTIVELY

Plumb Bob Check ☒ YES

Disk	Other Stations If Named	Other Stations Number TRSQ	GPS
SESSION A91.167	STATIC = GPS 0012 & 0138 PSEUDO KINEMATIC ROVERS = 0268, 0266, 0264, 0286		

JULIAN DAY FIRST DAY OF THE MONTH
Jan-001, Feb-032, Mar-060, Apr-091, May-121, June-152, July-182

26 Jan 89

Figure 15-14. Sample GPS field notes, short-form.

(SRPK), or no master pseudokinematics, is a very productive adaptation of pseudokinematics. The project referenced by Figure 15-18 partially utilized this field technique that resulted in the completion of a very large GPS control and cadastral survey project in less than one-half the time necessary for the usual static GPS procedures. SRPK is very effective with multiple rovers, three or more, each making the scheduled station occupations simultaneously. This not only results in the measured base lines between the rovers and reference station, but creates additional base lines between each of the rovers simultaneously. SRPK requires communication between all field personnel, either by cellular telephone or two-way radio, to coordinate the beginning and end of the common epochs. SRPK is a mission planning and session coordinating nightmare. This GPS technique requires that all rovers be on station, on time (to the minute), with setup times, tear-down times, travel times to multiple stations, etc., all with timing carefully orchestrated in order for the session to be a success. When the field work clicks, this technique is really worth the effort and complicated planning required. Production in terms of completed base lines is higher than with PK, but less than with K and rapid-statics.

In PK and SRPK, the antenna height on each rover antenna must be the same on the first circuit as it is on the return circuit at each point. The rover antennas can be of different heights, but they must not change between the first and second revolution. The use of fixed-height poles instead of tripods is strongly recommended. In the event that the rovers are shut down (not recommended) between moves, an additional 2 min, or more, must be allowed for warm-up to reestablish good satellite lock.

Figure 15-15 demonstrates the productivity possible with SRPK, even though in this example two of the five receivers remain fixed in location throughout the session. During this session, the four individual occupations resulted in the measurement of 37 base lines including both trivial and nontrivial vectors.

Had the session been PK, then only 25 baselines would have resulted during the same satellite window.

15-5-4. Kinematics

Kinematics (K) can approach the same accuracy and precision as 1-hr static GPS. Kinematics is based on solving the integer ambiguities (one ambiguity for each satellite) up front through a rapid initialization on a known base line. This base line can be known from a previous GPS survey, or it can be determined prior to a kinematic operation by an antenna swap, or a static observation. Once the integer ambiguities have been solved and tracking maintained to four or more satellites, the differential position of the rover antenna(s) relative to the fixed receiver antenna can be computed instantaneously.

The difference between pseudokinematic (fake kinematic) and kinematic GPS is the method of ambiguity resolution. Pseudokinematic is simply static processing of an hour of data with the middle portion of data missing. The ambiguities are solved during the processing just as if the receivers had been on station for the full time period. In kinematic processing, the ambiguities are solved up front by fixing a known base line. Once the ambiguities have been solved, they are carried forward throughout the survey. This is what requires the maintaining of lock to four satellites at all times. Without four satellites, the equations fall apart and the ambiguities cannot be carried forward. When this happens, the system must be reinitialized on another known base line, typically the last surveyed point.

The antenna swap in kinematics two receivers as shown in Figure 15-16. One receiver antenna setup is made on the base or reference station, the other is set up on an arbitrary point established at a convenient distance of approximately 2 to 10 m away (within a cable length of the reference station). The receivers record a specified number of epochs, usually numbering approximately 6 to 10; then

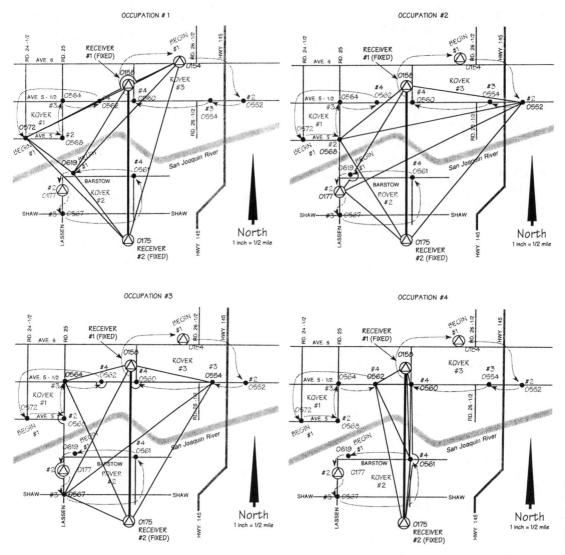

Figure 15-15. Session employing simultaneous rover pseudokinematics.

each antenna is moved to the opposite station to record the same number of epochs. This concludes an antenna swap.

Normally, two antenna swaps are performed as a safety measure. After the antenna swaps are completed, the antennas are returned to the home-leg position to initialize the kinematic survey. It is important to note that the only purpose of the antenna swap is to provide the required known base line. The antenna swap is not part of the kinematic survey. The kinematic survey is initialized by observing 6 epochs on the now known base line with the antenna of the stationary receiver on the base station. After the initialization, the roving antenna is removed to a fixed-height pole and placed on a vehicle to commence the kinematic survey proper. At this time, any other roving receivers would initialize on the known base line for 6 epochs at the swap point prior to being placed on their fixed-height poles and going mobile.

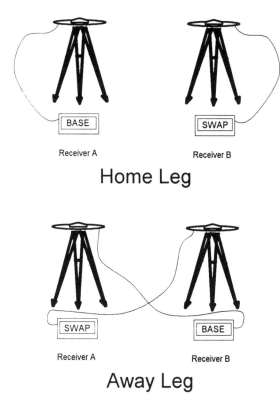

Home Leg

Away Leg

Figure 15-16. Antenna swap.

Should a loss of lock occur during kinematics, the rover simply returns to the last good station for additional measurements consisting of another 6 epochs, then continues as before, avoiding the problem that caused the loss of lock. It is not necessary to return to the location of the antenna swap to reinitialize. At the end of the session, each rover must return to the initializing point to close out the session.

An antenna swap may also be done at the end of a session as well as the beginning. This not necessary, as the check in observation on the known base line is quite sufficient to initialize the survey for reverse processing. If an antenna swap is performed at the end of a session, it must be done after all rovers have checked in on the known base line. If an antenna swap is done at the end, it may be performed between the base and any one of the rovers.

During the entire antenna swap and kinematic survey, the receivers must maintain lock on four satellites, or more. The receiver operators must be extremely careful when moving the antennas to keep them oriented to the sky, and out of the way of obstructions. Both power and antenna connections must be maintained throughout the session. It is a good precaution to connect two power sources to each receiver to ensure an unbroken power supply. Also, the antenna height of the base receiver and roving receiver must be the same when occupying the same point. The best way to perform the swap is to set up two tripods and tribrachs, attaching identical tribrach adapters to each of the antennas, When swapping, move only the antenna and adapter, thus ensuring that the antenna heights will remain constant at each point. In all kinematic GPS operations, it highly recommended that fixed-height poles be used on which to mount the antennas. Fixed-height poles will eliminate one of the most common causes of error, incorrect antenna height measurement.

If kinematic lock is lost, i.e., if any of the roving receivers should track fewer than four satellites, the kinematic survey can be reinitialized on any point that is well known in relation to the base station. Well known is defined as a point or station having been measured directly by GPS methods. Typically, this is the last station occupied prior to losing lock.

During the antenna swapping process, keep clear of the top of the antenna or loss of lock will occur. The antenna heights at the reference station and swap station must not change during either the swap or entire session. Antenna swaps yield very precise azimuths good to + or − 2 arc seconds. The receivers must maintain lock at all times; consequently, all receivers are left on during the entire session until after the last antenna swap is completed at the end of the session.

It is assumed that the tracking of four or more satellites will be maintained throughout the session without a loss of lock. Therefore, it is necessary to plan for a five-plus satellite window to allow for the occasional cycle slip.

When the operator anticipates a situation in which a loss of lock might occur, a temporary point such as a PK nail or hub and tack is set and a reading made at that location. This provides a point in the proximity to which the receiver operator can return should loss of lock occur.

Kinematic GPS, although a highly productive technique, is not in common usage because of the requirement for maintaining lock to at least four satellites at all times. For the best results, five or more satellites should be a requirement in mission planning. Kinematic GPS is second in productivity only to rapid-static. It is most demanding of the mission planner and receiver operators, and is very sensitive to route selection between stations due to the required avoidance of satellite signal interruptions from hazards to data collection such as trees and overhead obstructions. The avoidance of loss of lock is very difficult and virtually impossible in urban, wooded, or rough-terrain areas.

Work is proceeding on instantaneous ambiguity resolution that will allow reinitialization "on the fly" and provide robust kinematic operations in both real time as well as postprocessing. Results have been reported, and when the method is successfully implemented, kinematics will be a viable tool for not only the collection of data, but also real-time stakeout tasks

Refer to Table 15-2 for a comparison of the different GPS measurement techniques.

15-5-5. GPS Survey Party Staffing

GPS data must be processed on a daily basis. Neither the size of the project, nor the number of receivers really matter; postprocessing and network adjustment must proceed as the project progresses in order to spot possible areas of poor results or errors. On-site* post-

processing and network adjustment, mission and session planning, should all be a capability of the GPS survey party. Consequently, these requirements mandate participation in the survey by more than just receiver operators. A representative party size for four receivers on a campaign employing mostly static procedures, e.g., might be shown as in Figure 15-17.

Figure 15-17 does not necessarily suggest that four GPS receivers require eight persons. Some responsibilities can be combined to maximize cost efficiency without impacting productivity. For example, the project director, chief of party, and mission planner could be the same person. The tasks of the computer operator and those of the mission planner could be combined and made the responsibility of one person. This will only be successful provided that no complications or problems with either computations or mission planning arise requiring additional attention. Given the previous example of Figure 15-17 with mostly static procedures, this would (or could be) a workable solution. Computer operation and mission planning are two areas on which the progress of the campaign is crucially dependent.

Mission planning is done in near real time, preferably several hours before the actual mission day is scheduled to begin. With the evolution of more userfriendly, more automated postprocessing software in use at this time, the work of the computer operator has been reduced compared with several years ago. This is offset, however, by greater field productivity resulting from the use of rapid-statics, pseudokinematics, and kinematics, all doubling and redoubling the work of the mission planner.

Combining the responsibilities of computer operation and mission planning is a sensible solution for efficient and economical staffing for schedules and conditions where static procedures predominate, and two sessions per mission day are the norm. This combination

---

*For the purposes of discussion, "on-site" means in the proximity of the survey location, such as the field office or motel.

Table 15-2. A table of methods

Method	Observing Time	Accuracy	Productivity*	Required SVs	Notes
Static	45 min to 7 hr	cm, mm	9 new points	3 or more, recommend 4 plus	More time higher accuracy
Rapid-static	5–10 min	cm	27 new points	4 or more, the more the better	Typically uses P-code
Pseudo kinematic or pseudostatic	2 repeated 5–10 min obs.; total 10–20 min	cm	18 new points	4 or more, 4 common between repeated observations	Complete loss of lock is ok
Kinematic	Instantaneous epoch by epoch solutions; 1 min of obs. typically averaged for survey points	cm, highly subject to satellite geometry with only four satellites	54 new points (10,800 1-sec epoch per trajectory)	Must maintain lock to at least 4 SV's at all times, min. of 5 satellites for practical use	Most productive but most demanding, needs open terrain
Stop & go kinematic	1-min obs.; trajectory epochs ignored	Same as kinematic	54 new points	Same as kinematic	Same as kinematic

*Rough estimates if we assume a 3-hr window, 4 receivers, open terrain, 5 or more satellites above 15°, 1 base or "swing" station, normal static sessions, and accessible points spaced 1–2 km.

works for as many as possibly five or six receivers, but no more. Increases in productivity that result from using either rapid-static pseudokinematic, or kinematic procedures will quickly overload the person who must keep up with both computer operation and mission planning. The use of five or more receivers with mostly nonstatic procedures requires that the functions of mission planning and computer operation be separated and the responsibility of two different persons.

For high-productivity techniques using five or more receivers on a large project, combin-

ing static procedures with kinematic and pseudokinematic procedures, a different staffing strategy is more appropriate and necessary as the example in Figure 15-18 indicates. The best efficiency is realized here if the responsibilities of the project director and mission planner are combined. The computer operation is vested in one individual. The role of chief of party would fall on the most experienced of the receiver operators. This position would be a dual role. Although justified by special circumstances such as the level of experience of the operators, safety considerations, or unique problems in the field, it is necessary for the chief of party to be free to go wherever assistance is necessary. This requires another receiver operator. The survey from which Figure 15-18 was taken was of this nature. The party configuration and other details were as given there.

> 1 project director
> 1 chief of party
> 1 mission planner
> 1 computer operator*
> 4 receiver operators**

*Daily postprocessing and network adjustment.
**Note that safety, productivity, time constraints, or other factors might require more than one operator per receiver.
Figure 15-17. Proposed GPS party configuration for a medium-size network.

15-5-6. Vehicle Considerations

The selection of vehicles is less specialized and design-specific than in conventional terrestrial surveys. GPS does not require the transport of an inordinate amount of equipment inventory and, in most cases, would not exceed the capacity of the trunk of an automobile. In fact, many GPS surveys are successfully conducted with the receiver operators working out of ordinary rental sedans. Naturally, in these circumstances, the survey should not require off-road driving, or rapid GPS production techniques, unless special modifications are made to the vehicles. Use what is available and best-suited for the job at hand.

In the event that either rapid-statics, kinematics, or pseudokinematics are being employed, it is necessary to be as time-efficient as possible. It is not productive for the receiver operator to be required, due to space or transport limitations imposed by the type of vehicle, to assembly the array of equipment arrival at the site, then to disassemble the same equipment when the collection of data is complete. It is better to have a vehicle that allows for the transport of equipment as preassembled as possible to save time in setting up and

Project Details

Geographical area:	200 square miles, combination of city, suburban, rural, and remote
Terrain:	Rolling hills, river valley, river obstruction with limited locations for crossing
Ground cover:	Mature trees with large canopies over roads and access locations such as oaks, sycamore, conifer, agricultural tree crops such as almond, pistachio, citrus, etc.
Purpose of survey:	NAD 83 positioning of cadastral monuments, property corners, various BMs, and historical monuments
Minimum precision:	+ or − 1 cm or 1 part in 50,000, whichever is greater
Special considerations:	The majority of cadastral monuments are located within public rights of way, often in the pavement of busy roads and highways, requiring traffic control, signing, and traffic cone protection of the receiver operators
Number of stations:	Approximately 200
Average station spacing:	Approx. 0.75 m
Measurement techniques:	Static GPS 35% Simultaneous, rover pseudokinematic GPS 59% Conventional survey 5% Kinematics 1% due to tree canopies, obstructions, etc.
Campaign duration:	2 wk
Number of receivers:	5 single-frequency, 12-channel
Number of vehicles:	8 total (7 ea. 4 × 4's, all 2-way FM radio-equipped and 1 sedan)

GPS Party Configuration

1 project director:	Project supervision, mission planning, mission briefings and debriefings, public contact, public relations
1 computer operator:	assistant project director
1 chief of party:	Receiver 1 : 2 persons*
	Receiver 2 : 2 persons*
	Receiver 3 : 2 persons*
	Receiver 4 : 2 persons* (occasionally 1 person)
	Receiver 5 : 1 person
TOTAL STAFFING:	12 persons

*Two persons required due to safety considerations in high traffic areas.
Figure 15-18. Details of a large GPS campaign.

tearing down. The choice of two- or four-wheel capability is terrain- and weather-dependent.

The selection of most 4 × 4 types such as van, station wagon, utility body pickups, and other enclosed utility-type vehicles, all winch-equipped if possible, is the most common and universally successful for any type of GPS survey and practically any location where vehicle access to the station site is possible. These vehicles will be successful wherever the project is located. Some locations still require access by packing-in or airlift.

It is axiomatic that the vehicle must get the receiver operator to the station. Various functions of the vehicle, such as the electrical system, e.g., will be taxed, especially in cases where rapid GPS production techniques are employed. Proper tires for the terrain, a spare tire for the same application, accessories such as windshield wipers, a heater, and even air conditioning must all be in proper working order. In short, using an unreliable, poorly equipped vehicle will be counterproductive when the receiver operator must be on site, on station, and ready to begin data collection at a specific time.

The convenient and secure storage of GPS equipment is a prime requirement of the vehicle. There are two types of storage required. The first is for general transportation to and from the job site or from the main office to the field office. This type does not generally take productivity into account and the equipment is usually safely stored in shipping containers or cases. The second type is storage designed for the rapid deployment of the receiver gear from the vehicle to the station on site. This requires a different consideration. Time is crucial; often, the session planning and coordination allow only minutes for the receiver operator to unpack, set up the equipment, log on, and begin collecting data. This kind of scheduling mandates that maximum preassembly be utilized to save valuable time. There are commercial accessories available that allow the mounting of a GPS bipod or tripod on a universal roof assembly that can be transferred between different vehicle types.

Other custom variants might allow for the mounting of a GPS bipod or tripod rod on the front bumper of a vehicle. This type is a preference by many because the assembly can be viewed from the driver's seat while the vehicle is en route to the station. There are magnetic roof mounts available that permit the antenna to be transported on the roof between stations to help avoid the loss of lock during kinematic measurements. Keep these magnetic mounts away from the computer disks.

In the example shown in Figure 15-19, a 2.5-in. pipe flange has been bolted to the winch bumper. A galvanized pipe is screwed into the flange. The pipe is used as the GPS bipod (tripod) holder, and the GPS rod is inserted within for transport. The pipe is sleeved to provide padding for the protection of the GPS rod from vibration and abrasion during travel from station to station. Note that the antenna is attached to the GPS rod for transport, and the antenna cable to the receiver for quick deployment. The receiver, battery, and coiled antenna cable are housed

Figure 15-19.   Vehicle modification for high productivity.

within a foam-padded amo can bolted to the winch bumper. This assembly proved to be very safe and time-efficient. During the survey pictured in Figure 15-19, four of the five receiver vehicles were staffed by two field personnel due to the large geographical area being surveyed during this campaign. It proved to be more efficient, cost-effective, safer, and more productive to assign an extra person to assist in directing the driver to the station site while reading the to-reach, since it was not always possible for each receiver operator to memorize the route and on-site details of approximately 200 such stations. Time constraints and production schedules did not leave time for the receiver operator to visit the station prior to the days' mission. Once at the station, the extra person assisted in setting up and completing the GPS data log form. The typical elapsed time from arrival to log-on and beginning data collection was 2 min.

15-6. MISSION PLANNING

The term *mission* as used in GPS generally signifies a GPS work day. A mission is broken into several smaller parts called *sessions*. From GPS beginnings in the late 1970s, mission planning for GPS surveys has consisted of the same general and primary steps. The technology of field data collection has evolved from static procedures through several innovative techniques to improve efficiency such as kinematic, pseudokinematic procedures, to the latest, rapid-static. From the smallest to the very largest GPS survey, proper mission planning has not changed significantly, but has become more compacted due to more efficient GPS field techniques. The importance of proper planning cannot be overstressed, and it will help to guarantee the success of the project.

Geometry concerns have rotated to the zenith. Satellite geometry is more important than ground geometry in a well-configured network. Without good satellite geometry, the most ideally configured network would fail.

Measurements on the ground are indirect measurements made by differencing measurements to the satellites. Compared to doing a terrestrial resection, the measurements to the satellites are subject to the same geometrical constraints.

The satellites available for the session must not be bunched together in the sky, but be spread, ideally in four quadrants. This is not always possible, but using mission planning software available from several sources enables the mission planner to monitor an expression called the positional dilution of precision (PDOP). Generally, PDOP values of six or less are sufficient for GPS relative positioning techniques. Do not overlook the use of periods identified by the mission planning software in which the PDOP indicates a spike. This could indicate the rapidly changing geometry that can result in shorter occupation times, yielding the same accuracy as would result with longer occupation times.

The mission planner must have an intimate knowledge of the terrain. This requires a visit to each site and assurance that the recovery notes are still sufficiently accurate for the receiver operator to get to the station site and recover the station.

The mission planner must know the driving time required to access the station from any location in the project area. In the event packing-in is required, then the time necessary to reach the end of truck travel must be known, as well as the walking time necessary to reach the station. Most of the foregoing is information that should be available from recovery notes and to-reaches resulting from a thorough reconnaissance. Packing-in requires an awareness of the physical abilities of the receiver operator and the knowledge of possible dangers en route to the station before personnel assignments are made for the session. The personal strengths and weaknesses of each staff member must be given consideration when making session assignments.

Figure 15-20 shows a key route index map that was taken from an actual campaign; it details specified main access routes for the receiver operators and associated driving times between road intersections defining the geographical limits of the GPS network. The information was extracted from recovery notes prepared for the campaign, as well as other sources to aid in realistic mission and session planning for multiple sessions on the same mission day.

The selection of GPS field technique for the campaign can only be done after the reconnaissance is completed. If a station-to-station spacing represents, say, 5 min of travel or less, then rapid-statics, SRPK, PK, or K would be a strong consideration for high productivity.

The selection of GPS field methodology involves an intimate knowledge of the geography and transportation routes of the campaign area, as well as the capabilities of the various GPS field techniques, so that technique can be matched with area. The mission planner must have personal and first-hand information of the stations to be visited by the GPS crew, in addition to various access routes to the stations. Such an awareness does not result from the recovery notes alone, but from a station-by-station tour of the project for the purposes of familiarization, and formulation of the most favorable GPS technique for these specific locations.

After the agenda for a mission day has been designed by the mission planner, especially in those cases where high-productivity methods are to be employed, it is advisable to have the monuments in the schedule freshened before the session begins. This requires an available member of the party to proceed from station to station ahead of the GPS crew, uncovering the markers should they be buried, or opening well covers and bailing out water if necessary. Additionally, where monuments are located in a roadway, the setting of necessary traffic signs and cones in preparation for the arrival of the GPS observer is a significant time-saver.

As an aid to rapid recovery during the GPS measurement phase of the campaign, it would be very helpful if, during reconnaissance, the station is well lathed and flagged. If the monument is located in an asphalt roadway, then a unique GPS identity number can be painted on the pavement in a large enough size to be easily visible to the GPS observer from a vehicle approaching the station.

Initial advance preplanning should be for the sole purpose of estimating general campaign costs and scheduling, while providing sufficient padding to allow for contingencies. Actual mission planning should occur in near real time in order to be realistic enough to reflect the realities of present field conditions, allowing for flexibility in immediately responding and adapting to field, satellite, weather, personnel, equipment, and access difficulties that might exist. A survey will not be successful if inflexible mission planning was done 2 mon in advance of the project back in the home office.

Figure 15-21 shows an actual plan for an SRPK session prepared for the A session (there was also a B and C session on the same mission day) of Julian day 161 during 1991. The plan details a sketch of the geographical area, station assignments for each rover, start-up and shut-down times, move times, station identities, and more. This document was prepared at the end of the previous mission day (91.160), together with two other session plans for B91.161 and C91.161 (no figures shown). Prior to the beginning of mission day 91.161 the GPS field crew was briefed on the details of A91.161 (Figure 15-21), and the subsequent B and C sessions of the same day.

Although the example shown was for an SRPK session, it is recommended that this type of document be drawn regardless of the GPS technique selected. Figure 15-15 is a detailed breakdown of the four moves (setups) contained within session A91.161; it demonstrates the high-productivity potential of SRPK.

15-6-1. Premission Briefing

The premission briefing is an essential meeting between the mission planner and all GPS crew members and should occur immedi-

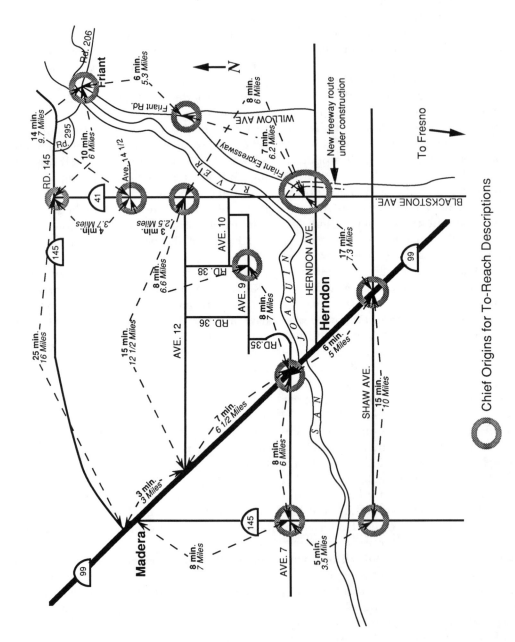

Figure 15-20. Key route index map.

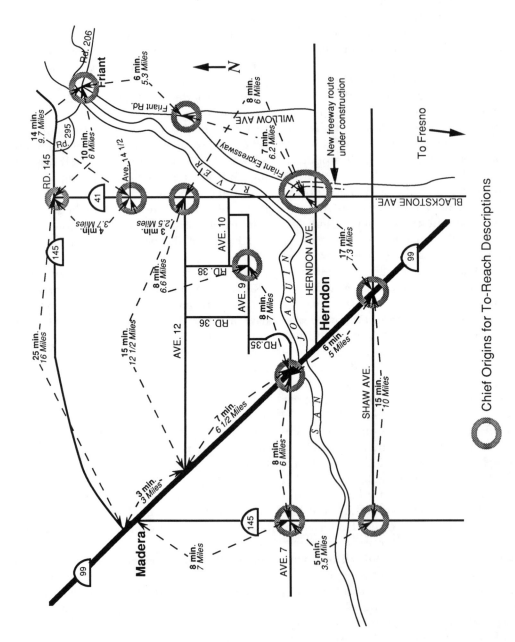

Chief Origins for To-Reach Descriptions

ately before the crew takes to the field. The briefing covers individual equipment assignments, vehicles, station assignments in each session, access to stations, observation schedules, receiver start-up and shut-down times, route changes that may differ from the to-reach, and any changes in conditions since reconnaissance that may impact access to station sites. If there is new information to be added to the recovery notes, this is the time to do it. Other factors to discuss include any new information such as new locks and keys, or combinations, bad dogs, new restrictions, or conditions imposed by property owners, and other practical information. This is the opportunity to comprehensively discuss the start-up and shut-down times as well as anything crucial to the success of the mission. Whatever has gone wrong during the last session should be discussed and critiqued by the receiver operators so that each can learn from the mistakes of others. Perhaps a better time for such input occurs at the time of the postsession debriefing. Feedback to the mission planner is crucial. For example, if insufficient time is being factored in for the moves in pseudokinematics, statics, or rapid-statics, or, on the other hand, the allotment of time is too generous, this would be the time to bring this matter to the attention of the mission planner so that adjustments can be made to the scheduled start-up and shut-down times to provide greater productivity.

Session plan distribution should be made to all receiver operators at the time of the pre-mission briefing to ensure the success of the session. Additionally, it is further recommended that each operator be given a folder for each station assignment that contains a copy of the following:

1. The to-reach
2. Station notes
3. Local map
4. Assessor's parcel map
5. Photo of station
6. Blank notes for the GPS observation, either long- or short-form.

15-7. SPECIAL EQUIPMENT

The importance of two-way radio communication can never be overstressed or overemphasized. At the very least, the receiver operators should be equipped with CB radios for inter-site communication. Ideally, the receiver vehicles should have two-way FM radios and walkies, and at least one of the vehicles in the field should have a cellular telephone if at all possible. The resulting close communication enables the coordination of start-up and shut-down times during the sessions and allows for variables in arrival times or setup times at the station sites due to wrong turns, delays, equipment malfunction, or similar misfortunes. It also allows the flexibility to extend data-collection times if necessary, or for earlier start-up time than scheduled by the mission planner, provided that proper satellite geometry is present. In the event of an emergency or any situation where assistance is necessary, communication is the key to response and should *never* be ignored.

15-7-1. Tripods and Tribrachs

Except for GPS hardware such as antenna, cables, receiver, etc., the normal equipment and accessories usually found in the inventory of most surveying offices are sufficient. GPS antenna assemblies are designed to fit on practically every type of tribrach. A tribrach and tripod are all that is required. Naturally, the calibration of the tribrach must be validated and the tripod in good repair. But why use a tripod and tribrach? The tribrach is expensive, subject to chronic centering (collimation) errors, is fragile, is not weatherproof, and with its sophisticated optics and precision machining, perhaps too refined and complex for the job at hand and maybe not essential, or even practical, for production-oriented GPS.

15-7-2. GPS Bipod-Tripod

As an alternative to the conventional tripod and tribrach, many governmental agencies, es-

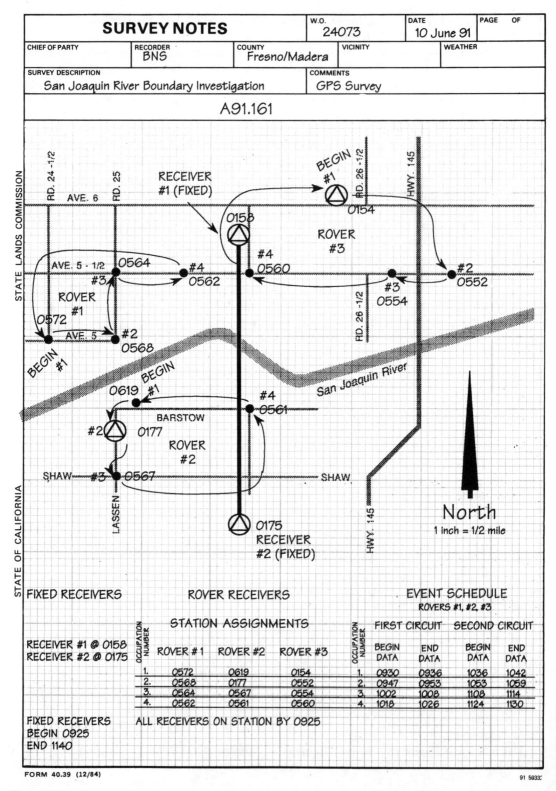

Figure 15-21. Typical session plan.

pecially federal and state, and private industry as well are switching to GPS bipod-type stakeout rods in preference to the conventional tripod and tribrach. For the price of a better-quality conventional tripod, a GPS stakeout rod can be purchased that is easier to set up, more suitable for GPS measurements, less subject to damage than the conventional tribrach, and about one-third the cost of a tripod and tribrach. Additionally, at least 2 to 3 min are cut from the setup time when a GPS bipod- or tripod-type stakeout rod is used. In conventional work, this may not seem like much, but when the session plan calls for a GPS receiver operator to break down the setup, pack up, and travel to a new station, where the equipment is again set up and the receiver logged on before data collection begins again, all in a time frame of 10 min, the shaving off of 2, 3, or 4 min from total operational time becomes significant. Time is of the essence; time saved in any GPS operation will yield greater productivity no matter what GPS field operation technique is utilized. Time is saved in setting up the antenna and time is saved during HI measurements, since it is only measured once, and in recording the data. The GPS rod assembly shown in Figure 15-22 is lighter in weight, more suitable for packing into a station, and occupies considerably less space, which is beneficial in those situations where storage capacity in a vehicle or aircraft is at a premium. Higher-precision-level bubbles are available that can yield centering of the 1-mm level. These assemblies make the GPS crew more mobile and flexible to individual site requirements and cut down on a bulky and unnecessary inventory. With the addition of screw-in rod extensions, the height of the antenna can be elevated several times beyond that of a tripod and tribrach, thus solving the problems associated with difficult setup locations where obstructions would have made observations impossible.

The GPS bipod or tripod is similar to the conventional stakeout rod, but with larger diameter for greater stability and strength. The length of the rod is adjustable by the selection

Figure 15-22.   GPS rod.

of appropriate lengths of screw-in sections that are available in various lengths, generally of 1, 2, or 3 feet. Once assembled, the rod is left in whatever length configuration is selected for the mission. The use of an adjustable-length (telescoping) rod is not a recommended alternative due to problems inherent with HI measurements. The goal is to reduce unnecessary field operations and eliminate the incidence of HI errors. A telescoping rod is subject to a change in the antenna height if the friction collar slips. Should this occur and go undetected, any GPS measurements made will be invalid. Leave it in the equipment room to keep company with the tripods and tribrachs. The use of uniform length should be a rule during the campaign. This does not mean that all the rods need to be exactly the same; they do not. However, to avoid errors in the HI measurement, this consistency should be maintained. The mismeasurement of HI is one of the more common mistakes made with GPS observations. If each GPS rod is premeasured for HI, this is one less task that the receiver operator must accomplish in his or her busy schedule and one less possible source of error.

All that needs to be recorded in the GPS field notes is the rod identification number. Since the rod is premeasured, the HI is already known to the mission planner and computer operator.

The choice between the two- or three-leg GPS stakeout rod is a matter of preference. However, there are advantages and disadvantages with either selection.

The bipod is more efficient to set up. How much more time-efficient is a matter of practice and technique but should not amount to more than 10 sec. The bipod is naturally somewhat lighter in weight than the tripod since it has less hardware and does not have that extra leg which sometimes gets in the way. The bipod will stand by itself, provided there are no gusts of wind or other conditions that could upset the balance. Consequently, the bipod must be attended throughout the occupation to prevent it from falling over.

The tripod version of the GPS rod is more stable in traffic, windy or gusty conditions and is recommended for most applications requiring unattended operations. Naturally, adequate carrying cases are advised to protect the GPS rod for shipment.

15-7-3. Multiple GPS Rod Bubbles

Centering (or plumbing) over the mark is accomplished by using the level bubble attached to the rod. GPS rods can be purchased with more than one bull's-eye level. It is recommended that an array of three such levels with a sensitivity of 10 min be installed on each rod. These rod bubbles shown in Figure 15-23 are arrayed at 120°, a convenient distance above the ground so as not to interfere with the operation of the bipod or tripod legs of the GPS rod. Make certain that they are in exact adjustment and firmly attached and mounted to the rod as the level bubbles may be subjected to unavoidably harsh handling. Why three bubbles? If only one bubble is used on the GPS rod assembly and it should go out of adjustment, the condition would probably

Figure 15-23.   GPS rod with three bubbles.

go unnoticed by the receiver operator and might result in good observations to the antenna, but bad observations relative to the mark since the antenna phase center is in a different plane. Two bubbles installed on the GPS rod and adjusted during plumbing calibration serve as a check on each other. Should one of the pair of bubbles go out of adjustment, the operator would not know which of the two is suspect, and would need to abort his or her session schedule until the problem is cured. The use of three bubbles is the practical limit and provides for a situation in which one of the levels has been knocked out of adjustment; the remaining two should have the same reading when the rod is set over the mark, thus reassuring the observer that collimation is still good. Thus, three bubbles would allow the session to proceed without interruption.

15-7-4. GPS Rod Collimation Adjustment: A Shortcut

Rod collimation can be accomplished using several different procedures. One of the simplest methods that does not require elaborate equipment can be done in any doorway. Simply suspend a plumb bob from a small nail driven securely into the top underside of a doorway. Mark the plumb point on the floor. Remove the plumb bob and set up the rod with the point of the rod on the plumbed point and the top of the rod plumbed underneath the nail (the plumb bob might be needed here) and secure the legs. Once this is accomplished, the rod is now vertical and in collimation with both points. Now adjust the rod bubbles so that they are centered. Check by slowly rotating the rod through 360°, noting any movement of the bubbles and any movement of the top of the rod away from its plumbed position under the plumb bob hanging from the nail. There should be no movement in either the bubbles or top of the rod. If such is not the case, the rod is either bent or the rod bubbles are at fault. Replace and/or repair and readjust as required.

15-8. HAZARDS TO SUCCESSFUL DATA COLLECTION

The interruption of satellite signals can be caused by a multitude of sources. Rain or condensation causing moisture on the electrical connections, or the head, hands, hard hat, etc., while servicing or maintaining the receiver and antenna can cause interference or blocking of the signal. Basically, anything that passes between the antenna and satellite can interrupt the signal. Keep away from the antenna while collecting data. The body can block signals from the satellite just as surely as tree limbs, power poles, and other inanimate objects. Avoid placing anything between the antenna and satellites or the signal will be blocked and a loss of lock will occur. Another classic is the ''experienced'' field hand who, during kinematics, automatically shoulders the GPS rod or tripod when picking up. This, of course, is guaranteed to cause a loss of lock and trip back to the last known position for reinitialization.

15-8-1. Multipath

Multipath is a condition in which a satellite signal arrives at the receiver antenna by way of several different paths. It is caused by reflected, indirect, signals from a satellite and can originate from a multitude of sources. Figure 15-9 is a good example. Bodies of water, structures, nearby vehicles (especially slab-sided semitrailers, or vehicles or similar design), freeway signs, chain-link fences, or similar reflective objects can cause a condition of multipath.

Sometimes, the use of a larger-accessory ground plain can reduce or eliminate this condition. Some antennas do not even have an integral ground plane. These antenna types should be avoided except for special applications. Avoid the problem whenever possible by locating the station away from possible interference. If the station is part of the primary net, then an eccentric might be considered, or possibly a higher setup to clear the possible reflection. Multipath is not detectable until post-data-reduction is done, so be observant. Proper reconnaissance will identify stations where this problem exists and precautions can be taken to minimize or eliminate the problem. During data collection, the survey vehicle must be parked far enough away from, or below, the antenna to eliminate the possibility of multipath from that source. Good site selection during reconnaissance will either minimize or eliminate problems due to multipath.

15-8-2.   Electronic Interference

Consult with the manufacturer regarding possible sources of electronic interference that could corrupt the signals of the satellites. The following electromagnetic energy sources will not necessarily affect all receivers. Some receivers are adequately shielded; however, it would be best to err on the side of caution if there is any doubt. In the design of any electronic circuit, one important factor is to separate the input signal from unwanted signals and amplify it in the required way without producing distortion beyond an acceptable degree. Effective passive filters are installed in the GPS receivers to filter out unwanted signals; however, if the unwanted signals are powerful enough, the filters are ineffective. Powerful signals cause the amplification to be adversely affected in the form of distortion and this can affect the GPS receiver performance. Some of the sources that may cause interference with GPS units are:

1. Vehicle detectors for actuating traffic lights
2. Portable transceivers for radio communication
3. Signals emitted from antennas generated by radio and television stations
4. Amateur ham band and citizen band transceivers
5. Microwave antennas and transmitters for equipment such as telephones
6. Radar installations

15-8-3.   Traffic Sensor Devices

Vehicle detectors come in two varieties: below-ground (loop detector) and above-ground (pole-mounted). The ground plane antenna on GPS units would shield the signal from the below-ground, or loop, detectors, that have a range of approximately 10 ft straight up. Therefore, they would present little risk of interference to the GPS receivers. The pole mounts have two operating frequencies: microwave (10.525 GHz) and ultrasonic. The microwave vehicle detectors have a greater possibility of affecting GPS receivers because of the

closeness of operating frequencies; also they are located above the GPS receiver antenna. On the other hand, the GPS receiver antenna would have to be relatively close to a vehicle detector because the low power (2.5 to 6.0 W) of the vehicle detector and limited range (60 ft more or less). Ultrasonic vehicle detectors have little or no effect on GPS receivers because of the difference in operating signal frequencies between the two units.

15-8-4.   Two-Way Radios

Portable transceivers such as handheld and mobile radios should be operated as little as possible and as far from the GPS unit as is convenient. Transmission is not recommended during data collection except with the manufacturer's OK. The concentrated signal in the vicinity of a portable transceiver can resemble a much stronger signal and possibly corrupt the satellite signal. Cellular phone transmissions, when done in the close proximity of the receiver, could interfere with satellite transmission reception as well.

15-8-5.   Radio, Television, Microwave Antennas

Radio, television, radar, and microwave antennas radiate a powerful signal and generally the antennas occupy high points such as the tops of mountains and buildings. These signals can also present a problem for GPS units. The signal from these antennas is so strong that it can affect a GPS signal a considerable distance away. Satellite dishes such as those for home use are for receiving only and present no problem unless the satellite dish is in the GPS signal path to the receiver antenna. The satellite dishes for commercial television stations both transmit and receive; however, they are very directional and so also should present no problem. The configuration is not necessarily an indication of the function of the antenna since antennas can be used for both transmitting and receiving.

Airports typically have more than a fair share of potential electronic interference.

Large airports always have a greater abundance of exotic electronic devices than any other location, most of which are capable of interference.

Other possible sources of problems such as high kv power transmission lines can cause problems during certain atmospheric conditions. Any time that buzzing or arcing can be heard, satellite signals could be subject to possible interference. Thunderstorms, even though miles away, are reported to be disruptive.

15-8-6.   Geomagnetic Disturbances

Geomagnetic disturbances are caused by solar flares, solar storms, and other similar natural solar phenomena. These solar disturbances release large amounts of energy in the ionosphere, an electrically conductive series of layers of the earth's upper atmosphere extending from 50 to 400 km above the surface.

Flares and the resulting geomagnetic storms that sometimes accompany them can disrupt low-frequency systems such as Loran C and GPS. Communication systems like television, radio, microwave, and short-wave ones are also impacted, sometimes to the point of total disruption. The current solar cycle is believed to be one of the highest ever and should continue into the mid–1990s.

Geomagnetic storms are sometimes accompanied by the appearance of the Aurora Borealis or northern lights, even in the lower mid-latitudes. In a time of major geomagnetic storms, such as the flares of March 1989, an aurora was visible in the Gulf states. The Aurora Borealis during this event was seen as far south in California as 35° north latitude.

The use of dual-frequency GPS receivers will minimize the effects of geomagnetic disturbance. The majority of receivers in use at this time are of the single-frequency variety that are susceptible to cycle slips caused by the electromagnetic noise associated with these disturbances.

The appearance of the aurora should be a visual warning to the project manager and mission planner to carefully view the post-data-reduction for signs of noisy data and possible cycle slips. Mission planning should include the monitoring of all geomagnetic and solar advisories that are available on the joint USAF/NOAA solar region summary bulletin board service. This bulletin provides values for the A-index and K-index and predicted values for A and K that can be used as an indication of solar activity levels. Generally, A-index values greater than 20 and K-index values greater than 5 are indicators of high geomagnetic disturbances, possibly contributing detrimentally to the collected data.

With single-frequency receivers, a few measures could be employed to a limited extent in an attempt to salvage a few mission days in a campaign unfortunate enough to be accidentally scheduled within a period of high solar activity. If there are short base lines in the project of 2 km or less, these short lines would be the preferred measurements to attempt. Longer base lines are typically out of the question and likely to fail. The order of preference for measurement techniques during high solar activity periods is

1. Statics
2. Kinematics
3. Pseudokinematics or SRPK

It is very possible that cycle slips will go unnoticed with the pseudokinematic technique during periods of high solar activity; increasing the data-collection time by 50% or more might help. It is reported that nighttime levels are somewhat lower than daytime levels and midday levels lower than early AM or late afternoon levels. Watch the resulting residual statistics with caution and modify the mission plans to include redundant vectors and independent checks on the observed stations. If unrepairable cycle ships occur in spite of these precautions, either obtain dual-frequency receivers or cancel the field operations until the geomagnetic disturbances subside.

15-8-7. Power Source

Any system that utilizes battery power is subject to a multitude of energy-related problems. Do not believe the sales representative when a claim is made that a brand A receiver will run all day on an AAA battery. Never underestimate the power consumption of the receiver system. Always allow for a comfortable, if not generous, safety margin. It does not matter how high-tech and exotic surveying technology becomes. The simple truth remains: We are slaves to our battery power source. The battery and two-bit battery connections remain the two most common causes of receiver failures. Always provide the receiver operators with backup battery reserves and extra cables and connectors, and insist that all batteries be topped off between missions.

The selection of battery capacity is dependent on the power requirements of the unique receiver, which can vary from manufacturer to manufacturer. A good guideline would be to select a battery with sufficient reserves for possibly 150% or more of the required power consumption for approximately two missions —i.e., two data-collection days or perhaps 10 hr of actual data collection, whichever is the greater. This would allow for the loss of battery efficiency from low temperatures or possibly a failure of the operator to recharge to a full 100% level. Naturally, pack-in situations require lightweight as well as ample reserves so allowances should be made to mix battery types—i.e., weight and capacity according to the specific session and station requirement —to cover long-term continuous data-collection sessions as well as situations in which a lightweight and free mobility are important.

15-8-8. Antenna Height Measurement

The importance of this measurement is often overlooked. GPS is a three-dimensional system and requires an HI to compute the position at the mark. Antenna height measurement is a common source of problems in the field. Without reliable antenna height measurements, the system will not accurately compute final position and elevation at the station. The position will be to some nonrepeatable point in space. The common error is either failure to measure the height or erroneous measurement of an HI. As a check against possibly flawed measurements, a good technique is to independently measure in two different units of length such as feet and meters. The use of fixed-height antennas is the greatest elixir for either bad or missing HI measurements. However, be advised in otherwise identical GPS rods that there can be variations in antenna height measurements. It is a good idea to individually identify each GPS rod with a unique name or number and HI. This identity is then incorporated as data entry on the GPS field notes and entered into the GPS receiver at the site. When conventional tripods and tribrachs are being employed, extra care must be exercised to ensure that the HI being measured is error-free. Once again, the best method involves measurement in two different systems, such as feet and meters, with an independent conversion of feet to meters or from meters to feet done in the field by the receiver operator as a check. It should be standard operational policy for field personnel to make this conversion as part of the data-collection process.

The use of a GPS rod, either bipod or tripod, practically eliminates this error due to missing HIs since the setups are done at a uniform height above the mark. The length of the rod from the point to the antenna needs to be measured and checked only once.

15-9. TYPICAL PROBLEMS ENCOUNTERED

No matter how detailed, complete, and intuitive the planning and preparations are, something will always occur to disrupt the survey.

GPS is subject to a host of familiar maladies common to conventional terrestrial survey operations and a few that are not. Most problems associated with delays and downtime are vehicle- and human-error-related. Table 15-3 is a tabulation of various things that went wrong during two actual campaigns, each survey consisting of approximately 200 stations. The first campaign employed static procedures with two, or more, nighttime sessions per mission during a 4-wk period in February and March 1989. The second survey employed a combination of static, pseudokinematic, and kinematic

procedures with two or more daytime sessions per mission during a 2-wk period in June of 1991. During each survey, premission briefings and postmission debriefings were conducted by the mission planner, with the chief of party, computer operator, and GPS receiver operators in attendance.

Items 1 through 9, and perhaps item 19, in Table 15-3 are, to the greatest extent, classified as human blunders and mostly preventable. Perhaps more emphasis can be placed on high-repeat problems at the time of the postmission debriefing or next premission

Table 15-3. Problems Encountered on Two Typical Jobs

Malady	Number of Instances
1. Receiver operator getting lost en route, wrong turn en route to a GPS station	6
2. Receiver operator on wrong GPS station	4
3. Bad setup (tribrach not leveled, GPS rod not plumb, etc.)	4
4. Receiver operator forgets to load an equipment component necessary for session	6
5. Operator error: Wrong data keyed, wrong epoch interval, etc.	4
6. Bad HI measurement	3
7. Cycle slip: Loss of lock due to carelessness	2
8. Vehicle stuck	2
9. Vehicle out of gas or lack of fuel delays or constrains planning	3
10. Vehicle will not start	4
11. Vehicle breakdown	4
12. Vehicle battery dead	2
13. Access to station not possible: Receiver operator locked out, new lock installed, no key, key not working, etc.	3
14. Traffic delays	4
15. Receiver battery problems: Battery failure, bad connections, etc.	11
16. GPS equipment failure, either receiver or antenna	2
17. Geomagnetic storm (causing unrepairable cycle slips): Dual-frequency receivers would not have been affected:	6
18. Bad weather: Thunder, lightning, snow, rain, etc.	1
19. Ground swing. Multipath (semitrailer parked near station, survey vehicle parked too close to the antenna, etc.)	2
SUMMARY of PROBLEMS by TYPE	
Human error, blunder-related	34
Vehicle, traffic, access	17
Unavoidable, would possibly occur again	9
GPS hardware	2
GPS power source, connectors	11
TOTAL	73
Summary	
GPS hardware-related	02%
GPS power source, connectors	15%
Unavoidable, usual delays	13%
Human error, non–GPS equip.-related	47%
Vehicle, traffic, access	23%

briefing. Some problems are due to operator fatigue; others are preventable by the application of good judgement, and some are going to occur no matter what preventative measures are taken.

Items 10 through 14 are vehicular or access-related problems: Some are preventable and some are not. Vehicle breakdowns are going to occur, but can be minimized with good maintenance and careful, nonabusive driving practices. Access problems can be reduced by revisiting the station (premission) to update the to-reach if time permits. Should there be sufficient time before a session begins, a luxury that does not occur very often, the receiver operator could tour assigned stations to become familiar with the routes to them. With good planning, traffic delays can be factored into most session plans.

Items 15 and 16 are GPS hardware-related. Item 15 is a familiar foe and will continue to be as long as a battery is a part of any technology. Batteries need to be recharged and connections are necessary to feed the energy to the electronics. This is nothing new as battery and connector difficulties have always been a problem with conventional survey instrumentation and will continue with GPS as well. With the implementation of standard procedural policy for field personnel, energy-related malfunctions can be minimized. Carrying spare batteries is essential; keeping both the main and spare battery fully charged is paramount. Policy should be that the receiver operator is responsible for topping off these power sources immediately after the mission day. If insufficient time exists between missions to completely recharge, then additional batteries must be issued to permit the cycling and charging of the battery inventory. Power interruptions due to faulty connectors are intolerable during data collection. Good manufacturer design should be insisted on; if problems occur, complain loudly. Keep the connections tight and tape them if necessary to prevent movement and moisture.

Item 16 is not supposed to happen, but does in spite of the best of design and manufacturing practices. The best advice is to buy reliable, robust equipment from manufacturers or firms with established postsale support and good performance history.

Items 17, 18, and 19 are random. Geomagnetic storms or ionospheric disturbances in the earth's upper atmosphere caused by solar disturbances are not predictable with any reliability. These disturbances can cause unrepairable cycle slips in single-frequency GPS receivers when levels are high. The use of dual-frequency receivers virtually eliminates this problem. Both the surveys referenced by Table 15-3 were impacted by solar disturbances that aborted several mission days. Had dual-frequency receivers been employed, the campaigns could likely have continued without interruption and costly crew downtime.

GPS is universally represented as an "all-weather system." This is not always true. GPS is subject to most of the same influences as conventional instrumentation, except those relating to line of sight. Thunderstorms and lightening can cause cycle slips, and the hazards of a strike from being near the tripod or GPS rod during such events are a reality. Rain can cause shorts, or open circuits in the power leads and connectors. Snow collecting on the antenna can cause a loss of lock, and wind can topple a tripod or GPS rod, possibly irreparably damaging equipment and causing downtime for repair or replacement. It would be advisable to carry as many spares as the budget and conditions warrant. An extra set of cables and power leads should be issued to each receiver operator. Rain, ice, snow, and mud can cause gross differences in the access times required to reach stations; this could impact mission planning to a substantial degree. Weather should always be factored in with all mission plans. Should bad weather set in, wait it out if economically possible. GPS is the most productive and efficient technology ever to impact the surveying profession. Lost time can be made up. Inclement weather is hard on equipment, results in undue hardship for field personnel, and because of hazardous driving and field conditions places the staff in possible danger.

15-10. NETWORK ADJUSTMENT

Network adjustments are performed for two reasons: (1) To detect and remove blunders, and (2) provide a best fit into the local datum. Blunder detection adjustments should be performed on the daily results throughout the project. Each day the daily results should be added to the previous vectors and a daily subtotal adjustment performed to verify data continuity. At the end of the project, it will then be a simple task to fix the control and perform a final constrained adjustment, fitting the data to the local control.

There are several excellent least-squares adjustment packages on the market for GPS observations. Most of these packages allow the inclusion of conventional data and use the information contained in the base-line processing solution files for weighting the lines. However, one commonly used package does not allow the mixing of conventional and GPS data and its expects the user to input a priori error estimates for weighting instead of using the solution information. There are pros and cons to either approach, both as to the inclusion of conventional data use of the base-line solution statistics. Either approach is correct and will produce excellent results if used properly. Following is a description of the process using least-squares adjustments. The method is applicable to either of the least-squares weighting schemes, the use of the base-line statistics or a priori error estimates, and based on realistic error estimates regardless of the weighting scheme used.

Most of the work in a network adjustment is performed as a *free adjustment*. A free adjustment is a minimally constrained adjustment where the latitude, longitude, and height components of a single station are held fixed. A free adjustment results in the adjustment of the data on itself without constraints as to the known positions of other stations or rotation angles between the GPS system and local datum. Free adjustments are for blunder detection and eliminating bad measurements. Once the data have been cleaned up, the control

station values can be held fixed and the final constrained adjustment performed.

Starting directly with a constrained adjustment results in greater difficulty in isolating any problems. If there are any problems in the adjustment, it would be extremely difficult to determine if the problem is actually in the GPS data or control itself. By cleaning up the data before fixing the control, it is easy to determine that the problem lies with the control. A properly performed network adjustment consists of daily free adjustments, daily subtotal free adjustments, a total free adjustment, and a total constrained adjustment. Each step of the adjustments, especially the total constrained adjustment, may take several iterations when vectors are deleted or deweighted. Daily processing and adjusting cannot be overstressed. It is important to find any errors as soon as possible so that they can be corrected immediately. Reobservation is a dirty word only if spoken at the wrong time.

15-10-1. Statistical Indicators

Before the steps in an adjustment are detailed, look at the apriori and a posteriori indicators that will determine the adjustment quality. These indicators are the *apriori error estimates*, either entered by the operator or taken from the base-line solution statistics; the *residuals*; the *normalized residuals* (standardized residuals); and the *standard error of unit weight* (variance of unit weight). To be accurate, all these indicators depend on realistic estimates of apriori error.

Apriori error estimates are generally created from the standard errors of the base-line components (sigma *X*, sigma *Y*, and sigma *Z*) and the correlation matrix. These are combined into the covariance matrix that is used for weighting the vectors. The apriori errors can also be empirically derived and specified manually as a base error (mm or cm) plus an allowable additional part-per-million error. Either way, it is important that the estimates of error be realistic.

The residuals are the amounts by which the adjusted vectors have been shifted in the adjustment. They are the differences between the ΔX, ΔY, and ΔZ components of the observed vector and adjusted vector. They are generally reported in meter units.

The normalized residuals are the residual divided by the apriori standard error. The normalized residual indicates outliers. A normalized residual of 2.0 indicates that the residual is twice as large as it should be based on the apriori errors. If the apriori errors are realistic and the standard error of unit weight is close to 1, statistical outliers will have a value of 3 or more.

The standard error of unit weight indicates the degree with which the data and apriori errors agree. The ideal, a standard error of unit weight equal to 1.0, indicates that the quality of the data exactly fits the model. The standard error of unit weight should approach 1.0 or less. Exceeding 1.0 by more than a very small amount indicates problems with the data, or overly optimistic error estimates—i.e., the apriori errors are too small. On the other hand, a very small standard of unit weight indicates pessimistic apriori errors, or extremely good data. If the estimates are pessimistic—i.e., if they can be realistically reduced—they should be. Pessimistic apriori errors can hide problem data. When the apriori errors are too large, the normalized residuals will be smaller than they should be, and data that might be an outlier with realistic error estimates can be hidden.

15-10-2. Network Adjustment Procedure

Starting with the first mission-day observations, a daily free adjustment is performed. The daily free adjustment is where the majority of work is done. It is at this stage that blunders are detected and removed, and final decisions as to the fixed or float solution selection are made.

Most processing software packages have a filter program that selects the best solution for each vector. This filter selects either the *float double difference solution* or *fixed double difference*

solution, depending on its particular algorithm looking at the base-line processing statistics in the solution files.

A suggested practice is to perform two adjustments if any float solutions are selected. One of the adjustments will contain the combination of fixed and float solutions selected by the filter, and the other all fixed solutions. A quick comparison of the standard error of unit weight will usually determine whether the fixed or float solutions are best.

Sometimes, it is necessary to go further and return one of the fixed solutions back to a float solution and perform a tired adjustment. Decisions like this are based on the SE of unit weight and values of the normalized residuals. The results of the daily free adjustment(s) should be the vectors in the adjustment with the lowest SE of unit weight. These would be the best fitting vectors and those to pass along as we build the total network.

Each day, the vectors of the final daily adjustment are added to the previously selected vectors, and a subtotal adjustment is performed. This is to verify that the day's sessions agree with those of other previous days. Although it is not common, a session not agreeing with others happens on occasion, due probably to bad broadcast ephemeris information. To find this out in a timely fashion, perform daily subtotal adjustments. Generally, the broadcast ephemeris is satisfactory for surveys up to a few parts per million. For surveys of 1 part per million or better, the use of precise orbits is mandatory.

The use of daily subtotal adjustments also provides an easy way of building the network using the best vector solutions available. It is virtually assured that the final free adjustment will be simply a matter of adding the final daily adjustment vectors to the subtotal and performing the final free adjustment.

When the final free adjustment has been completed, blunders have been removed and the data has been validated. At this point, the first task of the network adjustment has been accomplished: the removal of blunders. It is now time to constrain our free adjustment by the control and perform the final task, fitting to the local datum.

Generally, the free adjustments are performed using the WGS 84 latitude, longitude, and ellipsoidal heights derived from the baseline processing solutions. One of these positions is simply held in all three components for the free adjustments. To constrain the network, specify the datum and reference ellipsoid to which the free adjustment will be molded and fix the control stations by entering their positions relative to the datum selected.

For example, if it is desired to perform a constrained NAD 83 adjustment, make sure the adjustment program will use the GRS 80 reference ellipsoid parameters, then supply NAD 83 latitude and longitude, and ellipsoid eights (elevation plus geoidal separation) for the fixed positions. If NAD 27 positions are desired, make sure the Clark 1866 reference ellipsoid is used, and NAD 27 latitude ad longitude, and orthometric heights (sea-level elevations) for the control stations are entered for the fixed stations.

Once the control has been fixed, perform a constrained adjustment. It may be a minimally constrained adjustment or fully constrained adjustment. A minimally constrained adjustment has only enough control fixed to solve uniquely for scale and rotation biases between the GPS system and local datum. For example, two horizontal stations will provide one and only one solution for scale and azimuth rotation. Three vertical stations will provide one and only one solution for the gamma X and gamma Y rotations—i.e., the vertical tilts in the east-west and north-south directions. A fully constrained adjustment will provide redundant solutions for these bias parameters,—e.g., three or more horizontal control stations and four or more vertical control stations.

If the control values are good and estimates of error accurate, a standard error of unit weight close to 1.0 should result. If the standard error of unit weight is larger than 1.0, it indicates one of two things: Either the estimates of error are too optimistic, i.e., too

small, or the control does not fit well. Most commonly in an NAD 27 adjustment, it will be the control causing the problem. Unless there is a single station that can be proven to be in error, it will be necessary to mold the data to fit the local datum. This may take several iterations and involve deweighting of vector components and constraining, instead of fixing, some control values.

Avoid any temptation to leave the adjustment in the free adjustment stage, i.e., radial surveys. The data will fit best on itself, but will not fit the real world or local control. Any ties from the free adjustment to other national network stations could easily result in substandard closure statistics because the rotation biases were not solved and applied.

15-10-3. Dual Heights

Elevation, as measured with a level, is the height above the geoid, an irregular surface of equipotential gravity commonly associated with the mean sea level. Ellipsoid height is height above the reference ellipsoid, a smooth mathematical surface. GPS measures this ellipsoid height. The ellipsoid height used in the NAD 83 adjustments may be broken down into two elements: the *orthometric height* (commonly referred to as sea-level elevation) and *geoid height* (the height of the geoid above, or below the ellipsoid). Normally, the surveyor does not know the ellipsoid height of a station, but will know the elevation. The geoid height of any station can be interpolated from a tabular data set that has been created using a model such as the GEOID 90 model available from NGS. By adding the elevation and geoid height (a negative value in the continental United States), the ellipsoid height of the vertical control stations can be computed, and conversely the adjusted elevations of unknown stations by subtracting the geoid heights from the adjusted ellipsoid heights.

In Figure 15-24, the relationship between the geoid and ellipsoid in world terms can be seen. The WGS 84 or GRS 80 ellipsoids are mathematical surfaces defined to best fit the

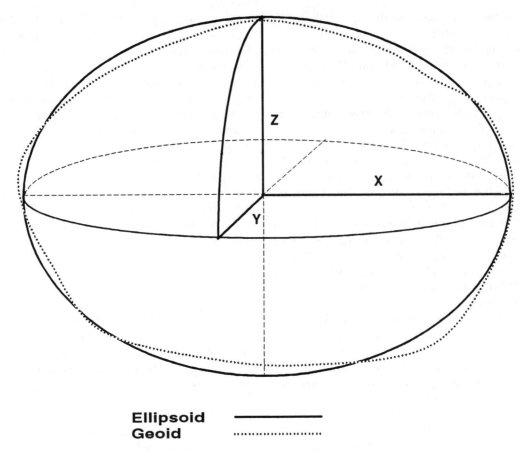

Ellipsoid ────────
Geoid ·····················

Figure 15-24. Ellipsoid and global geoid.

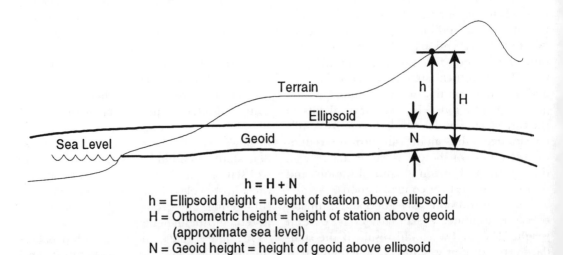

h = H + N
h = Ellipsoid height = height of station above ellipsoid
H = Orthometric height = height of station above geoid
       (approximate sea level)
N = Geoid height = height of geoid above ellipsoid
       (negative in continental U.S.)

Figure 15-25. Elements of height.

shape of the earth. The WGS 84 was intended to be the same ellipsoid as the GRS 80, but a mistake with precision caused a change in the inverse of flattening parameter.

In Figure 15-25, a detail section showing the relationship between the ellipsoid and geoid can be viewed. The calculation of the ellipsoid height h is shown as the addition of the elevation H and geoid height N. In the example shown, N is a negative number (the geoid is below the ellipsoid), resulting in h being smaller than H. This is the case in all of the continental United States. The term geoid height is somewhat confusing. Unlike the term ellipsoid height, or orthometric height, it does not refer to the station height in relation to the datum. It refers instead to the height of the geoid in relation to the ellipsoid. A more appropriate term would be geoid separation.

A good adjustment package will allow easy use of modeled geoid separations and easy conversion between ellipsoid heights and elevations. Ellipsoid heights are not of much use to the surveyor; water does not flow based on differences in ellipsoid height.

15-11. GPS AND THE FUTURE

A blending and blurring of differential GPS surveying and navigation will result. Work is being done to develop on-the-fly ambiguity resolution allowing for cm level for ships, aircraft, and most significantly the surveyor, and aircraft camera positioning to eliminate the need for costly ground control. On-the-fly ambiguity resolution will have other applications such as profiling tasks, and bathymetry without the need for static initialization and a host of others. For the average surveyor, the name of the game will be rapid-statics.

REFERENCES

ANANDA, M. P., ET AL. 1988. Global positioning system (GPS) autonomous user system. *Navigation, Journal of the Institute of Navigation 35*(2), Summer 1986.

BLACKWELL, E. G. 1985. Overview of differential GPS methods. *Navigation Journal of the Institute of Navigation 32*(2), Summer 1985.

CANNON, M. E. 1990. High-accuracy GPS semikinematic positioning modeling and results. *Navigation, Journal of the Institute of Navigation 37*(1), 26 Spring. 1990.

FERGUSON, K. E., ET AL. 1989. Kinematic and pseudo-kinematic surveying with the ASHTECH XII. Proceedings of ION GPS 1989. Colorado Springs, CO, Institute of Navigation, Sept. 27–29, pp. 35–37.

FERGUSON, K. E., and E. R. VEATCH. 1990. Centimeter level surveying in real time. GPS90, Ottawa, Canada, Sept. 3–7.

FREI, E., and G. BEUTLER. 1989. Some considerations concerning an adaptive optimized technique to resolve the initial phase ambiguities. Proceedings of the 5th International Geodetic Symposium on Satellite Positioning. DOD/DMA, NOAA/NGS, Las Cruces, NM, March 13–17, New Mexico State University, Vol. 2, pp. 671–686.

FREI, E., and G. BEUTLER 1988. Rapid static positioning based upon fast ambiguity resolution approach "FARA": Theory and results. *Manuscripta Geodaetica 15*, 325–356.

GEORGIADOU, Y., and A. KLEUSBERG. 1988. On the effect of ionospheric delay on deodetic relative GPS positioning. *Manuscripta Geodaetica 13*(1).

GPSIC. 1990. *GPS Information Center Users Manual.* U.S. Department of Transportation, U.S. Coast Guard.

Global positioning system papers. 1980, 1984, 1986. *Navigation*, Vols. I–3. Washington, D.C. Institute of Navigation.

GREEN, COLONEL G. B., ET AL. 1989. The GPS 21 primary satellite constellation. *Navigation, Journal of the institute of Navigation 36*(1), Spring.

GUNTER, H. W., ET AL. 1989. High-precision kinematic GPS differential positioning and integration of GPS with a ring laser strapdown inertial system. *Navigation, Journal of the Institute of Navigation 36*(1), Spring.

HATCH, R. 1989. Ambiguity resolution in the fast lane. Proceedings of ION GPS 1989. Institute of Navigation, Colorado Springs, CO, Sept. 27–29, pp. 45–50.

HATCH, R. 1990. Instantaneous ambiguity resolution. KIS Symposium, Baniff, Canada, September 11.

HOTHEM, L. D., C. C. GOAD, and B. W. REMONDI. GPS satellite surveying: A Practical aspects. *The Canadian Surveyor 38*(3), Autumn.

HOTHEM, L. D., and G. E. STRANGE. 1985. Factors to be considered in the development of specifications for geodetic surveys using relative positioning GPS techniques. Proceedings 1st International Symposium on Precise Positioning with the Global Positioning System, Vol. II. pg 87 Rockville, MD, April.

HWANG, P. Y. C. 1991. Kinematic GPS for differential positioning: Resolving integer ambiguities on the fly. *Navigation, Journal of the Institute of Navigation 38*(1), Spring.

KREMER, G. T. 1990. The effect of selective availability on differential GPS corrections. *Navigation, Journal of the institute of navigation 37*(1), Spring.

JORGENSEN, P. S. 1989. An assessment of ionospheric effects on the GPS user. *Navigation, Journal of the Institute of Navigation 36*(2), Summer.

KING, R. W., ET AL. 1985. Surveying with GPS. Monograph No. 9, School of Surveying, University of New South Wales, Kensington, NES Australia, Nov.

KUNCHES, J. M., and J. W. HIRMAN. 1990. Predicted Solar flare activity for the 1990's: Possible effects on navigational systems. *Navigation, Journal of the Institute of Navigation 37*(2), Spring.

LACHAPELLE, G., ET AL. 1988. Shipborne GPS kinematic positioning for hydrographic applications. *Navigation, Journal of the Institute of Navigation 35*(1), Spring.

LAPINE, COMMANDER L. A. 1990. Practical photogrammetric control by kinematic GPS. *GPS World 1*(3), May–June.

LEICK, A. 1990. *GPS Satellite Surveying*. New York: John Wiley & Sons.

Minutes of FGCC Meeting. 1988. Rockville, MD, May 27.

REILLY J. P. 1990. *Practical Surveying with GPS*. POB Publishing Company, Feb. Canton MI.

REMONDI, B. W. 1985. Performing centimeter accuracy relative surveys in seconds using carrier phase. Proceedings of the 1st International Symposium on Precise Positioning with the Global Positioning System. National Geodetic Information Center, NOAA, Rockville, MD, April 15–19, pp. 789–797.

REMONDI, B. W. 1985. Performing centimeter-level surveys in seconds with GPS carrier phase, initial results. NOAA Technical Memorandum NOS NGS-43, National Geodetic Information Center, NOAA, Rockville, MD.

REMONDI, B. W. 1988. Kinematic and pseudo-kinematic GPS. Proceedings of the Satellite Divisions International Technical Meeting. Colorado Springs, CO, Institute of Navigation, Sept. 19–23, pp. 115–121.

REMONDI, B. W. 1991. Pseudo-kinematic GPS results using the ambiguity function method. *Navigation, Journal of the Institute of Navigation 38*(1), Spring.

ROEBER, J. F. 1986–1987. Where in the world are we? *Navigation, Journal of the Institute of Navigation 33*(4), Winter.

TALBOT, N. C. 1991. High-precision real-time GPS positioning concepts: Modeling and results. *Navigation, Journal of the Institute of Navigation 38*(2), Summer.

VEATCH II, E. R., and J. OSWALD. 1989. The kinematic GPS revolution: Surveying on the move. ASPRS/ACSM Annual Convention, Baltimore, MD, Vol. 5, pp. 288–297.

WELLS, D., ET AL. 1987. *Guide to GPS Positioning*. Fredericton, New Brunswick, Canada: Canadian GPS Associates, May.

16

Survey Measurement Adjustments by Least Squares

Paul R. Wolf and Charles Ghilani

16-1. INTRODUCTION

The general subject of errors in measurement was discussed in Chapter 3, and the two classes of errors, *systematic* and *random* (or accidental), were defined. It was noted that systematic errors follow physical laws, and that if the conditions producing them are measured, corrections to eliminate these can be computed and applied; however, random errors will still exist in all observed values.

As explained in Chapter 3, experience has shown that random errors in surveying follow the mathematical laws of probability, and that any group of measurements will contain random errors conforming to a "normal distribution" as illustrated in Figure 3-5. With reference to that figure, it can be seen that random errors have the following characteristics: (1) small errors occur more frequently than large ones, (2) positive and negative errors of the same size occur with equal frequency, and (3) very large errors seldom occur. They must be avoided through alertness and careful checking of all measured values.

If proper procedures are used in surveying work—after eliminating mistakes and making corrections for systematic errors—the presence of remaining random errors generally should be evident. In leveling, e.g., as discussed in Chapter 7, circuits should be closed on either the starting bench mark or another of equal or higher reliability. Any misclosure in the circuit can then be computed, providing an indication of random errors that remain. Similarly, in an angle measurement as described in Chapter 6, the sum of all angles measured around the horizon at a point should equal 360°, and in plane surveying the sum of the angles in any closed polygon should equal $(n - 2)$ 180°, where n is the number of angles in the figure. Also, as discussed in Chapter 9, the algebraic sums of the latitudes and departures of a closed-polygon traverse must equal zero. After eliminating mistakes and correcting for systematic errors, any remaining deviations (misclosures) from these required conditions indicate the presence of random errors in the measured values.

In surveying, adjustments are applied to measured values to distribute misclosure errors and produce mathematically perfect geometric conditions; various procedures are used. Some simply apply corrections of the same size to all measured values, where each correction equals the total misclosure divided by the number of measurements. Others introduce corrections of varying size to certain values on

383

the basis of their suspected errors. Still others employ rules of thumb—e.g., the compass rule for adjusting latitudes and departures of closed traverses.

Because random errors in surveying are "normally distributed" and conform to the mathematical laws of probability, it follows logically that for the most rigorous adjustment procedure, corrections should be computed in accordance with that theory. The method of *least squares* is based on the laws of probability.

In the sections of this chapter that follow, the fundamental condition that is enforced in least-squares adjustment is described and an elementary example given. Then, systematic procedures for forming and solving least-squares equations are given, including the use of matrix methods. Following this, specific procedures for adjusting level nets, trilateration, triangulation, and traverses are described, and example problems solved.

16-2. THE FUNDAMENTAL CONDITION OF LEAST SQUARES

Making adjustments of measured values by the method of least squares is not new. It was done by the German mathematician Karl Gauss as early as the latter part of the 18th century. Until the advent of computers, however, least-squares techniques were seldom employed because of the lengthy calculations involved. Now the procedures are routinely performed.

Least squares is applicable for adjusting any of the basic measurements made in surveying, including observed differences in elevation, horizontal distances, and horizontal and vertical angles. Applying least squares for adjusting these observations in the commonly employed surveying procedures if leveling, trilateration, triangulation, and traversing is the thrust of this chapter. Least squares is also applied in photogrammetric, inertial, and GPS surveys, but these procedures are not described here.

For a group of equally weighted observations, the fundamental condition that is enforced in least-squares adjustment is that the sum of the squares of the residuals is minimized. This condition, which has been developed from the equation for the normal distribution curve (see Section 3-10), provides most probable values for the adjustment quantities. Suppose a group of m equally weighted measurements were taken having residuals $v_1, v_2, v_3, \ldots, v_m$. Then, in equation form, the fundamental condition of least squares is expressed as follows

$$\sum_{i=1}^{m} (v_i)^2 = (v_1)^2 + (v_3)^2 + \cdots$$

$$+ (v_m)^2 = \text{minimum} \qquad (16\text{-}1)$$

If measured values are weighted (see Section 3-22) in least-squares adjustment, then the fundamental condition enforced is that the sum of the weights p times their corresponding squared residuals is minimized, or in the following equation form:

$$\sum_{i=1}^{m} p_i(v_i)^2 = p_1(v_1)^2 + p_2(v_2)^2 + p_3(v_3)^2 + \cdots$$

$$+ p_m(v_m)^2 = \text{minimum} \quad (16\text{-}2)$$

Some basic assumptions underlying least-squares theory are that (1) mistakes and systematic errors have been eliminated, so only random errors remain; (2) the number of observations being adjusted is large; and (3) as stated earlier, the frequency distribution of the errors is normal. Although these basic assumptions are not always met, least-squares adjustment still provides the most rigorous error treatment available, and hence it has become very popular and important in modern surveying. Besides yielding most probable values for the unknowns, least-squares adjustment also enables (1) determining precisions of adjusted qualities, (2) revealing the presence of large errors and mistakes so steps can be taken to eliminate them, and (3) making possible the optimum design of survey procedures in the office before going into the field to take measurements. The latter topic is beyond the scope of this discussion but can be found in references cited at the end of this chapter.

16-3. LEAST-SQUARES ADJUSTMENT BY THE OBSERVATION-EQUATION METHOD

There are two basic methods of employing least squares in survey adjustments: (1) the *observation-equation* method and (2) *condition-equation* approach. The following discussion in this chapter concentrates on the former procedure.

In the observation-equation method, observation equations are written relating measured values to their residual errors and the unknown parameters. One observation equation is written for each measurement. For a unique solution, the number of equations must equal the number of unknowns. If redundant observations are made, then more observation equations can be written than are needed for a unique solution, and most probable values of the unknowns can be determined by the method of least squares. For a group of equally weighted observations, an equation for each residual error is obtained from each observation equation. The residuals are squared and added to obtain the function expressed in Equation (16-1).

To minimize the function in accordance with equation (16-1), partial derivatives of this expression are taken with respect to each unknown variable and set equal to zero. This yields a set of so-called *normal equations*, which are equal in number to the number of unknowns. The normal equations are solved to obtain the most probable values for the unknowns.

Example 16-1. As an elementary example illustrating the method of least-squares adjustment by the observation-equation method, adjust the following three equally weighted distance measurements taken between points *A*, *B*, and *C* of Figure 16-1:

$$AC = 431.71$$
$$AB = 211.52$$
$$BC = 220.10$$

In terms of unknown distances *x*, and *y*, the following three equations can be written:

$$x + y = 431.71 \text{ ft}$$
$$x = 211.52 \text{ ft}$$
$$y = 220.10 \text{ ft}$$

These equations relate unknowns *x* and *y* to the observations. Values for *x* and *y* could be obtained from any two of these equations so that the remaining equation is redundant. Notice, however, that values obtained for *x* and *y* will differ, depending on which two equations are solved. It is therefore apparent

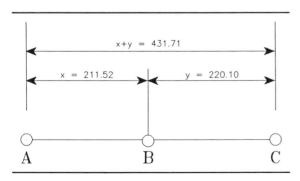

Figure 16-1. Measurements for least-squares adjustment of Example 16-1.

that the measurements contain errors. The equations may be rewritten as observation equations by including residual errors as follows:

$$x + y = 431.71 + v_1$$

$$x = 211.52 + v_2$$

$$y = 220.10 + v_3$$

To obtain the least-squares solution, the observation equations are rearranged to obtain expressions for the residuals; these are squared and added to form the function

$$\sum_{i=1}^{m} (v_i)^2$$

as follows:

$$\sum_{i=1}^{m} (v_i)^2 = (x + y - 431.71)^2$$

$$+ (x - 221.52)^2 + (y - 220.10)^2$$

This function is minimized, enforcing the condition of least squares, by taking partial derivatives with respect to each unknown and setting them equal to zero. This yields the following two equations:

$$\frac{\partial \Sigma v^2}{\partial x} = 0 = 2(x + y - 431.71)$$

$$+ 2(x - 211.52)$$

$$\frac{\partial \Sigma v^2}{\partial y} = 0 = 2(x + y - 431.71)$$

$$+ 2(y - 220.10)$$

These are normal equations. The reduced normal equations are as follows:

$$2x + y = 643.23$$

$$x + 2y = 651.81$$

Solving the reduced normal equations simultaneously yields $x = 211.55$ and $y = 220.13$. According to the theory of least squares, these values have the highest proba-

bility. Having the most probable values for the unknowns, the residuals can be calculated by substitution in the original observation equations, or

$$v_1 = 211.55 + 220.13 - 431.71 = -0.03$$

$$v_2 = 211.55 - 211.52 = +0.03$$

$$v_3 = 220.13 - 220.10 = +0.03$$

By substituting these adjusted values for x and y into the original observation equations, the following adjusted measurements result:

$$x + y = 431.71 - 0.03 = 431.68 \text{ ft} = AC$$

$$x = 211.52 + 0.03 = 211.55 \text{ ft} = AB$$

$$y = 220.10 + 0.03 = 220.13 \text{ ft} = BC$$

Note that the adjusted values are now consistent—i.e., $x + y = 431.68$—no matter which measurements are used. Whereas other adjustments could be made to achieve consistency, there is no other combination of residuals possible that will render the sum of their squares a smaller value. Thus, the condition of least squares is realized.

This simple example serves to illustrate the method of least squares without complicating the mathematics. Least-squares adjustment of large systems of observation equations is performed in the same manner.

16-4. SYSTEMATIC FORMULATION OF NORMAL EQUATIONS

In large systems of observation equations, it is helpful to utilize systematic procedures to formulate normal equations. Consider the following system of m linear observation equations of equal weight containing n unknowns:

$$a_1 A + b_1 B + c_1 C + \cdots + n_1 N - L_1 = v_1$$

$$a_2 A + b_2 B + c_2 C + \cdots + n_2 N - L_2 = v_2$$

$$\cdots \cdots \cdots \cdots \cdots \cdots \cdots \cdots \cdots \cdots \cdots \cdots \cdots \quad (16\text{-}3)$$

$$a_m A + b_m B + c_m C + \cdots + n_m N - L_m = v_m$$

In Equation (16-3), the a's, b's, c's, etc are coefficients of unknowns A, B, C, etc.; the L's constants and the v's residuals. By squaring the residuals and summing them, the function Σv^2 is formed. Taking partial derivatives of Σv^2 with respect to each unknown A, B, C, etc. yields n normal equations, After reducing and factoring the normal equations, we can obtain the following generalized system for expressing normal equations:

$$[aa]A + [ab]B + [ac]C + \cdots + [an]N = [aL]$$

$$[ba]A + [bb]B + [bc]C + \cdots + [bn]N = [bL]$$

$$[ca]A + [cb]B + [cc]C + \cdots + [cn]N = [cL]$$

. . . .

$$[na]A + [nb]B + [nc]C + \cdots + [nn]N = [nL]$$

$$(16\text{-}4)$$

In Equation (16-4), the symbol [ ] signifies the sum of the products; e.g., $[aa] = a_1 a_1 = a_2 a_2 + a_3 a_3 + \cdots + a_m a_m$; $[ab] = a_1 b_1 + a_2 b_2 + a_3 b_3 + \cdots + a_m b_m$; etc.

It can be similarly shown that normal equations may be systematically formed from weighted observation equations in the following manner:

$$[paa]A + [pab]B$$

$$+ [pac]C + \cdots + [pan]N = [paL]$$

$$[pba]A + [pbb]B + [pbc]C + \cdots + [pbn]N = [pbL]$$

$$[pca]A + [pcb]B + [pcc]C + \cdots [pcn]N = [pcL]$$

. . . .

$$[pna]A + [pnb]B$$

$$+ [pnc]C + \cdots + [pnn]N = [pnL] \quad (16\text{-}5)$$

In Equation (16-5), the terms are as described previously, except that the p's are the relative weights of the individual observations. Examples of the bracket terms are $[paa] = p_1 a_1 a_1 + p_2 a_2 a_2 + \cdots + p_m a_m a_m$; $[pbL] = p_1 b_1 L_1 + p_2 b_2 L_2 + \cdots + p_m b_m L_m$; etc.

16-5.   MATRIX METHODS IN LEAST-SQUARES ADJUSTMENT

It has been previously mentioned that least-squares computations are quite lengthy and therefore best performed on a computer. The algebraic approach—Equations (16-4) or (16-5)—for forming normal equations and obtaining their simultaneous solution can be programmed for computer solution; however, the procedure is much more easily adapted to matrix methods.

In developing matrix equations for least-squares computations, analogy will be made to the algebraic approach in Section 16-4. First, observation Equation (16-3) may be prepresented in matrix form as follows:

$$_m\mathbf{A}_n \, _n\mathbf{X}_1 = {}_m\mathbf{L}_1 + {}_m\mathbf{V}_1 \qquad (16\text{-}6)$$

where

$$\mathbf{A} = \begin{bmatrix} a_1 & b_1 & c_1 & \cdots & n_1 \\ a_2 & b_2 & c_2 & \cdots & n_2 \\ \cdot & \cdot & \cdot & \cdots & \cdots \\ \cdot & \cdot & \cdot & \cdots & \cdots \\ a_m & b_m & c_m & \cdots & n_m \end{bmatrix}, \quad \mathbf{X} = \begin{bmatrix} A \\ B \\ C \\ \cdot \\ \cdot \\ \cdot \\ N \end{bmatrix}^1$$

$$\mathbf{L} = \begin{bmatrix} L_1 \\ L_2 \\ L_3 \\ \cdot \\ \cdot \\ \cdot \\ L_m \end{bmatrix}^1, \quad \mathbf{V} = \begin{bmatrix} v_1 \\ v_2 \\ v_3 \\ \cdot \\ \cdot \\ \cdot \\ v_m \end{bmatrix}^1$$

On studying the following matrix representation, it will be realized that it exactly produces normal Equation (16-4):

$$\mathbf{A}^T\mathbf{A}\mathbf{X} = \mathbf{A}^T\mathbf{L} \qquad (16\text{-}7)$$

In this equation, $\mathbf{A}^T\mathbf{A}$ is the matrix of normal equation coefficients for the unknowns.

Premultiplying both sides of Equation (16-7) by $(\mathbf{A}^T\mathbf{A})^{-1}$ and reducing, we obtain

$$(\mathbf{A}^T\mathbf{A})^{-1}(\mathbf{A}^T\mathbf{A})\mathbf{X} = (\mathbf{A}^T\mathbf{A})^{-1}\mathbf{A}^T\mathbf{L}$$

$$\mathbf{I}\mathbf{X} = (\mathbf{A}^T\mathbf{A})^{-1}\mathbf{A}^T\mathbf{L} \qquad (16\text{-}8)$$

$$\mathbf{X} = (\mathbf{A}^T\mathbf{A})^{-1}\mathbf{A}^T\mathbf{L}$$

In this reduction, $\mathbf{I}$ is the identity matrix. Equation (16-8) is the basic least-squares matrix equation for equally weighted observations. The matrix $\mathbf{X}$ consists of most probable values for unknowns $A, B, C, \ldots, N$. For a system of weighted observations, the following matrix equation provides the $\mathbf{X}$ matrix:

$$\mathbf{X} = (\mathbf{A}^T\mathbf{P}\mathbf{A})^{-1}\mathbf{A}^T\mathbf{P}\mathbf{L} \qquad (16\text{-}9)$$

In Equation (16-9), the matrices are identical to those of the equally weighted equations, except that $\mathbf{P}$ is a diagonal matrix of weights defined as follows:

$$\mathbf{P} = \begin{bmatrix} p_1 & & & & & & \\ & p_2 & & & & & \\ & & p_3 & & & & \\ & & & \cdot & & & \\ & & & & \cdot & & \\ & & & & & \cdot & \\ & & & & & & p_m \end{bmatrix}^m$$

In the $\mathbf{P}$ matrix, all off-diagonal elements are zeros. This is proper when the individual observations are independent and uncorrelated—e.g., they are not related to each other. This is usually the case in surveying.

If the observations in an adjustment are all of equal weight, Equation (16-9) can still be used, but the $\mathbf{P}$ matrix becomes an identity matrix with ones for all diagonal elements. It therefore reduces exactly to Equation (16-8). Thus, Equation (16-9) is a general one that can be used for both unweighted and weighted adjustments. It is readily programmed for computer solution.

Example 16-2. Solve example 16-1 using matrix methods.

1. The observation equations of Example 16-1 can be expressed in matrix form as follows:

$$_3\mathbf{A}_2\,_2\mathbf{X}_1 = _3\mathbf{L}_1 + _3\mathbf{V}_1$$

where

$$\mathbf{A} = \begin{bmatrix} 1 & 1 \\ 1 & 0 \\ 0 & 1 \end{bmatrix}^2, \qquad \mathbf{X} = \begin{bmatrix} x \\ y \end{bmatrix}^1,$$

$$\mathbf{L} = \begin{bmatrix} 431.71 \\ 211.52 \\ 220.10 \end{bmatrix}^1, \qquad \mathbf{V} = \begin{bmatrix} v_1 \\ v_2 \\ v_3 \end{bmatrix}^1$$

2. Solving matrix Equation (16-7), yields

$$(\mathbf{A}^T\mathbf{A}) = \begin{bmatrix} 1 & 1 & 0 \\ 1 & 0 & 1 \end{bmatrix}\begin{bmatrix} 1 & 1 \\ 1 & 0 \\ 0 & 1 \end{bmatrix} = \begin{bmatrix} 2 & 1 \\ 1 & 2 \end{bmatrix}$$

$$(\mathbf{A}^T\mathbf{A})^{-1} = \frac{1}{3}\begin{bmatrix} 2 & -1 \\ -1 & 2 \end{bmatrix}, \quad \mathbf{A}^T\mathbf{L} = \begin{bmatrix} 643.23 \\ 651.81 \end{bmatrix}$$

$$\mathbf{X} + (\mathbf{A}^T\mathbf{A})^{-1}\mathbf{A}^T\mathbf{L} = \frac{1}{3}\begin{bmatrix} 2 & -1 \\ -1 & 2 \end{bmatrix}$$

$$\begin{bmatrix} 643.23 \\ 651.81 \end{bmatrix} = \begin{bmatrix} 211.55 \\ 220.13 \end{bmatrix}$$

Note that this solution yields $x = 211.55$ and $y = 220.13$, which are exactly the same values obtained through the algebraic approach of Example 16-1.

As previously stated, digital computers are normally used in least-squares adjustments due to the relatively lengthy nature of the calculations. Because the equations are so conveniently programmed and solved using matrix algebra, the balance of this chapter will stress this approach.

16-6. MATRIX EQUATIONS FOR PRECISIONS OF ADJUSTED QUANTITIES

The matrix equation for calculating residuals after adjustment, whether the adjustment is weighted or not, is

$$\mathbf{V} = \mathbf{AX} - \mathbf{L} \tag{16-10}$$

The standard deviation of unit weight for an unweighted adjustment is

$$S_0 = \sqrt{\frac{(\mathbf{V}^T\mathbf{V})}{r}} \tag{16-11}$$

The standard deviation of unit weight for a weighted adjustment is

$$S_0 = \sqrt{\frac{(\mathbf{V}^T\mathbf{P}\mathbf{V})}{r}} \tag{16-12}$$

In Equations (16-11) and (16-12), r is the number of degrees of freedom in an adjustment and equals the number of observations equations minus the number of unknowns, or $r = m - n$.

Standard deviations of the individual adjusted quantities are as follows:

$$S_{x_i} = S_0\sqrt{Q_{x_i x_i}} \tag{16-13}$$

In Equation (16-13), S_{x_i} is the standard deviation of the ith adjusted quantity—e.g., the quantity in the ith row of the $\mathbf{X}$ matrix; S_0 the standard deviation of unit weight as calculated by Equation (16-11) or (16-12); and $Q_{x_i x_i}$ the diagonal element in the ith row and the ith column of the matrix $(\mathbf{A}^T\mathbf{A})^{-1}$ in the unweighted case or the matrix $(\mathbf{A}^T\mathbf{PA})^{-1}$ in the weighted case. The $(\mathbf{A}^T\mathbf{A})^{-1}$ and $(\mathbf{A}^T\mathbf{PA})^{-1}$ matrices are the so-called *covariance* matrices.

Example 16-3. Calculate the standard deviation of unit weight and standard deviations

of the adjusted quantities x and y for the unweighted problem of Example 16-2.

1. By Equation (16-10), the residuals are as follows:

$$\mathbf{V} = \begin{bmatrix} 1 & 1 \\ 1 & 0 \\ 0 & 1 \end{bmatrix}\begin{bmatrix} 211.55 \\ 220.13 \end{bmatrix}$$

$$- \begin{bmatrix} 431.71 \\ 211.52 \\ 220.10 \end{bmatrix} = \begin{bmatrix} -0.03 \\ 0.03 \\ 0.03 \end{bmatrix}$$

2. By Equation (16-11), the standard deviation of unit weight is

$$\mathbf{V}^T\mathbf{V} = [-0.03 \quad 0.03 \quad 0.03]\begin{bmatrix} -0.03 \\ 0.03 \\ 0.03 \end{bmatrix}$$

$$= 0.0027$$

$$S_0 = \sqrt{\frac{0.0027}{3 - 2}} = \pm 0.052$$

3. With Equation (16-13), the standard deviations of the adjusted values for x and y are

$$S_x = \pm 0.052\sqrt{\frac{2}{3}} = \pm 0.042$$

$$S_y = \pm 0.052\sqrt{\frac{2}{3}} = \pm 0.042$$

In part 3, the numbers 2/3 under the radicals are the 1,1 and 2,2 elements of the $(\mathbf{A}^T\mathbf{A})^{-1}$ matrix of Example 16-2. The interpretation of the standard deviations computed under part 3 is that a *68%* probability exists the adjusted values for x and y are within ± 0.042 of their true values. Note that for this simple example, the three residuals calculated in part 1 were equal, and the standard deviations of x and y were equal in part 3. This is due to the symmetric nature of this particular problem (illustrated in Figure 16-1), but it is not generally the case with more complex problems.

16-7. ADJUSTMENT OF LEVELING CIRCUITS

The method of least squares is extremely valuable as a means of adjusting leveling circuits, especially those consisting of two or more interconnected loops that form networks. A simple example is illustrated in Figure 16-2. Here the objective was to determine elevations of *A*, *B*, and *C*, which were to serve as temporary project bench marks to control construction of a highway through the cross-hatched corridor. Obviously, it would have been possible to obtain elevations for *A*, *B*, and *C* by beginning at *BMX* and running a single closed loop consisting of only courses 1, 5, 7, and 4. Alternatively, a single closed loop could have been initiated at *BMY* and consist of courses 2, 5, 7, and 3. However, by running all seven courses, redundancy is achieved that enables checks to be made, blunders isolated, and precision increased.

Now that we have run all seven courses of Figure 16-2, it would be possible to compute the adjusted elevation of *B*, e.g., using several different single closed circuits. Loops 1-5-6, 2-5-6, 3-7-6, and 4-7-6 could each be used, but it is almost certain that each would yield a different elevation for *B*. A more logical approach, that will produce only one adjusted value for *B*—its most probable one—is to use all seven courses in a simultaneous least-squares adjustment.

In adjusting level networks, the observed difference in elevation for each course is treated as one observation containing a single random error. This single random error is the total of the individual random errors in backsight and foresight readings for the entire course. The table of Figure 16-2 lists the total difference in elevation observed for each course. In the figure, the arrows indicate the direction of leveling. Thus, for course number 1, leveling proceeded from *BMX* to *A* and the observed elevation difference was +5.10 ft.

Example 16-4. Adjust the level net of Figure 16-2 by least squares. Consider all observations to be equally weighted. Compute the precisions of the adjusted elevations.

1. First, observation equations are written relating each measurement of the difference in elevation of a line to the most probable values for unknown elevations *A*, *B*, and *C*

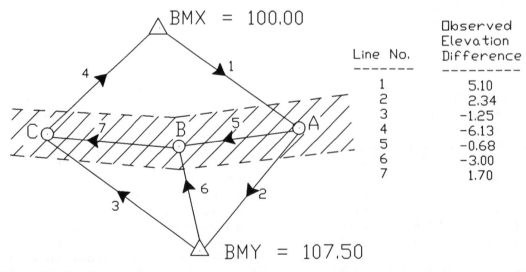

Line No.	Observed Elevation Difference
1	5.10
2	2.34
3	-1.25
4	-6.13
5	-0.68
6	-3.00
7	1.70

BMX = 100.00

BMY = 107.50

Figure 16-2. Leveling network.

and residual errors in the measurements, as follows:

$$A = BMX + 5.10 + v_1$$

$$BMY = A + 2.34 + v_2$$

$$C = BMY - 1.25 + v_3$$

$$BMX = C - 6.13 + v_4 \qquad (16\text{-}14)$$

$$B = A - 0.68 + v_5$$

$$B = BMY - 3.00 + v_6$$

$$C = B + 1.70 + v_7$$

2. Introducing the elevations of *BMX* and *BMY*, reducing, and rewriting observation Equations (16-14) in a form compatible with Equation (16-6) give the following:

$$A = 105.10 + v_1$$

$$-A = -105.16 + v_2$$

$$C = 106.25 + v_3$$

$$-C = -106.13 + v_4$$

$$-A + B = -0.68 + v_5$$

$$B = 104.50 + v_6$$

$$-B + C = 1.70 + v_7$$

3. The observation equations expressed in matrix form are as follows:

$$_7\mathbf{A}_{3\,3}\mathbf{X}_1 = {_7}\mathbf{L}_1 + {_7}\mathbf{V}_1$$

where

$$\mathbf{A}^T\mathbf{A} = \begin{bmatrix} 1 & 0 & 0 \\ -1 & 0 & 0 \\ 0 & 0 & 1 \\ 0 & 0 & -1 \\ -1 & 1 & 0 \\ 0 & 1 & 0 \\ 0 & -1 & 1 \end{bmatrix}, \quad \mathbf{X} = \begin{bmatrix} A \\ B \\ C \end{bmatrix},$$

$$\mathbf{L} = \begin{bmatrix} 105.10 \\ -105.16 \\ 106.25 \\ -106.13 \\ -0.68 \\ 104.50 \\ 1.70 \end{bmatrix}, \quad \mathbf{V} = \begin{bmatrix} v_1 \\ v_2 \\ v_3 \\ v_4 \\ v_5 \\ v_6 \\ v_7 \end{bmatrix}$$

4. The matrix solution for most probable values is

$$\mathbf{A}^T\mathbf{A} = \begin{bmatrix} 1 & -1 & 0 & 0 & -1 & 0 & 0 \\ 0 & 0 & 0 & 0 & 1 & 1 & -1 \\ 0 & 0 & 1 & -1 & 0 & 0 & 1 \end{bmatrix}$$

$$\times \begin{bmatrix} 1 & 0 & 0 \\ -1 & 0 & 0 \\ 0 & 0 & 1 \\ 0 & 0 & -1 \\ -1 & 1 & 0 \\ 0 & 1 & 0 \\ 0 & -1 & 1 \end{bmatrix}$$

$$= \begin{bmatrix} 3 & -1 & 0 \\ -1 & 3 & -1 \\ 0 & -1 & 3 \end{bmatrix}$$

$$(\mathbf{A}^T\mathbf{A})^{-1} = \frac{1}{21}\begin{bmatrix} 8 & 3 & 1 \\ 3 & 9 & 3 \\ 1 & 3 & 8 \end{bmatrix}$$

$$\mathbf{X} = (\mathbf{A}^T\mathbf{A})^{-1}\mathbf{A}^T\mathbf{L} = \frac{1}{21}\begin{bmatrix} 8 & 3 & 1 \\ 3 & 9 & 3 \\ 1 & 3 & 8 \end{bmatrix}$$

$$\times \begin{bmatrix} 210.94 \\ 102.12 \\ 214.08 \end{bmatrix}$$

$$= \begin{bmatrix} 105.14 \\ 104.48 \\ 106.19 \end{bmatrix}$$

$$\mathbf{A}^T\mathbf{L} = \begin{bmatrix} 1 & -1 & 0 & 0 & -1 & 0 & 0 \\ 0 & 0 & 0 & 0 & 1 & 1 & -1 \\ 0 & 0 & 1 & -1 & 0 & 0 & 1 \end{bmatrix}$$

$$\begin{bmatrix} 105.10 \\ -105.16 \\ 106.25 \\ -106.13 \\ -0.68 \\ 104.50 \\ 1.70 \end{bmatrix} = \begin{bmatrix} 210.94 \\ 102.12 \\ 214.08 \end{bmatrix}$$

Thus, the adjusted bench-mark elevations are $A = 105.14$, $B = 104.48$, and $C = 106.19$.

5. The residuals by Equation (16-10) are as follows:

$$\mathbf{AX} = \begin{bmatrix} 1 & 0 & 0 \\ -1 & 0 & 0 \\ 0 & 0 & 1 \\ 0 & 0 & -1 \\ -1 & 1 & 0 \\ 0 & 1 & 0 \\ 0 & -1 & 1 \end{bmatrix} \begin{bmatrix} 105.14 \\ 104.48 \\ 106.19 \end{bmatrix}$$

$$= \begin{bmatrix} 105.14 \\ -105.14 \\ 106.19 \\ -106.19 \\ -0.66 \\ 104.48 \\ 1.71 \end{bmatrix}$$

$$\mathbf{V} = \mathbf{AX} - \mathbf{L} = \begin{bmatrix} 105.14 \\ -105.14 \\ 106.19 \\ -106.19 \\ -0.66 \\ 104.48 \\ 1.71 \end{bmatrix}$$

$$- \begin{bmatrix} 105.10 \\ -105.10 \\ 106.25 \\ -106.13 \\ -0.68 \\ 104.50 \\ 1.70 \end{bmatrix}$$

$$= \begin{bmatrix} +0.04 \\ +0.02 \\ -0.06 \\ -0.06 \\ +0.02 \\ -0.02 \\ +0.01 \end{bmatrix}$$

6. Utilizing Equation (16-11), we obtain the estimated standard deviation of unit weight as

$$S_0 = \sqrt{\frac{0.0101}{7 - 3}} = \pm 0.050 \text{ ft}$$

where

$$\mathbf{V}^T \mathbf{V} = (0.04)^2 + (0.02)^2 +$$
$$(-0.06)^2 + (-0.06)^2 + (0.02)^2 +$$
$$(-0.02)^2 + (0.01)^2 = 0.0101.$$

7. By Equation (16-13), the estimated standard deviations of unknown elevations of A, B, and C are

$$S_A = S_0\sqrt{Q_{AA}} = \pm(0.05)\sqrt{8/21} = \pm 0.031 \text{ ft}$$
$$S_B = S_0\sqrt{Q_{BB}} = \pm(0.05)\sqrt{9/21} = \pm 0.033 \text{ ft}$$
$$S_C = S_0\sqrt{Q_{CC}} = \pm(0.05)\sqrt{8/21} = \pm 0.031 \text{ ft}$$

Note in these calculations that terms in the radicals are diagonal elements of the $(\mathbf{A}^T\mathbf{A})^{-1}$ matrix.

Example 16-5. Adjust the level net of Figure 16-2 by the method of weighted least squares. Use weights that are inversely proportional to the course lengths of 4, 3, 2, 3, 2, 2, and 2 m, respectively.

1. The $\mathbf{A}$, $\mathbf{X}$, $\mathbf{L}$, and $\mathbf{V}$ matrices are exactly the same as for Example 16-4. The diagonal elements of the P matrix are $1/4$, $1/3$, $1/2$, $1/3$, $1/2$, $1/2$, and $1/2$, respectively, and after we multiply each by 12, the $\mathbf{P}$ matrix becomes

$$\mathbf{P} = \begin{bmatrix} 3 & & & & & & \\ & 4 & & & & & \\ & & 6 & & & & \\ & & & 4 & & & \\ & & & & 6 & & \\ & & & & & 6 & \\ & & & & & & 6 \end{bmatrix}$$

2. The matrix solution of Equation (16-9) is

$$(\mathbf{A}^T\mathbf{PA})^{-1} = \begin{bmatrix} 0.0933 & 0.0355 & 0.0133 \\ 0.0355 & 0.0770 & 0.0289 \\ 0.0133 & 0.0289 & 0.0733 \end{bmatrix}$$

$$\mathbf{A}^T\mathbf{P} = \begin{bmatrix} 1 & -1 & 0 & 0 & -1 & 0 & 0 \\ 0 & 0 & 0 & 0 & 1 & 1 & -1 \\ 0 & 0 & 1 & -1 & 0 & 0 & 1 \end{bmatrix}$$

$$\times \begin{bmatrix} 3 & & & & & & \\ & 4 & & & & & \\ & & 6 & & & & \\ & & & 4 & & & \\ & & & & 6 & & \\ & & & & & 6 & \\ & & & & & & 6 \end{bmatrix}$$

$$= \begin{bmatrix} 3 & -4 & 0 & 0 & -6 & 0 & 0 \\ 0 & 0 & 0 & 0 & 6 & 6 & -6 \\ 0 & 0 & 6 & -4 & 0 & 0 & 6 \end{bmatrix}$$

$$A^T PA = \begin{bmatrix} 3 & -4 & 0 & 0 & -6 & 0 & 0 \\ 0 & 0 & 0 & 0 & 6 & 6 & -6 \\ 0 & 0 & 6 & -4 & 0 & 0 & 6 \end{bmatrix}$$

$$\times \begin{bmatrix} 1 & 0 & 0 \\ -1 & 0 & 0 \\ 0 & 0 & 1 \\ 0 & 0 & -1 \\ -1 & 1 & 0 \\ 0 & 1 & 0 \\ 0 & -1 & 1 \end{bmatrix}$$

$$= \begin{bmatrix} 13 & -6 & 0 \\ -6 & 18 & -6 \\ 0 & -6 & 16 \end{bmatrix}$$

$$A^T PL = \begin{bmatrix} 3 & -4 & 0 & 0 & -6 & 0 & 0 \\ 0 & 0 & 0 & 0 & 6 & 6 & -6 \\ 0 & 0 & 6 & -4 & 0 & 0 & 6 \end{bmatrix}$$

$$\times \begin{bmatrix} 105.10 \\ -105.16 \\ 106.25 \\ -106.13 \\ -0.68 \\ 104.50 \\ 1.70 \end{bmatrix}$$

$$= \begin{bmatrix} 740.02 \\ 612.72 \\ 1072.22 \end{bmatrix}$$

$$X = (A^T PA)^{-1} A^T PL$$

$$= \begin{bmatrix} 0.0933 & 0.0355 & 0.0133 \\ 0.0355 & 0.0770 & 0.0289 \\ 0.0133 & 0.0289 & 0.0733 \end{bmatrix}$$

$$\times \begin{bmatrix} 740.02 \\ 612.72 \\ 1072.22 \end{bmatrix}$$

$$= \begin{bmatrix} 105.15 \\ 104.49 \\ 106.20 \end{bmatrix}$$

In summary, the adjusted elevations for A, B, and C are 105.15, 104.49, and 106.20, respectively. Note that these differ slightly from the unweighted results of Example 16-4, as they should.

3. By equation (16-10), the residuals are as follows:

$$AX = \begin{bmatrix} 1 & 0 & 0 \\ -1 & 0 & 0 \\ 0 & 0 & 1 \\ 0 & 0 & -1 \\ -1 & 1 & 0 \\ 0 & 1 & 0 \\ 0 & -1 & 1 \end{bmatrix} \begin{bmatrix} 105.15 \\ 104.49 \\ 106.20 \end{bmatrix}$$

$$= \begin{bmatrix} 105.15 \\ -105.15 \\ 106.20 \\ -106.20 \\ -0.66 \\ 104.49 \\ 1.71 \end{bmatrix}$$

$$V = AX - L = \begin{bmatrix} 105.15 \\ -105.15 \\ 106.20 \\ -106.20 \\ -0.66 \\ 104.49 \\ 1.71 \end{bmatrix}$$

$$- \begin{bmatrix} 105.10 \\ -105.16 \\ 106.25 \\ -106.13 \\ -0.68 \\ 104.50 \\ 1.70 \end{bmatrix}$$

$$= \begin{bmatrix} +0.05 \\ +0.01 \\ -0.05 \\ -0.07 \\ +0.02 \\ -0.01 \\ +0.01 \end{bmatrix}$$

4. The estimated standard deviation of unit weight by Equation (16-12) is

$$S_0 = \pm \sqrt{\frac{0.046}{7 - 3}} = \pm 0.107$$

where

$$V^T PV = 3(0.05)^2 + 4(0.01)^2$$
$$+ 6(-0.05)^2 + 4(-0.07)^2$$
$$+ 6(0.02)^2 + 6(-0.01)^2 + 6(0.01)^2$$
$$= 0.0461.$$

5. By Equation (16-13), the estimated standard deviations in elevations of bench marks A, B, and C are as follows:

$$S_A = S_0\sqrt{Q_{AA}} = \pm(0.107)\sqrt{(0.0933)}$$

$$= \pm 0.033 \text{ ft}$$

$$S_B = S_0\sqrt{Q_{BB}} = \pm(0.107)\sqrt{(0.0770)}$$

$$= \pm 0.030 \text{ ft}$$

$$S_C = S_0\sqrt{Q_{CC}} = \pm(0.107)\sqrt{(0.0733)}$$

$$= \pm 0.029 \text{ ft}$$

Note that values under the radicals in the above calculations are diagonal elements of the $(\mathbf{A}^T\mathbf{PA})^{-1}$ matrix.

16-8.   ADJUSTMENT OF HORIZONTAL SURVEYS

In addition to level nets, which can be referred to as vertical surveys, another basic class that can be adjusted by least squares is *horizontal surveys*. They are run for the purpose of establishing horizontal positions of points, expressed either in terms of X and Y plane coordinates (usually state plane systems), or as geodetic latitudes and longitudes. These surveys are often executed as one of three specific types: (1) *trilateration*, (2) *triangulation*, or (3) *traversing*. The methods of running these surveys are described in Chapters 9, 11, and 12, respectively.

Trilateration consists exclusively of distance measurements, triangulation principally involves angle measurement with some base-line distances observed, and traverses contain both distance and angle measurements. Therefore, to perform least-squares adjustments of horizontal surveys by the method of observation equations, it is necessary to write observation equations for these two types of measurements. The equations are nonlinear, so they are first linearized using Taylor's theorem and then solved iteratively.

16-9.   THE DISTANCE OBSERVATION EQUATION

Distance observation equations relate measured lengths and their inherent random errors to the most probable coordinates of their end points. This procedure is often referred to as the method of the *variation of coordinates*. In this section, X and Y plane coordinates will be used. Referring to Figure 16-3, we may write the following distance observation equation for any line IJ:

$$L_{ij} + V_{L_{ij}} = \sqrt{(X_j - X_i)^2 + (Y_j - Y_i)^2} \quad (16\text{-}15)$$

In Equation (16-15), L_{ij} is the observed length of line IJ; $V_{L_{ij}}$ is the residual error in the observation; and X_i, Y_i, X_j, and Y_j are the most probable coordinates of points I and J. The right side of the equation is a nonlinear function of unknown variables X_i, Y_i, X_j, and Y_j. The equation may be rewritten as

$$L_{ij} + V_{L_{ij}} = F(X_j, Y_j, X_j, Y_j) \quad (16\text{-}16)$$

where

$$F(X_i, Y_i, X_j, Y_j) = \sqrt{(X_j - X_i)^2 + (Y_j - Y_i)^2}.$$

With the Taylor series, linearization of the function F, after we drop as negligible all

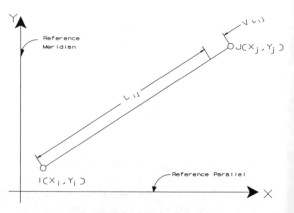

Figure 16-3.   Geometry of distance observation equation.

terms of order two or higher, takes the following form:

$$F(X_i, Y_i, X_j, Y_j) = F(X_{i_0}, Y_{i_0}, X_{j_0}, Y_{j_0})$$

$$+ \left[\frac{\partial F}{\partial X_j}\right]_0 dX_i + \left[\frac{\partial F}{\partial Y_i}\right]_0 dY_i + \left[\frac{\partial F}{\partial X_j}\right]_0 dX_j$$

$$+ \left[\frac{\partial F}{\partial Y_j}\right]_0 dY_j \qquad (16\text{-}17)$$

In Equation (16-17), X_{i_0}, Y_{i_0}, X_{j_0}, Y_{j_0} are initial approximations of unknowns X_i, Y_i, X_j, and Y_j; $(\partial F/\partial X_j)_0$ is the partial derivative of F with respect to X_i evaluated at the initial approximations, etc.; and dX_i, dY_i, dX_j, and dY_j are corrections to be applied to the initial approximations such that

$$X_i = X_{i_0} + dX_i, \qquad X_j = X_{j_0} + dX_j$$

$$Y_i = Y_{i_0} + dY_i, \qquad Y_j = Y_{j_0} + dY_j$$

Evaluating partial derivatives of the function F, substituting them into Equation (16-17), and then in turn substituting into Equation (16-16) and rearrange, we arrive at the following linearized observation equation for distance measurements:

$$K_{L_{ij}} + V_{L_{ij}} = \left[\frac{X_{i_0} - X_{j_0}}{IJ_0}\right] dX_i + \left[\frac{Y_{i_0} - Y_{j_0}}{IJ_0}\right] dY_i$$

$$+ \left[\frac{X_{j_0} - X_{i_0}}{IJ_0}\right] dX_j +$$

$$\left[\frac{Y_{j_0} - Y_{i_0}}{IJ_0}\right] dY_j \qquad (16\text{-}18)$$

where $K_{L_{ij}} = L_{ij} - (IJ_0)$, and

$$(IJ_0) = \sqrt{(X_{j_0} - X_{i_0})^2 + (Y_{j_0} - Y_{i_0})^2}$$

Evaluation of partial derivatives in the previous development is quite straightforward. However, to illustrate the procedure, the partial of F with respect to X_i is demonstrated. Having done this, we may easily visualize the

remaining partials without actually performing the steps.

$$F = \left[(X_j - X_i)^2 + (Y_j - Y_i)^2\right]^{1/2}$$

$$\frac{\partial F}{\partial X_i} = \frac{1}{2}\left[(X_j - X_i)^2\right.$$

$$\left. + (Y_j - Y_i)^2\right]^{-1/2}[2(X_j - X_i)(-1)]$$

Reducing gives

$$\frac{\partial F}{\partial X_i} = \frac{-X_j + X_i}{\left[(X_j - X_i)^2 + (Y_j - Y_i)^2\right]^{1/2}}$$

Rearranging and evaluating at initial approximations yield the following:

$$\frac{\partial F}{\partial X_i} = \frac{X_{j_0} - X_{j_0}}{(IJ_0)}$$

16-10. THE ANGLE OBSERVATION EQUATION

Angle observation equations relate measured angles and their inherent random errors to the most probable coordinates of the occupied station, backsight station, and foresight station. This is also termed the variation of coordinates method. Again, in this treatment X and Y plane coordinates are used. Referring to Figure 16-4, we may write the following observation equation for the measured angle at I between points J and K:

$$\theta_{jik} + V_{\theta_{jik}} = \text{azimuth}_{ik} - \text{azimuth}_{ij}$$

$$= \tan^{-1}\left(\frac{X_k - X_i}{Y_k - Y_i}\right)$$

$$- \tan^{-1}\left(\frac{X_j - X_i}{Y_j - Y_i}\right) + C \quad (16\text{-}19)$$

Equation (16-19) relates observed angle θ_{jik} and its residual error $V_{\theta_{jik}}$ to the most proba-

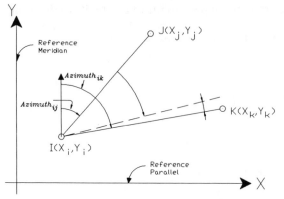

Figure 16-4.   Geometry of angle observation equation.

ble angle in terms of unknown variables X_j, Y_j, X_i, Y_i, X_k, and Y_k, the most probable coordinates of the points involved. The term C makes the equation general and accounts for the fact that the azimuths of lines IK and IJ can be in any directions. Figure 16-5 shows the 12 possi-

ble quadrant locations for stations J, I, and K for angles under 180°. For the six cases a through f, $C = 0$; for the six cases g through 1, $C = 180°$.

Equation (16-19) is also nonlinear and may be linearized using the Taylor series as follows:

$$\theta_{jik} + V_{\theta_{jik}} = U(X_j, Y_j, X_i, Y_i, X_k, Y_k) \quad (16\text{-}20)$$

where

$$U(X_j, Y_j, X_i, Y_i, X_k, Y_k)$$

$$= \tan^{-1}\left(\frac{X_k - X_i}{Y_k - Y_i}\right) - \tan^{-1}\left(\frac{X_j - X_i}{Y_j - Y_i}\right) + C$$

The Taylor series approximation for function U, after we drop as negligible all terms of

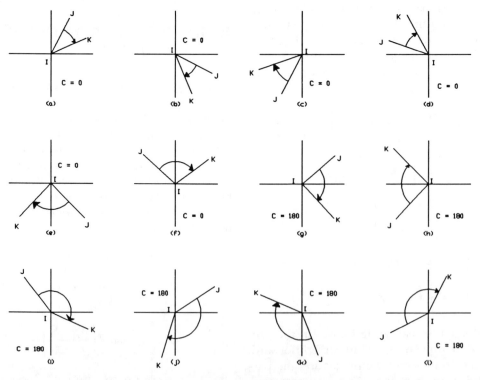

Figure 16-5.   Twelve different quadrant locations (a) through (l) for stations J, I, and K in angle measurement.

order two and higher, is

$$U(X_j, Y_j, X_i, Y_i, X_k, Y_k)$$

$$= U(X_{j_0}, Y_{j_0}, X_{i_0}, Y_{i_0}, X_{k_0}, Y_{k_0})$$

$$+ \left[\frac{\partial U}{\partial X_j}\right]_0 dX_j + \left[\frac{\partial U}{\partial Y_j}\right]_0 dY_j$$

$$+ \left[\frac{\partial U}{\partial X_i}\right]_0 dX_i + \left[\frac{\partial U}{\partial Y_i}\right]_0 dY_i$$

$$+ \left[\frac{\partial U}{\partial X_k}\right]_0 dX_k + \left[\frac{\partial U}{\partial Y_k}\right]_0 dY_k \quad (16\text{-}21)$$

In Equation (16-21), terms are defined as in Equation (16-17). Evaluating partial derivatives of the function U, substituting them into equation (16-21), and then in turn substituting into equation (16-20) and rearranging, we get the following linearized observation equation for angle measurements:

$$K_{\theta_{jik}} + V_{\theta_{jik}} = \left[\frac{Y_{i_0} - Y_{j_0}}{(IJ_0)^2}\right] dX_j + \left[\frac{X_{j_0} - X_{i_0}}{(IJ_0)^2}\right] dY_j$$

$$+ \left[\frac{Y_{j_0} - Y_{i_0}}{(IJ_0)^2} - \frac{Y_{k_0} - Y_{i_0}}{(IK_0)^2}\right] dX_i$$

$$+ \left[\frac{X_{i_0} - X_{j_0}}{(IJ_0)^2} - \frac{X_{i_0} - X_{k_0}}{(IK_0)^2}\right] dY_i$$

$$+ \left[\frac{Y_{k_0} - Y_{i_0}}{(IK_0)^2}\right] dX_k$$

$$+ \left[\frac{X_{i_0} - X_{k_0}}{(IK_0)^2}\right] dY_k \quad (16\text{-}22)$$

where $K_{\theta_{jik}} = \theta_{jik} - \theta_{jik_0}$, and

$$\theta_{jik_0} = \tan^{-1}\left(\frac{X_{k_0} - X_{i_0}}{Y_{k_0} - X_{i_0}}\right)$$

$$- \tan^{-1}\left(\frac{X_{j_0} - X_{i_0}}{Y_{j_0} - Y_{i_0}}\right) + C$$

$$(IJ_0) = \sqrt{(X_{j_0} - X_{i_0})^2 + (Y_{j_0} - Y_{i_0})^2}$$

$$(IK_0) = \sqrt{(X_{k_0} - X_{i_0})^2 + (Y_{k_0} - Y_{i_0})^2}$$

Equations (16-18) and (16-22) are linearized distance and angle observation equation, respectively, and a system of these may be formed and manipulated conveniently by matrix methods for adjusting horizontal surveys.

In equation (16-22), $K_{\theta_{jik}}$ and $V_{\theta_{jik}}$ are in radian measure. Since it is more common to work in the sexagesimal system in the United States and because the magnitudes of the angle residuals are generally in the seconds range, the equation's units may be converted to seconds by multiplying the right-hand side by ρ (rho) the number of seconds per radian, which is 206,264.8″/rad.

16-11. TRILATERATION ADJUSTMENT

As noted in Section 16-8, trilateration surveys consist of only distance measurements. The geometric figures used are many and varied. All are equally adaptable to the observation-equation method of adjustment, although each different geometric configuration poses a specific adjustment problem. Consider, e.g., the adjustment of simple Figure 16-6. Points A, B, and C are horizontal control points whose X- and Y-coordinates are known and fixed. The position of point U is to be established from the measurement of distances AU, BU, and CU. Obviously, any two of these distances would

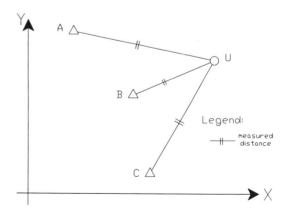

Figure 16-6. Simple trilateration example.

be sufficient to establish X- and Y-coordinates of U. The third distance is, therefore, redundant and makes possible the adjustment and calculations of most probable values of X_u and Y_u.

The observation equations are developed by substituting into prototype Equation (16-18). The equation for measured line AU, e.g., may be formed by interchanging subscript i with a, and j with u. For line BU the i and j subscripts of Equation (16-18) are replaced by b and u, respectively, ad for line CU, the subscripts i and j are replaced by c and u, respectively. It is to be noted that if one of the end points of a measured line is a control point, then its coordinates are invariant, and hence these terms drop out of the prototype observation equation. By using the procedures, described, the following three linearized observation equations result:

$$(L_{au} - AU_0) + V_{L_{au}} = \left[\frac{X_{u_0} - X_a}{AU_0}\right] dX_u$$

$$+ \left[\frac{Y_{u_0} - Y_a}{AU_0}\right] dY_u$$

$$(L_{bu} - BU_0) + V_{L_{bu}} = \left[\frac{X_{u_0} - X_b}{BU_0}\right] dX_u$$

$$+ \left[\frac{Y_{u_0} + Y_b}{BU_0}\right] dY_u$$

$$(L_{cu} - CU_0) + V_{L_{cu}} = \left[\frac{X_{u_0} - X_c}{CU_0}\right] dX_u$$

$$+ \left[\frac{Y_{u_0} - Y_c}{CU_0}\right] dY_u \quad (16\text{-}23)$$

where

$$AU_0 = \sqrt{(X_{u_0} - X_a)^2 + (Y_{u_0} - Y_a)^2}$$

$$BU_0 = \sqrt{(X_{u_0} - X_b)^2 + (Y_{u_0} - Y_b)^2}$$

$$CU_0 = \sqrt{(X_{u_0} - X_c)^2 + (Y_{u_0} - Y_c)^2}$$

Also, L_{au}, L_{bu}, and L_{cu} are the observed distances, and X_{U_0} and Y_{U_0} initial approxima-

tions of coordinates of point U, which can be obtained from a scaled diagram or computed from two of the observed distances.

This system of linear observation equations may be expressed in matrix form as follows:

$$_3\mathbf{A}_2\ _2\mathbf{X}_1 = {}_3\mathbf{K}_1 + {}_3\mathbf{V}_1$$

where $\mathbf{A}$ is the matrix of coefficients of unknown; $\mathbf{X}$ the matrix of unknown corrections dX_u and dY_u; $\mathbf{K}$ the matrix of constants, i.e., measured lengths minus lengths computed from initial approximate coordinates, and $\mathbf{V}$ the matrix of residuals in the measured lengths. Most probable corrections dX_u and dY_u, and hence the most probable coordinates X_u and Y_u, may be calculated by applying the least-squares matrix equation. If we consider equal weights for the observations the equation is

$$_2\mathbf{X}_1 = \left({}_2\mathbf{A}_3^T\ _3\mathbf{A}_2\right)^{-1}{}_2\mathbf{A}_3^T\ _3\mathbf{K}_1$$

The $\mathbf{X}$ matrix consists of corrections to be added to the initial approximations of the coordinates. Because Taylor series linearization drops terms higher than order one, an iterative solution is required. For the second iteration, the corrected coordinates are used to reformulate the $\mathbf{A}$ and $\mathbf{K}$ matrices. Then, these are used in another least-squares solution to obtain a new $\mathbf{X}$ matrix consisting of corrections. The process is repeated until the values of the computed $\mathbf{X}$ matrix become small enough to be considered negligible.

Example 16-6. Adjust the example of Figure 16-6 by least squares if measured distances AU, BU, and CU are 6049.00, 4736.83, and 5446.49 ft, respectively, and coordinates of the control points are as follows:

$$X_a = 865.40, \qquad X_b = 2432.55, \qquad X_c = 2865.22$$

$$Y_a = 4527.15, \qquad Y_b = 2047.25, \qquad Y_c = 27.15$$

A. First Iteration.

1. Calculate initial approximates for X_{u_0} and Y_{u_0}.

   (a) Calculate azimuth AB.

   $$\text{Azimuth}_{AB} = \tan^{-1}\left(\frac{X_b - X_a}{Y_b - Y_a}\right) + 180°$$

   $$\text{Azimuth}_{AB} = \tan^{-1}\left[\frac{2432.55 - 865.40}{2047.25 - 4527.15}\right]$$

   $$+ 180° = 147°43'34''$$

   (b) Calculate length AB.

   $$AB = \sqrt{(X_b - X_a)^2 + (Y_b - Y_a)^2}$$

   $$= \sqrt{\begin{array}{c}(2432.55 - 865.20)^2 \\ + (2047.25 - 4527.15)^2\end{array}}$$

   $$= 2933.58 \text{ ft}$$

   (c) Calculate azimuth AU_0, (from cosine law: $c^2 = a^2 + b^2 - 2ab \cos C$).

   $$\cos(\angle UAB)$$

   $$= \frac{6049.00^2 - 4736.83^2 + 2933.58^2}{2(6049.00)(2933.58)}$$

   $$\angle UAB = 50°06'50''$$

   $$\text{Az}_{AU_0} = 147°42'34'' - 50°06'50'' = 97°35'44''$$

   (d) Calculate X_{u_0} and Y_{u_0}

   $$X_{u_0} = 865.40 + 6049.00 \sin(97°35'44'')$$

   $$= 6861.324 \text{ ft}$$

   $$Y_{u_0} = 4527.15 + 6049.00 \cos(97°35'44'')$$

   $$= 3727.587 \text{ ft}$$

2. Calculate AU_0, BU_0, and CU_0. For this first iteration, AU_0 and BU_0 are exactly equal to their respective measured distances because X_{u_0} and Y_{u_0} were calculated from these measured values. Therefore,

$$AU_0 = 6049.00, \qquad BU_0 = 4736.83$$

$$CU_0 = [(6861.32 - 2865.22)^2$$

$$+ (3727.59 - 27.15)^2]^{1/2} = 5446.29 \text{ ft}$$

3. Formulate the matrices. (a) The **A** matrix. Observation Equations (16-18) may be simplified as follows:

$$a_{11}\, dX_u + a_{12}\, dY_u = k_1 + v_1$$

$$a_{21}\, dX_u + a_{22}\, dY_u = k_2 + v_2$$

$$a_{31}\, dX_u + a_{32}\, dY_u = k_3 + v_3$$

where

$$a_{11} = \frac{6861.32 - 865.40}{6049.00} = 0.991$$

$$a_{12} = \frac{3727.59 - 4527.15}{6049.00} = -0.132$$

$$a_{21} = \frac{6861.32 - 2432.55}{4736.83} = 0.935$$

$$a_{22} = \frac{3727.59 - 2047.25}{4736.83} = 0.355$$

$$a_{31} = \frac{6861.32 - 2865.22}{5446.29} = 0.734$$

$$a_{32} = \frac{3727.59 - 27.15}{5446.29} = 0.679$$

(b) The **K** matrix.

$$k_1 = 6049.00 - 6049.00 = 0.00$$

$$k_2 = 4736.83 - 4736.83 = 0.00$$

$$k_3 = 5446.49 - 5446.29 = 0.20$$

(c) The **X** and **V** matrices.

$$\mathbf{X} = \begin{bmatrix} dX_u \\ dY_u \end{bmatrix}, \qquad \mathbf{V} = \begin{bmatrix} v_{au} \\ v_{bu} \\ v_{cu} \end{bmatrix}$$

4. The matrix solution using unweighted least-squares Equation (16-7) is

$$\mathbf{X} = (\mathbf{A}^T\mathbf{A})^{-1}\mathbf{A}^T\mathbf{K}$$

$$\mathbf{A}^T\mathbf{A} = \begin{bmatrix} 0.991 & 0.935 & 0.734 \\ -0.132 & 0.355 & 0.679 \end{bmatrix}$$

$$\times \begin{bmatrix} 0.991 & -0.132 \\ 0.935 & 0.355 \\ 0.735 & 0.679 \end{bmatrix}$$

$$= \begin{bmatrix} 2.395 & 0.699 \\ 0.699 & 2.395 \end{bmatrix}$$

$$(\mathbf{A}^T\mathbf{A})^{-1} = \frac{1}{0.960} \begin{bmatrix} 0.605 & -0.699 \\ -0.699 & 2.395 \end{bmatrix}$$

$$\mathbf{A}^T\mathbf{K} = \begin{bmatrix} 0.991 & 0.935 & 0.734 \\ -0.132 & 0.355 & 0.679 \end{bmatrix}$$

$$\begin{bmatrix} 0.000 \\ 0.000 \\ 0.200 \end{bmatrix} = \begin{bmatrix} 0.144 \\ 0.135 \end{bmatrix}$$

$$\mathbf{X} = \frac{1}{0.960} \begin{bmatrix} 0.605 & -0.699 \\ -0.699 & 2.395 \end{bmatrix}$$

$$\begin{bmatrix} 0.144 \\ 0.135 \end{bmatrix} = \begin{bmatrix} -0.007 \\ +0.232 \end{bmatrix}$$

The revised coordinates of U are as follows:

$$X_u = 6861.324 - 0.007 = 6861.317$$

$$Y_u = 3727.587 + 0.232 = 3727.819$$

B. SECOND ITERATION.

1. Calculate AU_0, BU_0, and CU_0.

$$AU_0 = \sqrt{\begin{array}{l}(6861.73 - 865.40)^2 \\ + (3727.819 - 4527.15)^2\end{array}}$$

$$= 6048.963$$

$$BU_0 = \sqrt{\begin{array}{l}(6861.317 - 865.40)^2 \\ + (3727.819 - 2047.25)^2\end{array}}$$

$$= 4736.907$$

$$CU_0 = \sqrt{\begin{array}{l}(6861.317 - 2865.22)^2 \\ + (3727.819 - 27.15)^2\end{array}}$$

$$= 5446.443$$

2. Formulate the matrices. With these minor changes in the lengths, the **A** matrix (to three places) does not change. Hence, $(\mathbf{A}^T\mathbf{A})^{-1}$ does not change either. The **K** matrix does not change, however, as shown by the following computations:

$$k_1 = 6049.00 - 6048.963 = 0.037$$

$$k_2 = 4736.83 - 4376.907 = -0.077$$

$$k_3 = 5446.49 - 5446.443 = 0.047$$

3. The matrix solution.

$$\mathbf{A}^T\mathbf{K} = \begin{bmatrix} 0.991 & 0.935 & 0.734 \\ -0.132 & 0.355 & 0.679 \end{bmatrix}$$

$$\times \begin{bmatrix} 0.037 \\ -0.077 \\ 0.047 \end{bmatrix}$$

$$= \begin{bmatrix} -0.0008 \\ -0.0003 \end{bmatrix}$$

$$\mathbf{X} = \frac{1}{0.960} \begin{bmatrix} 0.605 & -0.699 \\ -0.699 & 2.395 \end{bmatrix}$$

$$\times \begin{bmatrix} -0.0008 \\ -0.0003 \end{bmatrix}$$

$$= \begin{bmatrix} -0.0003 \\ -0.0002 \end{bmatrix}$$

The revised coordinates of U are

$$X_u = 6861.317 - 0.0003 = 6861.317$$

$$Y_u = 3727.819 - 0.0004 = 3727.819$$

Satisfactory convergence is indicated by the very small size of the corrections computed in the second iteration. Having calculated most probable coordinates, we may then calculate residuals using Equation (16-10), followed by calculations for the standard deviation of unit weight and the standard deviations of adjusted coordinates using Equations (16-11) and (16-13), respectively. Of course, the measured lengths could be weighted, in which case the appropriate weighted least-squares equations would be used.

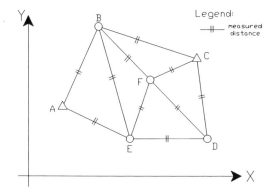

Figure 16-7. More complex trilateration network.

As was demonstrated in Example 16-6, *X*- and *Y*-coordinates of control stations can readily be held fixed in an adjustment. This is accomplished by assigning values of zero to their *dX* and *dY* terms, and hence those terms drop out of the equations. Note that in each observation equation of the example, only two unknowns appear, since in each case one end of those lines was a control station and thus was held fixed. Directions of lines can also be held fixed, and methods of doing so are described in references cited at the end of this chapter.

The procedure for adjusting larger trilateration networks, such as that shown in Figure 16-7, follows the same approach as that used for the small figure of Example 16-6. The basic difference is that the sizes of the matrices are larger, but the individual elements are still formed using the same prototype equation. Suppose in Figure 16-7 that *A* and *C* are control points whose coordinates are fixed. Thus, there are ten observations and eight unknowns. Points *A* and *C* in the network are held fixed by giving the terms *dX_a*, *dY_a*, *dX_c*, and *dY_c* zero coefficients. Hence, these terms drop out of the solution. The **A** matrix formulated from Equation (16-18) would have nonzero elements as indicated by the X's in Table 16-1.

Formulation of the matrices and their least-squares solution in an iterative manner

are readily programmed for computer solution.

16-12. TRIANGULATION ADJUSTMENT

Triangulation is a method of horizontal control extension in which angle measurements are the basic observations. Positions of widely spaced points are then computed based on these angles and only a minimum number of measured distanced called *base lines*. As in trilateration, the geometric figures used are many and varied.

Least-squares triangulation adjustment may be performed by condition equations, direction observation equations, or angle observation equations. In this section, the angle observation equation method is presented. This procedure is relatively simple and adaptable to any combination of angle measurements and geometric figures.

In formulating angle observation equations *I* is always assigned to the station of the angle's vertex. If we consider the angles turned clockwise, *J* is the backsight station and *K* the foresight station. This designation of stations must be strictly adhered to in forming observation equations from prototype Equation (16-22). The procedure is demonstrated in numerical examples in this chapter.

16-12-1. Adjustment of Intersections

Intersection is one of the simplest and sometimes most practical methods for locating the horizontal position of an occasional isolated point, if the point is visible from two or more existing horizontal control stations. Intersection is especially well-adapted over inaccessible terrain. For a unique position computation, the method requires that at least two horizontal angles be measured from two con-

Table 16-1.   Nonzero elements for **A** matrix from Equation (16-18)

Unknowns Distance	dY_b	dX_b	dY_d	dX_d	dY_e	dX_e	dY_f	dX_f
AB	X	X	0	0	0	0	0	0
AE	0	0	0	0	X	X	0	0
BC	X	X	0	0	0	0	0	0
BF	X	X	0	0	0	0	X	X
BE	X	X	0	0	X	X	0	0
CD	0	0	X	X	0	0	0	0
CF	0	0	0	0	0	0	X	X
DF	0	0	X	X	0	0	X	X
DE	0	0	X	X	X	X	0	0
EF	0	0	0	0	X	X	X	X

trol points, as angles θ_1 and θ_2 measured at control points A and B of Figure 16-8. If additional control is available, the position computation for unknown point U can be strengthened by measuring redundant angles θ_3 and θ_4 in the figure. When redundant measurements are taken, most probable coordinates of the point U may be calculated by the least-squares procedure.

Example 16-7.   Adjust the example of Figure 16-8 by least squares if the measured angles (equally weighted) and coordinates of control points are the following:

$$\theta_1 = 50°06'50''$$

$$\theta_3 = 98°41'17''$$

$$\theta_2 = 101°30'47''$$

$$\theta_4 = 59°17'01''$$

$$X_a = 865.40$$

$$X_b = 2432.55$$

$$X_c = 2865.22$$

$$Y_a = 4527.15$$

$$Y_b = 2047.25$$

$$Y_c = 27.15$$

1. Calculate initial approximations for AU_0, BU_0, CU_0, X_{u_0}, and Y_{u_0} as follows:

$$AB = \sqrt{\begin{array}{l}(2432.55 - 865.40)^2 \\ +(4527.15 - 2047.25)^2\end{array}}$$

$$= 2933.58 \text{ ft}$$

$$AU_0 = \frac{AB \sin \theta_2}{\sin(180° - \theta_1 - \theta_2)}$$

$$= \frac{2933.58 \sin 101°30'47''}{\sin 28°27'23''}$$

$$= 6049.00 \text{ ft}$$

$$\text{Azimuth } AB = \tan^{-1}\left(\frac{X_b - X_a}{Y_b - Y_a}\right) + 180°$$

$$= \tan^{-1}\left(\frac{2432.55 - 865.40}{2047.25 - 4527.15}\right)$$

$$+ 180° = 147°42'34''$$

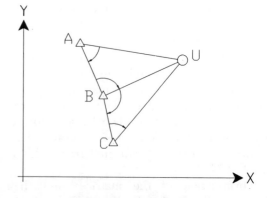

Figure 16-8.   Intersection example.

Azimuth $AU_0 = 147°42'34'' - 50°06'50''$

$$= 97°35'44''$$

$X_{u_0} = X_a + AU_0 \sin(\text{azimuth } AU_0)$

$$= 865.40 + (6049.00)\sin 97°35'44''$$

$$= 6861.35$$

$Y_{u_0} = Y_a + AU_0 \cos(\text{azimuth } AU_0)$

$$= 4527.15 + (6049.00)\cos 97°35'44''$$

$$= 3727.59$$

$BU_0 = \sqrt{\begin{array}{c}(6861.35 - 2432.55)^2 \\ + (3727.59 - 2047.25)^2\end{array}}$

$$= 4736.83$$

$CU_0 = \sqrt{\begin{array}{c}(6861.35 - 2865.22)^2 \\ + (3727.59 - 27.15)^2\end{array}}$

$$= 5446.29$$

2. Formulation of the matrices. As in trilateration adjustment, a control station's coordinate can be held fixed by assigning zeros to its dX and dY values, so that these terms drop out of the equations. In Figure 16-8, the vertex points of all angles are control stations; thus, their dX and dY values are zeros, and these terms do not appear in the observation equations. In forming the observation equations, I, J, and K are assigned as previously described. Thus for angle θ_1, i, j, and k are placed in prototype Equation (16-22) by a, u, and b, respectively. Angle θ_1 conforms to Figure 14-4b.

Referring to Equation (16-22), we may write the following observation equations for the four observed angles:

$$\left[\frac{Y_a - Y_{u_0}}{(AU_0)^2}\right](dX_u) + \left[\frac{X_{u_0} - X_a}{(AU_0)^2}\right](dY_u)$$

$$= \theta_1 - \left\{\tan^{-1}\left(\frac{X_b - X_a}{Y_b - Y_a}\right)\right.$$

$$\left. - \tan^{-1}\left(\frac{X_{u_0} - X_a}{Y_{u_0} - Y_a}\right) + 0°\right\} + v_1$$

$$\times \left[\frac{Y_{u_0} - Y_b}{(UB_0)^2}\right](dX_u) + \left[\frac{X_b - X_{u_0}}{(UB_0)^2}\right](dY_u)$$

$$= \theta_2 - \left\{\tan^{-1}\left(\frac{X_{u_0} - X_b}{Y_{u_0} - Y_b}\right)\right.$$

$$\left. - \tan^{-1}\left(\frac{X_a - X_b}{Y_a - Y_b}\right) + 0°\right\} + v_2$$

$$\times \left[\frac{Y_b - Y_{u_0}}{(BU_0)^2}\right](dX_u) + \left[\frac{X_{u_0} - X_b}{(BU_0)^2}\right](dY_u)$$

$$= \theta_3 - \left\{\tan^{-1}\left(\frac{X_c - X_b}{Y_c - Y_b}\right)\right.$$

$$\left. - \tan^{-1}\left(\frac{X_{u_0} - X_b}{Y_{u_0} - Y_b}\right) + 180°\right\} + v_3$$

$$\times \left[\frac{Y_{u_0} - Y_c}{(UC)^2}\right](dX_u) + \left[\frac{X_c - X_{u_0}}{(UC_0)^2}\right]$$

$$= \theta_4 - \left\{\tan^{-1}\left(\frac{X_{u_0} - X_c}{Y_{u_0} - Y_c}\right)\right.$$

$$\left. + \tan^{-1}\left(\frac{X_b - X_c}{Y_b - Y_c}\right) + 0°\right\} + v_4$$

3. Substituting control point coordinates and approximate coordinates X_{u_0} and Y_{u_0} into the linearized observation equations and multiplying by ρ, we can form the following **A** and **K** matrices. [Note that Equation (16-19) is used to calculate their **K** values.]

$$\mathbf{A} = \rho \begin{bmatrix} \dfrac{4527.15 - 3727.59}{6049.00^2} & \dfrac{6861.35 - 865.40}{6049.00^2} \\[2mm] \dfrac{3727.59 - 2047.25}{4736.83^2} & \dfrac{2432.55 - 6861.35}{4736.83^2} \\[2mm] \dfrac{2047.25 - 3727.59}{4736.83^2} & \dfrac{6861.35 - 24.32.55}{4736.83^2} \\[2mm] \dfrac{3727.59 - 27.15}{5446.29^2} & \dfrac{2865.22 - 6861.35}{5446.29^2} \end{bmatrix}$$

$$= \begin{bmatrix} 4.507 & 33.800 \\ 15.447 & -40.713 \\ -15.447 & 40.713 \\ 25.732 & -27.788 \end{bmatrix}$$

$$\mathbf{K} = \begin{bmatrix} 50°06'50'' - \left\{ \tan^{-1}\left(\dfrac{2432.55 - 865.40}{2047.25 - 4527.15} \right) - \tan^{-1}\left(\dfrac{6861.35 - 865.40}{3727.59 - 4527.15} \right) + 0° \right\} \\ 101°30'47'' - \left\{ \tan^{-1}\left(\dfrac{6861.35 - 2432.55}{3727.59 - 047.25} \right) - \tan^{-1}\left(\dfrac{865.40 - 2432.55}{4527.15 - 2047.25} \right) + 0° \right\} \\ 98°41'17'' - \left\{ \tan^{-1}\left(\dfrac{2865.22 - 2432.55}{27.15 - 047.25} \right) - \tan^{-1}\left(\dfrac{6861.35 - 2432.55}{3727.59 - 2047.25} \right) + 180° \right\} \\ 59°17'01'' - \left\{ \tan^{-1}\left(\dfrac{6861.35 - 2865.22}{3727.59 - 27.15} \right) - \tan^{-1}\left(\dfrac{2432.55 - 2865.22}{2047.25 - 27.15} \right) + 0° \right\} \end{bmatrix} = \begin{bmatrix} 0.00'' \\ 0.00'' \\ -0.69'' \\ -20.23'' \end{bmatrix}$$

Note that values in the **K** matrix for angles θ_1 and θ_2 are exactly equal to zero for the first iteration because initial coordinates X_{u_0} and Y_{u_0} were calculated using these two angles.

4. Matrix solution of Equation (16-7) for adjusted coordinates X_u and Y_u.

$$\mathbf{A}^T\mathbf{A} = \begin{bmatrix} 1159.7 & -1820.5 \\ -1820.5 & 5229.7 \end{bmatrix},$$

$$\mathbf{Q} = (\mathbf{A}^T\mathbf{A})^{-1} = \begin{bmatrix} 0.001901 & 0.000662 \\ 0.000662 & 0.000422 \end{bmatrix}$$

$$\mathbf{A}^T\mathbf{K} = \begin{bmatrix} -509.9 \\ 534.1 \end{bmatrix}$$

$$\mathbf{X} = (\mathbf{A}^T\mathbf{A})^{-1}(\mathbf{A}^T\mathbf{K})$$

$$= \begin{bmatrix} 0.001901 & 0.000662 \\ 0.000662 & 0.000422 \end{bmatrix}\begin{bmatrix} -509.0 \\ 534.1 \end{bmatrix}$$

$$= \begin{bmatrix} dX_u \\ dY_u \end{bmatrix}$$

and

$$dX_u = -0.62 \text{ ft}$$

$$dY_u = -0.11 \text{ ft}$$

$$X_u = X_{u_0} + dX_u = 6861.35 - 0.62 = 6860.73$$

$$Y_u = Y_{u_0} + dY_u = 3727.59 - 0.11 = 3727.48$$

Note that a second iteration produced negligible-sized values for dX_u and dY_u, thus, the solution converged after one iteration. The second iteration is not shown.

5. Matrix solutions for residuals and precisions by Equation (16-10) with **L** replaced by **K** is as follows:

$$\mathbf{V} = \mathbf{AX} - \mathbf{K}$$

$$= \begin{bmatrix} 4.507 & 33.80 \\ 15.447 & -40.713 \\ -15.447 & 40.713 \\ 25.732 & -27.788 \end{bmatrix}\begin{bmatrix} -0.62 \\ -0.11 \end{bmatrix}$$

$$\times \begin{bmatrix} 0.00'' \\ 0.00'' \\ -0.69'' \\ -20.23'' \end{bmatrix}$$

$$= \begin{bmatrix} -6.5'' \\ -5.1'' \\ +5.8'' \\ +7.3'' \end{bmatrix}$$

By Equation (16-11),

$$\mathbf{V}^T\mathbf{V} = \begin{bmatrix} -6.5 & -5.1 & 5.8 & 7.3 \end{bmatrix}$$

$$\times \begin{bmatrix} -6.5 \\ -5.1 \\ 5.8 \\ 7.3 \end{bmatrix} = 155.2 \text{ sec}^2$$

$$S_0 = \sqrt{\dfrac{\mathbf{V}^T\mathbf{V}}{m - n}}$$

$$= \sqrt{\dfrac{155.2}{4 - 2}} = \pm 8.8''$$

By Equation (16-13),

$$S_{X_u} = S_0\sqrt{Q_{X_uX_u}} = 8.8\sqrt{0.001901} = \pm 0.38 \text{ ft}$$

$$S_{Y_u} = S_0\sqrt{Q_{Y_uY_u}} = 8.8\sqrt{0.000422} = \pm 0.18 \text{ ft}$$

Finally, the standard deviation in the position of U, S_u, is the square root of the sum of the squares of and S_{X_u} and S_{Y_u} or as follows:

$$S_u = \sqrt{S_{X_u}^2 + S_{Y_u}^2} - \sqrt{0.38^2 + 0.18^2}$$

$$= \pm 0.42 \text{ ft}$$

16-12-2. Adjustment of Resections

Resection is a method that may be employed for locating the unknown horizontal position of an occupied theodolite station by measuring a minimum of two horizontal angles to a minimum of three stations whose horizontal positions are known. If more than three stations are available, redundant observations may be obtained and the position of the unknown occupied station can be computed by employing the least-squares procedure. Like intersection, this method is suitable for locating an occasional isolated point and is especially well-adapted for use over inaccessible terrain.

Consider the resection position computation for the occupied point U of Figure 16-9, having observed the three horizontal angles shown. Points A, B, C, and D are fixed control stations. Angles θ_1, θ_2, and θ_3 are ob-

served from station U, whose position is unknown.

Utilizing prototype Equation (16-22), we may write a linearized observation equation for each angle. These observation equations can be expressed in matrix notation as follows:

$$_3\mathbf{A}_2 \, _2\mathbf{X}_1 = {_3}\mathbf{K}_1 + {_3}\mathbf{V}_1$$

Application of the least-squares routine yields corrections dX_u and dY_u, the most probable coordinates X_u and Y_u, residuals, and the estimated standard deviation of the position of point U.

16-12-3. Adjustment of More Complex Triangulation Networks

The basic figure for triangulation is generally considered to be the quadrilateral, but frequently other geometrical figures, such as chains of quadrilaterals, central-point figures, etc., are used. Regardless of the geometrical shape of the triangulated figure, the basic least-squares approach is simply an extension of principles already discussed, which involves writing one observation equation for each measured angle in terms of the most probable coordinates of the points involved. The following example (see Figure 16-10) illustrates the adjustment of a quadrilateral.

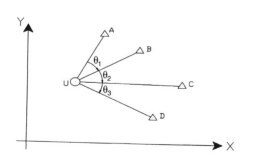

Figure 16-9.   Resection example.

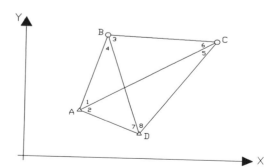

Figure 6-10.   Quadrilateral of Example 16-8.

Example 16-8. Adjust, by least squares, the quadrilateral of Figure 16-10, given the following observations and known data:

Observed angles (assume equal weights):

1 = 42°35′29.0″,	5 = 21°29′23.9″
2 = 87°35′10.6″,	6 = 39°01′35.4″
3 = 79°54′42.1″,	7 = 31°20′45.8″
4 = 18°28′22.4″,	8 = 39°34′27.9″

Fixed coordinates:

$$X_a = 4270.33, \qquad X_d = 7610.58$$

$$Y_a = 8448.90, \qquad Y_d = 4568.75$$

A computer program has been used to form the matrices and solve the problem. Equal weights were used. In the program, the angles were input in order 1 through 8. The **X** matrix in the solution is

$$\mathbf{X} = \begin{bmatrix} dX_b \\ dY_b \\ dX_c \\ dY_c \end{bmatrix}$$

The following computer-output listing, which is self-explanatory, gives the solution for Example 16-8. As shown, one iteration was satisfactory to achieve convergence because for the second iteration, the unknowns—corrections to the initial coordinates—are all zeros. Residuals, adjusted coordinates, their standard deviation, and adjusted angles are all given at the end of the listing.

Triangulation Example

Number of Observed Distances	≫ 0
Number of Observed Angles	≫ 8
Number of Observed Azimuths	≫ 0
Number of Unknown Stations	≫ 2
Number of Control Stations	≫ 2

Control Stations

Station	Northing	Easting
A	8448.900	4270.330
D	4568.750	7610.580

Initial Approximations of Unknown Station Coordinates

Station	Northing	Easting
B	16,749.769	5599.549
C	16,636.185	14,633.027

Angle Observations

Backsighted Station	Occupied Station	Foresighted Station	Angle
B	A	C	42°35′29.0″
C	A	D	87°35′10.6″
C	B	D	79°54′42.1″
D	B	A	18°28′22.4″
D	C	A	21°29′23.9″
A	C	B	39°1′35.4″
A	D	B	31°20′45.8″
B	D	C	39°34′27.9″

Weight Matrix
Dimensions: 8 × 8

1	0	0	0	0	0	0	0
0	1	0	0	0	0	0	0
0	0	1	0	0	0	0	0
0	0	0	1	0	0	0	0
0	0	0	0	1	0	0	0
0	0	0	0	0	1	0	0
0	0	0	0	0	0	1	0
0	0	0	0	0	0	0	1

Iteration No. 1
A Matrix
Dimensions: 8 × 4

− 24.2301	3.8804	9.6822	− 12.2549
0.0000	0.0000	− 9.6822	12.2549
16.2008	− 20.1080	0.2845	22.8298
7.7448	− 6.6023	0.0000	0.0000
0.0000	0.0000	− 3.0864	− 4.8244
0.2845	22.8298	− 9.9668	− 10.5750
16.4853	2.7219	0.0000	0.0000
− 16.4853	− 2.7219	12.7686	− 7.4305

K Matrix
Dimensions: 8 × 1

4.7351
− 4.7406
4.3168
− 4.3190
− 7.7651
4.1671
− 7.8755
8.5812

Covariance Matrix (Q_{xx})
Dimensions: 4 × 4

0.0016	0.0008	0.0018	− 0.0003
0.0008	0.0079	0.0077	0.0066
0.0018	0.0077	0.0106	0.0063
− 0.0003	0.0066	0.0063	0.0073

Covariance Matrix (Q_{xx})
Dimensions: 4 × 4

0.0016	0.0008	0.0018	− 0.0003
0.0008	0.0079	0.0077	0.0066
0.0018	0.0077	0.0106	0.0063
− 0.0003	0.0066	0.0063	0.0073

Unknowns (X) Dimensions: 4 × 1	Residuals (V) Dimensions: 8 × 1
0.0001	− 2.1010
− 0.0001	− 5.0324
0.0002	4.1834
− 0.0003	1.4172
	− 1.7581
	5.4004
	− 6.4838
	1.4744

Unknowns (X) Dimensions: 4 × 1	Residuals (V) Dimensions: 8 × 1
− 0.1909	− 2.1010
0.6449	− 5.0324
0.8542	4.1834
0.6987	1.4172
	− 1.7582
	5.4004
	− 6.4839
	1.4744

STANDARD ERROR OF UNIT WEIGHT = 5.606

Standard Deviations

SXB = ± 0.226
SYB = ± 0.498
SXC = ± 0.577
SYC = ± 0.478

Iteration No. 2
A Matrix
Dimensions: 8 × 4

− 24.2285	3.8793	9.6815	− 12.2538
0.0000	0.0000	− 9.6815	12.2538
16.2000	− 20.1054	0.2843	22.8272
7.7441	− 6.6012	0.0000	0.0000
0.0000	0.0000	− 3.0860	− 4.8235
0.2843	22.8272	− 9.9658	− 10.5734
16.4844	2.7218	0.0000	0.0000
− 16.4844	− 2.7218	12.7675	− 7.4303

Adjusted Angles

Backsighted Station	Occupied Station	Foresighted Station	Angle
B	A	C	42°35′31.1″
C	A	D	87°35′15.6″
C	B	D	79°54′37.9″
D	B	A	18°28′21.0″
D	C	A	21°29′25.7″
A	C	B	39°01′30.0″
A	D	B	31°20′52.3″
B	D	C	39°34′26.4″

K Matrix
Dimensions: 8 × 1

− 2.0977
− 5.0381
4.1798
1.4184
− 1.7575
5.3996
− 6.4825
1.4781

16-13. TRAVERSE ADJUSTMENT

Of the many methods for traverse adjustment, the characteristic that distinguishes traverse adjustment by least squares from approximate methods is that distance and direction observations are adjusted simultaneously in the least-squares approach. The result is an adjustment in which all geometrical conditions are satisfied; but more important, the values of the adjusted quantities are more probable with least-squares adjustment than with any other method because the sum of the weights times their corresponding squared residuals is minimized. In addition, least squares enables the assignment of relative weights to the observations based on their expected relative reliabilities.

In this section, least-squares traverse adjustment by the method of observation equations is presented in which an observation equation is written for each measured distance and angle. The basic observation equations for distance and angle measurements are Equations (16-18) and (16-22), respectively.

The apparent inconsistency arising from the fact that distance and angle observations with differing units are combined into one adjustment is resolved through the use of appropriate weights. In this procedure, a relative weight is assigned to each observation in accordance with the inverse of its variance (square of standard deviation). Based on variances, relative weights of distance and angle observations are given by

adjustment. The fundamental condition that is enforced in least-squares adjustment is the minimization of the pv^2 terms. Thus, it follows that if observations are weighted in accordance with the inverse of variance and the same units are used for the residual and standard deviation, the pv^2 terms will all be dimensionless and, therefore, compatible for simultaneous adjustment. The procedure is demonstrated in the example problem given in this section.

For an n-sided closed traverse, there are n distances and $n + 1$ angles, if we consider one angle for orientation. In Figures 16-11a and b, e.g., each closed traverse has four sides, with four distances, and five angles measured. Each new traverse station introduces two unknowns, their X- and Y-coordinates; thus, $2(n - 1)$ unknowns exist for any completely surveyed closed traverse. Therefore, for such a traverse, regardless of its number of sides, the number of redundant equations r (number of observations minus number of unknowns) is equal to $(2n + 1) - 2(n - 1) = 3$. Specifically, for the example traverses of Figure 16-11, $r = \{2(4) + 1\} - 2(4 - 1) = 3$.

Example 16-9.   Adjust by least squares the simple traverse shown in Figure 16-12   from

$$\text{Distance: } P_{L_{ij}} = \frac{1}{(S_{L_{ij}})^2} \qquad (16\text{-}24)$$

$$\text{Angle: } P_{\theta_{jik}} = \frac{1}{(S_{\theta_{jik}})^2} \qquad (16\text{-}25)$$

In Equations (16-24) and (16-25), $P_{L_{ij}}$ and $P_{\theta_{jik}}$ are weights, respectively, for distance and angle observations, and $S_{L_{ij}}$ and $S_{\theta_{jik}}$ are their respective standard deviations. Since standard deviations are generally not available before adjustment, they are usually estimated and referred to as a priori values. They are important because they significantly influence the

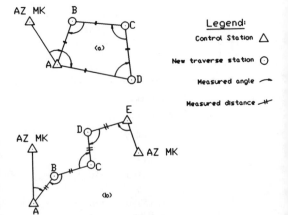

Figure 16-11.   Closed traverse examples.

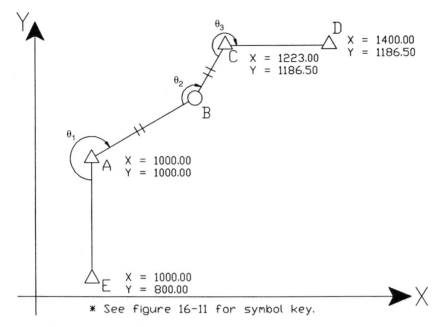

Figure 16-12. Traverse of Example 16-9.

the following measured survey data (coordinates of the control points are shown in the figure):

Observations	A priori Standard Deviations
AB = 200.00	±0.05 ft
BC = 100.00	±0.08 ft
θ_1 = 240°00′	±30 sec
θ_2 = 150°00′	±30 sec
θ_3 = 240°01′	±30 sec

1. Calculate initial approximate coordinates for station B.

$$X_{b_0} = 1000.00 + 200.00 \sin 60° = 1173.20$$

$$Y_{b_0} = 1000.00 + 200.00 \cos 60° = 1100.00$$

2. Formulate the **X** and **K** matrices.

$$\mathbf{X} = \begin{bmatrix} dX_b \\ dY_b \end{bmatrix}, \qquad \mathbf{K} = \begin{bmatrix} k_{L_{ab}} \\ k_{L_{bc}} \\ k_{\theta_1} \\ k_{\theta_2} \\ k_{\theta_3} \end{bmatrix}$$

$$\mathbf{K} = \begin{bmatrix} 200.00 - 200.00 \text{ ft} \\ 100.00 - 99.81 \text{ ft} \\ 120°00'00'' - 120°00'00'' \\ 150°00'00'' - 149°55'51'' \\ 119°59'00'' - 119°55'48'' \end{bmatrix} = \begin{bmatrix} 0.00 \\ 0.19 \\ 0'' \\ 249'' \\ 192'' \end{bmatrix}$$

The values in the **K** matrix are derived by subtracting computed quantities, based on ititial coordinates, from their respective observed quantities.

3. Calculate the **A** matrix. The **A** matrix is formed using prototype Equation (16-18) for the distances and Equation (16-22) for the angles. Note that angle coefficients are multiplied by $\rho(206'264.8''/\text{rad})$ to convert to seconds and be compatible with the values given in the **K** matrix.

$$A = \begin{bmatrix} \left(\dfrac{1173.20 - 1000.00}{200.00}\right) & \left(\dfrac{1100.00 - 1000.00}{200.00}\right) \\[2mm] \left(\dfrac{1173.20 - 1223.00}{100.00}\right) & \left(\dfrac{1100.00 - 1186.50}{100.00}\right) \\[2mm] \left(\dfrac{1000.00 - 1100.00}{200.00^2}\right)\rho & \left(\dfrac{1173.20 - 1000.00}{200.00^2}\right)\rho \\[2mm] \left(\dfrac{1000.00 - 1100.00}{200.00^2} - \dfrac{1186.50 - 1100.00}{100.00^2}\right)\rho & \left(\dfrac{1173.20 - 1000.00}{200.00^2} - \dfrac{1173.20 - 1223.00}{100.00^2}\right)\rho \\[2mm] \left(\dfrac{1100.00 - 1186.50}{100.00^2}\right)\rho & \left(\dfrac{1223.00 - 1173.20}{100.00^2}\right)\rho \end{bmatrix}$$

$$A = \begin{bmatrix} 0.866 & 0.500 \\ -0.498 & -0.865 \\ -515.7 & 893.1 \\ -2299.8 & 1920.3 \\ -1784.2 & 1027.2 \end{bmatrix}$$

4. Formulate the **P** matrix. Based on the given a priori standard deviations and if we use Equations (16-24) and (16-25), the **P** matrix is

$$P = \begin{bmatrix} \dfrac{1}{0.05^2} & & & & & \text{zeros} \\ & \dfrac{1}{0.08^2} & & & & \\ & & \dfrac{1}{30^2} & & & \\ & & & \dfrac{1}{30^2} & & \\ \text{zeros} & & & & \dfrac{1}{30^2} \end{bmatrix}$$

$$= \begin{bmatrix} 400.00 & & & & \text{zeros} \\ & 156.2 & & & \\ & & 0.0011 & & \\ & & & 0.0011 & \\ \text{zeros} & & & & 0.0011 \end{bmatrix}$$

5. Solving Equation (16-9) using these matrices yields the following values for the corrections to the initial coordinates (a second iteration, not shown, produced zeros for dX_b and dY_b):

$$dX_b = -0.11 \text{ ft}$$

$$dY_b = -0.01 \text{ ft}$$

The residuals, computed by Equation (16-10), are

$$v_{ab} = -0.10 \text{ ft}$$

$$v_{bc} = -0.12 \text{ ft}$$

$$v_{\theta_1} = 47.0 \text{ sec}$$

$$v_{\theta_2} = -17.4 \text{ sec}$$

$$v_{\theta_3} = -6.4 \text{ sec}$$

By Equations (16-11), (16-12), and (16-13), the standard deviation of unit weight and standard deviations of the adjusted coordinates are

$$S_0 = \pm 1.74$$

$$S_{X_b} = \pm 0.04 \text{ ft}$$

$$S_{Y_b} = \pm 0.05 \text{ ft}$$

6. The adjusted coordinates are

$$X_b = 1173.20 - 0.11 = 1173.09$$

$$Y_b = 1100.00 - 0.01 = 1099.99$$

7. Finally, the adjusted observations obtained by adding their residuals are

$$AB = 200.00 - 0.10 = 199.90$$

$$BC = 100.00 - 0.12 = 99.88$$

$$\theta_1 = 360° - (120°00'49'') = 239°59'13''$$

$$\theta_2 = 150°00'00'' - 0°00'17'' = 149*59'43''$$

$$\theta_3 = 360° - (119°59'00'' - 0°00'06'')$$

$$= 240°01'06''$$

16-14. CORRELATION AND THE STANDARD ERROR ELLIPSE

As given in Section 16-6, Equation (16-13) can be used to determine the standard deviations in a station's adjusted coordinates. These uncertainities are a reflection of the geometry of the problem and inexactness of the measurements used to determine the station's position. Standard deviations computed by Equation (16-13) are parallel to the X-Y adjustment axis. However, the station's largest positional uncertainties will not generally be aligned with these axes. For example, consider the uncertainties in the position of station U shown in Figure 16-13. Obviously, the uncertainties in this station's position due to the distance inexactness will not be aligned with the **X**- and **Y**-coordinate system, but rather they will vary along the direction of the line itself. For any horizontal adjustment problem, there is a correlation between the station's coordinate uncertainties. This correlation can be determined from elements of the covariance matrix by the equation

$$r_{xy} = \frac{Q_{xy}}{\sqrt{Q_{xx}} \sqrt{Q_{yy}}} \qquad (16\text{-}26)$$

where r_{xy} is the correlation coefficient between two unknown parameters; Q_{xy} the off-diagonal element in the xth row and yth column; Q_{xx} the diagonal element in the xth row and column; and Q_{yy} the diagonal element in the yth row and column. It should be noted that r_{xy} is always between zero and 1. When r_{xy} equals 1, a change in one of the unknowns directly influences the value of the other unknown. Zero indicates no correlation between the two unknowns, and thus changes in one unknown would not affect the other unknown. With Equation (16-26), the x and y coordinates of station C in Example 16-8 have a correlation coefficient of

$$r_{c_x c_y} = \frac{Q_{c_x c_y}}{\sqrt{Q_{c_x c_x}} \sqrt{Q_{c_y c_y}}}$$

$$= \frac{0.0063}{\sqrt{0.0106} \sqrt{0.0073}} = 0.716$$

From the standard deviations of a station's coordinates, the *standard error rectangle* can be drawn as shown in Figure 16-14. This rectangle has half-dimensions of S_x and S_y and encloses the so-called *standard error ellipse*. The semimajor and semiminor axes of this ellipse (the U-V axis) exist in the directions of the largest and smallest uncertainties of the point, respectively. The proper amount of angular rotation required to produce these axes is given by t in the equation

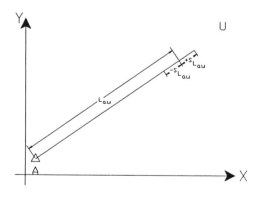

Figure 16-13. Uncertainty in station coordinates due to distance uncertainty.

$$2t = \tan^{-1}\left(\frac{2Q_{xy}}{Q_{yy} - Q_{xx}}\right) \qquad (16\text{-}27)$$

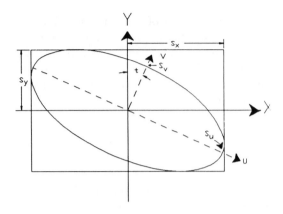

Figure 16-14. The standard error rectangle and ellipse.

where t is the rotation angle from the Y-axis to the semiminor axis of the ellipse, the V-axis. Note that $2t$ must be located in its appropriate quadrant according to the signs of Q_{xy} and $(Q_{yy} - Q_{xx})$ using the sign convention given in Table 16-2 before t is determined.

As noted previously, in this rotated system, the maximum positional uncertainty occurs along the rotated U-axis, and the minimum along the V-axis. The covariance matrix elements for the point rotated into these axes are

$$Q_{uu} = Q_{xx} \sin^2 t + 2Q_{xy} \cos t \sin t$$

$$+ Q_{yy} \cos^2 t \qquad (16\text{-}28)$$

$$Q_{vv} = Q_{xx} \cos^2 t - 2Q_{xy} \sin t \cos t$$

$$+ Q_{yy} \sin^2 t \qquad (16\text{-}29)$$

From these Q_{uu} and Q_{vv} elements, the resulting uncertainties in the U- and V-axis system are

$$S_u = S_0 \sqrt{Q_{uu}} \qquad (16\text{-}30)$$

$$S_v = S_0 \sqrt{Q_{vv}} \qquad (16\text{-}31)$$

where S_u is the maximum positional uncertainty of the station—at a confidence level of 39.4%— and S_v the axis perpendicular to the U-axis.

Equations (16-30) and (16-31) produce the semimajor and semiminor axes, for the *stan-*

Table 16-2. Sign conventions to determine the proper quadrant of $2t$

Sign Q_{xy}	Sign $(Q_{yy} - Q_{xx})$	Quadrant
+	+	I (0–90%)
+	−	II (90–180°)
−	−	III (180–270°)
−	+	IV (270–360°)

dard error ellipse. This ellipse can be modified to produce an error ellipse at any confidence level, or percent probability. The magnification factor is derived from F-statistics. These modifiers allow changing the standard error ellipse to an ellipse of any probability level. They give ratios of variances for varying degrees of freedom, for as the degrees of freedom increase, precision also can be expected to increase. The $F_{(2, \text{degrees of freedom}, \alpha)}$ statistics are shown in Table 16-3. It can be seen that for small degrees of freedom the F-statistics' modifier rapidly decreases and stabilizes for larger degrees of freedom. The confidence level of the error ellipse may be increased to any confidence level by the multiplier $c = \sqrt{2F_{\text{statistic}}}$. The resulting new uncertainties are

$$S_{u_\%} = S_u c = S_u \sqrt{2F_{\text{statistic}}}$$

$$\hspace{2.5cm} (16\text{-}32)$$

$$S_{v_\%} = S_v c = S_v \sqrt{2F_{\text{statistic}}}$$

Table 16-3. $F_{2, \text{degree of freedom}, \alpha}$ statistics for selected certainty levels

Degrees of freedom	90%	95%	99%
3	5.46	9.55	30.82
4	4.32	6.94	18.00
5	3.78	5.79	13.27
6	3.46	5.14	10.92
7	3.26	4.74	9.55
8	3.11	4.46	8.65
9	3.01	4.26	8.02
10	2.92	4.10	7.56
15	2.70	3.68	6.36
20	2.59	3.49	5.85
25	2.53	3.39	5.57
30	2.49	3.32	5.39
60	2.39	3.15	4.98

Example 16-10. Compute the standard and 95% error ellipses for the station C of Example 16-8.

1. Recall from Example 16-8 that $S_0 = 5.626$ with four degrees of freedom. Recall also that station C's covariance matrix elements were

$$Q_c = \begin{bmatrix} Q_{xx} & Q_{xy} \\ Q_{xy} & Q_{yy} \end{bmatrix} = \begin{bmatrix} 0.0106 & 0.0063 \\ 0.0063 & 0.0073 \end{bmatrix}$$

2. From Equation (16-27), two times the rotation angle t is

$$2t = \tan^{-1}\left[\frac{2(0.0063)}{0.0073 - 0.0106}\right] = -75°20'$$

Since $2t$ is negative with a positive numerator and negative denominator, it lies in quadrant II. Therefore, 180° must be added to $2t$ to obtain the proper value for t. Thus, $t = \frac{1}{2}(180° - 75°20') = 52°20'$.

3. From Equations (16-28) and (16-29), the station's rotated covariance matrix elements are

$$Q_{uu} = 0.0106 \sin^2(t)$$
$$+ 2(0.0063) \cos(t) \sin(t)$$
$$+ 0.0073 \cos^2(t) = 0.015$$
$$Q_{vv} = 0.0106 \cos^2(t)$$
$$- 2(0.0063) \sin(t) \cos(t)$$
$$+ 0.0073 \sin^2(t) = 0.002$$

4. From Equations (16-30) and (16-31), the semimajor and semiminor axes for the standard error ellipse are

$$S_u = S_0\sqrt{Q_{uu}} = 5.606\sqrt{0.015} = \pm 0.70 \text{ ft}$$
$$S_v = S_0\sqrt{Q_{vv}} = 5.606\sqrt{0.002} = \pm 0.29 \text{ ft}$$

5. If we use Equations (16-32) and Table 16-3, at a 95% level of confidence, the semimajor and semiminor axes are

$$S_{u_{95\%}} = \pm 0.70\sqrt{2 \times 6.94} = \pm 2.59 \text{ ft}$$
$$S_{v_{95\%}} = \pm 0.29\sqrt{2 \times 6.94} = \pm 1.08 \text{ ft}$$

16-15. SUMMARY

In this chapter, only a brief treatment of the subject of survey adjustments by least squares was presented, and emphasis has been placed on methods of adjusting the most commonly employed types of traditional ground surveys. These include level nets, trilateration, triangulation, and traverses. Example problems of each were given to clarify computational procedures. Many other more advanced topics in least squares are useful and important, but space limitations do not allow their discussion here. References cited pursue the subject in greater depth and should be consulted by those interested in further study.

REFERENCES

HIRVONEN, R. A. 1965. *Adjustment by Least Squares in Geodesy and Photogrammetry.* New York: Frederick Ungar Publishing Company.

LEICK, A. 1990. *GPS Satellite Surveying.* New York: John Wiley & Sons.

MIKHAIL, E. M. 1976. *Observations and Least Squares.* New York: Dun-Donnelly.

MIKHAIL, E. M., and G. GRACIE. 1981. *Analysis and Adjustment of Survey Measurements.* New York: Van Nostrand Reinhold.

RAINSFORD, H. F. 1957. *Survey Adjustments and Least Squares.* London: Constable.

WOLF, P. R. 1980. *Adjustment Computations: Practical Least Squares for Surveyors,* 2nd ed. Rancho Cordova, CA: Landmark Enterprises.

17

Field Astronomy for Azimuth Determinations

Richard L. Elgin, David R. Knowles, and Joseph H. Senne

17-1. INTRODUCTION

Surveying has been defined as the science of determining positions of points on the earth's surface. The four components of surveying measurements are: (1) vertical (elevations), (2) horizontal (distances), (3) relative direction (angles), and (4) absolute direction (azimuths). Due to recent developments in technology, the accuracy and efficiency of measuring these first three components have increased dramatically. This has resulted in accurate determination of the size and shape of figures. Unfortunately, determination of the orientation of figures, the fourth component, has not kept pace, even though inexpensive technology and equipment exist, such as precise timepieces, portable time signal receivers, ephemerides, programmable calculators, and computers. The purpose of this chapter is to provide sufficient theory, calculations, and field procedures so surveys in both the northern and southern hemispheres can be accurately oriented without significant increases in time and expense, in both the northern and southern hemispheres.

17-2. CELESTIAL SPHERE AND DEFINITIONS

To better visualize positions and movements of the sun, stars, and celestial coordinate circles, they are projected onto a sphere of infinite radius surrounding the earth. This sphere conforms to all the various motions of the earth as the earth rotates on its axis and revolves around the sun. Figure 17-1 illustrates the celestial sphere and principal circles necessary to understand celestial observations and calculations. Definitions of important points and circles on the sphere follow:

Great circle. A circle described on the sphere's surface by a plane that includes the sphere's center.

Zenith (Zn). The point directly overhead or where an observer's vertical line pierces the celestial sphere. Opposite zenith is the nadir.

Equator. A great circle on the celestial sphere defined by a plane that is perpendicular to the poles.

Horizon. A great circle on the celestial sphere defined by a plane that is perpendicular to an observer's vertical.

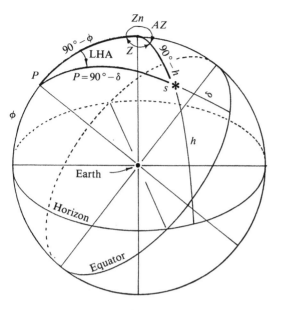

Figure 17-1. Celestial sphere.

Hour circle. A great circle that includes the poles of the celestial sphere. It is analogous to a line of longitude and perpendicular to the equator.

Vertical circle. A great circle that includes the zenith and nadir. It is perpendicular to the horizon.

Meridian. The hour circle that includes an observer's zenith. it represents north-south at an observer's location.

Altitude (h). The angle measured from the horizon upward along a vertical circle. It is the vertical angle to a celestial body.

Declination (δ). The angular distance measured along an hour circle north (positive) or south (negative) from the equator to a celestial body. It is analogous to latitude.

Prime meridian. Reference line (zero degrees longitude) from which longitude is measured. It passes through the Royal Naval Observatory in Greenwich, England; hence, it is also known as Greenwich meridian.

Longitude (λ). Angle measured at the pole, east or west from the prime meridian. Varies from zero degrees to 180°E and 180°W.

Latitude (φ). Angle measured along a meridian north (positive) and south (negative) from the equator. It varies from zero to 90°.

Greenwich hour angle (GHA). Angle at the pole measured westward from the prime (Greenwich) meridian to the hour circle through a celestial body. It is measured in a plane parallel to the equator and varies from zero to 360°. The GHA of a celestial body is always increasing—moving westward—with time.

Local hour angle (LHA). Angle at the pole measured westward from an observer's meridian to the hour circle through a celestial body. It differs from GHA by the observer's longitude.

Meridian angle (t). Equivalent to LHA, except it is measured either eastward or westward and is always less than 180°.

Astronomical triangle (PZS triangle) Spherical triangle formed by the three points: (1) celestial pole P, (2) observer's zenith Zn, and (3) celestial body S.

Polar distance (PS or p). Angular distance from the celestial pole P to a celestial body S. Known as codeclination: $p = 90° - \delta$.

Zenith distance (ZS or z). Angular distance along a vertical circle from an observer's zenith Zn to a celestial body S. Known as coaltitude: $z = 90° - h$.

Slide PZ. Angular distance from the pole P to an observer's zenith Zn. Known as coaltitude: $PZ = 90° - \phi$.

Angle Z. Angle measured at the zenith Zn, in a plane parallel with the observer's horizon, from the pole to a celestial body. Angle Z is denoted as AZ if it is measured as an azimuth, clockwise from zero to 360°.

Angle S. Angle at a celestial body between the pole P and observer's zenith Zn. Known as the parallactic angle.

Astronomical refraction. As light from a celestial body penetrates the earth's atmosphere, direction of the light ray is bent. Astronomical refraction is the angular difference between the direction of a light ray when it enters the atmosphere and its direction at the point of observation. Refraction causes celestial objects to appear higher than they actually are. Refraction corrections are required in some observation methods (altitude method).

Parallax. Apparent displacement of a point with respect to the reference system, caused by a shift in observation location. Celestial observations are considered to be made at the earth's center instead of on the earth's surface (a distance of approximately 3963 mi, or 6378 kn). Parallax corrections are required in some sun observation methods (altitude method).

Mean solar time. Uniform time based on a mean or fictional sun position. The mean sun is a point that moves at a uniform rate around the earth, making one revolution in exactly 24 hr.

Apparent sun time. Nonuniform time based on the apparent sun's position. Because the earth's orbit is eccentric and inclined to the equator, the apparent sun does not cross the observer's meridian exactly every 24 hr throughout the year.

Equation of time. Difference between apparent time and mean time.

Coordinated universal time (UTC). Uniform time based on mean-time at Greenwich. Since the earth's rotation is gradually slowing, approximately one leap second is added per year. UTC is broadcast by radio station WWV.

UT1 time. Mean universal time at the prime meridian obtained directly from the stars. It contains all the irregular motions of the earth and is corrected for polar wandering. UT1 is the time required for celestial observations.

DUT correction. Difference between UT1 time and UTC time.

17-3. AZIMUTH OF A LINE

Azimuth is defined as an angle measured clockwise from a reference meridian (north-south direction) to a line. Several types of meridians exist: astronomic, geodetic, grid, magnetic, record, and assumed. For field measurements, the most convenient, accurate, and retraceable reference is astronomic north. Once obtained, astronomic north can be converted to geodetic and grid north. Astronomic north is based on the direction of gravity (vertical) and axis of rotation of the earth. Geodetic north is based on a mathematical approximation of the earth's shape. The difference between astronomic and geodetic north is the LaPlace correction.

The term true north has frequently been used to indicate either astronomic or geodetic north. In central and eastern portions of the United States, LaPlace corrections tend to be small. Corrections in western states, however, may approach 20 arc-sec. Given present angle-measuring accuracies, LaPlace corrections may be significant, and use of the term true north should be discontinued. A direction determined from celestial observations results in an astronomic north reference meridian.

The azimuth of a line can be determined by measuring an angle from the line to a reference of known azimuth and computed from the following equation (see Figure 17-2).

$$AZL = AZ - \text{ang rt} \qquad (17\text{-}1)$$

This is a general equation. If *AZL* computes to be negative, 360° is added to normalize the azimuth.

For astronomic observations, a celestial body becomes the reference direction. If we know geographic location (latitude and longitude), ephemeris data, and either time or altitude, the azimuth of a celestial body can be computed. If time is used, the procedure is known as the *hour-angle method.* Likewise, if altitude is measured, the procedure is termed the *altitude method.* The basic difference between them is that the altitude method requires approximate

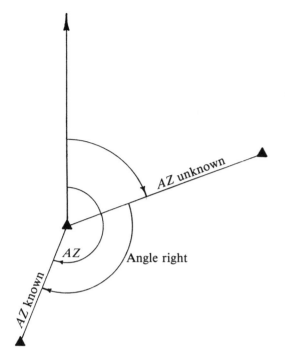

Figure 17-2. Azimuth/angle relationships.

time and an accurate vertical angle corrected for parallax and refraction, whereas the hour-angle method requires accurate time and no vertical angle.

In the past, the altitude method has been more popular for sun observations primarily due to the difficulty of obtaining and maintaining accurate time in the field (time accuracy requirements of the hour-angle method for the sun are greater than for Polaris). Recent developments of time receivers and accurate timepieces, particularly digital watches with split-time features, and time modules for calculators have eliminated this obstacle. The hour-angle method is more accurate, faster, requires shorter training for proficiency, has fewer restrictions on time of day and geographic location, is more versatile, and is applicable to the sun, Polaris, and other stars. Consequently, the hour-angle method is emphasized, and its use by surveyors is encouraged.

17-4. AZIMUTH OF A CELESTIAL BODY BY THE HOUR-ANGLE METHOD

Applying the law of sines to the *PZS* triangle shown in Figure 17-1, we obtain

$$\frac{\sin(90° - h)}{\sin \text{LHA}} = \frac{\sin(90° - \delta)}{\sin Z}$$

$$\sin(90° - h)\sin Z = \sin(90° - \delta)\sin \text{LHA}$$

$$\cos h \sin Z = \cos \delta \sin \text{LHA} \qquad (17\text{-}2)$$

From the five-part formula for spherical triangles,

$$\sin(90° - h)\cos Z = \sin(90° - \phi)\cos(90° - \delta)$$
$$- \cos(90° - \phi)$$
$$\times \sin(90° - \delta)\cos \text{LHA} \quad (17\text{-}3)$$

$$\cos h \cos Z = \cos \phi \sin \delta$$
$$- \sin \phi \cos \delta \cos \text{LHA}$$

Dividing Equation (17-2) by (17-3) yields

$$\frac{\cos h \sin Z}{\cos h \cos Z} = \frac{\sin \text{LHA} \cos \delta}{\cos \phi \sin \delta - \sin \phi \cos \delta \cos \text{LHA}}$$

Then

$$\tan Z = \frac{\sin \text{LHA}}{\cos \phi \tan \delta - \sin \phi \cos \text{LHA}} \quad (17\text{-}4)$$

Since

$$Z = 360° - AZ \quad \text{and} \quad \tan(360° - AZ) = -\tan AZ$$

then

$$\tan AZ = \frac{-\sin \text{LHA}}{\cos \phi \tan \delta - \sin \phi \cos \text{LHA}}$$

or $\qquad\qquad\qquad\qquad\qquad\qquad (17\text{-}5)$

$$AZ = \tan^{-1} \frac{-\sin \text{LHA}}{\cos \phi \tan \delta - \sin \phi \cos \text{LHA}}$$

Solving Equation (17-5) using the arctangent function on a typical calculator will result in a value between $-90°$ and $+90°$ and must be

Table 17-1. Correction to normalize azimuth

When LHA Is	Correction	
	If AZ Is Positive	If AZ Is Negative
0–180°	180°	360°
180–360°	0°	180°

normalized to between 0 and 360° by adding algebraically a correction from Table 17-1.

Rather than arctangent function, if the calculator has rectangular to polar conversion ($R \rightarrow P$), AZ can be directly computed without referring to Table 17-1. If we use an HP calculator, e.g., the numerator and denominator are reduced, placing the numerator value in the Y register and denominator number in the X register (display). Executing $R \rightarrow P$ will yield AZ (decimal degrees) in the Y register. Pressing the XY interchange key ($X \langle \rangle Y$) displays the azimuth. For most calculators, azimuths between 180° and 360° are displayed as minus values. Consequently, for a negative result simply add 360°. If the observation has been made in the southern hemisphere, the sign of ϕ should be negative. In this case, the azimuth is still measured clockwise from north.

Rotation of the celestial sphere increases the local hour angle (LHA) of a celestial body by approximately 15° per hour. (The average sun increases by exactly 15° per hour). Therefore, to calculate the LHA at the instant of observation, accurate time is required. In the United States, coordinated universal time (UTC) is broadcast by the National Bureau of Standards radio station WWV (WWVH in Hawaii) on 2.5, 5, 10, 15, and 20 MHz. Inexpensive receivers pretuned to WWV are available. Also, the signal can be received by calling (303) 499-7111. Calling 1-900-410-TIME gives accurate time, but does not provide DUT corrections. In Canada, Eastern standard time (EST) is broadcast on radio station CHU (3.33, 7.335, and 14.67 MHz). This can be converted to UTC by adding 5 hr to EST. GPS receivers are another source of precise time.

Time based on the actual rotation of the earth (UT1) is obtained by adding a correction

(DUT) to coordinated universal time (UT1 = UTC + DUT). DUT is obtained from WWV (WWVH and CHU) by counting the number of double ticks following any minute tone. These double ticks are not obvious and must be carefully listened for. Each double tick represents one-tenth of a second and is positive for the first 7 sec (ticks). Beginning with the ninth second, each double tick is a negative correction. The total correction, either positive or negative, will not exceed 0.7 sec. Although this DUT correction is very small, it is easy to apply and increases azimuth accuracy.

A stopwatch with a split (or lap) time feature is excellent for obtaining times of pointings. The stopwatch is set by starting on a WWV minute tone and then checked 1 min later with a split time. If a significant difference is observed, restart the stopwatch or apply this difference as a correction. Split times are taken for each pointing on the celestial body and added to the beginning UT1 time (beginning UTC corrected for DUT).

A calculator with a time module, such as the HP-41CX or HP48SX, is ideal for obtaining times. The module can be accurately set to the UTC, and the DUT applied using the $T + X$ function. UT1 will now be displayed by the calculator. Time (UT1) of each pointing on the celestial body can be stored and then recalled for subsequent calculations. To ensure that the timepiece has not gained or lost a significant amount of time, it should be rechecked with WWV after the observations. The key to accurate azimuths by the hourangle method is obtaining accurate time. Surveyors should develop skilled techniques for synchronizing starting time and obtaining split times on the celestial body.

To utilize ephemeris tables, the Greenwich date as well as the time of observation must be known. For observations in the western hemisphere, if UT1 is greater than local time, the Greenwich date is the same as the local date. If UT1 is less than local time, the Greenwich date is the local date plus 1 day. For the eastern hemisphere, if UT1 is less than local time (24-hr basis), Greenwich date is the same as the local date. If UT1 is greater than local time, the Greenwich date is the local date minus 1 day.

Both the observer's latitude and longitude are required for the hour-angle method. Usually, these values are readily obtained by scaling from a map such as a USGS 7.5-min quadrangle sheet. In general, to achieve equivalent azimuth accuracies, latitude and longitude must be more accurately determined for observations on celestial bodies close to the equator—e.g., the sun—than for bodies near the pole—e.g., Polaris.

Declination of celestial bodies—the sun, Polaris, and selected stars—is tabulated in most ephemeris tables for 0 hr UT1 of each day (Greenwich date). Linear interpolation for declination at the UT1 time of observation can be performed using the following equation:

$$\text{Decl} = \text{decl}\,0^h + (\text{decl}\,24^h$$
$$- \text{decl}\,0^h)\left(\frac{\text{UT1}}{24}\right) \qquad (17\text{-}6)$$

where decl 0^h is the declination at 0 hr for the Greenwich date of observation, and decl 24^h the declination at 0 hr for the next Greenwich day.

Linear interpolation of a nonlinear function results in an error. Except for the sun's declination, this error is insignificant. On sun observations, the contribution to total error in azimuth depends on numerous factors and usually is negligible. It can be eliminated, however, by using three-point interpolation or the special two-point nonlinear interpolation equation as follows:

$$\text{Decl} = \text{decl}\,0^h$$
$$+ (\text{decl}\,24^h - \text{Decl}\,0^h)\left(\frac{\text{UT1}}{24}\right)$$
$$+ (0.0000395)(\text{decl}\,0^h)\sin(7.5\,\text{UT1})$$
$$(17.7)$$

where declination is expressed in decimal degrees.

This equation is unique for the sun's declination. It should not be used for the declina-

tion of Polaris or any other star. A negative declination indicates that the celestial body is south of the equator and must be a negative value in Equations (17-5), (17-6), and (17-7).

The LHA at UT1 time of observation is necessary to compute the azimuth of a celestial body. It is defined as an angle measured westward at the north celestial pole, from the observer's meridian, to the celestial body's hour circle. hence, as can be seen from Figure 17-3, an equation for the LHA is

$$\text{LHA} = \text{GHA} - \text{W}\lambda \quad \text{(west longitude)} \quad (17\text{-}8)$$

or

$$\text{LHA} = \text{GHA} + \text{E}\lambda \quad \text{(east longitude)} \quad (17\text{-}9)$$

LHA should be normalized to between $0°$ and $360°$ by adding or subtracting $360°$, if necessary.

Greenwich hour angle (GHA) of celestial bodies—the sun, Polaris, and selected stars—is

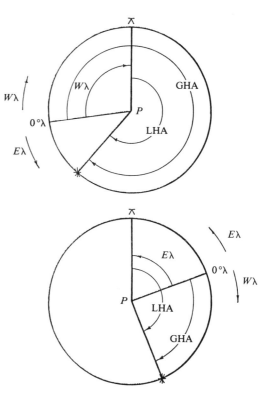

Figure 17-3. Relationship between LHA, GHA, and longitude as viewed from top of celestial sphere.

tabulated in some ephemeris tables for 0 hr UT1 time of each day (Greenwich date). Interpolation for GHA at the time of observation is required. Due to the earth's rotation, the GHA increases by approximately 360° during a 24-hr period, rather than changing by a degree or less as the ephemeris tables appear to indicate. Consequently, interpolation for GHA at UT1 time of observation can be performed using the following equation:

$$\text{HGA} = \text{GHA}\,0^h + (\text{GHA}\,24^h - \text{GHA}\,0^h + 360°)$$
$$\times \left(\frac{\text{UT1}}{24}\right) \qquad (17\text{-}10)$$

where GHA 0^h = GHA at 0 hr for the Greenwich date of observation, and GHA 24^h = GHA at 0 hr for the next day.

Except for the sun, once each year the tabulated GHA will increase past 360° from one day to the next. For observations on this one day only, 720° rather than 360° should be used in Equation (17-10).

Rather than listing the sun's GHA, some ephemeris tables provide the equation of time E. E is usually defined as apparent time minus mean time and can be converted to GHA using the following equation:

$$\text{HGA} = 180° + 15\,E \qquad (17\text{-}11)$$

where E is in decimal hours.

In those cases where E is listed as mean time minus apparent time, the algebraic sign of E should be reversed.

17-5. AZIMUTH OF A CELESTIAL BODY BY THE ALTITUDE METHOD

Applying the law of cosines to the PZS triangle in Figure 17-1 results in the following:

$$\cos(90° - \delta) = \cos(90° - h)\cos(90° - \phi)$$
$$+ \sin(90° - h)$$
$$\times \sin(90° - \phi)\cos Z$$
$$\cos Z = \frac{\sin \delta - \sin h \sin \phi}{\cos h \cos \phi}$$

Since $Z = 360° - AZ$ and $\cos(360° - AZ) = \cos AZ$, then

$$\cos AZ = \frac{\sin \delta - \sin h \sin \phi}{\cos h \cos \phi} \qquad \text{or}$$

$$AZ = \cos^{-1} \frac{\sin \delta - \sin h \sin \phi}{\cos h \cos \phi} \qquad (17\text{-}12)$$

Solving Equation (17-12) using the arccosine function on a typical calculator will result in a value between 0 and 180° and must be normalized to between 0 and 360°. A celestial body will be at equal altitudes with approximately equal angles from the meridian twice during a 12-hr period. Consequently, if the celestial body is east of the meridian (morning observation on the sun), AZ found from Equation (17-12) is the correct azimuth. If the body is west of the meridian (afternoon observation on the sun), AZ must be subtracted from 360° to obtain the correct azimuth. Normalizing the azimuth for celestial bodies other than the sun may be difficult or confusing.

Latitude is normally scaled from a topographic map as for the hour-angle method. Due to inaccuracies resulting from errors associated with altitude, declination at the precise time of observation is not so critical as in the hour-angle method. For sun observations, the zone time announced on radio or television corrected to Greenwich is sufficient when interpolating to obtain declination. For other celestial bodies, using declination at 0 hr without interpolating is usually sufficient.

Altitude h is a theoretical angle measured at the earth's center and assumes that light from a celestial body passes straight through the atmosphere (see Figure 17-4). It is normally determined by measuring a vertical angle near the earth's surface and correcting for parallax and refraction. In general, the earth's radius is insignificant compared with distances to celestial bodies, and parallax is considered to be zero. For the sun, however, parallax is significant and computed using the

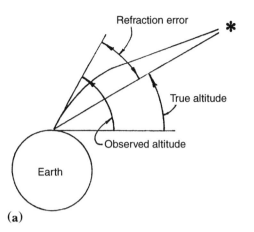

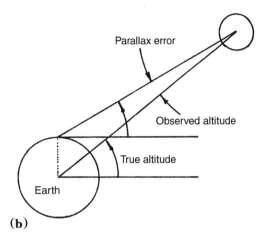

Figure 17-4. (a) Refraction. (b) Parallax.

following equation:

$$C_p = 0.002443 \cos V \qquad (17\text{-}13)$$

where C_p is in decimal degrees, and V the measured vertical angle.

Refraction is computed using the following equation:

$$C_r = \frac{0.273\, b}{(460 + F)\tan V} \qquad (17\text{-}14)$$

where C_r is in decimal degrees, and

b = absolute barometric pressure (not corrected to sea level) in inches of mercury

F = temperature in degrees Fahrenheit

V = measured vertical angle

If b is unknown, it can be calculated from the following:

$$b = \text{inverse}\log\frac{92{,}670 - \text{elevation}}{62{,}737} \qquad (17\text{-}15)$$

where elevation is in feet.

Refraction corrections are large for celestial bodies close to the horizon owing to the amount of atmosphere through which the light must pass. Since pressure and temperature are usually known only at the observation station, errors in refraction may become quite large and have a significant adverse effect on azimuth accuracies.

Applying parallax and refraction corrections to the measured vertical angle results in the following:

$$h = V + C_p - C_r \qquad (17\text{-}16)$$

17-6. SUN OBSERVATIONS (HOUR-ANGLE METHOD)

Sun observations, as compared with those on stars, provide surveyors with a more convenient and economical method of determining an accurate astronomic azimuth. A sun observation can be readily incorporated into a regular work schedule; requires little additional field time; and with reasonable care and proper equipment, an accuracy to within 6 arc-sec can be obtained. In the hour-angle method, a horizontal angle from a line to the sun is measured. Knowing accurate time of the observation and geographic position, we can compute the sun's azimuth. This azimuth and horizontal angle are combined to yield the line's azimuth.

Horizontal angles from a line to the sun are obtained from direct and reverse (face left and face right or face I and face II) pointings taken on the backsight mark and sun. It is suggested that repeating theodolites be used as directional instruments, with one of two general measuring procedures being followed: (1) a single foresight pointing on the sun for each pointing on the backsight mark or (2) multiple foresight pointings on the sun for each pointing on the backsight mark.

For the single foresight procedure, the sighting sequence is (1) direct on mark, (2) direct on sun, (3) reverse on sun, and (4) reverse on the mark—with times being recorded for each sun pointing. The two times and four horizontal circle readings constitute one data set. An observation consists of one or more sets, and a minimum of three sets is recommended. This procedure is similar to measuring an angle at a traverse station using a directional theodolite.

For the multiple foresight procedure, the sighting sequence is (1) direct on mark, (2) several direct on the sun, (3) an equal number of reverse on sun, and (4) reverse on mark—with times being recorded for each pointing on the sun. A minimum of 6 pointings (3D and 3R) on the sun is recommended. The multiple times, multiple horizontal circle readings on the sun, and two horizontal circle readings on the backsight mark constitute one observation.

In general, which measuring procedure to use is a matter of preference. The single foresight procedure is based on an assumption that sun pointings are of approximately equal precision as backsighting pointings. In turn, the multiple foresight procedure is based on an assumption that sun pointings are less precise than backsight pointings. Errors in accurately setting or synchronizing the timepiece have the same effect on both methods. The single foresight system lends itself to proper procedure for incrementing horizontal circle and micrometer settings on the backsight. The multiple foresight method permits a greater number of pointings on the sun in a shorter time span.

Since a large difference usually exists between vertical angles to the backsight mark and sun, it is imperative that an equal number of both direct and reverse pointings be taken. When reducing data to compute horizontal angles, direct readings on the backsight mark should always be subtracted from direct foresight readings on the sun and likewise for reverse readings. Add 360° if the resulting angle is negative.

Vertical angles to the sun are usually larger than those in typical surveying work, thereby increasing the importance of accurately leveling instruments. For a vertical angle of 5°, a leveling error of 10 arc-sec perpendicular to the direction pointed will result in smaller than a 1-arc-sec error in the horizontal circle reading. For a vertical angle of 45°, however, this error would be 10 arc-sec. Because of this and other errors, it is recommended that observations not be made when the sun's altitude is greater than 45°.

The sun cannot be observed directly through the telescope without using either an eyepiece or objective lens filter. In lieu of a filter, the sun's image and cross hair can be projected onto a white surface held approximately 1 ft behind the eyepiece. Both eyepiece and telescope focus must be adjusted to obtain a sharp image. Usually, only that portion of the cross-hair system situated within the sun's image is clearly visible. Observations with a filter are more convenient and slightly improve pointing accuracies. For total stations, an objective lens filter is mandatory to protect EDMI components. For the same reason, a telescope-mounted EDMI should be removed or covered with a lens cap before making observations.

The sun's image is large in diameter—approximately 32 min of arc—making accurate pointings on the center impractical. In lieu of pointing the center, both direct and reverse pointings may be taken on only one edge of the sun—usually the trailing edge (see Figure 17-5). The sun's trailing edge is pointed by allowing it to move onto the vertical cross hair. The leading edge is pointed by moving the vertical cross hair forward, until it becomes tangent to the sun's image. A correction to the sun's center is calculated from semidiameter and altitude. It is applied to the measured horizontal angle from a backsight mark to the

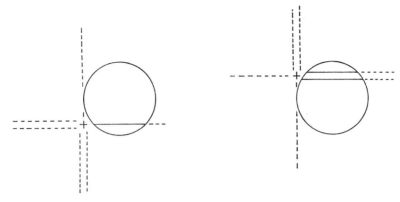

Figure 17-5.   Pointing the sun.

sun's edge. This correction *dH* is computed using Equations (17-17) and (17-18). The semidiameter of the sun is tabulated in most ephemeris tables.

$$h = \sin^{-1}(\sin \phi \sin \delta$$
$$+ \cos \phi \cos \delta \cos \text{LHA}) \qquad (17\text{-}17)$$
$$dH = (\text{sun's semidiameter})/\cos h \quad (17\text{-}18)$$

When pointing the left edge (left when facing the sun), *dH* is added to an angle right. When pointing the right edge, subtract. The left edge is always the trailing edge at latitudes greater than 23.5° N and the leading edge at latitudes greater than 23.5° S. An azimuth of the line *AZL* should be computed for each pointings on the sun.

After azimuths of the line have been computed, they are compared and, if found within acceptable limits, averaged. Azimuths computed with telescope direct and telescope reversed should be compared independently. Systematic instrument errors and use of an objective lens filter can cause a significant difference between the two. An equal number of direct and reverse azimuths should be averaged.

An alternate calculation procedure averages times and angles, or points on opposite edges of the sun and averages to eliminate semidiameter corrections. Due to the sun traveling on an apparent curved path and its changing semidiameter correction with altitude, this procedure usually introduces a sig-

nificant error in azimuth. Also, it does not provide a good check on the final azimuth. Averaging times and angles reduces the number of calculations; however, since modern calculators and software are readily available, accuracy should not be sacrificed by shortcutting calculations.

Errors in determining a line's azimuth can be divided into two categories: (1) measuring horizontal angles from a line to the sun and (2) errors in determining the sun's azimuth. Except when pointing on the sun, errors in horizontal angles are similar to any other field angle. The width of a theodolite cross hair is approximately 2 or 3 arc-sec. With practice, the sun's edge, particularly the trailing one, can be pointed to within this width. In many instances, pointing the sun may introduce a smaller error than pointing the backsight mark.

Total error in the sun's azimuth is a function of errors in obtaining UT1 time, and scaling latitude and longitude. The magnitude these errors contribute to total error is a function of the observer's latitude, declination of the sun, and time from local noon. The relationship of these parameters for selected latitudes and declinations is shown in Table 17-2, which illustrates the importance of input data accuracies at different times of day and year. It should be noted that 10 arc-sec of latitude are equivalent to approximately 1000 ft (300 m) on the earth's surface. Ten arc-sec of longitude are equivalent to approximately 880 ft (270 m) at 30° latitude and 500 ft (150 m) at 60° latitude.

As an example, for an early morning observation (time of 4 hr from local noon), during late fall (declination of $-23°$), at latitude of 30° N, assume that UT1 time is accurate within 0.3 sec, and scaling of latitude and longitude within 5″. From Table 17-2, the error due to time is 3″, those due to latitude and longitude are 1″ and 3″, respectively, and the total error is smaller than 5″ ($\sqrt{3^2 + 1^2 + 3^2}$). For this same example, if the observation was taken close to local noon, time and scaling longitude would be more critical.

Since errors in scaling latitude and longitude are constant for all data sets of an observation, each computed azimuth of the sun contains a constant error. Errors in time affect azimuth in a similar manner. Consequently, increasing the number of data sets does not appreciably reduce the sun's azimuth-error. The increase can, however, improve horizontal angle accuracy and therefore have a desirable effect.

17-7. EXAMPLE SUN OBSERVATION (HOUR-ANGLE METHOD)

Calculations are shown for the example field notes in Figure 17-6. An azimuth is computed for each pointing; however, in order to elimi-

Table 17-2. Azimuth errors related to time, latitude, longitude, and declination (Sun)

Latitude = 30° N		Declination = $+23°$		
Data Error		Time from Local Noon		
	0^h	2^h	6^h	6^h
1^s of Time	1′53″	11″	6″	6″
10″ of Latitude	0″	19″	7″	2″
10″ of Longitude	1′15″	7″	4″	4″

Latitude = 30° N		Declination = $-23°$		
Data Error		Time from Local Noon		
	0^h	2^h	4^h	6^h
1^s of Time	17″	14″	9″	7″
10″ of Latitude	0″	3″	2″	2″
10″ of Longitude	12″	9″	6″	4″

Latitude = 60° N		Declination = $+23°$		
Data Error		Time from Local Noon		
	0^h	2^h	4^h	6^h
1^s of Time	23″	19″	14″	12″
10″ of Latitude	0″	7″	7″	4″
10″ of Longitude	15″	13″	10″	8″

Latitude = 60° N		Declination = $-23°$	
Data Error		Time from Local Noon	
	0^h	2^h	4^h
1^s of Time	14″	13″	13″
10″ of Latitude	0″	3″	1″
10″ of Longitude	9″	9″	8″

For southern latitudes, similar errors are obtained by reversing the sign of the declination.

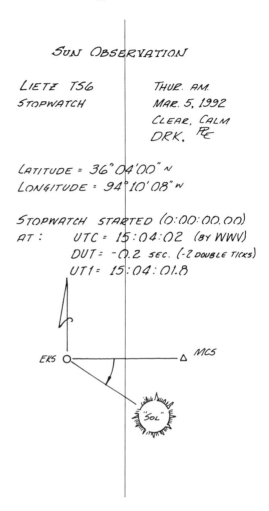

Figure 17-6 notes (field observation, left side):

SUN OBSERVATION

POINT.	TELE.	STOPWATCH TIME	CIRCLE READING
MCS	D		0-00-00
p	D	0:08:34.64	30-12-48
p	D	0:08:52.92	30-16-30
p	D	0:09:20.27	30-22-06
p	R	0:12:39.77	211-02-54
p	R	0:13:06.80	211-09-24
p	R	0:13:20.40	211-11-12
MCS	R		179-59-54

SUN OBSERVATION

LIETZ TS6   THUR. AM.
STOPWATCH   MAR. 5, 1992
                   CLEAR, CALM
                   DRK, FE

LATITUDE = 36°04'00" N
LONGITUDE = 94°10'08" W

STOPWATCH STARTED (0:00:00.00)
AT :   UTC = 15:04:02 (BY WWV)
         DUT = -0.2 SEC. (-2 DOUBLE TICKS)
         UT1 = 15:04:01.8

Figure 17-6. Example of sun observation field notes.

nate repetition, detailed calculation for the fourth pointing only will be demonstrated.

Example 17-1. Greenwich date of observation is March 5, 1992; latitude is 36°04'00", and longitude 94°10'08". From Table 17-3,

$$\text{GHA } 0^h = 177°06'30.1$$
$$\text{GHA } 24^h = 177°08'06''.3$$
$$\text{Decl } 0^h = -6°15'05''.9$$
$$\text{Decl } 24^h = -5°38'56''.7$$
$$\text{Semidiameter} = 0°16'09''.0$$

Fourth pointing (telescope reversed),

$$\text{UT1} = 15^h04^m01^s.8 + 0:12:39.77$$
$$= 15^h16^m4^s.57$$

From Equation (17-10),

$$\text{GHA} = 177°06'30''.1 + (177°09'55''.0$$

$$- 177°06'30''.1 + 360°)\frac{15^h16^m41^s.57}{24^h}$$

$$= 406.31780°$$
$$= 46.31780°$$

From Equation (17-8),

$$\text{LHA} = 46.28788° - 94°10'08''$$
$$= -47.85109°$$
$$= 312.14891°$$

Table 17-3. Ephemeris tables

TABLE A — SOKKIA

SUN, POLARIS

MARCH 1992

GREENWICH HOUR ANGLE FOR THE SUN AND POLARIS FOR 0 HOUR UNIVERSAL TIME

DAY	GHA (SUN)	DECLINATION	EQ. OF TIME APPT-MEAN	SEMI-DIAM.	GHA (POLARIS)	DECLINATION	GREENWICH TRANSIT
	o ' "	o ' "	M S	' "	o ' "	o ' "	H M S
1 SU	176 53 57.7	- 7 34 13.4	-12 24.15	16 10.0	123 20 35.8	89 14 08.10	15 44 03.
2 M	176 56 55.2	- 7 11 21.6	-12 12.32	16 09.7	124 20 06.5	89 14 07.94	15 40 05.
3 TU	176 59 59.9	- 6 48 23.6	-12 00.00	16 09.5	125 19 37.4	89 14 07.76	15 36 08.
4 W	177 03 11.7	- 6 25 19.8	-11 47.22	16 09.2	126 19 08.1	89 14 07.57	15 32 10.
5 TH	177 06 30.1	- 6 02 10.7	-11 33.99	16 09.0	127 18 38.0	89 14 07.35	15 28 13.
6 F	177 09 55.0	- 5 38 56.7	-11 20.33	16 08.7	128 18 06.7	89 14 07.11	15 24 16.
7 SA	177 13 26.2	- 5 15 38.1	-11 06.25	16 08.5	129 17 33.9	89 14 06.86	15 20 19.
8 SU	177 17 03.5	- 4 52 15.4	-10 51.77	16 08.2	130 16 59.2	89 14 06.61	15 16 22.
9 M	177 20 46.4	- 4 28 49.0	-10 36.90	16 08.0	131 16 22.9	89 14 06.36	15 12 25.
10 TU	177 24 34.9	- 4 05 19.3	-10 21.67	16 07.7	132 15 45.2	89 14 06.12	15 08 28.
11 W	177 28 28.7	- 3 41 46.6	-10 06.09	16 07.5	133 15 06.8	89 14 05.90	15 04 31.
12 TH	177 32 27.4	- 3 18 11.4	-09 50.17	16 07.2	134 14 28.4	89 14 05.69	15 00 34.
13 F	177 36 30.8	- 2 54 34.0	-09 33.95	16 07.0	135 13 50.8	89 14 05.50	14 56 37.
14 SA	177 40 38.6	- 2 30 54.8	-09 17.43	16 06.7	136 13 14.4	89 14 05.32	14 52 40.
15 SU	177 44 50.4	- 2 07 14.2	-09 00.64	16 06.5	137 12 39.6	89 14 05.13	14 48 43.
16 M	177 49 06.0	- 1 43 32.6	-08 43.60	16 06.2	138 12 05.8	89 14 04.92	14 44 46.
17 TU	177 53 25.0	- 1 19 50.3	-08 26.33	16 05.9	139 11 32.2	89 14 04.69	14 40 49.
18 W	177 57 47.0	- 0 56 07.6	-08 08.86	16 05.7	140 10 57.9	89 14 04.42	14 36 52.
19 TH	178 02 11.7	- 0 32 24.8	-07 51.22	16 05.4	141 10 21.8	89 14 04.13	14 32 55.
20 F	178 06 38.6	- 0 08 42.3	-07 33.43	16 05.1	142 09 43.7	89 14 03.83	14 28 58.
21 SA	178 11 07.4	0 14 59.6	-07 15.51	16 04.8	143 09 03.4	89 14 03.52	14 25 02.
22 SU	178 15 37.7	0 38 40.6	-06 57.48	16 04.6	144 08 21.7	89 14 03.22	14 21 05.
23 M	178 20 09.2	1 02 20.4	-06 39.38	16 04.3	145 07 39.1	89 14 02.94	14 17 09.
24 TU	178 24 41.6	1 25 58.4	-06 21.23	16 04.0	146 06 56.4	89 14 02.68	14 13 12.
25 W	178 29 14.4	1 49 34.6	-06 03.04	16 03.7	147 06 14.1	89 14 02.42	14 09 16.
26 TH	178 33 47.5	2 13 08.3	-05 44.84	16 03.5	148 05 32.6	89 14 02.18	14 05 19.
27 F	178 38 20.4	2 36 39.4	-05 26.64	16 03.2	149 04 51.9	89 14 01.93	14 01 22.
28 SA	178 42 52.9	3 00 07.3	-05 08.47	16 02.9	150 04 12.0	89 14 01.68	13 57 26.
29 SU	178 47 24.7	3 23 31.8	-04 50.35	16 02.6	151 03 32.6	89 14 01.41	13 53 29.
30 M	178 51 55.6	3 46 52.5	-04 32.29	16 02.3	152 02 53.5	89 14 01.14	13 49 32.
31 TU	178 56 25.2	4 10 09.0	-04 14.32	16 02.0	153 02 14.1	89 14 00.84	13 45 35.

APRIL 1992

GREENWICH HOUR ANGLE FOR THE SUN AND POLARIS FOR 0 HOUR UNIVERSAL TIME

DAY	GHA (SUN)	DECLINATION	EQ. OF TIME APPT-MEAN	SEMI-DIAM.	GHA (POLARIS)	DECLINATION	GREENWICH TRANSIT
	o ' "	o ' "	M S	' "	o ' "	o ' "	H M S
1 W	179 00 53.3	4 33 20.9	-03 56.44	16 01.8	154 01 33.9	89 14 00.53	13 41 39.
2 TH	179 05 19.7	4 56 27.9	-03 38.69	16 01.5	155 00 52.6	89 14 00.19	13 37 42.
3 F	179 09 44.1	5 19 29.7	-03 21.06	16 01.2	156 00 09.5	89 13 59.85	13 33 46.
4 SA	179 14 06.2	5 42 25.8	-03 03.59	16 00.9	156 59 24.6	89 13 59.51	13 29 49.
5 SU	179 18 25.9	6 05 15.8	-02 46.27	16 00.7	157 58 37.7	89 13 59.16	13 25 53.
6 M	179 22 42.9	6 27 59.6	-02 29.14	16 00.4	158 57 49.3	89 13 58.83	13 21 57.
7 TU	179 26 57.1	6 50 36.6	-02 12.20	16 00.1	159 57 00.0	89 13 58.52	13 18 01.
8 W	179 31 08.1	7 13 06.5	-01 55.46	15 59.8	160 56 10.4	89 13 58.23	13 14 05.
9 TH	179 35 15.8	7 35 28.9	-01 38.95	15 59.6	161 55 21.3	89 13 57.96	13 10 09.
10 F	179 39 19.9	7 57 43.6	-01 22.67	15 59.3	162 54 33.3	89 13 57.70	13 06 13.
11 SA	179 43 20.3	8 19 50.1	-01 06.65	15 59.0	163 53 46.6	89 13 57.44	13 02 16.
12 SU	179 47 16.7	8 41 48.2	-00 50.89	15 58.8	164 53 00.9	89 13 57.17	12 58 20.
13 M	179 51 08.7	9 03 37.4	-00 35.42	15 58.5	165 52 15.6	89 13 56.87	12 54 24.
14 TU	179 54 56.2	9 25 17.5	-00 20.25	15 58.2	166 51 29.7	89 13 56.56	12 50 27.
15 W	179 58 38.8	9 46 48.1	-00 05.41	15 58.0	167 50 42.5	89 13 56.21	12 46 31.
16 TH	180 02 16.4	10 08 09.0	00 09.09	15 57.7	168 49 53.2	89 13 55.86	12 42 35.
17 F	180 05 48.4	10 29 19.9	00 23.23	15 57.4	169 49 01.7	89 13 55.50	12 38 39.
18 SA	180 09 14.8	10 50 20.4	00 36.99	15 57.2	170 48 08.5	89 13 55.15	12 34 43.
19 SU	180 12 35.2	11 11 10.2	00 50.35	15 56.9	171 47 13.9	89 13 54.81	12 30 48.
20 M	180 15 49.3	11 31 49.1	01 03.29	15 56.7	172 46 19.0	89 13 54.50	12 26 52.
21 TU	180 18 56.9	11 52 16.7	01 15.80	15 56.4	173 45 24.2	89 13 54.20	12 22 56.
22 W	180 21 57.8	12 12 32.7	01 27.85	15 56.1	174 44 30.0	89 13 53.92	12 19 01.
23 TH	180 24 51.7	12 32 36.7	01 39.45	15 55.9	175 43 36.6	89 13 53.64	12 15 05.
24 F	180 27 38.5	12 52 28.4	01 50.57	15 55.6	176 42 44.1	89 13 53.36	12 11 09.
25 SA	180 30 17.9	13 12 07.5	02 01.20	15 55.3	177 41 52.3	89 13 53.07	12 07 13.
26 SU	180 32 49.9	13 31 33.6	02 11.32	15 55.1	178 41 00.7	89 13 52.78	12 03 17.
27 M	180 35 14.1	13 50 46.4	02 20.94	15 54.8	179 40 09.0	89 13 52.47	11 59 21.
28 TU	180 37 30.6	14 09 45.6	02 30.04	15 54.6	180 39 16.7	89 13 52.14	11 55 25.
29 W	180 39 39.1	14 28 30.9	02 38.61	15 54.3	181 38 23.3	89 13 51.80	11 51 30.
30 TH	180 41 39.6	14 47 01.8	02 46.64	15 54.1	182 37 28.3	89 13 51.45	11 47 34.

From Equation (17-7),

$$
\begin{aligned}
\text{Decl} = & -6°02'10''.7 \\
& + (-5°38'56''.7 + 6°02'10''.7) \\
& \times \frac{15^h 16^m 41^s.57}{24^h} \\
& + (0.0000395)(-6°02'10''.7) \\
& \times \sin(7.5 \times 15^h 16^m 41^s.57) \\
= & -5.78980° - 0.00022° \\
= & -5.79002°
\end{aligned}
$$

From Equation (17-5),

$$
AZ = \tan^{-1} \frac{-\sin(312.14891°)}{\cos(36°04'00'') \tan(-5.79002°) - \sin(36°04'00'') \cos(312.14891°)}
$$

$$
= \tan^{-1} \frac{0.74140323}{-0.47703565}
$$

Using $R \to P$, we obtain $AZ = 122.75815°$, or using $\tan^{-1}$, $AZ = -57.24185°$.

Since LHA is between 180° and 360° and *AZ* negative, the normalize correction from Table 17-1 equals 180°.

$$AZ = -57.24185° + 180°$$
$$= 122.75815°$$

Field R Ang Rt $= 211°02'54'' - 179°59'54°$
$$= 31.05000°$$

From Equation (17-17),

$$h = \sin^{-1}[\sin(36°04'00'')\ln(-5.79002°)$$
$$+ \cos(36°04'00'')\cos(-5.79002°)$$
$$\times \cos(312.14891°)]$$
$$= 28.7037°$$

From Equation (17-18),

$$dH = \frac{0°16'09''.0}{\cos(28.7037°)}$$
$$= 0.30688°$$

The left edge is pointed D & R; therefore, *dH* is positive.

Cor R Ang Rt $= 31.05000° + 0.30688°$
$$= 31.35688°$$

From Equation (17-1),

$$AZL = 122.75815° - 31.35688°$$
$$= 91.40127°$$
$$= 91°24'04''.6$$

Using the same calculation procedure for remaining pointings yields the following:

Direct		Reverse	
91°24'03".6	(1)	91°24'04".6	(4)
91°24'04".8	(2)	91°23'09".6	(5)
91°24'03".1	(3)	91°24'10".3	(6)

A comparison of direct and reverse azimuths indicates that the fifth value contains excessive error. Throw out

this azimuth and a direct azimuth (third) before averaging. (Average = 91°24'06")

17-8. SUN OBSERVATIONS (ALTITUDE METHOD)

Outdated and less accurate (approximately 1 min of arc, depending on numerous factors), the altitude method should only be considered when accurate time cannot be determined. Simultaneous vertical and horizontal angles to the sun's center are required. Therefore, a special sighting accessory must be used or both edges of the sun pointed simultaneously (quadrant method, see Figure 17-7). Due to the suns's large diameter, both edges cannot be accurately observed simultaneously using a filter. Since total stations require an objective lens filter, this essentially eliminates their use for this method.

The altitude method requires very accurate vertical angles that must be corrected for parallax and refraction. This is particularly critical when the sun is close to local noon because of rapid changes in azimuth, with little or no change in altitude. Therefore, observations should not be made within 2 to 3 hr of local noon. Refraction corrections are large and more difficult to accurately determine when the sun is close to the horizon. This restricts observations during the first hour or two after sunrise and before sunset.

Due to problems involved in obtaining horizontal and vertical pointings at the same instant, and the importance of vertical angle accuracy, a set of data should consist of several foresights on the sun for each backsight. A recommended sighting procedure is (1) direct on backsight mark, (2) three direct on sun, (3) three reverse on sun, and (4) reverse on backsight mark. The three direct and three reverse angles (horizontal and vertical) are averaged, with a single azimuth computed for each set. In order to minimize errors due to curvature of the sun's apparent path, time spans from first direct to last reverse pointings should be kept as short as possible.

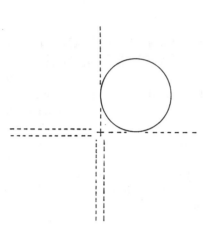

Figure 17-7.   Quadrant method.

17-9.   EXAMPLE SUN CALCULATION (ALTITUDE METHOD)

Example 17-2.

Local date = April 10, 1992
Avg. ang rt to sun's center = 326°47′30″
Avg. zenith angle to sun's center = 65°20′42″
Avg. time = 4:40 PM CST
Elevation = 1400 ft
Temperature = 70°F
Latitude = 36°04′00″
Greenwich date = April 10, 1986
From Table 17-3,

$$Decl\, 0^h = 7°57′43″.6$$

$$Decl\, 24^h = 8°19′50″.1$$

$$UT = 4^h40^m + 12^h + 6^h$$

$$= 22^h40^m$$

From Equation (17-6),

$$Decl = 7°57′43″.6$$

$$+ (8°19′50″.1 - 7°57′43″.6)\frac{22^h40^m}{24^h}$$

$$= 8.3101°$$

Vert ang = 90° − 65°20′42″

$$= 24.6550°$$

From Equation (17-13),

$$C_p = 0.002443 \cos(24.6550°)$$

$$= 0.0022°$$

From Equation (17-15),

$$b = \text{inverse log} \frac{92,670 - 1400}{62,737}$$

$$= 28.50 \text{ in. of Hg}$$

From Equation (17-14),

$$C_r = \frac{0.273(28.50)}{(460 + 70)\tan(24.6550°)}$$

$$= 0.0320°$$

From Equation (17-16),

$$h = 24.6550° + 0.0022° - 0.0320°$$

$$= 24.6252°$$

From Equation (17-12),

$$AZ = \cos^{-1}\frac{\sin(8.3101°) - \sin(24.6252°)\sin(36°04′00″)}{\cos(24.6252°)\cos(36°04′00″)}$$

$$= 97.8830°$$

For an afternoon observation (sun west of meridian),

$$AZ = 360° - 97.8830°$$
$$= 262.1170°$$

From Equation (17-1),

$$AZL = 262.1170° - 326°47'30''$$
$$= -64.6746°$$
$$= 295.3254°$$
$$= 295°20'$$

17-10. POLARIS OBSERVATIONS (HOUR-ANGLE METHOD)

In most land surveying situations, determination of the astronomic azimuth by sun observations is satisfactory. However, for direction accuracy requirements of approximately 6 arcsec or fewer, a star observation is required. At middle latitudes of the northern hemisphere, Polaris is preferred. It moves very slowly as seen by an observer on earth and is easily located. At near-pole and near-equator latitudes, a star other than Polaris should be selected. If close to the equator, Polaris may not be visible, and horizontal refraction can be a problem. When near the pole, time and leveling become very critical in azimuth determinations.

Several observation methods and calculation procedures can be applied to determine astronomic azimuth from Polaris. For all practical purposes, however, the hour-angle method is the only one that should be considered. Figure 17-8 depicts the apparent motion of Polaris to an observer on earth. The relationship between the north celestial pole, azimuth of Polaris, and horizontal angle from line *AB* to Polaris is shown.

Four important positions of Polaris during its daily rotation around the pole are (1) upper culmination (UC), (2) western elongation (WE), (3) lower culmination (LC), and (4) eastern elongation (EE). If an observer sights Po-laris exactly at upper or lower culmination, its azimuth is zero, which of course simplifies computing the line's azimuth. This method of observation is not practical in that culmination occurs for only an instant, and movement is most rapid in either an east or west direction. Consequently, accurate direct and reverse pointings on Polaris are not possible. Likewise, observation procedures and computations can be simplified by making observations at either eastern or western elongation. These also have distinct disadvantages, since elongation normally occurs only once each day during hours of darkness and possibly at an inconvenient time of night. Instead of observing Polaris at culmination or elongation, surveyors should be prepared to make observations at any time (hour angle) and perform the necessary calculations.

Figure 17-9 can assist in locating Polaris. It is the end star in the handle of constellation Ursa Minor (Little Dipper) located between the constellation Ursa Major (Big Dipper) and Cassiopeia. Two stars of the Big Dipper's cup point to Polaris. Finding Polaris in the instrument's field of view for the first time can be exasperating, but it need not be if approached in a systematic manner. The instrument can be prefocused on some distant object, such as a bright star or the moon. Horizontally, Polaris can be located with the telescope sight. Approximate vertical angles to Polaris can be computed by estimating expected UT1 times of observation and applying the following equation:

$$h = \phi + (90° - \delta)$$
$$\times \cos[\text{GHA } 0^h - W\lambda + (\text{UT1})(15°02')]$$
$$(17\text{-}19)$$

As an example, assume a Polaris observation is to be taken at 7:15 PM CST, December 7, 1992 (local date), at longitude 91°46′ W and latitude 37°57′ N. UT1 time and Greenwich date of observation would be 1^h15^m December

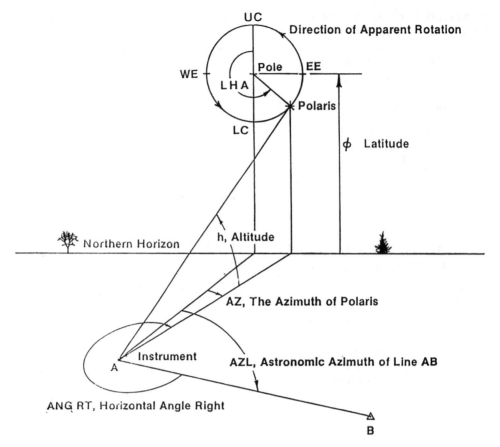

Figure 17-8. Polaris movement and relationship of horizontal angle, azimuth, and local hour angle.

8. GHA 0^h on this date is 40°31′ and declination 89°15′.

$$h = 37°57' + (90° - 89°15')$$
$$\times \cos[40°31' - 91°46' + (1.25^h)(15°02')]$$
$$= 38°35'$$

Converting this approximate vertical angle to zenith angle yields 51°25′. Since the altitude of Polaris is approximately equal to the observer's latitude, precise leveling of the theodolite is extremely critical for mid and high northern latitudes.

Polaris observations made at night present problems not encountered in survey work performed during daylight hours. In particular, illumination is necessary to see the cross hairs against the night sky. Theodolites usually have lighting systems as accessories. If a lighting system is not available, point a light into the hole located at the theodolite's reflecting mirror to illuminate both horizontal circle and cross hairs, or use a flashlight to illuminate the cross hairs by reflecting light at an angle into the objective lens.

Horizontal angles from a line to Polaris are obtained from direct and reverse pointings taken on the backsight mark and star. The single foresight for each backsight procedure (as discussed under sun observations) is recommended with at least three sets being taken. For high-order surveys, horizontal circle and micrometer settings should be incremented between sets.

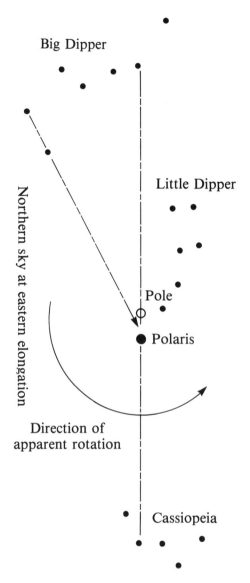

Northern sky at upper culmination

Big Dipper

Little Dipper

Pole

Polaris

Cassiopeia

Northern sky at eastern elongation

Northern sky at western elongation

Direction of apparent rotation

Northern sky at lower culmination

Figure 17-9. Chart to locate Polaris.

for hand computations, averaging angles and times for each set is normally sufficient. Azimuths for each set should be compared and, if found within acceptable limits, averaged. A slightly more accurate procedure is computing an azimuth for each pointing as recommended for sun observations.

Excluding instrument and personal errors, accuracy of an azimuth determination is affected by errors in obtaining UT1 time of pointing Polaris and scaling latitude and longitude. Determining specifications for time, latitude, and longitude in order to meet a prescribed azimuth accuracy is somewhat complicated since they vary with the observer's latitude and LHA at time of pointing. Table 17-4 lists maximum azimuth errors resulting from inaccuracies in time, latitude, and longitude at several latitudes. In general, for any observations made on Polaris in the continental United States, if pointing times are obtained to 1 sec and latitude and longitude are scaled to 1000 ft (300 m), the resulting azimuth accuracy is within 0.5 arc-sec (if we disregard errors in the horizontal angle from mark to star).

17-11. EXAMPLE POLARIS CALCULATION

Calculations are detailed for the example field notes in Figure 17-10. These notes illustrate standard procedures to increment horizontal circle and micrometer settings for a directional theodolite. An azimuth is computed for

If the time span between direct and reverse pointings on Polaris is kept below 4 min, an azimuth computed from averaged horizontal angles and UT1 times would contain a maximum error of only 0.2 arc-sec at 60° latitude and smaller at lower latitudes. Consequently,

Table 17-4. Maximum azimuth error related to time, latitude, and longitude (polaris)

Data Error	Latitude 20°	Latitude 40°	Latitude 60°
1ˢ of Time	0″.23	0″.28	0″.42
10″ of Latitude	0″.05	0″.15	0″.49
10″ of Longitude	0″.15	0″.18	0″.28

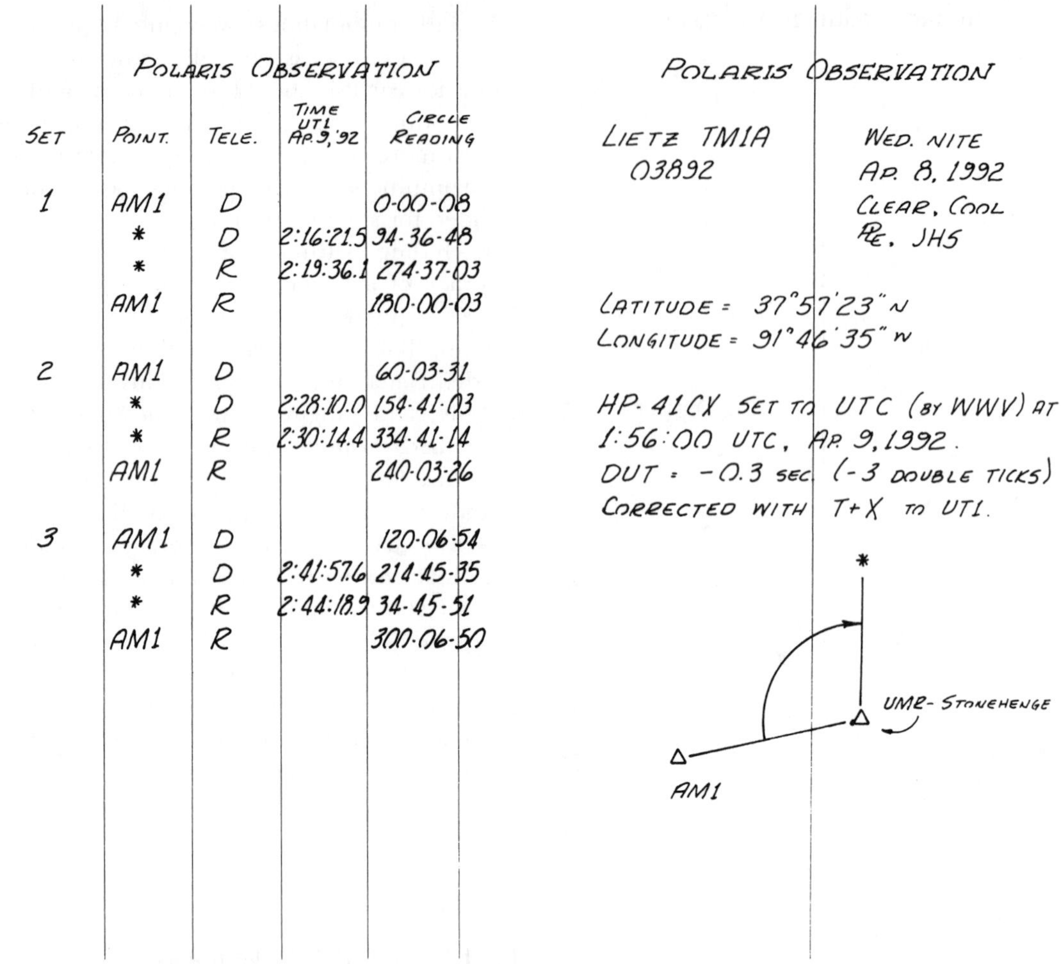

POLARIS OBSERVATION

SET	POINT.	TELE.	TIME UT1 AP. 9, '92	CIRCLE READING
1	AM1	D		0-00-08
	*	D	2:16:21.5	94-36-48
	*	R	2:19:36.1	274-37-03
	AM1	R		180-00-03
2	AM1	D		60-03-31
	*	D	2:28:10.0	154-41-03
	*	R	2:30:14.4	334-41-14
	AM1	R		240-03-26
3	AM1	D		120-06-54
	*	D	2:41:57.6	214-45-35
	*	R	2:44:18.9	34-45-51
	AM1	R		300-06-50

POLARIS OBSERVATION

LIETZ TM1A
03892

WED. NITE
AP. 8, 1992
CLEAR, COOL
RE. JHS

LATITUDE = 37°57'23" N
LONGITUDE = 91°46'35" W

HP. 41CX SET TO UTC (BY WWV) AT
1:56:00 UTC, AP. 9, 1992.
DUT = -0.3 SEC (-3 DOUBLE TICKS)
CORRECTED WITH T+X TO UT1.

UMR- STONEHENGE

AM1

Figure 17-10. Example of Polaris observation field notes.

each pointing; however, in order to eliminate repetition, detailed calculations for the first pointing only will be shown.

Example 17-3. The local date of observation is April 8, 1992. Greenwich date of observation is April 9, 1992.
From Table 17-3,

$$\text{GHA } 0^h = 161°55'21".3$$

$$\text{GHA } 24^h = 162°54'33".3$$

$$\text{Decl } 0^h = 89°13'57".96$$

$$\text{Decl } 24^h = 89°13'57".70$$

First pointing (telescope direct),

$$\text{UT1} = 2^h16^m21^s.5$$

From Equation (17-10),

$$\text{GHA} = 161°55'21".3$$
$$+ (162°54'33".3 - 161°55'21".3 + 360°)$$
$$\times \frac{2^h16^m21^s.5}{24^h}$$
$$= 196.10560°$$

From Equation (17-8),

$$\text{LHA} = 196.10560° - 91°46'35"$$
$$= 104.32921°$$

From Equation (17-6),

$$Decl = 89°13'57''.96$$
$$+ (89°13'57''.70 - 89°13'57''.96)$$
$$\times \frac{2^h16^m21^s.5}{24^h}$$
$$= 89.232760°$$

From Equation (17-5),

$$AZ = \tan^{-1} \frac{-\sin(104.32921°)}{\cos(37°57'23'')\tan(89.232760°) - \sin(37°57'23'')\cos(104.32921°)}$$

$$= \tan^{-1} \frac{-0.96888969}{59.03056203}$$

Using $R \to P$, we obtain

$$AZ = -0.94033°$$
$$= 359.05967°$$

or using $\tan^{-1}$,

$$AZ = -0.94033°.$$

Since LHA is between 0 and 180° and AZ is negative, the normalize correction from Table 17-1 is 360°.

$$AZ = -0.94033° + 360°$$
$$= 359.05967°$$
$$\text{Field D ang rt} = 94°36'48'' - 0°00'08''$$
$$= 94.61111°$$

From Equation (17-1),

$$AZL = 359.05967° - 94.61111°$$
$$= 264.44856°$$
$$= 264°26'54''8$$

Using the same calculation procedure for remaining pointings yields the following:

Set	Direct	Reverse
1	264°26'54".8	264°26'47".9
2	264°26'53".7	264°26'47".5
3	264°26'55".3	264°26'48".5

Avg. AZL = 264°26'51".3.

17-12. STARS OTHER THAN POLARIS

At times, neither Polaris nor the sun is suitable for azimuth observations. Polaris obviously cannot be observed south of the equator and may not produce accurate results at high northern latitudes. In some cases, sun observations may not produce sufficient accuracy for the work desired, or a cloudy day can require a night observation if a time schedule is to be met. Also, Polaris may be cloud covered while the southern sky is clear. There is a fifth magnitude star, Sigma Octantis, within one degree of the south celestial pole that can be observed in the southern hemisphere using the same procedures as Polaris. Of course, both latitude and declination are negative, and the azimuth will, as usual, be measured clockwise from north. Since the star is nearly invisible to the naked eye, accurate star maps or precomputation are necessary for location.

In general, pointings to stars can be made more accurately than those on the sun since the image is precisely defined, atmospheric turbulence is lower, and thermal expansions of the theodolite and tripod are eliminated. In addition, bright stars are visible shortly after sundown, thus permitting observations to be made in twilight.

If we assume that GHA and declination ephemeris data are available for other stars, calculations are identical to those for Polaris. However, their apparent motion is rapid, and for stars close to the equator, the movement is similar to the sun's. Consequently, observations should be avoided when the vertical angle is above 45°—i.e., their directions change rapidly with time, particularly if near the observer's meridian.

Some stars are not visible during the entire night, and their locations change throughout the year. Before attempting any observations, it is necessary to become familiar with a star's position. This involves determining when it will be above the horizon and identifying its location. Many publications, such as Sokkia's *Celestial Observation Handbook* and *Ephem-*

erist, provide visibility charts to aid in this procedure.

17-13. SUMMARY

Astronomic azimuth provides an accurate and efficient means of orienting field surveys. Textbooks and surveying courses have covered celestial observations for many years, but until recently this aspect of surveying measurements has been essentially academic and not employed in surveying practice. This is due in part to emphasis on the altitude method for sun observations and the inherent associated problems. Surveyors are encouraged to take advantage of available technology and use the hour-angle method for all azimuth determinations, whether the sun, Polaris, or selected stars are observed.

Ample software is available to perform all necessary calculations for the hour-angle method. Included are modules for hand-held calculators that not only serve as timepieces, data collectors, and computers, but also generate ephemeris data. As professionals, however, surveyors must have a basic understanding of the underlying theory of celestial observations and be able to test and verify accuracies of the software used.

In the near future, it is hoped that *direction*, the fourth component of surveying measurements, will no longer be a stepchild to the other three. As a result, all field surveys will be accurately oriented to a retraceable reference.

REFERENCES

BUCKNER, R. B. 1984. *A Manual on Astronomic and Grid Azimuth*. Rancho Cordora: Landmark Enterprises.

ELGIN, R. L., D. R. KNOWLES, and J. H. SENNE. *Celestial Observation Handbook and Ephemeris*. 1985–1993 Overland Park: Sokkia/Lietz.

ELGIN, R. L., D. R. KNOWLES, and J. H. SENNE. 1989. *Practical Surveying for Observations*. Canton: P.O.B. Publishing.

ELGIN, R. L., D. R. KNOWLES, and J. H. SENNE. 1986–1989. The tech on celestial observations. *P.O.B. Magazine* 12 1 (Oct.–Nov. 1986 through June–July 1990).

MACKIE, J. B. 1978. *The Elements of Astronomy for Surveyors*, 8th ed. London: Charles Griffin.

MUELLER, I. I. 1969. *Spherical and Practical Astronomy as Applied to Geodesy*. New York: Frederick Ungar Publishing Company.

NASSAU, J. J. 1948. *Practical Astronomy*. New York: McGraw-Hill.

18

Map Projections

Porter W. McDonnell

18-1. INTRODUCTION

The earth is round; maps are flat. If a particular map is to show only a very small portion of the earth, such as a few city blocks, the roundness of the earth is insignificant. On the other hand, if a map is to show the western hemisphere, the roundness presents a major problem—i.e., some kind of deformation will be necessary. To illustrate, a large section of orange peel can only be flattened if it is stretched and torn.

A map depicting only a small area is often called a *plan*. A map showing a large portion (or all) of the earth, where curvature of the surface becomes a factor, is called a *map* or *chart*, the latter term being used for a map designed for navigational purposes. Preparation of a plan involves a simple rectangular grid, whereas a map or chart commonly requires the selection of a suitable map projection to deal with the earth's shape.

Although the roundness of the earth is not a factor in drawing a plan—a map of limited area—the topic of map projections is nevertheless important to land surveyors. Increasingly, land surveyors are making use of plane coordinate systems (Chapter 19) that extend over hundreds of kilometers (or miles) even though the job at hand is small. When a plan shows a limited land area and is drawn as if the earth were flat, the data shown may be so precise that a knowledge of map projections is needed in the survey computations.

For the purposes of this chapter, and for small-scale mapping generally, the earth can be considered a sphere with a radius of 6370 km (or about 3960 mi). Actually, the dimension is greater across the equator than from top to bottom (pole to pole).

For large-scale mapping and in geodetic surveying, the earth's true shape has to be considered (see Chapter 19). In these cases, the earth is assumed to be a spheroid instead of a sphere. Figure 18-1 is a flowchart in which the choice of a datum (sphere or spheroid) is shown as the first step in evolving a map projection.

This chapter is based on the author's textbook, *Introduction to Map Projections* (New York: Marcel Dekker, 1979), with permission from the publisher. (Second edition available from Landmark Enterprises, Rancho Cordova, CA, © 1992.)

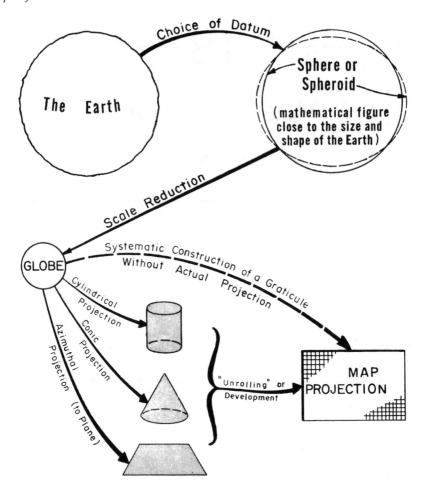

Figure 18-1.   Evolution of a map projection. In some cases, a geometric projection to a developable surface is involved, but usually the term *cylindrical, conic,* or *azimuthal* is used to classify a projection that only resembles such a case. The dashed arrow shows this possibility. (From P. W. McDonnell, Jr., 1979, New York: Marcel Dekker, with permission.)

18-2.   PROJECTION

If all points within some large portion of the earth—e.g., the western hemisphere—are to be represented on a flat map, *two* transformations of the sphere or spheroid are necessary. First, there must be a scale reduction to make sure a huge area fit into the limits of a sheet. Second, there needs to be a systematic way of deforming the rounded surface of the sphere or spheroid to make it flat (see Figure 18-1).

It is very useful to think of these operations as always being done in two steps, in the order mentioned. First, the full-sized sphere is greatly reduced to an exact model called a *globe*. Second, a map projection is generated to convert all or part of the globe into a flat map. There are an infinite number of ways, literally, of accomplishing this second step.

Certain reference lines and points have been established on the earth. The equator and two poles are known to all (see Figure 18-2). Lines running north and south, from pole to pole, are *meridians*. One of them, passing through Greenwich, England, has been chosen arbitrarily to be the *prime meridian*. It is assigned an angular value of 0°. Each other meridian is identified by its angular distance

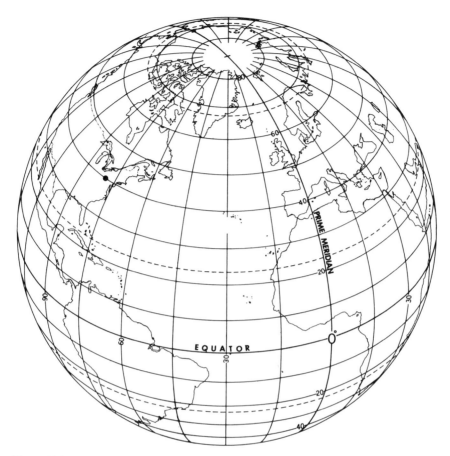

Figure 18-2. Network of meridians and parallels. Pittsburgh is located at $\phi = 40°$ N and $\lambda = 80°$ W.

east or west of the prime meridian. The meridian through Pittsburgh, e.g., is 80° west of Greenwich and said to have a *longitude* of 80° W. (The line itself is a meridian, and its spherical coordinate the longitude). The Greek letter lambda (λ) is used for longitude. Lines crossing all meridians at right angles and running parallel to the equator are called *parallels*. Each parallel is identified by its angular distance north or south of the equator, known as its *latitude*. The parallel passing through Madrid, Pittsburgh, and Peking is at 40° N and is often called the 40th parallel if there is no chance of confusing it with the parallel at 40° S. The Greek letter phi (ϕ) is used for latitude.

A network of meridians and parallels is called a *graticule*. When the sphere or spheroid is reduced in size, as shown in Figure 18-1, the graticule becomes, of course, a reference network for all points on the *globe* just as it is on the earth itself.

The map being produced is two-dimensional. Points on the map sheet have x and y positions based on some rectangular system of reference. The map projection process is the systematic transformation of all spherical coordinates ϕ and λ of the globe into corresponding rectangular coordinates x and y on the map. Mathematically, $x = f_1(\phi, \lambda)$ and $y = f_2(\phi, \lambda)$, meaning that x and y positions on the map are functions of ϕ and λ. These functions must be: (1) unique, so a particular point will appear at only one position on the map; (2) finite, so a particular point will not appear at infinity and be unplottable; and (3) continuous, so although stretching or shrinking of features may occur, there will be no gaps. Projections do exist in which the functions are not finite for the entire globe.

In some cases, the *x* and *y* positions may be obtained by imagining an intermediate step involving a cylinder, cone, or plane as shown in Figures 18-1 and 18-3. To illustrate this type of projection, imagine a ray of light projected radially from the globe's center to a tangent surface (Figure 18-3). A point on the globe having a certain φ and λ can be transferred to the surrounding surface, which then is "unrolled" or *developed* to form a plane map. More commonly, the functions used to get *x* and *y* positions are purely mathematical and do not involve a developable surface. Many of these mathematical concoctions bear some resemblance to the geometrically projected cases shown in Figure 18-3. The cylinder, cone, and plane thus provide a convenient basis for classifying a large number of projections. A projected graticule is classified as (1) *cylindrical* if it takes on a rectangular appearance (Figure 18-3a), (2) *conic* if it looks fan-shaped (Figure 18-3b), and (3) *azimuthal* if its resembles a map projected directly to a plane (Figure 18-3c). The term azimuthal refers to the property that azimuths (or bearings or directions) from the central point to other points are not deformed during the projection process. This term is discussed further in Sections 18-9 and 18-12. An example of the cylindrical group is the well-known Mercator projection.

If a projected graticule has only a slight similarity to geometrically projected cases, it may be classified as pseudocylindrical, pseudoconic, or pseudoazimuthal. The pseudocylindrical projection is not rectangular in appearance, but the parallels are all horizontal, suggesting a relationship to the cylindrical group (see Section 18-12).

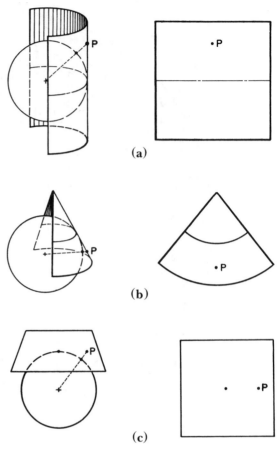

Figure 8-3. Developable projection surfaces. (a) Cylinder tangent to a globe at the equator and developed or unrolled map. (b) Cone tangent to a globe along a parallel and developed or unrolled map. (c) Plane tangent to a globe at the North Pole and part of resulting map.

18-3. SCALE

When a large area, such as the western hemisphere, is shown on a small sheet of paper, the result is said to be a small-scale map. A large-scale map, of course, is the opposite; an example would be the map of a few city blocks on a large sheet. The large-scale map is generated from a relatively large globe. For a plane coordinate system to be used in surveying computations, a full-size globe is used (see Chapter 19).

The usual way of expressing scale in numerical terms is by a dimensionless *ratio* or *representative fraction* (RF).

$$\text{RF scale} = \frac{\text{globe distance}}{\text{earth distance}} \quad (18\text{-}1)$$

If 200 km (124.3 mi) are represented on a globe as 1 cm, the RF scale is

$$\frac{1.00 \text{ cm}}{200 \text{ km}} = \frac{0.0100 \text{ m}}{200,000 \text{ m}} = \frac{1}{20,000,000}$$

Distances on the globe are 20,000,000 times smaller than on the earth itself. (Two-dimensional surfaces or areas are reduced in both dimensions and thus are smaller by a factor of $20,000,000^2$ or 4.00×10^{14}, but only the linear scale is usually stated.) This scale, often called the principal scale, may be written for convenience as 1 : 20,000,000. If the radius of the generating globe R is known, the RF scale is equal to $R/6370$ km converted to a dimensionless ratio with a numerator of unity.

Scale also may be expressed in *unit equivalents*. In the case mentioned, 1 cm represents 200 km or 1 cm = 200 km, in which it is understood that the smaller unit is a globe distance, and the larger unit an earth distance. It is standard practice to assign unity to the smaller unit rather than to say 1 km = 0.005 cm or 1 km = 0.05 mm. Other examples of unit equivalent scales are 1 in. = 300 mi and 1 in. = 2000 ft.

Scale may be shown graphically, with convenient multiples of earth distances marked off along a bar. In the first example (1 cm = 200 km), scale divisions equivalent to 100 or 500 km might be used. The size of a 500-km division would be 500,000 m/20,000,000 = 0.025 m or 2.5 cm. Graphic scales, being pictorial, are very helpful to a map user.

18-4. SCALE FACTOR

All dimensions of the earth are reduced proportionately when it is reduced to a globe. Some dimensions do not undergo any further change as the surface of the generating globe is projected to become a map. Figure 18-3a shows the case of a cylinder tangent to the globe at the equator. As the cylinder is unrolled, or developed, the equator maintains its original length. Such a line is called a *standard line* or line of exact scale. It is said to have a *scale factor* or "particular scale" of 1.000. If a certain line is doubled in length during the projection process, it is said to have a scalar factor of 2.000. In equation form,

$$\text{Scale factor} = \frac{\text{map distance}}{\text{globe distance}} \quad (18\text{-}2)$$

The scale factor on any map will vary from point to point and may vary in different directions at the same point, being 1.000 along only standard lines or at a standard point. No map has a uniform scale. An RF, such as 1 : 20,000,000 applies to the generating globe itself and is correct for the map only when the scale factor is 1.000.

18-5. MATHEMATICS OF THE SPHERE

Before proceeding with this discussion of map projections, it is important to review the geometry of spheres.

If the radius of the globe is R, the circumference is $2\pi R$. That, of course, is the length of the equator and any meridian circle. The various parallels are shorter in circumference than the equator. The North and South poles are really the 90th parallels in the northern and southern hemispheres, but they have zero lengths.

The length of a particular parallel can be calculated by multiplying length of the equator by cosine of latitude. If we use ϕ for latitude, the relationship is

Length of parallel

$$= (\text{length of equator})(\cos \phi) \quad (18\text{-}3)$$

Figure 18-4 shows why Equation (18-3) is correct. The figure depicts a cross section of a sphere in which the radius of any parallel, in its own plane, varies with cos ϕ. The circumference of any parallel, in turn, is $2\pi R \cos \phi$. Obviously, partial lengths of parallels, falling between any two meridians, also vary with cos ϕ, becoming zero at the poles where cos ϕ is zero.

To illustrate this relationship, consider the meridians through Pittsburgh (80° W) and Denver (105° W). The distance between these two meridians is a maximum at the equator —namely,

$$\frac{25}{360} \, 2\pi R = 2779 \text{ km}$$

using 6370 km as the radius of the earth. Between the two cities, as measured along the 40th parallel, the distance is

$$\frac{25}{360} \, 2\pi R \cos 40° = 2129 \text{ km}$$

If a map having an RF scale of $1 : 20,000,000$ shows the 40th parallel as a standard line, the distance between the cities, measured along the parallel, will be

$$\frac{25}{360} \left(\frac{2\pi R \cos 40°}{20,000,000} \right) = 0.000106 \text{ km} \quad \text{or} \quad 10.6 \text{ cm}$$

If another map of the same scale has a scale factor along the 40th parallel of 1.15, the distance will be (10.6) (1.15) = 12.2 cm. The ratio of $2\pi/360$ can be viewed as a conversion factor for degrees to radians.

The surface area of a sphere is $4\pi R^2$, exactly equal to that of a cylinder having the same diameter and height. The circumference

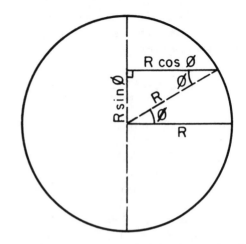

Figure 18-4. Cross section of a globe, showing that the radius of any parallel of latitude is $R \cos \phi$ and the distance between its plane and that of the equator $R \sin \phi$.

of such a cylinder is $2\pi R$ and the height $2R$. The area is $(2\pi R)(2R) = 4\pi R^2$, as stated for the sphere.

The surface area of a zone between any two parallels on a sphere may be found in a similar manner. It is equal to a strip on the surrounding cylinder having the same height (see Figure 18-5). The height of the strip shown is $R \sin \phi$ and the area, therefore, is

Area, equator to parallel

$$= (2\pi R)R \sin \phi = 2\pi R^2 \sin \phi \quad (18\text{-}4)^*$$

This formula may be used, e.g., to find what fraction of the earth's surface lies within the Arctic Circle (66.5° N). The plane of the Arctic Circle is parallel to the equator. The distance between the two planes is $R \sin 66.5°$. This is a height like that of the gray strip shown in Figure 18-5. The area within the Arctic Circle

---

* Equation (18-4), for the area of the zone between the equator and any parallel, also may be found by integration. A narrow zone located at any latitude will have a width $R \, d\phi$ and radius $R \cos \phi$. Its area is $(2\pi R \cos \phi)(R' d\phi)$ and the total area desired therefore is

$$\int_0^{\leq} 2\pi R^2 \cos \phi \, d\phi = 2\pi R^2 \sin \phi$$

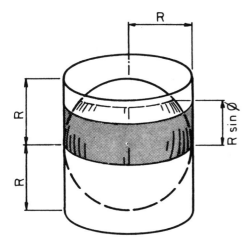

Figure 18-5. Globe surrounded by a cylinder of same height $2R$. Surface area of shaded zone on the sphere is equal to the gray strip or band on a cylinder having the same height.

may be found by subtraction from the area of a hemisphere.

$$2\pi R^2 - 2\pi R^2 \sin 66.5° = 2\pi R^2(1 - \sin 66.5°)$$

This may be seen as a circumference $2\pi R$ times a strip height of $R - R\sin 66.5°$. The fraction of surface area within the Arctic Circle is found by dividing this by the area of a sphere. The answer is independent of R, as follows:

$$\frac{2\pi R^2(1 - \sin 66.5°)}{4\pi R^2} = \tfrac{1}{2}(1 - \sin 66.5°)$$

$$= 0.0415 \quad \text{or} \quad 4.15\%$$

The term *great circle* refers to any arc on the earth, or globe, formed by a plane containing the center of the sphere. Each meridian is a great circle (or actually half of one, running only from pole to pole); the equator is another example. The shortest distance between any two points on the earth's surface is a great circle route. The shortest route between Pittsburgh and Denver is not the one discussed earlier, but rather a great circle route running slightly above the 40th parallel.

The shortest distance between two points may be found by first calculating the central angle subtended by the two points—measured in the plane of the great circle—using an expression from spherical trigonometry. In Figure 18-6, if D is the central angle between points A and B, ϕ_a the latitude of A, ϕ_b the latitude of B, and $\Delta\lambda$ the difference of longitude between A and B, the expression is

$$\cos D = \sin\phi_a \sin\phi_b$$
$$+ \cos\phi_a \cos\phi_b \cos\Delta\lambda \quad (18\text{-}5)$$

Note that ϕ_a and ϕ_b must be expressed as plus or minus (north or south of the equator), but $\Delta\lambda$ may be the longitudinal difference in either direction—not necessarily smaller than 180°. The latter statement is correct because $\cos\Delta\lambda$ is the same either way, e.g., $\cos 20° = \cos 340°$ and $\cos 200° = \cos 160°$. Central angle D can be converted to a surface distance or arc distance by assuming that the earth is spherical. The length of each degree of a great circle is, of course, $2\pi R/360°$.

The shortest distance between Pittsburgh and Denver is found as follows, using latitudes

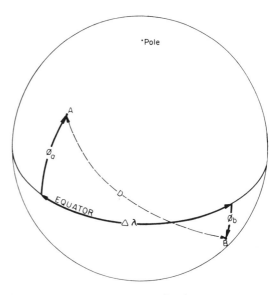

Figure 18-6. Terms in Equation (18-5) include desired central angle D between points A and B, latitudes ϕ_a and ϕ_b of the two points, and difference in longitude $\Delta\lambda$.

and longitudes from the previous example:

$$\cos D = \sin 40° \sin 40°$$
$$+ \cos 40° \cos 40° \cos 25°$$
$$D = 19.09°$$
$$\text{Surface distance} = (2\pi R/360°)(19.09°)$$
$$= 2122 \text{ km}$$

This figure should be rounded to 2120 km.

Although a great circle provides the shortest possible route between two points, it may be a difficult one to follow if navigating manually by compass. In flying from Pittsburgh to Peking, it is simpler to go over Denver, due west all the way, than to follow a route with a constantly changing bearing—i.e., the great circle route would be northwest at first and southwest later. The route of constant bearing, or constant azimuth (in this case, the 40th parallel), is called a *loxodrome* or *rhumb line*. Of course, parallels and meridians are loxodromes, but in general a loxodrome is a route that crosses every meridian at the same angle (see Figure 18-7). The trip from Pittsburgh to Peking could follow a series of loxodromes that together approximate a great circle route. The pilot would change his or her bearing a

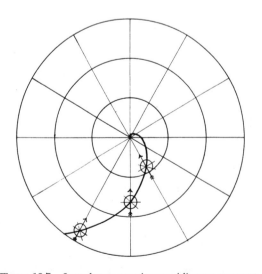

Figure 18-7. Loxodrome crossing meridians at constant angle. (From E. Raisz, 1962, New York: McGraw-Hill, *Principles of Cartography*, with permission.)

number of times, but not continually. There is a map projection with the valuable characteristic that straight lines drawn on it are great circle routes (the gnomonic, see Figure 18-24) and another map projection on which straight lines are loxodromes (the Mercator, see Figure 18-19).

Another useful bit of geometry relates to the trapezoid. Its area is

$$\text{Area} = \frac{b_1 + b_2}{2} h \qquad (18\text{-}6)$$

A 1° quadrangle on a globe (bounded by parallels and meridians 1° apart) resembles a trapezoid but is on a curved surface instead of a plane. The two bases and height are slightly curved. A 1-*min* quadrangle is even more like a trapezoid because there is much less curvature involved; this idea is used in Section 18-10.

18-6.   CONSTANT OF THE CONE

On some map projections, including one used in the state plane coordinate system (Chapter 19), a parallel of latitude is drawn as part of a circle. Longitudinal coverage $\Delta\lambda$ represented by the circular arc may be a full 360° or any smaller number, such as 16° for a map of France.

Figure 18-8 shows a central angle L that is related to $\Delta\lambda$ but generally not equal to it. For example, a map may show all 360° of the 40th parallel as a circular arc in which $L = 231°$. Central angle L is related to $\Delta\lambda$ by a constant k as follows:

$$L = k\,\Delta\lambda \qquad (18\text{-}7)$$

In the case cited, $L = (0.643)(360°) = 231°$. For reasons explained in Section 18-12 and Chapter 19, k is called the *constant of the cone*.

In dealing with circles, it is useful to remember that a chord length can be calculated as follows, referring again to Figure 18-8:

$$\text{Chord} = 2r \sin \tfrac{1}{2}L \qquad (18\text{-}8)$$

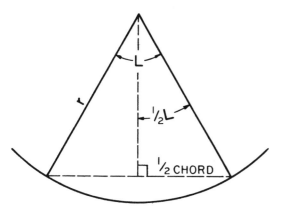

Figure 18-8. Central angle L in conic projections is not equal to $\Delta\lambda$ on globe, and radius r must be derived. (It is not equal to that of globe or of parallel itself, as given in Figure 18-4).

This expression is applied in route surveying as well as the study of map projections.

Often, there is a need to calculate x- and y-coordinates (or tangent offsets) of several points along the circular arc (see Figure 18-9). This computation is useful in plotting, route surveying, and state plane coordinate computations. Assume that a particular parallel has a radius of 100.0 cm and the meridians will cross it at intervals of 5.60 cm. (Figure 18-9 is not drawn to scale.) The central angle for each 5.60-cm arc is

$$L = \frac{5.60}{100.0} \text{ rad}$$

or

$$\left(\frac{5.60}{100.0}\right)\left(\frac{360}{2\pi}\right) = 3.2086°$$

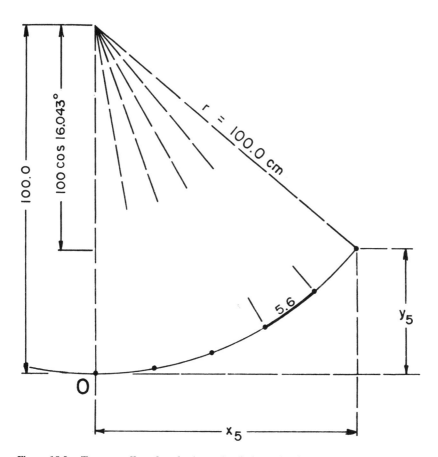

Figure 18-9. Tangent offsets for plotting point 5 along circular arc.

A plotting table may be prepared giving x- and y-coordinates from the point 0, where the central meridian meets the parallel. If five graticule points are needed, the last one will be as follows:

$$x_5 = 100.0 \sin(5 \times 3.2086°)$$

$$= 27.64 \text{ cm}$$

$$y_5 = 100.0 - 100.0 \cos(5 \times 3.2086°)$$

$$= 3.89 \text{ cm}$$

If the meridians are spaced 4° apart in longitude, the constant of the cone must be $3.2086/4.00 = 0.8022$.

The next five sections describe some characteristics found in various map projections.

18-7. STANDARD LINES

In the discussion of scale factor (Section 18-4), standard lines were defined as lines that do not change length when projected from a generating globe to a map. Some projections have only one standard line, such as the equator in Figure 18-3a, but others have many. Several conic projections, e.g., have two standard parallels.

18-8. EQUIDISTANT PROJECTIONS

Although no map projection can offer a uniform scale, some have it in one direction. A projection may have a scale factor of 1.000 in the north-south direction (all meridians are standard; see Figure 18-15), or in the east-west direction (all parallels are standard; see Figure 18-14). Such projections may be described as equidistant.

18-9. AZIMUTHAL PROJECTIONS

Section 18-2 mentioned the characteristic that all azimuthal projections share, namely, that directions to all points with respect to a central point are not deformed during projection from globe to map. (The direction to a distant point is important in the operation of airports, seismographs, radio stations, etc.) An example is shown in Figure 18-18.

18-10. EQUAL-AREA PROJECTIONS

If the *relative size* of all features on a generating globe is maintained during the process of projection to a map, the projection is said to be *equal-area* (also equivalent or equiareal).

It has been pointed out that no map has a scale factor of 1.000 everywhere. If area is to be preserved but scale cannot be, then a given feature on the globe, such as a state, will have to be plotted with a scale factor greater than 1.000 in one direction and smaller than 1.000 in another. It can be shown that such "compensatory scale factors" on an equal-area projection occur in perpendicular directions, often called the *principal directions*. If a circle with a radius of 1 cm is drawn on the surface of a large generating globe, it appears as an *ellipse* when projected to an equal-area map (see Figure 16-10). If one semiaxis is reduced to 0.5 cm, the other will be increased to 2.0 cm to make the ellipse and circle contain the same area.

18-11. CONFORMAL PROJECTIONS

Although equal-area projections have compensatory scale factors and allow a tiny circle, as just described, to be distorted in order to avoid a change in its area, *conformal* projections have equal scale factors in all directions at any one point. A tiny circle is not distorted at all but becomes simply a larger or smaller circle, depending on its location on the map. Instead of preserving size, conformal projections preserve *shape* (see Figure 18-11). They also are called orthomorphic projections.

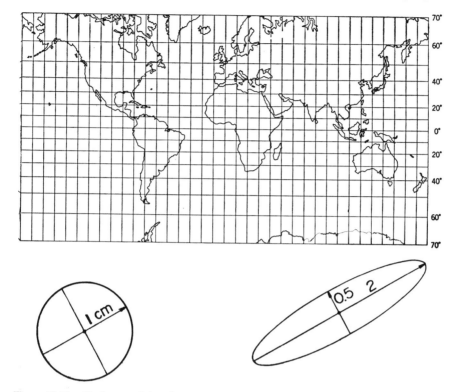

Figure 18-10.   Circle on a globe, shown at left, projects as an ellipse on a map. Example from an equal-area projection is shown.

If the shape of everything on a globe could be preserved on a map, as for the tiny circle in the preceding paragraph, the map would have a uniform scale, which is impossible. Thus, it is correct to say that shapes of small features will be preserved in the course of projection to the conformal map. Conformal projections are ideal for setting up plane coordinate systems for use in surveying, because a surveyor's transverse is small in comparison to the portion of the earth on a particular system. Its angles will be the same when placed on the

Figure 18-11.   In conformal projections, a circle on a globe, at left, will project as another circle (a special kind of ellipse). A scale factor of 0.6 is illustrated.

coordinate system as when they were measured in the field. A single scale factor (generally not 1.000) will apply to all distance measurements, unless the survey is of unusually great size of precision (see Chapter 19).

A map certainly cannot be both equal-area and conformal. It cannot have minimum and maximum scale factors at a point that are compensatory and yet equal. Some projections are neither equal-area nor conformal. The tiny circle referred to earlier is then distorted in both size and shape (see Figure 18-12).

It can be shown that for any projection, such a circle is invariably transformed into an ellipse of some size and has a pair of axes (the principal directions) that remained perpendicular during projection. Unless the projection is conformal, other angular relationships at the point are disturbed. Angular deformation at a point is zero for the principal directions and reaches some maximum value for another pair of lines (see Figure 18-13). Figure 18-14 shows how this maximum angular deformation varies over an equal-area projection of the world. Clearly, the meridians and parallels do not always meet at right angles as they did on the globe. It is evident in the figure that the pair of perpendiculars that remain perpendicular after projection is not necessarily the pair of graticule lines at a point.

The mathematics of how the tiny circle is deformed into an ellipse was developed by M. A. Tissot in 1881. Further discussion appears in *Introduction to Map Projections*.[1]

18-12. SIMPLE EQUIDISTANT PROJECTIONS

If we use the three basic projection surfaces —cylinder, cone, and plane—it is possible to generate three very simple equidistant projections. They are equidistant in the sense that all meridians are standard (the scale factor is 1.000 in the north-south direction). In each case, there is one standard parallel.

All three projections are examples of an idea mentioned in Section 18-2; they are not *literally* projected to a cylinder, cone, or plane but rather are designed mathematically to have a desirable property. They may be thought of as true projections on which the spacing of parallels has been later modified to match their spacing on the globe, making them equidistant.

The first two in the group—cylindrical and conic—are not very important in themselves, but serve to introduce more complex projections that have great value to surveyors.

Cylindrical Equidistant Projection

This projection is also called *plane chart, plate carrée, simple cylindrical,* or the *cylindrical equal-spaced projection.*

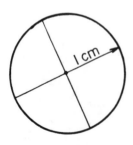

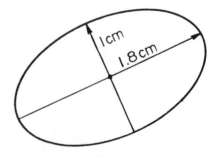

Figure 18-12. In a projection that is neither equal-area nor conformal, a circle on a globe, at left, will project as a nonlinear ellipse of different size.

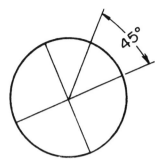 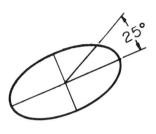

Figure 18-13. When a circle is projected as a noncircular ellipse, the angle between two radial lines will be deformed, except for the two perpendicular that become the ellipse axes.

A cylinder wrapped all the way around the generating globe touching the equator will have the same circumference as the sphere —namely, $2\pi R$. If the whole world is shown on this projection, construction is begun by drawing the equator as a straight line of this length (the equator is a standard line). The meridians are standard also and drawn as straight vertical lines with a length πR. Figure 18-15 shows the resulting graticule, consisting of perfect squares. They are standard in their north-south dimension but, except at the equator, are wider than the corresponding "squares" or quadrangles on the globe. The

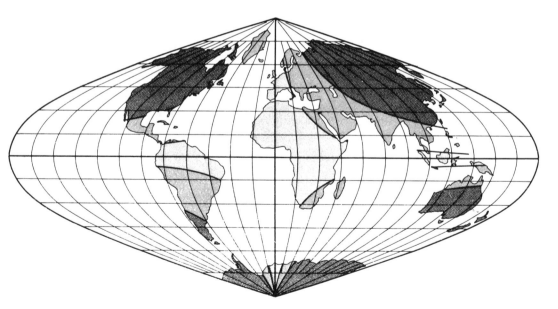

Figure 18-14. World map on a sinusoidal projection, showing lines of equal maximum angular deformation (10 and 40°). Projection is equal-area. Standard lines include the central meridian and all parallels. (From A. H. Robinson and R. D. Sale, 1984, *Elements of Cartography*, 5th ed., New York: John Wiley & Sons, with permission.)

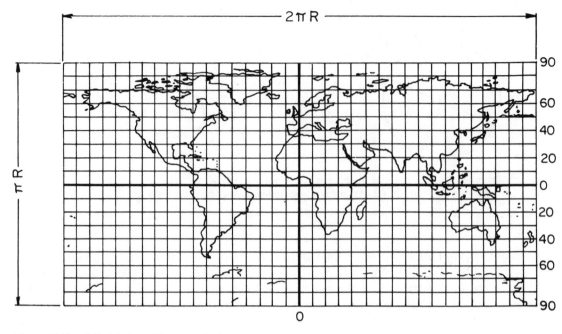

Figure 18-15. Cylindrical equidistant projection.

meridians fail to converge, resulting in the north and south poles appearing as lines the length of the equator instead of points.

By definition, the scale factor is 1.000 along the standard lines. It is greater than 1.000 along the parallels. The 60th parallel, e.g., has a "globe distance," or length, of $2\pi R \cos 60°$ (see Section 18-5) but a "map distance" equal to $2\pi R$—the same as the equator. The scale factor is the ratio of these lengths.

$$\text{Scale factor} = \frac{\text{map distance}}{\text{globe distance}}$$

$$= \frac{2\pi R}{2\pi R \cos 60°} = \sec 60° = 2.000$$

The east-west scale factor varies with $\sec \phi$, being 1.000 on the equator and infinity at the poles.

This projection is so easy to construct, there is little need to think in terms of x- and y-coordinates being functions of ϕ and λ. The relationship, however, is $x = C\lambda$ and $y = C\phi$, meaning λ and ϕ are plotted to some scale as if they were rectangular coordinates.

Conical Equidistant Projection

The projection just discussed was classified as cylindrical, even though spacing of the parallels was determined by the requirement that it be equidistant rather than by any actual geometric projection to a cylinder. The *conical equidistant* is designed in exactly the same way.

A conical equidistant projection is best suited for mapping areas in the vicinity of one standard parallel just as the cylindrical equidistant is appropriate for areas near the equator. This projection, as well as the other conics, is generally chosen for an area lying entirely on one side of the equator.

A cross-sectional view of a globe and tangent cone is depicted in Figure 18-16. The apex is at A and the point of tangency at T. The angle at the apex between globe axis and cone element AT is seen to be equal to the latitude of the standard parallel. In triangle ATO, the tangent of ϕ is R/AT and

$$AT = \frac{R}{\tan \phi} = R \tan \phi$$

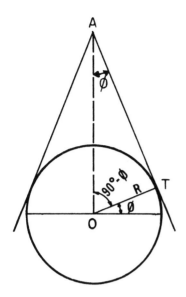

Figure 18-16. Cross-sectional view of a globe and tangent cone.

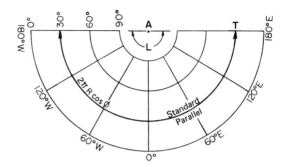

Figure 18-17. Developed cone for the northern hemisphere, conical equidistant projection, with standard parallel at 30°.

Distance AT may be used as a compass setting r for drawing the standard parallel on a map.

The radius of the standard parallel *on the globe is $R \cos \phi$*, as illustrated in Figure 18-4. Its length, of course, is $2\pi R \cos \phi$. *On the map*, after the cone has been "unrolled," the parallel has the same length but the radius used in drawing it, AT or $R \cot \phi$, makes it appear as less than a full circle (see Figure 18-17). The central angle at A called L, in radians, is equal to the arc length divided by the radius.

$$L = \frac{2\pi R \cos \phi}{R \cot \phi} = 2 \cos \phi \tan \phi$$

$$= 2\pi \cos \phi \frac{\sin \phi}{\cos \phi} = 2\pi \sin \phi$$

This angle may be converted to degrees (if we multiply by $360°/2\pi$).

$$L \text{ (in degrees)} = 2\pi \sin \phi \frac{360°}{2\pi} = 360° \sin \phi$$

The L angle is independent of scale (R was cancelled out of the expression). If $\phi = 30°$, then $L = 180°$, as shown in Figure 18-17. A full 360° of longitude is shown within a semicircle. The "constant of the cone" k, defined in Section 18-6, is $\sin \phi$ in the case of a conical equidistant projection.

The standard parallel is divided into equal parts and the meridians drawn as straight radial lines of standard length. As in the cylindrical equidistant, the North Pole will be a line instead of a point. For $\phi = 30°$, distance AT is $1.732R$, while the distance from T to the pole is $(60/180)\pi R$ or only $1.04R$.

The cylindrical equidistant projection, covered previously, is really only a special case of the conical equidistant in which the standard parallel is at $\phi = 0°$, radius $AT = $ infinity, the constant of the cone $k = \sin \phi = 0$, and central angle $L = 0°$ (the meridians being parallel). In other words, a cylinder is merely a special kind of a cone having its apex at infinity.

The conical equidistant projection is known also as the "simple conic." It could reasonably be chosen for a map covering only a few degrees of latitude, such as a tourist map of the Trans-Canada Highway. In that case, the standard parallel might be 50° N; however, better conic projections are available.

Azimuthal Equidistant Projection

It has been pointed out that a cylinder is really a special cone with its apex at infinity. A plane that is tangent at the pole may be viewed as a special cone also. It has an altitude equal to zero and its standard parallel is at 90° N or S. (It is just a little bit flatter than a cone made tangent to 80° N, e.g.) Distance *AT* is zero or *R* cot 90°. The constant of the cone *k* = sin 90° = 1.000, meaning that the central angle *L*, in degrees, for a full 360° of longitude is 360°. A graticule has a fan-shaped appearance like regular conics if the fan is thought of as being wide open (see Figure 18-18, p. 450). Meridians radiate like spokes of a wheel and are separated by the same angles as on the globe. The projection is called the polar azimuthal equidistant. *Introduction to Map Projections* discusses the nonpolar, or oblique, case where the plane may be tangent to any selected point.

In the polar case, the azimuthal property requires meridians to be drawn with their actual differences in longitude. If the projection is to be equidistant, all of them will be standard lines; parallels will be equally spaced concentric circles. The opposite pole will be a large circle drawn with a radius of πR.

The oblique case is often centered at an airport, radio station, or seismograph because it correctly presents both directions and distances from that point.

18-13. PROJECTIONS FOR PLANE COORDINATE SYSTEMS

The three simple projections described in Section 18-12 are neither equal-area nor conformal. Variations exist that do have one or the other of these properties. For surveying purposes, the three projections discussed are modified in the following ways:

1. They are made conformal by sacrificing the equidistant property—i.e., the scale factor is allowed to vary along the line formerly held standard.

2. A spheroid is adopted instead of a sphere because there is to be no scale reduction; field work and computations are done with an RF scale of 1:1 (full size). The dimensions are not purely for plotting purposes, as in cartography.

3. In the cases of the cylindrical and conic projections, *two* standard parallels are used instead of one. This serves to keep the scale factor closer to 1.000 over a wide region. Instead of having all scale factors equal to or greater than 1.000 at all points, as was true in the simple cases of the previous section, projections with two standard parallels include values slightly smaller than 1.000 between the parallels and greater than 1.000 beyond them.

4. For the cylindrical projection, the supposed cylinder is turned 90°, running transversely to the earth's axis. The two standard lines thus are not parallels of latitude but parallel lines adjacent to a selected central meridian. Only a limited area—a "zone" near the central meridian—is included in the coordinate system.

Among cylindrical projections, the conformal one is the Mercator, shown in Figure 18-19 (p. 451). The meridians fail to converge just as they did in Figure 18-15; therefore, the east-west scale factor again varies with sec ϕ.

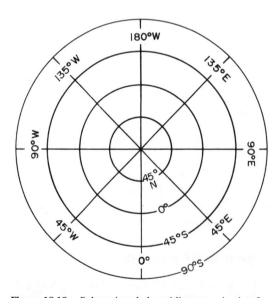

Figure 18-18. Polar azimuthal equidistant projection for the entire world.

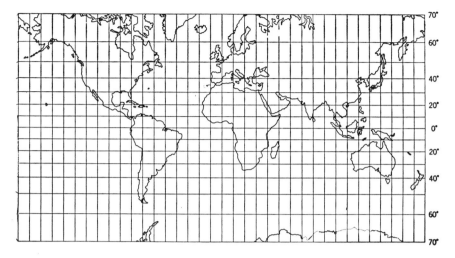

Figure 18-19. Mercator projection. Scale factor in all directions increases with sec ϕ, becoming 2.00 at 60° and infinity at the poles. (From A. H. Robinson and R. D. Sale, 1984, *Elements of Map Projections*, 5th ed., New York: John Wiley & Sons, with permission.)

Because the Mercator is conformal, the scale factor along the meridians must vary in the same way. The two poles, therefore, fall at infinity. The *transverse* Mercator is shown in several figures of Chapter 19.

Among conic projections, the conformal one is the Lambert conformal conic. It is used in the state plane coordinate system for more than half of the states. In a particular state, such as Connecticut, a best-fitting "cone" was selected having its central meridian about midway across the state and its standard parallels just inside the north and south borders (see Chapter 19).

Among azimuthal projections, the conformal one is the stereographic. When based on a sphere, the stereographic is formed as a true projection to a tangent plane from a diametrically opposite point (see Figure 18-23). The polar case is used in the military grid systems (Chapter 19). The oblique case has also been chosen for plane coordinate systems in a few places, including New Brunswick and Prince Edward Island.

Figures 18-20, 18-21, 18-22, 18-23, and 18-24 illustrate five map projections and show how a particular triangle projects on each. The an-

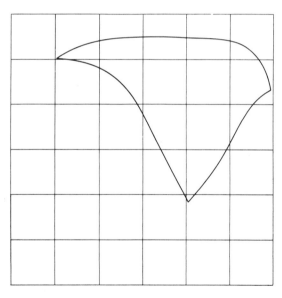

Figure 18-20. Cylindrical equidistant projection of the western hemisphere. Great circle routes joining Anchorage, Madrid, and Buenos Aires are plotted on this and the following four figures; in each case, the central meridian is 90° W. (From P. W. McDonnell, Jr., 1979, *Introduction to Map Projections*, New York: Marcel Dekker, with permission.)

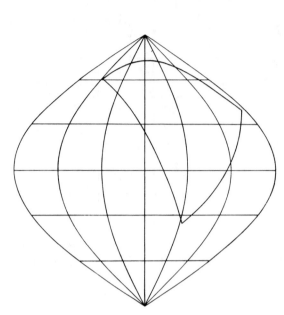

Figure 18-21.   Equal-area sinusoidal projection of the western hemisphere.

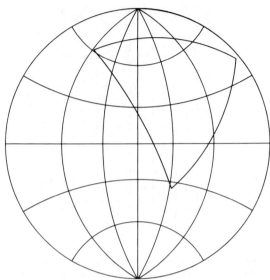

Figure 18-23.   Conformal stereographic projection centered on the equator; meridians and parallels are perpendicular.

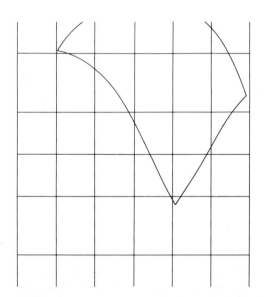

Figure 18-22.   Conformal Mercator projection. Angles formed where great circles meet are equal to corresponding angles on earth.

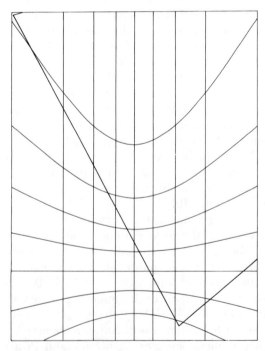

Figure 18-24.   Gnomonic projection centered on the equator and 90° W, with graticle lines at 15° intervals. The projection extends to infinity in all directions, but it offers the unique advantage of displaying any great circle as a straight line.

gles at each apex are the same (not deformed) in the conformal projections shown in Figures 18-22 and 18-23.

NOTE

1. McDonnell, P. W., Jr. 1992. *Introduction to Map Projections*, 2nd ed. Rancho Cordova, CA: Landmark Enterprises.

REFERENCES

DEETZ, C. H., AND O. S. ADAMS. 1944 *Elements of Map Projection*. Special Publication No. 68, U.S. Coast & Geodetic Survey.

ROBINSON, A. H., AND R. D. SALE. 1984 *Elements of Cartography*, 5th ed. New York: John Wiley & Sons.

19

Plane Coordinate Systems

R. B. Buckner

19-1. INTRODUCTION TO PLANE COORDINATE SYSTEMS

It has been common practice for surveyors to use rectangular (*X*- and *Y*-) coordinates in surveys. Coordinates are useful in the design and layout of subdivisions, layout of construction control, computation and plotting of traverses, surveying of land boundaries, establishment of mapping control, etc. All too often, however, surveyors have used arbitrary coordinate systems, resulting in thousands of different surveys referred to thousands of unrelated origins. These independent systems are useless for any purpose other than the original one— i.e., the data cannot be conveniently plotted on a map made by others or stored in a data bank or land information system. They cannot be used to relate land boundary corners to each other in adjacent areas, to tie together public works geometrically over large areas, or as control survey reference systems for any purpose except the single one for which they were created.

The systems described in this chapter were designed to provide single systems for very large regions and overcome the disadvantages of independent systems. The described systems

were designed to eliminate the need for a knowledge of geodesy, despite the fact that the systems cover very large regions. The state plane coordinate system, in particular, was intended to be adopted by private surveyors and government agencies for horizontal control as used in land, construction, topographic, and other surveys.

Three systems are discussed: (1) the state plane coordinate system (hereafter referred to as SPCS or SPC), (2) the universal transverse Mercator system (UTM), and (3) the universal polar stereographic system (UPSS). These can be considered actually only two systems —namely, the SPCS and UTM-UPSS.

The SPCS divides the United States into large zones, each one covering an entire state or a proportion of a state. Zones are bounded by either state lines or county (parish in Louisiana) lines within states. The UTM-UPSS divides the entire world into 62 zones, each of them much larger than those of the SPCS. Zone boundaries are lines of latitude and longitude. This latter system was not developed for surveying purposes. It was originally developed as a military grid system following World War II, but has been widely used for civilian purposes by many countries, including the

United States. The main focus in this chapter will be the SPCS, it being more directly used by surveyors, but a general description of the UTM-UPSS will also be given.

Material in Chapter 13, especially as related to ellipsoids and the geoid, will be helpful in the study of this chapter. The information material in Chapter 18 related to projection surfaces and Chapter 23 related to horizontal control surveys and datums will also be helpful. Some sections of Chapters 11 and 12 and Chapter 17 also relate to the subject matter of this chapter.

19-2. STATE PLANE COORDINATE SYSTEMS

19-2-1. Development of the State Plane Coordinate Systems

The State Plane Coordinate System of 1927

Prior to the 1980s, the SPCS was referenced to latitudes and longitudes based on the *North American Datum of 1927* (NAD27), which are positions resulting from mathematical adjustments of the North American Datum, made around 1927. At that time, a reference ellipsoid called the Clarke Spheroid of 1866 was used. This ellipsoid had been in use for many years and was devised with parameters (semimajor and semiminor axes) and a physical center designed to closely fit the North American continent.

The SPCS began in 1933 at the suggestion of the North Carolina Highway Department. Until then, the triangulation stations of the U.S. Coast and Geodetic Survey (USC and GS), established at the taxpayers' expense, were not serving a sufficiently broad purpose. Their positions were expressed solely in spherical coordinates (latitudes and longitudes) and a highway department or other surveying and mapping organization could only tie its work

to the high-order control network by becoming involved in the complexities of geodesy. The USC and GS proceeded to design a rectangular system for each state. O. S. Adams, a mathematician with the USC and GS, is credited with developing the mathematics for the first systems.

With plane coordinate systems developed for them, many states enacted statutes that legally adopted the mathematical definition and name of their system. In these states, the coordinates may be used in a land description merely by referring to the system by name and stating *X*- and *Y*-values. However, such laws simply made it possible to use coordinates as part of a land description; they did nothing to encourage or require their use in descriptions or as mathematical reference ties to preserve corner evidence, or employment of the system for mapping control or other general use as a measuring and computational tool. Although the SPCS has been used very little in property surveying, it has been used more extensively in applications such as highway engineering, photogrammetric control for mapping, general control surveying, and mining surveying.

The State Plane Coordinate System of 1983

During the 1960s, with electronic distance measurements becoming more common, ties to SPC monuments also became more common. These surveys, with their wider extent and accuracy, began to detect inaccuracies in the published positions of the official control monuments. Because of this and other reasons such as the need for higher accuracy of a worldwide scope for satellite tracking and missile ranging, the USC and GS began to discuss a readjustment of the existing control and adoption of a different reference ellipsoid. The project began officially in 1974 by the National Geodetic Survey (the NGS evolved in 1970 from the USC and GS) and was completed in 1986. It involved the solution of 928,735 simultaneous equations using over 266,000 control stations in the United States,

Canada, Mexico, and Central America. Not only were the data readjusted for a better mathematical "balance," but also a new ellipsoid was adopted, more suited to worldwide use than the Clarke Spheroid of 1866. In comparison to the Clarke Spheroid, the newer ellipsoid, called the *Geodetic Reference System of 1980* (GRS80), has slightly different parameters (value of major and minor axes) and it is *geocentric* (earth's center of mass is its center). The differences between these ellipsoids are discussed in Chapter 13. The readjustment, along with adoption of the GRS80 ellipsoid, has now created the *North American Datum of 1983* (NAD83).

The combined effects of the readjustment and change in reference ellipsoid caused shifts in the latitude and longitude of existing control points and any other points referenced to the system. Naturally, the *X*- and *Y*-coordinates of the points also changed since plane coordinates are mathematically related to geodetic positions. Typically, the shifts in coordinates have been under 50 m in most of the United States, but range as high as 100 m in the extreme western states and over 100 m in parts of Alaska.

Utilizing Both SPCS27 and SPCS83

Surveyors must recognize that since differences exist between NAD27 and NAD83, there are now two state plane coordinate systems, designated as SPCS27 and SPCS83, each being based on the respective adjustment and ellipsoid used at the two different points in time. Ideally, a complete shift should eventually be made to NAD83. But, the shift may never be complete because of the very large number of old coordinates in files that would need to be converted to NAD83.

Although the theory, mathematics, projection surfaces, and basic nature of SPC are the same in both SPCS27 and SPCS83, and learning one or the other is not particularly complicated, the differences between the two must be understood. Besides the fact that the geo-

graphic positions of all points shifted, there are some minor changes in a few terms, symbols, and other details. One immediate difference that will be noted is that SPCS83 coordinates are published in meters, whereas SPC27 coordinates were given in feet. Since there was a movement in this country during the 1980s to gradually shift to the metric system and the meter has been the official standard of measurement in the United States since the 19th century anyway, it was decided that new coordinates should be published in meters. Another difference is that azimuths were reckoned from the south end of the meridian in NAD27, but are reckoned from north in NAD83. Other important differences between SPC27 and SPC83 will be revealed as the systems are described in the following sections. Transformations between NAD27 and NAD83 coordinates are covered in Section 19-2-8.

Impact of Global Positioning Systems on SPC

During the late 1980s, global positioning systems (GPS) became more widely used for the accurate positioning of new control stations. These systems have begun to detect small inaccuracies in the 1983 readjusted values of existing control stations, much as EDM found errors in NAD27 values during the 1960s. Although the errors are much smaller than those causing concern prior to NAD83, various programs have begun, some on a state level, to use GPS to set a standard. The result is that values of existing stations may be shifted in these areas from their NAD83 adjusted values. Called "supernet," these 1 : 1,000,000 precision networks utilize existing high-order control stations when accurate GPS positions have been determined and adjust other control stations within the region, based on these positions. For example, Tennessee has now been covered by such a network of GPS stations, creating what has been called SPC83/90 coordinates in Tennessee. Similarly, Florida has such a system. Maryland's is called SPC83/91, the last number in the designation being the year of completion of the revision to SPC83. Thus, we actually have three sets of coordinates to consider in some areas: SPC27, SPC83, and SPC83/9? These programs were ongoing at the time of writing and their impact was not fully realized. Typically, the shifts to SPC83 positions so far have been in the range of 0.4 to 0.7 m using the GPS supernets.

19-2-2. Applications and Advantages of SPCS

State plane coordinates can be used in geodetic control surveys, property surveys, engineering and construction surveys, mining surveys, topographic and other mapping surveys, land subdivision design, land information systems, hydrographic surveys, highway and street design, utility and infrastructure design and maintenance, and general urban and regional planning. Whenever the property corners, control corners of transportation and utility systems, prominent cultural and natural features, and any underground workings have all been accurately located on the same reproducible coordinate system, all can be mapped using one mapping base and interrelated on the ground for location and design purposes. Positioning underground features (such as buried gas mains and mining shafts and tunnels) on the SPCS add a safety feature since accurate positioning with respect to the ground surface gives information on where to dig or avoid digging. The SPCS has a significant application as a measuring and control survey tool for land boundary analysis and other purposes, as will be discussed as one of its advantages.

It is ironic that the original supposition that SPC would be used to describe property corners and be the basis for legislation on SPC has not been accepted by land surveyors as an advantage of the system. Rejection or low acceptance on this basis, however, should not preclude recognition of the many other advantages of the system in land surveying and other applications. Some of these important advantages of the SPCS are summarized as follows:

1. After SPCs have been established for a particular property corner, the corner has a higher degree of permanence. If the monument making the corner becomes list, it can be reestablished from any nearby points having accurate and known coordinates. Thus, SPCs provide an essentially indestructible reference tie (as distinguished from the concept of coordinates as part of the land description), the usefulness as such being proportional to the density of monumentation and accuracy of the system and the measurements made to reference the corners to the system.

2. In a manner similar to the preservation of corner positions by making SPC reference ties, use of the grid meridian of SPC provides an accurate and reproducible reference meridian for surveys. This aids specifically in the retracement of boundaries and overall permanence. This advantage exists even without making ties to control monuments since an SPC grid meridian can be determined through the use of astronomy.

3. When SPCs have been determined on property corners and/or the grid meridian used as a reference on property surveys in an area, all land boundaries in adjacent areas are easier to relate to each other. This aids in performing boundary retracements or describing land.

4. The system is useful as a measuring tool over large regions, since coordinate geometry is useful in the analysis and layout of land boundary and other positions. Several types of indirect measurements can be made employing a combination of traverse, intersection, resection, and other triangulation-trilateration schemes.

5. Long traverses, such as those for route surveys, may be closed on distant control stations rather than by doubling back for loop closures, saving time and increasing overall efficiency.

6. SPCs are useful as photogrammetric and other mapping control systems, as well as for locating planimetric details for such maps

and for the grid lines used on the maps for scaling purposes.

7. Public works agencies, planning commissions, utility companies, and others who deal with surveying and mapping can operate more efficiently if all survey work is tied to a common horizontal datum. This advantage results in much the same way that having a common vertical datum for elevations assures higher efficiency in dealing with design and other projects.

8. As the systems are utilized and densified, they can evolve to a point of even higher efficiency and usefulness as related to the development of multipurpose cadastral or land information systems for general planning, large-scale mapping, design and maintenance of the infrastructure, and other purposes related to land parcel and other records.

19-2-3. The Basics of State Plane Coordinate Systems

Each zone of a SPCS is an area covering an entire state or part of a state. Within each zone, surveyed points located on the surface of the earth are projected to "developable" surfaces—i.e., surfaces that can be "rolled out" into a plane. The two surfaces are a cone and cylinder. The conical projection used is called the Lambert conformal conic projection. The cylindrical projection is termed the transverse Mercator (see Chapter 18 for a discussion of these projections). Typically, the conical projection is used in states that are long in an east-west direction because the *standard lines* (Section 18-7) are parallels of latitude. The cylindrical projection is used largely in states that are long in a north-south direction since standard lines run nearly north-south (Section 18-13).

Figures 19-1a and 19-2a show the basic concepts of these two projections. Conformal projections are actually mathematically derived but the figures are sufficient for explanation. One zone (for southeast Alaska) uses an oblique Mercator projection that is not discussed here. *Conformality* is described in Section 18-11.

Figures 19-1 and 19-2 show a cone and cylinder covering a large portion of the earth rather than actual zones that are, as stated earlier, limited to a single state or region. In order to create each zone, the cones and cylinders are placed in many different locations such that standard lines (lines *AB* and *CD* in Figures 19-1 and 19-2) are centered on each respective zone and slightly within it. In the figures, the circle depicts the reference ellipsoid.

Zones vary somewhat in width, and thus the conical and cylindrical parameters vary. The

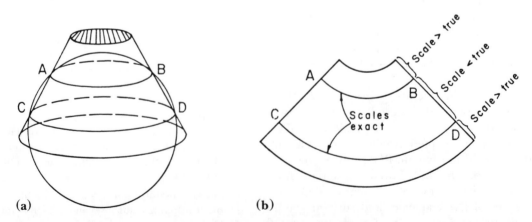

(a) **(b)**

Figure 9-1.   (a) Conic surface on earth. (b) Developed cone.

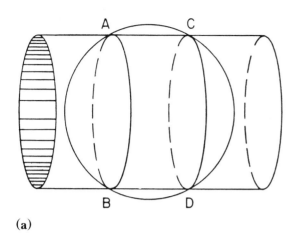
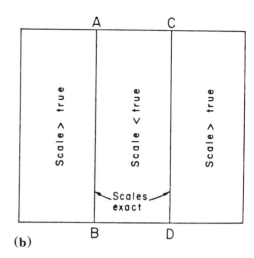

(a)   **(b)**

Figure 19-2.   (a) Transverse cylinder on earth. (b) Developed cylinder.

cone and cylinder both intersect the ellipsoid slightly. This allows zones to be wider than if the figures were made tangent to the ellipsoid. This point will perhaps be more clear when scale factors are discussed.

Distances lying along the standard lines are the same on both the ellipsoid and projection surface (cone or cylinder). Within the standard lines, a distance projected from the ellipsoid to the projection surface becomes smaller if the line lies between the standard lines. This is what is meant by "scale < true" in the figures. For the small portion of a zone lying outside of the standard lines, the opposite is true; thus, "scale > true."

Both the developed cone and developed cylinder have a *central meridian* (not shown in the figures) that constitutes the *Y*-axis of a plane coordinate system and is assigned a large *X*-coordinate (*Easting*) so that negative *X*-coordinates are avoided at the western extremes of each zone. The Lambert system of SPCS27 usually has a value of 2,000,000 ft assigned to this meridian, whereas 500,000 ft is most common in the TM system. Commonly used values for this meridian in SPC83 are 600,000 m in Lambert zones and 200,000 m in TM zones, although much variation exists in these values among the various states. The *X*-axis is placed well to the south of a zone and

assigned a *Y*-coordinate (*Northing*) of zero. Thus, a rectangular grid is simply superimposed on the developed cone or cylinder.

A grid superimposed on the developed plane, shown in Figures 19-1b and 19-2b, creates a reference direction called grid north that differs from geodetic north, except on the central meridian. The "true" (the term geodetic or astronomic is preferred) meridians converge toward the central meridian as they go north in both kinds of zones. Convergence will be discussed further in later sections of this chapter.

In the original design of the SPCS, it was decided that no distance as measured and projected to the ellipsoid should differ from the distance as projected to the projection surface by more than 1 part in 10,000. This design was apparently decided based on the assumption that practicing surveyors, using transit and tape, would normally not do work better than about 1 part in 5000, and if they neglected the "scale factors" (ratio of the two mentioned distances), no serious errors would be introduced. This meant that (as can be proved mathematically) zones in either the Lambert or TM systems could not have a total width of more than about 158 mi. In fact, if it is assumed that the appropriate corrections would always be made and users could deal with larger differences between actual ground positions and projected positions, zones could have been made much wider, thus reducing the number of zones in each state. Apparently assuming that the mathematics and discrepancies would be handled properly, three states (Montana, Nebraska, and South Carolina) changed their SPCS27 from multizone to only one zone in each state for SPC83.

19-2-4.   Referencing Surveys to the State Plane Coordinate System

When very low-order accuracy might be sufficient, surveys could be tied to SPCS by the same methods used with the more limited single-survey control systems. For most work, however, there are a few additional refinements to consider. (1) The measured horizontal distances must be projected to the reference ellipsoid, and (2) the ellipsoid distances must next be projected to the projection surface. In addition, whether these two refinements are neglected or not, grid direction must be employed in order to have any reasonable approximation of SPC after the survey is completed. The care used in determining each of these three factors is dictated by the accuracy required in the resulting coordinates. In any case, it should be recognized that after attending to these variables, computations for coordinate differences (departures and latitudes), coordinates, areas, and other coordinate geometry are no different than those on any plane surface.

Variable 1 cited above results in what was formerly called a *sea-level factor* (SPCS27) and is now called an *ellipsoidal factor* (SPC83). A general term used to describe either of these is "elevation factor." Variable 2 results in a *scale factor* (see Section 18-4). The elevation factor multiplied by the scale factor results in a *grid factor*, sometimes called *grid combination factor* or simply "combination factor." These concepts are depicted in Figure 19-3a. In this figure, the projection surface (either a cone or cylinder) contains the plane coordinate grid.

It is shown passing through two standard lines, where the scale factor is 1.00000, and cutting the earth's surface, where the grid factor is 1.00000. The grid factor is smallest for a surface distance measured in the center of the zone. This distance must be reduced to the ellipsoid and then further to the projection surface (grid). The grid factor is shown as 1.00000 at a point where a surface distance must first be reduced because of its elevation but then increased by a like amount because it lies beyond the standard parallel. In most states, the zone is limited by the initial design decision to keep scale factors between 0.99990 and 1.00010. At the ocean shore, just beyond the zone in Figure 19-3, the elevation factor would be 1.0000.

Horizontal distances measured at the local elevation must each be multiplied by their respective grid factors so as to have each distance on the projection surface for computations on a plane. It should be realized that this factor is very close to 1.0000 and varies with the ground elevation of the line above the ellipsoid and location of the line with respect to the standard lines.

Besides correcting distances to their equivalent length on the grid, azimuths must be oriented to the grid system of the zone where the survey lies. This involves either using azimuth marks placed in conjunction with the SPC monuments or determining azimuths astronomically and correcting them for convergence and other factors.

The above factors and how to determine them are further explained in the following sections.

Elevation Related Factors

As can be seen from Figures 19-3a and b, there is a difference between the way elevation is handled in SPCS27 and SPCS83. The ellipsoid used for NAD27 was designed to fit North America closely, resulting in a very small difference between the position of the ellipsoid and geoid at any location. At Meade's Ranch, Ks, the *geoid height* (height of the geoid above

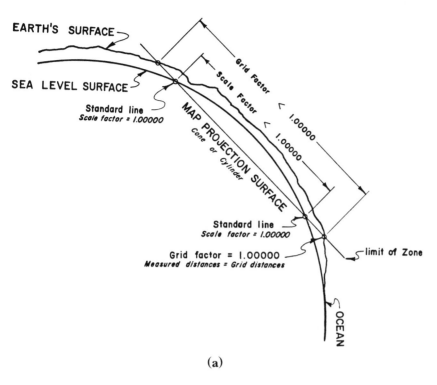

(a)

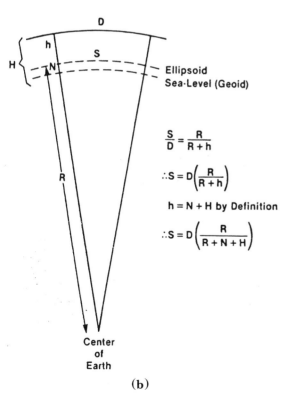

$$\frac{S}{D} = \frac{R}{R+h}$$

$$\therefore S = D\left(\frac{R}{R+h}\right)$$

$h = N + H$ by Definition

$$\therefore S = D\left(\frac{R}{R+N+H}\right)$$

(b)

Figure 19-3. (a) Cross section of a state coordinate zone. *Note:* This surface is more properly called the "geoid." Furthermore, the geoid height, which is too small to illustrate clearly in the sketch, should be considered when using SPC83. See the text and Figure 19-3b for more of an explanation. (b) Reduction to the ellipsoid (shown with a negative geoid height). (NOAA Manual NOS NGS-5 State Plane Coordinate System of 1983.)

the ellipsoid) was set to zero. It was never above 12 m anywhere in the United States. Thus, the ellipsoid and geoid were considered to coincide, for all practical purposes, and the surface was simply called "sea level." With the GRS80 used in NAD83, the geoid heights are more significant across the United States, so they should be considered. In essence, elevations being referred to the geoid (sea level) should undergo another small correction to the ellipsoid when using SPCS83.

Figure 19-3b shows the geometry and equations. In this figure, h is simply the average ground elevation when using SPCS27, but it is $H + N$ when using SPCS83, as shown. Since the geoid height in the conterminous United States is negative, the figure is drawn to show the geoid below the ellipsoid, depicting the situation for most of the country. In Alaska, the ellipsoid is below the geoid, resulting in positive geoid heights. Geoid heights are given in control station description sheets for SPC83.

Scale Factors

Scale factors change with changes in latitude in Lambert zones and with *easting* (X-coordinate) or longitude in TM zones. Actually, scale factors are constant only at specific points and so changes along lines must be considered. Usually, the mean latitude of a line can be used in Lambert zones and the approximate mean easting in TM zones. Scale factors were formerly listed as a function of these variables in SPC projection tables published by the NGS for each state, applicable to SPCS27. As of this writing, the NGS had not published tables for NAD83 scale factors on the TM system, but had done so for the Lambert system. New publications accompanying SPCS83[1] suggest another means, using equations instead of interpolation from tables, to determine scale factors in both systems. Although the equations are simple, computer programs have been made available from NGS for determining these factors. Except in the case of the three states that changed zone systems, the NAD27 tables are still close

enough, in most cases, for extraction of scal factors for the Lambert system. This point wi be further clarified in Section 19-2-5. Exam ples later in this chapter will explain the calcu lation and use of scale factors.

Grid Factors

The grid factor for a survey line is simpl the product of the elevation factor and scal factor for the line. It is a matter of preferenc as to whether elevation factors and scale fac tors are considered separately or combine into one grid factor. The results on distance are the same.

Grid Directions and Convergence Angles

As was discussed in Section 19-2-3, geodeti ("true") north and grid north coincide onl along the central meridian in each zone. A other locations east or west of the centra meridian, geodetic north converges toward th central meridian. The amount of convergenc is proportional to the distance from the cen tral meridian. Grid meridians are paralle whereas geodetic meridians are not. Tru north cannot be used in plane surveying sys tems without resulting errors due to meridia convergence. Figure 19-4 depicts the concep of convergence.

If a survey starts and ends on a SPC monu ment having a reliable nearby azimuth mar with a published, accurate azimuth, or sight ings can be made to another SPC station fro the starting station, convergence does not hav to be considered. Angles are then simply bal anced between the known grid azimuths an accurate, adjusted grid azimuths computed o all survey lines between the azimuth check points. But, if neither a terrestrial azimut mark nor another SPC monument can b sighted for azimuth control at the desired lo cations, grid azimuths must be determine astronomically using the sun or a star, or gyr system, and the astronomic azimuths cor rected to grid azimuths. The relationship among astronomic, geodetic, and grid az

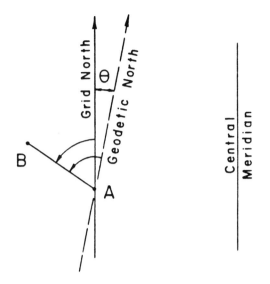

Figure 19-4. Traverse line with a negative convergence.

imuth are as follows:

Grid azimuth = geodetic azimuth

$$- \text{ convergence } \Delta + (t - T) \quad (19\text{-}1a)$$

Geodetic azimuth = astronomic azimuth

$$+ \text{ Laplace correction} \quad (19\text{-}1b)$$

The *convergence* or *mapping angle* was called θ in the Lambert system and $\Delta \alpha$ in the TM system in SPCS27, but is now referred to as γ in both the Lambert and TM systems for SPCS83. It will be generally termed "convergence Δ" in this discussion. In the Lambert system,

$$\text{Convergence } \Delta = l \times \Delta \lambda \quad (19\text{-}1c)$$

where l is a constant for a zone and $\Delta \lambda =$ the longitude of the central meridian minus the longitude of the station. In the TM system,

$$\text{Convergence } \Delta = \Delta \lambda \times \sin \phi + g \quad (19\text{-}1d)$$

where $\phi =$ the latitude of the station, $\Delta \lambda$ is as defined above, and g is a value that varies with $\Delta \lambda$ and ϕ. Since g is less than 0.1", even for lines several miles long, it will not be considered further in the discussion in this chapter.

The $t - T$ correction (otherwise known as the *arc-to-chord* correction or *second difference*)

is an angular correction to the geodetic azimuth, being the difference between the pointing observed that on the ellipsoid (generally the same as on the ground) and on the grid. Figure 19-5 depicts the $t - T$ correction. The magnitude of $t - T$ is greatest for long lines near the edges of SPC zones. It is about 1" of arc for a line approximately 3 mi long, near the edge of a zone and oriented parallel to the standard lines. Its importance, in terms of overall survey accuracy, is discussed in Section 19-2-5. Its evaluation is explained in Section 19-2-9.

The Laplace correction is caused by deflection of the vertical, which is the angular deflection between the plumb line and a line perpendicular to the ellipsoid at a point (see Chapter 13 for a discussion of deflection of the vertical). More specifically,

$$\text{Laplace correction} = \eta \tan \phi \quad (19\text{-}1e)$$

where

$\eta =$ deflection of the vertical in the prime vertical (east-west) component

$\phi =$ the latitude of the point

A Manual on Astronomic and Grid Azimuth[2] lists values for Laplace corrections at approximately 3000 stations throughout the United States. Estimates can be made from this listing or data provided by the NGS. An NGS computer program ("Deflect 90"), e.g., yields Laplace corrections with input of latitude and longitude. Control station descriptions for

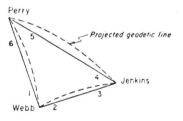

Figure 19-5. Effect of $t - T$ correction. Solid lines are grid lines and appear straight when plotted on grid. Dashed lines are projections of geodetic lines and correspond to lines of sight. (Reprinted with permission).

SPCS83 contain either an estimate of the Laplace correction at the station or other astro-geodetic deflections (i.e., deflections of the vertical), from which Laplace corrections can be calculated using Equation (19-1e). If such a station is within a few miles of the place where an astronomic azimuth is being determined, the value given on the station description is probably close enough to use in the nearby areas. Additional comments are made on this point in Section 19-2-5.

Techniques for the determination of astronomic azimuth are covered in Chapter 17 and also the manual cited above.

Types and Uses of SPC Control Data

The NGS frequently modifies the format and content of SPC control data as SPCS83 evolves in its uses by the surveying and mapping community. For this reason, the exact format and availability of some types of data may vary somewhat from what is described here. The data are of several types. As of this time of writing, SPC information can be obtained from the National Geodetic Information Center (see Appendix 2).

Anyone can subscribe to National Oceanographic and Atmospheric Administration's (NOAA) automatic mailing service by completing a form that can be requested by writing to the address given in the appendix. Subscribers receive periodic notices of availability of new or revised data for their particular area of interest. The important types and forms of information are as follows.

1. GEODETIC CONTROL DIAGRAMS. These diagrams show geodetic control established by federal surveying and mapping agencies and various state and local governments. The most common diagrams are depicted as overlays on USGS topographic maps and cover the conterminous United States and Hawaii. Each of these maps covers an area of 1° latitude by 2° longitude, at a scale of 1:250,000. Another series, at a scale of 1:500,000 is produced for Alaska. Other control diagrams, at larger

scales, have been produced for the coast regions of the United States.

Control diagrams are useful for identifying the names and general locations of triangulation and traverse stations in the region covered on each map, and for planning the layout of local horizontal control systems, and thus are an important starting place for referencing local surveys to SPC.

2. GEODETIC CONTROL DATA. The principal data given on these data sheets are the final adjusted SPC83 coordinates for each named station, including latitude, longitude and the plane X- and Y-coordinates (listed as easting and northing, respectively, on the data sheets). Also listed for each station are the name of the zone in which the station lies, the convergence angle and scale factor at the station, and the elevation and geoid height of the station. These data sets are published for 1° by 2° blocks, corresponding to the areas depicted on the geodetic control diagrams.

These data sheets are useful for making computations related to horizontal control in a region, since they list the coordinates and other numerical data pertinent to each station. But, they do not contain the details necessary to find or fully utilize the stations in the field. This information is found in the station data sheets.

3. STATION DATA SHEETS. A station description is printed for each horizontal control station listed in the geodetic control data. These sheets contain the information necessary to find each station and other descriptive physical aspects of the monuments and the azimuth mark and reference marks associated with them. They also contain station mark history and station recovery information. The data given on the geodetic control sheets are all repeated on these data sheets, but more complete data are given. These data include reference to the USGS quadrangle sheet where the station lies, distance and direction to each reference mark and azimuth mark from the control monument, method of establishing the control and order of accuracy, the shift in

position between NAD27 and NAD83, both the SPC and UTM data, deflection of the vertical, standard errors estimated for most variables listed, and possibly other data that might be available for a particular station.

The coordinates have begun to be listed as northings and eastings, a departure from the traditional, mathematical designation of Cartesian coordinates as X and Y, and the Y is listed before the X. This change may cause some confusion for individuals who have been accustomed to working with coordinates in a mathematical (algebraic and analytical geometry) sense and who recognize the more logical alphabetically ordered listings. For purposes of consistency with the world of mathematics and clarity in the equations utilized in algebra and geometry when referring to plane coordinate systems, this author emphasizes the $X - Y$ designations in this chapter.

These data sheets are the primary reference sheets needed for utilizing horizontal geodetic control. From these, the surveyor should be able to determine all necessary information to reference local surveys to a particular control station.

4. PROJECTION TABLES. The NGS publishes plane coordinate projection tables for each SPC zone. The tables give ellipsoidal constants and projection-defining constants for the zone, and other information. The tables for SPCS83 differ from those of SPCS27. Here we will not go into an in-depth discussion of how to use the tables, the user being referred to previously cited *NOAA Manual NOS NGS-5* for an explanation of the listings and their use. The tables for the Lambert system list scale factors and other values for each zone in a manner similar to what was formerly done for SPC27. However, the tables for the TM system do not list these values in the same straightforward manner and the notation is not explained. The above-mentioned manual must be consulted in order to use the tables.

5. MANUALS AND REFERENCES. In order to use SPC monumentation and data, various other references need to be consulted. The above-mentioned manual is most important because it contains tables that supplement the projection tables specific to each zone and various equations and theories that go beyond the explanations in this short treatise. Its counterpart for SPCS27 is Special Publication No. 235, *The State Coordinate Systems*.[3] Some other references, listed at the end of this chapter, may be helpful in understanding SPCS and its use.

19-2-5. Precision and Accuracy Needed in SPC Variables

The precision and accuracy of the elevation factors, scale factors, and other variables associated with SPCS naturally affect the precision and accuracy of coordinates, just as random or systematic errors in measurements would. The specifications for each survey dictate the care that must be used in determining these factors for each survey. Each variable should be investigated and error propagation theories applied, the results of such investigation being tested against the precision and accuracy requirements of the survey. In many cases, an average grid factor can be used for an entire survey, especially when the survey area is relatively small and the elevations are nearly constant. In other instances, an average scale factor might be used, but elevation factors may have to be changed with each line as topography changes. Each situation is different and the surveyor, as measurement analyst, must decide what is required. For a complete understanding of the concepts of measurement analysis, *Surveying Measurements and their Analysis* should be consulted.

Although the following suggestions are admittedly not based on a rigorous error analysis, they do have a good theoretical basis, in relationship to "ordinary" surveying work.

Precision and Accuracy of Elevation Variables

Geoid heights are around -30 to -36 m in the states east of the Mississippi, vary to a minimum of about -10 m in the Rocky

Mountain region, and are somewhat larger in the remainder of the western conterminous states. A good approximation in most of the eastern states would be -100 ft. Simply subtracting 100 ft from ground elevations in these states would shift the ellipsoid and geoid to an average separation of no more than what existed with NAD27 (separations varying from 0 to 12 m) and closer in most instances.

For most survey work, the elevation can be determined by interpolation between contour lines on a USGS or larger-scale map. Normally, the mean elevation of the two ends of a survey line is sufficiently accurate for determination of the elevation factor.

A combined error of 20 ft (6 m) in the elevation and geoid height causes an error of only 1 part in 1,000,000 or about 0.01 ft in a 10,000-ft length. The surveyor should check the geoid heights in the region of the survey and further investigate the precision needed in elevation and geoid height for various accuracy requirements before deciding on any approximations.

The mean radius of the earth R can be approximated as 20,906,000 ft or 6,372,150 m for both SPCS27 and SPCS83 with no significant error resulting in elevation factors.

The recent adjustment to the *National Geodetic Vertical Datum* (called NGVD88) does not affect the computation of ellipsoid factors or anything else related to SPC. The maximum differences between elevations of bench marks before and after this adjustment are about 1.5 m, and generally the differences are much less. This amount, as can be seen from the above discussions, has negligible effects on elevation factors.

Care in Determining Scale Factors

The shifts in positions between NAD27 and NAD83 seldom exceed 1 second of arc in either latitude or longitude. Such shifts do not affect the scale factors by more than 1 part in 10,000,000 or even 1 part in 1,000,000 in the worst shot situations in Alaska and the extreme western part of the conterminous United

States, near the edges of a zone. For this reason, the SPCS27 tables for scale factors can be used for SPCS83 (Lambert system only) without any appreciable error. The round-off error due to the fact that these tables list scale factors to seven decimal places is comparable to the errors due to a shift in datum (seven decimal places for a number close to 1.0000 translates to about 1 part in 10,000,000). It would seem that this precision is sufficient for most survey work. When the equations listed in the previously cited *NOAA Manual NOS NG-5* are used, all variables should be to enough significant figures to prevent intolerable round-off error in grid distances. The use of these equations generally results in at least seven decimal places in scale factors.

Scale factors change in a north-south direction in Lambert zones and east-west direction in TM zones. Scale factors change most rapidly near the edges of a zone, the rate of change gradually increasing from the zone's center toward its edges. Except for very long lines oriented in the direction of the changes in scale factor, however, the location and orientation of a line within a zone are relatively unimportant. For example, an error of 1 part in 1,000,000 is introduced near the edges of most zones for a length oriented in the direction of the changes.

Besides line length and the above variables, the accuracy in lengths and general scope of the survey are other variables affecting the precision required in both scale and elevation factors.

Precision and Accuracy Requirements Related to Directions

Considering the above discussion on the $t - T$ correction and in view of the short-range nature of EDM used by most surveyors, we realize that this correction would not generally need to be considered. It is usually negligible for most "ordinary" survey work.

In most instances, particularly in mountainous areas, the Laplace correction should *not* be considered negligible. Its magnitude should always be investigated. It is commonly under 5″ in flat areas near the coast and in the plains states, but often exceeds 5″ in hilly country. In mountainous areas, it often exceeds 10″ but rarely 20″. However, it has been found to approach 1′ in certain areas in Alaska and can change rather abruptly in many areas.

It may be realized that the consideration of only the convergence angle is necessary in situations where azimuth accuracies need to be no better than about ±20 to ±30″. That is, the $t - T$ and Laplace corrections might both be ignored if this is the specified accuracy of azimuths. In such situations, Equations (19-1a) and (19-1b) reduce to grid azimuth = astronomic azimuth − convergence Δ. However, when azimuths must be at least ±5 to ±10″, the Laplace correction should be considered. When azimuths with an accuracy of ±1 to ±2″ are desired, the $t - T$ correction should be applied.

19-2-6. Sample Traverse

This example illustrates how SPC are used in a typical traverse. The field procedure is the same in either the Lambert or TM system and for SPCS27 and SPCS83. The differences between computational procedures using SPCS27 and SPCS83 are minor and have been noted here. The following example uses the SPCS27 in a Lambert zone.

The traverse in Figure 19-6 consists of two sides, beginning at NGS triangulation station Bank and closing at Tower, a station established by a county road department in Pennsylvania. The objective is to determine the SPC of station *A*. Elevations are as follows:

Bank = 1763 ft according to NGS data

  A = 1340 ft estimated from USGS quad sheet

Tower = 1542.77 ft according to county records

The county falls in Pennsylvania's north zone. Data for station Bank include its latitude of 41°48′30.000″. Figure 19-6 shows the traverse extending generally to the south of station Bank. Scaling vertically down the page reveals

that station Tower is about 5000 ft south of station Bank. This corresponds to 50 sec of latitude (actually 1 sec of latitude equals approximately 101 ft). Thus, the average latitude of the traverse, to the nearest minute, is 41°48′. Another common way to obtain this figure is by scaling on a USGS quad sheet using the $2\frac{1}{2}$-min tick marks along the margins and interior of the sheet.

Table 19-1 lists excerpts from projection tables for Pennsylvania that provide data for this traverse problem. Using a latitude of 41°48′ in the north zone, we find the scale factor to be 0.9999791. If we assume 1500 ft as an approximate average elevation, the sea-level factor is given in Table 19-2 as 0.9999283. The grid factor (Section 19-2-4) is the product of these two numbers—namely, 0.9999074. The significance of the last digit (4) is questionable because elevation was only approximate, but that digit is not needed anyway. Measured distances contain only six digits, and even the last of these is uncertain. A grid factor of 0.999907 is adopted.

Figure 19-6 shows an azimuth line at station Bank existing traverse line at station Tower, both as solid lines. Azimuths of these two fixed lines (from grid north) are shown using a solid clockwise arc. Dashed lines and arcs indicate *measured* quantities (two distances and three angles). The published data on stations Bank and Tower are as follows:

Triangulation station Bank:
  X = 2,567,887.24 ft
  Y = 602,126.54 ft
  Azimuth to Bank azimuth mark
   = 291°11′07″ from grid south
  (converted to north basis on drawing)
  Approximate elevation = 1763 ft

Station Tower:
  X = 2,564,481.50 ft
  Y = 597,001.53 ft
  Azimuth to Station Pin
   = 298°23′09″ from grid north
  (as published by the county)
  Elevation = 1542.77 ft

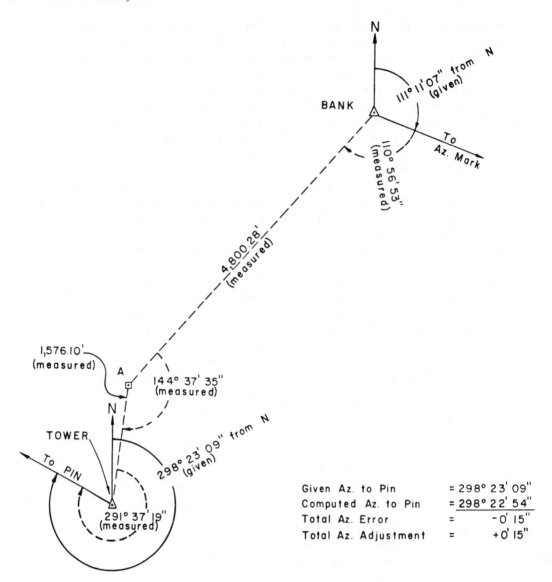

Given Az. to Pin	= 298° 23' 09"
Computed Az. to Pin	= 298° 22' 54"
Total Az. Error	= −0' 15"
Total Az. Adjustment	= +0' 15"

Figure 19-6.   Traverse to establish state coordinates of station *A*.

The traverse computation has been carried forward in Tables 19-3 and 19-4 using the compass rule, but a least-squares approach could have been employed. For the compass rule, the steps involved are as follows:

1. Distribution of the angular error of closure. This provides corrected grid azimuths for all traverse sides. Figure 19-6 illustrates that angles to the right were measured. The misclo-

sure (or the difference between the *given* azimuth at station Tower and azimuth *computed* by using the three measured angles and starting azimuth at Bank) was found to be −0°00′15″. This error (misclosure) was distributed equally to the three angles, as noted in Table 19-3. Final grid azimuths are listed in column 6.

Table 19-4 completes the traverse computation, using grid azimuths from the north.

Table 19-1. Excerpts from Pennsylvania projection tables

Lambert Projection for Pennsylvania—North (Table I)

Latitude	R (ft)*	y' y Value on Central Meridian (ft)	Tabular Difference for 1 sec of Latitude (ft)	Scale in Units of 7th Place of Logs	Scale Expressed as a Ratio
41°46'	23,628,092.47	582,957.90	101.21800	− 106.9	0.9999754
47	23,622,019.39	589,030.98	101.21850	− 99.0	0.9999772
48	23,615,946.28	595,104.09	101.21900	− 90.8	0.9999791
49	23,609,873.14	601,177.23	101.21933	− 82.2	0.9999811
50	23,603,799.98	607,250.39	101.22000	− 73.2	0.9999831

Note: R as listed in these tables is *not* the radius of the earth. See Figure 19-7 for an explanation.
Source: NGS Special Publication No. 267. 1968. Washington DC.

Such grid azimuths are recommended over grid bearings because signs of the trig functions will, in every case, give signs of latitudes and departures. This is a convenience if trig functions are generated by a pocket calculator.

2. Reduction of measured distances to grid distances. This was done in Table 19-4 by multiplying the values in column 3 by those in column 4. The resulting correction is 0.093 ft per 1000 ft (smaller than 1 part in 10,000).

3. Computation of latitudes and departures, columns 7 and 8.

Table 19-2. Sea-level factors

Elevation in ft (m)	Sea-Level Factor
500 (152)	ᴜ.9999 761
1000 (305)	0.9999 522
1500 (457)	0.9999 283
2000 (610)	0.9999 043
2500 (762)	0.9998 804
3000 (914)	0.9998 565
3500 (1067)	0.9998 326
4000 (1219)	0.9998 087
4500 (1372)	0.9997 848
5000 (1524)	0.9997 608
5500 (1676)	0.9997 369
6000 (1829)	0.9997 130
6500 (1981)	0.9996 891
7000 (2134)	0.9996 652
7500 (2286)	0.9996 413
8000 (2438)	0.9996 173

4. Computation of preliminary grid coordinates. These are listed in Table 19-4, columns 9 and 10. The preliminary X-coordinate of A, e.g., is 2,564,667.15. Misclosures in the X and Y directions are found by subtracting the given or fixed coordinates of station Tower from those obtained by traversing. The relative misclosure or ratio of precision is found to be 1 : 10,200.

5. Coordinate adjustment by compass rule. Adjustments of the preliminary coordinates are computed in proportion to the accumulated traverse distance up to any given station. The adjustment of the preliminary X-coordinate of A, e.g., is (−0.012)(4.80) = −0.06 ft, where 4.80 is the distance traversed to A in thousands of feet. These corrections were applied in columns 9 and 10 to obtain final state plane coordinates for station A. The full corrections (−0.08 ad −0.62) applied at station Tower will, of course, give the fixed values written in previously.

Note that the misclosures and adjustments are in the *coordinate* columns, whereas in a loop traverse (closing on itself), it is usual to list the misclosures and adjustments in the *latitude* and *departure* columns. (The algebraic sums of latitudes and departures are made to equal zero.) Incidentally, it is possible to apply the coordinate adjustment method to a loop

Table 19-3. Angular closure and adjustment

1 Station Occupied	2 Backsight	3 Azimuth to Foresight Azimuth to	4 Prelim. Azimuth Angle to Right Prelim. Azmith	5 Adjust. Cumul. Adjust.	6 Final Azimuth Angle to Right Final Azimuth
		Bank azimuth	111°11′07″		111°11′07″
Bank	Bank azimuth	A	110°56′53″	+5	110°56′58″
		A	220°08′00″	+5	222°08′05″
		Bank	42°08′00″		42°08′05″
A	Bank	Tower	144°37′35″	+5	144°37′40″
		Tower	186°45′35″	+10	186°45′45″
		A	6°45′35″		6°45′45″
Tower	A	Pin	291°37′19″	+5	291°37′24″
		Pin	298°22′54″	+15	298°23′09″*

*Given.

traverse as a special case in which the first and last stations in Table 19-4 are the same.

A traverse in a transverse Mercator zone is handled similarly except that scale factors are tabulated for every 5000 ft of X distance from the central meridian, rather than every minute of latitude. This X distance, called X', is zero at the central meridian. For any survey, the X' used is simply the difference between the average X-coordinate (found from coordinates of the control points or a USGS quad sheet) and the X-coordinate of the central meridian. The latter is a constant of the zone called C, usually set at 500,000.00 ft (prior to adoption of the *1983 Datum*). See Table 19-6, discussed later.

19-2-7. Conversions Between Plane and Geodetic Coordinates

Boundries between zones were placed at state and county lines to minimize the occasions when a survey would involve more than one zone, the assumption being that most surveys stay confined within political boundaries. On a design project such as a bridge between Ohio and Kentucky, it is necessary to use the same plane coordinate zone for the surveying, for closure and adjustment purposes. The coordinate system of either zone can be adopted. Station descriptions of NGS stations lying near zone boundaries usually contain the plane coordinates of all nearby zones, which avoids any need to make conversions. If such coordinates are not given, a conversion is necessary. The conversion starts with the latitude and longitude that are, of course, the same regardless of the zone. For the Ohio-Kentucky example, the plane coordinate projection tables can be used to convert the latitude to a Y-coordinate and the longitude to a X-coordinate in the zone corresponding to the tables used.

The conversion problem can exist in either direction—i.e., a plane coordinate of a point might be known and a precise latitude and longitude desired for that point. Of course, all such conversions can be made approximately by using grid ticks on USGS maps, but if high precision is needed, computations must be made. Because of the large numbers involved, at least 10 significant figures must be used in all variables to avoid round-off errors.

Prior to SPCS83 and the increased use of computers and programmable calculators, conversions were commonly made using fewer automated methods. Currently, software is

Table 19-4. Traverse closure and adjustment (compass rule) using azimuths from north

1	2	3	4	5	6	7	8	9	10
				Grid Distance	Sine			Preliminary X Adjustment	Preliminary Y Adjustment
Station	Grid Azimuth from North	Measured Distance	Grid Factor	Cumulative Distance	Cosine	Departure	Lattitude	Final X	Final Y
Bank							(Given) →	2,567,887.24	602,126.54
	222°08'05"	4800.28	0.999907	4799.83	−0.6708761	−3220.09	−3559.41	2,564,667.15	598,567.13
				(4800)	−0.7415694			Adj. = −0.06	Adj. = −0.46
A								2,564,667.09	598,566.67
	186°45'45"	1576.10	0.999907	1575.95	−0.1177540	−185.57	−1564.98	2,564,481.58	597,002.15
				(6376)	−0.9930428			Adj. = −0.08	Adj. = −0.62
Tower							(Given) →	2,564,481.50	597,001.53
							Misclosure =	+0.08	+0.62

Linear misclosure = $\sqrt{0.08^2 + 0.62^2} = 0.62$ ft

Relative misclosure = $0.62/6376 = 1/10,200$

Compass rule:

X adjustment = $-0.08/6.376 = -0.012$ ft per 1000 ft of cumulative distance

Y adjustment = $-0.62/6.376 = -0.097$ ft per 1000 ft cumulative distance available from NGS and other sources to make the conversions. For purposes of explanation and illustration of the variables involved, a "long-hand" solution has been retained here. The problem will be presented by two examples. Both will be solved using projection tables published for SPCS27, which seems to afford the maximum explanation for understanding the theories involved.

Example 19-1 (Lambert). A project was being planned in 1980 in Carbon County, PA (in the north zone) by the AAA Mapping Company. In 1975, the Acme Survey Company ran a long traverse for the Lehigh County Development Authority (south zone). They established station K that is in Lehigh County but only 1000 ft south of the county line, making it advantageous to the AAA Mappers as a control point for work in Carbon County. South zone coordinates for K according to the 1975 survey were $X_s = 2,586,745.20$ ft and $Y_s = 515,262.32$ ft. Find the north zone coordinates for station K.

Table 19-5. Excerpts from north zone and south zone tables for Pennsylvania

Lambert Projection for Pennsylvania—South (Table I)

Latitude	R (ft)	y' y Value on Central Meridian (ft)	Tabular Difference for 1 sec of Latitude (ft)	Scale in Units of 7th Place of Logs	Scale Expressed as a Ratio
40°41'	24,493,042.22	491,784.21	101.19783	− 140.2	0.9999677
42	24,486,970.35	497,856.08	101.19833	− 134.9	0.9999689
→ 43	24,480,898.45	503,927.98	101.19866	− 129.2	0.9999703
44	24,474,826.53	509,999.90	101.19917	− 123.2	0.9999716
45	24,468,754.58	516,071.85	101.19950	− 166.8	0.9999731

Lambert Projection for Pennsylvania—North (Table I)

Latitude	R (ft)	y' y Value on Central Meridian (ft)	Tabular Difference for 1 sec of Latitude (ft)	Scale in Units of 7th Place of Logs	Scale Expressed as a Ratio
40°41'	24,022,804.71	188,245.66	101.20467	+ 166.3	1.0000383
42	24,016,732.43	194,317.94	101.20467	+ 150.4	1.0000346
→ 43	24,010,660.15	200,390.22	101.20467	+ 134.9	1.0000311
44	24,004,587.87	206,462.50	101.20467	+ 119.8	1.0000276
45	23,998,515.78	212,534,78	101.20450	+ 105.1	1.0000242

	North Zone	South Zone
C	2,000,000.00 ft	2,000,000.00 ft
Central meridian	77°45'00".000	77°45'00".000
R_b	24,211,050.37 ft	24,984,826.43 ft
y_0	455,699.10 ft	407,025.76 ft
l	0.66153 97363	0.64879 31668
$\dfrac{1}{2\rho_o^2 \sin 1''}$	2.357×10^{-10}	2.358×10^{-10}

Source: NGS Special Publication No. 267, Washington DC 1968.

There is no direct way to go from south zone coordinates to north zone coordinates (except approximately). The latitude and longitude of station K must be found as an intermediate step. This initial computation (to obtain ϕ and λ) and the final step (to obtain X_n and Y_n) involve only right-triangle trigonometry and the tables for both zones. Excerpts from the tables are shown in Table 19-5.

The small angle in the right triangle ABP of Figure 19-7 is θ; both the convergence and hypotenuse R appear in the tables. The "op-posite side" AP is $X' = R \sin \theta$. The "adjacent side" $AB = R \cos \theta$. The X-coordinate of any point P (station K in the example) is $R \sin \theta + 2,000,000.00$ ft because the central meridian has been assigned that large number to avoid negative coordinates anywhere in the zone. The Y-coordinate of point $P = R_b - R \cos \theta$, where R_b is the radius from apex B to the lowest parallel listed in the table, said parallel being actually somewhat south of the zone. R_b is a constant in any one zone and a 10-digit number if carried to hundredths of

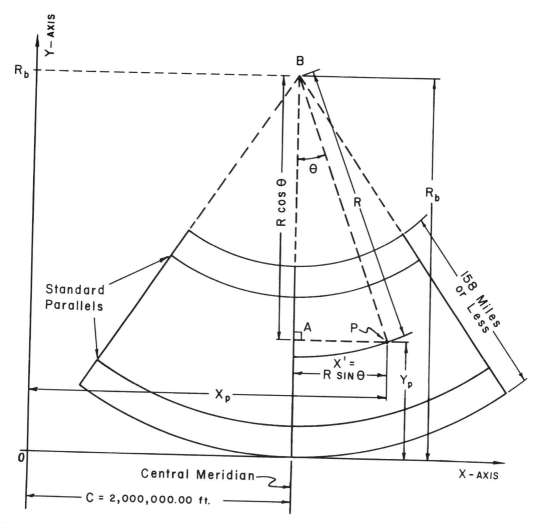

Figure 19-7.   Relationships for any point P.

a foot. When that accuracy (hundredths) is desired in the coordinates, the sines and cosines must also be carried to 10 places.

To obtain ϕ and λ from X_s and Y_s,

$$X' \quad \text{or} \quad R \sin \theta = X_s - 2,000,000.00$$

$$= 586,745.20$$

$$\tan \theta = \frac{\text{opp.}}{\text{adj.}} = \frac{X'}{R_b - Y_s}$$

$$= \frac{586,745.20}{24,984,826.43 - 515,262.32}$$

$$= \frac{586,745.20}{24,469,564.11} = 0.0239785718$$

$$\theta = +1.373607741°$$

$$R = \frac{R_b - Y_s}{\cos \theta} = \frac{24,469,564.11}{0.9997126380}$$

$$= 24,476,597.75 \text{ ft}$$

With convergence θ and radius R now known, the longitude λ and latitude ϕ can be obtained from the south zone tables. The plus value of θ indicates station K is east of the central meridian. This is also evident from the general location of Lehigh County and by the positive value of X'. The longitude corresponding to a calculated value of θ may be interpolated from a table (not shown), but it is easier to recognize that K is east of $77°45'00''$ (the central meridian) by an angle

$$\Delta \lambda = \frac{\theta}{0.6487931668} = 2.117173564°$$

where the denominator is the constant l (the rate of change of θ for a unit change in longitude for the south zone. Longitude can now be obtained directly by subtraction, but in this case, where both zones have the same central meridian, only the difference in longitude, just found, is needed. The decimal form is convenient for the north zone computation.

Latitude is obtained from the south zone projection table, shown in the upper half of

Table 19-5, by interpolation. Look for an R value in the second column that is the next larger number than the one calculated. In this case, the value found is 24,480,898.45, which corresponds to a latitude of $40°43'$. Additional seconds of latitude are found by using the tabular difference in the fourth column as an aid to interpolation.

Additional seconds

$$= \frac{24,480,898.45 - 24,476,597.75}{101.19866}$$

$$= 42.4976''$$

$$\phi = 40°43'42.4976''$$

To obtain X_n and Y_n from λ and ϕ, the first step is to find the north zone convergence θ. This could be done from tables, but as before, it is easier to use the constant rate of change of θ in the zone.

$$\theta = (0.6615397363)(2.117173561°)$$

$$= 1.400594441°$$

The radius R is found in the north zone projection table by interpolation between $\phi = 40°43'$ and $\phi = 40°44'$.

$$R = 24,010,660.15$$

$$- (42.4976'')(101.20467) = 24,006,359.19 \text{ ft}$$

$$X_n = R \sin \theta + 2,000,000.00$$

$$= 2,586,776.63 \text{ ft}$$

$$Y_n = R_b - R \cos \theta$$

$$= 211,863.41 \text{ ft}$$

where R_b is the north zone constant, 24,211,050.37 ft. Differences of a few hundredths can occur rather easily in such a problem. For example, if the latitude had been rounded off to $40°43'42.498''$, Y_n would be increased by 0.04 ft. However, the positions of survey points are generally uncertain in the hundreths place, anyway. The sample problem in Section 19-2-6 had compass rule corrections for station A of 0.06 and 0.47 in the X and Y directions and would have had a correction greater than the 0.04 mentioned, even if the misclosure had been 1 : 50,000. Also, of course, the given coordinates of both stations Bank and Tower in that example, and of station K in this one, contained original survey errors.

Example 19-2 (Mercator). A photogrammetric mapping project is being planned near the southern border of Idaho. The area to be covered lies mostly in Owyhee County (west zone) but extends into Twin Falls County (central zone). The entire map will be plotted on west zone coordinates. Station Rock in Twin Falls County was part of an earlier control survey and has the following central zone coordinates that now must be transformed into west zone coordinates: $X_c =$ 224,662.81 ft and $Y_c = 131,703.72$ ft.

Transverse Mercator zones are more complex than Lambert zones. The meridians are not straight lines; the parallels are not circular arcs (see Figure 19-8. The physical significance of some numbers taken from the tables will not be apparent in the following solution to Example 19-2 (see Table 19-6).

As in a Lambert zone, $X' = X - C$, the distance from the central meridian to a station. In this type of zone, C usually is 500,000.000 ft. The computation is shown in Table 19-7 on a form provided in the projection tables. The Y_0 term is the Y-coordinate of the point where the parallel through Station Rock crosses the central meridian. This coordinate will always be smaller than Y because of the curvature of the parallel. P is taken from the tables by interpolation given the Y-coordinate, and if we know X', d is found on the same page.

The Y_0 value just found is used to compute the latitude ϕ and then the quantity H, from

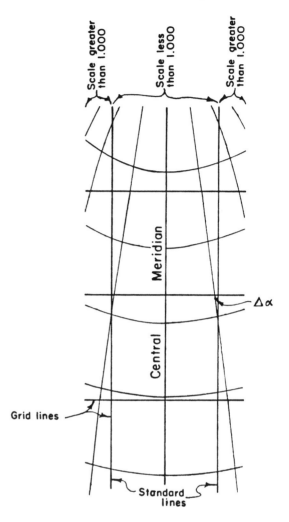

Figure 18-8. Parallels and meridians on transverse Mercator grid.

another page of the tables, as follows:

$$\phi = 42°01' + \left(\frac{130{,}073.60 - 127{,}532.48}{101.21950} \right)''$$

$$= 42°01'25''.105$$

$$H = 75.484549 - (25.105)(0.00032858)$$

$$= 75.476300$$

The small quantity a is obtained from the last column as -0.888.

Next, "approx. $\Delta\lambda$" is calculated to the nearest second as shown ($X' \div H$). This is

Table 19-6. Excerpts from Idaho projection tables

Y Correction for Computation of Geographic Positions from Plane Coordinates, Transverse Mercator Projection, Idaho, East and Central Zones

$$P(x'/10{,}000)^2 + d = V(\Delta^2/100)^2 + c$$
$$P \text{ taken out for } y\text{-coordinate}$$
$$d \text{ taken out for } x'$$

	y	P	ΔP	x'	d
	0	2.12307	2053	0	0.00
→	100,000	2.14360	2071	50,000	+0.01
	200,000	2.16431	2089	100,000	+0.02
	300,000	2.18520	2106	150,000	+0.04
	400,000	2.20626	2126	200,000	+0.07
	500,000	2.22752	2144	250,000	+0.07
	600,000	2.24896	2164	300,000	+0.06 ←

Transverse Mercator Projection Idaho, East and Central Zones

Latitude	Y_0 (ft)	ΔY_0 per sec		H	ΔH per sec	V	ΔV per sec	a
42°00'	121 459.32	101.219 33		75.504 258	328.48	1.224 910	1.26	−0.895
→ 42°01'	127 532.48	101.219 50	→	75.484 549 →	328.58	1.224 986	1.25 →	−0.890
42°02'	133 605.65	101.219 83		75.464 834	328.70	1.225 061	1.25	−0.885
42°03'	139 678.84	101.220 00		75.445 112	328.80	1.225 136	1.25	−0.880
42°04'	145 752.04	101.220 50		75.425 384	328.90	1.225 211	1.23	−0.875

Transverse Mercator Projection, Idaho, East and Central Zones

	$\Delta \lambda''$	b	Δb	c
→	2600	+1.484	+0.007	−0.111
	2700	+1.491	+0.001	−0.116
	2800	+1.492	−0.005	−0.121
	2900	+1.487	−0.011	−0.125
	3000	+1.476	−0.017	−0.130
→	3600	+1.270	−0.059	−0.131
	3700	+1.211	−0.067	−0.128
	3800	+1.144	−0.075	−0.124
	3900	+1.069	−0.084	−0.120
	4000	+0.985	−0.091	−0.115

Transverse Mercator Projection, Idaho, West Zone

Latitude	Y_0 (ft)	ΔY_0 per sec		H	ΔH per sec	V	ΔV per sec	a
42°00'	121 457.62	101.217	83	75.503198	328.48	1.224 892	1.26	−0.895
→ 42°01'	127 530.69	101.218	00	75.483489	328.58	1.224 968	1.25	−0.890
42°02'	133 603.77	101.218	50	75.463774	328.68	1.225 043	1.25	−0.885
42°03'	139 676.88	101.218	67	75.444053	328.80	1.225 118	1.25	−0.880
42°04'	145 750.00	101.219	00	75.424325	328.90	1.225 193	1.23	−0.875

Continued

Table 19-6. *Continued*

Table for e

y \ x'	100,000	200,000	↓ 300,000	400,000
→ 0	0.0	0.1	0.3	0.7
500,000	0.0	0.1	0.3	0.8
1,000,000	0.0	0.1	0.4	0.9
1,500,000	0.0	0.1	0.4	1.0
2,000,000	0.0	0.2	0.5	1.1
2,500,000	0.0	0.2	0.5	1.3

	East and Central Zones			West Zone	
y	M	↓	ΔM	M	ΔM
0	0.008	7592	847	0.008	7593	847
→ 100,000	0.008	8439	854	0.008	8440	854
200,000	0.008	9293	862	0.008	9294	862
300,000	0.009	0155	870	0.009	0156	870
400,000	0.009	1024	877	0.009	1026	877

Table 19-6. Excerpts from Idaho Projection Tables

Table for g $\quad \Delta \alpha'' = \sin \phi (\Delta \lambda'') + g$

				$\Delta \lambda''$			
Latitude	0''	1000''	2000''	3000''	↓ 4000''	5000''	6000''
41°	0.00	0.00	0.02	0.08	0.19	0.37	0.63
42°	0	0	0.02	0.08	0.18	0.36	0.63
→ 43°	0	0	0.02	0.08	0.18	0.36	0.62
44°	0	0	0.02	0.08	0.18	0.35	0.61
45°	0	0	0.02	0.08	0.18	0.35	0.60

Source: NGS Special Publication No. 306, 1975. Idaho Projection Tables; Washington D.C.

used to obtain b from the tables by interpolation as follows:

$$b = +1.270 + (0.48)(-0.059) = +1.242$$

The final value of $\Delta \lambda$ in seconds $= (X' \pm ab) \div H$, in which the sign of ab depends on the sign of X'. When X' is minus, as in this example, the sign of ab is unchanged from what the table indicates (also minus in this example). When X' is plus, the indicated sign of ab is reversed. The footnote in Table 19-7 explains this in an equivalent way. The $\Delta \lambda$ term is converted to degrees and subtracted from the longitude of the central meridian to obtain λ. This concludes the *inverse computation* (ϕ and λ from X and Y) using central zone tables. The *forward computation* will use west zone tables (see Table 19-6).

Table 19-8 shows the computation on a form provided in the same government booklet. Referred to the central meridian of the west zone, $\Delta \lambda$ is plus (to the east) and in seconds equals $+2651.990$. For later use, $\Delta \lambda''/100^2$ is filled in. The Y_0, H, V, and a terms are interpolated, being tabulated for given latitudes. The b term, as in the inverse computation, is obtained from another part of the tables, listed for given values of $\Delta \lambda''$. The small quantity c is found there also but not entered on the form.

The X' distance equals $H \cdot \Delta \lambda'' \pm ab$. When the sign of the $H \cdot \Delta \lambda''$ term is minus, the sign of ab is reversed from that given by

Table 19-7. Calculating ϕ and λ for Example 19-2

Station Rock (Central Zone, ID)

X	224,662.81	Y	131,703.72
C	$-500,000.00$	$P\left(\dfrac{x'}{10,000}\right)^2 + d$	-1630.12
X'	$-275,337.19$	Y_0	130,073.60
P	2.15017	Approx. $\Delta\lambda = X' \div H$	$-3648''$
d	$+0.06$	$\Delta\lambda = (X' \mp ab) \div H$	$-3648''.010$
H	75.476300	$\Delta\lambda$	$-1°00'48''.010$
a $\quad$ b	$-0.888 \quad +1.242$	Central meridian	$114°00'00.000$
ϕ	$42°01'25''.105$	$\lambda = $ C.M. $- \Delta\lambda$	$115°00'48''.010$

When ab is $(+,$ decrease$/-,$ increase$)X'$ numerically.

Table 19-8. Calculating X and Y for Example 19-2

	West Zone	
Station		Rock
ϕ		42°01'25".105
λ		115°00'48".010
C.M.		115°45'00".000
$\Delta\lambda$ = Central mer. $-\lambda$		+0°44'11.990
$\Delta\lambda''$		+2651.990
$\dfrac{\Delta\lambda''^{\,2}}{100}$		703.305
H		75.475240
V		1.224999
a $\qquad b$		−0.888 $\qquad$ +1.488
$x' = H \cdot \Delta\lambda \pm ab$		+200,158.26
$V\left(\dfrac{\Delta\lambda''}{100}\right)^2 \pm c$		861.43
Tabular y or Y_0		130,071.77
x		700,158.26
y		130,933.20
$\Delta\alpha''$		
$\Delta\alpha$		
Geod. az. to az. mk.		
Grid az. to az. mk.		

$x = x' + 500,000$

$y = \text{tab. } y + V\left(\dfrac{\Delta\lambda''}{100}\right)^2 \pm c$ When ab is $\dfrac{-\text{decrease}}{+\text{increase}}$ numerically.

H and V = tab. H and tab. V

$H \cdot \Delta\lambda$

$\Delta\alpha'' = \Delta\lambda'' \sin\phi + g$ $\qquad$ g increases $\Delta\lambda'' \cdot \det\phi$ numerically.

Grid az. = geod. az. $- \Delta\alpha$

the tables. In this example, ab retains its minus sign. The X-coordinate is $X' + 500,000.000$, and the letter may be written X_w to indicate the west zone. The Y_0 term, obtained earlier, is called "tabular y" on the form in Table 19-8 and in the footnote. The Y-coordinate or Y_w is calculated using the equation in that footnote.

19-2-8. Transformations Between NAD27 and NAD83 Coordinates

A transformation would be desirable whenever a surveyor had control data on both NAD27 and NAD83 in an area and wanted to utilize stations from both systems in a survey. The typical case would be when SPC27 coordinates had previously been determined for a survey and it is desired to place these stations on SPCS83 for incorporation with other surveys referenced to SPC83.

The following comparison of positions of NGS triangulation station "Smathers," located in Johnson City, might help us to understand the rudiments of transformations:

	NAD27	NAD83
ϕ	36°18'28.877"	36°18'29.24041"
λ	82°20'49.760"	82°20'49.21519"

	SPC27 (ft)	SPC83 (m)	SPC83 (ft)
X	3,076,094.78	927,999.661	3,044,612.22
Y	717,533.73	225,226.210	738,929.66

Note that the shift is $+0.362''$ in latitude and $-0.545''$ in longitude. This is about 37 ft in northing and -40 ft in easting. The published plane coordinates for SPC83 represent the values in meters. The values in feet for SPC83 were determined from the metric values using the pre−1960 definition of the foot (see the second subsection under Section 19-2-9). The large differences in values in feet have nothing to do with the datum readjustment. These discrepancies are caused by changes in the origin values. Had the central meridian and origin for X-coordinates not been assigned new values, the coordinate shifts in northing and easting would have been only 37 and -40 ft, rather than several thousand feet. These comparisons will illustrate, among other things, it is hoped, that transformations between NAD27 and NAD83 must start from geodetic coordinates. If values are known only in plane coordinates, a conversion to geodetic coordinates must be made (see Section 19-2-7). It may also be realized that simple conversions from meters to feet is not the problem when considering transformations, and that neither the numerical value for the origins nor the units used for plane coordinates have anything to do with transformations.

As with other surfaces, coordinate transformations can be made mathematically between different ellipsoids without any errors other than the usual computational (round-off) errors. Uncertainties much larger than this occur in transformations between NAD27 and NAD83 because of "distortions" in the readjustment (nonuniformity caused by the variations in weighting the data, relative accuracies of the data used, etc.). Except at the stations used in the readjustment (as with the comparison in the above tabulation), any transformation has uncertainties larger than most surveying projects would tolerate. The best procedures available cannot be executed without uncertainties on the order of 10 cm or more for points located anywhere between control stations used in the readjustment.

The best way to arrive at SPC83 coordinates is to use the field data (distances, angles, azimuths, etc.) for the survey and calculate coordinates as explained in Section 19-2-6. This avoids the transformation errors caused by distortions in the system. For survey purposes, this may be the only recourse. For mapping and other purposes requiring less accuracy than land or control surveys, some of the transformation systems mentioned in the next paragraph may be useful.

Since the mid–1980s, the NGS and others have devised various methods for the transformation of coordinates. One of the first was called LEFTI ("leftie"). It is a four-parameter similarity transformation, said to be accurate to about 1 m. The latest system suggested by the NGS is called NADCON (North American Datum conversion). It relies on a simultaneous model of the shift values for a large region and local modeling. The accuracy is said to be around 10 to 15 cm. Another recent conversion package is called CORPSCON. Details on this method are not yet available, but it is said to be comparable to NADCON. The mathematics of these transformations will not be presented here. The user does not perform any calculations, but only keys values of latitude and longitude into the program. The NGS sells the NADCON program as one of several in its library of available programs related to SPC.

A discussion of these and other ideas on transformations can be found in *NAD 83 State Plane Coordinates and Datum Transformations*[5] and *North American Datum of 1983*.[6] It is likely that other methods will be devised in the future, but it is unlikely that any will yield accuracy in results sufficient for use in most surveys. It is reiterated here that the best way to place points on either SPC27 or SPC83 is to start and end a traverse or other control survey on stations having the coordinates known in the respective system, then use the appropriate, respective projection constants and values to place measurements on the desired ellipsoid and grid.

19-2-9. Miscellaneous Problems Concerning State Plane Coordinates

Plat Distances and Areas When SPC Are Used

Except when the grid factor happens to be 1.00000, ground distances will not agree with grid distances and therefore lengths of property lines shown on a plat will not be consistent with distances determined from the SPC. Also, the areas computed using SPC will be the areas projected to the grid, not those of the ground surface. For example, an "inversed" distance of 2643.46 ft using SPC when the grid factor is 0.9999675 is consistent with a ground distance of 2643.55 ft, since the grid length is simply divided by the grid factor to obtain ground length. Similarly, an area of 1,735,900 sq ft as computed using SPC would be divided by the square of the grid factor to determine the ground area, resulting in 1,736,013 sq ft.

These relationships should cause no problem for the knowledgeable surveyor. Other than the surveyor, nobody else would probably make use of the coordinates anyway, so misuse or confusion would be unlikely. A simple note somewhere on the plate giving the average grid factor is all that the surveyor needs to relate ground measurements to grid data. For other possible users who may understand co-ordinate geometry, but not SPC, a note on the plat would be useful, such as "to calculate ground distances from lengths computed using SPC, divide the computed lengths by the grid factor," and "to calculate ground areas from grid areas computed from SPC, divide the computed areas by the square of the grid factor."

Conversion of Metric Coordinates to Feet

Since many surveyors want to continue using feet instead of meters, a conversion must be made from published NAD83 metric coordinates. This would be a simple matter if it were not for the fact that there are still two definitions of the foot in existence. Prior to 1960, there was only one definition of the foot, based on the meter. By international agreement, the foot (actually the inch) was redefined in 1960 so as to arrive at a value mutually compatible between certain countries of the world. The difference is slight, but is important when SPC conversions are made.

The historic reason why the former definition was retained by the USC and GS in 1960, while the rest of industry adopted the international definition, was that SPCs were at that time published in feet (based on the meter as a standard of measure) and changing their values in feet based on a new definition would have meant that all coordinate values would have changed. To prevent confusion, the USC and GS was given special permission to retain the old definition. This then became known as the "survey foot." With a complete readjustment and change of ellipsoid as per NAD83, where coordinate values would change anyway, and because it was decided that plane coordinates would be listed in meters and not feet, many users felt that the old definition would logically be discarded. Ultimately, probably because of nostalgia and perhaps a lack of historical perspective and understanding of the original reasons for retaining the earlier defi-

nition, a movement began to retain the survey foot in making conversions of SPC83 values. The NGS, not wanting to disregard the desires of users of the system, did not dictate either definition.

Despite the logic of discarding it, many states have retained the older definition. As of this writing, approximately 40 states had either enacted legislation concerning SPC83 or were in the process of doing so. Of these states, 15 of them have specified that the former definition of the foot would be retained, 5 have specified that the newer definition should be used, and 20 have not addressed this question at all. It is likely that the lack of mention of the definition of the foot in most of the laws on SPCS83 is because it was commonly understood at one time that the obsolete survey foot would be discarded after NAD83 was completed and the newer internationally agreed on definition adopted. The two definitions are

Pre−1960 Definition (survey foot)	1960 Redefinition (international foot)
1 m = 39.37 in.	1 in. = 2.54 cm

From the above exact definitions, the following can be derived:

$$1 \text{ m} = (3937/1200) \text{ ft}$$

$$= 3.280833333 \text{ ft}$$

$$1 \text{ m} = (1250/381) \text{ ft}$$

$$= 3.280839895 \text{ ft}$$

There is no difference between the results of conversions until the digits reach the 6th or 7th significant figure, depending on the number being converted. A conversion of 1,000.00 m would be 3,280.83 ft according to the old system and 3,280.84 ft according to the new system, the number differing by one unit in the 6th place, but a value of 4,000.000 m would convert to 13,123.36 ft according to the new system and 13,123.33 ft according to the old system, the number differing by three units in the 7th place. When converting numbers as large as SPC, the difference in results can be several feet. For example, the Smathers station in Johnson City, TN has coordinates as follows, as published and converted using the two definitions:

	Former definition	New definition
SPC83 (m)	SPC83 (ft)	SPC83 (ft)
X 927,999.661	3,044,612.22	3,044,618.31
Y 225,226.210	738,929.66	738,931.14

As can be seen, the difference is nearly 6 ft in the X-coordinate and 1.5 ft in the Y-coordinate. The differences are proportional to the numerical value of the coordinate, which is the largest on the easterly side of a Lambert zone and the northerly side of a TM zone. The differences also depend on the size of values assigned to the central meridian (origin of eastings) and the origin for northings.

Arc-to-Chord Correction

The arc-to-chord $(t - T)$ correction was discussed in subsections under Sections 19-2-4 and 19-2-5. Figure 19-5 illustrates the concept. When high-order precision is required, this correction should be considered. The longer the survey line, the more it deviates from a straight line joining the two stations. The azimuth obtained directly from the X- and Y-coordinates of the ends of the line can be converted by adding the $t - T$ correction. The result, designated here as the *terminal grid azimuth*, is the value to use for referencing traverses or other control surveys to an existing long line, as well as for converting from grid to geodetic azimuth using Equation (19-1).

The $t - T$ correction varies with the direction of a line and its location in the zone, as well as its length. Fomerly, with SPC27, the formula recommended for Lambert zones, was

$$t - T = -\frac{X_2 - X_1}{2\rho_o^2 \sin 1''}$$

$$\times \left(Y_1 - Y_o + \frac{Y_2 - Y_1}{3} \right) \quad \text{(19-2a)}$$

where $1/(2\rho_o^2 \sin 1'')$ is a constant for a zone. The result is in seconds of arc.

Subscripts 1 and 2 refer to the near and far ends of the line, respectively, and Y_0 is the Y value on the central meridian located roughly halfway between the two standard parallels. The value of Y_1 determines the algebraic sign of $t - T$ since Y_0 is subtracted from it. It is noted that $t - T$ approaches zero as X_1 and X_2 become closer in value (north-south orientation).

For the TM system, using the equations cited in publications for SPC27, we obtain

$$t - T = \frac{(Y_2 - Y_1)(2X_1' + X_2')}{6\rho_o^2 \sin 1''} \quad \text{(19-2b)}$$

where $1/(6\rho_o^2 \sin 1'')$ is a constant for a zone. The result is in seconds of arc.

The X' values are computed from $X' = X - X_{cm}$, where X is the coordinate of the respective station (1 or 2) and X_{cm} the X value of the central meridian for the zone. For a line running east-west (Y_1 and Y_2 being equal), $t - T$ is zero.

Publications for SPCS83 recommend one equation for both the Lambert and TM systems, which is

$$\delta = -(2.36 \times 10^{-10}) \Delta X \Delta Y$$

for coordinates in ft or   (19-3a)

$$\delta = -(25.4 \times 10^{-10}) \Delta X \Delta Y$$

for coordinates in m   (19-3b)

Note here that $t - T$ is called δ in SPCS83 publications. In this equation, the result is in seconds of arc. The variables are defined differently for the two systems. They are

$$\Delta X = X_2 - X_1$$

and

$$\Delta Y = Y_1 - Y_0$$

for the Lambert system. For the TM system,

$$\Delta X = (2X_1' + X_2')/3$$

and

$$\Delta Y = Y_2 - Y_1$$

As an example for the Lambert system, suppose that a line about 5 mi long in Pennsylvania has the following coordinates for its ends:

$$X_1 = 1,403,740.00 \text{ ft},$$

$$X_2 = 1,429,740.00 \text{ ft}$$

$$Y_1 = 742,370.00 \text{ ft},$$

$$Y_2 = 747,470.00 \text{ ft}$$

From SPC27 projection tables for Pennsylvania, the North zone,

$$Y_0 = 455,699.10 \text{ ft},$$

$$\frac{1}{2\rho_o^2 \sin 1''} = 2.357 \times 10^{-10}$$

Substituting these values into Equation (19-2a) yields $t - T = -1.767''$. Substituting them into Equation (19-3a) yields $\delta = -1.759''$. This is the amount that would be algebraically added to the azimuth computed from the plane coordinates for starting a survey from station 1. This is what is called the *terminal grid azimuth*.

As can be seen, the differences between using these two equations is negligible, a suitable value for the correction for most surveys being -1.8 or just an even $-2''$. The magnitude of the correction further illustrates why this correction is usually negligible. Most lines measured in surveying are much shorter than 5 mi.

As an example for the TM system, suppose the following line, approximately 5 mi long, near the west side of the west zone in Illinois has coordinates of the line ends as follows:

$$X_1 = 125,150.00 \text{ ft},$$

$$X_2 = 126,450.00 \text{ ft}$$

$$Y_1 = 945,810.00 \text{ ft},$$

$$Y_2 = 974,010.00 \text{ ft}$$

From SPC27 projection tables for Illinois, west zone, we obtain

$$\frac{1}{\rho_o^2 \sin 1''} = 0.7861 \times 10^{-10}$$

Substituting these values into Equation (19-2b) yields $t - T = -2.490''$. Substituting them into Equation (19-3a) yields $\delta = -2.492''$. Thus, the azimuth computed using the coordinates given would be reduced by about 2.5'' for a terminal grid azimuth.

The difference in using these equations is insignificant. Furthermore, the magnitude of the correction, as with the Lambert system, is small enough that it could ordinarily be ignored for the line lengths normally encountered in surveying.

It is noted that Equation (19-3) is derived using average values of the constants for the zones in each system. These values are close enough to use in all zones unless the highest orders of accuracy are required. Equation (19-3) can be used whether coordinates originate from SPC27 or SPC83. However, the appropriate choice must be made according to the units used (meters or feet).

19-3.   THE UTM-UPS SYSTEM

19-3-1.   The 62 Zones

The UTM is often confused with the TM system used in SPC. Although the two can be interrelated mathematically since both are tied to geodetic coordinates, this is seldom done except for academic purposes, SPCs are not based on UTM. The explanation here has one purpose, clearing up any confusion between the TM used in SPC and the UTM system.

At the end of World War II, the U.S. Army Map Service devised the universal transverse Mercator and universal Polar stereographic grids to divide the entire world into 62 zones. There were about 100 heterogeneous grids then in use. The new system, incorporating some of the earlier ideas, has gained wide acceptance around the world. The two polar zones are not described here. Readers are referred to *Introduction to Map Projections*[7] for a detailed explanation of these systems.

The populated regions of the earth, from 80° S to 84° N, are divided into 60 zones in the UTM system, each of which is 6° wide in longitude. The zones are numbered from 1 to 60 beginning at 180°W as shown in Figure 19-9. Zone 11, e.g., from 120°W to 114°W includes Los Angeles, CA. Zone 18 contains New York City, and Tokyo is in zone 54.

Grid zone designations, such as 15°S in Figure 19-9, are part of a military grid reference system, which is explained further in the book cited above. The civilian system is entirely numerical. Each zone is similar to a transverse Mercator zone of the state plane coordinate system. The central meridian is assigned an easting of 500,000 m. For the northern half of the zone, above the equator, a northing of zero is assigned to the equator, as shown in Figure 19-10. For the southern half of the zone, below the equator, the equator is assigned a northing of 10,000,000 m. Thus, the horizontal position of a certain point on earth may be uniquely stated as follows:

Northern hemisphere, zone 13

506,021.43 − 4,385,107.38

in which it is understood that the easting is given first, as is most commonly done when listing or considering *X*- and *Y*- (Cartesian)

coordinates. If it is known that the point is in eastern Colorado, the hemisphere and zone need not be stated. The same coordinates occur in adjacent zones about 320 mi to the east or west, in Kansas or Utah.

The scale factor varies from 0.99960 at the central meridian to 1.00000 at a distance of 180,000 m to the east and west, and larger than 1.00000 beyond that. UTM zones are wider than state plane coordinate zones and therefore involve a greater range of scale factors. In most states, zones of the SPCS are limited to 158 mi in the direction of varying scale factors to keep factors within a range of 0.99990 to 1.00010 (1 part in 10,000). The value at the central meridian of a UTM zone is smaller than 1.00000 by 1 part in 2500.

19-3-2. Principal Digits

Topographic maps, such as those of the U.S. Geological Survey, often show UTM grid ticks in blue along the edge or *neatline*. If desired, opposite grid ticks may be joined to form grid lines. In labeling the grid ticks, it is customary to print two of the digits in larger type and omit trailing zeros except once on each neatline. Where a map shows grid ticks every 1000 m, they would be labeled as follows:

^{2}12$^{000m.}$E ^{20}84$^{000m.}$N
^{2}13 ^{20}85
^{2}14 ^{20}86
^{2}15 ^{20}87

On a smaller scale map, showing grid ticks at 10,000-m intervals, four trailing zeros are dropped and only one of the principal digits is shown.

^{2}10$^{000m.}$E ^{20}80$^{000m.}$N
^{2}2 ^{20}9
^{2}3 ^{21}0

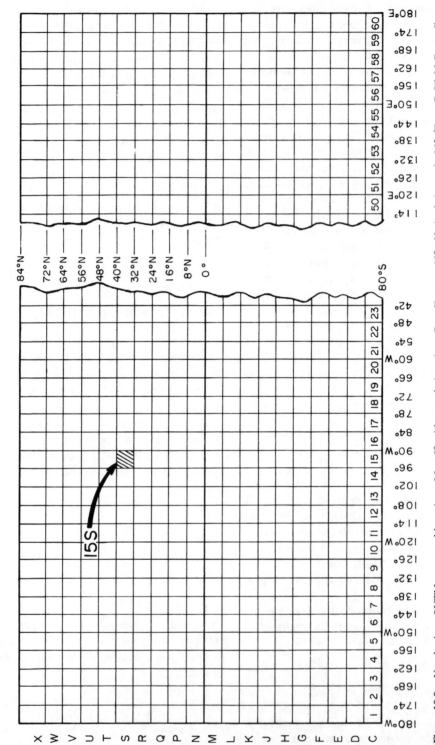

Figure 19-9. Numbering of UTM zones and lettering of 6 × 8° grid zone designations. Row X covers 12° of latitude instead of 8°. (From P. W. McDonnell, Jr., 1992, *Introduction to Map Projections*, 2nd ed., Rancho Cordova, CA: Landmark Enterprises, with permission.)

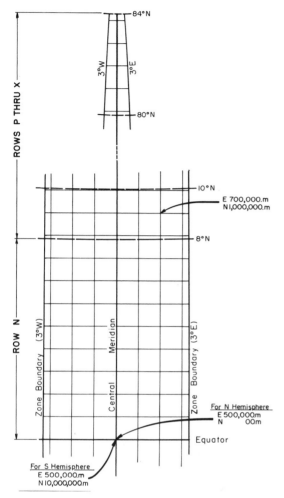

Figure 19-10. Rectangular grid superimposed on projection of UTM zone. (From P. W. McDonnell, Jr. 1992, *Introduction to Map Projections*, 2nd ed., Rancho, Cordova, CA: Landmark Enterprises, with permission.)

Principal digits are always in the thousands and ten-thousands places.

19-3-3. Convergence

In a UTM zone, as in a transverse Mercator state plane coordinate zone, the approximate expression for convergence is $C = \Delta\lambda \sin\phi$, where C and $\Delta\lambda$ must be in the same units (degrees, minutes, or seconds). As with SPC $\Delta\lambda$ is the difference in longitude between the location and central meridian and ϕ is the latitude. The formula is correct within 1 sec.

Further details about the UTM and UPS systems can be found in the end-of-chapter notes and reference section.

NOTES

1. J. E. Stem. 1990. *State Plane Coordinate System of 1983*. Manual NOS NGS-5. Rockville, MD: NOAA.

2. R. B. Buckner. 1984. *A Manual on Astronomic and Grid Azimuth*. Rancho Cordova, CA: Landmark Enterprises.

3. H. C. Mitchell, and L. G. Simmons, 1974. *The State Coordinate Systems*. Special Publication No. 235. Washington DC: U.S. Government Printing Office.

4. R. B. Buckner. 1983. *Surveying Measurements and Their Analysis*. Rancho Cordova, CA: Landmark Enterprises.

5. D. Doyle. 1991. *NAD83 State Plane Coordinates and Datum Transformations* (workshop notes). Rockville, MD: NOAA.

6. C. L. Schwarz, Ed. 1989. North American Datum of 1983. Professional Paper NOS-2, NOAA, Rockville, MD.

7. P. W. McDonnell, Jr. 1992. *Introduction to Map Projections*, 2nd ed. Rancho Cordova, CA: Landmark Enterprises.

REFERENCES

STOUGHTON, H. W., Ed. *The North American Datum of 1983, AAGS Monograph No. 2* (a collection of papers describing the planning and implementation of the readjustment of the North American horizontal network). Bethesda, MD: ACSM (undated).

20

Photogrammetry

Andrew Kellie
Wayne Valentine

20-1. INTRODUCTION

Photogrammetry has influenced survey practice drastically since its introduction as a mapping tool. Aerial photographs specifically for mapping purposes were first taken in 1913, although ground photography had been used to a limited extent in field surveys as early as 1894.[1] The first major mapping project in the United States was conducted by the U.S. Geological Survey and the Tennessee Valley Authority in the 1930s. At that time, some 40,000 sq mi (103,500 sq km) of the Tennessee River basin were mapped. At present, aerial photogrammetry is used for virtually all small-scale mapping done in the United States and is seeing increasing use for large-scale mapping as well.

Applications of photogrammetry in survey practice include topographic mapping, site planning, earthwork volume estimation for proposed roads, and overburden estimates for surface mines. Also, photogrammetry is being used for boundary surveys in the western United States. In addition to its mapping role, aerial photography is used for planning ground surveys, estimating timber classifications for taxation purposes, and in the conduct of boundary retracement.

20-2. PHOTOGRAPHS AND MAPS

Photogrammetry is the process of preparing accurate maps or obtaining precise measurements from photographs. These data may be presented as a map or stored in digital form, depending on their intended purpose.

Photogrammetric mapping may employ photos taken either from the ground or the air. Photography obtained on the ground is termed *terrestrial* photography. It has been used for surveys of traffic accident sites and to obtain elevations of historic buildings or structures. Terrestrial photogrammetry is discussed beginning with Section 20-3. A terrestrial camera system capable of stereo photography is show in Figure 20-1. *Aerial* photography requires the use of cameras mounted in aircraft. Aerial photos are the most widely used type of photography employed for photogrammetric mapping. For mapping purposes, aerial photography is usually obtained with the camera axis vertical, although oblique photos have been used as well. Examples of vertical and oblique aerial photography are shown in Figures 20-2 and 20-3.

Vertical photographs have the appearance of a map. There are two reasons, however, why a photograph should not be used as one. First,

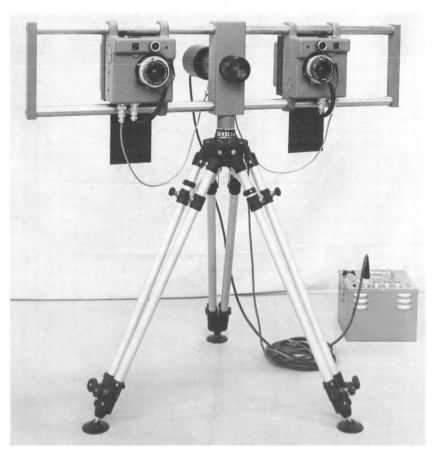

Figure 20-1. Kelsh K-460 Universal stereometric camera mounted on tripod. Electronic shutters and surface contrast optical projector (SCOP) configured with single film cassettes. (Courtesy of Danko Arlington, Inc.)

a map presents an orthogonal projection of the features shown; a photograph shows a central perspective projection. This is noticeable on vertical aerial photography in which features are displaced radially outward from the photo center. Second, the scale of a photograph is not constant. For vertical aerial photos, the photo scale depends on the ratio between camera focal length and height of the aircraft above the terrain. Unless the ground is perfectly level and the camera axis perfectly vertical, this ratio is not constant and photo scale varies accordingly. As a result, when precise measurements are required, a photographic mapping process is used to eliminate radial displacements present in mapping pho-

tography and bring features shown on the photograph to a constant scale.

One characteristic of mapping photography is that photographs overlap each other both at the ends and on the sides. This overlap is necessary to stereoscopically study them. Stereoscopic (or three-dimensional) vision is possible when the same object is viewed from two different positions. Separation of the human eyes serves this function, allowing three-dimensional vision. The principle of separate viewing positions is also used in photogrammetric operations and achieved by taking photographs of the same scene from two different photo stations. With the photographs placed under a stereoscope, or mounted in a stereo-

Figure 20-2.   Vertical photography. (Courtesy of Photo Science, Inc., Lexington, KY.)

plotter, the operator views the scene imaged within the area of overlap on the photographs in three dimensions. The standard endlap used in photogrammetric mapping is 60%, sidelap 30%. The overlap and sidelap on a series of photos taken for mapping are shown in Figure 20-4.

20-3.   THE PHOTOGRAMMETRIC MAPPING PROCESS

A typical photogrammetric mapping project requires four stages: (1) obtaining mapping photography, (2) completing ground control surveys for the mapped area, (3) compiling map data from the photos, and (4) checking the finished product for compliance with project specification.

The decision to use photogrammetric methods to complete a mapping project is primarily economic, although workload and project deadline requirements have to be considered as well. In general, as the area mapped increases, the cost per acre (hectare) for photogrammetric mapping decreases. One reason is that the fixed costs of photo acquisition are prorated on the basis of area. Also, as the area mapped increases, use of analytic photogrammetry to extend ground control becomes more feasible, thereby reducing ground control costs.

It is frequently assumed that the mapped area must exceed 50 acres (20 hectacres) if

Figure 20-3. Oblique photography. (Courtesy of Photo Science, Inc., Lexington, KY.)

photogrammetric mapping costs were to be competitive with ground techniques. Photogrammetric mapping of much smaller areas may be competitive, however, if suitable photography of the project site is already available. Prior to contracting for new photography, it is essential to check with photogrammetric mapping firms working in the project area. It may be possible to purchase existing photography and thereby decrease mapping costs.

20-4. AERIAL CAMERAS

Mapping photography is obtained using *metric* cameras having a focal length, lens distortion curve, and internal geometry determined from calibration by the National Bureau of Standards. The procedure used is beyond the scope of this book, but all government mapping contracts, and many private ones, require camera calibration so the extent of distortion and aberration present is known.

Distortion results in image displacement on the photograph and is important in mapping because it affects a photograph's geometry. *Radial* distortion results in image displacement symmetrically toward or away from the lens axis, and its magnitude is determined by calibration. *Tangential* distortion results in image displacement at 90° to a radial line. It is caused by improper lens centering, but usually is small enough to be negligible in modern mapping cameras.

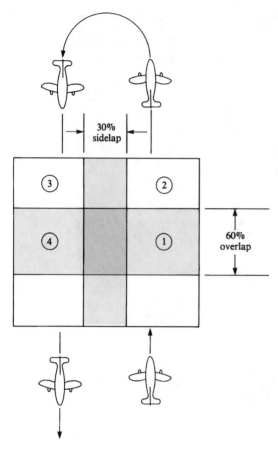

Figure 20-4. Overlap and sidelap on a single vertical aerial photograph.

Lens aberrations are of two primary types: (1) *spherical* and (2) *chromatic*. Spherical aberration results when all rays striking a lens do not form a plane image, but rather one that is convex or spherical in shape. Chromatic aberration results from the unequal bending of light rays of different wavelengths (colors) so an image of incorrect hue appears. Aberrations can be controlled by using lenses of different types—e.g., convex and concave lenses in series—and glass types with different indices of refraction.

The lens in a modern mapping camera may contain a dozen or more separate elements that are designed to minimize the distortions just described. Typically, a between-the-lens-type shutter is used, having speeds of 1/250 to 1/500 sec. Lens apertures vary, with $f/6.3$ and $f/4.2$ being common.

Four different types of cameras can be used for aerial photography and classified as (1) single-lens frame, (2) multilens, (3) strip, and (4) panoramic cameras. Of these, the single-lens frame model is by far the most common for mapping. The panoramic camera, which is actually a scanner, has been used in the LANDSAT satellite series, whereas the multilens design (of which the Trimetrigon camera is an example) is primarily of historical interest. Strip cameras, which do not contain a conventional lens system, are not generally used for mapping photography.

Single-lens frame cameras are classified by both their focal length and angular field of view. This relationship is shown in Figure 20-5. A *normal-angle* mapping camera has a focal length of $8\frac{1}{4}$ in. (0.210 m) and subtends a 75° field of view. The *wide-angle* mapping camera has a 6-in. (0.153-m) focal length and subtends a 93° field of view. The superwide-angle camera has a $3\frac{1}{2}$-in. (0.089-m) focal length and 122° angular field. Of these, the wide-angle camera with its 6-in. (0.153-m) focal length is the most common mapping camera at present. It is not as prone to the lens distortion problems encountered with a superwide-angle lens, and the 93° field of view requires less ground control than needed for a normal-angle mapping camera.

In modern frame mapping cameras, the format is 9 × 9 in. (23 × 23 cm) between side fiducial marks, which are precisely measured lines in the camera that image on the exposed film. They are used as references in the photogrammetric process to establish the photo principal point and measure film distortions. Modern cameras have at least four fiducial marks, and some have four additional ones in the corners.

The scene imaged on the format is easily computed by knowing the photo scale. For example, the side dimension of ground coverage of 1:12,000 scale aerial photography with a

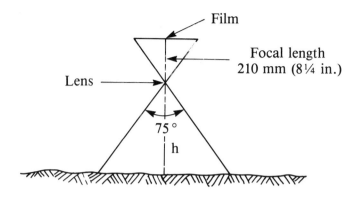

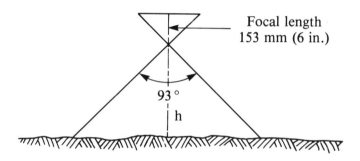

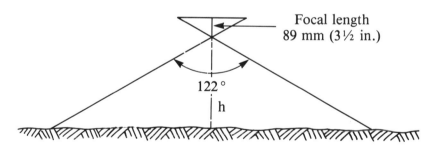

Figure 20-5. Ground coverage at constant altitude of normal, wide-angle, and superwide-angle mapping cameras.

9 × 9 in. (23 × 23 cm) format is

$$\frac{9 \text{ in.} \times 12,000}{12} = 9000 \text{ ft} \quad (2743 \text{ m})$$

Camera focal length influences both photo scale and vertical exaggeration. The scale of a vertical aerial photo is determined by the mathematical relationship between camera focal length and altitude above ground level at which the photograph is taken. This relationship is shown in Figure 20-6, where *ab* represents the photographic film on which ground

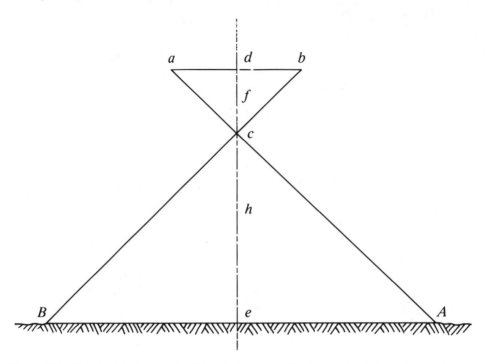

Figure 20-6. Relationship between focal length f and altitude above ground h on photo scale.

image BA is projected by lens c. Lines ab and BA are parallel because each is perpendicular to the camera axis de. As a result, triangles abc and ABC are similar ($AAA = AAA$). It then follows that the camera focal length cd is proportional to the altitude above ground ce. Further, photo scale ab is proportional to ground distance BA. Mathematically,

$$\frac{ab}{BA} = \frac{dc}{ce} \qquad (20\text{-}1)$$

Let dc equal f, the focal length of the camera and ce equal h, the height above ground level. Expressing photo scale as a rational fraction, we can rewrite Equation (20-1) as

$$\frac{1}{x} = \frac{f}{h} \qquad (20\text{-}2)$$

The focal length f is fixed by the camera being used, and the denominator of the photo scale fraction established by scale require-

ments, hence, selection of photo scale and camera focal length determine flying height.

Example 20-1. Assume that project requirements dictate a photo scale of 1:12,000. Camera focal length is 6 in. (153 cm). Altitude above ground is then found by substitution in Equation (20-1a).

$$1/12{,}000 = \frac{0.5}{x}$$

$$x = 6000 \text{ ft} \quad (1828 \text{ m})$$

As an alternative, an $8\frac{1}{4}$-in. (0.210-m) focal length might have been used, in which case substitution in Equation (20-1a) would show the altitude above ground to be 8250 ft (2515 m).

Whereas photo scale is a function of both focal length and altitude, vertical exaggeration depends on both the altitude above ground and *air base*, the horizontal distance between camera exposures. Its relationship to the altitude above ground is shown in Figure 20-7. The altitude/air base relationship is expressed mathematically as the *base-height ratio.*

The base-height ratio is important in mapping photography because as it increases so do the vertical exaggeration and photogrammetrist's ability to measure elevations within the stereomodel. The importance of camera focal length to vertical exaggeration is that photo scale is fixed by stereoplotter and map-scale requirements. A small focal length gives a large base-height ratio; a longer focal length produces a small base-height ratio with correspondingly less vertical exaggeration.

Example 20-2. Assume that 1:12,000 photography is to be obtained using a 6-in. (0.153-m) focal-length camera. Sixty percent overlap is required between adjacent photos. The air base would be $(1 - 0.6)(9000 \text{ ft}) = 3600$ ft (1097 m), and the corresponding base-height ratio is

$$\frac{3600}{6000} = \frac{1}{1.7}$$

For a normal-angle mapping camera with an $8\frac{1}{4}$-in. (0.210-m) focal length used to obtain the same scale photography, the air base remains the same, but the base-height ratio becomes the following:

$$\frac{3600}{8250} = \frac{1}{2.3}$$

A typical aerial camera system includes (1) the camera itself, (2) a film magazine, (3) the viewfinder, and (4) the intervalometer. The camera is mounted in a gimbal frame over a hole in the aircraft fuselage or a pod located beneath the aircraft. The camera can be rotated within the frame to compensate for crab (or drift) of the aircraft along the flight line. During exposure, the film is held flat by means of a vacuum system.

Following exposure, an automatic drive mechanism advances the exposed frame, and new film is fed from the magazine. Film magazines hold from 200 to 400 ft (60 to 120 m) of film.

The intervalometer is an electronic device that triggers the camera shutter at a predeter-

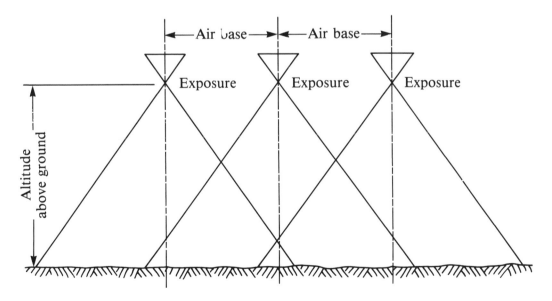

Figure 20-7. Relationship between air base and altitude.

mined time interval. The interval between exposures depends on the air base required for the photography and speed of the aircraft. Exposures can also be obtained manually by means of different air-base measurement devices, such as etched lines built into the viewfinder. Mapping cameras in current use include Fairchild, Zeiss, and Wild camera systems. A typical mapping camera is shown in Figure 20-8.

Mapping photography can employ any or all four types of photographic emulsions: (1) panchromatic, (2) natural color, (3) panchromatic infrared, and (4) color infrared. At present, panchromatic photography is probably the most widely used film type for mapping, although natural color photography is becoming increasingly popular. Infrared films, both panchromatic and color, are employed extensively for vegetation and hydrographic studies. When specifying the film type for a mapping project, possible future uses of the photography should be considered. For example, although panchromatic photography might be entirely suitable for mapping, natural color photography—because of its increased interpretative value—may be more satisfactory than panchromatic in preparing

an environmental impact statement for the mapped area. Regardless of the emulsion selected, it must be fine-grained and have high resolution.

Flying height is another consideration in selecting the best photographic emulsion. As flying height increases, the effects of atmospheric haze and pollution, which scatter light in the blue wavelength, become more pronounced. These effects can be minimized on panchromatic photography by the use of a yellow (minus blue) filter. Alternatively, color infrared film that is not sensitive to the blue light wavelengths can be substituted. Substantial filtering, however, is not possible when a natural color emulsion is specified.

20-5. FLIGHT PLANNING

After the mapping specifications have been determined, a photographic flight plan is prepared prior to the photo mission, so the aerial photography obtained will meet mapping specifications and stereoplotter requirements. Flight planning must ensure photographic coverage of the entire area to be mapped in a minimum air time.

Figure 20-8. Wild RCIOA aerial mapping cameras. (Courtesy of Wild, Inc.)

Mapping or measurement accuracies need to be determined before intelligent flight planning can proceed. For topographic mapping, the minimum contour interval is determined by the C factor of the stereoplotter and flight height h, according to the relationship $Cl = h/C$.

C factors range from about 1000 to 1200 for projection type (e.g., Kelsh) plotters to 1800 for optical-mechanical (e.g., Wild B-8) plotters to over 2000 for analytical plotters. Spot heights on well-defined points can be determined with much greater accuracy, typically 1/4000 to 1/10,000 of the flight height for optical-mechanical plotters. Planimetric (xy) position-measuring capability for control densification typically ranges from 1/8000 flight height (RMS) for universal analogue plotters using semianalytical techniques to 1/40,000 flight height (and smaller) for fully analytical methods using self-calibration and bundle adjustment. The phot-to-map enlargement ratio is commonly five times for projection plotters and up to 10 times for other plotters.

The first step in preparing a flight plan involves marking the area to be photographed on a base map—usually a U.S. Geological Survey quadrangle sheet. Second, flying height above the ground is computed from the known camera focal length and specified photo scale. Third, photo overlap requirements and flying speed fix the time interval at which exposures must be made, as well as the horizontal spacing between flight lines.

Example 20-3. Aerial photography is to be acquired over an area 3.5 mi (5.6 km) wide and 7.2 mi (11.6 km) long. From map-scale and accuracy requirements, it has been determined that photography must be at an average scale of 1 : 3600. Photo format is 9 × 9 in. (23 × 23 cm). Endlap and sidelap are 60 and 30%, respectively. The aircraft to be used for photography has a ground speed of 100 mph (160 kph).

To begin planning, the area to be covered is marked on a suitable map sheet. Flight lines are oriented along the major axis of the mission area to minimize air time. The major axis true azimuth is scaled from the quad sheet to determine the aircraft headings.

Next, the height above terrain is found from Equation (20-2) (h being the height above ground, f the camera focal length, and s the photo scale). Here, $h = 0.5$ ft $(3600) = 1800$ ft (549 m).

Examination of a quadrangle sheet shows the average elevation for the mission area is 150 ft (46 m) above the National Geodetic Vertical Datum. Barometric altitude (required by the pilot) is then found from the following relationship:

$$A = h + e \qquad (20\text{-}3)$$

where A is the barometric altitude, h the height above the terrain, and e the average elevation over the mission area. By substitution, $A = 1800$ ft $+ 150$ ft $= 1950$ ft (595 m).

With the flight-line azimuth and altitude known, the intervalometer setting can be determined. After computing the air base, we can find this from the following relationship:

$$P = 100(G - B)/G \qquad (20\text{-}4)$$

where P is the percent endlap, G the ground distance of the photo along the flight line, and B the air base.

By substitution,

$$60 = \frac{(9 \text{ in.})(300/\text{in.}) - B}{(9 \text{ in.})(300/\text{in.})}100$$

and

$$B = 1080 \text{ ft} \quad (329 \text{ m})$$

The intervalometer setting can now be computed from

$$S = B/V \qquad (20\text{-}5)$$

where S is the intervalometer setting (in sec), B the air base (in ft), and V the ground speed (in ft/sec). Therefore, $S = 1080 \text{ ft}/147 \text{ ft/sec}$.

$$S = 7.35 \text{ sec}$$

This must be rounded to 7 sec because fractional second timing cannot be set on most intervalometers. Recomputation of the air base follows this rounding, and $B' = 147 \text{ ft/sec} \times 7 \text{ sec} = 1029 \text{ ft}$ (314 m).

The spacing between flight lines along the minor axis of the mission area can be found from

$$L = W/w \qquad (20\text{-}6)$$

where L is the number of flight lines needed, W the width of mission area, and w the width of the area covered by each photo. Here, the width of each photo is 9 in. $\times$ 300 ft/in. = 2700 ft (823 m). Because a 30% sidelap is specified, this width must be accordingly decreased and

$$w^1 = 2700 - (0.30)(2700) = 1890 \text{ ft} \quad (576 \text{ m})$$

Substituting, we obtain $L = 18,480 \text{ ft}/1890$ ft = 9.8. Therefore, 10 flight lines will be used.

Finally, the total number of photos for the mission is computed from the following:

$$N = L(l)B' \qquad (20\text{-}7)$$

where N is the number of photos; L the number of flight lines; l the length of the flight line, in ft (m), and B' the revised air base, in ft (m). In this example, $N = 10(38,016 \text{ ft})/1029 = 370$.

To ensure complete photographic coverage of the mission area, it is prudent to have two photos outside the mission area on both ends of each flight line. This increases the number of photos needed to 410.

20-6. GROUND CONTROL FOR PHOTOGRAPHIC MAPPING

Field surveys for photo control constitute an important part of the photogrammetric mapping process. The ground control network is used to fix the map scale and level the model in a stereoplotter.

The extent of photo control surveys required for a particular mapping project depends on the number of stereomodels needed for mapping. Each stereomodel requires a minimum of two horizontal and three vertical control points, but for redundancy five each are suggested. Total control requirements are lessened somewhat, however, because control points are visible in more than one model. Also, if the area mapped is very large, ground control may be extended mathematically through several models by a technique known as *analytic bridging*. When this is done, ground control requirements are reduced accordingly.

Control surveys can be conducted by a photogrammetric mapping company or the consulting firm ordering photogrammetric mapping services. In either case, the actual ground points used for control are selected by a photogrammetrist because control points must appear in certain parts of a stereomodel to be helpful in the mapping process.

Field control surveys can either precede or follow acquisition of aerial photography. When photography is taken specifically for a mapping project, it is possible to place ground control points in advance of the photo flight and *premarked* to make them visible on the aerial photography itself.

The field crew conducting the survey is provided by the photogrammetrist with a topographic map showing ideal locations for control. The stations shown can serve for both horizontal and vertical control or only one such function.

Control points are marked on the ground by a target large enough to be visible at the scale of photography desired. Targets may be

X- or *T*-shaped or occasionally circular. The control station itself is the target center. Materials used for targeting include cotton cloth, wooden panels, or plastic sheeting. Since such targets are subject to vandalism, targets painted on pavement are preferred. Placement of the targets is critical. They must be set out of shadow, on the side of hills facing the flight line, in areas providing sufficient contract between the target and background. Targets should be located on ground as flat and level as possible.

The target itself must take a minimum 0.01-in. (0.0003-m) image on the photo. Therefore, the total target size across in feet should be about 1/1000 of the photo-scale numerator, and the width of each leg in inches should also be 1/1000 of the photo-scale number. For example, if the photo scale is 1:12,000, the total target length is 1/1000 × 12,000 or 12 ft (3.6 m), and the width of the leg 1/1000 × 12,000 or 12 in. (0.3 m). Increasing leg length and width is recommended in situations of poor contrast.

Postmarking refers to control surveys undertaken after aerial photography has been obtained. Field crews are supplied with annotated photographs showing physical features to be used as control points, which can serve for either horizontal or vertical control or both. Horizontal and vertical control surveys then locate the control points to be targeted. Caution by the field crew is necessary when postmarking is used to ensure correct ground identification of points set for control purposes.

The precision required for both horizontal and vertical control surveys depends on a number of factors. If the control survey can be designed to serve purposes other than photo control, costs will be reduced. For example, if photo control surveys are also used in construction, the cost can be prorated between two required operations. Photo scale must be considered, since for small-scale maps, lower control survey precision is acceptable than that required for large-scale maps. Finally, the areal extent of a control net and long distances between control stations may require greater precision.

For purposes of photo control only, in which a map is the end product, horizontal and vertical control should be at least third-order. To meet this precision, a simple traverse using theodolites and EDMIs is practical when areal extent and terrain permit. Vertical control can consist of differential or trigonometric leveling, depending on the precision needed and contour interval to be mapped. When the aerial extent of a project is large, horizontal control stations of the state plane coordinate system can often be used to advantage, with a horizontal control traverse beginning and ending on such stations. Bench marks referenced to the National Geodetic Vertical Datum are similarly helpful.

On completion of the control surveys, positional data in terms of x-, y-, and z-coordinates of the control stations are furnished to the photogrammetrist.

20-7. STEREOPLOTTING INSTRUMENTS

Three types of stereoplotting instruments are in current use for photogrammetric mapping: (1) optical projection, (2) mechanical projection, and (3) analytic plotting machines.

The optical-projection plotter recreates the recorded scene by projecting light through photo diapositives (transparent positive prints) onto a platen. The image so projected can then be viewed stereoscopically by any of three viewing systems. The first of these is known as the *anaglyphic* method. The left- and right-hand projection lamps of the plotter are covered by blue and red filters, respectively. The plotter operator wears glasses of corresponding colors to view the stereomodel in three dimensions.

The second viewing system is known as the *polarized-platen viewing* (ppv) method. This design requires polarizing filters on the projector lamps. Polarized glasses worn by the operator provide stereoviewing of the model.

The *stereo image alternator* system has a rotating shutter interposed between the operator and platen that alternately exposes the image formed by left projector and the corresponding image formed by the right. Thus, the operator can view the stereomodel without using spectacles. An example of a projection-type stereoplotter is shown in Figure 20-9.

The primary advantages of the projection-type stereoplotter are its simplicity and low cost. It has a disadvantage that for direct plotting of a stereomodel, the map is compiled at the model scale. Hence, map scale is tied to projector magnification of the diapositives and ultimately to photo scale. This disadvantage can be remedied by installing a digitizing table and an encoder-equipped platen. Triaxial coordinates of image features can then be stored on magnetic tape or disks and used to plot the map at any desired scale.

Projection plotters are limited by the ability of the human eye to resolve detail as it appears in the stereomodel. This problem can be overcome by employing a mechanical plotter constructed so that an operator views the diapositives directly through a microscope to magnify the stereomodel. A system of rods mechanically recreates the stereomodel. Movement of the floating mark seen by the operator is duplicated by the mechanical rods and transferred to a tracing table, where the map is plotted. An advantage of the mechanical plotter is that it can accept photography of the focal length of any commonly used aerial cameras. Also, the output format can be varied to provide a one to 10 times enlargement or reduction of photo scale. As a result, it is possible to use higher-altitude photography than would be possible with a projection plotter and still produce a map at the same scale. A typical mechanical plotter is shown in Figure 20-10.

The third kind of stereoplotter in common use is the analytic type. The analytic plotter does not use optical or mechanical systems to restitute photogeometry, but relies instead on

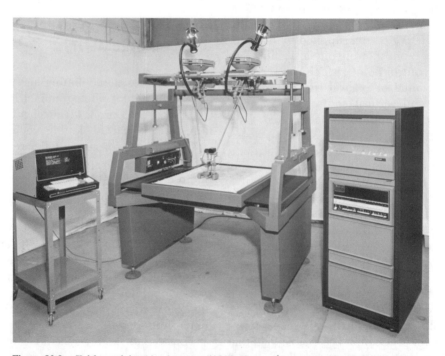

Figure 20-9.   Kelsh model projection-type 5030B plotter. (Courtesy of Danko Arlington, Inc.)

Figure 20-10.   Wild B-8 mechanical projection plotter. (Courtesy of Wild Inc.)

computations using coordinates measured on photo points to define the map positions of features. An operator views the stereomodel through a variable magnification system, and the plotter provides a microprocessor system with coordinates of the features being mapped. A computer system then directs map plotting on a separate table. Usually, an analytic plotter incorporates a closed-circuit television system that permits the operator to monitor map preparation. A typical analytic plotter system, including the components just mentioned, is shown in Figure 20-11.

20-8.   STEREOPLOTTER OPERATIONS

Stereoplotter operations include making relative and absolute orientations to tie the mapping photography to ground control and plotting the map manuscript itself. These two

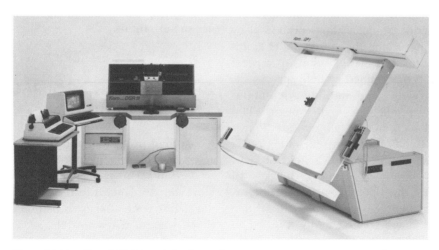

Figure 20-11.   Kern DSR-11 analytic plotter. (Courtesy of Kern Instruments, Inc.)

operations produce a map that is accurate in scale and correctly portrays an area's topographic features.

Prior to stereoplotter orientation, the aerial photography to be used is developed as a diapositive, rather than a paper print, and mounted in the plotter. Glass or film diapositives are employed in stereoplotters because these materials are dimensionally stable, whereas differential shrinkage of paper prints would cause distortions to appear in the resulting map.

Diapositive mounting systems permit rotation about three axes, referred to as the *omega*, *phi*, and *kappa* axes (lateral tilt, longitudinal tip, and azimuth) respectively. In the *relative orientation* process, the motions are alternately adjusted and readjusted to recreate on the stereoplotter the same geometric relationship between the photographs that existed when they were taken. The relative orientation process is iterative and continues (except in analytical plotters) until this relationship is created. When obtained, Y parallax has been removed from the stereomodel. Y parallax is the apparent displacement of the two measuring marks in the Y direction with respect to each other.

Following relative orientation, *absolute orientation* of the stereomodel matches photo scale to map scale by orienting the stereomodel to the horizontal control points plotted on the map manuscript. Model scaling is followed by model leveling, which the stereomodel is adjusted to conform to the elevations of vertical control points determined by field surveys.

With the relative and absolute orientations complete, mapping can finally begin. The stereomodel seen by the plotter operator is three-dimensional. It contains an index point at the center of the operator's field of view. The index point or floating mark can be moved by the operator about the x, y, and z stereoplotter axes within the three-dimensional image created by the plotter either horizontally, vertically, or both. All stereoplotters permit indexing the z-axis in the stereomodel against the vertical control points; some allow indexing the x- and y-axes also.

The actual plotting of planimetric features within the stereomodel is done by coupling the x and y movement of the floating mark to a drawing or scribing device. Alternative, planimetric features can be described by the x- and y-coordinates, and the assignment of numerical codes to the points so defined facilitates computer plotting of the map features.

Topographic features can be shown by contours or as profiles taken across the stereomodel. Contour points are located by setting the floating-mark index to the desired elevation and then moving it in the stereomodel until it appears to rest on the ground surface. The plotter operator then shifts the floating mark, being careful to keep it in contact at all times with the apparent model surface. Movement of the floating mark in this manner draws the contour locations. As the contour is traced, the linkage system between the floating mark and a plotting device draws or scribes the contour on the map manuscript. If an automated mapping system is used, x- and y-coordinates at the contour elevation being traced are sensed and recorded at predetermined intervals of time or distance and transferred to a magnetic tape or disk for later computer plotting. The sequence of operations just described parallels the plotting operation as practiced in most medium- and large-size organizations.

20-9. MAPPING ACCURACY EVALUATION

A mapping project should include both map accuracy requirements and the procedures to be used in accuracy testing. The specifications most commonly used in the United States are the National Map Accuracy Standards (NMAS), which require maps at scales of 1 : 20,000 or smaller to have 90% of the positional features tested within 1/50 in. at map scale of their true ground position. For scales larger than 1 : 20,000, the positional tolerance is 1/30 in. Further, 90% of all elevations tested must be within one-half of the contour interval shown on the map, regardless of map scale. The Engineering Map Accuracy Standards (EMAS) promulgated by the American Society of Civil Engineers are appropriate for large-scale maps used for engineering design and measurement. The EMAS provide for specifying limiting standard and mean-absolute errors in three dimensions and can be tailored to meet accuracy requirements.[2]

Proper conduct of the test survey is essential in map accuracy evaluation. First, there should be commonality of the control used in preparing the map and that used in accuracy evaluations. If map control is to be checked, the verification should be independent of the map accuracy evaluation process. Second, the evaluation survey must be of higher precision than the techniques used in preparing the map itself. This ensures that any error detected is a function of a photogrammetric mapping mistake and not a result of the evaluation survey. Third, the need for closed traverses and level circuits must be recognized. Preferably, traverses should begin and end on established control to eliminate the effect of scale error. Fourth, if discrepancies are detected, it is important that the field data involved be rechecked before transmitting it to the photogrammetrist. In addition, possible causes for the error might also be examined. For example, if an error in contour location is detected, a recheck of the control on which the contours are based may reveal an excellent control survey, but transposed figures in control point elevations furnished to the photogrammetrist.

NMAS requirements are standard on governmental mapping projects. On private photogrammetric projects, the map accuracy specifications themselves, the conduct of the evaluation surveys, and the designation of who must conduct the surveys are all matters to be addressed in the mapping contract.[3]

20-10. SIMPLE PHOTO MEASUREMENTS

It is possible to make simple measurements of distance, area, azimuth, and elevation from paper prints of aerial photographs. The results will not approach the accuracy attainable with advanced stereoplotter or analytic photogrammetric systems, but are useful in forestry, geology, and boundary retracement applications.

Basic measurement of distance, azimuth, and area can be made from a single photograph. They are less subject to error from photographic distortion if made on the *effective area* of the photo rather than elsewhere. The effective area is defined as the part of a photo that is closer to the center of the photo being used than to any other photo center. On a 9 × 9 in. (23 × 23 cm) aerial photograph, the effective area measures roughly $4\frac{1}{2}$ × $6\frac{1}{2}$ in. (10 × 16 cm) about the center of the photo, if we assume 60% endlap and 30% sidelap (see Figure 20-12).

Photo measurements begin by determining the photo scale. This can be done by comparing distances on the photo with the same

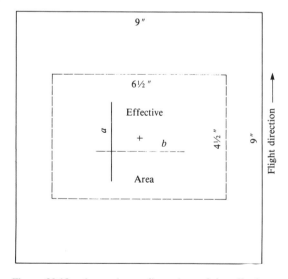

Figure 20-12. Approximate dimensions of the effective area of a 9 × 9-in. photo.

lengths measured either on the ground or from a reliable map.

Example 20-4. Assume that the distance between ground points is 1427 ft (435 m). The corresponding photo distance is 2.47 in. (62 mm). Photo scale is

$$\frac{0.062}{435} = \frac{1}{X} = \frac{1}{7000}$$

With the photo scale determined, the lengths of other lines visible in the photograph can be secured. In making photo measurements, the most finely divided scale available is used. For example, at a photo scale of $1:12,000$, $1/60$ in. is equal to approximately 17 ft (5 m) on the ground.

Area can be measured from single photos by summing the areas of component geometric figures. When irregular areas are to be determined, either a dot grid or polar planimeter can be used. The latter provides direct measurement of photo areas for conversion to corresponding ground areas. A dot grid contains a known number of dots per square inch. Counting the number of dots included within tract boundaries and dividing by the number of dots per square inch give photo area, so the ground area can be computed.

To determine the azimuth of lines on aerial photos, at least one line of known azimuth must be visible on the photo. This can be observed in the field or scaled from a map. When section lines are visible, their cardinal orientation can be used as a reference. It is helpful to use a line of known orientation and protractor to draw a north arrow on the photo used for measurements.

In addition to measurements of distance, direction, and area, heights of objects visible on photos can be determined by stereoscopic viewing. Stereoscopic or three-dimensional viewing of pairs of aerial photos is possible in overlapping parts of the photos. Measurements are made using a *parallax bar* on photos viewed through a stereoscope (see Figure 20-13).

To measure parallax: (1) Locate the *principle point* on each photo at the intersection of lines drawn between the *fiducial marks* on opposite sides of the photo. (2) Mark each principal point by punching a small hole in the photo itself. (3) Tape one photo to the table and observe both photos through the stereoscope. (4) Rotate the second photo until a stereoscopic image appears and tape the second photo to the table. (5) Transfer the position of the principal point of the first photo onto the second photo and mark the *conjugate principal point* with a pinhole. Similarly, trans-

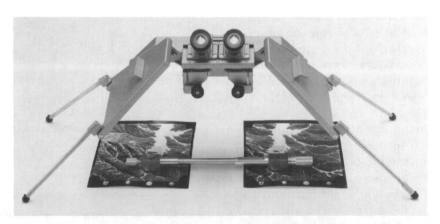

Figure 20-13. Parallax bar and stereoscope. (Courtesy of Topcon, Inc.)

fer the conjugate principal point of the second photo onto the first.

If the photos are correctly oriented, the principal and conjugate principal points of both photos will fall on a straight line. The line connecting these points is the x-axis used in parallax measurements.

Next, it is necessary to determine the air base of the photography, which can be done in a number of ways. Perhaps the simplest, however, is to measure the distance between the principal and conjugate principal points on photo number one.

If the photography scale and focal length of the camera used to obtain it are known, then it is possible to find the height of objects visible on aerial photos by Equation (20-8), as follows:

$$h = \frac{(P_2 - P_1)A'}{B} \qquad (20\text{-}8)$$

where

$$h = \text{height of the object}$$

$$P_1 \text{ and } P_2 = \text{the parallax bar readings}$$

$$\text{at the top and bottom}$$

$$\text{of the object, respectively}$$

$$B = \text{the air base}$$

$$A' = \text{elevation above terrain}$$

$$\text{of the photo}$$

Example 20-5. Aerial photography at a scale of $1:12,000$ has been obtained with a 6-in. (153-mm) focal-length camera. The measured air base on the photography is 91.5 mm, and the height above terrain at which the photos were taken was 6000 ft. A tree has a parallax bar measurement of 133.62 mm on the base and 132.7 mm on the bottom.

$$h = \frac{(133.62 - 132.70)}{91.5} 6000$$

$$= 60 \text{ ft} \quad (18 \text{ m})$$

This formula is simple and its use limited to gentle terrain and average flying heights.[4]

20-11. SPECIAL PRODUCTS

In addition to preparing line maps, photogrammetric techniques can be used to advantage for alternative map products. Perhaps the best known of these is an *orthophoto* as a substitute for a map. It shows the same detail provided by a photo and is an orthogonal projection suitable for scaled measurements. It is possible to superimpose map data, such as highway route numbers, place names, and contour lines on the orthophoto and produce an orthophoto map. The cost of constructing special-purpose, large-scale orthophoto maps is presently significantly higher than that for comparable line maps. An ortophoto plotting machine is shown in Figure 20-14.

Another map product, available from analytic plotting systems, or from any system that can output map data in digital form, is an isometric projection of a mapped area. Typically, the mapping software used permits changing the horizontal and vertical scales and rotation of the projection axes. Such projections are pertinent in site planning, in studies of road location, and overburden volume distribution.

A third product obtainable from the photogrammetric process consists of in situ volume estimates prepared without the intermediate step of map preparation if the stereoplotter used is equipped with a digitizing system. The procedure employed consists of obtaining triaxial coordinates of points on the stereomodel corresponding to ground features that have a proposed design elevation. Model and design data for these points are then used with appropriate computer software to provide volume estimates.

Another product that is beginning to have a major impact on mapping sciences is the *digital terrain model* (DTM), which is a data file of

Figure 20-14.   Kelsh K-320 Orthoscan with OTM converter. (Courtesy of Danko Arlington, Inc.)

the terrain, created from a stereomodel with a digitized stereoplotter. Thousands of points are measured with the plotter, and the x-, y-, z-coordinates of the points stored in computer-compatible form. Many programs have been developed that manipulate data and create a variety of graphical and statistical outputs; e.g., contour plats at various contour intervals and scales can be extracted from the DTM. Slope maps, aspect plots, elevation zones, seen areas, profiles, cross sections, overburden volumes, etc. are other examples.

A DTM is created in the orthophoto production process. The USGS is saving these digital models for $7\frac{1}{2}$-min quads to produce orthophotos. Tapes of these digital elevation models (DEMO) in 7- or 15-m (RMS) accuracy can be ordered from the USGS through the

National Cartographic Information Center, 507 National Center, Reston, VA 22092.

20-12.   PHOTO INTERPRETATION

Photo interpretation provides the surveyor with information useful for boundary retracement, control station recovery, and survey planning. The high cost of field operations justifies the time spent in photo examination of the project site, because it may indicate omitted deed calls, possible encroachments, or existing control adjacent to the project site.

Boundary retracement is facilitated by plotting deed calls to photo scale on clear acetate and then superimposing it on the photo. Coincidence with boundary indicators on the phot is readily apparent, and stereoscopic study unaffected by the transparent overlay. It must be noted, however, that absolute geometric registration is unlikely, due both to photo distortions and field measurement errors. When used in this manner during the planning of a survey, utility easements, existing rights of way, and cemeteries—which may require further deed research for accurate location—become apparent.

Boundary indicators on aerial photos include lines of trees, fence rows or fences, land use changes, forest type or age changes, roads, railroads, streams, and ridges. Although many of these indicators can be seen on single photos, additional data are available when stereoscopic examination is used. When boundary identifiers are visible on aerial photography, approximate measurements of distance, direction, and area made as previously described are useful to either reject evidence shown on the photo or to accept it for ground study.

Site drainage information to aid boundary retracement is best obtained using stereoscopic study. In addition to streams and visible drainage ways, low areas can be indicated by topographic relief or vegetation types. For example, in the northeasthern United States, a change from spruce fir forest cover to alder, red maple, and cedar usually indicates the existence of a poorly drained area, whereas in the Midwest, grassed waterways in fields indicate drainage sites.

All the boundary indicators noted vary with time. As a result, photography taken at different times and seasons will indicate the location of different or additional boundary lines. Multidate photography is available to surveyors at many Agricultural Stabilization and Conservation Service and Soil Conservation Service offices. In addition, a listing of all the mapping photography for a specific area held by the federal government can be obtained from the National Cartographic Information Center.

Aerial photos are also useful in planning and recovering control station location. Control station descriptions, obtained from the National Geodetic Survey, can be plotted on a photo and taken to the field by the survey crew to aid in recovery. The location of intervisible control can be estimated from stereoscopic examination of control station locations and intervening land cover and topography.

20-13. CLOSE-RANGE PHOTOGRAMMETRY

By common definition, the term *terrestrial photogrammetry* is applied to projects in which photographs are taken from the ground, and the object distance is about 1000 ft (300 m) or greater. *Close-range photogrammetry* (CRP) applies to projects in which the object distance is shorter than 1000 ft. As generally applied, terrestrial photogrammetry is used as a terrain mapping tool, whereas close-range photogrammetry can be employed to "map" or measure practically any surface or object.

20-14. APPLICATIONS

Surveyors, as expert measurers, are often called on to make measurements of unusual shapes and objects. The expert measurer must be aware of measuring tools that can be applied for unusual applications such as close-range photogrammetry. This method has been used successfully to model surfaces and shapes such as ship hulls and propellers, aircraft fuselages and wings, and architectural works. It has been used to monitor retaining wall movement, deformation of large structures, errosion from earth slopes, and to "map" sculptures, the anatomy of live bodies, and the surfaces of teeth. The potential of close-range photogrammetry to map, model, or measure shapes and surfaces appears to be virtually unlimited. Of course, several of the examples cited are primarily of academic interest, but they indicate the potential. Prime applications are those concerned with structural, industrial, and architectural fields. Techniques for such applications will be discussed here.

20-15. PHOTOGRAMMETRIC SYSTEMS

Before deciding which camera and restitution systems will be utilized, a careful assessment of the required measurement accuracy must be made. The total "error budget" can be allocated to three components—i.e., control, camera, and restitution errors. If we consider control errors separately, camera-restitution components are the key elements of a photogrammetric system. Essentially three combinations are available: (1) a metric camera with analogue restitution, (2) nonmetric camera with analytical restitution, and (3) metric-analytical camera. Compromises on the camera side of the combination require higher-order restitution equipment and vice versa. If measurements are to be made in only two dimensions versus three dimensions, the system and control requirements can be substantially modified.

20-19. CAMERA SYSTEMS

Generally speaking, any high-quality camera can be used for close-range photogrammetry. However, the use of nonmetric cameras requires fully analytic restitution techniques to achieve any level of confidence in the accuracy of results. In the absence of analytical capability, metric cameras are a practical must (Figure 20-1). There are several single and dual metric camera systems on the market, with a variety of formats ranging from $6\frac{1}{2} \times 9$ cm to 23×23 cm.[5]

Several integrated systems comprising stereometric (dual) cameras mounted on a bar with a precisely known base and matching stereoplotter specifically made for close-range work are also available. These systems offer the advantage of quicker photo acquisition and reduced object-space control requirements. They are particularly suitable for applications involving numerous repetitions of similar type. For example, these systems are extensively used in Europe for automobile accident documentation and investigation.

Camera format and focal length must be considered in relation to object size, number of stereomodels required for coverage, control and accuracy requirements, and plotting equipment available. Obviously, single model coverage is desirable for reasons of control economy, ease of planning, plotting, etc. Model coverage is a function of object distance, focal length, and format. Accuracy is a function of focal length and camera base, but focal length may be limited by the stereoplotters.

Phototheodolites are useful in certain applications where object-space control is difficult to achieve. Rotations of the camera and theodolite axes can be recorded to facilitate plotter setup. However, since small rotation errors will be magnified in the plot, such control systems are not capable of the highest accuracies.

Conventional analogue topo-mapping plotters are designed for 9×9 in. format photography. They have a finite range of focal lengths and may put restrictions on camera bases and z range. If the camera focal length cannot be reproduced in the plotter, three-dimensional plotting is not possible without special techniques, either by enlarging the photo to stimulate a photo of acceptable f or using affine restitution. In both cases, accuracy suffers somewhat because of the additional steps required. However, two-dimensional plotting is possible so long as the model is leveled well to make the plotting plane parallel with the film plane. Thus, close-range photogrammetry may be limited with conventional analogue stereoplotters, and the user must understand the restrictions. The restrictions do not affect those plotters specifically designed for close-range work, nor do they apply to analytical plotters, since solutions to the space relationships are solved mathematically rather than mechanically.

20-17. PLANNING

The normal case assumes that camera axes are nominally horizontal and parallel, and on a common base line (see Figure 20-15). Other geometrical relationships are possible, such as convergent photography, particularly with analytical plotters. However, by far the greatest number of structural and industrial uses can be accommodated with typical geometry. For planning purposes, the size of an object to be mapped and the clear unobstructed space in front if it must be determined. The key objective in planning is to achieve complete coverage with as few models as possible, consistent with accuracy requirements. Format, focal length, and object distance all figure in this objective. The maximum object distance D theoretically possible to achieve a specified accuracy in the XZ-plane is determined from

Equation (20-9), as follows:

$$dx = dz = \pm D\, dp / f \qquad (20\text{-}9)$$

where dx, dz are errors on the vertical plane, D is the object distance, f the focal length, and dp the pointing error on the measuring mark (dp typically ranges from 2 to 20 μm RMS). A conventional mapping coordinate system in the XZ-plane is assumed. Some plotters permit interchanging the Y and Z motions, so if the camera axes are parallel, a stereogram can be treated like conventional vertical photos.

The vertical unobstructed coverage of a model can now be computed by the known geometric relationships (Figure 20-16). Note that all the vertical format may not be usable, since the foreground may occupy a large percentage of the scene unless the camera is

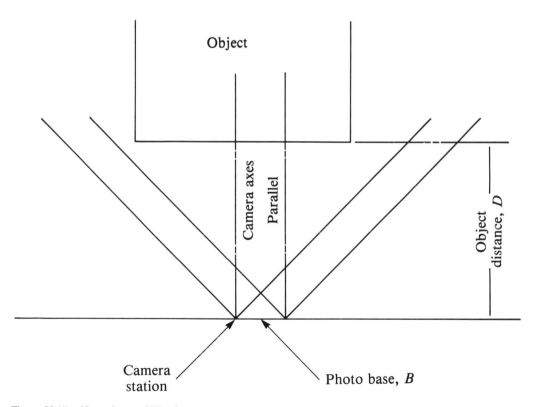

Figure 20-15. Normal case, CRP, plan view.

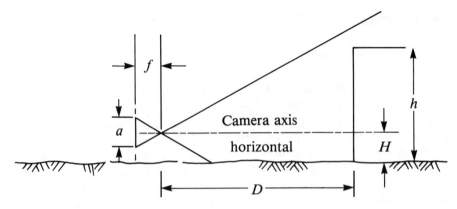

Figure 20-16. Normal case, CRP, profile view.

elevated with respect to it. Assuming flat terrain and the camera at ground level, we can compute the *minimum* object distance in which the whole object can be photographed in one frame as follows:

$$D_{min} = \frac{(H - h)f}{1/2a} \qquad (20\text{-}10)$$

where H is the height of an object, h the camera height, and a the vertical format dimension between the fiducial marks. For tall objects, D_{min} may be greater than D_{max}. In this case, the camera must be elevated to use the entire frame, or perhaps multiple levels of photography will be required to assure at least 20% overlap.

The base-to-object distance ratio (B-D ratio) should be greater than about 1/3 to 1/4 (0.33 to 0.25) for stereoscopic viewing, and no lower than about 1/20 (0.05) for geometric and photogrammetric strength of figure. The range of acceptable B-D ratios may be further modified by the plotting equipment and cameras used. For example, the "8" range on one popular mapping plotter requires the B-D ratio to be greater than 1/5.6 (0.18).

For many applications, accuracy in the Y direction (parallel with the camera axis) is not critical. But when it is, accuracy can be calculated by

$$dy = \pm \frac{D^2}{f(B)} \cdot p. \qquad (20\text{-}11)$$

A B-D ratio near the upper limit may be necessary for specified accuracies in the Y direction. For usual situations, a B-D ratio of about 1/10 represents a good compromise, accuracy and equipment limitations permitting.

20-18. CONTROL

Control is required to establish model scale and define the origin and orientation of the three-axis coordinate system in which the stereomodel will be plotted. These requirements can be satisfied by using either control in model space, or by surveys of the camera locations, object distances, and orientation of the camera axes. Since small measurement errors in camera orientation are magnified in plotting, the latter technique is suitable only in situations where high accuracies are not of primary concern. The first method, however, is completely similar to aerial photogrammetric techniques and therefore readily understood and capable of the greatest precision.

To achieve the highest accuracies, several well-distributed, carefully coordinated, suitably marked points must be established in the model. For metric cameras and analogue plotters, four points are needed; six points are required for nonmetric cameras and analytical plotters. The survey of these control points is achieved by standard methods, with special care to assure necessary precision.

A much simpler technique involving only basic surveying operations is suitable for many applications. One or more plumb lines are established in the object space, either with an actual line, targets on a plumb line, or a leveling rod held vertically. A horizontal line is established parallel with the plotting plane, either in the same plane as the plumb line or on a parallel one at a known distance from the plumb line. Known distances are introduced on both horizontal and vertical lines. If Y-axis measurements are to be taken, a third horizontal, orthogonal line in the Y-direction with known distances is established. For many architectural applications, the building itself may be used to establish horizontal and vertical lines and the known horizontal distances can be measured.

20-19. PHOTOGRAPHY

Depending on the camera, roll film, cut film, or glass plates may be used. For the latter two, some foolproof system of separating the exposed from the unexposed stock has to be devised and strictly followed. In addition, some means of scene and exposure identification is necessary. A notebook can be kept, recording the exposure number and scene. Far better is a title placed in the scene giving the date, location, and other pertinent information. It must show images on both stereopairs for an unambiguous record.

A high-quality exposure meter is a *must*. It is recommended that two extra exposures be taken, one stop over and one under the indicated setting to permit selecting the exposure with the best contrast and detail for plotting.

20-20. SUMMARY

Photogrammetry provides surveyors with a mapping and estimating tool that produces decided cost benefits for certain projects, when compared to conventional field methods. The economic advantages offered by photogrammetric techniques depend on the project size and accuracy requirements.

NOTES

1. J. E. Colcord. 1981. Photogrammetry as an aid to triangulateration traverse reconnaissance. Proceedings of the 41st ACSM Annual Meeting. Washington, D.C., pp. 483–491.
2. ASCE. 1983. *Map Uses, Scales, and Accuracies for Engineering and Associated Purposes*. New York.
3. For an at-length discussion of photogrammetric mapping contracts, see C. C. Slama, ed., 1980. *Manual of Photogrammetry*, 4th ed. Falls Church, VA: American Society of Photogrammetry.
4. For more precise methods of measuring height and elevation, see P. R. Wolf. 1983. *Elements of Photogrammetry*, 2nd ed. New York: McGraw-Hill.
5. For a list of camera systems available, see *Handbook of Non-Topographic Photogrammetry*. 1985. Falls Church, VA: American Society of Photogrammetry.

REFERENCES

American Society of Photogrammetry. 1985. *The Handbook of Close Range Photogrammetry and Surveying*. Falls Church, VA.

Moffitt, F. H., and E. M. Mikhail. 1980. *Photogrammetry*, 3rd ed. New York: Harper Collins.

Thompson, M. M., and G. H. Rosenfield. 1971. Map accuracy specifications. *Surveying and Mapping 31*(1), 57.

U.S. Department of Transportation. 1968. *Specifications for Aerial Surveys and Mapping by Photogrammetric Methods for Highways*. Washington, DC.

21

Compass Surveying

F. Henry Sipe

21-1. INTRODUCTION

Compasses discussed in this chapter are devices used to determine direction, with a magnetized needle balanced on a pivot. The needle and pivot are housed in a box containing a circular ring divided into degrees and/or half-degrees. When a compass is held steady and the needle swings freely, it points in a northerly-southerly direction, and the degree mark to which it points can be read on the circle.

Ancient civilizations did not have a compass; they navigated chiefly by stars, winds, waves, and landmarks. Discovery of the compass needle opened the universe to exploration. Colombus used a compass on his voyage to America in 1492, and he may have been one of the first long-distance travelers to verify a difference between magnetic and true north.

A forerunner of the compass needle was a mineral called "lodestone." About the 10th century A.D., it was discovered that a piece of lodestone floating in water on a reed—or perhaps suspended from a thread—always came to rest in a certain position. Although it is now popularly believed that the needle points toward a "magnetic pole," this is wrong. Its direction depends on the horizontal com-

ponent of the magnetic field where the compass is located.

Lodestone has an attraction for iron and can transfer this quality to iron itself. Legends say that the Chinese knew about magnets 2500 B.C. However, it was not until about the 11th century A.D. that the practical use of a magnetized needle for navigation became known. A compass needle on a pivot in a circular housing whose rim was graduated was described by a Frenchman named Peregrinus in 1269 A.D. The circle was first divided into 8 parts, then later 16, 32, 64, and now 360 (or 720 if half-degrees).

Progress in applying the compass needle was slow. The compass we know today is not an instrument of the highest accuracy, but it does have the advantages of simplicity, low cost, and ease of use.

Most early land surveys in the United States, including those by George Washington, were made "by the needle." It was an essential tool for exploring and mapping America. And despite the development of transits, theodolites, and electronic measuring devices, the compass (often still part of transits and theodolites) remains a valuable instrument.

This chapter describes various compasses and their uses for persons who have a limited background in them. It can help foresters, geologists, archaelogists, builders, real estate people, land developers, and lawyers. Many engineers and surveyors do not receive adequate compass instruction, and they should find this chapter helpful.

21-2. COMPASSES AND THEIR USE

In a broad sense, surveying includes exploration or *orienteering*, making or interpreting maps, and locating property boundaries, roads, bridges, buildings, or other structures. The word "survey" is said to come from the French and means to "view" or "oversee." Only those uses in which the compass is the main instrument are discussed in this chapter. Simpler compasses will be described first and then more sophisticated ones.

"*Pocket*" or *hand-held compasses* are small compasses not designed to be mounted on a tripod and should not be used to mark property boundaries. Instructions come with the better ones. One well-known type is shown in Figure 21-1. Graduation marks may be in 90° quadrants or in 260° azimuths. Some are liquid-filled to allow the needle to come to rest

quickly. The Suunto compass is held close to the eye; by an optical illusion, the vertical sight line appears to project above the compass.

Tripod compasses can be mounted on a tripod or Jacob staff. The latter is a wooden pole sharpened at the bottom (often tipped with metal) and tapered at the top to fit into a ball-joint socket. The staff was commonly used for early surveys. It cannot stand on hard surfaces, lacks stability, and is not recommended for property boundary surveys. It may be employed to run timber cruise lines or for reconnaissance with greater accuracy than a hand compass.

One of the best-known tripod compasses is the *Brunton* (Figure 20-2). It is graduated in quadrants or azimuths and can be tilted to read vertical angles. With a needle about 2 in. long, it can be read to the nearest degree, although some models have digital readouts. The Brunton is used for preliminary surveys, road layouts, timber cruising, reconnaissance, and topographic work.

Other *open-sight compasses* are made by Keuffel & Esser, Warren-Knight (Figure 21-3), and formerly Gurley, and have these features: (1) tripod or Jacob staff mount, (2) needles 3 to 5 in. long, (3) graduations to degrees or half-degrees in quadrants or azimuths, and (4) folding sights—some with verniers and declination adjustments. Bull's-eye levels and ball-

Figure 21-1. Silva-type 15 compass. (Courtesy of Silva Compass.)

Figure 21-2. Brunton compass. (Courtesy of Forestry Suppliers.)

Figure 21-3.   Surveying compass. (Courtesy of Warren-Knight Co.)

joint sockets are used for leveling the compass. Cardinal directions are reversed to provide direct reading.

A good open-sight compass is suitable for surveys requiring accuracies of 1:2000 to 1:3000. A proficient operator should get 1:2000 at a lower cost than with telescopic instruments for woods surveys. An important advantage of a compass is that errors tend to compensate, whereas errors with plate instruments can accumulate.

Woods surveys are sometimes made with compasses built into or affixed to transits or theodolites with a telescope replacing an open sight. These *telescopic instruments with compasses* are heavier, more awkward to handle, take longer to use, and cost more than an open-sight compass. Their use is not justified for surveys requiring an accuracy lower than 1:2000. They will not be described in this chapter.

21-3.   MAGNETIC DECLINATION

Little is known about the earth's magnetism. There is a concentration of magnetic force in northern Canada and south of Australia. In Canada, the "center" is about 75° north latitude and 101° west longitude, but it moves and changes intensity. The north end of a compass needle dips more and more when nearing the North Pole, so a counterbalance is placed on the south end. In the south magnetic field, the situation is reversed. Near both pole areas, the needle loses its sense of direction and remains in any set position.

Because the center of greatest attraction is not at the true geographic poles, a magnetized needle points to true geodetic north in only a few places. Figure 21-4 shows the U.S. Government Isogonic Map of Declination in the United States for 1985. The line of zero declination is called the *agonic* line. In the United States, declination is *west* on the east side of this line and *east* on the west side. An isogonic map may be purchased from the Arlington and Denver offices of the U.S. Geological Survey; ask for the latest map—it is printed about every five years (1985, 1990, etc.).

The term *declination* should not be confused with *variation*. Declination is the angle between true north and magnetic north at a given time and place. Variation is the *change* in declination of the needle caused by the passage of time and other disturbances. In Colonial times, surveyors often used the word variation when they meant declination as used today.

21-4.   HOW TO DETERMINE DECLINATION

1.   The most accurate declination at a given location can be obtained from the National Geophysical Data Center (NGDC), National Oceanic and Atmospheric Administration (NOAA), E 62, 365 Broadway, Boulder, CO 80303 (phone 303-497-6478 or 6222). Give the point's latitude and longitude to the nearest minute or two, or the exact state, county, city, township, and/or location. If the latitude and longitude are given, the declination can often be obtained during the original phone call or a return phone call.

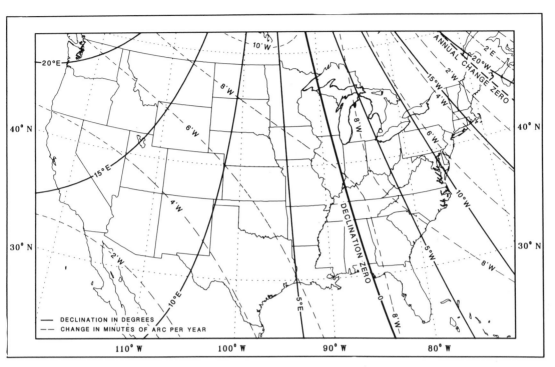

Figure 21-4. Isogonic chart for 1985. (Courtesy of U.S. Geological Survey.)

2. A second method to interpolate the declination for your location is from an isogonic map; update it by adding to a west declination the annual west change multiplied by the number of years since the map was printed. Subtract from an east declination the annual west change. The agonic line has been generally moving westward for many years.

3. Another procedure is to calculate it from the declination shown on a $7\frac{1}{2}$-min government quadrangle covering the place of survey. This figure is for the quad center and year of publication in the lower right-hand corner. It must be updated by the method described under secular variation.

4. The U.S. Geological Survey has established magnetic stations at several thousand places in the United States. Reference marks are visible from the stations unless foliage or new structures intervene. A description of the Elkins magnetic station is given in Figure 21.5.

   The description of any magnetic station can be obtained from the NGDC. The decli-

nation is calculated by reading the bearing of a reference mark and subtracting the smaller bearing from the larger one. Government azimuths are referred to *south*.

5. Declination can be determined at a national geodetic marker. The National Geodetic Survey (NGS) has installed many markers in the United States. Read the magnetic bearing from the station to one or more reference marks, calculate the line's true bearing from its azimuth and subtract the smaller bearing from the larger. A description of any station can be obtained by writing to the Director, NGS, Information Center C18, Rockville, MD 20852 (phone 301-443-8631). Give the latitude and longitude of the marker or place and ask for a copy of the horizontal control data sheet for that station. An NGS marker can have several drawbacks—e.g., it may be inaccessible or obliterated.

6. A sixth method involves the establishment of your own meridian station. Select a convenient place at the county seat, preferably in an open spot on public property or on a school campus. Do not set both a north and

STATE: West Virginia	COUNTY: Randolph	STATION: Elkins 1924

FORM C&GS-769 (10-67)

U.S. DEPARTMENT OF COMMERCE
ENVIRONMENTAL SCIENCE SERVICES ADMINISTRATION
COAST AND GEODETIC SURVEY

LATITUDE	38°	55.8
LONGITUDE	79	50.9

DESCRIPTION OF MAGNETIC STATION

STATION MARKER: Drainpipe, filled with concrete, set in ground with standard bronze disk on top.

APPROX. DISTANCE	TRUE AZIMUTH†		NO.	PROMINENT OBJECTS VISIBLE FROM THE STATION
350 yds.	66°	48.1	1	Spire on residence of H. B. Martin
200 yds.	111	03.6	2	Spire on house of Mrs. Arthur Lee (Graceland)
150 yds.	153	43.9	3	Top of tower on home of Senator S. B. Elkins (Halliehurst)

The station is on the campus of Davis & Elkins College, in the open space west of Science Hall, at the foot of the slope which extends up to the Elkins residence.

It is 236.9 ft. from the N.W. corner of the low stone wall in front of Science Hall, 10.7 ft. S. of the N. line of that building. It is 251.3 ft. from the S.W. corner of the Liberal Arts Building and is 23 paces from a large clump of shrubbery and trees. It is marked by a south meridian drainpipe set in the ground and filled with concrete 365.5 ft. due S. of the station, 141.8 ft. from the stone fence along Sycamore St. and 22 ft. S. of the S. back stop of the tennis courts.

Date of last report: 1966

The present (1983.0) estimated value of magnetic declination at this station is 6° 57.4' west. This value should be used only at the station; for other points in the area either chart or tabular values should be used.

†From South around by West: 0°= South; 90°= West; 180°= North; 270°= East.

CHECKED BY	DATE	APPROVED BY	DATE

Figure 21-5. Description of Elkins magnetic station. (Courtesy of National Geophysical Data Center.)

south station. Set one marker, make a true meridian observation, and read a true bearing to three or four prominent points several hundred feet away. Then, set a compass over the station, sight on a reference mark, and read the magnetic bearing. Declination at that time, date, and place for a particular compass is found by subtracting the smaller from the larger bearing. Do this on three or four days at about 6 PM and average the readings. Government quads show whether the declination is east or west.

21-5. MAGNETIC VARIATION

Magnetic needles change direction because of secular, daily, and irregular variations. *Secular variation* is the change in declination over long periods of time. For many years, the U.S. government has operated observatories to record these shifts. Available tables show movement over the last 200 or more years for latitudes and longitudes at 5- to 10-yr intervals. Tables for an entire state are reasonably priced and can be obtained from the NGDC; a computer-derived table for a given location can also be purchased from the NGDC. If there are no natural or artificial disturbances, the table's accuracy in recent decades is probably within 2 min for a 10-yr period. Secular-change data are less reliable for early years.

Chart and tabular values of magnetic declination, computer-derived from a magnetic field model, are considered to be accurate within $\frac{1}{2}°$. Data from repeat stations are observed readings updated using secular change from the model and accurate enough for a surveyor to get a reliable compass index correction.

For many years, surveyors in the eastern United States used a *rule of thumb* of 1° variation for every 20 yr. It should be discarded in favor of government tables or actual ground observations. Future variations cannot be accurately predicted.

Since government tables are only to the nearest degree of latitude and longitude and

in 5- or 10-yr intervals, it is necessary to find variation by interpolation for a more specific location and year (see Figure 21-6). The figure is based on an original survey made in 1890 and a resurvey made in 1983 at latitude 38°40' and longitude 79°20'. A circle locates the survey. The 1890 declination is given at four 1° intersection lines, interpolated to 1980 as 4°17' for point of survey. (It is immaterial whether interpolation is up-down or left-right). The change to 1980 must then be brought up to 1983. The 1980 isogonic map shows an annual change of 8.1' at the survey location. Assuming a uniform continuation from 1980 through 1983, add $(8.1 \times 3) + 4°17' = 4°41'$. Although these calculations appear very accurate, in reality they are merely the best approximation obtainable from presently available ways to handle data on a continental basis.

Secular variation can often be found in a more practical way. If a line is resurveyed between two authentic markers and its present bearing determined, the difference between the present bearing and that of a former survey is the variation.

Secular variation of one line does not necessarily agree with variation of an adjoining line in the same deed description; the two lines may have been surveyed at *different* times.

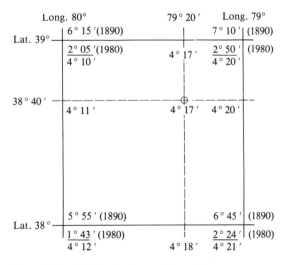

Figure 21-6. Calculation for secular variation.

This is a common cause of failure of "calls" (bearings and distances) to close when plotted on paper. Also, because land has been subdivided and transferred many times, landowners and lawyers have arbitrarily selected calls originating many years apart and placed them in new deed descriptions, not realizing that magnetic bearings of lines change as years pass.

Misunderstanding secular variation can cause boundary disputes. For example, a line whose original bearing was N 50° E may now be N 54° E. If a surveyor now runs the line as N 50° E, it will encroach on the owner to the left (northwest) side of the correct line.

A compass needle also changes direction throughout the day because of *daily variation*. During morning daylight hours, the north end of the needle is east of its mean position for the day; in the afternoon, it is west of its mean. Table 21-1 from U.S. Government Publication No. 40-1 shows daily variation at three places. Movement may be related to the sun and ionosphere electric currents; it can be allowed for with a vernier and is not significant on small surveys.

A needle does not follow the chart in Table 21-1 on *stormy* days not related to visible wind or rainstorms. *Magnetic storms* of sufficient magnitude to affect surveyors' field work schedules are rare, perhaps one day in five years. Magnetic conditions and forecasts are available from NOAA, Space Environment Service Center, R432, Boulder, CO 80303. Magnetic storms are seldom a factor in the day-to-day use of compasses.

Deposits of iron ore in parts of the United States seriously affect compasses. An isogonic map shows areas where actual declination has been observed to vary as much as several degrees from normal. For example, in the Mesabi Iron Range of northern Minnesota, the map shows a difference of $11\frac{1}{2}°$. Other ores and geological formations may cause lesser disturbances. Using a compass near *magnetic ore deposits* requires special study and experience.

Every piece of *survey equipment* and clothing containing ferrous metal must be tested to see if it affects the needle and, if so, kept at a safe distance or a nonferrous item substituted. Some questionable articles are eyeglass frames, the eyelet on a magnifier, belt buckle, wristwatch, or a notebook and pencils.

Surveyors must be alert to detect and avoid or compensate for *miscellaneous disturbances* on or near the line of sight the might affect the

Table 21-1.   Average daily variation of needle for 10 "quiet" days of month, rounded to nearest minute*

Standard Time	January and February November and December			March and April September and October			May and June July and August		
	Sitka, AK	Cheltenham, MD	Tucson, AZ	Sitka, AK	Cheltenham, MD	Tucson, AZ	Sitka, AK	Cheltenham, MD	Tucson, AK
8 AM	+2	+2	+2	+5	+5	+4	+8	+6	+5
9 AM	+2	+3	+3	+5	+4	+3	+8	+5	+3
10 AM	+2	+2	+2	+4	+2	+1	+5	+1	0
11 AM	+1	0	+1	+2	-1	-1	+1	-2	-2
Noon	0	-2	-1	-1	-4	-2	-2	-5	-3
1 PM	-1	-3	-2	-2	-5	-3	-4	-6	-4
2 PM	-2	-3	-2	-3	-5	-3	-5	-5	-3
3 PM	-2	-2	-2	-3	-4	-2	-6	-4	-2
4 PM	-2	-2	-1	-3	-2	-1	-5	-2	-1
5 PM	-1	-1	0	-3	-1	-1	-4	-1	-1
6 PM	-1	0	0	-2	-1	-1	-2	0	0

*A plus sign means the north end of the needle is east of its mean position for the day; a minus sign, the needle west of its mean position.

Source: Adapted from J. H. Nelson, L. Hurwitz, and D. Knapp, 1962, *Magnetism of the Earth*, U.S. Government Publication No. 40-1, U.S. Coast and Geodetic Survey, Washington: U.S. Government Printing Office, p. 43.

compass needle. The longer the needle, the more sensitive it is. The distances shown here are the closest a Warren-Knight surveyor's compass model 40-1300 should come to certain objects:

1. Rusty barbed wire fence on ground: little effect

2. Rusty barbed wire fence standing: 15 to 20 ft

3. New wire fence: 20 to 25 ft

4. Automobile (compact): 50 ft

5. Railroad track (90-lb rails): 75 ft

6. Gas pipelines (26 to 36 in. in diameter): 100 ft

7. Culverts and pipes: variable

There is widespread belief that electric lines interfere with compass needles. U.S. Government Publication No. 40-1 states: "Fortunately, *alternating* currents have no perceptible effect on the ordinary compass, since direction of the resulting magnetic field changes so rapidly the compass cannot respond to it."[1] However, *direct* current will affect a needle. Metal in most electric wires is nonferrous, but steel towers supporting them will have a serious effect.

The best way to avoid magnetic objects found on the ground, such as junk piles, is to set a station safely beyond the disturbance or project the line from the backsights.

21-6. ACCESSORY EQUIPMENT

A 200-ft Lietz utility fiberglass *tape*, graduated in feet and tenths, is preferred for a compass survey. It is inexpensive, easy to read and see in the woods, does not break if kinked, and can be "done up" in a loop from hand to elbow like a clothesline. A 200-ft Lietz utility tape stretches about $\frac{3}{4}$ in. more than the same length steep tape in hot weather if equal tension is applied, so a pull about 5 lb lower than for a steel tape is recommended. Too tight a pull can, however, help compensate for sag or tape being off-line.

One surveyor suggests that a cut fiberglass tape can be repaired by peeling off about $\frac{3}{4}$ in. of the outer coat on each end. Cut two thin metal shims about $1\frac{1}{2}$ in. long and a little narrower than the tape. Punch a small hole through them at each end. Place the tape ends together, place a shim on each side with the burred side next to the tape, and tap a $\frac{1}{4}$-in. brad through each set of holes and tape. Glue shims to the tape, cut of the brads and batter them, cover the splice with fiberglass glue or sealer, and let dry; file off any projections. An emergency splice can be made by punching a hole through the two ends, inserting a 2-in. paper clip through the holes, and wrapping the splice with friction tape, wire, string, or strapping tape.

Range poles in common use are too narrow for good compass surveys: They are hard to see in the woods. Three kinds of larger poles are available: (1) $1\frac{1}{2}$-in. ID PVC white drain, waste and vent pipe; (2) aluminum tubing, and (3) 2.375-in. OD white PVC pipe. They are available at many builders' supply houses. Some surveyors prefer a 2.375-in. pole with rod level and red enamel painted 12-in. band beginning 12 in. back from each end. A $6\frac{1}{2}$-ft length is good because it places the red band at about eye level for an average-height person, where it can be used for sighting vertical angles. If the diameter is too large to fit a hand, a screen-door handle can be screwed to the pole.

Vertical angles may be read with a Suunto *clinometer* or Abney level (see Figure 21-7). The Suunto is preferred because it is faster in woods. The surveyor and frontperson should read their clinometers to the nearest $\frac{1}{4}°$ to a point on the rod or on the other person at the observer's eye height—i.e., parallel to the ground. If readings differ by more than $\frac{1}{4}°$, the operation must be repeated until they agree within that limit. If unequal readings result using one versus two eyes, *heterophoria* may exist. (Heterophoria is "insufficient action of one or more muscles of the eye, so one eye

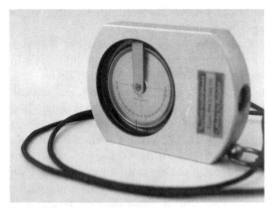

Figure 21-7.   Suunto clinometer.

deviates from the correct direction.'') If it does, a comparison of vertical angles read with a transit will indicate whether one or two eyes is better. There can be a slight index difference between Suuntos, so they should be tested with a transit.

An anvil-type shrub *pruner*, carried in a hip pocket, is a good substitute for a machete in many surveys. It is lightweight, not dangerous, and it can cut branches up to $\frac{3}{4}$ in. The average frontperson tends to cut too much brush, especially with a machete; most of the cut brush is uselessly off the line of sight, thereby slowing the work. If a pruner cannot do the job, use a hand ax.

21-7. SURVEYING WITH AN OPEN-SIGHT COMPASS

This section describes how to survey with a Warren-Knight surveyor's compass model 40-1300. In a flat field devoid of magnetic disturbance, establish a closed traverse with five sides of 175 to 200 ft each. Set a hub and center tack at each corner and measure between them to the nearest 0.01 ft. Measure all interior angles to the nearest minute with a transit or theodolite. Set the compass over the westernmost tack with the plumb bob, vernier at zero, and sight on the hub and tack to the northeast. Read the bearing of the line to the

nearest minute and repeat for the remaining lines. Employ the compass bearing of the first line as a transit bearing and convert transit angles to bearings. If the compass survey has been well-done, compass bearings should be within 5 to 10 min of the transit bearings. Using taped distances and observed compass bearings, calculate the error of closure by latitudes and departures. If the error is poorer than 1 in 2500, repeat the compass survey until it is consistently better.

21-7-1. The Vernier

To operate the vernier on the Warren-Knight 40-1300 compass, set up at the first station with the plumb bob and set the vernier to zero. Sight on the next marker, clamp the spindle, and bring the vertical thread on line with the hub, using the ball-joint tangent screw. Next, read the bearing of the line, estimating minutes. Turn the vernier control knob (with the vernier clamp loose) under the south vane until the needle rests at the next *lower* graduation on the outer circle. For example, if the bearing is estimated as N 54°10' E, rotate the circle until the needle reads N 54° E and read the vernier to determine the extra minutes. The answer should not be far from 54°10'.

The *Y* (zero) of the inner vernier arc will be to the *left* of the outer ring zero for NE and SW quadrant lines; to the *right* in NW and SE quadrants. A similar procedure is followed for reading minutes, when running a line with a bearing predetermined to the nearest minute. For example, a trial line is run and misses the objective by too great a distance for easy offsetting to the true line. Calculate the correct bearing to the nearest minute and rerun the line on that bearing.

21-8. PRELIMINARY PROCEDURES

Before starting a boundary-line field survey, study all documents relating to it. A record search must include deeds of adjoiners, boundary agreements, court decrees, previous surveys, road plats, etc. The surveyor should inform adjoiners of plans and perhaps inspect the area. Secular variation and the estimated present bearing must be ascertained.

A job often overlooked is plotting calls on transparent paper to a scale of 1 in. = 2000 ft. By tracing over those lines with a sharp red pencil and superimposing the tracing on a $7\frac{1}{2}$-min USGS quadrangle, it is possible, especially on large surveys or those calling for natural or cultural features, to see how lines match culture and topography, and to detect serious mistakes in calls. Calls may also be plotted to scale of an aerial photo, which may be found in county offices of the Soil Conservation Service or Agricultural Stabilization Conservation Service. The oldest flights show fields, roads, fences, streams, etc., as they were many years ago. Contact prints are preferred over enlargements for stereoscopic viewing.

21-9. THE TRIAL OR RANDOM LINE

A crew may consist of only a surveyor (observer) and front rod-chainperson (frontperson). Extra people may be assigned to cut brush, carry supplies, etc., but should be kept out of the line of sight. If possible, begin at the south end of an undisputed line to avoid sighting into the sun and make reading the needle's north end easier. If there is only one authentic marker, start from it with variation applied.

21-9-1. If Starting Point Can Be Set Up On

A compass is placed over the mark and leveled. If the line is short and daily variation is negligible, keep the vernier at zero and run on the degree or half-degree graduation nearest the calculated bearing. This avoids mistakes in setting off minutes on the vernier, and the line will end close to the next point. For example, if N 54°10′ E is run as N 54° E, the line will miss a point $\frac{1}{4}$ mi away by only 3.8 ft and offsets to the correct line are small. Since

calculations for secular variation may not be accurate within 10 min, often trial lines need not be run on specific minutes.

The observer now directs the frontperson (who carries a hand ax, rangle pole, shrub pruner, clinometer in degrees and percent, and a roll of flagging) to proceed on the line of sight. Tape is passed around a small sapling or other object near to and rear of the observer to prevent it from rubbing a tripod leg when pulled toward, and to allow the observer to pick up the tape without changing viewing position. This eliminates the risk of disturbing the needle by bumping the tripod. As the frontperson goes forward, the observer holds the tape loosely. If the frontperson diverges more than a foot or two off line, tape is snubbed, and the frontperson brought back on line. When tape is extended 200 ft (or some lesser distance in a multiple of 25 ft because the pole can no longer be seen easily), the observer calls "chain" and snubs the tape. A last check is made of the needle position, tapping the tripod to see if the needle shimmies properly.

The frontperson places the ax on the ground and turns and faces the observer, with tape in one hand and pole in the other. With zero held at the pole center, the tape and pole are moved right or left, forward or back, until the pole is at the correct distance and direction. In wooded areas, the tape is held by both persons at a uniform height above ground, say, about $2\frac{1}{2}$ ft. In cleared areas, place the tape on or close to the ground. Proper tension should be applied and the tape must not be out of alignment over 2 ft. When all requirements are met, the observer calls "OK." The frontperson drops the tape, holds the pole steady and vertical, and both persons read their clinometers. The observer records the average angle.

Meanwhile, the frontperson marks the pole position with a small stick or nail and cuts a stake about 3 ft long and $1\frac{1}{2}$ in. in diameter, sharpens one end and flattens the other, and drives it at the pole point.

The observer now records more data in the notebook. Date, weather, names, and perhaps addresses of the persons present and the job each performs should be recorded at the top of the right-hand page.

21-9-2.   If Starting Point Cannot Be Set Up On

Set the compass 3 or 4 ft away, on line to be run, by first estimating direction. If the needle shows that the compass is off-line, move it until the backsight hits the middle of the marker—this is called "jiggling in." Distance is measured from the side of the marker approximating its center (see Figure 21-8).

21-9-3.   If There Is Magnetic Interference at Starting Point

If the beginning point is a wire fence intersection and the line runs with or near a fence, a different tactic is required. The compass must be offset 20 to 25 ft at a right angle to the line being run; a cut stake (station O) is driven at that point. If this point also is in a wire fence, a further adjustment is employed. The compass is placed 20 to 25 ft away from the first stake O but in the opposite direction from the line to be run. The compass is jiggled in until the selected bearing hits stake O. Distance is, of course, measured from station O (see Figure 21-9).

If a bearing has not been preselected and the line to be run is parallel to the first 100 or 200 ft of fence (not always a good practice because a fence may be crooked or go in the

wrong direction), a temporary cut stake is set by the frontperson at a point the same distance from the fence as is station O, and up to 200 ft forward. A multiple of 25 ft is selected if possible. Bearing to it is read to nearest minute. A trial line can be run on that bearing or the nearest $\frac{1}{2}°$ left or right.

21-9-4.   Continuing the Line

The observer, assuring that the station O stake is reasonably vertical, raises the needle off pivot, releases the spindle clamp, folds vanes (front one first), lifts the legs, brings them together, and carries the instrument to the next station. The compass should not be detached from the tripod.

En route between stations, the observer watches for evidence of previous line marks, corner markers, and possession. Any seen must be tied to and noted, such as "at Sta. ‾ + 62 ft pass 22 ft Rt. of 20″ white oak with ‾2 old blazes on south side." Such entries are usually placed on the right-hand page at an appropriate place. Distance to centers of streams, roads crossed, etc. are recorded. If the line is run along or near a fence, at each station a right-angle offset to it is measured and the distance recorded.

Slope station numbers are marked on the flattened top of each stake and a ribbon tied near its top. Fluorescent orange plastic flagging is suggested. Either person may flag and number. This process is repeated as the line is projected.

21-9-5.   Obstacles on Line

In wooded areas, trees and other obstacles are encountered on the line of sight. These are avoided in several ways. Leafy twigs (the most common obstacle) can be snipped with the pruner. The line can be projected beyond small trees or saplings by tilting the pole from side to side, or the observer's head can be moved a little from side to side and an average extension made. It is seldom desirable to cut trees or saplings because work is delayed with-

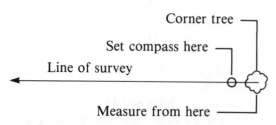

Figure 21-8.   'Jiggling in.'

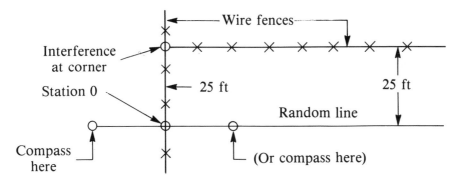

Figure 21-9. Random line to avoid fence.

out increasing accuracy, and the surveyor is exposed to claims for damages. If the front-person holds the pole at the middle of the body, the latter makes a good target. Use imagination: Try a venetian-blind slat (marked at the midpoint and stored in one end of the range pole) held horizontally like a cross, equidistant on each side of the pole.

Eventually, an obstacle too large to be seen past will be met. The usual way to avoid it is by offsetting the compass, at either the station from which the obstacle is first seen or a new station near it. The offset is made by raising the needle and moving the instrument to a point measured at a right angle to the sight line. On offsets of a foot or two, the right angle is estimated; the 90° offset sights are used for longer ones. They are measured horizontally, even if short broken shots (breaking chain) are required.

After the compass is offset, a shot is taken past the obstacle to a temporary point, where an offset is made back to the original line being run. A ribbon placed on or around the obstacle helps when backsighting. Short offsets need not be recorded, but longer ones should.

If a fence comes within 15 or 20 ft of the compass, an offset should be made away from it; this is preferred to changing the bearing of the line being run. Several offsets may be needed on a line, some to the right, others to the left.

21-9-6. Backsight Control

After the compass is set up, leveled, and oriented, a backsight is taken to the previous stake, sighting from the front threaded vane through the rear slit. Three black lines (optical illusions) may be seen, the heaviest in the middle being the correct one. Turn the compass by hand until the heavy black line hits the previous stake. Then, move to rear of the compass, level it, and set the correct bearing at the north end of the needle. Go back to the front vane side and sight the previous stake again. If the line of sight is more than an inch or two off, the reason should be sought and the matter corrected. Among possible reasons are:

1. Previous stake not vertical.
2. Compass not set vertically over point where stake enters ground.
3. Slight vanes not vertical.
4. Needle not riding free.
5. Magnitude attraction at either station.
6. Stake may have been set at wrong place.

In this situation, the ability and judgment of an observer are tested severely. A projected backsight may be used if attraction is found. However, it should seldom be projected more than one shot or two without needle verification, even through fields. The practice of "leapfrogging" (setting over every other station and using backsights for alternate calls) is not good surveying.

21-9-7. Nearing End of a Line

When a survey is nearing the distance called for, the marker called for should be sought first at the slope distance and to either side of the line. A fence corner may be significant if a marker is not found.

If the marker is not found at slope distance, the horizontal length of each shot is computed with a pocket calculator or taken from a table. Table 21-2 is a portion of a slope reduction

Table 21-2.   For reducing slope distances to horizontal

			Slope Reduction Table			
Degree	2.000	1.750	1.500	1.250	1.000	Degree
3	1.997	1.748	1.498	1.248	0.999	3
$\frac{1}{2}$	96	47	97	48	98	$\frac{1}{2}$
4	1.995	1.746	1.496	1.247	0.998	4
$\frac{1}{2}$	94	45	95	46	97	$\frac{1}{2}$
5	1.992	1.743	1.494	1.245	0.966	5
$\frac{1}{2}$	91	42	93	44	95	$\frac{1}{2}$
6	1.989	1.740	1.492	1.243	0.995	6
$\frac{1}{2}$	87	39	90	42	94	$\frac{1}{2}$
7	1.985	1.737	1.489	1.241	0.993	7
$\frac{1}{2}$	83	35	87	39	91	$\frac{1}{2}$
8	1.981	1.733	1.485	1.238	0.990	8
$\frac{1}{2}$	78	31	84	36	89	$\frac{1}{2}$
9	1.975	1.728	1.482	1.234	0.988	9
$\frac{1}{2}$	73	26	79	33	86	$\frac{1}{2}$
10	1.970	1.723	1.477	1.231	0.985	10
$\frac{1}{2}$	67	21	75	29	83	$\frac{1}{2}$
11	1.963	1.718	1.472	1.227	0.982	11
$\frac{1}{2}$	60	15	70	25	80	$\frac{1}{2}$
12	1.956	1.712	1.467	1.223	0.978	12
$\frac{1}{2}$	53	08	64	20	76	$\frac{1}{2}$
13	1.949	1.705	1.462	1.218	0.974	13
$\frac{1}{2}$	45	02	59	16	72	$\frac{1}{2}$
14	1.941	1.698	1.455	1.213	0.970	14
$\frac{1}{2}$	36	94	52	10	68	$\frac{1}{2}$
15	1.932	1.690	1.449	1.207	0.966	15
$\frac{1}{2}$	27	86	45	04	64	$\frac{1}{2}$
16	1.923	1.682	1.442	1.202	0.961	16
$\frac{1}{2}$	18	78	38	198	59	$\frac{1}{2}$
17	1.913	1.674	1.434	1.195	0.956	17
$\frac{1}{2}$	07	69	31	92	54	$\frac{1}{2}$
18	1.902	1.664	1.427	1.189	0.951	18
$\frac{1}{2}$	897	60	22	85	48	$\frac{1}{2}$
19	1.891	1.655	1.418	1.182	0.946	19
$\frac{1}{2}$	85	50	14	78	43	$\frac{1}{2}$
20	1.879	1.644	1.410	1.175	0.940	20
$\frac{1}{2}$	73	39	05	71	37	$\frac{1}{2}$
21	1.867	1.634	1.400	1.167	0.934	21
$\frac{1}{2}$	61	28	396	63	30	$\frac{1}{2}$

Source: F. Henry Sipe, Elkins, WV Aug. 1971.

table. If shot lengths are in multiples of 25 ft, the measured distance and vertical angle are merely intersected in the table; the number found is entered in the horizontal distance column ($\frac{1}{4}°$ distances are interpolated). The table is adapted for any unit by moving decimal points. Odd distances can be calculated and multiplied by the 1.000 (cosine) column.

When all horizontal distances are entered, they should be added, subtotals carried to the next page, and a final total found. Slope station numbers will help to detect major errors in addition. If the total horizontal distance run is slightly different from that called for, a station is set at the proper horizontal point, numbered, and notes corrected.

Diligent search must now be made to retrieve the marker called for. If found, the trial line is extended or shortened to a place at a right angle to the marker, and the last shot length adjusted accordingly. A right angle is determined by trial and error. Set and number a stake at the proper place. Bearing and length of the tie to the marker must be recorded.

21-9-8. Reference Objects

Measure the bearing and distance to two or three reference trees or objects to help perpetuate the location of the markers, especially if they are deteriorating or subject to easy removal.

21-10. CORRECT BEARING BETWEEN MARKERS

Most projected trial lines will not hit the marker called for. The correct bearing between markers can be found in several ways.

First, if only a quick approximation is needed, remember that a 1° angle causes an offset of 92 ft in 1 mi. If this formula is proportioned with the actual distance run and distance missed, the correction angle is ob-

tained. For example, 12 ft in 1320 ft are equivalent to 31 min. If a line run N 60° E is 12 ft to the right of the marker, the correct bearing is N 59°29′ E.

The angle tangent of miss is the offset divided by the distance run. In this example, $12/1320 = 0.0090909$, and the tangent inverse is 0.508563, which multiplied by 60 is 31 min.

A rule of thumb used by older surveyors for small angles of miss was to multiply the feet of "miss" by 57.3, divide the result by the distance run, and multiply by 60. The result is 31. Succeeding lines can be run $\frac{1}{2}°$ counterclockwise.

However, before calculations for corrected bearings are made, all recorded offsets (as shown in Figure 21-10) must be balanced out. There are two 25-ft left offsets and one 70-ft right offset, leaving a net 20-ft right offset. But the marker was found 5.6 ft left, or 14.4 ft right of the trial line if it had been projected without offsets. For a trial line 2728 ft long, the angle of miss is 18 min. If the trial line bearing had been changed in midstream instead of offsetting, a right triangle could not have been used in the calculations. Therefore, 90° offsets are preferred over bearing changes.

21-11. CORRECT DISTANCE BETWEEN MARKERS

The length of the line between markers can be found in several ways. For example, square the distance run, add it to the square of the distance of miss, and take the square root of the total. To illustrate, 1320 squared is 1,742,400, +12 squared is 144 = 1,742,544. The square root of 1,742,544 is 1320.05.

The best trig formula is to divide the distance run by the cosine of the angle of miss. Cosine of 31 min is 0.9999593, which divided into 1320 = 1320.0537. Rounding off for rural work gives 1320.

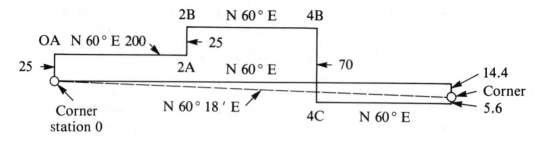

Figure 21-10. Offsets from random line.

21-12. WHEN SECOND MARKER IS NOT FOUND

Trial lines must be continued until an authentic marker is found. Apply the same variation to each line unless some good reason is found to change. If a called for marker is not found until the fourth corner, as in Figure 21-11, lines *A-B-C-D* are trial lines with distances called for, and the bearing and length of *D-D*1 is measured. Distribute that distance among trial lines in the proportion that the length of each line is to the total length of the three trial lines. Corners *B*1 and *C*1 are located on the same bearing from *B* and *C* as the bearing of line *D-D*1.

Total distance run is 2225 ft and closing tie N 11° E 37 ft. The tie for every foot of trial line run is 37/2225 or 0.0166 ft. The second corner, therefore, is 0.0166 × 750 or 12.45 ft N 11° E from *B*. The third corner is 0.0166 ×

1050 or 17.43 ft, but to this is added the tie to the second corner or 29.88 ft N 11° E from *C*.

21-13. CORRECT BEARING AND DISTANCE OF FINAL LINES

In Figure 21-11, correct measurements are determined in two ways. First, triangle *A-B-B*1-*A* can be plotted with a drafting machine on a scale large enough to measure *A-B*1 from the plat. At 1 in. = 50 ft (about the smallest that can be read by eye), line *A-B*1 is N 76°10′ E 755 ft.

A second way is to calculate line *A-B*1 as an omitted measurement (see Table 21-3). Tangent of the missing bearing is 733.2/180.9 = 4.05307, so the bearing is N 76°08′ E.

The length of line *A-B*1 is best found by dividing its departure (733.2) by the sine of its bearing (0.97087) = 755.2 ft. Distances in

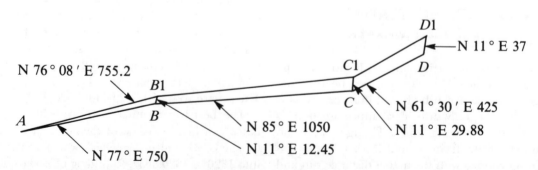

Figure 21-11. Diagram of offsets when second or third corner is not found.

Table 21-3. Calculation for omitted measurement

Bearing	Distance	Cosine	Sine	Latitude (cos × dist) N	S	Departure (sine × dist) E	W
N 77° E	750.0	0.22495	0.97437	168.7		730.8	
N 11° E	12.4	0.98163	0.19081	12.2		2.4	
			Subtotal =	180.9		733.2	

woods are not normally measured closer than 0.1 ft, so four- or five-place trig tables are satisfactory. No other trig formulas for solving oblique triangles are required for compass surveys.

Two good ways to solve for the bearing and length of lines *B*1-*C*1 and *C*1-*D*1 are by scaling from a plat or making the omitted measurement. Often, a shorter time is needed to measure a line in the field than to calculate it in the office.

21-14. SURVEYING BY TRUE BEARINGS

Occasionally, true bearings are to be retraced. A surveyor may plan to set off a declination on the compass believing that procedure is easier and more accurate. The writer recommends against this because the declination may be set off on the wrong side, which doubles the error. Also, declination at best is only an estimate and will vary with different instruments, time of day, etc. It is better to allow for declination by calculating a magnetic bearing for the line to be run.

For example, a line N 60° E true bearing with a present estimated declination of 6°40″ W would be run N 66°40′ E (or S 66°40′ W) magnetic. West declination is subtracted in NE and SW quadrants, added in NW and SE quadrants, and vice versa for east declinations. Plot true bearings near cardinal directions, N, E, S, and W to prevent errors. For example, a true bearing of N 88° E with 6°40′ west declination would be run S 85°20′ E magnetic. New or original compass surveys should usually be

run by magnetic bearings with declination shown on the plat.

21-15. SURVEYING ALONG ROADS

If traffic is light and safety rules followed, the road center can be traversed with short shots. The compass is set up in the middle of the road over a mark or PK nail. On sharp curves, shots may be as short as 50 ft, but longer sights are desirable. Backsights may be taken if a small stone pile is left.

This method leaves small segments on either side of the road center. Their area can be calculated from formulas and tables if the "middle ordinate" is noted on each shot. Or plot each segment on a large scale and planimeter its area. If segments are small and compensating, they may be ignored for most rural work.

When traffic is heavy, survey on the shoulder or edge, with offsets measured to the road center at each setup. Road bearings when no cars are passing. Traverse and road center are plotted; road alignment drawn with a French or railroad curve; chord bearings and distances are scaled from the plat.

21-16. MEANDERING STREAMS

If streams can be waded, survey in the center with temporary stakes in small mounds of stones for markers. Segment areas often compensate and may be unimportant. At selected intervals such as sharp bends, tie to reference trees. Preface your description of the meander

lines, "Thence with the meanders of...."
The word meander should not be confused
with meander lines of public-land surveys. The
latter denotes metes and bounds surveys along
the mean high-water marks of permanent natural bodies of water and usually does not
define ownership lines.

21-17. ACCURACY OF THE WORK

After the trial survey, error of closure may be
found in two ways to see if it meets a standard
desired. In the first method, plot traverse bearings and distances with a drafting machine to
a large scale. The distance between the end
and beginning is scaled from the plat and
divided into total perimeter, giving error of
closure.

For the second method, use latitudes and
departures like omitted measurements. In
Table 21-4, cosines and sines are omitted. Differences between north and south latitudes
(7.2) squared (51.84) are added to the difference between east and west departures (21.8)
squared (475.24) = 527.08. The square root of
527.08 is 22.96, which divided into a traverse
perimeter of 12471 = error of 1 in 553. As this
is generally below a satisfactory error for compass surveys, field or office mistakes must be
found and corrected.

Beginners may have poor closure errors
with open-sight compasses. Taping distances
and reading vertical angles in rural rough-
terrain wood surveys are a source of more
errors than bearings. Practice should improve
precision until satisfactory accuracy is attained
consistently. Davis and Foote say,

> There can be no rules for the relative accuracy
> of different classes of surveys.... Each surveyor
> must establish the limit of error using his own
> judgement.... The best surveyor is not one who
> is extremely accurate, but one who makes a
> survey with sufficient accuracy to serve its pur-
> pose without waste of time or money.[2]

Error of closure is usually calculated only
for the trial traverse. It is best not to balance
courses to attain perfect mathematical calls.
Error should be left in to show future surveyors how precise the former survey was, and
how much leeway is available to reestablish
missing markers. Balancing courses in compass surveys in a purely mathematical way can
put errors at the wrong places.

21-18. ACREAGE OF THE SURVEY

There are several ways to find the area of a
traverse. To begin, plat on a large scale and
divide into right triangles. Figure 21-12 shows

Table 21-4. Calculation for error of closure

Bearing	Distance	Latitude (cos × dist) N	S		Departure (sin × dist) E	W
N 27-45 E	1279	1131.9			595.5	
N 76-08 E	3020	723.8			2932.0	
S 46-30 E	2362		1625.9		1713.3	
S 51-30 W	2540		1581.2			1987.8
N 67-41 W	3540	1344.2				3274.8
	12,741	3199.9	3207.1		5240.8	5262.6
			−3199.9			−5240.8
			7.2			21.8

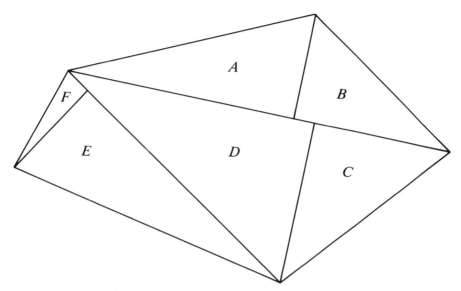

Figure 21-12.   Traverse divided into triangles.

the traverse used for Table 21-4 divided into six triangles. With a 50-scale ruler, measure two shorter sides of each triangle, multiply them and divide by 2, which gives units of area in each triangle. Add the units of all triangles, giving total units. Convert the units to acreage based on the scale of the ruler used. If the plat scale is 500 ft per in., each square covers 250,000 sq ft or 5.739 acres—i.e., 5.739 × number of square inches = acreage. Or multiply the number of square inches by 250,000 and divide by 43,560.

An alternate method involves using a planimeter. It shows the number of square inches, which can be multiplied and divided as in the previous example. To use a planimeter, assemble and set the pivot at a point (usually to left of traverse) so the tracer pin can reach all corners. The pin is place over a corner with the measuring wheel at zero and then moved clockwise along lines by hand. When the circuit is ended, the number is recorded, the tracer point moved around a second and third time, and all numbers recorded. If the third number is not close to three times the first number, go around again until a consistent

average is attained. Lower-price planimeters are satisfactory for compass surveys.

The most accurate method is by double meridian distances (DMDs) or coordinates. These are explained in other chapters.

Finally, bearings and distances are punched into programmed calculators or computers, and error of closure, acreage, etc. are printed on paper or screen.

21-19.   OFFSETTING

After the final corner locations are decided on, the surveyor locates and marks the final lines on the ground. Seldom are calculated lines rerun on new bearings because (1) offsetting trial stakes to new lines is cheaper, unless offsets are long and on steep hillsides; (2) new lines would meet obstacles that require new offsets—cutting down trees to preserve projected lines is not wise because trees make good boundary markers, cost is higher, and property owners may object; and (3) new compass lines could also miss corners, requiring more offsets ("bending" the last part of a line to hit a corner is improper).

21-19-1. If There Are No Recorded Offsets

Trial line stakes are merely offset to the final lines. The offset distance by which the trial line missed the corner is prorated between the trial line stakes. Tables 21-5 and 21-6 show how this is done.

A 547.6-ft trial line missed the corner by 12.6 ft. Each foot run therefore is entitled to an offset of 0.002301 ft. Leave this number in calculator and multiply it by horizontal length of each shot; round off to the nearest hundredth. Enter each offset in the offsets column with its direction and add cumulatively. Compare the total with the final tie for a check.

Although the accuracy of a compass survey does not justify offsetting to the nearest hundredth foot, they are entered in notes to prevent accumulation of small errors. If final offsets are not within a tenth of a foot of the measured tie, find the error.

21-19-2. If There Are Recorded Offsets

They may be in both directions (see Table 21-6 and Figure 21-10). The net offset must be calculated first. Add left NW offsets (55.6) and add right SE offsets (70.00). Subtract 55.6 from

70.0, leaving a net offset of 14.4 ft. Then 14.4/547.6 = 0.026966 offset per foot run. Multiply this factor by each horizontal distance and enter it in the offset column.

Despite the apparent simplicity of this calculation, make a scale plat in the office, like Figure 21-9. Exaggerate the offset scale to help measure small offsets.

21-19-3. Offsetting

One person can make short offsets as noted in Table 21-5. Start at the beginning of the trial line. Pass a stick or survey pin through the end loop of a 50-ft engineer's tape and insert it beside the trial stake. Measure offsets to the nearest tenth of a foot at a right angle to the trial line and set a nail, small stick, or other marker there. Pull up the trial stake and drive it at the offset point. Estimate right angles by standing over the trial stake, with one arm extended toward the previous station and the other arm directed to the next station. Bring the arms together in front of your body for a pointer. An exact right angle is not critical until the offsets become long. Then, use a pocket compass.

If the offsets are on a down slope, a plumb line can be used or the verticality estimated. For very steep slopes, offsets can be made in several level segments. If offsets are uphill, set the zero end of the tape at a point a foot or two farther out than the actual offset. Com-

Table 21-5.   A trial line with no recorded offsets

Station	Magnetic Bearing	Slope Distance	Vertical Angle	Horizontal Distance	Offsets
0 = stone pile					to NW
	N 60° E	200.0	0	200.0	4.60
2					
	N 60° E	200.0	+8	198.1	4.56
4					9.16
	N 60° E	150.0	+4½	149.5	3.44
				547.6	12.60
	N 30° W		0	12.6 =	12.60
16-in. white oak corner					

Table 21-6. A trial line with recorded offset in both directions

Station	Magnetic Bearing	Slope Distance	Vertical Angle	Horizontal Distance	Offsets
0 = stone pile					
	N 30° W		0	25.0	
0a					SE 25.00
	N 60° E	200.0	0	200.0	+5.26
2a					SE 30.26
	N 30° W		0	25.0	+25.00
2b					SE 55.26
	N 60° E	200.0	+8	198.1	+5.21
4b					SE 60.47
	S 30° E		0	70.00	−70.00
4c					NW 9.53
	N 60° E	150.0	+4½	149.5	−3.93
5.5c				547.6	5.60
	N 30° W		0	5.6	= 5.60
16-in. white oak corner					

pare the offset distance to the recorded one and place a mark at the difference. For example, if the trial offset is 20 ft and recorded as 18.6, move the stake to the 1.4-ft mark. Two persons are more efficient for long offsets.

21-20. MARKING LINES

Marking lines in woods is very desirable and has been ruled mandatory in some court cases. It allows fences to be built on line, provides safer timber cutting or land clearing, and trespass is less likely. It is disturbing for clients to spend time and money for a survey, and then need it done again in 10 or 15 yr to satisfy a new owner or divide the property.

21-20-1. Kinds of Marks

There is no generally approved method for marking private boundary lines, nationwide. A good standard is to make two hacks about 6 to 8 in. apart on trees within arm-reaching distance from the lines; if trees are scarce, mark trees farther away. Make a hack by striking trees at about breast height with the hand ax held at a slight angle above horizontal, hitting hard enough to go through the bark into wood. If the tree is on line, two hacks are given it fore and aft.

The Bureau of Land Management's manual suggests "notches" (which it calls hacks) or blazes on public-land survey lines. A blaze removes bark and part of the wood about 6 in. long and 2 to 4 in. wide. Notches and blazes are harmful on some species of trees because rot may enter, whereas a single-stroke hack usually heals in 1 yr and takes less time. Signs placed on boundaries at key points are desirable. On lines in open areas, durable monuments should be installed at intervisible points.

21-20-2. Field Procedures

Two persons can offset and mark. The crew chief offsets a stake and stands astride it. The marker stands over the previous stake. They view a straight line between them and agree on any line trees. The marker comes forward, marking en route, at first guided by the crew chief. As soon as the marker sees the next stake, the chief moves forward to the next stake, and the operation is repeated.

If they cannot see each other—often because random shots were too long—they move toward each other by sound or trial and error, snipping brush or removing obstacles until they are intervisible. If a small hill intervenes, one or both persons can hang a cap or kerchief on a pole over the station and wave it. At times, a three-point system helps: They move toward each other until they are intervisible *and* each can see the other's stake. Then, they line each other in alternately, like jiggling in an instrument.

21-21.   PAINTING LINES

Lines should be painted when marking because they can be seen more easily, trespass is discouraged, and it costs less than painting later (paint may be brushed or sprayed on). A three-person crew (offsetter, marker, and painter) can cover 1 to $1\frac{1}{2}$ mi a day with a gallon of paint. Orange or yellow implement enamel is recommended.

The painter follows the marker and brushes enamel on hacks over an area of about 3×5 in. Corner marks and reference objects are also painted. The painter should carry a gallon can not over half full (to avoid spilling) and use a built-up bail for easier holding. Carry extra paint in plastic bottles. Scrape off rough or loose bark with a paint scraper. Although a three-person crew is efficient, an ambidextrous marker can also paint.

21-22.   MONUMENTATION

Proper selection, marking, and description of markers for property corners is a very high-priority job of land surveyors. Courts recognize monuments as important evidence. Markers should be *unique* to prevent confusion with any other nearby object.

21-22-1.   Trees

If a tree is called for and found, it must be used. If there is a choice, select one that will likely live long—one a logger might not cut. Three hacks are placed on corner tress on sides the line enters and leaves. Hacks should be 6 to 8 in. apart, about breast height. Reference objects should be described in field notes by species and size—i.e., diameter, breast height, and DBH. They should be given three hacks facing the corner. If some abnormality exists, describe it as leaning, forked, broken top, etc. Record the bearing and distance (usually horizontal) from the corner to the center of reference objects (at about stump height for trees), Make two or three references at each corner.

21-22-2.   Stones

A stone, large end down, sunk in the ground for one-half to two-thirds its length, and surrounded by a stone pile will last indefinitely. It should be regular in shape with 3- or 4-in. letters chiseled in a suitable place near the top. Grooves should be deep enough so that they can be felt with a finger and not be confused with natural groovings. Other marks can be placed on stones, such as the year of the survey. Bottles or bits of glass in the bottom of a hole will provide uniqueness and certainty of identity. Notes must show what was used: dimensions, shape, marks, perhaps approximate weight or size, and the depth in ground.

21-22-3.   Other objects

Pipes, rods, axles, car springs, gun barrels, etc. make good markers. They should be 30 to 36 in. long, project 8 to 10 in. above ground, and set in a mound of stones. Nails or spikes are sometimes left as markers, but unless they are driven into roots or solid material, they may not last. The bigger the spike the better. Do not leave wood stakes as final markers. Surveyors should avoid the word "pin" because it is too vague.

If fence posts must be used as markers, reference them with two or three good objects. If a corner falls on solid rock, chisel a letter on it. If a marker cannot be set or made at a corner, or it may be disturbed, make a "witness" corner, usually in the property line coming to that corner. Edge-of-road rights of way are good locations.

Markers may be set at intervals along boundary lines. A marker set 100 or 200 ft from a corner, in the property line, makes an excellent line pointer for future surveyors. Markers in mowed fields must be set below cutting height. Follow state or federal rules.

21-23. DESCRIPTION OF SURVEY

A good property boundary survey description includes (1) the *preamble*, the place or jurisdiction where the survey is located; (2) the *body*, corner markers, reference objects, bearings, distances, acreage, surveyor's name and address, date of survey, reference to an attached plat, how lines are marked and painted, orientation meridian, etc.; and (3) the *being clause*, what previous deed or document the survey represents or is a part of, grantor, grantee, date and place of record.

A hypothetical description for the traverse in Table 21-4 follows. The calls include many conditions found on rural woods compass surveys. Most of the reference trees are on the client's land and were selected to make good intersection angles for measurements if the corners must be relocated later.

A tract of land in Beverly District Randolph County West Virginia on waters of Files Creek about 4 miles south of Elkins, more particularly described as follows:

Beginning at a stone pile as called for, corner to William Smith and John Brown, from which a 10″ red oak bears N 30 E 12.3 ft., an 8″ hemlock bears S 73 E 7.6 ft. and a 14″ sugar maple bears N $62\frac{1}{2}$ W 16.8 ft., thence with two lines of said Brown.

N 27-45 E, at 239 ft. crossing Laurel Run, at 282 ft. crossing public Rt. 69, in all 1279 ft. to a 30″ black gum with three old marks on southwest side, from which an 8″ gum is N 82 E 11.7 ft. and a 6″ red maple is S 11 W 11.9 ft., thence

N 76-08 E, at 1240 ft. crossing a ridge, in all 3020 ft. to a point in rail fence, locust called for but not found, corner to said Brown and Harry

Johnson, set a 4 × 6 × 20″ stone 12″ in ground in mound of stones, chiselled a 3″ H in top of said stone, from which an H chiselled on a 3-cubic yard boulder bears N $32\frac{1}{2}$ W 18 ft. and a 10″ locust bears S 52 W 14.5 ft., thence with or along said rail fence and line of said Johnson.

S 46-30 E 2362 ft. to a rotten 20″ white oak stump at intersection of old rail fences on a ridge, white oak called for, corner to said Johnston and in line of Henry Jones, drove a 1″ × 30″ iron pipe in center of said stump, from which a 6″ crooked beech is S 17 W 4.0 ft. and a 6″ iron wood is N 61 W 13.5 ft., thence leaving said fence and with said Jones

S 51-30 W, at 1080 ft. passing under an electric line serving a residence on property surveyed, at 1090 ft. crossing said Rt. 69, in all 2540 ft. to a point in center of Laurel Run as called for, corner to William Smith set a 1″ × 30″ iron pipe witness marker in a mound of stones on northeast bank of said run N $51\frac{1}{2}$ E 12.0 ft., said pipe being capped with an aluminum cap marked John Campbell LLS 314, thence with meanders of said run and line of said Smith.

N 67-41 W, at 422 ft. leaving center of said run, in all 3540 ft. to the beginning, containing_____acres more or less, as surveyed by John Campbell of Parsons, West Virginia in October 1982 and shown on a plat attached hereto and made a part of this description;

All lines being marked with two hacks, reference trees being marked with three hacks; hacks and markers painted with yellow implement enamel; bearing magnetic; made with Warren-Knight surveyor's compass #17841 with an error of closure of 1 in._____;

Being the same land conveyed by Peter Doe and Helen Doe his wife to George Harrison by deed dated January 14, 1923 and recorded in the Office of the Clerk of_____County West Virginia in Deed Book 216 at page 17.

This description written by John Campbell LLS on November 27, 1984.

21-24. REPORT OF SURVEY

Finally, a certificate or report should be written for most surveys. It may include significant facts not shown on plats or indescriptions such as:

1. What records were examined
2. Who was contacted and what was said, furnished, or showed
3. Problems encountered and how they were resolved
4. Comments on occupancy, possession, encroachments, gaps, etc.
5. Methods and equipment used
6. Uncertainties of position and error of closure
7. Distribution of documents

An actual report of survey follows:

Report of Survey for Richard W. Wilson

I certify that I did, in 1975 resurvey a tract of land in Preston County West Virginia conveyed by Wm. M. O. Dawson, Commissioner of School Lands and Charles C. Craig and Elizabeth Craig his wife to Marshall G. Wilson and Sarah F. Wilson his wife, by deed dated November 16, 1897, recorded in the Office of the Clerk of Preston County West Virginia in Deed Book 83 at page 474;

That before making the survey I examined the following documents:

DEEDS: 28/66, 31/258, 35/348, 49/280, 53/16, 53/67, (etc.).

PLATS: Julius Monroe's and Tom Clark's plats of Tannery Tract.

WILL BOOK: 22/427.

That of 10 markers called for in 83/474 only three were found—stone pile at SW corner and 2 chestnuts on south lines; that there were several overlaps on south lines between Wilson, Preston Tannery and Filsinger (the latter involving an 1888 survey by Julius K. Monroe of the "old Mountain road"); that these overlaps were explained to Filsinger and Clark (who represents the Tannery) by plat of 12/1/1975 and report of 2/17/1976 and my proposal to compromise them based partly on following the old Crab Orchard (Pine Swamp) road; that said compromise was not objected to by them;

That in addition to an overlap on south lines, there was a deed error of closure of about 106

feet in a north-south direction; that because neither of the north stones was found, the west line was shortened by 53 ft. and the east line lengthened by 53 ft.;

That one other overlap was found on the west line as shown on the plat;

That Richard W. Wilson and all adjoiners were sent a copy of my preliminary report and plat of 2/17/1976 and that no objections thereto were received;

That I made a final plat and description of survey which are attached hereto and made a part of this report;

That all lines have been marked with 2 hacks, and witness trees with three hacks; all painted with Persian Orange Enamel;

That copy of the final plat has been sent to Richard W. Wilson, Bertus M. Craig, Thomas L. Craig, Joseph C. Filsinger, Ronald White, Ed Smith, Thomas G. Clark and surveyor Robert A. White;

The plat and description should be recorded in the County Clerk's Office; that this survey is now complete, having been made with Warren-Knight surveyor's compass #17841 with an error of closure of 1:1516.

October 4, 1975   (seal)   (sgd) F. H. Sipe, L.L.S.
Elkins, WV

NOTES

1. J. H. Nelson, L. Huritz, and D. Knapp. 1962. *Magnetism of the Earth*. U.S. Government Publication No. 40-1, U.S. Coast and Geodetic Survey. Washington, D.C.: U.S. Government Printing Office, p. 15.

2. R. E. Davis, and F. S. Foote. 1968. *Surveying*, 6th ed. New York: McGraw-Hill, p. 34.

REFERENCES

HOWE, H. H., AND L. HURWITZ. 1964 *Magnetic Surveys*, Serial No. 718, National Geodetic Survey. Washington, D.C.: U.S. Government Printing Office.

KJELLSTROM, B. 1976 *Be Expert with Map and Compass*. New York: Charles Scribner's Sons.

SIPE, F. H. 1979 *Compass Land Surveys*. Philadelphia, PA: Warren-Knight.

22

The Planetable: Instruments and Methods

Robert J. Fish

22-1. INTRODUCTION

The planetable alidade is an articulated hybrid surveying and drafting instrument system that enables a topographer to simply and directly communicate with the map sheet that, in a very real sense, is an extension of the topographer's perception of a landscape. The instrument is capable of performing all the usual survey functions with the exception of field astronomy. It is also unsurpassed as a basic instrument for teaching fundamental concepts of surveying because the geometric principles are readily grasped, and the objective of a survey operation is clearly indicated on the map sheet.

22-2. PRACTICAL APPLICATIONS

On balance, advantages of the planetable-alidade system often outweigh any limitations. Integration of the "obsolete" planetable method can often be advantageously applied to contemporary problems as a valuable adjunct to the spectrum of current advanced technological instrumentation.

Photogrammetry is not always effective. Dense conifer forest, deciduous forest in leaf, swampland of tall grasses, and featureless terrain such as sand dunes can be mapped in more detail by ground survey techniques. There is no substitute for on-site inspection of terrain features, especially in large-scale mapping of built-up areas, where a considerable volume of underground detail and paved surfaces is encountered.

The planetable method can also be used for some types of construction layout staking, such as embankments, cuts, ramps, berms, or other similar situations where earthwork slopes are best determined graphically. Nuances of gradients at paved intersections for vehicular traffic or taxiways and/or runways can be solved and staked by expendient application of the planetable method.

22-3. PLANETABLE METHOD

Map manuscripts can be produced by various techniques using ground surveying methods. Three main ones considered in this chapter are (1) the level and tape, (2) planetable alidade, and (3) transit stadia.

The procedure using an automatic level and tape has some applications in mapping terrain strips for roadways or other similar routes. Detail is field-plotted on a prepared sheet mounted on the planetable board but without the alidade.

The usual method for capturing terrain details is to locate features with a tape and right-angle prism directly from a premarked ground system, such as a centerline, base line, or in the case of a site plan, a suitable grid. Plotting the map manuscript is done on prepared field sheets concurrently with the field readings, thus obviating keeping field notes. Elevations are determined by a self-reducing rod. The field team is normally composed of three to five people, depending on the density of detail to be mapped.

Productivity is measured by the output of a topographer or draftsperson. If the site is densely cluttered with much small detail, proportionally more drafting time than measuring time is required, whereas a sparser terrain needs less drafting but added personnel to do the increased legwork. To render the technique cost-effective, judgment is required to balance field-measuring time versus drafting time.

If a telescopic alidade is selected, the same general process is used. However, certain accuracy aspects are forfeited as the alidade is purely graphic and the level-tape method semigraphic. The planetable-alidade procedure is best suited to smaller-scale ratios and areas or site work typically running lower than 1:600 (1 in. = 50 ft) with the upper limit of perhaps 1:62,500 (or about 1 in. = 1 mi). For some landscape applications or special engineering works, scales as large as 1 in. = 8 ft are used, obviously for very small sites.

Perhaps the most cost-effective, versatile, and flexible technique is the transit-stadia method. Recent advances in total station types of EDMIs permit all aspects of fieldwork to be accomplished simultaneously. If the theodolite/transit is equipped with a sensitive control level on the telescope, third-order bench leveling can be combined with the normal capture of terrain detail. If a second-order theodolite is employed, along with an EDMI, an acceptable vertical accuracy result is also possible. A direct/reversed pointing on the retroprism or

target along each traverse leg for both the fore and backsights is recommended.

Field reduction of such observations is not necessary and can be a part of the office reduction procedure. Measurements of distances and directions between instrument stations are obtained in the usual way by whatever technique is necessary to fulfill the mapping project accuracy specification. Optionally, stadia measurements locating discrete terrain features of minor importance, or where a lower accuracy is acceptable, will be both faster and more blunder-free than the newer EDMI technique. In any case, concise, legible, compact unambiguous field notes are essential for subsequent office plotting of details. Field personnel can be reduced to two, but a more cost-effective approach utilizes a three-member team. Since the ordinary drafting table in an office is substituted for planetable board in the field, observations can be performed under weather conditions not possible with other procedures.

22-4. LIMITATIONS AND DISADVANTAGES OF THE METHODS

A fourth method, not previously discussed, is considered by many practitioners to be the standard, basic, or "normal" technique. For obvious reasons, it has serious shortcomings and will not be discussed in detail here. Essentially, it consists of capturing terrain detail by the level-tape method, except that field notes are kept by the expedient use of sketches for subsequent office plotting and evaluation. A premarked ground system, usually a grid, centerline, or base line, is set out to locate details along with a system of temporary bench marks. A comprehensive and complete detailed sketch requires considerable time, which reduces the efficiency of the entire survey field team. It is not drawn to scale, so a distorted perception of terrain features may result in omissions and blunders. The situation can be compounded by a draftperson's "interpretation."

The level-tape procedure is limited to more gentle gradients and not particularly adaptable to terrain where large differences of elevation are encountered. Water courses, ravines, canyons, cliffs, dense brush, or heavy traffic, etc. render the technique cost-prohibitive. It is imperative that all areas of the survey theater be accessible by foot. It is not readily adaptable to large open areas but very practical on strip-type mapping.

The planetable-alidade system produces a graphical analog or scale model of the terrain. Analysis is limited to the terrain model only. The sheer bulk of ancillary equipment, such as a sun umbrella, drafting gear, map case, slide rule, or calculator, along with the six basic components (Section 22-5) limits survey team mobility. All field observations must be simultaneously reduced and plotted on the field sheet. Since the telescope is positioned nearer to the ground, obstructions such as brush, shrubbery, fences, etc. become troublesome.

The transit-stadia method has the fewest functional defects. Perhaps the weakest link of the system is the notekeeping aspect. Any type of manual notekeeping is subject to both blunders and omissions. Aside from some rather minor arithmetic operations on field notes to reduce measurements to true horizontal distances and elevations of terrain details, the field notes function only to transcommunicate field date to office plotting of the manuscript, after which they have little value.

The remainder of this chapter will be limited to these two general methods that are about equally cost-effective, produce a similar comparable result, and from a functional standpoint are very nearly interchangeable.

22-5. DESCRIPTION OF THE PLANETABLE AND ALIDADE

There are six main components of the planetable system.

1. A short stout tripod of special design
2. Modest-size drawing board
3. Telescopic alidade (newest state of the art, integrated with an EDMI)
4. Planetable sheet of suitable drafting media
5. Specially designed stadia rod or retroprism
6. Ancillary equipment with field cases and expendable supplies

Figure 22-1 shows some necessary ancillary equipment for planetable operation (more items are listed in Section 22-11).

22-6. TRIPOD

Much of the accuracy or inaccuracy in planetable work is attributable to the tripod. The moving parts for leveling and azimuth control are an integral part of the tripod function. It must be designed as a geometrically rigid support that will maintain the board in a fixed azimuth, parallel with the horizon, while drafting operations are performed, thereby impart-

Figure 22-1. Planetable with two old-style alidades and some ancillary equipment.

ing an intermittent and variable torque to the tripod and head.

Two main styles of mounting and movement are available: (1) the geodetic survey type and (2) Johnson model. The first style has leveling screws, a clamp, and tangent screw similar to the lower motion of an old-type transit. The Johnson movement is composed of a double ball-and-socket arrangement on a coaxial clamp stem, so the board can first be leveled and then fixed in any desired azimuth. The geodetic-type has some advantages for graphical triangulation over the Johnson model, because it can be set more accurately at a given azimuth using the tangent screw. Flexure of the tripod and board is inherent in both styles. Since the Johnson-type is simpler to use, quicker in setup, has fewer parts, and for most purposes is equally satisfactory, it is the more popular of the two.

22-7. BOARD OR TABLE

Various board designs have been constructed of wood, metal, plastic, or combinations thereof. Most older boards were fabricated of well-seasoned pine about $\frac{3}{4}$ in. thick and ranging from 15×15 in. to the more popular 24×31 in. size. Individual planks are joined by tongue-and-groove joints, with an end piece fitted across the grain along the shorter sides to minimize a tendency to warp. These types of boards should always be stored on edge. When in use, the alidade is moved to different positions on the board, thus preventing it from retaining a perfect plan due to flexure. It is therefore a waste of time to level the board with very high precision.

Some boards are rounded on the long edges to mount standard-size planetable sheets, 31 in. wide and any length, by rolling the unused portion of the sheet and fastening it with spring clips to the board's underside. Other boards have eight flush-headed thumb screws for attaching a 24×31 in. sheet stock that has been appropriately perforated. Each screw

head is recessed below the board's surface to allow the alidade blade freedom of movement over the screws. A thread mounting plate that mates with the threaded stud of the tripod head is attached to the board's underside. Sometimes, the board is stiffened with sleepers radiating from the mounting plate.

22-8. ALIDADE

Only the telescopic alidade will be described in detail. Figure 22-1 shows two old-style models. The nearer one is the "exploration" version, smaller and more portable than the pedestal "geological survey" type. The so-called "peepsight" alidade has a use limited generally to military reconnaissance purposes in conjunction with a smaller 15×15 in. board.

A telescopic alidade consists of a base or blade composed of either a plain or articulated fiducial straightedge that rests directly on the planetable sheet. It has (1) a pylon or pedestal extending from the straightedge base to a pair of standards—very short in the exploration-type alidade; (2) a trunnion axis; and (3) a sleeve in which a telescope of rather modest magnification fits. On some styles, the telescope can be rotated 180° between stops on its longitudinal axis. This arrangement permits adjusting the alidade by a procedure similar to that for the old-style wye level. If the alidade is self-indexing (automatic), follow the adjustment method recommended by the manufacturer. As typical for all leveling instruments, the classical peg-check is appropriate, regardless of the reading system.

The telescope eyepiece is generally swivel-mounted, so the observer's eye need not be aligned with the telescope's longitudinal axis. Some older-style alidades, not so equipped, forced the observer to look directly into the barrel, which could be awkward or even impossible on very steep shots. Most bases or blades have a slight arch to distribute the weight toward the tips, thereby increasing the turning moment to increase the alidade's stability.

All alidades are equipped with a partial circle for measuring vertical angles. Various methods for reading the circles employ (1) a fixed or nonadjustable index—the index correction must be read at every observation, (2) an adjustable index controlled by an attached level vial, (3) automatic indexing with a pendulum-actuated compensator, and (4) electronic sensing as part of the EDMI system. Depending on the instrument, the vertical angle may be read in the customary unit of degrees and minutes, or by an auxiliary scale to simplify reduction of the vertical and horizontal components.

Figure 22-2 displays a Kern self-reducing alidade with its eyepiece fixed at a 30° angle to the horizontal for comfortable reading, ever on steeply inclined sights. The precise and rugged parallel arm can be moved over a wide range, thus simplifying the sighting and plotting operations. A pin at the end of the scale pricks the point into the paper at the push of a knob. Plotting accuracy is increased by about 25%.

Figure 22-3 shows the first and smallest EDMI mountable on an alidade and facilitates planetable operations. The minisize Benchmark Surveyor PT-1 EDMI mounted on an

Figure 22-3. Survey PT-1 EDMI mounted on an alidade. (Courtesy of Benchmark Co.)

alidade measures just $2 \times 2 \times 4$ in., weighs only 1.6 lb, has automatic slope reduction and an accuracy of $\pm(5$ mm $+$ 10 ppm).

22-9. PLANETABLE SHEET

Material for an ideal planetable sheet has the following properties:

1. Moisture-proof and nonhygroscopic
2. Dimensionally stable under variable conditions of temperature and humidity
3. Able to take ink or pencil
4. Capable of withstanding repeated erasures
5. Stiff enough to resist fluttering in a light breeze
6. Rollable
7. Tough enough to resist surface scuffing
8. Transparent for production of preliminary prints
9. Suitable for long-term archival quality storage
10. Reasonable in cost and availability
11. Nonglaring (low reflectivity)

Figure 22-2. Self-reducing alidade. (Courtesy of Kern Instruments, Inc.)

Older material were made of cloth-backed detail paper 31 in. wide and up to about 10 ft long. These sheets were rolled and kept in waterproof cylindrical maps cases during transportation. Some flat sheets were laminated detail papers with their individual grains oriented at right angles. Still others were three layers thick, the middle being a thin sheet of aluminum forming a "sandwich."

For small projects, ordinary colored poster board may be used. For larger more important work, mylar works well if the board is covered by a light green or buff vinyl material available for that purpose. A double matte tends to adhere to the board, whereas a single matte may "skate."

If a plan is to be edited on the planetable, a mylar reproduction is used instead of an ordinary ammonia print and exposed in a vacuum frame to protect scale integrity.

22-10. STADIA RODS

Only rods specifically designed and manufacturered for use in stadia work should be em-

ployed. Self-reducing alidades require a special rod matched to the particular instrument.

Figure 22-4a and 22-4b show the face and back of duplex rods graduated in feet and meters: (1) for use on short sights and (2) for long shots. Both were hand-built and the stadia faces handpainted with a black and white flat paint for a nonglare surface. The graduated coaser stadia faces lies in a single unbroken plane permitted by the nonfolding design. When employed in heavy brush, their bold open graduations can be read more accurately than other designs. Complicated marking systems and/or fine graduations should be avoided for long shots, because it is difficult to discern small details. The foot rod has been read to about 2200 ft under ideal conditions.

The other faces are commercially available metal tapes fastened only at the bottom, run in guides, and spring-loaded at the top. It is essential to equip the rod with an adjustable level as shown in the figures, since accuracy in stadia measurements depends on the rod being held vertically. Note its low placement on the rod face where it can be readily checked if

Figure 22-4. Observe and reverse sides of hand-built stadia rods, graduated in feet and meters.

the rod is held on a TBM fire hydrant or other high mark.

22-11. ANCILLARY EQUIPMENT AND EXPENDABLE SUPPLIES

These include the following:

1. A bull's-eye level for the board.

2. A box compass, also called a trough compass or declinator.

3. Spring clips or clamps for attaching rolled sheets.

4. Pressure-sensitive drafting tape for mounting the planetable sheets and other uses.

5. Tubes for storing and transporting rolled sheet stock.

6. Drafting scales.

7. Stadia slide rule, ordinary slide rule, or pocket calculator.

8. Smooth paper or a plastic sheet for cutting a mask to protect portions of the map sheet not in use.

9. Drafting instruments and a set of technical pens.

10. Straightedge 18 or 20 in. long, assorted triangles, and curve templates.

11. Several lead-chucks or wood pencils from about 2H through 9H, with a small emery board or file for generating conical or chisel points.

12. Erasers for ink or pencil and for cleaning the map sheet. If mylar is used, a small can of lighter fluid ("white" gas can be used) for cleaning drafting tape residue, removing unwanted pencil lines, or "dry cleaning" when the sheet becomes soiled from the hands or the alidade blade. Caution should be used when applying the lighter fluid to the sheet and rubbing with a small cloth or tissue, because it removes all pencil marks, but ink is not affected.

13. A sharp knife such as a Swiss army knife.

14. Signal flags and/or ratios.

15. Tape and plumb bobs.

16. Hubs, lath, hammer, nails, guy wire, target faces, signal cloth, light lumber, saw, shovel, brush-cutting tools.

17. Retroprism and plumb pole if EDMI alidade is used.

18. Transit or theodolite for supplemental control work.

19. Loose-leaf notebook with a supply of field forms.

20. Sun umbrella, folding camp stool.

21. Plumbing "fork" for transferring the position of a ground station to the map sheet or vice versa on large-scale mapping.

22. Variable scale or spacing divider for interpolating contours.

23. Small plastic squeeze bottle of detergent for washing hands, tools, and the alidade blade. Paper towels, water jug, or canteen.

24. Suitable cases for transport of all equipment.

22-12. ADJUSTMENTS OF INSTRUMENTS AND EQUIPMENT, AND DETERMINATION OF CONSTANTS

These important factors in doing accurate surveying should be checked at suitable intervals. A baseline and several bench marks established conveniently near the office—if not already available—will save time and bolster confidence. A schedule of dates for checking individual pieces of equipment can be prepared and supplemented as intuition and/or unsatisfactory checks surface.

22-13. STRAIGHT EDGES

An alidade fitted with plain fiducial edges must be checked to determine whether both edges are straight and mutually parallel. To test for straightness, place the alidade on a light table and bring it into contact with a high-grade steel straightedge. A damaged or worn edge can be trued by rubbing it with a large file. Lay the file down and stroke the alidade edge cautiously while in full contact.

To check for parallelity, set up the planetable, insert two push pins into its surface so the blade ends make contact. Align the vertical hair on a distant well-defined object. Bring the other edge into contact with the push pins and the vertical cross hair should again cut the object. Caution: A swivel prismatic eyepiece erects the· image, but it is reversed, left to right. It is a great convenience to have both alidade edges parallel.

If the alidade is equipped with an articulated fiducial edge similar to the linkage of a drafting machine or a parallel rule, first check the rule for straightness as before. Sight the vertical hair on a distant well-defined target, with the rule in an extended position. Draw a fine line with a chisel-pointed 9H pencil the full length of the ruler. Shift the alidade toward the line to nearly its closed position; again sight the distant object. The ruler should register on the pencil line as previously drawn. If there appears to be an error of unacceptable magnitude, the instrument should be sent for repair. It is not necessary to check the difference in direction between the fiducial edge and collimation axis of the telescope with any style alidade.

22-14. TRIPOD

Check the tripod for loose shows and tighten if necessary. If the wood has shrunk beyond the range of accommodation, remove the shoe, clean the mating surfaces, and reset the epoxy putty.

Examine the tripod for smooth functioning. If the leveling motion or azimuth clamp does not function perfectly when the board is mounted, disassemble the head. Burnish all mating surfaces with steel wool to remove fine-grained deleterious material. Wipe the surfaces with a small cloth saturated with carbon tetrachloride, naphtha, or 'white' gas. Re-lubricate with a good grade of powdered graphite, including the mating surfaces between the wood and metal parts. Satisfactory work rests largely on smooth positive functioning of the tripod head.

22-15. PLANETABLE BOARD

Boards should always be stored on edge—a warped board is useless. It cannot be used with a self-indexing alidade because the compensator range may be exceeded.

To check a board for warp, place a steel drafting straightedge diagonally across corner to corner, then about 4 in. from each edge. If a space of more than about 0.1 in. is detected, corrective action should be initiated. High spots can be relieved with a carpenter's plane and sander or the board dampened and loaded to remove gross warpage; however, this procedure could take several days. If the board appears useless for immediate use, a reasonably good substitute can be fabricated from a sheet of $\frac{3}{4}$-in. plywood. Simply cut a suitable size, bevel and sand the edges, seal with shellac or varnish, and install the mounting plate in the center of the bottom surface.

22-16. LEVELING DEVICE

An omnidirectional (bull's-eye) level is usually mounted directly on the blade or base. Some styles have a separate appliance consisting of a conventional level vial set in a small housing, with a base or foot plane on the bottom. To check and adjust the level, place it at the board center. Level the board in the usual way and then turn the level end for end. It will now indicate twice any error of adjustment. Before adjusting, shift the level to a corner of the board and note the run due to flexure in the board and tripod. Return the level to the board's center and complete the adjustment if it is grossly off.

22-17. MAGNETIC COMPASS

To check the compass functioning, lower the needle to the pivot. It should swing freely without sluggishness and sense the magnetic meridian with a maximum error of about 0.1°. First align the fiducial index with the needle's north end. Draw a line on a mounted sheet along the box edge or alidade. Swing the box several degrees and compare with the line as drawn. The needle should settle on the fiducial index mark. Repeat the trial several times to verify the function. If the needle comes to rest in several positions and any one is far off, the compass should be sent to a repair shop.

If the needle does not settle with the ends at the same level, the cover glass can be removed and the needle lifted from its pivot. Simply slide the counterbalance along the needle until the balance is obtained. Replace the needle with care as the pivot point is a finely honed steel pin that is easily damaged, particularly if carried without the needle being raised and locked off the pivot.

22-18. STRIDING LEVEL

Older-style alidades are equipped with a striding level that rests in cylindrical collars on the telescope barrel. It should be checked by reversals in the same manner as the board level: (1) Place the alidade on the board in the usual way; (2) bring the bubble to the center of its run with the tangent screw; (3) reverse the level, adjust half the apparent error by the means provided and the other half with the tangent screw; and (4) repeat the test.

22-19. PARALLAX

Parallax simply means that the conjugate focus of the eyepiece and objective do not coincide at the plane of the cross-line reticle. If the eyepiece is not properly focused on the reticle,

movement of the cross lines relative to the background occurs as the observer's eye is moved from side to side or up and down. This condition precludes obtaining accurate stadia intercepts. The adjustment is personal in nature and must be made by every individual observer. The adjustment should be checked several times each day as tired eyes might have a slightly different focal length than rested ones.

To adjust for parallax, move one eye relative to the eyepiece while slowly rotating the eyepiece in its spiral track, with the objective lens focused on a distant object. The adjustment is complete when the cross line is in sharp focus without any apparent movement. If the eyepiece is equipped with a resetting diopter scale, note the reading for each eye of the observer.

22-20. RETICLE

To check the reticle for coincidence in the plane of rotation of the telescope, sight a well-defined target with the cross line near the field of view center. Clamp and then with the tangent screw, track an object across the field. If the deviation is less than about 10 widths of the cross line, no change is needed. To adjust a gross error, loosen any two adjacent capstan retaining screws and very gently tap on the screw head to rotate the entire reticle cell inside the telescope. Tighten the same two screws just enough to resist transport vibration.

If the telescope is mounted on a sleeve, the reticle can also be checked for collimation. Sight the cross lines on a distant target and rotate the telescope 180°. The apparent deviation is twice the error. Loosen the capstan reticle screws and bring the cross lines halfway back to the target. Repeat the test.

The adjustment procedure should now include the familiar peg-check. Use the same test as for a dumpy level.

22-21. CONTROL LEVEL TO THE VERTICAL ARC

After the peg-check on the striding level, bring the bubble to the center of its run. With the tangent screw to the vernier frame, set the index to read zero. Being the control level to its center of run with the adjusting nuts on the mounting tube. The index should now read zero, and both level bubbles stand at the center of their runs.

22-22. DETERMINATION OF THE STADIA CONSTANT

The usual alidade telescope is equipped with a cross-line reticle composed of a central vertical line, one horizontal line, and two or more lines parallel with the central horizontal line. These auxiliary lines project a nearly constant angle of reference for any objects sighted in the field of view. The interval or angle of interception is carefully set by the manufacturer to a specific ratio or submultiple thereof, usually 1:100 between the extreme lines. Some instruments have additional lines set at ratios

1:400, with other older instruments providing ratios as high as 1:800. The so-called self-reducing alidades have a set of curved lines on a cam and gear-actuated reticle. In any case, a correction for the intercept interval should be determined empirically for each individual instrument.

On a cloudy day, set up the alidade at a nearly level site. Plumb an initial point to the ground with the plumbing fork directly under the trunnion axis. From this point, accurately lay out a taped base line setting points at 10, 20, 30, 50, 75, 100, and 200 ft. Hold a stadia rod vertically on each point and carefully read the respective intercepted distances on the rod. Plot differences between the true known lengths and observed stadia distances, and connect them by a smooth curve, as in Figure 22-5.

Most test results reveal that the largest and most rapid changes occur at the nearer distances. Manufacturers of modern equipment approach a zero correction for all lengths, but due to the nature of an internal-focusing telescope, the correction actually follows a slight curve. It may be extrapolated without an error of consequence beyond a 200-ft distance. The resultant increase in accuracy, particularly for short distances, is well worth the effort.

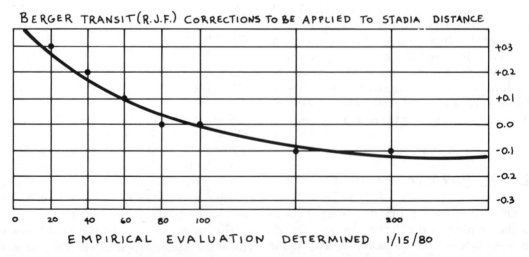

Figure 22-5.   Determining stadia constant.

22-23. STADIA RODS

All graduations must be clean and sharp. Rod levels should be sensitive enough to permit the rod to be plumbed with an error not exceeding about $\frac{1}{2}$ in. on a 12-ft rod. Improvise a test stand at the angle of the main and porch roofs. Simply mark the plumb-bob strike at a ground station directly under the roof angle. Then, register a corner of the rod at the ground station and press the rod to the roof angle with a slight pressure. Adjust the attached level. Repeat at intervals not exceeding 3 mon, or whenever the rod or level might have been damaged through accident or dampness.

22-24. FIELD OPERATIONS WITH THE PLANETABLE

Preparatory activities for fieldwork will ordinarily include some or all of the following jobs:

1. Establishing suitable vertical and horizontal control stations in and around the area to be mapped
2. Testing and checking all components of the system, such as tripod head and shoes, testing adjustment of the alidade, checking the stadia rod levels or plumbing pole for an EDMI, charging all battery-actuated equipment, sharpening brush-cutting tools, etc.
3. Drafting and plotting the control stations on planetable sheets *in ink*, or otherwise preparing them for field use, depending on whether the mapping project is fill-in mapping to enhance a photo job; revising and updating existing maps, developing and drafting models for construction layout; etc.

22-25. CONTROL

As on any survey operation, adequate vertical and horizontal control are essential. The use of planetable methods to map large areas is almost outdated. Economic factors make pho-

togrammetric procedures the preferred mode. To map an area photogrammetrically, the plotter operator must "see" the ground. Impacted areas of sand dunes, swamps, forests, and high-rise buildings can be filled in by using features that appear on the manuscript sheets as control. Large-scale mapping requires discrete ground stations related to a reference surface by survey methods generally employed for third-order control.

Bench marks are set for vertical control, and horizontal control stations are established by traverse, triangulation, intersection, resection, and trilateration. All should be referred to a specific datum surface—the reference spheroid commonly used for geodetic purposes, a state plane coordinate system, or usually a local coordinate network for small isolated projects.

Survey control is usually a separate field operation performed to a higher degree of accuracy than is possible with just a planetable. For small sites, such as a park of not more than perhaps 5 acres, exclusive use of a planetable may be feasible. Larger areas must be more closely controlled, and the multiple sheets registered exactly at the margins.

Actual maximum positional errors in the horizontal control should not exceed 25% of that for a plotted position's error. Thus, if a point mapped at a scale of 1 in. = 100 ft can be plotted with an error not exceeding 1 ft at publication scale, the maximum positional error of the nearest controlling ground station should not exceed 0.25 ft on the ground. Vertical control should be run to at least third-order standards, and the height of the trunnion axis determined with its error not exceeding 0.1 of the contour interval.

Small site work may require elevations to very close limits if paved areas, sewers, floor elevations of buildings, etc. are required. If a planetable is used for the construction of paved areas, provision should be made to limit the gross vertical error to 0.02 ft.

Horizontal control is generally marked on the ground by concrete monuments, iron pins, or tacked hubs. Intersection stations can be any high structure that occurs naturally in the landscape. Outstanding signals are commercial-type radio or TV antennas, elevated water tanks, church spires, or other landmark buildings. Even small trees that have been flagged, insulators on utility poles, steel transmission towers, flagpoles, etc., might be used. In planetable work, intersection stations are more valuable than ground stations. All signal objects should be intersected from ground stations that have been determined by the typical survey methods discussed.

All vertical and horizontal control stations must be plotted on the planetable sheet. A description of the station, its elevation, and horizontal position coordinates should be noted on the planetable sheet or in a list or catalog arranged for quick field reference.

22-26.   THE SETUP

On arrival at a site, the planetable board is screwed firmly to the tripod head, and the prepared sheet, carefully mounted with spring clips, thumb screws, or drafting tape. A planetabler works bent at the waist, so the board's top surface should be set up about belt high or perhaps an inch or two below the elbows. Care must be taken to avoid bodily contact with the board's edge.

If particularly high setups are necessary to clear obstructions such as tall grass, shrubbery, brush, or fences, the table may be mounted in the box of a pickup truck. Snub the tripod shoes in a "spider," provide a catwalk for the operator, and stabilize the vehicle with a jack under each side. The flat roof of a nearby building can be a potential setup station. A catwalk of planks for the observer will minimize flexure of the roof structure as the planetabler's weight is shifted. Protect the roof from damage with a plywood overlay.

Sites to be avoided are steep, rough, boulder-strewn places, spongy or swampy areas if footing is not fair, and walkways or paved stretches subject to pedestrian or vehicular traffic. Tracts where moisture might be a problem should be shunned because unexpected "rain" from an irrigation system could ruin an unprotected sheet. Or worse, muddy puddle water splashed by passage of a vehicle might damage the alidade and ancilliary equipment. Consideration must be given to potential problems from trees bearing ripe fruit, bird droppings from nests on buildings or trees, crop-dusting operations, airborne pollution from industrial plants and refineries, etc.

Ideally, a setup station should be selected at random to maximize command of the terrain to be mapped or laid out. The footing should be stable so the shoes can be firmly planted. If the ground is unstable, drive two-by-two wood hubs flush with the surface for the tripod shoes to rest on. Sites should be reasonably level, not in conflict with any traffic, and in natural shade provided by trees or buildings, if possible. To satisfy these conditions, the board is only infrequently set over a known ground control station. Once the most advantageous spot has been selected, the position must somehow be accurately plotted.

The classical procedures for locating a random setup station are (1) resection (the three-point problem) and (2) the two-point problem (inaccessible base). Other solutions use the principles of triangulation, trilateration, intersection, and side-section. Convenience or necessity dictates an analytical approach using a transit, theodolite, EDMI, or GPS. Once the position has been determined by application of either graphical or analytical procedures, it is customary to locate topographic details to be mapped or laid out by taking side shots using radiation and intersection.

22-27.   RESECTION

Simply stated, resection is the accurate determination of a discrete position occupied by a survey instrument from angle observations at only the unknown setup station. The technique depends on sighting no fewer than three suitably situated signals (see Figure 22-6) whose locations are known, either from coordinates (digital) or by their plotted positions on a map sheet (analog).

The position of a randomly selected setup station can probably be defined most readily using a transit or theodolite in conjunction with a programmable calculator—e.g., an HP 97 calculator and the "canned" resection program. Two advantages are (1) The setup station is ascertained with greater accuracy and (2) intersection control stations that may fall outside the sheet limits can be employed, the usual condition for large-scale mapping. The setup position is then plotted on the sheet utilizing the derived coordinates, pricked with a needle, circled with a 3×0 pen, and labeled for identification. It should be suitably marked on the ground, referenced, and described in the field notebook for future recovery.

The graphic method of resection has been used for a long time, and several general solutions have been developed. The first method to be described is simple to understand but inconvenient because extra sheets of tracing paper or mylar are required. The procedure is satisfactory on a calm day, but even a light breeze can frustrate the operation. To perform it, simply mount an appropriate piece of tracing paper temporarily over the planetable sheet, using draft tape. A point is selected, needle-pricked near the expected final position, and a ray drawn from this trial point toward each of the three known stations plotted previously on the sheet. Use a hard chisel-pointed pencil that will accurately register the fiducial edge of the alidade blade.

With the board now approximately oriented, leveled, and clamped in azimuth, carefully align the vertical cross line with each of the distant known signals while passing over the trial point. If carefully done, each ray will not be in error by more than 1 min of arc. The rays represent true horizontal angles between the control stations.

Loosen the drafting tape and shift the tracing paper about on the planetable sheet by

trial until each ray is made to "cut" its corresponding control signal plotted on the sheet. The trial point previously needle-marked is then pricked through, circled, and labeled for identification as before. The tracing paper is rolled, returned to the field storage tube for several more field solutions, and then discarded. The azimuth clamp is loosened, the alidade blade registered along any ray, and the board rotated until the cross line cuts the signal, and the azimuth motion reclamped. When sighting other signals as a check against any gross blunders, the alidade blade should cut both the needle-pricked point and the corresponding signal. The magnetic compass should register north, as previously plotted on the sheet.

The easiest and quickest resection solution by an analog procedure is a modified *Lehmann's method*. The concept is simple and the solution based on the following principles. As demonstrated in the tracing-paper method, when the board is properly oriented and the alidade sighted to each of the control signals *A*, *B*, and *C* (see Figure 22-6), rays drawn from their respective signals will intersect at a *unique* point. Also, when rays are drawn from the control signals, *the angles at their intersections are true angles*, whether or not the board is properly oriented.

To execute, set up the planetable at a selected random station. Orient the board by estimation or with the box compass. Draw rays from the control signals through the as yet undetermined location directly on the planetable sheet (rays *a*, *b*, *c*). Loosen the azimuth clamp and rotate the table either clockwise or counterclockwise until a second trial produces a second set of rays (*a'*, *b'*, *c'*) on the other side of the expected location, generally determined intuitively by inspection. Then, lines drawn from the intersection of *ab* to *a'b'*, *ac* to *a'c'*, and *bc* to *b'c'* represent small portions of three arcs of circles that pass through the unknown station setup and the control stations *A*, *B*, and *C*. The point so determined at the arc intersections is needle-pricked and the construction lines erased, or if on a mylar map

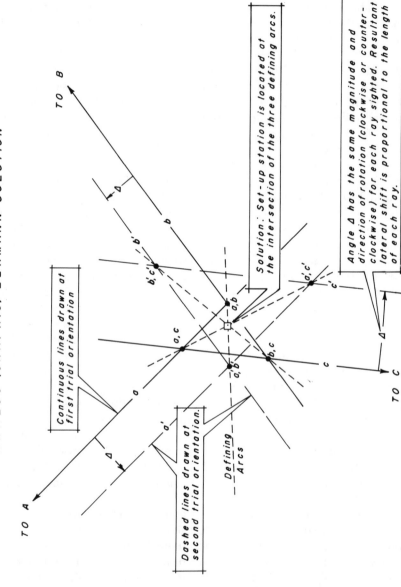

THE 'THREE POINT' PROBLEM

ANALOG (GRAPHIC) LEHMANN SOLUTION

Continuous lines drawn at first trial orientation

Dashed lines drawn at second trial orientation.

Defining Arcs

TO A

TO B

TO C

Solution: Set-up station is located at the intersection of the three defining arcs.

Angle Δ has the same magnitude and direction of rotation (clockwise or counter-clockwise) for each ray sighted. Resultant lateral shift is proportional to the length of each ray.

Figure 22-6. Solution of three-point problem.

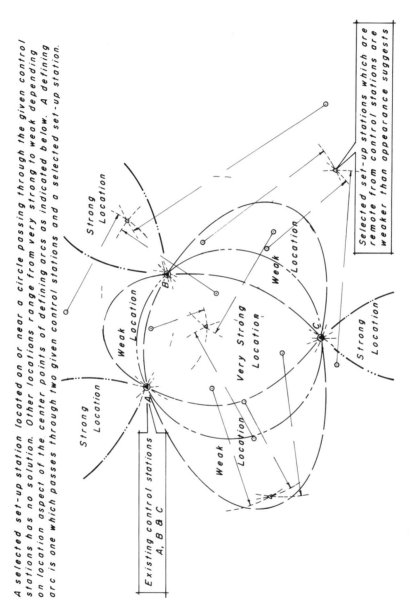

A selected set-up station located on or near a circle passing through the given control stations has no solution. Other locations range from very strong to weak depending on location aspect of the center points of defining arcs as indicated below. A defining arc is one which passes through two given control stations and a selected set-up station.

Strong Location

Strong Location

Weak Location

Weak Location

Very Strong Location

Weak Location

Strong Location

Existing control stations A, B & C

Selected set-up stations which are remote from control stations are weaker than appearance suggests

Figure 22-6. (*Continued*).

547

sheet, washed off with some white gas, and the point circled and labeled for identification. The azimuth clamp is loosened, the alidade blade aligned on any ray leading to a control signal, and a cross line set on the signal by rotating the board then clamped in azimuth, as for the tracing-paper method. Strength of the solution can be gauged by how strong an intersection the arcs make.

Two other solutions—Bessel's and Collin's—require additional extended figures to be drafted and fail in function if some of the necessary plotting falls beyond the board edges.

22-28. TWO-POINT PROBLEM

Resection can also be performed when only two control signals are identifiable, and neither can be physically occupied, as in Figure 22-7. The same general concept applies to the two solutions previously discussed. Any two-point solutions requires an auxiliary setup station and is therefore not recommended for economical operation or cost-effectiveness.

As in the three-point resection problem, the best solutions in most cases is an analytical one, with preliminary observations made with a transit or theodolite. If these are made in advance of planetabling, the solution then becomes an office procedure, so the setup stations can be plotted on a sheet in the same way as for other control stations.

To proceed, an unmeasured base line composed of two convenient mutually visible ground stations is chosen at random, marked with hubs or in other temporary ways, and so situated that control stations A and B can be observed from both (Figure 22-7). Occupy each ground station with a transit or theodolite and take azimuth cuts to control signals A and B using an appropriate procedure. The resultant intersection figures are solved by plane trigonometry (coordinate geometry), using an assumed length and azimuth for the base line. An inverse is computed for both the assumed

and actual coordinate positions of control signals A and B. The distances calculated from the assumed data are scaled up or down in a simple ratio, and an azimuth equation is applied to the assumed azimuths. The figures are then recomputed using the true azimuths and distances between all stations to get known coordinate positions of the selected ground stations that are plotted on the sheet. Caution should be exercised in selecting auxiliary stations. The solution will be weak if any intersection angles at the known control signals are smaller than about 35° or larger than 145°.

Several graphic or analog solutions have been developed. As demonstrated in the three-point problem, the solution is complete when the map sheet is correctly oriented. This is achieved when, on an accurately oriented board, intersecting rays are drawn *from* the direction of the control signals *to* the vicinity of the setup station.

Orientation is readily made by selecting a station on the line joining the two control signals. Accurately orient the board along that line and draw a ray toward the second, or "working" setup. This ray is at a *true angle* with the line joining the two control signals. Setup at the second station, backsight on the first, and clamp the board in azimuth. Then sight the two control signals, cutting their plotted positions, and draw rays from them. The setup lies at their intersection.

The more general field condition is not quite as simple. As in the three-point problem, the tracing-paper solution is easier to understand and parallels the analytical method. Steps to be taken are: (1) Mount a sheet of tracing paper over the map sheet. (2) From each end of the chosen base line, sight both signals using the base line as a reference. (3) Join the intersected stations with a straight line. (4) Loosen the tracing paper and superimpose that line over the one joining the two signals on the map sheet and temporarily resnub. (5) Loosen the azimuth clamp and sight the alidade to either control station, clamp in azimuth. The board is now correctly oriented. (6) Sight the other control signal, draw a ray *from* the plotted position. Their intersection is the board's position.

In more frequently occurring situations, one or both of the control stations are accessible. Again, the best solution is an analytic one and essentially a variation of an eccentric station case.

Proceed by selecting a setup point where both control stations can be seen. Measure the distance to the accessible ground station by stadia, tape, or an EDMI and observe the angle between the two control stations. Inverse the coordinates between them and solve the triangle by the sine law. Compute the triangle's two legs as a traverse to determine the setup station's position, and plot and identify it in the usual way.

From an analog solution, draw an arc having its center at the control station and a radius equal to the measured distance. With the board approximately oriented, intersect a ray from each control station. Loosen the azimuth clamp and rotate the board slightly. Again, draw rays from each control station to intersect. As in the three-point problem, the two arcs pass through the unknown station, so their intersection is the point sought.

With the addition of an EDMI to the planetable alidade, solutions are shortened and more accurately fixed. If two ground stations are accessible and distances to them are measured, the position sought is at the intersection of the two arcs. Where an analytical solution is necessary, measure the two accessible distances, inverse the coordinates of the known stations, solve the triangle (three known sides), compute the two accessible legs as a traverse, and plot the coordinates as usual.

Caution is required when the minimum number of control stations is employed in any of the location procedures because no redundant checks are possible. Regardless of how erroneously or carelessly the signals have been plotted, coordinated, identified, sighted, and the intersecting rays drawn, *an apparent solution always follows*. A prudent operator *always* makes a redundant check of some kind —orientation with the box compass, an extra

sighting to an additional control station or previously plotted features. A blunder in the basic location of a setup station or telescope HI makes all data obtained at the station worthless. If the planetable is being used in the reverse mode as a layout instrument, the consequences can be *very costly*, far more expensive than just a few hours of nonproductive crew time. A control scheme must always offer some provision for a redundant check.

22-29. ORGANIZATION AND QUALIFICATIONS OF THE FIELD PARTY

The minimum number of persons on a topographic crew can be reduced to only two, but progress in most cases will be painfully slow. Adding a notekeeper who calculates the observations to the minimum-size party of a planetable operator and rodperson expedites progress. If field work is remote from highways, a truck driver doubling on camp chores is an undeniable necessity. Also, if the average distance to be walked between side shots is shorter than about 120 ft, addition of a second rodperson is justified. The cost-effective ratio is enhanced if the truck driver also serves as a rodperson when the situation warrants, clears brush, transports equipment from station to station, and/or helps set up the camp and does the cooking.

Survey personnel qualifications should be optimized based on experience, job requirements, functions, etc. One untrained or incompetent person in a topographic crew handicaps a team that must function in a highly cooperative effort. If the team is to be cost-effective and the work attain professional quality, *all* members must be efficient.

A notekeeper/computer can be less experienced in field procedures than other members of the team. If also a proficient draftsperson, he or she can take over when the operator must be away from the table. An almost ideal situation results when all members of a topo-

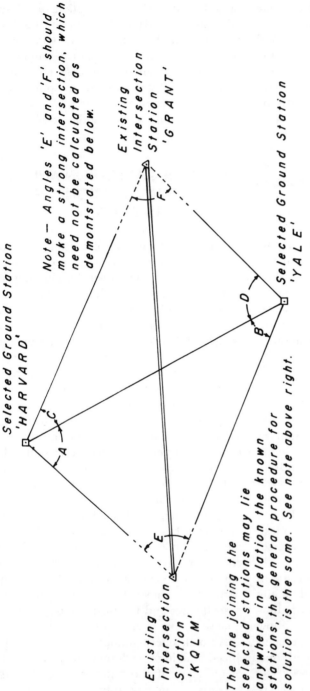
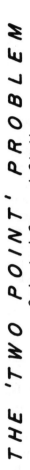

THE 'TWO POINT' PROBLEM

Selected Ground Station
'HARVARD'

Note — Angles 'E' and 'F' should make a strong intersection, which need not be calculated as demonstrated below.

Existing Intersection Station 'GRANT'

Selected Ground Station 'YALE'

Existing Intersection Station 'KQLM'

The line joining the selected stations may lie anywhere in relation the known stations, the general procedure for solution is the same. See note above right.

KNOWN DATA

	East	North
Δ 'KQLM'	319 412.63	522 620.70
Δ 'GRANT'	325 620.04	523 102.20

Coordinates, State Plane

A = 55° 10' 14" } Measured Angles
B = 44° 03' 10" }

C = 33° 19' 21" } Measured Angles
D = 70° 00' 54" }

Figure 22-7. Two-point problem.

550

STEP ① Assume azimuth & distance 'HARVARD' to 'YALE', 150°00'00", 5000.00 Ft.

STEP ② Solve triangles 'KQLM', 'HARVARD', 'YALE' and 'GRANT','YALE','HARVARD', by 'SINE LAW', or by intersections as follows.

(Parenthesis indicates solved-for quantities)

	East	North		
'HARVARD'	10 000.000	10 000.000	150°00'00"	5000.000
'YALE'	12 500.000	5 669.873	285°56'50"	(4158.038)
'KQLM'	(8 501.983	6 812.301)	25°10'14"	(3522.141)
'HARVARD'	10 000.000	10 000.000		

AND:

'YALE'	12 500.000	5669.873	40°00'54"	(2822.894)
'GRANT'	(14 315.087	7831.860)	296°40'39"	(4829.162)
'HARVARD'	10 000.000	10 000.000		

STEP ③ Inverse between 'KQLM' and 'GRANT' on both datums, assumed & SPCS

'KQLM'	8 501.983	6812.301	(80°03'07.5"	5901.837) (assumed)
'GRANT'	14 315.087	7831.860		

'KQLM'	319 412.63	522 620.70	(85°33'52.3"	6226.057) (SPCS)
'GRANT'	325 620.04	523 102.20		

(+5°30'44.8") Rotate all assumed azimuths

(1.054935) Expand all assumed distances

STEP ④ Traverse from 'KQLM' through 'HARVARD', 'GRANT', 'YALE' to 'KQLM'

'KQLM'	319 412.63	522 620.70		
'HARVARD'	(321 308.671	525 816.156)	30°40'58.8"	3715.631
'GRANT'	325 620.040	523 102.200	122°11'23.8"	5094.454
'YALE'	(323 495.002	521 015.930)	225°31'38.8"	2977.971
'KQLM'	319 412.630	522 620.700	291°27'34.8"	4386.462

STEP ⑤ An inverse from 'HARVARD' to 'YALE' should be the same as obtained from applying the azimuth rotation and scale factor, namely:
150°00'00" +5°30'44.8" = 155°30'44.8" (SPCS)
5000.000 x 1.054 935 = 5274.677 (SPCS)
A small round-off error should be expected.

Figure 22-7. (Continued).

graphic crew can satisfactorily run the plan-etable, keep notes, and act in a utility capacity. Turnabout is particularly expedient when the planetabler and rodperson alternate duties on sequential setups. In addition to the obvious relief from working steadily in a bent-over position, the operator can complete the field plotting, interpolate contours, and draft de-tails of features in the area just traversed while they are fresh in mind. This activity is per-formed in the short time it takes the rodper-son to walk from shot to shot.

Progress and cost-effectiveness are mea-sured by the number of points plotted and processed per day and depend on the se-quence and interaction of team members dur-ing observance of a spot elevation. All must be alert, and individual effort is expected to maintain the edge required for uninterrupted progress, so innovative procedures are called for. If some doubt exists about the board's ground position, orientation, or HI, the side-shot positions can be plotted tentatively on an overlay sheet. After the ambiguity is clarified, the side shots can be integrated on the main sheet as an office job. Obviously, this speeds the field work at the expense of some addi-tional inside time.

22-30. MAPPING

Actual selection of side shots depend on the mapping project requirements. Scale of the map governs spacing of side shots on a plan-etable sheet. More than about 10 shots per square inch become difficult to process manu-ally. Large, wide flat areas with a low rough-ness factor are deceiving, and there is a ten-dency to spread the shots too far apart. This psychological factor must be resisted and a fair number of side shots taken in terrain that is comparatively plain and uninteresting. The rodperson should walk a fairly closely spaced interval or pattern, so all the ground is in-spected at close range; otherwise, important small features may remain unseen and there-fore not plotted.

The planetable is admirably suited for map-ping some features and information that can-not normally be detected on aerial pho-tographs. These include (1) the size and type of culvert pipe, (2) depths and sizes of sewers and storm inlets, (3) size and species of indi-vidual trees and shrubs, (4) floor levels of buildings, (5) details of pipelines and irriga-tion systems, and (6) buried power or tele-phone lines.

Elevations of barren sand dunes and fields with crops such as corn, sorghum, and cotton also present problems for photogrammetrists. An example is an irrigated citrus orchard that was mapped when a dense canopy hid the fact that the only ground a camera could "see" was on the elevated farm roads between tree rows. Extensive plan revision was necessary after construction started to compensate for "missing" material in depressions, where or-chard trees once stood.

A similar mistake occurred in 1952 on the Ohio Turnpike. A preliminary design required radical revision when "ground truth" on-site surveys revealed that contours had been drafted on top of the forest canopy, in some places more than 100 ft above the ground! *There is no substitute for a close inspection of the ground by a well-trained competent rodperson.*

22-31. DEPICTION OF GROUND CONFIGURATION, TOPOGRAPHIC FEATURES

Selecting a ground point to be mapped that represents a surface is very important. An ex-act representation of warped ground forms is not possible, but a close approximation is at-tainable from the plotted locations/elevations of ground shots. Proper procedures produce results far superior in both accuracy and econ-omy of operation to the much-abused "grid" or "checkerboard" arrangement.

The controlling-point method is based on two facts: (1) Three points in space determine a plane that can be passed through any three points and (2) only a plane surface can be analyzed in a practical way on the planetable sheet.

To accurately portray the ground surface, a rodperson visualizes the surface to be mapped as being subdivided into triangles defining a plane that does not deviate from any point on the ground surface by more than one-half the contour interval. Any three spot elevations control the size, location, orientation, and differences of elevation enclosed by the triangle.

When plotted on a planetable sheet, sequential points are connected by drawing a straight line. The triangle side having the greatest difference in elevation is divided proportionately or interpolated with an appropriate device such as a spacing divider or variable scale. An additional tick mark is included at the same elevation as the third point of the triangle. Using a pair of small triangles, draw lines through each interpolated point parallel with the line considered level, terminating at the defining triangle limits. If the rodperson's perception of the defining plane is substantially correct, it follows that the contours so drafted are an acceptable approximation of the actual ground surface, limited by one-half the contour interval.

In the interest of economy, contours are drawn where some confusion may exist about which points are needed to clarify a particular plane area. Unambiguous portions are deferred for office work. With a little experience, an operator or draftsperson can intuitively select only those line segments requiring interpolation, because in most instances a few interpolated lines along the steeper gradients are sufficient to define an entire hillside.

The sequence or pattern in which a rodperson chooses side shots is also critical. The number of them required is evident if features such as buildings, trees, walks, curbs, gutters, drives, fences, manholes, utility poles, etc. are located first. Those features having continuity —i.e., curbs, fences, and building lines—can be side shot first, sequentially, and immediately sketched in. For example, if the outlet

end of a culvert is a side shot, the following sight should be on the inlet end. Size, type of material, and any special items are noted on the sheet. When a considerable number of details and features must be labeled, brief notes are essential. When a point is plotted, it becomes the elevation's decimal point. For clarity, *never* note the elevation beside the digits by a small cross. For brevity, truncate the elevation figure as 32.45 instead of 1032.45.

22-32. OBSERVING, PLOTTING, NOTEKEEPING

Observing and reading a stadia rod with a telescopic alidade is a quick and seductively simple procedure that results in a product superior in accuracy and lower in cost than more conventional methods. The basic sequence for taking side shots is as follows:

1. The rodperson holds the stadia rod at a point or station to be mapped.

2. The operator aligns the alidade on a rod using the telescope's open rifle sight, with the fiducial edge of the alidade straightedge nearly cutting the plotted setup station on the map sheet.

3. Focus the telescope and align the vertical cross line on the rod.

4. Slide the alidade laterally until the straightedge cuts the setup station as plotted.

5. Level the telescope and read the rod where the level sight line strikes it.

6. Turn the tangent screw until the lower stadia line cuts the rod exactly at a foot mark and note the rod reading on the upper cross line. The difference between the two readings is the stadia intercept.

7. The intercept ratio customarily set by the manufacturer is almost exactly 1 to 100; therefore, multiply the stadia intercept by 100.

8. Apply a scale along the fiducial straightedge and plot the point.

9. Subtract the level rod reading from the HI above the datum plane and note the eleva-

tion. If a level sight line does not strike the rod, modify the procedure with an inclined stadia observation.

10. Read the stadia intercept first, then sight anywhere on the rod and record the reading.

11. Reduce the inclined sight to its horizontal and vertical components, plot the location, and compute the elevation as before.

In the procedure outlined, the vertical angle must be precisely determined. An operator may opt to read it in the standard degree-and-minute units and then compute the horizontal and vertical components using tables, a stadia slide rule, or a programmable pocket calculator. As an alternative, an auxiliary scale, the Beaman stadia arc, can be employed. In both cases, before the vertical angle is read, the control level on the vernier frame must be brought to the center of its run.

A Beaman arc has two variable-spaced scales that are read against fixed indexes, so the graduations represent whole convenient ratios of the vertical and horizontal components on inclined sights. It is not necessary to know the derived mathematical theory. The fiducial point on most Beaman arcs' vertical scale is numbered 50, a constant to be subtracted from all readings. Sight the rod as before and read the stadia intercept to be recorded by the notekeeper. Bring the control level to the center of its run and with the tangent screw, *set to the nearest graduation on the vertical arc exactly at the index* and note the horizontal cross-line rod reading. The difference in elevation between the trunnion axis and rod reading is the product of the vertical arc reading and stadia intercept.

On some models, the horizontal scale of a Beaman arc is numbered zero when the telescope is level and goes up to 20 or more. On others, the numbers begin at 100 for a level reading and move down to 80. The incremental spacing of both styles is exactly the same. The first numbering system indicates the percentage of correction to be applied to the stadia distance (always minus). The second

graduated scale gives the factor as a percentage of the stadia distance. Vertical angles smaller than about 3° need not be corrected.

A self-reducing alidade has an articulated cam and/or gear-actuated curved-line reticle interposed at the conjugate focus appearing in the field of view to reduce inclined sights to horizontal distances. Hence, the distance is read directly on a vertical rod without calculation. The vertical component is obtained similarly with the rod intercept indicated by a ratio to the vertical component. The so-called self-indexing alidade is convenient, quick, and an improvement on the older-style instruments.

Figure 22-8 depicts a different planetable project. Note the lath nailed to the board's corner that greatly adds to the system's stability. So does having the rodperson stand directly behind the rod. Staking for initial grading operations on site work is well within the accuracy limits of the planetable.

Figure 22-9 shows a loose-leaf noteform with several side shots reduced preparatory to plotting. Loose-leaf forms are preferred because the book is less bulky than its bound counterpart, and as field work progresses, the notes can be removed and correlated with the planetable sheets in the office. Active notes remain in the field for reference. Since all operations are manual in nature and performed by people, certain calculations, simple as they may be, are subject to blunders. Experience shows that most mistakes are arithmetic; few if any readings are sour. If a plotted point does not fit, usually it can be found quickly, identified in the field notes, and the reduction checked. When the project is completed and all pertinent information placed on the finished map sheet, the field notes retain very little value as a permanent record.

22-33. PRACTICAL SUGGESTIONS: AVOIDING PITFALLS

Obscured sight lines caused by trees, brush, shrubs, fences, buildings, vehicles, etc. are

Figure 22-8.   Using the planetable in layout mode to stake a curved entrance drive.

(DIST.)	(BEA. ARC)	(DIFF)	(ROD)	(DIFF)	(ELEV.)	(GRADE)	
			1349.57				
221	—	—	7.20			1342.37	T.B.M.
135	—	—	5.0	—	44.6	43.09	C 1.5 ₵ INJECT. WELL
177	—	—	5.42	—	44.15	43.85	CO.3 N.W. COR PARK. LOT.
152	—	—	5.37	—	44.2	43.14	C1.1 ₵ VALLEY GUT.
147	—	—	5.22	—	44.35	43.90	CO.45 ₵ ISLAND
82	—	—	5.37	—	44.2	43.99	CO.2 " "
118	53	+ 3.5	3.5	-5.4	44.2	43.15	C1.1 ₵ VALLEY GUT.
140	—	—	5.72	—	43.85	43.86	CO.0 ₵ ISLAND
157	—	—	6.02	—	43.55	44.17	FO.62 GRADE BREAK

T.E.S.T. / SITE GRADING @ VIA RIO VERDI   9/28/84 (R.J.F. ☆
85° CLEAR) B.C. ∅
INSTRUMENT ON DIAGONAL LINE BETWEEN   JOB #6016
I.P.³ @ N.W. ₵ S.E. CORNERS / 140.07 FEET
FROM I.P. @ N.W. CORNER. (T.B.M. BRASS CAP @ ₵X ELEV.= 1342.37)

COPYR[ ] BY LEFAX, [ ]ADELPHIA, [ ] 9132   PR[ ]ED IN U.S.A

Figure 22-9.   Field notes for initial site grading at a small shopping mall.

some of the most pervasive obstacles in field work. Carefully selecting the setup stations can eliminate at least part of that problem. Good reconnaissance will determine whether certain areas hidden below low ridges or buildings can be "seen" without extensive brush cutting. Setup stations should be tentatively marked with lath to check sight lines at critical locations to be mapped, and for obstructions between the stations. About 20 min are lost in nonproductive crew time whenever the table must be moved because a few side shots cannot be observed. If the party is staffed with three or more people, some forward progress is always possible by only an operator and rodperson. Planetable work can be suspended entirely if additional control work such as bench leveling is done by the observer and rodperson while the notekeeper plots contours, sketches details, or mounts a new sheet.

If the stadia rod is partially obscured, the observer can traverse the rod, searching for enough showing through "the hole" to obtain a reading for the stadia intercept, and then complete the shot. Generally, the rod can be shifted a little to the left or right to take advantage of small openings. It may be waved through a *small* arc of 2 or 3 in. if two openings are not directly in a vertical line, one above the other. Sometimes, the rod can be read by setting the upper cross line (instead of the lower one) at a full foot mark. Or failing that, any two rod graduations can be read. For a noncritical side shot on the ground, a half- or quarter-interval can be used. If the obstruction is not too high, the rod can be "booted" up to perhaps 6 ft—not a generally recommended practice but effective for ground shots of not more than a few hundred feet. Obviously, considerable teamwork is required between the planetabler and rodperson to successfully map an area where terrain features severely limit visibility.

Frequently, in locating or mapping buildings, trees, branches, overhanging eaves, or other obstacles prevent the rodperson from setting the rod close to the proper point. The best approach has the rodperson move at right angles to the sight line from alidade to the desired point, so the stadia distance is the same as if held at the building corner or tree. The alidade is aligned to the object before plotting the point, if it can be seen. Otherwise, the rodperson measures the offset, which the operator uses to plot the true point.

For a circular standpipe tank, a single side shot at the tank's base is taken on its center and ray lines drawn tangent to the edges. By trial, fit a circle tangent to the two rays, while cutting the single point, using a template or spring compass. Short circular arcs, such as curb returns, can be located with three side shots defining the arc, or by a single sight to the PI plus the rodperson's measurement of the external. Walkway widths, small buildings, driveways, and other small narrow features can also be quickly measured by the rodperson without assistance.

When mapping in cropland, it is advantageous to use the rows as guides, spacing shots for optimum coverage along each row traversed, thereby causing little or no damage to crops. Small grain crops, such as wheat, oats, and barley, are damaged minimally because the rodperson moves over the ground just once. The board can be located at the field's edge, so it is not necessary for other crew members to enter the field. Grid surveys require two or three people to traverse and retrace the grid lines at least twice.

To map swampland or bogs where footing is soft, a satisfactory setup can generally be made by driving wood hubs to support the tripod shoes and providing a 1 × 6 in. board about 6 ft long for the operator to stand on. In the northern states, swamp and lake areas are best surveyed during or after freezing weather.

Some forests are easily mapped because dense undergrowth does not flourish in deep shade. Sight lines are limited, however, so relatively more horizontal and vertical control is required than in open terrain.

When working in streets with relatively heavy traffic, the notekeeper can be a guard to protect the rodperson. Standard control devices, such as signs and traffic cones, may be necessary. If traffic is extremely hazardous, work must be delayed until a Sunday morning or holiday. Equipping the survey truck with a revolving light provides a temporary barricade. As a last resort, contract with the local police to assign one or more officers for traffic control.

Sometimes, it is advantageous to check or edit topographic surveys that have been made photogrammetrically. Use two boards: Number one in the normal way and on number two mount the contact prints as a stereo pair to be viewed with a pocket stereoscope. Terrain details that may be indistinct if put in a plotter can be given on-site enhancement. Occasionally, additional terrain features are side shot to improve the photo control—e.g., a sidewalk shot made if a truck was directly over an aerial panel painted on the street at the instant of film exposure. Terrain details can be fuzzy in the deep shade of multistory buildings or hidden beneath bridges and overhanging trees. Brush or tall grass can conceal ditches and outlets of drainage structures.

In breezy weather, vibration of the planetable is damped by applying a strut under slight compression between a board edge and ground. Drive a lath beside the board's edge and cut it off about $\frac{1}{4}$ in. above the board's bottom surface. Bend the lath into a slight curve and tuck it beneath the board so a slight upward force is applied, or nail the lath to the board edge with a small box nail not more than 1 in. long.

22-34. PLANETABLE APPROACH, BUT WITHOUT A PLANETABLE

Most of the advantages of planetable mapping can be realized by substituting a transit or theodolite for the planetable alidade. The objective, general procedure, and result are essentially the same; namely, to produce by ground survey methods and manual plotting a topographic plan of the earth's surface to an appropriate scale showing the relief, culture, and terrain features by means of symbols. A transit or theodolite is employed in the tacheometric mode.

This technique is more tolerant of weather conditions than the planetable-alidade method. Light, misty rain (use Rite-in-the-Rain field books), intermittent showers, light snow, moderate wind, or temperature extremes may to some extent impede field progress, but these conditions do not completely deter a resolute survey party. In reasonably good weather, with typical relatively unobstructed features to be mapped, an experienced crew composed of an operator and rodperson can book about 350 to 400 side shots s day. Production can be increased to perhaps 700 to 900 per day by adding a notekeeper. If the work area warrants the addition of a second rodperson to make a four-member crew, the total may reach 1500 side shots per day. If we assume gently rolling terrain with about 20 side shots per acre, it translates to approximately 75 acres mapped per day.

Major disadvantages compared with the planetable method are (1) an increase in the volume of notekeeping and number of side shots required, (2) more office work to prepare the plotted map sheet, and (3) a reentry on the job site is necessary to ground check the plotted map sheet.

Some advantages include (1) not much ancillary equipment to carry; (2) the survey party become more mobile; (3) vertical and horizontal control work can be combined concurrently with procurement of terrain features; (4) the telescope is nearer eye level, thus more comfortable for the operator and better for seeing over obstructions; (5) higher magnification enables the instrumentperson to take rod readings through smaller holes in the brush and at longer sight lengths, resulting in fewer setups; and (6) ground elevations are determined with higher accuracy because of the greater stability of the mounting base and improved reading system on the vertical circle.

22-35. MORE PRACTICAL SUGGESTIONS: AVOIDING PITFALLS

Note that reductions, plotting, and contouring should be carried out by the same people who did the field work. Familiarity with the job site, control scheme, and field notes expedites preparation of the map sheet. It is a serious mistake if management precludes field personnel from office work and vice versa because the surveying, plotting, and drafting are one continuous phased project. Considering every phase a distinct operation that requires specialists for each one results in blunders, increased man-hours, and a work product of inferior quality.

Plotting is somewhat simplified and mistakes avoided if the horizontal circle's zero point is always oriented to the north. Easy verification is provided by checking the zero against the compass magnetic needle at each setup. On small site surveys at large scales, the circle should be oriented in register with the magnetic compass at the first setup station. At subsequent stations, zero on the circle is aligned parallel with the first setup.

For example, assume that the instrument is at station *A*, with the circle clamped at zero and the magnetic compass reading north. Station *b* is then sighted at a clockwise azimuth of 30°00′. Then, set up at *B* with the circle clamped at 30°00′ plus 180°00′ and point to *A*. This procedure handles the orientation in a purely mechanical way. As a variation, set or check the circle azimuth reading at *B*, invert the telescope before backsighting *A*, which is equivalent to adding or subtracting 180°00′.

If the horizontal control was run previously as a traverse and adjusted for misclosure, the control data should be expressed in the usual *XY*-coordinates, and horizontal distances and *azimuths* between sequential stations derived therefrom. *Never use bearings*. When occupying any station, sight the adjacent ones with the circle clamped at the azimuth or back azimuth

as pertinent, turn the instrument to zero, and let the needle swing free. It should settle very near a north reading.

If the vertical control was run by differential leveling, the station hubs could be included in the bench loops as turning points. To get the HI at any setup, measure with a pocket tape from the hub to the trunnion axis. Checking readings on TBMs or adjacent stations will avoid blunders in computing the HI.

A recommended standard operating method reserves the first few lines of the field noteform, at all new or reoccupied stations, for entry of the check readings. It is not always convenient to take them first and the tendency to record these essential observations in the main sequence of side shots must be resisted. Also, some prominent well-defined terrain feature should be selected at each station setup to serve as an azimuth mark. Its azimuth is entered in the noteform to guard against the natural proclivity of a transit or theodolite to drift slightly in azimuth. Orientation of the circle can be quickly checked at any time, without the assistance of a rodperson.

The detail sheet is prepared the same way as a normal planetable map. Generally, the control grid is drafted to the required accuracy and setup stations plotted by coordinates from the adjusted traverse or other control scheme. The drafting media—vellum, mylar, etc.—must be suitable for tracing. A flat scale is less awkward than the ordinary triangular type.

If is essential that a large protractor printed on paper, bristol board, or mylar and graduated to quarter degrees be used (see Figure 22-10). Plastic types are too thick and drafting machines too slow. A protractor graduated clockwise through 360° is centered *under* the plotted station and accurately oriented across a whole diameter, using a meridian line drafted through the plotted station. A needle (or better, a push pin) that has been honed to a small diameter is inserted at the exact station and covered protractor center to form a socket for a repeat register.

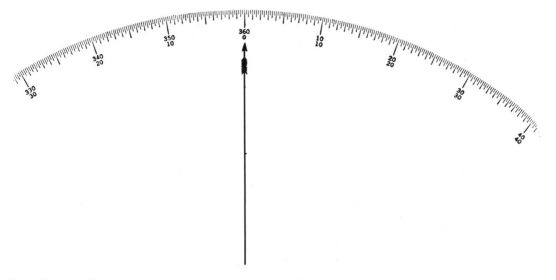

Figure 22-10. Full-scale protractor to be used under plotting sheet.

The flat plotting scale is prepared by affixing a small $\frac{1}{4}$ in. strip of ordinary drafting tape at the zero point (see Figure 22-11, top panel). About one-third of the tape is attached to the underside and pierced at the fiducial zero point. Fold the tape approximately $\frac{1}{8}$ in. from the scale edge and press into contact with the upper surface. Register the zero point from the bottom side by again piercing the push pin (Figure 22-11, bottom panel). Then, place the zero point over the plotted station mark and gently probe for the socket made earlier. Reinsert the push pin to place the scale in near-perfect register with the plotted station and protractor center.

In Figure 22-12, the plotting scale is pinned through the protractor's center, which is underneath the mylar sheet. Some details have already been inked and the pencil notes washed off. Individual points can be plotted to an accuracy of about 0.02 in. for ordinary work with higher accuracy attainable at a slower pace. Normal plotting rate is about 400 shots per hour if a "reader" is available. The field notes are arranged to be read from left to right and plotted in the order recorded in the field (1) azimuth, (2) distance, (3) elevation, and (4) description.

22-36. EARTHWORK VOLUMES BY TACHEOMETRY

Cross sections in borrow pits can be developed quickly and easily by a small field survey party using the stadia method. At the site for a borrow (or pit) area, select a base line, preferably along the longitudinal axis. Terminal points of the base line should be referenced to points in sheltered areas. From the terminals, sight one or two objects for azimuth marks that are unique and easily described. Hub and profile the base line in the usual way: 30-ft (10-yd) intervals simplify the calculations.

Then (1) sequentially set a transit/theodolite over each hub and profile the original ground surface at right angles to the base line; (2) instruct the rodperson to side shot every break in the gradient, no matter how slight; (3) extend the profile beyond the expected limits of the borrow area/pit; (4) after the material has been removed and dressed to a smooth contour, again hub the base line from the terminal points and cross-profile as before; (5) the resultant cross sections can be plotted if required, or the end areas are to be checked by planimetering; and (6) computing end

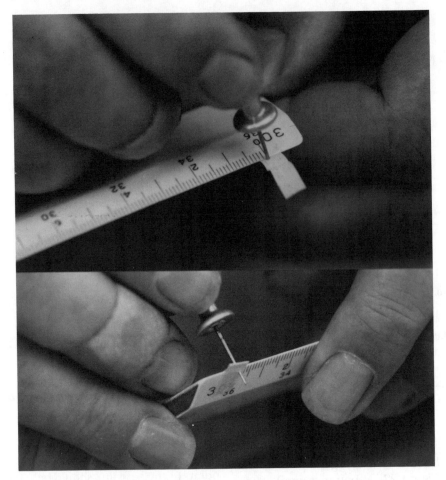

Figure 22-11.   Attaching drafting tape to zero point of plotting scale.

areas by the coordinate method (see Chapter 9) is recommended, since many more points have been sighted and overall accuracy is greater. The technique can be used to advantage for monthly estimates, where equipment is working the pit and stakes have a very short life.

This is a very cost-effective and accurate method. A minimal field staff suffices (the rodperson can be unskilled) and all cross files are quickly located by the stadia method. After the rodperson completes a profile, both the instrument and rod are moved to the next profile where the operator sets up, sights along the base line, turns 90°, and signals the rodperson on to line. He or she then proceeds

toward the instrument, holding the rod at random intervals on *every* discernable grade break. On arrival at the base-line hub, he or she assists the operator in measuring the HI, gives a check reading on an adjacent base-line hub, and proceeds *away* from the instrument on a prolongation of the first profile segment to the borrow-pit limit. The instrument and rod *both* move to the next base-line, where the process is repeated in the reverse direction.

The method functions very well for borrow pits of any length and up to about 1200 ft in width, with differences in elevations of 50 ft or more. There is some loss in accuracy on the longer sight lengths. If the borrow area/pit requires sight lines longer than 600 ft for any

Figure 22-12. Plotting side shots.

significant segment, select two parallel base lines so stationed that the resultant cross profiles are continuous. The range can be increased with improved accuracy if a theodolite/EDMI/prism is substituted for the stadia method. This cross-profile technique can also be employed advantageously in strip-type situations, such as roadways in rugged terrain.

22-37. PLANETABLE USE IN CONSTRUCTION WORK

A new approach employing the planetable and contoured models for construction layout of earthwork and paving will be discussed. In this context, a contoured model simply means the contoured plan of a construction site is drawn on a conventional planetable sheet or other dimensionally stable media to a suitable scale and contour interval, instead of preparing a three-dimensional architectural model. It must be accurately drafted to fit the exact geometry of contract drawings. The maximum allowable error in all horizontal and vertical positions mandated by the project specifications deter-

mines the model's horizontal scale and contour interval.

This technique is based on the concept that if a portion of the earth's surface can be mapped by planetable methods, then the process can be reversed. Therefore, the same equipment can be used for layout if a suitable site model is available. As usual in any layout situation, horizontal and vertical control are necessary.

In practice, the planetable is set up and oriented in the customary way at a convenient location for the layout requirements. Accurate correlation is essential between the ground features to be constructed and the survey control system.

The operator selects points to be set—e.g., at the toe of a slope, top of cut, intersection of two slopes at a valley or ridge line, and curved surfaces below bridge abutments, ditches, etc. Distances from the setup station to selected points are accurately scaled from the model and recorded in the noteform. The alidade fiducial edge is placed beside the plotted setup station mark and aligned on the chosen point. A rodperson is directed along the line; the stadia distance to a trial point is read; then, the rodperson is guided by hand signals forward or backward to the selected point (a tracking-mode instrument is helpful). After the proper location has been accurately established and a hub or lath set, the elevation is computed and recorded. The cut or fill is lettered on the lath for the contractor's use.

Some advantages of this technique are as follows:

1. The geometry shown on the contract drawing is checked and by abnormal conditions or blunders detected.
2. The stake-setting process is fast and requires only two people.
3. An operator can see the objects without abstract visualizations.
4. The procedure can be used regardless of traffic conditions or heavy equipment around and between members of the layout party.
5. Moderate differences in elevation of about 50 ft, up or down, are not a deterrent.
6. Since only certain points are set, it is not necessary to stake or replace the project survey centerline or base line.
7. In complex configurations, such as interchange ramps or channels relocations, the slopes can be readily checked during progress of the construction, thus allowing the contractor an option to make midcourse corrections.
8. If a site is prone to vandalism or any other conditions where stakes may have a rather short life, they can be economically reset.

22-38. CONTOURED MODELS

Most engineering plans that include construction elements of paving, earthwork, channel relocations, etc are in some respects deficient in geometrically defining the surfaces to be constructed. For example, a long straight street with a constant gradient can be satisfactorily manifested by the profile grades and a typical section. However, a curved superelevated highway with traffic islands and tapered curb lanes crossing an intersection, with valley gut-

ters, can be a layout surveyor's nightmare. Runway and taxiway intersections, notorious for their flat gradients, can be prebuilt with the aid of a scale model to detect gradients that may be too flat for drainage and bumps where crown lines intersect. Street crossing having steep gradients that must transition to more gentle ones, and cul-de-sacs in steep terrain can be checked for the best geometrical development of their warped surfaces.

The procedure is very simple. Select an appropriate scale and contour interval for optimal effect on the site under investigation. A larger scale and close contour interval permit more detail to be drafted. The elevation of a discrete point is readily estimated to one-tenth of any contour-interval, and on a scale of 1 in. = 20 ft, it can be located within a horizontal error of perhaps 0.2 ft. These limits are satisfactory for checking paved areas in parking lots, intersections, runways, etc. Earthwork ramps in an interchange require a 1-ft interval and scale of 1 in. = 50 ft.

Select the proper media on which to draft the grid and plan details. Using basic information from the contract drawings, compute coordinates for stations on the project survey centerline, base line, or other reference lines; also, for the coordinate positions of center points on curb returns, bridge abutments, drives, valley gutters, storm inlets, etc. Plot all the various construction details as they are built. Interpolate the contours, *scrupulously solving every line*. The resulting plan will show what the site should look like if built exactly according to plan.

Minor adjustments can be made at this time. If something that appears to be a major blunder in the contract drawing becomes evident, clarification should be sought from the design or project engineer. Readily detected are gradients on paved surfaces that are too flat or do not bend at intersections, closed contours indicating potential ponding, reversed flow, "dry" inlets where storm water will possibly run past, bone-jarring reverse gradients across valley gutters, and manhole elevations that are too high.

23

Control Surveys

Carlos Najera

23-1. INTRODUCTION

Control surveys provide horizontal and vertical positions of points to which supplementary surveys are adjusted. Control surveys provide the standard of accuracy for subsequent and subordinate surveys to attain. All projects, including route surveys, photogrammetry, and topographic mapping, are made up of a series of vertical and horizontal field surveys. These secondary surveys are dependent on control for position and relative accuracy.

23-2. BASIC CONTROL NETWORKS

The U.S. Department of Commerce is responsible for establishing and maintaining basic control networks for the nation. Through its office of the National Oceanic and Atmospheric Administration (NOAA), and NOAA's subordinate offices of the National Ocean Survey (NOS) and National Geodetic Survey (NGS), horizontal and vertical geodetic control networks are surveyed, adjusted, and the results published. Geodetic surveys are affected by and take into account the curvature of the earth, astronomic observations, and gravity determinations.

As part of the control program, the Federal Geodetic Control Committee (FGCC) prepared classifications and standards for geodetic control surveys. The specifications were reviewed, in part, by the American Society of Civil Engineers (ASCE), American Congress on Surveying and Mapping (ACSM), and the American Geophysical Union. In 1974 the FGCC published "Classification, Standards of Accuracy, and General Specifications of Geodetic Control Surveys." To support the requirements the FGCC published in 1975, "Specifications to Support Classification, Standards of Accuracy, and General Specifications of Geodetic Control Surveys." In 1984 "Standards and Specifications for Geodetic Control Networks was published to replace parts of the previous publications. Because requirements and methods for acquisition of geodetic control are changing rapidly, provisions were made to provide upon request, updated information as it was released. The address to obtain current information or be placed upon a mailing list is National Geodetic Information Branch, NCG174, NOAA, Rockville MD 20852. All Federal agencies, are required to comply with the latest standards.[1,2]

Table 23-1 outlines the FGCC general requirements for horizontal and vertical control.

Table 23-1. Standards for the classification of geodetic control and principal recommended uses

Horizontal Control

Classification	First-Order	Second-Order Class I	Second-Order Class II	Third-Order Class I	Third-Order Class II
Relative accuracy between directly connected adjacent points (at least)	1 part in 100,000	1 part in 50,000	1 part in 20,000	1 part in 10,000	1 part in 5000
Recommended uses	Primary national network. Metropolitan area surveys. Scientific studies.	Area control that strengthens the national network. Subsidiary metropolitan control.	Area control that contributes to, but is supplemental to, the national network.	General control surveys referenced to the national network. Local control surveys.	

Vertical Control

Classification	First-Order Class I	First-Order Class II	Second-Order Class I	Second-Order Class II	Third-Order
Relative accuracy between directly connected points or bench marks (standard error)	$0.5 \text{ mm } \sqrt{K}$	$0.7 \text{ mm } \sqrt{K}$	$1.0 \text{ mm } \sqrt{K}$	$1.3 \text{ mm } \sqrt{K}$	$2.0 \text{ mm } \sqrt{K}$
			(K) is the distance in kilometers between points.)		
Recommended uses	Basic framework of the national network and metropolitan area control. Regional crustal movement studies. Extensive engineering projects. Support for subsidiary surveys.		Secondary framework of the national network and metropolitan area control. Local crustal movement studies. Large engineering projects. Tidal boundary reference. Support for lower-order surveys.	Densification within the national network. Rapid subsidence studies. Local engineering projects. Topographic mapping.	Small-scale topographic mapping. Establishing gradients in mountainous areas. Small engineering projects. May or may not be adjusted to the national network.

Source: FGCC 1984. *"Standards and Specifications for Geodetic Control Networks,"* Rockville MD, Reprinted August 1993.

The recommended uses range from local control surveys to the national control network. Control surveys used to support the national network or for densification of the net must be coordinated with the NGS. Special procedures, methods of recording, instrumentation, and other requirements must strictly conform to them. See the reference section at the end of this chapter for a list of publications related to geodetic surveying.

23-3. HORIZONTAL CONTROL SURVEYS

The NGS, formerly the U.S. Coast and Geodetic Survey (USC and GS), has established a high-precision horizontal traverse network across the United States. The traverse lines run generally north and south, and east and west, to form a rectangular grid. The traverses consist of extremely precise length, angle, and azimuth determinations. The adjusted traverses are used to provide scale for the worldwide satellite triangulation network and upgrade the triangulation network established by USC and GS. Table 23-2 defines the classifications and standards of accuracy for traverse. The specifications for triangulation and trilateration are given in Tables 11-1 and 12-2, respectively.

The NGS publishes the horizontal control data of the geodetic networks. The data sheets contain, in part, station description, order of accuracy, latitude and longitude, and the state plane coordinate values. Data sheets also contain a physical description of the station and directions to find it.

The reference datum for horizontal control data has been the *North American datum of 1927*, referenced to the Clark Spheroid of 1866. However, the geodetic control networks have been readjusted and referenced to a more earth-mass-centered ellipsoid. The new datum is known as the *North American Datum of 1983* (NAD '83). Stations existing prior to implementation of NAD '83 will have two values: (1)

the *North American Datums of 1927* and (2) NAD '83. Care should be taken to ensure that all control data for a project are on the same datum.

National network geodetic control data are widely distributed, without cost, to all federal, state, and local government organizations. Other users are charged a nominal fee. Contact the NGS for these data and to obtain copies of the publications listed in the references at the end of this chapter. Also available are control diagrams, which are quadrangles, usually 1° latitude by 2° longitude, that have the horizontal control net superimposed on a topographic map. The control diagrams provide the location, extent of control stations, and station names.

The advantage to using the national geodetic control networks is that high standards of accuracy are available. Employing state plane coordinates greatly facilitates computations and adjustments. A wide variety of programs is available for computers, from portable hand-helds to large mainframes, to manipulate data referenced to state plane coordinates.

The type of survey required to extend horizontal control depends on a particular project. Route surveys may be controlled by traverse; large-area mapping by triangulation or trilateration. Factors to consider are (1) the size and configuration of the project area, (2) terrain, and (3) degree of accuracy required. Accuracy requirements for various projects have been outlined by the FGCC. Table 23-3 offers standards for a variety of projects.

The American Congress on Surveying and Mapping ACSM adopted *Technical Standards for Property Surveys* on June 28, 1964.[3] These standards require linear measurements to have an accuracy of 1 part in 10,000 and angular closure of 5″ per angle within a closed traverse. This is approximately third-order traverse as defined by FGCC. ACSM also encourages the use of the NGS control networks.

In 1988, the American Title Association (ATA) and ACSM adopted *Minimum Standard Detail Requirements for Land Title Surveys*.[4] These minimum standards require closure of 1 part

Table 23-2. Classification, standards of accuracy, and general specifications for horizontal control

Traverse Classification	First-Order	Second-Order Class I	Second-Order Class II	Third-Order Class I	Third-Order Class II
Recommended spacing of principal stations	Network stations 10–15 km. Other surveys seldom less than 3 km.	Principal stations seldom less than 4 km except in metropolitan area surveys, where the limitation is 0.3 km.	Principal stations seldom less than 2 km except in metropolitan area surveys, where the limitation is 0.2 km.	Seldom less than 0.1 km in tertiary surveys in metropolitan area surveys. As required for other surveys.	
*Horizontal directions or angles**					
Instrument	0".2	1".0 or 0".2	1".0 or 0".2	1".0	1".0
Number of observations	16	12** or 8	8** or 6 ; 6	4	2
Rejection limit from mean	4"	5" or 4"	5" or 4"	5"	5"
Length measurements					
Standard error†	1 part in 600,000	1 part in 300,000	1 part in 120,000	1 part in 60,000	1 part in 30,000
Reciprocal vertical angle observations‡					
Number of and spread between observations	3 D/R—10"	3 D/R—10"	2 D/R—10"	2 D/R—10"	2 D/R—20"
Number of stations between known elevations	4–6	6–8	8–10	10–15	15–20
Astro azimuths					
Number of courses between azimuth checks‖	5–6	10–12	15–20	20–25	30–40
No. of obs./night	16	16	12	8	4
No. of nights	2	2	1	1	1
Standard error	0".45	0".45	1".5	3".0	8".0

Azimuth closure at azimuth checkpoint not to exceed#	1".0 per station or 2"$\sqrt{N}$.	1".5 per station or 3"$\sqrt{N}$. Metropolitan area surveys seldom to exceed 2".0 per station or 3"$\sqrt{N}$.	2".0 per station or 6"$\sqrt{N}$. Metropolitan area surveys seldom to exceed 4".0 per station or 8"$\sqrt{N}$.	3".0 per station or 10"$\sqrt{N}$. Metropolitan area surveys seldom to exceed 6".0 per station or 15"$\sqrt{N}$.	8" per station or 30"$\sqrt{N}$.
Position closure §,#					
After azimuth adjustment	0.04 m$\sqrt{K}$ or 1:100,000	0.08 m$\sqrt{K}$ or 1:50,000	0.2 m$\sqrt{K}$ or 1:20,000	0.4 m$\sqrt{K}$ or 1:10,000	0.8 m$\sqrt{K}$ or 1:5000

$$\sigma_m = \sqrt{\frac{\Sigma y^2}{n(n-1)}}$$

*The figure for "Instrument" describes the theodolite recommended in terms of the smallest reading of the horizontal circle. A position is one measure, with the telescope both direct and reversed, of the horizontal direction from the initial station to each of the other stations. See FGCC "Detailed Specifications" for the number of observations and rejection limits when using transits.

**May be reduced to 8 and 4, respectively, in metropolitan areas.

†The standard error is to be estimated by

where σ_m is the standard error of the mean, ν a residual (i.e., the difference between a measured length and the mean of all measured lengths of a line), and n the number of measurements. The term "standard error" used here is computed under the assumption that all errors are strictly random in nature. The true or actual error is a quantity that cannot be obtained exactly. It is the difference between the true and measured value. By correcting each measurement for every known source of systematic and constant errors, however, one may approach the true error. It is mandatory for any practitioner using these tables to reduce to a minimum the effect of all systematic and constant errors so that real accuracy may be obtained. (See p. 267 of U.S. Coast and Geodetic Survey Special Publication No. 247, *Manual of U.S. Geodetic Triangulation*, revised edition, 1959, for the definition of "actual error.")

‡See FGCC "Detailed Specifications" on "Elevation of Horizontal Control Points" for further details. These elevations are intended to suffice for computations, adjustments, and broad mapping and control projects, not necessarily for vertical network elevations.

§Unless the survey is in the form of a loop closing on itself, the position closures would depend largely on the constraints or established control in the adjustment. The extent of constraints and actual relationship of the surveys can be obtained through either a review of the computations, or a minimally constrained adjustment of all work involved. The proportional accuracy or closure (i.e., 1/100,000) can be obtained by computing the difference between the computed value and fixed value and dividing this quantity by the length of the loop connecting the two points.

‖The number of azimuth courses for first-order traverses are between Laplace azimuths. For other survey accuracies, the number of courses may be between Laplace azimuths and/or adjusted azimuths.

#The expressions for closing errors in traverses are given in two forms. The expression containing the square root is designed for longer lines where higher proportional accuracy is required. The formula that gives the smallest permissible closure should be used. N is the number of stations for carrying azimuth, and K the distance in kilometers.

Source: FGCC, (1975) 1980, *Classification, Standards of Accuracy, and General Specifications of Geodetic Control Surveys*, Silver Spring MD (reprint).

Table 23-3. Synopsis of horizontal control classifications

Order, Class	Superior	First-Order	Second-Order Class I	Second-Order Class II	Third-Order Class I, II
General title	Transcontinental control.	Primary horizontal control.	Secondary horizontal control.	Supplemental horizontal control.	Local horizontal control.
Purpose	Transcontinental traverses. Satellite observations. Lunar ranging.	Primary arcs. Metropolitan area surveys. Engineering projects.	Area control. Detailed surveys in very-high-value land areas.	Area control. Detailed surveys in high-value land areas.	Area control. Detailed surveys in moderate- and low-value land areas.
Network design	Control develops the national network.		Control strengthens the national network.	Control contributes to the national network.	Control referenced to the national framework.
Accuracy	1:1,000,000	1:100,000	1:50,000	1:20,000	1:10,000 1:5000
Spacing	Traverses at 750 km. Spacing: stations at 15–30 km or greater. Satellite as required.	Arcs not in excess of 100 km. Stations at 15 km. Metropolitan area control 3–8 km.	Stations at 10 km. Metropolitan area control at 1–2 km.	As required.	As required.
Examples of use	Positioning and orientation of North American continent. Continental drift and spreading studies.	Surveys required for primary framework. Crustal movement. Primary metropolitan area control.	Metropolitan area densification. Land subdivision. Basic framework for densification.	Mapping and charting. Land subdivision. Construction.	Local control. Local improvements and developments.

Source: FGCC, (1975) 1980, *Classification, Standards of Accuracy, and General Specifications of Geodetic Control Surveys*, (reprint). Silver Spring, MD.

in 10,000 or approximately third-order accuracy. The FGCC, as shown in Table 23-2, recognizes that high land values require higher standards of accuracy.

23-4. RECONNAISSANCE

Reconnaissance for a control survey begins with the research of existing stations. The basic net is used by not only federal agencies but also state, county, and city governments, and utility companies. State and local agencies should be contacted to determine the extent of control in the general project area. County surveyors and recorders are the sources for mapping based on state plane coordinates. Only stations that meet the minimum standards of the designed project should be considered for the main scheme.

Existing control stations should be plotted on a map of suitable scale to depict the project area and surroundings. For large projects, topographic maps published by the U.S. Geological Survey (USGS) in the 7.5-min series make excellent base maps for control layout. After existing control stations and original lines of sight are plotted, the purposed control survey scheme is plotted in a tentative position. Triangulation or traverse schemes should be laid out in nearly perfect geometric configurations. In triangulation, the strength of figures is of prime importance; in a traverse, the routes between controlling stations are more-or-less straight lines, with the distances between new stations about equal in length.

Reconnaissance in the field begins with recovery of existing stations and checking original lines of sight. Over the years, trees grow and structures are erected, blocking lines of sight.

Recovery of stations that are part of the national network can be reported to the NOS. "Report on Conditions of Survey Mark," NOAA Form 76-91, is used to update the horizontal control data sheets. The report is in preaddressed postcard form with prepaid postage and the use of it is encouraged. Permission from landowners should be obtained before entering their property, whether recovering existing stations or positioning new station sites. Perpetuation of the new stations can be enhanced by establishing good relations with the landowner. Landowners may also inform you of plans for future development that may threaten the existence of present or proposed stations.

Control surveys for a construction project must have station sites located outside of the proposed construction area. New station sites have to be situated to provide clear lines of sight to other selected monuments and also maintain the desired geometric configuration of the control figures. Proposed structures should be considered when selecting new station sites.

Final location of the new station site is a compromise between clear lines of sight to adjoining stations, permanence, strength-of-figures consideration, and cooperation of the landowner.

23-5. MONUMENTATION

Monumentation, as prescribed by the FGCC, requires that all first- and second-order, class I horizontal control stations be monumented and described. The monumentation includes the station mark, an underground mark (if possible), two or more reference marks, and an azimuth mark. These specifications also apply to second-order, class II stations although temporary points, particularly in traverses, are permitted. Third-order monumented stations should be placed in protected areas and described with sufficient detail so that unmistakable recovery can be made in the future. It should be emphasized that the continued value of a control survey is dependent on permanence of the station marks.

Station marks set by the USC and GS or the NGS are bronze disks generally set either in rock outcroppings or on concrete monuments that will have a subsurface mark. If the original mark is destroyed or distributed, the subsurface mark may be intact. Disks set in rock outcroppings are countersunk and concreted in a drill hole.

Reference marks (RMs) are also bronze disks set in concrete, but their markings are different, and there is no subsurface monument. The RMs are located to form approximately a 90° angle with the station mark, usually within 100 ft. Azimuth marks, constructed in the same manner as RMs, are located at least $\frac{1}{4}$ mi from the station. A clear line of sight at tripod height is required between the station and azimuth mark.

Detailed methods and specifications for FGCC-approved monumentation are found in the USC and GS *Manual of Geodetic Triangulation*, Special Publication No. 247[5], and *Specifications for Horizontal Marks*,[6] ESSA Technical Memorandum C and GSTM-4.

Not all control survey projects require monumentation to these rigid standards. Surveys of lesser standards can have commensurate monumentation.

23-6.   INSTRUMENTATION

Instruments required for measuring horizontal directions must be of high quality. First-order instruments are defined as optical-reading theodolites with micrometer readings smaller than 1 sec. The FGCC has recognized the Kern DKM-3 and Wild T-3 as representative of this quality. A second-order instrument is an optical-reading theodolite with micrometer readings of 1 sec. The Askania A2, Kern DKM-2, Wild T-2, and Zeiss Th-2 are examples

of suitable theodolites for second-order surveys. Good-quality transits and repeating theodolites are acceptable for third-order, although they are not recommended. The extra effort required to obtain the specified accuracy with this type of instrument makes the 1-sec direction theodolite much more desirable.

In recent years, electronic distance-measuring instruments (EDMIs) have been improved to provide high accuracy at relatively moderate costs. The availability of EDMIs is such that if purchase is not possible, rentals are an alternative for a project. Their use facilitates establishing base lines in triangulation and greatly improves the speed and accuracy of trilateration and traverse surveys.

Three types of EDMIs predominate: (1) electro-optical devices that use visible light as the carrier wave, (2) those that employ an infrared light source, and (3) others that operate with microwaves. EDMIs using visible light and microwaves can measure distances of 100 km and are generally considered to be long-range instruments. Infrared equipment may have either medium- or short-range measuring capabilities. Distances over 10 m can be read with some infrared equipment.

Technology in this field is continually improving and capabilities for each type of EDMI may change (see Chapter 5 for a detailed discussion of this equipment).

23-7.   LENGTH MEASUREMENTS

The FGCC-established requirements for length measurements on different types of surveys are listed in Table 23-4.

Table 23-4.   Commensurate length measurements for designated surveys. (Letters A to G represent the various classes of accuracy described in the narrative portion of the table.)

Type of Survey	First-Order Specifications	Second-Order Specifications		Third-Order Specifications	
		Class I	Class II	Class I	Class II
Base lines	A, B	B	C	D	E
Trilateration	B	C	D	E	F
Traverse	D	E	F	G	

All EDMIs should be serviced regularly, calibrated at least annually, and checked over lines of known distances at shorter intervals. Calibration base lines have been established by the NGS to test EDMIs; for additional information, contact the NGS. Technical memorandums NOS NGS-8[7] and NOS NGS-10,[8] also provide additional information.

Horizontal control surveys of high-order accuracy require precise methods of execution. Detailed descriptions of rigid procedures to measure distances and horizontal directions are found elsewhere in this book, along with the mathematical requirements for triangulation, trilateration, and traverse.

Horizontal control surveys, to this point, have been discussed only with reference to the national control net. Control surveys need not be referenced to the NAD '27 or NAD '83: They can be referenced to any datum. Some local governments have established datums that can be used as the basis for high-order surveys. Although not recommended for very precise work, an assumed datum may be arbitrarily established for either horizontal or vertical control.

The FGCC has defined the accuracies of first-, second-, and third-order surveys but do not consider those of lower accuracy. In actual practice, especially on large construction projects, control surveys may be on three levels: primary, secondary, and auxiliary. By FGCC standards, the control could be second-, third-, and fourth-order, if a fourth order existed. Whether a survey is based on the national network geodetic control or an assumed datum, the FGCC standards of accuracy can still be used.

23-8. VERTICAL CONTROL SURVEYS

Vertical control surveys provide elevations of reference to a single vertical datum. In the United States, the *National Geodetic Vertical Datum of 1929* (NGVD29) is the reference vertical datum used by the NOS. Vertical control data published by the NGS is referenced to NGVD29, formerly the *Sea Level Datum of 1929* (SLD29). The 1929 adjustment of the vertical net was made to the mean sea-level elevations of 21 tide gages in the United States and five in Canada. The NGS is redefining the 1929 datum; it will be known as the *North American Vertical Datum of 1988* (NAVD88).

To understand some of the adjustments required in geodetic leveling, the relationship of gravity and a level surface is defined.

Sea-level datum is an ellipsoid that approximates the geoid, earth.

A *level surface* is everywhere perpendicular to the direction of gravity as defined by the plumb line; it is also called an *equipotential surface*, meaning that every point on it has the same gravity potential.

Gravitation is an attraction force exerted toward the center of the earth's mass. *Centrifugal force* is generated by the earth's rotation and exerted perpendicular to the axis of rotation. Centrifugal force increases as altitude increases and varies from zero at the poles to a maximum at the equator. *Gravity* is the resultant force of gravitation and centrifugal force.

The variable effect of centrifugal force causes level surfaces to be nonparallel (Figure 23-1). In geodetic leveling, elevations established at altitudes other than mean sea level will require an *orthometric correction* to compensate for the nonparallel surfaces. The correction is applied to level lines running in a north-south direction to rectify the northward convergence of level surfaces. The computation can be performed by one of two methods.

The first method is described in the USC and GS *Manual of Leveling Computational and Adjustment*, Special Publication No. 240.[9] The orthometric correction is computed from three variables: (1) measured height, (2) scaled latitude, and (3) a constant factor that assumes normal gravity at the given latitude. The formula and tables given in Special Publication No. 240 are represented in Figure 23-2.

The second method differs from the first in that gravity is measured at bench marks at

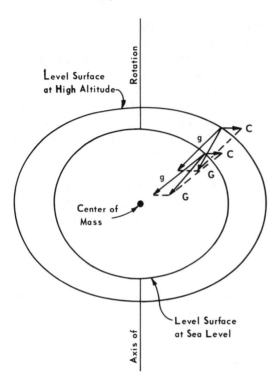

Figure 23-1. Gravity *G*: the result of gravitation *g* and centrifugal force *c*.

designated intervals. Gravity anomalies are detected with frequent gravity determinations and its effects minimized. The resulting adjusted elevations give the true relationship (hydrostatic head) between bench marks.

The USC and GS used the first method in the 1929 adjustment. Gravity determinations were very time-consuming and expensive. Currently, the NGS is releveling the national vertical control net and adjusted elevations will consider gravity anomalies. Modern *gravimeters* are capable of quick, accurate, and economical observations.

The national vertical control network is composed of a series of interrelated nets. Basic net *A* is composed of circuits, with an average grid size of 100 to 300 km. Basic net *B* is a subdivision of circuits, with an average grid size of 50 to 100 km. A densification net with a

line spacing of 10 to 50 km provides additional vertical control points, as required. Table 23-5 showed specifications for basic vertical control nets and other surveys. The classification and specifications have been prepared by the FGCC. Relative accuracy for each classification is shown in Table 23-1. Published vertical control data are available in the same format and from the same source as horizontal control data.

Crustal motion and extraction of subterranean natural resources are known causes of vertical bench-mark movement. To keep elevations updated and accurate, the NGS has established a program that requires a resurvey and readjustment approximately every 15 yr. Effort must be made to assure that the most current bench lists are being used.

Reconnaissance for a new vertical control line or net begins with the recovery of existing bench marks within the project limits. The existing bench-mark elevations are required to be at least equal to the classification of accuracy for the proposed vertical control. Table 23-6 shows the number of marks to be checked for each classification. The NGS or originating agency should be contacted if checks are not within acceptable units.

Recovery of bench marks that are part of the national vertical control net should be reported to the NOS. NOAA Form 76-91, "Report on Condition of Survey Mark," is used to update the vertical control data sheets. The report is in preaddressed, postcard form with prepaid postage.

The proposed level survey should be tentatively plotted on a map of suitable scale. Existing vertical control should also be plotted. Unless there are extraordinary circumstances, the same map for horizontal control should be used. In addition, it may be more economical to employ new horizontal control monuments also as bench marks. Horizontal control is limited in position due to line of sight and balance of lines, whereas vertical control is not so limited. Permission from landowners should

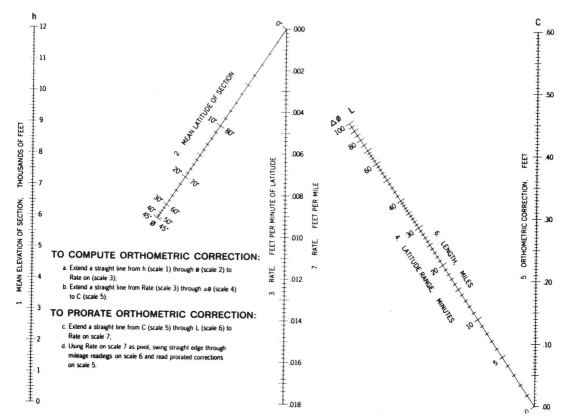

Figure 23-2. Nomogram for orthometric correction. (Courtesy of California State Lands Commission.)

be obtained prior to entering the property, whether recovering existing bench marks or establishing new ones. Landowner cooperation is very important in perpetuating monuments.

Generally, NGS marks are bronze disks set in bedrock or massive concrete structures, such as building foundations or bridge abutments. More recently, marks have been put on steel rods driven to refusal and capped with a bronze disk. Monuments set for vertical control must be absolutely stable vertically.

The final location of marks should conform to the line spacing and interval selected for the project. On construction jobs, monuments and marks are so located that they will not be destroyed during construction. Stability and permanence are key elements for bench marks.

Instruments suitable for high-precision leveling and to fit the various classifications are listed in Table 23-6. Metal turning pins should be used if possible. When leveling over sandy or marshy terrain, wooden stakes with a double-headed nail driven into their tops provide turning points. A turning plate is appropriate on concrete or other hard-packed surfaces. Asphaltic concrete surfaces should be avoided for turning points and instrument setups, especially on hot days.

Vertical control surveys can be accurate to the specifications given without reference to NGVD29. Cities, counties, and sanitation and drainage districts may have other established

Table 23-5. Classification, standards of accuracy, and general specifications for vertical control

Classification	First-Order	Second-Order		Third-Order
	Class I, Class II	Class I	Class II	
Principal uses Minimum standards; higher accuracies may be used for special purposes	Basic framework of the national network and metropolitan area control. Extensive engineering projects. Regional crustal movement investigations. Determining geopotential values.	Secondary control of the national network and metropolitan area control. Large engineering projects. Local crustal movement and subsidence investigations. Support for lower-order control.	Control densification, usually adjusted to the national net. Local engineering projects. Topographic mapping. Studies of rapid subsidence. Support for local surveys.	Miscellaneous local control; may not be adjusted to the national network. Small engineering projects. Small-scale topo, mapping drainage studies and gradient establishment in mountainous areas.
Recommended spacing of lines National network	Net *A*; 100–300 km, *class I* Net *B*; 50–100 km, *class II.*	Secondary net; 20–50 km.	Area control: 10–25 km.	As needed
Metropolitan control; other purposes	2–8 km As needed.	0.5–1 km As needed.	As needed. As needed.	As needed. As needed.
Spacing of marks along lines	1–3 km	1–3 km	Not more than 3 km.	Not more than 3 km.
Gravity requirement	0.20×10^{-3} gpu	—	—	—
Instrument standards	Automatic or tilting levels with parallel-plate micrometers; Invar scale rods.	Automatic or tilting levels with optical micrometers or three-wire levels; Invar scale rods.	Geodetic levels and Invar scale rods.	Geodetic levels and rods.

Field procedures	Double-run; forward and backward, each section. 1–2 km.	Double run; forward and backward, each section. 1–2 km.	Double or single run. 1–3 km for double run.	Double or single run. 1–3 km for double run.
Section length	1–2 km.	1–2 km.	1–3 km for double run.	1–3 km for double run.
Maximum length of sight	50 m *class I*; 60 m *class II*	60 m	70 m	90 m
Field procedures*				
Max. difference in lengths				
Forward and backward sights				
Per setup	2 m *class I*; 5 m *class II*	5 m	10 m	10 m
Per section (cumulative)	4 m *class I*; 10 m *class II*	10 m	10 m	10 m
Max. length of line between connections	Net A; 300 km Net B; 100 km	50 km	50-km double run. 25-km single run.	25-km double run. 10-km single run.
Maximum closures†				
Section. Forward and Backward	3 mm $\sqrt{K}$ *class I*; 4 mm $\sqrt{K}$ *class II*	6 mm $\sqrt{K}$	8 mm $\sqrt{K}$	12 mm $\sqrt{K}$
Loop or line	4 mm $\sqrt{K}$ *class I*; 5 mm $\sqrt{K}$ *class II*	6 mm $\sqrt{K}$	8 mm $\sqrt{K}$	12 mm $\sqrt{K}$

*The maximum length of line between connections may be increased to 100 km for double run for second-order, class II, and to 50 km for double run for third-order in those areas where the first-order control has not been fully established.

†Check between forward and backward runnings, where K is the distance in kilometers.

Source: FGCC (1975) 1980, *Classification, Standards of Accuracy, and General Specifications of Geodetic Control Surveys*, (reprint). Silver Spring MD.

Table 23-6. Minimum requirements for connections to previous vertical control surveys

New Survey Order	Previous Survey Order		
	First	Second	Third
First	Check 3 marks. Contact originating agency after 4.	Check 2 marks. Contact originating agency after 3.	Check 2 marks.
Second	Check 2 marks. Contact originating agency after 3.	Check 2 marks. Contact originating agency after 3.	Check 2 marks.
Third	Check 2 marks.	Check 2 marks.	Check 2 marks.

Source: FGCC, (1975) 1980, *Specifications to Support Classification, Standards of Accuracy, and General Specifications of Geodetic Control Surveys,* Silver Spring, MD (reprint).

vertical datums. The stated standards of accuracy are applicable to any datum.

23-9. SUMMARY

Control surveys require superior methods, equipment, and execution. High-accuracy results can only be attained by strict adherence to established procedures.

The surveyor in responsible charge should have expertise and experience in control surveys. Many decisions are based on judgement, and mistakes are very costly. Personnel assigned to high-precision surveys must be conscientious, capable, and knowledgeable of task and equipment. Properly executed surveys provide high-accuracy control and facilitate subsequent work.

NOTES

1. FGCC. (1975) 1980. *Classification, Standards of Accuracy, and General Specifications of Geodetic Control Surveys.* Silver Spring MD (reprint).

2. _____ . (1975). 1980. *Specifications to Support Classification, Standards of Accuracy, and General Specifications of geodetic Control Surveys.* Silver Spring MD (reprint).

3. ACSM. 1946. *Technical Standards for Property Surveys.*

4. ATA and ACSM. 1988. *Minimum Standard Detail Requirements for Land Title Surveys.*

5. F. R. Gossett. (1950) 1959. *Manual of Geodetic Triangulation.* Special Publication No. 247, U.S. Coast and Geodetic Survey, (revised). Washington DC.

6. _____ . 1958. Specifications of Horizontal Marks. ESSA, Washington, DC.

7. C. T. Whalen. 1978. Control leveling. NOAA Technical Report NOS-73 NGS-8. Silver Spring MD.

8. C. J. Fronczek. 1977. Use of calibration base lines. NOAA Technical Memorandum NOS NGS-10. Silver Spring MD.

9. H. S. Rappleye. 1948. *Manual of Leveling Computation and Adjustment.* Special Publication No. 240, U.S. Coast and Geodetic Survey. Washington DC.

REFERENCES

ADAMS, O. S. 1915. *Application of the Theory of Least Squares to the Adjustment of Triangulation.* Special Publication No. 28, U.S. Coast and Geodetic Survey. Washington DC.

DRACUP, J. F., C. J., FRONCZEK, and R. W. TOMLINSON. 1977. Establishment of calibration base lines. NOAA Technical Memorandum NOS HGS-8. Silver Spring MD.

FLOYD, R. P. 1978. *Geodetic Bench Marks*. NOAA Manaual NOS NGS-1.

HOSKINSON, A. J., and J. A. DUERKSEN. 1952. *Manual of Geodetic Astronomy*. Special Publication No. 237, U.S. Coast and Geodetic Survey. Washington DC.

MITCHELL, H. C., and L. G. SIMMONS. (1945) 1977. *The State Coordinate Systems (A Manual for Surveyors)* Special Publication No. 235, U.S. Coast and Geodetic Survey (revised). Washington DC.

MUSSETTER, W. (1941) 1959. *Manual for Reconnaissance for Triangulation*. Special Publication No. 225, U.S. Coast and Geodetic Survey (revised). Washington DC.

SHOMAKER, M. C., and R. M. BERRY. 1981. *Geodetic Leveling*. NOAA Manual NOS NGS-3. Rockville, M.D.

SPENCER, J. F., JR., J. E. STEM, and W. W. WALLACE. 1983. *Products and Services of the National Geodetic Survey*. NOAA. Rockville MD.

WHALER, C. T., and E. I. BALAZS. 1975. Test results of first-order class III leveling. NOAA Technical Report NOS-68 NGS-4. Silver Spring MD.

24

Construction Surveying

Boyd L. Cardon and Edward G. Zimmerman

24-1. INTRODUCTION

Construction surveying operations comprise approximately 60% of all surveying work being performed and should be considered a definite specialty of the modern surveyor. Three basic objectives of construction surveying are (1) providing layout stakes, located both horizontally and vertically, that construction personnel can utilize in an accurate and efficient manner to position structures or earthwork projects; (2) ongoing replacement of layout stakes as a project progresses toward completion, along with periodic checking of projects to ensure compliance with design dimensions; and (3) providing a map at the completion of a project, showing the final project location and configuration, incorporating any changes or modifications in project design—an "as-built" map.

Other than the correct application of basic surveying principles and following the objectives stated, no strict procedure governs construction surveying. The type of project, environment of the project site, requirements of the construction force, and economic realities must all be considered before adopting any particular procedure for executing a construction survey.

Many construction surveying procedures depend on the motivation and resourcefulness of an individual surveyor in providing sufficient information to the builder at a minimum expense to the client. Discussions in this chapter recognize the fact and are geared to provide general principles, rather than a specialized methodology for various operations.

24-2. PROJECT CONTROL SURVEYS

Prior to design and construction of a particular project, extensive surveying must be performed. Property boundaries of a project site must be mapped and monumented for acquisition purposes, and a topographic map of the area has to be prepared for visibility studies and to aid the engineers in developing project designs. The survey network established to complete the property and topographic work will contain horizontal and vertical control monuments that, when supplemented by additional monuments, can ultimately control the construction surveys. Care should be taken by surveyors to properly locate these monuments for maximum use, not only for initial surveys, but through the completion of ensuing construction surveys.

It is important for a surveyor in charge of a project to describe and reference all major horizontal control monuments. Methods illustrated in Figure 24-1 can be used. To preserve vertical control monuments (bench marks), it is recommended that an adequate number of differential level circuits be run to establish supplementary bench marks removed from areas of construction and possible displace-ment, yet close enough for efficient use by construction personnel.

Throughout the duration of the construction, all reference monuments and supplemental bench marks should be periodically checked to detect and correct, if necessary, any displacements that may occur. Following completion of construction, all control monuments that can be readily replaced should be,

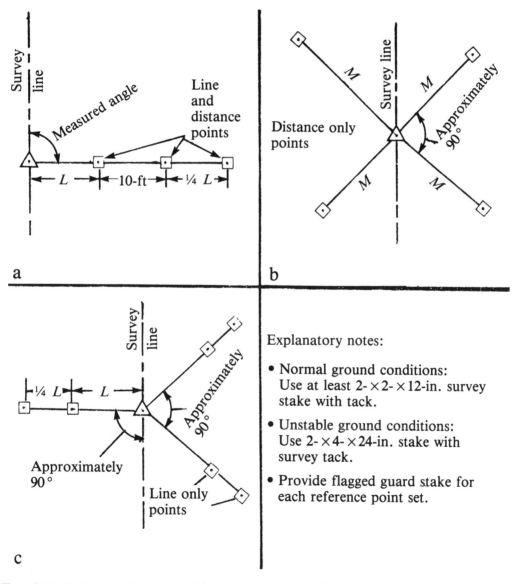

Figure 24-1. Reference staking methods. (a) Line and distance ties. (b) Distances-only ties. (c) Line-only ties.

in order to perpetuate the original survey for use in future expansion or modification of the project.

Surveyors should brief construction workers concerning control monuments and reference stakes and obtain their cooperation in preserving as much of the control system as possible. Destruction of monumentation is unavoidable sometimes, but close communication between builder and surveyor should be maintained to minimize loss of control points.

24-3. GRID NETWORK FOR CONSTRUCTION CONTROL

For most large structural projects, such as water treatment facilities or manufacturing plants, it is common engineering practice to establish a rectangular grid system. Such a system can be based on a local coordinate system or incorporated in a state plane coordinate system (SPCS). Establishing a local system requires less preliminary surveying, but connecting a grid system to the SPCS provides the obvious advantage of integrating it with the same horizontal reference datum for all surveys in a large area (see Chapter 19).

The ends of major x and y grid axes should be monumented with poured-in-place concrete monuments supporting a metal disk. Axes intersection monuments of lesser permanence, perhaps 2×4 in. stakes, should be set at even 50- or 100-ft intervals along each grid line. The system should be aligned in cardinal directions and, if not assigned SPCS values, be given large enough x- and y-coordinates to prevent negative values from developing if the system is extended to the west or south.

Any point located on a grid system has a mathematical relationship to all other points on the system. All major construction features of a project will have an assigned coordinate value, allowing a particular feature to be located from any convenient control monument, through the inverse calculation process. The availability of electronic survey and calcu-

lation equipment makes this type of construction control highly practical.

24-4. HORIZONTAL CONSTRUCTION LOCATION

Stakes are set to mark the horizontal locations of a structure's key features, such as corners, intersections, etc. The initial stakes are temporary in nature and usually destroyed during excavation and rough grading. These stakes are controlled by permanent stations previously set outside the immediate area of construction. The permanent stations, in turn, are an extension of original horizontal control surveys and should be referenced as described in Figure 24-1.

The location of a structure itself is accomplished by setting stakes opposite all important points and at sufficient intervals to define alignment of the structure's vertical planes. The stakes should be located at some convenient even-numbered horizontal distance from the actual construction surface (4 ft, 6 ft, 8 ft). The actual offset distance can be determined by discussion with the construction force, taking into consideration the following factors: (1) adequate clearance space to prevent destruction of stakes during construction operations and (2) close enough to allow builders to make simple measurements, using short rules and carpenter levels.

In the construction of lengthy facilities or linear structures such as highways, a survey line can be established parallel to and at a convenient horizontal offset distance from the pavement edge or building face. Stakes can be set at intervals to provide convenient measurements for the builders but not so close that they cause confusion and waste the surveyor's effort.

Clearly defined and marked stakes, easily understood by the builder, are a necessity along with adequate guard stakes to provide visibility, identification, and protection. The recommended type and size of stakes used for construction purposes are as follows:

1. Normal ground conditions: $2 \times 2 \times 8$ in. wood stakes

2. Unstable ground conditions: $2 \times 4 \times 18$ in. wood stakes

3. Abnormally hard ground: $\frac{1}{2} \times \frac{1}{2} \times 4$ in. wood stakes or large nail or bridge spike

4. Asphalt pavement: concrete or PK nail

5. Marker or guard stakes: $1 \times 2 \times 18$ in. wood stakes or $\frac{1}{2} \times 2 \times 48$ in. wood lath

Table 24-1. Recommended Staking Accuracies

Type of Construction	Horizontal Accuracy	Vertical Accuracy
Vegetation clearing	± 1 to 2 ft	N/A
Excavation/embankment	± 0.1 ft	± 0.1 ft
Pipelines (gravity flow)	± 0.1 ft	± 0.01 ft
Pipelines (pressure flow)	± 0.1 ft	± 0.1 ft
Street or highway	± 0.01 ft	± 0.01 ft
General structures	± 0.01 ft	± 0.01 ft
Prefabricated steel structures: base plates pumps, valves, weirs, etc.	± 0.005 ft	± 0.005 ft

Note: If inches and fractions are required, use the following relationship: 1 in. $= 0.08333$ ft or 0.1 ft $= 1\frac{13}{64}$ in. (approx.).

24-5. VERTICAL CONSTRUCTION LOCATION

Horizontal construction planes are located by grade stakes—usually the same stakes driven for horizontal location. After we set a grade stake, the elevation of its highest point is determined. The difference in elevation between the stake top and construction grade is calculated and translated into a "cut" (design elevation lower than the stake) or "fill" (design elevation higher than the stake). The cut C or fill F is marked on the guard stake to guide the builder. This difference should also be noted in the surveyor's field book for future use in replacing destroyed guard stakes or in the event of a dispute.

When conditions may dictate, it is acceptable to use a line drawn on an adjacent structure, the top of a curb, or a nail driven into a nearby tree or pole to provide a grade reference. Whatever mark or object is used, it is vital to tell the constructor the method of setting the horizontal and vertical references and the way they are marked. This will not only avoid confusion and errors, but also ensure that the particular survey system is compatible with the builder's needs.

After a surveyor has completed his or her layout and marked the stakes, it becomes the builder's responsibility to devise measurement methods to transfer alignment and grade from the surveyor's stakes to the construction plans. This operation is generally accomplished by using grade-alignment wires or strings, batter boards, laser devices, or other suitable means. Applications of the methods noted to a particular type of construction operation will be given in appropriate sections of this chapter.

24-6. PIPELINE CONSTRUCTION STAKE OUT

Pipeline construction can be divided into two classifications: (1) gravity-flow lines and (2) pressure-flow lines. A gravity-flow pipeline utilizes the force of gravity to move liquids through a line on an engineered slope. Thus, the pipe grade must be closely controlled during layout and construction to maintain design elevation and horizontal location. (See Table 24-1 for recommended staking accuracies.)

Examples of gravity-flow pipelines include surface-water drainage pipes, culverts, sewage, and some irrigation systems. In contrast, pressure-flow lines generally depend on a pump to provide movement of liquids through the line; hence, they can be designed without providing constant slopes. Exceptions to this principle apply to pressure lines that may be prefabricated, and construction in locations where climatic conditions require installation of drains at low points along the line to prevent freezing damage. When accurate pressure-line elevation is not required, only horizontal location is necessary.

Except for conditions similar to those stated, a constructor can usually determine pressure pipeline depths from construction plans and specifications. For example, residential water systems require only enough depth to adequately protect the pipe. This allows the contractor to follow a predetermined depth from the existing ground profile.

24-6-1. Gravity-Line Construction Layout

Principal points, such as manhole locations, and the beginnings and ends of curves, are established on the ground along the design pipeline centerline location. An offset line parallel to the pipe centerline, and far enough from it to prevent displacement during excavation and construction, is established. A line of stakes is placed in the ground at a usual interval of 50 ft, where the grade line is constant, and as close together as 10 or 20 ft on a vertical-curve grade line. A marker stake is placed behind the grade stake (the side of the grade stake opposite the pipe centerline). A flat side of the marker stake faces the construction and provides a surface to mark "*C*" to the invert or pipe flow-line (Figure 24-2), and the grade stake's station location. The reverse side of the marker stake is used to display the horizontal offset from the pipe center.

If the abnormally hard ground or pavement along the offset line precludes driving wooden stakes, it is recommended that 20p nails or bridge spikes be used in place of stakes. Paint marks or a chisel cut can also serve the same purpose on concrete surfaces. After all stakes or marks have been established, the elevations of all stations are determined by leveling. The elevation for each station is recorded in a field book, along with the design elevation of the

pipe invert and the resulting calculated *C*. A suggested field noteform is shown as Figure 24-3.

After the cuts are calculated, they are marked at the appropriate space on the guard stakes. A copy of the cuts should be made and given to the contractor as a reference cut-sheet for use if the guard stakes are destroyed during excavation.

Pipeline excavation is usually performed with a trenching machine that is guided by a horizontal wire or string suspended on the stakes set by the contractor. Guidelines are set directly over the surveyor's offset stakes, at a predetermined elevation above the pipe invert.

For example, in Figure 24-4, the *C* is 7.36 ft to the invert at station 12 + 50. Therefore, the guideline is set 2.64 ft above the grade stake, resulting in an even 10.00-ft vertical difference to the pipe invert. The guideline will be set on both the horizontal alignment and vertical slope.

Layout of the guideline is the contractor's responsibility. Because the *C* is to the pipe invert, a contractor must also compensate to allow sufficient overexcavation space for bedding material in the trench bottom and also for the proposed pipe-shell thickness (see Figure 24-5).

After the excavation is completed, the contractor erects boards across the trench opposite each grade stake. Grade boards are fabricated by driving a 2 × 4 in. upright on each side of the trench and nailing a 1 × 6 in. board to them. With the surveyor's stake and its *C*, the top edge of the cross or grade board is set to an even-foot vertical interval above the pipe invert. A nail is driven into the top edge of the grade board to mark horizontal alignment and a string or wire stretched between each alignment nail to provide a checking line for pipe installation. Each pipe section is set in place and checked by comparison with the guideline. A long wooden pole with a right-angle foot on its bottom end and marked at the vertical interval transfers the invert grade

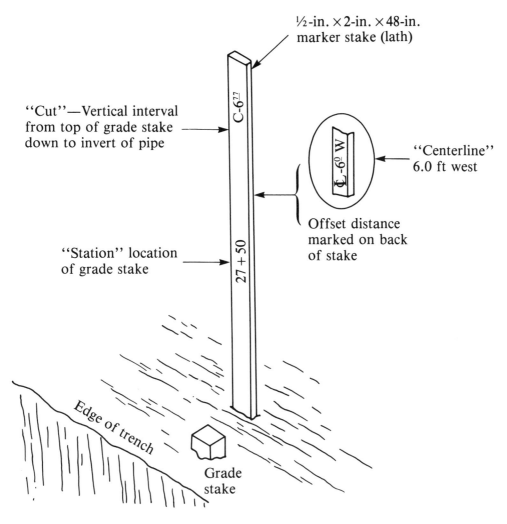

½-in. × 2-in. × 48-in.
marker stake (lath)

"Cut"—Vertical interval
from top of grade stake
down to invert of pipe

C-6⁷⁷

"Centerline"
6.0 ft west

₵-6⁰ W

Offset distance
marked on back
of stake

"Station" location
of grade stake

27 + 50

Edge of trench

Grade
stake

Figure 24-2. Marking construction stakes for underground work.

from guideline to pipe invert, as shown in Figure 24-6.

24-6-2. Pressure Pipelines

Construction layout for pressure lines does not require the same rigid vertical control as gravity-flow lines. It may be unnecessary to set any cuts at all or perhaps very few, whereby existing or proposed underground facilities demand a precise vertical location of a pressure line. Horizontal location is important, though, to maintain design alignment and location.

24-6-3. Use of Laser Beams

A contractor may find it is more efficient to use one of the variety of laser-equipped survey instruments to replace the guideline/grade-board method of providing grade and alignment. The instrument can be set up at a principal control station and sighted on a target site over the next control point. The laser

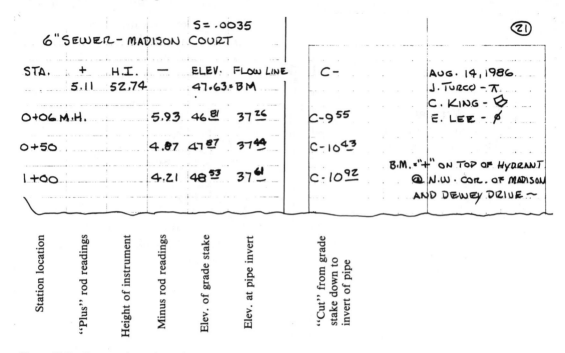

Figure 24-3. Suggested noteforms for underground construction stakes.

beam thus oriented replaces the guideline as shown in Figure 24-7. The laser beam is visible on the face of a measuring rule intercepting it —hence, furnishes a horizontal and vertical reference.

Another specialized laser instrument, shown in Figure 24-8, can be placed directly on the pipeline invert and aligned along the centerline and required slope. The laser beam is projected onto a target held at the end of each pipe section as it is laid, aligning the pipe horizontally and vertically.

A laser instrument can also be used to guide a trenching machine by providing an intense spot of light on a target attached to its digging arm. By observation, the trencher operator has a continual guide for maintaining the alignment and correct depth of the excavation.

A electronic laser-sensing unit can be installed on a trencher to automatically keep it aligned. Sensor cells detect variations in the light when the machine wanders from the grade or alignment, and it actuates servocon-

trols to restore the trencher to proper alignment and grade.

Although the laser replaces guidelines and eliminates much grade setting and checking work, this instrument still requires conventional surveying methods for correct positioning and alignment. It is emphasized that most laser devices are rather limited in application and lack the adaptability of a transit or theodolite. Application of laser techniques is beneficial for large, high-production projects.

Benefits of using lasers include (1) reduction of required labor; (2) a more accurate guide for line and grade; (3) ease of checking ongoing work; and (4) the fact that the ditch may be backfilled as soon as the pipe is installed.

24-7. BUILDINGS AND STRUCTURES

An important requirement for a surveyor providing structural layout is to ensure correct location of the structure within the building

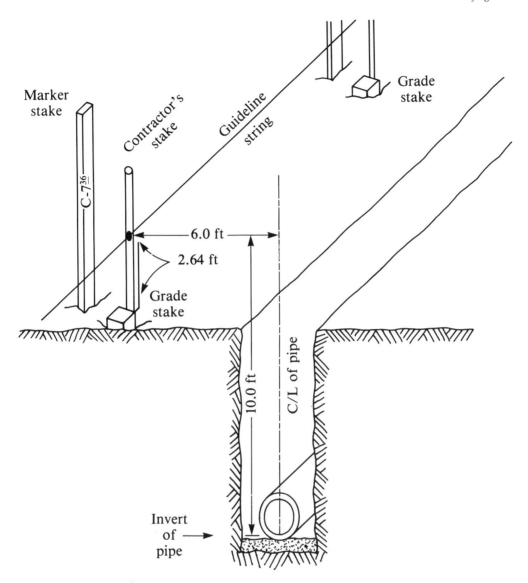

Figure 24-4.   Relationship of grade stakes, guideline, and pipe invert.

site. The ownership line of a project site must be located prior to any construction staking, to both provide a base line for layout and verify that the proposed building does not encroach on adjoining properties.

24-7-1. Batter Boards

All corners and key positions of the proposed building are located and staked on the ground to provide a dimensional check of the layout. These points are then referenced by constructing batter boards, as shown in Figure 24-9. A batter board is set approximately 3 to 8 ft from each end of the intersecting building lines. Tops of the cross pieces are set whole numbers of feet above or below the main floor elevation or some other horizontal plane of the proposed building. If possible, all batter boards should be set at the same elevation,

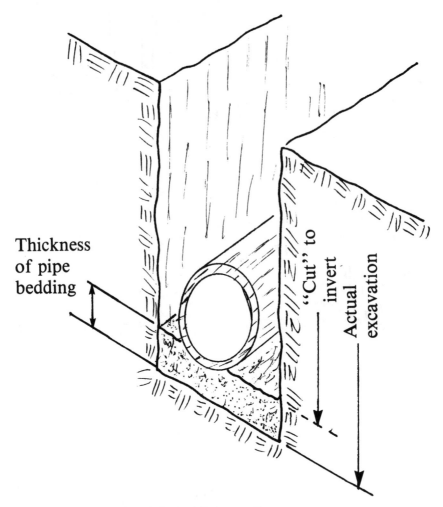

Thickness of pipe bedding

"Cut" to invert

Actual excavation

Figure 24-5.   Cross section of pipe and ditch excavation.

thereby creating a level line along strings or wires stretched between the boards. The lines are attached to nails driven into the tops of the cross pieces at the intersection of the building faces.

After erection of the batter boards, excavation for the structure's footings or basement can begin. Batter boards are protected and preserved to provide grade and alignment throughout excavation operations and construction of the first floor. It is advisable to set points on building lines, at locations removed from construction activities, in order to furnish reference directions if a batter board is destroyed.

When the footing or basement excavation is completed, additional batter boards can be placed inside the excavation or survey stakes established by using batter boards set originally to align footings, floor slabs, and the base of exterior walls. Individual locations for features such as column footings, interior wall corners, and anchor bolts can be marked by survey stakes or scratches on the concrete

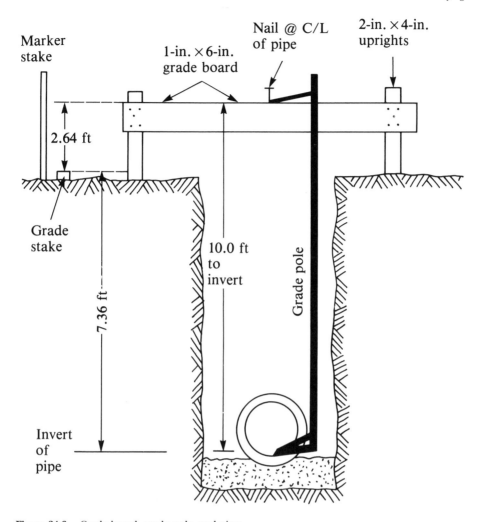

Figure 24-6. Grade board, grade pole, and pipe.

floor. Grades for such features are provided by stakes set to finished elevations of the floor or footing.

Once the footing or floor slab of a building has been built and checked, layouts for the walls, columns, and structural members are located directly on the slab or footing. Forms for concrete pours, prefabricated walls, or steel members can be aligned directly to the surveyor's marks that provide a direct check of the construction grade and alignment. On multistoried buildings, as each floor is completed, the alignment and grade controls are transferred up to the next higher floor, main-

taining plumbness and level of the building. Controls are transferred by plumb lines, an optical-plummet theodolite, vertical laser beams, or combinations of these methods.

24-8. HIGHWAYS AND STREETS

Staking for a highway construction project is usually controlled from survey points remaining in place from earlier topographic and right-of-way acquisition surveys. Surveyors preserve the control stakes by placing reference

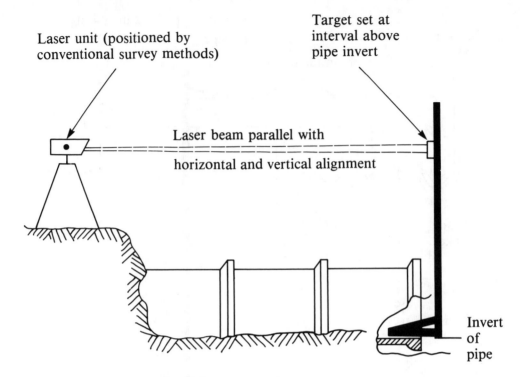

Laser unit (positioned by conventional survey methods)

Target set at interval above pipe invert

Laser beam parallel with horizontal and vertical alignment

Invert of pipe

Figure 24-7. Laser beam used as guideline.

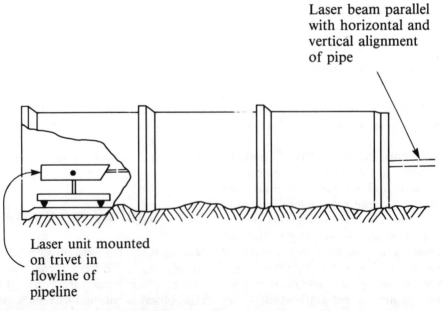

Laser beam parallel with horizontal and vertical alignment of pipe

Laser unit mounted on trivet in flowline of pipeline

Figure 24-8. Use of laser beam inside pipeline.

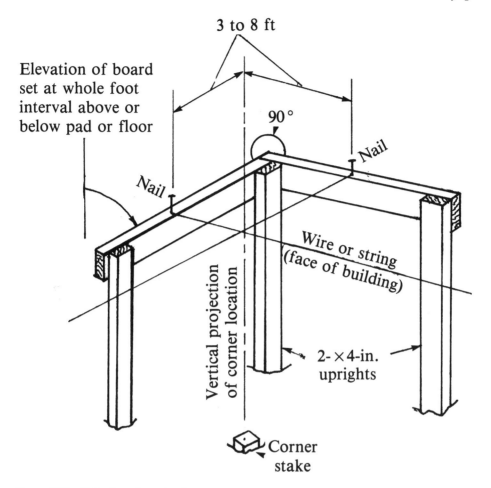

3 to 8 ft

Elevation of board
set at whole foot
interval above or
below pad or floor

90°

Nail

Nail

Wire or string
(face of building)

Vertical projection
of corner location

2-×4-in.
uprights

Corner
stake

Figure 24-9. Batter board construction.

points well clear of the construction area, using the methods shown in Figure 24-1. All references should be recorded in a field book to provide permanent data for checking and replacement of displaced or destroyed control stations.

24-8-1. Clearing Stakes

The first construction stakes provided for the contractor are points marking limits of the construction project. These stakes will be used to guide clearing of brush and vegetation prior to grading and are generally known as "clearing stakes." They are placed at the sidelines of a project, usually at 100-ft stations. Wooden

lath, $\frac{1}{2} \times 2 \times 48$ in., located to the nearest foot, are generally used.

24-8-2. Rough-Cut Stakes

After a construction area has been cleared of brush, the contractor needs stakes to guide the subgrade or "rough-cut" grading operations. These stakes are set (1) along the project centerline on 50-ft intervals, (2) at the beginning and end of all horizontal or vertical curves, and (3) at any other grade or alignment transition. The points are located to the nearest 0.1 ft only, usually by a $1 \times 2 \times 18$ in. stake that serves as both the locator and marker. After setting, an elevation is taken on

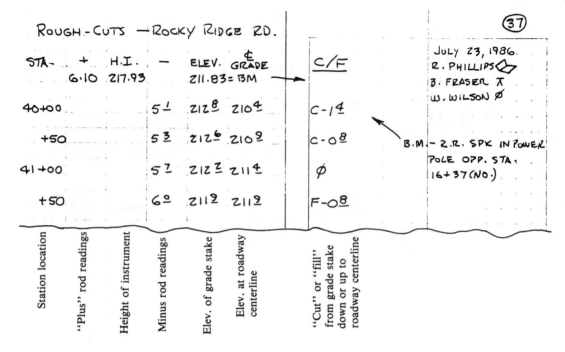

Figure 24-10. Suggested noteform for roadway rough cuts.

the ground beside the point and the C or F determined and recorded in a field book (Figure 24-10). The C or F is then marked on the stake. In rough terrain, it is advisable to set a $2 \times 2 \times 4$ in. stake in addition to the marker. Figure 24-11 shows markings to be made on the stake. Centerline marks are generally adequate for rough-cut grading because the contractor will set additional ones as needed to show the sidelines of a project.

24-8-3. Random Control Stationing

In the preceding discussion of highway and street staking, it was assumed that the surveyor has been using the project centerline (centerline-offset method) to locate and place construction stakes—i.e., location by construction station and offset distance from centerline. This method of surveying has definite advantages, such as simplicity of calculation, because most angles are right angles, and the ease of checking the work. However, there are several

disadvantages in using the offset method—e.g., most stakes set initially are lost during rough grading and have to be reset for final construction. In addition, this method requires a lot of time to complete and is more expensive than other available choices.

An alternate layout method that should certainly be considered is to employ electronic survey systems (or total stations of the type discussed in Chapter 5 and 6 in combination with coordinate geometry, to provide staking from a few select control stations. To use this method (random control survey), the project must be mapped and designed on either the applicable SPCS or an assumed local coordinate datum.

In addition to bearing and distances, coordinate-based construction drawings show grid values for all major construction features and all survey control points. Supplemental layout stations may be set, their positions and density governed by intervisibility and convenience, along with actual or offset stakes for

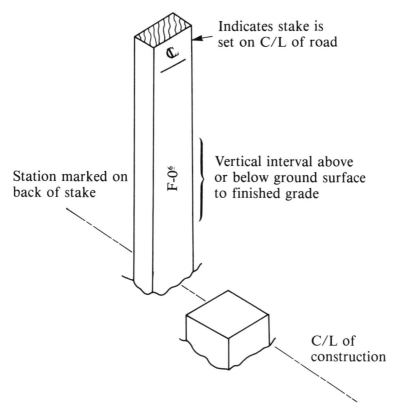

Indicates stake is set on C/L of road

Station marked on back of stake

F-0⁶

Vertical interval above or below ground surface to finished grade

C/L of construction

Figure 24-11. Marking rough-grade stakes.

construction features. The offset points can be placed directly from occupied control stations and require no intermediate mark. To provide direction and distance data from control stations to offset points, it is necessary to calculate a coordinate inverse solution (see Chapter 19). This solution can be made in the field using a hand-held electronic calculator, but for an average-size construction project requiring hundreds of points to be located, most of the surveyor's time would be spent calculating rather than construction staking.

Virtually all surveying firms should have computer equipment that accepts given data, addressing the points to be staked, printout listings of the control stations occupied, and the azimuths and distances to the required layout points. For grade stakes, a vertical angle, slope distance, and elevation can also

be calculated. An entire project may be computed and printed out on data sheets and given to the surveyor for his or her layout work.

In staking operations, a control station is occupied, the total station oriented to a preselected backsight, and the required angle and distance set in the instrument to lay out a selected stake. It is possible to set a large percentage of the construction stakes for a project from a single random control point. As a check on calculations and field work, it is advisable to verify the location of a few stakes from a different random control point or measure between a few set construction stakes. It is recommended that a total station or at least a theodolite equipped with an EDMI unit be used for this type of project. However, a transit-tape survey can be adapted to random control work, if allowances are made for the lesser accuracy of transits and inconvenience of direct taping.

24-9. SLOPE STAKES

Slope stakes are driven to define cut and fill areas in the construction of highways, railroads, canals, dams, and other projects requiring earth movement. Volumes of excavation and fill, waste and/or borrow, can then be computed to aid in estimating construction cost and time.

24-9-1. Slope-Stake Definitions

Basic terms in slope staking are defined and illustrated in Figures 24-12 and 24-13.

Slope. The inclined surface of an excavation or embankment.

Slide slope. The slope of the sides of a canal, dam, or embankment; custom has sanctioned naming the horizontal distance first as 1.5 to 1 (or frequently $1\frac{1}{2}$: 1), meaning a horizontal distance of $1\frac{1}{2}$ to 1 ft vertical. A better form not subject to misinterpretation by thoughtless transposition is 1 on 1.5.

Slope intercepts. Intersection of the original ground and each side slope.

Slope stakes. Stakes set at slope intercepts. They are usually offset a distance of 5 to 10 ft to be safe during construction and marked as shown in Figure 24-12.

Slope-stake accuracy. Standards acceptable for setting slope stakes (see Table 24-2).

Clearing-line stakes. Stakes or lath that mark construction area limits.

Grade. The slope of a road surface with a vertical rise or fall expressed as a percentage of the horizontal distance—e.g., a 3% upgrade means a rise of 3 ft per 100 ft of horizontal distance.

Grade points. Points that have the same ground and grade elevations—i.e., there is no cut or fill.

Grade stakes. Stakes set at grade points. Three transition sections occur when going from all cut to all fill. Three grade stakes must be located as in Figure 24-12. They are marked C 0.0/a, C 0.0/0.0, and F 0.0/b, where a and b are one-half the roadway widths in cut and fill, which may

differ. A line connecting the three grade stakes marks the change from cut to fill. *Grade stakes* may also mark a desired elevation or grade and are called *blue tops* or *red tops*.

Grade rod. The vertical distance from instrument HI to the finished grade. It is positive if the HI is above the ground surface and subgrade (finished grade), and negative if the HI is below the proposed roadbed or dike.

Ground rod. The vertical distance from the instrument HI to the existing ground surface.

Reference stake. Stake set at a distance and difference in elevation from a slope stake, to provide the information needed if the slope stake is destroyed.

24-9-2. Slope-Staking Cuts and Fills

Two basic methods will be discussed here: (1) when a computer printout with road width, centerline, and side-slope data is given (see Tables 24-3 and 24-4) and (2) when profile levels have not been run, so cuts and fills must be computed from station to station.

The equipment needed is a leveling instrument (hand level, Abney level, automatic level, engineers' level, Rhodes arc, or theodolite); a 100- or 50-ft cloth tape or EDMI; level rod, right-angle prims, marking crayon, stakes and bag, hammer, notebook, and 4H pencil. A minimum of two people is needed. A four-person crew works well by assigning one each as notekeeper, rodperson, instrumentperson, and stake driver. The steps utilized in setting slope stakes for the two methods will be listed.

METHOD 1.   A computer printout gives the data for centerline and slope intercepts from precalculated cross-section information. Steps to be taken in slope staking, in sequence, are listed below:

1. Visually inspect ground profile from station to station to check for congruency with the computer printout. Gross errors found in the centerline profile must be checked and corrected.
2. Mark centerline stake with cut or fill on one side, the station on the other, and drive it on centerline.

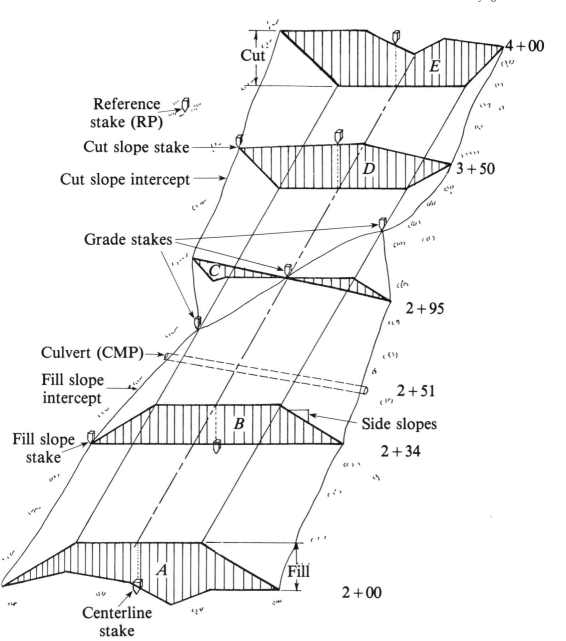

Cut

Reference stake (RP)

Cut slope stake

Cut slope intercept

Grade stakes

Culvert (CMP)

Fill slope intercept

Fill slope stake

Centerline stake

Side slopes

Fill

4+00

3+50

2+95

2+51

2+34

2+00

E

D

C

B

A

Figure 24-12. Typical road diagram with types of road cross sections.

3. At right angles to the centerline, the rodperson goes out a distance called for on the computer printout for left- and right-hand slope stakes. With a level, the difference in ground elevation between the trial point and centerline is read. From the computer printout, mentally calculate the difference in elevation between centerline and left-hand slope stake and the centerline and right-hand slope stake. If differences in ground elevations and computer elevations are within desired accuracy, mark cut or fill with the

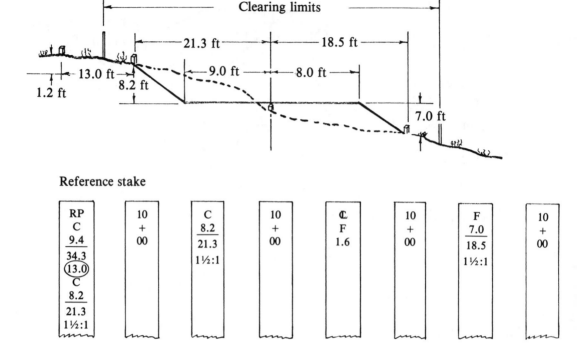

Figure 24-13.   Slope stakes.

distance from centerline and side slope on the stake front facing the centerline. Drive stake aslant so equipment operators can read the data from construction machinery. Reference stakes are marked and set at the

required distance, usually 5 to 10 ft. Record all information for the centerline and sideslope stakes in a notebook.

4. If computer printout is in error at centerline or left or right slope stakes, use method 2.

Table 24-2.   Slope-Stake Accuracy

	Survey Class						
	A	B	C	D	E	F	G
Allowable deviation of slope-stake line projection from a true perpendicular to tangents and a true bisector of angle point.	1°	2°	2°	3°	3°	4°	4°
Horizontal and vertical accuracy for slope-stake references and clearing limits. Use larger value for all slope stakes, clearing limits, and reference stakes.	0.2' or 0.5%	0.2' or 1%	0.2' or 1%	0.2' or 1.5%	0.2' or 1.5%	0.3' or 2%	0.4' or 3%

Table 24-3. U.S. Forest Service Slope-Stake Computer Printout

		Bear Gulch Spur 3		
LT Slope		Template		RT Slope
2.00:1	14.0	0	14.0	1.5:1
Station				
416 + 50	Ground elevation 5288.74			
	Grade elevation 5301.53			
	Left	Slope Stakes	Right	
	(−16.0/46.0)	−12.8	(−15.3/37.0)	

LT Slope		Template		RT Slope
1.00:1	14.0	0	14.0	1.00:1
Station				
425 + 00	Ground elevation 5216.59			
	Grade elevation 5214.49			
	Left	Slope Stakes	Right	
	(4.3/28.3)	2.1	(1.0/15.0)	

METHOD 2. Profile levels have not been run, so cuts and fills must be computed at each station, and the centerline elevation must be carried from station to station. Sequenced steps in slope staking are given in the following description:

1. At centerline, compute the difference between ground-surface and grade elevations.

Mark the cut or fill on one side of the stake, the station on the other, and drive it on centerline.

2. The left-side slope stake is set by trial and error. Estimate the difference in elevation between the center stake and left trial point.

3. Mentally calculate the distance out to the slope stake. For beginners, the formula $d = \frac{1}{2}b + s$ (cut or fill), where d is the horizon-

Table 24-4. Federal Highway Administration Slope-Staking Computer Printout

403 + 50.00		RP	RP/SS Diff.	SS	SS	RP/SS Diff.	RP
		5289.8	−0.1	5289.6	5321.2	−9.6	5330.9
		−22.9	−3.8	−23.1	9.0	15.2	18.7
		63.4		67.2	32.0		47.2
Grade	5313.71						
Surf	0.58						
VC Corr.	0.00						
Cl Corr.	0.00						
Elev.	5321.46	312.7	312.8	5313.1	5312.8	5312.7	5312.2
GL FR	5313.13	−0.4	−0.3	0.0	−0.3	−0.4	−0.9
CL Rod	8.33	21.0	14.0	0.0	14.0	21.0	23.0
PR Off	0.0	HP		GP		SH	HP
Gradient	−0.007						
Slope LT	−2.0						
Slope RT	1.0						

RP = reference point
HP = hinge point
SS = slope stake

tal distance from centerline; b is the road width or template; and s, the side slope, may be useful in locating the trial point.

4. Measure out this distance perpendicularly from the centerline.

5. Read rod with level at this point and calculate the difference in the grade rod and ground rod.

6. Mentally compute the actual distance from the centerline that the rodperson should be.

7. Compute the difference between the actual and estimated distances.

8. Differences of a few tenths are usually acceptable (0.1 to 0.3 ft). On steep hillsides and rough terrain, more is allowed. Mark the cut or fill with distance from centerline and side slope on one side and station on reverse side of stake. Drive the stake aslant so that the cut or fill data face the centerline, allowing heavy-equipment operators to read data from construction machinery. Set reference stakes are required and record information in notebook.

9. If the distance has been missed badly, a new trial point is selected by moving *in* or *out* using a more exact estimate, and another reading is taken. Repeat the process until accuracy standards are met. On *steep* slide slopes, make small corrections in or out.

10. The right-side slope stake is set by the same trial-and-error procedure. The rodperson lines up with centerline and left-hand slope stake. When the slope-stake position is found, the stake is marked, referenced, and recorded.

Example 24-1. Use the trial distances for the left- and right-hand slope stakes given in the computer printout in Table 24-3 to set slope stakes for station 416 + 50.

1. Mark the centerline stake with F 12.8 on one side and 416 + 50 on the other. Drive it into the ground with fill facing beginning of project.

2. The rodperson goes 37.0 ft to the right, perpendicular to the centerline. With a level, check the ground elevation between the

centerline and trial point—perhaps 2.4 ft. Difference in computer elevations is $-12.8 - -15.3 = 2.5$, close enough.

3. Mark the slope stake and drive it aslant with data facing the centerline.

4. For the left slope stake, the rodperson goes 46.0 ft from the centerline for trial point and aligns with the center and right-hand slope stakes. Read ground elevation difference between the centerline and trial point with a level—perhaps 3.0. Printout calls for $3.2 (-12.8 - -16.0)$, which is good enough. Mark and drive the slope stake aslant facing the centerline.

5. Set reference stakes and record the data in a notebook.

6. If the difference is badly missed, use method 2 to set the stakes.

Example 24-2. Use Table 24-3 to set slope stakes for station 403 + 50.

1. Mark C 8.3 on one side and station on the other side of the centerline stake and drive it into the ground with fill data facing decreasing stations.

2. The rodperson goes 15.2 ft toward the centerline for the right-side reference-point (RP) stake and lines in with the left-hand RP. Difference in ground elevation of RP and rodperson should be 9.6 ft.

3. Check to see if the difference meets standards; if not, use method 2.

4. Mark C 9.0/32.0, 1 : 1 on the stake and drive it aslant into the ground, facing the centerline.

5. The rodperson goes 3.8 ft away from the centerline for the left RP stake in line with the right RP stake. Difference between ground elevation of the left RP and rodperson should be 0.1 ft.

6. Check the difference for accuracy and use method 2 if adjustments are needed.

7. Write F 23.1/67.2, 2 : 1 on one side of the stake and station 403 + 50 on the other; drive the stake aslant into ground, facing the centerline.

8. Record the information in a field book.

Example 24-3. Use method 2 in Figure 24-14, to set the centerline and slope stakes at station 1 + 00 in building a dike or embankment.

1. Difference between the ground rod and grade rod at the centerline is 29.4 ft. Mark *F* 29.4 on one side of the stake and station on the other side and drive it in.

2. Estimate the difference in elevation between the center stake and left-hand trial point. Suppose that the ground rises from the centerline 1.1 ft in about 60 ft. Difference between the left trial point and centerline grade is 28.3 ft.

3. Mentally calculate the distance out to the slope stake, 10 + 2(28.3) = 66.6 ft, since the side slope is 2:1 and road width 20 ft.

4. The rodperson goes out 66.6 ft, perpendicular from the centerline.

5. Read the rod with the level and get fill from the grade minus ground rod—perhaps −21 − 7.3 = −28.3.

6. Actual distance from the centerline should be 10 + 2(28.4) = 66.8 ft, close enough. Mark and reference the stake and drive it in aslant.

7. For the right slope stake, suppose that the ground rises about 4 ft in 60 ft. Difference

between the right trial point and centerline grade is 25.4 ft.

8. Mentally calculate the distance out to the slope stake, 10 + 2(25.4) = 60.8 ft.

9. The rodperson goes out 60.8 ft, perpendicular from the centerline to the right.

10. Read the rod with the level and get fill from the grade rod minus ground rod—perhaps −21 − 4.1 = −25.1.

11. Actual distance from the centerline should be 10 + 2(25.1) = 60.2. If the ground is level, move in 0.7 ft, mark, reference, and drive in the stake. Otherwise, make a better estimate, compute a new distance, and take a reading to repeat the trial-and-error procedure until the difference between actual and estimated distances is acceptable.

Example 24-4. Use method 2 in Figure 24-15 to set the centerline and slope stakes for a cross section in cut. Assume a level roadway of 20-ft width for ideal conditions, with no ditches.

1. The grade rod and ground rod difference at the centerline is 11.2 ft. Mark *C* 11.2 on the stake with the station on the opposite side and drive it.

2. For the left trial point, estimate the difference between the center stake and trial

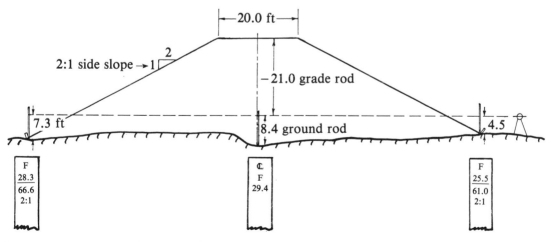

Figure 24-14. Cross section (embankment) at station 1 + 00.

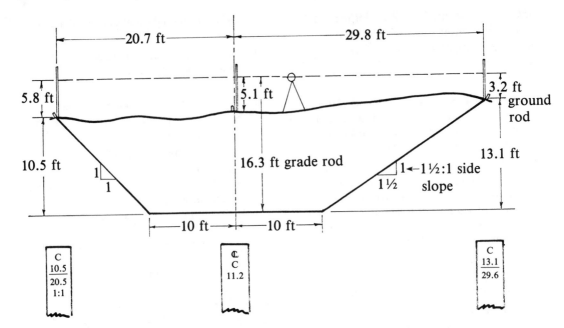

Figure 24-15.   Slope stakes for cross section in cut.

point. In about 20 ft, the ground elevation drops 1 ft. The elevation difference between the trial point and centerline grade is 10.2 ft.

3. Mentally calculate the distance out to the slope stake, 10 + 1(10.2) = 20.2 ft, since the side slope is 1 : 1 and one-half the road width is 10.

4. The rodperson goes out 20.2 ft, perpendicular from the centerline to the left.

5. With the level, read the rod to get a cut by subtracting ground rod from grade rod—perhaps 16.3 − 5.6 = 10.7.

6. Actual distance from the centerline should be 10 + 1(10.7) = 20.7 ft, not close enough. Always move toward the actual distance for a new trial point.

7. Estimate a new trial point. Suppose that the ground drops 0.2 ft in 0.5 ft. The rodperson holds 20.7 ft from the centerline.

8. Take a new reading with a level, say 5.8, to get a cut of 10.5 (16.3 − 5.8).

9. Repeat the calculation of actual distance from the centerline 10 + 1(10.5) = 20.5 ft, close enough. Mark, reference, and drive the stake.

10. The right trial point is found by estimating

the difference between its elevation and centerline, say, a 2-ft difference in 30 ft out, so the elevation difference between centerline grade and trial point is 13.2.

11. Mentally compute the distance out to the slope-stake, 10 + 3/2(13.2) = 29.8 ft, since the side slope is $1\frac{1}{2}$: 1 and one-half the road width is 10.

12. The rodperson holds rod 29.8 ft out, perpendicular from the centerline.

13. With a level, read the rod and get cut from the grade rod minus the ground rod—perhaps 16.3 − 3.2 = 13.1.

14. Actual distance from the centerline should be 10 + 3/2(13.1) = 29.6 ft, close enough to the estimated figure. Mark, reference, and drive in the stake.

15. Record this information in a notebook.

24-9-3. Slope-Staking Culverts

Culvert lengths are adjusted to fit on-the-ground conditions with a minimum flow grade of 2%. Inlets and outlets are carefully placed to avoid erosion and clogging. Pipe diameters are determined by a small area nomograph or approximation. A minimum of a 1-ft cover at subgrade is required. Consult the project engineer for exceptions in the standards. Method 1 is used to set culverts when fills are at each slope stake; method 2 is employed when cuts are at each slope stake.

METHOD 1.

1. By the trial-and-error procedure or computer printout, set the left- and right-hand slope stakes.

2. With a level, determine the difference in elevation between the left and right slope stakes.

3. Compute flow grade by dividing the elevation difference by the horizontal distance between slope stakes. Check to see if the minimum is met; find the length of pipe by measurement or the Pythagorean theorem.

4. Compute the subgrade cover by subtracting the culvert diameter from fill at the centerline.

5. Multiply flow grade by one-half the road width. Subtract this quantity from the subgrade cover; the value should be 1 ft or more.

6. Mark C 0.0 to the invert, diameter, and length of pipe, and station on the stakes and

drive them aslant at the two ends of the culvert.

7. Reference both ends and record information in a notebook.

8. Catch basins should be staked as required.

METHOD 2. For simplicity, this method assumes *academic conditions* of a level roadway.

1. Calculate the distance from the centerline to each end of the culvert by multiplying the shoulder side slope by the pipe diameter + 1 ft of cover at the road edge and adding this to one-half the road width.

2. Add the width of the flat-bottom ditch to the distance obtained in step 1.

3. Rodperson goes out the required distance in step 2 from the centerline.

4. Compute the cut at this point by adding the centerline cut, diameter of the pipe, feet of cover at roadway edge, and elevation change from centerline.

5. Treat the rodperson's position in step 3 as the centerline. Use cut data from step 4 and by trial-and-error methods set the slope stakes for the left- and right-hand sides.

6. Mark the stake with cut, culvert size, and catch-basin dimensions.

7. Record this data in a notebook.

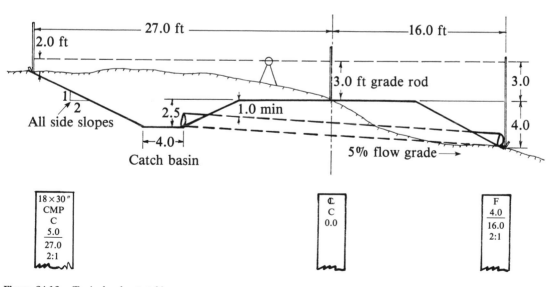

Figure 24-16. Typical culvert staking.

Example 24-5. Stake the culvert in Figure 24-16.

1. Set the right-hand slope stake by trial and error.
2. Mark state *F* 4.0/16.0 2 : 1 slope.
3. Calculate the distance to the left-hand culvert end—perhaps 8 + 2(2.5) = 13.0 ft, since the shoulder side slope is 2 : 1 and pipe diameter plus 1 ft of cover 2.5 ft.
4. The rodperson goes 17.0 ft left from the centerline so as to include 4.0 ft of the catch basin (flat-bottom ditch).
5. Compute the cut at this point by adding the cut at centerline (0.0), diameter of pipe (1.5 ft), cover at roadway edge (1.0 ft), and elevation change from centerline (2.5 ft) to get 5.0 ft.
6. Treat the rodperson's position as if at centerline. Use trial and error to set the left slope stake.
7. Mark on stake 18″ × 30′ CMP, *C* 5.0/27.0 2 : 1 slope.
8. Check flow grade—in this case, 1.5/30 = 5%, which is good.
9. Set the reference stakes and record all data in a notebook.

24-10.   EARTHWORK DEFINITIONS

Measurement of volumes of earth, rock, concrete, stockpiles of material, and capacities of reservoirs, bins, and tanks is often required. These volumes are found by measuring lines and areas that have a relationship to the desired volume. Indirect methods are employed rather than direct methods. Volumes are approximated by utilizing pyramids, prisms, or prismoids. Three major methods are applied: (1) the cross-section method, (2) borrow-pit method, and (3) contour-area method.

Earthwork measurements are obtained by field surveys, topographic maps, and photogrammetry. Normally, cuts and fills are balanced (except for embankments, dams, highways on low ground, etc.) to minimize the cost. On large projects, the analysis of quantities of earth movement is done by *mass diagrams*.

They are plotted with stations as abscissas versus volume as ordinates (see books on route surveying for more details).

The following is a list of basic terms and definitions:

Borrow. Quantity of earth, rock, or other fill material brought in from outside sources.

Waste. Quantity of earth, rock, or material hauled away from a construction site.

Haul. Distance over which material is moved or the product of volume and distance it is moved.

Free-haul. Distance the contractor will move a cubic unit of excavated material and place it in fill without added cost.

Cross-section. A vertical section of the ground surface or underlying strata, or both, measured at right angles to the centerline or across a stream. Also, a horizontal grid system laid out on the ground for determining contours, quantities, or earthwork, etc. by means of grid-point elevations. Several types of cross sections are illustrated in Figure 22-12: (1) an *irregular section* (area labeled a) used in rough topography; (2) a *level section* (area labeled b) used in flat terrain; (3) a *side-hill section* (area labeled c) used in passing from cut to fill or on side-hill locations; (4) a *three-level section* (area labeled d) used where ordinary ground conditions exist; and (5) a *five-level section* (area labeled e) used in rough terrain.

24-10-1.   Cross-Section Method

This method is employed for computing volumes on linear construction projects such as canals, highways, and railroads. Ground profiles are secured at right angles to the centerline of a project at specified intervals or stations by field methods, photogrammetry, or topographic maps. A *design template* (outlines of planned embankment or excavation) can be superimposed on the plot of each cross section to obtain the area of excavation or embankment. These are called *end areas* and can be determined by computation or planimetering. Volumes are calculated by the average-end-area, prismoidal, or pyramid formula.

24-10-2. Average-End-Area Formula

For most earthwork, the solid between two cross sections can be considered a prism whose right cross-sectional area is the average of the two end areas. The volume for two sections with areas A_1 and A_2 separated by a horizontal distance L (Figure 24-17) is

$$V_{\text{AEA}} = \frac{A_1 + A_2}{2} \times \frac{L}{27} \text{ cu yd} \qquad (24\text{-}1)$$

The exact volume of the prism is given when $A_1 = A_2$. Equation (24-1) is approximate and gives answers that are generally slightly larger than the true prismoidal volume. In practice, it is used because of its simplicity—and the contractor is happy since generally a little extra payment results. To increase accuracy, cross sections must be taken closer together. Computation of earthwork quantities by this method was speeded up by using a nomogram before the advent of calculators and computers. In general, average-end-area volumes are satisfactory and legal for most volume calculations unless noted otherwise.

For a pyramid (see Figure 24-18),

$$A_2 = 0 \quad \text{and} \quad V = \frac{AL}{3 \times 27} \text{ cu yd} \qquad (24\text{-}2)$$

Example 24-6. Compute the volume of excavation between station $2 + 00$ with an end area of 567 sq ft, and station $2 + 50$ with an end area of 392 sq ft.

By Equation (24-1),

$$V_{\text{AEA}} = \frac{567 + 392}{2} \times \frac{50}{27} = 890 \text{ cu yd}$$

Example 24-7. Compute the volume of a pyramid having $L = 100$ and a base area $A = 478$ sq ft.

$$V = \frac{(478)(100)}{3 \times 27} = 590 \text{ cu yd}$$

24-10-3. Prismoidal Formula

A prismoid (Figure 24-19) is a solid bounded by planes whose end faces are parallel and have the same number of sides. Most earthwork volumes are prismoids, but relatively few of them warrant the precision of this formula. It usually gives a volume smaller than the average-end-area formula but is inconvenient to use. The prismoidal formula, a result of Simpson's rule from calculus, is given by the following:

$$V_p = \frac{L(A_1 + 4A_m + A_2)}{6 \times 27} \text{ cu yd} \qquad (24\text{-}3)$$

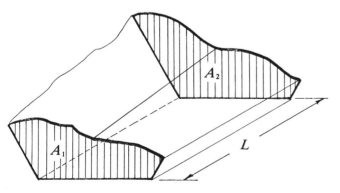

Figure 24-17. Volume by average-end-area method.

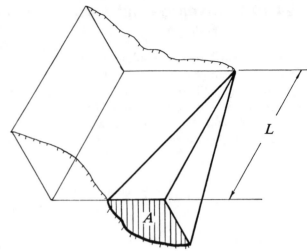

Figure 24-18.   Volume of pyramid.

where L is the horizontal distance between A_1 and A_2, A_1 and A_2 are the end areas, and A_m is the area of the plane section midway between A_1 and A_2. The formula can be used for figures whose bounding surfaces are warped and continuous. Note that $A_m \neq (A_1 + A_2)/2$.

To use the formula, A_m must be found. If the coordinates of the end faces A_1 and A_2 are known, the *coordinates of A_m are found by averaging corresponding coordinates of the two end faces*. A_m can be computed by the coordinate method; otherwise, field methods are required for find A_m.

In concrete work and rock excavation, the precision and labor of the prismoidal formula are justified.

Example 24-8. Use the prismoidal formula to find the volume of earthwork between stations 5 + 00 and 6 + 00 with end areas 231 and 152 sq ft and A_m = 189 sq ft.

$$V_p = \frac{100[231 + 4(189) + 152]}{6 \times 27} = 703 \text{ cu yd}$$

Prismoidal volumes can be obtained by applying corrections to the average-end-area volume. Tables and formulas for these corrections are found in most route surveying books. One such correction formula, accurate for three-level sections and close enough for most

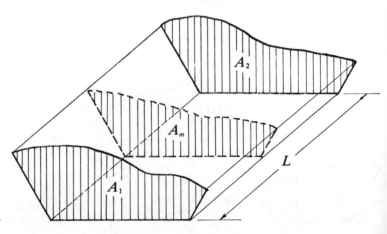

Figure 24-19.   Volume of prismoid.

others, is

$$C_p = \frac{L}{12 \times 27}(c_1 - c_2)(w_1 - w_2) \text{ cu yd} \quad (24\text{-}4)$$

where $C_p = V_{AEA} - V_p$, c_1 and c_2 are center-line heights in cut or fill, and w_1 and w_2 are distances between outer slope stakes (from slope intercept to slope intercept), and L is the horizontal distance between cross sections. Thus, an alternate way of finding V_p is

$$V_p = V_{AEA} - C_p \text{ cu yd} \quad (24\text{-}5)$$

For projects requiring only a few cross sections, routine manual computations can be reduced by using either the average-end-area formula or prismoidal formula. For large projects, computers and programmable calculators are used to determine quantities. In open terrain free of obstructions, photogrammetric methods will generate earthwork data (see photogrammetry books for more details).

Example 24-9. The three-level cross-section notes for two stations 50 ft apart in Table 24-5, are plotted in Figure 24-20. The road base is 30 ft wide and side slopes are 2:1. Use the average-end-area formula to find the volume between stations.

Figure 24-20 has the origin for cuts at the centerline and coordinates assigned in the

Table 24-5. Three-Level Cross-Section Notes for 30-ft-Wide Road with 2:1 Slide Slopes

Station	Left Slope Stake	₵	Right Slope Stake
6 + 50	C8.0	C7.0	C3.0
	31.0		21.0
7 + 00	C10.0	C8.0	C6.0
	35.0		27.0

conventional manner. The numerator is a vertical distance above the centerline. The denominator is horizontal distance from the centerline. By the coordinate method and matrix system, areas A_1 and A_2 are computed. Areas are positive, so ignore the algebraic sign. Start by listing Y- and X-coordinates of each corner in columns in succession as the figure is traversed counterclockwise, with the coordinates of the starting point repeated as the last column. In matrix format, to find the area: (1) Add the product of the diagonals from left to right; (2) add the product of the diagonals from right to left; and (3) take one-half the difference of (1) and (2).

$$A_1 = \frac{1}{2}\begin{vmatrix} 0 & 0 & 3 & 7 & 8 & 0 & 0 \\ 0 & 15 & 21 & 0 & -31 & -15 & 0 \end{vmatrix}$$

$$= \frac{1}{2}[7(-31) + 8(-15) - 3(15) - 7(21)]$$

$$= 264 \text{ sq ft (note that the matrix has}$$

two rows and seven columns)

$$A_2 = \frac{1}{2}\begin{vmatrix} 0 & 0 & 6 & 8 & 10 & 0 & 0 \\ 0 & 15 & 27 & 0 & -35 & -15 & 0 \end{vmatrix}$$

$$= 368 \text{ sq ft}$$

Then

$$V_{AEA} = \frac{A_1 + A_2}{2} \times \frac{L}{27} = \frac{264 + 368}{2} \times \frac{50}{27}$$

$$= 585 \text{ cu yd by Equation (24-1)}$$

Example 24-10. Again using Table 24-5 and Figure 24-20, find the volume by the prismoidal formula.

In Figure 24-20, the coordinates of A_m are found by averaging the respective end-area coordinates. The far-left coordinate of A_m is

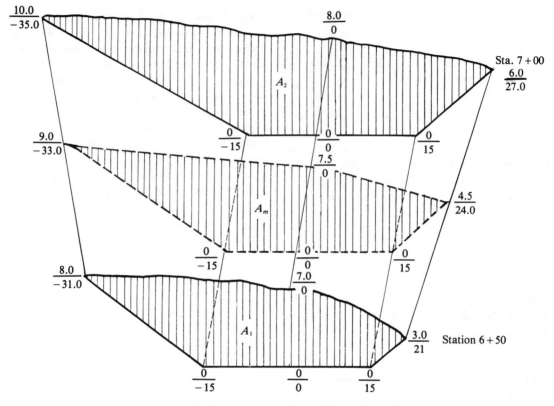

Figure 24-20.   Mass diagram for three-level cross section using prismoidal formula.

$\frac{1}{2}(10 + 8) = 9$ for the numerator and $\frac{1}{2}(-35 + -31) = -33$ for the denominator. Other coordinates are found in a similar manner. The area of A_m by coordinates is

$$A_m = \frac{1}{2} \begin{vmatrix} 0 & 0 & 4.5 & 7.5 & 9 & 0 & 0 \\ 0 & 15 & 24 & 0 & -33 & -15 & 0 \end{vmatrix}$$

$$= 315 \text{ sq ft}$$

Substituting in Equation (24-3) yields

$$V_p = \frac{L(A_1 + 4A_m + A_2)}{6 \times 27}$$

$$= \frac{50[264 + 4(315) + 368]}{6 \times 27} = 584 \text{ cu yd}$$

Note that V_p is smaller than V_{AEA}.

Example 24-11. As an alternate method and check on V_p, use Equation (24-5) for Table 24-5 and Figure 24-20.

By Equation (24-4), compute C_p using Table 24-5, $c_1 = 8.0$, $c_2 = 7.0$, $w_1 = 62$, $w_2 = 52$, and $L = 50$, where units are in feet.

$$C_p = \frac{L}{12 \times 27}(c_1 - c_2)(w_1 - w_2)$$

$$= \frac{50}{12 \times 27}(8 - 7)(62 - 52) = 2 \text{ cu yd}$$

Substituting C_p and V_{AEA} (from Example 24-9) in Equation (24-5), we obtain

$$V_p = V_{AEA} - C_p = 585 - 2 = 583 \text{ cu yd}$$

In practice, the average-end-area formula is adequate for most volume calculations. If high precision is needed, the prismoidal formula should be used.

24-10-4. Borrow-Pit Method

Quantities of material excavated or filled on a construction project and the number of cubic yards of loose material such as gravel, coal, etc. in stockpiles can be found by borrow-pit leveling. The area to be measured is staked in squares of 10, 20, 25, 50, or more feet, with the choice depending on the accuracy desired and size of the project area. Layout may be done by using a transit and/or tape. Outside the construction site, a bench mark of assumed or known elevation is established and preserved. In addition, the grid of squares should be referenced to points outside the construction area. Elevations are determined by sighting on all corners of the squares (grid points) before and after construction. The volume of a rectangular prism is the base area times the average of corner heights. Total volume is found by adding all rectangular prism volumes using the formula

$$V = \frac{A}{4 \times 27}(\Sigma h_1 + 2\Sigma h_2 + 3\Sigma h_3 + 4\Sigma h_4) \text{ cu yd}$$

$$(24\text{-}6)$$

where h_1, h_2, h_3, and h_4 are corner heights common to exactly one, two, three, or four squares, respectively, and A is the area of one square.

Marginal areas may be best approximated by triangles. The truncated triangular prism volume is found by the following formula:

$$V = \frac{A}{3 \times 27}(h_1 + h_2 + h_3) \text{ cu yd} \quad (24\text{-}7)$$

where A is the area of the triangle, and h_1, h_2, and h_3 are the triangle corner heights.

The total is found by summing all volumes of rectangular and triangular prisms.

Example 24-12. use Table 24-6 and Figure 24-21 to calculate the volume by the borrow-pit method.

Table 24-6. Borrow-Pit Tabulation for Figure 24-21

Point	h_1	h_2	h_3	h_4
A1	3.0			
B1		6.1		
C1		4.2		
D1	5.1			
A2		4.3		
B2				5.5
C2				4.8
D2		5.0		
A3	4.7			
B3			5.7	
C3				5.0
D3		4.7		
B4	5.2			
C4		4.8		
D4	4.7			
Totals	22.7	29.1	5.7	15.3

From Tables 24-6, the sum of corners common to one square is $h_1 = 3.0 + 5.1 + 4.7 + 5.2 + 4.7 = 22.7$. The sum of corners common to two squares is $h_2 = 6.1 + 4.2 + 4.3 + 5.0 + 4.7 + 4.8 = 29.1$. Similarly, the sums of corners common to three and four squares are $h_3 = 5.7$ and $h_4 = 15.3$, respectively. Since each square is 25 by 25 ft, $A = 625$ sq ft. Substituting in Equation (24-6), we obtain

$$V = \frac{625}{4 \times 27}[22.7 + 2(29.1) + 3(5.7) + 4(15.3)]$$

$$= 920 \text{ cu yd}$$

Example 24-13. Use Table 24-7 and Figure 24-22 to find the volume of earth removed from squares 1, 10, 11, 12, 13, and 14.

Sides of squares are 50 ft; volumes of squares 10, 11, 12, 13, and 14 are found by Equation (24-6).

$$V = \frac{2500}{4 \times 27}[25.1 + 2(18.8) + 3(7.0) + 4(6.2)]$$

$$= 2500 \text{ cu yd}$$

Volume of the triangular prism in square 1 is given by Equation (24-7)

$$V = \frac{1250}{3 \times 27}(4.0 + 5.2 + 6.0) = 230 \text{ cu yd}$$

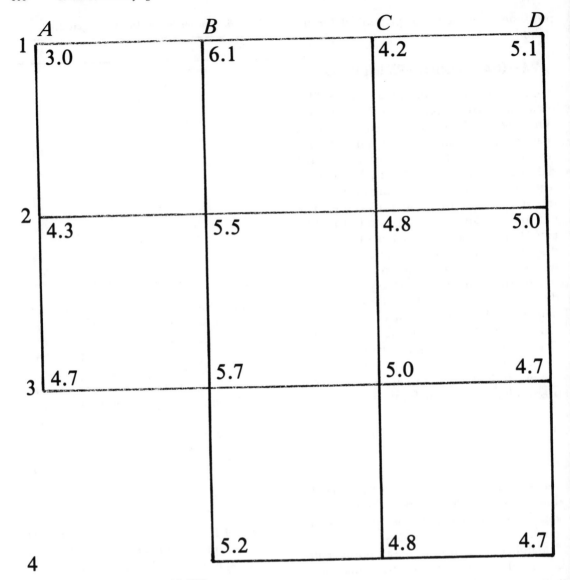

Figure 24-21. Borrow-pit cuts with 25-ft squares.

where $B1$ has a cut of 6.0 ft. The total volume is 2730 cu yd, obtained by adding triangular and square prisms.

24-10-5. Contour-Area Method

Whenever an accurate contour map of the project area exists, it can be planimetered. U.S. Geological Survey topographic maps are readily available for most areas. The volume between adjacent contour lines is computed by the average-end-area formula and written as follows:

$$V = \frac{A_1 + A_2}{2} \times \frac{CI}{27} \text{ cu yd} \qquad (24\text{-}8)$$

where CI is the contour interval and A_1 and A_2 are planimetered areas (see Figure 24-23).

Table 24-7. Borrow-Pit Tabulation for Figure 24-22

Point	h_1	h_2	h_3	h_4
A2	4.0			
B2		5.2		
C2		3.1		
D2	4.8			
A3		5.5		
B3				6.2
C3			7.0	
D3	5.5			
A4	6.3			
B4		5.0		
C4	4.5			
Totals	25.1	18.8	7.0	6.2

Example 24-14. The capacity of a reservoir will be computed from given planimetered areas and contour intervals utilizing Equation (24-8).

Elevations	4820	4825	4830	4835
Areas in sq ft	630	350	1010	1180

Since the two middle areas are added twice and the contour interval is 5 ft.

$$V_{AEA} = (630/2 + 852 + 1010 + 1180/2) \times \frac{5}{27}$$

$$= 512 \text{ cu yd}$$

24-10-6. Mass Diagrams

A continuous graph of cumulated cut and fill volumes plotted as ordinates versus stations along the centerline as abscissas is called an *earthwork mass diagram*. To analyze earthwork distribution, the mass diagram is usually plotted below or above the centerline profile. In the plot, cut is positive and fill negative, with an appropriate allowance for shrinkage and swell. Shrinkage occurs when fill material is compacted and occupies a smaller volume.

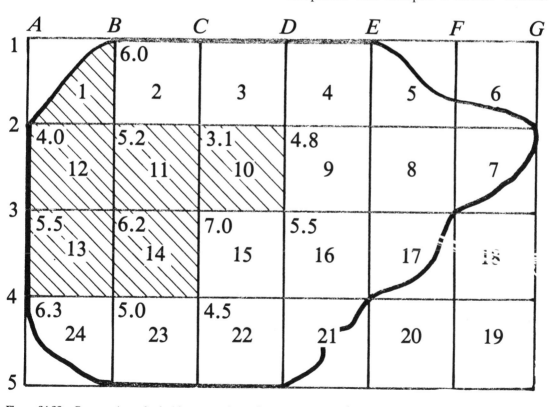

Figure 24-22. Borrow-pit method with rectangular and triangular prisms (50-ft squares).

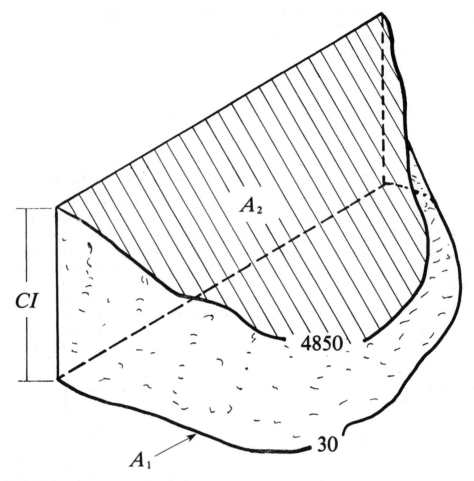

Figure 24-23. Volume by contour-area method.

Swell occurs when excavated material such as shattered rock occupies more volume. The graph indicates to a contractor where cuts should go and borrow or waste occur (see Figure 24-24). It is used by contractors to plan their work and prepare a bid (lump sum or price per cubic yard).

In Figure 24-24, an increasing or rising curve as from A to C indicates excavation or cut, the descending or decreasing curve from C to G signifies a fill or embankment. Generally, the maximum or minimum points on the mass diagram are located at grade points on the centerline profile (points C, G, I, and J). An exception occurs when there are extensive side-hill cuts and fills at the same centerline station. Horizontal lines BD and FH are called *balance lines*. Total cuts and fills between the two end points of the balance line will be equal. If the final ordinate equals the initial ordinate, fills and cuts balance in the mass diagram. Material must be borrowed if the final ordinate is smaller than the initial ordinate or obtained by "daylighting" cut slopes on curves. Material is wasted when the final ordinate is greater than the initial ordinate and can be used to flatten fill side slopes.

The economics of doing earthwork can be analyzed by using a mass diagram, given the

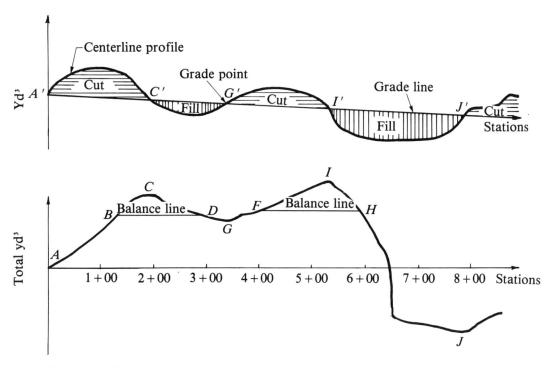

Figure 24-24. Centerline profile and mass diagram.

cost of excavation, borrow, overhaul (volumes moved greater than free-haul distances), and free-haul distance are known. The limit of economic haul (LEH) is the distance beyond which it is cheaper to borrow or waste material and is determined by the following:

$$LEH = \text{free-haul distance}$$
$$+ \frac{\text{cost of excavation}}{\text{cost of overhaul}} \quad (24\text{-}9)$$

Example 24-15. Use Figure 24-25 to find (1) balance lines for the material between stations 0 + 00 and 20 + 00, (2) quantities of excavation, (3) total overhaul, and (4) cost of earthwork. Assume that the free-haul distance is 600 ft (6 stations), the cost of excavation $3/cu yd, and overhaul $1.50/station yd.

By Equation (24-9), the limit of economic haul is

$$LEH = 6 + \frac{\$3.00}{\$1.60} = 8 \text{ stations}$$

1. For balance lines, consider the first loop from station 0 + 00 to 5 + 00. It falls within the limits of free-haul distance FH. The cut volume from *B* to *C* is used to fill from *A* to *B*. In the second loop, balance lines of length equal to FH and LEH are drawn horizontally on the graph. The overhaul area is represented by triangular areas *EDQ* and *GRH*. Excavation from *E* to *D* wasted is 2000 cu yd; overhaul from *D* to *E* is spread between *G* and *H*. Approximately 1500 cu yd must be borrowed for fill in *H* to *I*. For the two small consecutive loops, balance line *IKM* is placed horizontally making *IK* = *KM* (both *IK* and *KM* must be shorter than the free-haul distance). Excavation from *J* to *L* is put in *I* to *J* and *L* to *M*. From *M* to *N*, about 1500 cu yd of borrow are needed.

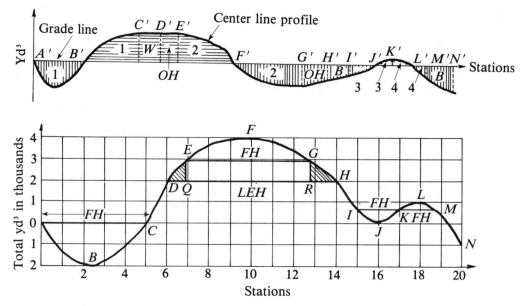

Figure 24-25. Centerline profile and mass diagram for small project.

2. Total excavation taken from the graph is

B to C	2000 cu yd
C to F	4000 cu yd
J to L	1000 cu yd
	7000 cu yd

Total borrow from the graph is 3000 cu yd; hence, total excavation is 10,000 cu yd.

3. Volume of overhaul is determined by planimetering or approximating triangular areas *EDQ* and *GRH*, perhaps 1000 station yd.

4. Total cost is

Excavation	10,000 cu yd × $3/cu yd	= $30,000
Overhaul	1000 × $1.50	= 1500
	Total	$31,500

24-10-7. Digital Elevation Models and Aerial Photographic Methods

A network of three-dimensional coordinates (x, y, z) of surfaces can be obtained by using modern photogrammetric techniques. These are used to create a data file in a computer system. Computers can then be programmed to generate both graphical and quantitative data. It is beyond the scope of this chapter to discuss or develop methods covered in photogrammetric textbooks. However, surveyors should know that aerial photography can be used to measure stream channels, reservoir volumes, stockpile quantities, etc. As always, economics comes into play, so a cost analysis will be needed before deciding to employ conventional or photogrammetric methods.

Pre- and posteruptions digital elevation models were made of Mount St. Helens in Oregon by the Western Mapping Center of the National Mapping Division, U.S. Geological Survey. On May 18, 1980, the volcanic reaction displaced approximately 3.5 billion cu yd of material, devastated more than 123,500 acres of land, removed the top 1300 ft of the mountain, and left a crater about 2450 ft deep. The St. Helens DEM model gave immediate, inexpensive, and accurate data. In the future, satellites and spaceships will provide topographic maps and surface measurements of the earth and other celestial bodies by employing photogrammetric methods and digital elevation models.

24-10-8. Earthwork Computational Devices

Hand-held programmable calculators and computers are readily available to compute volumes by the various methods discussed in this chapter. Hand-held programmable calculators utilize programmed chips for earthwork calculations and have peripheral portable printers. Surveyors can write their own programs or input selected ones—this tool is obviously indispensable for today's surveyor.

24-10-9. Cost Estimation

Payment for slope staking is on a per-mile basis and estimated by considering the type of terrain, accuracy desired, weather, and number of stations occupied. Many road projects are let for bid to small businesses by federal agencies such as the U.S. Forest Service and Federal Highway Administration. Experience warns: *Allow margin on jobs bid.* It is best to visit the actual construction site and walk the proposed road PI line so that a reasonable bid can be made. Many firms have suffered from biddings jobs unseen. Some projects are best bid by an hourly rate rather than a lump-sum bid, especially for inexperienced contractors.

24-11. TUNNEL SURVEYS

Although tunnel surveys and mine surveys appear to be identical operations, they differ in some important aspects because of their individual nature and objectives. Comparison between this chapter and Chapter 29 will more clearly define their relationship.

24-11-1. Surface Survey

Tunnel surveys are based on an aboveground survey that connects points representing each portal of a tunnel. A traverse connecting the portal points determines the azimuth, distance, and differences in elevation of each end of the proposed tunnel. Local conditions and length will influence the choice of methods used in the connecting survey. It is highly recommended that the survey be based on the applicable state plane coordinate system or adapted to the suitable local coordinate system. Data from the initial survey are vital to the design of alignment and grade of the proposed tunnel (see Figure 24-26).

24-11-2. Control Monuments

A permanent system of monuments is established within an area outside each tunnel portal to provide reference for alignment. If local conditions permit, monuments are set on an extension of the tunnel centerline to establish its direction outside each portal. Transfer of centerline alignment is accomplished by occupying an aboveground point and sighting ahead or double centering on required centerline points into the tunnel, as shown in Figure 24-26.

During construction, all worker and equipment movement in the tunnel will be along the centerline. Therefore, it is essential to establish and maintain co-operation with the work force in scheduling surveying activities at mutually convenient times.

24-11-3. Construction Control

Centerline and grade stakes within the tunnel are usually set in the roof to avoid displacement and destruction by the constant flow of people and machinery as construction proceeds. If the stakes are set on the floor, they should be offset into an area along the tunnel's edge. The size and cross section of the tunnel will be a major consideration in the location and density of construction stakes.

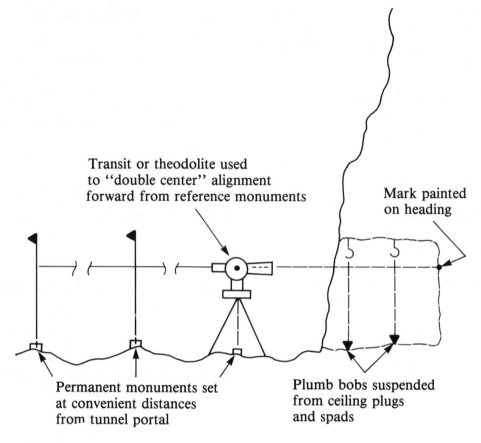

Transit or theodolite used
to "double center" alignment
forward from reference monuments

Mark painted
on heading

Permanent monuments set
at convenient distances
from tunnel portal

Plumb bobs suspended
from ceiling plugs
and spads

Figure 24-26. Centerline produced into tunnel.

The methods and machinery used for tunnel excavation also influence the choice of staking methods. A small tunnel being excavated by air hammers and blasting from a single heading can be guided by a centerline set into the tunnel roof. The contractor can make necessary measurements from this line to control the direction and grade of the tunnel heading.

24-11-4. Vertical Shaft Control

Highway and railroad tunnels may be several miles in length. Tunnel excavations are carried inward from both portals and in both directions from vertical shafts (adits) that are dug to the grade and alignment of the tunnel at intermediate locations along its route.

24-11-5. Vertical Alignment by Plumb Bob

The required adits are located in the field by reference to the aboveground survey. As an adit is being excavated, it is extremely important to make frequent checks to maintain plumbness of the shaft. A common method suspends a heavy plumb bob (10 to 15 lb) on either a wire or heavy twine. Oscillations of the bob can be dampened by suspending it in a tray containing high-viscosity oil. The bob can be hung from a removable bracket attached to the surface side of the shaft. The horizontal location of the attachment point, hence the plumb wire, is set by appropriate measurements from the survey control line. Throughout the excavation operations, a plumb bob is periodically suspended, providing a reference for measurements to shaft sides as a check on plumbness.

24-11-6. Vertical Alignment by Optical Collimator

An optical vertical collimator is shown in Figure 24-27. This Topcon Auto-V-Site instrument allows users to establish and check vertical lines of sight, both up and down, with the same eyepiece, and has a centering function provided. It is useful for vertical collimation at construction sites, mines, shipyards, and other types of vertical surveying.

A vertical collimator (1) may be more convenient than a plumb bob; (2) can be used to set marks directly on the floor of a completed shaft; and (3) permits direct measurements to be made to the shaft walls, instead of using plumb-bob wires. A sturdy bracket, built out from the shaft edge, will support a tripod or trivet on which to mount the instrument.

24-11-7. Vertical Alignment by Laser

A laser-equipped instrument is employed like an optical collimator to make measurements directly by observing where the light beam intersects a scale or rule. After it is set up and aligned, the laser has the advantage of being left in place to furnish a visible line for constant reference. A laser-equipped unit should be located unobtrusively along one side of the shaft and not obstruct excavation operations. Targets, generally a flat piece of steel or wood with a $\frac{1}{2}$- to $\frac{3}{4}$-in. hole drilled in it, are set at intervals along the shaft wall. As long as the laser beam passes through the target holes, it remains plumb and in correct orientation.

24-11-8. Transfer of Elevation

Establishing correct elevations for each shaft and tunnel section is best accomplished through direct measurement down the shaft from a surface bench mark, utilizing either a steep tape or EDMI unit. When using a steel tape, it is freely suspended from a surface point, high enough to be observed with a surveyor's level. The tape should have a 20-lb weight at its bottom end to provide a correct calibrated length; temperature corrections must also be considered. Simultaneous observations are made on the tape with surveyors' levels at the surface and on the shaft floor. The observed difference, after proper corrections are applied, is the vertical difference between the HI of the level at the surface and the HI at the shaft floor (see Figure 24-28).

An EDMI may be used to transfer elevation, provided the favorable visibility conditions are presented. The EDMI unit and reflectors must be mounted on a vertical line and the height of both instrument and reflector centers determined by spirit leveling. It is necessary to devise a mounting system for the EDMI and reflector to ensure a stable vertical alignment. After an observation is made, the corrected distance is the vertical interval between instrument and reflector centers, as shown in Figure 24-29.

24-11-9. Transferring Alignment in a Shaft

Maintaining the horizontal alignment of the tunnel segments is a demanding task, requir-

Figure 24-27. Topcop VS-A1 (Auto-V-Sight) optical collimator. (Courtesy of Topcon.)

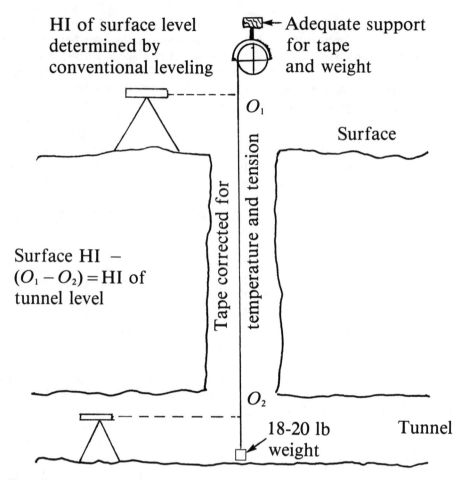

HI of surface level
determined by
conventional leveling

Adequate support
for tape
and weight

O_1

Surface

Tape corrected for

temperature and tension

Surface HI −
$(O_1 − O_2) =$ HI of
tunnel level

O_2

18-20 lb
weight

Tunnel

Figure 24-28. Vertical distance in shaft by steel tape.

ing the combined applications of surface and tunnel surveying techniques, plus ingenuity by the surveyor. Vertical transfer of horizontal control through a shaft is primarily to link the aboveground survey with the working surface of the tunnel, thereby enabling construction to follow design alignment. Transfer of surface control to the shaft bottom places a point on both shaft sides. From these points, the alignment is carried forward until it meets the alignment from an adjacent shaft or side tunnel. Obviously, the limited shaft diameter along with the lack of checks to surface alignment make this a very demanding operation.

24-11-10. Transfer by Plumb Bobs

Points are located at both sides of the shaft top at the intersection of the surface survey line. An appropriate structure is erected to anchor two plumb bobs suspended on piano wires at the surface point locations. Thus, the plumb lines recreate the horizontal location and alignment of the surface points. An instrument set up on the tunnel floor approximately on line with the plumb lines is carefully brought into correct alignment by lateral adjustment, averaging any swing of the plumb

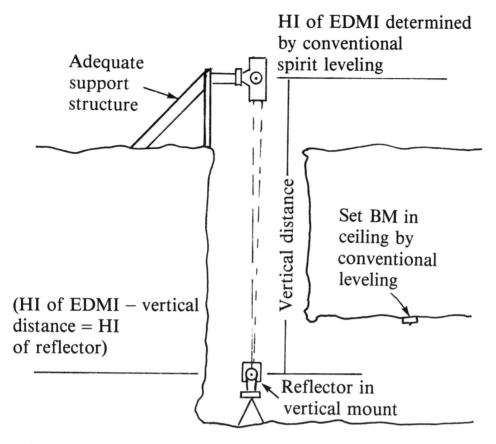

Figure 24-29. Vertical distance in shaft by EDMI.

lines. Control points are then established on both sides of the shaft by at least four direct and four inverted observations. Alignment can be extended in either direction horizontally by a very careful process of double centering, using a laterally adjustable target and scale (Figure 24-30).

24-11-11. Transfer by Vertical Collimator

The same basic procedure employed to check plumbness by vertical collimation is used, with the exception that two points at the top of the shaft, horizontally controlled, are occupied by the collimator. Marks are then made, or points set at the shaft bottom, where

vertical lines of sight are intersected by the floor surface. The resultant points are utilized in a precise double-centering operation to prolong the tunnel alignment. Laser devices, projecting a vertical light beam, can be used following the same general procedure.

24-11-12. Azimuth by Gyroscopic Theodolite

Another method of establishing and maintaining azimuth in tunnel construction is afforded by the gyro-theodolite (Figure 24-31). This instrument is equipped with an electric gyro that rotates at approximately 22,000 RPM. The axis of gyro rotation is in a horizontal

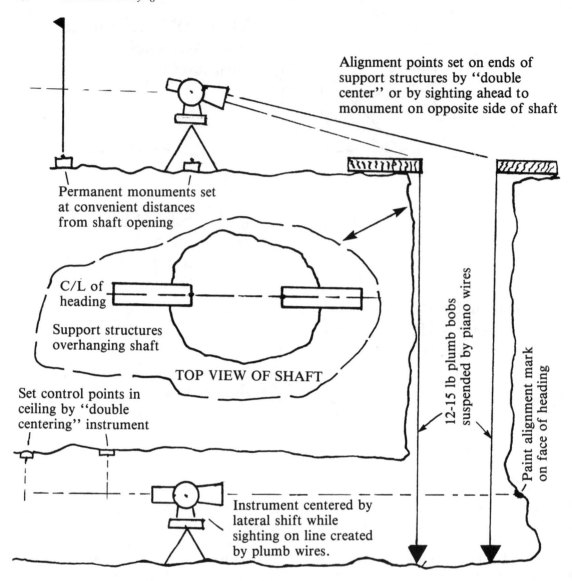

Alignment points set on ends of support structures by "double center" or by sighting ahead to monument on opposite side of shaft

Permanent monuments set at convenient distances from shaft opening

C/L of heading

Support structures overhanging shaft

TOP VIEW OF SHAFT

Set control points in ceiling by "double centering" instrument

12-15 lb plumb bobs suspended by piano wires

Paint alignment mark on face of heading

Instrument centered by lateral shift while sighting on line created by plumb wires.

Figure 24-30. Transfer of alignment down shaft by plumb bobs.

plane; therefore, the earth's rotation causes the spin axis to turn toward the meridian plane. When this happens, inertia carries the spin axis somewhat beyond the meridian; the axis then proceeds back toward, and a bit beyond, the meridian. This movement is referred to as *precession* and is more or less a continuous oscillation to either side of the meridian. An average pointing of the instrument can be achieved by observing the reversing points of the precession cycle. The procedure may take from 20 to 30 min, but yields an azimuth with a standard deviation of 20″. Although this is an expensive instrument, the accuracy and convenience realized by use of this theodolite will balance out the cost factor, particularly on a large project.

Azimuth determined by this instrument is a true azimuth, since it is based on the earth's axis of rotation. Therefore, the user should know the effect of meridian convergence when a project survey is based on the SPCS or an arbitrary grid system. Close attention must be given to angular adjustment applications, if the observed true azimuth is to provide the correct grid direction.

24-11-13. Use of Laser Beams in Tunnels

The laser beam has been successful for tunnel construction, since it furnishes a readily accessible guide for both grade and alignment. Such an instrument can be mounted a foot or two from the tunnel wall and 3 or 4 ft above the floor. When aligned parallel with the centerline and slope of the tunnel, the highly visible laser beam provides a reference line for the bore's elevation and alignment. As shown in Figure 24-32, mounts and setups for the laser unit and its target are placed and kept in correct pointing by conventional survey procedure.

A ceiling-mounted laser can be used to guide a tunneling machine, as illustrated in

Figure 24-31. Gyro-theodolite GAK1/T16. (Courtesy of Wild Heerbrugg, Ltd.)

Figure 24-33. The tunnel machine has two targets attached to its front and rear. When the machine is correctly positioned for line and grade, the laser beam passes through a hole in the rear target and strikes the center of the front (opaque) target. An operator can observe the spot of light and make necessary adjustments to keep the machine aligned.

24-12. BRIDGES

A control survey and topographic site map are generally required except for very small bridges (see Figure 24-34). The resulting data are used for design and in the permit process with governmental agencies. A control survey of this nature should conform to high-order accuracy standards and employ theodolites and EDMIs.

Geographic configuration of the site and length of the proposed bridge should be considered in the design of a control survey. For most control systems, a quadrilateral presents a suitable figure. Conveniently located stations for construction control staking should be provided in the control figure that is based on an appropriate SPCS. When completed, the control survey can be expanded to provide additional monuments as required for use in both offshore and onshore construction locations.

Piers and other offshore structures are located by either simultaneous observations from two control points or set with a total station unit. Regardless of the instrument used in offshore operations, it is of critical importance to select control stations that present the strongest geometrical figures, including the point being located. After a pier location is established, the contractor installs a sheet-piling cofferdam and excavates to bedrock for the pier footing, and then places the foundation. The location survey extended to the footing top guides pier construction.

As work progresses, periodic checks are made to verify plumbness of the pier struc-

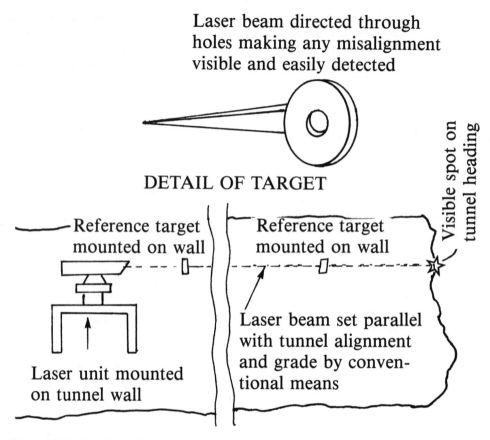

Laser beam directed through holes making any misalignment visible and easily detected

DETAIL OF TARGET

Reference target mounted on wall

Reference target mounted on wall

Visible spot on tunnel heading

Laser unit mounted on tunnel wall

Laser beam set parallel with tunnel alignment and grade by conventional means

Figure 24-32.   Use of laser in tunnel.

tures and their elevations. Final operations in pier construction are setting anchor bolts and beam sockets to anchor the bridge spans. All these components must be located by triangulation or trilateration from the onshore control stations. If possible, reference marks made on adjacent piers should be traversed to provide additional positional checks.

Onshore staking will furnish line and grade for the abutments, wing walls, approach structures, and roadways. The latter structures can be located from the road construction or control surveys. The face lines of abutment structures are set and referenced to either batter boards or points located beyond the construction limits. If portions of an abutment will extend into the water, an offset line established on the riverbank and measurements made by the contractor locate the structure's waterward face.

As in pier construction, when footing and structure bases have been completed, survey control is transferred to their tops and used to guide construction of the facilities upward to completion. Again, it is usually the surveyor's responsibility to make periodic checks to ensure compliance with specified alignment and elevation. When abutment construction has been completed, a precise survey from the control system provides locations for the installation of anchor bolts and other span structure items. After this hardware has been placed, it is checked for position relative to the control system and abutments on the op-

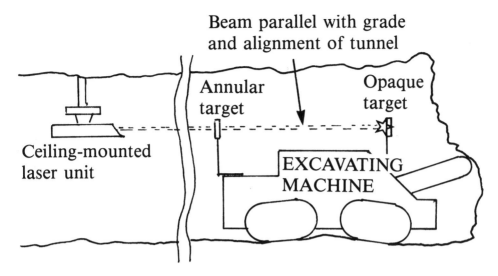

Figure 24-33. Tunneling machine direct by laser.

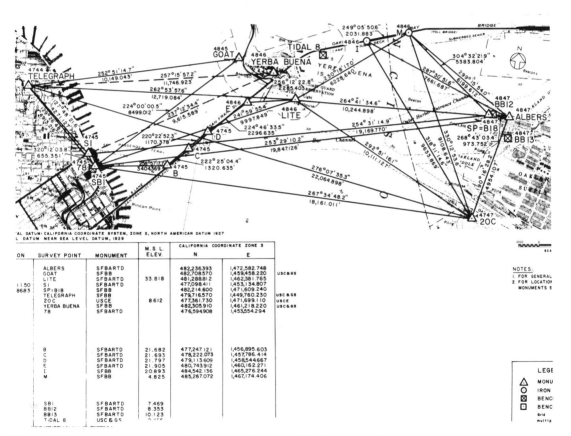

ON	SURVEY POINT	MONUMENT	M.S.L. ELEV.	CALIFORNIA COORDINATE ZONE 3 N	E	
	ALBERS	SFBARTD		482,236.393	1,472,582.748	
	GOAT	SFBB		482,708.570	1,459,458.220	USCB@S
	LITE	SFBARTD	33.818	481,288.812	1,462,381.765	
11.50	SI	SFBARTD		477,098.411	1,453,134.807	
86.83	SP=BI8	SFBB		482,214.600	1,471,609.240	
	TELEGRAPH	SFBB		479,716.570	1,449,760.230	USC & GS
	20C	USCE	8.612	477,361.730	1,471,699.110	USCE
	YERBA BUENA	SFBB		482,305.910	1,461,218.220	USC&@S
	78	SFBARTD		476,594908	1,453,554.294	
	B	SFBARTD	21.682	477,247.121	1,456,895.603	
	C	SFBARTD	21.693	478,222.073	1,457,786.414	
	D	SFBARTD	21.797	479,113.609	1,458,544.667	
	E	SFBARTD	21.905	480,743.912	1,460,162.271	
	I	SFBB	20.893	484,542.136	1,465,276.244	
	M	SFBB	4.825	485,267.072	1,467,174.406	
	SBI	SFBARTD	7.469			
	BBI2	SFBARTD	8.353			
	BBI3	SFBARTD	10.123			
	TIDAL 8	USC & GS				

AL DATUM: CALIFORNIA COORDINATE SYSTEM, ZONE 3, NORTH AMERICAN DATUM 1927
L DATUM: MEAN SEA LEVEL DATUM, 1929

NOTES:
1. FOR GENERAL
2. FOR LOCATION
   MONUMENTS S

LEGE
△ MONU
○ IRON
⊠ BENC
☐ BENC
   Grid
   multip

Figure 24-34. Portion of survey map for design and control of bridge construction.

posite shoreline. As the span is being constructed, a surveyor may be required to conduct various vertical and horizontal alignment checks of the structure.

Permanent monuments are set on the completed bridge structure following the completion of construction. They should be conveniently located and well-referenced, since they provide a basis for future deformation surveys monitoring any motion in the bridge and pier structures.

24-13. DAMS

As in most construction projects, a boundary and topographic survey must be made of the dam site prior to undertaking even the design work. A dam/reservoir is unique in that large areas will be inundated by the proposed reservoir, so surveys may extend over 100 mi. This type of survey is utilized for many purposes: (1) geological exploration, (2) hydrography of existing water bodies, (3) property acquisition for reservoir and dam facilities, (4) topography for site design work, and (5) data for access road construction. Also to be considered are power-line routes, future recreational facilities encircling the reservoir, and provision for a deformation/subsidence survey around the completed dam/reservoir area.

A survey of this scope must, by necessity, be based on a SPCS. The primary control surveys, both horizontal and vertical, should meet first-order standards (see Figure 24-35). Secondary controls for topography, etc. can be performed to second-order specifications. Several methods of surveying may be employed depending on the project's magnitude and geographic conditions. Among various methods to consider are conventional triangulation, traverse, trilateration, Doppler, or inertial positioning. The topography work should be done by photogrammetric methods.

When the dam location has been selected, a very dense system of horizontal and vertical control stations is established in the area. Per-

manent monuments are constructed in convenient locations on both up- and downstream sides of the dam. A construction control survey is based on the primary control survey and should meet first-order standards (see Figure 24-36).

Dam construction may extend over several years; therefore, it is essential that the construction control system be designed for permanence and adaptability to changing requirements as construction progresses. Design consideration must also be given to ease of checking the overall system and for efficient replacement of points that are destroyed or displaced.

Construction of a dam will require survey control to excavate diversion tunnels, construct cofferdams up- and downstream from the dam location, and locate the dam base at bedrock. Numerous other peripheral construction projects will be undertaken in addition to the dam work, such as cableway crossings, on-site railways, access roads, rock quarries, fabrication shops, concrete-mixing plants, and utility lines. All these facilities require location and construction staking prior to, and simultaneous with, construction of the dam. It is obvious that careful scheduling is required to meet the project's surveying needs by employing an economical number of personnel to avoid delaying construction.

The control survey and all major points of the dam are assigned x-, y-, and z-coordinate values, so azimuth and distance can be developed between selected control stations and points of construction by coordinate inverse procedure. Thus, horizontal positions for concrete forms or other features of construction can be established by simultaneous observations from two theodolites occupying control stations. The instruments are backsighted along fixed control lines, then aligned to the computed azimuth of a desired point. An intergrated EDMI and the theodolite of adequate precision may be used for the same position.

Provision of vertical control is accomplished by direct differential leveling or trigonometric

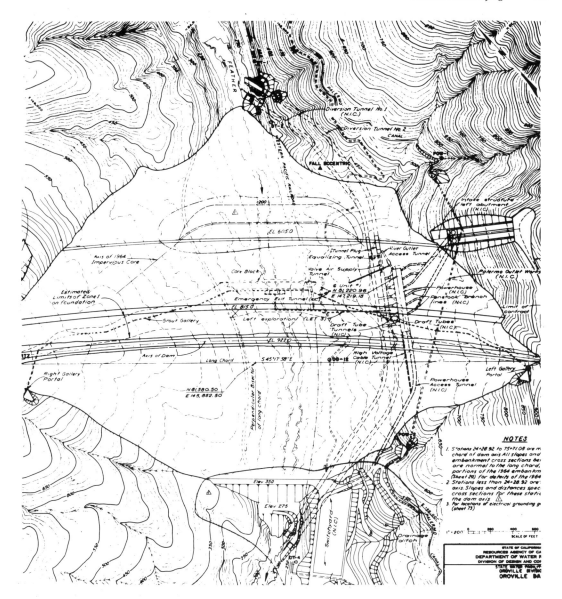

Figure 24-35. Portion of map for design and control of reservoir and dam.

leveling using a total station of suitable precision. In addition to controlling major features of a dam, a surveyor may be required to lay out internal structures such as penstocks, turbine and generator bases, and other various machinery. These operations require a wide variety of surveying techniques along with a great deal of ingenuity on the surveyor's part. It is advisable to consider using laser-equipped survey instruments to facilitate horizontal and vertical alignment.

Along with providing ongoing construction control, survey requirements will often include periodic checks of finished construction to monitor settlement or displacement of the dam. On completion of the dam, a precise system of horizontal and vertical survey stations is established on the dam itself and at

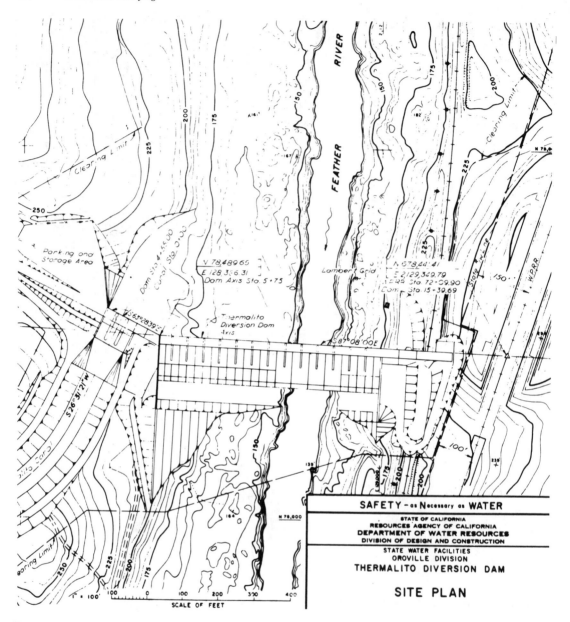

Figure 24-36. Portion of map showing construction points for dam structure and site.

locations around the reservoir to provide a basis for deformation surveys. These will generally be carried on for years after completion of construction to provide a program of safety analyses through confirmation of design-predicted movements.

24-14. AS-BUILT SURVEYS

The final surveying job after construction is finished involves preparation of an "as-built" survey and map. All primary control stations for the project should be rechecked and any missing or displaced monuments replaced or readjusted. Likewise, highway and street centerlines or other easily destroyed control marks must have adequately established reference points. Permanent monumentation of all major control points is necessary for possible use in any modification or expansion of the project.

With the control system lines, all major changes in the completed project are mapped. Few large facilities have been finished without making a field adjustment or modification to the original design. Therefore, it is essential to prepare a detailed map showing all changes that may have been made during construction, in relation to the completed project. Measurements from an as-built survey often are the basis of payment for actual quantities of work and materials used in construction.

The complete as-built survey should depict the primary control system, references to all major survey control points, locations and dimensions of the completed project, and notations of any modifications to the original design. This map becomes a permanent record and must be preserved for many uses in future engineering or surveying relative to the completed project.

REFERENCES

DAVIS, R. E., F. S. FOOTE, J. M. ANDERSON, and E. M. MIKHAIL. 1981. *Surveying Theory and Practice*, 6th ed. New York: McGraw-Hill.

KAVANAGH, B. F., and S. J. GLENN BIRD. 1992. *Surveying*, 3rd ed., Englewood Cliffs, NJ: Prentice Hall.

MEYER, C. F., and D. W. GIBSON. 1980. *Route Surveying and Design*, 5th ed. New York: Harper Collins.

MOFFITT, F. H., and H. BOUCHART. 1992. *Surveying*, 9th ed. New York: Harper-Collins.

WOLF, P. R. and R. C. BRINKER. 1994. *Elementary Surveying*, 8th ed. New York: Harper Collins.

25

Route Surveys

David W. Gibson

25-1. INTRODUCTION

Highways, railways, canals, tunnels, dams, pipelines, and transmission lines are constructed works having linear shapes classified as *routes*. A unique system for expressing route geometry has developed that, once learned, can be applied without fail to this broad range of projects by all surveyors, designers, and constructors.

25-2. THE ROUTE COORDINATE SYSTEM

Positions of features along a route are expressed by three coordinates: (1) *station* S, (2) *offset* OS, and (3) *elevation* Z (see Figure 25-1).

Station S is the horizontal distance along a preselected *base line*. Due to a route's linear form, a single line is chosen to be the base of a geometry system, and its choice is extremely important. The selected base line is usually the center of pavement for highways, the center of rails for railroads, or another feature that parallels the route's linear form.

The base line is *stationed*, first in concept and later physically, on the ground during project layout. Station are points separated by a fixed horizontal distance, such as 100 ft for traditional U.S. routes or 1000 m for the metric expression of route geometry. Stations are numbered from zero at the initial point to 1, 2, 3, etc. throughout the length of the route. For 100-ft stationing, station 25 is therefore 2500 ft from station 0, measured horizontally along the base line. Points between stations are designated by a *plus* distance—the distance past the previous full station. For example, a point at station 25 + 45.12 is 45.12 ft beyond station 25, or 2545 ft from station 0 + 00.

Under the metric system, stations have assumed different definitions in various countries. In Canada, station 2 + 151.262 is 2 km plus 151.262 m from station 0 + 000. The *layout interval* for setting base-line stakes is chosen as 10, 20, or 50 m depending on the route's geometric complexity. Therefore, with a 20-m layout interval, 50 stakes will be set from station 1 + 000 to 2 + 000. Calculations are carried to the millimeter.

An *offset* OS is measured as the minimum perpendicular horizontal from the base line, either right (plus) or left (minus), referenced to a person standing on the base line facing up-station.

Elevation Z has the same meaning in route geometry as for other surveys and is refer-

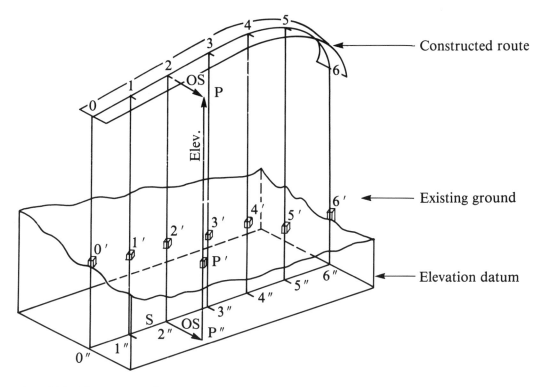

Figure 25-1. The route coordinate system.

enced to a recognized vertical datum such as the NGVD.

25-3. HORIZONTAL ALIGNMENT PARTS AND NOMENCLATURE

Base-line horizontal geometry consists of *tangents, simple curves, simple spirals, compound curves, reverse curves, broken-back curves,* and *combining spirals.* As shown in Figure 25-2, tangents are straight portions of base line and when extended meet at a *point of intersection,* also termed *vertex* V. The angle each tangent makes with the local direction of north is the *tangent bearing* or *azimuth.* The *intersection angle* Δ (or I) between tangents equals the difference between tangent azimuths. *Simple curves* are arcs of circles providing a constant rate of direction change from tangent to tangent. The

curve starts at the *point of curvature* PC and ends at the *point of tangent* PT (also called BC and EC for the beginning and end of the curve or TC and CT for tangent to curve and curve to tangent).

A *compound curve* provides an instantaneous change in curvature at a *point of compound curvature* PCC, where an initial curve leads into a second curve of different radius but curving in the same direction. At a *point of reverse curvature* PRC in a *reverse curve,* curvature changes magnitude and direction instantaneously. *Broken-back* compound or reverse curves replace the PCC or PRC with a short length of tangent.

Spirals are curves of variable radius used to replace instantaneous curvature changes at a PC, PT, or PCC with a gradual easement of curvature change. *Simple spirals* join a tangent portion of roadway with a circular curve, while

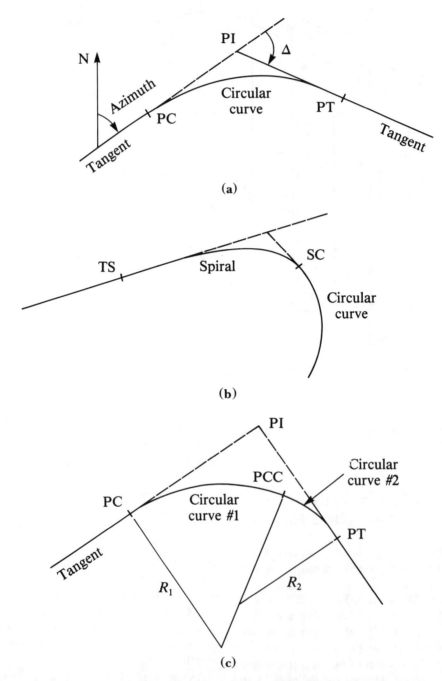

Figure 25-2. Base-line shapes and nomenclature. (a) Simple curve. (b) Simple spiral. (c) Compound curve. (d) Reverse curve. (e) Combining spiral. (f) Broken-back compound curve.

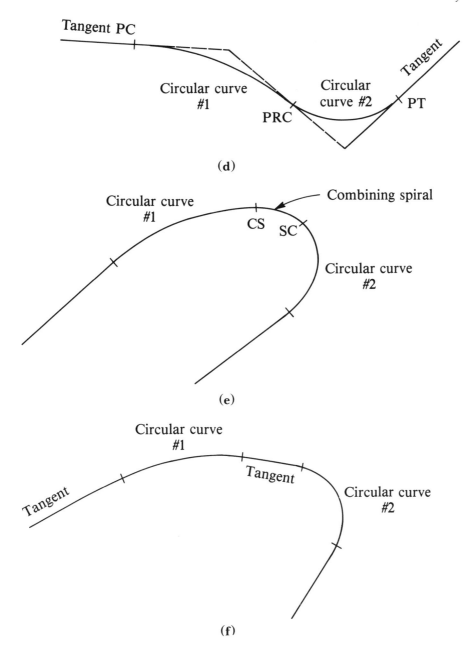

Figure 25-2. (*Continued*)

combining spirals connect two circular curves of differing radius. A *tangent to spiral* juncture on the incoming tangent is labeled TS, while *spiral to curve* SC denotes the union with the circular curve. CS and ST denote similar points but of higher stationing on the leaving spiral, where the curve joins the spiral and the spiral meets the tangent. A *combining spiral* provides a gradual curvature change at a compound curve PCC.

25-4. VERTICAL ALIGNMENT PARTS AND NOMENCLATURE

In addition to having a particular horizontal geometry, the base lines's *vertical alignment* consists of distinct components: (1) *vertical tangents* and (2) *vertical curves*. For vertical tangents, the base line has constant slope, termed *percent grade*, which expresses the rise or fall of the base line per 100 units of run. A positive 3% grade rises 3 ft per 100 in an up-station direction.

Vertical tangents meet at a *point of vertical intersection* PVI. To allow the smooth transition of grade from one tangent to another, a vertical curve is used, consisting of a portion of a parabola. Whereas circular horizontal curves give a constant rate of change of azimuth, parabolic vertical curves yield a fixed rate of change of grade. The beginning of a vertical curve, where the parabola meets the incoming tangent, is termed *point of vertical curvature* PVC, while its juncture with the outgoing tangent is the *point of vertical tangent* PVT. In some references, these points are also termed the *begin vertical curve* BVC and *end vertical curve* EVC.

25-5. CROSS-SECTION ALIGNMENT PARTS AND NOMENCLATURE

Cross sections of a route often have repetitive geometry from place to place, leading to the concept of a *typical cross section*, often called a *template*. For example, a highway cross section may have constant shape consisting of a *crown*, *side slope*, *superelevation*, and *standard offsets*.

A crown may be expressed as a percent, rise-per-foot ratio, or height difference. For two 12-ft traffic lanes, a 2% crown equals a cross slope of $\frac{1}{4}$ in. per ft and height difference of 0.24 ft between the pavement center and edges. For earthwork cuts and fills, it is customary to express side slopes as a ratio of run : rise—e.g., a side slope of 2 : 1 indicates a 1-ft rise or fall for each 2 ft of run. Side slopes

range from vertical of $\frac{1}{2}$: 1 in rock cuts to 6 : 1 for gradual swales and shoulders.

In horizontal curves, the pavement is built at a constant cross slope called superelevation, usually expressed as a percent. Superelevation raises the outside edge of the pavement to aid a vehicle in negotiating a curve by providing better balance of friction and centrifugal forces. A typical superelevation rate is 10%, meaning that the outside edge of a 12-ft traffic lane is 1.2 ft above the inside.

Near the PC of a circular curve, a *transition* is employed to remove the incoming crown and rotate the pavement into superelevation. Likewise at the PT, a second transition is needed to rotate the pavement back to a normal crown.

Typical cross sections also show standard offset to roadway features, such as the edges of the pavement, shoulders, and side ditches. Elevation of the base line at a particular station is termed the *profile grade*, where the term "grade" refers to elevation, not percent slope.

25-6. DEGREE OF CURVE

Degree of curve D is a statement of curvature or "tightness" of a horizontal curve. Mathematically, degree of curve expresses the azimuth change in degree experienced by a vehicle in one station (100 ft) of travel. Figure 25-3 presents two curves with a vehicle traveling one station on each. In Figure 25-3a, the vehicle's azimuth change and therefore degree of curve is small, whereas in Figure 25-3b, a large change in direction gives a large D.

Route designers choose a safe D according to the specified velocity of travel. Table 1 gives typical maximum safe values of D for highways in the United States, indicating that values of D seldom exceed 25° on a very tight curve. Flat curves have a D of 1° or smaller.

An inverse mathematical relationship exists between D and R—i.e., as one decreases the other increases—but given D or R, the other can be calculated. Figure 25-3 indicates that D

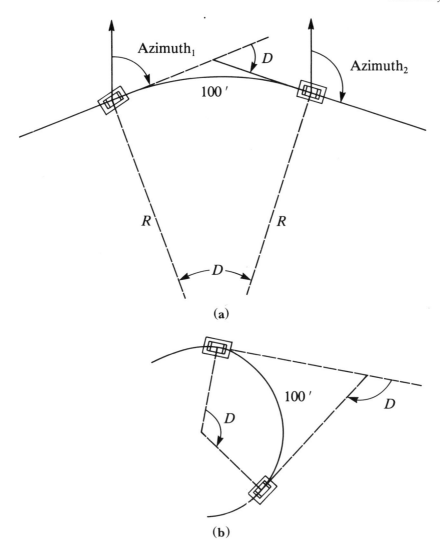

Figure 25-3.   Degree of curve. (a) Flat curve. (b) Tight curve.

Table 25-1.   Safe degrees of curve

Design Velocity (MPH)	Maximum Safe D (deg)
30	25
40	13
50	8
60	5
70	4
80	2.8

also equals the central angle subtending 100 ft of arc. Using the following relationship between the central angle in the radians, arc length, and radius, we obtain

$$100 = RD \tag{25-1}$$

with R in ft and D in rad. Solving for R gives

$$R = \frac{100}{D} \tag{25-2}$$

With D in decimal degrees and R in ft, combining constants yields

$$R = \frac{5729.573}{D} \qquad (25\text{-}3)$$

and

$$D = \frac{5729.573}{R} \qquad (25\text{-}4)$$

Early railroad practice used a *chord definition* D_c for degree of curve, because railroad stations were set 100 ft apart on the chord, not the arc. D_c results in a slightly larger numerical value than D for the same curve. Figure 25-4 indicates that 100 ft of chord extends farther along the curve, thereby making D_c larger than D. By solving the cross-hatched triangle,

$$50 = R \sin \tfrac{1}{2} D_c$$

or

$$R = 50/\sin \tfrac{1}{2} D_c \qquad (25\text{-}5)$$

and D_c can be easily calculated from D by

$$D_c = 2 \sin^{-1} \tfrac{1}{2} D \qquad (25\text{-}6)$$

25-7. COMPUTING CURVE PARTS

The intersection angle Δ and D (degree of curve) are basic parts of a horizontal curve, usually selected during design. From D the radius R is easily calculated, and from R and Δ the remaining curve parts are determined.

Arc length L can be calculated knowing R and Δ in either decimal degrees or rad. With Δ in rad,

$$L = R \Delta \qquad (25\text{-}7)$$

with Δ in degrees,

$$L = 2\pi R \left(\frac{\Delta}{360} \right) \qquad (25\text{-}8)$$

where π is 3.14159.

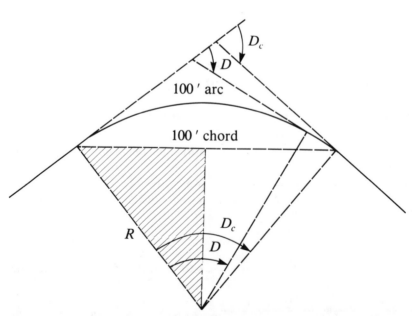

Figure 25-4.   Chord definition of degree of curve.

Tangent distance T and the *external E* may be calculated from triangle *cdb* in Figure 25-5. The *intersection angle* Δ is duplicated at the radius point as the *total central angle* and bisected by the line joining the PI with the radius point. By right-triangle trigonometry,

$$T = R \tan \tfrac{1}{2}\Delta \qquad (25\text{-}9)$$

The external equals the distance from the PI to the radius point minus the radius

$$E = \frac{R}{\cos \tfrac{1}{2}\Delta} - R \qquad (25\text{-}10)$$

The *long chord LC* joins the PC with the PT and may be calculated along with the *middle ordinate M* from triangle *cab*. Point *a* is the long chord's midpoint; therefore,

$$LC = 2R \sin \tfrac{1}{2}\Delta \qquad (25\text{-}11)$$

M is the difference between *R* and distance *ab_r*, giving

$$M = R - R \cos \tfrac{1}{2}\Delta \qquad (25\text{-}12)$$

Other useful curve relationships are as follows:

$$\text{versine } \tfrac{1}{2}\Delta = 1 - \cos \tfrac{1}{2}\Delta \qquad (25\text{-}13)$$

$$\text{exsecant } \tfrac{1}{2}\Delta = \frac{1}{\cos \tfrac{1}{2}\Delta} - 1 \qquad (26\text{-}14)$$

$$M = R \text{ versine } \tfrac{1}{2}\Delta \qquad (25\text{-}15)$$

$$E = R \text{ exsecant } \tfrac{1}{2}\Delta \qquad (25\text{-}16)$$

$$LC = 2T \cos \tfrac{1}{2}\Delta \qquad (25\text{-}17)$$

$$M = E \cos \tfrac{1}{2}\Delta \qquad (25\text{-}18)$$

$$M = \tfrac{1}{2}LC \tan \tfrac{1}{4}\Delta \qquad (25\text{-}19)$$

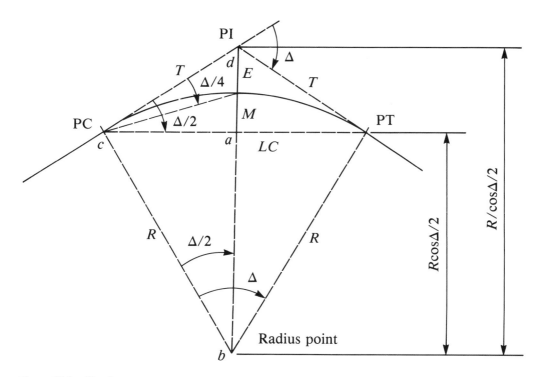

Figure 25-5. Circular curve parts.

25-8. LAYOUT OF HORIZONTAL CURVES

To lay out an alignment, stakes must be set at each base-line station (the PI stakes usually have already been set). Therefore, along tangents, a theodolite gives line from PI to PI, and a tape is used to measure distances for stations, such as 1, 2, 3, and 4 in Figure 25-6.

But when a curve is encountered, special techniques must be employed to set the stakes along the circular curve. The most common method is *deflection angles* and *chords* from a theodolite set on the PC. A theodolite at the PC sights the PI, and deflection angles d are turned to desired points on the curve, such as stations 5, 6, and 7 in Figure 25-6. Chords are slightly shorter than the arcs they subtend and are measured as either *total chords* from the PC to curve points or *subchords* between successive stations on a curve.

Deflection angles equal half the angular size of the total arc subtended. In Figure 25-7, *total arc x* is the length from the PC to a point on the curve and has an angular value θ_x, the central angle it subtends. Therefore,

$$d_x = \tfrac{1}{2}\theta_x \qquad (25\text{-}20)$$

where d_x is the deflection angle for arc x and θ_x the central angle subtending arc x. The chord c_x is found by solving the cross-hatched

triangle, giving

$$c_x = 2R\sin\tfrac{1}{2}\theta_x \qquad (25\text{-}21)$$

and since $\theta_x/2$ equals d_x,

$$c_x = 2R\sin d_x \qquad (25\text{-}22)$$

Equation (25-22) indicates that a chord for arc x is twice the radius multiplied by the sine of the corresponding deflection angle.

Example 25-1. For the curve in Figure 25-6, assume that $R = 400$ ft, $\Delta = 70°$, and the PC station $= 4 + 30$. Calculate the total deflection angles and chords required to lay out each full station from a theodolite on the PC.

Station 5 is 70 ft into the curve. The angular equivalent in rad for 70 ft of arc is 70/400; d_x is half of that or 0.087500 rad, and the decimal equivalent is found by the standard conversion to be $5°00'48''$. The chord for 70 ft of arc is found by Equation (25-22) to be 69.910 ft.

Station 6 information is found in a like manner. The total arc is 170, making d_x equal to $1/2(170/400)$ or 0.21250 rad, which converts to $12°10'31.3''$. The required chord measured from the PC is found by Equation 25-22 to be $2(400)\sin d_x = 168.72$ ft. The total solution is shown in Table 25-2.

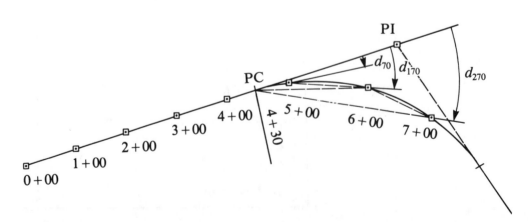

Figure 25-6. Deflection angles.

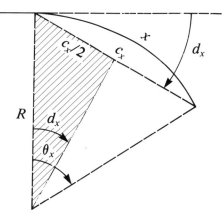

Figure 25-7. Deflection angle nomenclature.

An alternate approach for the layout of a curve uses deflection angles and subchords, whereby curve points are located by taped subchords from the previously set station instead of total chords from the PC. Also, it is customary to calculate the required deflection angles by an additive process using subdeflection angles instead of the direct calculation of total deflection angle, as in the previous example. This procedure is especially recommended when calculations are being done by hand or in the field, because blunders are more easily found and avoided. *Subdeflection* angles equal the incremental deflections between sightings. Since the degree of curve D is the central angle for 100 ft of arc, then the

subdeflection angle for 100 ft of arc, d_{100}, is half of D. A subdeflection angle for an odd value of arc length x is then found by the following simple proportion:

$$d_x = \left(\frac{x}{100}\right) d_{100} = \left(\frac{x}{100}\right)\frac{1}{2}D \quad (25\text{-}23)$$

Corresponding subchords are then computed by Equation (25-22) from the subdeflection angles. This method reduces computational time because each curve usually has only three different subdeflection angles, from the PC to the first full station, between full stations, and from the last full station to the PT. As shown in Figure 25-8, the curve of the previous example has three subdeflection angles: d_{70}, d_{100}, and $d_{18.69}$; and three corresponding subchords: c_{70}, c_{100}, and $c_{18.69}$. Total deflection angles can then be computed by simple addition. This technique will be shown in the following example.

Example 25-2. Using the same curve data as in Example 25-1, tabulate the computations in Table 25-3.

25-9. LAYOUT OF A CURVE BY COORDINATES

With a total station instrument, layout of a circular curve is facilitated by using radial-

Table 25-2. Curve layout data by total deflection angles and total chords

Curve data: $L = R \Delta$ (rad) = 488.692
PT station = (4 + 30) + 488.692 = 9 + 18.69

Station	(Total Arc) from PC, x	Angular Value of x, θ_x (rad)	Deflection Angle $d_x = \theta_x/2$ (rad)	(deg)	Total Chord from PC, c_x (ft)
PC = 4 + 30	0	0	0	0	0
5 + 00	70	70/400	0.08750	5°00'48"	69.91
6 + 00	170	170/400	0.21250	12°10'31"	168.72
7 + 00	270	270/400	0.33750	19°20'14"	264.90
8 + 00	370	370/400	0.46250	26°29'57"	356.95
9 + 00	470	470/400	0.58750	33°39'40"	443.42
PT = 9 + 18.69	488.692	488.692/4004	0.61086	35°00'00"	458.86

Check: The last deflection angle should equal half of Δ.

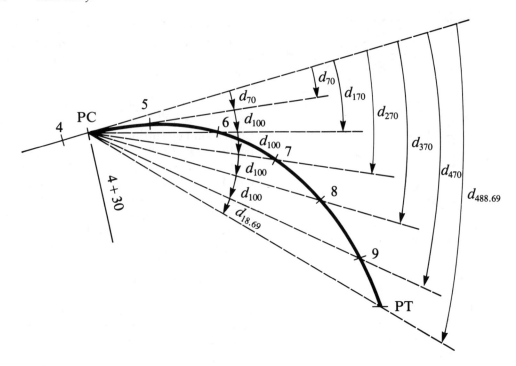

Figure 25-8. Subdeflection angles and total deflection angles.

stakeout techniques. As shown in Figure 25-9, the instrument can be set at any point whose coordinates are known and the desired curve points visible. Coordinates for instrument locations are established previously by control traversing. With the instrument on a point,

say, A, a backsight can be taken to a point of known coordinates, say, B, and the reference azimuth can be calculated as follows:

$$\alpha_{AB} = \tan^{-1} \frac{E_B - E_A}{N_B - N_A} \qquad (25\text{-}24)$$

Table 25-3. Curve layout data by subdeflection angles and subchords

Station	Arc	Total Deflection Angle Subdeflection	Subchord	Computed Data
PC = 4 + 30		0°00′00″		$D = 100/R = 0.25$ rad
	70	5°00′48″	69.91	$d_{100} = \frac{1}{2}D$
5 + 00		5°00′48″		$= 0.12500$
	100	7°09′43″	99.74	$= 7°09′43″$
6 + 00		12°10′31″		$d_{70} = (70/100)d_{100}$
	100	7°09′43″	99.74	$= 5°00′48″$
7 + 00		19°20′14″		$d_{18.69} = (18.69/100)d_{100}$
	100	7°09′43″	99.74	$= 1°20′19″$
8 + 00		26°29′57″		$c_{100} = 2R \sin d_{100}$
	100	7°09′43″	99.74	$= 99.74$
9 + 00		33°39′40″		$c_{70} = 2R \sin d_{70}$
	18.69	1°20′19″	18.69	$= 69.91$
PT = 9 + 18.69		34°59′59″		$c_{18.69} = 2R \sin d_{18.69}$
				$= 18.69$

Check: The last deflection angle equals $\frac{1}{2}\Delta$ within roundoff tolerances.

Coordinates for curve points are readily determined using precomputed deflection angles. Given the PI coordinates (N_{PI}, E_{PI}), azimuth of the incoming tangent α_1, intersection angle Δ, and curve radius R, curve-point coordinates are calculated by first finding coordinates for the PC.

$$N_{PC} = N_{PI} - R \tan \tfrac{1}{2}\Delta \cos \alpha_1 \qquad (25\text{-}25)$$

$$E_{PC} = E_{PI} - R \tan \tfrac{1}{2}\Delta \sin \alpha_1 \qquad (25\text{-}26)$$

Using precomputed total deflection angles, we can calculate the curve-point coordinates by

$$N_x = N_{PC} + |2R \sin d_x| \cos(\alpha_1 + d_x) \quad (25\text{-}27)$$

$$E_x = E_{PC} + |2R \sin d_x| \sin(\alpha_1 + d_x) \quad (25\text{-}28)$$

where x is the total arc from PC to curve point, and d_x the corresponding total deflection angle, positive for clockwise (CW) and negative for counterclockwise (CCW) deflections.

The required layout angle and distance for the curve point are then determined from coordinates by

$$\text{Layout angle} = \tan^{-1}\left(\frac{E_x - E_A}{N_x - N_A}\right) - \alpha_{AB} \quad (25\text{-}29)$$

where the measurement is clockwise from the reference backsight. The horizontal distance is found from coordinates.

$$\text{Layout dist.} = \left[(N_x - N_A)^2 + (E_x - E_A)^2\right]^{1/2}$$
$$(25\text{-}30)$$

25-10. TRANSITION SPIRALS

As vehicles travel circular curves, they exert outward centrifugal force on the roadbed in direct proportion to the degree of curvature present and velocity of travel. Since a vehicle on a straight tangent produces no centrifugal force and a finite amount of curvature abruptly begins at the PC of the circular curve, the required centrifugal force builds up instantaneously, causing a distinct thrust on the roadbed, vehicle, occupants, and cargo. Riders experience discomfort, the cargo shifts, and the vehicle's undercarriage may be damaged.

The effect is particularly serious on fixed-guideway rail systems because the vehicle must follow the route geometry exactly. On highways, the driver anticipates upcoming curvature and begins turning the vehicle slightly

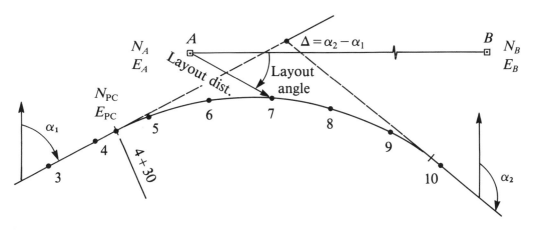

Figure 25-9. Radial layout.

before the PC, as in Figure 25-10. If we assume that the steering wheel is turned at a constant rate, the car's curvature will increase linearly until the full degree of curvature equals that of the circular curve. During this period the vehicle is following a *spiral*, a geometric shape of constantly changing curvature in which the centrifugal force increases linearly over a length of roadbed, thereby reducing the thrust of a sudden increase. But an automobile following a natural spiral deviates from the traffic-lane center and drifts toward the inside, causing a potentially unsafe condition, particularly on multilane highways. Therefore, on highways a spiral is inserted to put the pavement on this natural spiral and permit the

vehicle to maintain a position in the traffic-lane center.

25-11. SPIRAL GEOMETRY

As shown in Figure 25-11, the tangent-spiral juncture is termed TS, spiral to curve point SC. Constructing a local tangent at the SC and producing it back to the tangent give the *spiral* PI. The deflection angle at the PI is Δ_s, the *spiral angle*. To make room for the spiral, the circular curve's PC is shifted inward by the *throw o*. The *spiral's length* L_s is measured along the curve from the TS to the SC. A

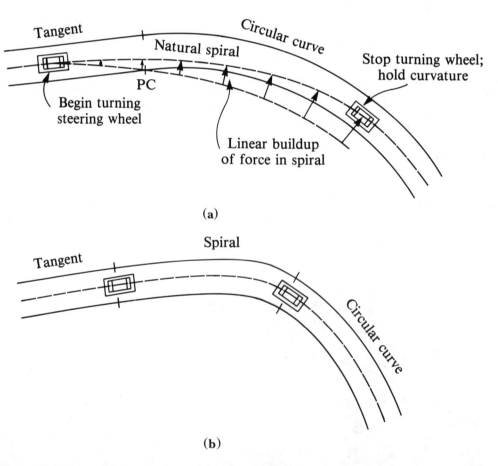

(a)

(b)

Figure 25-10.   Spirals. (a) Nonspiraled curve. (b) Spiraled curve.

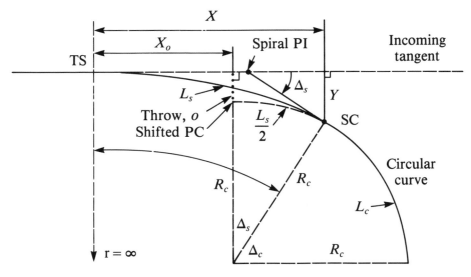

Figure 25-11.   Parts of a spiral.

special rectangular coordinate system is set up in which Y is the *tangent offset to the SC* and X the *distance along the tangent to the offset's foot*.

A route designer usually chooses R_c and a spiral's length L_s. From these, the remaining parts of the spiral can be determined.

The spiral angle Δ_s is calculated based on two geometric properties of the spiral: (1) The length of spiral L_s equals twice the length of circular curve from the shifted PC to SC and (2) the spiral angle Δ_s is duplicated at the circular curve radius point, as shown in Figure 25-11. Therefore,

$$L_s = 2R_c\,\Delta_s \qquad (25\text{-}31)$$

Solving for the spiral angle gives

$$\Delta = \frac{1}{2}\left(\frac{L_s}{R_c}\right) \quad (\Delta_s \text{ in rad}) \qquad (25\text{-}32)$$

The decimal degree value of Δ_s is found by the standard conversion, multiplying by $180/\pi$.

Spiral coordinates X and Y are found by an infinite-series calculation, as follows:

$$X = L_s\left[1 - \frac{\Delta_s^2}{5(2!)} + \frac{\Delta_s^4}{9(4!)} - + \ldots\right] \qquad (25\text{-}33)$$

and

$$Y = L_s\left[\frac{\Delta_s}{3} - \frac{\Delta_s^3}{7(3!)} + \frac{\Delta_s^5}{15(7!)} - + \ldots\right] \qquad (25\text{-}34)$$

where X or Y will have the same units as L_s, and Δ_s is in rad.

The X-coordinate of the throw x_o and the throw itself are calculated by right-triangle trigonometry using triangle *abc* of Figure 25-12, giving

$$x_o = X - R_c \sin \Delta_s \qquad (25\text{-}35)$$

$$o = Y - R_c(1 - \cos \Delta_s) \qquad (25\text{-}36)$$

Example 25-3. A route designer chooses a spiral length of 200 ft and circular curve ra-

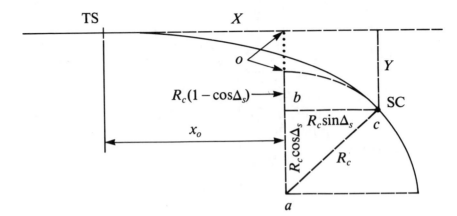

Figure 25-12. Spiral geometry.

dius of 1000 ft. Find spiral quantities Δ_s, X, Y, o, and x_o.

$$\Delta_s = (L_s/R_c)/2 = (200/1000)/2 = 0.10000 \text{ rad}$$

$$= 5.7296°$$

$$X = 200\{1 - 0.1^2/[5(2!)] + 0.1^4/[9(4!)]$$

$$-0.1^6/[13(6!)]\} = 199.80 \text{ ft}$$

$$Y = 200\{0.1/3 - 0.1^3/[7(3!)] + 0.1^5/[11(5!)]$$

$$-0.1^7/[15(7!)]\} = 6.662 \text{ ft}$$

$$o = 6.662 - 1000[1 - \cos(0.1)] = 1.666 \text{ ft}$$

$$x_o = 199.80 - 1000 \sin(0.1) = 99.967 \text{ ft}$$

25-12. INSERTION OF SPIRALS IN AN ALIGNMENT

To fit spirals into an alignment, designers first choose a circular curve radius R_c and an intersection angle Δ, then compute the tangent distance for a simple circular curve $T = R\tan(\Delta/2)$, and a station value for the unshifted PC.

To make room for a spiral of chosen length L_s, the circular curve must be shifted along the angle bisector as shown in Figure 25-13. The component of shift measured along the

tangent is $o/(\tan \frac{1}{2}\Delta)$. This is then added to x_o to determine the distance along the tangent from the unshifted PC to the TS. The unshifted PC station may then be adjusted by this amount to find the TS station

$$\text{TS station} = \text{unshifted PC station}$$

$$-\left(x_o + \frac{o}{\tan \frac{1}{2}\Delta}\right)$$

The SC station is found by adding L_s, as follows:

$$\text{SC station} = \text{TS station} + L_s$$

L_c, the reduced length of the circular curve, is determined by first calculating Δ_c. Since two values of Δ_s added to Δ_c must equal the total intersection angle Δ,

$$\Delta_c = \Delta - 2\Delta_s$$

The length of the circular curve from SC to CS is therefore

$$L_c = R_c \Delta_c \quad (\Delta_c \text{ in rad}) \qquad (25\text{-}37)$$

and

$$\text{CS station} = \text{SC station} + L_c$$

Finally, the ST station is found by incrementing the CS station by L_s.

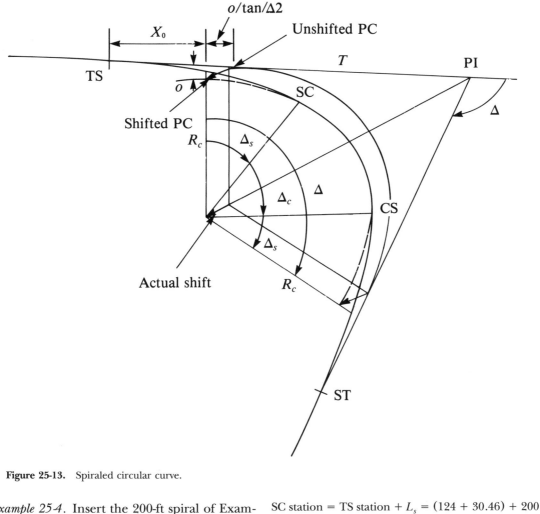

Figure 25-13. Spiraled circular curve.

Example 25-4. Insert the 200-ft spiral of Example 25-3 into an existing horizontal alignment. The unshifted circular curve of radius 1000 ft has a PC station of 125 + 34 and an intersection angle Δ of 50°.

Using previously calculated values for x_o, o, and Δ_s, we obtain

TS station = unshifted PC station

$$-[x_o + o/\tan(\Delta/2)]$$

$$= (125 + 34)$$

$$- (99.967 + 1.666/\tan 25°)$$

$$= 124 + 30.46 \text{ ft}$$

SC station = TS station + L_s = (124 + 30.46) + 200

$$= 126 + 30.46$$

$$\Delta_c = \Delta - 2\,\Delta_s = 50° - 2(5.7296)$$

$$= 38.5408° = 38°32'37''$$

$$= 0.67266 \text{ rad}$$

$$L_c = R_c\,\Delta_c = 1000(0.67266) = 672.66 \text{ ft}$$

CS station = SC station + L_c

$$= (126 + 30.46) + 672.66$$

$$= 133 + 03.12$$

ST station = CS station + L_s = (133 + 03.12) + 200

$$= 135 + 03.12$$

25-13.   LAYOUT OF SPIRALS

The usual method of spiral layout is by deflection angles and chords, with a theodolite at the TS (or ST). Deflection angles are computed by one of three available methods: (1) exact deflection angles by spiral coordinates, (2) approximate deflection angles by geometry, or (3) exact deflection angles by a computed correction to the approximate value.

Chords are either (1) total chords from the TS (or ST) or (2) subchords from station to station. A total station theodolite will be best suited to exact angles by coordinates and total chords from the TS. A more traditional layout using a theodolite and tape will employ methods (2) or (3) for angles and subchords for distances.

Spirals can also be established by a Cartesian coordinate method, whereby the theodo-

lite is located at any point of known coordinates, a backsight is taken to another, and a layout angle and distance are computed for each point to be set.

Regardless of the method, the *local spiral angle* δ must be computed for each desired point on the spiral. In Figure 25-14, consider a point a distance l from the TS. If a local tangent is constructed and extended back to an intersection with the incoming main tangent, the local spiral angle is created. This angle is computed from Δ_s by knowing that *spiral angles are proportional to the square of their respective distances from the TS*. Since Δ_s is the angle for a point L_s from the TS, the relation can be written

$$\delta_l = \Delta_s \left(\frac{l}{L_s}\right)^2 \qquad (25\text{-}38)$$

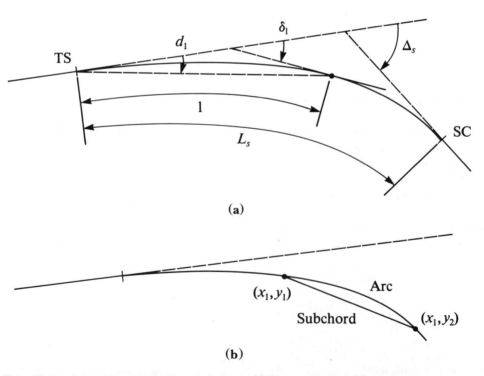

Figure 25-14.   Spiral deflection angles and subchords. (a) Deflection angles. (b) Subchords.

25-14. EXACT DEFLECTION ANGLES AND TOTAL CHORDS BY SPIRAL COORDINATES

From the computed local spiral angle, *local spiral coordinates* x_l and y_l can be found using Equations (25-33) and (25-34), substituting l for L_s and δ_l for Δ_s. The exact *deflection angle* d_l and *total chord* c_l are then established by trigonometry, as follows:

$$d_l = \tan^{-1}\left(\frac{y_l}{x_l}\right) \qquad (25\text{-}39)$$

and

$$c_l = (x_l^2 + y_l^2)^{1/2} \qquad (25\text{-}40)$$

25-15. SPIRAL SUBCHORDS

As shown in Figure 25-14b, exact spiral subchord lengths may be calculated by applying the distance formula to *local spiral coordinates* (x_1, y_1) and (x_2, y_2) for the subchord's endpoints

$$c_{12} = \left[(x_2 - x_1)^2 + (y_2 - y_1)^2\right]^{1/2}$$

Coordinates are first computed using Equations (25-33) and (25-34).

25-16. SPIRAL DEFLECTION ANGLES BY GEOMETRY

From calculus, *the deflection angle is slightly larger than one-third the corresponding local spiral angle*, giving a very useful relation, as follows:

$$d_l = \frac{\delta_l}{3} \quad \text{(approx.)} \qquad (25\text{-}41)$$

For the layout of tight spirals or applications requiring more accuracy, a small correction must be subtracted to get an exact relation

$$d_l = \frac{\delta_l}{3} - C'' \qquad (25\text{-}42)$$

where C'' is a correction expressed in seconds and found by

$$C'' = 0.00309\,\delta^3 + 0.00228\,\delta^{-5} \qquad (25\text{-}43)$$

with δ in decimal degrees. The correction is small and can be neglected on most work. Typical values are shown in the following tabulation:

δ (deg)	C''
5	0.4
10	3.1
15	10.46
20	24.83
25	48.58
30	84.11
35	133.88
40	200.39

C'' is a nonlinear function of δ, increasing at a greater rate for larger δ's.

Example 25-5. For the 200-ft spiral of Example 25-3, compute the approximate and exact angles to lay out each full 50-ft station.

Determine distance l by subtracting the TS station from the desired spiral point to be set. For example, for station $125 + 50$

$$l = (125 + 50) - (124 + 30.46) = 119.54$$

Use Equation (25-38) to calculate the local spiral angle $\delta_{119.54}$

$$\delta_l = \left(\frac{l}{200}\right)^2 \Delta_s = \left(\frac{119.54}{200}\right)^2 (5.7296)$$

$$= 2.04687° = 2°02'48.7''$$

From Equation (25-41), the approximate deflection angle

$$d_l = \frac{\delta_l}{3} = \frac{2.04687}{3} = 0.68229 = 0°40'56.2''$$

and by equation (25-43), the correction is $C'' = 0.03''$. Subtracting for the exact deflection angle, we obtain

$$d_l = d_l \text{ approx.} - C'' = 0°40'56.2'' - 0.03''$$
$$= 0°40'56.2''$$

For other points, calculations are repetitive as shown in Table 25-4.

25-17. LAYOUT OF SPIRALS FROM CURVE SETUPS

At times, the total spiral is not visible from the TS, requiring instrument setups on the spiral to set the remaining points. For example, in Figure 25-15, most of the spiral is obstructed from the TS. In this case, remaining points should be "backed in" from an instrument at the SC.

The SC may have already been set, but if not, it can be located by using a spiral's long and short tangents. Once X and Y have been calculated, the *long tangent LT* from *TS* to spiral PI and *short tangent ST* from spiral PI to SC can easily be found by right-triangle trigonometry, as shown in Figure 25-15, giving the following:

$$LT = X - Y \tan \Delta_s \qquad (25\text{-}44)$$

and

$$ST = \frac{Y}{\tan \Delta_s} \qquad (25\text{-}45)$$

With the instrument on the SC, a backsight to the spiral PI provides proper orientation along the local tangent. At each point on a spiral, a *local circle* can be constructed having a common tangent with the spiral and radius equal to the spiral's at that point. At the TS, radius is infinite and decreases to R_c at the SC. Between these points, the radius decreases in inverse proportion to length from the TS. For a point a distance l from the TS,

$$\frac{r_l}{R_c} = \frac{L_s}{l} \qquad (25\text{-}46)$$

where r_l is the instantaneous spiral radius and hence the radius of the local circle.

Spiral deflections are based on the principle that *a spiral departs from any local circle at the same rate as from the tangent at the TS*. For Example, in Figure 25-16, the offset a_1 at a point 50 ft down the tangent equals the offset from the local circle to the spiral 50 ft from the place where the circle is constructed. Likewise, offsets a_2 and a_3 are equal. The angular size of the offsets is readily calculated by the

Table 25-4. Exact spiral deflections for Example 25-5

Station	l	δ_s	Approximate Deflection	C''	Exact Deflection
TS = 124 + 30.46	0.00	0°00'00''	0.0	0.0	0.0
124 + 50	19.54	0°03'16.9''	0°01'05.6''	Neg.	0°01'05.6''
125 + 00	69.54	0°41'33.6''	0°13'51.2''	Neg.	0°13'51.2''
125 + 50	119.54	2°02'48.7''	0°40'56.2''	0.03	0°40'56.2''
126 + 00	169.54	4°07'02.1''	1°22'20.7''	0.22	1°22'20.5''
SC = 126 + 30.46	200.00	5°43'46.5''	1°54'35.6''	0.58	1°54'34.9''

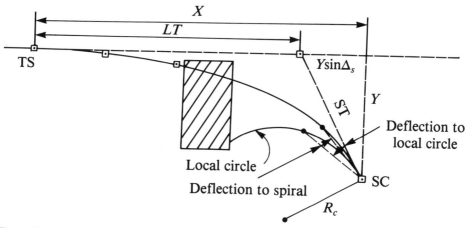

Figure 25-15. Spiral layout from SC.

standard deflection angle relations for a spiral, as follows:

$$d_{50} = \frac{\delta_{50}}{3} - C''$$

$$d_{100} = \frac{\delta_{100}}{3} - C''$$

$$d_{150} = \frac{\delta_{150}}{3} - C'', \text{ etc.}$$

Angular deflections are equal whether measured from a tangent or local circle.

In Figure 25-15, the local circle at the SC has radius R_c. To set a point distance x from the instrument, first turn a deflection angle d_x from the local tangent to the local circle. Then, sight back to the spiral using a deflection angle computed as if the instrument were at the TS. The total angle to turn is the difference between these two deflections, as follows:

$$d_x = d_x(\text{local circle}) - d_x(\text{spiral}) \quad (25\text{-}47)$$

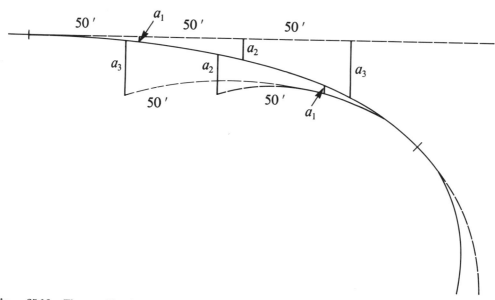

Figure 25-16. Theory of local circles.

25.18. VERTICAL ALIGNMENT

Figure 25-17 has two plotted profiles: (1) the *original ground* along a route's base line and (2) *profile grade* of the proposed base line itself.

After station stakes are set on the proper horizontal alignment, a *profile survey* determines ground elevation at each station stake and distinct breaks in grade between stations. Differential scales are chosen, such as 1 in. = 50 ft for stationing and 1 in. = 5 ft for elevations, giving *vertical exaggeration* to show elevation changes more clearly.

Using the original ground profile, a designer then lays down trial vertical tangents for the proposed base line in an attempt to optimize construction costs versus serviceability of the route in terms of percent grade, earthwork cuts and fills, and sight distances at crests. Vertical tangents meet a PVs, *points of vertical intersection*, with each having a chosen PVI station and elevation when the design is finalized. The plus and minus vertical tangent grades are then computed. At each PVI, a *vertical curve* is chosen to provide gradual transition of the grade as a vehicle travels from one vertical tangent to the next. The main design choice is curve *length L*, the horizontal distance from the *point of vertical curvature* PVC to the *point of vertical tangent* PVT.

25-19. VERTICAL CURVE PROPERTIES

Figure 25-18 indicates four types of vertical parabolic curves: (1) sag, (2) summit, (3) rising, and (4) falling. Grades of vertical tangents are termed g_1 for the incoming grade and g_2 for outgoing, usually expressed in percent. The curve's horizontal projection, length L, is in either ft or stations, but stations will be used in this discussion for computational convenience.

At the PVI, A represents the *total tangent deflection* expressing the *total change in grade between tangents* found by the algebraic expression

$$A = g_2 - g_1 \qquad (25\text{-}48)$$

Positive A indicates an increase in positive grade as a vehicle travels the curve, whereas a negative A denotes an increased negative grade. For positive A's, the curve is above the tangents and vice versa.

Most curves are *equal-tangent vertical curves* consisting of arcs of vertical-axis parabolas. As

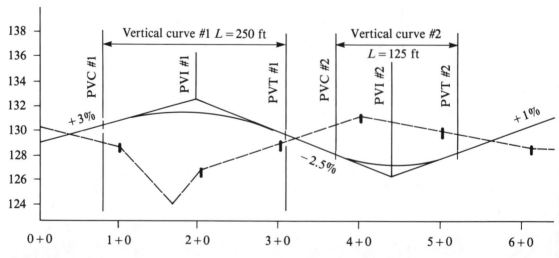

Figure 25-17. Profiles.

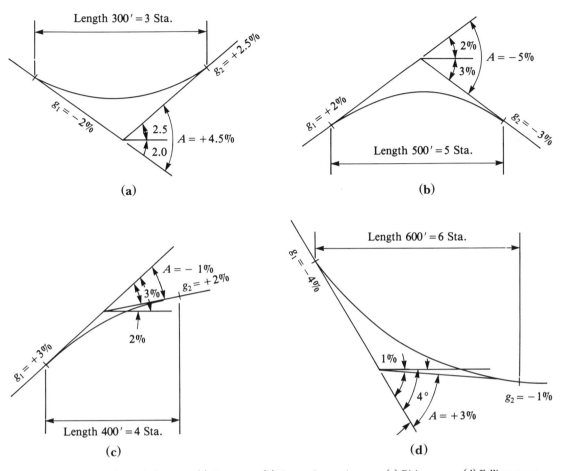

Figure 25-18.   Example vertical curves. (a) Sag curve. (b) Crest of summit curve. (c) Rising curve. (d) Falling curve.

shown in Figure 25-19, the horizontal distance between projections from a and b is the curve's length L, and the PVI's projection splits L into "equal tangents" of $L/2$.

Another important property of parabolas is that *the rate of change of grade is constant.* For example, in Figure 25-20, a vehicle enters a five-station (500-ft) vertical curve at $+2\%$ and leaves at -3%, thereby experiencing a total change of -5%. Since the curve is five stations long, the vehicle changes its grade by -5% in five stations or -1% per station. This *grade change per station*, termed r, is calculated by

$$r = \frac{A}{L} \qquad \text{(25-49)}$$

with A being the curve's total grade change in percent and L in stations.

An additional important property states that a parabolic curve's *external distance E* always equals its *middle ordinate M*. As shown in Figure 25-21, E is the vertical distance from the PVI to the curve and its middle ordinate M terminates on the long chord's midpoint. Since the chord midpoint's elevation is the average between the PVC and PVT elevation, E is calculated from the following:

$$E = \frac{1}{2}\left(\text{PVI elevation} - \frac{\text{PVC elevation} + \text{PVT elevation}}{2} \right) \qquad \text{(25-50)}$$

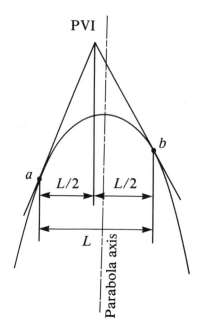

Figure 25-19. Equal-tangent property of curves.

25-20. COMPUTING CURVE ELEVATIONS BY FORMULA

Elevations of points on vertical curves can be calculated directly by algebraic formula. In Figure 25-22, a coordinate system is established with the Y-axis placed through the PVC and X along the elevation datum. Y-coordinates denote elevations and the X values represent horizontal distances from the

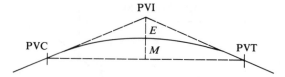

Figure 25-21. External equals middle ordinate.

PVC in a general formula, as follows:

$$Y = \tfrac{1}{2}rX^2 + g_1X + \text{PVC elevation} \quad (25\text{-}51)$$

where Y is the curve point elevation in ft, X the number of stations from the PVC to the curve point, r equals the rate of grade change $= A/L = (g_2 - g_1)/L$, and g_1 is the incoming tangent percent grade.

The second term g_1X is the rise (or fall) of the incoming tangent in ft over the distance spanned by X, and when added to the PVC elevation (third term) gives the tangent elevation above (or below) the curve. The first term $\tfrac{1}{2}r X^2$ represents a vertical offset from the tangent to the curve.

Two alternate sets of units can be used in the calculation: (1) X and L in ft, with grades as decimals or (2) X and L in stations, with grades in percent. The latter is used in this presentation.

Example 25-6. For a vertical curve, five stations in length, PVI station $56 + 25$, PVI elevation 155.00 ft, $g_1 = +2.00$, and $g_2 =$

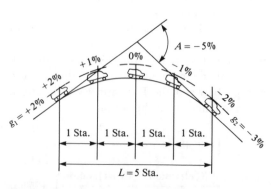

Figure 25-20. Constant rate of grade change.

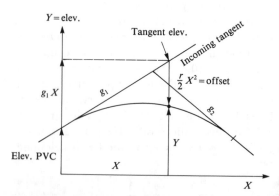

Figure 25-22. Vertical curve coordinate system.

−3.00, compute the curve elevations for each full station on the curve.

Using the equal-tangent property, we can calculate the PVC and PVT stations by decreasing and increasing the PVI station by $\frac{1}{2}L$.

$$\text{PVC station} = \text{PVI station} - \tfrac{1}{2}L$$
$$= 56 + 25 - 250 = 53 + 75$$
$$\text{PVT station} = \text{PVI station} + \tfrac{1}{2}L$$
$$= 56 + 25 + 250 = 58 + 75$$

Elevations of the PVC ad PVT are now calculated, as follows:

$$\text{PVC elev.} = \text{PVI elev.} - g_1(\tfrac{1}{2}L)$$
$$= 155.00 - (2.00)(2.50) = 150.00 \text{ ft}$$
$$\text{PVT elev.} = \text{PVI elev.} + g_2(\tfrac{1}{2}L)$$
$$= 155.00 + (-3.00)(2.50) = 147.50 \text{ ft}$$

The rate of grade change is

$$r = \frac{g_2 - g_1}{L} = (-3.00 - 2.00)/5.00$$
$$= -1.00\% \text{ per station}$$

For a desired curve point, say, station 54 + 00, X is first determined

$$X_{54} = (54 + 00) - (53 + 75) = 0.25 \; stations$$

Then, Y is found by the following formula:

$$Y_{54} = \tfrac{1}{2}(-1.00)(0.25^2) + (+2.00)(0.25) + 150.00$$
$$= 150.469 \text{ ft}$$

The remaining calculations are summarized in Table 25-5.

25-21. CURVE ELEVATION OFFSETS BY PROPORTION

When computations are done with a hand-held calculator, the offsets from a tangent to curve can also be computed by proportion, based on the property of a parabola: *Offsets from a tangent to a parabola are proportional to the squares of the distances from the point of tangency.* Since the offset to the curve's center is E at a distance of $L/2$, other offsets are computed by this proportion

$$\text{Offset}_x = E\left(\frac{X}{\frac{1}{2}L}\right)^2 \qquad (25\text{-}52)$$

For example, the offset for station 54 + 00 in Example 25-6, which the formula gives as −0.031, could also be calculated by first finding the external E

$$E = \frac{1}{2}\left(155.00 - \frac{150.00 + 147.500}{2}\right) = 3.125 \text{ ft}$$

and then using Equation (25-52) for an X of 0.25, as follows:

$$\text{Offset}_{54} = 3.125(0.25/2.50)^2 = 0.031 \text{ ft}$$

Table 25-5. Calculation of curve elevations for Example 25-6

Point Station	X(sta.)	Elevation PVC	$g_2 X$	Tangent Elevation	Offset	Curve Elevation
PVC = 53 + 75	0.00	150.00	0.00	150.000	0.000	150.000
54 + 00	0.25	150.00	0.50	150.500	−0.031	150.469
55 + 00	1.25	150.00	2.50	152.500	−0.781	151.718
56 + 00	2.25	150.00	4.50	154.500	−2.531	151.969
57 + 00	3.25	150.00	6.50	156.500	−5.281	151.219
58 + 00	4.25	150.00	8.50	158.500	−9.031	149.469
PVT = 58 + 75	5.00	150.00	10.00	160.000	−12.500	147.500

25-22. UNEQUAL-TANGENT VERTICAL CURVES

When design requirements indicate that a single equal-tangent curve will not suffice, an unequal-tangent curve may be used. Tilting a full parabola's axis will shift the PVI away from the midpoint between the PVC and PVT, giving unequal "tangents." However, doing so will complicate the mathematics considerably.

A true unequal-tangent parabola can be closely approximated by *compound vertical curves*, whereby two equal-tangent curves are joined at the PVI station at a *point of compound vertical curvature* (PCVC). At that point, r changes from the first vertical curve's value to that of the second curve.

Figure 25-23 shows a compound vertical curve fitted between the same tangents of Example 25-6. Points A and B are the PVIs of equal-tangent curves and located at midpoints of their respective tangents. The PCVC is vertically below the PVI and has an elevation found by proportion between A and B. The grade from A to B is the same as that from the PVC to PVT, which may be easily calculated.

Two equal-tangent curves can now be computed, one between PVC and PCVC, and the other between PCVC and PVT.

25-23. HIGH OR LOW POINT OF A VERTICAL CURVE

Vertical curve high or low points, sometimes termed "turning points" occur when the curve's tangent has zero grade. To calculate the turning-point station, use the constant grade change property of a parabola. A vehicle enters a curve at grade g_1 and must rotate that amount to become level at the turning point. Since it rotates at constant rate $r\%$ per station, dividing r into g_1 gives the number of stations required to level the vehicle, as follows:

$$X_{tp} = \frac{-g_1}{r} \qquad (25\text{-}53)$$

where X_{tp} is the number of stations beyond the PVC to the turning point, g_1 the incoming

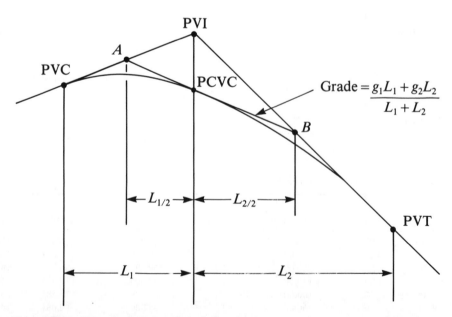

Figure 25-23.   Unequal-tangent vertical curve.

percent grade, and r the rate of grade change in percent per station. The negative sign will make the sign of X_{tp} positive if the proper algebraic values of other terms are used.

Example 25-7. For the previous five-station curve with $g_1 = +2.00\%$ and $g_2 = -3.00\%$, find the station of the turning point.

First, calculate r.

$$r = \frac{A}{L} = [-3.00 - (+2.00)]/5.00$$

$$= -1.00\%/\text{station}$$

Then,

$$X_{tp} = -2.00/-1.00 = 2.00 \text{ stations}$$

X_{tp} is added to the PVC station for a turning-point station of $55 + 75$. Its elevation can be found by another application of the curve elevation formula, Equation (25-51).

25-24. PASSING A VERTICAL CURVE THROUGH A FIXED POINT

In some design situations, a curve with tangents of fixed percent must pass through a point P of fixed station and elevation. To find the needed length L, the proportional offsets property of the parabola presented in Section 25-17 is employed. The property holds whether offsets are measured from the same tangent or different ones as long as horizontal distances are measured from the proper tangent point. Applied to this problem, the relation gives

$$\frac{o_1}{o_2} = \frac{\left(\dfrac{L}{2} - m\right)^2}{\left(\dfrac{L}{2} + m\right)^2} \qquad (25\text{-}54)$$

where o_1 equals the offset from the g_1 tangent, o_2 is from g_2, m is the horizontal distance from the PVI to P, $L/2 - m$ equals the distance from the PVC to P, and $L/2 + m$ is from the PVT to P. Distance m is positive when P is left of the PVI and negative right. The proper algebraic sign will result from the following:

$$m = \text{PVI station} - \text{P station} \qquad (25\text{-}55)$$

Offsets o_1 and o_2 are easily calculated before applying Equation (25-54) by determining the two tangent elevations at P. Subtraction of these from P's elevation will give the offsets, as follows:

$$o_1 = \text{P elevation} - \text{tangent elevation}_1$$

$$o_2 = \text{P elevation} - \text{tangent elevation}_2$$

Solving Equation (25-54) gives the unknown L directly

$$L = 2m \frac{1 + \left(\dfrac{o_1}{o_2}\right)^{1/2}}{1 - \left(\dfrac{o_1}{o_2}\right)^{1/2}} \qquad (25\text{-}56)$$

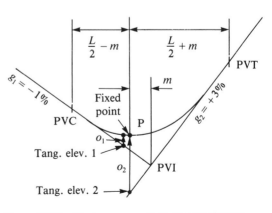

Figure 25-24. Passing a curve through a fixed point.

Example 25-8. In Figure 25-24, incoming and leaving tangents of -1.00% and $+3.00\%$ meet at a PVI of station $150 + 00$ and elevation 531.00 ft. Find the length L of the vertical curve that passes through point P of station $148 + 00$ and elevation 535.00 ft.

Calculations follows these steps

$$m = (150 + 00) - (148 + 00) = +2.00 \text{ stations}$$

$$\text{tangent elevation}_1 = \text{PVI elevation} - g_1 m$$

$$= 533.00 \text{ ft}$$

$$\text{tangent elevation}_2 = \text{PVI elevation} - g_2 m$$

$$= 525.00 \text{ ft}$$

$$o_1 = \text{P elevation} - \text{tangent elevation}_1 = 2.00 \text{ ft}$$

$$o_2 = \text{P elevation} - \text{tangent elevation}_2 = 10.00 \text{ ft}$$

$$\text{offset ratio} = o_1/o_2 = 0.2000$$

$$\text{square root of offset ratio} = 0.2000^{1/2} = 0.4472$$

and

$$L = 2m(1.4472/0.5528) = 10.473 \text{ stations}$$

REFERENCES

KAVANAGH, B. F. 1989. *Surveying with Construction Applications*, 3rd ed. Englewood Cliffs, NJ: Prentice Hall.

MEYER, C. F., and D. W. GIBSON, 1980. Route Surveying and Design, 5th ed. New York: Harper Collins.

26

Hydrographic Surveying

Capt. Donald E. Nortrup

26-1. INTRODUCTION

Hydrographic surveying is the wet equivalent of topographic surveying. Its objective is to delineate the shape of a portion of the earth's surface concealed by water. The surface being mapped cannot be observed directly or occupied, so it is necessary to infer topography from depth measurements. Simply, it is the process of deducing underwater topography from numerous discrete observations of depth at positions throughout the survey area. The quality of its product depends on the accuracy and density of these observations.

Hydrographic survey operations are undertaken to:

1. Provide basic data for nautical charting
2. Obtain site detail for alongshore or offshore construction
3. Assess condition of port and marina facilities
4. Measure quantities in dredging projects
5. Determine extent of siltation and for numerous other reasons

Regardless of the purpose, magnitude, or scale of the survey, the principles remain the same. Identical basic measurements are required and topography is inferred.

As a source of data for nautical charting, hydrographic surveying is generally undertaken by agencies of the world's maritime nations. Standards and procedures for these surveys are codified in manuals in the U.S. National Ocean Survey's (NOS) *Hydrographic Manual*, 4th edition[1]. Although the following discussion draws heavily on nautical charting procedures, treatment is rudimentary and intended as a guide for occasional hydrographic surveyors. Anyone undertaking such surveys on a regular or extensive basis should obtain one of the charting agency manuals.

26-2. PHASES OF HYDROGRAPHIC SURVEYING

The execution of a hydrographic survey can be divided into four components: (1) preliminary office preparations, (2) preliminary field work, (3) sounding operations, and (4) data preparation.

26-2-1. Preliminary Office Preparations

Prospective hydrographic surveyors should gather as much information as possible regarding the survey area. It is essential to obtain descriptions and positions of all horizontal control in the vicinity. Other valuable materials include topographic maps, prior hydrographic surveys, aerial photography, meteorological characteristics, etc. Project specifications will dictate survey scale, minimum sounding density, required positioning frequency, vertical datum, and horizontal control references system.

Survey requirements must be thoroughly examined and understood. The area to be surveyed has to be clearly defined, particularly the inshore limits. It is far more expensive and time-consuming to survey to the shoreline than to some depth contour at which a sounding vessel can safely navigate. If streams enter the survey area, determine the upstream extent of

the project tract. It is beneficial to obtain a relatively large-scale map depicting the entire survey area and mark on it the survey area limits. This will be a valuable reference throughout the planning process.

Field plotting sheets used during sounding operations are prepared at survey scale using a convenient projection or plane coordinate grid. The NOS, prior to the advent of computer-generated projections, utilized the polyconic projection because of its relative ease of manual construction and low distortion. Several plotting sheets will normally be required to depict an entire survey area. The location and coverage of each sheet are determined by developing a sheet layout, as depicted in Figure 26-1. A sufficient overlap between adjacent sheets should be provided to

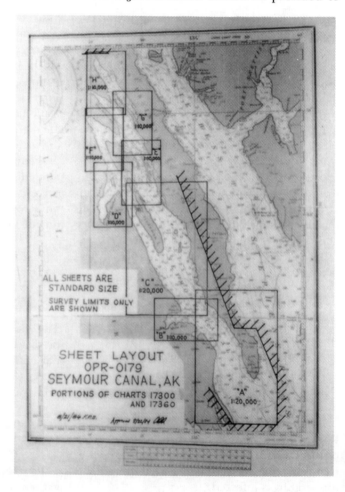

Figure 26-1. Sheet layout on planning base map.

ensure sounding and depth-contour continuity across junctions. Typical survey scales for nautical charting surveys range from 1 : 50,000 and smaller in open coastal waters, to 1 : 10,000 in channels and waterways, to 1 : 5000 and larger in harbors.

Preparation of a comprehensive sounding plan will expedite field work. Typically, sounding operations involve measuring depths along lines traversed by a sounding vessel, each line yielding a bottom profile. The sounding plan consists of a predetermined set of sounding lines. The lines are oriented to cross bottom contours at between 45 and 90° to facilitate definition of bottom topography. Required sounding density is normally specified in terms of line spacing, generally 1 cm at survey scale —e.g., 50-m line spacing at 1 : 5000 scale. Soundings are plotted along lines at intervals of one-half their spacing. The plan can be transferred to the field sheets to guide boat crews.

Positioning afloat is relative and transient. It is relative to the extent that all positioning methods involve observations on, or measurements from, reference points of known location. Positioning is transient in that, once determined, the position cannot be precisely reoccupied, as in land surveying. The selection of a positioning method determines the number and type of shoreside reference points needed to support a survey. A well-prepared positioning plan is critical to the expeditious execution of sounding operations. The plan should include (1) the method of positioning throughout the survey area, (2) identification of existing horizontal control points, (3) sites for establishing additional control points, and (4) an observation scheme to locate them. Positioning afloat is treated as a separate topic later.

As in any survey activity, a logistics plan is necessary. Support for the shoreside portion of a survey—i.e., recovering and extending horizontal control—is similar to that of other land surveys. In addition, a hydrographic survey requires the service and support of a sounding vessel that is (1) large enough to

safely accommodate the needed personnel and equipment including, if possible, space to plot on the field sheet as operations proceed; (2) small enough to maneuver in restricted portions of the survey area; (3) able to provide adequate electrical power to operate the survey instruments; and (4) possessed of seakeeping and endurance capabilities commensurate with the survey area. Sounding vessels range from open skiffs operating around piers and wharfs, to 30-ft launches sounding in channels and bays, to 200-plus-ft ships in offshore ocean operations.

It is necessary to arrange berthing or launching accommodations for the sounding vessel and ensure a fuel supply. To the extent possible, berthing should be secured close to the survey area. Unnecessary transit to the working area wastes both time and fuel.

A preliminary survey area inspection, after initial plans have been developed, but before actual field work begins, will reveal planning weaknesses while the cost of making changes is minimal. During this inspection:

1. Water-level monitoring sites are visited and their suitability evaluated for installation of intended gages. If tidal bench marks exist, they are recovered.

2. Horizontal control marks critical to the positioning plan are recovered to ensure their availability.

3. Sites intended for use as reference points are visited and their suitability evaluated. Visibility into the survey area is checked to ensure that installation will provide the intended coverage. Availability of electrical power, if needed, is assessed.

4. Access to needed sites ashore is obtained from property owners.

5. Logistical support arrangements are made.

26-2-2. Preliminary Field Work

Before beginning sounding operations, it is necessary to establish some shoreside reference points and install the water-level monitoring facility. Magnitude of the horizontal control effort is dependent on the availability and proximity of existing control and the positioning system to be used. For large survey areas, it may be advantageous to concentrate the initial control work and provide sufficient reference points for only a portion of the survey area. This will allow sounding operations to begin while the control work continues.

Recording water-level gages are installed as soon as practicable, providing an opportunity to ensure their proper operation prior to commencing sounding. A backup gage in the survey area ensures against gaps in the water-level record should either gage malfunction. If the survey area is nontidal, water-level monitoring only during period of sounding operations is adequate.

26-2-3. Sounding Operations

Water depth measurements are made as a vessel transits lines designated by the sounding plan. Depending on the positioning and sounding systems in use, the vessel may stop at designated points to observe its location and the depth, or it may proceed along the line at a constant speed obtaining both soundings and positions at predetermined time intervals.

Sounding vessel supplies include (1) depth measurement and positioning equipment, (2) the field plotting sheet, (3) plotting devices, (4) a clock, and (5) the ubiquitous record book (sounding volume). The field sheet should depict the projection graticule, sounding plan, survey limits, all positioning reference points with their identifiers and descriptions, and an electronic positioning lattice, if appropriate.

The simplest sounding plan consists of a set of discrete observation points combined with positioning parameters for each. The sounding vessel simply proceeds from point to point, recording at each the time, measured depth, and position parameters.

Continuous sounding operations begin by bringing the vessel near the starting point of a sounding line. This may require observing and

plotting several "trial" positions until the vessel is properly located on the line. Sounding operations begin when vessel speed is constant. Each sounding line begins with a position observation. The first record book entry reflects the observation time, measured depth, and position parameters. It is helpful if the vessel's speed and heading or some form of line identifier, are also noted. Sounding data are indexed by assigning a "position number" to each observation of position parameters. In addition to recording observations, the vessel's position is plotted on the field sheet and identified by position number (Figure 26-2). If depths are being measured by a recording echo sounder, it is imperative that each position event be marked on the sounder record and labeled by position number and time.

Events are repetitive as the vessel proceeds along the sounding line. As each time-dependent sounding interval elapses, a depth observation is made and recorded along with its time. Positions may be observed for each sounding or at some appropriate multiple of the sounding interval. A position determination is generally required at least every 2 to 4 cm along the plotted line. Each position is assigned a number and recorded and plotted. This process continues until the line is completed. A position is observed at the end of each line, as at the beginning. Course and speed changes may be made as the vessel proceeds along a line. It is imperative that all such changes be noted in the record book along with the time.

The sounding vessel continues from one sounding line to the next in accordance with the prepared plan. A surveyor continuously checks the plot to ensure that the sounding and position intervals meet requirements. On-line adjustments of these intervals are possible, but changing speed may be more efficient.

It is extremely difficult to position and maneuver a vessel onto and along predesignated sounding lines without some form of guidance to assist the ship operator. This can be provided by following temporary ranges, main-

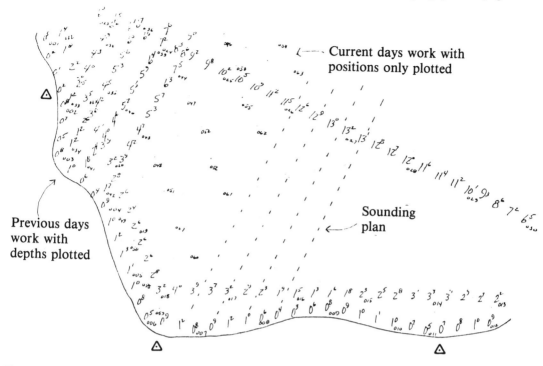

Current days work with positions only plotted

Sounding plan

Previous days work with depths plotted

Figure 26-2. Field plotting sheet.

taining constant range from an electronic ranging station, or by a number of other methods that depend on the positioning method in use. Possible methods of track control are discussed later.

At the end of each sounding day, plotted positions are checked for accuracy, soundings added to the field sheet, and preliminary depth contours constructed. Spacing between successive positions is checked for consistency and reasonableness. For observations made at regular intervals with uniform vessel speed, spacing between plotted positions will be constant. Conversely, any variation in spacing should correspond to a change in vessel speed or time between observations. Errors in observing or plotting result in inconsistent or unreasonable relationship with preceding and succeeding positions. The record book is checked for completeness; the entire plot should be reproducible from the record book alone.

Observed depths are corrected, as nearly as possible, to the vertical datum reference prior to plotting. Depths observed between positions are located along sounding lines by time interpolation between positions.

It is important to construct preliminary depth contours on a day-by-day basis. In this early phase, the contour interval should be sufficiently small to bring all plotted soundings into consideration. Any anomalous appearing contours indicate errors in the sounding data and/or the need for additional field observations.

If additional sounding operations are needed to adequately delineate a portion of the survey area, lines may be added to the sounding plan between the existing ones. This process of "splitting" sounding lines can be repeated until their spacing is sufficiently close to provide confidence in the contours. If a small and/or isolated feature is to be investigated, it may be advantageous to run sounding lines forming rays that intersect at the feature's location. This method concentrates soundings at the point in question.

In the final analysis, the quality of a hydrographic survey is a function of the confidence with which the hydrographer is able to construct depth contours.

26-2-4. Data Presentation

When field work is complete, results of the survey must be communicated to the user in an understandable, comprehensive, and reliable data package. This package will be comprised of four components: (1) a final graphic, (2) sounding records, (3) ancillary records, and (4) a narrative report.

The final graphic depicts survey data in a form that allows the user to visualize underwater topography. It corresponds in both scale and coverage with the plotting sheet used during field operations and depicts essentially the same information. Whereas the field sheet is a working document, the final graphic is "for the record."

Except when data density precludes clear representation, all recorded depth measurements, with final correctors applied, are plotted. When clear representation is impracticable, only critical depths, usually least depths, are included. Contours are constructed to facilitate visualization of topography without cluttering the presentation, thus allowing greater intervals between them than on the field sheet. Although the locations of positioning reference points should be shown, the inclusion of electronic control lattices is not necessary. Shoreline details may be included if available. Depending on the purpose of the survey, other items that might be depicted are (1) floating aids to navigation (buoys); (2) hazards to navigation (rocks, reefs, wrecks, etc.); and (3) bottom material (sand, mud, etc.)

Sounding records include measurements and observations made in the process of actually acquiring depth information. Such data are contained in the survey record books and include depth measurements with associated time of observation and position data, positioning system initialization comparison data and resultant correctors, and any other information that could affect the final depth and/or position information. Sounding records should be so complete that the final graphic could be reproduced by an independent party from the records alone.

Ancillary records are those documents from which portions of the sounding records are derived. Included in this component are analog echo-sounder traces, analog water-level gage traces, horizontal control records supporting reference-point locations, computations of sound velocity and transducer depth correctors, and other supporting materials. All records must be unambiguously cross-referenced, by time or position number, to the sounding records.

A narrative report is prepared for each survey to document the methods, equipment, and qualitative observations. This report describes the following:

1. Sounding vessels used
2. Depth-measuring equipment employed and methods of corrector determination
3. Horizontal control scheme and procedures used to locate reference points
4. Vessel-positioning system, initialization/calibration methods, and corrector computation routine
5. Survey area characteristics that might be of value to a user but are not depicted on the final graphic

26-3. POSITIONING AFLOAT

Sounding vessel position is determined by measurements made from or to reference points (RPs) of known location. Such measurements define lines of position (LOPs) along which the vessel is located. The point of intersection of two or more simultaneously determined LOPs defines the vessel's location.

Measurement of distance to an RP or the angle between two RPs defines a circular LOP. Measurement of bearing to, or azimuth from, an RP or observations of two RPs on range yield straight LOPs. Measurement of the difference between ranges to two RPs produces hyperbolic LOPs.

Dissimilar as well as similar LOPs may be combined to define a position. In the interest of accuracy, LOPs with intersection angles smaller than 30° should be avoided. At small intersection angles, minor errors in measurement cause relatively large displacements in plotted positions.

26-3-1. Positioning by Tagline

Tagline positioning, illustrated in Figure 26-3, has limited application but can be quite expedient when appropriate. It is used almost exclusively in delineating water depths alongside vessel mooring facilities—e.g., piers, wharfs, and bulkheads.

Positioning is accomplished by projecting a LOP, corresponding with a sounding line, into the survey area. A sounding vessel then proceeds along the line measuring depths at predetermined distances from an RP.

This method requires establishing a base line along a structure adjacent to the survey area. RPs are located along the base line at intervals corresponding to the desired line spacing. Distance measurements to the sounding vessel and projection of the sounding line/LOP are made from these RPs. Depending on the survey's purpose, the base line may serve as a stand-alone reference for the survey or be connected to an extended horizontal control network.

Several methods exist for projecting sounding line/LOPs into a survey area. Lines beginning at RPs A and B in Figure 26-3 are projected by providing artificial ranges AL and BM. Each of the RPs requires marking with objects clearly visible over the entire line.

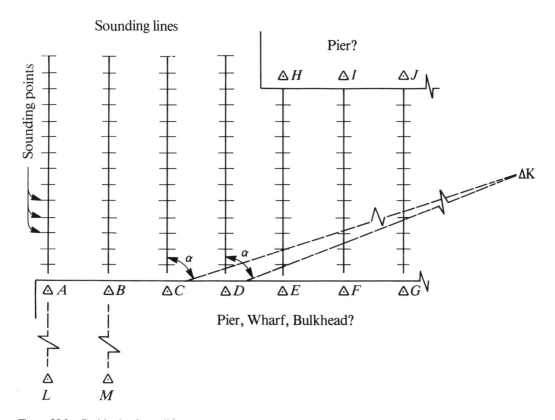

Figure 26-3. Positioning by tagline.

When viewed from a vessel, the objects representing points *A* and *L* will be coincident or exhibit no horizontal displacement if the vessel is on the desired line. When using ranges, it is desirable to have relatively large horizontal displacement between the range markers, preferably equal to or greater than the sounding line length.

Lines beginning at RPs *C* and *D* are projected by occupying the points with an angle-measuring instrument (a sextant is appropriate) and setting off a predetermined angle from a remote RP. The vessel is then guided onto and along the line by an observer. The remote RP should be relatively distant to keep resulting lines virtually parallel.

Lines beginning at RPs *E*, *F*, and *G* are projected by simply stretching a guideline between the end points of the lines. The vessel then moves along it.

Regardless of the line-projecting method, distance measurements are similar. Using a cloth tape or similar device, a vessel measures the distance from the base line as it proceeds along the sounding line.

Record books should be set up so that each position is defined by line parameters and distance from the base line. Because the sounding vessel occupies predesignated points,

the position plot and sounding plan are the same.

26-3-2. Three-Point Sextant Fix

The three-point sextant-fix method of positioning, illustrated in Figure 26-4, is broadly applicable, accurate, reliable, and requires a minimum investment in equipment. The method can be used in almost all waters within visual range (approximately 5 km) of the shoreline.

A circular LOP is defined by measuring the angle between two RPs from a vessel, using a sextant in the horizontal position. Simultaneous observations of two such angles, sharing a common center RP, yield two LOPs whose intersection defines the vessel's position. This positioning method requires a large number of RPs ashore to provide good-fix geometry throughout a survey area. Reference points may be required at as small as 100-m intervals along the shoreline. It is imperative that each RP be located accurately, relative to all the others. An erroneously located RP results in anomalous shifts in plotted positions along sounding lines, as combinations of RPs are shifted from set to set.

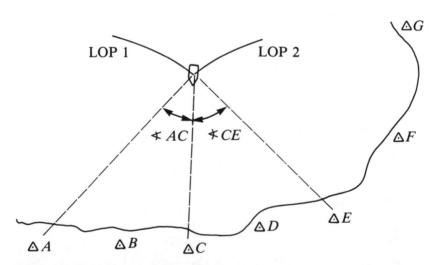

Figure 26-4. Positioning by three-point sextant fix.

To make RPs visible from a vessel, it is necessary to construct signals over them. These may vary from banners of a few square feet of brightly colored cloth to tripods or tepees similarly covered and as high as 16 ft. Existing landmarks can be located and serve as RPs along with intersection stations having known positions.

For each position observed by this method, the record book should reflect the position number, time, identity of RPs, and angles observed (see Figure 26-5). Reference points and angles are always recorded in the same sequence—i.e., if RPs are listed from left to right as viewed from the vessel, then angles are recorded in that order.

The three-point sextant fix is a counterpart to resection in horizontal control surveying. As such, vessel position can be computed analytically. A much more expedient field plotting method is provided by the three-arm protractor (see Figure 26-6). This device is made of transparent plastic with one fixed and two movable arms radiating from the center of a circular disk. Each arm bears a finely etched line along its length, intersecting at the protractor's center. The protractor's arc is scaled left and right of the fixed arm from 0 to 180°, and each movable arm carries a small vernier.

The observed angles of a three-point fix are set, left and right, respectively, on the protractor by positioning the movable arms on the arc. Use of the vernier permits adjustment to within 1' of arc. The protractor is placed on the plotting sheet and adjusted until the fixed-arm etched line passes through the plotted location of the center RP, and the movable arms pass through the left and right RPs, respectively. When properly positioned, the protractor center represents a vessel's position. A hole is provided at the center of the protractor so its position can be marked on a plotting sheet.

Properly executed, the three-point sextant-fix method is capable of producing positions with an accuracy of approximately 1 m per km

Posn #	Time 75° W	Sndg		Corrections			Depth		Position Data		Remarks
		Mtrs		T. D.	Tide	Velo.	Mtrs	¹⁄₁₀			
Date: 5/21/86			Boat: *Whaler A*						Project: *Lake Wobegon*		
0205	081530	1	4	+0.5	-0.6	0.0	1	3	012	60-13	*Begin Line B7*
									046		*Cvs 090°*
	45	3	7	/	0.6	0	3	6	087	48-52	*Spd ¹⁄₂*
	1600	5	4		0.6	0	5	3			
	15	8	2		0.7	0	8	0			
0206	30	10	5		0.7	0.1	10	2	012	68-42	
									046		
	45	15	9		0.7	0.1	15	6	112	57-20	
	1700	16	3		0.7	0.1	16	0			
	15	17	2		0.7	0.1	16	9			

Figure 26-5.   Record book.

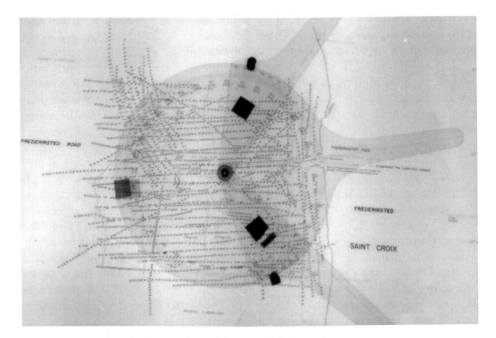

Figure 26-6.   Three-arm protractor.

of range from the RPs. Sextants, like all other instruments, must be properly adjusted as described in standard navigation references. Observers should stand as close together as possible when measuring angles and make observations simultaneously. Reference points should be at approximately the same elevation.

The intersection angle of the LOPs is a function of the RP configuration chosen. Choosing which RPs to observe for each position is a critical element of the method. In the worst case, when the three RPs and vessel lie on a common circle, an indeterminate solution results. Although judicious selection of RPs is basically an acquired skill, a few generalizations are helpful. Good RP configuration results when the following occur.

1. The vessel lies inside a triangle formed by the three RPs.
2. The center RP lies between the vessel and a line drawn between the other two.
3. One angle changes rapidly as the vessel progresses.

4. The distance between the center RP and each of the others exceeds the distance between the vessel and center point.

Except in the case where two RPs form a range, angles smaller than 30° should be avoided.

Control of the vessel track is difficult with this method. It is theoretically possible to maneuver a vessel along a circular LOP by holding one angle constant. A more practical means of track control is to follow the boat compass and timing turns, although the result is guidance rather than control.

The lack of practical track control is one weakness of the method. Among others are its high manpower requirement and vulnerability to human error, the time required to obtain and plot each position, and dependence on visibility.

26-3-3.   Azimuth Intersection

This positioning method, illustrated in Figure 26-7, can be used in areas within visible

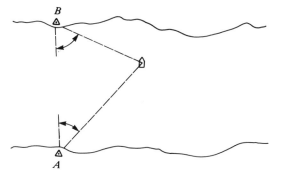

Figure 26-7. Positioning by azimuth intersection.

range of the shoreline. Like the three-point sextant fix, it is reliable, accurate, and likely to require little additional investment in equipment.

Straight LOPs are defined by measuring the angle at an RP between a second RP and the vessel. Two such LOPs intersecting at an angle between 30 and 150° are required to define the vessel's position. Angles are measured using simple theodolites to within 1' of arc.

The method requires additional logistical support to occupy the RPs and reliable, interference-free radio communication between the vessel and observers. Measurements are made simultaneously at a time tick, or mark, radioed from the vessel and observed directions then radioed back to and recorded on the vessel.

Because RPs are occupied, changing them becomes more time-consuming and cumbersome. Pairs of RPs should be carefully picked to provide geometrically strong positioning through the survey area during development of the positioning plan. Fewer RPs will normally be needed than for the sextant-fix method.

For each series of positions obtained from a pair of RPs, the record book should identify the occupied RPs, the RP on which each initials his or her instrument, and the initial plate settings. Initial settings should be checked regularly and recorded.

For each position observed, the record book should reflect the position number, time, RP

identity, and the instrument readings. Records should be set up so there is no doubt in correlating readings and RPs. Observers may record the instrument readings on site to back up the record book.

Azimuth intersection positions can be computed analytically or plotted manually with an azimuth template and straightedge. A relatively large circular template, graduated from 0 to 360°, is laid under the transparent plotting sheet. It is centered at the RP of the observer and oriented so that the template's initial setting is on the line connecting the observer's RP and on which the initial sighting was made. Each LOP is then represented by a straight line from the observer's RP through the instrument-observed direction on the template. It is necessary to repeat the process for the other LOP.

This method suffers the inefficiency of not providing a system of track control. Although its accuracy potential exceeds that of a sextant fix, it is relatively inflexible and requires more logistical support. Both plans demand substantial manpower and are limited by visibility, and quite vulnerable to human error. In choosing one method over the other, the deciding factor may be a difference in the necessary RP location effort versus the logistical support required by the azimuth intersection method.

26-3-4. Range / Range Electronic

Nearly all hydrographic surveys for nautical charting use electronic positioning methods. Most such systems operate in the range-measuring mode illustrated in Figure 26-8. Although expensive, electronic systems provide increased usable range, visibility independent, reduced manpower requirements, and convenient track control.

Circular LOPs are defined by measuring range (distance) between the vessel and RP. Two LOPs, resulting from simultaneous measurements, intersecting at an angle between 30 and 150°, establish a vessel's position.

Ranges can be derived by timing the travel

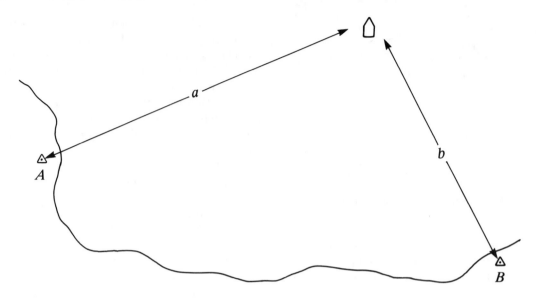

Figure 26-8.   Positioning by electronic ranging.

of electromagnetic pulses from the vessel to the RP and back, then converting time to distance. Some systems are based on phase-comparison principles similar to those employed in geodetic EDMIs. Because phase-comparison systems do not measure total range, it is necessary to initialize them at a point with predetermined system coordinates. These systems indicate range in terms of lanes (equal to one-half wavelength). Only fractions of lanes are actually measured, while the whole lane count, once initialized, is accumulated.

Pulse signal systems operate at superhigh frequencies (SHF), limiting operating range to approximately 15 km. Line-of-sight conditions are required. Phase-comparison systems operate at medium frequencies (MF), with operat-

ing ranges in excess of 150 km. Shallow LOP intersection angles preclude the use of range/range systems along the base line between the RPs and base-line extensions (see Figure 26-9).

Relatively open survey areas may call for as few as two or three RPs to support electronic positioning. Complex or restricted areas can demand numerous RPs and frequent relocation of electronic shoreside equipment. RPs should provide a clear transmission path into the survey area, with no intervening land masses. Since each RP will be occupied by an electronic shore station, accommodation of the station and electrical supply become factors in RP selection. Medium-frequency equipment requires substantial antenna systems,

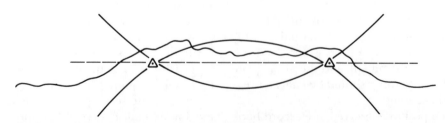

Figure 26-9.   Range/range geometry.

whereas SHF devices suffice with tripod-mounted electronics packages. A shoreside power supply is convenient, but batteries can be used at the cost of increased logistical support.

Additional RPs may be needed to initialize phase-comparison systems. Initialization can be accomplished by intercomparison with sextant fixes, azimuth intersections, or properly operating SHF positions. Once the system is correctly initialized, including the partial lane count, it should be checked regularly to ensure proper maintenance of the whole-lane count. For each position observed, the record book should reflect the position number, time, RP identity, and range (or lane count).

Range/range positions are computed analytically or plotted manually using a range lattice and an interpolating protractor. The range lattice is plotted on the field sheet in advance of survey operations. A lattice consists of concentric circles around each RP at intervals of 10 to 12 cm, representing a fixed number of meters or lanes. An interpolating pro-

tractor is a series of closely spaced concentric circles etched on a transparent medium at intervals corresponding to a fixed number of meters or lanes for the scale of the plotting sheet (see Figure 26-10). The protractor should be at least twice the diameter of the lattice interval. In use, the protractor is laid over the field sheet and maneuvered until the sum of the lattice-ring value and that on the protractor ring coincident with it equals the observed values for both ranges.

For example, to plot the coordinates 270, 130 the protractor is placed to make the 20 ring coincide with the 250 lattice ring, and the 30 ring to coincide with the 100 lattice ring of the other RP. The vessel's position is represented by the center of the protractor. The protractor is used to either add or subtract from lattice values.

Although the potential for human error is much lower than in visual positioning methods, it is higher for systematic errors. Inaccuracies in initializing phase-comparison systems are a common source of such errors. The best,

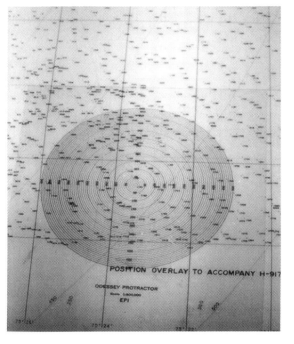

Figure 26-10. Interpolating protractor.

albeit expensive, means to avoid systematic errors is the use of redundant—i.e., three or more—LOPs at each position.

26-3-5. Range/Azimuth

This widely used method is applicable within visible range of the shoreline and particularly useful in complex areas, where sight distances are limited and full electronic positioning net deployment is impractical. It is also frequently used in those areas along the base line of range/range systems, where their accuracy is degraded.

The method combines one circular LOP from an electronic ranging system, with an azimuth observed on shore (Figure 26-11). The observer and electronic shore station are typically colocated at the same RP, which has the salutary effect of producing LOPs that always intersect at 90°.

Since the RP is occupied, this method requires additional logistical support to get equipment and personnel to the site. Changes from one RP to another involve relocation. Since one observation is made on the vessel (range) and the other (azimuth) ashore, reliable communications are mandatory.

Reference-point requirements are relatively light. The entire survey area must be visible from at least one RP, and each RP must be intervisible with a minimum of one other.

For each series of positions from an RP, the record book should identify the RP of the observer/range station, the RP on which the instrument is initialed, and the initial plate setting. This setting must be checked and recorded regularly. Each position record should include the position number, time, range, and observed direction.

Plotting positions can be accomplished by preplotting range rings around the RP and intersecting them with azimuths, using a template as previously described. Interpolation of range is made using a metric scale. Among the strengths of this method are excellent track control and consistently good intersection geometry.

26-4. DEPTH MEASUREMENT

Measurements of depth at known positions provide data points from which topography is inferred by construction of depth contours. They are made either directly by a sounding

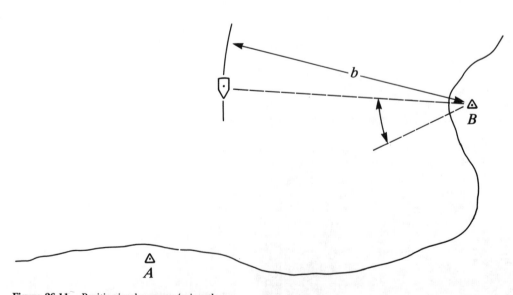

Figure 26-11. Positioning by range/azimuth.

pole or leadline or indirectly by an acoustic echo sounder.

26-4-1. Direct Measurement

A sounding pole is analogous to a ruler since it is a wooden staff, approximately 4 cm in diameter and 5 m long, graduated and marked in units of length. The pole ends are metal-capped to compensate for buoyancy; plates may be attached to them to prevent penetration into soft bottom material. Increments of length are marked by notches cut in the pole, and a distinctive color scheme is used to facilitate reading.

In use, the pole is simply lowered vertically into the water until it touches bottom and is read at the water's surface. Although crude, the sounding pole is a cheap, unambiguous measuring device for depths up to about 4 m.

Like a cloth tape, the lead line is a stretch-resistant cord, graduated and marked in units of length, with an attached lead weight. A line with a corrosion-resistant and flexible wire (phosphor-bronze) core is preferred, although other materials may perform acceptably. Lead weights vary from 5 to 15 lb.

Length increments are marked by attaching distinctively colored bunting and/or shaped leather strips to the line. The line is graduated and marked by comparison with a steel tape while thoroughly soaked and under tension equivalent to the lead weight. Daily comparisons with a steel tape should be made.

For other than a wire-core line, more frequent comparison with a tape or other standard is required. If stretching is detected, a table of correctors must be developed and applied

In use, the lead line is lowered into the water until the weight touches bottom, and a reading is taken at the water's surface. Care must be exercised to ensure that the line is vertical when it is read.

26-4-2. Indirect Measurement

Depth measurement using an acoustic echo sounder is accomplished by transmitting an acoustic pulse vertically into the water column and measuring the elapsed time until receipt of the echo reflected by the bottom. Elapsed time is converted to depth, based on the speed of sound in the water.

Echo sounders register depth data either digitally as an analog record or by an instantaneous "flash" against a graduated circular depth scale. Flasher instruments are the least expensive but require manual reading at each sounding point and do not provide a permanent record. The analog recording instruments produce a graphic profile along each sounding line from which representative and critical depths can be selected for plotting. Digital output may be provided to supplement a recording instrument.

All recording instruments provide an operator the facility to superpose an "event" mark on the record. When conducting survey operations, it is imperative that the vessel's position be correlated in time with the sounding record. This is accomplished by marking each position determination as an event on the sounding record and labeling it with the position number.

All echo sounders are inherently ambiguous. They do not measure depth to a point vertically beneath the vessel, except coincidentally. The combination of signal frequency and transducer geometry determines the shape and size of the acoustic wave form through the water. The resultant depth measurement is the shortest travel path to the bottom and back within the wave form.

If the bottom is anything other than perfectly flat, the shortest distance is unlikely to be a vertical path. Further, any vessel motion deflects the wave-form axis from the vertical, thus amplifying ambiguity. It should be noted that errors introduced by these factors nearly always result in echo-sounding measured depths slightly smaller than vertical depths.

Depths measured by echo sounder must be corrected for transducer depth, and any difference between the assumed and actual speed of sound in water, to obtain actual depth. An echo sounder measures the distance between the transducer and bottom. Because transducers are mounted well below the water surface, the distance between the water surface and transducer face must be added to each depth measurement. It may be possible to physically measure transducer depth on small vessels while they are out of the water.

All echo sounders convert signal travel time to depth, based on a design (sometimes adjustable) speed of sound in water. Any difference between the design and actual speed introduces measurement error. The speed of sound in water is a function of temperature, density, and salinity, and varies vertically within the water column, horizontally with location, and temporally. It is possible to periodically measure water column parameters, compute actual sound velocities, and calculate depth correctors.

Transducer depth and velocity correctors can be determined collectively by direct comparison of an echo sounder to a bar-check or vertical (lead-line) cast. This method is valid only when direct comparisons can be made to depths at least 75% of the deepest survey depth measurements.

A bar check is a flat, acoustically reflective, negatively buoyant surface equal in length to the sounding vessel's beam. A stretch-resistant line graduated and marked in the sounding units of the echo sounder is attached to each end. The bar check is lowered beneath the vessel (Figure 26-12) to a specific depth and maneuvered directly beneath the transducer. An echo sounder will indicate the distance between the transducer and bar. With the bar relatively near the transducer, the difference between bar depth and indicated depth is the transducer corrector. As the bar is lowered and comparisons are made at deeper depths, changes in the difference results from sound-velocity variation. Comparative readings are made as the bar is lowered and again while being raised. In ideal conditions, bar-check comparisons can be made to depths approaching 30 m.

Depth corrector values are based on averages of repeated comparisons. Typically, bar checks are conducted in the survey area at the beginning and end of each day's sounding operations. If it becomes apparent that velocity correctors are changing, either with time or operating area, it is necessary to block the comparative data by area or time when averaging and applying correctors.

Plotting comparison differences as a function of indicated depth allows the selection of a range of depths to which each increment of correction is to be applied—e.g., +0.6 m between 27.6 and 33.3 m.

Acoustic echo sounders record echoes from any reflective surface, not just the bottom. The most common and bothersome of these "false echoes" are reflections from kelp or weed beds. Higher-frequency instruments (100 + KHz) are particularly susceptible to false echo indications. Any doubt about the legitimacy of an echo can be resolved by performing a lead-line comparison.

26-5. WATER-LEVEL MONITORING

Depths measured from a floating vessel are necessarily relative to the water surface at the time of measurement. Water surface elevation, or water level, is not a stable reference datum but varies from minute to minute in tidal areas and from season to season in nontidal bodies of water. To accurately infer the bottom shape from discrete measurements of depth, it is essential that survey depths be relative to a common and stable reference surface—i.e., a vertical datum. Measured depths are converted to survey depths by applying correctors equal to the difference between the water level and vertical datum at the time of each measurement.

Some form of mean water level is generally used as the vertical datum. Mean low-tide level might be appropriate in a tidal area, while the average water level observed during the survey may be best for a manmade lake. In either case, it is necessary to observe water levels over a period of time in order to define the datum.

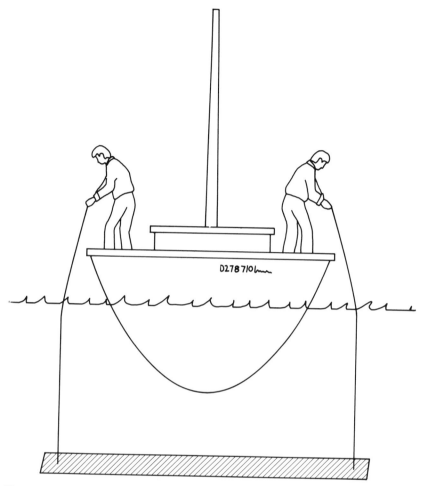

D2787/0

Figure 26-12. Bar check.

Once established, water-level variations relative to the datum can be quantified and depth correctors computed.

The simplest water-level monitoring device is a wooden staff, graduated and marked in units of length. When used as a stand-alone device, the staff should be sufficiently long and installed so that it extends over the entire range of anticipated water levels. Whenever possible, the staff should be affixed to a permanent structure or feature in the survey area. It must be free of any vertical movement (see Figure 26-13).

Bench marks (BMs) are an integral part of any water-level monitoring installation. A level

circuit run from the staff to the BMs determines their elevation relative to staff zero. Bench marks serve multiple purposes—e.g., (1) provide a check on the staff's vertical stability, (2) supply reference elevations should it become necessary to replace a staff, and (3) monument the datum level once it is defined. Levels should be rerun if any staff movement is suspected and also at the survey's conclusion.

A recording water-level gage is commonly installed at a monitoring station to eliminate the need for an observer and provide a continuous record. The gage zero can be correlated to the staff, and ultimately to the BMs, by

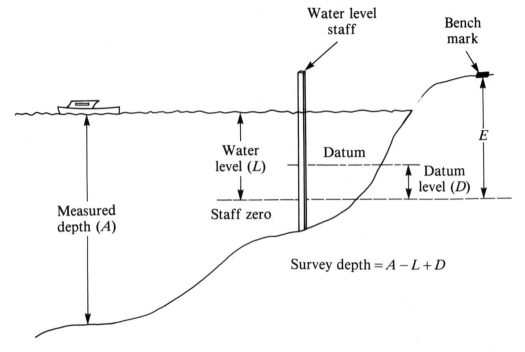

Figure 26-13.   Water-level monitoring system.

taking simultaneous readings of the gage and staff. Such comparative readings should be made regularly, preferably daily, throughout the survey period to ensure that the gate is functioning properly and the gage zero remains constant.

Water-level observations, whether read by an observer or recorded, must be accurately time-correlated. Time is the index by which water-level correctors are tied to depth measurements.

Because the vertical datum often is not defined until a survey is completed, it may be impossible to make actual water-level corrections as the survey progresses. Preliminary corrections can be derived by assuming a datum —e.g., staff or gage zero—and comparing water-level observations to it. This provides corrected depths relative to a common and stable reference surface, although not the desired one. After the datum has been defined, correctors are recomputed and applied to the final survey data.

26-6. STATE OF THE ART

The methods and procedures addressed in this chapter are conceptually valid, and a small survey could indeed be carried out by applying them. They are primitive, however, and a surveyor desiring to enter hydrographic surveying on a regular and extensive basis should be aware of the technology and costs of performing competitively in today's market.

Computers and plotters pervade the process from the preparation of plotting sheets through the real-time acquisition of data to the production of the final graphic. They serve as principal components of automated acquisition systems and elements of sophisticated positioning and depth-measuring systems.

Although conventional ground survey methods remain indispensable in the effort to locate reference points ashore, aerotriangulation and satellite positioning systems are regularly applied to support hydrographic surveying. The total station concept has been applied to short-range positioning afloat (range/azimuth) and refined to the extent that autotracking, autoreading, autotransmitting devices are available. Electronic positioning systems provide multiple LOPs and least-squares fitting. The advent of the global positioning system promises to revolutionize positioning afloat, even at survey accuracies.

Acoustic echo sounders vary from high-frequency, narrow beam to multiple/selectable frequency to beam-forming swath sounding systems. Airborne sounding systems using a laser-light medium have been developed, and an electromagnetic medium device is being tested.

Regardless of all the technology, the basic measurements remain the same: depth and position.

NOTES

1. NOS. *Hydrographic Manual*, 4th ed. (1976) 1981. Washington DC.

REFERENCES

HYDROGRAPHIC SURVEYING, 3rd ed., 1992, Ingham, A. and Abbott, V., Blackwell Scientific, Cambridge, MA 02142.

27

Boundary Surveys

Donald A. Wilson

27-1. INTRODUCTION

Webster's Dictionary defines boundary as any line or thing making a limit, whereas *Corpus Juris Secundum* states that a boundary is a line or object indicating the limit or furthest extent of a tract of land or territory; a separating or dividing line between counties, states, districts of territory, or tracts of land.[1]

The term *boundary surveying* is used synonymously with other terms, such as *land surveying* and *property surveying*. Basically, each one is attempting to identify the same type of surveying, i.e., surveying of property ownership, or rights.

Ownership and rights in land cannot only be defined, but since land itself is a physical entity, it can be measured, marked, and described in words and on drawings. The definition of land ownership of rights gives rise to boundaries, and it is these boundaries with which the land surveyor is concerned and licensed to deal.

27-2. OWNERSHIP, RIGHTS AND INTERESTS IN LAND

By law, an *estate* is the interest a person has in real or personal property. In general, real estates are classified according to the time of enjoyment, as (1) estates in fee, usually termed *fee simple*, (2) life estate, (3) estate for years, and (4) estate at will. It is important to understand what interest, or rights, a person or entity has in a parcel of land, since it governs what they may do with that interest. A person may only transfer what they own or can only be affected according to that interest.

Understanding the nature of estates becomes extremely critical when several persons or entities are involved, each with different interests.

When we refer to land ownership, it is generally meant that an estate in fee simple is involved. There are numerous situations where the ownership is complicated, with, or by, other interests, such as easements, encumbrances, encroachments, agreements, and the like. They must all be identified before their boundaries can be determined and ultimately surveyed.

27-3. TRANSFER OF TITLE

Title to property is the means whereby the owner has the just possession of the property, or the means whereby a person's right to property is established.[2] Whatever title a person has in property may be conveyed, acquired, transferred, or lost. The means by which this is accomplished may be categorized as follows:

1. Title by public grant (e.g., a patent from the United States)
2. Title by private grant (such as a deed)
3. Title by will (from the decedent)
4. Title by descent (intestate succession)
5. Title by involuntary alienation (bankruptcy or foreclosure)
6. Title by adverse possession or unwritten agreement (prescription in case of easements)
7. Title by eminent domain (public taking with compensation)
8. Title by escheat (property reverting to state)
9. Title by dedication (e.g., dedicating property to public use)
10. Title with the element of estoppel entering (reliance on former acts or assertions)
11. Title by accretion (land built up by the action or water)
12. Title by parol gift (when followed by adverse possession or appropriate action of the parties)
13. Title by operation of law (statutory)

It is essential to determine which of these affect the parcel(s) in question, since they will dictate what records must be examined and what other evidence may be important or available.

27-4. BOUNDARY SURVEYING

Although boundary surveying usually denotes a survey of boundaries of land ownership, it can also mean other things, since there are other kinds of boundaries. There may be boundaries defining, or delineating:

1. Rights other than strict ownership—e.g., a restricted area, a common area, or an easement
2. An area of encroachment by one person on the record owner, or title holder
3. An area of land use
4. Zoning
5. An administrative district, such as for a school
6. An area of jurisdiction, such as a wetland
7. A political or government entity, such as national, state, county or municipal lines, or urban areas

In order to survey, or measure, a boundary, it must first be identified. When it comes to land ownership, it is a matter of law as to what the boundary is. It may be a deed line, line of occupation, grant line, line of agreement, line set by the court, or some other line based on the law or by the action of the parties. Once identified, the line may be measured, or surveyed, computed, and shown on a drawing.

In surveying, the boundary line must be found, or located, on, under, or over the surface of the earth, depending on what it is and where it is. Its location is based on *evidence*.

The land surveyor assembles the available evidence, then by applying rules of law and evidence, and through experience, reaches a conclusion as to what is the best, or most reliable, evidence from which to define the property line, or the lines that define the parcel in question. Goals sought in a land survey may be, or include:

1. Where a parcel is located on the surface of the earth, or elsewhere
2. Location of one or more lines
3. Road frontage
4. Acreage
5. Set back lines
6. Encroachments
7. Subdivision into smaller units
8. Peripheral, such as easements, access, or other attached, or inherent, rights

27-5. EVIDENCE

Property boundaries are determined through the interpretation and use of *evidence*. Evidence is distinguished from proof in that the former is the medium or means by which a fact is proved or disproved, whereas the latter is the result of effect of evidence.[3] Evidence is not proof; evidence leads to proof.

Evidence may be found in three basic forms, any or all of which may have an effect on the survey, or identification of the boundaries and corners of the parcel being surveyed:

1. *Documentary.* Also called *record evidence*, it comprises all those writings and documents that relate to land, title, or boundaries: deeds, mortgages, wills, maps and plats, photographs, and the like.

2. *Physical.* Also called *real evidence*, this category includes those objects, natural and artificial, that mark or indicate the location of boundary lines and corners. Natural objects often called for are trees, waterbodies, stones, ridge tops, and the like, while artificial objects frequently encountered are stone and concrete bounds, iron pins and pipes, wood stakes, buildings, and roads. These are termed *monuments* when called for in title documents.

3. *Parol.* Sometimes called *testimonial evidence*, it consists of what persons knowledgeable about the title or location of a tract of land have to say about it. Their explanations or descriptions may lead to other information or explain inconsistencies or ambiguities and aid in the resolution of conflicts between other forms of evidence.

Some type of evidence is used in the determination and location of property boundaries. Those lines and corners established few to many years ago have been perpetuated by the evidence left behind or created subsequent to the origin of the tract or title. Secondary evidence is inferior to primary or original evidence, but will suffice when determined to be the best available evidence remaining. It is the surveyor's responsibility to seek out all the immediate evidence, analyze it, and reach a conclusion or determination on the best evidence available.

27-5-1. Collection of Evidence

Collecting evidence is an art as well as a science. Determining how much research is necessary, how much field search is required, recognizing when additional experts are needed, all come with experience and are governed by the nature of the problem.

It is the preponderance of evidence that indicates the location of property, property lines, and property corners, as well as many other features relating to the property. The surveyor collects evidence bearing on property location to do the survey. All available evidence should be examined and evaluated, or weighed, to reach a proper conclusion. This part of the survey process is the detective work and demands skill as well as patience in order to insure a complete job. Once the evidence is known, measurements may be taken for verification and to obtain data for drawing plats and maps.

Conclusions based on scant or inadequate evidence frequently are incorrect or faulty. As A. C. Mulford wrote in 1912:

> It is far more important to have a somewhat faulty measurement of the spot where the line truly exists than to have an extremely accurate measurement of the place where the line does not exist at all.[4]

There often is a question of how much research should be undertaken. The only proper answer to the question is, "enough to do the job." Budget should not be a controlling factor, but sometimes is, and time constraints often dictate limits. Both of these factors should be eliminated, or at least controlled, whenever possible.

Deeds, or other source(s) of title, should be traced to the origin of description , whenever possible. Without knowing the date a parcel was created, the conditions and circumstances at the time cannot be evaluated, so it would be impossible to consider the intent of the parties at the time the description originated. Senior-junior rights cannot be determined without knowing the order of conveyancing and corrections, for the change in magnetic declination cannot properly be made without knowledge of the date when the observations were made. A 50- or 100-yr difference could be very significant, particularly when dealing with a long line. And finally, as descriptions are repeated over time in subsequent documents, numerous scrivener's errors arise, and sometimes critical information, such as the recitation of easements and other rights affecting the parcel, are excluded.

Abutting parcels must be examined as well, for the boundary lines are common ones, and frequently additional information, sometimes even a survey, will appear in an abutting chain of title that would otherwise be overlooked.[5]

Anything referenced in documents should be examined. If referenced documents are not a matter of public record, a search must be made elsewhere.

Comprehensive field investigation must also be undertaken, to (1) find all features mentioned in the documents, if still remaining, (2) discover any features that affect the property but are not mentioned in any documents, such as property corners set by landowners or subsequent survey(s), (3) identify any encroachments affecting properties or property lines, and (4) discover other features affecting the title, such as wells, burial grounds, prescriptive easements, other interests not of record, and the like.

27-5-2. Analysis and Evaluation of Evidence

This stage of the survey is an extremely critical one. Evidence collected or discovered must be carefully evaluated to insure its applicability and correctness. Courts consider the admisibility and relevancy of evidence before weighing it for its influence in leading to a conclusion. Some of the considerations in a typical survey are as follows:

1. *Roads*. Is the road in existence the same as the one called for in the description; is it in its original location or has it been altered?

2. *Trees*. Is the tree being considered the correct species, and is it old enough to be the one called for in the *original* description?

3. *Fences*. Is the fence found of the correct type and style, and how long has it been there? Very little wire existed prior to the 1880s when most barbed wire types were invented.

4. *Monuments*. Is the one found the original monument; has it been moved?

5. *Survey*. Is the survey found correct, is it complete, how was it performed, and on what is it based?

6. *Distances*. What units and what basis (type of chain, measurement technique, corrections, etc.)?

7. *Directions*. How observed, to what degree of accuracy, true or magnetic?

8. *Area*. How derived, what units?

The procedure of analyzing the evidence may be a highly complex one. There are numerous rules of survey and of law that must be considered to resolve the conflicts which are invariably found. Exercising judgment and applying the appropriate rules are what make the surveyor's job less than routine. Frequently encountered problems may be categorized as follows:

1. Conflicts in title documents
2. Conflicts between title documents
3. Conflicts between title documents and physical evidence
4. Conflicts between items of physical evidence
5. Conflicts between surveyors' opinions/ interpretations
6. Conflicts in law[6]

    (a) Between decisions

    (b) Between decision and litigant

27-5-3. Perpetuation of Evidence

After the evidence of property location and its lines and corners have been determined, it must be perpetuated. Missing corners are set, deteriorating corners renewed, and fading marks rejuvenated. The record must be made current and information may be perpetuated through survey plats, land descriptions, sketches, formal and informal reports, a letter to the client or the client's agent, and notes or memoranda to the file. Public record affords some protection for the work, but there are times when alternative means will suffice, or are even superior.

Without some report of the survey, the work becomes increasingly less valuable as marks wane, monuments disappear or are moved, memories fade, and property changes in ownership. Without usable and available information, encroachments may occur, and boundary disputes are more likely. If this can be avoided, or minimized, the surveying profession has provided a service to the public.

27-6. PRESUMPTIONS

A *presumption* is a statement, sometimes of fact, sometimes of law, and sometimes mixed, that can be considered as true without further proof. Presumptions may be classified as *rebuttable* or *irrebuttable*. A rebuttable presumption is one assumed true until proven otherwise. Most rules of law that control the location or other aspects of real property are of this type. An irrebuttable presumption is one that is conclusive and absolute. There are few presumptions that fall into this category and most authorities consider them to be substantive rules of law and not rules of evidence.

An *inference* is not the same as a presumption. It is a deduction of fact that may logically follow, or be drawn, from another fact or set of facts.

Presumptions are not evidence; they are substitutes for evidence. There are many presumptions that apply to the survey or property and the interpretation of evidence, particularly the interpretation of documents and descriptions. Examples of presumptions are ownership to the centerline of a street, and knowing that a survey is performed correctly, the lines were actually run, measurements were made horizontally, and a survey is referenced to the magnetic meridian. All these presumptions may be rebutted by proof to the contrary.

27-7. JUDICIAL NOTICE

Because some facts material to litigation are matters of common knowledge, they need not be proved through the presentation of evidence. Judicial notice is a mechanism whereby formal proof may be dispensed with because the fact is known to the court as being a matter of general, or common, knowledge.[7]

Judicial notice may be defined as the cognizance of certain facts that a judge or jury may properly take or act on without proof because they are already known to them, or because of that knowledge which the judge or jury is assumed to have.

Judicial notice is not the same as judicial knowledge, although they are used interchangeably. The former refers to things that are commonly known, whereas the latter refers to things that courts are deemed to know by virtue of their office. In a broad sense, the term *judicial notice* is used to denote both judicial knowledge (which courts possess) and common knowledge (which every individual possesses).[8]

Courts may be requested to take judicial notice of facts in many areas. Matters that are commonly judicially noticed are subjects from encyclopedias and textbooks, matters of law, matters of survey, custom and usages, geography and geographical facts, language, meanings of words and phrases, laws of nature and weights and measures, among many other categories. Courts may also take judicial notice on their own, of matters of common knowledge, at their discretion.

27-8. RULES OF CONSTRUCTION

Every deed, otherwise valid, will be considered to have intended to convey an estate of some nature.[9] Therefore, every attempt should be made to uphold the deed whenever possible.

When descriptions set forth in deeds are not ambiguous, they must be followed. When, and only when, the meaning of a deed is not clear, or is ambiguous or uncertain, will a court of law or equity resort to established rules of construction to aid in the ascertainment of the grantors' intention by artificial means whereby such intention cannot otherwise be ascertained. Unlike a settled rule of property that has become a rule of law,[10] rules of construction are subordinate and always yield to the intention of the parties, particularly the intention of the grantor,[11] whereby such intention can be ascertained.

The rules for construction of deeds are essentially those applicable to other written instruments and contracts generally. Although, as a general consideration, the same broad rules govern the construction of both deeds and wills, in some jurisdictions deeds are more strictly construed than wills.[12]

When the intention of the parties is uncertain, resort must be had to well-settled but subordinate rules of construction to be treated as such and not as rules of positive law.[13] All rules of construction are but aids in arriving at the grantor's intention and they may be applied only when the application of the rule with respect to the intent of the parties does not banish all doubt concerning the conclusions to be drawn from the language of the conveyance and the circumstances attending its formulation.

27-8-1. In General

Words contained within a deed are presumed to have a purpose.[14] And documents are to be viewed in light of the surrounding circumstances.[15] This includes the conditions at the time,[16] as well as the law in existence at the time of the conveyance.[17] In interpreting a document, the court will place itself as nearly as possible in the situation of the parties at the time of the conveyance.[18]

Documents are to be construed according to their plain terms[19] and words given ordinary meaning.[20]

27-8-2. Sufficiency of Description

A deed will not be avoided because some particulars of the description of the premises are false or inconsistent, so long as it is sufficient to identify the premises.[21] However, if the description is so imperfect that it is impossible to know what land was intended to be conveyed, a deed may be void.[22] The basic rule is, *if a description is certain enough to enable a person to locate the land, it is sufficient.*[23]

A description is adequate so long as it allows the property to be located, even if an actual survey is required in order to do so.[24] Courts do not want to defeat a description, or render an instrument void, if it is at all possible to sustain it. Courts have consistently ruled on the sufficiency of a description when lands can be located by a surveyor[25] or extrinsic evidence.[26]

Generally, a deed will not be declared void for uncertainty in a description of land conveyed, if it is possible, by any reasonable rules of construction, to ascertain from the description, aided by extrinsic evidence, what property is intended to be conveyed.[27] Under the maxim that that is certain which can be made certain,[28] courts properly lean against striking down a deed for uncertainty and generally will adopt liberal rules of construction to uphold a conveyance.[29] *Descriptions are not to identify land, but to furnish the means of identification.*[30]

Only when it remains a matter of conjecture what realty was intended to be conveyed by the deed, after resorting to such extrinsic evidence as is admissible, will the deed be held void for uncertainty of description of realty.[31] A deed, devise, or a reservation of real estate, the description of which is impossible of ascertainment, is void.[32]

27-8-3. Intention of the Parties

The main object in construing a deed is to ascertain the intention of the parties from the language used and to effectuate such an intention when not inconsistent with any rule of law. It is not what the parties meant to say, but the meaning of what they did say that is controlling.[33]

27-8-4. Four Corners of the Instrument

The intention of the grantor, gathered from the four corners of the instrument, is the controlling principle, and the court will enforce that intention, no matter where in the instrument it appears.[34] It is the grantor's intention, expressed in the instrument and not as shown by extrinsic evidence, that governs in determining the title conveyed.[35]

27-8-5. Contemporaneous Instruments

A general rule of construction in ascertaining the intention of the parties is that separate deeds or instruments executed at the same time and in relation to the same subject matter, between the same parties, may be taken together and construed as one instrument.[36] The rule will not be applied, however, to allow an unambiguous conveyance to be modified by contemporaneous instruments that are not a part of the identical transaction in which the deed was given.[37]

When a deed almost identical with the one before the court was executed at the same time by the grantor but to a different grantee, the other deed may be used to aid in construing the ambiguous one.[38]

When a deed and written agreement were executed at the same time, the agreement referring to the deed, the two instruments would be construed together to determine the intention of the parties.[39]

27-8-6. Consideration of Entire Instrument

A deed must be construed as a whole, and a meaning given to every part of it.[40] In construing a deed, each and every word must be given a meaning, if possible,[41] and the instrument construed within the limits of the four corners.[42] This rule applies to not only deeds, but also all other instruments.[43]

The legal effect of a deed is not determined from a single word, or part, or relative position of different parts, but the entire instrument.[44] All parts must be considered and, unless conflicting, given effect.[45] And when conflicts arise, or appear, other rules are resorted to for their resolution.

27-8-7. References Part of the Description

Any items, referenced in a description, are part of the description,[46] with as much effect as if copied into the description.[47] One of the reasons for employing references and not copying other information is so not to encumber the description, the reference being sufficient.[48] In fact, not just *any* reference, but all instruments in a chain of title when referred to will be read into it.[49]

27-8-8. Specific Description Controls General Description

When particular and general descriptions conflict and are contradictory to one another, the particular will control unless the intent of the parties is otherwise apparent on the face of the instrument.[50] It does not matter which one comes first in the deed.[51]

27-8-9. False Description May be Rejected

A plainly erroneous description will be rejected and reasonable meaning given to the deed that will conform to the intent of the parties.[52] Also, any particular of a description may be rejected, if it is manifestly erroneous, and enough remains to identify the land intended to be conveyed.[53]

When the land in a deed is so described that it can be ascertained, it will pass, even though some part of the description is false.[54] The principle is that a mistake in the description of lands in a deed will not void the deed, or defeat the legal title of the grantee or any one claiming under his or her title.[55] When property intended to be conveyed can be ascertained from such parts of description in deed as are found correct, the property will pass, and the incorrect pairs of the description will be disregarded.[56]

27-8-10. Construe Description Against Grantor; In Favor of Grantee

Generally, a grantee in a deed or other instrument, who accepts the instrument, is bounded by the recitals contained there, even though he or she does not sign it.[57] However, the general rule is that all grants, deeds, and leases are to be most strongly construed against the grantor if there is any doubt or uncertainty as to the meaning of the grant.[58] The decisions are many and the rule is a prevalent one. But there must be doubt, or ambiguity, for a grantee is not entitled to the benefit of an alternative interpretation favoring him or her when ambiguity is not established.[59]

27-8-11. Meaning and Intending Clause

A clause in a deed, at the end of a particular description of the premises by metes and bounds, "meaning and intending to convey the same premises conveyed to me," etc., does not either enlarge or limit the grant.[60] The Maine court also said, in Brown v. Heard,[61] that such wording was "merely a help to trace the title." The court stated in Sinford v. Watts[62] that the clause "meaning to convey lot known as," etc., following a particular description by metes and bounds, did not enlarge the grant, unless the contrary appears, because such a clause is ordinarily intended as a help to trace the title.

27-9. RELATIVE IMPORTANCE OF CONFLICTING ELEMENTS*

Courts make determinations between conflicting calls. They have generally agreed on a classification and gradation of calls in a grant or survey of land, by which their relative importance and weight are to be determined. Although the rules of comparative dignity of types of calls have been said to be not artificial rules built on mere theory, but the true results of human experience, they are not conclusive, imperative, or universal, but are called rules of construction, adaptable to circumstances, or only rules of evidence, or merely helpful in determining to which of conflicting calls controlling effect shall be given. Therefore, so a call that would defeat the parties' intention will be rejected regardless of the comparative dignity of the conflicting calls, and when calls of a higher order are made by mistake, the calls of a lower order may control, as most clearly indicating the intention of the grant.[63]

The general order of precedence of guides in determining boundaries is as follows.

1. CONTROL OF CORNERS AND MARKED LINES. A call for an established corner may, unless uncertain or mistaken, control other conflicting calls.[64] Lines marked or surveyed, when

---

*Reprinted with permission from 11 C.J.S. Boundaries, copyright © 1954 by West Publishing Co.

found, constitute the true boundaries and control any less certain matter of description or identity, provided that such lines were intended by the parties as lines of the land to be conveyed, and reference is made in the deed to the lines of the survey.[65]

2. NATURAL MONUMENTS.   Natural or permanent objects or monuments, definitely located, generally control other and conflicting calls, unless a different intention is indicated, or the call for such monument is clearly erroneous or less certain. The reason for this rule is that natural monuments or objects afford greater certainty than computations of course or distance and are less subject to error in calls. Further, the true intention of the parties will more probably be ascertained by adopting the call for natural monuments.[66]

3. ARTIFICIAL MONUMENTS.   Calls for artificial monuments, objects, marks, or established corners will, unless made by mistake or not mentioned in the deed, generally control other and conflicting calls, except those for natural objects. However, the rule does not apply to a call for a monument that is false or mistaken, or to monuments not mentioned in the deed, but a monument not existing at the time a deed is made, and afterward erected by the parties with intent to conform to the deed, will control. The reason for the rule is that there is less likelihood of a mistake in a call for an artificial monument than one for course and distance, the former being the more reliable call. Further, the parties are presumed to have taken note of the monument in viewing the premises.[67]

4. MAPS, PLATS, AND FIELD NOTES.   Maps, plats, or field notes referred to in a grant or conveyance are generally regarded as incorporated into the instrument and furnishing the true description of the boundaries of the land. Further, they have been held to stand on the same footing as monuments.

A plat many control notes, lines, descriptions, and landmarks, but whether or not a plat will have a controlling effect depends on the particular facts of the case, as the description that best identifies the land in accordance with the interest of the parties is controlling.[68]

5. ADJOINERS.   In absence of calls for other monuments, calls for adjoiners will, as a rule, control other and conflicting calls, since when they are certain, they are monuments of the highest dignity. Adjoiners, in order to have a controlling effect, must be established and well known, and must be called for in the conveyance; when such calls are manifestly erroneous, they will be disregarded.

Even the unmarked lines of an adjacent tract, if it is well established and its position can be ascertained with accuracy, will control a call for courses and distances. In the application of this rule, such a line may be given the dignity of an artificial object.

When a call for an adjoiner is made under a mistaken belief as to its true location, it may be rejected, and courses and distances held controlling, even though the line or corner called for was marked, at least where such a construction is most consistent with the intention to be derived from the entire description. However, it does not always follow that courses and distances will control, since a mistake does not reverse the general rule, but rather leaves the court free to construct the survey in such a manner as will best give effect to the intention to be determined from the entire instrument.[69]

6. METES AND BOUNDS.   Metes and bounds in the description of property granted, if established, always control courses and distances.[70]

7. COURSES AND DISTANCES.   Although courses and distances have been held unreliable, they will ordinarily govern in the absence of located calls of a superior type. However, the rule is not absolute, and in some circumstances they may be controlled by other evidence.

Because of the liability of chain carriers to error, courses and distances have been declared to be among the most unreliable calls, and the most unsatisfactory calls in a survey. Nevertheless, when there are no located calls of a type ordinarily considered superior to courses and distances, such as for monuments or lines marked and surveyed, courses and distances called for in the description of the boundaries should be given considerable weight and will ordinarily govern.[71]

8. COURSES OVER DISTANCES. Although there is some authority to the contrary, the general rule is that when a departure from either course or distance becomes necessary, the distance must yield. In other words, courses prevail over distances if they do not agree, with distances regarded as more uncertain than courses.[72]

9. QUANTITY. Quantity is ordinarily the least certain element of description, of little importance, and the last element to be considered in determining boundaries. However, in special circumstances it may become more valuable, or control—e.g., when superior calls are lacking or leave the boundaries doubtful, or if there is a clear intention to convey a certain quantity.[73]

27-10. RULES OF SURVEY

In locating and running the boundary lines of lots or tracts of land of private owners, reference should be made to the calls in the grant and the field notes carried in the grant or the map or plan with reference to which the conveyance was made. If there is no ambiguity, the land must be located and the lines run according to the description in the conveyance. None of the calls should be rejected or disregarded if they can be harmonized and applied in any reasonable manner. When, however, the descriptions given are conflicting, the courts have established an order of precedence among the several calls in the conveyance. Accordingly, in restoring lost lines and corners, visible and actual landmarks are

to be preferred, but if they cannot be ascertained, resort must then be made to courses and distances. Courses and distances in a call surrender to natural objects and judicially determined corners, and all calls are to be made straight between corners, unless a different intention appears from the description in the muniments of title.[74]

27-10-1. Footsteps of Surveyor

A line is a boundary or division only to the extent to which it is definitely described.[75] The lines of a survey as actually run and marked on the ground are controlling as to boundaries fixed with reference to such a survey.[76] Or as sometimes expressed, the rule is that the tracks of the surveyor, so far as is discoverable on the ground with reasonable certainty, should be followed.[77]

When title to land has been established under a previous survey, the surveyor's duty is to solely locate the lines of the original survey.[78] He or she cannot establish a new corner, nor can he or she even correct the erroneous surveys of earlier surveyors. The surveyor must track the footsteps of the first.[79]

27-10-2. Straight-Line Presumption

When a line is described as running from one point to another, it is presumed, unless a different line is described in the instrument, or marked on the ground, to be a straight line, so that by ascertaining the points at the angles of a parcel of land, the boundary lines can at once be determined. The rule of surveying, as well as law, is to reach the point of destination by the line of shortest distance, and lines should never be deflected, except in order to conform to the intention of the parties.[80]

27-10-3. Reversal of Calls

Courses may be run in a reverse direction when, by doing that, a difficulty can be overcome and the calls harmonized. Doing so should only be as a last resort, and ordinarily done only when running forward will not result in a closed figure. In addition, such a procedure can only be performed when lines of survey are actually measured and run on the ground, and monuments and boundaries in the deed must be followed when the courses and distances are reversed. Never is it permissible to disregard natural objects, either as corners or lines.[81]

27-10-4. Magnetic North Presumption

Courses in a deed are to be run according to the magnetic meridian, unless something appears to show that a different method was intended.[82]

27-10-5. Horizontal Versus Slope Measurement

Ordinarily, it is presumed that measurements of, and distances along, property lines are horizontal. However, in some cases surface measurement is found to be the method employed, particularly when such a method is the custom of the locality, or is dictated by circumstances.[83]

27-10-6. Lines of Ancient Fences

In the absence of natural boundaries or monuments, and of monuments or stakes set in the course of the original survey, the lines of ancient fences and long-continued occupation of adjacent lots and blocks in the same plat, if evidently intended to mark the true lines of such lots and blocks, have greater probative force than mere measurements of courses and distances.[84]

27-10-7. Relocation of Lost Monument

When monuments designating the boundaries of land are obliterated and cannot be

found, they are to be relocated by the field notes and plats of the original survey.[85]

27-10-8. Relocation of Lost Line

In relocating or reestablishing the lost lines of an old survey, the tracks of the original survey should be followed so far as it is possible to discover them, and the purpose of a resurvey is to find where the original lines ran. All locations, calls, and distances must, if found, be followed.[86]

27-10-9. Conflicting Grants

When there is a conflict of boundaries in two conveyances from the same grantor, the title of the grantee in the conveyance first executed is, to the extent of the conflict, superior, even though the conveyances were made with reference to a map or plat.[87]

27-10-10. Conflicting Surveys

The lines and field notes of a valid senior survey control over those of a junior survey, particularly when the junior is bounded with express reference to the elder. The field notes of a junior survey ordinarily cannot, but in some circumstances may, be looked to to establish the boundary of the senior survey.[88]

27-11. EASEMENTS AND REVERSIONS

An easement may be defined as a right, privilege, or liberty that one has in land owned by another; it is a right to a limited use in another's land for some special or definite purpose. An easement owner or holder does not own the land itself; the easement holder merely has the right to do certain things on the land of another. It is more than a mere personal privilege, however. It constitutes an actual interest in the land, and as such is treated as realty.

Easements are transferred the same as real estate, by deed, will, intestate succession, through adverse possession (prescription), etc. They may be created and terminated by the parties, or as a result of their actions.

Easements that benefit other land are called *appurtenant* and run with the land. Those that exist independently and do not benefit other land are easements *in gross*. The land benefited by an easement is the *dominant tenement* or *dominant estate*, whereas the land burdened by the easement is the *servient tenement* or *servient estate*. Easements may be *affirmative*, whereby the owner is allowed to do certain things on the land of another, or *negative*, whereby an owner is prevented from doing certain things on his or her own land. Easements may also be *apparent* or *nonapparent*, and may be *continuous* or *noncontinuous*.

There are nearly an infinite number of kinds of easements. Some of the more common are as follows:

1. Right of way
2. Flooding or flowage
3. Avigation
4. Use of water
5. Drainage
6. Overhanging eaves
7. Light and air
8. View

27-11-1. Creation of Easements

An easement may be created by any one of the following ways.

1. EXPRESS GRANT. This is usually called a "deeded easement," although an easement may be created with documents other than deeds, particularly in probate proceedings.

2. RESERVATION OR EXCEPTION. Although these terms are sometimes thought to be the same and are often used together, a technical distinction does exist. An exception is the process by which a grantor withdraws from the conveyance land that would otherwise have been included; the grantor merely retains or keeps the part excepted. A reservation vests in the grantor a new right or interest that did not exist before; it operates by way of an implied grant.

3. AGREEMENT OR CONVENANT. An agreement or covenant operates the same as a grant and is construed the same as an express grant. Whether it runs with the land depends on whether the act that it embraces concerns or relates to the land.

4. IMPLICATION. The general rule of law is that when an owner of a tract of land conveys part of it to another, he or she is said to grant with it, by implication, all easements that are apparent and obvious, and reasonably necessary for the fair enjoyment of the land granted. Easements by *necessity* also fall under this category.

5. ESTOPPEL. The word "estop" means to stop, prevent, or prohibit. Legally, an estoppel is a bar raised by the law that precludes a person *because of his or her conduct* from asserting rights that he or she might otherwise have —rights as against another person who in good faith relied on such conduct and was led thereby to change his or her position for the worse.

6. PRESCRIPTION. Long continued use of another's land for purposes in the nature of an easement may create permanent rights in the user. Such an easement stands in all respects on the same footing as an easement acquired by grant.

In most states, the requirements for establishment of an easement are the same as those for acquiring title by adverse possession. The difference is that in adverse possession, possession ripens into title, whereas in prescription, use develops into an easement. The minimum requirements are as follows:

(a) Use must be adverse.
(b) Open and notorious.
(c) Continuous.
(d) Exclusive.
(e) Under claim of right.
(f) For the statutory period.

For prescription, the requirements for continuity and exclusivity are more relaxed than those for adverse possession.

7. DEDICATION. Easements may be created by dedicating land to the public use, such as streets on a plan of a subdivision or development. For public rights to be created, however, there must be an acceptance of the dedication by the appropriate public authority.

8. EMINENT DOMAIN. Eminent domain is the power of the sovereign to take land or rights through the process of condemnation by paying just compensation.

27-11-2. Termination of Easements

Easements may also be terminated by a variety of means.

1. EXPIRATION. When established for a fixed or limited amount of time, an easement will expire according to its own terms.

2. RELEASE. An easement may be transferred (conveyed) from the dominant owner to the servient one.

3. MERGER. if the owner of the servient estate acquires the dominant estate, the easement will terminate through merger of title, since a person cannot have an easement in his or her own land. Whether the easement revives on a redivision of the land depends on the circumstances.

4. ABANDONMENT. Ordinarily, land and rights in land cannot be abandoned. However, certain types of easements may be terminated through the process of abandonment. Mere nonuse does not constitute abandonment, and the burden of proof is on the person claiming the abandonment.

5. ESTOPPEL. An easement may be terminated by the conduct of the easement holder, even though he or she may have had no intention of giving up the easement.

6. PRESCRIPTION. An easement may be extinguished by prescription if there is an interference with the use of the easement, satisfying the same requirements as for the creation of an easement by prescription.

7. DESTRUCTION OF THE SERVIENT ESTATE. There can be no easement in a servient estate if it no longer exists.

8. CESSATION OF NECESSITY. A way of necessity ceases when the necessity for it ceases.

9. EMINENT DOMAIN. Proceedings may terminate an easement, particularly if the result of the proceedings is to sever, or block, the easement.

27-11-3. Reversion

When an easement is terminated, it reverts to the land that by the right(s) was taken. This is true whenever an estate less than fee simple absolute terminates, whether it is an easement, a life estate, or some other interest. When reversionary clauses exist, their terms must be satisfied. Otherwise, the rights revert to the owner of the burdened real estate. Reversion is automatic and takes place at the point in time when the encumbrance is extinguished or terminates.

Recognizing reversion is critical when it comes to highways, railroads, flooded areas, and the like. For years, large areas have been encumbered and for the most part unusable, then all of a sudden, they are no longer burdened. The highway, railroad, or waterbody no longer exists in a legal sense, although it may in a physical sense. This usually gives rise to new boundary questions and only by knowing the history and pattern of ownership can such questions be properly addressed.

Easements burden land, but do not, by themselves, alter existing boundaries. Therefore, when an easement is extinguished, it is necessary to determine the existence and location of boundaries prior to its creation. Any land parcels created after the easement result in additional boundaries that are determined by appropriate rules. One of those rules is the presumption that land bordering on a highway or street extends to its center, provided, however, that the grantor owned that far and unless the contrary appears (Figure 27-1). It must be stressed that although this rule has extensive application, it is a presumption and not an absolute rule of law.[89]

Since easements and their associated reversion rights are treated similarly to real property, a survey of the same should be approached in the same manner as a survey for a parcel of land. Each has a history of title and boundaries the same as any other parcel of real estate.

27-12. ADVERSE POSSESSION

Title may be acquired, or lost, through *adverse possession*. That is, possession by a person or group of persons against the true or record owner of such a nature that, after the passage of a specified amount of time, such possession will ripen into title. In most states, these requirements are as follows:

Open. Visible and not hidden or concealed.

Continuous. Not intermittent or interrupted.

Exclusive. Against the true owner, meaning that no one else can share it.

Adverse. Against the true owner, meaning that no agreement exists, or is permissive.

Notorious. Sufficient to put the person of ordinary prudence on guard.

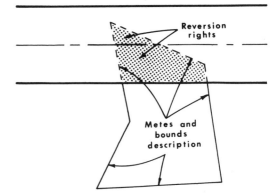

Figure 27-1. Reversion rights are defined by property lines as they existed before the creation of the easement —in this case, the street.

Statutory period. The number of years through which the possession must take place. It varies considerably from state to state, and may vary according to other factors, or conditions.

In some states, there are additional requirements, such as a requirement of good faith, payment of taxes, and the like.

As soon as all the requirements have been satisfied, *the possession ripens into title*, and the possessor becomes the new owner with a title that is as good as if received by grant or deed. This title may not be a *marketable title* under the marketability standards because there is no chain of record title behind the new owner connected to him or her, but it may be made marketable through court action. Adverse possession is sometimes difficult to prove and the burden of proof is on the adverse possessor to prove his or her title.

Some state courts require that the proof be of a *clear and convincing* nature, although some merely require that the possession be shown by a *preponderance of the evidence*, or the *balance of probabilities*.

27-12-1. Color of Title

A few states require that the adverse possessor have *color of title*—i.e., something that on its face appears to be title, but in reality is not. An often recited example is a faulty tax deed.

Even in those states that do not require color of title, having it may be significant to the possessor. A few states have a shorter time requirement with color of title than without. And in a number of states, whether color of title exists has an effect on how much land actually passes title.

With color of title, when possession ripens into title, the entire parcel described in whatever constitutes the color of title passes, whereas without color of title, only the amount of land actually possessed, and used, will pass. In some jurisdictions, when no color of title exists, there is an additional requirement that the land be occupied and enclosed.

27-12-2. Prescriptive Easements

Even though adverse possession and *prescription* are sometimes used synonymously, in most jurisdictions there is a subtle difference. Prescription relates to the *use* of a parcel of land, whereas adverse possession relates to its possession. The former results in an easement in the property, whereas the latter results in ownership of the fee.

Easements are created by prescription through the use of the property, satisfying the same requirements as for adverse possession. The requirements of continuity and exclusivity may be somewhat relaxed because of the nature of an easement as opposed to fee ownership, but basically the requirements are the same and both result in the creation of a new title in the adverse user, or possessor.

27-12-3. Surveyor's Role

Surveyors need to be on the alert for evidence that may indicate third party rights. Evidence in the form of use or encroachments should be taken note of and reported, since they may have an effect on the title, or ownership, of the parcel in question. Recently, some courts have stated that it is part of the surveyor's duty and responsibility to report such activity by others that is in conflict with the record title, whenever it is apparent.

27-13. UNWRITTEN RIGHTS

Rights may also be acquired, or boundaries established, without writings through four other legal doctrines—namely, *acquiescence, estoppel, parol agreement*, and *practical location*. The difference between these four and adverse possession is that they result in the *establishment of boundaries*, as opposed to the transfer of title to a parcel of land. The four doctrines do not involve a transfer of ownership; they are merely forms of agreement, expressed or implied, by the parties *in an attempt to fix that which is uncertain*. Because of this theory, such activity is not contrary to the statute of frauds, which requires that, to be enforceable, contracts for the transfer of land must be in writing. Consequently, the courts have long been in favor of parties agreeing on the location of their common boundary.

Most jurisdications require that a boundary be *in dispute* to qualify for an agreement. Stated another way, if a boundary is not in dispute and the parties desire a different location, they cannot relocate it by agreement since that would result in a transfer of title to a parcel of land, and therefore be contrary to the statute of frauds. For a boundary to be considered in dispute, courts have stated that it must fall into one of the following categories:

1. In actual dispute.
2. Be unknown or uncertain.
3. Be unascertainable.

Most courts require that at least a reasonable investigation be made before concluding that one of the foregoing is the case. They have also stated that any of the following may give rise to at least one of the above:

1. The boundary in question is unmarked and unknown.
2. There are two different locations desired by the abutting landowners.

3. There is one location desired by one owner, but the other owner does not agree with it.

4. There is an ambiguous and irreconcilable description of the boundary.

27-13-1. Acquiescence

Also called *recognition and acquiescence*, this doctrine involves the establishment of a boundary by abutting owners who have recognized a line as their true boundary. No agreement is involved, merely an acquiescence in the existence of the boundary line for the requisite period of time, which is usually equivalent to the statute of limitations. In most states, the time requirement is the same as for adverse possession. The requirements necessary for the doctrine to take effect are (1) occupation to a visible line, (2) mutual acquiescence in the line for a period of time, and (3) by adjoining landowners.

27-13-2. Estoppel

The doctrine of boundary by estoppel, like its related doctrines of boundary agreement, results in the establishment of a boundary by the parties in a place other than the true boundary. The requirements for this to take place are (1) a representation by one owner, or a failure to assert the facts, (2) a reliance on the representation or lack thereof, resulting in (3) injury or damages to the owner in reliance. All the elements must be present, and when they are, the owners have established a mutual boundary in a new location.

27-13-3. Parol Agreement

In many jurisdictions, parties may agree on the location of their mutual boundary by parol. Courts have long recognized parol agreements as binding on the parties and their successors. Parties cannot relocate a known line for their convenience because that would be a transfer of title to a parcel of land and therefore contrary to the statute of frauds, but they may agree to fix that which is unknown, or uncertain.

Written agreements are favored over oral agreements, but the latter do persist and numerous boundaries have been established in the past through the express verbal agreement of both parties. Even in those states that today require agreements to be in writing, past agreements may still be binding on all parties, particularly if they occurred prior to the current statute.

27-13-4. Practical Location

Practical location is the term used to describe the type of agreement in which the parties apply a construction to their descriptions and determine a line in the case of disagreement, doubt, or uncertainty. It is simply an actual designation on the ground, by the parties, of the monuments and bounds called for in their conveyances, or a location of their common boundary through a practical interpretation.

Boundaries by agreement, as well as those created through adverse possession, are binding on the parties and their successors in title. The courts have held to a standard that it is the responsibility of the surveyor to recognize and report such evidence whenever it is known or apparent.

Great case must be taken not to base other boundaries on distances from boundaries established by adverse possession or agreement. The parties' actions affect one line and one line only, and do not affect boundaries or titles of other persons who are not parties thereto.

27-14. APPORTIONMENT OF EXCESS AND DEFICIENCY

it is important to know when lots owe their origin from a common grantor, and what kind of conveyancing took place, whether it was *sequential*, *simultaneous*, or a combination of both. The general rule, called the *apportionment rule*, is that when a tract of land is subdivided into parts or lots, the title to which becomes vested in different persons, none of the grantees is entitled to any preference over the others on the discovery of an excess or deficiency in the quantity of land contained in the original tract. Additionally, the excess must be divided among, or the deficiency must be borne by, all the parts or lots in proportion to their areas.[90]

27-14-1. Sequential Conveyancing

In sequential conveyancing, senior-junior rights must be taken into account. When one or more parcels of land is conveyed from a parent tract, subsequent parcels cannot include part of them, since they are no longer owned by the grantor, who can only convey that which remains. This becomes of great concern when there is more or less land, or frontage, in the parent tract than supposed. For example, if lots were conveyed according to their numbers as shown in Figure 27-2, the first lots would be entitled to what their deeds state, whereas later lots, particularly the last one(s), could only receive what was left to convey.

Six 100-ft lots are sold out of a parcel believed to be 600 ft in width, when in reality it is not. In this example, lots 1 through 5 would each be entitled to 100 ft of frontage, while lot 6, being a remainder lot, is entitled to the remainder, whether it is more or less. The one exception occurs when there is excess, and it can be shown that the grantor intended to retain it, such as for access to back land.

The same rule applies to area. If a grantor conveys, in sequence, five lots from a 500-acre tract, the last lot is entitled to whatever acreage remains, whether more or less. Again, the one exception occurs when the grantor intended to retain excess. See Figure 27-3.

This becomes a very critical issue in surveying original range lots, or lots defined by early surveys, since they intentionally were made to "overrun" in area. The early surveyors did that to allow for poor-quality land and surveying errors such as "swag of chain." In cases such as this, entire blocks often have to be surveyed in order to account for outsales and properly determine what footage or acreage remains.

27-14-2. Simultaneous Conveyancing

When parcels in a group are created in a simultaneous manner, no one parcel is given preference over another, and each must bear a portion of the excess or deficiency. This applies in cases of subdivision, in which lots were created by the acceptance of a plan, and in cases of partition, in which a scheme of land division is approved, usually by the court, but in some cases may be agreed to by co-owners and accomplished through the use of one or more deeds.

Taking the previous example, we arrive at Figure 27-4. The 600-ft block was found upon survey to have only 594 ft of frontage. Therefore, each lot is only 99 ft long instead of 100. Each lot must bear the deficiency in propor-

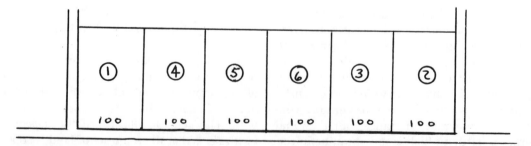

Figure 27-2. Example of sequential conveyances. Lots were conveyed in order of numbering.

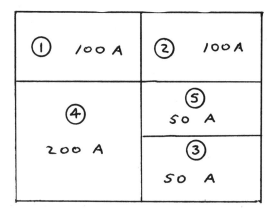

Figure 27-3. Five lots sold from a 500-acre tract.

tion to its frontage, and the rule applies without regard to the order of the conveyances of the lots by the common grantor.

27-14-3. Combination

Only through adequate study of the conveyances from a common grantor in a block can one know what type of conveyancing has taken place. Frequently, the case is a combination of both of the foregoing. For example, consider the following. In Figure 27-5a, lots 1 through 4 have been conveyed in a sequential manner, later followed by the 4th lot being divided in a simultaneous fashion. Lot 4 in Figure 27-5a will receive the remainder; then the frontage for lot 4 in Figure 27-5b is a result of the apportionment of the error over lots 1 through 4.

Lot 4 in the first example (Figure 27-5b) and lot 4 in the second example (Figure 27-5b) will not have the same frontage even though, if there had been no errors, or no deficiency, they would be the same.

27-14-4. Remnant Rule

Some state courts have applied what is known as the *remnant rule* whereby, even though parts or lots are sold according to a plan, the plan shows a number of the lots regular in shape and size and a remnant lot that is irregular. The grantee or grantees of the remnant take whatever is left, whether it is of greater or less area than shown on the plat. The reason for the rule is that when a tract is platted into lots, all regular in form except one, it is presumed that the subdivider intended to lay as many uniform lots as possible, leaving the residue in one lot, which should absorb any surplus or deficiency because of a mistake of measurement in the platting. The rule has been held not applicable, however, when the plat dimension of the irregularly shaped lot or lots is as definitely fixed as that of the regular lots.

State rules should be examined in detail before deciding which rules to use, especially if the remnant rule is considered.[91]

In Figure 27-5b, the total frontage for lots 1 through 4 will be short, and out of that, number 4 will receive only whatever remains after lots 1 through 3 have been conveyed.

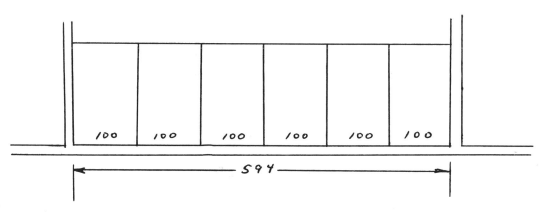

Figure 27-4. Lots created all at one time.

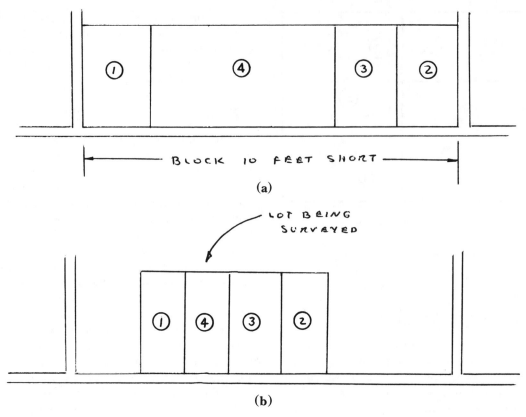

Figure 27-5. (a) is the simultaneous creation of four lots, whereas (b) represents the subsequent sequential creation of four lots in lot 4 from (a).

Although these rules are universal and must be considered whenever multiple sales are made by one grantor, care must be exercised in the determination of boundaries. Conditions change, subsequent transfers occur, and improvements are made. Occupation and possession may influence the ownership of a lot and ultimate location of its boundaries after a lapse of time.

NOTES

1. 11 C.J.S. Boundaries, §1.
2. *Black's Law Dictionary.*
3. *Black's Law Dictionary.*
4. A. C. Mulford. 1912. *Boundaries and Landmarks. A Practical Manual.* New York: D. Van Nostrand Company.
5. N. J. Sup. 1859. Evidence of corners and distances in prior deeds or maps of the same or adjoining lands, and of other evidence aliunde competent to prove any actual fact, is competent to confirm or control other courses and distances in a deed, on a question of boundary. Opdyke v. Stephens, 28 N.J. Law, 83.
6. While the surveyor does not practice law, there must be sufficient familiarity with rules and principles of law in order to make proper decisions in accordance with guidelines from the courts.
7. 29 Am. Jur. 2d, §14.
8. 29 Am. Jur. 2d, §14.
9. Penienskice v. Short, 194 A. 409 (Del.).
10. Unlike a rule of construction, a settled rule of law or rule of property is one that fastens a specific import and meaning on particular language employed in a deed and states arbitrarily the legal effect that such language will have, attaching to it a specific and unimpeachable intention, even though the parties em-

ploying the language may have had and may have envinced quite a different intention. Such rules therefore ingraft certain meaning on language employed in a deed and determine what effect is given to such language in law. In other words, a rule of property is to be applied automatically as a result of the language used, and the court will not refuse to apply such a rule merely on the surmise that the grantor did not intend that his or her phraseology operate in the way that the rule makes it. 23 Am. Jur. 2d, Deeds, §224, Settled rules of property.

11. Alabama Medicaid Agency v. Wade, 494 So.2d 654 (Als. 1986). See also Kennedy v. Rutter, 6 A.2d 17 (Vt., 1939).

12. 23 Am. Jur. 2d, Deeds, §223.

13. 26 C.J.S. Deeds, §82.

14. Wittmeir v. Leonard, 122, So. 330 (Al. 1929). See also Rhomberg v. Texas Co., 40 N.E.2d 526 (Ill. 1942).

15. Freeman v. Affiliated Property Craftsmen, 72 Cal. Rptr. 357 (1968). See also State v. Ladd, 268 A.2d 894 (N.H.) and Wilson v. DeGenaro, 415 A.2d 1334 (Conn. 1979).

16. Montgomery v. Central Nat. Bank & Trust Co. of Battle Creek, 255 N.W. 274 (Mich. 1934).

17. Elder v. Delcour, 269 S.W.2d 17 (Mo. 1954). In U.S. Trust Co. of N.Y. v. Boshkoff, 90 A.2d 713, 148 Me. 134 (1952), the court said, "The law of the state in which land is situated controls its descent, devise, alienation, and transfer, and the construction of instruments intended to convey it."

18. Sanborn v. Keroack, 171 A.2d 25 (N.H. 1961).

19. Faison v. Faison, 31, S.W.2d 828 (1930).

20. Weber v. Grover, 291 P.2d 173; Raritan State Bank v. Huston, 161 N.E. 141; Bradshaw v. Bradbury, 64 Mo. 334 (1876). See also Wood v. Mardilla, 140 P. 279 (Cal. 1914). In Franklin Flourspar Co. v. Hoseck, 39 S.W.2d 665, 329 Ky 454, the court said, "Although words may change in meaning, they must be read in deed as parties understood them and as they were commonly used where deed was written."

21. Wing v. Burgis, 13 Me. 111 (1836).

22. McChesney's Lessee v. Wainwright, 5 Ohio 452. See also Carter v. Barnes, 26 Ill. 454 (1861).

23. Cunnigham's Lessee v. Harper, 1 Wright 366 (Ohio 1833).

24. Town of Brookhaven v. Dinos, 431 N.Y.S.2d 567.

25. Brunotte v. DeWitt, 196 N.E. 489; see also Smiley v. Fries, 104 Ill. 416 (1882) and McCullough v. Olds, 41 P. 420.

26. Persinger v. Jubb, 17 N.W. 851 (Mich. 1883); see also Smith v. Crawford, 81 Ill. 296 (1876).

27. City of North Mankato v. Carlstrom, 2 N.W.2d 130 (Minn).

28. City of Brookhaven v. Dinos, 431 N.Y.S.2d 567.

29. City of North Mankato v. Carlstrom, 2 N.W.2d 130 (Minn.).

30. Daly v. Duwane Construction Co., 106 N.W.2d 631 (Minn.).

31. State v. Rosenquist, 51 N.W.2d 767 (N.D.).

32. Edens v. Miller, 46 N.E. 526.

33. Urban v. Urban, 18 Conn. Sup. 83.

34. Davidson v. Davidson, 167 S.W.2d 641 (Mo. 1943).

35. Rummerfield v. Mason, 179 S.W.2d 732 (Mo. 1944).

36. See Rudes v. Field, 204 S.W.2d 5, 146 Tex. 133.

37. 23 Am. Jur. 2d, Deeds, §237.

38. Hacker v. Carlisle, 388 So.2d 947 (Ala. 1980).

39. Brock v. Brock, 16 So.2d 881 (Ala. 1944).

40. Lyford v. City of Laconia, 72 A. 108 (N.H. 1909). See also 26 C.J.S. Deeds, §84.

41. Daniels Gardens v. Hilyard, 49 A.2d 721 (Del. Ch. 1946).

42. Schreier v. Chicago & N.W. Ry. Co., 239, N.E.2d 281 (Ill. App. 1968).

43. Ott v. Pickard, 237 S.W.2d 109 (Mo. 1951).

44. Reynolds v. McMan Oil & Gas Co., 11 S.W.2d 778 (1928).

45. Spiller v. McGehee, 68 S.W.2d 1093 (1934).

46. Perry v. Buswell, 94 A. 483 (Me. 1915).

47. Jacobs v. All Persons, etc., 106 P. 896 (Cal. App. 1910).

48. Cragin v. Powell, 128 U.S. 691.

49. Scheller v. Groesbeck, 231 S.W. 1092; Hathaway v. Rancourt, 409 A.2d 209 (Me. 1979).

50. 72 A.L.R. 410. In Perry v. Buswell, 94 A. 483 (Me. 1915) the court said, "Of all rules of construction, none is more rigid than the one that, where the language describing the grant is specific and definite, as, for instance, by metes and bounds, the grant cannot be en-

larged or diminished by a later general description, or by mere reference to deeds through which title was obtained. And this rule holds because the specific description is necessarily more indicative of intention than the general one.''

51. City of Rome v. Vescio, 58 A.D.2d 990, 397 N.Y.S.2d 267.

52. Wadleigh v. Cline, 108 A.2d 38 (N.H. 1954).

53. Lane v. Thompson, 43 N.H. 320 (1861).

54. Scofield v. Lockwood, 35 Conn. 425 (1968).

55. Bevans v. Henry, 49 Ala. 123 (1873).

56. McCausland v. York. 174 A. 383; 133 Me. 115.

57. Webb v. British American Oil Producing Co., Civ. App., 281 S.W.2d 726.

58. Treharne v. Klint, 58 N.E.2d 638 (Ill. 1945).

59. Foslund v. Cookman, 211 A.2d 190; 125 Vt. 112.

60. Smith v. Sweat, 38 A. 554 (Me. 1897).

61. 27 A. 182, 85 Me. 294 (1893).

62. 122 A. 573; 123 Me. 230 (1923).

63. 11 C.J.S. Boundaries, §47.

64. 11 C.J.S. Boundaries, §47.

65. 11 C.J.S. Boundaries, §49.

66. 11 C.J.S. Boundaries, §50.

67. 11 C.J.S. Boundaries, §51.

68. 11 C.J.S. Boundaries, §52.

69. 11 C.J.S. Boundaries, §53.

70. 11 C.J.S. Boundaries, §54.

71. 11 C.J.S. Boundaries, §55.

72. 11 C.J.S. Boundaries, §56.

73. 11 C.J.S. Boundaries, §57. This points out a very interesting conflict in philosophy. Landowners consider frontage and acreage to be of the utmost importance (they are the basis of taxation and selling price), whereas the courts have historically felt otherwise.

74. 12 Am. Jur. 2d, Boundaries, §55.

75. Fuller v. Fuller, 146 N.E. 174, 315 Ill. 214.

76. Settegast v. Meyer, 257 S.W. 343 (Texas).

77. Ballard v. Stanolind Oil & Gas Co., 80 F.2d 588 (Texas). See also Fleming v. Atlas, Tex. Civ. App. 51 S.W.2d 632. In Ralston v. Dwiggins, 225 P. 343, 115 Kan. 842, the court said, ''The footsteps of the surveyor may be traced backwards as well as forwards and any ascertained moment may be adopted as a starting point where difficulty exists in ascertaining the lines actually run.'' The courts will follow the footsteps of a surveyor if they can be traced. See Livingston Oil & Gas Co. v. Shasta Oil Co., 114 S.W.2d 378 (1938).

78. Williams v. Barnett, 287 P.2d 789, Cal. App.2d 607.

79. DeEscobar v. Isom, 245 P.2d 1105, 112 Cal. App.2d 172; Giles v. Kietzmeier, 239 S.W.2d 706. See also Diehl v. Zanger, 39 Mich. 601, in which the court stated, ''Object of a resurvey: A resurvey, made after monuments of the original survey have disappeared, is for the purpose of determining where they were, and not where they should have been.''

80. 11 C.J.S. Boundaries, §9(1).

81. 11 C.J.S. Boundaries, §9.

82. Wells v. Jackson Iron Mfg. Co., 47 N.H. 235 (1866).

83. See 12 Am. Jur. 2d, Boundaries, §58.

84. Galesville v. Parker, 83 N.W. 646, 107 Wis. 363.

85. Hiltscher v. Wagner, 273 P. 590, 96 Cal. App. 66. See also Houston Oil Co. of Texas v. Choate, 215 S.W. 118 (Texas).

86. 11 C.J.S. Boundaries, §15.

87. 11 C.J.S. Boundaries, §60.

88. 11 C.J.S. Boundaries, §61. See also Shackelford v. Walker, 160 S.W. 807, 156 Ky. 173.

89. 12 Am. Jur. 2d, Boundaries, §39.

90. 97 A.L.R. 1227. *Rights as between grantees in severalty of lots or parts of some tract, where actual measurements vary from those given in the deeds or indicated on the map or plat.*

91. See 97 A.L.R. 1227.

REFERENCES

BROWN, C. M., W. G. ROBILLARD, and D. A. WILSON. 1986. *Boundary Control and Legal Principles*, 3rd ed. New York: John Wiley & Sons.

BROWN, C. M., W. G. ROBILLARD, and D. A. WILSON. 1981. *Evidence and Procedures for Boundary Location*, 2nd ed. New York: John Wiley & Sons.

DEAN, D. R., JR., and J. G. MCINTIRE. 1982. *Establishment of Boundaries by Unwritten Methods and the Land Surveyor*. I.S.P.L.S. Surveying Publication Series 6, Indiana Society of Professional Land Surveyors and School of Engineering, Purdue University.

WILSON, D. A. 1991. *Easements and Reversions*. Rancho Cordova, CA: Landmark Enterprises.

28

Boundary Location Along Waterways

Roy Minnick

28-1. INTRODUCTION

Property boundary location along waterways is probably the most complex of all boundary locations. Water lines naturally fluctuate and so any boundary dependent on a fluctuating line must change. This often leads to litigation among litoral owners since a changing property line may cause one person to lose and another to gain. As a result, courts rooted in common law have often established precedents for riparian boundary location based on equity rather than more precise and scientific location principles. In recent years, courts have modified some viewpoints and reversed themselves on others. As a result, boundary location along waterways is fraught with uncertainties, not only in location, but also title. Matters are made worse by unclear laws. What does all of this mean to the surveyor? How does the surveyor begin to understand riparian boundary location? Most important, what does the surveyor need to know to perform the duties and discharge the responsibilities associated with surveying a boundary along a waterway or shoreline.

Fortunately, some basic principles are available to all, and with these, the surveyor need not be greatly concerned even though the legal situation may appear confusing. Generally speaking, the law defines the boundary, and the surveyor locates the boundary. When the boundary is not defined, the surveyor can be a key investigator and finder of fact to aid the property owners and the court arrive at a definition of the boundary and then locate the boundary by map and description.

Before the discussion of boundary begins, some aspects of ownership and title must be considered.

28-2. SOVEREIGN LANDS AND THE PUBLIC TRUST

Sovereign lands are those lands beneath waterways that existed naturally when the state was admitted to the Union, either as one of the original colonies, or one of those states admitted on "equal footing" with the original colonies. Sovereign land title is vested in the state at statehood. Sovereign lands are comprised of the beds of navigable waterbodies. The extent of the sovereign lands is not defined at the time of statehood. In general, sovereign lands are not subject to adverse claims, prescriptive rights, or other unwritten rights. The navigable sovereign lands are intended for general public use as highways for commerce, navigation, fisheries, and, more recently in some states, other public uses such as recreational fishing and boating. The sovereign's right, exercised on behalf of the public, is considered paramount, or senior, to the privately owned upland and takes the bed of the waterbody, even though it may be in conflict with land descriptions for the uplands.

The extent of upland ownership—i.e., the littoral ownership—is limited to the shoreline and does not include the bed of the waterbody, if the waterbody is navigable.

The sovereign traditionally held the lands in trust for the general public who could use the waterbodies for "commerce, navigation and fisheries." Sovereign navigable waterways are inseparable from the public trust doctrine. Each state, on admission to the Union, became the trustee. The states have discharged these trusts in various ways, and since the public trust question is a matter of state law, a number of approaches have been used, and the law from state to state is inconsistent.

U.S. Chief Justice Taney, speaking for the Supreme Court, explained the ownership of such lands as follows:

> For when the Revolution took place, the people of each state became themselves sovereign, and in that character hold the absolute right to all their navigable waters and the soils under them for their own common use, subject only to the rights since surrendered by the Constitution to the general government.[1]

Even within a state, public trust law may be inconsistent and uncertain since the courts take a case-by-case approach within broad principles of law. The legal results in a given situation, therefore, are hard to predict.

Sovereign title may be affected by a number of factors, including lands patented by the U.S. government during the time of territorial status and lands granted by treaty, such as grants to Mexican citizens confirming their title claims at the close of the Mexican War in 1848. Some states also vary the application of the general public trust doctrine by enacting special statutes dealing with specific sites, or particular problems.

The public trust doctrine, when applied to sovereign lands in the beds of navigable waterbodies, has several elements, including but not limited to the following:

1. The physical character of the land
2. Source of title to the land
3. Changes of the character or boundaries of the land

Other elements can be considered, such as revocation of the trust, land exchanges, or estoppel, but these are beyond the scope of this chapter.

28-2-1. Physical Character of the Land

In general, the public trust doctrine applies to tidelands, submerged lands, and the beds of navigable streams, rivers, and lakes. In contrast, under English law, the trust extended only to lands subject to the ebb and flow of the tides that, in the small island, included in essence all navigable waters. With its long rivers and large interior lakes, however, America expanded the trust definition to include all navigable waters even if nontidal.

28-2-2. Source of Title

By virtue of its admission to the Union as a sovereign state, each state acquired title to all the *ungranted* trust lands within the boundaries of the new state. The extent of the grant of sovereign trust land is determined by its character on the data of admission to the Union, not its character at a later date. Hence, considerable difficulty surrounds proving the actual character of the land at statehood inasmuch as accurate surveys representing the grants do not exist. However, evidence of its character at a later date has been admissible in a court of law to show its character at statehood, and scientific studies usually try to shed light on the issue.

28-2-3. Changes of the Character or Boundaries of the Land

Changes in the extent and boundaries results from shoreline processes that build up, or eat away the land along the waterbody. The boundary line separating the uplands from the waterbody then changes. Sometimes, these changes are quite dramatic over a number of years, even going so far as to completely eliminate a parcel, or double its size. In other instances, submerged lands may become tidelands, or the reverse.

28-3. CLASSIFICATIONS OF LANDS INVOLVED IN WATER BOUNDARY LOCATION

There are various classes of lands invoked in water boundary situations: tidelands; submerged lands; navigable lakes, streams, and rivers; and swamp and overflowed lands after 1850.

28-3-1. Tidelands

Tidelands are lands subject to the public trust lying between the lines of mean high and low tide covered and uncovered successively by the ebb and flow of the tides, including the shores of every bay, inlet, estuary, and navigable stream as far up as tidewater goes and until it meets the lands made swampy by the overflow and seepage of freshwater streams.

28-3-2. Submerged Lands

Submerged lands are also subject to the public trust. Submerged lands are those covered by water at any stage of the tide and would include the bed of the sea and of bays and inlets.

Ownership of the submerged lands between the low water mark and 3-mi limit became important when oil was discovered in this area and the necessary technical advances made

extraction feasible. In 1947, the U.S. Supreme Court decided that these lands were held by the federal government in trust for the people. In 1953, however, Congress passed the Submerged Lands Act that conveyed whatever interest the federal government had in these lands to the states, with the exception of those lawfully held by the federal government. Congress also specifically reserved its paramount regulatory power under the commerce clause of the U.S. Constitution.

28-3-3. Navigable but Nontidal Waterbodies

A third category of lands that may be subject to the common-law public trust is navigable nontidal waters, technically distinguished from tidelands and submerged lands although both are navigable for title purposes. Again, the "navigability" of waters for determining whether the public trust doctrine applies is ascertained by the uses made while in a natural state. The federal test of navigability is usually used, and the basic question is to determine whether the waters were navigable in fact at the time of statehood. The standard is whether

> ...They are used, or are susceptible of being used, in their natural and ordinary condition, as highways for commerce, over which trade and travel are or may be conducted in the customary modes of trade and travel are or may be conducted in the customary modes of trade and travel on water...navigability does not depend on the particular mode in which such use is or may be had...nor on an absence of occasional difficulties in navigation, but on the fact, if it be a fact, that the stream in its natural and ordinary condition affords a channel for useful commerce.[2]

28-3-4. Swamp and Overflowed Lands (Swamplands)

Unlike tidelands, submerged lands, and navigable waters, "swamp and overflowed" lands are not generally considered to be subject to the public trust. In 1850, Congress passed the Swamp Lands Act that granted to each state all the unsold swamp and overflowed lands within its border. Swamp and overflowed lands are those that were "...unfit for cultivation by reason of their swampy character and requiring drainage or reclamation to make them available for beneficial use" (*Black's Law Dictionary*) and could be sold by states to private persons without restraint as to ownership or use.

28-3-5. The Difference Between Swamplands and Sovereign Lands

Thus, the lands belonging to a state along or within a waterbody may be divided into two categories: those that it received from the federal government by grant—the swamp and overflowed lands, which are not subject to the common trust, and those that it received by virtue of its sovereignty—the tidelands, submerged lands, and lands under navigable waters, which are subject to the public trust. The all-important distinction to be made between the two is, in practice, not an easy one. For example, coastal lagoons are difficult to specifically state, as of statehood, whether the land was properly characterized as swamp and overflowed lands, or as tidelands. A solution usually involves litigation, or a boundary agreement and/or exchange, after a thorough study has been conducted and the determination of one or the other is uncertain.

28-4. PUBLIC TRUST LANDS HELD BY PRIVATE PERSONS

Disposition of these lands were subject to state law, and some states immediately began conveying tidelands and sometimes adjoining submerged lands into private hands. Since tidelands were held by the state in trust for the people, there were always some who doubted the validity of sales to private individuals. Some early law cases held that such sales were void or voidable. In 1892, however, the U.S. Supreme Court decided the landmark federal public trust case, Illinois Central R. R. v. Illinois, which involved a grant of waterfront on Lake Michigan to a corporation, which the Illinois legislature later wanted to revoke. The court, in speaking about the status of the title, described it as:

> ...A title held in trust for the people of the State that they may enjoy the navigation of the waters, carry on commerce over them, and have the liberty of fishing therein free from the obstruction or interference of private parties. The interest of the people in the navigation of the waters and in commerce over them may be improved...for which purpose the State may grant parcels of the submerged lands.... [3]

The court then went on to say that the state could not abdicate its duty and control over the trust, nor could the trust be extinguished or impaired by a transfer of title. However, the trust could be revoked if the state relinquished title to particular parcels without any substantial impairment of the public interest in the lands and waters remaining.

Taking this cue, the many state courts resolved the same uncertainty about sales of tidelands by holding that, although the state could convey tidelands to private persons, the title so conveyed was subject to the public trust rights and powers of the state and public.

Exceptions to the general statement that a private tidelands' owner holds subject to the public trust are found when the sale was made under a special statute clearly evidencing the legislative intent to alienate tidelands free of the public trust and when the state has otherwise revoked the trust or would be precluded from asserting it. In general, however, privately owned tideland is subject to the public trust.

No statutory authorization to sell submerged lands and lands under navigable waters appears to have ever existed, in contrast with the statutes authorizing the sale of tide-lands. Such sales likewise may be subject to attack, and if void, title would remain with the state as trustee.

28-5. NONNAVIGABLE WATERBODIES

Nonnavigable waterways are not owned by the state as sovereign. Landowners adjacent to nonnavigable waterways own the bed of the waterways. If there is only one owner around the entire body of water, than the bed is owned by that party. If there are several owners along the shore, then each is entitled to a proportionate share of the underlying lands. In this case, the portions are divided proportionately among the littoral owners by agreeable and sensible means. Several methods are covered in the second edition of *Boundary Control and Legal Principles*, 3rd ed.[4] The most commonly used are the "pie method" for dividing round lakes and "long lake method" for dividing long lakes. Thread of stream or center of stream is customary along rivers and other methods are used in bays and coves. In situations where the surveyor is called on to proportionately divide nonnavigable waterways, any method that is agreeable to the parties concerned, and is within the surveyor's authority, is acceptable.

A surveyor attempting to proportionately divide a lake, e.g., needs to obtain agreement from all parties, preferably by acknowledgement on a record map, even if some parties are not clients.

28-6. DETERMINING NAVIGABILITY

Navigability determination is a difficult and lengthy assignment. Navigability means different things to different people, and it lacks a precise scientific definition. Generally, waters are navigable in fact when they are used or are susceptible for being used in their ordinary conditions as highways of commerce, of which trade and travel are conducted in the customary modes of trade and travel on water. Some states have passed laws declaring the factors necessary to constitute navigability. Others totally ignore the situation. In Minnesota, a stream capable of floating a canoe may be considered navigable. In many states, the ability of a stream to carry logs may be a factor in declaring a stream to be navigable. State laws and courts, when they address the issue, generally consider a waterbody navigable when it is suitable for use as a public way. The state, as "sovereign," has a duty to protect the rights of the public from abuse or infringement unless a law to the contrary has been acted and upheld. The state as the owner of the bed of navigable waterways is at least coequal and, in some instances, superior to adjacent upland owners.

28-6-1. Riparian Rights

The upland owners themselves are coequal and have certain property rights entitled to protection. These rights are usually called *riparian rights* and accrue to an adjacent upland owner. Riparian rights may include such things as the right to "wharf-out" into the water body to construct bulkheads, enjoy the fruits of accretion, and suffer the flip side, erosion. An upland owner, in order to have riparian rights, must share a common boundary with the owner of the bed of the waterway. Any intervening ownership, no matter how infinitesimally small, can deprive a person of these riparian rights.

Surveyors must carefully examine a chain of title histories of property to be certain that no intervening interest exists. It is common to find that even though an original deed was riparian, the present vesting document does not include riparian rights. These usually occur when a strip of land lies between a meander line or private subdivision meandered line. The strip is "dropped out" scriveners, or perhaps inadvertently.

28-6-2. Federal and State Law Governing Navigability

There are two broad categories of law governing navigability: state and federal law. State law considers navigable waterways to be generally a public highway. The public may use the waterway as well as the adjacent upland owner. In some jurisdictions, navigability is based on limited tests such as commerce, navigation, and commercial fishing. In other states, this narrow traditional definition of navigability has been broadened to include whatever uses the general public may wish to make of the waterway. This includes recreational fishing, water skiing, and other such "noncommercial" uses.

Federal law is based on the commerce clause in the U.S. Constitution that defines the authority of the federal government to control the use of navigation on interstate waterways. Although the federal government is expanding its control of navigable waterways and even includes some waterways that were naturally navigable, by and large their efforts relate to land use regulation and navigation on the water surface.

Generally, water boundary disputes are settled in state courts, between the state and its citizens. An exception occurs when the federal government is an adjacent upland owner and a dispute arises between the federal government and the state. In this case, federal courts are used. Federal courts are also used to settle water boundary disputes between various states.

From many different standpoints such as permissible uses, property ownership, title and boundaries, the determination of navigability is perhaps the most important distinction that must be made along waterbodies. While navigability disputes are usually settled in courts or by law, the surveyor may find him or herself gathering facts and evidence that will support the contention of either navigability or nonnavigability. In fact, the surveyor may well be instrumental in the determination of navigability, provided that the surveyor is aware of the implications and impact of his or her actions.

28-6-3. Navigability by Statute

Each state address navigability in its own way. Some use statutes, others court cases, and some a combination of the two. A list of statutes has been compiled and is listed as a reference at the end of the chapter.

28-6-4. Navigability in Fact

The second method for determining navigability is to determine in fact that the waterway was used at the time of statehood as a public highway for commerce, navigation, and fisheries or other purposes allowed under state law. Sometimes, several months are spent attempting to locate historical evidence of use in an effort to determine the navigability of waterbodies. Rainfall records may be checked to determine if sufficient water existed for navigation.

In some western states, waterbodies that were navigable in fact at the time of entry into the Union and for years thereafter in their natural condition are now dry as a result of irrigation diversion, or some other artificial cause. One such case is the San Joaquin River in the central valley of California. It once carried steamboats up into the foothills of the Sierra Mountains, but now is dry most of the year. It is hard for landowners to understand that the river, in a land title sense, is navigable.

28-6-5. Susceptibility to Navigation

The third method for determining navigability involves waterbodies in existence at the time of statehood, still in existence, and large enough to be susceptible to navigational uses as may be defined by the state in which the waterbody is located. Waterbodies that fall into this category may be relatively small mountain lakes that are only now becoming valuable because of the public's unquenchable thirst for recreational land; they occurred naturally at statehood, and are still in place, and susceptible to use.

Generally, waterbodies, in order to be navigable for title purposes, had to exist as a natural body of water either navigable in fact or susceptible to navigability under its ordinary condition at the time the state entered the union. The mere fact that the body of water dried up periodically or will be dry for periods of time each year does not in itself prevent a declaration of navigability. For example, in 1908, a Supreme Court case in Florida considered a case involving title to the bed of Lake Jackson in Leon County. Most of the bed during ordinary water levels could be navigable only by flat-bottomed boats drawing no more than 6 in. of water. Large portions of the lake bottom were dried out for such long periods of time that crops were harvested on the lake beds. The court held Lake Jackson to be navigable. The fact that the lake went dry at times did not strip of navigability since in its ordinary state it was navigable.

28-7. BOUNDARIES ALONG TIDAL WATERBODIES

Boundaries along tidal waterbodies are usually tide lines. For example, along the western coast of the United States, the ordinary high water mark is normally considered to be the boundary between the state and upland owner. In areas where the shoreline is natural, the *ordinary high water mark* may be defined as the mean high tide line averaged over a substantial period of time. The *ordinary high water line* is the level that the water reaches in ordinary stages, not in a period of drought, and not in a period of flood.

Tidal riparian boundaries often rely on tidal observations over an 18.6 year period. The 18.6 year period is a full cycle of the varying relationship among the positions of the earth, sun, and moon. Various methods have been described over the years for locating mean

high water line boundaries. In general, tidal observations by the federal government are used, and an elevation for the mean high water is determined. With this elevation, a contour line is run, and the mean water located for the moment. The surveyor, when platting this line, enters the data and time, as well as the elevation of the contour. The line is not permanent, but it existed at the location at that point in time.

From the surveyor's standpoint, it is necessary to first ascertain what the boundary may be between the bed of the water and the upland owner. This is generally a matter of law and for attorneys. Where the boundary may ultimately be located is normally within the province of the surveyor. For example, a court may determine that the boundary between two points on a shore is the mean high tide line. In the event that the court does not define the elevation of the boundary, it is the surveyor's task to determine the elevation of the mean high tide line in a professional manner and then locate this line on the ground. For example, the mean high water line in the given stretch may be determined by observation to be 2.5 ft above the mean lower water. The surveyor locates this contour line with reference to mean lower low water by the usual methods of survey.

In other cases, as in the state of Washington, the vegetation line may be the boundary established by law. On nontidal water boundaries, a variety of techniques may be used to determine the legal boundary. For example, in Clear Lake, CA, the largest lake in California with over 100 mi of shoreline had a low water mark boundary definition of zero on the Rumsey gauge. The gauge was established by Rumsey in the late part of the 19th century and daily observations on a water-level staff have been made ever since. As a result, the surveyors interested in mapping the line between the upland property owners and the state of California can use this contour line that was defined by survey leveling to be a contour elevation of 1318.26 ft above sea level.

28-8.   WATER BOUNDARIES IN PUBLIC LANDS

The federal government patented public lands to the ordinary high water line, even though the survey was run along a meander line. Meander lines are set as part of the original survey to segregate bodies of waters from the lands to be patented to the upland owner. These lines are not usually boundary lines, unless specifically indicated. They were used to close the upland survey and determine the area to be purchased. The title of the upland extends to the boundary of the water as it may be determined. Some general comments taken from the various U.S. manuals of survey instructions are:

1. All manuals call for meander corners to be established at all those points where the lines of the public surveys intersect the "banks" of bodies of water, waterways, or islands that are to be separated for the public lands or acreage.

2. All manuals refer to the position of the banks of a river or stream as, facing downstream, the bank on the left-hand side is the "left bank," and the bank on the right-hand side the "right bank."

3. The manual of 1930 defines a meander line as "the traverse of the margin of a permanent, natural body of water."

4. Tidewaters are not mentioned in the manuals of 1851 to 1881, but are mentioned after 1881.

5. The manual of 1902 states that lands bounded by waters are to be meandered at a mean high water mark (Section 154).

6. The manual of 1902 states that unless an irregular or sinuous line closely follows a stream or body of water, it is not entitled to be called a meander line (Section 153). (See also Sections 108 and 151.)

The Manual of United States Surveying[5] adds:

1. Large lakes, navigable rivers, and bayous are by the law of Congress made public highways, and as the government surveys progress, they are mentioned and segregated from the public lands.

2. Wide "flats" having wide, irregular expansions occur in rivers that are not navigable. Such expansions are permanent bodies of water, the area of which is more than 40-acres and embraces more than one-half of a legal subdivision of 40 acres, they should be meandered of both banks.

28-9.   EFFECT OF NATURAL SHORELINE PROCESSES ON BOUNDARIES

After defining the boundary, by law and fact, it is necessary to consider some shoreline processes before location of the line can be commenced. Notably, accretion, erosion, avulsion, and the sundry acts of humans in building dams, bulkheads, carving cutoffs, and building groins and jetties.

Accretion is the gradual imperceptible deposit of buildup of land along the shore of a waterbody. The accretion results in the placement of alluvion that belongs to the upland owner or owners. In this case, the surveyor is called on to first of all identify the amount of accretion and then apportion that accretion as necessary and in a manner recognizing the coequity of the upland owners. The accretion must be apportioned from the last deed line using one of a number of methods, two of which follow:

1. Project the lot lines from the old shoreline to the new.

2. Extend the lot lines at an angle to provide a share of area of alluvion proportionate to the amount of frontage the owner had prior to accretion.

Erosion is the gradual and imperceptible washing away or reduction of land along the water boundary. In this case, the owner is losing area. The boundary is shifting to take more and more of the upland property and in fact the upland owner may lose the entire parcel to erosion.

Avulsion is the sudden and perceptible separation of land by the violent action of water. In this last case, property boundary lines sometimes remain fixed in place, just prior to the avulsive action. For example, a stream that during a flood stage suddenly adopts a new channel does not alter the ownership of the abandoned river bed. If the waterway is navigable, it so happens that the owner of the channel will end up owning the abandoned bed of the river that now cuts the parcel in two parts. Accretion and erosion, on the other hand, result from natural causes and the property lines shift as the shoreline shifts.

28-10. ARTIFICIAL CHANGES TO SHORELINES

Other factors to be considered in locating boundary lines along waterways are the physical and legal effects of artificial changes. Examples of artificial changes are dams, groins, piers, and other shoreline structures that prevent the body of water from moving naturally. In most states, artificial accretion resulting from interruption of the natural processes belongs to the owner of the shore; this is also true of lands subject to federal rules of interpretation.

28-11. OPERATIVE WORDS: SUDDENLY AND GENERALLY

In all of the above cases, the words "suddenly," "generally," and other operative words are subjective. For example, in the definition of avulsion we used the word suddenly. Various contenders in litigation will argue over the definition of suddenly. Viewpoints may range from considering suddenly as spanning a period of time overnight to as long as 7 yr.

28-12. WHEN ARTIFICIAL IS NATURAL

On the Colorado River, the Boulder Dam has been generally construed by several federal courts to have no effect on downstream property boundaries. Although it is clearly an artificial influence, the court has held that it has acted primarily as a control to prevent both floods and drought and the resultant regulation of the flow does not inhibit the gradual erosion and accretion of the river, although the court admits that the range is greatly diminished by the dam controlling the flows of water.

28-13. TITLE: EXTINGUISHED VERSUS REEMERGENCE

Other problems arise when considering situations where entire parcels have been eroded away and then processes have reversed themselves and accretion occurs. In this situation, is the title extinguished and the accretion divided between the adjacent upland owners at the time the process reversed or does the title reemerge?

28-14. OWNERSHIP OF ISLANDS

Surveyors may face other situations involving islands in waterways. The ownership of an island in a navigable waterbody may depend on whether the island grew from the bed of the river since statehood or whether it existed prior to statehood. If the island formed after statehood, then it probably will belong to the state; if it existed before statehood, then it is subject to survey and patent, and does not belong to the owner of the bed of the waterbody. Islands formed in nonnavigable streams usually belong to adjacent upland owners. They are divided according to where the division line of the stream is located.

28-15. WATER BOUNDARY DETERMINATION

A variety of legally acceptable evidence may be utilized to establish water boundaries. At the outset, keep in mind that the boundary may be the last natural condition, or may have been altered by accretion or erosion. Therefore, as to each of the following tests, you may have add the caveat that the appropriate test is to be applied as to the last natural condition of the waterbody.

In order to understand the legal test of a tidal water boundary, it is first necessary to understand certain terms or words. The definitions here are those adopted by the U.S. Coast and Geodetic Survey [now the National Ocean Survey (NOS)]:

High water. The maximum height reached by a rising tide. The height may be due solely to the periodic tidal forces or it may have superimposed on it the effects of prevailing meteorological conditions.

High water line. The intersection of the plane of mean high water with the shore. The shoreline delineated on the nautical charts of the U.S. Coast and Geodetic Survey is an approximation to the high water line.

Higher high water. The higher of the two high waters of a tidal day.

Mean high water. The average height of the high waters over a 19-yr period. For shorter periods of observations, corrections are applied to eliminate known variations and reduce the result to the equivalent of a mean 19-yr value. All high water heights are included in the average, where the type of tide is either semidiurnal or mixed. Only the higher high water heights are included in the average, where the type of tide is diurnal. So determined, mean high water in the latter case is the same as mean higher high water.

Mean higher high water The average height of higher high waters over a 19-yr period. For shorter periods of observations, corrections are applied to eliminate known variations and reduce the result to the equivalent of a mean 19-year value.

Low water. The minimum height reached by a falling tide. The height may be due solely to the periodic tidal forces or it may have superimposed on it the effects of meteorological conditions.

Lower low water. The lower of the low waters of any tidal day.

Mean low water. The average height of low waters over a 19-yr period. For shorter periods of observations, corrections are applied to eliminate known variations and reduce the result to the equivalent of a mean 19-yr value. All low water heights are included in the average, where the type of tide is either semidiurnal or mixed. Only the lower low water heights are included in the average, where the type of tide is diurnal. So determined, mean low water in the latter case is the same as mean lower low water.

Mean lower low water is frequently abbreviated lower low water on U.S. Coast and Geodetic Survey charts, the average height of lower low waters over a 19-yr period. For shorter periods of observations, corrections are applied to eliminated known variations and reduce the result to the equivalent of a mean 19-yr value.

Generally speaking, state law applies to the location of water boundaries. However, federal law will apply if the boundary is determined by a federal patent or for locating the boundaries of navigable waters of the United States under federal law.

A water boundary for the purposes of federal law is the ordinary high water mark, or mean high water. This has been defined in *Borax Consolidated, Ltd. v. Los Angeles*,[6] as the average of all the high waters occurring over a tidal epoch of 18.6 yr. This includes averaging both high waters that occur daily on the Pacific coast.

28-15-1. Nontidal Waterbodies

Water boundaries on nontidal waterbodies are generally determined by state law. On nontidal nonnavigable bodies, the upland owners take to the middle, whatever that may be.

How are the low water mark and high water mark determined on nontidal navigable waterbodies? Although at first glance this question seems to have an easy answer, it does not, in fact. Nontidal waterbodies are not subject to any sort of regular fluctuation. Substantial seasonal changes occur, and they differ from year to year depending on many factors such as rainfall. Hence, there is no mean water mark in the same sense as that found along tidal bodies of water. In addition, a great majority of the waterbodies in this category have been subjected to artificial influences, such as dams, or diversion for irrigation.

One Attorney General's opinion best summarized the definition of low water mark as:

The elevation of water in the non-tidal navigable lake or stream at its low point during a normal year, not affected by floods, droughts, or other special circumstances.[7]

The opinion further advised that any competent evidence may be used to establish the low water mark on a nontidal navigable lake, including but not limited to maps, historical data, testimony, and physical characteristics of the lake bed, or adjacent terrain. This approach has generally been confirmed in case law. Some of the sources that provide the information needed are listed in the reference section here.

28-16. SUMMARY

Water boundary location is difficult because of the wide variety of interests in land and types of land, that are involved. Further complications arise because of the ambulatory nature of water boundaries. Even the stability associated with written land titles must be set aside.

Land descriptions along waterways may be inconsistent with the facts as indicated on the ground, as a result of natural processes. Sovereign lands are seldom even described and derive from the state's act of admission to the United States, or from its status as one of the original 13 states. A search of records will not disclose the existence of sovereign lands, and state legislatures have not provided much guidance for the surveyor.

NOTES

1. Martin v. Waddel, 41 U.S. 349 (1842).
2. U.S. v. Utah, 283 U.S. 64 (1931); Utah v. U.S., 403 U.S. 9 (1971).
3. Illinois Central R.R. v. Illinois, 146 U.S. 307 (1892).
4. C. M. Brown 1988. *Boundary Control and Legal Principles*, 2nd ed. New York: John Wiley & Sons.
5. J. H. Hawes. 1882. *The Manual of United States Surveying*. Philadelphia, PA: Lippincott & Co.
6. Borax Consolidated, Ltd. v. Los Angeles, 296 U.S. 10 (1935).
7. 430 Ops. Calif. Atty. Gen., 296.

REFERENCES

MINNICK, R. 1994. *Water Boundaries for Land Surveyors*. Rancho Cordova, CA: Landmark Enterprises.

1992. *Clark on Surveying and Boundaries*, 6th ed. Charlottesville, VA: Walt Robillard Michie Co.

COLE, G. 1994. *Water Boundaries*, 2nd ed. Rancho Cordova, CA: Landmark Enterprises.

MINNICK, R., ed., 1989. *Glossaries for Surveyors*, 2nd ed. Rancho Cordova, CA: Landmark Enterprises.

BLM. 1993. *Manual of Instructions for Survey of the Public Lands of the United States*, Washington, DC: U.S. Government Printing Office (also Rancho Cordova, CA: Landmark Enterprises).

MINNICK, R., ed. 1991. *Compilation of State Water Boundary Statutes*. Rancho Cordova, CA: Landmark Enterprises.

29

Mining Surveys*

E. Franklin Hart and John S. Parrish

29-1. UNDERGROUND†

29-1-1. Mine Surveying Practice

Surveying and mapping practice in the mining industry encompasses most surveying fields; its main difference is that it has a direct effect on the safety of people working in the mines. Accurate surveys and reliable maps are a prerequisite to a successful mining operation. If an accident occurs, such as a roof fall or an onrush of water or oxygen-deficient air, surveying operations must be performed immediately to aid in rescue efforts. Time is important, and confidence in the surveying and mapping system of the mine is essential. If a rescue borehole is needed, a spatial position must be promptly established on the surface or in an adjacent mine; a good surface-underground three-dimensional coordinate system is required.

In the deep-mining industry, conventional plane surveying practice is used to accomplish most surveying and mapping needs. The environment in which surveying operations are performed is different. Because surveying is usually carried out in the dark, the cross hairs and micrometer of the theodolite, and the plumb-bob string or object sighted, must be illuminated. Other adverse environmental problems include dripping water, noise, mine gases, dust, high-voltage electricity, high traffic density, and air velocity. In extremely deep mines, surveyors also encounter high temperatures, and work must be performed in an air-conditioned environment. Small openings for traversing usually mean difficult setups and short traverse lengths. To accomplish a good deep-mine survey under these conditions requires experienced personnel and well-designed survey systems.

In the surface-mining (open-pit) industry, surveying operations are much like those in

---

*Analysis, procedures, and suggestions presented here are solely the responsibility of this author and *no liability or responsibility* shall be incurred by the government agency with whom this author is employed.

†Underground section written by E. Franklin Hart.

the heavy-construction industry. Mine surveyors work closely with mining engineers, geologists, soil scientists, and photogrammetrists in developing an operation plan that can result in efficient mining operations. Surveyors must then perform the operations' stakeout and monitor their results. Layout includes setting slope stakes, locating blast holes, and setting markers to implement the operation plan. Monitoring includes deformation measurements for checking pit-wall stability and indirect quantity measurements for payments to contractors. Much of the monitoring is done photogrammetrically after semipermanent photo panels have been established; aerial photogrammetry is accomplished on a weekly or monthly basis. In some cases, terrestrial photogrammetry, combined with indirect conventional surveying measurement, is used on a daily basis for quantity surveys, particularly when soil and rock are moved more than once. Because of surveyor involvement in surface-mining operations, preparation of plans and permits has evolved as his or her responsibility. Often, the surveyor is the owner's (company) representative in communication with regulatory agencies and contractors.

A common task for mine surveyors is to determine whether or not the necessary work can be performed with company surveyors. Frequently, they must let for contract geodetic, mineral (boundary), and photogrammetric surveying services because of the expert nature of these services and their cost-effectiveness. It is important that a mine surveyor be proficient in writing specifications for contract services and requires inspections before accepting the contracted work. Both education and experience are necessary in developing competent mine surveyors able to handle the broad professional scope of services that must be rendered. These persons are highly respected as members of the mine management team.

Operational deep-mine surveying will be discussed here. Mine surveyors must be innovative in applying appropriate technology and equipment to solve the unique problems encountered in practice.

29-1-2. Definitions

Figure 29-1 illustrates some common glossary terms used in mine surveying and mapping. A brief definition of each term follows.

Adit. An opening, usually horizontal, driven from the surface to a working area of the mine.

Back. Top or roof of an underground opening.

Cropline. Intersection of an ore body and the earth's surface.

Cross cut. Horizontal opening (breakthrough) between entries.

Drift. Horizontal or nearly horizontal opening in or into an ore deposit.

Entry or *Heading*. An opening driven in or into an ore deposit for use as haulways, ventilation, and/or access.

Face. End wall of an entry or opening where ore is being extracted.

Level. Working section designated by an elevation difference from the surface.

Open-pit mine. Large open excavation developed for extracting deep or steeply pitching ore deposits.

Pillar. Block of ore/rock between the entry and cross cut used to support the overburden.

Projection. Direction for advancement of the mine.

Raise. Vertical or sloped opening in the ore from a level.

Rib. Wall of an entry.

Roof. Top of the room or entry.

Room. Area from the last crosscut to the face.

Shaft. Vertical or sloped opening in or into a mine used for a haulway, ventilation or access.

Slope. Inclined opening driven to the ore deposit.

Spad. Metallic marker usually set in a wood plug in the roof for traverse and projection control.

Stope. Working section or room from which ore is extracted.

Strip mine. Mine developed for excavating ore, usually coal, along the contour of the surface.

Top. Top of the room or opening.

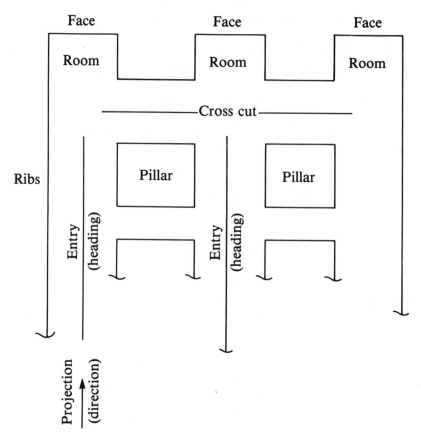

Figure 29-1.   Top view of a mining section.

29-1-3. Deep-Mine Maps

A requirement of the Mine Safety and Health Act of 1977 is that maps of all coal mines be certified by a registered engineer or registered surveyor of the state in which the coal mine is located.[1] A summary of state and federal laws and regulations governing mine maps has been listed by Williams.[2] Laws and regulations for all types of mining vary from state to state, and in some cases, they are more rigorous than federal laws and regulations. A section of the West Virginia underground coal mine map requirements is quoted as an example of deep-mine requirements.

The operator of every underground coal mine shall make, or cause to be made, an accurate map of such mine, on a scale of not less than one hundred, and not more than five hundred feet to the inch. The map of such mine shall show:

(1) Name and address of the mine;

(2) The scale and orientation of the map;

(3) The property or boundary lines of the mine;

(4) The shafts, slopes, drifts, tunnels, entries, rooms, crosscuts and all other excavations and auger and strip mined areas of the coalbed being mined;

(5) All drill holes that penetrate the coalbed being mined;

(6) Dip of the coalbed;

(7) The outcrop of the coalbed within the bounds of the property assigned to the mine;

(8) The elevations of tops and bottoms of shafts and slopes, and the floor at the entrance to drift and tunnel openings;

(9) The elevation of the floor at intervals of not more than two hundred feet in:

(a) At least one entry of each working section, and main and cross entries;

(b) The last line of open crosscuts of each working section, and main and cross entries before such sections and main and cross entries are abandoned; and

(c) Rooms advancing toward or adjacent to property or boundary lines or adjacent mines;

(10) Contour lines passing through whole-number elevations of the coalbed being mined, the spacing of such lines not to exceed ten-foot elevation levels, except that a broader spacing of contour lines may be approved for steeply pitched coalbeds by the person authorized to do so under the Federal act; and contour lines may be placed on overlays or tracings attached to mine maps;

(11) As far as practicable the outline of existing and extracted pillars;

(12) Entries and air courses with the direction of airflow indicated by arrows;

(13) The location of all surface mine ventilation fans, which location may be designated on the mine map by symbols;

(14) Escapeways;

(15) The known underground workings in the same coalbed on the adjoining properties within one thousand feet of such mine workings and projections;

(16) The location of any body of water damed in the mine or held back in any portion of the mine, but such bodies of water may be shown on the overlays or tracings attached to the mine maps used to show contour lines;

(17) The elevation of any body of water damed in the mine or held back in any portion of the mine;

(18) The abandoned portion or portions of the mine;

(19) The location and description of at least two permanent base-line points coordinated with the underground and surface mine traverses, and the location and description of at least two permanent elevation bench marks used in connection with establishing or referencing mine elevation surveys;

(20) Mines above or below;

(21) Water pools above;

(22) The location of the principal streams and bodies of water of the surface;

(23) Either producing or abandoned oil and gas wells located within five hundred feet of such mine and any underground area of such mine;

(24) The location of all high-pressure pipelines, high-voltage power lines and principal roads;

(25) The location of railroad tracks and public highways leading to the mine, and the mine buildings of a permanent nature with identifying names shown;

(26) Where the overburden is less than one hundred feet, occupied dwellings; and

(27) Such other information as may be required under the Federal act or by the department of mines.

The operator of every underground coal mine shall extend, or cause to be extended, on or before the first day of March and on or before the first day of September of each year, such mine map thereof to accurately show the progress of the workings as of the first day of July and the first day of January of each year. Such map shall be kept up to date by temporary notations, which shall include:

(1) The location of each working face of each working place;

(2) Pillars mined or other such second mining;

(3) Permanent ventilation controls constructed or removed, such as seals, overcasts, regulators and permanent stoppings, and the direction of air currents indicated; and

(4) Escapeways designated by means of symbols.

Such map shall be revised and supplemented at intervals prescribed under the Federal act on the basis of a survey made or certified by such engineer or surveyor, and shall be kept by the operator in a fireproof repository located in an area on the surface chosen by the operator to minimize the danger of destruction by fire or other hazard.[3]

In most states, mine laws require that these maps be referenced to a rectangular coordinate system and either two or three permanent survey monuments on the surface of the mine property. These monuments should be described and shown on the mine map so that they can quickly be found for use by any surveyor having permission to utilize them.

They are particularly helpful for tying together adjacent mine workings. However, having all maps and surveys connected to the National Geodetic Survey (NGS) network is desired.

Underground mine maps are used by all mining personnel. Not only are they employed to comply with mine laws, they also provide relative location for anyone utilizing them. They are necessary to determine volumes of ore and rock extracted so quantity control can be checked and royalty payments calculated. Almost all deep-mine engineering work performed after production begins depends on the maps.

29-1-4. Surface-Mine Maps

Surface-mine maps, open-pit maps, and strip-mine maps differ from underground maps and are very dependent on the production operation plan of an individual mine. They resemble site plans or heavy-construction operations. Quantity determination is critical; thus, surface depiction is updated either daily or weekly. Often, payment is made on the basis of material moved by the contractor.

A reclamation map is required for surface mining; it is produced using topographic maps and the assistance of other scientists. It shows both the premining and postmining plan of the land, including land use.

Surveying control for surface mining usually does not need to be as accurate as for deep mining. It is an advantage to have permanent survey control, so construction stakeout can be accomplished according to the plans and photo panels for aerial photography can be tied to a permanent system. This enhances regular updates of maps by aerial photogrammetry, keeping quantity surveys both accurate and timely.

29-1-5. Horizontal and Vertical Control for Mine Surveys

Because of the hazards affecting miner safety, the cost to develop underground mines,

and absence of the advantages in conventional surveying practice, a comprehensive surveying system must be established for each mining operation. Its base should be the NGS network for both horizontal and vertical control. Deep-mine surveys are performed three-dimensionally. It is frequently necessary to correlate surveys on the earth's surface with those underground. If a rescue borehole is requested between the surface and a position in a mine, time and a tested survey system are of the essence.

Important surface features, such as highways, power lines, gas lines, gas wells, and other structures, require system coordinates. Adjacent mine workings must be made part of the same system to eliminate encroachments and enhance mine planning. Without a good survey system, the three-dimensional working environment becomes cumbersome for an engineering department. All surrounding mine and other maps must be reviewed and fitted to the master map of the mining operation. Too often, different coordinate systems exist for adjacent mines, the highway department, railroad company, gas company, surface landowner, mineral landowner, or a mineral seam holder at a different level. This problem needs to be solved at the project's beginning.

Chrzanowski and Robinson recommend that three levels of surveying control be established.[4] The first level should be for basic control in the mining area and control in the permanent workings; the second for surveying the mine headings and development areas; and a third level should be for short traverses necessary to map and make daily projections for production.

In the United States, the first level is generally a second- or third-order NGS network tie performed using NGS specifications. Often, this tie is contracted to consultants as part of the original surface surveying and mapping project. Figure 29-2 shows surveying monumentation for an underground mine.

The second level of control is usually surveyed by mine company surveying personnel with theodolites, EDMIs, tapes, sometimes

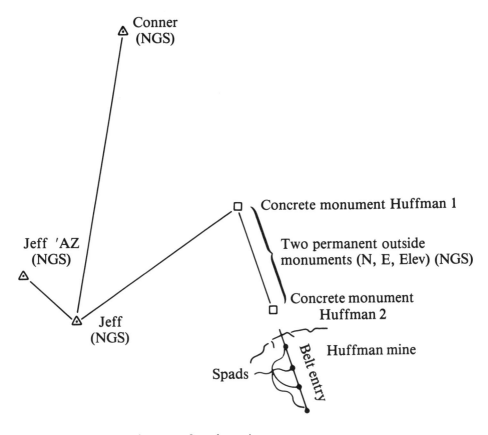

Figure 29-2. Monumentation system for a deep mine.

complemented by gyro-theodolites (Chapter 24), automatic nadir plummets (Figure 29-3), and levels. The same personnel using theodolites and tapes perform the necessry work for third-level control. Direction is particularly difficult to maintain underground. For this reason, it is advisable to employ good-quality theodolites that are maintained and checked at regular intervals. When a projection is being advanced over a large area, a gyro-theodolite essentially enhances accurate direction.

For good-quality surveying control, the chief mine surveyor must attempt to retain experienced surveying personnel and provide modern and efficient equipment. Company-adopted standards of practice can prove very helpful in developing the mine surveying control system, particularly in large mines.

29-1-6. Underground Surveying Equipment

Equipment for underground surveying is not too different from that used on the surface. Theodolites are the primary angle-measuring instruments; for underground use, they should be protected from dust and moisture. Most theodolites produced today have been sealed at the factory. Theodolite micrometers and cross hairs are illuminated by either an attached light or cap lamp.

Special attachments are sometimes added when very steep slopes are encountered, but less frequently now because of an increasing shift to automatic nadir plummets and gyro-theodolites, along with new technology.[5]

Tripods differ if height restraints are met; sometimes, a special tripod has to be made for

Figure 29-3.   ZNL automatic nadir plummet. (Courtesy of Wild Heerbrug Instruments, Inc.)

a particular need. When an EDMI is utilized, it is preferable to send signals transmitted through the telescope. Lower height restraints also work better with trigonometric levels. Since direction control is so critical, it is important that a reliable and accurate theodolite be employed on underground control traverses.

For distances, a steel tape is probably still used for 60% of the measurements. However, for higher-order underground and check surveys, an EDMI is best. Use of EDMIs underground is now very prevalent and significantly

improves survey system accuracy. One constraint exists in that, for safety reasons, an EDMI cannot be operated in some underground environments. Some EDMIs have been made explosion-proof, but the extra weight and cost for this have not justified its use.

A standard plumb bob is the most common target. It is illuminated by a cap lamp and hung from a spad in the roof (see Figure 29-4). Small reflectors are used underground with EDMIs and conventional reflectors on forced-centering traverses. Special explosion-proof illuminated reflectors can be attached to the spad (Figure 29-5).

With the increased use of more-precise theodolites that provide accurate vertical or zenith angles, trig levels are common. Automatic levels are still employed but on a much reduced scale. Level rods used underground are likely to be a fold-up engineer's rule or roll-up rule. Sometimes, a sawed-off conventional rod is substituted, but less often, because of the different heights encountered.

Total stations with data collectors are now being introduced underground but may be restricted for safety reasons. They must be capable of withstanding rough handling due to underground transportation, but they offer advantages by expediting data collection.

29-1-7. Underground Traversing

From permanent monuments outside the mine, control has to be transferred under-

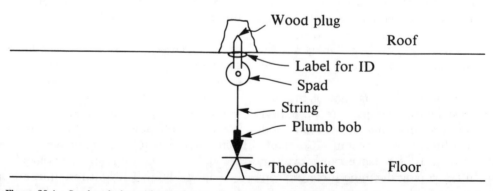

Figure 29-4.   Spad and plug with plumb bob.

Figure 29-5. Illuminated mining prism attached on a spad. (Courtesy of Seco Manufacturing Co., Inc.)

attached to the spad for station identification. The identification can be a sequential number, conventional stationing number, or code. Figure 29-6 demonstrates traversing underground. The theodolite setup underground is made by first suspending a plumb bob from an opening in the spad and then positioning the theodolite under the bob. A small centering point is drilled on the theodolite scope for this purpose. A telescope roof plummet offers an alternative for positioning under a spad. A backsight is taken to a plumb bob suspended from another spad, using the cap lamp to observe the string or target. The cross hairs and micrometer must also be lighted by a cap lamp unless a battery source is available for internal lighting. Figure 29-6 shows the data that must be collected at each station.

Sight spads are frequently set to guide the production crews for alignment. These spads are usually located within 10 ft of the traversing spad on the projection for advancement and generally set perpendicular to it for cross cuts. These are only for use in production and should not be confused with traverse stations.

ground by traversing or plumbing down a shaft for coordinates, using a gyro-theodolite for direction control. Chrzanowski et al. have published good reference material on the transfer of control down shafts.[6] Discussion in this section will focus on underground traversing technology.

After traversing to the mine site, by using conventional techniques, a position is established on a spad in the roof underground, usually on a defined projection. A metal tag is

After setting the spads for advancement and looping the traverse for closure, mapping is accomplished by measurements to pillar corners using pluses and offsets or in large rooms by angles and distances. These data are necessary to map the underground workings. Thus, the first traverse run underground serves two purposes: (1) to extend control using a designed alignment and operating plan and (2) serve as a reference for mapping.

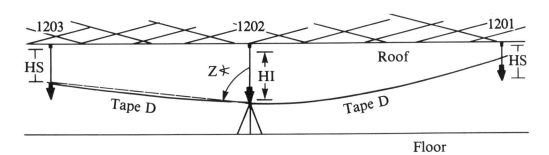

Figure 29-6. Traverse station underground.

After advancing a distance of 1000 to 1500 ft, a check survey of higher order is performed using an EDMI with forced-centering techniques. Longer distances are available for sights; thus, better direction accuracy is expected. The main purpose for this survey is to transfer the best coordinates and direction to the last two spads set. This can be accomplished using a gyro-theodolite to establish a direction on the last two spads and by measuring the distances across the other spads independently. In advancing an underground working a long distance, direction is much more critical to the traverse than distance. When a shaft is encountered, another check is made on the traverse, and the coordinates and direction of the underground survey are adjusted to fit the surface system.

Experience is important in traversing underground and often demands innovation when short sights cannot be avoided. To extend the projection direction, multiple setups under the same spad are made, and the average of the position ahead used for final location of the spad. Surveyors must meet accuracy requirements because, ultimately, a test of the survey system will be made.

29-1-8.   Underground Leveling

When leveling underground using conventional surface surveying equipment, procedures are the same except that stations are on the roof, either as a spad or roof bolt. Because the rod is positioned from above, backsights are considered negative and foresights positive. Time constraints have caused much underground leveling to be performed while traversing with the theodolite and using trigonometry. Since the accuracies of vertical and zenith angles are adequate on theodolites, good leveling results are achieved by trig levels. The 1-min mining transits did not offer acceptable precision levels to achieve the same results, and it has been difficult to convince some surveyors of leveling exactness, when using modern theodolites. Before deciding leveling methodology for underground surveys, the maximum allowable error should be established. Trig levels usually produce the desired results.

29-1-9.   Computer Usage in Underground Surveying and Mapping

The process of reducing field notes, plotting the results, and determining the quantities of ore and rock is a routine and tedious task best performed with computers and plotters. Not only is valuable time saved, accuracy is also improved in mapping and quantity determinations. A software package has been developed by Howard to use a low-cost microcomputer interfaced with an interactive digital plotter to map underground coal workings.[7] Table 29-1 lists the pillar location commands and shows the proper sequence of data and codes in the pillar location command line. Braces { } indicate that one of the enclosed codes must be used. Brackets [ ] indicate the enclosed codes or data are optional. If an item stands alone, an entry is necessary.

An example of command use to input field notes is shown in Figure 29-7, and a sample portion of a mine map prepared using this software depicted in Figure 29-8. For large mines, an integrated collecting, computing, and plotting system can be set up employing computers.

29-1-10.   Safety Considerations

The surveyor, an important member of a mine management team, is responsible for miner safety, particularly with respect to the maps and surveys required in the operations. To produce the necessary maps and surveys, equipment such as EDMIs and gyro-theodolites can cause serious safety problems if not properly used. The ambient working conditions encountered in mines pose difficult working environment problems associated with high-voltage electricity, mine gases, dripping water, poor visibility, noise, and heavy traffic. Surveyors must understand the safety implica-

Table 29-1. Pillar location commands

HDn:	Identifies the heading (numbered serially from 1, left to right) where n is the heading number.
PLn:	Identifies the next pillar to be located on the left side of the heading, where n is the pillar number.
PRn:	Identifies the next pillar to be located on the right side of the heading, where n is the pillar number.
FS:	Identifies the spad from which the pillars are measured.
TS:	Identifies the spad sighting on for azimuth determination.
AZ:	Defines the azimuth for the heading. Azimuth determination is made by this command or a pair of FS/TS commands.

Pillar corner location command format

$$\text{Plus}\begin{Bmatrix} \text{First} \\ \text{Second} \\ \text{Mid pt.} \\ \text{Wall} \\ \text{Top} \\ \text{Bottom} \end{Bmatrix} \begin{Bmatrix} \text{Left} \\ \text{Right} \end{Bmatrix} \left[\begin{Bmatrix} \text{F} \\ \text{S} \\ \text{M} \\ \text{W} \\ \text{T} \\ \text{B} \end{Bmatrix}\begin{Bmatrix} \text{L} \\ \text{R} \end{Bmatrix}\right] \text{ offset [offset] } \left[\begin{Bmatrix} \text{Coal} \\ \text{Shale} \\ \text{Slate} \\ \text{Sandstone} \end{Bmatrix}\text{Height}\right]$$

tions of working with electronic equipment in this environment; its improper use in a gassy area mine can cause an explosion. Inaccurate measurements could lead to mining into old workings, possibly filled with water- or oxygen-deficient air. Accurate maps correlated with good surface and underground coordinate systems are a must in the mining industry, particularly during rescue operations.

Surveyors must know more than measurement technology when working in deep-mine environments and should be trained in safety and first-aid. They need a fundamental knowledge of mining technology, especially ventilation and mine gas-handling practices. At least one survey crew member should be certified to use the mine safety lamp and methanometer. For the safety of the miners and the survey crew, good communication is necessary between all those working underground.

29-2. ABOVEGROUND*

29-2-1. Background

A variety of textbooks is available containing opinions, instructions, and legal citations dealing with retracements, resurveys, restoration of original corners, and subdivision of sections of the rectangular survey system. Additional texts cover commercial subdivisions dealing with survey, resurvey, and restoration principles of lot, easements, air rights, rights of way, and other related urban land development practices. Riparian law fills numerous volumes and restoration of lost corners on many types of metes-and-bounds [homestead entry surveys (HES), desert land claims (DLC), small holding claims (SHC), and land grants] parcels have been challenged in the court, resulting in some well-established common law practices being upheld and published.

In spite of the ever-increasing library of information being written as a guide for surveyors, there remains a critical void concerning resurveys of a patented mineral survey. No established common law can be found to draw on for principle or instruction; the Bureau of Land Management's (BLM) *Manual of Instructions*[8] (hereafter referred to as the 1973 manual) contains almost nothing on the subject and the *Mineral Survey Procedures Guide*[9] has only one-and-a-half pages on resurveys of

---

*Aboveground section written by John S. Parrish.

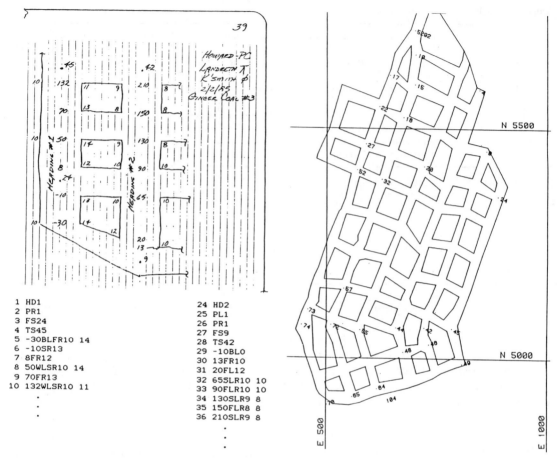

```
 1 HD1
 2 PR1
 3 FS24
 4 TS45
 5 -30BLFR10 14
 6 -10SR13
 7 8FR12
 8 50WLSR10 14
 9 70FR13
10 132WLSR10 11
   .
   .
   .
```

```
24 HD2
25 PL1
26 PR1
27 FS9
28 TS42
29 -10BL0
30 13FR10
31 20FL12
32 65SLR10  10
33 90FLR10  10
34 130SLR9  8
35 150FLR8  8
36 210SLR9  8
   .
   .
   .
```

Figure 29-7. Example of typical underground coal mine survey notes and the pillar location commands to enter into the computer. The command number is generated by the program; only the command is entered by the user. (Courtesy of Jack Howard, Jr.)

Figure 29-8. Sample mine map generated by plotter. (Courtesy of Jack Howard, Jr.)

patented mining claims. Hereafter, any reference to "mineral survey" will mean "patented mineral survey."

There are three typical forms of mineral surveys: (1) lodes, (2) millsites, and (3) placers. In most cases, lost corners of mill sites and placers are restored by the more commonly known "grant boundary" procedure. A grant boundary solution is generally acceptable when there is an insignificant deviation found between the corner monuments remaining and record plat/field notes. Most difficulties arise when the recovered corner monuments differ

considerably from the record, and/or unrecorded gaps/overlaps are identified during the retracement phase of the resurvey work. The following discussion will center primarily around the patented lode mineral survey.

29-2-2. Historical Development of Mineral Surveying Procedures

Mineral surveys started in the early to mid-1800s in the southeast portion of the United States. During the Gold Rush days of the mid-1800s, surveys were made under various rules established by local mining districts or political entities. Survey procedures, monumentation, and recordation varied dramatically, often lacking in quantity and quality. Beginning in 1865, several federal mining laws were enacted, the most significant being the Act of May 10, 1872. This 1872 act provides the General Mining Laws that are still in force today, with the addition of certain amendments. Specific standards and procedures for surveying mineral claims were detailed and have remained essentially unchanged. Chapter 10, "Mineral Surveys," of the 1973 manual consists of eight pages that briefly guide the deputy mineral surveyor on survey requirements and provide only hints for a retracement surveyor to follow.[10]

Special instruction issued to the deputy mineral surveyor by the former General Land Office (GLO) and today's BLM contains detailed instructions for surveying a mineral claim for patent. Additionally, several books have been written and published providing detailed information on mineral surveys. Most of them contain similar information. Underhill, *Mineral Land Surveying*,[11] published in 1906, typifies this information and will be referenced on occasion throughout the discussions to follow.

Some significant and specific procedures govern surveys of mining claims that differ from the rectangular system or metes-and-bounds surveys such as homestead entry surveys, desert land claims, or land grants. Survey closure is required to be 1/2000, whereas lode claims may not exceed 1500 ft in length nor more than 300 ft in width on each side of the center lode line. Measurements to bearing trees are made to a cross X on the face of the tree (as opposed to the center). The survey is paid for by the claimant and only duly authorized deputy mineral surveyors can perform mineral surveys for patent. Other important aspects involve the survey's intent with respect to discovery points, center lode lines, parallelity of end lines, and extralateral rights, to name a few.

29-2-3. Common Surveying Practices

Gathering and correctly interpreting all previous survey notes are imperative. Of equal importance is a knowledge of the actual field procedures and equipment used, though in conflict with that actually recorded. Experienced surveyors recognize differences and utilize their knowledge to recover difficult original evidence and resolve record/physical conflicts in harmony with the intent of original monumentation. It is important that surveyors make note of actual or apparent original surveying practices that conflict with recorded field notes and prescribed procedures.

Underhill's opening paragraph of Chapter 4 of *Mineral Land Surveying* offers some insight on common practices and anticipated results. As he states:

> About the simplest survey that the western surveyor is called on to make is that of a lode location. It is, however, somewhat complicated by the fact that as a rule he is assisted by the claimant himself in the work and thus often lacks an efficient assistant, with the result that the character of the results suffer.[12]

It is important to note that Chapter 5, "Surveying for Patent," contains the following: "The deputy surveyor then surveys the claim exactly as described for the location survey, except that the work is done more carefully, and with greater safeguards."[13] Forward-thinking deputy surveyors would usually survey a location as though for patent, thus saving efforts in redoing a survey after the patent survey request was authorized. The only additional work remaining was to mark the corners and accessories with the assigned mineral survey number prior to completion of "running" field notes.

Examination of a typical set of mineral survey field notes of a lode will indicate (1) starting at the discovery point, (2) running along the center lode line to a center end-line point (monumentation optional depending on locale), (3) measuring to corner no. 1, (4) courses and distances between each successive corner (including the opposite center end-line point), and (5) physical closure back to corner no. 1. If the field notes read differently, they would be rejected and the surveyor asked to rewrite them. In reality, the actual field procedure was probably as follows: beginning at the discovery point; then along the center lode line to a center end-line point; then offsetting left and right to the lode corners; then again from the discovery point measuring the calculated distance in an opposite direction not to exceed 1500 ft along the center lode line to the center end-line point; then offsetting (parallel to the opposite end line) left and right to the lode corners.

In many cases, the side lines were never actually run as recorded in the field notes. A noticeable lack of topographic calls along side lines over difficult terrain is a positive indicator of that common practice. Additional support for this method is found in *Mineral Land Surveying*, where Underhill discusses survey procedures for lode locations:

> ... At which point the claimant having desired to end his claim, a right angle is turned off (from the center lode line), and the stakes set... on each side of the center line.... We now extend the line through No. 5 and No. 6 [tangent points on the center lode line] and here knowing that the survey can be completed with another sight, the previous measurements are reduced to horizontals, the total subtracted from 1500 and the result laid out. Corners No. 3 and No. 4 are then set as for the westerly end.[14]

Retired and active deputy mineral surveyors confirm these procedures and common-sense analysis of existing conditions found during survey retracements verifies the practice.

When multiple side-by-side lodes were run, the center lode line was seldom traversed and

corners set by the shortest procedures available. Underhill suggested that

> In the case of groups of claims, the surveying may be often greatly simplified by a little forethought. This is evident in the case of those locations which lie side by side when one surveyed center line may serve for the whole group, the end lines being run from its two ends.[15]

29-2-4. Surveying for Patent

It is unnecessary to recite the many requirements of surveying for patent. It is appropriate, however, to strongly recommend that all practicing land surveyors possess and become familiar with the 1973 BLM manual, the BLM's *Mineral Survey Procedures Guide*, and at least one or more textbooks similar to Underhill's *Mineral Land Surveying*. These publications outline the minimum requirements for surveying mining claims and provide today's surveyor with valuable insight in understanding the principles, procedures, and intent of a mineral survey.

29-2-5. Intent: Parallelism, Discovery, Monumentation

Land surveying is more an art than a science. Recognizing the difference is important if valid and successful professional surveying services are to be provided to a client. Some simple decisions, when only one lode corner appears lost, are to apply a "grant boundary" solution; reestablish by record courses/distances; or reset from record calls to nearby lode corners.

Lode sidelines are often reestablished on the ground from calls to points on lines of adjoining lodes. Referring to Figure 29-9, a surveyor may be asked to locate the line between corners 1 and 2 of lode *A*. Corner 2 of lode *A* is assumed lost and a tie to line 1-2 of lode *A* contained in the field notes for lode *B*. These notes state that line 1-2 of lode *A* is intersected at a point 500 ft along the course

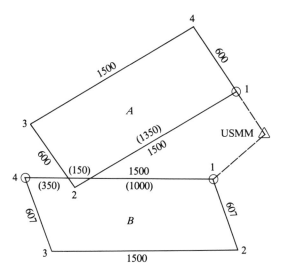

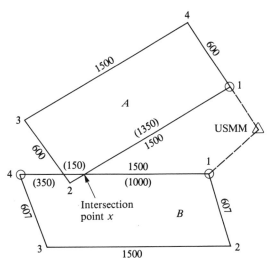

Figure 29-10. Intersection point *x* established from record ties of survey *B* to line 1-2 of survey *A*. Corner 2 *A* was assumed lost.

Legend: △ USMM found
○ Lode corner found

Figure 29-9. Senior survey *A* and junior survey *B* showing tie between line 1-2 of *A* and line 1-4 of *B*.

from corner 4 to corner 1 of lode *B* and is 1350 ft from corner 1 of lode *A* (or called 150 ft from corner 2 of lode *A*). Clients desiring the least expensive survey possible, coupled with some surveyors' blatant acceptance of only record ties, often result in erroneous establishments of desired deed lines.

Figure 29-10 illustrates that point *x* has been established strictly from record and field ties to only one corner 1/lode *A* and corners 1-4 lode *B*. Figure 29-11 shows that corners 3 and 4 of lode *A* were eventually located, and lode *A* actually set differently on the ground than the record indicates. The initial blunder appears to be a "computed" tie for an intersection of lodes *A* and *B* after having tied only to the U.S. Mineral Monument (USMM). Such computed ties are common, and a junior lode surveyor is required to note conflicts with adjoining lodes whether or not actual corners are located. After corners 3 and 4 of lode *A* were recovered, it was a simple matter to search at record distances from corners 1 and 3 to find corner 2.

A well-established principle of the configuration for a lode is that end lines are intended to be parallel. Figures 29-9 through 29-11 demonstrate that each lode must stand on its own merits, both in monument control and

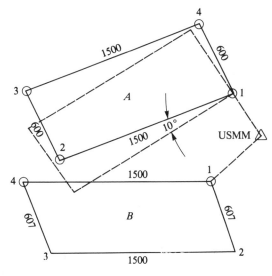

Figure 29-11. Corners 3 and 4 of survey *A* were located, which led to the location of corner 2*A*. A 10° bearing blunder with survey *A* is noted. Line 1-2 of survey *A* was erroneously reestablished by using only the computed ties from survey *B*.

intent of patent for a specific claim. Adjoining surveys can be used as secondary control, when physical evidence of a specific lode is totally lost. Actual existence of parallel end lines is similar to the supposition that all regular $\frac{1}{4}$ corners are on line and at midpoint between section corners on either side. Seldom is this the case, but end lines are usually found to be substantially parallel. This principle, or intent, must be carefully considered during resurvey work on a lode claim.

Another important element of lode surveys is the significance of discovery points. Without a discovery point, the lode could never have existed. Claimants must have worked their discoveries sufficiently to validate a lode claim, and the location survey had to begin at the discovery point. Before a mineral claim is assigned a number and an order for survey issued with the intent for patent purposes, the discovery will be witnessed on several occasions. If there is a total lack of corner evidence, a discovery point cannot be found, improvements are nonexistent, and the ground has not been significantly disturbed by man or nature, it is suggested that a search for the lode be made elsewhere. It is not uncommon to eventually locate an actual claim several miles away from map-projected locations. Discovery points alone, if verified beyond reasonable doubt, can be used to reestablish lode locations in lieu of erroneous ties to secondary monuments, lines, or topographic features.

The most significant components of a mineral survey are its monuments. These controlling points form the extent and terminus of the connecting lines for a lode. Acreage is subserviant to actual monuments and the area they contain. U.S. Code Annotated, Title 30, *Mineral Lands and Mining* states:

> ... The said monuments shall at all times constitute the highest authority as to what land is patented, and in case of any conflict between the said monuments of such patented claims and the descriptions of said claims in the patents issued therefor the monuments on the ground shall

govern, and erroneous or inconsistent descriptions or calls in the patent descriptions shall give way thereto.[16]

Corners and lines of adjoining surveys should be used as a last resort and only when all evidence of controlling corners for the lode being resurveyed are determined lost.

29-2-6. Retracing the Patent

Once equipped with a better understanding of the surveying procedures and intent of a patented mineral claim, a retracement can be approached with sharpened logic and greater success. Searching through and collecting all written field notes pertaining to the patented lodes are essential. Plats are part of the notes and require a thorough search for amended, revoked, and adjoining surveys. Note-keeping styles varied greatly among deputy mineral surveyors, and each state accepting differing formats.

Be sure to acquire all the original field notes. Often, only notes containing descriptions around lodes are secured, while pertinent calls to accessories, man-made structures, and natural features are recorded in the "general description" following the lode traverse notes. When reviewing notes of multiple side-by-side claims, read them carefully to determine to which lode corner the call relates. Occasionally, accessories are recorded on the face of a plat in lieu of field notes. When dealing with a corner common to two or more claims, be sure all claim descriptions are secured and carefully read. It is not uncommon to record a bearing tree with one claim and delete the reference in the notes of an adjoining claim. Species, diameters, markings, bearings, and distance sometimes conflict between notes of two adjoining surveys. If only one set of notes is used and the correct information is contained in the second (unused) set of notes, a surveyor could fail to recover evidence of an original monument.

Many mineral surveys were done with intent but later revoked for various reasons. Monu-

ments were seldom destroyed on these revoked surveys, and valuable ties may exist to
aid in recovering evidence of adjoining
patented claims. Though a corner of a revoked survey does not control a patented
claim, it may prove valid for locating or
reestablishing corners of an adjoining claim
called for in its notes.

Never attempt to reestablish a missing lode
claim corner with searching out the evidences
of all lode corners. A blunder along one end
line or on the center lode line may be overlooked if all corners are not searched for.

Many mineral survey corners are overlooked because a surveyor has not measured
from the correct point on a bearing tree.
Sections 10-34 and 10-38 of the 1973 manual
provide instructions for measuring distances to
bearing trees: "The exact point on the
tree . . . to which connection is made is indicated by a cross or other unmistakable
mark."[17] Measurement to a cross on the face
of the blaze or to the blaze without a cross has
been standard practice for mineral surveys.
Figure 29-12 illustrates the differences between distances measured from the centers of
two bearing trees instead of to the face. Tree
diameters and steepness of the hillside where
the corner was set contribute to the success or
failure in finding a rotted wood post below
ground level, if distances are measured from
the wrong points on the bearing trees. Bearing
differences often are slight, providing little
awareness of error by a retracement surveyor.

Figure 29-13 shows only one original bearing tree. Since bearings were usually less accurate than measurements, it is necessary to
search left and right of the reestablished bearing off the bearing tree. With larger-diameter
bearing trees, a crucial mistake could be
made if distances are measured from the tree
centers.

Occasionally, a mineral surveyor employed
crew members with experience in rectangular
survey systems. These employees, through
habit, may have measured distances to the
centers of bearing trees, even though their
instructions were to measure to the face or

cross. When such a situation is suspected, it
must be evidenced by existing corner to bearing tree measurements of the lode corners
being retraced.

29-2-7. Conflicting Patent Locations

Not uncommon to mineral surveys are erroneous locations of claims on maps, in description, or by field-note ties included with
subsequent surveys (either rectangular or
metes and bounds). Searches for patented
claim locations must be in an area of known
mining activity. Placers are seldom found on
ridges; lodes will often provide two or more
discoveries; and millsites are usually located
where buildings can be erected and access is
easy. When all signs of such conditions are
absent, start looking elsewhere for the claims.

Government surveyors (GLO, BLM, and
deputy mineral surveyors) were and are instructed to make actual ground ties to corners
of conflicting mining claims and note crossing
of claims along section lines. If these instructions were followed explicitly, such record ties
could be used with confidence to reestablish
apparent lost corners of mineral claims. Unfortunately, short cuts were and are still being
made, leaving a challenge when locating corners of mining claims. Figures 29-14 and 29-15
are actual, characterizing situations discovered
during subsequent field investigations of lode
surveys.

Five adjoining millsites were surveyed in
1883 and tied to their parent lode and a
USMM on a ridge high above the millsites.
(Figure 29-14). In 1922, a GLO survey of the
line between sections 21 and 22 stated:

> . . . 47.19 chs. Intersect line 3-4 of Allen Millsite,
> Survey No. 19B, 88 lks. S. 49° 15' W., from cor.
> No. 3, which is a decayed pine post, 4 ins.
> square, set in mound of stone, mkd. and wit-
> nessed as described by the surveyor general.[18]

This call placed corner 3 east of the section
line. A tie was also called to line 1-2 of the

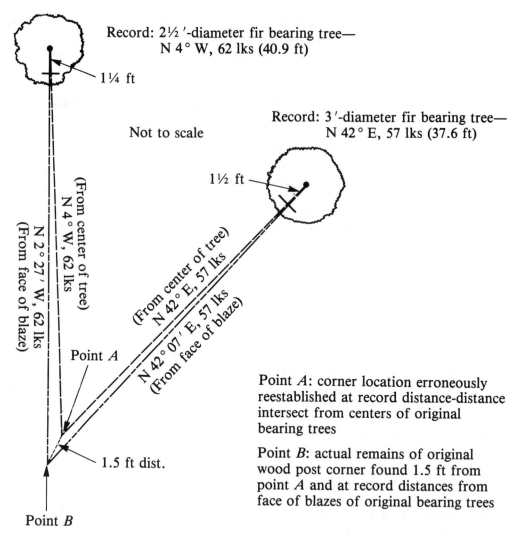

Record: 2½'-diameter fir bearing tree—
N 4° W, 62 lks (40.9 ft)

1¼ ft

Record: 3'-diameter fir bearing tree—
N 42° E, 57 lks (37.6 ft)

Not to scale

1½ ft

(From center of tree)
N 4° W, 62 lks

N 2° 27' W, 62 lks
(From face of blaze)

(From center of tree)
N 42° E, 57 lks

N 42° 07' E, 57 lks
(From face of blaze)

Point A

Point A: corner location erroneously reestablished at record distance-distance intersect from centers of original bearing trees

1.5 ft dist.

Point B: actual remains of original wood post corner found 1.5 ft from point A and at record distances from face of blazes of original bearing trees

Point B

Figure 29-12. Two original bearing trees. Original corner monument is wood post rotted portion remaining below ground level.

millsite, and at the northwest corner of section 21, a tie was made to USMM no. 5. The 1975 dependent resurvey of the millsites and section line found corner 3 of Allen millsite west of the section line, with no similarity to the 88 link call.

Field ties were made to the USMM, west ¼ and northwest corner of section 21, and 9 of the 12 corners of the millsites (3 lost by erosion and road construction). A 30-ft error from record was found between the USMM and

millsites and a 10-ft error between the northwest corner of section 21 and the USMM. The parent lode for Allen millsite contained the USMM, with ties to an adjoining lode to the south and millsites east. It is evident that ties from the millsites to the USMM were computed through the lodes in 1883. Likewise, intersections of the section line with the millsite were computed through 1883 record ties, after making a tie to USMM no. 5. Consider the erroneous relocation of the millsites if

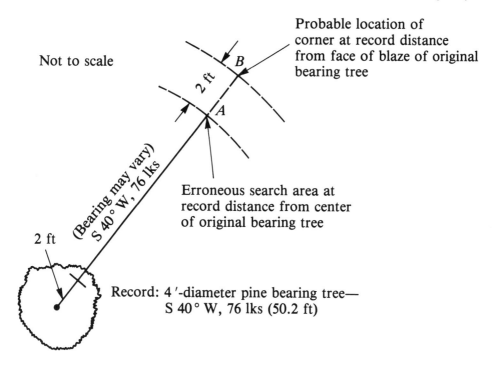

Not to scale

Probable location of
corner at record distance
from face of blaze of original
bearing tree

2 ft

B

A

(Bearing may vary)
S 40° W, 76 lks

Erroneous search area at
record distance from center
of original bearing tree

2 ft

Record: 4′-diameter pine bearing tree—
S 40° W, 76 lks (50.2 ft)

Figure 29-13. A thorough search at point *B*, left and right of the record bearing from the bearing tree, is the most likely location to find the original corner.

their original corners had not been found, and 1922 GLO ties were used to relocate "lost" corners of these millsites.

In 1904, the Interior Board of Land Adjustments (IBLA) rendered an opinion concerning the locations of two patented lode claims. A patent had been used for the Emma Nevada Lode (Survey No. 4348) on December 14, 1886. The Silver Monument Lode (Survey No. 15,714) received patent on April 28, 1902. Both were tied to the southwest corner of section 7, with the Silver Monument Lode tied also to the south $\frac{1}{4}$ of section 7 (Figure 29-15). Later in 1902, the grantee of the patented Emma Nevada Lode filed a protest against the patent application for the Silver Monument Lode, claiming a conflict existed in the field. The protest was dismissed on the grounds that no evidence of such a conflict existed by examination of notes and plats for the two lode claims. The Emma Nevada Lode grantee appealed, and a field investigation was eventually

made, confirming the existence of a conflict between the two lodes.

Defense for the patentee of the Silver Monument Lode stood firm on the premise that no conflict existed, due to the fact that the records did not conflict and ties to the section and quarter-section corner were an integral part of the locus (location) of a patented claim. Argument was finally reduced to the physical locations of the corners of each lode, without regard to their ties to corners of the public survey or USMM. The principle of "monument control" was upheld, with the patent being rejected to any portion of the Silver Monument Lode actually in conflict with the Emma Nevada Lode.

Examination of the Emma Nevada/Silver Monument Lode conflict reinforces the fact that care must be used when reestablishing lost mineral claims from corners of the public survey, USMMs, or other recorded calls to adjoining claims. Similar situations are more

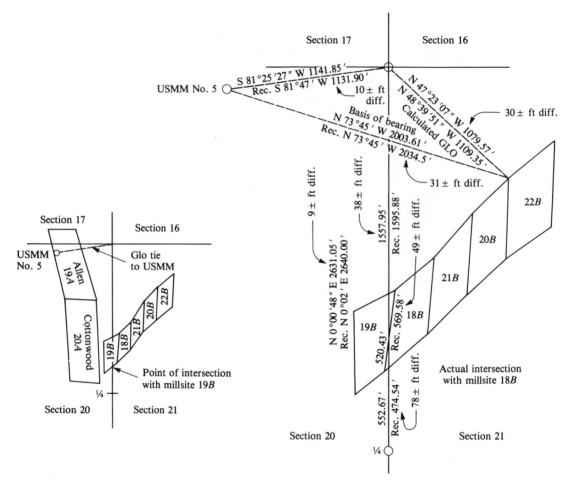

Figure 29-14. Left, 1922 GLO survey with ties to millsite 19*B* and USMM no. 5. Right, dependent resurvey showing numerous discrepancies in record ties between GLO and mining claim. Section line intersection with millsite 19*B* was calculated through GLO tie to USMM no. 5 and parent lodes.

the exception than the rule, but exist frequently enough to warrant special attention when no physical evidence of the lode is locatable.

29-2-8. Restoration of Lost or Obliterated Mineral Survey Corners

Legal precedence concerning reestablishment guidelines for lost mineral survey corners is lacking, so careful consideration of the intent and original survey procedure must in-

fluence resurvey decisions. Information shown in Figure 29-16 provides three solutions: (1) parallel end line/sideline, (2) grant boundary, and (3) record end line/sideline. No solution will restore the lode close to its record size.

Solutions 1 and 2 provide essentially the same results, but 1 maintains the strict parallelity of end lines with slightly more acreage than 2. Bearings along the end and sidelines are close to record and either solution 1 or 2 would be difficult to argue. Solution 3, however, least resembles the intent of the original survey, resulting in substantially nonparallel

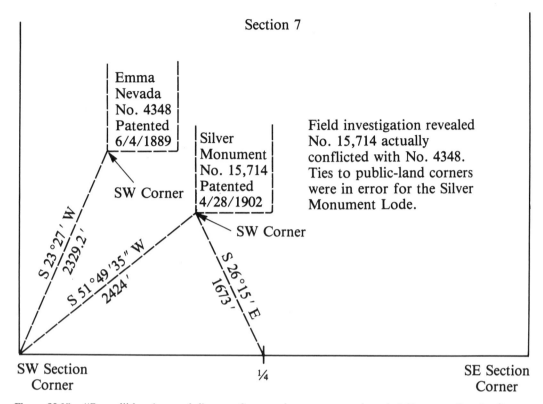

Figure 29-15. "Record" bearings and distances from southwest corner and south 1/4 corner of section 7 to southwest corners of surveys no. 4348 and 15,714. Plats and notes did not indicate a conflict between lodes.

end lines, notably disproportionate sidelines, and an acreage slightly larger than patented. Parallel end lines should be the decisive factor with the given conditions.

Figure 29-17 adds a more difficult conflict to the same lode, with an apparent 2° blunder along the westerly end line. When both distances and bearings depart significantly from record, a grant boundary solution (solution 2) provides a poor resemblance to the original record intent. The record distance-distance solution 3 still violates the intent of parallel end lines. Again, the recommended solution is 1.

Figures 29-17 and 29-18 were purposely selected with final acreage being less than that contained in the patent to emphasize the principle that acreage is nearly always the last consideration. If the lode had been long on the sideline, then excess acreage would have been enjoyed by the patentee.

The 1980 *Mineral Survey Procedures Guide* provides brief direction for reestablishing missing lode corners when one, two, or three corners are missing. The opening paragraph appropriately states:

> There is no hard and fast rule for establishing missing corners of mining claims. The method should be selected that will give the best results, bearing in mind that end lines should remain substantially parallel.[19]

Many surveyors agree with solutions to situations A, B, and C of Figure 7. The 1980 guide recommends using the broken boundary (nonriparian) or grant boundary method in many cases, but caution is again recommended if significant differences are found between the actual and record bearings and distances.

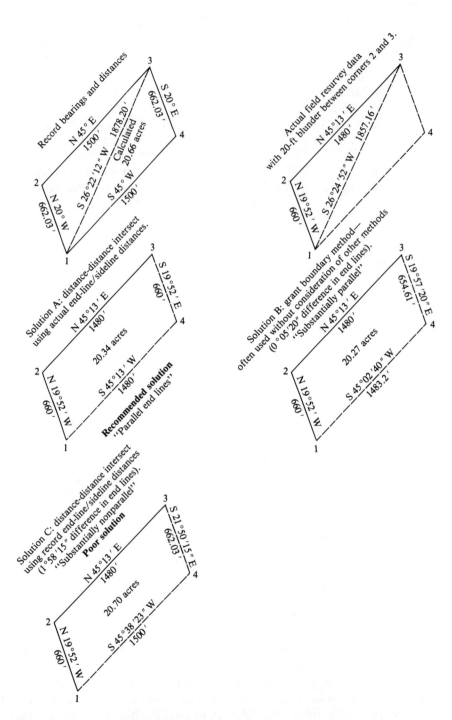

Figure 29-16. Lode corners 1, 2, and 3 located. Corner 4 missing. Recovered end line/sideline close to record bearing but 20-ft blunder found on sideline.

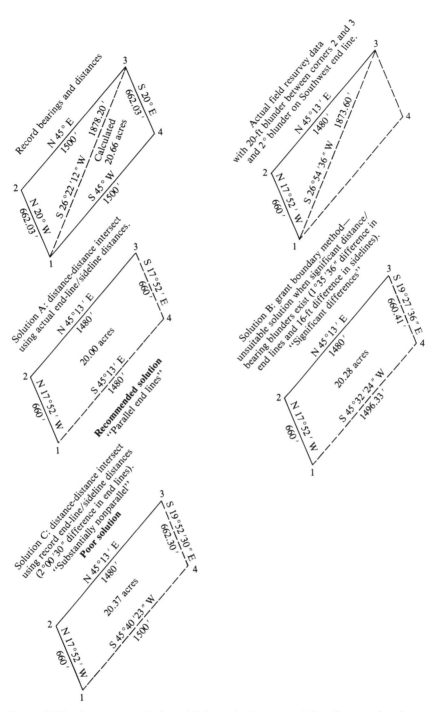

Figure 29-17. Lode corners 1, 2, and 3 located. Corner 4 missing. Recovered end line/sideline substantially different from record in both distance and bearing.

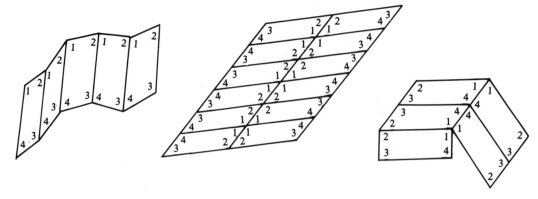

Figure 29-18.   Corners common to several claims. No gaps or overlaps exist.

29-2-9. Gaps and Overlaps Not of Record

Contained among the sporadic and overlapping mineral survey complexes are thousands of platted gaps and overlaps. Since mineral surveys are predicted on a claim of mineral deposit (and associated millsites), there exists no need or provision to patent all the surface area into neat manageable blocks of land. Seldom are platted gaps or overlaps actually the same dimensions as recorded, and frequently they are much larger or nonexistent. In addition to the inequities found with the platted gap or overlap is the reality that thousands of gaps or overlaps not officially platted are discovered during resurveys of mineral claims.

Mineral claims were often surveyed adjoining each other, with corners common to two or more claims. There is no doubt that all claim lines are common (Figure 29-18). When claims are offset with corners called along the lines of adjoining claims, a potential for gaps or overlaps exists. Referring to Figure 29-19, that the record notes and plat for lode *B* call a portion of line 1-4 (lode *A*) common to line 2-3 (lode *B*). Bearing and distance from corner 2 (lode *B*) are called to corner 1 (lode *A*); line 2-3 (lode *B*) is called along line 1-4 (lode *A*); corner 4 (lode *A*) is called along line 2-3 (lode *B*). With all these record calls, one would expect to find common lines between lodes *A*

and *B*. Unfortunately, such common lines seldom exist. Many discrepancies are found with the record, and it becomes the responsibility of a surveyor to identify the lines of specific claims.

A tendency may exist to "fill in the gap" and ignore the "overlap." Analysis of conditions leading up to the development of unrecorded gaps or overlaps is necessary at this time. Lode *A*, in Figure 29-19, was surveyed prior to lode *B* and a patent issued, with the location and acreage strictly dependent on the positions of corners 1 through 4. If no end line or sideline monuments were set, then lode *A* would be delineated by straight lines between the successive corners. Lode *B* was surveyed several years later and a patent also issued, based strictly on the corners of lode *B*. If lode *B* actually conflicts (overlaps) with lode *A*, then surface ownership will remain with lode *A* for that portion of conflict. When lode *B* is discovered a distance away from lode *A*, then a gap is identified and title to the strip of land vested in the owner of record adjoining lodes *A* and *B*.

Another example of a gap or overlap situation can be found in Figure 29-20, where corner 6 of the Abundance Lode is found 27 ft on either side of line 1-4 of Abundance No. 2 Lode. A field survey reveals that line 5-6 is substantially parallel with line 2-3 of the Abundance Lode. There is no legal justification for

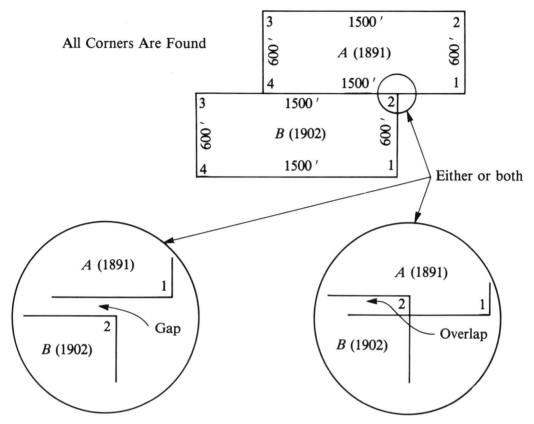

All Corners Are Found

Figure 29-19. Monuments of the specific patented lode control bounds for surface ownership. Gaps and overlaps are often discovered (not created) during resurvey.

moving corner 6 onto 1-4 of Abundance No. 2 Lode. Conversely, creating an angle point at corner 6, along line 1-4 of Abundance No. 2 Lode, would increase or decrease the acreage of Abundance No. 2 Lode without defensible reason.

When a mineral survey corner is lost, such as corner 4 of lode *B* (Figure 29-21), and called for on the line of an adjoining claim, it becomes imperative to retrace all existing corner locations prior to reestablishing the lost corner. When a blunder is identified in distance and/or bearing, the claim boundaries must be reestablished with closest conformity to the original location of the claim corners. With the discovery of a 2° blunder between lodes *A* and *B* and the relatively accurate lengths of the known end and sidelines of lode

B, missing corner 4 cannot logically be reestablished on line 1-2 of lode *A*.

29-2-10. Summary

Examples have already been discussed in which ties to USMMs and public-land corners have been determined erroneously. Distant ties are more likely to have been computed, less accurate in measurement, and thus not as valuable in reestablishing missing mineral claims or claim corners. When physical evidence conflicts with record ties and no reasonable solution is apparent, it may simply evolve to the principle of "closest and best." Courts usually decide on the validity of a survey procedure based on a preponderance of evidence. If surveyors can demonstrate reason-

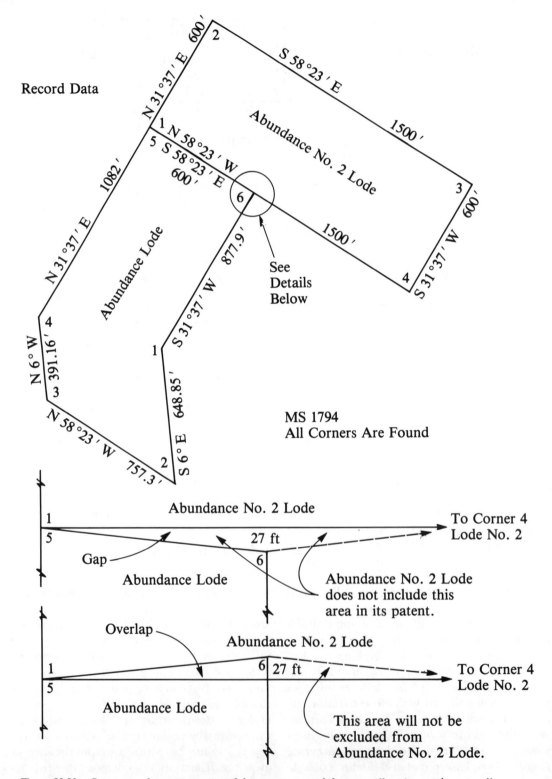

Figure 29-20. Gaps or overlaps: monuments of the patent, control the patent (location on the ground).

Missing Corner: Called for on "Senior Line"

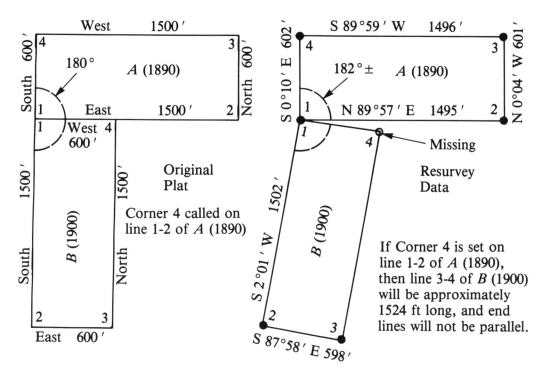

Legend: ● Found Corner

Figure 29-21. Missing corner 4 of *B* (1900) is reestablished by a distance-distance intersect from corners 1 and 3 of *B* (1900), using existing end line/sideline distances of *B* (1900). A gap will exist between *A* and *B*.

able effort in having gathered all the evidence, logically explain their solutions, express a good working knowledge of the background and intent of an original survey, then they will likely be successful when survey projects are challenged in court.

NOTES

1. Federal Mine Safety and Health Act of 1977. Public Law 91-173, U.S. Government Printing Office, Washington, DC. As amended by Public Law 95-164.

2. W. R. Williams. 1983. *Mine Mapping and Layout*. Englewood Cliffs, NJ: Prentice-Hall.

3. Coal Mine Safety Law of 1971. West Virginia Department of Mines, Charleston, WV.

4. A. Chrzanowski, and A. J. Robinson. 1981. Mining surveys. *Surveying Theory and Practice* (R. E. Davis, F. S. Foote, M. James, and E. M. Mikhail). New York: McGraw-Hill.

5. W. W. Staley. 1964. *Introduction to Mine Surveying*. Stanford, CA: Stanford University Press.

6. See Chrzanowski and Robinson, *Mining Surveys*, and Stanley, *Introduction to Mine Surveying*.

7. E. F. Hart, and J. D. Howard, Jr. 1983. Use of microcomputer for plotting and quality determinations of underground coal workings. American Society for Photogrammetry/ACSM Fall Meeting, Salt Lake City, UT.

8. BLM. 1973. *Manual of Instructions for Survey of the Public Lands of the United States*. Washington, DC: U.S. Government Printing Office.

9. _____ . 1980. *Mineral Survey Procedures Guide*. Washington, DC: U.S. Government Printing Office.

10. _____ . Mineral Surveys. *1973 Manual*. Chap. 10.

11. J. Underhill. *Mineral Land Surveying*. Denver, CO: The Mining Reporter Publishing Company. Reprinted with preface by Uzes, F. D., by Landmaric Enterprises, Rancho Cordova CA, 1982.

12. Underhill. Chap. 4, p. 52.

13. _____ . Surveying for patents. Chap. 5, p. 87.

14. _____ . Chap. 4, pp. 52, 54.

15. _____ . p. 90.

16. U.S. Code Annotated, Title 30, *Mineral Lands and Mining*. 1942. St. Paul, MN and Brooklyn, NY: West Publishing Co. and Edward Thompson Company, Subsection 34, p. 307.

17. BLM. *1973 Manual*. Sect. 10-34 and 10-38, p. 223.

18. BLM. *1980 Guide*. p. 57.

REFERENCES

CHRZANOWSKI, A., and E. F. HART. 1983. Role of the surveyor in North America in studies of ground movements in mining areas. Proceedings of American Society for Photogrammetry IACSM. Annual Meeting. Washington, DC.

FRUSH, C. G. 1973. Mine surveying. *SME Mining Engineering Handbook*, Vol. 2. New York: Society of Mining Engineers, AIME.

HART, E. F., and R. H. OWENSBY, JR. 1984. Safety considerations when surveying in deep mining operations. Proceedings of American Society for Photogrammetry/ACSM. Annual Meeting. Washington, DC.

30

Public-Land Surveys*

Roy Minnick and John S. Parrish

30-1. ORIGINAL SURVEYS†

30-1-1. Introduction

The public-land survey system covers all of the continental United States except the original 13 colonies, Texas, and Hawaii. It is a system begun by federal law in 1785 and was largely put in place by 1805. Most changes since that time have been refinements to the system, rather than major revisions. The system established the concept of surveying and marking the land prior to the government's selling the land to its citizens. Though a novel concept in the world of 1785, it was successful in encouraging rapid settlement of the continent by placing most land in private ownership.

The monuments and records left by the original public-lands surveys provide the basis for land titles in the public-land states, and locating, relocating, restoring, or interpreting traces of the surveys is a difficult and complex task that often results in widely varying interpretations that often must be settled by legal means.

History of U.S. Public-Land Surveys

On May 20, 1775, the Continental Congress passed an ordinance calling for some mode of locating and disposing of lands in the western portion of the United States. Under this ordinance, the first public-land surveys were made. On May 18, 1796, Congress provided for the appointment of surveyor general and directed the survey of the lands northwest of the Ohio River and above the mouth of the Kentucky River. By an act of Congress, approved April 25, 1812, all matters connected with the surveying and sale of the public lands, or in any way relating to those lands, were placed under the jurisdication of the General Land Office (GLO), the chief officer of which was the commissioner of the GLO. Originally, this office was in the Department of the Treasury, but it was later transferred to the Department of the Interior, where it remained. The Act of July

---

†Original Surveys section written by Roy Minnick.

*Retracement Surveys section written by John S. Parrish, and begins on page 770. Analysis, procedures, and suggestions presented here are solely the responsibility of this author and *no liability or responsibility* shall be incurred by the government agency with whom this author is employed.

16, 1946, abolished the GLO and its duties were assigned to the newly created Bureau of Land Management (BLM).

All executive duties relating to the public lands were performed by the commissioner of the GLO, under the direction of the Secretary of the Interior. After the GLO was abolished, these duties were performed by the direction of the BLM. The work of carrying out the field surveys and preparing and maintaining the office records for U.S. public-land surveys is the function of a group of civil-service employees, and the work in each surveying district is in direct charge of one person. The execution of U.S. public-land surveys must follow the instructions issued in the *Manual of Instructions for Survey of the Public Lands of the United States*[1], which has been published at various times by the Department of the Interior and printed by the U.S. Government Printing Office. It is supplemented by (1) standard field tables that contain tables particularly useful to U.S. public-land surveyors and (2) an ephemeris that contains data relating to the daily positions of the sun and Polaris.

From the time when the U.S. public-land surveys were first authorized until 1910, all federal land surveys were carried out by private surveyors working under a contract with the government. By an act of Congress, approved on June 25, 1910, provision was made for the establishment of a permanent corps of surveyors, employed under civil-service regulations, but the office of surveyor general was retained. The old contract system of surveying was ended, and the direct system of surveying by employees of the government was instituted and remains to this day.

Public Lands Defined

The original public domain included the lands that were turned over the federal government by the Colonial states, areas acquired later from native Indians or foreign powers, or after private acquisition, have been returned by law to public ownership and the status of public lands. After admission of states into the Union, the federal government has continued to hold title to and administer the unappropriated lands. Various acts expressly provide that the title to unappropriated lands within these states shall be retained by the United States. Lands in the territories not appropriated before they were acquired are the exclusive property of the United States, to be administered, or for disposal by such titles as the government may deem most advantageous. Congress has the power, derived from Article IV, Section 3, of the Constitution, of disposing of the public domain and making all rules and regulations.

The director of the BLM determines what are public lands, what lands have been surveyed, what are to be surveyed, what have been disposed of, what remains to be disposed of, and what are reserved. The United States, through the Department of the Interior, has the authority to extend the surveys as may be necessary to include lands erroneously omitted from earlier surveys. There are provisions for administrative and legal relief should controversy arise between the United States and landowners who may be affected by Department of Interior decisions.

Beds of navigable bodies of water are not public domain and not subject to survey and disposal by the United States. Sovereignty is in the individual states. Under the laws of the United States, the navigable waters have always been common highways. This includes all tide-water streams and other important permanent bodies of water, whose natural and normal condition at the date of the admission of a state into the Union was such as to classify it as navigable water. Tidelands that are covered by the normal daily overflow are not subject to survey as public land.

Swamp and overflowed lands in Alabama, California, Florida, Illinois, Indiana, Iowa, Louisiana, Michigan, Minnesota, Mississippi, Missouri, Ohio, Oregon, and Wisconsin, through public domain, pass to the states upon identification by public-land survey, and approved selection, the title being subject to the disposal by the states. The Act of March 2,

1849 (9 Stat. 352) granted to the state of Louisiana all its swamp and overflowed lands for the purpose of aiding in this reclamation. The Act of September 28, 1859 (9 Stat. 519), extended the grant to other public-land states then in the Union. The grant was also extended to Minnesota and Oregon by the Act of March 12, 1860 (12 Stat. 3). These various grants were carried over into R.S. 2479 (43 U.S.C. 982). A notable exception to the swampland laws is found in the Arkansas Compromise Act of April 29, 1898 (30 Stat. 367; 43 U.S.C. 991), by which all right, title, and interest to the remaining unappropriated swamp and overflowed lands reverted to the United States.

Swamp and overflow grants apply to elevations below the uplands where, without the construction of levees or drainage canals, the areas would be unfit for agriculture. The grants apply to all swamp and overflowed lands unappropriated at the dates of the granting acts, whose character at that time would bring them within the provisions of the grant. Figure 30-1 shows the states that were created out of the public domain. In 1983, 2,265,144,960 acres remain under administration of the federal government.

Laws Relating to the Public Lands

Principal early laws are found in the following:

1. "An ordinance for ascertaining the mode of locating and disposing of lands in the western territory, and for other purposes therein mentioned," passed by the Continential Congress on May 20, 1785.

Figure 30-1. States created out of the public domain. (*Manual of Instructions for the Survey of the Public Lands of the United States—1973.* 1975. Reprint. Rancho Cordova CA: Landmark Enterprises.)

2. The Acts of May 18, 1796 (1 Stat. 464); May 10, 1800 (2 Stat. 73); February 11, 1805 (2 Stat. 313); April 25, 1812 (2 Stat. 716); April 24, 1820 (3 Stat. 566); April 5, 1832 (4 Stat. 503); July 4, 1836 (5 Stat. 107); and March 3, 1849 (9 Stat. 395).

Based on these early laws, that part of the Northwest Territory that became the State of Ohio was tue experimental area for the development of the rectangular system. Here the plans and methods were tested in a practical way. Notable revisions of the rules were made as surveys progressed westward, until the general plan was complete.

Adoption of the rectangular system marked an important transition from the surveying practice that generally prevailed in the Colonial states, where lands were described by irregular metes and bounds, each parcel depending more or less on the description of its neighbors. All the early laws were incorporated in the revised statutes and Title 43 of the U.S. code.

The federal government never enacted a unified land law. Instead, federal legislation was enacted to meet the immediate needs of a nation possessing large tracts of unmapped and vacant land. Much of the early legislation related to the survey and distribution of the public domain. More recently, the legislation has centered around regulation, use, and resource management.

General Rules for Public-Land Surveys

Generally, from examination of federal legislation, the following synopsis is prepared:

1. The boundaries and subdivisions of the public lands are unchangeable after the United States passes title. Title may be passed after the public lands have been surveyed by official surveyors operating under approved instructions. The survey consists of establishing monuments on the ground, and the preparation of field notes and plats approved by the legally constituted authority. Currently this is the chief of cadastral surveys in the BLM.

2. The original township, section, quarter-section, and other original monuments must stand as the true corners of the subdivisions they were intended to represent and will be given controlling preference over the recorded directions and lengths of lines.

3. Quarter-quarter-section corners not established during the original survey shall be placed on the line connecting the section and quarter-section corners, and midway between them, except on the last half-mile of section lines closing on the north and west boundaries of the township, or on other lines between fractional or irregular sections.

4. The centerlines of a regular section are to be straight, running from the quarter-section corner on one boundary of the section to the corresponding corner on the opposite section line.

5. In a fractional section where no opposite corresponding quarter-section corner has been or can be established, the centerline of such section must be run from the property established quarter-section corner as nearly in a cardinal direction to the meander line, reservation, or other boundary of such fractional section, as due parallelism with section lines will permit.

6. Whenever possible, lost or obliterated corners of the approved surveys must be restored to their original locations.

30-1-2. The Manual of Instructions for Survey of the Public Lands

The *Manual of Instructions for Survey of the Public Lands of the United States*, hereafter referred to as the 1973 manual, describes methods approved by the federal agency responsible for surveys of the public lands. The 1973 manual's purpose is to ensure that the surveys are made in conformance to statutory and its federal judicial interpretation. It is, in modern terminology, a procedures manual for use of the BLM personnel. Contrary to popular belief, it has not binding effect on other surveyors, unless specifically required under state laws. The 1973 manual is, however, studied carefully by most property surveyors in public-land states, since it provides insight into survey techniques and provides the basis for retracing original survey lines.

From time to time, when special situations occurred that were not covered in the general manual, special instructions were issued. These are usually available in the survey records offices of the BLM. These special instructions form a part of the survey record for the public lands.

30-1-3. Development of the Manual

Surveying, in general, is the art of measuring and locating lines, angles, and elevations on the surface of the earth, within underground workings, and on the beds of bodies of water. A "cadastral survey" creates (or reestablishes) marks and defines boundaries of tracks of land. In the general plan for survey of the public lands, this includes a field-note record of the observations, measurements, survey marks placed in the ground, and a plat that represents the cadastral survey, all subject currently to approval of the director, BLM.

The surveys of public lands have been conducted since 1785. The first surveys, covering parts of Ohio, were made under the supervision of the geographer of the United States, in compliance with the Ordinance of May 20, 1785. Detailed instructions were not needed in these initial surveys, because only the exterior lines of the townships were surveyed and only mile corners established. Township plats were marked by subdivisions into sections or "lots" 1 mi square, numbered from 1 to 36, commencing with no. 1 in the southeast corner of the township, and running from south to north in each sequence to no. 36 in the northwest corner of the township.

The Act of May 18, 1796 provided for the appointment of a surveyor general, whose duty was to survey the public lands northwest of the Ohio River. Half the townships were to be subdivided into 2-mi blocks, and the rule for numbering of sections within the township was

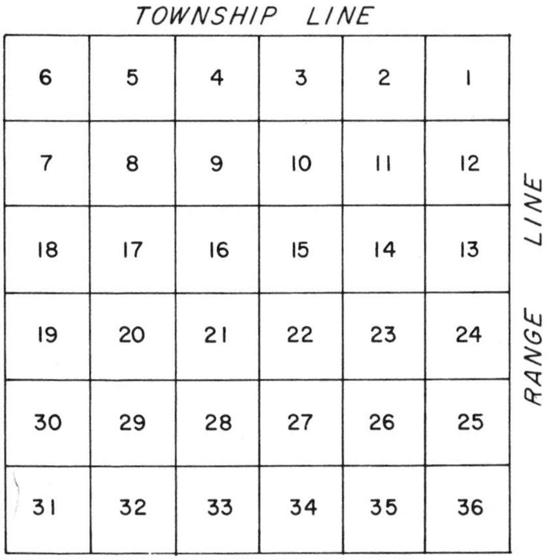

TOWNSHIP LINE

6	5	4	3	2	1
7	8	9	10	11	12
18	17	16	15	14	13
19	20	21	22	23	24
30	29	28	27	26	25
31	32	33	34	35	36

RANGE LINE

Figure 30-2.   Regular township showing current method of numbering sections. (*Manual of Instruction for the Survey of the Public Lands of the United States—1973.* 1975. Reprint. Rancho Cordova CA: Landmark Enterprises.)

changed to that practiced today (see Figure 30-2). Subsequent laws called for additional subdivision, and the system of surveys was gradually refined to its present form. General instructions were given to the surveyor general by the Secretary of the Treasury, then in charge of land sales, and later by the commissioner of the GLO. Instructions to deputy surveyors, those actually in charge of the field work, were issued by the surveyor general. A surveyor of the lands of the United States south of Tennessee was appointed in 1803, with the same duties as the surveyor general, and eventually a surveyor general was appointed for each of many public-land states and territories.

In 1831, the commissioner of the GLO issued detailed instructions to the surveyor general concerning surveys and plats. The applicable parts were incorporated by individual surveyors general in bound volumes of instructions suitable for those in the field by deputy surveyors. From these directions evolved the Manual. The Act of July 4, 1836 placed the overall direction of the public-land surveys under the principal clerk of surveys in the GLO. The immediate forerunner of the manual was printed in 1851 as *Instructions to the Surveyor General of Oregon; Being a Manual for Field Operations.*[2] Its use was extended at once to California, Minnesota, Kansas, Nebraska, and New Mexico. A slightly revised version of these instructions were issued as the 1855 manual. Source for the current 1973 manual and reprints of the older manuals are listed at the end of this chapter.

30-1-4. Manual Supplements

Two supplements are available for the 1973 manual, as follows:

1. *The Ephemeris of the Sun, Polaris and Other Selected Stars with Companion Data and Tables,*[3] which has been published annually, in advance, since 1910.

2. *Restoration of Lost or Obliterated Corners, and Subdivision of Sections, a Guide for Surveyors.*[4] The subject matter under this title first appeared in the decisions of the Department of the Interior, 1 L.D. 339; 2d ed., 1 L.D. 671 (1883). There have been several revisions and extensions, the latest in 1973. Providing an introduction to the rectangular system of public-land surveying and resurveying, with a compendium of basic laws relating to the system, it answers many common questions arising in practical work. It is intended especially for persons outside the BLM, who are interested in former or present public lands.

30-1-5. Distance Measurement

By law, the chain is the unit of linear measure for survey of the public lands. The general exception is the survey of town sites, where the measurement is in feet. Other exceptions are made to meet special requirements. Measurements recorded on the public-lands system are supposed to be horizontal distances, given in miles, chains, and links. The following units of measure are encountered in the so-called "public-land states":

$$1 \text{ chain (ch)} = 66 \text{ ft}$$
$$1 \text{ ch} = 100 \text{ links}$$
$$80 \text{ ch} = 1 \text{ mi}$$
$$1 \text{ ch} = 4 \text{ rods}$$
$$1 \text{ ch} = 4 \text{ poles}$$
$$1 \text{ ch} = 4 \text{ perches}$$
$$1 \text{ link} = 0.66 \text{ ft}$$
$$1 \text{ rod} = 16\tfrac{1}{2} \text{ ft}$$
$$1 \text{ mi} = 5280 \text{ ft}$$
$$1 \text{ mi} = 80 \text{ ch}$$
$$1 \text{ sq mi} = 640 \text{ acres}$$
$$1 \text{ acre} = 43{,}560 \text{ sq ft}$$
$$1 \text{ acre} = 10 \text{ sq ch}$$
$$1 \text{ sq rood} = \tfrac{1}{4} \text{ acre}$$
$$1 \text{ vara av} = 33.372 \text{ in. (Florida)}$$
$$1 \text{ vara av} = 33.333 \text{ in. (Texas)}$$
$$1 \text{ arpent} = 0.8507 \text{ acres}$$
$$\text{(Arkansas and Missouri)}$$
$$1 \text{ arpent} = 0.84625 \text{ acres}$$
$$\text{(Mississippi, Alabama,}$$
$$\text{and Florida)}$$
$$1 \text{ arpent} = 0.845 \text{ acres (Louisiana)}$$
$$1065.75 \text{ ft} = 1000 \text{ French ft}$$
$$\text{(Louisiana)}$$
$$\text{Side of a sq arpent} = 192.50 \text{ ft}$$
$$\text{(Arkansas and Missouri)}$$
$$\text{Side of a sq arpent} = 191.994 \text{ ft}$$
$$\text{(Mississippi, Alabama,}$$
$$\text{and Florida)}$$

Currently, distance measurements are usually made with steel tapes, varying in length from 1 to 8 ch. Graduations are in chains and links and sometimes tenths of links. Earlier surveys were usually measured with a device that actually resembled a chain (see Figure 30-3). In recent years, the use of electronic measurement devices has become more commonplace.

Figure 30-3. Surveyor's chain. (Courtesy of F. D. Uzes.)

30-1-6. Direction Measurement

The direction of each line of the public-land surveys is determined with reference to the true meridian defined by the axis of the earth's rotation. Bearings are stated in terms of angular measure referred to the true north or south.

Prior to the issuance of the 1890 manual, the surveyor's compass (Figure 30-4), utilizing a magnetic needle, was frequently used to obtain the direction of a line. After 1890, the magnetic compass could be used only in subdividing sections and meandering, if the area was known to be free of local attraction. After the 1894 manual was issued, all directions of lines had to be surveyed with reference to the true meridian, independent of the magnetic needles.

A field-note record is required of the average magnetic declination over the area of each

Figure 30-4. Surveyor's compass. (Courtesy of F. D. Uzes.)

survey. The value is shown on the plat and in the field notes. The principal purpose of this record is to provide an approximate value for use in local surveys and retracements, where a start is to be made by the angular value of the magnetic north in relation to the true north.

30-1-7. Establishing Direction

Current practice is to determine true azimuth by one of the following methods: (1) direct observations of the sun, Polaris, or other stars; (2) observations with a solar attachment; or (3) the turning of angles from triangulation stations of the horizontal control network. Use may also be made of a gyro-theodolite, properly calibrated and previously checked on an established meridian. Detailed information on methods of survey currently authorized is found in the current 1973 manual. For earlier methods of survey, refer to the manual in force at the time period under study.

30-1-8. The System of Rectangular Surveys

The underlying principle of the rectangular system of public-land surveys is to provide a simple and certain form of real property boundary identification and a certain land description of the parcels of public-lands. The survey has been in progress since 1785. Although most original public land surveys no longer cover extensive land areas, a knowledge of the original survey methods is essential to retrace survey lines and locate the landmarks and monuments that identify public-load property boundaries. Federal law provides for the general methods of survey. In accordance with the law, various specific instructions have been issued for the conduct of the actual field surveys to lay out and mark the lines on the ground.

Initial points for the commencement of the public-land surveys have been selected whenever needed, under special instructions from the GLO. Each initial point is a well-marked point perpetuated by a permanent monument, whose latitude and longitude are determined and recorded. The positions of the initial points are shown in Table 30-1, and the areas governed by each meridian in Figure 30-5.

The rectangular system of surveys, when it began in present-day Ohio in 1785, was experimental, and the first initial point was not selected until 1805. The first survey, e.g., used the west boundary of Pennsylvannia. Other references were used in Ohio and are listed in Table 30-2. For additional reading on Ohio surveys, see Peter's *Ohio Lands and Their History*.[5]

Principal meridians and base lines are established through each initial point to provide a reference for the survey (see Figure 30-6). A principal meridian is run on a true meridian extending north and south. Regular quarter-section and section corners are established alternately at intervals of 40 ch and regular township corners at intervals of 480 ch. Meander corners are established at the intersection of the line with meanderable bodies of water. The base line is extended east and west from the initial point on a true parallel of latitude. Standard quarter-section and section corners are established alternately at intervals of 40 ch and standard township corners at intervals of 480 ch. Meander corners are established where the line intersects meanderable bodies of water.

Each parallel of latitude is perpendicular at every point to the true meridian passing through that point. Since the surface of the earth is curved, the true meridians through different points on a true parallel of latitude are not parallel, but converge toward the North and South Poles. Because the meridians converge, the base line is curved slightly. Since the survey instrument line of sight is straight, survey techniques had to be developed to place the parallels of latitude on the ground, with sufficient accuracy to maintain the standards of alignment. Three methods explained in the current manual are named the solar method, tangent method, and secant method.

The solar method utilizes an attachment on the survey instrument to observe the sun. If such an instrument, in good adjustment, is employed, the true meridian may be determined by observation with the solar unit at each transit point. A turn of 90° in either direction then defines the true parallel, and if sights are taken not longer than 20 to 40 ch distant, the line so established does not appreciably differ from the theoretical parallel of the latitude. The resulting line is a succession of points, each one at right angles to the true meridian at the previous station.

The tangent method for determination of the true latitude curve consists in establishing the true meridian at the point of beginning, from which a horizontal deflection angle of 90° is turned to the east or west as needed; the projection of the line is the tangent and projected 6 mi in a straight line. As measurements are completed for each corner point, offsets are measured from the tangent to the parallel, on which line the corners are established.

After a 6-mi length of the parallel has been run, another tangent is located at the last point thus set on the parallel, and similar offsets are measured from that tangent. Figure 30-7 illustrates the establishment of a parallel at latitude 45°34′5″ N. Azimuths of the tangent to the parallel and the offsets to be measured from the tangent to the parallel must be calculated or determined from tables and vary with the latitude.

The secant method of laying out a parallel of latitude resembles the tangent method, but the offsets are measured from a secant 6 mi long. The secant is a great circle intersecting a parallel of latitude at the first- and fifth-mile points and tangent to a parallel of latitude at the third-mile point. The secant is run as a straight line, and its bearing at any point is the angle that it makes with the true meridian at that point. The secant is projected 6 mi in a straight line, and as the measurements are completed for each corner point, proper offsets are measured north or south from the secant to the parallel, on which parallel the corners are established.

Table 30-1.   Meridians and Base Lines of the U.S. Rectangular Surveys.

Meridian	Adopted	Governing surveys (wholly or in part) in States of	Initial Points Latitude °	′	″	Initial Points Longitude °	′	″
Black Hills	1878	South Dakota	43	59	44	104	03	16
Boise	1867	Idaho	43	22	21	116	23	35
Chickasaw	1833	Mississippi	35	01	58	89	14	47
Choctaw	1821	do	31	52	32	90	14	41
Cimarron	1881	Oklahoma	36	30	05	103	00	07
Copper River	1905	Alaska	61	49	04	145	18	37
Fairbanks*	1910	do	64	51	50.048	147	38	25.949
Fifth Principal	1815	Arkansas, Iowa, Minnesota, Missouri, North Dakota, and South Dakota	34	38	45	91	03	07
First Principal	1819	Ohio and Indiana	40	59	22	84	48	11
Fourth Principal	1815	Illinois	40	00	50	90	27	11
do	1831	Minnesota and Wisconsin	42	30	27	90	25	37
Gila and Salt River	1865	Arizona	33	22	38	112	18	19
Humboldt	1853	California	40	25	02	124	07	10
Huntsville	1807	Alabama and Mississippi	34	59	27	86	34	16
Indian	1870	Oklahoma	34	29	32	97	14	49
Kateel River†	1956	Alaska	65	26	16.374	158	45	31.014
Louisiana	1807	Louisiana	31	00	31	92	24	55
Michigan	1815	Michigan and Ohio	42	25	28	84	21	53
Mount Diablo	1851	California and Nevada	37	52	54	121	54	47
Navajo	1869	Arizona	35	44	56	108	31	59
New Mexico Principal	1855	Colorado and New Mexico	34	15	35	106	53	12
Principal	1867	Montana	45	47	13	111	39	33
Salt Lake	1855	Utah	40	46	11	111	53	27
San Bernardino	1852	California	34	07	13	116	55	48
Second Principal	1805	Illinois and Indiana	38	28	14	86	27	21
Seward	1911	Alaska	60	07	37	149	21	26
Sixth Principal	1855	Colorado, Kansas, Nebraska, South Dakota, and Wyoming	40	00	07	97	22	08
St. Helena	1819	Louisiana	30	59	56	91	09	36
St. Stephens	1805	Alabama and Mississippi	30	59	51.463	88	01	21.076
Tallahasee	1824	Florida and Alabama	30	26	03	84	16	38
Third Principal	1805	Illinois	38	28	27	89	08	54
Uintah	1875	Utah	40	25	59	109	56	06
Umiat‡	1956	Alaska	69	23	29.654	152	00	04.551
Ute	1880	Colorado	39	06	23	108	31	59
Washington	1803	Mississippi	30	59	56	91	09	36
Willamette	1851	Oregon and Washington	45	31	11	122	44	34
Wind River	1875	Wyoming	43	00	41	108	48	49

*USC and GS station "Initial, 1941" is located S 66° 44′ E, 2.85 ft distant from the initial point of the Fairbanks Meridian. The geodetic station (latitude 64° 51′ 50.037″ N, longitude 147° 38′ 25.888″ W) was inadvertently used as the origin from which to compute positions on the Fairbanks Meridian protractions diagrams.

†The Kateel River initial point is identical with USC and GS station "Jay, 1953."

‡The Umiat initial point is identical with USO and GS station "Umiat, 1953," positions are as published by the U.S. Coast and Geodetic Survey.

Source: BLM, *Manual of Instructions for Survey of the Public Lands of the United States*, 1973. Washington, DC: U.S. Government Printing Office.

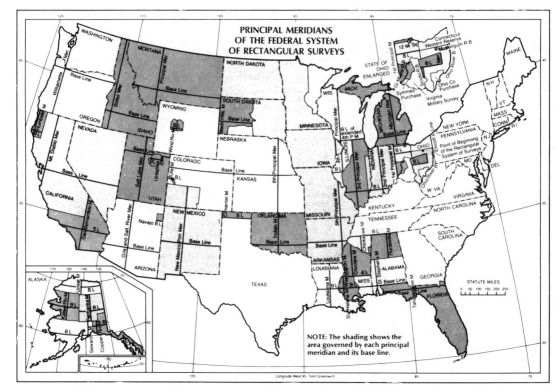

Figure 30-5.   Principal meridians of federal system of rectangular surveys. (*Manual of Instructions for the Survey of the Public Lands of the United States—1973*. 1975. Reprint. Rancho Cordova CA: Landmark Enterprises.)

Table 30-2.   Public-Land Surveys Having No Initial Point as an Origin for Both Township and Range Numbers.

Survey (and Year Commenced)		Townships Numbered	Ranges Numbered
Ohio River Survey (Ohio)	1785	North from Ohio River	West from west boundary of Pennsylvania
U.S. Military Survey (Ohio)	1797	North from south boundary of military grant	West from west boundary of the Seven Ranges
West of the Great Miami (Ohio)	1798	North from Great Miami River	East from Ohio-Indiana boundary
Ohio River Base (Indiana)	1799	North from Ohio River	From Ohio-Indiana boundary and its projection south
Scioto River Base (Ohio)	1799	North from Scioto River	West from west boundary of Pennsylvania
Muskingum River Survey (Ohio)	1800	1 and 2	Ten
Between the Miamis, north of Symmes Purchase (Ohio)	1802	East from Great Miami River	North from Ohio River (continuing numbers from Symmes Purchase)
Twelve-Mile-Square Reserve (Ohio)	1805	1, 2, 3, and 4	None

Source: BLM, *Manual of Instructions for Survey of the Public Lands of the United States*, 1973. Washington, DC: U.S. Government Printing Office.

As in the tangent method, the bearing angles or azimuths of the secant, and the offsets must be determined by calculators or from a table. Figure 30-8 illustrates the establishment of a standard parallel in latitude 45°34.5′ N by the secant method. Note that the bearings angles are converted from azimuths referred from true north for the first 3 mi and true south for the last 3 mi.

The various lines in the rectangular system of surveys are extended from or tied in with the principal meridian and base line. These lines are classified as follows: (1) standard parallels and guide meridians, (2) township exteriors, (3) section lines, (4) subdivision-of-section lines, and (5) meander lines.

Standard parallels, also called correction lines, are extended east and west from the principal meridian at intervals, currently, of 24 mi north and south of the base line, in the matter prescribed for the survey of the base line. Under some earlier instructions, the standard parallels have been placed at different intervals. In these instances, present practice requires additional standard parallels. These

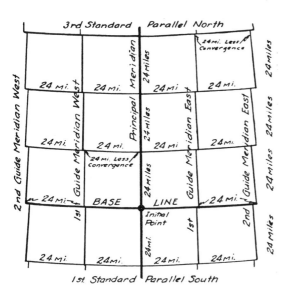

Figure 30-6. Township grid system. (From L. Stewart, 1975, *Public Land Surveys; History, Instructions, and Methods*, Rancho Cordova, CA: Carben Survey Reprints.)

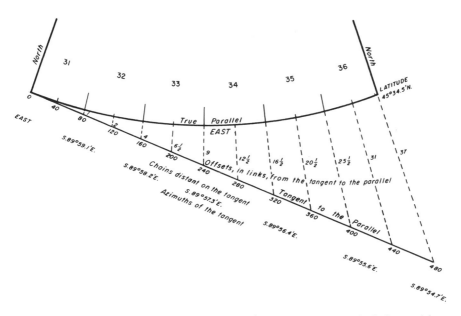

Figure 30-7. Tangent method to lay off a base line. (*Manual of Instructions for the Survey of the Public Lands of the United States—1973*. 1975. Reprint. Rancho Cordova CA: Landmark Enterprises.)

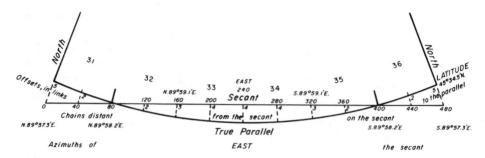

Figure 30-8. Secant method to lay off a base line. (*Manual of Instructions for the Survey of the Public Lands of the United States—1973.* 1975. Reprint. Rancho Cordova CA: Landmark Enterprises.)

are referred to as intermediate correction lines or auxiliary standard parallels. They may be given local names, but they are surveyed like regular standard parallels.

Guide meridians are extended north from the base line, or standard parallels at intervals of 24 mi east and west from the principal meridian, in the same way as the principal meridian is run. The guide meridians are terminated at their intersections with the standard parallels. Guide meridians are projected on the true meridian, and the fractional measurement is placed in the last half-mile. At the intersection of the guide meridian with the standard parallel, a closing township corner is established. The parallel is retraced between the first standard corners east and west of the point for the closing corner in order to determine the exact alignment of the line closed upon. The distance is measured and recorded only to the nearest corner on the standard parallel.

When guide meridians are run south from the base or correction lines, they are initiated at the theoretical point for the closing corner of the guide meridian, calculated on the basis of the survey of the line from south to north, initiated at the proper standard township corner. At the theoretical point of intersection, a closing township corner is established. In some instances, guide meridians have been placed at greater intervals, and in the event additional guide meridians are to be established, they are called auxiliary guide meridians or

sometimes given local names. They are surveyed like regular guide meridians.

Under ideal conditions, the order of survey is to (1) set the initial point, (2) extend the principal meridian and base line, and (3) establish the standard parallels and guide meridians to form quadrangles. The corners are established at 40-ch intervals, except for the corners marking lines that close upon standard parallels or base lines. Generally, excess or deficiency resulting from survey or convergency of meridians is added to or subtracted from the last half-mile of the line.

The subdivision of quadrangles into townships are surveyed, whenever practicable, successively through a quadrangle in ranges of townships, beginning with the townships on the south. The meridional township boundaries have precedence in the order of survey and are run from south or north on true meridians. Quarter-section and section corners are established at intersections of the line with meanderable bodies of water. A temporary township corner is set at a distance of 480 ch, pending the determination of its final position. The temporary point is then replaced by a permanent corner in proper latitudinal position.

The corners that are set on a base line or standard parallel at the time the line is run are called standard corners. They refer only to the land just north of the base line or standard parallel. The corners that are afterward set on the base line or standard parallel, during the process of subdividing the land to the south of the parallel, are called closing corners. For example, a standard parallel may have both standard corners controlling surveys to the north and closing corners ending survey lines coming from the south.

A meridional exterior is terminated at the point of intersection with a standard parallel. The excess or deficiency in measurement is placed in the north half-mile. A closing corner is established at the point of intersection. The parallel is retraced between the nearest standard corners to east and west to find the exact alignment, and the distance to the nearest corner is measured and recorded.

The latitudinal township boundary is run first as a random line, setting temporary corners, on a cardinal (true) course from the old toward the new meridional boundary, and is corrected back on a true line if conditions are ideal. When both meridional boundaries are new lines or both have been established previously, then a random latitudinal boundary is run from east to west. In either case, if defective conditions are not met with, the random is corrected back on a true line. Regular quarter-section corners and section corners are established at intervals of 40 ch, alternately, counting from the east, and meander corners are set where the true line intersects meanderable bodies of water. The fractional measurement is placed in the last half-mile.

The bearing of the true line is calculated from the falling of the random. The falling is the distance, on normal, by which a line falls to the right or left of an objective corner. The temporary points on any random line are replaced by permanent corners on the true line. The true line is blazed through timber, and distances to important items of topography are adjusted to correct true line measurement.

Often, actual conditions dictate that the ideal procedure be modified. In this case, special instructions are prepared for the surveyor and made a part of the public record of the survey. The variety of possibilities is almost limitless, and so study of the specific condition is the best approach.

Once the quadrangle has been subdivided into townships, the townships are identified as indicated in Figure 30-9. When townships are to be subdivided into sections, the corners on the township boundaries are located or otherwise reestablished. In irregular situations, special instructions are prepared.

The subdivision lines that form the boundaries are called section lines. Those running north and south are meridional section lines, and those running east and west are called latitudinal section lines. When townships are subdivided into sections, the meridional section lines have precedence. They are initiated at the section corners on the south boundary of the township and run north parallel to the governing east boundary. Meridional lines are numbered counting from the east and surveyed successively in the same order. If the east boundary is within limits, but has been found by retracement to be imperfect in alignment, the meridional section lines are run parallel to the mean course. Regular quarter-section and section corners are established alternately at intervals of 40 and 80 ch as far as the northernmost interior section corner.

A meridional section line is not continued north beyond a section corner until after the connecting latitudinal sectional line has been surveyed. In the case of the fifth meridional section line, both latitudinal section lines connecting east and west are surveyed before continuing with the meridional line beyond a section corner. The successive meridional lines are surveyed as convenient, but none should be carried beyond uncompleted sections to the east.

The last mile of a meridional line is continued as a random line, each successive random line being parallel to the true east boundary of the section to which it belongs. A temporary quarter-section corner is set at 40 ch, the distance measured to the point of intersection of the random line with the north boundary of the township, and the falling of the random line east or west of the objective section corner noted. The random is then corrected to a true line by blazing through timber and per-

TOWNSHIP GRID

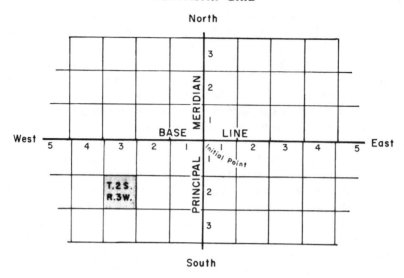

TOWNSHIP 2 SOUTH, RANGE 3 WEST

SECTION 14

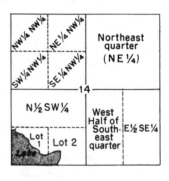

Figure 30-9. Quadrangle diagram showing system of numbering townships and relationship of parallels and meridians. (*From BLM*, *Restoration of Lost or Obliterated Corners and Subdivision of Sections, a Guide for Surveyors*, Rancho Cordova, CA: Carben Survey Reprint.)

manently establishing the quarter-section corner on the true line at a distance 40 ch from the south, placing the fractional measurement in the north half-mile.

When the north boundary of the township is a base line or standard parallel, the last miles of the meridional section lines are continued as true lines parallel to the east boundary of the township. Permanent quarter-section corners are established at 40 ch from the south, and closing corners at the points of

intersection with the north boundary. The distance is measured to the nearest standard corner in each case. New quarter-section corners for the sections of the township being subdivided are established at mean distances between closing corners or at 40 ch from one direction, depending on the plan of the subdivision of the section.

The latitudinal section lines are normally run on random lines from west to east, parallel to the south boundaries of the respective sections. The distances are measured to the points of intersection of the random lines, with the north and south lines passing through the objective section corners; bearings of true lines are calculated on the basis of the fallings. Each random line is corrected to a true line by blazing and marking between the section corners, including the permanent establishment of quarter-section corners at the midpoints on the true lines.

In the west range of sections from the random latitudinal section, lines are run from east to west, parallel to the south boundaries of the respective sections. On the true lines, the permanent quarter-section corners are established at 40 ch from the east, placing the fractional measurements in the west half-miles.

When surveying of the township is complete, the sections are identified by numbers from 1 to 36. The present numbering system is shown in Figure 30-9. The order of survey of the section lines is also indicated in Figure 30-10.

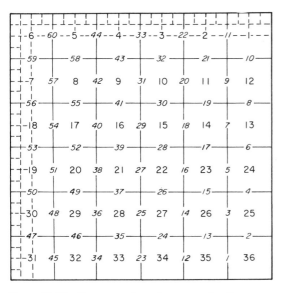

Figure 30-10. Sequence of numbers on section lines shows normal order of subdivision of township into sections. (*Manual of Instructions for the Survey of the Public Lands of the United States—1973.* 1975. Reprint. Rancho Cordova, CA: Landmark Enterprises.)

As in the subdivision of quadrangles into townships, the inevitable excess or deficiency resulting from acceptable survey practices is placed in the northern and western half-mile of the section. Closing section corners are set at the intersection of the section line and either base lines or standard parallels.

Sections of land in the public-land survey systems are not subdivided by survey in the field unless special instructions have been issued. However, some subdivision of section lines is usually drawn on the township plat without being located on the ground by survey. These lines are said to be protracted on the official plat.

30-1-9. Subdivision by Protraction

When the subdivision of section lines are protracted on the plat, all regular sections are shown by broken straight lines connecting the opposite quarter-section corners. The sections bordering the north or west boundary of a normal township, except section 6, are further subdivided by protraction into parts containing two regular half-quarter-sections and four lots. Section 6 has lots protracted against both the north and west boundaries, and so contains two regular half-quarter-sections, one quarter-section section, and seven lots. The position of the protracted lines and regular order of lot numbering as shown in Figure 30-11. The lots are numbered in a regular series progressively from east to west or north to south in each section. The lots in section 6 are numbered commencing with no. 1 in the northeast, then progressively to no. 4 in the northwest, and south to no. 7 in the southwest fractional quarter-quarter section.

The regular quarter-quarter-sections are aliquot parts of quarter-sections, based on midpoint protraction. These lines are not indicated on the official township plat (see Figure 30-9).

Sections invaded meanderable bodies of water, or approved land claims that prevent regular subdivisions, are subdivided by protraction into regular and fractional parts as needed to form a suitable basis for the administration of the public lands and to describe the public lands separately from the segregated areas.

Examples of sections subdivided by probation that show the effect of segregating land claims and meanderable bodies of water are shown in Figures 30-11a, b, and c.

The meander line of a body of water and boundary line of private claims are platted using information obtained from the field survey. When a land grant or meanderable body of water falls within a section, the section is subdivided as nearly as possible in conformity with the regular plan. Note the example of the meanderable river in Figure 30-11b. The dashed lines are arranged so that the maximum number of aliquot parts can be ob-

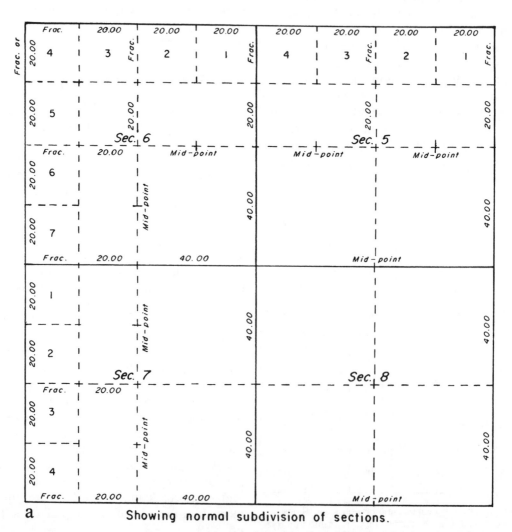

a   Showing normal subdivision of sections.

Figure 30-11. Examples of subdivision by protraction. (a) Normal subdivision of sections. (b) Meanderable river. (c) Meanderable lake. (d) Mineral or land claims. (*Manual of Instructions for the Survey of the Public Lands of the United States—1973*. 1975. Reprint. Rancho Cordova, CA: Landmark Enterprises.)

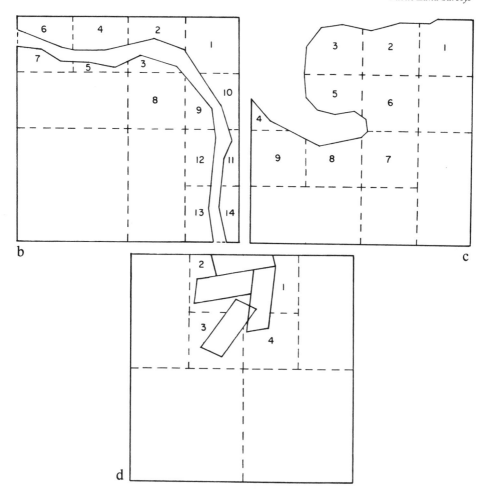

Figure 30-11. (*Continued*).

tained. The remainder are irregular parcels called lots or fractional lots. Lots generally do not lie within two sections, nor do they cross from one fraction quarter-section to another.

A uniform system for numbering lots in fractional sections has been devised. Begin with the eastern lot of the north tier, call it no. 1, and continue the numbering west through the tier, then east in the second, west in the third, east in the fourth tier, until all the fractional lots have been numbered. These directions are maintained even though some of the tiers contain no fractional lots. This method applies to any part of a section. If a section has been partly surveyed at different times, lot numbers should not be duplicated. If the souther part of a section is surveyed, and the lots are numbered 1 through 4, and later the northern part is surveyed, the first lot in the northern part should be numbered 5.

30-1-10. Subdivision by Survey

Occasionally, sections are subdivided by the BLM. These subdivisions are based on laws governing public-land surveys and often special instructions have been issued. The laws can be found in the U.S. codes, and any special instructions obtained from the bureau.

Private surveyors are often called on to subdivide sections to show the aliquot parts or fractional lots that were sold by the government but were not surveyed by the original government surveyor at the time the section lines were run. The procedure is to first identify or reestablish the monuments along the section boundaries, which are usually the section corners and quarter-corners. The steps to be taken to reestablish the boundaries of the section are as follows:

1. Obtain field notes, township plats, and special instructions from the BLM. If the possibility of litigation exists, obtain certified copies.

2. Visit the site and recover the monuments and other original survey evidence.

3. Restore lost or obliterated corners that are necessary for the subdivision of the section.

Restoration procedures are described in Section 30-1-17.

30-1-11.   Subdivision of Sections into Quarter-Sections

Run straight lines from the established quarter-section corners to the opposite quarter-section corners. The point of intersection of the lines thus run will be the corner common to the several quarter-sections or center of the section.

Upon the lines closing on the north and west boundaries of a regular township, the quarter-section corners were established originally at 40 ch to the north or west of the last interior section corners. The excess or deficiency in measurement was thrown into the half-mile next to township or range line, as the case may be. If such quarter-section corners are lost, they should be reestablished by proportionate measurement based on the original record.

When there are double sets of section corners on township and range lines, the quarter-section corners for the sections south of the township line and east of the range line

usually were not established in the original surveys. In subdividing such sections, new quarter-section corners are required, so placed as to suite the calculations of the areas that adjoin the township boundary as indicated on the official plat, adopting proportional measurements, where the new measurements of the north or west boundaries of the section differ from the record distances.

30-1-12.   Subdivisions of Fractional Sections

The law provides that when opposite corresponding quarter-section corners have not been or cannot be fixed, the subdivision-of-section lines shall be ascertained by running from the established corners north, south, east, or west, as the case may be, to the watercourse, reservation line, or other boundary of such fractional section, as represented on the official plat.

In this, the law presumes that the section lines are due north-and-south or east-and-west lines, but usually this is not the case. Hence, in order to carry out the spirit of the law, it will be necessary in running the centerlines through fractional sections to adopt mean courses, where the section lines are not on due cardinal, or to run parallel to the east, south, west, or north boundary of the section as conditions may require, where there is no opposite section line.

30-1-13.   Subdivision of Fractional Quarter-Sections

The subdivisional lines of fractional quarter-sections will be run from properly established quarter-quarter- or sixteenth-section corners, with courses governed by the conditions represented on the official plat, to the lake, watercourse, reservation, or other irregular boundary that renders such sections fractional.

Reasonable discrepancies between former and new measurements may generally be expected when retracing the section boundaries. The shortage or surplus is distributed by proportion in establishing a sixteenth-section corner. For example, the length of the line from the quarter-section corner on the west boundary of section 2 to the north line of the township was reported by the official survey as 43.40 ch, and by the county surveyor's measurement it was found to be 42.90 ch. The distance that the sixteenth-section corner should be located north of the quarter-section corner would be determined by proportion—i.e., as 43.40 ch (the official measurement of the whole distance) is to 42.90 ch (the county surveyor's measurement of the same distance) so is 20 ch (the original measurement) to 19.77 ch (the county surveyor's measurement). By proportionate measurement in this case, the sixteenth-section corner should be set at 19.77 ch north of the quarter-section corner, instead of 20 ch north of said corner, as represented on the official plat. In this manner, the discrepancies between original and new measurements are equitably distributed.

30-1-14. Meandering and Meander Corners

The traverse of the margin of a permanent natural body of water is termed a meander line. All navigable bodies of water and other important rivers and lakes are segregated from the public lands at mean higher-water elevation. In original surveys, meander lines are run for the purpose of ascertaining the quality of land remaining after segregation of the water area. Meander lines are not boundary lines, unless specific action has been taken to make them so.

The running of meander lines has always been authorized in the survey of public lands fronting on large streams and other bodies of water. But the mere fact that an irregular or sinuous line must be run, as in the case of a reservation boundary, does not entitle it to be called a meander line except where it closely follows the bank of a stream or lake. The riparian rights connected with meander lines do not apply in the case of other irregular lines, as the latter are strict boundaries.

A meander corner is established at every point where a standard, township, or section line intersects the bank of a navigable or other meanderable body of water. No monument should be placed in a position exposed to the beating of waves and action of ice in severe weather. In such cases, a witness corner should be established on the line at a secure point near the true point for the meander corner. The distance across a body of water is ascertained by triangulation or direct measurement, and the full particulars are given in the field tables.

The surveyor commences at one of the meander corners, follows the bank or shoreline, and determines the length and true bearing of each course, from the beginning to the next meander corner. All meander courses refer to the true meridian and are determined with precision. When it is impossible to survey the meander line along the mean high-water mark, the notes should state the distance therefrom and the obstacles that justify the deviation.

The field notes of meanders show the corners from which the meanders commenced and upon which they are closed, and exhibit the meanders of each fractional section separately. Following, and composing a part of the notes, should be a description of the adjoining land, soil, and timber, and the estimated depth of innundation to which the bottomland is subject.

Rivers

Facing downstream, the bank on the left-hand side is termed the left bank and that on the right-hand side the right bank.

Navigable rivers and bayous, as well as all rivers not navigable, the right-angle width of which is 3 ch and upward, are meandered on both banks at the ordinary mean high-water mark by taking the general courses and distances of their sinuosities. Rivers not classed as navigable are usually not meandered above the point where the average right-angle width is less than 3 ch.

Shallow streams and intermittent streams without well-defined channels or banks are not meandered, even when more than 3 ch wide. Tidewater streams are meandered at ordinary mean high tide as far as navigable, even when less than 3 ch wide. Tidewater inlets and bayous are recorded and meandered if more than 3 ch in width, but when nonnavigable are not meandered when less than 3 ch wide.

Lakes

All lakes of the area of 50 acres and upward are meandered. In the case of lakes that are located entirely within the boundaries of a section, a quarter-section line—if one crosses the lake—is run from one of the quarter-section corners, on a theoretical course to connect with the opposite quarter-section corner, to the margin of the lake, and the distance is measured, At the point thus determined, a "special meander corner" is established.

When one or both of the opposite quarter-section corners cannot be established, and in all cases where the distance across a lake exceeds 40 ch or the physical crossing is difficult, a temporary special meander corner is established at the computed intersection with the centerline of the section, when surveying the meander line. The temporary point is later corrected to the true centerline position for monumentation, at midpoint in departure (or latitude), or at proportionate distance in a fractional section.

If a meanderable lake is found to be located entirely within a quarter-section, an "auxiliary meander corner" is established at some suitable point on its margin, and a connecting line run from the monument to a regular corner on the section boundary. A connecting traverse line is recorded, if run, but it is also reduced to the equivalent direct connecting course and distance, all of which is stated in the field notes. Only the course and length of the direct connecting line are shown on the plat of the survey.

The meander line of a lake lying within a section is initiated at the established special or auxiliary meander corner and continued around the margin of the normal lake at its mean high-water level to a closing at the point of beginning.

Artificial lakes and reservoirs are not usually segregated from the public lands, but the true position and extent of such bodies of water are determined in the field and shown in the plat. Other exceptions to the general rule are shallow or poorly defined "lakes" that are actually pools that collect because of permafrost and lack of drainage or that are ephemeral desert playas formed seasonally or in wet years. These lakes should be meandered even when larger than 50 acres.

Even though the United States has parted with its title to the adjoining mainland, an island in a meandered body of water, navigable or nonnavigable, in continuous existence since the data of the admission of the state into the Union, and omitted from the original survey may still be public land of the United States. As such, the island is subject to survey. The riparian right that attaches to the lottings along the meander line of the mainland pertains only to the bed of the stream and such islands as may form within the bed subsequent to the disposal of the title.

Any township boundary or section line that will intersect an island is extended as nearly in accordance with the plan of regular surveys as conditions permit, and the usual township, section, quarter-section, and meander corners are established on the island. If an island falls in two sections only, the line between the sections should be established in its proper theoretical position, based on suitable sights and calculations. If an island falls entirely in one section and is large enough to be subdivided (over 50 acres in area), a suitable sight or calculation is made to locate on the margin of the island an intersection with the theoretical position of any suitable subdivision-of-section line. At the point thus determined, a special meander corner is established. In the case of an island falling entirely in one section and too small to be subdivided, an auxiliary meander corner is established at any suitable point on its margin that is connected with any regular corner on the mainland. The direct course and length of the connecting line are given in the field notes and shown on the plat.

30-1-15. Marking Lines Between Corners

The survey is marked on the ground in the following ways:

1. The regular corners of the public-land surveys are marked by fixed official monuments.

2. The relation to natural topographic features is recorded in detail in the field notes.

3. The locus of the lines is marked on forest trees by blazing and hack marks.

A blaze is a smoothed surface cut on a tree trunk at about breast height. The bark and a small amount of the live wood tissue are removed with an ax or other cutting tool, leaving a flat surface that forever brands the tree.

A hack is a horizontal notch cut well into the wood, also made at about breast height (see Figure 30-12). Two hacks are cut to distinguish them from other, accidental marks.

The blaze and hack mark are equally permanent, but so different in character that one mark should never be mistaken for the other. The difference becomes important when the line is retraced in later years.

Trees intersected by the line have two hacks or notches cut on each of the sides facing the line, without any other marks whatever. These are called sight trees or line trees. A sufficient number of other trees standing within 50 links of the line, on either side of it, are blazed on two sides quartering toward the line in order to render the line conspicuous and readily traced in either direction. The blazes are made opposite each other toward the line, the farther the line passes from the blazed trees (Figure 30-13).13

30-1-16. Identification of Existent Corners

The terms "corner" and "monument" are not interchangeable. A corner is a point determined by the surveying process. A monument is the object or physical structure that marks the corner point.

The corners of the public-land surveys are those points that determine the boundaries of the various subdivisions represented on the official plat—i.e., the township corner, section corner, quarter-section corner, subdivision corner, or meander corner. The "mile corner" of a state, reservation, or grant boundary does not mark a point of a subdivision; it is a

Figure 30-12. Line tree with hack marks. (*Manual of Instructions for the Survey of the Public Lands of the United States—1973*. 1975. Reprint. Rancho Cordova, CA: Landmark Enterprises.)

Figure 30-13. Line tree with blaze. (*Manual of Instructions for the Survey of the Public Lands of the United States—1973*. 1975. Reprint. Rancho Cordova, CA: Landmark Enterprises.)

station along the line, however, and long usage has given acceptance to the term. An "angle point" of a boundary marks a change in the bearing, and in that sense it is a corner of the survey.

Surveys of the public-land surveys have included the deposit of some durable memorial —e.g., a marked wooden stake or post, a marked stone, an iron post having an inscribed cap, a marked tablet set in solid rock or a concrete block, a marked tree, a rock in place marked with a cross X at the exact corner point, and other special types of markers —some of which are more substantial; any of these is termed a monument. The several classes of accessories, such as bearing trees, bearing objects, reference monuments, mounds of stone, and pits dug in the sod or soil, are aids in identifying the corner position. In their broader significance, the accessories are a part of the corner monument.

An existent corner is one whose position can be identified by verifying the evidence of the monument or its accessories, by reference to the description in the field notes, or located by an acceptable supplemental survey record, some physical evidence, or testimony. Even though its physical evidence may have entirely disappeared, a corner will not be recorded as lost if its position can be recovered through the testimony of one or more witnesses who have a dependable knowledge of the original location.

The recovery of previously established corners is simplified by projecting retracements from known points. The final search for a monument should cover the zone surrounding one, two, three, or four points determined by connection with known corners. These corners will ultimately control the relocation in case the corner being searched for is declared lost.

The search for the original monument should include a simultaneous search for its accessories. The evidence can be expected to range from that which is least conclusive to that which is unquestionable; the need for corroborative evidence is therefore in direct proportion to the uncertainty of any feature in doubt or dispute. The evidence should agree with the record in the field notes of the original survey, subject to natural changes. Mounds of stone may have become embedded, puts may filled until only a faint outline remains, blazes on bearing trees may have decayed or become overgrown.

After due allowance has been made for natural changes, there may still be material disagreement between the particular evidence in question and the record calls. The following considerations will provide useful in determining which features to eliminate as doubtful:

1. The character and dimensions of the monument in evidence should not be widely different from the record.
2. The markings in evidence should not be inconsistent with the record.
3. The nature of the accessories in evidence, including size, position and markings, should not be greatly at variance with the record.

Allowance for ordinary discrepancies should be made in considering the evidence of a monument and its accessories. No set rules can be laid down as to what is sufficient evidence. Much must be left to the skill, fidelity, and good judgment of the surveyor, bearing in mind the relation of one monument to another and the relation of all to the recorded natural objects and items of topography. No decision should be made in regard to the restoration of a corner until every means have been exercised that might aid in identifying its true original position.

An obliterated corner is one at whose point there are no remaining traces of the monument or its accessories, but whose location has been perpetuated, or the point may be recovered beyond reasonable doubt by the acts and testimony of the interested landowners, competent surveyors, other qualified local authorities, witnesses, or by some acceptable record evidence. A position that depends on the use of collateral evidence can be accepted only as duly supported, generally through proper relation to known corners, and agreement with the field notes regarding distances to natural objects, stream crossing, line trees, and off-line tree blazes, etc., or unquestionable testimony.

A corner is not considered lost if its position can be recovered satisfactorily by means of the testimony and acts of witnesses having positive knowledge of the precise location of the original monument. The expert testimony of surveyors, who may have identified the original monument prior to its destruction and recorded new accessories or connections, is by far the most reliable, though landowners are often able to furnish valuable testimony. The greatest case is necessary in order to establish the bona fide character of the record intervening after the destruction of an original monument. Full enquiry may bring to light various records relating to the original corners and memoranda of private markings, and the surveyor should make use of all such sources of information. The matter of boundary disputes should be carefully looked into, insofar as adverse claimants may base their contentions on the evidence of the original survey. If such disputes have resulted in a boundary suit, the record testimony and court's decision should be carefully examined for information that may shed light on the position of an original monument.

Testimony

The testimony of individuals may relate to the original monument or accessories, prior to their destruction, or to any other marks fixing the locus of the original survey. Weight will be given such testimony according to its completeness, its agreement with the original field notes, and the steps taken to preserve the original marks. Such evidence must be tested by relating it to known original corners and other calls of the original field notes, particularly to line trees, blazed lines, and items of topography.

There is no clearly defined rule for the acceptance or nonacceptance of the testimony of individuals. It may be based on unaided memory over a long period or definite notes and private markers. The witness may have come by his or her knowledge casually, or he or she may have had a specific reason for remembering. Corroborative evidence becomes necessary in direct proportion to the uncertainty of the statements advanced. The surveyor should bear in mind that conflicting statements and contrary views of interested parties may lead to boundary disputes.

The surveyor will show in the field notes, or report of a field examination, the weight given testimony in determining the true point for an original corner. The following points will serve as a guide:

1. The witness (or record statement) should be duly qualified: The knowledge or information should be first-hand, not hearsay; it should be complete, not merely personal opinion.

2. The testimony (or record statement) should be such that it can stand an appropriate test of its bona fide character.

3. The testimony (or record) must be sufficiently accurate, within a reasonable limit, for what is require in normal surveying practices.

Topographic Calls

The proper use of topographic calls of the original field notes may assist in recovering the locus of the original survey. Such evidence may merely disprove other questionable features, or it may be a valuable guide to the immediate vicinity of a line or corner. At best, it may fix the position of a line or corner beyond reasonable doubt.

Allowance should be made for ordinary discrepancies in the calls relating to items of topography. Such evidence should be considered more particularly in the aggregate; when it is found to be corroborative, an average may be secured to control the final adjustment. This will be governed largely by the evidences nearest the particular corner in question, giving the greatest weight to those features that agree most closely with the record and to such items as afford definite connection.

A careful analysis should be made by the surveyor before using topographic calls to fix an original corner plot. Indiscriminate use will lead to problems and disputes when two or more interpretations are possible. Close attention should be given to the manner in which the original survey was made. Instructions for chaining in the earlier manuals indicate that memory was an important factor in recording distances to items of topography. Early field notes often appear to have shown distances only to the nearest chain or even a wider approximation.

In comparing distances returned in the original field notes with those returned in the resurveys, gross differences appear in a significant number of instances. In some cases, the original surveyor apparently surveyed a line in one direction, but then reversed the direction in his or her record without making corresponding changes in distances to items of topography. These facts have sometimes caused distrust and virtual avoidance of the use of topography in corner restoration, when proper application might be extremely helpful. Misapplication usually may be avoided by applying the following tests:

1. The determination should result in a definite locus within a small area.

2. The evidence should not be susceptible to more than on reasonable interpretation.

3. The corner locus should not be contradicted by evidence of a higher class or by other topographic notes.

The determination of the original corner point from even fragmentary evidence of the original accessories, generally substantiated by the original topographic calls, is much stronger than determination from topographic calls alone. In questionable cases, it is better practice in the absence of other collateral evidence to turn to the suitable means of proportionate measurement.

Witness Corners

Ordinarily, a witness corner (WC) established in the original survey will fix the true point for the corner at record bearing and distance. Where the witness corner was placed on a line of the survey, if no complications arise, it will be used as control from that direction in determining the proportionate position of the true point. Thus, the record bearing and distance would be modified, and the witness corner would become an angle point. Unfortunately, the factual statements of the original field notes are not always clear. The record may indicate that the witness corner was established on a random line, or there may be an apparent error of calculation for distance along the true line. The monument may not have been marked WC, either plainly or at all. In these instances, or where there is extensive obliteration, each corner must be treated individually. The important consideration is to locate the true corner point in its original position.

Line Trees

A line tree or definite connection to readily identified natural objects or improvements may fix a point of the original survey. The *mean* position of a blazed line may help to fix a meridional line for departure or a latitudinal line for latitude. Such blazed lines must be carefully checked, because corrections may have been made before final acceptance of the old survey, or more than one line may have been blazed.

Under the law, a definitely identified line tree is a monument of the original survey and is used properly as a control point in the reestablishment of lost corners by the appropriate method of proportionate measurement. In this case, it is treated just as is a recovered corner, and it becomes an angle point of the line.

30-1-17. The Restoration of Lost Corners

A lost corner is a point of a survey whose position cannot be determined beyond reasonable doubt, from either traces of the original marks or acceptable evidence or testimony that bears on the original position, and whose location can be restored only by reference to one or more interdependent corners.

The rules for the restoration of lost corners should not be applied until all original and collateral evidence has been developed. When these means have been exhausted, the surveyor will turn to proportionate measurement, which harmonizes surveying practice with legal and equitable considerations. This plan of relocating a lost corner is always employed unless outweighed by conclusive evidence of the original survey.

The preliminary retracements show the discrepancies of courses and distances between the original record and findings of the retracement. The retracement is based on the courses and distances of the original survey record, initiated and closed upon known original corners. Temporary stakes for future use in the relocation of all lost corners are set when making retracements.

Existing original corners may not be disturbed. Consequently, discrepancies between the new measurements and those shown in the record have no effect beyond identified corners. The differences are distributed proportionally, within the several intervals along the line between the corners.

The retracements will show various degrees of accuracy in lengths of lines, where in every case it was intended to secure true horizontal distances. Until after 1990, most of the lines were measured with Gunter's link chain. Such a chain was difficult to keep at standard length, and inaccuracies often arose in measuring steep slopes by this method.

All discrepancies in measurement should be carefully verified, with the object of placing each difference where it properly belongs. Whenever it is possible to do so, the manifest errors in measurement are removed from the general average difference and placed where the blunder was made. The accumulated surplus or deficiency that then remains is the quantity that is to be uniformly distributed by the methods of proportionate measurement.

A proportionate measurement is one that gives equal relative weight to all parts of the line. The excess or deficiency between two existent corners is so distributed that the amount given to each interval bears the same proportion to the whole difference as the record length of the interval bears to the whole record distance. After the proportionate difference is added to or subtracted from the record length of each interval, the sum of the several parts will equal the new measurement of the whole distance.

The type of proportionate measurement to be used in the restorative process will depend on the method that was followed in the original survey. Standard parallels will be given precedence over other township exteriors, and ordinarily the latter will be given precedence over subdivisional lines; section corners will be relocated before the position of lost quarter-section corners can be determined.

Double Proportionate Measurement

The term *double proportionate measurement* is applied to a new measurement made between four known corners, two each on intersecting meridional and latitudinal lines, for the purpose of relating the intersection to both.

In effect, by double proportionate measurement the record directions are disregarded, except only when there is some acceptable supplemental survey record, some physical evidence, or testimony that may be brought into the control. Corners to the north and south control any intermediate latitudinal position. Corners to the east and west control the position in longitude. One identified original corner is balanced by the control of a corresponding original corner on the opposite side of a particular missing corner that is to be restored. Each identified corner is given a controlling weight inversely proportional to its distance from the lost corner. Lengths of proportioned lines are comparable only when reduced to their cardinal equivalents. The method may be referred to as a "four-way" proportionate measurement and is generally applicable to the restoration of lost corners of four townships and lost interior corners of four sections.

In order to restore a lost corner of four townships, a retracement will first be made between the nearest known corners on the meridional line, north and south of the missing corner and on that line a temporary stake will be placed at the proper proportionate distance; this will determine the latitude of the lost corner. Next, the nearest corners on the latitudinal line will be connected and a second point will be marked for the proportionate measurement east and west; this point will determine the position of the lost corner in departure (or longitude). Then, through the first temporary stake, run a line east or west, and through the second temporary stake a line north or south, as relative situations may determine; the intersection of these two lines will fix the position for the restored corner.

Figure 30-14 illustrates the plan of double proportionate measurement. Points *A*, *B*, *C*, and *D* represent four original corners that will control the restoration of the lost corner *X*. On the large-scale diagram, point *E* represents the proportional measurement between *A* and *B*, and similarly, point *F* represents the proportional measurement between *C* and *D*. Point *X* satisfies the first control for latitude and second control for departure.

A lost township corner cannot safely be restored, nor the boundaries ascertained, without first considering the field notes of the four intersecting lines. It is desirable also to examine the four township plats. In most cases, there is a fractional distance in the half-mile to

the east of the township corner and frequently in the half-mile to the south. The lines to the north and west are usually regular, with quarter-section and section corners at normal intervals of 40.00 and 80.00 ch, but there may be closing-section corners on any or all of the boundaries, so it is important to verify all distances by reference to the field notes.

Lost interior corners of four sections, when all the lines from there have been run, will also be reestablished by double proportionate measurement. The control for such restoration will not extend beyond the township boundary. If the controlling corner on the boundary is lost, that corner must be reestablished beforehand. When the line has not been established in one direction from the missing township or section corner, the record distance will be used to the nearest identified corner in the opposite direction.

Thus, in Figure 30-14, if the latitudinal line in the direction of point *D* has not been established, the position of point *F* in departure would have been determined by reference to the record distance from point *C*; point *X* would then be fixed by cardinal offsets from points *E* and *F*, as already explained.

When the intersecting lines have been established in only two of the directions, the record distances to the nearest identified corners on these two lines will control the position of the temporary points; then, from the latter, the cardinal offsets will be made to fix the corner point.

Record Distance

What is intended by record distance is the measure established in the original survey. Experience and good judgment are required in applying the rules. If the original survey was carelessly executed, no definite standard can be set up as representing that survey. On the other hand, the work may have been reasonably uniform within its own limits, yet inaccurate with respect to exact base standards. It is the consistent excess or deficiency of the original work that is intended here, if that can be

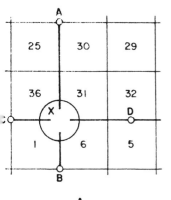

Lost township corner in vicinity of X

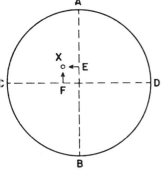

A,B,C,D—Control corners
  E—Proportionate point for X in latitude between A and B
  F—Proportionate point for X in departure between C and D
Correct position of X is at intersection of lines extended East or West
  from E, North or South from F.

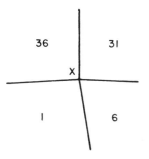

Restored corner showing true direction of township lines

Figure 30-14. Double proportionate measurement. (*Manual of Instructions for the Survey of the Public Lands of the United States—1973.* 1975. Reprint. Rancho Cordova, CA: Landmark Enterprises.)

determined within practical limits. Otherwise, the only rule that can be applied is that a record of 80.00 ch in distance means just that by exact standards—true horizontal measurement.

Single Proportionate Measurement

The term *single proportionate measurement* is applied to a new measurement made on a line to determine one or more positions on that line. By single proportionate measurement,

the position of two identified corners controls the direction of that line. The method is sometimes referred to as a "two-way" proportion, a north-and-south proportion or an east-and-west proportion. Examples are a quarter-section corner on the line between two section corners, all corners on standard parallels, and all corners occupying intermediate positions on a township boundary line.

In order to restore a lost corner on a line by a single proportionate measurement, a retracement is made connecting the nearest identified corners on the line. These corners control the position of the lost corner. Control corners are usually corners established in the original survey of the line. The lost corner is then reestablished at a proportionate distance on the true line connecting the recovered corners. Proper adjustment is made on an east-and-west line to secure the latitudinal curve. Any number of intermediate lost corners may be located on the same plan.

Restoration of lost corners of standard parallel is controlled by the regular standard corners. These include the standard township, section, quarter-section, and sixteenth-section corners, and meander corners, and also closing corners that were originally established by measurement along the standard line, as points from which to start a survey.

Lost standard corners will be restored to their original positions on a base line, standard parallel, or correction line by single proportionate measurement on the true line connecting the nearest identified standard corners on opposite sides of the missing corner or corners, as the case may be.

Corners on base lines are regarded the same way as those on standard parallels. In former practice, the term "correction line" was used for what has later been called the standard parallel. The corners first set in the running of a correction line are treated as original standard corners. Those that were set afterward at the intersection of a meridional line are regarded as closing corners.

All lost section and quarter-section corners on the township boundary lines will be restored by single proportionate measurement between the nearest identified corners on opposite sides of the missing corner, north and south on a meridional line, or east and west on a latitudinal line.

Two sets of corners have been established on many township lines and some section lines. Each set applies only to sections on its respective side of the line. Which corners control in the restoration of a lost corner will depend on

how the line was surveyed. Three common cases are discussed, as follows:

1. When both sets of corners have been established by measurement along the line in a single survey, each corner controls equally for both measurement and alignment (see Figure 30-15a).

2. When a single set of corners was established in the survey of the line and closing corners were subsequently established at the intersection of section lines on one side, the corners first established control both the alignment and first proportional measurement along the line. The original quarter-section corners nearly always referred to sections on only one side of the line, after the closing corners were established on the other side. The quarter-section corners for sections on the side to which the closing corners refer were not established in older surveys. The correct positions are as protracted on the plat of those sections (see Figure 30-15b).

3. Sometimes, one set of corners was established for one side of the line, and a second

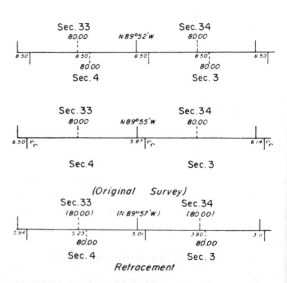

Figure 30-15. Single proportionate measurement. (*Manual of Instructions for the Survey of the Public Lands of the United States—1973.* 1975. Reprint. Rancho Cordova, CA: Landmark Enterprises.)

set of corners for the other side in the course of a later retracement (see Figure 30-15c).

The line is regarded as having been fixed in position by the senior survey, unless the survey was officially superseded. If both sets of corners are recovered, a junior corner lying off the line is treated in the same manner as a closing corner insofar as the alignment is concerned. Since it was established in the course of a retracement reporting the most recent measurement of the line, a junior corner properly can be used for control in restoring a lost corner of the line insofar as measurement is concerned. This procedure is not advisable when the corner is far off-line, because a bearing in the connecting section line would change its true position relative to other corners of the line. That condition can only be shown by retracing enough of the connecting section line to determine its bearing.

All lost quarter-section corners on the section boundaries within the township will be restored by single proportionate measurement between the adjoining section corners, after the section corners have been identified or relocated.

In those cases where connections from the lost quarter-section corner to other regular monuments of the line nearer than the section corners have been previously noted, these will ordinarily assume control in the restoration. Such monuments may include another quarter-section corner, an angle point, or a line tree—any of which may have been established when the line was previously surveyed or resurveyed (see also Section 30-1-8).

Half-Mile Posts, Alabama and Florida

In early practice in parts of Alabama and Florida, so called "half-mile posts" were established at distances of 40 ch from the starting-section corner. The term was applied when the line might be more or less than an exact 80 ch in record length, and when by later

methods the latitudinal lines have been run as "random and true." The practice contemplated that in some cases these subdivisional lines be run in cardinal directions to an intersection, where the next section corner would be placed, and either or both lines might be more or less than 80 ch in length. In some cases, the section corners were placed across the township at intervals of 80 ch on one of the cardinal lines, and the other lines were run on random only. On the first plan, the half-mile post would not be at midpoint unless the line turned out to be 80 ch in length. On the second plan, the half-mile post on the lines first run would be in true position for the quarter-section corner, but on the lines last run, they would usually not be on true line, nor at midpoint.

The rules for restoration of half-mile posts may be stated specifically as follows:

1. In case the half-mile post and quarter-section corner are recorded as being at a common point, the identified half-mile post will be restored as the quarter-section corner.

2. If there is evidence of the position of the section corners in both directions and the record leaves doubt as to the establishment of the half-mile post on the true line, the quarter-section corner will be monumented at midpoint on the true line, disregarding the record of the half-mile post.

3. In the absence of evidence at one or both section corners and when the record leaves doubt regarding the running and marking of the true line, the half-mile post will be employed on a north-and-south line for the control of the latitude of the quarter-section corner, or on an east-and-west line for control of its position in departure, using the record correction for distance. The alignment of the section boundary and position of the quarter-section corner on the true line will be adjusted to the location of the two section corners, after the double proportionate measurements have been completed.

4. When the field notes show proper location for alignment and record correction for distance, the half-mile post will be employed for the full control of the position of the quarter-section corner and restoration of the lost section corners. The position of the quarter-section in latitude on a north-and-south line, or in departure on an east-and-west line, will be ascertained by making use of the record correction for the distance from the half-mile post. The alignment from the position of the half-mile post to the point for the quarter-section corner will be determined by the position of the section corner to the south, if the record correction for distance is to be made to the north; the section corner to the north will be used if the record correction for distance is to be measured to the south; and similarly on east-and-west lines.

Meander Corners

Lost meander corners, originally established on a line projected across the meanderable body of water, usually will be relocated by single proportionate measurement.

Occasionally, it can be demonstrated that the meander corners on opposite banks of a wide river were actually established as terminal meander corners, even though the record indicates that the line was projected across the river. If the evidence outweighs the record, a lost meander corner in such a case will be relocated by single point control.

A lost closing corner will be reestablished on the true line that was closed on, and at the proper proportional interval between the nearest regular corners to the right and left. In order to reestablish a lost closing corner on a standard parallel or other controlling boundary, the line that was closed on will be retraced, beginning at the corner from which the connecting measurement was originally made. A temporary stake will be set at the record connecting distance, and the total distance and falling noted as the next regular corner on the line on the opposite side of the missing closing corner. The temporary stake will then be adjusted as in single proportionate measurement.

A recovered closing corner not actually located on the line that was closed on will determine the direction of the closing line but not its legal terminus. The correct position is at the true point of intersection of the two lines.

The new monument in those cases where it is required will always be placed at the true point of intersection. An off-line monument in such cases will be marked as an amended monument (AM) and connected by course and distance. The field notes of the closing line will include a full description of the old monument as recovered and clear statement that the new monument is set at the true point of intersection.

When an original closing corner is recovered off the line closed on and the new monument established at the true point of intersection, the original position will control in the proportionate restoration of lost corners, dependent on the closing corner. In a like manner, the positioning of sixteenth-section corners or lot corners on the closing line, between the quarter-section corner and the closing corner, will be based on the measurement to the original closing corner.

The foregoing are the general rules for the restoration of lost or obliterated corners. The special cases that are hereafter cited, with respect to broken boundary lines and limited control, do not have wide application and similar importance, except under those conditions and as explained in the succeeding text.

The preceding instructions will be applicable in the large majority of cases. If there seems to be some difficulty or inconsistent result, a careful check should be made of the record data. The special instructions for the original survey, the plat representation, or some call of the field notes may clarify the problem on further study. This research assumes a large importance in the more difficult problems of the recovery of an old line or boundary.

Broken Boundaries: Angle Points of Nonriparian Meander Lines

In some cases, it is necessary to restore—or possibly to locate for the first time—the angle points within a section of the record meander courses for a stream, lake, or tidewater, which may be required under the special rules that are applicable to nonriparian meander lines.

In these cases, the positions of the meander corners on the section boundaries are determined first. The record meander courses and distances are then run and temporary angle points marked. The residual error is shown by the direction and length of the line from the end of the last course to the objective meander corner. The residual is distributed on the same plan as in balancing a survey for the computation of the area of the lottings, as represented on the plat.

The general rule is that the adjustment to be applied to the latitude or departure of any course is to the resolved latitude or departure of the closing error as the length of all the courses. Each adjustment is applied in a direction to reduce the closure. If the northings are to be increased, then the southings will be decreased. A line due east would then be given a correction to the north (in effect, to the left); a line due west, also to the north (in effect, to the right). Each incremental correction is determined and applied in proportion to the length of the line.

The field adjustments for the positions of the several angle points are accomplished simply by moving each temporary point on the bearing of the closing error on amount that is its proportion of that line, counting from the beginning. The particular distance to be measured at any point is to the whole length of the closing error as the distance of that point from the starting corner is to the sum of the lengths of all the courses (Figure 30-16).

The same principle is followed to plot lottings of dependently resurveyed sections in their true relative positions, when the record meander line and true shoreline differ greatly because of distortion.

Grant Boundaries

In many of the states, there are irregular grant and reservation boundaries that were established prior to the public-land subdivisional surveys. In these cases, the township and section lines are regarded as the closing lines. The grant boundary field notes may call for natural objects, but these are often supplemented by metes-and-bounds descriptions. The natural calls are ordinarily given precedence; next, the existent angle points of the metes-and-bounds survey. The missing angle points are then restored by uniformly orienting the record courses to the left or right and adjusting the lengths of the lines on a constant ratio.

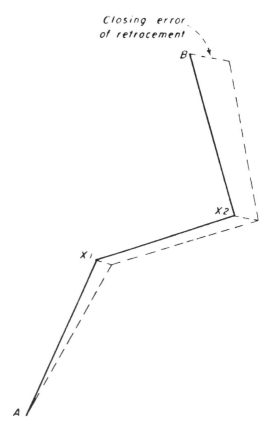

Figure 30-16. Adjusting angle points on riparian meander line. (*Manual of Instructions for the Survey of the Public Lands of the United States—1973*. 1975. Reprint. Rancho Cordova, CA: Landmark Enterprises.)

Both angular and linear corrections are made in the direction needed to reduce the falling of the trial lines laid down according to the record.

The retracement of the grant boundary is begun at an identified corner. Calls for natural objects are satisfied and the existent angle points recovered. Then, between the identified or acceptable points, the position of missing angle points is determined by these following steps:

1. Reduce the record courses and distances to the total differences in latitude and departure. Compute the direction and length of a line connecting the identified points.

2. Determine the actual differences in latitude and departure between the same identified points by retracement. Compute the direction and length of the connecting line, based on these figures.

3. The angular difference of direction between the connection lines computed in 1 and 2 gives the amount and direction of the adjustment to apply to the *record* bearing of each intermediate course.

4. The ratio of the length of the line computed in 2 to that computed in 1 gives the coeffi-

cient to apply to the record length of each intermediate course.

After the adjustments are completed, an additional search for evidence of the record markers should be made. The adjusted locations for the angle points are in the most probable original position, and a better check of collateral evidence is possible. If no further evidence is recovered, the adjusted points are then monumented.

In Figure 30-17, *A* and *B* are identified points of the original boundary. It is desired to restore intermediate points *T*, *S*, *R*, *J*, *I*, *H*, and *G*, which have been temporarily marked at T_t, S_t, R_t, J_t, I_t, H_t, and G_t in conformance with the original record starting from point *A*. The record position of point *B* in relation to point *A* is designated B_t. The adjustment has been made in the four steps already described.

The same procedure may be followed whenever it is desired to retain the *form* of the traverse being adjusted, since the interior angles are unchanged, and the increase or decrease in lengths of lines is constant. The adjustment may be likened to a photographic enlargement or reduction. Mechanically, this process requires that the record distances of

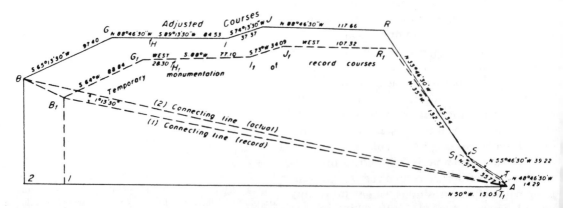

Figure 30-17. Adjusting grant boundary. (*Manual of Instructions for the Survey of the Public Lands of the United States—1973.* 1975. Reprint. Rancho Cordova, CA: Landmark Enterprises.)

the traverse legs between identified points be reduced or increased simultaneously with the rotation of the record bearings until the two identified points coincide.

Original Control

When a line has been terminated with measurement in one direction only, a lost corner will be restored by record bearing and distance, counting from the nearest regular corner—the latter having been duly identified or restored. Examples will be found where lines have been discontinued at the intersection with large meanderable bodies of water or at the border of what was classed as impassable ground.

An index correction for average error in the original measurement should be used, if appropriate. Additionally, in cases where a retracement has been made of many miles of the original lines, between identified original corners, and there has been developed a definite angle from cardinal that characterizes the original survey, it is proper to make an allowance for the average difference.

30-1-18. Resurveys

A *resurvey* is a reconstruction of land boundaries and subdivisions accomplished by rerunning and remarking the lines represented in the field-note record or on the plat of a previous official survey. The field-note record of the resurvey includes a description of the technical manner in which the resurvey was made, full reference to recovered evidence of the previous survey or surveys, and a complete description of the work performed and monuments established. The survey, like an original survey, is subject to approval of the directing authority.

A dependent resurvey is a retracement and reestablishment of the lines of the original survey in their true original positions, according to the best available evidence of the positions of the original corners. The section lines and lines of legal subdivision of the dependent resurvey in themselves represent the best possible identification of the true legal boundaries of lands patented on the basis of the plat of the original survey. In legal contemplation and in fact, the lands contained in a certain section of the original survey and those in the corresponding section of the dependent resurvey are identical.

An independent resurvey is an establishment of new section lines and often new township lines, independent of and without reference to the corners of the original survey. In an independent resurvey, it is necessary to preserve the boundaries of those lands patented by legal subdivisions of the sections of the original survey that are not identical with the corresponding legal subdivisions of the sections of the independent resurvey. This is done by surveying out by metes and bounds and designating as tracts the lands entered or patented on the basis of the original survey. These tracts represent the position and form of the lands alienated on the basis of the original survey, located on the ground according to the best available evidence of their true original positions.

A retracement is a survey that is made to ascertain the direction and length of lines and identify the monuments and other marks of an established prior survey. Retracements may be made for any of several reasons. In the simplest case, it is often necessary to retrace several miles of line leading from a lost corner, which is to be reestablished to an existent corner that will be used as a control. If no intervening corners are reestablished, details of the retracement are not usually shown in the record, but a direct connection between the two corners is reported as a tie. On the other hand, the retracement may be an extensive one made to afford new evidence of the character and condition of the previous survey. Recovered corners are rehabilitated, but a retracement does not include the restoration of lost corners or reblazing of lines through the timber. The retracement may sometimes be complete in itself, but usually it is made as an early part of a resurvey.

Bona Fide Rights of Claimants

Bona fide rights are those acquired in good faith under the law. A resurvey can affect bona fide rights only in the matter of position or location on the earth's surface. The surveyor will be concerned only with the question of whether the lands covered by such rights have been actually located in good faith. Other questions of good faith, such as priority of occupation, possession, continuous residence, value of improvements, and cultivation, do not affect the problem of resurvey except as they help to define the position of the original survey.

The basic principles of protecting bona fide rights are the same in either the dependent or independent resurvey. Each is intended to show the original position of entered or patented lands included in the original description. The dependent resurvey shows them as legal subdivisions; the independent resurvey as segregated tracts. Each is an official demonstration by the BLM, according to the best available evidence of the former survey. There is no legal authority for substituting the methods of an independent resurvey in disregard of identified evidence of the original survey.

The position of a tract of land, described by legal subdivisions, is absolutely fixed by the original corners and other evidences of the original survey and not by occupation or improvements or by the lines of a resurvey that do not follow the original. A conveyance of land must describe the parcel to be conveyed, so that it may be specifically and exactly identified, and for that purpose the law directs that a survey be made. Under fundamental law, the corners of the original survey are unchangeable. Even if the original survey was poorly executed, it still controls the boundaries of land patented under it.

The surveyor should neither rigidly apply the rules for restoration of lost corners with-

out regard to effect on location of improvements nor accept the position of improvements without question, regardless of their relation or irrelation to existing evidence of the original survey. Between these extremes will be found the basis for determining whether improved lands have been located in good faith or not. No definite set of rules can be laid down in advance. The solution to the problem must be found on the ground by the surveyor. It is his or her responsibility to resolve the question of good faith as to location.

It may be held generally that the entryperson has located his or her lands in good faith if such care was used in determining the boundaries as might be expected by the exercise of ordinary intelligence under existing conditions. The relationship of the lands to the nearest corners, existing at the time the lands were located, is often defined by fencing, culture, or other improvements. Lack of good faith is not necessarily chargeable if the entryperson has not located himself or herself according to a rigid application of the rules laid down for the restoration of lost corners when (1) complicated conditions involve a double set of corners, both of which may be regarded as authentic; (2) there are not existing corners in one or more directions for an excessive distance; (3) existing marks are improperly related to an extraordinary degree; or (4) all evidence of the original survey that have been adopted by the entryperson as a basis for location have been lost before the resurvey is undertaken.

In cases involving extensive obliteration at the date of entry, the entryperson of his or her successors in interest should understand that the boundaries of the claim will probably be subject to adjustment in the event of a resurvey. A general control applied to the boundaries of groups of claims must be favored as far as possible in the interest of equal fairness to all and the simplicity of resurvey. A claim cannot generally be regarded as having been located in good faith if no attempts have been made to relate it in some manner to the original survey.

The Dependent Resurvey

The dependent resurvey is designed to restore the original conditions of the official survey according to the record. It is based on, first, identified original corners and other acceptable points of control, and, second, the restoration of lost corners by proportionate measurement in harmony with the record of the original survey. Some flexibility is allowable in applying the rules of proportionate measurement in order to protect the bona fide rights of claimants.

The dependent resurvey is begun by making a retracement of the township exteriors and subdivisional lines of the established prior survey within the assigned work. Concurrently, a study is made of the records of any known supplemental surveys, and testimony obtained from witnesses concerning obliterated corners. The retracement leads at once to identification of known and plainer evidence of the original survey. A trial calculation is made of the proportionate positions of the missing corners, followed by a second and more exhaustive search for the more obscure evidence of the original survey. If additional evidence is found, a new trial calculation is made. Corners still not recovered are marked only as temporary points that may be influenced by acceptable locations. These steps give the basic control for the resurvey. The surveyor then weighs the less certain collateral evidence against the proportionate positions so obtained.

A comparison of the temporary points with the corners and boundaries of alienated lands often helps in determining how the original survey was made, how the claims were located, or both. In analyzing the problem of a particular corner's location, it is often helpful to determine where the theoretical corner point would fall if a three-point control was used. In extreme cases, the collateral evidence may be weighed against the position obtained by the

use of two-point control, particularly when supported by well-identified natural features. It may then provide that the original corner, which would otherwise be lost, has been perpetuated by an acceptably located claim.

Ordinarily, the one-point control is inconsistent with the general plan of a dependent resurvey. The courts have sometimes turned to this as the only apparent solution of a bad situation, and unfortunately this has been the method applied in many local surveys, thus minimizing the work to be done and the cost. Almost without exception, the method is given the support that "it follows the record." This overlooks the fact that the record is equally applicable when reversing the direction of the control from other good corners, monuments, or marks. The use of one-point control is only applicable when the prior survey was discontinued at a recorded distance or it can be shown conclusively that the line was never established. If the line was discontinued by the record, the field notes may be followed explicitly. If it was discontinued by evidenced unfaithfulness in execution, its use would be limited to the making of a tract segregation where the claimant has given confidence to the so-called field notes.

Once it is accepted, a local point of control has all the authority and significance of an identified original corner. The influence of such points is combined with that of the previously identified original corners, in making final adjustments of the temporary points. The surveyor must therefore use extreme caution in adopting local points of control. These may range from authentic perpetuations of original corners down to marks that were never intended to be more than approximations. When a local reestablishment of a lost corner has been made by proper methods without gross error and officially recorded, it will ordinarily be acceptable. Monuments of unknown origin must be judged on their own merits, but they should never be rejected out of hand without careful study. The age and degree to which a local corner has been relied on by *all* affected landowners may lead to its adoption as the best remaining evidence of the position of the original corner. The surveyor must consider all these factors. However, he or she cannot abandon the record of the original survey in favor of an indiscriminate adoption of points not reconcilable with original evidence.

The field-note record of the resurvey should clearly set forth the reasons for the acceptance of a local point when it is not identified by actual marks of the original survey. Recognized and acceptable local marks will be preserved and described. When they are monuments of a durable nature, they are fully described in the field notes and a full complement of the required accessories recorded, but without disturbing or remarking the existing monument. New monuments are established if required for permanence, in addition to—but without destroying—the evidence of the local marks.

While still in the field, the surveyor should make certain that he or she has noted complete descriptions of all identified or accepted corners for entry in the official record of the resurvey, so that the record will embrace the following:

1. A complete description of the remaining evidence of the original monument

2. A complete description of the original accessories as identified

3. A concise statement relating to the recovery of a corner based on identified line trees, blazed lines, items of topography, or other calls of the field notes of the original survey, in the absence of evidence of the monument or its accessories

4. A statement relating to the relocation of an obliterated monument or a statement of the determining features leading to the acceptance of a recognized local corner

5. A complete description of the new monument

6. A complete description of the new accessories

The Independent Resurvey

An independent resurvey is designed to supersede the prior official survey only insofar as the remaining public lands are concerned. The subdivisions previously entered or patented are in no way affected as to location. All such claims must be identified on the ground and then protected in one of two ways. Whenever possible, the sections in which claims are located are reconstructed from evidence of the record survey, just as in a dependent resurvey. When irrelated control prevents the reconstruction of the sections that would adequately protect them, the aliened lands are segregated as tracts. A particular tract is identical with the lands of a specific description, based on the plat of the prior official survey. The tract segregation merely shows where the lands of this description are located with respect to the new section lines of the independent resurvey. In order to avoid confusion with section numbers, the tracts are designated beginning with no. 37. The plan of the independent resurvey must be such that no lines, monuments, or plat representations duplicate the description of any previous section where disposals have been made.

The statutory authority to review the effect of an independent resurvey on the boundaries of privately owned lands rests in the courts. A decision of the court is binding in fixing a boundary between private lands. It would be contested in fixing a boundary between public and patented lands only if monuments of the official survey have not been considered, the court having no authority to set aside the official survey.

The independent resurvey is accomplished in three distinct steps, as follows:

1. The reestablishment of the outboundaries of the area to be resurveyed, following the methods of a dependent resurvey

2. The segregation of lands embraced in any valid claim, based on the former approved plat

3. The survey of new exterior, subdivisional, and meander lines by a new regular plan

30-1-19. Special Instructions

The detailed specifications for each survey are set out by the officer in administrative charge of the work in a written statement entitled "Special Instructions." The special instructions are an essential part of the permanent record of the survey, both as historical information, and because they show that the survey was properly authorized. The immediate purpose is to outline the extent of the field work and the method and order of procedure. Coupled with the 1973 manual, the special instructions contain the technical direction and information necessary for executing the survey. Emphasis is given to any procedure unusual in application, but no lengthy discussion is required of procedures that are adequately covered in the 1973 manual. The special instructions are written in the third person.

30-1-20. Special Surveys

Special surveys are surveys that involve unusual applications of or departures from the rectangular system (see Figure 30-18). They often carry out the provisions of a special legislative act. A particular category of special surveys has to do with various types of water boundaries. In some cases, the special instructions merely expand the regular methods. In complicated special surveys, the methods are carefully detailed. Examples are:

1. Tracts and lots
2. Subdivision of sections, special cases
3. Metes-and-bounds surveys
4. Town-site surveys
5. Small tract surveys
6. Mineral segregation surveys
7. Mine surveys
8. Water boundaries

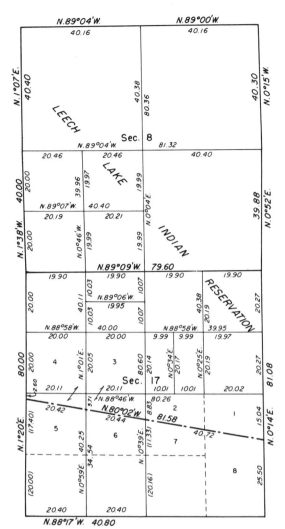

Figure 30-18. Example of a departure from the rectangular system. (*Manual of Instructions for the Survey of the Public Lands of the United States—1973*. 1975. Reprint. Rancho Cordova, CA: Landmark Enterprises.)

30-1-21. Field Notes

The field notes are the written record of the survey. This record identifies and describes the lines and corners of the survey and procedures by which they were established or reestablished. The new subdivisions to be platted (or replatted in the case of some resurveys) and quantity of land in each unit are derived from the field notes. The laws governing surveys of the public lands have required the return of field notes from the beginning.

The initial notes are kept in pocket field tablets. The final field notes for filing are transcribed from the field tablets and typewritten on regulation field-note paper. It is desirable that the final field notes conform to the general arrangement and phraseology set out in the 1973 manual. A large part of the final field notes must be extended from an abbreviated field record. At the same time, much of the minute detail of the initial notes may be summarized into a form of record that refers directly to the completed survey. This distinction in the two stages of the record is carried through the text. The subject in hand is the transcribed field notes, the record that is extended from the field tablets; this record is termed the *field notes* (see Figure 30-19).

All appropriate notes of the method and order of the survey procedures are entered in the field tablets. The tablets should show the dates on which each part of the field work was done. The field tablet record should supply the information needed for a complete preparation of the final record. Because of the great variety of survey types, the surveyor must plan carefully how the notes in the field tablet are to be arranged. The chief of field party is responsible for the accuracy and sufficiency of this record.

The work of transcribing the record usually receives the personal attention of the surveyor, but it is important that the arrangement of notes in the tablets and use of abbreviations are such as to be readily understood by others who are familiar with the technical processes. Due regard should be given to the 1973 manual requirements and form, though it intended that set forms of expression be used flexibly and modified when necessary to conform to the survey procedure. The work of the reviewing officers is directed to the fundamental requirements of the 1973 manual and written special instructions. Comments *as to the form* of the transcribed field notes are based

55.72	A bench mark of the U.S. Geological Survey, published elevation 7,946.987 ft. above mean sea level, bears South, 5.62 chs. dist.; a brass tablet seated in a sandstone boulder, conforming to the Geological Survey record.
80.00	Point for the stan. cor. of secs. 32 and 33, South 2 lks. from the secant.
	Set an iron post, 28 ins. long, 2½ ins. diam., 18 ins. in the ground to bedrock, encircled by a mound of stone, 3 ft. base to top of brass cap, mkd.

SC

T 13 N | R 21 E
S 32 | S 33

1972

from which

    A yellow pine, 9 ins. diam., bears N. 43 3/4° E., 27 lks. dist., mkd. T13N R21E S33 SC BT.

    A large sandstone outcropping, the highest point of which bears N. 57°35' W., 87 lks. dist., mkd. X BO.

Land, rolling west of creek; level table land above top of slope east of creek.
Soil, rich sandy loam and rocky loam.
Timber, mostly juniper, with some yellow pine and blue spruce; undergrowth, sagebrush.

    NOTE. ___ The field notes of the survey of the S. bdy. of secs. 33, 34, and 35 continue on the same form, and are omitted. The field notes of the survey of the S. bdy. of sec. 36 have been varied in order to show certain other forms of record.

	East, along the S. bdy. of sec. 36, on a transit line describing the secant, which bears S. 89°58.7' E.
	Over level land, through dense undergrowth.
40.00	Point for the stan. ¼ sec. cor. of sec. 36, North 2 lks. from the secant.
	Set a sandstone, 24 x 10 x 6 ins., 16 ins. in the ground, mkd. SC¼ on N face.
	Raise a mound of stone, 4 ft. base, 2 ft. high, N of cor.
45.00	Begin gradual descent.
48.92	Bank of Crystal Lake, bears N. 42° E. and S. 37° W.; point for the meander cor. of sec. 36, North 2.4 lks. from the secant.
	Set a sandstone, 27 x 8 x 8 ins., 18 ins. in the ground, mkd.

    6 grooves on N,
    MC on E, and
    6 grooves on W face.

Figure 30-19. Specimen field notes. (*Manual of Instructions for the Survey of the Public Lands of the United States—1973.* 1975. Reprint. Rancho Cordova, CA: Landmark Enterprises.)

on broad grounds, but it is necessary that the notes follow a standard form.

Random lines with fallings are shown in the field tablets but are omitted from the transcribed field-note record, except when some special purpose is served by showing the detail of a triangulation, offset, or traverse.

The township is considered the unit in compiling field notes. Normally, the field notes of all classes of lines pertaining to a township, when concurrently surveyed and not previously compiled, are included in a single book. In the survey of a block of exterior lines only, all the field notes may be placed in one book.

The field notes and plat are considered the primary record of any survey, and on their approval and acceptance the responsibility for the survey vests in the accepting authority. After the final record has been prepared, accepted, and officially filed, the field tablets and related field data are disposed of.

Field-Note Abbreviations

The following abbreviations, especially suited to field notes of surveys, are permitted in the final transcript record and used when repetitions in the form of the record and the expressions used are such as to make the abbreviations readily understood. Some of these abbreviations, as *appropriate*, are employed on the township plat. All abbreviations will be given capital or lowercase letters—the same as would be proper if the spelling were to be completed.

Abbreviation	Term
A	acres
alt.	altitude
a.m.	forenoon
Am.	amended
app. noon	apparent noon
app. t	apparent time
asc.	ascend
BM	bench mark
bet.	between
bdy., bdrs.	boundary, boundaries
ch, chs	chain, chains
cor., cors.	corner, corners
corr.	correction
decl.	declination
dep.	departure
desc.	descend
diam.	diameter
dir.	direct
dist.	distance, or distant
E	east*
e.e.	eastern elongation
elev.	elevation
ft	foot, feet

Abbreviation	Term
frac.	fractional
Gr.	Greenwich
GM	guide meridian
hor.	horizontal
h or hr	hour, hours
h.a.	hour angle
in., ins.	inch, inches
lat.	latitude
lk., lks.	link, links
l.m. noon	local mean noon
l.m.t.	local mean time
log.	logarithmic function
long.	longitude
l.c.	lower culmination
m or min	minute, minutes (time)
meas.	measurement
mer.	meridian
Mi. Cor.	mile corner
mkd.	marked
MS	mineral survey
N	north*
NE	northeast
NW	northwest
No.	number
obs.	observe
obsn.	observation
orig.	original
p.m.	afternoon
pt.	point
Prin. Mer.	principal meridian
R., Rs.	range, ranges
red.	reduction
rev.	reverse
s	second, seconds (time)
sec., secs.	section, sections
S	south*
SE	southeast
SW	southwest
sq.	square
Stan. Par.	standard parallel
sta.	station
temp.	temporary
t.	time
T., Tp., Tps.	township, townships
u.c.	upper culmination
USLM	U.S. Location Monument
USMM	U.S. Mineral Monument[†]
vert.	vertical
W.	west*
w.e.	western elongation
x	separating dimension values

[†]Discontinued in favor of the preferred term "U.S. Location (USLM)."
*Optional use of period.

30-1-22. Plats

The plat is the drawing that represents the lines surveyed, established retraced, or resurveyed, showing the direction and length of each line; the relation to the adjoining official surveys; the boundaries, description, and area of each parcel of the land; and as far as practicable, the topography, culture, and improvements within the limits of the survey. Occasionally, the plat may constitute the entire record of the survey (see Figure 30-20 for a sample of a township plat).

Ordinarily, an original survey of public lands does not ascertain boundaries: It creates them. The running of lines in the field and platting of townships, sections, and legal subdivisions are not alone sufficient to constitute a survey. Although a survey may have been physically made, if it is disapproved by the authorized administrative officers, the public lands that were the subject of the survey are still classed as unsurveyed.

The returns of a survey are prepared in the state survey office or service center and transmitted to the director, BLM, by the state or service center director for consideration as to acceptability. The survey only becomes official when it is accepted on behalf of the director by the officer to whom he or she has delegated this responsibility. Any necessary suspension or cancellation of a plat or survey must be made by the same approving authority.

The legal significance of plat and field notes is set out in Alaska United Gold Mining Co. v. Cincinnatti-Alaska Mining Co., 45 L.D. 330 (1916).

It has been repeatedly held by both State and Federal courts that plats and field notes referred to in patents may be resorted to for the purpose of determining the limits of the area that passed under such patents. In the case of *Cragin* v. *Powell* (128 U.S. 691, 696), the Supreme Court said: "It is a well settled principle that when lands are granted according to an official plat of the survey of such lands, the plat itself, with all its notes, lines, descriptions and landmarks, becomes as much a part of the grant or deed by which they are conveyed, and controls so far as limits are concerned, as if such descriptive features were written out upon the face of the deed or the grant itself."[6]

These legal principles apply to subsequent deeds of transfer related to the official plat.

The public lands are not considered surveyed or identified until approval of the survey and filing of the plat in the administering land office by direction of the BLM. The subdivisions are based on and defined by the monuments and other evidences of the controlling official survey. As long as these evidences are in existence, the record of the survey is an official exhibit and, presumably, correctly represents the actual field conditions. If there are discrepancies, the record must give way to the evidence of the corners in place.

In the absence of evidence, the field notes and plat are the best means of identification of the survey, and they will retain this purpose. In the event of a resurvey, they provide the basis for the dependent method and the control for the fixation of the boundaries of alienated lands by the independent method.

30-1-23. Current Information About the Public Lands

Current information about public-lands surveys may be found in the previous cited 1973 *Manual of Instructions for the Survey of the Public Lands of the United States*. This manual is available from the U.S. Government Printing Office. Earlier versions of the manual are helpful to learn the survey practices in force at the time of a particular survey activity. Manuals were issued in 1855, reprinted as the 1871 manual, and again in 1881, 1890, 1894, 1902, 1930, and 1947. Prior to 1855, special instructions were issued for various territories or states. All the manuals were prepared to guide the surveyors who were actually conducting the public-lands surveys. The manuals de-

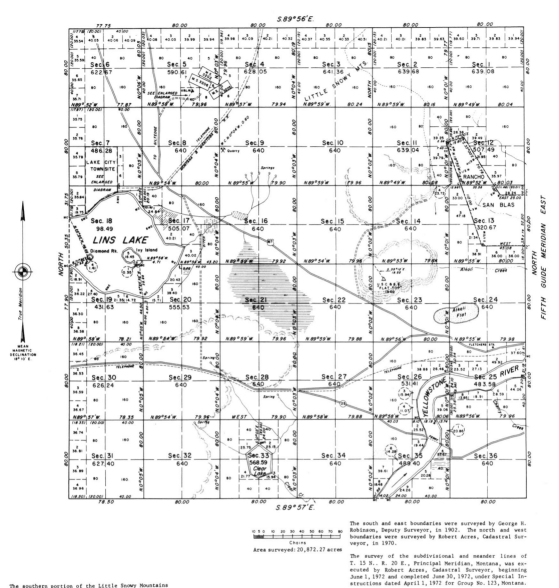

The south and east boundaries were surveyed by George H. Robinson, Deputy Surveyor, in 1902. The north and west boundaries were surveyed by Robert Acres, Cadastral Surveyor, in 1970.

Area surveyed: 20,872.27 acres

The survey of the subdivisional and meander lines of T. 15 N., R. 20 E., Principal Meridian, Montana, was executed by Robert Acres, Cadastral Surveyor, beginning June 1, 1972 and completed June 30, 1972, under Special Instructions dated April 1, 1972 for Group No. 123, Montana.

The southern portion of the Little Snowy Mountains is covered with a moderately heavy stand of pine, oak, and fir timber.

Figure 30-20. Sample of a portion of a township plat. (*Manual of Instructions for the Survey of the Public Lands of the United States—1973*. 1975. Reprint. Rancho Cordova, CA: Landmark Enterprises.)

scribed the methods to be used in conducting cadastral surveys, so that they would conform with federal statutes and the judicial interpretation that has occurred from time to time. The various manuals are the best place to begin any investigation of public-lands surveys.

Various supplements to the manual have been issued to cover specific topics. The one most often used by persons outside government service is the previously cited *Restoration of Lost or Obliterated Corners and Subdivision of Sections, a Guide for Surveyors.* The present edition is a 40-page pamphlet and provides an introduction to the public-lands survey and resurvey system and lists some of the basic laws on which the system is based. The pamphlet answers many common questions arising in the practical work of surveyors, land title specialists, attorneys, and others who have an interest in public-land systems. The pamphlet originally went into effect on June 30, 1883, and revisions appeared as shown in the following list:

1. First revision in effect: October 16, 1896

2. Second revision in effect: June 1, 1909

3. Third revision in effect: April 5, 1939

4. Fourth revision in effect: May 8, 1952

5. Fifth revision in effect: June 3, 1963

6. Sixth revision in effect: 1974

The standard field tables were issued as supplements for use by cadastral surveyors. The tables were heavily used to avoid lengthy calculations in the field, but with the arrival of hand-held programmable calculators and easy access to larger office calculators, the demand for the tables diminished, and they were allowed to go out of print a few years ago.

The previously cited *The Ephemeris of the Sun, Polaris and Other Selected Stars with Comparison Data and Tables* has been published since 1910. It was used by surveyors in the field who wanted to determine latitude and longitude.

All current public-lands surveys are under the jurisdiction of the BLM in the U.S. Department of Interior. The BLM was established on July 16, 1946. At that time, the functions of the GLO and U.S. supervisor of surveys were transferred to it. With that bureau, the public-lands survey continues in the Division of Cadastral Survey. This division operates through state and service center directors.

Information about specific surveys are readily available in the state offices of the BLM. Addresses for various offices are located in Appendix Z.

30-2. RETRACEMENT SURVEYS*

30-2-1. Introduction

The case studies presented in this section are not intended to insult or degrade any professional surveyor connected with the survey/resurvey work. If there is room for improvement by the reader (and there always is, no matter how extensive a surveyor's experience or knowledge), then a new idea or approach may have been presented. Comments to the contrary are welcomed by this author.

Practical experience coupled with almost limitless research form a strong foundation and defense for the professional land surveyor engaged in the illusive task of retracement/resurvey work. Land surveying will never be void of erroneous approaches (either past or present) and varied conclusions, but the frequency of such errors can be greatly reduced with conscientious research in both field and office settings. It is significant whether survey work is in the "metes-and-bounds states" or the "rectangular states"; of significance is a surveyor's *dedication, determination,* and *imagination.*

In order to adequately follow in the footsteps of previous surveyors, it is necessary to become familiar with the techniques, bias, language, instructions, obstructions, and related data pertinent to the survey project. Information resources seem endless at times, but shortcuts will invariably result in shortcomings, with the final results often discovered at a later and more costly time. The surveyor who states that he or she cannot "afford to do a complete job" (for the price being paid) cannot afford to do the job at all. Money "saved" doing an inadequate job will be spent many times over by irate clients, subsequent landowners, attorneys, and future surveyors. Land surveyors are professionally obligated to provide the most accurate and legal retracement/resurvey/survey product possible.

Case studies discussed in this chapter are derived from real situations. True names and locations have been deleted for reasons of privacy and simplicity of explanation. Examples include metes-and-bounds and rectangular surveys to emphasize the commonality of research and effort in both. The necessity to search "beyond" that point normally accepted by an "average" surveyor is demonstrated.

Case studies no. 1, 3, 4, 7, and 8 involve the rectangular system; studies no. 2 and no. 6 are metes and bounds. Case study no. 5 includes both survey systems.

Each geographic area has peculiar surveying challenges. A trip across the United States is not needed to discover these differences. They are usually present right next door. A tactful and professional surveyor will step carefully into new territory and thoroughly research the project prior to making survey decisions.

*Evidence—written, physical and testimonial—*is the common link between all surveys. It is the single largest responsibility and challenge for all professional surveyors. It is not necessary to be overly fearful in changing survey locations; rather, caution must always be exercised so that evidence is seldom overlooked.

Case studies that follow are located in the western states. However, the principles expressed are applicable to surveys and resurveys in most areas of the United States.

30-2-2. Types of Land Surveys

Considerable discussion can arise over definitions of land surveys and public land surveys. A distinction becomes meaningless when the task at hand is to locate and describe a client's property, usually based on previous deeds. While squabbling over such minor definitions, a surveyor may overlook the importance of carefully analyzing patents and deeds, plats/notes, occupancy, testimony, aerial photographs, and myriads of other resource information. These essential research elements will provide a more accurate location and description of a client's property. The courts determine property ownership, but surveyors can

influence court decisions by analysis, location, delineation, and documentation of evidences and deeds.

There are two basic types of land surveys most recognized by today's professional land surveyor: (1) metes-and-bounds and (2) rectangular. Most eastern states plus Texas and Hawaii are commonly referred to as "metes-and-bounds states," but similar surveys can be found throughout the remaining "rectangular" states. Some common metes-and-bounds public-land surveys may be as follows:

1. Homestead entry surveys (HES)

2. Donation land claims (DLC)

3. Exchange surveys (ES)

4. Mineral surveys (MS)

5. Reservation surveys

6. Military surveys

7. Town site surveys

8. Land grants

9. Ranchos

Thousands of surveys have been placed on the ground—i.e., monumented, platted, and described—but never patented or recorded. These elusive surveys are a continual source of trouble for the surveyor, when overlooked during the research/survey stage of the project. Though often unrecorded or difficult to retrieve from the public record, such surveys can be a source of important information, detailing the more accurate location of a particular land parcel. Regardless of the type of land survey requested, it becomes evident that *local* research, *public* contact, diligent *retracement*, and professional *documentation* are minimum requirements for all land surveyors.

30-2-3. Research

The part of land surveying operations least understood and seldom willingly paid for by a client is the research required for a survey.

After all, a client cannot hold a "conversation" in his or her hand, build a fence with "deeds" of adjoining landowners, or sell the 0.65 acres of land "lost" on the final plat description due to a senior adjoining (overlapping) survey. It becomes even more difficult to be the second or third surveyor in a community to establish a second or third set of monuments intended to define limits of the "same" parcel of land. Facetiously stated "more research means more problems." Realistically stated, "more research means better results."

Research must gather all the *written, physical*, and *testimonial* elements surrounding a survey project. Prior to accepting a job, a client must be informed of the potential research necessary to provide him or her with an accurate product. Contacts with adjoiners are imperative. A partial list of additional survey information sources follows:

1. BLM
2. Engineering/surveying companies
3. Real estate/title insurance companies
4. Transportation (local, state, or federal)
5. County records (registry of deeds/probate)
6. Forest service, Park Service, and like agencies
7. Utilities (electrical, gas, sanitation, telephone, etc.)

There is really no limit to the origin of possible survey information that can be uncovered with forethought and imagination. Only when surveyors feel comfortable that those following will not uncover new details can they cease to search for additional evidence of all survey problems.

30-2-4. Retracement/Resurvey Techniques

Eight case studies will describe some retracement/resurvey techniques.

Case Study No. 1: Missing 1897 Quarter-Corner Stone Between Sections 34 and 35

BACKGROUND.   A rectangular township was originally surveyed in 1897. Only three monuments within the township were marked stones, the remainder were wood posts. Most of the monuments had two or four bearing trees as accessories. Three forest fires had covered the township since the original survey, leaving dense undergrowth and a 6- to 8-in. layer of duff on the ground. The survey project required locating 24 original section and/or quarter-section corners. The missing quarter for sections 34/35 was described as a stone marked with "$\frac{1}{4}$" on the west face.

SURVEY RESULTS.   In 1975, a four-person survey crew spent approximately four months and successfully located 20 of the required 24 original monuments and/or their accessories. Of the four "lost" corners, three were wood posts with very small original bearing trees; the fourth was a stone monument for the quarter-corner of sections 34/35. Dependent resurvey results were provided to private contract surveyors for subdivision of certain sections by aliquot part. The contractors were required to delineate patent boundaries on the ground. Additional time was spent with the contractors searching for all lost corners. One hundred person-hours were spent searching for the quarter-corner of sections 34/35. Resurvey results indicated that the original surveyor commonly stubbed quarter-corners on east-west lines. Variations from the original record ranged from 2 to 288 ft per half-mile and up to 2° off cardinal directions. All persons connected with the survey knew that the quarter-corner of sections 34/35 physically existed, but efforts had to be abandoned at some point in time. Figure 30-21 illustrates the situation with record and resurvey results.

AFTERSHOCK RESULTS.   About 1979, a third surveyor began retracement work in section 27 and decided to search again for the "lost"

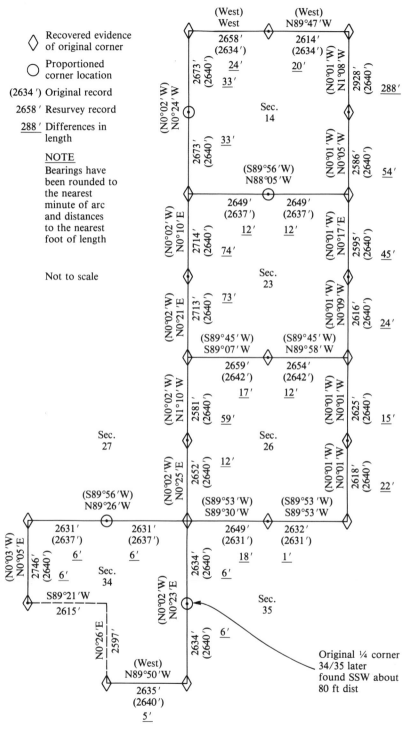

Recovered evidence of original corner

Proportioned corner location

(2634') Original record

2658' Resurvey record

288' Differences in length

NOTE
Bearings have been rounded to the nearest minute of arc and distances to the nearest foot of length

Not to scale

Figure 30-21. Case study no. 1.

quarter-corner of sections 34/35. The original marked stone was found several feet under a road fill and 80 ft SSW of the "single proportioned" quarter-corner of sections 34/35. This precipitated the potential for resubdivision of section 34. Fortunately, lands were exchanged in the area negating the need to resubdivide section 34 for present ownership purposes. Section 34 will have to be corrected before future disposal of lands there can occur. The status of an original corner in place was reemphasized, and a need to continue searching when optimism still exists concerning corner evidence was reinforced.

Case Study No. 2: Homestead Entry Survey of 1912

BACKGROUND.   Prior to 1913, the GLO retained the authority for surveying homestead entry surveys (HESs). To a minor extent in 1912 and totally after March 1913, the Forest Service was authorized, under special instructions from the GLO, to use its own surveyors to survey HESs. As a carryover from some of the "stubbing" techniques customary to early GLO and contract surveyors, many of the pre–1913 HES surveys were not actually closed on the ground, as their official notes indicated. Practices varied from leaving the last leg of the traverse open, computing corner locations from previously accepted nearby surveys, and/or stubbing off from random line through the body of the HES.

RESURVEY RESULTS.   Figure 30-22 depicts the dependent resurvey results of this 1912 HES. Corners 1, 3, 4, and 5 were recovered in their original positions and corner 2 reestablished from its original bearing trees. An astronomic observation was taken to determine the basis of bearing. Distances between corners 1/2 and 4/5 were relatively close to record, but distances between corners 2/3, 3/4, and 5/1 varied significantly from the original record.

DISCUSSION.   Several suppositions may be made from these resurvey results. (1) The original surveyor ran a random traverse through the center of the HES survey and stubbed corners off this random line. An error of approximately 100 ft was made along the random line and remained undetected because the survey was not actually closed. (2) The HES lines were actually run from corner 1 through 5, with a major error of approximately 117 ft between corners 2 and 3. A smaller mistake (or sloppy chaining) between corners 3 and 4 resulted in a final (undetected) blunder of approximately 98 ft between corners 5 and 1. Or (3) an unlikely 100-ft compensating chaining error was made between corners 2/3 and 5/1.

Specific calls were made to a road, fence, and creek between corners 1/2 and 4/5. Two very general calls were noted between corners 3/4, but none recorded between corners 1/5. Observation (1) is most logical though observation (2) could easily be urged. The end result is much the same, demonstrating the practice of "stubbing" as opposed to the close traverse required in survey instructions and returned in approved notes/plat.

PRACTICAL APPLICATION.   Suppose that corner 3 had been destroyed. The most common technique for reestablishing it would be by "grant boundary proportioning" between original corners 2 and 4, as shown in Figure 30-23. A few calculations show that a grant boundary proportioning solution to lost corner 3 would place the new corner position 99 ft from where corner 3 is actually located. Do not blindly enter a proportioning situation without gathering more information. A prudent surveyor will always gather additional supporting evidence, when a major blunder of this nature is uncovered. Clients often request only partial surveys of their deeded lands to enable them to develop and sell a few lots at a time. An incompetent surveyor, asked by a client to establish lot corners between corners 1, 2, and 3, could likely survey from corner 1 to 2 and erroneously reestablish corner 3 at

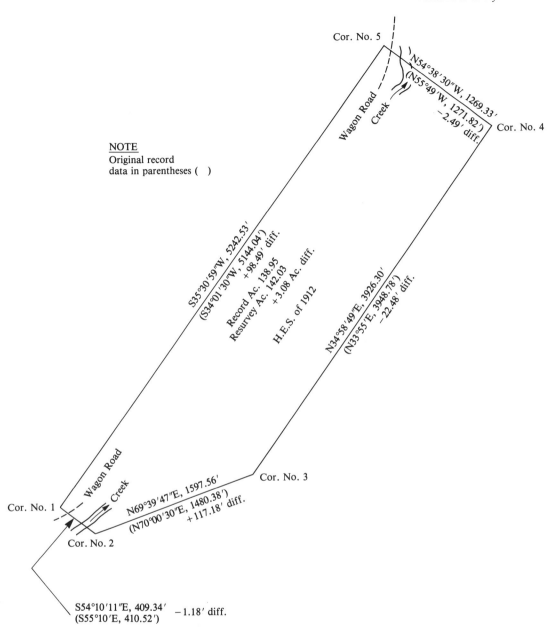

Figure 30-22.   Case study no. 2.

record bearing and distance from corner 2. Numerous combinations of errors could be made through failure to retrace and analyze the "entire" HES claim.

A positive note to this survey exists. The claim is totally encompassed by federal land, and due to the "excess" nature of the original survey blunder, there are approximately 3 acres more than contained in the original patent description. The surveyor has little difficulty explaining excess acreage to a client. Would a client be as easy to negotiate with if

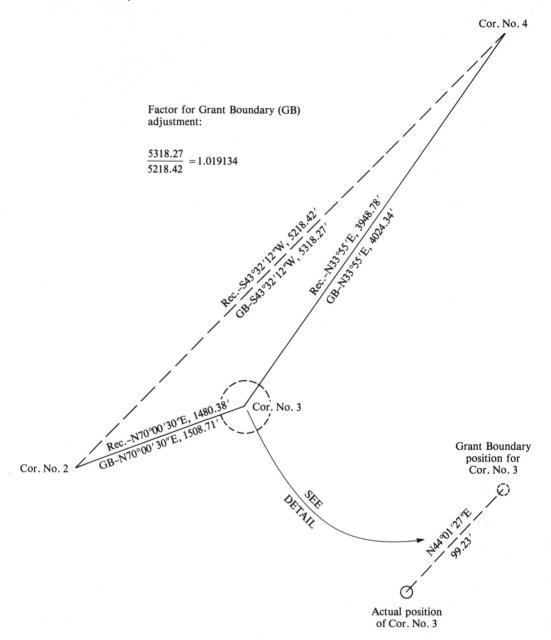

Figure 30-23. Grant boundary comparison of corner no. 3 in case study no. 2.

survey results show 3 acres fewer than the patent description indicates?

Case Study No. 3: Fractional Section 9

BACKGROUND. Refer to Figure 30-24. Sections 4, 5, and 8 of a fractional township were originally surveyed in 1891. A completion survey of the southeasterly position of said township was done in 1917. A sectional correction line was established in 1917 to return as many regular sections as possible from the remaining portion of this township. Fractional sections were created when the 1917 survey closed

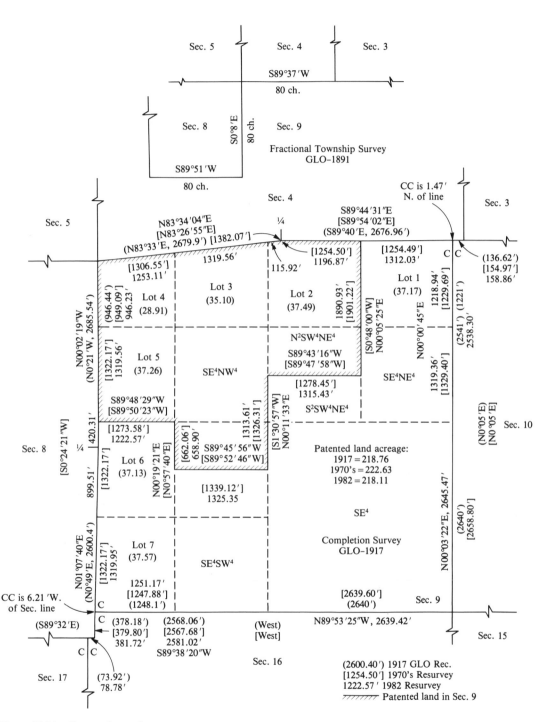

Figure 30-24. Case study no. 3.

against the earlier one. Patent was issued to about 219 acres in Section 9, after the 1917 survey, and described by government lots and aliquot parts. Three previous attempts to subdivide all or portions of fractional section 9 were made by registered surveyors prior to a dependent resurvey in 1982. None of the prior surveys was done correctly, and one had resulted in establishing "property corners" at all angle points of the 219 more-or-less private acres in section 9. The most recent and damaging survey had been made in the 1970s.

RESURVEY RESULTS.   A thorough search for corners controlling section 9 resulted in recovering all but the east quarter-corner of section 8. The north and west quarter-corners of section 9 were not set in 1917 and are not common with the south quarter-corner of section 4 or east quarter-corner of section 8, respectively. Several errors were made by surveyors since the 1917 survey: (1) A straight line was established between the southeast and northeast corners of section 8. The 1917 survey indicated an angle point at the east quarter-corner of section 8. A prior surveyor probably referred only to the 1891 survey, which returned the east line of section 8 as "S 0°08′ E." (2) The south quarter-corner of section 4 was assumed to be common with the north quarter-corner of section 9. Based on this erroneous assumption, the east sixteenth-corner (common to lots 1 and 2) on the north line of section 9 was placed at midpoint between the south quarter-corner of section 4 and northeast closing corner (CC) of section 9. Government lot 2 is patented private land, and lot 1 reserved public domain. (3) A private surveyor of the 1970s failed to locate any corners along the east or south boundaries of section 9. The 1917 record bearings were used for the east and south lines of section 9. Subdivisional corners within section 9 were then established by calculation from the northeast and southwest CCs of section 9. The 1970s private survey

showed about 223 acres within the described patent.

DISCUSSION.   Completion of the 1982 dependent resurvey revealed previously set subdivisional monuments up to 67 ft inside federal land and a total return of only 218 acres for the described patent—only $\frac{1}{3}$ of 1% (0.003%) smaller than the original patented acreage. Unfortunately, the landowners resisted the idea that they had only "1" acre less than their original patent and insisted they had been cheated out of "5" acres, the approximate difference between the 1970s and 1982 survey. Lengthy discussions, considerable patience, polite listening, and a confession to the landowner of erroneous survey technique by the 1970s surveyor finally brought the landowner to an agreement with the 1982 dependent resurvey.

In this case, the merits of a proper and accurate resurvey are self-evident. Later conversations revealed that survey costs were the determining factor for "shortcuts" in the 1970s survey. Resurvey and subdivision of section 9, largely in mountainous terrain, grossed only $750.00 for the 1970s surveyor. The client probably received what was paid for.

Case Study No. 4: Subsidence and the Shifting Quarter-Corners

BACKGROUND.   Figure 30-25 illustrates a peculiar situation that was overlooked by two previous land surveyors in their attempts to locate the common line between sections 1 and 6. Section 1 is government land (reserved public domain), and section 6 has private ownership, with lots being sold and homes built in the northwest quarter of said section.

In 1891, the west line of section 6 was originally surveyed on a "north" bearing and returned as 1 mi (80 ch) in length with the quarter-corner at midpoint. In a GLO survey of 1919, the township to the west was completed. The west side of T—N, R20E was re-

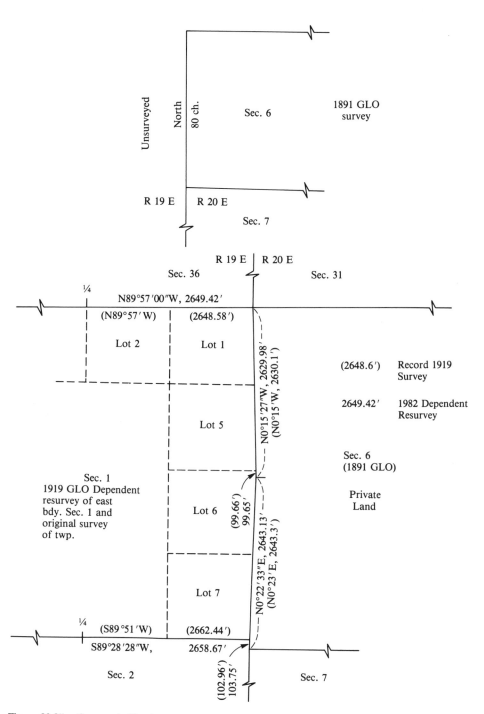

Figure 30-25. Case study No. 4.

traced and found to be out of closure limits, so CCs and offset quarter-corners were established adjoining the said township from the west. The original west quarter-corner of section 6 was recovered and remonumented by the 1919 GLO surveyor. A corner for the east quarter of section 1 was established about $1\frac{1}{2}$ ch southerly from the west quarter-corner of section 6. An angle point of 38 min was noted at the west quarter-corner of section 6, and each "half"-mile differed slightly from the original record.

The first private surveyor merely ran a straight line between the corner common to sections 6/7 (southwest corner section 6) and the northwest corner of section 6. Stakes were set for a landowner to build a "section-line" fence near the two quarter-corners. An expensive log-cabin-type home with eloquent landscaping was developed within 75 ft and east of the newly built fence line. The landowner stated that he and his surveyor had found the two quarter-corners to be considerably east of where they "should" be and ignored their position and any reason for their possible mislocation.

A second private land surveyor was later asked to subdivide most of the northwest quarter of section 6 into building lots. He failed to note the 1919 GLO survey; duplicated the "straight-line" relocation of the west line of section 6; did not question the log cabin's owner about the existence of the two quarter-corners; and apparently assumed that they were "lost." Neither of the two surveyors set monuments for the two "missing" quarter-corners. The later surveyor left several lot corner monuments along the north half-mile of the west line of section 6.

RESURVEY RESULTS. Details are shown in Figure 30-26. During a dependent resurvey in 1982, both surveyors were contacted, plats of their work acquired, and landowners along the common section lines interviewed. The log cabin's owner, near the two quarter-corners,

escorted the 1982 surveyor to two iron pipe/brass-capped quarter-corner monuments. They had been set by the GLO in 1919 during its dependent resurvey of the west line of section 6. The monuments were about 120 ft east of a straight line between the southwest and northwest corners of section 6.

It is "unusual" to have GLO iron pipe corners so far from "record" location. An investigation of possible reasons behind their present position was begun. Standing near the quarter-corners, it was evident that the area had been changed by a landslide. Large rocks were on end, and a very broken ground surface surrounded the immediate area. Several hundred feet away, the ground was rolling with a smoother surface, indicating more stable conditions. A deep, barren wash was noted just below an old irrigation canal northwesterly of the area. The landowner's father explained that the canal was originally used for ore-milling purposes on a placer claim, located in the northwest portion of section 6.

About 1930, a large quantity of water had entered the canal during spring runoff, causing the canal to breach just above the quarter-corners. Examination of the area by a soil scientist/hydrologist indicated that a bentonite layer had been supersaturated by the excessive water flow. This caused a massive but very localized landslide that encompassed both quarter-corners, moving them almost straight east from their original positions. Observation from a quarter-mile away vividly shows the landslide results.

DISCUSSION.   The two previous land surveyors had not considered the angle point recorded in the 1919 GLO notes and plat. Unfortunately, the direction of this angle point did not favor the log cabin's owner. After the two quarter-corner positions were properly (legally) relocated, the log cabin was only about 15 ft east of the common line between sections 1 and 6. Considerable landscaping had been done in section 1. Fortunately, no lots had

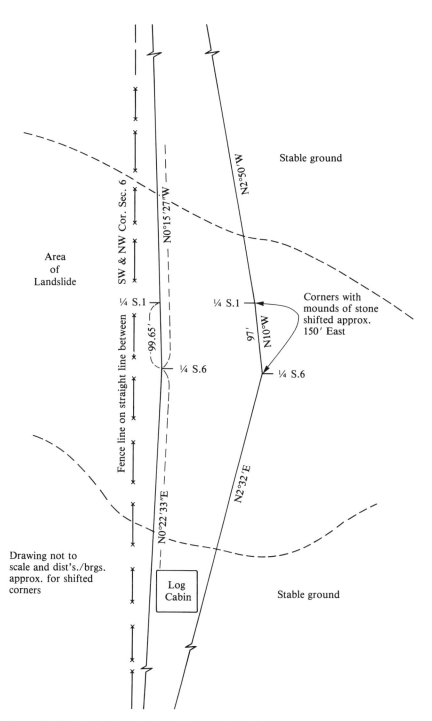

Figure 30-26. Further investigation of case study no. 4.

been developed along the northerly portion of the section line. Original corner monuments were reset based on their locations from the 1919 survey. Negotiations were begun between administrators of the government land and the owner of the expensive log cabin.

CONCLUSION. The log cabin had been located based on the first private-land surveyor's position for the common line between sections 1 and 6, but both surveyors failed to recognize the 1919 GLO resurvey/survey. Records were readily available from a local government surveyor's office and the BLM state office public room. There is no defense for failure to secure and utilize all government-generated survey information prior to initiating field work. It is a more difficult task to locate privately generated survey information.

Case Study No. 5: The Wandering Corner and Transposed Notes

BACKGROUND. A request was made in late fall of 1976, just after the first threatening snowstorm of a new winter season, to remonument a very important section corner controlling privately and federally owned land. Recovery notes of the corner described a properly marked stone, firmly set in the ground, with one bearing tree. *Without* reference to copies of the original notes, the cornerstone was remonumented. It had the correct number of grooves on two sides and a plainly scribed bearing tree northwest of the cornerstone. The stone was referenced, its markings and position noted (one groove on the south and three grooves on the west faces?), and a standard iron pipe with brass cap set in its place, bearing the proper stamping and surveyor identification. The cornerstone was buried alongside the iron pipe. A tie was made to the northwest bearing tree snag, but no other accessories were found at the time.

Winter immediately grasped the countryside, rendering it impractical to continue perpetuation of this corner before the spring thaw. Shortly thereafter, copies of the original 1896 survey notes were secured and they described *four* bearing trees. Additionally, there were notations by GLO surveyors in 1904, 1910, and 1917. Following are extractions from the original records:

[8-20-1896]...Set a granite stone, 14 × 12 × 6 ins., 9 ins. in the ground for cor. of secs. 13, 14, 23 and 24, marked with 3 notches on S., and one notch on E. edges, from which: A fir, 14 ins. diam., bears N.54°E., 247 lks. dist., marked T.—N.R.-E.S.13B.T. A fir, 14 ins. diam., bears S.38°E., 222 lks. dist., marked T.—N.R.-E.S.24B.T. A fir, 16 ins. diam., bears S.27°W., 199 lks. dist., marked T.-N.R.-E.S.23B.T. A fir, 24 ins. diam., bears N.41°W., 87 lks. dist., marked T.°N.R.-E.S.14B.T.

[9-24-1904]...at the cor. of secs. 13, 14, 23 and 24, which is stone, marked and witnessed as described by the Surveyor General.

[5-18-1910]...I began at cor. of secs. 13, 14, 23 and 24 where I find corner stone missing. At point for cor. as determined from the bearing trees which are plain and in good condition, I set a granite boulder, 18 × 12 × 10 ins., 12 ins. in the ground, for cor. of secs. 13, 14, 23 and 24, mkd. with 3 notches on the S., and 1 notch on the E. edge, witnessed as described by the Surveyor General.

[8-11-1917]...Survey of H.E.S.—...Retracement and subdivision of Sec. 13, T.—N., R.-E. From the Cor. of Secs. 13, 14, 23 and 24, which is a granite stone, 6 × 11 × 10 ins. above ground, marked and witnessed as described by the Surveyor General.

In April of 1977, the corner was revisited and a thorough search made for the other three bearing trees. Two more were found, but bearings and distances did not concur with the original notes. The standing bearing tree to the northwest was about 29 ft farther than record from the corner position. It was evident that this corner may have been prematurely remonumented, and further research was necessary.

PROBLEM ANALYSIS. Refer to Figure 30-27. Findings and conclusions noted on a state corner recordation form are as follows:

[4-21-1977] I found the original corner stone firmly set with a cross on the top and 1 groove on the south and 3 grooves on the west faces. After carefully tying in the three remaining original bearing trees it was noted that no correlation to the recorded bearing tree ties could be made. A careful search for another monument was made at recorded positions off each bearing tree with negative results. The chain of original recordation for this section corner is noted above. In an attempt to resolve the bearing tree and corner stone discrepancies I located the nearest 4 original corners of HES—along with a well imbedded mound of stone believed to be the position for the S-S 1/64 corner for sections 13 and 14. I tied these points, along with the corner stone and the three original bearing trees, with the use a a Wild T-2 theodolite and a Hewlett-Packard 3800 distance meter. After analyzing the relative positions of each found corner monument and the bearing tree accessories with respect to their original records I determined a position for the corner of sections 13, 14, 23 and 24 based on the following:

1. The corner position was first determined by using the record bearings and distances as recorded in (the) 1917 HES survey. This position was found to closely concur with the record bearings to the three remaining original bearing trees but only concurred with the record distance to the northwest bearing tree. However, by exchanging the distances (as recorded in the 1896 . . . record) between the northeast and the southeast original bearing trees all three distances concurred closely to the original 1896 record.

2. It is apparent that a distance recording error was either made in the field at the time of the original survey or during one of the transcriptions of the original surveyors notes. With this conclusion it followed that the corner position would have to be reestablished from distance-distance intersects off the original bearing tree positions.

3. Therefore, I reversed the record distances between the northeast and the southeast original

bearing trees and meaned the two values obtained from the distance-distance intersect positions between the northwest/northeast and the northeast/southeast original bearing trees.

4. Finally, I noted that the original corner stone was firmly imbedded in the ground but was turned 90 degrees clockwise from its recorded position (i.e., the 3 grooves are on the west and the one groove is on the south faces of the stone) indicating it having been disturbed and moved some time after the 1917 HES survey by . . .

LOCAL SURVEYOR CONTACTS. Further research uncovered a subdivision plat in section 23. A copy of said plat was secured and the responsible surveyor, who resided in an adjoining state several hundred miles away, contacted. Explanations and conclusions were forwarded to the surveyor with a request for comment and/or concurrence. The following letter was received very soon after:

To Review: I was hired by a (local) firm to lay out the roads and survey some fronts on the Plat of [No Name] Creek. May stamp was used, that makes me responsible.

[Mr. John Doe] was hired to survey all section breakdowns and present to [the firm]. [John] was cutting lines by himself with stobs for line and bringing up the chaining on the weekends with his young son.

This "unorthodox" method was a little alarming to me, so for a check I started at the corner in question, west to Forest Service Road, southwesterly along road to transmission line, easterly to corner, northerly along Forest Service Road and across the creek to starting corner. This tied in only two corners and I missed them about 40.00 feet in closing. So we went to the North quarter and again missed [John] 20.00 feet or more.

I put this information to (the firm) to review with (John Doe).

The (local) firm had a reputation of high-ball subdivisions and after I had placed the road pattern and some front corners, I again related

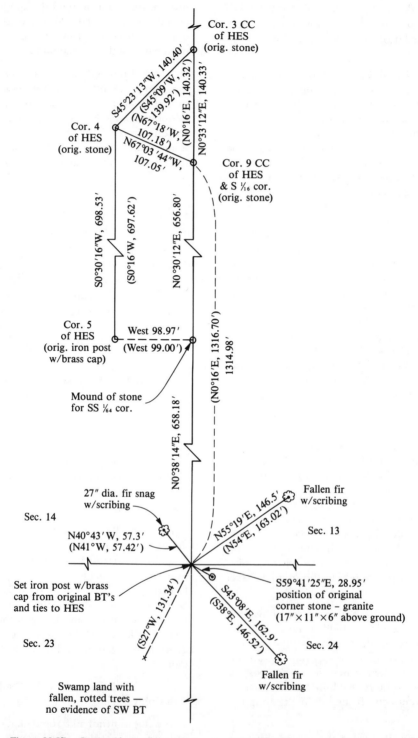

Figure 30-27. Case study no. 5.

to [the firm], the comp. sheet could be in error. At this point, I left the project and was told all final staking would be done by "others" and I learned these "others" were nonlicensed L.S.'s.

I remember the corner from my first visit and one large yellow pine bearing tree was in evidence.

I am sure I could place the corner within a foot or so in the position of its original.

I can only wonder of the terrible possibility of this corner being moved to conform with the original breakdown, which was in gross error.

Seems hard to believe, but a possibility.

I would be glad to assist in any manner possible, and for you to remove the stone and put it back in its original position would be a service to our profession.

Yours truly, . . . [9]

CONCLUSION.   It is a very difficult decision (at best) to move an original corner monument, when it (1) was found firmly set in the ground, (2) had obviously been witnessed in its position for more than 20 yr, and (3) had been used to subdivide local properties. The principle of locating the best available evidence, analyzing conflicting evidence, and formulating a legally supportable corner location is important and essential to the success of a professional surveyor's work. It is imperative that all involved parties be contacted and provide adequate time to respond to a surveyor's proposals and conclusions; of most importance is the manner and permanency of documentation of the existing circumstances surrounding surveying decisions.

If adjoiners agree to move a corner to its "original" location, the problem is easily resolved. If not, then further negotiations must follow. One solution may be maintaining two corner positions—i.e., one *section corner* and one *property corner*. A "cookbook" answer does not exist for every survey problem. Each case has its own peculiarities, and a legally acceptable solution remains the biggest challenge for

the professional surveyor. Occasionally, the courts are required to make a final "legal" decision.

Case Study No. 6: Grant Boundary Relocation of an HES Corner

BACKGROUND.   When we refer to Figure 30-28, a chronological listing of events best explains this case, as follows:

1. Original survey of the HES made about 1915.

2. Occupancy and building construction near the lines between corners 1/2 and 2/3 in place.

3. Retracement and plating of lines 1/2 and 2/3 with ties to encroached buildings and roads accomplished about 1955.

4. Lot purchased near corner 2, by present landowner, with expansion of existing home and improvements to surrounding land. Further encroachment occurred at this time.

5. Government surveyor resurveyed from corner 1 through 3, noted corner 2 had been destroyed, reset corner 2 by grant boundary method between corners 1 and 3, and posted the common property line between the HES and federal land.

6. Landowner hired his own surveyor, who agreed with the government surveyor. Dissatisfied, the landowner hired a second surveyor who told the story "another way." Second surveyor found barbed wire imbedded in a stump at a "convenient" position that would lessen the extent of his client's encroachment.

7. Differing opinions remained the point of heated discussion between land-owners, attorneys, and public-land administrators for several years.

ADDITIONAL RESEARCH.   With the conflicting information noted, meetings were initiated with *all* parties involved. After the inspection of numerous documents, the disputed corner location area was visited, and the existing and proportioned corners of the HES photoidentified. A search through old aerial photo files

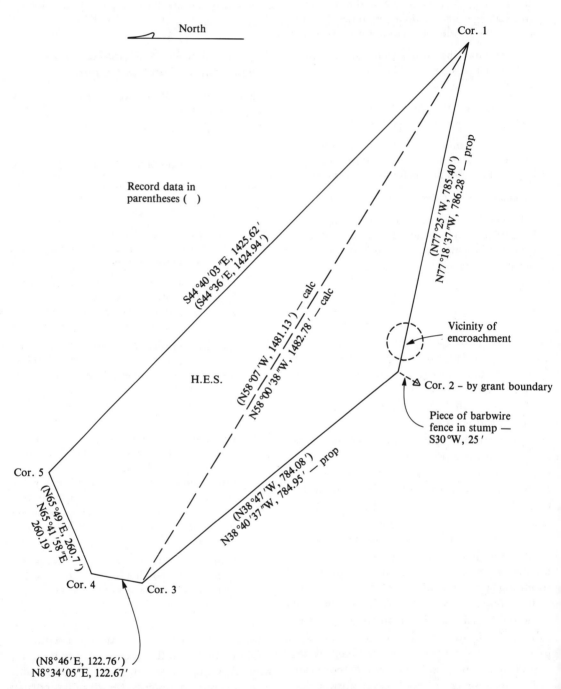

Figure 30-28.   Case study no. 6.

disclosed flights over the project site in the mid-1950s. Comparison of early and more recent aerial photographs substantiated the existence of two large crowned conifer trees at the locations of the original bearing trees for corner 2. The corner position had been disturbed and about 4 ft of soil including the two original bearings, were removed from the immediate area.

During meetings with the landowner's attorney and surveyor, procedural elements of the survey were reviewed, and the aerial photographs discussed. The surveyor acknowledged an understanding of photogrammetry and recognized its use as a legal and viable tool for substantiating original corner evidence. He agreed to go over the new findings and indicated a response would be forthcoming.

RESULTS AND CONCLUSIONS. Unfortunately, several months lapsed without a response from the surveyor. The attorney was again contacted and informed that the government would proceed based on its surveyor's conclusions unless a response was received immediately. The attorney "strongly" warned the government surveyor that reluctance to respond was not sufficient reason to proceed as proposed, and that to do so would be ill-advised on the government's part. Pressure often precipitates results, so the opposing surveyor was reminded of the government's intent to contact the state surveyor's board and also request an official BLM dependent resurvey of the HES. With this added fuel, the surveyor finally agreed to accept the government surveyor's findings and advised his client and attorney accordingly.

It is unfortunate that power and money speak so loudly, as evidenced in this case. When land boundary disputes linger between opposing parties and their legal counsel, the only benefiting party is the attorney. If formal courtroom litigation results, additional money is often wasted and the legally verifiable corner and property line evidences will generally prevail. The key to successfully staying out of

court is communication, professionalism, and education on the part of *all* parties involved. Preceding these elements is the responsibility of the land surveyor involved to gather all possible evidence and present his or her conclusion to all parties in the most concise and understandable manner. Litigation can often be avoided with round-table discussions based on legally supportable proposals and conclusions.

Case Study No. 7: Following in the Surveyor's Real Footsteps

This case and accompanying Figures 30-29 and 30-30 are vivid examples of the shortcuts and random surveying techniques so often suspected of the early GLO surveyors. The original survey was made in 1875 at elevations from 5000 to 10,000 ft above sea level, and terrain ranging from mild to "very difficult" and impassible at times. A considerable effort had been expended over a 5-yr period to locate the original corners controlling private and federal lands interspersed throughout the township.

Seven of the original 85 internal township corners and 12 original exterior boundary corners were found. When survey connections were made between these 19 corners and previously found corners along the north and east boundaries, the total township was reasonably regular compared with the original platted bearings and distances. A couple of striking differences—not unusual for an 1875 survey —were noted. The bewildering aspect of the resurvey efforts was the apparent absence of corners inside the township.

The task was assigned to investigate the existing corner evidences and attempt to locate additional evidence of original corners. After several hours of unsuccessful search near the quarter-corner of sections 19 and 20, all previous input and inferences were abandoned and the basics used over the past 23 yr applied. Corner search has *some* consistencies but demands creativity and insight in cases where the original surveyors deviated from the

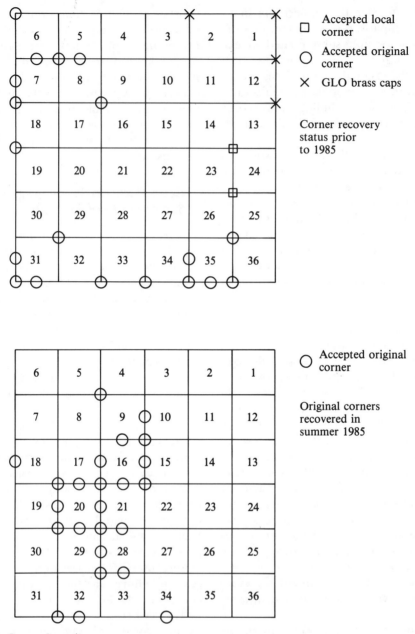

Figure 30-29. Case study no. 7.

mandated procedures that they were under oath to follow. A new start was made, armed with copies of the original field notes/plats, 7.5-min topographic maps of the area, aerial photos, and appropriate scales and marking pencils.

Within 15 min, three record calls were correlated with the topographic quandrangle map and transferred to aerial photography. The search area was localized on the ground, and a blazed tree found within $\frac{1}{2}$ ch of the photo plot. A search northerly from this blazed tree

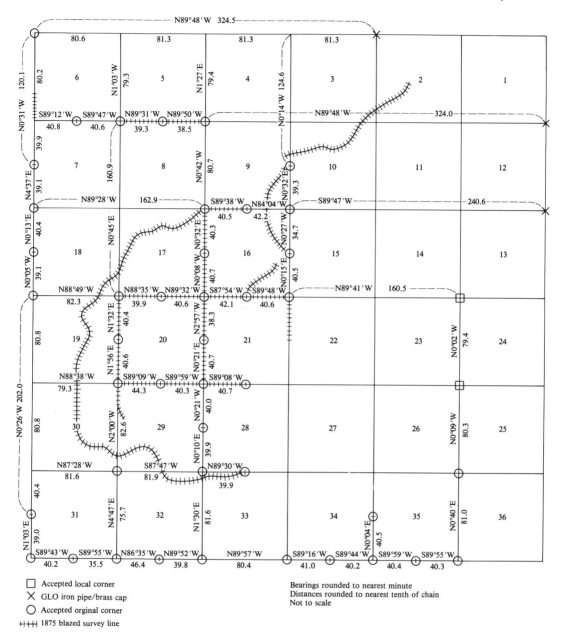

Figure 30-30. Data for case study no. 7.

uncovered several more similarly blazed trees. After it was determined that the search area was too far north, new efforts were begun by working south from the initial blazed tree. An original bearing tree was located, then a second, and some interesting discoveries followed.

Subsequent searches at various times over the next 12 wk netted an additional 23 original corners. The enlightening aspect of this research came when a "random" line was discovered through the northerly portion of section 17. The blazed trees were bored with an increment bore and dated back 110 yr

(1875) to the initial face of the blaze. This discovery led to the location of about 12 mi of random blazed survey line, marked by the original surveyor. These blazed lines were considerable distances from the projected section lines.

Figure 30-30 illustrates the approximate locations of these random blazed lines. It was evident that the surveyor had carried his "lats and deps" along these lines and offset to his corners as he neared their calculated positions. More than 50% of the original corners in the west half of the township were found, but no additional original corners in the township's east half could be located. A prime factor contributing to success in the west half was the absence of logging and forest fires since the original survey. The east half had been heavily logged and burned since the 1875 survey. Also, there were numerous calls of record that fit in the west half, whereas they were severely lacking in compatibility over most of the township's east half. The most difficult terrain was in the east half. Many more corners were set in the original survey than could be located, but imaginative approaches to discover more of them ran out.

Figure 30-30 shows the differences and comparable accuracies found throughout the township. It was fairly simple to determine which corners were stubbed. Locations of the random blazed lines indicate that the 1875 surveyor was a good "ridge-runner" and readily sought points of visibility and "paths of least resistance." He was actually quite creative in his approach, but this creativity was a hindrance to successfully locating more of his original corners. Some corners were thoroughly searched on four different occasions, with new ideas and approaches, but remained undiscovered. More corners are definitely in existence. It is difficult to place limits on search time and area, but efforts must terminate at some date. Perhaps other corners will be found later.

RETROSPECT. What about the efforts of surveyors for several years prior to this recent involvement? Some merely lacked time and experience in identifying original evidences. Others failed to recognize or acknowledge the presence of "110-yr old" blazed trees left by the original surveyor. Some were led from more logical search areas by "local" erroneous corners set about 1935. Still other had relied on the unsuccessful efforts of previous surveyors and failed, themselves, to make additional searches prior to their final resurvey efforts. Each new project demands consideration of all previous evidence and must include an independent and unbiased search for additional record and/or physical information that may have been overlooked. A valuable maxim states: "Search until you find the original evidences or until you feel certain that no one will find them after you."

Case Study No. 8: Illusive 1866 Corner Monuments

BACKGROUND.   An original GLO survey was conducted between sections 11 and 12 in 1866, including the west quarter-mile of the north line of section 12. The township was completed by another GLO surveyor in 1875. Dependent resurvey work was accomplished between 1948 and 1950. The 1866 survey plat and field notes described a stone monument at each corner location, with bearing tree accessories scribed for the two section corners and 1/16 corner. The 1875 GLO survey connected with the 1866 survey but never overlapped in section 12. Figure 30-31 illustrates the relationship between the 1866 and subsequent surveys/resurveys.

Several private and Forest Service resurveys were accomplished from about 1900 through 1984. Prior to 1985, none of the resurveys had located evidence of the 1866 corners between sections 11 and 12. During the 1985 field season, a contract awarded to a private surveyor required subdivision and posting of federal boundaries in and around section 12.

RESURVEY ANALYSIS.   Several inconsistencies became apparent when the various surveys/resurveys were compared and lines of section 12 retraced. A composite of retracement data is shown in Figure 30-32. Early in the 1985

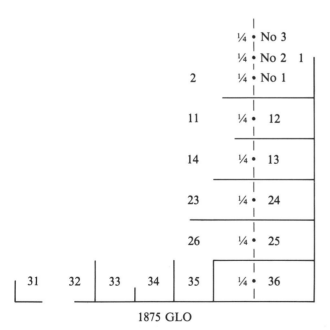

6	5	4	3	2	1
7	8	9	10	11	12
18	17	16	15	14	13
19	20	21	22	23	24
30	29	28	27	26	25
31	32	33	34	35	36

1866 GLO

2	1
11	12
	13
	24
	25

1948–50 BLM

¼ • No 3
¼ • No 2   1
2   ¼ • No 1

11   ¼ •   12

14   ¼ •   13

23   ¼ •   24

26   ¼ •   25

31 32 33 34 35   ¼ •   36

1875 GLO

The 1875 GLO survey completed the township lines and established ¼ corners ''inside'' sections 1, 12, 13, 24, 25 and 36

Figure 30-31.   Case study no. 8.

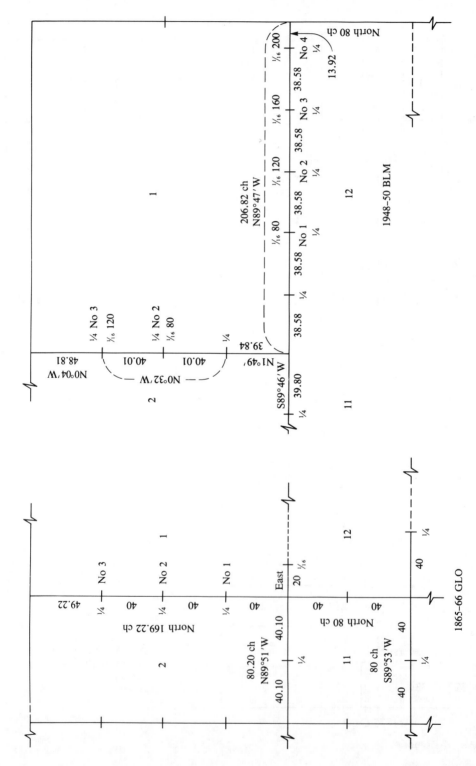

Figure 30-32. Details for case study no. 8. (See Figure 30-33 for details of ties to 1866 corners.)

Figure 30-32. *(Continued).*

793

resurvey, 5 of the 1875 corners were recovered, and all 8 of the 1948 to 1950 corners around elongated section 12 were found. The 1948 to 1950 field notes described a properly marked stone for section corner 1, 2, 11, and 12 that was found set firmly in the ground. The same field notes further explained that the found stone had been locally accepted and adjudicated by a 1915 court. A final comment stated that "no evidence of the original (1866) bearing trees were found."

EXTRA EFFORT PAYS OFF! Each recovered corner involved with section 12 was reviewed, including several corners of adjoining sections. Analysis of topographic and man-made calls in the 1866 field notes correlated very closely with existing features on the ground. A mill race had been called in both the 1866 and 1948 to 1950 field notes as being north and east of section corner 1, 2, 11, and 12. Figure 30-33 illustrates the location of the mill race with respect to each survey.

The mill race is easy to identify on the ground and meanders in a northwesterly direction. The contractor's random traverse line intersected the mill race several hundred feet east from the 1948 to 1950 GLO monument for section corner 1, 2, 11, and 12. From this point of intersection, the 1866 record distance was measured west and fell in the bottom of a 5-ft-deep ravine running southwesterly. Bearing trees for the original 1866 corner were described as a 9-in.-diameter pine and 36-in.-diameter fir. Inspect of the area quickly netted a 26-in.-diameter fir bark shell and pine bark crown at the record bearings and distances as recorded in the 1866 field notes. If these suspicious remains of fir and pine proved to be the original 1866 bearing trees, the 1948 to 1950 "corner" would be about 135 ft too far west.

Positive results fell quickly into place. The contractor reevaluated corners on the west line of section 12 and searched about 135 ft east from the existing fence line. With proficiency gained from professional experience and judgment, the contractor found the original marked stone, with a mound of stone, for the quarter-corner of sections 11 and 12. Soon after, he uncovered, from beneath several inches of soil, the original marked stone monument for section corner 11, 12, 13, and 14. To complete his search, the original stone monument for section corner 1, 2, 11, and 12 —properly marked—was uncovered in the bottom of the shallow ravine. The previously located fir and pine bark remains were verified as the original 1866 bearing trees.

DISCUSSION. A problem now existed. Two locally accepted corner monuments had been used to subdivide "expensive" properties. Permanent residences and vacation homes had been built on the subdivision lots. Numerous survey plats, both private and government-generated, had been filed in the county records. The whole community had been relying on two monuments that were not controlling points, as called for in the original patents to private lands in sections 1, 2, 11, and 12. Regardless of the existing circumstances, two *original* monuments had been found in direct conflict with "local" monuments. A third original monument was found in conflict with a long-standing fence line.

As of January 1986, agreements do not exist between adjoiners of properties affected by these multiple corners. Additional resurvey and investigation are needed, along with meetings between government and private home and lot owners who have occupied their space for many years since the original patents. One solution may be to hold the "local" corners as control for existing subdivisions and occupancy, while using the "original" corners for control of future land divisions. One basic principle must hold—i.e., controlling corners of an original survey, found undisturbed, will retain their intended identity and control descriptions of patents issued from their locations.

It will become necessary to redescribe existing land parcels in conformance with metes and bounds *within* the rectangular system rather than aliquot part descriptions *of* the

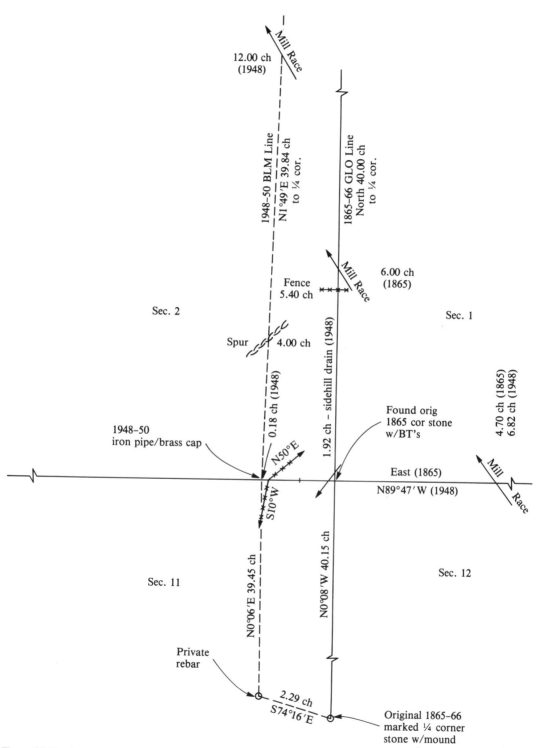

Figure 30-33.   Location of mill race in case study no. 8.

rectangular system. Professional surveyors will play an important role in explaining the preceding situation to landowners and redescribing their land parcels.

SUMMARY.   An objective observation of case study no. 8 leaves a serious question as to why the original evidences of the 1866 corners were not discovered before 1985. Surely, evidences must have been less deteriorated in 1948 or 1915, when much of the resurveying activity was happening. Research, thus far, has failed to uncover records of any "1915 court case," so its significance is as yet unkown. As to establishment of the unmarked stone monument found in 1948 to 1950 for section corner 1, 2, 11, and 12, it is possible that it was stubbed "$\frac{1}{2}$ mi east" off the original quarter-corner of sections 2 and 11. Retracements of other lines between original corners of the 1866 GLO survey show evidence of stubbing, resulting in several hundred feet and a few degrees of discrepancy with the same survey of record.

CONCLUSIONS.   Surveying advice is never complete without a reminder from A. C. Mulford, who was well ahead of his time. In the preface of his book *Boundaries and Landmarks*, he states:

...No attempt is made to describe how the lines should be measured; the intent is rather to furnish suggestions as to the method of locating the line to be measured—in short, finding it. It is far more important to have faulty measurements on the place where the line truly exists, than an accurate measurement where the line does not exist at all.[7]

In this book's last chapter, "Responsibilities of the Surveyor," the profession's duty is defined:

The watchwords of the surveyor are Patience and Common Sense.
Curiously enough the Surveyor is isolated in his calling, and therein lie his responsibility and his temptations. The lawyer comes nearest to understanding the work, yet of the actual details of a survey most lawyers are woefully ignorant.... To the skilled accountant of the bank the traverse sheet is a closed book. Dishonesty in ordinary business life cannot long be hid and errors in accounts quickly come to light, but the false or faulty survey may pass unchallenged through the years, for few but the Surveyor himself are qualified to judge it.... Therefore I believe that to every Surveyor who values his honor and has a full sense of his duty the fear of error is a perpetual shadow that darkens the sunlight.
Yet it seems to me that to a man of active mind and high ideals the profession is singularly suited; for to the reasonable certainty of a modest income must be added the intellectual satisfaction of problems solved, a sense of knowledge and power increasing with the years, the respect of the community, the consciousness of responsibility met and work well done. It is a profession for men who believe that a man is measured by his work, not by his purse, and to such I commend it.[8]

NOTES

1. U.S. Department of the Interior, BLM. 1973. *Manual of Instructions for Survey of the Public Lands of the United States*. Washington, DC: U.S. Government Printing Office.

2. 1857. *Instruction to the Surveyor General of Oregon; Being a Manual for Field Operations*. GLO: Washington D.C.

3. *The Ephemeris of the Sun, Polaris, and Other Selected Stars with Companion Data and Tables*. GPO: Washington D.C. (published annually)

4. *Restoration of Lost or Obliterated Corners and Subdivision of Sections, a Guide for Surveyors*. 1883–1974. Rancho Cordova, CA: Carben Survey Reprints.

5. 1980. W. E. Peters. *Ohio Lands and Their History*. New York: Ayer.

6. Alaska United Gold Mining Co. v. Cincinatti-Alaska Mining Co. 1916. 45 L.D. 330.

7. A. C. Mulford. 1912. *Boundaries and Landmarks*. New York: Van Nostrand Reinhold, Preface. 1974 Rancho Cordova CA Carben Survey Reprints.

8. _____ . Responsibilities of the Surveyor. *Boundaries*. pp. 88–89.

REFERENCES

BURT, W. A. (1854) 1978. *A Key to the Solar Compass —a Surveyors Companion*. Rancho Cordova, CA.

CAZIER, L. 1978. *Surveys and Surveyors of the Public Domain*. Washington, DC: U.S. Government Printing Office.

CLEVENGER, S. (1874) 1978. *A Treatise on the Method of Government Surveying*: Rancho Cordova, CA: Carben Survey Reprints.

A Collection of Original Survey Instructions: 1815–1881. 1978. Rancho Cordova, CA: Landmark Enterprises.

DONALDSON, T. (1884) 1970. *The Public Domain*. New York: Johnson Reprints.

DORR, B. F. (1886) 1978. *The Surveyors Guide and Pocket Table Book*. Rancho Cordova, CA: Carben Survey Reprints.

GATES, P., ed. 1974. *Public Land Policies; Management and Disposal*. New York: Ayer.

HAWES, J. H. (1868) 1977. *Manual of United States Surveying*. Rancho Cordova, CA: Carben Survey Reprints.

MCENTYRE, J. G. 1986. *Land Survey Systems*. Rancho Cordova, CA: Landmark Enterprises.

STEWART, L. (1935) 1975. *Public Land Surveys; History, Instructions, and Methods*. Rancho Cordova, CA: Carben Survey Reprints.

WHITE, A. 1985. *The History of the Rectangular Survey System*. Washington, DC: U.S. Government Printing Office.

31

Optical Tooling

James P. Reilly

31-1. INTRODUCTION

Occasionally, surveyors may be requested to provide very precise dimensional control in the assembly and alignment of aircraft jigs, automobile and farm implement manufacturing, and other machine elements. The term optimal tooling refers to surveying techniques that have been introduced into the aircraft and other industries to make accurate dimensional layouts possible. Many people refer to this process as *optical alignment*.

Optical tooling is not surveying in the classical sense. During World War II, it became necessary to manufacture airplanes and ships in very large numbers, and the story goes that standard surveyors' levels and transits were used to align fuselags and wings in British manufacturing facilities. Prior to that time, all alignment was done with a taut wire in laying out keels for large ships, etc. Since World War II, several instrument manufacturers working with U.S. aircraft companies have designed instruments specifically for optical tooling.

With limited space available for so many topics in a handbook, only a short treatise on optical tooling can be presented. The objective here is to give important information on the "basics." Many books and reference manuals have been written on the subject, and some are listed at this chapter's end.

Instruments used today by optical-tooling specialists will be mentioned, along with the new computer-controlled real-time systems that have gained a strong position in an industry still using some instruments and technology developed in World War II and the Korean Conflict days.

31-2. OPTICS, COLLIMATION, AND AUTOCOLLIMATION

Since optical instruments are the backbone of optical tooling, it is necessary to understand the basics of light passing through glass and reflected off mirrors. Principles of collimation and autocollimation will also be explained.

31-2-1. Refraction

Refraction changes the direction and bends a ray of light as it passes obliquely from one medium to another of different density.

Light passing from a medium like air into a dense transparent material such as glass has its speed reduced. Transparent material can be rated by the speed of light in a vacuum divided by the speed of light through the material. This number is called the material's index of refraction, universally referred to by the letter n. The index of refraction for ordinary glass is about $n = 1.5$, but of course, manufacturers of optics do not use ordinary glass. For the flint glass utilized in some theodolites, $n = 1.62$. The index of refraction varies slightly for colored glass.

31-2-2. Reflection

When a light ray strikes almost any surface, it is partially absorbed, and some of its colors are reflected in every direction as diffused light. The colors reflected are the means by which the color and shape of an object can be seen.

Smooth surfaces, like glass or highly polished metals, absorb very little light and reflect most of it in a definite direction—the basis of a mirror. A ray that strikes a mirror surface is reflected so the *angle of reflection* equals the *angle of incidence* (see Figure 31-1).

One characteristic of glass is that for a large angle of incidence, all the light will be reflected back into the glass. The angle of incidence at which this occurs is called the *critical angle*. Incident angles still larger are termed *angles of total reflection*. This phenomenon is depicted in Figure 31-2.

The critical angle can be computed by Snell's law. It is defined as the angle whose sine is the ratio of the two indices of refraction. As an example, where $n = 1.00$ for air and $n = 1.60$ for optical glass, sine (critical angle) $= 1.00/1.60 = 0.625$ and critical angle $= 38°41'$.

The ordinary mirror in every home is a glass plate with silvering on the back. A rear-surface mirror produces unwanted reflections that result in a less sharp image (Figure 31-3). If the mirror is not flat, again the reflected image will not be sharp (Figure 31-4).

Mirrors for optical tooling must be optically flat and silvered on the front surface. A person hearing the word "mirror" automatically thinks of glass. This is not necessarily so; polished metal surfaces can be employed, provided that they are perfectly flat.

31-2-3. Collimation

In optical tooling today, essentially all instruments used have one or more telescopes,

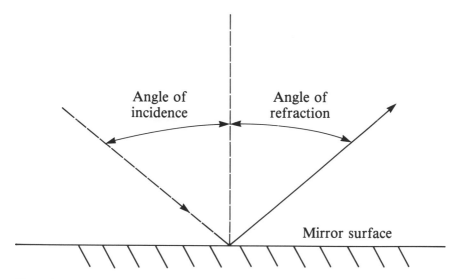

Figure 31-1. On mirror surface, angle of incidence equals angle of reflection.

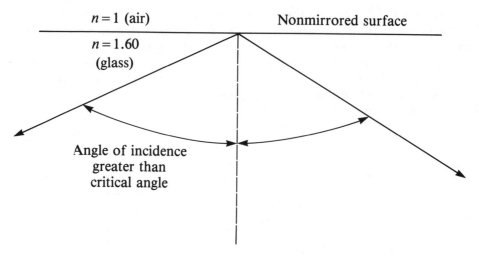

n = 1 (air)   Nonmirrored surface

n = 1.60
(glass)

Angle of incidence
greater than
critical angle

Figure 31-2.   On nonmirrored glass surface, angle of incidence equals angle of reflection, when angle of incidence exceeds critical angle.

each with one or more reticles (cross lines). The telescopes have glass optics (lens), and the light that passes through the telescope, from an eyepiece ad the objective lens, is controlled by the design characteristics of the different lenses in the optical path.

In optics, collimation means producing parallel rays. In the optical-tooling field, it also signifies the lines of sight of two instruments parallel. A collimator is a telescope focused to infinity and the reticle illuminated. Rays of light emanating from the reticle will be parallel when they leave the objective lens. If another telescope focused to infinity is pointed at the collimator, the collimator reticle will be seen as a perfect target at infinity.

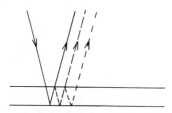

Figure 31-3.   Rear-surface mirror produces unwanted reflections that result in reflected image not being sharp. (Courtesy of Wild Heerbrugg, Ltd.)

If the line of sight of a telescope is parallel to the line of collimation, the telescope reticle will be seen to coincide with that in the collimator, regardless of whether there is any small displacement up or down, right or left, between the two objectives. Since parallel rays are involved, the distance between the objectives is not important. The telescope can be placed directly in front of the collimator, as in Figure 31-5. When sighting distances are limited, collimators are used as references for angle measurements.

31-2-4.   Autocollimation

Autocollimation is the process of sighting with a telescope focused to infinity on an optically flat mirror. The reticle, which is in the focal plane of the telescope, must be brightened from the eyepiece side. Rays of light emanating from the illuminated reticle leave the objective as a parallel beam. If the mirror is exactly at 90° to the line of sight, the rays will be reflected back along their own paths to form an image in the focal plane at the reticle position. When looking through the eyepiece, the telescope reticle will be seen

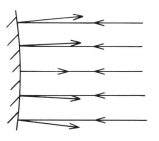 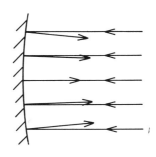

Figure 31-4. If mirror is not flat, rays of light will not be reflected back along their paths, and reflected image will not be sharp. (Courtesy of Wild Heerbrugg, Ltd.)

to coincide with its own reflected image, which is inverted (see Figure 31-6).

If the mirror is now turned through a small angle δ, the reflected rays will be moved through an angle 2δ. If at least some reflected rays still enter the telescope objective—i.e., provided that the mirror has not been turned too far—the cross-lines' image remains formed in the focal plane, but it will be displaced with respect to the telescope cross wires (see Figure 31-7).

With an autocollimation telescope on a theodolite, it is very simple to determine the angle of tilt. Read the horizontal and vertical circles before turning the mirror, then bring the cross lines into coincidence with their reflected images, and again read the two circles. To compute the angles turned, use the following equations:

$$\Delta H = H_1 - H_2 \qquad (31\text{-}1)$$

$$\Delta V = V_1 - V_2 \qquad (31\text{-}2)$$

where H_1 and H_2 are the first and second horizontal circle readings, V_1 and V_2 the first and second verticle readings, and ΔH and ΔV the horizontal and vertical angles through which the mirror has turned.

31-3. INSTRUMENTATION

Optical tooling has been accepted as the only accurate method of making measurements on subjects too large to be ascertained by mechanical instruments such as comparators. Optics provide a line of sight that is absolutely straight, has no weight, and serves as a perfect base from which to make accurate measurements.

Most optical-tooling measurements in shop construction are very basic and made to determine if a tool (and the object being worked on) is (1) aligned, (2) flat or level, (3) plumb, and (4) square. Exact alignment, precise linear and angular measurements, and perfect level and plumbness for vertical control can be obtained with a properly adjusted surveyor's level and theodolite. When measurements to thousandths of a foot were sufficient for shop practice, standard surveying instruments could meet the requirements. Today, however, tolerances of less than one-thousandth of an inch and decimals of a second of arc have resulted in refinements of even the better surveying instruments.

With the advent of modern digital computers and rapid advancements made in CAD/CAM systems (computer-aided design/computer-aided manufacturing), there is a

Collimator

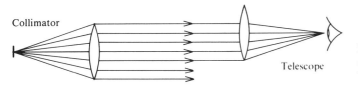

Telescope

Figure 31-5. Sighting with a telescope into a collimator. (Courtesy of Wild Heerbrugg, Ltd.)

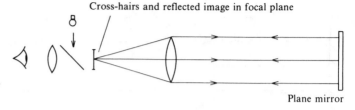

Cross-hairs and reflected image in focal plane

Figure 31-6. Telescope schematic with autocollimation eyepiece reflecting image of cross lines back into focal plane of cross lines (reticle). (Courtesy of Wild Heerbrugg, Ltd.)

Plane mirror

greater demand for X-, Y-, Z-coordinate determination instrumentation. With X, Y, Z known accurately, the four required conditions can be checked mathematically.

31-3-1. Alignment

In assembling large structures, such as aircraft fuselages, ships, etc., it is necessary to establish a line of sight from which the subcomponents are positioned. This line is usually defined by a mirror or optical instrument (the alignment telescope) (Figure 31-8). It can be a permanent fixture on the tool or jig, or attached to optical-tooling bars outside the tool or structure being assembled. The alignment telescope provides a long, absolutely straight, permanent optical reference line. It can be focused for any distance from practically zero (the point sighted being actually in contact with the front end of the telescope) to infinity. The magnification varies from 4× at zero distance to 47× at infinity. Figure 31-9 shows an alignment telescope being used for the alignment of compressors.

31-3-2. Level

Surveyors need no further explanation of the word *level*. In optical tooling, tolerances

fall within a narrow range, and the distances observed are much shorter than those experienced in outdoor surveying.

The levels used are generally tilting or automatic instruments equipped with a parallel-plate micrometer (optical micrometer) reading directly to 0.001 in. or smaller. They can be focused for any distance from the front end of the telescope to infinity; the magnification varies from about 20× at short distances to 30× or more at infinity. The level used can be either the three- or four-level screw type. Observations are normally made on optical-tooling scales specifically manufactured for optical-tooling applications.

Optical micrometers were originally manufactured as attachments for levels and transits to be used for precise leveling and alignment. Today, several European manufacturers of precise surveying instruments are incorporating the optical micrometer inside the instrument (see Figure 31-10).

An optical micrometer consists of a disk of optical glass with flat, parallel faces called a planoparallel plate. It is designed to permit precise tilting by moving a graduated drum, located as shown in Figure 31-11. The device illustrated is mounted on the instrument in place of a sunshade, with the plate in front of

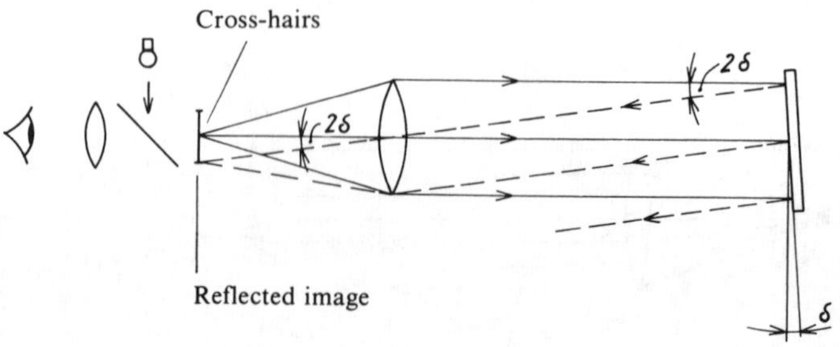

Cross-hairs

2δ

2δ

Figure 31-7. Mirror tilted through angle δ; rays reflected back are shifted through 2δ. Cross lines and reflected images no longer coincide. (Courtesy of Wild Heerbrugg, Ltd.)

Reflected image

δ

Figure 31-8.   Alignment telescope. (Courtesy of Cubic Precision, K & E Electro-Optical Products.)

Figure 31-9.   Optical alignment of compressors. (Courtesy of Cubic Precision, K & E Electro-Optical Products.)

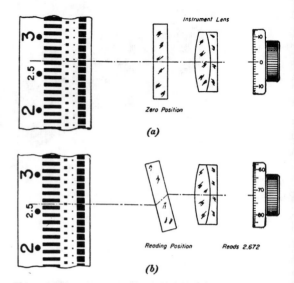

Figure 31-12. Principle of optical micrometer. (Courtesy of Cubic Precision, K & E Electro-Optical Products.)

Figure 31-10. Wild N-3 tilting level. (Courtesy of Wild Heerbrugg Instrument, Inc.)

the objective lens. A lower right-hand screw clamps the micrometer in position. A knob and drum control the planoparallel plate. Figure 31-11 shows the line of sight moved 0.081 in. to the right, as read on the drum.

When the optical micrometer is mounted on a precise level, it would be rotated 90° from that shown in Figure 31-11, so the movement of the planoparallel plate will be up and down. Figure 31-12 illustrates schematically the use of an optical micrometer for measuring a vertical distance. The sights are taken on an accurately graduated steel scale divided into tenths of an inch. Figure 31-12a shows the line of sight falling between 2.5 and 2.6 in. for the zero position of the micrometer. The drum is graduated in both directions from 0 to 100. To

Figure 31-11. Brunson optical micrometer. (Courtesy of Brunson Instrument. Co.)

avoid turning in the wrong direction, the drum is first set to zero and then rotated to make the line of sight move toward the graduation, with the lesser value on the steel scale (2.5 in. in Figure 31-12b). The drum always records movement of the line of sight from its zero position. Figure 31-12b shows a movement of 68.0 thousandths, thus making the reading 2.500 + 0.068 = 2.568 in.

31-3-3. Square

Several special instruments designed primarily for optical-tooling jobs include adaptions of the higher-accuracy theodolites familiar to land and geodetic surveyors.

A jig transit is an optical instrument with a telescope mounted so that it can be rotated about a horizontal and vertical axis. Like most optical instruments, it can be focused from the front end of the telescope to infinity, with magnification varying from 20 × at short distances to about 30 × at infinity. Figure 31-13 shows a jig transit mounted on a special stand. It does not have a horizontal or vertical circle and is used mostly to establish vertical reference planes. It is equipped with an optical micrometer and a front-surface mirror on the

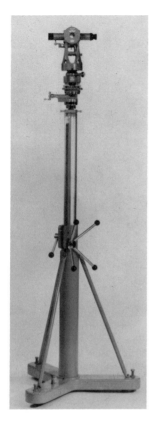

Figure 31-13. Jig transit and lateral slide attached to instrument stand. (Courtesy of Brunson Instrument Co.)

ment to be moved a short distance without disturbing the instrument stand. The process, called *bucking-in*, is exhibited schematically in Figure 31-15.

Theodolites may be the conventional optical type used by surveyors to measure horizontal and vertical angles to 1 sec of arc or better. Some, however, are manufactured specifically for optical tooling and equipped with an autocollimation eyepiece and internal wiring (see Figure 31-16). Their shortest focusing distance is usually 1 m or more, and the magnification about 30 ×.

A theodolite can autocollimate to a reference mirror, then turn a 90° angle to establish a plane perpendicular to the reference line. If a mirror is moved along a flat surface or rotated about an axis, any change in flatness is evident by the cross-lines displacement. The size of movement can be measured with the horizontal and vertical circles, as discussed earlier.

The telescopic transit square is an instrument similar to a jig transit, but it has a second telescope perpendicular to the main one through the instrument's hollow axis. This

end of the horizontal (trunion) axis. The mirror is perpendicular to that axis, so when it is autocollimated with a reference telescope, the plane established by the line of sight is perpendicular to the reference line of sight. Figure 31-14 illustrates this concept. A lateral slide just below the jig transit allows the instru-

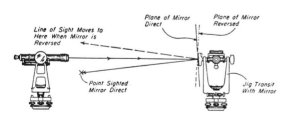

Figure 31-14. Setting jig transit at right angles to sight line. (Courtesy of Cubic Precision, K & E Electro-Optical Products.)

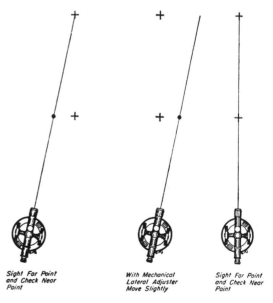

Figure 31-15. Bucking-in. (Courtesy of Cubic Precision, K & E Electro-Optical Products.)

Figure 31-16. Wild T-3A autocollimation theodolite. (Courtesy of Wild Heerbrugg Instruments, Inc.)

Figure 31-17. Brunson 76R telescope transit square. (Courtesy of Brunson Instrument Co.)

"cross telescope" is normally equipped for both collimation and autocollimation (see Figure 31-17). On some transit squares, the main telescope is identical to that of the jig transit, but the cross telescope has a fixed magnification of about 30 × and the focus is fixed at infinity.

31-4. COORDINATE DETERMINATION INSTRUMENTS

These devices have made their appearance in the optical-tooling equipment market. They operate as two electronic theodolites connected directly to a microcomputer (see Figure 31-18). The computer has special software that take the horizontal- and vertical-circle readings from the two theodolites and generate X-, Y-, and Z-coordinates of points on the tool or jig. With the three coordinates of many points now determined, it is simple for a computer to examine the four critical parameters —i.e., aligned, flat or level, plumb, and square. The computer also performs very sophisticated analyses of the data, such as checking if all points lie on a cylinder, sphere, paraboloid, etc.

31-5. ACCESSORIES

The accessories available to optical-tooling specialists can easily number 100. They include such items as scales, scale holders, targets, etc. One of the more important ones is the optical micrometer (parallel-plate micrometer). Another accessory used extensively is the autocollimation prism. It is a 90° prism (roof prism) mounted in a housing that allows the prism to be rotated horizontally and vertically (see Figure 31-19). Generally, there is a relatively sensitive plate level in the housing to facilitate leveling the prism precisely, thus setting its roof edge horizontal.

In the horizontal plane, the 90° prism functions as a mirror. In the vertical plane, it performs much like a retroreflector used for electronic distance measurements—i.e., an in-

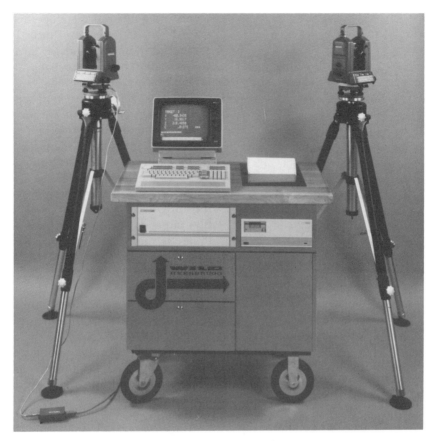

Figure 31-18. Coordinate-analyzing theodolite system. (Courtesy of Wild Heerbrugg Instruments, Inc.)

coming ray of light is turned through 180° and returned parallel to itself.

31-6. APPLICATIONS

Once the basics of optical tooling are understood, the number of applications is infinitely large. It is a matter of marrying mechanical knowledge to optical experience, tackling each problem as it arises. A list of some common optical-tooling applications found in industry follows.

31-6-1. Checking the Parallelism of Rollers

Rollers in various types of mills must be exactly parallel with each other and at correct heights. To check for parallelism, establish a control line parallel to the machine axis, in a location where you can have a clear line of sight for the entire length of the machine. Because of the high accuracies required, this control line should be close to the machine, say, within 10 to 20 ft. The method used to establish this line will vary from one location to another. In most cases, the line would have been established when the machine was installed. If not, conventional surveying methods are employed to define it.

After the control line has been fixed, the procedure for checking the roller depends on the equipment used. The method will be discussed for the two most commonly used instruments: (1) the transit square and (2) theodolite.

Figure 31-19. Wild GAP1 autocollimation prism. (Courtesy of Wild Heerbrugg Instruments, Inc.)

Transit-Square Procedure

Position the transit square on an instrument stand at one end of the control line. At the opposite end of the line, place an autocollimation mirror on another stand at approximately the same height as the transit-square's telescope axis. Point the main telescope of the transit square at the mirror and have a second person rotate the mirror until the observer can see the cross-hair image reflected back into the telescope. The observer then directs the person at the mirror how far and in which direction it has to be rotated to achieve autocollimation—i.e., exactly superimpose the reflected cross lines on the instrument cross lines.

After autocollimation has been completed, a line is projected from the surface of each roller out to the control line. This can be done crudely, using a taut string. A control point must then be established approximately 6 in. in front of the intersection of the string line and control line, in the direction of the transit square. The observer at the transit

square now tilts the telescope axis down, focuses in on the area, and directs the person setting the point to its exact location on the control line. This process is repeated opposite every roller on the machine.

The final step in checking the parallelism of the rollers begins with the autocollimation mirror remaining fixed in position. The transit square and instrument stand are moved and positioned over the control point on the line opposite the first roller to be checked. The transit-square's main telescope is pointed at the mirror with the cross telescope pointing in the direction of the machine. The transit square is autocollimated to the mirror; a scale held against the back edge of the roller; a reading taken through the cross telescope; the scale moved to the front edge of the roller; and the scale read again. If the roller is perpendicular to the control line, as it must be, the two scale readings will be the same. If they differ, the roller must be adjusted until the readings are equal. The transit square is moved along the control line and the procedure described repeated.

Theodolite Procedure

The only difference in the method when using a theodolite is that after autocollimation to the mirror, a 90°00′00″ angle is turned to the roller.

31-6-2. Checking Levelness of a Table or Plate

The simplest method to determine levelness is with a level on an instrument stand, taking readings to a scale or leveling staff that is moved over the surface. The general approach positions an instrument stand where the entire table or plate surface can be seen. If the surface to be checked is a table, place the level just a few inches in height above it. Mount an optical-tooling scale vertically in its holder and read the scale at different positions on the table. Adjustments can then be made so that when all scale readings are equal, the table is level.

To check the surface of a plate, the optical scale used generally is much longer than the one employed on tables. If the plate is recessed below the floor surface, an Invar staff similar to those utilized in precise leveling is suitable.

31-6-3. Checking Flatness of a Rail

To check for flatness, mount a mirror on a special base tripod. Let d equal the distance between the feet. Place the base at the beginning of the rail with the feet at points 0 and 1 (see Figure 31-20). With an autocollimation theodolite, point to the mirror, autocollimate, and read vertical circle V_1.

Move the base through d so the feet are now at points 1 and 2, autocollimate, and read the vertical circle V_2. Continue in this manner, moving the base in steps of length d until the rail end is reached.

Points 0 and 1 are assumed to be the datum points—i.e., height of 0 = height of 1 = 0. The vertical-circle reading V_1 is the datum reading. Subtract all vertical-circle readings from V

$$\alpha_{1-2} = V_1 - V_2, \qquad \alpha_{2-3} = V_1 - V_3 \qquad (31\text{-}3)$$

$$\Delta H_{1-2} = d \sin_{1-2}, \qquad \Delta H_{2-3} = d \sin_{2-3} \qquad (31\text{-}4)$$

$$\text{Height of 2} = \text{height of 1} + \Delta H_{1-2}$$

$$\text{Height of 3} = \text{height of 2} + \Delta H_{2-3}$$

These measurements can be highly accurate. As an example, if distance d between the feet on the base is 200 mm (approximately 8

in.), an error of ± 1 sec in the vertical angle will result in an error of only ± 0.001 mm (± 0.00004 in.) in height difference. This small disparity would be almost impossible to find by direct reading on a scale.

31-6-4. Transferring a Line to a Different Level

The shipbuilding industry provides a good example of this operation. A line from the ship's bow to its stern must be transferred to a lower level inside the vessel. The easiest way to perform this task (see Figure 31-21) is with an autocollimation prism and optical instrument —e.g., a jig transit, transit square, or theodolite. Line 1-2 being transferred to a lower level must be over a staircase or hatch, and points 1 and 2 must be close enough to this opening so the next lower surface can be seen without any obstruction.

The optical instrument is set up on a stand at point 1 and the autocollimation prism on its stand at point 2. Point the instrument to point 2 so the line of sight is in the plane of the line to be transferred. Focus the instrument to infinity and sight on the prism. A second person then turns the prism in the horizontal plane with the slow-motion screw until autocollimation is obtained.

Move the optical instrument to point 3 on a lower level. The exact location of point 3 is not known at this time, but temporarily select a position from which the autocollimation prism can be seen on the upper level. Tilt the autocollimation axis downward (it must not be rotated in the horizontal plane) and view it with the instrument. Move the instrument laterally until autocollimation is completed by using the instrument's slow-motion screws. After final autocollimation, the line of sight is moved down and point 4 established on the deck.

The procedure just described to transfer a line downward, while maintaining the required direction and same vertical plane, must go one step farther. A visible graduated scale should be on the autocollimation prism. Then,

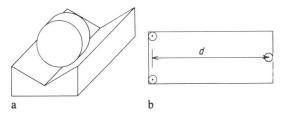

a   b

Figure 31-20. (a) Mirror in circular mount can be placed in base with V-notch. (b) d is distance between the feet.

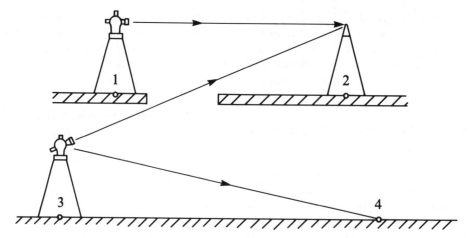

Figure 31-21. Transferring a line to a different level.

when the optical instrument at point 1 auto-collimates to the prism at point 2, the scale reading on the vertical cross line can be recorded. After the instrument is positioned at point 3 and autocollimation finished, it has to be moved laterally until the vertical cross line is at the same position on the scale as it was at point 1. This process can be repeated to transfer the line to lower levels.

31-7. SUMMARY

Optical tooling is a highly specialized and fascinating area, whose theory and application have just been touched on here. Further reading will be needed in a glossary of optical-tooling terms, in addition to a study of pertinent books and instrument manufacturers' literature listed in the bibliography at the chapter's end (manufacturers' catalogs describe and offer for sale targets, cup mounts, reticle projectors, and numerous other items).

REFERENCES

Aerospace Industries Association of America. 1981. *NAS Optical Tooling Standards*. Washington. D.C.

Bod, D., and D. Wright. 1981. *Optical Alignment Reference Manual*. Seattle: WA Boeing Commercial Airplane Company.

Brinker, R. C. 1969. *Elementary Surveying*, 5th ed. Scranton: PA International Textbook.

Brunson Instrument Company. *Optical Tooling for Optical Alignment, Precise Optical Measurement and Three-Dimensional Control of Planes and Angles by Optical Coordinates*. Kansas City, MO (undated).

Jackson, P. 1979. *Wild Instruments for Optical Tooling and for Precise Measurements in Industry and Laboratories: with Special Emphasis on Autocollimation Equipment*. Heerbrugg, Switzerland: Wild Heerbrugg.

Keuffel & Esser Company. *Optical Alignment Manual*. Morristown, NJ (undated.)

Kissam, P. 1962. *Optical Tooling for Precise Manufacture and Alignment*. New York: McGraw-Hill.

McGrae, J. D. 1964. *Optical Tooling in Industry*. New York: Hayden.

32

Land Descriptions

Dennis J. Mouland

32-1. INTRODUCTION

Land descriptions are prepared for use in various types of deeds or documents transferring interests in land. They should be clear, complete, and concise identifications of particular pieces of land. A properly written description describes land in a way that avoids any confusion between the intended parcel and other parcels. Unfortunately, every professional land surveyor has encountered many land descriptions (also called *legal descriptions*) that leave much to be desired. In fact, the history of legal descriptions is one of constant turmoil and chaos.

Modern-day scriveners—i.e., writers of deeds—have a wealth of reference material available for use in preparing good descriptions. However, mistakes are repeated and ambiguous legal descriptions filed into the records systems. There is ample help available, but the majority of description authors are either inattentive or simply undertrained. It is therefore imperative that all persons involved in preparing and reviewing legal land descriptions be familiar with the basics of structure, content, and the legal principles that surround the words used in new descriptions. This chapter presents the principles of description writing and lists a few of the many published references that are helpful.

32-2. BASIC DESCRIPTION STRUCTURE

Legal land descriptions consist of three basic parts: (1) the caption; (2) the body; and (3) exceptions and/or reservations (if any), also called qualifying statements. Depending on the method employed to describe a parcel of land, one or more of these basic parts may be omitted, but some form of "caption" is always required.

The caption is a type of introduction and "specific-purpose statement" that sets the stage for a complete description. Basic background information and certain parameters are given; any calls in the body that conflict with the caption are nullified. For example, if a description reads "a parcel of land lying within the southwest quarter of section 10," then no portion of the ensuing content can go beyond that limiting factor. Statements within the description might erroneously guide the reader outside the southwest quarter of section 10, but no title is passed on those portions beyond the limits given in the caption.

The caption also acts as a statement on what the description is intended to do. It may say this is a parcel of land or an easement for a road or perhaps a three-dimensional chunk of airspace. The caption also usually helps identify which type of description system the body uses to describe the land. An example is a parcel based on the public-land survey system (PLSS) that recites the section, township, range, and meridian in the caption, and then uses the PLSS in the body. A metes-and-bounds description refers to some identification of the area, and then says "more particularly described as follows." This statement leads you into the body. There are many combinations of usable methods and words. These will be discussed in Section 32-11.

Suffice to say the caption is an extremely important portion of the description and quite often the most ignored. All statements made in a description must be "filtered" through the caption to qualify as binding calls on the interest being passed on the land.

The body of a description must be (1) a clear recital of all the pertinent facts to describe the land and (2) complete, without contradictions to the caption or any other factors influencing the location. A description body can take on many differing forms, primarily depending on the description method employed.

The exceptions and/or reservations should close the description. These items are sometimes referred to as "qualifying statements," because they can cast a completely different meaning on the deed's total intent. There is a definite difference between an exception and a reservation, although these words were used interchangeably in the past and caused a great deal of confusion regarding the real intent of the parties. The basic rule to remember is that an exception cuts out or removes subsequent statements from the entire intent of the deed, whereas courts have held a reservation to be a portion of land included in the general transfer of "fee title," but with some type of condition or special case placed on a portion of the transaction. Some descriptions will never need any qualifying statements, but writers should always read carefully to determine the possible extent of interests in land.

An example of an exception is as follows: "Lot 5, section 12, except the southerly 25 feet." Here, the fee title of lot 5 is transferred to the buyer, with the exception of that southerly 25 ft. The 25 ft still belong to the "grantor." In contrast, a reservation would be "Lot, 5, section 12, reserving therefrom the southerly 25 feet for an easement." This statement now conveys fee title to all of lot 5, but places a condition on that 25 ft. The buyer or "grantee" must pay taxes on all of lot 5, but will have only restricted use of the southerly 25 ft and must continue to allow access for the parties involved.

Occasionally, a description may have an addendum of some kind after the body or other statements, usually called a conclusion. The conclusion contains less important data that are to be given the reader. An example is a statement about the acreage, or perhaps the county and state in which the parcel is located. One method used in the midwest and west is

> Said described parcel containing 20.50 acres, more or less, all in Morgan County, MO.

Three sample descriptions, with the basic structural parts identified, are as follows:

Example 32-1.

A parcel of land lying within the Sangre de Cristo Land Grant, more particularly described as follows: } caption

Beginning at the intersection of U.S. highway 64 and State highway 38, said point being marked by a brass cap; thence north 100 feet; thence west 100 feet; thence south 100 feet; thence east 100 feet to the point of beginning; } body

Excepting therefrom the easterly 35 feet that are within the highway right of way; } exception

Containing 0.15 acres, more or less, all in Taos County, NM. } conclusion

Example 32-2.

A parcel of land lying in section 14, township 12 north, range 8 east, $\rbrace$ caption
6th PM, described as follows:

The northwest quarter of the southwest quarter of said section 14;$\rbrace$ body

Reserving therefrom an easement for ingress and egress for the $\rbrace$ reservation
grantor across the the westerly 25 feet thereof.

Example 32-3.

A parcel of land lying within the Fort Hall Indian Reservation in $\rbrace$ caption
Bingham County, ID, described as follows:

Beginning at a point that lies S21-42'-35"W, 405.22 feet from
USGS Triangulation Station "Hall"; thence N 15-21-30 E, 208.71 feet; $\rbrace$
thence N 74-38-30 W, 208.71 feet; thence S 15-21-30 W, 208.71 feet; body
thence S 74-38-30 E, 208.71 feet to the point of beginning.

32-3. DESCRIPTION SYSTEMS

In the realm of legal land descriptions, there are many avenues used to describe a parcel of land. Often referred to as "systems" or "methods," they consist of the various legal survey processes employed in a local area, usually within a state. Scriveners should be thoroughly familiar with the methods available locally in past years to describe land, as they will influence the method used for revising a description today. When writing a new description, a surveyor is not at liberty to make a capricious decision. Quite often, confusion has resulted from using a different system than that governing the parcel's history. Thus, if a parcel that has been described since patent by aliquot parts (legal subdivisions) is now to be rewritten for a sale of a portion thereof, a change to the metes-and-bounds system will probably be required. However, the scrivener must be certain that the calls used in the description do not conflict with any elements of the original description.

A basic discussion of each major description system is necessary to help define their parameters and limitations. Surveyors should realize that some systems are not available in certain areas of the country. The seven major description systems are (1) metes and bounds; (2) lot and block; (3) portion of another parcel("ly");

(4) legal subdivisions; (5) reference to a map, plat, or deed; (6) coordinates; and (7) strip.

32-4. METES-AND-BOUNDS DESCRIPTIONS

The metes-and-bounds system is the most widely used method for describing land. The terms come from Old English law and custom. "Metes" refers to the directions and distances that were measured (or meted) out around the parcel. "Bounds" refers to a call in the description for a certain adjoiner. Such a call could be for another landowner, an adjoining deed, a previously established line, or a physical monument that helps to define the parcel. A metes-and-bounds description usually takes the reader around a piece of land by citing a bearing (or direction) and then a distance, and will call for various important facts along the way to aid in clarifying where that boundary is truly located. As an example, Smith currently owns all three parcels shown in Figure 32-1, which she purchased from Childers. The description that was used in the warranty deed between them reads as follows:

A parcel of land within the City of Anywhere, FL described as follows; BEGINNING at a city brass cup marking the intersection of the centerlines of Main Street and First Street; thence N38W along the centerline of Main Street 105 feet;

thence N52E 25 feet to the northerly line of Main Street, being the TRUE POINT OF BEGINNING: Thence continue N52E 100 feet to a half-inch iron bar; thence N38W 225 feet to a 1-inch iron pipe; thence S52W along the Ott property as shown in deed recorded in Docket 175, page 201, a distance of 100 feet to a cross ("+") on the sidewalk, said point being on the northerly line of Main Street; thence along the northerly line of Main Street 225 feet to the TRUE POINT OF BEGINNING.

There are several important facts to be considered while analyzing this deed. Certain calls and facts included in this description help to ensure that future problems will not arise. First, there are specific bearings and distances all the way around the property. Second, whenever a monument is available, it is called for in the description. This maintains the description's integrity even if a future survey finds the bearings, distances, or both to be in error. The deed is very clear that the parties involved intended the parcel to go to the monuments, and not just held at a distance that may have been arrived at without benefit of a survey. Finally, the "bounds" calls for "the northerly line of Main Street" and "along the Ott property..." also aids in avoiding conflicts with the neighbors. (The "neighbors" can be a person or a city street.) In eastern states and some other areas to a certain extent, a metes-and-bounds land description may not contain any bearing and distance information. They were very simply written descriptions, but sometimes are extremely difficult to locate on the ground or determine whether any conflicts exist. If the same parcel in Figure 32-1 had been written in this manner, it would read:

Bounded on the southeast by Sutton, on the northeast by Lopez, on the northwest by Ott, and on the southwest by Main Street.

(It is highly recommended that modern-day use of this pure bounds or adjoiners type of description be avoided. Voluminous research is required of the everyday user, who is usually untrained in such matters.)

Continuing a discussion of Figure 32-1, assume that Smith decides to sell the southeasterly "third" to Roeder and describes a parcel 75 by 100 ft, using the same elements as noted in the deed from Childers. Next, suppose Smith decides to sell another piece, this time to Jones. The scrivener must remember that the new description to Jones should not in any way conflict with the deed description Smith originally obtained from Childers. Further, there should not be any conflicts with the land already sold to Roeder. If there are specific places or monuments that control the line, they should be called for—i.e., a 24-in. oak tree at a corner. If the northeasterly line is to be a specific distance, then do not refer to the oak tree. If the oak tree was mentioned in the deed to Roeder, then it must be specified in the deed to Jones or else a conflict could arise, especially if a future survey revealed that the 24-in. oak tree was not 75 ft from the $\frac{1}{2}$-in. iron bar or even on the line against Lopez.

A surveyor with proper training, paying strict attention to historical details of the adjoiners' land descriptions, can use the metes-and-bounds system to prepare very clear and specific descriptions. Further study in the fine art of these descriptions is advised. Important basics to remember regarding a metes-and-bounds descriptions are as follows:

1. Research the parcel's "description history."
2. Research all adjoiners' description histories.
3. Start from a well-described and recoverable point.
4. Avoid "bounds only" descriptions.
5. Use bearings and distances, if known.
6. Call for monuments when available and applicable.
7. Call for adjoiners, if needed to avoid conflicts.
8. Use clarity but avoid verbosity.
9. Add qualifying statements, if any.
10. Give acreage to reasonable accuracy, if required and known.

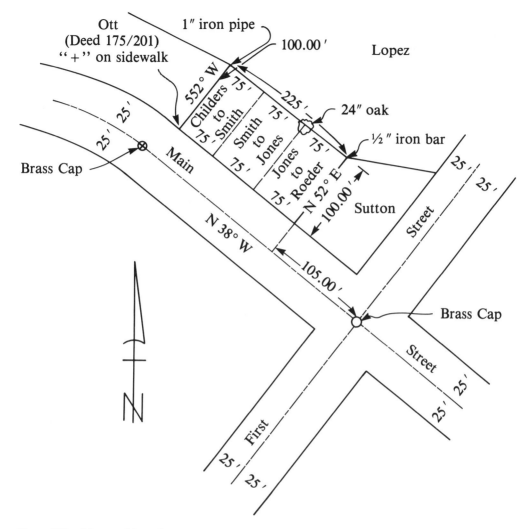

Figure 32-1. Metes and bounds.

Also, remember that (1) the caption controls the remainder of a description, (2) portions of the metes and bounds that go beyond the limitations set forth in the caption will be nullified, (3) the body also has some control over any qualifying statements and the conclusions, (4) acreage calls usually give way to the actual area described, and (5) reservations and exceptions not related to the parcel described will also be nullified.

Well-written metes-and-bounds descriptions are not an accident, nor a product to be mass-produced by untrained aides, but rather are a well thought out and researched docu-

ment giving all pertinent information in an organized manner. Descriptions must identify only one piece of land, to the exclusion of all others. They cannot conflict with any other parcels and must faithfully serve the public throughout the future.

32-5. LOT-AND-BLOCK DESCRIPTIONS

Lot-and-block descriptions come in many forms. Most metes-and-bounds states have some type of lot-and-block identification peculiar to the local area. Modern-day subdivisions rely on a modified system, sometimes calling only for a lot number and the subdivision name or condominium project name. In portions of the country where the French have had an influence, the "arpents" may be designated by a lot numbering system. The New England states have "lots and ranges," whereas the old Veterans' tracts were laid out with other forms of lot numbering. Even in the "rectangular" west, town sites were established with varying forms of lot-and-block numbers assigned.

One common denominator of all lot-and-block description variations is the need for a map, plat, or lotting plan used in their layout. When preparing a new description for parcels located in an area that employs this system, a scrivener should consult the appropriate plan or document and be absolutely certain that the lot and/or block number used is correct. A check should be made for any duplication in the numbering systems that might cause confusion in the future. Problems can also arise if there is more than one copy or revision to a lotting plan or plat. Reference must always be made to the recording office, a date on the plat or plan, or even a reference to another fact given on the map to ensure the accuracy of the parcel being described.

Lot-and-block descriptions can be very simple in form, if the precautions mentioned are taken. Two examples of lot-and-block descriptions are as follows:

A tract of land in Wyoming County, NY, described as follows: lot 10, range 5, and the northern half of lot 11, range 5; containing 120 acres more or less.

Lot 108, Shady Deal Estates, as shown on plat recorded in Book 4 of Maps, page 45, Garfield County Records, CO.

At times, a lot-and-block description can be supplemented by a metes-and-bounds description to help identify the parcel. But caution is required in the particular calls to be used, as the lot-and-block portion of the description will act as a caption and could override some of the metes-and-bounds calls.

32-6. PORTIONS OF ANOTHER PARCEL DESCRIPTIONS

This is a catch-all name for a description system that can take on many differing styles. G. Wattles, the dean of legal descriptions, called these "ly" descriptions, for the simple reason that almost all of them have a directional word ending with those two letters. An example of such a description would be, "the northerly 100 feet of section 12," or "the easterly half of lot 5." When properly researched, "ly" descriptions can be very simple and still clear.

However, using "ly" descriptions can cause many problems for future surveyors and title insurers. Problems can arise in subsequent surveys if the status of the adjoining deeds is not researched.

Consider the situation shown in Figure 32-2a. Lot 10, range 4 was owned by A, who in 1978 sold the "southerly half of lot 10, range 4." The record dimensions of the lot, based on the lotting plan, are 100 by 200 ft. In 1983, A decides to sell the remainder of lot 10. However, in 1982, the new owner of the southerly half of lot 10 had a legal survey made that resulted in the dimensions shown as measured *M* (Figure 32-2b). When A decides to sell the remainder of lot 10, he has the title company write a description that reads:

The northerly 100 feet of lot 10, range 4 . . .

The writer of the new description had failed to consider the resurvey's effect. The westerly line of lot 10 was measured to be 198.00 ft, rather than the 200 ft of record. Therefore, the "southerly half" in the 1978 description will be placed at midpoint, or 99.00 ft from either corner. But the 1983 deed sold the "northerly 100.00 feet," which actually do not exist! An overlap of 1 ft on the westerly line

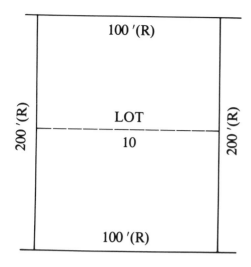

(R) = Record distances from lotting plan

a

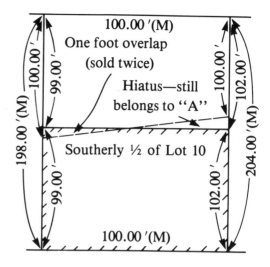

(M) = Measured distances using found monuments

b

Figure 32-2. Conflicting "ly" descriptions.

dividing the two buyers has been created. Basic land-law principles in this country will use the "first in time, first in right" approach to this situation. Therefore, the second buyer loses out. Title was already passed on that small sliver of land to the first buyer.

The east line of lot 10 is another problem. Record information places the overall length of the boundary at 200 ft, but the survey revealed that it is actually 204.00 ft. The original buyer (in 1978) acquired the southerly half of lot 10, which thus would be 102.00 ft in length. The 1983 buyer has a deed reading the "northerly 100 feet," so an opposite situation exists from that on the west. A gap or hiatus has been created: a sliver of land 2 ft wide and tapering to nothing about midway through the lot. Title never passed on that sliver, so in reality it still belongs to A, who may never know what has happened. The situation depicted in Figure 32-2 occurs often in the real world of surveys and description interpretation. What should have been the description of the second sale? One approach would be:

The northerly half of lot 10...

In the case just discussed, this would be an adequate solution. A good way to describe a remainder in a parcel is to refer to the original sale, such as:

Lot 10, range 4, EXCEPT the southerly half thereof...

It is evident that the seeming simplicity of the "ly" description can introduce many problems if it is not researched before preparing a final draft. Numerous other situations can arise when employing this method. An apparently clear representation of the parties' intent at the time of the sale can become very unclear a short time later.

Figure 32-3 exhibits five different lots for which descriptions were written referring to a portion of the larger parcel, but the intent is not clear:

1. Case a calls for the "northerly 50 feet," but the intent is not obvious. Perhaps a metes-and-bounds description would be better.
2. Case b supposedly describes the "westerly 35 feet." But were the intentions of the parties

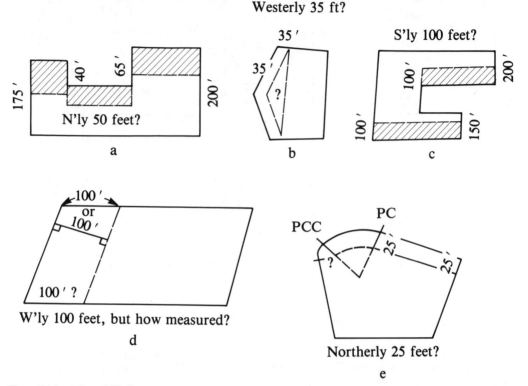

Figure 32-3. Misused "ly."

just a 35-ft strip or 35 ft on both the north and south boundaries, with a straight line between them? Other evidence or testimony, if available, might be required before a boundary could be determined on the ground. However, all the confusion could have been eliminated by a more strict adherence to the facts and willingness to use a different method for description.

3. Case c calls for the "southerly 100 feet" and is obviously confusing.

4. Whenever a scrivener is working with non-rectangular parcels of land, such as in case d, the line of reference must be clearly defined. The "westerly 100 feet" can be measured in at least two ways. First, the call for 100 ft might be measured along both the north and south boundaries, then connected. Second, the 100-ft call may be parallel to the west boundary. Again, the parties' intentions must be clearly expressed in the description.

5. Finally, case e shows northerly 25 ft or a lot, but does not explain how to deal with the curve. Should the line follow the curve and thus become a portion of the westerly 25 ft also? Or was the intent to run the new line parallel to the northerly line, and extend it through the curve to an intersection with the exterior lot line? These factors and other possible ramifications must be considered.

Land descriptions containing "ly" wordings can also be compounded. Figure 32-4 shows a parcel effectively described by compounding, as follows:

A parcel of land located in the city of Prescott, AZ, more particularly described as follows.

The north 100 feet of the east 100 feet of tract 4, block 5, of the original townsite of Prescott, containing 0.23 acres.

Again, in Figure 32-4, assume that the owner of the 0.23 acres decides to sell the north half thereof. The scrivener can go back to the previous deed and simply add to it by writing, "the northerly half of the north 100 feet of the east 100 feet of tract 4, block 5,..." Another variation sometimes employed refers to another deed or a plat and compounds thereon. For example, a description might read, "the northerly 200 feet of the westerly 50 feet of that parcel described in the deed recorded in page 10094 of the County Records."

This is convenient, but all pertinent facts must be considered. Some scriveners do not like this method, because it requires a reader to depend on another document that may not be available or is difficult to attain. The availability of "referred to" documents should be considered before employing this method. Anyone not familiar with the use of any variation of "ly" descriptions should consult the reference sources at this chapter's end.

32-7. LEGAL SUBDIVISIONS

Since the late 1700s, a system of land survey and description, the public-land survey system (PLSS), has helped to shape the majority of the United States. The method of describing land within that system is referred to as "legal subdivisions," "aliquot parts," or "rectangular" descriptions. The PLSS is used in all but the 20 metes-and-bounds states: Maine, New Hampshire, Vermont, Massachusetts, Connecticut, Rhode Island, New York, New Jersey, Pennsylvania, Maryland, Delaware, Virginia, West Virginia, Kentucky, North and South Carolina, Tennessee, Georgia, Texas, and Hawaii.

The states listed primarily rely on metes and bounds, lot and block, and "ly" descriptions. Texas was under the influence of the Spanish for many years, and at the time of acquisition by the United States, the private lands had already been substantially divided and described under a Spanish version of the metes-and-bounds system. Therefore, Texas was not included within the PLSS. Some portions of the state are divided into a rectangular system similar to the PLSS, but many legal aspects differ from the PLSS. Sections based on railroad surveys and the like should not be confused with the system being described in this chapter. Such descriptions are closer to the lot-and-block descriptions discussed earlier.

Ohio was the original testing ground of the PLSS, so there are several differing systems within the state. This situation requires the surveyor and scrivener to be very well versed in the system applicable to that particular area. Many significant changes were made in the rectangular methods during the early years in Ohio. This portion of the chapter deals only with the other rectangular (PLSS) states. The varying methods used in Ohio are complex, and the reader is advised to seek other references on the subject.

The public-land survey system is based on a rectangular grid extended out over a large

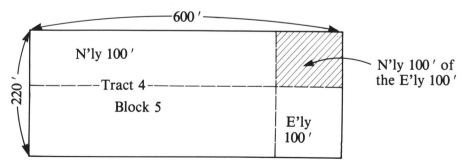

Figure 32-4. Compound "ly."

area, sometimes more than one state. Chapter 30 describes the basic system, so this discussion assumes that the reader is familiar with PLSS terminology. Laws and regulations governing the PLSS have been spelled out over the years in various government manuals, the most recent being the *Manual of Instructions for the Survey of the Public Lands of the United States*,[1] hereafter referred to as the 1973 manual. This publication is produced by the Department of the Interior, Bureau of Land Management (BLM), the present-day successor to the General Land Office (GLO). A thorough understanding of PLSS methods and terms enables a scrivener to easily describe land.

Figure 32-5 depicts a normal township of land with 36 sq mi (sections). Notice that calling for "all of section 12, township 21 north, range 7 east, Gila and Salt River meridian" identifies a section of land that cannot be confused with any other. Figure 32-6 shows section 26 and its principal corners in that same township. The section is divided by east-west and/or north-south lines into approxi-

mately equal (or aliquot) parts. Each of these aliquot parts is now also easily described. Parcel A is properly described as the "southwest quarter" of section 26; parcel B, the "northwest quarter of the southwest quarter of section 26." The system can be broken down further into parcel C, which is the "north half of the northeast quarter of section 26," etc. This method of using quarters of quarters, and so on can be employed to describe a parcel of land down to $2\frac{1}{2}$ acres. Parcel D is "the southeast quarter of the southeast quarter of the northwest quarter of the northwest quarter of section 26, T.21 N., R.7 E., GSRM." The complete legal description of parcel E would read as follows:

> The south half of the northeast quarter of the southeast quarter of the southeast quarter of section 26, township 21 north, range 7 east, Gila and Salt River meridian.

This last description identified a 5-acre parcel of land, without conflicts with any adjoiner.

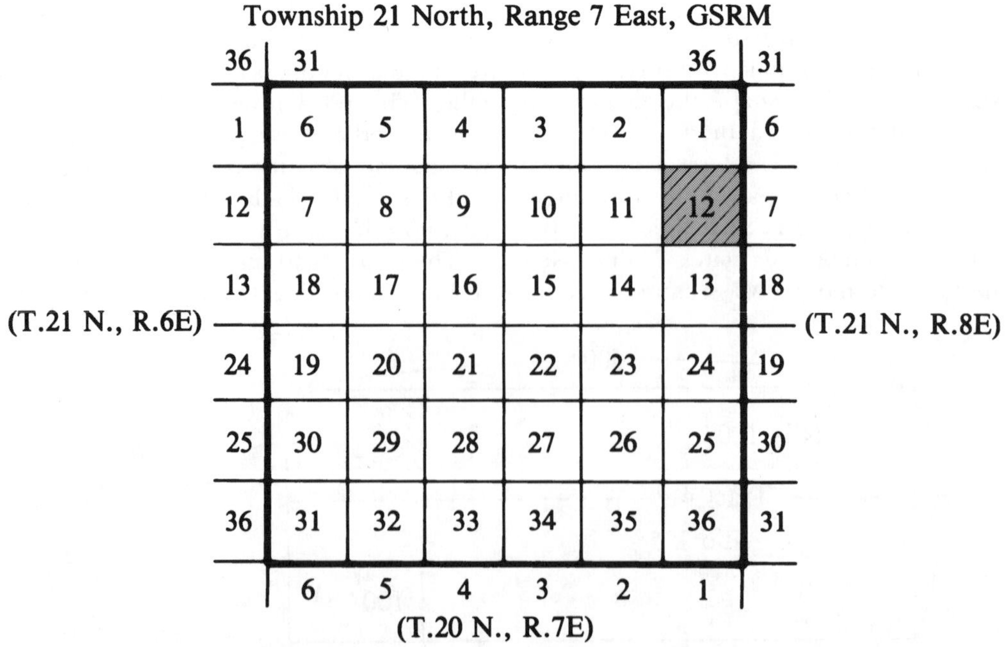

Township 21 North, Range 7 East, GSRM

36	31						36	31
1	6	5	4	3	2	1	6
12	7	8	9	10	11	12	7
13	18	17	16	15	14	13	18
24	19	20	21	22	23	24	19
25	30	29	28	27	26	25	30
36	31	32	33	34	35	36	31

(T.21 N., R.6E)   (T.21 N., R.8E)

6   5   4   3   2   1

(T.20 N., R.7E)

Figure 32-5.   Typical township.

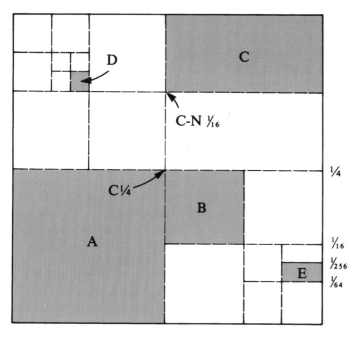

Figure 32-6.   Sec. 26, T.21N, R.7E, GSRM.

32-7-1. PLSS Abbreviations

A method of abbreviating those somewhat long and monotonous descriptions has been developed. Again if we refer to Figure 32-6, parcels A through E, governed by the Gila and Salt River meridian, are properly described as follows:

Parcel	Legal Description
A	SW1/4 Sec. 26, T.21N., R.7E., GSRM.
B	NW1/4 SE1/4 Sec. 26, T.21N., R.7E., GSRM.
C	N1/2 NE1/4 Sec. 26, T.21N., R.7E., GSRM.
D	SE1/4 SE1/4 NW1/4 NW1/4 Sec. 26, T.21N., R.7E., GSRM.
E	S1/2 NE1/4 SE1/4 SE1/4 Sec. 26, T.21N., R.7E., GSRM.

An important word of caution: Misplacement of commas in these abbreviated forms of legal subdivisions can be disastrous. The rule is as follows: *Use of a comma ends the land unit being described, and starts another unit.*

If parcel B in Figure 32-6 is written as "NW1/4,SE1/4," this includes both the northwest quarter and southeast quarter of section 26, a total area of 320 acres. However, the intent was to sell off only a 40-acre parcel, the NW$\frac{1}{4}$ of the SE$\frac{1}{4}$. If Smith decided to sell Jones both parcels A and B, the description would read:

> The southwest quarter, the northwest quarter of the southeast quarter, in section 26...

This could be abbreviated as follows:

> SW1/4, NW1/4 SE1/4 Sec. 26, T.21N., R.7E., GSRM.

Another point to remember is that mentioning the meridian on which the description is based is always required. Many states under the PLSS have more than one meridian, so the scrivener must always check and include the meridian name or number to avoid creating an erroneous deed.

32-7-2. Irregularities in the PLSS

The public-land survey is extremely complex and not completely understood by many practicing land surveyors. Several irregularities exist that must be dealt with to avoid future errors in deed and description interpretation. Whenever any rectangular parcel of land was "invaded" by a nonrectangular boundary, parcels were created that could not be described by aliquot parts. At least one boundary of the parcel would not conform to the rectangular scheme. These parcels are called "lots" or "government lots," and they differ from any other form of lot mentioned earlier in this chapter. The terminology must be kept clear when dealing with more than one type of lot in a single description. Some of the situations that create lots are listed as follows:

1. Aliquot parts invaded by nonrectangular parcels:
   (a) Land grants
   (b) Military reservations
   (c) Indian reservations
   (d) State boundaries
   (e) National Park boundaries
   (f) Mineral surveys (lode, placer, mill site)
   (g) Town sites
   (h) Homestead entry surveys
   (i) Small holding claims
2. Meander lines:
   (a) Rivers
   (b) Lakes and islands
   (c) Oceans
   (d) Some swamps and tidelands
   (e) Other fractional lines within a township
3. Certain township exteriors:
   (a) North and west exteriors in normal township
   (b) Any exterior, if township surveyed irregularly
4. Anywhere in township after a dependent resurvey:
   (a) Lines found to be more than 21 min from "cardinal"
   (b) Distances found to be "out of limits"
   (c) Area of parcel found to be "out of limits"
5. Anywhere in a township invaded by a tract created in an independent resurvey

The preceding list proves that the author of a new PLSS land description must be very

familiar with the survey history of the township and section in which a parcel is located. All records of original and subsequent surveys of the GLO and BLM are available in a "public room" at each BLM state office. Occasionally, these records are housed by other federal or state offices or possibly the county surveyor's office. The PLSS record system is also a rather complex system, so be sure all the records available for a certain township have been consulted. The GLO/BLM often created —and still does create—supplemental plats that may rename certain parcels or create new lots that differ from the original survey plat. The effects of dependent and independent resurveys are discussed later.

The most common location of lots is along township exteriors, normally along the north and west boundaries. The reason for these lots was to throw any excess or deficiency in the surveys against these boundaries. Hence, a lot is not truly an aliquot part and should not be descried as such. Government lots were assigned numbers at the time of their creation and follow certain rules in their numbering pattern.

Figure 32-7 shows the northern tier of sections in a normal township. Sections 1 through 5 contain four lots each, with the numbering system starting over again in each section. A lot was never supposed to be part of more than one section. The numbers "80" in the southerly portions of the northwest and northeast quarters of these sections serve as reminders that those parcels are aliquot. But the lots are not aliquot and can vary in acreage significantly from an aliquot part.

Section 2 does not technically contain a northwest quarter of the northwest quarter. There is no such aliquot part in section 2. This parcel of land should be referred to as "lot 4, section 2." It is correct to say that section 2 has a NW$\frac{1}{4}$. The south half of the NW$\frac{1}{4}$ exists and contains 80 acres. Section 6 has a different situation because it lies against both the north and west boundaries of the township. Lot 4, section 6 could vary significantly from 40 acres. Lots generally can range anywhere

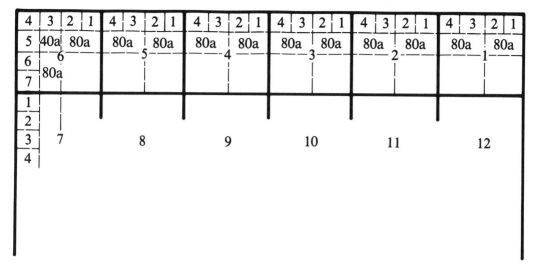

Figure 32-7. North boundary of a township.

from 5 to 50 acres. Therefore, it is not proper to call for the $NW\frac{1}{4}$ $NW\frac{1}{4}$ of section 6. Figure 32-8 presents several plats with other lotting situations.

Surveyors in Ohio must be aware that lotting was not used in most areas of the state. It was a later invention of the GLO. Description preparers should be familiar with the time period under which the original government surveys were performed in their particular locale. The original manual underwent many changes and variations through the years. Surveys and plats based on one manual will differ from those done under the next. An excellent discussion of the changes through the years between manuals was written by J. G. McEntyre in *Land Survey Systems.*[2]

The GLO and later the BLM conducted many dependent resurveys to reconstruct the original survey onto the ground, replacing lost corners according to the manual, and resurveying the lines. Plats were created that displayed the new measured bearings and distances of the section lines. Once land has gone to patent (the government term for "deeded to a private citizen"), the government cannot change the location or relative size of the patent. In federal circles, this is called "protecting the patent" or "bona fide rights."

When a normal 640-acre section was found to be significantly different from the record, the GLO/BLM would lot the remaining government land in a section. The basic rules are that if any boundaries of the township or section are found to be "out of limits," a lot would be created in place of the aliquot part previously shown. Figure 32-9 illustrates such a case. Section 3-34 of the 1973 manual discusses this principle.

32-7-3. Independent Resurveys

Independent resurveys are being treated separately here to emphasize that whenever a description is being written in a township where any such resurvey has been conducted, there is a good possibility the parcel being described has received a new "name" from the government. This does not mean that the location or relative size of the patent has changed, but the description has. This entire subject is a great mystery to most land surveyors. Actually, many cases have arisen where private lands within an independent resurvey have been erroneously run by a private land

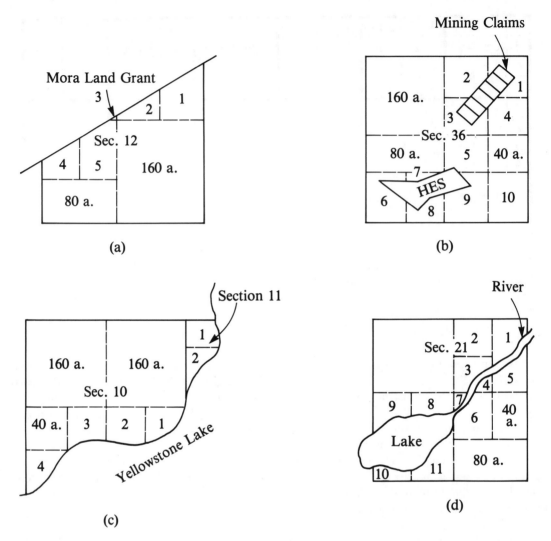

Figure 32-8. Lotting situations.

surveyor, thereby creating significant trespass and liability.

The purpose of an independent resurvey is to restructure the sections within a given township, without regard to the original survey. The need arises when the government discovers a township was originally poorly surveyed and the existing lines are very distorted. Also, when insufficient evidence remains from the original survey, independent resurveys provide a solution to land locations.

Before an independent resurvey is carried out, the government is required to protect any patents or other private land interests within the township. These parcels are segregated from the new section patterns. It would be unfair and illegal for the government to survey the patents according to their original descriptions but then ignore the original survey. Therefore, the process includes surveying all private land in-holdings according to survey evidence under which the patent was issued. This procedure is called "tracting out."

Figure 32-10 shows a portion of a township where an independent resurvey has created tracts from lands previously described as

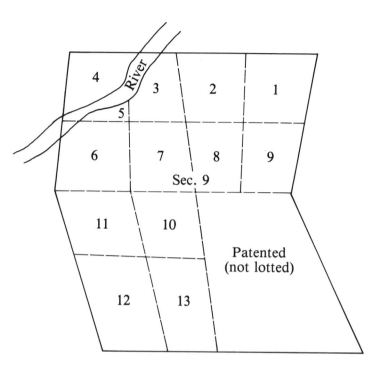

Figure 32-9. Lotting distorted sections.

Same piece of land, only the sections have been "restructured"

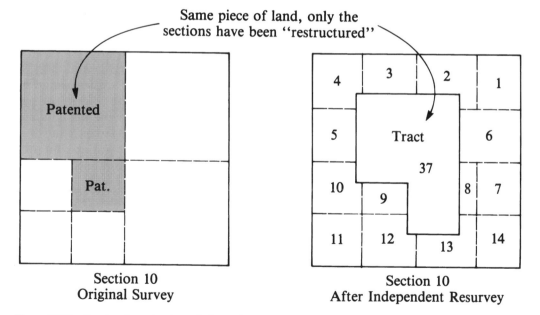

Section 10
Original Survey

Section 10
After Independent Resurvey

Figure 32-10. Results of tracting in an independent resurvey.

aliquot parts. Plat A depicts the original survey. A patent was issued on the NW$\frac{1}{4}$ and NE$\frac{1}{4}$ SW$\frac{1}{4}$ of section 10. The independent resurvey will first locate the patent by the original survey and description, using the remaining original evidence. After surveying and monumenting this tract, the original survey is ignored and a completely new network of sections created. Again, the patent has not changed location on the surface of the earth, nor has it changed in relative size from what it always was intended to be. Tract numbers begin after the highest section number in the township. In most cases, the first tract will be 37, as there are 36 sections is a normal township. Note that tracts are usually restricted to one township only.

A problem arises when a description is being prepared or a survey is to be run for a piece of land that has been tracted, and the scrivener must research the survey history to avoid writing an incorrect description. If we refer again to Figure 32-10, to write a new description for a portion of the NW$\frac{1}{4}$ of section 10, the dates of the patent and resurvey must be compared. On virtually all patents issued by the federal government, the patent description includes the words, "according to the official plat of the survey of the said lands, returned to the General Land Office by the Surveyor General. ..." Therefore, the plat and survey cannot be ignored. If a portion of the NW$\frac{1}{4}$ of section 10 is being sold and a description is drawn up reading, "the N$\frac{1}{2}$ NW$\frac{1}{4}$ section 10 ... ," this description probably refers to the old section 10, which would lie within the created tract 37. Thorough research will help to identify such potential cases and resolve them. Because the remaining tracts do not conform to the newly created rectangular scheme, lots will appear in the areas affected.

Two methods are available to properly describe such a parcel: (1) Use a metes-and-bounds description to describe any subsequent divisions of the parcel or (2) add the words, "according to the official plat dated May 4, 1881." The latter allows a writer to employ the same aliquot parts description used in the

patent to avoid confusion. The second method creates one problem by not giving signals that the area has been affected by an independent resurvey. The independent resurveys create unusual cases and scriveners must be aware of the nature of the surveys and lands being presently described.

32-7-4. Nonrectangular Units in the PLSS

Although the PLSS is basically a rectangular system, a few units of land that were patented by the government did not conform to the rectangular grid. Included are the items mentioned in Section 32-7-2. Discussing each of them will clarify procedures on writing descriptions for lands that are subsequently sold within these nonrectangular parcels.

Land grants, or private land claims, were usually large units of land on which the government issued patents for special purposes. In the east and midwest, many military or veterans' grants were created; the west contained Spanish or Mexican land grants (sometimes referred to as ranchos). Descriptions within these grants are usually in the metes-and-bounds style, except in New England, where lots and blocks were devised.

Basically, the same history lies behind military and Indian reservations. Large parcels of land were set apart from the public domain and excepted from the rectangular survey net. Some reservations in the west are now being divided into townships and sections by the BLM.

Two other nonrectangular land units appearing frequently in the west are the homestead entry survey (HES) and mineral survey (MS). The HESs were surveyed independently of the rectangular grid and therefore cannot be described by aliquot parts. The HES was not supposed to exceed 160 acres. When land transactions are made within the HES, the official HES number should always be called for, and then a metes-and-bounds description prepared for the new unit. When describing an entire HES, the HES number is sufficient, but citing the township and range, if established, and the meridian is helpful. There are a few cases where the same HES number was issued to two different parcels of land.

There are a number of cases where HES lands partly or entirely conformed to the rectangular system. The scrivener should be aware of this from the records research and watch for any conflicts made in the past by adjoining deeds, whose authors were not aware of the facts.

Mineral surveys are a different matter. Surveyors should become familiar with the basics of mineral surveys covered in Chapter 29. Patent was issued by an MS number, with the official plat referred to in the patent. Mineral surveys do not usually conform to the rectangular system. Lode claims were to be parallel to the discovered vein of ore. Placer claims normally run parallel to the stream involved. Mill sites could be just about anywhere. If a description is being written for any mineral survey in total, reference to the MS number is adequate.

As with the HES, a reference to the township, range, and meridian, if known, further defines the parcel. In numerous cases, the same MS number was issued to more than one claim at opposite ends of the state. A new description written for the sale of a portion of a mineral survey written by metes and bounds will suffice. Care is required, as with any metes-and-bounds description, to make the correct calls for lines, monuments, and any other factors that will help to properly define the parcel. In areas of concentrated mineral surveys, surveyors and title officers must recognize the possibility of overlapping claims. Research of patent dates may help to resolve which claim has senior rights to patent, but not necessarily rights to the minerals. That subject requires a great deal of expertise in mineral law.

In summary of the PLSS as a whole, a description writer should realize that the apparent simplicity of the system is misleading. The PLSS is very complex and still has some questions waiting to be answered by the courts.

Research of the survey and description history is essential. Familiarity with the terms and regulations in the 1973 manual is vital. Every BLM state office is available for assistance.

32-8. REFERENCE TO A MAP, PLAT, OR DEED

This chapter has already briefly mentioned these types of legal description. Many times, a scrivener can describe a parcel of land by simply referring to another document of public or quasipublic record. A basic principle of land law states that whenever a deed calls for a survey, map, plat, or another deed, the called for item becomes a part of the new deed, as if written on the deed itself. This method is safe if the record item referenced is a correct call and there are no conflicting calls in the item. Again, the duty of a historian—to research the descriptions and surveys of lands adjoining this parcel—is assigned to be description writer. A part of the duties of a description author is to avoid the perpetuation of error. If an error was made, such as a bad call in another deed or document, it should not be carried on into the future by referencing that error in the new document. Three examples of descriptions are presented that refer to other documents. The scrivener has already made a reasonable attempt to research the facts and determined that no errors or conflicts exist in the record instruments.

Example 32-4. "All of lots 3, 4, 5, and 9 of the Greenlee addition to the town of Springerville, AZ, as shown on the map thereof recorded in the Apache County Records."

Example 32-5. "The northerly 25 feet of lot 27 of the El Dorado subdivision, Santa Fe County, NM, as shown on the official recorded plat thereof."

Example 32-6. "All that land described in deed recorded in Book 25, page 99 of the

Beaverhead County Records, EXCEPT the southerly 100 feet thereof; and RESERVING therefrom the westerly 50 feet for an easement to the grantor.''

32-9. COORDINATE DESCRIPTIONS

Parcels of land can also be described by using various forms of coordinate descriptions. A rectangular coordinate pair is assigned to each corner of the parcel. Many towns, cities, and counties have established local coordinate systems that are well established and monumented, and of course should be of public record. In the western United States, some large tracts of land—i.e., Indian reservations and land grants—do not have a local system of coordinates for ties, but the state plane coordinate system can be utilized. Basic factors to remember in using coordinates in a legal description are as follows:

1. Define the coordinate system clearly.

2. Always give coordinates in the same order (north, east).

3. Describe coordinates in sequential order around the parcel.

4. Use other calls as required.

5. State whether the values given, if any, are grid or ground distances.

There are some variations available to the scrivener. A parcel can be described by metes and bounds tying only the point of beginning to a coordinate system. Bearings and distances can be called for, as well as monuments, plus the coordinates. Make certain that the critical call is not being supplanted by other calls intended only as secondary information. Coordinate systems can give way to other calls, especially calls for monuments.

One drawback to this method is the lack of expertise in geometry (or spherical geometry) on the part of nonsurveyors. Although coordi-

nates are becoming more fashionable, a scrivener should consider all available methods and select one that will best serve the public. Two sample descriptions using the coordinate method follow:

A parcel of land located within zone 5, of the California State Plane Coordinate System, more particularly described as follows:
The point of beginning having coordinates of N 605,456.890, E 235,998.450; thence to a point having coordinates N 605,288.004, E 235,999.852; thence to a point having coordinates N 605,285.354, E 236,544.597; thence to the point of beginning.
A parcel of land located in the west zone of the Arizona State Plane Coordinate System, more particularly described as follows:
BEGINNING at a point with values of 855,984.221 N, and 345,562.553 E; thence north 1000 feet to a point with values of 856,982.008 N, and 345,562.578 E; thence west 250.00 feet to a point with values of 856,982.015 N, and 345,312.657 E; thence S14-02-10E 1030.77 feet to the point of BEGINNING.
The above description using ground distances, and containing 2.86 acres.

32-10. STRIP DESCRIPTIONS

Strip descriptions are used for parcels of land running in a "strip," usually controlled by centerline data. This is particularly suited for rights of way and easements. A scrivener dealing with strips of land for these purposes must know whether the acquisition is going to be in "fee" or just a limited interest. This factor should also be considered when researching description history. Often, descriptions of adjoiners ignore a strip easement by using calls to the centerline, thereby perpetuating confusion. An example of a strip description follows:

A strip of land 50-feet wide, 25 feet on each side of the following described centerline:
Beginning at a point on the south line of section 4, township 15 north, range 6 west, PMM, being 258.91 feet west of the southeast corner of said section 4; thence N15°25'W 1000 feet;

thence N27°16′E 431.11 feet to a point on the north line of the SE 1/4 SE 1/4 of said section 4, being 320.00 feet from the south 1/16 corner of sections 3 and 4.

Because of certain situations, or as a result of the land ownership pattern, a strip description may use a "survey line" rather than centerline for the metes-and-bounds portion of the body. For example,

A strip of land lying 20 feet north and 35 feet south of the following described line: . . .

A precaution should be noted if the strip being described changes directions significantly. A call for "north and south" of the line described may no longer be true if the referenced line turns to the north or south. Another method used in strip descriptions places the entire strip on one side of the referenced line. An example of this would be:

A strip of land 30 feet wide, the northerly and northeasterly line of which is herein described; . . .

Occasionally, a strip description may refer to stationing (see Chapter 25) by relating to various factors along the line. Stationing could be employed in the following manner:

A strip of land 66 feet in width, 33 feet on each side of the following described line;

Beginning at a point on the westline of Homestead Entry Survey No. 401, being S15°-37′W, 245.88 feet of corner no. 5 of said HES, and being given an engineer's station of 0 + 00; thence N88°55′30″E to station 4 + 53.77; thence through a tangent curve, concave to the southwest, having a central angle of 30°, and a radius 4969.77 feet to station 7 + 28.90; thence S61°04′30″E to station 14 + 34.21, being on the east line of said HES No. 401, and bearing S4°34′E 100.00 feet from corner no. 2 of HES 401.

Stationing in a description is acceptable but sometimes not desirable, because the average person reading a deed will not be familiar with it. Subtracting one station from the other does not always give the described distance—use of negative stations and/or station equation can change the situation significantly.

Several problems can arise in using the strip description. Figure 32-11 presents two cases: (1) when a curved centerline intersects

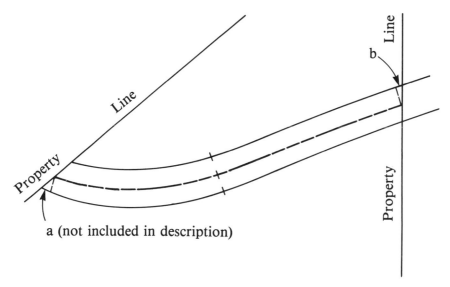

Figure 32-11. Strip descriptions.

the property line on which the strip is to terminate and (2) when the described center-line ends on a line that is not perpendicular thereto. The strip description caption says the strip is on each side of this line, but the line described ends at the property line as shown. The small tracts noted as *a* and *b* have been omitted from the description but were intended to be included. One common method includes a statement at the end of the metes-and-bounds description of the centerline, such as:

> ...The sidelines of said described strip being shortened or elongated to meet the property lines of the grantor.

When a strip description makes angle point at a property line and then continues into another parcel, the use of exceptions is advisable to be certain that the desired parcel is properly described. Figure 32-12 shows such a case. The areas designated *b* and *c* must be dealt with. Area *b* will probably be needed for the strip, but might not be acquired from the correct owner. On the other hand, *c* really is of no use to the strip but will be included unless excepted. Be certain that these parcels are being excepted or acquired from the correct owner.

As was mentioned, strip descriptions can be written to cover a right of way or an easement across multiple parcels with differing owners. Often, the caption will read, "any portion of lot 12, range 4, of the Warren Military Heroes Grant, which lies within the following description, which is a strip of land 30 feet wide, 15 feet on each side ... ." Highway departments as well as utility companies resort to such methods so they can use the same strip description wording on dozens of deeds, without having to specifically address how the strip intersects with each intermediate property line. Although this procedure can save many hours of computations during the acquisition stage, it does not give the future surveyor or title agent a definite idea of how much land is involved, or where the strip is precisely located on any one parcel.

One other important point should be made regarding "blanket easements." These are very

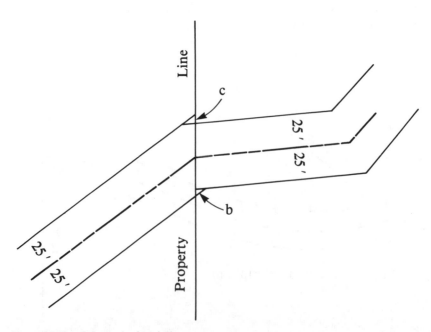

Figure 32-12.   Other strip descriptions.

loosely written easements to identify which parcel of land is subject to the acquisition. The actual strip location is not given.

> A strip of land 200 feet in width for the purpose of an electric transmission line, over and across the NW 1/4 of section 18, T.1S., R.29W., 6th PM.

The use of such a nebulous strip description should be avoided, again because of the great disservice to the public and future land owners. The blanket easement should be employed only in rare cases, where there is absolutely no survey control available, and the boundaries of the land being crossed are completely indiscernible.

32-11. COMBINING DESCRIPTION SYSTEMS

Notice that throughout the explanations of the various description systems, a scrivener can decide to describe land using more than one method. However, it will require the writer to be even more careful of the manner in which calls are used. Remember the caption's "power." By using one method in the caption and another in the body, certain conflicts can arise that may change the parties' original intent. For instance, a description may use the PLSS in the caption but then employ the metes-and-bounds system for the body. Any portion of the body that in fact strays beyond the aliquot parts mentioned in the caption would be voided.

Another consideration is in order: Mixing systems within the body can be dangerous. Although information available may be fragmented and from several types of description, the scrivener should strive to employ the same description method in the body. Jumping back and forth between metes and bounds, the PLSS, coordinates, or references to other deeds can be extremely confusing. Some important data may be omitted from the de-

scription. A well-written metes-and-bounds description can employ varying methods in very limited ways and still succeed when the scrivener uses good judgment.

When researching description histories, find out how the land was described previously in order to avoid creating any new conflicts. For example, assume that a parcel of land was originally patented as the $SE\frac{1}{4}$ $SE\frac{1}{4}$ of section 34. When the owner decides to sell the entire parcel, in most cases it is best to describe the land in exactly the same way. If the scrivener were to rewrite the description in a metes-and-bounds format, based on the original plat, it would probably read as follows:

> Beginning at the southeast corner of section 34, T. 10, N., R.5W., PMM, thence north 1320 feet; thence west 1320 feet; thence south 1320 feet; thence east 1320 feet to the point of beginning.

To the novice, this may seem acceptable, but many problems were introduced. Assume that Figure 32-13 is the result of a survey made 25 yr after the metes-and-bounds description was created and the section found to be irregular. The law says the original undisturbed monuments will forever fix the section limits. The metes-and-bounds description does not agree with the actual shape of dimensions of the $SE\frac{1}{4}$ $SE\frac{1}{4}$. It would be easy to decide that the intent was to follow the lines of the aliquot parts, but the deed does not say that. Perhaps the same grantor owned some adjoining lands, and his or her intent was not to follow the section lines. Any assumptions made as to the intent of parties long gone are merely guesses.

Confusion could have been eliminated if the new description followed the same system used to describe the patent, which was the aliquort parts of a section, *if* this was the intent of the parties. Also, other calls in the metes-and-bounds portion of the body could have prevented the chaos. The words, "thence north, along the section line, 1320 feet to the south 1/16th corner..." would have made the intent obvious. Here, calls for bounds clarify the intent. The section line and 16th cor-

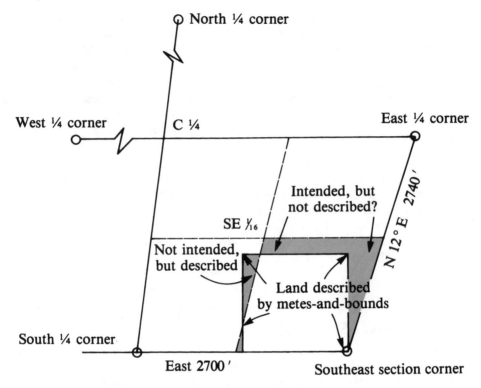

Sec. 34, T. 10 N., R. 5 W., PMM.

Figure 32-13. Combining description systems.

ner are called as bounds and therefore control over the distances and bearings given.

The opposite situation can also result when changing a metes-and-bounds description into an aliquot parts description. This is rarely successful, but often attempted by deed preparers willing to assume intents. The main point of this discussion on changing description systems in the course of the parcels description history is that unless the scrivener is fully versed in the relationships of calls and evidences on the ground, an erroneous deed is in the making.

32-12. EFFECT OF STATE LAWS

After discussing the many systems or methods available to describe land, a general statement regarding the effect of myriad volumes of state laws is pertinent. When lands passed from the "sovereign," or the government, the laws and rules controlling the survey of private lands changed. The federal government no longer had jurisdiction over those lands, except where they were intermingled with remaining federal lands having public-domain status. Some states use the 1973 manual as an integral part of their laws. Others created their own sets of laws to set a lost corner or determine the manner in which a section was to be subdivided. States have varying rules for the priorities of calls on a deed. Placement of errors of closure in a metes-and-bounds description also varies between states.

As mentioned earlier in this chapter, Texas established a rectangular system having little relationship with the PLSS. State laws differ greatly on riparian boundries. A hotly contested question still exists within the federal government regarding the manner in which reacquired lands are to be surveyed and described. Adverse possession normally does not run against the federal government, but the effect use lines have on lands the government has bought from a private citizen is vague. Be aware of state laws and never make hasty assumptions regarding the similarities to federal laws.

32-13. SUPERIORITY OF CALLS

With a few exceptions, the order in which legal description calls can control each other is similar. This is an important subject for the writer of a new description. Certain calls will outweigh others; therefore, it is crucial to know how and in what order to write a new description, so the parties' real intent is forever noted on the deed. The following list is basically accepted in most of the country.

1. Calls for monuments. These can be natural, such as lakes, streams, or trees. Physical monuments include any type of marker set by a surveyor or referenced objects that somehow coincide with the boundary.
2. Secondary monuments. These are the same type objects but were not specifically called for. With reference to other maps or documents, they are found to be acceptable evidence for locating the parcel.
3. Reference to a record boundary, such as a map, plat, or deed.
4. Calls for ties, usually by bearing and distance, to other monuments or boundary lines of record, but not in direct relationship to the lines being described.
5. Distances on the boundary.
6. Bearings on the boundary. This can also include angular relationships between two lines.
7. "Areas" when not used as the specific call. Descriptions that give an area at the conclusion are not controlled by the area, but rather the caption and body.

Studying a few brief examples of these calls will be useful. A very basic rule in land law is the superiority of a call for a monument. In almost every case where any other element within a legal description seems contradictory, a call for a monument will hold. For instance, if a distance is given in a metes-and-bounds description, but later is found to be in error, the call for a monument (if any) will govern. The deed may call for "100 feet to a 24′ oak tree... ," but what if the oak tree is 110 ft away? The oak tree is a superior call. Mentioning it in the description shows it to be the intended corner, and the distance is a secondary call to find the correct area or tree. A call for another boundary or other line of record shows an intent of the parties to avoid any overlap of hiatus. The distance call may not be correct, but the call for the adjoining parcel to share a common boundary cannot be mistaken as to intent.

When not being used with monuments or calls to adjoiners, distances are superior to bearings. This principle comes primarily from the belief that the measurement of a distance is more accurate than a bearing. This was true for most of history, but has changed in the last 20 yr with the common use of more accurate angle-measuring equipment. Nevertheless, the legal principle still holds because the concept remains that ordinary landowners can determine a distance more accurately than a bearing.

Surprising as it may seem, area given in a description is usually the last call to be considered. The average citizen believes the exact opposite. A person having a deed with an area stated thereon is adamant about having at least that much land. But the entire structure of land law and surveying sciences leans away from this theory. The federal government recognized the danger of relying solely on acreage calls in patents. Almost every government patent issued over the last 200 yr called for an acreage and then stated, "according to the official plat. ... " Remember, a call for a map, plat, survey, or the field notes on a deed is treated as if those items were actually printed directly on the deed. Therefore, monuments and other evidences of the survey on which the patent was based are called for and will be the superior calls. When a quarter-section is found to have only 150 acres, instead of the 160 stated in the patent, monuments are the controlling factor. The acreage is simply an addendum to the more superior calls in the patent. This same rule holds as long as the description is not written in a way to make it obvious that the acreage is a superior call. An example of this would be "the north 30 acres of the Smith Ranch, as described in . . ."

When a call for acreage is made in this manner, it is actually a dimension call that will control location of the land. However, when an acreage call is given and then followed by a metes-and-bounds description, the acreage call gives way to the bearings and distances called for. Reasons for the inferiority of acreage calls are numerous, but acreages have traditionally been somewhat "floating" values. In the northeastern states, acreages frequently were dramatically exaggerated to make up for poor-quality land. Wisdom on the part of early surveyors and scriveners showed that the acreage could change significantly, depending on the method of survey. In reality, if our entire system was based on acreages alone, no amount of time could ever resolve the chaos and lawsuits arising therefrom!

Modern-day description preparers must be familiar with the superiority of calls. The duty of today's surveyors is to "retrace the footsteps of the original surveyor." In far too many cases, those footsteps are very difficult to find or follow. Descriptions must be written in a way to ensure that future surveyors, title officers, lawyers, and especially landowners can clearly follow our footsteps.

32-14. DESCRIPTION TERMINOLOGY

A final area to be explored in legal land descriptions is terminology. All too often, a scrivener capriciously employs terms and phrases that sound normal and common but are vague to those who attempt to use descrip-

tions. Every office that prepares land descriptions must have a law dictionary. Many are available, but the most widely employed in the surveying field is *Black's Law Dictionary*. The 10 most commonly used and misunderstood terms are discussed briefly in the following listing:

Adjacent. This word is often used incorrectly instead of adjoining. Adjacent is defined as lying near, close to, or in the vicinity of but not necessarily touching. Although thought to mean touching, this term indicates only that a person is in the right neighborhood, not on the right line.

Adjoining. A much-preferred term when saying a boundary actually is common with another.

Along. A call for a boundary to be along a line implies that a person is moving with the call, and the line itself is the route of motion. The term is not interchangeable with by a line, or on a line, or with a line. When referring to a strip description or road, along refers to the centerline, unless otherwise called for.

Contiguous. According to *Black's Law Dictionary*, the term contiguous is somewhat vague and should be avoided. In close proximity, near though not in contact, and adjoining all fall within the meaning. Other terms that do not carry such ambiguities are preferable.

Due (as in direction). Courts have determined that due north refers to astronomic north, not as assumed north. Geodetic north is not included in this term. The use of true instead of due does not improve clarity. These terms should be used only when the bearings or azimuths are actually based on astronomic observations. Statements regarding the basis of bearings in a deed can be made in the description. In the absence of a stated basis of bearings, bearings called for are considered to be relative to each other, but not necessarily to bearings in other deeds.

Either. This term is assumed to mean the same as on both sides when describing a strip. "Thirty-three feet on either side of the following described centerline" is an incorrect use. Either implies that there is a choice of one side or the other of the centerline, but not both sides, which was the intent. Scriveners should employ the word *each* instead.

Excepting (as opposed to *reserving*). As discussed in Section 32-2, these terms are not interchangeable. Older deeds, especially in the northeast, did use them in tandem. However, the word excepting means omitted or cut off, whereas the term reserving indicates that only a portion of something is being withheld. When reserving an easement from a foregoing description, the implication is that the fee title to the land has been transferred, but a partial interest in the form of an easement is to be created and held for the grantor. When an "exception" is made, anything described in the exception that was also described in the body will be removed from the transaction.

More or less. A most commonly used term in land descriptions, more or less is sometimes misunderstood. This phrase indicates that the reader of a call should be cautious. It indicates a certain amount of uncertainty in the accuracy of a call. The term should be reserved for situations in which the quantity is questionable. When used at the end of an acreage call, more or less implies that the area is not to be taken as any real influence on the description; i.e., the body controls the land, and the area is simply given as a supplemental call for information only.

Parallel to. For two lines to be parallel to each other, they must by mathematics be equidistant from each other at any point. Lines or objects cannot be parallel "with" each other, but must be parallel "to." Although parallel implies straight lines, this term is often used with curves. The principle indicates that the curves are concentric, which means they share a common radius point and are also equidistant from each other at any given point.

Tract (as opposed to *parcel*). These two terms are the most frequently used in captions in referring to the land being described. The term parcel is much preferred because it does not have any other meanings with which it can be confused. Tract is a term used within the PLSS for lands segregated during an independent resurvey. Similarly, using "a section of land" can be confused with a 640-acre entity of the PLSS. The term lot does have more than one official meaning, and it should be made clear which system is being referred to in a deed.

32-15. SUMMARY

The subject of legal land descriptions is very complex and far more detailed than this single chapter can demonstrate. An attempt has been made to present a skeletal review of the subject. Surveyors, engineers, architects, lawyers, realtors, and others dealing with land use and transfer should refer to detailed textbooks on the subject. The single most important factor to remember is that intent of the parties is always the most controlling factor. Therefore, any new description created must clearly indicate what those intentions are, without hasty assumptions. Carefully choose the system to describe the parcel. Use words and phrases that truly describe the intent. Write each description with calls for records, bounds, and other details tailored to avoid future conflicts. Be concise and to the point. And above all, research the history of descriptions for a parcel and the adjoiners thereto. A cautious and professional attitude toward this subject will greatly improve the quality of descriptions within our record system.

NOTES

1. BLM. 1973. *Manual of Instructions for Survey of the Public Lands of the United States*. Washington, DC: U.S. Government Printing Office.
2. J. G. McEntyre. 1986. *Land Survey Systems*, 1st ed. Rancho Cordova, Landmark Enterprises.

REFERENCES

BLACK, H. C. 1979. *Black's Law Dictionary*, 6th ed. St. Paul, MN: West Publishing Co.

BROWN, C. M. 1986. *Boundary Control and Legal Principles*, 3rd ed. New York: John Wiley & Sons.

CUOMO, P. A. Advanced Land Descriptions 1994, Rancho Cordova CA: Landmark Enterprises.

GRIMES, J. S. 1987. *Clark on Surveying and Boundaries*, 6th ed. New York: Bobbs-Merril.

WATTLES, G. H. 1979. *Writing Legal Descriptions in Conjunction with Survey Boundary Control*, 2nd ed. Orange, CA: Gurdon Wattles.

WATTLES, W. C. 1974. *Land Survey Descriptions*, 10th ed. Orange, CA: Gurdon H. Wattles (revised).

33

The Role of the Surveyor in Land Litigation: Pretrial

John Briscoe

33-1. INTRODUCTION

The expert in any of a number of technical fields is frequently, as a witness or an adviser, drawn into the vortex of land litigation. The expert in the land case may be a surveyor, title abstractor, property appraiser, engineer, geologist, hydrologist, oceanographer, biologist, or geographer. But there is one denominator that, in land cases, each of these experts bears in common with the other. It lies in the role each plays in a land dispute, whether it concerns title, political or property boundaries, water rights, or other aspects of natural resource use. In these cases, the expert's role is much more like the lawyer's than the part of an expert in virtually any other kind of law case.

In fact, the role of the land expert is unique —particularly the role of the land surveyor in relation to the law. This is perhaps best shown by contrast. A neurosurgeon does not excise a brain tumor by reference to a rule of law. Nor does an engineer design a bridge according to any laws except those of physics. By contrast, the land expert, in making a survey or forming an opinion as to the title, value, or location of property, applies law in the same way as a lawyer who is drafting deeds or subdivision

restrictions. In turn, the expert's professional ancestors often played a major role in shaping that law. Presidents Washington and Lincoln were land surveyors, as was Thomas Jefferson, who persuaded Congress to establish the U.S. Coast Survey in 1807 and is reputed to have been the originator, in 1785, of the rectangular system of surveys that has fostered many of the principles of boundary location.[1] In the main, the law of boundaries and surveying has developed through government manuals, special instructions written by surveyors, and through surveying customs and usages that were, in time, adopted by the courts as law.

It is perhaps because of his or her close relation to the law that the expert in a land case—more often than experts in other kinds of cases—asks to be initiated in the mysteries of the legal process. It is hoped that this chapter will help dissipate some of the mystery that shrouds the rules of procedure and evidence in civil cases. Two points of unpatronizing caution, however, are in order. First, these vast subjects, taught over a period of years in law schools and mastered only with years more practice, can be only superficially treated in this chapter. Second, it is one thing, say, to read Hoyle on bridge, and another to apprehend the subleties of the game. This chapter's objective, then, is to impart to the reader a nodding acquaintance with the principles of evidence and procedure.

33-2. RULES OF EVIDENCE

To a large extent, a land surveyor's contribution to land litigation is governed by the rules of evidence. Any expert advice or testimony that he or she may offer can only be used within the limits prescribed by these rules. The land surveyor contemplating serving on a litigation team should therefore be aware of the extent to which his or her information can be used at the time of trial. By using the evidence that is most likely to be accepted by the court, an expert can avoid complications at trial and, as a result, greatly improve both individual effectiveness and that of the rest of his or her litigation team. To this end, some basic rules of evidence will be discussed.

33-3. RELEVANCE

Relevance is the fundamental principal of the law of evidence. In a sense, all other rules of evidence in one manner or another address the question of relevance by focusing on the reliability or credibility of evidence, which are, on reflection, components of relevance. The less reliable or credible an item of evidence is, the less relevant it is. McCormick wrote that relevance is logically the first rule of evidence,[2] and this principle has been codified in many jurisdictions:

> All relevant evidence is admissible, except as otherwise provided. ... Evidence which is not relevant is not admissible.[3] Except as otherwise provided by statute, all relevant evidence is admissible.[4]

Although the concept of relevance may seem self-evident, it is useful to examine the definitions that have been formulated. "Relevance," wrote McCormick, "is probative worth."[5] The California rule, for use in California state courts, reads:

> "Relevant evidence" means evidence having any tendency to make the existence of any fact that is of consequence to the determination of the action more probable or less probable than it would be without the evidence.[6]

The federal rule, used in federal courts, provides:

> "Relevant evidence" means evidence, including evidence relevant to the credibility of a witness or hearsay declarant, having any tendency in reason to prove or disprove any disputed fact that is of consequence to the determination of the action.[7]

Both the California and federal definitions of relevance have two essential components: (1) The proferred evidence must tend to prove or disprove a proposition and (2) that proposition must be of consequence to the action. Thus, if the location of a section corner, e.g., is of no consequence to the action—i.e., if its location would in no manner affect the outcome of the lawsuit—evidence of its location would be irrelevant. In addition, if the location of the corner were of consequence, but the offered evidence did not tend to establish its location, it too would be irrelevant.

Although the general rule is that relevant evidence is admissible, the trial judge may in his or her discretion exclude evidence that is admittedly relevant (and not objectionable under the other rules of evidence) in certain well-established circumstances. These circumstances occur when admitting the proferred evidence would cause (1) a danger of unduly prejudicing the jury, (2) a danger of confusing the jury, and (3) an undue consumption of the court's time.[8] Explicit photographs of a maiming victim are sometimes excluded, e.g., even though relevant (to show the victim was in fact maimed), for the photographs may so inflame the jury that it loses sight of the more central issue—whether the defendant was in fact the attacker. The rule that relevant evidence may be excluded if it might confuse the jury is infrequently invoked today. Modern trial judges are far more willing to allow complex scientific, economic, or other technical information to go to the jury than their predecessors.

Relevant evidence may also be excluded if it serves only to beat a dead horse. After three or four eyewitnesses testify to the setting of the disputed corner monument, for instance, the trial judge may refuse to allow more to take the stand. Also, if an inordinate amount of time would be consumed establishing a minor or tangential point, one might expect the judge to draw the line. As Justice Oliver Wendell Holmes once aptly remarked, "... that objection [that the offered evidence would require far too much court time] is a purely practical one—a concession to the shortness of life."[9]

Finally, in addition to the matters of prejudice, danger of confusion, and undue consumption of time, relevant evidence is typically excluded in other well-defined situations, for the purposes of "public policy."

As ought to be apparent from this brief discussion of the concept of relevant evidence, what is relevant and, thus, admissible depends in large measure on the substantive rules governing the dispute. For example, in a lawsuit in which the sole issue is whether one party has acquired title to land by adverse possession, what evidence is relevant is a function of the elements needed to establish title by adverse possession in that state. If payment of property taxes is not a necessary element as it is in a few states, such as California, then evidence of payment of taxes is, strictly speaking, irrelevant.

"Procedural" rules also can determine what is relevant evidence and what is irrelevant. For example, one such rule governing the trial of quiet title actions provides that the plaintiff must prevail on the strength of his or her own title and may not prevail merely by showing infirmities in the defendant's title.[10] Thus, un-til the plaintiff in a title case has introduced sufficient evidence that, if unrebutted, would entitle the plaintiff to a decree quieting his or her title, any evidence he or she may offer that is directed solely at attacking the defendant's claim is irrelevant.

33-4. DOCUMENTARY EVIDENCE

33-4-1. Authentication

Before a document may be received in evidence at trial, it must first be shown to be what the party offering it claims it to be. Authentication is the process of establishing this. The California definition of authentication, representative of other jurisdictions, is this:

> Authentication of a writing means (a) the introduction of evidence sufficient to sustain a finding that *is the writing that the proponent of the evidence claims it is* or (b) the establishment of such facts by any other means provided by law.[11]

Thus, even though a writing is relevant and not subject to any exclusionary rule, a foundation must be laid by authentication before it can be introduced into evidence.[12] The federal rule respecting authentication, Rule 901 of the Federal Rules of Evidence, is very similar to the California rule.

Presumptions, which are considered in a later section of this chapter, frequently help in authenticating documents. In California, e.g., a "book, purporting to be printed or published by public authority, is presumed to have been so printed or published."[13] Similarly, in common with many other states, California law presumes the authenticity of "ancient documents," providing four conditions are met. The document must be at least 30 yr old; it must be in a condition that raises no suspicion concerning its authenticity; it must have been kept or found where one would expect it to be; and last, it must have been generally treated as authentic by persons having an interest in the matter.[14] Another presumption that can

aid in authenticating a document holds that a writing is presumed to have been truly dated.[15] All such presumptions respecting the authenticity of documents, however, may be rebutted by showing that the document is fake or forged. These presumptions are not conclusive, but their effect is that unless contrary evidence
is shown, the document will be presumed authentic.

In the case of government documents, the process of authentication is greatly hastened by the use of certified copies. (Two other evidentiary hurdles to the introduction of documents, the best-evidence rule and part of the hearsay rule, are also overcome by the use of certified copies.) Virtually all states, and federal law, provide that properly certified copies are prima facie evidence of the fact that the original is on file with the public entity, and of what the original contains. By prima facie it is meant that the opposing party may seek to show that there is no original, or the original is not genuine, or the purported copy is not a true copy, however, if he or she cannot, the certification establishes the authenticity of the document. Thus, a certified copy of the field notes of a government survey will be treated as authentic as if one produced the original and authenticated it. It should be apparent then that using a certified copy is preferable even to using the original. If we assume that one could obtain an original government document, it would still be necessary to establish *its* authenticity, which usually is done by calling as a witness the custodian of the document. Using a certified copy dispenses with the need to call a "live" witness for authentication.

In addition to the presumptions discussed, there are other commonplace methods of establishing the authenticity or genuineness of a document. A common method of proof is by comparison of the offered writing with some other admittedly genuine writing, called an "exemplar." The comparison may, of course, be made by expert witnesses in order to form their opinion.[16] But the two writings—the disputed and exemplar—may be offered in evidence and considered by the jury or other trier of fact as circumstantial evidence to prove genuineness or lack thereof.[17] The foundation for such evidence is proof of the preliminary fact of genuineness of the exemplar. The court must find that it was "admitted or treated as genuine" by the adverse party, or it must be "otherwise proved to be genuine to the satisfaction of the court."[18] The exemplar must also be one that was written naturally and independently of the purposes of the litigation. A specimen writing or signature prepared specially for the use of an expert ("post litem motam") is not admissible. And, defendants cannot offer exemplars of their handwriting, made at the trial or after arrest, because of the possibility of fraud in allowing them to corroborate their own testimony by a prepared specimen.[19]

Finally, the authenticity of an instrument relied on by the plaintiff or defendant also may be established *prior to the trial*, in discovery proceedings or at a pretrial conference. These matters are considered later in this chapter.

33-4-2. Best Evidence or Original Document Rule

The best-evidence rule may be the most egregiously misnamed rule in the law. Prosecutors may speak of the smoking gun as the "best evidence" of guilt, and surveyors may debate what constitutes the best evidence of an obliterated corner. But the best-evidence rule is not a rule of general application that admits only the most probative or persuasive evidence. It applies only to documents, and it would be more aptly called the "original document" rule, which is the name that will be used here.

Succinctly stated the rule is this: In proving the terms of a writing, when the terms are material to the case, the original writing must be produced unless it is shown to be unavailable for some reason other than the serious fault of the one who wishes to prove the writing's terms.[20] The California rule reads:

> Except as otherwise provided by statute, no evidence other than the original of a writing is admissible to prove the content of a writing.[21]

The federal rule is distinctive in one significant respect. It is not called the best-evidence rule but rather "requirement of original." Like the California rule, the federal rule has been expanded to cover not only writings but also recordings and photographs.[22]

The purpose of the rule is plain. If a question arose at trial concerning an entry in the field notes of an original government survey, the most reliable evidence of what was entered would be the original field notes themselves. A handwritten or typewritten transcription of the original notes—not to mention oral testimony of their contents—would be too susceptible to errors, whether inadvertent or purposeful. This problem is presented less by modern photocopying methods but nonetheless persists.

> The reasons for the rule are thus stated in [Section 602] of the Model [Evidence] Code: "Slight differences in written words or other symbols may make vast differences in meaning; there is great danger of inaccurate observation of such symbols, especially if they are substantially similar to the eye. Consequently there is opportunity for fraud and likelihood of mistake in proof of the content of a writing unless the writing itself is produced. Hence it should be produced if available.[23]

It should be noticed that the statement of the rule given above specifies that the original is required only when the terms of the writing are material to the case. Stated in reverse fashion, if the writing is "collateral" to the principal issues of the case, the original need not be produced. Suppose a witness testifies that he first observed the blaze tree on December 4. When asked how he knows it was December 4, he replies that he was carrying that day's newspaper at the time and made a mental note of the date given below the paper's banner. In such an instance, it would

hardly be necessary to produce the original of the newspaper to prove the date.

The original document rule comes into operation when an attempt is made to offer *secondary* evidence—e.g., a copy or oral testimony, to prove the contents of an original writing. Sometimes, however, it may be difficult to determine whether a writing is an original document or secondary evidence for purposes of the rule. If there is in a practical sense no single original but several duplicates of equal reliability, any one could be considered primary evidence, and the policy of the rule would be satisfied. This theory is approved by modern authorities, but the cases still make conflicting and confusing distinctions between different kinds of duplicates.[24] Carbon copies are now generally regarded as duplicate originals and may be introduced without showing the unavailability of the original.[25] Printed and other mechanically reproduced copies should, like carbons, be regarded as originals for purposes of the rule. However, except as to business records under the new California statute, photographs and photostats are still generally considered secondary evidence.[26]

The original document rule is inapplicable if the writing "was not reasonably procurable by the proponent by use of the court's process or by other available means."[27] Prior California cases reached the same result by treating the writing as "lost."[28] Even though a writing is outside the reach of process, the original document rule applies if other means of procurement are available. But when the writing is in the hands of a third person, whose attitude is hostile and who refuses to give it up, some authorities hold that compliance is excused.[29]

It should be borne in mind that the three points treated in this section of the chapter are by no means the only evidence issues that must be considered when seeking to introduce—or to prevent the introduction—of a document. The relevance issue, discussed in the preceding section, pertains with respect to all evidence; the document may constitute hearsay; and there may be a question of whether the document can be judicially noticed. The hearsay rule is taken up next.

33-5. THE RULE AGAINST HEARSAY OR THE RULES PERMITTING HEARSAY

What is hearsay? You are urged to put out of mind any present ideas of what hearsay is and to consider the following variously formulated definitions thoughtfully. McCormick defines it thus:

> Hearsay evidence is testimony in court or written evidence, of a statement made out of court, such statement being offered . . . to show the truth of matters asserted therein, and thus resting for its value upon the credibility of the out-of-court asserter.[30]

The federal rules provide:

> "Hearsay" is a statement [either written, oral, or non-verbal conduct, if intended as an assertion], other than one made by the declarant while testifying at the trial or hearing, offered in evidence to prove the truth of the matter asserted.[31]

The California definition is:

> "Hearsay evidence" is evidence of a statement that was made other than by a witness while testifying at the hearing and that is offered to prove the truth of the matter stated."[32]

These definitions show that there are two essential components of hearsay evidence. The first is almost self-evident: It is evidence of a statement made out of court. The witness on the stand, e.g., relates what someone told him or her during a meeting. It should be recalled that the "statement" may be written as well as oral. Thus, field notes of a survey when introduced at trial are evidence of a statement made out of court. (In most jurisdictions, nonverbal conduct, if intended to assert something, may also constitute a "statement"; a deaf-mute's sign-language statement is an example.) It is the second component of the hearsay definition, however, that is elusive; the statement must be offered at the trial or hearing for the purpose of proving the truth of the matter asserted in it. If it is offered for another purpose, it is *not hearsay*. Some examples may illustrate.

In a murder trial, suppose that the defendant testifies that just before he killed the victim, the victim raised a gun and shouted, "I intend to kill you." If the victim's statement were offered to prove the truth of what was stated—i.e., that he intended to kill the defendant, it would constitute hearsay. It is a statement made out of court, offered for the truth of the matter asserted. If, however, it were offered to show simply that the victim had in fact said the words—irrespective of their truth—and thus had put the defendant in fear of his life (which would help establish self-defense), the evidence would not constitute hearsay.

Suppose a witness testifies that another person told her the defendant was driving 70 mi an hour. Again, the statement is one made out of court, but is it hearsay? If it is offered to show that the defendant was in fact driving at that speed, it is. But it may be offered for other purposes. If it is offered to show that the one reporting the speed of the defendant (the "declarant") could speak English, it is not hearsay. Similarly, if it is offered to prove that the declarant was conscious at the time, it is again not hearsay.

If a map is offered to show the truth of what is depicted on it, as for example the location of monuments, it is hearsay, for it is a "statement" made out of court and offered to prove the contents of the statement. If the question in the case is imply whether the map exists—regardless of what it depicts—it does not constitute hearsay.

The critical factor in these examples is whether the evidence of the out-of-court statement is offered to prove the truth of what was said or merely show the making of a statement. This is why an out-of-court statement may be objectionable when offered to show one thing but not when offered to show another; the trier of fact may or may not need to consider the truthfulness of the out-of-court declarant. If the jury must decide whether the car was being driven at 70 mi an hour, to use one of the examples, it needs to know whether the out-of-court declarant was telling the truth. She was neither placed under oath nor subject to cross-examination when she made the statement, and thus two of legal system's most revered guarantors of truth are lacking. If the jury, on the other hand, is not concerned with the speed of the car, but only with whether the declarant was conscious, it has no need to know whether the declarant was speaking the truth about the speed of the car. It need only know whether the witness on the stand is telling the truth when she testifies that the declarant spoke; the witness on the stand has taken the oath and may be cross-examined.

The general rule, of course, is that hearsay evidence is not admissible,[33] but many exceptions have been made. The exceptions to the hearsay rule are fairly uniform among the states and in the federal system. Some exceptions deal specifically with criminal matters, and others would rarely be encountered by the reader. Accordingly, the following list is not intended as exhaustive, but emphasizes those that may arise in land and boundary cases. Also, many of the exceptions do not operate unless certain rather technical matters are first shown. These "foundational" matters will be discussed only sparingly.

33-5-1.   Admissions of a Party

Evidence of a statement made out of court by a party to a lawsuit may be introduced, for the purpose of proving the truth of the statement, by an opposing party.[34] It may not be used, however, by the party who made the statement. Suppose the plaintiff in an adverse possession case once told a neighbor that her occupation of the property had been only intermittent. The defendant may call the neighbor to the stand and have here testify to what the plaintiff told him, and it is admissible for the truth of the matter asserted—i.e., it may be considered as evidence that the plaintiff's possession had not been continuous. The reason for this exception is apparent. The plaintiff is a party to the action and, thus, available to explain or deny the statement.

33-5-2.   Declarations Against Interest

The hearsay exception for admissions is so frequently confused with the exception for declarations against interest that the erroneous expression "admissions against interest" is often heard. Although the same hearsay statement may constitute both an admission and a declaration against interest, the two have different elements. Essentially, a declaration against interest must be one that, when made, was antithetical to the declarant's pecuniary or proprietary interest, and the declarant must be unavailable to testify at trial. An admission, on the other hand, need not have been against the declarant's interest at the time it was made, although it usually is. Moreover, the exception for admissions applies only to statements by parties to the lawsuit, whereas the exception for declarations against interest applies to any declarant.

Examples in the real property field of declarations against interest are common. Statements by one in possession of real property that he is not the true owner, that she has fenced beyond her true boundaries, or that the property had been purchased with trust funds, have been traditionally considered declarations against interest.[35] Examples in other areas of law are admitting liability for an automobile accident, acknowledging a debt, or acknowledging a breach of contract. The belief that such statements would not be made unless true overcomes the infirmities of the lack of oath and opportunity for cross-examination.

In all such cases, however, the declarant must be unavailable to testify at trial. What "unavailable" means varies from jurisdiction to jurisdiction. If, however, he or she is dead, beyond the subpoena power of the court, or physically incapacitated, all courts would consider this person unavailable.

33-5-3. Business Records

Certain entries in business records are considered admissible evidence in most jurisdictions, notwithstanding that they are hearsay.[36] Some such entries, of course, might come in under other exceptions to the hearsay rule. If the entries were made by a party opponent, they may be admissible as admissions; perhaps the entries, whether made by a party or not, were declarations against interest at the time they were made. Because of the general reliability of records kept in the regular order of business, however, a specific exception to the hearsay rule for such records has been made. The rule no doubt originated with the admission of a shopkeeper's books to show the making of a debt; without his or her books, the shopkeeper could not testify as to the creation of the debt except by independent recollection.[37]

Today, evidence of entries in business records is admissible so long as certain matters are shown. (These requirements may vary somewhat from jurisdiction to jurisdiction.) One requirement is that the records must bear the earmarks of reliability. The federal rules, e.g., specify that business records are inadmissible if "the source of information or the method or circumstances of preparation indicate lack of trustworthiness."[38] It is difficult to generalize on circumstances that indicate a lack of trustworthiness, but self-serving declarations certainly fall within this category.[39] Erasures or other alterations of records, or mutilations such as torn-out pages, also cast suspicion on the reliability of the records. If the erasure, alteration, or mutilation can be explained to the satisfaction of the court, it may nonetheless be received into evidence, but these conditions make more difficult the task of introducing such records.[40]

In this respect, two well-known authors have some sound suggestions on the making of survey field notes, to which the land surveyor should adhere rigidly.

When a survey is performed for the purpose of gathering data, then the field notes become the record of the survey. If the notes have been carelessly recorded and documented, falsified, lost, or made grossly incorrect in any way, the survey or a portion of the survey is rendered useless. Defective notes result in tremendous waste of both time and money. Furthermore, it will become obvious that, no matter how carefully the field measurements are made, the survey as a whole may be useless if some of those measurements are not recorded or if the meaning of any record is ambiguous.

The keeping of neat, accurate, complete field notes is one of the most exacting tasks. Although several systems of note-keeping are in general use, certain principles apply to all. The aim is to make the clearest possible notes with the least expenditure of time and effort.

A hard pencil—4H or harder—should be used to prevent smearing. The notebook should be of good quality, since it is subjected to hard usage. *No erasures should be made*, because such notes will be under suspicion of having been altered. If an error is made, a line should be drawn through the incorrect value and the new value should be inserted above. In some organizations the notes are kept in ink, but this is rather inadvisable unless a waterproof ink is used or unless there is no possible chance of the notes becoming wet.

Clear, plain figures should be used, and the notes should be lettered rather than written. *The record should be made in the field book at the time the work is done*, and not on scraps of paper to be copied into the book later. Copied notes are not original notes, and there are too many chances for making mistakes in copying or of losing some of the scraps. All field computations should appear in the book so that possible mistakes can be detected later.[41]

33-5-4. Certain Official Documents

An exception to the hearsay rule, akin to the exception for business records, has been made for certain official documents.[42] Such matters as court records, records of acts of the legislature, observations recorded by the National Weather Service, and land office records fall within this exception. The requirements for the exception are similar to those for the business records exception.

33-5-5. Past Recollection Recorded

Another exception to the hearsay rule, which in many respects resembles the exception for business records, addresses what is generally called "past recollection recorded."[43] According to this exception, evidence of what is contained in a memorandum or other writing may be admitted into evidence if certain elements are shown. First, the writings must have been made by the witnesses themselves, or someone under their direction, or some other person for the purpose of recording the witnesses' statement at the time it was made as, e.g., a shorthand reporter taking the statement of a witness after a crime. Second, the writing must have been made when the fact recorded in the writing actually occurred or was fresh in the witnesses' memory. Generally, this may be shown simply by having the witnesses testify that the statement they made was a true statement of the fact recorded. Third, the witnesses must presently have insufficient recollection to testify fully and accurately to the facts recorded in the writing.

From the foregoing discussion, it should be apparent that a surveyor's properly recorded field notes are generally admissible into evidence as past recollection recorded and business records as well.

33-5-6. Reputation Concerning Land Boundaries and Land Customs

The California rule respecting land boundaries and customs is typical of those found in most jurisdictions today.

Evidence of reputation in a community is not made inadmissible by the hearsay rule if the reputation concerns boundaries of, or customs affecting, land in the community and the reputation arose before controversy.[44]

Formerly, it was a requirement for the admissibility of reputation evidence concerning land boundaries or customs that the reputation must have been "ancient"—i.e., it must have arisen during past generations.[45] Today, however, the generally accepted principle is simply that the reputation must have arisen in the community before the present controversey arose; there is no requirement that the reputation be ancient.

The general principle for the rule admitting reputation evidence is this:

> [T]he fact sought to be proved . . . [being] [o]f too ancient a date to be proved by eye-witnesses, and not of a character to be made a matter of public record, unless it could be proved by tradition, there would seem to be no mode in which it could be established. It is a universal rule, founded in necessity, that the best evidence of which the nature of the case admits is always receivable.[46]

This principle has special application in boundry cases:

> It must be obvious, that when the country becomes cleared and in a state of improvement. It is oftentimes difficult to trace the lines of a survey made in early times.[47]

> Questions of boundary, after the lapse of many years, become of necessity questions of hearsay and, reputation. For boundaries are artificial, arbitrary, and often perishable; and when a generation or two have passed away, they cannot be established by the testimony of eye-witnesses.[48]

It should be emphasized that this exception to the hearsay rule does not concern individual statements respecting land boundaries (which statements are treated separately below), but rather the prevalent brief of community reputation. The reputation may be proved in several ways. Individual witnesses may testify that, according to reputation in the community, a certain ditch or road is the boundary of the tract in question. In addition, old maps and old surveys,[49] if they are shown to have been consulted by the community in its dealings with the land, may be admitted as evidence of reputation in the community concerning boundaries. In the same manner, old deeds and leases may be admissible as evidence of community reputation.[50] Books concerning matters of local history may also recite facts that, if shown to reflect community reputation, would be admissible.[51] Although boundaries may be proved by reputation evidence, *title* on the other hand may not be so proved.[52]

In addition to reputation concerning boundaries, the hearsay exception extends to evidence of reputation concerning a myriad of customs affecting land. The elements of the doctrine of custom, a principle of substantive law (evidence is procedural law), were taken from the 18th century writings of Sir William Blackstone. Among the requirements are that the custom must have been used so long that the memory of man runneth not to the contrary. Also, the custom must have been uninterrupted, it must have been peaceable, and acquiesced in, and it must be reasonable. "Customs ought to be *certain*. A custom that lands shall descend to the most worthy of the owner's blood is void: for how shall this worth be determined?" In addition, the custom when established must be compulsory, and it cannot be repugnant to another.[53]

Other recent cases have applied the common law doctrine of custom in land cases.[54] In such cases where rights in land are asserted to exist by virtue of custom, the reputation of the custom may be the only evidence that can establish it, making this exception to hearsay a quite valuable rule.

33-5-7. Statements of Deceased Persons

To be distinguished from the hearsay exception for community reputation concerning boundaries and land customs is the distinct exception for certain individual statements respecting boundaries. Although the two hearsay exceptions have common historical roots in many jurisdictions, today the two are usually treated as discrete exceptions to the general rule forbidding the introduction of hearsay evidence.

The general requirements for admitting such declarations are, first, that the declarant is dead or unable to testify and, second, that the circumstances indicate no reason for the declarant to have misrepresented the truth.

33-5-8. Ancient Documents

An exception to the hearsay rule found in virtually all jurisdictions applies to what are, for convenience, called ancient documents. Typically, if a writing is more than 30 yr old, contains nothing that casts suspicion on its reliability, and if the writer is shown to have been a qualified witness—e.g., having firsthand knowledge, the writing may be received for the truth of the matter asserted within it.

The federal and California rules for ancient documents are structured somewhat differently from these general principles. Federal Rule of Evidence 803(16) provides that "[s]tatements in a document in existence twenty years or more the authenticity of which is established" are admissible as an exception to the hearsay rule. The question of the authenticity of the ancient document is separately treated in Rule 901(b)(8). That rule provides that authentication is established by:

> Evidence that a document or data compilation, in any form, (A) is in such condition as to create no suspicion concerning its authenticity, (B) was in a place where it, if authentic, would likely be, and (C) has been in existence 20 years or more at the time it is offered.

The California rule contains a requirement not imposed by other jurisdictions:

> Evidence of a statement is not made inadmissible by the hearsay rule if the statement is

contained in writing more than 30 years old and the statement has been since generally acted upon as true by persons having an interest in the matter.[55]

The California Law Revision Commission, which drafted the California Evidence Code (the code was enacted in 1965), noted that the ancient documents concept had been extended from the area of authentication to the hearsay rule. The commission's view was that the age of a document alone is not a sufficient guarantee of the trustworthiness of the statement contained in it. For that reason, it added the requirement that the statement must have been generally acted on as true by people having an interest in the subject.

33-6. THE RULE REQUIRING PERSONAL KNOWLEDGE

Thus far, it should be apparent that no matter how contrived or obscure the rules of evidence may seem, they are unmistakably the product of the steadfast purpose of the common law and American courts to receive only the most reliable evidence. This purpose also accounts for the fundamental rule requiring that witnesses testify only to matters of which they have personal knowledge.[56] A customary statement of the rule is that a witness who testifies to a fact that could have been perceived by the senses must have had an opportunity to observe and also must have actually observed the fact.[57] Federal Rule of Evidence 602, e.g., provides

A witness may not testify to a matter unless evidence is introduced sufficient to support a finding that he has personal knowledge of the matter. Evidence to prove personal knowledge may, but need not, consist of the testimony of the witness himself. This rule is subject to the provisions of Rule 703, relating to opinion testimony by expert witnesses [discussed later in this chapter].

The essential concept of the rule requiring personal knowledge is that, with the exception of opinion testimony given by experts, a witness may testify only to facts, and only when he or she has personal knowledge of those facts. Thus, we find the Supreme Court of Alabama in 1848 writing:

The general rule requires, that witnesses should depose only to facts, and such facts too as come within their knowledge. The expression or opinions, the belief of the witness, or deductions from the facts, however honestly made, are not proper evidence as coming from the witness; and when such deductions are made by the witness, the prerogative of the jury is invaded.[58]

As early as 1349, it was held that witnesses, as opposed to jurors, were not challengable

Because the verdict will not be received from them, but from the jury; and the witnesses are to be sworn "to say the truth," without adding "to the best of their knowledge," for they should testify nothing but what they … know for certain, that is to say what they see and hear.[59]

Although not nearly so formally categorized, as in the case of the hearsay rule where logic, the presence of other indicia of reliability, and sheer need for the testimony are present, exceptions to the rule requiring personal knowledge have been formulated. It may also be said that some of these exceptions have been created to prevent ludicrous results. For example, competent witnesses are invariably allowed to testify to their age and data of birth.[60] Similarly, witnesses have been allowed to testify as to their kinship to others.[61] The same experience holds true when the required evidence as a practical matter can be found only in a body of records, whether public or private. It is simply impractical, if not impossible, in such cases to require the summoning to the witness stand of each person having personal knowledge of the events recorded in such records. Similarly, testimony respecting data collected from the use of scientific instruments, formulas, and tables invariably entails a dependence on the statements of other persons, and even of unknown inventors, computer programmers, etc. Yet it is clearly impracticable to adhere rigidly to the firsthand knowledge requirement, when the sole source of the necessary firsthand knowledge "connection" cannot feasibly be produced.

33-7. BURDEN OF PROOF

The expression "burden of proof" conveys an idea that may be instinctively understood—that in an adversary proceeding, one side must bear the onus of proving its position is correct. Put another way, if the evidence on each side of a case is equally persuasive, in "equipose," one side nevertheless must prevail. In a championship prize fight, e.g., the challenger must take the crown from the champion. If they draw, the champ keeps the title.

The common law has refined the concept of burden of proof to a remarkable sophistication. As refined, the expression burden of proof actually denotes two distinct burdens, for convenience called the "burden of producing evidence" and "burden of persuasion." The burden of producing evidence is sometimes referred to as the "burden of production," or the "duty of going forward."[62] The burden of persuasion is occasionally called the "risk of nonpersuasion,"[63] or, unhappily, "burden of proof."[64]

The burden of production is best understood in the context of a jury trial. If one party has the burden of production on an issue—e.g., whether a particular survey was fraudulent—he or she must produce sufficient evidence from which it might reasonably be concluded that the survey was in fact fraudulent. If this is not done, the judge may direct a finding against him or her and refuse to allow the jury to decide the issue. If, however, sufficient evidence is introduced—say, the expert testimony of a surveyor who has analyzed the

survey and concluded it is fraudulent—the burden is discharged. The party may still carry the burden of persuasion, however.

The burden of persuasion is the burden on a party to "prove"—i.e., to convince the trier of fact, of the existence or nonexistence of a fact essential to its claim for relief or to its defense. The plaintiff in a quiet-title action, e.g., claims to own the disputed property, and it is incument on the plaintiff to convince the trier of that fact and not rely simply on the defendant's inability to show title. Likewise, a defendant who asserts in defense that a deed to the plaintiff was not executed by the purported grantor, but was forged, has the burden of convincing the trier of fact of that point.

The burden of persuasion is quite different from the burden of production. Unlike the burden of production, the burden of persuasion is not assigned until all the evidence has been received.[65] Moreover, the burden of persuasion does not "shift" during the course of the trial. There is one similarity, however: The burden of persuasion is allocated by the same principles that allocate the burden of production in the first instance—i.e., before any "shift" in that burden.[66]

The word "prove," as used above, means different things in different contexts. In general, it means that the party having the burden of persuasion on an issue must induce a *particular degree* of conviction in the mind of the trier of fact (the jury or, when there is no jury, the judge). If the required degree of conviction is not achieved, the trier of the fact must assume that the fact does not exist.[67] The degree, or quantum, of proof varies. In criminal cases, the prosecution must prove each element of the crime charged "beyond a reasonable doubt."[68] The expression "beyond a reasonable doubt" is readily understandable. One judge has remarked that it requires a remarkable skill in language to make it plainer by explanation.[69] In civil actions, on the other hand, the degree of proof required of the party bearing the burden of persuasion is ordinarily "the preponderance of the evidence."

To prove a matter by a preponderance of the evidence is not simply to produce more witnesses or introduce more documents than the opponent.[70] Simply stated, parties meet this burden of persuasion when they produce evidence, together with inferences that the jury can reasonably make from the evidence, that is more convincing than their opponent's evidence. Since this concept has been treated in rarified detail elsewhere,[71] this discussion should suffice for present purposes.

33-8. PRESUMPTIONS

It was suggested earlier that the hearsay rule is perhaps the one rule of evidence most recognized by name among laymen. Few non-lawyers, though, would profess much knowledge of the arcane complexities of the hearsay rule. The evidence rules called "presumptions" may be only slightly less known by name, but in contrast to the hearsay rule are widely assumed to be understood. Generally speaking, though, presumptions are grossly misunderstood. A leading writer on the principles of evidence characterizes presumption as

> ... The slipperiest member of the family of legal terms, except its first cousin, "burden of proof," .... Agreement can be secured to this extent, however: a presumption is a standardised practice, under which certain oft-recurring fact groupings are held to call for uniform treatment whenever they occur, with respect to their effect as proof to support issues.[72]

The reasons for the creation and use of presumptions in the law include such factors as probability and judicial economy. If the issue in an action is whether a letter giving notice of a matter was received, and the burden is on the sender to prove its receipt, experience tells us that if the letter was duly posted, it was probably delivered. Thus, the presumption that a letter duly mailed was re-

ceived was created for its probability, not to mention the saving of the court's time in hearing additional evidence that would tend to prove the letter's receipt.

An additional consideration in the creation of presumptions is fairness to the party with the burden of producing evidence on the issue. One who posts a letter, unless it is returned as undeliverable, ordinarily has little means of ascertaining whether the letter was in fact received. If it is his or her burden to show the letter's receipt, it is only fair that he or she enjoy this presumption.

Thousands of words have been spent in refining the definition and role of presumption in the law.[73] To understand the modern and generally held concept of presumptions, it is instructive to examine the work of the drafters of the California Evidence Code, which became law in 1967. California Evidence Code, Section 600(a) defines a presumption as an "an assumption of fact that the law requires to be made from another fact or group of facts found or otherwise established in the action." Section 600(a) also states "a presumption is not evidence." Also, Section 140 of the California code limits "evidence" to matters "offered" as proof, thus reinforcing the notion that presumptions are not evidence.

Just what is the effect of a presumption? On reflection, it should be plain that there are several ways a presumption can operate. It can serve to foreclose the issue, permitting no evidence to be received that would tend to dispute the presumed fact. This is called a "conclusive" presumption. The prevailing views, however, hold that a presumption is "rebuttable" and operates either to shift the burden of production or burden of persuasion on an issue.

As the passages set forth below indicate, the reason for presumptions affecting the burden of production is merely to further the policy of expediting the trial of cases, and presumptions are selected for this policy only if they serve no loftier policy, such as the policy of promoting the stability of titles. The California statute provides:

> A presumption affecting the burden of producing evidence is a presumption established to implement no public policy other than to facilitate the determination of the particular action in which the presumption is applied.[74]

To advert to McCormick's rationale for presumptions, those affecting the burden of production only are simply designed to dispense with superfluous proof. The drafters of the California Evidence Code adopted this reasoning:

> Typically, such presumptions [affecting the burden of production] are based on an underlying logical inference. In some cases, the presumed fact is so likely to be true and so little likely to be disputed that the law requires it to be assumed in the absence of contrary evidence. In other cases, evidence of the nonexistence of the presumed fact, if there is any, is so much more readily available to the party against whom the presumption operates that he is not permitted to argue that the presumed fact does not exist unless he is willing to produce such evidence. In still other cases, thee may be no direct evidence of the existence or nonexistence of the presumed fact; but, because the case must be decided, the law requires a determination that the presumed fact exists in light of common experience indicating that it usually exists in such cases.[75]

On the other hand, a presumption affecting the burden of persuasion (or "proof") is, in the words of the California Evidence Code:

> A presumption established to implement some public policy other than [merely] to facilitate the determination of the particular action in which the presumption is applied, such as the policy in favor of establishment of a parent and child relationship, the validity of marriage, the stability of titles to property, or the security of those who entrust themselves or their property to the administration of others.[76]

The Official Comment to California Evidence Code, Section 605 helps to explain the nature of a presumption affecting the burden of persuasion, and those characteristics of policy that distinguish it from a presumption affecting merely the burden of production:

> Frequently, presumptions affecting the burden of proof [persuasion] are designed to facilitate determination of the action in which they are applied. Superficially, therefore, such presumptions may appear merely to be presumptions affecting the burden of producing evidence. What makes a presumption one affecting the burden of proof is the fact that there is always some further reason of policy for the establishment of the presumption. It is the existence of this further basis in policy that distinguishes a presumption affecting the burden of proof from a presumption affecting the burden of producing evidence. For example, the presumption of death from seven years' absence (Section 667) exists in part to facilitate the disposition of actions by supplying a rule of thumb to govern certain cases in which there is likely to be no direct evidence of the presumed fact. But the policy in favor of distributing estates, of settling titles, and of permitting life to proceed normally at some time prior to the expiration of the absentee's normal life expectancy (perhaps 30 or 40 years) that underlies the presumption indicates that it should be a presumption affecting the burden of proof.[77]

Thus, the furthering of some policy other than the policy favoring expeditious resolutions of lawsuits is the distinctive characteristics of a presumption affecting the burden of persuasion.

In addition to the larger classification of presumptions called "rebuttable presumptions," a separate but small category of "conclusive presumptions" is recognized in California and several other jurisdictions. Conclusive presumptions, again, are those that simply establish something as a given fact regardless of conflicting evidence. In effect, such presumptions are really substantive rules of law that the parties cannot contradict by introducing evidence.

33-9.   PRIVILEGES

The rules of privilege, like the rules excluding hearsay and irrelevant evidence, operate to exclude evidence from introduction in the courtroom. Unlike the rules of relevance and hearsay, however, these rules were not developed for the purpose of assuring the admission of only probative and reliable evidence. On the contrary, evidence excluded because of a privilege often constitutes the most trustworthy and probative evidence on its point. Rather, the rules of privilege were shaped to protect certain confidential relationships from breach. In general, it may be said that the law deems the need for confidentially in certain relationships—marriage, for one—more important than the need for probative evidence at trial.

When a privilege not to disclose information exists, it applies to all phases of litigation. A client, e.g., who exercises the privilege not to disclose confidential advice from his or her lawyer cannot be compelled to divulge the information while testifying at trial, nor in response to interrogatories, nor during deposition.

It is useful at this point to characterize some of the specific privileges, such as the work-product and attorney-client privilege, which are common to most American jurisdictions today, and describe the manner in which they are invoked.

33-9-1.   The Attorney-Client Privilege

This privilege, long recognized by the common law, is not uncommonly interpreted by laymen too broadly, chiefly by the failure to distinguish between a client and consultant. In brief, it is the privilege that attaches to confidential communications between lawyer and client. It is necessary to examine each element of that short definition. California Evidence Code, Section 952 contains a typical definition

of a "confidential communication between client and lawyer":

> [I]nformation transmitted between a client and his lawyer in the course of that relationship and in confidence by a means which, so far as the client is aware, discloses the information to no third persons other than those who are present to further the interest of the client in the consultation or those to whom disclosure is reasonably necessary for the transmission of the information or the accomplishment of the purpose for which the lawyer is consulted, and includes a legal opinion formed and the advice given by the lawyer in the course of that relationship.[78]

Thus, information conveyed by clients to their attorney in the presence of opposing parties to a negotiation is not privileged, nor is advice given by the lawyer in the presence of such persons. On the other hand, the presence of the attorney's secretary or clerk, who may be there for a purpose of assisting the lawyer while such communications are made, does not vitiate the privilege.

Because of the difficulty of having to prove affirmatively that a communication was made in confidence, most jurisdictions indulge in a presumption that a communication made between an atttorney and client was a confidential one.[79] It is then incumbent on the party who would compel disclosure to establish that no confidentiality attached to the communication.

But it should not be thought—and this is a somewhat subtle point—that simply because a client communicates certain information to his or her attorney in confidence and in the course of the attorney-client relationship, the client is thereby immunized against having to testify about that information. A person's knowledge of certain facts, which is not privileged, does not subsequently become so simply by being communicated to a lawyer. Land surveyors who perform certain operations in the course of conducting a survey and are later sued for negligence should, to assure the best possible representation from their attorney, explain thoroughly their actions in conducting the survey. But by so disclosing these matters to their attorney, they are not thereby privileged from testifying about them when called as a witness. Surveyors must testify to their operations in the field and office, for in those respects they are merely testifying to matters of which they have firsthand knowledge. They may not be compelled to testify as to what they *told* their lawyer—for that is protected by the attorney-client privilege—but if we assume that they told their lawyer the truth, it becomes simply a matter of the form of the question. The examiner may properly ask, "Please explain how you determined that this is the true location of the quarter-corner." The examiner may *not* ask, however, "Please tell the court what you told your lawyer about your determination of the true location of the quarter-corner."

The concept of the "privilege-holder" is important in understanding the nature of a privilege. The holder of the attorney-client privilege, e.g., is the client, not the lawyer. If the client asserts his or her privilege, the lawyer may not breach the confidence and disclose the communication unless the legal services were sought or obtained to enable or aid someone to commit or plan a crime or fraud.[80] Conversely, if the client voluntarily waives his or her privilege, the attorney no longer has ground to refuse to disclose the matter in question.

It should be plain that experts who are consulted by an attorney do not thereby have an attorney-client relationship, such that their communications are privileged. The "client" in such a situation is the person on whose behalf the attorney consulted the experts, not the experts themselves. These communications may nonetheless be protected by the attorney-client privilege, notwithstanding that the experts are not the clients of the attorney. If the true client has conveyed certain information to his or her attorney that meets the requirements of the privilege, and the attorney in turn discloses this information to the experts, and if that disclosure "is reasonably necessary for the . . . accomplishment of the purpose for which the lawyer is consulted," many jurisdictions hold that it is protected every bit as much as the original communication from the client to the attorney.[81] Thus, if a client has consulted an attorney with respect to a property-boundary problem, and the attorney requires the advice or services of a land surveyor to properly counsel his or her client, such a communication would presumably be privileged. It should be pointed out that such communications between an attorney and expert are generally also protected by the attorney's "work-product" privilege, discussed below. One salient characteristic distinguishing the work-product privilege from the attorney-client privilege, it should be noted, is that the holder of the work-product privilege is generally the attorney, not the client.

33-9-2. The Privileges for an Attorney's Work Product: An Absolute and a Qualified Privilege

The privilege for an attorney's work product, longer recognized by the English courts[82] than the American courts, is too often confused with the attorney-client privilege. One helpful way for experts to keep the two discretely in mind is to recall that this privilege, not the attorney-client privilege, *may* make their consultations to a lawyer privileged against disclosure. Also, as mentioned above, another distinguishing characteristic of the privilege is that its holder is the lawyer, not the client. But the point that ought to be impressed is the reason and nature of this privilege: to protect from disclosure the attorney's efforts in preparing cases so that clients have the benefit of thorough, considered preparation. Imagine the furtive doings of lawyers who feared that they could be deposed to tell all of the weaknesses in their cases.

33-9-3. The Privilege Against Self-Incrimination

It has long been established in America constitutional law that a defendant in a criminal case has a privilege (1) not to be called as a witness and (2) not to testify.

33-9-4. Other Privileges

There are a number of other privileges, some created by statute and some by court decisions, that arise out of family or professional relationships. Among these are fundamental privileges, recognized in most jurisdictions, that arise from the marriage relationship, confidential communications between a patient and his or her physician, a clergyman and a penitent, and other relationships. Privileges have also been recognized for such diverse matters as how one voted at a public election by secret ballot[83] and for trade secrets. [84] Due to their limited applicability to land cases, these privileges are not discussed in detail here.

33-10.   JUDICIAL NOTICE

It should, on reflection, seem self-evident that certain matters ought to require no formal proof. There are 365 days in a year; February has 28 days except in leap years; an event that occurred on December 25 occurred on Christmas. These facts are said to be "judicially noticed"; their notoriety is such that the court will find them to be so, or direct the jury to, dispensing with the need for proof.

But what are the limits of what a court may judicially notice, and what predicates are there for its taking notice of certain facts, without the usual strict requirements of proof? This subject, while seemingly simple, is nonetheless one of the most elusive subjects of evidence law. If this section of the chapter confounds more than it enlightens, perhaps it does so because the law of judicial notice is more perplexing than illuminating. So if the following remarks raise more questions than they answer, perhaps they are apt.

A beginning must be found somewhere, and perhaps the standard legal dictionary is as good a place as any. The definition of judicial notice contained in *Black's Law Dictionary* gives a glimpse of the concept of judicial notice, but as discussed below, is perhaps too restrictive by modern standards:

Judicial Notice. The act by which a court, in conducting a trial, or framing its decision, will, of its own motion, and without the production of evidence, recognize the existence and truth of certain facts, having a bearing on the controversy at bar, which, from their nature, are not properly the subject of testimony, or which are universally regarded as established by common notoriety, e.g., the laws of the state, international law, historical events, the constitution and course of nature, main geographical features, etc. [Citations.] The cognizance of certain facts which judges and jurors may properly take and act upon without proof, because they already know them. [Citation.]

The true conception of what is "judicially known" is that of something which is not, or rather need not be, unless the tribunal wishes it, the subject of either evidence or argument. [Citation.] The limits of "judicial notice" cannot be prescribed with exactness, but notoriety is, generally speaking, the ultimate test of facts sought to be brought within the realm of judicial notice; in general, it covers matters so notorious that a production of evidence would be unnecessary, matters which the judicial supposes the judge to be acquainted with actually or theoretically, and matters not strictly included under either of such heads. [Citation.][85]

Some of the respects in which this definition is today unduly restrictive are these: (1) Judicial notice of certain facts may be obtained upon motion of a party, and not merely upon the court's own motion; (2) frequently even the production of evidence is required before judicial notice will be invoked; and (3) facts that are (a) susceptible of immediate and accurate determination and (b) not subject to reasonable dispute may be judicially noticed in many jurisdictions, as much as facts that are already notorious.

An 1896 California case, reported in the official judicial reports of that state, contains an account of the use of judicial notice. People v. Mayes[86] was a prosecution for the theft of a steer. A witness for the defendant testified that he observed someone other than the defendant steal the steer on the night in question "betwixt 9 and 10, I suppose." The judge in his charge to the jury instructed them that as a matter of judicial knowledge, the moon on that night rose at 10:57 PM. The defendant, trying to place the time of moonrise as early as possible, had sought by affidavit to show that the moon on the night in question rose at 10:35 PM. The California Supreme Court, however, held that the judge, by consulting reliable calendars and almanacs, had properly taken judicial notice of the correct time of the rising of the moon.[87]

This principle of judicial notice, as you have it, certainly seems so appropriate, so self-evidently necessary for the efficient trial of cases, that its limits could be dictated simply by the exigencies of the individual case. This should be plainly seen, from the following example, not to be the case. A quite comparable principle was invoked by the ecclesiastical jurists who presided at the trial of Galileo; then it was also invoked on a matter of astronomy. From *The Times of London*, October 23, 1980, we have the report headlined "Vatican Is To Reexamine Galileo's Heresy." The short item reads:

> The Pope has called for a fresh inquiry "with full objectivity" into the case of Galileo, the great astronomer and mathematician imprisoned by the Catholic Church in the 17th Century for being "vehemently suspected of heresy" [for having agreed with Copernicus that the earth revolves around the sun].[88]

The ecclesiastical court had but taken what we would call judicial notice that the sun revolves around the earth.

This story should illustrate then not only the potential usefulness of judicial notice, but also the need for caution in its invoking. The determination of a court to take judicial notice of facts is a matter of some discrimination, entailing distinctions not unlike those found in the formulation of the hearsay rule and its exceptions, but more often overlooked.

Those who are reflected on the rules of evidence thus far will be acutely aware of the need for a principle such as judicial notice to establish critical points of a case. The principle is elusive, however, and may be more readily understood from actual examples than abstract statements of principle. With respect to the surveys and methods of disposal of the public lands, a broad variety of records of the General Land Office (GLO) and the Bureau of Land Management (BLM), including plats, field notes, etc., are subject to judicial notice.[89] Other documents of relevance to land and boundary cases that courts have traditionally noticed include topographic maps of the Department of Interior,[90] reports of the California Debris Commission and the U.S. Bureau of Reclamation,[91] segregation surveys of swamp and overflowed lands (the government's official record of those lands granted to the states by the Arkanas Swamp Lands Act of 1850),[92] and records, including hydrographic and topographic surveys, for the National Ocean Survey (NOS) and its predecessor organizations (the U.S. Coast Survey and, later, the U.S. Coast and Geodetic Survey).

Having considered the kinds of documents of which the courts have taken judicial notice and bearing in mind the rationale of the hearsay rule, we may now appropriately consider whether the *truth* of matters asserted in such documents is judicially noticed equally with the documents themselves. That is to say, it is one thing to take judicial notice of the charts and descriptive reports of the U.S. Coast Survey; it is another to take judicial notice of the *truth* of all matters asserted within those documents. In Beckley v. Reclamation Board, the California Court of Appeals in 1962 wrote:

> It is, of course, true that the courts take judicial notice of all matters of science and common knowledge, and of the reports of the California Debris Commission referred to above; that is to say, we take judicial notice of the fact that the reports were made, and of their contents. We do not, however, take judicial notice that everything said therein is true. These reports are based upon studies made by engineers with opinions and conclusions drawn from those studies. But engineers are not infallible, nor are all statements contained in the reports, even those stated as facts, irrefutable. (If they were, then there could have been no justification for the Grant Report which modified the Jackson report, corrected errors therein, and changed the physical plan [for a flood control project] in several material respects where experience and further studies have proved earlier engineers' assertions of "fact" and their conclusions faulty....) To assert the immutableness of statements in official documents would constitute abdication by the courts in favor of adjudication by engineering fiat. [Citations omitted.][93]

Needless to say, the distinction between taking judicial notice of an official document and its contents and taking notice of the truth of matters asserted there is frequently lost. Courts frequently make findings of fact based solely on items contained in judicially noticed official documents. Thus, as mentioned above, courts have found that the moon arose at a particular time because it was so reported in an official document. Curiously, it has been noted that "nowhere can there be found a definition of what constitutes competent or authoritative sources for purposes of verifying judicially-noticed facts."[94]

33-11. THE OPINION RULE

Ordinarily, witnesses must be shown to have personal knowledge of the matter they are to testify about (this personal knowledge is some- times called the "foundational fact"), and they must confine their testimony to what they have perceived. Inferences from those perceptions are for the jury to draw. Stated conversely, the general rule is that witnesses may not testify about their opinions. This rule is an ancient one; at common law witnesses were to testify only to "what they see and hear." Wigmore quotes Lord Coke in 1622 writing, "[i]t is no satisfaction for a witness to say that he 'thinketh' or 'persuadeth himself'."[96]

Legal commentators have written a great deal on what constitutes an "opinion," by which is meant a conclusion of inference, and of course on whether there ought to be a rule excluding such evidence. Is it an opinion, e.g., to say that a person was drunk? Is it an opinion to say that a fence was not fit to keep stock off of land?[97] The distinction between fact and opinion is a rarified one and, on reflection, appears perhaps to be purely artificial. To say there were "remnants" of a fence post is to draw a conclusion from certain observations; so is it to say that the fence post was of redwood; or that it was in a state of decomposition from water saturation and not having been painted with preservative; or that the monument set in 1922 was to mark a property corner? In formulating a conclusion in each of these examples, the speaker considers not only what he or she actually saw, but also additional data from a personal store of knowledge. In the first example, the speaker recalls and considers how fence-post remnants appear; in the second, considers what redwood looks like; in the third, considers the effects of not treating buried wood; and in the fourth, takes into account the acts of a certain land surveyor in 1922. With the possible exception of what partly rotted wood looks like, much of this knowledge may not be shared by the population at large. That is to say, were another to see what the witness saw, the testimony of that person might well be different.

In the last statement of the hypothetical witness given above, he or she is uttering what most people "intuitively" know to be "opinion" as opposed to fact. But to truly relate mere fact, in the sense of only what was observed, the speaker would have to convey raw sensory data—perhaps not even in words. For there are no words to convey only what was perceived, unencumbered by data learned elsewhere—e.g., that an object with certain visual and palpable characteristics is wood. Until we learn to transmit raw sensory data directly from the observer to the trier of fact, it is perhaps helpful to bear in mind that the distinction between fact and opinion is artificial, though useful.

33-12.   EXPERT TESTIMONY

The traditional exception to the common law's prohibition on opinion evidence was the testimony of experts, which was originally permitted only in the rarest of circumstances, when the expert's opinion was grudgingly felt essential to a decision. Today, the rules respecting the admissibility of an expert's testimony, whether in the form of an opinion or otherwise, are codified in most jurisdictions. It is convenient then to consider two codifications of these principles. The Federal Rules of Evidence, e.g., provide:

> If scientific, technical, or other specialized knowledge will assist the trier of fact to understand the evidence or to determine a fact in issue, a witness qualified as an expert by knowledge, skill, experience, training, or education, may testify thereto in the form of an opinion or otherwise.[98]

California Evidence Code, Section 801, which is similar to the rules of most states, provides:

> If a witness is to testify as an expert, his testimony in the form of an opinion is limited to such an opinion as is:
>
> (a) Related to a subject that is sufficiently beyond common experience that the opinion of an expert would assist the trier of fact....[99]

33-12-1.   Expert Nonopinion Testimony

One noteworthy aspect of the federal rule is that it concludes the expert "may testify...in the form of an opinion *or otherwise*." Certainly, land surveyors, when they have been qualified as an expert and other predicates have been shown, may testify that in their opinion the section corner in question was originally set at a certain location. It is certainly proper for surveyors, on other occasions, to testify to the customary surveying methods in seeking to recover original corners, the contents of applicable instructions from the GLO to U.S. deputy surveyors, etc. Notwithstanding the nearly metaphysical problems associated with the distinction between opinion and fact, such matters are not what the law considers "opinions." However, they may be of even more value in enabling the trier of fact to reach a decision than the ultimate opinion of an expert, such as a land surveyor. In many cases, it may be unnecessary for the land surveyor even to render an opinion; it may suffice simply to relate to the court some segment of his or her store of specialized knowledge.

33-12-2.   A Precondition: the Need for Expert Testimony

Whether an expert will be permitted to testify depends in the first instance on the need of the trier of fact to hear such testimony:

> There is no more certain test for determining when experts may be used than the common sense inquiry whether the untrained layman would be qualified to determine intelligently and to the best possible degree the particular issue without enlightenment from those having a specialized understanding of the subject involved in the dispute.[100]

Both Federal Rule 702 and Section 801 of the California Evidence Code contain this requirement. When expert testimony, whether in the form of opinion or not, is excluded, it is usually because it is unhelpful and therefore a waste of the court's time.[101]

33-12-3. A Foundational Fact: The Expert's Qualifications

To testify as an expert, whether to matters of opinion or merely matters within his or her specialized knowledge, such as scientific procedures, a witness must first be shown to be "qualified" to give such testimony. In a typical formulation of the required qualifications of an expert, Federal Rule 702 specifies that the requisite expertise may be the product of "knowledge, skill, experience, training or education." Most states describe expert qualifications in similar terms.[102] Under these criteria, it is certainly not a difficult matter to qualify a witness as an expert.[103]

As a matter of procedure, the qualification of an expert is a "preliminary" or "foundational" fact that must be determined by the judge, who is vested with a broad discretion in making such determinations.[104] The party offering the witness as an expert has the burden of proving that the expert is qualified (i.e., unless the opposing party has acquiesced to the witness's testimony by failing to object). Ordinarily, the jury will not be informed of the fact that the court has determined the witness to be qualified as an expert.[105] Obviously, a jury is not bound to accept the opinion of an expert, and the weight to be given to the opinion is solely within the jury's province.[106]

In determining whether a witness is qualified to testify as an expert, the trial judge typically considers the following kinds of factors: (1) the number of experts in the field; (2) whether the type of testimony involves generally recognized and related scientific fields; (3) whether claims of accuracy by the witness have been substantiated by accepted methods of scientific verification; (4) whether the expert's background, whether based on education,

practical experience, etc., is sufficient to enable him or her to testify; and (5) whether the expert's background relates sufficiently to the type of testimony to be given.[107] Proffered testimony of a land surveyor as to the customary practices of lending institutions, e.g., will most likely be found to be beyond his or her experience and, thus, inadmissible.

An expert need not be academically trained in the subject matter of his or her expertise to be qualified to testify; the expertise may be the product of skill, experience, or training obtained outside of an academic environment. In one California case, the testimony of an FBI expert who did not have a university degree, but did have considerable experience in fingerprint identification, was allowed.[108] In another, a product salesman was permitted to testify as to the value of the product, even though he was not an appraiser by profession.[109] In addition, there is no requirement that one be licensed to engage in the business in which he or she is expert in order to give expert testimony. It is conceivable, e.g., that one for whatever reason has never troubled to obtain a state license to practice land surveying may be an expert in corner recovery.

33-12-4. The Bases for an Expert's Opinion

Another topic of expert testimony that has received much attention is: What matters may experts properly use as the *basis* for their opinion testimony? Federal Rule 703 provides an answer:

> The facts or data in the particular case upon which an expert bases an opinion or inference may be those perceived by or made known to him at or before the hearing. If of a type reasonably relied upon by experts in the particular field in forming opinions or inferences upon the subject, the facts or data need not be admissible in evidence.[110]

Essentially, then, experts may base their opinion on three sources of information. First, they may consider what they have perceived with their own senses. A treating physician testifying at a trial is an example of an expert relying on what he or she has perceived. Likewise, a land surveyor who has made a field survey has personal observations on which to base testimony. The second basis consists of facts made known to experts at or before the hearing. This kind of information is frequently presented to the exert in the form of the much maligned "hypothetical question." In this situation, expert witnesses are asked to assume as true a series of facts (evidence of which must, as a general rule, have been received in court) and then to render their opinion on the basis of those assumptions. Finally, the experts may rely on matter that may be "reasonably relied on" by other experts in their area. Physicians, e.g., may rely on statements made to them by their patients concerning the history of their condition.[111] In appropriate cases, the physician may also rely on the reports and opinions of other physicians.[112] Experts on real property values may rely on inquiries they have made of other people, commercial reports, market quotations, and relevant sales of which they have learned, for the purpose of rendering an opinion on the value of property being condemned.[113]

It should be noted that, under the latter category, experts may base their opinion testimony on matters that, under other circumstances, would not be admitted in evidence because they constitute hearsay. For example, all the information obtained in the preceding example by the expert on real property values from other persons, commercial reports, etc. is hearsay. Are there any restrictions on experts' use of hearsay as a basis of their testimony? There is no general rule that governs this issue, except the test whether it is acceptable practice in the expert's field to rely on such statements.[114]

On what kinds of matter may land surveyors reasonably rely in their practice of land surveying and thus use as a basis for their testimony in court? There are certain indispensible mat-

ters, such as plats, field notes, contracts, and special instructions for original government surveys, that are considered reasonable bases of opinions. Recognized texts (such as that by Bouchard and Moffitt) likely fall into this category as well. In most circumstances, however, the determination must be made on a case-by-case basis. If the surveyors attempt to locate a bearing tree noted in the field notes of an 1869 survey and find nothing but new-growth trees in the terrain, may they rely on the community "reputation" of a devastating forest fire in 1940? May the rely on newspaper accounts of such a fire to account for the inability to find the bearing tree? Similarly, in an accretion-avulsion case, may they rely on newspaper or diary accounts of a great flood to form a basis for their opinion as to the manner in which a river changed its course. These are the kinds of questions that need to be pondered in advance by both the attorney and expert witnesses preparing to testify. It is largely a question of custom among experts in the particular field, and it is impossible to dogmatize about the kinds of matter that will be permissible bases for opinions in any particular case.

33-12-5. Miscellaneous Observations on Expert Testimony

It should be added that, contrary to an ancient and discredited, yet frequently repeated dictum, it is permissible for experts to give their opinion on an "ultimate issue" in the case. Thus, if the question in a trial is the true location of the original corner between sections 1, 2, 11, and 12, a qualified land surveyor may offer his or her opinion as to its location, notwithstanding that the issue is the ultimate question in the case.[115]

Questions asking, "Is this possible that...," while once of dubious propriety when posed to an expert, are now accepted in most jurisdictions as proper.[116]

Many court decisions have addressed the proper scope of expert testimony to be given in land cases by surveyors and civil engineers.[117] In a California case requiring the location of the "high water mark" boundary, the expert testimony of land surveyors that the water of the river had washed a smooth, visible line along the base of the riverbank, below which it was devoid of vegetation, was held competent evidence.[118] On the other hand, it has been held that a trial court erred in admitting the testimony of a surveyor that an arch projected less than 3 in. over the boundary line of a lot, when the survey was not actually made by the witness but rather by his employees, and when he was present only for a few minutes at the beginning of the survey, which took 6 or 7 hr.[119] It is not possible to generalize about the circumstances in which an expert may testify on the basis of work performed by subordinates, but this situation is always potentially troublesome and should be considered in advance of trial. Broadly speaking, the practice of having such work done by subordinates and the degree of supervision are salient factors in determining the admissibility of the testimony.

In general, the testimony of a competent surveyor, who in surveying the location of a line followed the rules of the federal land office, is admissible to show the location of a disputed boundary line of a government section.[120] Also, as may seem obvious, the testimony of a surveyor concerning the location of lands within a section is not made inadmissible simply because the surveyor did not survey the entire section, when no such survey was essential to locate the lands in question.[121]

There are, as in all fields of law, anomalous cases. One court remarked, erroneously it seems clear, that opinion evidence of the location of a boundary line is inadmissible. (That court did, however, hold that allowing such testimony was not a sufficient reason for reversal.[122]) Another court has similarly held that it was "harmless error" to allow the admission of testimony of a surveyor that his or her survey was correct and another was not.[123]

33-13. PRETRIAL: THE DISCOVERY PHASE

With this theoretical background, it is useful to look at some of the procedures leading up to the trial of a land case, focusing particularly on the "discovery phase" of the pretrial. It is frequently during this phase that land experts are consulted by one of the parties and become participants in the discovery process.

Discovery is a generic term used to describe pretrial procedures that enable each party to learn a number of things about their opponent's case. Through discovery, a party can learn the contentions an opponent makes, facts and witnesses known to the opponent, and documents in the opponent's possession. What follows is a brief discussion of the most commonly used discovery devices. It should be emphasized that to be effective, discovery must be conducted meticulously. Several examples will help to show how land experts, working with their attorney, can help to assure that discovery yields truly helpful information and is not squandered.

33-13-1. Interrogatories

Interrogatories are written questions one party asks of (or "propounds to") another party, which must be answered under oath. (They may not be sent to a witness who is not a party.) They may inquire whether the party makes a certain contention: "Do you contend that the corner common to township 3 south, range 2 west, and township 4 south, range 3 west, Mt. Diablo base and meridian, as set by the Deputy U.S. Surveyor, has been obliterated?" Interrogatories may ask what facts the opponent relies on for making a particular contention: "If your answer to the preceding interrogatory is in the affirmative, please describe all facts that you contend support your contention that the corner has been obliterated." Interrogaories may inquire whether the opponent knows of any documents pertinent to the lawsuit: "Please identify all documents that you contend support your contention that the corner has been obliterated." Interrogatories are also used to learn the names and addresses of witnesses known to the other party: "Please identify by name and last known address all persons with knowledge of the sale of Blackacre by Mr. Green to Mr. Snowden."

In land cases, the expert is frequently asked to help draft interrogatories to the opposing party and to respond to interrogatories served ("propounded" again is the usual expression) by the opposing party. A cooperative effort by both the lawyer and expert generally will produce the best results; the lawyer will be most familiar with the legal issues and principles, whereas the land expert will know more of the technical, scientific, and practical aspects of the particular subject matter. Both are needed in preparing useful interrogatories. Take the interrogatory concerning the township corner used in the preceding example and suppose it read, "Do you contend that the corner common to [the description should be precise so that there is no question what corner is referred to] is lost or obliterated?" Your opponent responds, "Yes." At this point, you do not know whether the opponent contends the corner is obliterated or he or she contends it is lost. This problem may be the result of poor drafting (usually the fault of the attorney) or ignorance of the distinction between a lost and an obliterated corner (about which the expert could have educated the attorney).[124] In other instances, the land expert may be able to suggest whole lines of inquiry that, otherwise, would remain unknown to the lawyer.

Of interrogatories and discovery mechanisms in general, it can be said that the scope of permitted inquiry is far broader than the limits of relevance at trial. At trial, evidence and testimony must be "relevant," which for *that* purpose is defined as tending to prove or disprove a fact in issue. During discovery, on the other hand, a matter is "relevant" if it pertains "to the subject matter of the action" or is "reasonably calculated to lead to the discovery of admissible evidence."[125] The matter inquired into during discovery need not be

relevant to "the issues" so long as it is relevant to the "subject matter" or is reasonably designed to lead the inquiring party to evidence that he or she could introduce at trial. Also, it is permissible in discovery to elicit hearsay evidence, whether there is an applicable exception or not.[126] To object to an interrogatory of deposition question on the ground that it asks for hearsay (and further, to refuse to answer it for that reason) is to invite a judge to impose "sanctions"—a monetary penalty—as well as award your opponent his or her costs and attorneys fees in securing an answer. However, while it is no objection to a discovery inquiry (such as an interrogatory) that the question calls for hearsay, information that is protected by a privilege is protected from disclosure during discovery to the same extent as during trial.

33-13-2. Depositions

A deposition is a discovery device by which the oral testimony of a witness, under oath, is taken before a shorthand reporter, who later transcribes the testimony. A deposition may be taken of a party or nonparty witness, so long as the witness is within the geographic limits of the court's subpoena power. It is usually taken in the office of the attorney who wishes to take the deposition. Unlike trial, there is no judge present to rule on objections during a deposition.

A deposition, like interrogatories, also differs from courtroom testimony in that the scope of relevance is much broader than at trial. The witness may even give hearsay testimony that would be inadmissible at trial, so long as the testimony is "relevant to the subject matter of the action," or "calculated to lead to the discovery of admissible evidence."[127]

Occasionally, an impasse develops during a deposition, and when the witness's attorney instructs the witness not to answer a particular question, and the attorney asking it insists on an answer. Since no judge is present, such disputes must be taken to court for a decision whether the witness must respond. Occasionally, the deposition is interrupted at this point to allow the attorneys to argue the matter before a judge promptly, sometimes in a matter of days. If the question is not critical to the balance of the testimony, however, the deposition may continue on other points, with the dispute to be argued later, or the disputed question may be ignored. The witness, of course, should always follow his or her attorney's instruction when told not to answer.

It is not often that experts being deposed will be instructed by their attorney not to answer a question. Such an instruction not to answer is usually based on the assertion that the information asked for is privileged, and ordinarily the only such privilege attaching to information in the expert's possession is the attorney's work-product privilege. But for most purposes, that privilege is waived when the experts are identified as potential witnesses for trial, and their discussions and correspondence with their attorney are usually then subject to discovery. It is more often the case that witnesses are instructed not to answer when the witnesses are the clients themselves, and the questions concern information protected by the attorney-client privilege. On rare occasions, witnesses will be instructed not to answer because the inquiry has strayed from the bounds of discovery relevance.

Invariably, surveyors and other experts who are to testify in a land case will be deposed before trial. (Obviously, it can be rather hazardous for an expert to testify at trial if the corresponding expert who will testify for the opposing party has not been deposed. The best experts in any field can never be certain they have overlooked nothing, that there is no plausible way other than their own of interpreting the evidence.) The deposition of an expert is taken—i.e., the questions are asked by the opposing attorney. The surveyor's own attorney will seldom ask questions during a deposition, for he or she has nothing to "discover."

The deposition of a land surveyor, or any expert witness for that matter, generally covers a predictable range of topics, although not necessarily in predictable order. The witnesses are asked their qualifications: education, membership in professional associations, number of years in practice, work experience, professional journals subscribed to, etc. If the case concerns the location of a water boundary, the deponents may be asked how many such problems they have encountered during their career. If the witness is a civil engineer, he or she is frequently asked what percentage of the practice is devoted to land surveying.

Next, the witnesses will typically be asked about their substantive preparation for trial: (1) Specifically, what have they been asked to do in connection with the case? (2) In carrying out their assignment, what materials have they reviewed and what investigations made? (3) What are their conclusions (opinions)? And (4) what reasons do they have for their conclusions?

Obviously, in a complicated case, a thorough deposition of a land expert, which fully covers these four broad areas, may take days. The response to the simple question, "What materials have you reviewed in the course of your work on this case?" may alone take many hours. Almost invariably, the witness is required to being his or her deposition all such materials (or copies of them). The attorney deposing the witness, if thorough, will ask the witness to identify each document, map, etc. and explain its contents and relevance to the case. Each item will be given an exhibit number (plaintiff's exhibit 1, etc.) and made part of the record of the deposition.

Preparing to testify at a deposition is as essential as preparing to testify at trial. Experts will want to adduce every fact and document they have examined and be able to articulate clearly the reasons for their opinions, for if the attorneys have taken the depositions skillfully, they will take care to press the witnesses at every point for a definitive answer: "Are there *any other* materials you have reviewed in the course of your investigation, which you have not identified? . . . Are you certain? . . . If you later recall any you may have forgotten today, will you notify me promptly?" Or, "You have stated three bases for your conclusion. Do you have any other reason for your conclusion? Be certain now—these three only?" If at trial, surveyors produce a survey that they did not have at their deposition or state an additional reason for their conclusion, they may be in for some rough moments. The opposing attorney may try to make it appear the new information was deliberately withheld or recently fabricated.

On the other hand, the deposition may be conducted in an unskillful or unthorough manner. In this regard, the witnesses should recall that they are obligated only to answer the questions asked. If the attorney neglects to inquire into a critical point or press for a witness's reasoning, the witness has no obligation to volunteer information. For this reason, most attorneys instruct their own witnesses who are about to be deposed to listen carefully to the question asked and answer *that question* only.

Land experts are frequently—and prudently—asked to help their attorney prepare to take the deposition of the opposing party's expert. The attorney may have little or no experience with land or boundry problems and, thus, may need a healthy does of education. On the other hand, the attorney may be well versed in the mathematics entailed, but have little knowledge of the historic practices of the GLO, e.g., or rules respecting restoration of lost corners, or the method of capitalizing income. Even when knowledgable about the subject, the attorney will usually find suggestions of his or her expert helpful, much as the suggestions of a law partner on procedural or tactical aspects of the case. The expert should always suggest that the attorney inquire further into those areas that the expert finds troubling or unclear. The deposition will probably be the only occasion before trial to learn what the opponent's expert has found or concluded on the subject.

Depositions, of course, are primarily a discovery device—i.e., a technique to learn what

a witness, whether a party or not, may know about the case, or about matters affecting the case. They do have other uses, however. Ordinarily, depositions may not be read at trial, since the witnesses themselves should testify. But if witnesses contradict their deposition testimony, they may be confronted with it. Also, if the witnesses have died or are otherwise unable to be produced for trial (as when they live beyond the court's subpoena power and refuse to come voluntarily), on proof of their unavailability their deposition may be used at trial.

33-13-3. Requests for Admissions

The request for admissions is another discovery tool the expert may be asked to help prepare. It is frequently the last-used discovery device and is designed to eliminate uncontroverted issues of fact and expedite the conduct of trial. Like a set of interrogatories, a request for admissions is a written document sent to the opposing party. It may request that the opponent admit the genuineness of a document (recall that a document must be shown to be authentic before it may be introduced into evidence) or truth of the statement.

As was explained earlier, the first requirement for the admissibility of a document is a showing that it is genuine—e.g., that it in fact is the plat of the survey in question, the letter it purports to be, etc. Even in a relatively uncomplicated boundary case, there may be scores of documents that a land surveyor and his or her attorney will seek to introduce at trial. The request for admissions may obviate the need for proof of the genuineness of these documents at trial.

The party to whome the request is sent has a specified time to admit or deny the genuineness of each document, usually 30 days.[128] If he or she admits the genuineness of a particular document, there will be no need to establish the document's authenticity at trial. If the genuineness is denied and the document is shown at trial to be authentic, the party who denied the authenticity may be held liable for the cost of establishing it, even though he or she may have won the lawsuit. These costs can be substantial, particularly when proving the authenticity of a document requires producing a witness who lives a great distance from the site of the trial.

The request for admissions can also be used to establish matters of fact. Suppose one issue in a lawsuit is the correct location of a section corner, another the date of death of a certain man, and another the prestatehood status of the state of California. An attorney seeking to streamline the case presentation might propound requests for admission in the following form:

Request for Admission of Fact No. 1. Do you admit that the correct location of the northwest corner of section 1, T.1S., R.24W., Gila and Salt River Meridian, as set by United States Deputy Surveyor John A. Barry in 1902, is as shown on Exhibit A attached hereto? [Exhibit A is a plat of a survey conducted by the surveyor for the propounding party, showing his placement of the corner in question.]

Request for Admission of Fact No. 2. Do you admit that Gideon Lightfoot, whom you assert to be a predecessor in interest of Plaintiff, died on August 4, 1912?

Request for Admission of Fact No. 3. Do you admit that prior to its admission to statehood on September 9, 1850, the State of California had never enjoyed the status of a territory of the United States of America?

If a request for admission is admitted, there will be no need to establish the fact at trial; the attorney will instead read the admission to the jury or judge. On the other hand, if the request for admission of fact is denied, the truth or falsity of the fact in question will be resolved at trial based on the evidence presented. And, as mentioned above, if the fact is established at trial, the denying party may be charged with the cost of proving the fact. This economic incentive naturally tends to cause a party to take seriously a request for admission.

33-13-4. Inspection of Land

Another discovery device that is frequently employed in land cases is the request to inspect property. In many states, land surveyors do not require permission of landowners to enter their property when the surveyors are conducting a survey.[129] In jurisdictions where a surveyor does not enjoy this privilege, it will be necessary for the attorney to request that the surveyor be allowed onto the property. If the request is denied, the court will invariably order that the surveyor be allowed to enter the property, unless the request is clearly frivolous.

NOTES

1. 28 Journs of Cong. 375–381 (1785). Jefferson was chairman of the committee appointed by the Continental Congress to prepare a plane for the survey and disposition of the public lands. Because of this, he is often mentioned as the inventor of the rectangular system. The prevailing opinion is that the system was not the result of any single individual's thinking. Patton, 1 Land Titles 289 (2nd ed. 1957).

2. McCormick on Evidence, §151, at 314 (1954).

3. Fed. R. Evid. 402.

4. Cal. Evid. Code, §351.

5. McCormick on Evidence, §151, at 314 (1954).

6. Fed. R. Evid. 401.

7. Cal. Evid. Code, §210.

8. See, e.g., Cal. Evid. Code, §352; Fed. R. Evid. 403.

9. Reeve v. Dennett, 145 Mass. 23, 11 N.E. 938, 943 (1887).

10. See, e.g., Ernie v. Trinity Lutheran Church, 51 Cad, 2d 702, 706 (1959).

11. Cal. Evid. Code, §1400.

12. Cal. Evid. Code, §1401(a).

13. Cal. Evid. Code, §644.

14. Cal. Evid. Code, §643.

15. Cal. Evid. Code, §640.

16. Cal. Evid. Code, §1418.

17. Cal. Evid. Code, §1417; Castor v. Bernstein, 2 Cal. App. 703, 705, 84 P. 244 (1906); People v. Storke, 128 Cal. 486, 488, 60 P. 1090 (1900); People v. Gaines, 1 Cal. 2d 110, 115, 34 P. 2d

146 (1934); People v. Davis, 65 Cal. App. 2d 255, 257, 150 P. 2d 474 (1944) (prosecution for maintaining illegal betting establishment; comparison of questioned betting marker with admittedly genuine loan application).

18. Cal. Evid. Code, §1417.

19. People v. Briggs, 117 Cal. App. 708, 711, 4 P. 2d 593 (1931).

20. McCormick on Evidence, §196, at 409 (1954).

21. Cal. Evid. Code, §1500. California Evidence Code, Section 250 in turn defines "writing" to include "handwriting, typewriting, printing, photostating, photographing, and every other means of recording upon any tangible thing any form of communication or representation including letters, words, pictures, sounds or symbols. ... "
    The exception for "collateral writings" is found in Evidence Code, Section 1504, which makes a copy admissible "if the writing is not closely related to the controlling issues and it would be inexpedient to require its original."

22. The general rule is Federal Rule of Evidence 1002; the exception for collateral writings, recordings, or photographs is contained in Rule 1004(4).

23. Model Code of Evidence, Rule 602, comment at 300 (1942); see also McCormick on Evidence, §197, at 410 (1954); 4 Wigmore on Evidence, §1179 (rev. ed. 1974); Cal. Evid. Code, §1500 comment.

24. See 64 Harv. L. Rev. 1369 (1951); McCormick on Evidence, §206, at 419 (1954); 4 Wigmore on Evidence, §1232 (rev. ed. 1974).

25. Pratt v. Phelps, 23 Cal. App. 755, 757, 139 P. 906 (1914); Edmunds v. Atchinson etc. Ry. Co., 174 Cal. 246, 247, 162 P. 1038 (1917); Hughes v. Pac. Wharf Co., 188 Cal. 210, 219, 205 P. 105 (1922); People v. Lockhart, 200 Cal. App. 2d 862, 871, 19 C.R. 719 (1962).

26. Cal. Evid. Code, §§1500, 1550; see Hopkins v. Hopkins, 157 Cal. App. 2d 313, 321, 320 P. 2d 918 (1958); Annot. 142 A.L.R. 1270 (1942); Annot. 76 A.L.R. 2d 1356 (1958).

27. Cal. Evid. Code, §1502.

28. Zellerbach v. Allenberg, 99 Cal. 57, 73, 33 P. 786 (1893); see also Heinz v. Heinz, 73 Cal. App. 2d 61, 66, 165 P. 2d 967 (1946) (original

negatives and prints in photographer's studio in New York).

29. McCormick on Evidence, §202, at 414 (1954); 4 Wigmore on Evidence, §1211 (rev. ed. 1974).

30. McCormick on Evidences, §225, at 460 (1954).

31. Fed. R. Evid. 801(c).

32. Cal. Evid. Code, §1200(a).

33. Fed. R. Evid. 802; Cal. Evid. Code, §1200(b).

34. Cal. Evid. Code, §1200. The federal rules interestingly treat admissions as nonhearsay, as opposed to treating them as exceptions to the rule. Fed. R. Evid. 801(d)(2).

35. Lamar v. Pearre, 90 Ga. 377, 17 S.E. 92 (1892); Smith v. Moore, 142 N.C. 277, 55 S.E. 275 (1906); Carr v. Bizzell, 192 N.C. 212, 134 S.E. 462 (1926); Barlow v. Greer, 222 S.W. 301 (Tex. Civ. App. 1920).

36. Cal. Evid. Code, §1271; Fed. R. Evid. 803(6).

37. Wigmore on Evidence, §1518 (rev. ed. 1974); Radtke v. Taylor, 105 Or. 559, 210 P. 863 (1922).

38. Fed. R. Evid. 803(6).

39. See Cummins v. Pennsylvania Fire Ins. Co., 153 Iowa 579, 134 N.W. 79 (1912), in which the court held that memorandum entered by an agent of the insurance company, since deceased, in his "policy register," were self-serving statements and hence inadmissible.

40. See Annot., 142 A.L.R. 1406 (1939).

41. Bouchard and Moffitt 1959. *Surveying*, 4th ed. New York: Harper and Cullins, p. 2.

42. See, e.g., Cal. Evid. Code, §§1280–1281; Fed. R. Evid. 803(8) and (9).

43. Cal. Evid. Code, §1237; Fed. R. Evid. 803(5).

44. Cal. Evid. Code, §1322; see also Fed. R. Evid. 803(20).

45. See, e.g., Shutte v. Thompson, 82 U.S. (15 Wall.) 151 (1873); Dawson v. Town of Orange, 78 Conn. 96, 61 A. 101 (1905); Mechanics' Bank & Trust Co. v. Whilden, 175 N.C. 52, 94 S.E. 723 (1917).

46. McKinnon v. Bliss, 21 N.Y. (7 Smith) 206, 218 (1860).

47. Montgomery v. Dickey, 2 Yeates 212 (Pa. 1797).

48. Harriman v. Brown, 35 Va. (8 Leigh) 697, 707 (1837).

49. Taylor v. McConigle, 120 Cal. 123, 52 P. 159 (1898); Seaway Co. v. Attorney General, 375 S.W. 2d 923 (Tex. Civ. App. 1964); Adams v. Stanyan, 24 N.H. 405 (1852).

50. Sasser v. Herring, 14 N.C. (3 Dev. L.) 340, 342 (1832); Weld v. Brooks, 152 Mass, 297, 25 N.E. 719 (1890).

51. Morris v. Lessee, 32 U.S. (7 Pet.) 554, 558–559 (1833).

52. Crippen v. State, 80 S.W. 372 (Tex. Crim. 1904); School District of Donegal Township v. Crosby, 112 A. 2d 645 (Pa. Super. 1955); Henry v. Brown, 39 So. 325 (Ala. 1905); Howland v. Crocker, 89 Mass. (7 Allen) 153 (1863).

53. W. Blackstone. 1969. One commentaries. Reprinted from 1899 Cooley edition. (St. Paul, MN: West Publishing, pp. 66–72.

54. See United States v. St. Thomas Beach Resort, Inc., 386 F. Supp. 769, 772–773 (D. Virgin Islands 1974); Application of Ashford, 440 P. 2d 76 (Hawaii 1968); State Highway Commission v. Fultz, 491 P. 2d 1171 (Or. 1971); City of Daytona Beach v. Tona-Rama, Inc. 294 So. 2d 73, 81 (Fla. 1974) (Boyd, J. disenting).

55. Cal. Evid. Code, §1331.

56. The obverse of the rule requiring personal knowledge is one of the older common law rules, i.e., that witnesses generally may not testify as to their opinions. See generally 2 Wigmore on Evidence, §§650–670; 97 C.J.S. *Witnesses*, §52 (1957); 58 Am. Jur. *Witnesses*, §§113–114 (1971).

57. For representative cases, see Barnett v. Aetna Life Ins. Co., 139 F. 2d 483 (3rd Cir. 1943); State v. Dixon, 420 S.W. 2d 267 (Mo. 1967); State v. Johnson, 92 Idaho 533, 447 P. 2d 10 (1968).

58. Donnell v. Jones, 13 Ala. 490, 510 (1848). In the Donnell case, the court held that the opinion of a witness familiar with business matters whether a levy of attachment had destroyed the credit of a business was held properly excluded.

59. Anon. Lib. Ass. 110, 11 (1349), quoted in Phipson, Evidence 398 (19th ed. 1952).

60. See, e.g., Antelope v. United States, 185 F. 2d 174 (10th Cir. 1950); Hancock v. Supreme Council Catholic Benevolent Legion, 69 N.J.L. 308, 55 A. 246 (1903) (witness allowed to testify to the age of an elder brother).

61. Brown v. Mitchell, 88 Tex. 350, 31 S.W. 621, 623 (1895) (the witness was allowed to testify that he was the child of the deceased, notwithstanding that he lacked firsthand knowledge of the sole event that could have provided such knowledge).

62. McCormick on Evidence, §336, at 783, fn. 3 (2nd ed. 1972). Interestingly, assigning the task of producing evidence to the parties, rather than to the judge, is a peculiar characteristic of the common law of England not found in the civil law systems of continental Europe. 9 Wigmore on Evidence, §2483, at 266–267 (3rd ed. 1940).

63. 9 Wigmore on Evidence, §2485, at 271 (3rd ed. 1940).

64. Cal. Evid. Code, §500.

65. McCormick on Evidence, §337, at 788 (2nd ed. 1972).

66. McCormick on Evidence, §337, at 788 (2nd ed. 1972).

67. Morgan, Basic Problems of Evidence 19 (1954). See also: The Murders in the Rue Morgue, *The Complete Tales and Poems of Edgar Allan Poe* (New York: The Modern Library, 1965), p. 159. Mr Dupin says, "You will say, no doubt, using the language of the law, that 'to make out my case, I should rather undervalue than insist upon a full estimation of the activity required in this matter.' This may be the practice in law, but it is not the usage of reason. My ultimate object is only the truth."

68. In *re Winship*, 397 U.S. 358, 364 (1970).

69. Newman, J., in Hoffman v. State, 97 Wis. 571, 73 N.W. 51, 52 (1897).

70. See Livingston v. Schreckengost, 255 Iowa 1102, 25 N.W. 2d 126, 131 (1963).

71. See Burch v. Reading Co., 240 F. 2d 574 (3d Cir. 1957), *cert. den.* 353 U.S. 965 (1957); Sargent v. Massachusetts Accident Co., 29 N.E. 2d 825, 827 (1940); McDonald v. Union Pac. R. Co., 109 Utah 493, 167 P. 2d. 685, 689 (1946); 9 Wigmore, *supra*, §2498, at 326.

72. McCormick on Evidence, §308, at 639 (1954).

73. See, e.g., McBaine, Burden of Proof: Presumptions, 2 UCLA L. Rev. 13 (1954); Lowe, The California Evidence Code: Presump-

tions, 53 Calif. L. Rev. 1439 (1965); Mc-Cormick on Evidence, §309, at 639–643 (1954).

74. Cal. Evid. Code, §603; for discussion, see 53 Calif. L. Rev. 1445 (1965).

75. Official Comment to Cal. Evid. Code, §603.

76. Cal. Evid. Code, §605; see also 53 Calif. L. Rev. 1447 (1965)

77. *Ibid*.

78. Cal. Evid. Code, §952.

79. See, e.g., Cal. Evid. Code, §917.

80. For the crime or fraud exception, see, e.g., Cal. Evid. Code, §956.

81. See, e.g., Cal. Evid. Code, §952.

82. See, e.g., 8 Wigmore on Evidence, §2319, pp. 618–622 (3rd ed. 1940).

83. Cal. Evid. Code, §1050.

84. Cal. Evid. Code, §1060.

85. *Black's Law Dictionary*, 4th ed. 1951. St. Paul, MN: West Publishing, Co., p. 986.

86. 113 Cal. 618, 45 P. 860 (1896).

87. *Ibid*. at 624–625.

88. *The Times of London*, Oct. 23, 1980, col. 5, p. 6.

89. Livermore v. Beal, 18 Cal. App. 2d 535, 541 (1937).

90. Newport v. Temescal Water Co., 149 Cal. 531 (1906); Union Transp. Co. v. Sacramento County, 42 Cal. 2d 235, 239 (1954).

91. Gray v. Reclamation Dist. No. 1500, 174 Cal. 622 (1917).

92. Foss v. Johnstone, 158 Cal. 119, 110 P. 294 (1910).

93. 205 Cal. App. 2d 734, 741–742. For the same proposition, see People v. Long, 7 Cal. App. 3d 586, 86 C.R. 590 (1970); Marocco v. Ford Motor Co., 7 Cal. App. 3d 84, 86 C.R. 526 (1970); B. Witkin, California Evidence, §180, at 167 (2d ed. 1966).

94. Comment, "The Presently Expanding Concept of Judicial Notice," 13 Vill. L. Rev. 528, 545 (1968).

95. Phipson, *Evidence*, 398 (9th ed. 1952), quoting from Anon. Lib. Ass. 110, 11 (1349).

96. 7 Wigmore on Evidence, §1917, at 2 (3rd ed. 1940) quoting from Adams v. Canon, Dyer 53b.

97. Such testimony was held inadmissible in Baltimore & O.R. Co. v. Schultz, 43 Ohio 270, 1 N.E. 324 (1885).

98. Fed. R. Evid. 702.

99. Cal. Evid. Code, §801.

100. Ladd, Expert Testimony, 5 Vand. L. Rev. 414, 418 (1952).

101. See 7 Wigmore on Evidence, §1918.

102. See, e.g., Cal. Evid. Code, §720.

103. The relative ease with which such qualifications can be established is shown by the California cases of Moore v. Belt, 34 Cal. 2d 525 (1949); Hyman v. Gordon, 35 Cal. App. 3d 769, (1973); and Brown v. Colm, 11 Cal. 3d 639 (1974).

104. Cal. Evid. Code, §405; see Putensen v. Clay Adams, Inc., 12 Cal. App. 3d 1062, 1080 (1970); People v. Murray, 247 Cal. App. 2d 730, 735 (1967).

105. Cal. Evid. Code, §405.

106. See, e.g., California Jury Instructions, Civil, 5th ed., No. 2. 40.

107. See People v. King, 266 Cal. App. 2d 437, 443–445 (1968), in which voice prints and testimony relating to them were rejected.

108. People v. Stuller, 10 Cal. App. 3d 582, 597 (1970).

109. Naples Restaurant, Inc. Corberly Ford, 259 Cal. App. 2d 881, 884 (1968).

110. Fed. R. Evid. 703.

111. People v. Wilson, 25 Cal. 2d 341 (1944).

112. Kelley v. Bailey, 189 Cal. App. 2d 728 (1961).

113. Betts. v. Southern Cal. Fruit Exchange, 144 Cal. 402 (1904).; Hammond Lumber Co. v. County of Los Angeles, 104 Cal. App. 235 (1930); Glantz v. Freedman, 100 Cal. App. 611 (1929).

114. That these determinations must be made on a case-by-case basis was noted in Board of Trustees v. Porini, 263 Cal. App. 2d 784, 793 (1968).

115. See Fed. R. Evid. 704; Cal. Evid. Code, §805.

116. See, e.g., Bauman v. San Francisco, 42 Cal. App. 2d 144 (1940); in *Re J.F.*, 268 Cal. App. 2d 761 (1969); Cullum v. Seifer, 1 Cal. App. 3d 20, 26 (1969).

117. See, e.g., Richfield Oil Corp, v. Crawford, 39 Cal. 2d 729 (1952).

118. Mammoth Gold Dredging Co. v. Forbes, 39 Cal. App. 2d 739 (1940).

119. Hermance v. Blackburn, 206 Cal. 653 (1929).

120. Porter v. Counts, 16 Cal. App. 241 (1911).

121. Heinlen v. Heilbron, 97 Cal. 101 (1892).

122. Andrews v. Wheeler, 10 Cal. App. 614, 618 (1909).

123. Tognazzini v. Morganti, 84 Cal. 159 (1890).

124. See generally, State of California v. Thompson, 22 Cal. App. 3d 368 (1971).

125. See, e.g., Cal. Civ. Proc. Code, §2030; Fed. R. Civ. P. 33.

126. Cal. Civ. Proc. Code, §§2016(b); 2030(c); Fed. R. Civ. P., §26(b).

127. See, e.g., Cal. Civ. Proc. Code, §§2019, 2030.

128. See, e.g., Calif. Civ. Proc. Code, §2033; Fed. R. Civ. P. 36(a).

129. See, e.g., Cal. Civ. Code, §846.5.

REFERENCES

BRISCOE, J. 1984. *Surveying the Courtroom*. A guide to evidence and civil procedure. Rancho Cordova, CA: Landmark Enterprises.

BROWN, C. 1994. *Evidence and Procedures for Boundary Location*. 5th ed. New York: John Wiley & Sons.

34

Courtroom Techniques

Walter G. Robillard

34-1. INTRODUCTION

After all phases of surveying, mapping, and other related activities are completed, individuals who are engaged in surveying and mapping sciences may find that their work is just commencing. Whether the individuals conducted a boundary survey, a building stakeout, or prepared a topographic or planimetric map, the product may be questioned by the client or a third party who had no pecuniary interest in the original job. As a result, professionals may be embroiled in a legal action, either as a party or witness defending their work for the client or explaining their actions. In either situation, their professional qualifications and capabilities will be questioned and "hung out to air" for all to see and evaluate.

The proper preparation for courtroom techniques begins with the start of the initial survey or job. No amount of courtroom maneuvering or tactics can explain away a poor survey or partially completed job, yet those who work will at one time or another be called on to explain their work before a tribunal. In any completed job, surveyors will be either defending their work, using the legal system as a shield to defend themselves from others, or using the system as a sword to thrust at their opponent.

This chapter will look at two phases of courtroom techniques.

34-2. PRETRIAL INVOLVEMENT

The question of pretrial involvement depends on which party the surveyor has been asked to represent or the nature of the involvement. If the litigation is initiated as a result of a survey conducted by the surveyor and his or her client initiates the cause of action, the surveyor will appear for the plantiff. All states and the federal code state, "A civil action is commenced by filing a complaint with the court."[1]

Although the complaint is a legal action, it is to the client's benefit that the attorney and surveyor work as a team. The original complaint should contain all the causes of action. Although it is compiled and filed by the attorney, this does not preclude the surveyor from having input into its preparation. Any knowledgeable attorney knows that all the major issues should be addressed, because it may be difficult to amend a complaint at a later date.

867

34-3. IN REM

Since many of the actions are *in rem*,[2] against the land, it is important that the land in question be properly described. Then, if there is any question as to title, the plantiff must recover on the strength of his or her title and not the weakness of the defendant's title. Or the plaintiff may seek equity.[3] It is important that the surveyor understand the elements of partitioning, adverse possession, agreement, acquiescence, and riparian. These should be properly raised in the complaint.

Since land litigation is initiated in the situs or location, usually county, where the land is located, it becomes important that the surveyor inform the attorney of any unique situations, such as the land being located in two or more counties. This situation provides the attorney to *forum shop* to seek the best county for litigation.[4]

34-4. COMPLAINT

The original complaint must contain a brief statement of all possible claims. Once the initial complaint is filed and served, the discovery process is initiated. It is at this time that the defendant may seek to file a cross action or counterclaim. Since a cross action is an action brought by the defendent against the party who is plantiff and on a cause of action growing out of the same transaction that is in controversy,[5] the defendant's or plaintiff's surveyor may once again find his or her work or decisions being questioned. The surveyor for the defendent should be in a position to advise his or her client and the client's attorney relative to descriptions, surveys, and title.

Hand in hand, surveyors may find themselves a party to a counterclaim. The counterclaim's sole requisite is to diminish or defect the plaintiff's demands or claims.[6] This could result with the surveyor becoming a party in his or her own right. If the action were initiated by the plaintiff as a result of a survey performed for the defendant by the surveyor, the defendant in turn can say, "If I am held responsible for legal actions or for damages as a result of work performed for me, by you, I will seek any damages the plaintiff collects from me, for you." The defendant will seek to collect any reward to the plaintiff from the one who performed the survey.

Precluding any cross claims or counterclaims, the surveyor can be very helpful to the defendant in drafting his or her answer. Here the surveyor should be asked to read and determine the adequacy of any descriptions in the complaint for completeness and adequacy. The surveyor's comments should be sought as to lines and evidence of possession.

34-5. DISCOVERY

Discovery is defined as "the ascertainment of that which is previously unknown; the disclosure or coming to light of that previously hidden; the acquisition of knowledge or facts."[7]

Federal Rule 26(b)(1) provides that discovery is

> ...Any matter, not privileged, which is relevant to the subject matter involved in the pending action....

This provides that one party can seek all available information or evidence as possible from the other party. The same rule provides for the production and inspection of evidence, will give access to documents, and the power to test and sample and to enter on land for inspection and to survey.

Along these lines, you may encounter the deposition. Plainly put, you may be deposed to find out what you known, what you did, what you do not know, and what you did not do. It becomes a legal fishing trip. A deposition may be written or oral. Usually it is oral, but it is part of the complete discovery process. If properly used, it can be one of the most effective discovery tools available to the attorney. A deposition may be taken from any person—e.g., a party to the legal action or even a potential witness. Depositions can serve a twofold purpose. They may be used to discover, but can also be an effective tool to preserve testimony. Since depositions are considered as extrajudicial, they usually can not be used as evidence at the trial, except in certain instances when the witness is not available or his or her testimony is to be impeached.

34-6. DEPOSITIONS

If you are asked to give a deposition, it is highly recommended that an attorney accompany you. In most instances, attorneys will try to "beat" their opponents to depositions. This can be good or bad. If your attorney, on the other hand, is to depose a second surveyor, your attorney should seek your input on forming suitable questions. A deposition can be a success or failure. If you give a deposition or even help to frame questions, there should be defined goals. What is it you want to keep the other party from discovering; what is it you want to discover; and what is your opponent trying to discover? When an attorney deposes a surveyor, he or she should have the following five goals or objectives in mind:[8]

1. To gain information, including the identity and location of sources of information
2. To preserve testimony for possible use during trial
3. To find out the depth of the information the witness knows, both about his or her own case and yours
4. To commit the deponant to a version of the facts from which he or she cannot later deviate so as to adapt to a changed complexion of the case
5. To confront the party to be examined with damaging evidence, or otherwise show the party the weakness in his or her case, thereby inducing the party to drop the case or seek a favorable settlement

In giving depositions, surveyors may find that their attorney may devastate them; the attorney preserves his or her objections i.e., puts them off until the end. Considerable time and trouble may be saved by attorneys if they request certain procedures, but they in turn may cause future harm to the surveyor by:

1. Stipulation on signature. This provides for waiving of the witnesses' signature. *Never do this. Always read and then sign. Never waive.*
2. Stipulation of fact. Think before you stipulate any fact. You may lose a lot and gain nothing.
3. Exhibits. You may use exhibits. Do not stipulate exhibits. Have them identified for future use.

Item 2 can have serious implications here at trial if the attorney states, "We wish to stipulate that J. Smith is an expert witness in surveying." This can have a twofold effect. One stipulation that the surveyor is an expert will keep any of his or her qualifications from the record. In doing this, the judge and/or jury will never hear the surveyor's qualifications in relation to the other witnesses. Of course, this could help in that if your qualifications are stipulated, the opposite party cannot object at a later date. The proper methods must be at the discretion of your attorney. By stipulating you give up, but get nothing in return. You are limiting your potential to do future damage.[9] In item 3, if you stipulate as to exhibits and their meaning, you can offer testimony, and the individual who reads your deposition will be able to correlate your exhibits to your testimony.

A competent attorney will prepare you for a deposition. This should include an explanation as to who will be present and the role of each individual. During a deposition, your attorney may object to a question asked of you and will direct you to either answer it or remain quiet and not answer. If directed to remain quiet, *do not answer the question.* The opposing attorney may become demanding, *but do not answer*. Follow the instructions of your attorney. In this "play," your attorney is the director. You are simply the actor.

Your attorney will be permitted to cross-examine you, but for the most part it will only be to clarify some vague point or expand on some chain of thought that was omitted.[10]

Some suggestions for a deposition include:

1. Rely on your attorney. Do not do it alone.
2. Know the ground rules for the deposition.
3. Tell the truth. Even when asked if you have talked with your attorney or client.
4. Answer only the questions asked. Do not offer.
5. Relate only to facts. Do not speculate, give opinions, or make inferences. Do not use "I think," "I guess," or "maybe."
6. Do not interject humor.
7. Watch for trick questions of opinion. Always see the documents.
8. Understand the case. Read the complaint before you come.
9. Be professional at all times.
10. Rely on your attorney. Do not do it alone.

The depth and scope of questions in a deposition are usually more broad than those asked at trial. Since a major purpose of a deposition is for discovery that will lead to evidence, deposition questions are not trial questions for purposes of proving one's case. When you give a deposition, do not try to prove your position or argue as you would in court. The attorney simply wants to probe, punch, and probe again in an attempt to find any weaknesses in your case and your testimony.

34-7. THE TRIAL

In any trial, the winning team is the one in which the attorney, as captain, knows each player's strengths and weaknesses. The attorney capitalizes on the strengths and protects the weaknesses. He or she should draw a game plan, placing you in position where you would best fit in.

Regardless of your position, you, the surveyor, must be used by the attorney, and you must remember that you are playing the attorney's game—according to the attorney's rules —and in the attorney's ballpark. You will not be permitted to inject any of your rules or even complain of a foul.

34-8. EXPERT CATEGORIES

An attorney who uses an expert will place him or her into one of three categories, as follows:[11]

1. The expert who will be used at trial
2. The expert who is a consultant, but who will not be used at trial
3. The expert who has relevant knowledge or he was an actor or viewer, or she may have been involved, but will not testify

If you are to be used as an expert witness at trial, your attorney may be requested to "list" your name or identify you during discovery but is under no obligation to identify you as an expert if not requested to do so. Your attorney must be asked. Since discovery can be limited to include trial witnesses, it is important that each witness to be used at trial be identified at an early stage. Only a poor attorney will attempt to establish or build a case on the opponent's experts.

Once you are identified as an expert witness, then the opposition can commence discovery on you. Your attorney or client may have spent large sums of money and time in conducting surveys, then under the rules of discovery, all this information is discoverable and available to the opposition for a very small cost on their part: the cost of the deposition and reproduction of copies.

When an attorney desires more than just basic facts and opinions, he or she may resort to Federal Rule 26 (b)(4)(A)(ii). The court will require a fair return for your time,[12] and the wise deponent will require this in advance of the deposition. The opposing attorney will use many methods to limit the cost of the deposition to his or her client, while charging a fee for time.

In Lewis vs. United Air Lines Transportation Corp., 32 F. Supp. 21, 23, the court stated:

> To permit a party by deposition to examine an expert of the opposite party before trial, to whom the latter has obligated himself to pay a considerable sum of money, would be equivalent to taking property without making any compensation. . . .

Your attorney may decide to seek your help as an expert, but not use you at the time of trial. As those of an expert, your view may be sought for advice or to interpret data.

However, the opposing counsel may ask the court to permit you to "speak." The court will look to the aspect that the party seeking the information cannot obtain the facts, opinions, or evidence relative to the subject matter from any other means. If this testimony is permitted, the party seeking discovery must pay you as an expert and should pay your client a fair portion of the fees and expenses that are incurred due to the discovery.

In this category will fall that expert who is (1) informally consulted, as distinguished from one "who has been retained or specially employed by another party in anticipation of litigation or preparation for trial"; (2) a regular employee of the party; or (3) formally retained to help in preparation of the trial, but whose testimony will not be used at trial.

The 10th Circuit Court in Ager v. Jane C. Stormont Hospital 622 F.2d 496 (1980), established guidelines when it wrote:

> In our view, the status of each expert must be determined on an ad hoc basis. Several factors should be considered: (1) the manner in which the consultation was initiated; (2) the nature, type and extent of information or material provided to or determined by, the expert in connec-

tion with his review; (3) the duration and intensity of the consultative relationship; and (4) the terms of the consultation, if any (e.g., payment, confidentiality of . . . data or opinions, etc.) Of course, additional factors bearing on this determination may be examined if relevant.

34-9. SURVEYOR'S PREPARATION

As no player would consider playing any game of combat without adequate practice, no surveyor should consider offering testimony without the same. Depending on the magnitude and complexity of the case, certain basic fundamentals must be observed. We must realize it is the attorney who leads the presentation, yet the surveyor should insist that at least one meeting be held prior to trial to discuss the facts and strategy. A courtroom is no place to hold a "surprise party." The attorney should explain the trial plan and how testimony will be used in the overall plan. This will also assure that the attorney understands your testimony. This visit should include a complete examination of all the exhibits and record documents. Make certain that you understand the documents you will use. Once you testify in error to one of your own documents, your credibility is certainly questioned by the jury and you will provide the opposition additional cross-examination impeachment data. You should feel comfortable with all your exhibits and documents when your refer to them in the course of the trial, Most attorneys compile a trial notebook in which they identify their plan, the witnesses, exhibits, including anticipated problems of authentication and admissibility. As a witness, you should also prepare your own list of exhibits and keep it handy for reference and any problems you anticipate. Uncertainty and unfamiliarity with your exhibits and documents may weaken and discredit your testimony, and will make you appear unprofessional.

In most instances, testimony relative to past surveys and past events will be the norm of the trial. These events may have occurred well in the past. In order to make your testimony more creditable, a visit to the area just prior to trial is a necessity. This visit should be made with all the documents, in hand, and closely examined. Such a visit will serve to refresh your memory as to the small facts that you may have forgotten. With the documents in hand, close your eyes and place yourself in the time frame when the notes were made. Make a sincere effort to picture the scene, as it was, noting the relationship of all the objects, people, events, and conversations. If other people are involved it would be logical to invite them to accompany you. When two witnesses relate the exact same story as to dimensions of monuments, etc., possibilities exist. They have agreed on the same story or all were present when the measurements were made. More credibility is engendered when all tell the same story, but it is tempered in the minor aspects. This could occur when three individuals testify as to the diameter of a tree. One states it is 9 in., a second, 8 in., and a third, 10 in. The general size is important, not the exact size. If all testified it were 10 in., each would have had to measure it with a diameter tape. Credibility is present when one witness states, "It is a red oak 35 inches in diameter," another stating "It is a red oak 3 feet in diameter." This is credibility. Regurgitation diminishes credibility.

34-10. THE SURVEYOR ON THE WITNESS STAND

Physical characteristics depict the credibility of a witness. You must believe your own testimony to make others believe it. Speech, body language, mannerisms, and *you* play an important role. When you sit in the witness chair, all eyes are on you. Your responsibility to the jury is to convey, both in speech and non-speech, those mannerisms that convey the credibility of your story.

Eye contact is a must. One who avoids eye contact with the jury is questioned for his or her honesty, for people know when one is less than truthful. He or she will not look at you. Eye contact should be extended to the judge, the attorneys, and the jury, more particularly to that one strong individual who sits on the jury. Your story must be told in such a way that you are believed.

Attorneys believe it is necessary and surveyors thrive on the honor to act as an expert witness in land-related cases. However, as a witness, you may be just as important as a lay witness. Your attorney must look first to credibility. Will the individual be a creditable witness? Will he or she be believed? If the individual lacks credibility, believability is also lacking.

Credibility goes to the foundation of the entire testimony or story that is being told. The attorney should weight the credibility of the witness against the entire story the witness has to relate and must decide when a simple, down-home story about monuments and the survey will engender more trust than the relation of technical terms and theories. This is one advantage local witnesses have over the imported "hired gun," who is brought in to add weight. The decision as to in what capacity a witness will act and be utilized will affect the usefulness and degree of cross-examination, which includes whatever impeachment to which he or she will be subjected.

In analyzing the story, the attorney should consider if your testimony is from personal actual experience or was it obtained vicariously through others. This situation can occur when a supervisor of surveys reviews a crew's work without being an actual member of the survey party. The supervisor is then called in to testify as to what was done, what was seen, and what was decided. The survey party did the work, they recovered the evidence, they completed the field books and set the monuments, yet the supervisor reviewed the data, made the decisions, and gave the instructions —far from the job. The modern trend is to require those individuals who actually accomplished the work to testify. You will be asked if your conclusions and decisions are based on your own actual experiences and observations or those of others. The most convincing testimony is that based on personal experience and not that of others.

For both expert and lay witnesses, credibility of testimony goes hand in hand with consistency. However, inconsistency can be expected in testimony, for the human mind can "play tricks" on memory and the ability to recall events, as well as the fact that each individual sees the same event from a different vantage. Attorneys will use inconsistencies to advantage and will try to discredit otherwise acceptable testimony. Such words as "lies," "cover-ups," and "old age" may be tossed at the witness and jury. The judge and jury should have it pointed out by your attorney that your testimony is consistent with established facts accepted by other witnesses and the profession in general.

How does your testimony relate to published standards of practice, professional treatises and texts, or other known facts presented by the witnesses? As a witness, you have an obligation to point out to your attorney any inconsistencies in the evidence and your testimony. At times, an opposing attorney will dwell on an insignificant inconsistency to cloud the issue. If possible, avoid these areas of conflict.

The witnesses' credibility is intimately linked to the testimony credibility, and the matter of personal credibility is dependent on the personal characteristics of the person who will relate the story. When a witness is called on to testify, it is assumed that the testimony will be unbiased and have a clean neutral approach, free from suspicion. Regardless of what is said or written, expert witnesses are interested in the results and outcome of the litigation. The age-old statement that a "witness is unbiased" helps to offset some of the stigma, in that the witness is only interested in relating the "true facts." As a whole, society views the testimony of a biased witness with distrust. Biased testimony can be both positive and negative. As a person can slant testimony in favor of an individual, so can he or she slant the same testimony in a negative manner. Testimony is the

perception of the evidence. It is not what you actually see, but what you think you see, for we all "see" the same events from different vantage points. This is why two surveyors will relate the markings seen on a wooden post in different terms. The basic presumption is that all individuals are honest in their testimony. If this were so and two individuals saw the same evidence in the same manner, then there would be no discrepancies in testimony. The mind will play tricks. It can take random scratches on a stone and make identifiable markings that take on form.

Since each of the surveyors on both sides is telling the truth, your attorney must make the jury believe that you are telling the "whole truth."

After direct examination, you will be subject to cross-examination by the opposing attorney, who will attempt to show with clarity that bias does exist, that you are incompetent, a gun for hire and not worthy of being called professional. Some attorneys make a general practice of cross-examination, but most do not use this technique to their fullest advantage. Once cross-examination is complete, your attorney will be given an opportunity to rehabilitate you in the eyes of the jury and judge. There are times that cross-examination can work to your advantage when it is poorly used or the attorney engenders hostility from the jury because of the way in which he or she handled the witness.

If your attorney is an experienced trial lawyer and knowledgable of this potential for bias prior to the trial, he or she can limit its effect or neutralize cross-examination in several ways. Your attorney can work it into the direct testimony in such a manner as the following questions and answers illustrate:

Q. It is possible that a second surveyor could see these numbers or letters in a different light?

A. The second is a matter of trial tactics. Some attorneys believe it is wise to give the opposition all of the "log chain" they can swim the river with, or "all they will get and fight for the rest." This will include giving any damaging testimony up front in the direct testimony and catch the opposition off guard. This will take some of the force from the opposition, and in most instances it is well received by the jury.

An example is when your client's deed does not cover and describe all the lands your client claims, yet your remaining testimony could be suspicious as being biased on his or her behalf. Your attorney should realize that a knowledgeable surveyor or attorney can come to the same conclusion and testify to these facts. It would be wise, in this situation, at the first part of your testimony to ask that question.

Q. Does your client's deed cover and describe all the property your client claims?

A. Of course, the answer is no, and the facts are out. Now you can get on with your case and leave the damaging testimony behind. On closing, he or she can state and argue that your testimony cannot be considered as biased, yet having come at first, the jury may have forgotten its impact.

Many times an expert is called in to testify "after the fact." That is, all of the field work has been completed, and it was not until litigation that you were consulted, when it was considered that your testimony could better the case. The attorney's first responsibility is to establish and build up your credentials and experience and make your testimony necessary and creditable in the eyes of the jury. Professional credibility is dependent on personal credibility: Looks, speech, stature, education, experience, professional associations, and many other elements are considered.

It is a "knee jerk" reaction for attorneys to use you as an expert witness, rather than a lay witness. This is normal, yet it may be more of an ego trip for the surveyor than a benefit for the client or attorney. The client and attorney should closely examine the capacity in which you will be used, deciding which will be more beneficial: lay or expert? An expert's appearance in court may be a sword or shield—i.e., you can be used to attack his or her cause or defend the attorney's client. The attorney is under no obligation to use a qualified expert witness in that capacity. It may be to the attorney's greatest advantage to have you available simply as an advisor. Your presence may tend to swing the balance, neutralize the opponent, and keep the opponent's testimony "clean."

Although all jurisdictions have their own rules as to the use of witnesses, most have adopted or closely relied on the *Federal Rules of Evidence*, as practiced in the federal courts, addressing the basic use of witnesses, as follows.

Rule 601: General Rule of Competency

Every person is competent to be a witness except as otherwise provided in these rules. However, in civil actions and proceedings, with respect to an element of a claim or defense as to which state law supplies the rule of decision, the competency of a witness shall be determined in accordance with state law.

In the decision rendered in U.S. v. Lightly, 677 F.2d 1027, the 4th Circuit Court indicated that every witness is presumed competent to testify unless it can be shown that the witness does not have personal knowledge of the matters he or she will testify to or that the witness does not have the ability to recall events or information. The ultimate decision as to whether an individual can or cannot appear as a witness is the question of competency, and the final ruling is left to the decision of the judge, who has the power to exclude any witness he or she determines legally incompetent.

Surveyors may find that their competency to testify is questioned by the opposition. This has happened in several situations where the work was accomplished by a field party, yet the client engaged you to testify as a witness. Although you have the necessary credentials and present the proper image before the court, you have no personal knowledge of the survey other than what you were told by the field-persons and what you read in the field books. A wise attorney will realize that you are incompetent to testify as to your personal knowledge. The attorney now has a hearsay problem or will have to prove that you are the best available witness.

The greater the position of responsibility in your firm and the more removed from the field work aspects, the less the possibility of your having personal knowledge of the field facts. A good opposition attorney will make this work to his or her advantage. This was also considered by the federal courts in Rule 602.

Rule 602: Lack of Personal Knowledge

A witness may not testify to a matter unless evidence is introduced sufficient to support a finding that he has a personal knowledge of the matter. Evidence to prove personal knowledge may, but need not consist of testimony of the witness himself. This rule is subject to provisions of Rule 703 relating to opinion testimony by expert witness.

Once a surveyor serves as a witness, he or she can see the importance of these two rules relative to work accomplished by field crews but not personally examined by the principal or crew chief in the field. The above two rules can also be applied to lay witnesses. If lay witnesses are to testify, they can testify only to those facts within their personal knowledge and experience. You can testify as to what you saw, what you observed, and of any evidence or statements made to you. In some instances, the court may let you express an opinion.

Traditionally, the courts have let lay witnesses express opinions in some areas that address themselves to the sense—e.g., speed, "The car was going fast"; time, "She was gone a long time"; distance, "It was about a quarter of a mile"; weights, "The stone was heavy"; sanity, " He acted crazy"; and sobriety, "He was drunk." The courts permit this because these opinions are based on perception and may be helpful to the finder of fact. For a professional and technical person, being a lay witness can cause difficulties in that the individual is programmed to give opinions.

The role of the expert witness is discussed in Rule 702.

Rule 702: Testimony by Experts

If scientific, technical, or other specialized knowledge will assist the trier of fact to understand the evidence or to determine a fact at issue, a witness qualified as an expert by knowledge, skill, experience, training, or education, may testify thereto in the form of an opinion or otherwise.

The exact position of an expert witness was stated by F. Lee Bailey in *The Defense Never Rests*. Although he was speaking on behalf of the cross-examination of an expert, his words are true regardless of the capacity in which the expert is used:

The cross-examination of an expert witness, which I was about to do, poses added problems. He's a professional who understands the trial process; he knows how much he can get away with; he knows how to answer questions. And usually, though not always, he is thoroughly versed in his field. Give an expert a broad enough question, and he may bury you. That's why a trial lawyer has to be a crammer; he has to know his stuff well enough to catch any weakness on the part of the witness.[13]

The sole purpose of having experts is to permit them an expression of their opinion of the evidence they recovered or what was presented and then be able to even answer hypothetical questions. This is an important field and will be discussed later. Once you are qualified as an expert, or if you consider yourself an expert and hold your qualifications out as those of an expert, it should be stressed to your attorney that a bland, rote, uninteresting presentation of your qualifications, just to meet the intent of the court and requirements for legal sufficiency, will fail to help you in the eyes of either jury or judge. You and the attorney now have the opportunity to "toot your horn" and tell it all. The attorney's responsibility is to establish and build your credibility with the jury. The attorney must give considerable thought and care to this matter. The surveyor has a responsibility to aid the attorney. An experienced individual will maintain a biographical outline and make certain that the attorney has a copy before the trial, so it can be studied in depth. The attorney's job is to present your qualifications to the jury, so that they will accept you as the most qualified individual in your field. It is helpful if the court is presented with a copy of you biographical outline; the attorney has an absolute right to place your qualifications before the court.

When you, as an expert, determine conclusions based on information at hand, the courts are ambivalent as to what is correct and proper. A court that accepts the common law rule will only permit you to base an opinion on those facts and evidence that were presented at the trial in chief. As an expert, if your opinion is based on facts X and Y, then facts X and Y must be placed in evidence by one of the witnesses. Under the more liberal federal rules, the expert's opinion may go beyond the evidence as long as it is reasonable to do so. Thus, you may form an opinion to a hypothetical question based on facts X, Y, and Z, even if fact Z was never introduced in evidence or given in testimony. The federal rules address this quest for information in the fact that other experts in the field customarily rely on these facts for their opinions. However, all must realize that this testimony is still an opinion and not evidence.

Since an expert's opinion may or may not be based on introduced evidence, and applying the laws of probability (for no reasonable experts will ever consider themselves 100% certain), both the attorney and surveyor should be careful in the use of words. A good attorney can make your words ineffective. Never tell the jury what their verdict must be or what the ultimate issue is, unless you are asked by the judge. Such phrases as "She owns the land," "He had adverse possession," or "That

is a boundary by agreement," should be avoided. That is what the jury is to decide. Many experts toss the words "possible" and "probable" in their testimony with little regard to their meaning. At law, the word possible is nothing more than a 50-50 chance or less. On the other hand, probable carries a legal interpretation of a "fair degree of certainty" or to betting people, a probability of 75% or more.

In most instances, the direct examination of an expert witness is routine and without its problems, unless both the witness and attorney failed to set out their plan and approach. The cross-examination of an expert is another story. The old law school examination question is, "When do you cross-examine a witness?" The answer is *never!* We know this is not true or feasible. In cross examination, the expert witness is like any other witness and all the usual rules of impeachment apply.

The expert will probably be asked to submit exhibits during the direct examination. It is essential that the proposed exhibits show exactly what you want to depict and illustrate and no more. The expert must be prepared to meet any and all challenges to qualifications, work, and even dress. The attorney may be smiling at you, but in reality is trying to stay ahead of you so that you will take that fatal step and be tripped up and discredited. One of the easiest areas in which this is done is in the use of exhibits. Have complete control over all the exhibits you use, from the initial preparation to courtroom use. Make the exhibits work for you. Maps, field books, illustrations, descriptions, and affidavits are but a few that you may use. Present an outline to your attorney, indicating what the exhibits will illustrate, pointing out each exhibit's strengths and weaknesses. The attorney must plan how to introduce the evidence, its authentication, and what it will do for the case. A foundation must be set for all evidence.[14] A joint decision needs to be made prior to trial whether the documents and exhibits will be introduced in evidence, or if they will be only for the purpose of refreshing your memory.

34-11. EXHIBITS USED IN COURT

To introduce photographs requires that you state that you have personal knowledge that the picture is a true and correct representation of the scene. You do not have to have taken the photograph yourself. Relevancy of the photograph must be shown. This can be for the simple purpose of showing markings on a tree. The courts have considered slides and contact prints to be included in this category.

Charts, maps, models, and graphs make excellent exhibits and evidence. They can be used to explain your position or the opponent's position. The court usually does not hold to "absolute accuracy" for such exhibits if you prepare them yourself, but any errors will certainly be highlighted on cross-examination. Show only that information you wish to show and no more. Do not confuse the jury.

Reports and field books are a common form of evidence in land cases. Usually, they play an important role in refreshing memory as to past events, or to show evidence recovered by earlier field crews that are no longer available. The attorney must determine if these books and records will be used simply as a memory refresher or whether they will be introduced in evidence. Unless adequately prepared, the attorney may become embroiled in the middle of a procedural battle. In order to be able to use records to refresh memory, four elements must be proven: (1) It is in writing, (2) the writing was made at or near the time of the event by you or an individual acting under your supervision, (3) you are unable to remember the events depicted in the field book, and (4) you must vouch that you are certain that the facts were correct when they were made.

Every attorney and expert witness goes to court armed with their learned texts, which usually are carried in a bag or under the arm, but they are visible to both jury and judge. Security is the reason. Individuals expect to be able to point to the accepted test to prove their point of view. Once experts refer to a passage in any accepted reference, they "open" the entire book for examination. Experts will be expected to know all of the information that is in the book.

In litigation, public documents may fill boxes. In most land-related cases, every party may show up with the same documents to prove different points of view. The documents may consist of originals, certified copies, xerox copies, carbon copies, or hand copies. Land cases depend on documents. Problems may be encountered when your entire case is predicated on a single document that is found to be inadmissible into evidence, and although the document is in the courtroom, it is unusable. Experienced attorneys anticipate problems of introduction or admissibility, and they will work around them and have back-up alternatives. Certified documents stand on their own; when possible, work with *certified* copies of public documents. If the original document is not available and a copy is to be introduced, account for the loss of the original. Some documents may meet the criteria as ancient documents; if they do, use this alternative, for they are an exception to the hearsay problem. If documents cannot be introduced for one purpose, a wise attorney may be able to introduce them for a second purpose. Determine in the discovery process if the other side is going to introduce the document, and if it does, use the document to your advantage.

In working closely with the attorney, the survey can dramatize his or her evidence by preparing colorful exhibits. If these are completed in a professional manner, the opposing attorney may be tempted to use your exhibits. It is always helpful for a juror to have a copy in hand, where it can be closely examined.

34-12. CROSS-EXAMINATION

At the start of cross-examination, you will have just completed the direct examination and will in all probability still be on the stand. Your mouth is dry, the palms of your hands are wet and cold, and your stomach is in turmoil. The attorney who now stands before you feels exactly the same, knows all of the law school professors said not to cross-examine—but they really did not mean him or her. By common law right, any witness who is permitted to testify on direct must submit to cross-examination. The party may waive the right to cross-examine.

The purpose of cross-examination is to probe deeper into your testimony and discover facts that may be favorable to the cross-examiner. To be subject to cross-examination is difficult but not impossible if you understand it and use it to your advantage. Keep in mind that this is the attorney's game, field, and forum, and you must play according to the attorney's rules. The rigid rules that apply to direct examination do not apply here. The attorney may ask leading questions, complex questions, and just about anything with which he or she can get away.

Preparation for cross-examination should have commenced with the very first question of direct examination. The attorney must have all the facts of the case, pro and con. Hold nothing back to your attorney. Unfavorable facts can be addressed on direct examination. The more your attorney is able to address, the less there will be for cross-examination. If unfavorable facts are revealed on direct examination, the jury will feel your credibility.

In most jurisdictions, the scope of cross-examination is limited to those facts that were presented on direct examination. Keeping in mind that the cross-examiner has but two purposes,[15] to minimize or destroy your direct testimony and to develop independent evidence that will aid the case (at your expense), you will have an advantage, for you may be able to second guess him or her.

The cross-examiner may be able to beat you into submission by using time-tried tactics,[16] probing and pitching, using such elements as prior inconsistent statements, implausibility of your testimony, your inability to remember, biased testimony, a lack of truth and veracity, and your basic qualifications, either too educated or not enough education.

Each of these is used to engender anger and bait you to play the cross-examiner's game. The cross-examiner wants to destroy you and has the tools and means to make your testimony ineffective, if you permit it. "Hired gun," "You can be bought," and "They paid you to testify." Such phrases and more will be said in attempts to discredit you.

34-13. PAYMENT FOR SERVICES

As the cost of expert witnesses have increased, some attorneys have attempted to get as much expertise as cheaply as possible. In order to do this, they have used the system in order to limit the amount of payments to experts. Using the subpoena power of the courts, some attorneys may attempt to force you to testify by seeking your knowledge under the guise of a lay witness, and then use you as an expert as the trial progresses. Many surveyors will play this game. Under the common law, experts cannot be forced to testify against their will. Of course, this cannot be said for a lay witness, for the courts have held that lay witnesses have a public duty to testify.

If expert witnesses are forced to testify against their will, it would be a violation of their constitutional rights, for their property will be taken without due process of law. You can sell it or give it away, but you cannot be forced to part with it for free. Unfortunately, this common law rule is starting to erode. In Kaufman v. Edlestein, 539 F.2d 811 (1976), the court realized that the expert is no longer living in a simple world. Courts have realized that experts are needed in today's complex world and must complement the legal world. In its decision, the court identified certain parameters in order to change the age-old custom. It stated that experts are entitled to receive reasonable and adequate compensation, that they cannot be compelled to do any work in preparation for their testimony, and if they have performed work, they cannot be compelled to do additional work if they already formulated their views on the subject.

Their opinions must relate to those facts that they already have gathered.

After the testimony of each witness and other legal procedures, the case will go to the jury or judge for a decision. There is a winner and a loser. No matter whether you won or lost, the case will be tried and retried for years, and you will wonder what was right and what was wrong. You will find that there were actually three testimonies: (1) the one you prepared for, (2) the one you actually presented, and (3) the one you wished you had given. The last one was the best.

After reading the preceding pages, the expert could have eliminated all of it by simply reading the following checklist. By gleaning the main points identified in the text, the expert witness should consider the following as a minimum to be followed when called on to testify:

34-14. CHECKLIST FOR THE EXPERT WITNESS

1. Be honest. Tell the truth no matter what, whether it helps your case or harms it. You have no obligation to introduce evidence but if asked, tell the truth. If you made an error in earlier testimony, correct it as soon as possible.

2. Do not guess. If you do not know the answer, say so. When you answer and can give positive answers, do so. If you cannot, do not use "I guess," "I believe," or "Maybe." They still mean you do not know.

3. Maintain a professional stature. This will include dressing properly and being polite, especially to the judge. "Yes, Judge," and "Yes, Your Honor" are words of respect.

4. Speak in your own words. Do not use words on the stand that you do not use in your daily speech. Do not memorize testimony.

5. Do not volunteer information. This applies on direct, but especially on cross-examination. It is the attorney's job and responsibility to extract the information. Help your attorney only.

6. Avoid hearsay. Do not open your testimony for objections. Never relate what other people told you—a certain objection.

7. Talk to someone. Direct your answers to the jury or judge. They are the ones you want to convince—*not* the attorneys. Talk so you can be heard.

8. Watch the attorneys and judge. You can see when objections are coming and keep from disclosing information if the objection is "sustained."

9. Control your documents and exhibits. Do not let them control you. Work with your attorney before trial. Know what your attorney will do. It will impress the jury.

10. Keep cool. Do not let anyone lead you into anger. If you do, you lose.

11. Do not be funny. The court is no place for humor. Sit-down comedy is unnecessary.

12. Guard your speech. Discuss the case only with your attorney and client. *No* other person. This may save you a court reprimand.

13. Expect the unexpected. Never be complacent. Keep on guard at all times.

14. Trust your attorney. You were taught to trust your compass needle. Trust the attorney.

NOTES

1. Federal Rules of Civil Procedure. Rule 3
2. *Black's Law Dictionary*. 1968. St. Paul. MN: West Publishing Co.
3. W. Barthold. 1968. *Attorney's Guide to Effective Discovery Techniques*. Englewood Cliffs, N.J.: Prentice Hall, p. 64.
4. _____ . p. 88.
5. _____ . pp, 112–118.
6. J. Gelin, and D. W. Miller. 1982. *The Federal Law of Eminent Domain*. Charlottesville, VA: Michie, pp. 371–372.
7. _____ . p. 371.
8. J. W. McElhaney. 1981. *Trial Notebook*. Chicago, IL: American Bar Association, Chap. 3.
9. F. L. Bailey, 1971. *The Defense Never Rests*. New York: New American Library, p. 25.
10. R. A. Givens. 1980. *Advocacy*. Colorado Springs, CO: McGraw-Hill, p. 216.
11. McElhaney. p. 107.
12. *Ibid.*
13. Bailey. p. 25.
14. Givens. p. 216
15. McElhaney. p. 107.
16. *Ibid.*

35

Land and Geographic Information Systems

Grenville Barnes

35-1. INTRODUCTION

Over the past decade, the interest in land and geographic information systems (LIS/GIS) has been overwhelming. Surveyors, geographers, engineers, landscape architects, environmentalists, planners, and professionals from a number of other related disciplines have embraced this technology and begun to build information systems to assist them in their work. In the United States, the development of integrated information systems has focused to a large extent on the local level where most land-related information is acquired and maintained. This has been particularly true for highly urbanized areas (cities) where most of the information is related to the property parcel.

Fundamental to this movement has been a belief that better information leads to better decisions. It is argued that the most effective way of improving information is to manage it as a corporate resource. For urban governments, this entails forging a unified approach to land information management so that data maintained by different agencies can be shared and integrated across agency boundaries. This will not only minimize data inconsistencies and redundancies, but also by combining information resources across several agencies will produce a synergism that will significantly improve current, isolated information systems.

The previous edition of this book contained a chapter on land information systems (LIS) written by J. McEntyre. McEntyre began the chapter with a general history of surveying in the United States, as well as current practices with regard to the description of land parcels. This was followed by a discussion of the need for a multipurpose cadastre based on the shortcomings of current land data systems. Different components of an LIS were examined and a system was proposed that would promote the integration of different types of information through a common state plane coordinate system and the introduction of a registration of title system. The current chapter has been updated to reflect the changes that have occurred since this earlier chapter was written. There is less emphasis on the history of surveying and description of land parcels (dealt with adequately elsewhere in this book) and more emphasis on the evolution of LIS/GIS and what role GIS technology plays in this development.

The LIS concept has evolved from a focus on a parcel-based information system, originally termed multipurpose cadastre,[1-3] into a broader and more integrated network of information systems, often termed GIS/LIS. This rapid evolution and the integration of several systems and disciplines have made it extremely difficult to define terms such as land data systems, LIS, GIS, AM/FM, multipurpose cadastre, and cadastre in a clear and permanent manner. However, by examining how these terms have evolved over time, we can begin to understand more about the concept itself and what general trends are emerging in this field. Such an examination is included in the early part of this chapter.

In order to narrow the scope of this chapter, the discussion focuses primarily on parcel-based information systems (or LIS), as the traditional role of the land surveyor places the surveying profession (at least potentially) at the helm of LIS development. The GIS literature is diverse and multidisciplinary and the author has purposely included a broad array of citations in the hopes that this will assist the reader in negotiating his or her way through the myraid of publications dealing with this broad subject.

Defining LIS/GIS in a succinct and distinctive manner is extremely difficult partly because of the rapid development of these systems over a relatively short period. The problem of defining these terms has sometimes been likened to that of catching a cricket: You no sooner have it in your hand than it has jumped away from you again. As a preface to defining these terms, it should be pointed out that professionals from different disciplines are likely to adopt different definitions and to view these systems and related issues from slightly different perspectives. Since the LIS/GIS field is inherently multidisciplinary, these different perspectives serve to give us a fuller understanding of this broad field.

LIS can be defined as an information system that focuses on data and information that is referenced to the property parcel. This includes data relating to the spatial location and dimensions of the parcels (usually from property surveys), land tenure (right holders, nature of land rights and restrictions), land use

(regulations, area zoning), and land administration (political and administrative units). The principal components of an LIS are (1) people (users, producers, managers, etc.), (2) an information base, (3) technology (computer mapping systems or GIS), and (4) procedures, standards and protocols that facilitate the exchange of information (see Figure 35-1).

GIS can be defined in a similar manner to LIS, with the exception that the information in the system will generally not be parcel-based but related to some natural resource polygon or linear feature (road, river, etc.). However, a review of papers published on the topic shows that the overwhelming majority focus on GIS as a technology with a secondary emphasis on the people or institutional components. In other words, GIS is seen as a tool for managing and analyzing spatial data as opposed to a resource.

One of the reasons LIS is more centrally concerned with institutional or "people" issues than GIS is that parcel-based information describes people's rights to land, and in most cases, the boundaries of parcels are not defined simply by physical boundaries, but by an array of information (relating to monuments, measurements, deed descriptions, etc.) that, when interpreted according to certain predefined rules, will define the invisible line between property corners. This can be contrasted against the boundaries of a natural resource such as a forest, where the boundary

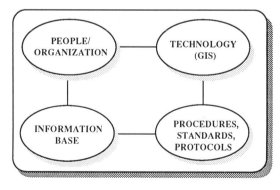

Figure 35-1. Principal components of an LIS.

is simply determined by the physical manifestation of the forest.

The debate on the exact difference between LIS and GIS is thankfully decreasing in intensity as those involved realize that such arguments produce diminishing returns in our understanding of LIS/GIS issues. There are so many commonalities and overlaps which inextricably link these two systems that it is clearly more productive to concentrate on the major issues of joint concern than on defining "turf" boundaries. Both LIS and GIS are concerned with the integration and analysis of spatial data for addressing certain issues and problems. Today's environmental, infrastructure, land tenure, and socioeconomic problems demand the integration of a variety of spatial data. In the remainder of this chapter, the term LIS/GIS will be used to address issues of joint concern. When GIS is used independently, this refers to the GIS hardware/software package described more fully in the section on GIS. The term LIS should be interpreted in a broader manner as defined in Figure 35-1.

The chapter is organized into seven related sections. The section on the evolution of LIS begins with the early development of fiscal cadastres in England and Europe and traces the evolution of juridical/fiscal cadastres into multipurpose cadastres and finally LIS/GIS networks. The following section on LIS within the land administration framework positions LIS in the center of this framework, thereby demonstrating how LIS related to other land administrative functions (surveying, registration, regulation, etc.) and why it is important to view land records modernization efforts in relation to this broad framework. Subsequent sections on geodetic reference framework, cadastral overlay, and land registration deal more specifically with the role of each of these components in an LIS. The section on GIS discusses the two major structures (vector and raster) for storing spatial data in a GIS. Methods used to relate and integrate graphic and attribute data are explained and various GIS applications listed. Finally, the section on case studies gives practical examples of LIS implementation in the United States.

35-2. THE EVOLUTION OF LAND INFORMATION SYSTEMS

35-2-1. Early Fiscal Cadastres

The origins of today's LIS can be traced back several hundred years to cadastral initiatives in England and Europe. In England, The Domesday Book, compiled in the 11th century, consisted of a textual record containing a comprehensive inventory of the real and personal property of each landholder. This inventory is often regarded as the origin of the modern cadastre concept.[4] In the early 1800s, Napoleon I of France ordered a survey of some 100 million parcels with the express purpose of collecting and spatially referencing fiscal (taxation) data to individual parcels. These early initiatives were essentially concerned with the development of a fiscal cadastre for supporting the land taxation system.

As surveying and mapping technologies improved, the definition of parcels, originally delineated for tax purposes, became sufficiently accurate to support the legal descrption of property rights. Accurate cadastral maps showing the parcellation of land in a specific community were used as a means of describing the spatial extent of property rights reflected on officially registered deeds or titles. This development gave rise to the juridical cadastre.

Although the use of a cadastral map became common practice as early as 1800 in many western European countries, this has not generally been the case in North America. With few exceptions, the only community-wide (county-wide) parcel maps available in the United States are those created for assessment/taxation purposes. However, this situation is being rectified as counties begin to computerize their land records and develop integrated LIS.

35-2-2. The Multipurpose Cadastre

In the mid-1970s, the concept of a multipurpose cadastre (MPC) was advanced by McLaughlin in his previously cited 1975 paper and others' work,[5] in North America. Interest in this concept was fueled by an increasing need to integrate fiscal and legal land tenure data with other land-related data. The previously mentioned 1980 report by the National Research Council (NRC), entitled *Need for a Multipurpose Cadastre*, documented this need and attempted to define the role that should be played by the federal government in the development of the MPC. The approach followed in this study was explained as follows:

> Rather than attempting to resolve all land-information systems problems, it was decided to consider the basic components (reference frame, base map, and cadastral overlay) of a multipurpose cadastre, which, if properly established and maintained, would provide the common framework for all land-information systems. In the

process of defining a federal role, the roles of the state and local governments and those of the private sector (companies and citizens) were considered germane. With these roles established and the MPC conceptualized, the relationship of land data files to the multipurpose cadastre was considered.[6]

At this time, the MPC was viewed as a parcel-based system that incorporated natural resource, land tenure, fiscal, administrative, and other land-related data as illustrated in Figure 35-2. The central elements of this model were a geodetic reference framework, base map, and cadastral map or overlay.

The problem of treating natural resource and other nonparcel data as attributes to the property parcel was subsequently recognized and the MPC model modified. In the previously cited 1983 follow-up NRC report, entitled *Procedures and Standards for a Multipurpose Cadastre*, the nonparcel elements were separated from the MPC. This new model (see Figure 3) contained many of the same components as the earlier 1980 model, but the natu-

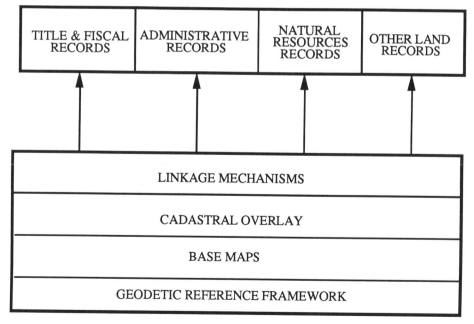

Figure 35-2. The 1980 multipurpose cadastre model.

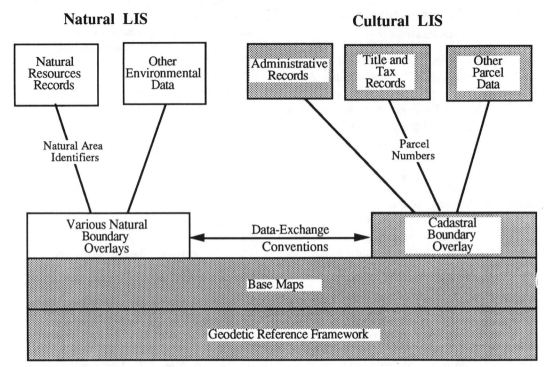

Figure 35-3.   The 1983 multipurpose cadastre model.

ral resource component was incorporated in a "natural LIS" that was linked to the MPC, now called a "cultural LIS." The geodetic reference framework, base map, and cadastral overlay still constituted the fundamental components of this cultural LIS (or MPC). In this report, the MPC was defined as:

> ... The core module of a large-scale community-oriented information system designed to serve both public and private agencies and individual citizens, by (1) employing the proprietary land unit (cadastral parcel) as the fundamental unit of spatial organization of land information and (2) employing local government land-record offices as the fundamental unit for information dissemination.[7]

A number of counties have based their LIS design on the recommendations made in the two NRC reports and the contribution this work made toward focusing attention on cadastral issues and problems in the United States should not be underestimated.

35-2-3.   Multipurpose LIS

Environmental concerns and increasing pressure on land resources have led to a more integrated approach toward land management and development. The protection of valuable natural resources like wetlands requires a knowledge of the location of these resources, as well as information on who holds rights to the underlying land. In the case of wetland, federal regulations have effectively removed the development right from the owners of this land. Other environmental concerns like soil erosion, groundwater pollution, and deforestation also require knowledge of land rights in order to identify offenders as well as landholders who will be affected by stricter public land-use controls. These situations underline the need to integrate parcel-based information, typically provided by an MPC, with natural resource and other nonparcel information. This need has led to a broader and more integrated LIS model.

In 1985, a model for a multipurpose LIS (MPLIS) was proposed (see Figure 35-4) and tested by researchers at the University of Wisconsin.[8, 9] Proponents of the MPLIS model felt that the MPC model formulated in the late 1970s and early 1980s was too strongly focused on the parcel-based aspects of an LIS. In order to promote a broader approach that was truly multidisciplinary, the MPLIS was designed to remove this parcel-based bias. Different types of information (soils, land use, land ownership, etc.) were stored on different layers and the concept of a base map was dispensed with as it was felt that the base information on the

map would depend on the purpose and discipline of the user. Instead, information typically found on maps such as the USGS 7.5 min. quad sheets (roads, rivers, contours, etc.) could be incorporated as individual layers. In this way, additional layers could be incorporated into the system whenever the need for that information arose.

This layer-based approach is also more consistent with the approach used in modern computerized mapping systems and GIS. Some of the differences between this model and the MPC models advanced in the NRC reports are undoubtedly due to an attempt to cater to both computerized and manual environments in the latter. Note, however, that the MPLIS still depends on a geodetic reference framework to provide spatial compatability so that the different layers can be properly registered.

35-2-3.   LIS /GIS Networks

Although the MPLS provides a general model for an integrated information system, it does not provide guidance on how individual information systems can be networked together. The model assumes that the different players will resolve questions associated with standards and policies for exchanging and integrating data from different agencies. However, this integration between local, state, and federal agencies is often regarded as the ultimate challenge in LIS/GIS development. There is consequently a need for a statewide initiative that brings the land information management community together in one forum so as to forge a common LIS/GIS strategy. Efforts in Wisconsin[10] and North Carolina[11] were early examples of such initiatives, but similar efforts are now underway in a number of states throughout the United States and Canada.

The approximately 82,000 local agencies (below state level) that make decisions with regard to land use,[12] make it extremely difficult to promote standard policies and approaches on a statewide basis in the United

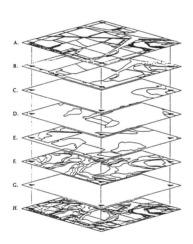

**Concept for a
Multipurpose Land Information System**

Section 22, T8N, R9E, Town of Westport, Dane County, Wisconsin

Data Layers:	Responsible Agency:
A. Parcels	Surveyor, Dane County Land Regulation and Records Department.
B. Zoning	Zoning Administrator, Dane County Land Regulation and Records Department.
C. Floodplains	Zoning Administrator, Dane County Land Regulation and Records Department.
D. Wetlands	Wisconsin Department of Natural Resources.
E. Land Cover	Dane County Land Conservation Committee.
F. Soils	United States Department of Agriculture, Soil Conservation Service.
G. Reference Framework	Public Land Survey System corners with geodetic coordinates.
H. Composite Overlay	*Layers integrated as needed, example shows parcels, soils and reference framework.*

Land Information and Computer Graphics Facility,
College of Agricultural and Life Sciences, School of Natural Resources
UNIVERSITY OF WISCONSIN-MADISON

Figure 35-4. The multipurpose LIS model. (Courtesy of Land Information and Computer Graphics Facility, College of Agricultural and Life Sciences, School of Natural Resources. University of Wisconsin-Madison.)

States. In countries like Australia, where most of the decision-making power is at the state level, progress in this area has been more rapid. In the state of South Australia, e.g., institutional reform and LIS development have followed a nodal approach.[13] This approach identifies certain primary nodes (information systems) that have a very close relationship and uses these as focal points in a larger LIS network. For example, property assessment, land registration, and surveying are incorporated in a legal-fiscal (cadastral) node. Similarly, information systems concerned with natural resource, utilities/infrastructure, and socioeconomic information can be associated with the environmental, utilities, and socioeconomic nodes, respectively. Figure 35-5 represents a nodal model that has been adapted

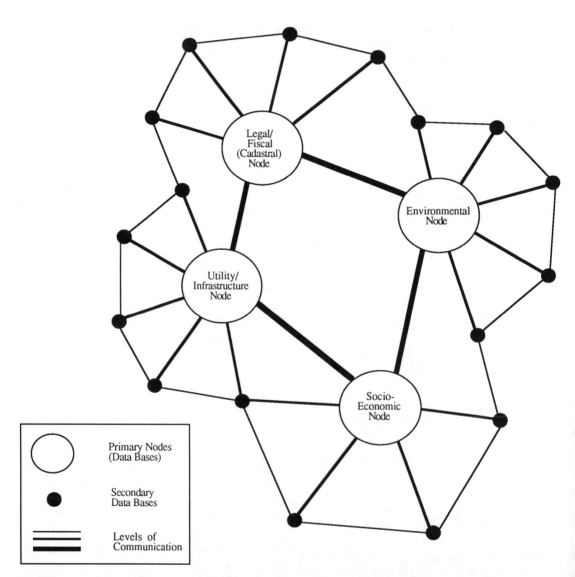

Figure 35-5. The nodal LIS/GIS network. (Adapted from M. E. Sedundary, 1984, LOTS: A South Australian approach to a total, fully integrated LIS, Proceedings of FIG Symposium, Edmonton, Alberta, Canada.)

from the model being followed in South Australia. The advantages of a nodal approach have been described as follows:

> This approach allows these nodes to be developed either in sequence or in parallel, subject to need and separate cost justification. It also allows the choice of the most suitable hardware/software for the data management needs of the particular function. Each individual development must, however, accord with a published development policy to ensure compatibility and the required level of data integration.[14]

This nodal model is an attractive compromise between a highly centralized model (single state-level LIS agency with broad powers) and completely decentralized one in which it is extremely difficult to reach consensus on a common LIS development policy.

Although the LIS nodal network can be viewed as the natural evolution of the earlier MPC models, implementing such a model is particularly challenging in the United States. The democratic principles that led to the decentralization of decision making have multiplied the number of players with some interest in LIS development. Also, the fact that many of the utilities are owned by private companies means that not only must means be found to vertically integrate information and policies across different levels of government, but private-public sector partnerships must also be created. It is therefore not surprising that many people concerned with LIS view these institutional problems as being far more challenging than those of a purely technical nature.

In summary, the trend in LIS development is toward networked solutions that focus initially on connecting those agencies that are linked through common land data and information needs. Instead of focusing entirely on parcel-based systems (MPC), models are emerging that place these systems in a larger LIS/GIS context incorporating land-related information associated with natural resources, utilities, and socioeconomic conditions.

35-3. FUNDAMENTAL COMPONENTS OF LIS

While there is an increasing trend toward the integration of information into broad-based multipurpose systems, it is essential that the subsystems within this network are designed to meet their primary needs first. From a surveying perspective, the legal/fiscal node (LIS) is the subsystem most relevant to the general functions of the profession. This section describes two extremely important components of an LIS, both of which are integrally involved and dependent on surveying. These are the geodetic reference framework (GRF) and cadastral overlay.

35-3-1. Geodetic Reference Framework

An examination of the multipurpose cadastre and MPLIS models described in the previous section shows that the GRF is regarded as a fundamental layer in all these models. The definition and function of a GRF have been described as follows:

> A geodetic reference framework forms the spatial foundation for the creation of any Land-Information System (LIS). Consisting of monumented points whose locations have been accurately determined with respect to a mathematical framework, this system permits the spatial referencing of all land data to identifiable positions on the Earth's surface. A geodetic reference framework provides not only an accurate and efficient means for positioning data,
> but it also provides a uniform, effective language for interpreting and disseminating land information.[15]

The same publication reports that in 1983, the GRF in the United States consisted of approximately 750,000 control points. These points were established by the National Geodetic Survey (NGS) to support various geodetic, mapping, defense, and other functions. Since this time and particularly the advent of GPS, many local governments have established control networks based on the NGS network. In some cases, these points have been established to provide photo control points for mapping from aerial photography, and in other cases as a means of controlling local property surveys. Generally, the coordinates of these points are determined on the state plane coordinate systems (SPCS) that have been defined for each state.

These mathematically defined control points should be distinguished from the physically defined points of the public-land survey system (PLSS). Mathematical positions do not depend on the physical demarcation of points, as those positions can always be relocated even if the markers disappear. Mathematically defined points provide information about absolute locations on earth. PLSS points, on the other hand, are defined principally through physical monuments placed at these points, but also through relative measurements to surrounding points on the rectangular "network." Determining points on a mathematical coordinate system, such as the SPCS, places these points in a common spatial framework that facilitates data integration. If, e.g., two subdivisions at opposite ends of a county are connected by survey to mathematically defined geodetic control points, they can easily be integrated into one consistent database or parcel map since they are on the same coordinate system. If they are not connected to a GRF, they remain as two free-floating maps. The problems this produces (gaps and overlaps) when attempts are made to integrate them are well known to surveyors and others concerned with LIS implementation.

Dale and McLaughlin[16] list several required capabilities of a GRF:

Identify points, lines, and areas (and hence parcels).

Indicate size and shape of features.

Permit calculation of areas.

Show relative and absolute positions.

Help in the location and relocation of boundaries.

Aid subdivision work and setting out new estates.

Aid engineering and underground services work.

Support land information system development.

This list emphasizes the role of a GRF for defining property parcels, the basic reference unit for LIS. Unfortunately, in North America the importance of connecting property surveys to a GRF has not been fully appreciated until recently. The fact that coordinates are generally placed at the low end of the hierarchy of boundary evidence shows that they are still not a primary means of defining boundaries. Some surveyors argue that the additional efforts of connecting to the GRF will raise the cost of surveys in spite of evidence to the contrary. A 1970 Canadian study found that when the costs of using an "isolated" (unconnected) survey approach versus a coordinated control network approach for boundary restoration were compared, the latter was 25% cheaper.[17] As this study was done prior to the development of GPS, a coordinated approach can be expected to become even more cost-effective as GPS technology becomes more commonplace in the surveying profession.

In another study that compared the costs and benefits of a GRF, a positive benefit/cost ratio of between 1.7 and 4.5 was found.[18] This analysis was not approached from a purely surveying perspective, but instead focused on all potential users (or consumers) of land information. It is interesting to examine the major features of a GRF as identified in this study:

A collection of permanently marked and maintained points

Coverage of an extensive area

A spatial relationship of known accuracy

Relationships expressed in a common mathematical language

Universal availability of geodetic information

Epstein and Duchesneau also correctly pinpoint universal compatibility as the primary contribution, or benefit, of a GRF. In their analysis, they compared costs required to create data compatibility (which are avoided because of the GRF) with costs to create, maintain, and use the GRF in order to make data compatibility possible.

Three important design criteria need to be considered when designing a GRF:

Density

Accuracy

Cost

In the 1983 NRC report, densities of 0.2 to 0.5 mi were recommended for urban areas and 1 to 2 mi for rural areas. In addition, it recommended that "the entire area concerned should be covered at a uniform density with a simultaneously adjusted network of control survey stations."[19] These density recommendations were obviously made after considering such factors as intervisibility (between control points) and the density of points in the PLSS. Since the land records in most states refer to the PLSS as a means of defining parcels (the PLSS covers approximately 80% of the land in the United States), it makes sense to use PLSS points (when possible) for the GRF. This does, however, involve additional work as many of these points are not demarcated and will require remonumentation. If the PLSS points are used, density will be dictated by this framework. For example, if section corners are selected for the GRF, this will result in 1-mi density. On the other hand. if township corners are selected, the geodetic control points will have a spacing of about 6 mi.

The recommendation of "uniform density" is one which the author finds difficult to justify, particularly from a cost and time perspective. The density of a GRF should be based on the need for such points. If the GRF is to be used only for cadastral purposes and, for arguments sake, the parcels in an area are 10,000 acres in size, the justification for a 1-mi density

network would be extremely difficult. However, if the GRF is to be used to spatially define other features such as roads and natural resources, this problem starts to disappear. In urban areas, where parcel sizes tend to be similar, a uniform density may also be more justifiable.

In deciding on the density for a GRF, it is therefore extremely important first of all to consider the uses of the framework. The use of the framework for cadastral (property) surveying and mapping purposes is an obvious one. Less obvious is the use of the GRF for mapping noncadastral features (roads, buildings, fields, land cover, land use, etc.) and surveying such features as wells, soil sample sites, utilities, and wetlands. Since these requirements will vary across a jurisdiction, the condition of uniform density runs counter to a needs-based approach. In other words, making a distinction between urban and rural requirements is not sufficient. We need to consider the variations within these broad areas and base density on the requirements in these smaller subareas. For example, areas that are likely to be developed will generally require a higher density than those that are stable and unlikely to change in the near future.

The accuracy standards for horizontal and vertical control in the United States are well defined and accepted by the majority of users. Horizontal control standards based on the relative accuracy between adjacent points are as follows:

First-order (1 part in 100,000)

Second-order

  Class I (1 part in 50,000)

  Class II (1 part in 20,000)

Third-order

  Class I (1 part in 10,000)

  Class II (1 part in 5000)

In order to maximize the utility of the network, it has been recommended that control surveys be done to third-order, class I or higher.[20,21] Further details on these standards and vertical control standards are given in

Surveying Theory and Practice[22] and elsewhere in this book.

The cost of creating a GRF is frequently the most important design criterion and one that often dictates the density and accuracy of the framework. As Dale and McLaughlin point out, "the need is not necessarily for high precision surveys, but rather for the best value for money."[23] Unrealistically high requirements in terms of density and accuracy will invariably lead to unaffordable costs, thereby jeopardizing the successful implementation of an LIS. In any event, the cost of creating an adequate GRF for LIS development will be substantial. Based on the experience of the Southeastern Wisconsin Regional Planning Commission, the relocation, monumentation, and control survey work amounted to approximately 20% of the total cost of implementing an LIS in local government. This translated into $3600 per square mile in 1985 dollars.[24] The percentage cost was larger than any of the other cost components of the project, such as the preparation of the topographic base map and cadastral base sheets (11%), cadastral map preparation (7%), digitization (7%), management and overhead (16%), hardware and software (15%), support (13%), and hardware and software maintenance (6%). With the advent of GPS, the per-point cost of control has dropped substantially and is currently estimated at $500 to $600 per point.

The challenge in designing a GRF is therefore to strike a balance between density, accuracy, and cost requirements. As GPS technology becomes less expensive and the constellation of satellites increases to provide 24-h coverage, it is likely that the cost factor will become less prohibitive, allowing the implementation of a more accurate and denser framework. This, in turn, should promote more reliable and consistent data for inclusion in a broad-based LIS/GIS.

35-3-2. Cadastral Overlay

One of the most challenging components of an LIS in the United States is the cadastral overlay. This overlay may be viewed as a digital cadastral map that depicts the size, shape, and location of all parcels within a jurisdiction.

Ideally, the cadastral overlay should be based on field survey data, some of which are conventionally reflected in deeds, subdivision plats, official maps, etc. Achieving this ideal is, however, extremely difficult due to inconsistencies in the legal descriptions of land parcels across a jurisdiction like a county. Not only are these parcels defined to different standards of accuracy, but they are also defined by different surveyors using different technology and techniques at different times, and recorded in different documents and formats. In many cases, parcels may not have been defined by survey but simply deed description (e.g., north half of lot 12). These unsurveyed parcels are particularly problematical as the errors in the original survey of the parent parcel (such as a section) are propagated until such time as a proper survey is carried out. The implications of this can be appreciated if one considers a 98-acre parcel that is incorrectly recorded as 100 acres. This error will probably not become significant until the parcel is subdivided into 5- to 10-acre lots or smaller. Originally, the missing 2 acres represent only 2% of the parcel area, but this increases to 20% when the parcel size is 10 acres. The cadastral mapping process forces one to address these problem areas, since all parcel data across a jurisdiction must be taken into consideration.

Manual cadastral maps, when such maps exist, are generally used as index maps. Indexing is either done by reflecting the deed number (vol. page) and survey record number for each parcel (or group of parcels) on the map, or assigning a unique parcel identifier (PID) to individual parcels. The PID acts as a common reference for all survey records and deeds associated with a particular parcel. Since the map is only an index, accurate and precise information at the parcel level must be obtained from the associated records referred to on the map. Bearings, distances, and monument description information, e.g., are generally not included on the cadastral map. From a surveying perspective, a cadastral index map is really only useful as a means of locating sources of more accurate and precise boundary data. However, in a broader LIS/GIS environment, there is a danger that a cadastral overlay will be regarded as far more than a general index.

Since the cadastral overlay is essentially a computerized version of the cadastral map, the integrity of an overlay that is merely an index must be questioned. This is particularly relevant when the cadastral overlay is merged with other layers to form a multipurpose LIS/GIS. Utility line data on the utility layer, e.g., may be spatially related to property boundary lines (e.g., 5 ft from the property line). Therefore, any error in the cadastral overly will be propagated in the utility layer and any other layer that is spatially referenced to property boundaries. Another potential problem arises from the fact that the cadastral overlay is the de facto representation of cadastral information in a multipurpose LIS/GIS, even though the overlay may only be a general index. It is common knowledge in surveying that the true location of a boundary is determined by considering physical, record, survey, and other evidence. Monuments in particular are important indicators of the boundary position. The cadastral overlay and its accompanying attribute data do not generally contain monument data. As a result, the usefulness of an LIS/GIS for supporting property surveying has recently been questioned and debated at some length[25-27]. The inclusion of monument and other property survey data that are not included in a cadastral overlay leads to the creation of a database that is sometimes termed a digital cadastral database (DCDB) or cadastral evidence database. This approach specifically accommodates such conditions as multiple monuments and coordinate values for a single corner location.[28] In some instances, this database has been designed to be object-oriented, whereby cadastral "objects" (control, measurements, deed descriptions, etc.) are related in a hierarchical fashion (see Kjerne's,[29] and Kjerne and Dueker's papers[30] for a more detailed description).

In order to clarify the questions surrounding the usefulness of the cadastral overlay, it is important to remember that the overlay is an abstract representation of the real situation. Ideally, this abstraction should be based on actual field measurements to the original positions of property corners and these measurements connected to a GRF so that coordinates on a state- or county-wide basis are determined. Only then will this become a true and consistent representation of the cadastral situation. In the meantime, the cadastral overlay should represent our best approximation of the true legal situation given the data, funding, technology, and personnel available for the project.

There are several different approaches that can be used to compile and maintain a cadastral overlay. These range from simply digitizing existing tax maps to the creation of an expert system. In the list below, 10 different approaches are distinguished:

1. *Digitized tax maps in AutoCad environment.* Tax maps are simply digitized and stored as individual AutoCad drawings without any connection to a GRF or one another.

2. *Digitized tax maps "zipped" together in a GIS environment.* Tax maps are digitized, the resulting coordinates transformed into a countywide coordinate system (preferably SPCS), edge-matched, and stored in a seamless database.

3. *Photogrammetric mapping.* This involves the use of aerial photography for identifying and mapping (digitally) parcel boundaries. The effectiveness of this approach depends to a large extent on the visibility of physical features along the boundaries. This method can be used to good effect in countries like England where physical features (fences, hedges, walls, etc.) are used to demarcate boundaries.

4. *Combination of photogrammetric mapping and field measurement.* Field measurements are used as a complement to photogrammetry in order to define parcel boundaries that are not visible on the aerial photography (or orthophotography).

5. *Coordinate geometry (COGO) approach based on data from subdivision plats, deeds, and other documents.* The coordinates (preferably SPCS) of all parcel corners are computed from the bearing, distance, and angle data available from plats, deeds, and other sources. Inconsistencies are either retained as attributes to the boundary or parcel, or resolved through field survey or photogrammetric evidence.

6. *Coordinates obtained directly from field surveys connected to a GRF.* All property surveys are connected to a GRF so that field-derived state plane coordinates are determined. Coordinates can thus be entered into the database via the keyboard.

7. *Measurement management approach.*[31] This approach makes use of actual field measurements that are stored in the database and adjusted periodically to determine parcel coordinates. By focusing on measurements, as opposed to coordinates, this approach provides the flexibility to include new field measurements in subsequent adjustments in order to upgrade the coordinate values. The effectiveness of this approach, however, depends on surveyors sharing measurement data with the agency maintaining the system. For more details on this approach, see the papers by Hintz and Onsrud[32] and Buyong et al.[33]

8. *Creating a topologically structured database for parcel data in a GIS.* This involves the creation of point-line polygon topology in the database so that adjacency, intersection, and other nonmetric relationships are captured (see the section on vector-based GIS for more details).

9. *Creation of a digital cadastral database (DCDB).* This database not only includes topological relationships between points, lines, and polygons, but also relationships between other cadastral "entities," such as monuments, fences, control points, etc. that are required to make decisions about the true (original) location of boundaries.

10. *An expert system built around a DCDB.* This approach involves the addition of professional rules, norms, and conventions to a DCDB. This might include such rules as

those used to deal with conflicting title or boundary elements. This system should be viewed as a decision-support system that organizes the boundary/title evidence and aids the decision maker in asking the right questions. It is not a replacement for the surveyor or any other decision maker. For a general discussion on expert systems, see Waterman's *A Guide to Expert Systems.*[34]

From a surveying perspective, a DCDB is certainly a desirable approach, but for the purpose of supporting LIS development, it may be difficult to justify. The author believes that professional surveying organizations should initiate an effort to design and implement a DCDB. Although surveyors stand to benefit the most from such an effort, this will also serve to upgrade the cadastral overlay and therefore the LIS/GIS as a whole.

The approaches to cadastral overlay creation and management may be characterized in terms of different levels of sophistication, data, cost, accuracy, and effort. An attractive compromise between these various alternatives is to begin with an approach that is relatively quick and inexpensive, such as digitizing tax maps, and then incrementally upgrade this database by imposing basic geometric constraints and including selected measurement data.[35] In this way, a basic cadastral overlay is created with a minimum amount of effort and cost, and a mechanism provided to incrementally improve the accuracy of these data.

35-4. LIS WITHIN THE LAND ADMINISTRATION FRAMEWORK

The major responsibility of land surveyors in the United States and elsewhere is the demarcation and recordation of the spatial extent of rights to land. The kind of rights that exist in a society, and the manner in which they are described, are a function of the land tenure system operating in that community, state, or country. In other words, the land tenure system controls the way in which land is divided and allocated to individuals or groups. In the United States, the concept of private ownership (fee simple) of small, well-defined parcels of land is predominant, whereas in the traditional native American land tenure system, such ownership and division would have been a foreign and unacceptable idea.

Although the land tensure system is incorporated to a large extent in society's land policy, the operational component of land tenure is encapsulated in the land administration framework of a country. According to McLaughlin and Nichols, "land administration provides the mechanism for allocating and enforcing rights and restrictions concerning land."[36] The various subcomponents of the land administration framework, identified in Figure 35-6, are discussed below.

The land survey subcomponent deals with the demarcation and delineation of land rights. In the United States this entails the definition of the size, location, and shape of the parcel over which these rights are held, as well as the documentation of the results in a subdivision plat, survey plan, and/or deed description. Land registration deals with the definition of the nature of the land rights and the identification of the holder(s). This information is contained in a deed that is generally "registered" in a deeds registry, recorder's office, or title office for the purpose of publicly documenting the current land tenure status. In the United States this office is generally at the county level, but in other countries, it may be at the state or even federal level. The assessment of land parcels is undertaken to determine the tax liability of individual land owners. Property taxes are a major source of

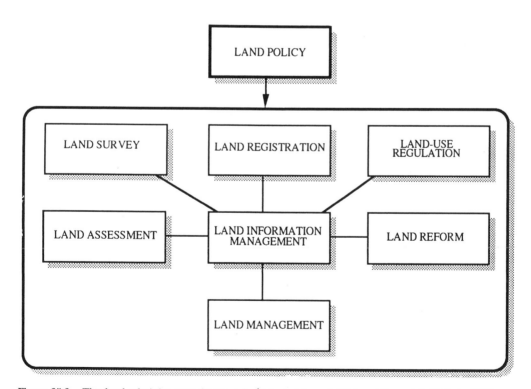

Figure 35-6. The land administration framework. (Adapted from J. McLaughlin and S. Nichols, 1989, Resource management: the land administration and cadastral systems component, *Surveying and Mapping* 49(2).

funding for most local governments, allowing them to fulfill their obligations with regard to public safety, welfare, education, health, and infrastructure.

Land reform is particularly important in countries where there is an inequitable distribution of land or fragmented rural parcels are consolidated and redivided into more economic farming units. The latter activity has been carried out extensively in such European countries as Switzerland and Austria. Land management in the context of land administration refers to practices and procedures followed on publicly owned land. In the United States, the largest landholder is, in fact, the federal government that manages large tracts of forest and parklands through agencies like the Bureau of Land Management (BLM) and Forestry Service. A subcomponent of land administration that has been receiving increasing attention is the area of land-use control or regulation. As our society has recognized the need to protect public interests in the environment and plan for more harmonious urban settlements, so the regulation of private land use (through a mechanism such as zoning and building regulations) has grown.

A common thread between all these subcomponents is their need for land information in order to carry out their respective functions in an equitable, efficient, and effective manner. From a broader land administration perspective, it is desireable to integrate the information collected and managed on a daily basis in each subcomponent. The management of this information is done by LIS/GIS that increasingly provide linkages between survey, registration, regulation, taxation, and other data in an effort to improve our land administration capabilities. Although the technology tends to pull the focus toward the hardware and software, it is important to view GIS technology in this broader land administration framework in order to appreciate the ultimate goals of such systems. It is equally important to view surveying activities in the same light, as the need for such services is ultimately driven by the broader land administration needs.

34-4-1. Land Registration Systems

No discussion of LIS is complete without a discussion of land registration systems. Together with cadastral (property) surveying, land registration forms the backbone of any juridical cadastre and is therefore an important component of a parcel-based LIS.

Most publications on land registration systems[37, 38] identify three distinct registration systems: private conveyancing, registration of deeds, and registration of title. Private conveyancing is a system that was popular in England some time ago and will not be discussed here as it is not viewed as being relevant to LIS development in North America. Before we discuss the other two systems, it should be noted that these systems are not the only alternatives open to a jurisdiction. There are many systems that have been improved in some way or another, and it is the principles behind these systems that are important as opposed to the finer details of each individual system.

35-4-2. Registration of Deeds: Rudimentary Systems

The recording or registration system used in most jurisdictions in the United States is known as a rudimentary registration of deeds (ROD) system. Under this system, the instrument used to convey land is a deed drawn up by a competent person, or the parties to the transaction. There is no legal requirement to register the deed in the local registry or recorder's office, but many of these documents are, in fact, registered because of mortgage loan requirements. Typically, a deed will describe parties to the transaction (grantor and grantee) and their details (address, marital status, etc.); signature of person who prepared the deed; signature of grantor, and a legal description that may be a simple metes-and-bounds description. Further details can be obtained from any of a number of books published on the subject of real estate transactions[39] or legal principles in surveying.[40]

When deeds are archived in the deeds registry, they are generally indexed by entering the names of the parties in a grantor/grantee index and grantee/grantor index. The former are arranged alphabetically according to the last name of the grantor and the latter according to the last name of the grantee. In order to prove that a person has clear title to a piece of land, it is necessary to search back through all the deeds affecting that parcel. The documents uncovered in this search form what is commonly known as a "chain of title." Often, this search must stretch back to the time that the land was first transferred from the public domain, which in many areas may be well over a hundred years. The following extract demonstrates the extremes to which these chains of title can stretch:

A New Orleans attorney called at the RFC office regarding a loan for his Louisiana client. After exhaustive work, the attorney ran the chain of title back to 1803 and sent the report to RFC as instructed. Presently, he received a letter from the RFC complimenting him on the able manner in which the title abstract was prepared, but stating that approval could not be given for the loan until title was searched for the period to 1803. The attorney's reply was a classic and is reproduced herewith:
Your letter regarding titles in CASE No. 802649 received. I note you wish titles to extend further than I have presented them. I was unaware that any educated man in the world failed to know that Louisiana was purchased by the United States from France in 1803. The title to land was acquired by France by right of conquest from Spain. The land came into possession of Spain by right of discovery made in 1492 by a Spanish-Portuguese sailor named Christopher Columbus, who had been granted the privilege of seeking a new route to India by then reigning monarch, Queen Isabella.
The good Queen, being a pious woman and careful about titles (almost as careful, I might say, as the RFC) took the precaution of securing the blessing of the Pope of Rome upon the voyage before she sold her jewels to help Columbus. Now, the Pope, as you know, is the emissary of Jesus Christ, who is the son of God, and God,

it is commonly accepted, made the world. Therefore, I believe it is safe to presume that he also made that part of the United States called "Louisiana," and I hope to hell you're satisfied.[41]

In almost all states (excluding Iowa where the code specifically prohibits in-state sales of title insurance; see Wall's paper,[42] landowners are required by lending institutions to take out a title insurance policy in order to provide some security against "loss or damage" arising from valid claims on the property. It should be pointed out that in most cases, this policy merely provides financial support for loss or damage, costs, and legal fees in the case of a suit. It does not prevent a claimant bringing a successful suit against the holder and dispossessing that person of their title. This is not the case in a registration of titles system in which the state guarantees title.

4-3. Improved Registration of Deeds System

In many jurisdictions, the rudimentary system discussed in the previous section has been modernized and improved to deal with the modern-day demands on the system. Generally, the motivation for such improvement comes from increases in the frequency of transactions and the realization that the registry is, in fact, a very valuable information system and not merely a passive archive for storing documents. Listed below are examples of improvements that have been made to rudimentary deeds systems in North America and elsewhere:

Requiring that all deeds be registered in a public deeds registry (compulsory registration)
Giving registered deeds a higher legal priority than unregistered deeds
Requiring that all deeds refer to a corresponding survey plan
Examination of deeds to check for consistency with parent deeds
Creation of a parcel-based or tract index that contains one page per parcel with each page

reflecting transaction details for each individual parcel

Establishment of a register organized on a parcel-by-parcel basis that contains certain details abstracted from the deed (grantor, grantee, deed references, plan references, etc.)

Limiting the "chain-of-title" search to a certain period (such as 20 yr) and discounting all documents prior to that date

Computerizing all indices in the registry to make searching more efficient

Microfilming all deeds and documents for safer archiving

The trend is toward systems in which the data are complete, reliable, consistent, and easily accessible. Improved systems operate in France, South Africa, Holland, Brazil, and many other countries. Although these systems differ quite significantly from one another, they all contain certain improved elements like those listed above. In some cases, these so-called improved systems may even resemble registration of title systems. In discussing the South African registration system, Simpson remarks that:

> Since there is no country in which land parcels are more closely defined than in South Africa, and since no transaction in land there can be affected without being registered under the parcel to which it relates, the only reason for classifying the South African system as deeds rather than title registration would appear to be that, technically, it is not the fact of registration which proves title but the document, if duly registered.[43]

Once again, the important aspect of this discussion is not the definition of certain registration systems, but rather the principles that characterize these systems. Modernization of land records systems should not only be seen as the computerization of land-related data, but also as an opportunity to improve the land registration system through incorporating one or more of the above mentioned principles, or by moving to a registration of title system.

35-4-4. Registration of Title

The Torrens registration of title (ROT) system is one of the earliest and certainly one of the most popular title registration systems in the world. While serving as Registrar General in South Australia in 1858, R. Torrens designed and implemented the system that has been exported to many countries around the world. Torrens argued that a land registration system should be organized around the units of registration (land parcels), not around the persons who were purchasing or selling those units. He rightfully pointed out that a parcel of land is far more permanent than any grantor or grantee and is therefore a much more convenient unit of reference for ownership and other land tenure data. Furthermore, it can be argued, the rapid accumulation of documents, and the need to search through all these documents to prove title, will make searching increasingly more cumbersome and less efficient. Torrens asserted that if the most recently registered document could portray the current status of a parcel of land, then there would be no such accumulation of deeds and no need to carry out time-consuming retrospective searches.

Most discussions of Torrens' ROT system focus on three central principles:

Curtain principle

Mirror principle

Title guarantee principle

The curtain principle can be described as the drawing of a curtain behind the current instrument of transfer. This means that all prior documents are ignored and it is not necessary to carry out a chain-of-title search to prove title. The current instrument reflects the current legal status with regard to the associated parcel of land. This is the so-called mirror principle. In the Torrens system, the instruments of transfer are known as certificates of title (see Figure 35-7) and one certificate is created for each parcel of land. These certificates are bound together in a register. In order to determine who owns a piece of land, one need only refer to the relevant page in the register.

The importance of the register is further enhanced by a state guarantee that protects all owners shown on the register. If a valid claim is made against an owner on the register, the state will compensate the claimant, thereby

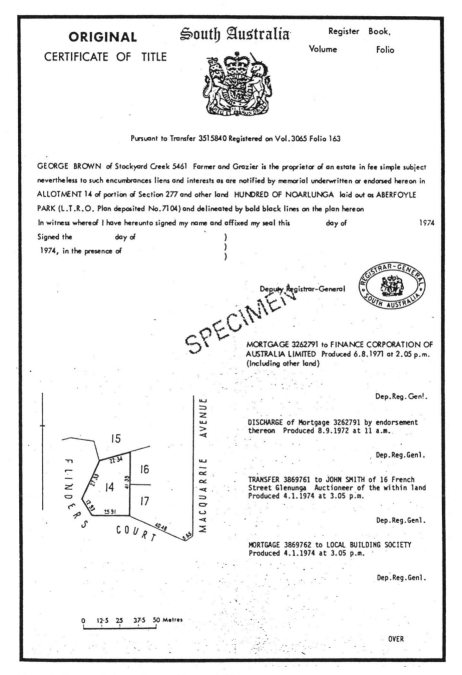

Figure 35-7.   South Australian certificate of title.

protecting the validity of the register. In other words, title under this system is indefeasible. This state guarantee of title should not be confused with title insurance that does not provide indefeasible title. Under the Torrens system in Australia, an assurance fund was set up to provide the means for compensating claims against the register. Initially, adverse possession was also not recognized as it was thought that this would undermine the veracity of the register. In recent times, this thinking has been reversed and adverse possession is now accepted in many Torrens jurisdictions.[44]

Since there has been a great deal of debate and criticism of Torrens ROT systems in the United States, it is useful to reexamine the original principles that Torrens viewed as being important. The list given below is a summary version of the principles advanced by Torrens in a Publication entitled *An Essay on the Transfer of Land by Registration*:[45]

> Title should not be retrospective: It should not be necessary to search back through multiple documents in order to establish title.
>
> An estate should only be transferred once the public register has been updated.
>
> Registered title should be indefeasible.
>
> Registration should be compulsory.
>
> Once a parcel of land is registered and placed on the register, it may not be "deregistered."
>
> Parcels of land should be defined by ground survey and the resulting map or plat shown on the certificate of title.

When an encumberance, such as a mortgage, is extinguished, the details reflected on the certificate are simply cancelled. The certificate also reflects many secondary rights or encumberances, such as easements, leases, or charges, that may be associated with an individual parcel. Most land rights can therefore be ascertained by locating the relevant page in the register and examining the certificate that constitutes that page.

Given the many benefits that the Torrens ROT system appears to offer, why has this system not caught on in the United States? At one stage, Torrens legislation existed in over 20 states, but there have been virtually no success stories. Even in Cook County, IL, which had an ideal opportunity to introduce the system when the great Chicago fire destroyed almost all the publicly recorded documents, this has not met with much success. In fact, there has been a surprising amount of opposition to this system, primarily because the system labeled as Torrens in the United States did not include some of the fundamental principles originally advanced by Torrens.[46] Some of the criticisms that have been leveled against the Torrens system include (1) title insurance is required for many Torrens titles, (therefore, they cannot be regarded as more secure than title under a ROD system), (2) ROT does not clear up title defects or prevent them from arising, (3) initial registration is an expensive process, (4) certificates of title do not reflect all possible claims on a parcel of land, (5) state guarantee of title does not provide adequate protection for individuals with valid claims against registered parcels, (6) maps used to support ROT are "cumbersome" and there are now more "efficient means" of relocating property corners.[47, 48]

A brief examination of the basic Torrens principles listed previously will reveal that the systems being called "Torrens" in the United States differ quite significantly from the system envisaged by the originator. Title insurance is an American institution and outside of the United States and parts of Canada does not exist. Title insurance companies have been in operation since the late 19th century and the requirement for such insurance has become part of any land conveyance that uses the parcel of land as collateral for a bank loan. Therefore, it is not surprising that this is still required by banks who do not understand the benefits or workings of an ROT system.[49] In the United States these policies are more a part of the banking/credit system than they are a necessary part of the registration system. The fact is that thousands of ROT, as well as ROD, systems around the world operate quite securely and effectively without any form of title insurance.

Before a piece of land is placed on the Torrens register, every effort is made to locate individuals or groups who have an interest in the land. In many instances, the public record is also searched, but in some developing countries, initial registration has been based solely on the documentary evidence presented by landholders. In this initial registration process, every effort is made to resolve conflicts and existing or potential title defects, particularly when the ROT is implemented on a systematic basis throughout a jurisdiction. It is true that this process may be expensive, but to date no studies have been conducted that prove this assertion or compare this cost with the expense of continuing with an inefficient system. Notable studies that have advanced our knowledge in this regard include those by Larsen[50] and Jeffress.[51]

Another criticism leveled at the Torrens system is that rights such as liens are not reflected on the certificate of title. This assertion is true, but whenever there is some secondary claim against the title, a caveat is placed over the property and all transactions involving that parcel are frozen. This is designed to take care of minor claims that do not qualify as high-frequency transactions.[52]

State guarantees of title have been severely criticized by those opposing Torrens ROT in the United States. In particular, they point to the need to set aside an assurance fund and the treatment of unregistered rights. It is rather strange that so much attention is given to a feature that, although a major component of the original Torrens system, has ceased to be an active part of many of these systems. For example, during the first 18 yr of the Torrens system in the state of New South Wales (Australia), no claims were made against the state. The infrequency of claims is actually a measure of the success of the Torrens system and some states in Australia, e.g., have even closed their assurance funds.[53]

The notion that old "cumbersome" maps or plats no longer have a role in the relocation of original boundaries because more efficient methods are now available is tantamount to withdrawing a major source of boundary evidence. In some instances, this may be the only source of evidence. Cadastral maps or plats, regardless of the registration system in use, provide vital information that is essential for the accurate relocation of property boundaries on the ground. This is particularly true when all physical evidence (monuments, fences, etc.) has disappeared. The emergence of accurate positioning technologies, such as GPS, may change boundary definition in the future, but historical research of old documents and maps will always be a part of a land surveyor's job.

Williams throws further light on the question of the Torrens system's dismal performance in the United States:

> It [Torrens ROT] failed in numerous states, most notably in California, because the statutes were poorly written and do not make the certificate conclusive in all cases except fraud. . . . Opposition by vested interests, most notably title insurers, has no doubt played a large role in discouraging the use of [Torrens] registration, as has the banking industry's reluctance to finance a mortgage on a certificate . . . The real problem is the abysmal unfamiliarity with
> Torrens registration among lawyers, abstracters, banks, and insurers.[54]

Another problem has been the deviation from some of Torrens' original principles, particularly the principle of compulsory registration. Given the lack of success of the Torrens systems in the United States, a more viable approach may be to improve the existing ROD systems. This is already occurring in many counties around the country through the creation of a tract index, placing limits on chain-of-title searches and other measures listed previously in the section on the improved registration of deeds. The advantage of such an approach is that it represents an incremental improvement over time, not the drastic changes that would accompany the introduction of a true Torrens system. However, opportunities to modernize the land registration system as part of the LIS development process are not being adequately exploited in the United States and future efforts should actively involve registrars in the land records modernization process.

35-4-5. The Role of Surveying in Land Registration

Since the primary unit of registration is the land parcel, the land surveyor plays a major role in defining the location, extent (area), and dimensions of the parcel. As registration systems move from a person-based (grantor-grantee) system to a parcel-based system, one can expect this role to increase.

Rudimentary metes-and-bounds descriptions (e.g., bounded on the north by Jones and the east by Smith) are an effective means of describing land parcels if boundaries are well known and there are relatively few land transfers. However, they contribute very little in terms of information on absolute location, dimensions, and area, all basic requirements for defining a parcel. Thankfully, these rudimentary descriptions are increasingly being replaced by proper surveys. Although many of these surveys are not connected to any geodetic reference framework, there is increasing pressure for surveyors to tie into these networks. This is important for adequately defining the spatial attributes of parcels, but it is becoming equally important for the development of parcel-based LIS.

Simpson asserted that there are few places in the world where parcels are more accurately defined than in the South African system. Accuracy is clearly a desirable characteristic as far as land registration and LIS are concerned, but perhaps more important, the South African system defines all spatial attributes of parcels in a concise and complete manner. These data are represented on a survey diagram that is attached to every deed concerned with the transfer of land rights.

Figure 35-8 shows a sample diagram that contains the following spatial attributes:

Shape and dimensions (sides and azimuth)

Coordinates of all parcel corners relative to a country-wide coordinate system (parcel identifiers)

Unique parcel identifier

Administrative district

Physical description of monuments ("beacons")

Geodetic control points to which survey was connected

Whenever a deed refers to rights or restrictions over a part or the whole parcel, it refers directly to the attached survey diagram. In this way, land surveyors deal with the spatial aspects of land registration and lawyers (conveyancers) with the legal, nonspatial aspects. The quality of the product (deed/diagram) is such that no state guarantee or title insurance is required to provide security to the system.

Although deeds in the United States increasingly refer to survey plans, particularly subdivision plats, opportunities for the creation of country-wide cadastral overlays are being missed by the surveying profession. If land surveyors wish to enter the mainstream of LIS development, they will have to adopt a broader outlook toward property surveys and move away from a purely measurement-oriented role. This is particularly true in the light of advances in geopositioning technologies like global positioning system (GPS) that simplify the measurement process and therefore make measurement more accessible to untrained individuals outside the surveying profession.

Advances in information systems technologies, such as GIS, provide new opportunities for integrating surveyed spatial data with the attribute data that are traditionally captured in land registration systems. In fact, GIS offers many other opportunities for those of us that deal with spatial information. Before examining potential applications of this technology it is necessary to understand more about the technology itself.

35-5. GEOGRAPHIC INFORMATION SYSTEMS

The following GIS definitions, taken from several of the foremost GIS specialists, describe the principal components of a GIS. Marble[55] explains that a GIS is composed of the following components:

Data input component for collecting and processing spatial data

Data storage and retrieval component that organizes the spatial data so that they can be efficiently retrieved and edited

Data manipulation and analysis component, which reconfigures (generalizes, aggregates, etc.) data for specific purposes

Data-reporting component for displaying data in a tabular or map form

Other definitions treat GIS as a set of tools for managing and analyzing spatial data:

...A powerful set of tools for collecting, storing, retrieving at will, transforming, and displaying spatial data from the real world.[56]

...Computer assisted systems for the capture, storage, retrieval, analysis and display of spatial data.[57]

GIS rely on the integration of three distinct aspects of computer technology: database management (of graphic and nongraphic data); routines for manipulating, displaying, and plotting graphic representations of the data; and algorithms and techniques that facilitate spatial analysis.[58] Increasingly, GIS are being seen as systems for integrating, managing, and analyzing spatial data from a diverse range of sources. These data may be tabular (e.g., tables containing parcel identifiers, owners' names, street addresses, assessed value), graphic (e.g., maps, plats, drawings), image

(e.g., scanned images of deeds and aerial photographs), raster (e.g., Landsat and SPOT images), or video data. Two general approaches, raster and vector, have been used to organize the data in a GIS.

35-5-1. Raster-Based GIS

Raster-based GIS make use of a series of grid cells or pixels as a spatial framework for referencing data. The advantage of this approach is that it provides a simple method for investigating the spatial relationships between different features as each feature is referenced to a particular cell(s), and each cell is addressable, as shown in Figure 35-9. A few decades ago, organizing data in this manner would require a manually drafted raster transparency that was then overlaid on the map and the relevant attributes of each cell noted. To compare data from another map of the same area, the same procedure would be followed and the relevant data recorded against each cell. The two sets of attributes for each individual cell could then be compared to reveal the status of that particular area on the ground.

Present-day technology has significantly improved our ability to capture raster data more efficiently. Most notably, advances in satellite-borne scanners, such as the multispectral scanner (MSS) and thematic mapper (TM) of LandSat, and the high-resolution visible (HRV) instruments of the SPOT system now enable us to capture large sets of data on an ongoing basis. These systems produce raster data in digital form, with pixel values representing reflectance values of particular wavebands. The resolution, or area of the pixel, in the LandSat and SPOT systems varies from 80 × 80 m down to 10 × 10 m (SPOT panchromatic). Although these scanners do not have a high enough resolution for capturing parcel boundary data, they are extremely useful for capturing land cover and land-use data over large areas. Further details on remote sensing can be found in the work of Lillesand and Kiefer,[59] the American Society of Photogrammetry,[60] and Campbell.[61]

SIDES METRES		DIRECTIONS		CO-ORDINATES Y SYSTEM Lo 29° X			S.G. No.
			Constants	− 91 000,00	+3 070 000,00		
AB	44,33	322 27 20	A	− ... 727,98	+ 2 026,07		*Approved*
BC	22,91	52 25 00	B	− 754,99	+ 2 061,22		
CD	44,32	142 27 10	C	− 736,84	+ 2 075,19		
DA	22,91	232 24 40	D	− 709,83	+ 2 040,05		
DE	6,10	232 24 40	E	− 714,66	+ 2 036,33		*Surveyor-General.*
EF	8,63	7 25 50	F	− 713,54	+ 2 044,88		
FD	6,10	142 27 10					
			11 M8 ⊕	− 572,68	+ 1 878,22		
			54 M8 ⊕	− 818,36	+ 2 198,17		

The figure D E F represents a Road Servitude.

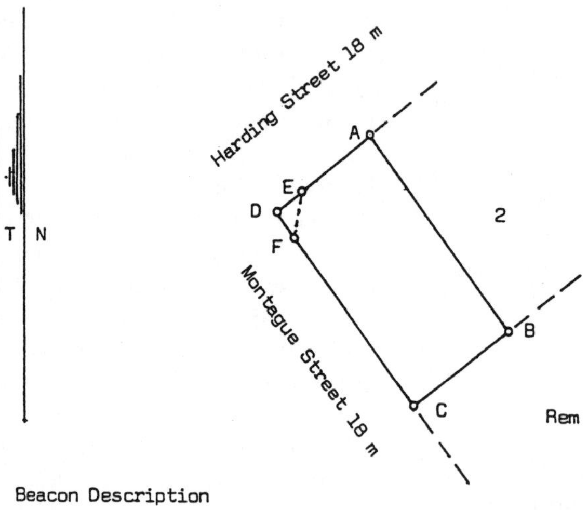

Beacon Description

A,B,C,E,F − 12 mm iron pegs

D   − 12 mm iron pipe 0,1 m west of wooden corner post

SCALE: 1 : 1 000

The figure   A B C D

represents   1 015 square metres   *of land, being*

SUB 1 OF LOT 599 NEWCASTLE TOWNSHIP

situate in the Borough of Newcastle, County of Klip River   *Province of Natal*

surveyed in November 1973

by me
                                                        Land   Surveyor.

This diagram relates to	*The original diagram is* Gr.	*File No.*
No.	Vol. *No.* 59 Fol.101	*S.R. No.*
	~~*Transfer*~~/ Grant	*Comp.* HS—8B—4B—3
Registrar of Deeds	*No.* 3100	*Degree Sheet* 14

Figure 35-8.   South African property diagram.

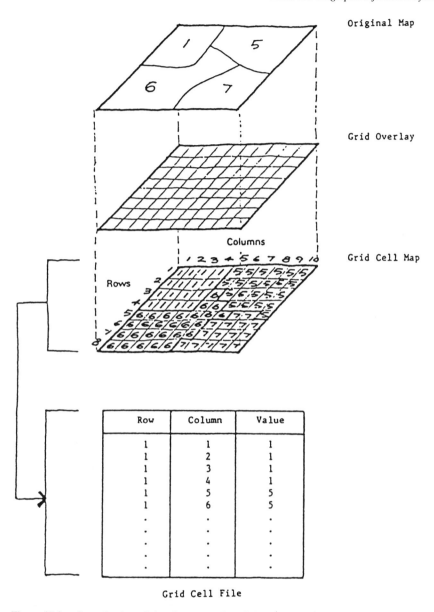

Original Map

Grid Overlay

Grid Cell Map

Columns

Rows

Grid Cell File

Figure 35-9. Organization of data in a raster-based GIS. (From J. Dangermond, 1983, A classification of software components commonly used in geographic information systems, *Design and Implementation of Computer-Based Geographic Information System* (D. Peuquet and J. O'Callaghan, Eds.), New York: IGU Commission on Geographical Data Sensing and Processing.)

An area that is receiving increasing attention and holds great promise for GIS is digital photogrammetry. Digital cameras make it possible to capture terrestrial and aerial data directly in digital form. Although this is clearly another form of remote sensing and the digital products do not equal the quality of traditional aerial photography, digital cameras have several advantages over satellite-based remote sensing. Most important, the pixel resolution is higher (pixel size varies from 5 to 60 μ) and there is more flexibility in the selection of resolution as this is to a large degree a function of the flying height. In addition, the geometric accuracy of the data obtained via digital photogrammetry is superior to that of satellite images.

Most of the efficient map-scanning devices produce digital raster data by scanning the manual map and storing the result in an array of pixels. The initial capture of these data is certainly much more efficient than manual digitizing methods; however, the editing that is required subsequent to capturing the data tends to make scanning somewhat less attractive. The complexity of the original map (different line weights, shading, stains) also influences the decision of whether to use scanning or manual digitizing. Current research into the automation of the editing process and more effective raster to vector conversion routines will no doubt make scanning an increasingly attractive option for the acquistion of map data.

Raster-based GIS have limited applicability to parcel mapping and most of the types of data that land surveyors both produce and use. Since LIS are essentially parcel-based, vector-based GIS that record linear data more precisely are generally selected to support LIS.

35-5-2. Vector-Based GIS

Vector-based GIS generally treat map data as point, line, and polygon entities. Points are defined by X- and Y-coordinates, preferably in a "real-world" coordinate system (e.g., state plane, UTM), but local coordinate systems as well as digitizer table coordinates are also commonly used. Lines are defined by pairs of specific points, and polygons by a series of lines. The interrelationship of these three entitites in a GIS database is central to the database structure in a vector-based GIS.

When humans look at a map, we can automatically, or implicitly, tell which points are the corners of a particular parcel and which lines constitute the boundaries of that parcel (polygon). In a computer environment, such implicit relationships must be explicitly defined because computers cannot intuitively derive these relationships. In most GIS, point-line-polygon topology is used to define these relationships. Topology in this context refers to nonmetric relationships between different entities and is essential for describing conditions of connectivity, intersection, and adjacency. Figure 35-10 shows an area covered by five parcels that will be used to illustrate how basic topological relationships are incorporated in a database. The state plane coordinates for all property corners and boundary distances are as shown.

The point table (see Table 35-1) defines the spatial location of all point entities. It does this by associating a unique point identifier with its relevant X- and Y-coordinates. Generally, the type of point (e.g., property corner, centroid) is included in the table so that points may be separated into thematic layers. Other data items, such as monument description, date set, surveyor's name, date, etc., can be added to this table or created in a separate table and related to the point table through the point identifiers. For a more detailed discussion of cadastral database design issues, readers are encouraged to refer to von Meyer's paper.[62]

The line table (see Table 35-2) not only defines the two points defining the terminals of the line, but defining one of these points as a beginning point and the other as an endpoint, a direction is imparted to the line. This facilitates the inclusion of adjacency conditions that are defined by means of the polygon identifiers situated to the left (LPOLY) and right (RPOLY) of the line. As in the point table, the type of line (e.g., boundary, street centerline, powerline, pipeline, etc.) is included so that different types of lines can be distinguished and dealt with separately when required. Attribute data that are not part of the topology can also be included in the table. These may include distances (from a deed, plat, or current survey), dates (e.g., date of survey, date pipe laid), physical description

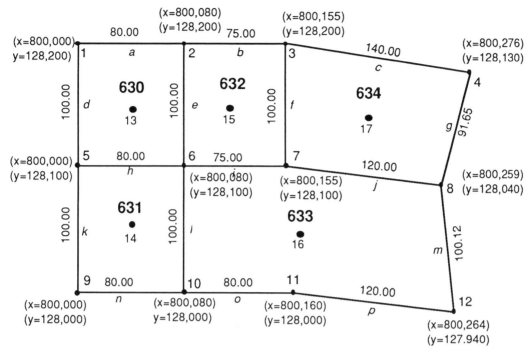

Figure 35-10.   Parcel map showing point, line, and parcel identifiers (not to scale).

(e.g., diameter of pipe, voltage, material of pipe), and many other types of attributes.

The polygon table (see Table 35-3) lists the polygon identifiers and centroid of each polygon. The term centroid is often used loosely in the GIS literature to mean any point that is more or less in the center of the polygon and is therefore more accurately regarded as a pseudocentroid. This point is useful for labeling polygons or displaying certain attribute

Table 35-1.   Point table

Point ID	X-Coordinate	Y-Coordinate	Type
1	800,000.00	128,200.00	Parcel corner
2	800,080.00	128,200.00	Parcel corner
3	800,155.00	128,200.00	Parcel corner
4	800,276.24	128,130.00	Parcel corner
5	800,000.00	128,100.00	Parcel corner
6	800,080.00	128,100.00	Parcel corner
7	800,155.00	128,100.00	Parcel corner
8	800,258.92	128,040.00	Parcel corner
9	800,000.00	128,000.00	Parcel corner
10	800,080.00	128,000.00	Parcel corner
11	800,160.00	128,000.00	Parcel corner
12	800,263.92	127,940.00	Parcel corner
13	800,040.00	128,150.00	Parcel centroid
14	800,040.00	128,050.00	Parcel corner
15	800,117.50	128,150.00	Parcel corner
16	800,171.96	128,080.00	Parcel corner
17	800,206.96	128,120.00	Parcel corner

Table 35-2.   Line table

Line ID	Begin Pt.	End Pt.	L. Poly.	R. Poly.	Type	Distance
a	1	2	—	630	Boundary	80.00
b	2	3	—	632	Boundary	75.00
c	3	4	—	634	Boundary	140.00
d	1	5	630	—	Boundary	100.00
e	2	6	632	630	Boundary	100.00
f	3	7	634	632	Boundary	100.00
g	4	8	—	634	Boundary	91.65
h	5	6	630	631	Boundary	80.00
i	6	7	632	633	Boundary	75.00
j	7	8	634	633	Boundary	120.00
k	5	9	631	—	Boundary	100.00
l	6	10	633	631	Boundary	100.00
m	8	12	—	633	Boundary	100.12
n	9	10	631	—	Boundary	80.00
o	10	11	633	—	Boundary	80.00
p	11	12	633	—	Boundary	120.00

data (e.g., area, assessed value, owner's name) in a graphic form. In many instances, the centroid is used for referencing attribute data and may even be used to graphically represent the parcel when the boundaries of the parcel are unimportant. In Milwaukee, this approach has been used to analyze the distribution of building permits granted during a specific period, as well as spatial patterns of assessed property values, age of buildings, liquor license applications, property tax delinquencies, and many other citywide applications.[63] Due to the vast amount of data needed to topologically define every single property parcel in a jurisdiction such as a county, this topology is excluded from the database and all parcel-related data are associated with the parcel centroid. Huxhold explains:

Many large jurisdictions, for example, do not create parcels as individual polgons, but choose

Table 35-3.   Polygon table

Poly ID	Centroid	Type	Vertices
630	13	Parcel	1,2,6,5,1
631	14	Parcel	5,6,10,9,5
632	15	Parcel	2,3,7,6,2
633	16	Parcel	6,7,8,12,11,10,6
634	17	Parcel	3,4,8,7,3

rather to create boundaries as separate lines that, when displayed on a map, correctly portray the parcels. Parcel identifiers [centroids] are then usually digitized on a separate overlay that can be combined with the parcel boundary overlay to display the parcels with their identifiers. Without explicit polygon definitions, these parcels cannot be manipulated as geometric entities (to perform area calculation, area shading, point-in-polygon determination, etc.). They can, however, be displayed and viewed by the map user for many valuable applications (parcel map retrieval, parcel attribute display, parcel map overlay with other geographic features, etc.) as long as unique parcel identifiers are included in the base map information. These identifiers should be located inside the parcel so that their *XY* coordinates can be used to associate attribute data with the correct location of the parcel when its boundaries are displayed.[64]

It should be pointed out that from a surveying standpoint (as opposed to a more general GIS viewpoint), the inclusion of parcel topology is highly desirable since many surveying applications focus on the parcel level.

Many GIS make use of a relational database structure for organizing and managing attribute data (including topological data). This structure essentially consists of a series of tables that are linked or related by means of common data fields as illustrated in Figure 35-11. The flexibility and minimization of data redundancies give relational databases an advantage over those structured on a hierarchical or network basis. Conceptually, any number of related tables may be created, but memory and processing efficiency will clearly constrain the size of the database. In Figure 35-11, five related tables containing parcel, assessment, owner, census, and deed data are shown. As the arrows indicate, the parcel table can be related to the assessment, owner, and deed tables via the parcel identifier. Likewise, the parcel table can be linked to the census table through the census tract number.

Although there appears to be increasing debate about the role of GIS in managing data from property surveys (see the section on the cadastral overlay), it should be clear from the above discussion of relational databases that this structure is flexible enough to handle most data. In the context of property surveys, different tables could be created for different types of evidence (record, measurement, phys-

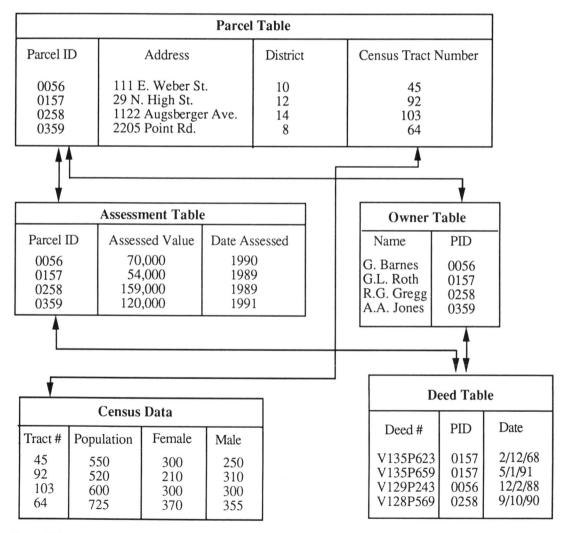

Figure 35-11.   Relational database structure for storing attribute data.

ical, etc.). In deciding on the exact location of the original boundary, surveyors will compare evidence from various sources and arrive at a decision even if this evidence is conflicting. In most retracement surveys, such conflict will exist, but this does not mean that any of the evidence should be ignored and therefore left out of the database. Rather it should be included in the database and "related" to associated evidence data.

35-5-3. Geocoding

One of the most valuable assets of a GIS is its capability to integrate data from different sources. These data must, however, be spatially referenced. In other words, all attribute data must be associated with or referenced to a spatial unit that can be mapped. Examples of such reference units include parcels, blocks, districts, subdivisions, zoning districts, census tracts, garbage collection districts, land-use units, forestry units, townships, sections, easements, health districts, emergency response zones, incorporated areas, and many other units. In Milwaukee, over 200 of these reference units were found to be used for the various urban management and administration activities.[65] In cases where these units are components of a larger area (e.g., sections and townships, parcels and blocks), data from the smaller units can simply be aggregated up to the larger units. However, in many cases, these units do not share any boundaries and were chosen without any consideration of the other units. In urban environments, addresses may be used as a bridge between these units and the attribute data referenced to these units. Huxhold describes the situation as follows:

> Because local government has such a direct impact on individuals and their properties, most of its functions require an address as part of their information base, and these addresses can be located in each of the 200 different geographic areas. A given address, for example, could be located in Census Tract 43, Aldermanic District 9, Housing Code Inspection District 213, Solid Waste Collection Route 3A, Police Squad Area

C23, and so on. Conceivably, then, each address could have over 200 geocodes assigned to it in order to be able to summarize data by any geographic area... .

> The assignment of geocodes to records in nongraphics data bases is called *geocoding*. Geocoding is a process that is performed either manually or with a computer to assign location identifiers to data for further geoprocessing of the data for analysis.[66]

In order to perform geocoding in a computerized mode, a geographic base file is created that links addresses with the various spatial units so that the units associated with each address (or address string) are explicitly included in a single table or set of related tables. To some extent, the parcel file shown in Figure 35-11 acts as a geographic base file by cross-referencing parcel, district, and census tract units. The most well-known geographic base files are the GBF/DIME and TIGER files created by the U.S. Census Bureau for the 1970 and 1990 population censuses. These files make use of street segments, usually the line between two street intersections, as a basis for connecting addresses to other units. Although the precision of the graphic data in the DIME and TIGER files is inadequate for cadastral purposes, it is extremely useful for regional and network analysis as well as infrastructure management. For more information on these files, see the work of Marx,[67] the U.S. Bureau of the Census,[68] and Sobel.[69]

35-5-4. GIS Applications

GIS are used in a number of different disciplines for a broad range of applications. Any discipline, or interdiscipline, concerned with spatially related data can benefit from this technology. In a paper on the current and potential uses of GIS, Tomlinson specifically mentions the following application areas: forestry; property and land parcel data; utilities; transport, facility, and distribution planning; civil engineering; agriculture and environment. GIS is also being used extensively for modernizing land records, particularly in highly urbanized areas. Since this application leads directly to the creation of LIS, a more detailed discussion of three counties that have incorporated GIS in their modernization efforts is given in the next section.

35-6. CASE STUDIES

Three case studies are included in the final section of this chapter. They are all county-level initiatives that, because of their dominant urban character, were developed with a strong focus on parcel-level data. In addition to addressing the various technical components of these systems, we have attempted to portray the motivation behind each of these projects.

35-6-1. Wyandotte County, KS

Wyandotte County is a 159-sq-mi area situated in the Kansas City metropolitan area. The county has been described as "an aging industrial center experiencing problems typical of older, blue-collar communities in the United States."[70] The primary motivation for the development of an LIS in Wyandotte County was to improve the assessment records and collection of delinquent property taxes. Problems in the assessment system were initially identified in 1971 when a new computerized billing system was introduced. The large amount of delinquent taxes subsequently became a serious political issue, particularly after the county surveyor estimated that as much as $9 million was owed to the county. Having identified the dimensions of this problem, he suggested that the inadequate assessment system could be improved through the development of an LIS based on an accurate parcel map. Much of the early success of this LIS can therefore be attributed to the county surveyor, who became known as the "$9 million man!"[71]

Like many county systems, the Wyandotte County LIS, known as LANDS, began with a major base mapping program that was to serve as a basis for the cadastral overlay. This base map was compiled stereo-photogrammetrically from 1:1200 scale aerial photography and produced at an identical scale. The map shows planimetric details (such as buildings, roads, and utility poles) as well as contours at a 2-ft contour interval. The Kansas state plane coordinate system is indicated at 500-ft intervals, and horizontal and vertical accuracies are estimated to be in the range of 3 ft and 1 ft, respectively.[72] Two series of maps are maintained within LANDS. The 1:2400 scale assessment parcel section maps (see Figure 35-12) are derived from the quarter-section base maps (see Figure 35-13).

The cadastral overlay was added to the base map by referring to deeds, assessment data, and subdivision plats and fitting these data to the graphic planimetric details of the base map. Rhodes described the process as follows:

> Because Kansas is part of the United States Public Land Survey, section corners and lines are the basis for all parcel descriptions. In Wyandotte County, however, a precise and accurate location for these corners has not been available. Monumentation has either been physically obliterated, lost, or made ambiguous by perpetuation of multiple positions for the same corner. Subsequently, deeded property descriptions are not definitive references to parcel location. Even subdivision plats could not be plotted accurately in many cases. Many gaps and overlaps were noted as parcel boundaries were added to the maps, although, the anomalies are not shown.

> The bottom line in all this is that accurate parcel maps are essential to a modern land records system, however, the cost of remonumentation was prohibitive. Therefore, an incremental approach was taken ensuring an adaptable mapping process in which new, more precise location data could be incorporated as it became available.[73]

Although gaps and overlaps are not shown graphically on the cadastral overlay, conflicting data are retained and stored as attributes to the relevant parcel(s).

The fact that there is no compulsory registration of deeds in the county created further

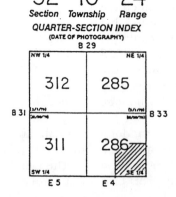

Figure 35-12. Wyandotte County assessment parcel section map.

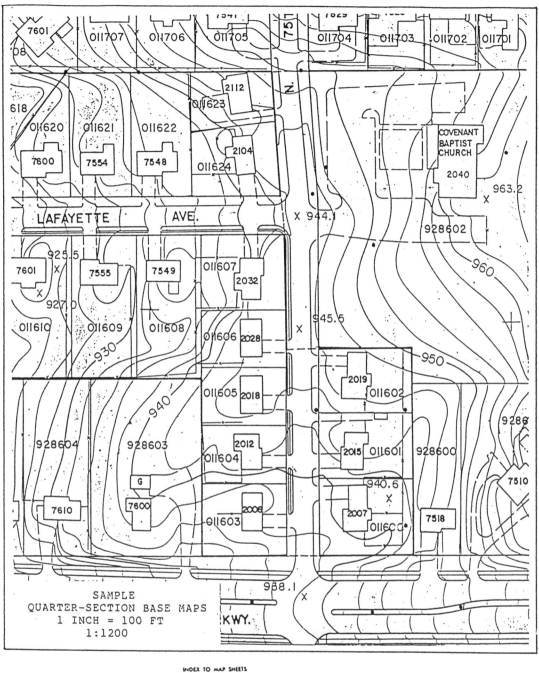

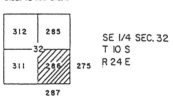

Figure 35-13. Wyandotte County quarter-section base map.

problems as it meant that the public record was incomplete. All of the 66,000 parcels in the county were manually digitized using customized software and parcel topology added to these data. The inclusion of this topology is considered an essential part of the cadastral overlay. A six-digit parcel identifier (e.g., 931631) has been designed based on the number of the subdivision or tract of land (first four digits) and an actual parcel number (last two digits).

Although the need for a geodetic reference framework was recognized early on in the project, it was only in 1988 that resources were made available for this purpose. The network was designed to have a uniform density of approximately 2 mi, with some additional control in areas where future development was anticipated. The establishment of the network (monumentation, observations, adjustment, etc.) was undertaken as a cooperative project involving federal government agencies, county agencies, and private surveyors. GPS technology was used to determine positions for the 67 points in the network and the observing work was completed in a mere 3 wk. Field closures achieved ranged from 1 : 240,000 to better than 1 : 3,000,000.[74] A series of azimuth marks were also established approximately halfway between the GPS points. These positions will be determined by traverse to second-order, class II standards, thereby providing an accurate geodetic framework at 1-mi density. The final component of this cooperative project was a 5-day workshop offered to local land surveyors on GPS, state plane coordinates, coordinate transformations, and other related topics. This should be part of any LIS project as it not only provides education, but also promotes the cooperation of local professionals who are vital to the sustainability of the entire effort.

In 1989, a new set of aerial photography was flown in order to update the planimetric database. All of the GPS points and 135 section corners were targeted prior to flying so that coordinates for the latter could be determined photogrammetrically.

The major node in the Wyandotte County LIS/GIS is a parcel-based database known as LANDS. Crane described LANDS as follows:

> LANDS, an acronym for Land Data System, has been operational since 1982. LANDS is a multi-purpose land information system consisting of large-scale planimetric base maps, a consolidated data base of land information, and automated mapping capabilities; all tied together by a 6-digit unique parcel ID . . . . The entire system is maintained daily in an online, real-time, interactive environment, and data entry is distributed among the offices responsible for their respective data.[75]

The LANDS database consists of 86 primary data elements that are arranged into 12 hierarchical segments as shown in Table 35-4.[76] The system is based on an IBM 4381 mainframe computer and the data are stored in a hierarchical database called DL/I.

LANDS is closely linked to an appraisal information system, or node, called CAMA (computer-aided mass appraisal). This system contains detailed appraisal data and is used to support the property appraisal and taxation functions. The county has also installed an ARC/INFO GIS on a new PRIME computer that is used primarily for managing environmental data (e.g., soils and geology) and carrying out spatial analysis. Recently, a database containing video imagery was added to the system. This consists of a high-quality image of each property stored on a laser disk and 70 "moving video" cassette tapes showing the general neighborhood around individual parcels.[77] These data have proved to be extremely useful in the appraisal process.

The greatest financial gains from the project have come through recouping delinquent real estate taxes. In 1983, for instance, revenues from this source exceeded the 10-yr budget for the entire project. The availability of topographic and planimetric data means planners, engineers, and surveyors do not have to acquire new data for every new development project. This amounts to substantial savings in dollars, time, and effort as those data

Table 35-4. LANDS Database data elements by segment

Parcel Segment	**Environment Segment**
Parcel no.	Land use
Taxable status (exempt/active)	Zoning
Tax volume	Master plan code
Tax district	Appraisal class
Tax book	**Remark Segment**
Tax reference	Sequence no.
Land value	Remark type
Improvement value	Remark text
Total value	**Tax Roll Segment**
Mortgage company	Tax year
Loan no.	Item no.
Map no. (quarter section)	Land value
Plat name/section-township-range	Improvement value
Coded for deletion—date	Total value
Low-high block/lot no.	Tax rate
Name Segment	General tax
Last name	Special assessment
First name	Advertising fee
Middle initials	Total tax
Name suffix	Tax penalty date
Name type (owner or billing agent)	Payment status
Name sequence no.	**Payments Segment**
Phonetic name code	Payment date
Address Segment	Payment type
Address type (property, owner, billing)	Batch no.
Direction	Tax amount
Street no.	Interest amount
Street name	Payment code
Street suffix (St., Dr., Ave.)	Refund type
City	**Billing Description Segment**
State	Legal description
Zip code	**Geo Segment**
Miscellaneous address (nonstandard)	Census block
Address sequence no.	Census tract
Legal Description Segment	Tract suffix
Block	Voting ward
Lot (from)	Voting precinct
Lot (to)	Precinct suffix
Metes-and-bounds description (unformatted)	School district
Special Assessment Segment	Drainage district
Special code	Political districts
Payout year	Service districts
Tax year	**Audit (for Each Segment)**
Tax year amount	Transaction ID code
Tax year item	Initials
Envelope Segment	Department
Centroid *X-Y*	Form no.
Delta *X*	Form type
Delta *Y*	Date
Frontage	
Area	

are freely available from the county at a nominal cost of $5 per quarter-section. This can be compared with an estimated $5000 to gather the same data for a 40-acre site.[78] Other applications include cut/fill analysis, accident site reconstruction, hydrographic studies, site and development planning, assessment, and legal projects. It is estimated that the system saves an estimated $100,000 per year just on these activities.

Having received awards from both the Urban and Regional Information Systems Association (URISA) and National Association of Counties (NACO), the Wyandotte County LIS/GIS represents an exemplary system. It is appropriate to conclude this case study with the wisdom that has been gained by the county surveyor who championed this system to its award-winning status:

> Cooperation, communication, and sharing are perhaps the three most essential ingredients in the building and establishment of land information systems in Wyandotte County. Economically accessible geodetic control is changing our lives. However, we have found that any successes in the past and the goals we are striving toward today hinge greatly on these three ingredients. Without the benefit of cooperative effort, new techniques and new technology have little more than a minuscule impact on the ways of the world.[79]

35-6-2. Franklin County, OH

Franklin County envelopes the Columbus Metropolitan area that has been cited as the fastest growing urban center in the northwest quadrant of the United States. The 543-sq-mi area is composed of 17 townships, 26 incorporated cities, and approximately 330,000 parcels of land. It is further complicated by the four different systems used for the original subdivision of this land. These consist of U.S. military lands (5 × 5 mi townships), Virginia military lands (metes and bounds), the refugee tract, and U.S. Congress lands. The latter two tracts of land were subdivided into conventional 6 × 6-sq-mi townships and 1-sq-mi sections.

The LIS project was initiated by the Franklin County Auditor's Office primary as a means of upgrading its property appraisal capabilities. The motivation for the project was described as follows in the Project Overview Document:

> Under the Ohio Revised Code (ORC), county auditors are permitted to prepare appraisal maps that contain information on traditional tax maps prepared by the county engineer *and also additional data*, such as building diagrams and locations, and other appraisal information maintained by the county auditor. A modern, computerized appraisal mapping system would allow the auditor to easily access property record data (now stored on property record cards) with the appraisal map. The preparation of these modern appraisal maps is clearly part of the appraisal and valuation process and also is part of the auditor's duties as assessor of all real property in the county.[80]

Funding for the project was obtained from the auditor's real estate assessment fund that is set aside to cover costs related to the appraisal process. For this reason, the auditor's office has been the key agency in this project.

The project for developing what was termed a computerized appraisal and mapping system (CAMS) began in 1987 with the establishment of a 96-point geodetic reference framework. The positions of all points were determined by a GPS to first-order accuracy (1 : 100,000). This network was densified by adding 47 azimuth points. The total cost to create this network (including monumentation) was $234,500, which meant an approximate $1600 per point in 1987 dollars.[81] Photography at a scale of $1'' = 1320$ ft (1 : 15,840) was used to densify the control network through aerial triangulation in order to provide photo control for mapping planimetric details from larger-scale photography.

The digital base map, or planimetric database, was derived from $1'' = 660$ ft (1 : 7920) scale aerial photography and plotted at a scale of $1'' = 100$ ft (1 : 1200). The resulting map is important as a means of controlling the quality of the database and as a product for external users. This was particularly true in the case of Franklin County, as almost all of the database creation work was undertaken by outside contractors and subsequent quality checks were done in-house by examining line maps produced from these data. However, the database that is used to create the map must be viewed as the most important outcome of this "mapping" effort. Once the computerized system is fully operational at the auditor's office, the use of manual products like maps will undoubtedly decrease and be replaced by the more comprehensive and flexible digital database. The major data layers in the planimetric database together with their respective data elements are summarized in Table 35-5. Initially, contour data were not included in the database, but they were later added after it was established that there was a significant demand for such data. A contour interval of 2 ft was selected for this dataset.

The cadastral overlay was created by researching all relevant deeds, subdivision plats, and other available plans depicting property boundaries. Once again, this work was contracted out and the resulting cadastral map delivered to the auditor's office for quality control. This map consisted of boundary lines drawn on $1'' = 100$ ft $(1:1200)$ scale orthophotography derived from $1'' = 600$ ft $(1:7200)$ scale aerial photography flown specifically for the project. Once the cadastral maps had passed certain internal quality checks, they were sent to another contractor for digitizing. The resulting digital data are checked once more in the auditor's office to ensure that all data elements are included in the respective data layers.[82]

The data described above reside on a DEC 6310 and Synercom's GIS, INFORMAP III, is used for spatial queries and managing the graphic data. ORACLE is used as a database management system for the attribute data. Since a significant amount of real estate appraisal data already existed on another system (IBM 3090), it was necessary to network these systems together by means of an SNA gateway so that data could be passed back and fourth

between the two databases. This network is now operational and personnel at the auditor's office estimate that the system will reduce the reappraisal process by about a year. In Franklin County, properties are reappraised every 6 yr, and under the old system this process would require 3 yr to complete.

Besides the obvious benefits to the auditor's office, various local authorities will gain a current digital database as a foundation for their own systems. For the proper local areas, this represents an inexpensive opportunity to improve their land information management capabilities. In many cases, they have neither the funds nor the personnel to create such a database for their jurisdictions. Since the database creation process has been controlled by a single county office, the quality standards of the data should be consistent throughout the county. This will facilitate data integration among the various incorporated areas and other local authorities across the county by placing them on a common foundation.

This project recently came under the political spotlight when the incumbent auditor came up for reelection. He was challenged by a political opponent who claimed that the LIS/GIS was "an expensive gas guzzler" and the funding for the program should instead be directed toward the school districts in the county.[83] The auditor won the election handily, but continued investigations of the project eventually unearthed certain irregularities with regard to campaign funds received from several of the companies contracted to develop the LIS/GIS and he has been forced to resign. It is, perhaps, unfortunate that this LIS/GIS program became a political "football," but the reality is that such programs generally require large sums of money ($18 million in the case of Franklin County) that inevitably draw them into the political spotlight. The challenge is to develop an information system that is recognized as a community resource. Furthermore, it is important to emphasize benefits to the public in the form of improved decision making, which will hopefully lead to more effective governance.

Table 35-5. Franklin County appraisal GIS data items and layers

DATA ITEM (Layer #)	DATA ITEM (Layer #)
Title Block Data (1–3)	**Roadway Centerline Symbols** (20)
	Roadway centerline intersections
Primary Transportation (8)	**Utilities (Water)** (21)
Airfield pavement (runway)	Water manhole
Airfield pavement (apron)	Fire hydrant
Airfield pavement (unpaved runway)	Water valve
Roadway (paved)	**Utilities (Storm)** (22)
Roadway (unpaved)	Storm manhole
Alley (paved)	Catch basin/curb inlets
Alley (unpaved)	End of pipe (storm)
Overpass/bridge	**Utilities (Storm Lines)** (23)
Roadway under construction	Headwall (storm)
Secondary Transportation (9)	Paved storm drainage ditch
Major walkways	Misc. storm drainage structures
Parking lot (paved)	**Utilities (Sanitary)** (24)
Parking lot (unpaved)	Sanitary manhole
Drives (paved)	**Utilities** (25)
Drives (unpaved)	Utility pole
Highway berm (paved)	Electric manhole
Transportation Annotation (10)	**Utilities (Telephone)** (26)
Parking lot labels	Telephone manhole
Supplemental trans. annotation	**Utilities (Gas)** (27)
Supplemental roadway annotation	Gas manhole
Primary Structures (11)	Gas valve
Building outline	**Misc. Symbols** (28)
Dam	Misc. poles
Secondary Structures (12)	Misc. manholes
Building outline (secondary)	Satellite dishes
Building foundations	Antennas
Major retaining walls	Misc. utility structures
Docks, slabs, courts, towers	Swamp
Storage tanks	**Utilities** (29)
Misc. structures	Transmission line
Roof separations	Pipelines
Substations	**Area Lines** (30)
Structure Annotation (13)	County boundary line
Structure annotation	**Misc. Annotation** (31)
Misc. structure annotation	Swamp annotation (1″ = 100′ scale)
Misc. Structure Symbols (14)	Misc. annotation (1″ = 100′ scale)
Monuments	County boundary annotation
Swimming pools	(1″ = 100′ scale)
Primary Hydrography (15)	**Misc. Annotation** (32)
Streams, rivers, creeks, lakes, etc.	Misc. annotation (1″ = 400′ scale)
Secondary Hydrography (16)	County boundary annotation
Drainage ditches, swamp outlines	(1″ = 400′ scale)
Hydrography Annotation (17)	**Misc. Annotation** (33)
Hydrography annotation	Misc. annotation (1″ = 1000′ scale)
Railroad (18)	Municipality annotation (1″ = 1000′ scale)
Railroad	County boundary annotation
Railroad (abandoned)	(1″ = 1000′ scale)
Centerline (19)	
Roadway centerline	

DATA ITEM (Layer #)	DATA ITEM (Layer #)
Misc. Annotation (34)	**Parcels** (47)
Misc. annotation (1″ = 2000′ scale)	Parcel outline
Municipality annotation	Railroad rights of way
(1″ = 2000′ scale)	Vacation line
County boundary annotation	**Parcel Annotation** (48)
(1″ = 2000′ scale)	Parcel dimensions
Misc. Area Line (35)	**Parcel Centroids** (49)
Paved recreational surfaces	Centroid
Unpaved recreational surfaces	**Parcel Symbology** (50)
Playing fields	One-half land hook
Golf courses	Full land hook
Small cemetery fences	Dimension marker
Control Symbols (36)	**Parcel Leader Lines** (51)
Horizontal control	Leader line
Combination horizontal/vertical control	**Subdivision Lot Lines** (52)
Vertical control	Lot line
Hydrography Intersections Symbol (37)	**Subdivision Lot Numbers** (53)
Hydrography intersections	Lot numbers
Tax District Centroids (41)	Vacation labels
Centroid symbol	**Easement Lines** (54)
Subdivision Boundaries (42)	Cross-county easement
Subdivision boundary	**Easement Annotation** (55)
Subdivision Centroids (43)	Cross-county easement dimensions
Subdivision centroids	**Parcel Condominium Centroids** (56)
Subdivision Index Annotation (44)	Centroid symbol
Annotation	**Condominium Centroid** (57)
Rights-of-Way (45)	Centroid symbol
Right-of-way line	**Appraisal Centroid** (58)
Right-of-Way-Annotation (46)	Centroid symbol
Right-of-way dimensions	**Right-of-Way Centerline** (59)
	Centerline

Source: M. Rhodes and E. Crane. 1984. Lands—A Multipurpose Land Information System, Proceedings of FIG Symposium. Edmonton, Alberta, pp. 86–87.

Although the Franklin County program coordinates its efforts to some extent through an advisory committee composed of interested individuals from various city and county agencies, the implementation of the project was controlled by the auditor's office. Ideally, the funding and control of the program should be broad-based, either by forming a close consortium of various county and city agencies, or creating a new agency that is specifically charged with the implementation of an LIS/GIS (see the next section). A well-designed system is often regarded as one that can withstand major administrative and political changes. Given the high political profile of the Franklin County system, it will be interest-ing to observe the direction of this program under the leadership of a new auditor.

35-6-3. Prince William County, VA

Prince William County (PWC) is located approximately 35 mi southwest of Washington, DC, in an area that is experiencing rapid population growth and development pressure. The county is 350 sq mi in area and supports a population of almost 230,000 people. The county experienced a population increase of 45% during the 1980s and it is anticipated that the 83,000 parcels in the county will increase by over 6500 per year.[84]

The development of the PWC LIS can be traced to the establishment of a Land Use Information Committee (LUIC) in the early 1980s. This committee, comprised of department heads from 15 different county agencies, was given the responsibility of "developing a concept and devising a means of more efficiently managing the County's land information."[85] The following problems were identified by the LUIC and a subcommittee appointed to deal with technical issues:

Lack of a model or unifying concept for linking geographic information in the county

No unique parcel and structure identifiers that would facilitate data integration and sharing between different county agencies

Updating problems due to information redundancies and inconsistencies

Data quality problems due to lack of standards, data inconsistencies, and procedural deficiencies

No standardized and reliable method for retaining historical data on parcel changes and administrative actions that affect parcels

Difficulties in upgrading existing automated systems because they were designed for specific functions[86, 87]

The LUIC recommended that a separate Office of Mapping be created in order to promote a unified approach toward land information management in the county. Musselman explains how this was achieved:

In 1985 the first county-wide department-level was established in Prince William County. Parts of several departments were put together to form the Office of Mapping, the base mapping was collected from Public Works, the parcel mapping from Assessments and street naming and addressing from the Planning Office.[88]

Using the multipurpose cadastre as a model, the Office of Mapping focused its initial efforts on the control and base mapping required to support such a model. The initial geodetic reference framework consisted of 83 points whose positions were determined to second-order, class I accuracy by means of GPS. The network was designed to provide a uniform density of points spaced approximately 2 mi apart. Funding for the subsequent densification of this network down to a half-mile density is derived from a per acre development fee that is assessed on every new development in the county.

The next phase in the LIS development process was to acquire color aerial photography at a scale of 1 : 14,400 for the purpose of creating a digital base map. All photo identifiable topographic and planimetric data were plotted at a scale of $1'' = 200'$ (1 : 2400). The initial database was based on 1987 photography and the county plans to update these data by acquiring new color aerial photography every 2 yr. The cadastral overlay was created primarily from available deeds and subdivision plats.

The LIS has been designed to include two major components. The administrative information system (AIS) deals primarily with attribute or tabular data that are used heavily for day-to-day functions. Marshall describes the rationale behind the AIS as follows:

In developing the AIS, land information was viewed as a "corporate resource." We recognized that fundamental changes were necessary to our management of information—not only to enable more effective sharing of data, but more importantly to better operate the various administrative functions for which the County had responsibility... the land information system had to be a management tool that could improve the flow of information and make administrative functions more effective and efficient.

We knew that there was a core of information about parcels, and the structures and units that occupied them, that must be readily available to all users—items such as ownership, zoning, correct address, etc. Thus, at the heart of the AIS would be a "corporate database" that would be updated as administrative functions occurred in agencies. All application systems would eventually be designed to update data in the corporate base as well as draw from it. The capabilities that could be provided by a relational data base management system were essential to making this concept work by enabling flexible structuring of data relationships. A variety of agency keys, such as address, map sheet number, or file number could be related and indexed to the parcel identifier, making retrieval of information across agencies feasible.[89]

ORACLE is used as a database management system for these data and an HP 3000 computer was purchased to support all central AIS requirements.

The second LIS component is known as the GIS. This system deals primarily with locational data and is based on ARC/INFO running on a PRIME 6450 in the Office of Mapping. Major layers or coverages in the GIS include soils, parcels, floodplains, wetlands, zoning, and landuse.[90] The functions and significance of the GIS have been outlined as follows:

> The GIS provides the geographical or locational aspect to the LIS, enabling users to retrieve data for specific parcels of land. In addition to supporting the LIS, the GIS provides flexibility in mapping, enabling numerous types of geographic data to be mapped at any scale quickly and efficiently. The GIS also provides for more accurate geographic data maintenance and update. Our GIS uses automated processes to integrate, analyze, and display geographic data which are stored in the ARC/INFO format.... The GIS is also a tool for efficiently storing, maintaining, and accessing land information. Through graphic presentation and data analysis, it can be used to make better-informed decisions on land use issues facing the County. The impact of GIS can be significant, considering that 85% of the issues addressed by the Board of County Supervisors are land-related.[91]

The model that emerges from the PWC LIS, as well as the systems in Wyandotte and Franklin County, is one in which the acquisition and maintenance of graphic (map) data are concentrated in one office and the attribute data maintained by the various offices who make use of these data. The graphic (planimetric, topographic, and cadastral) database provides a common spatial platform that facilitates the sharing and exchange of attribute data between various local agencies.

35-7. SUMMARY

In this chapter, LIS/GIS have been discussed by examining the historical roots of LIS in the early fiscal and juridical cadastres and tracing their evolution into the multipurpose cadastre and ultimately LIS/GIS networks. The geodetic reference framework and cadastral overlay were identified as components that are fundamental to an LIS and most relevant to surveyors. Since surveyors are integrally involved with these two components, these activities should be used as a springboard into the boarder LIS/GIS environment. The section on land registration systems focused on rudimentary and improved registration of deeds systems, as well as the Torrens registration of title system. Until relatively recently, the Torrens system has been seen by many as the best alternative to the deeds recordation systems commonly found in the United States and Canada. Given the failure of the Torrens-type systems in the United States and the fact that similar benefits can be derived from an improved registration of deeds system, much more attention is now being paid to the incremental improvement of existing deeds systems. The section on GIS emphasized the mechanisms for integrating different data elements within a digital environment. Raster- and vector-based GIS were described briefly and several GIS application areas listed together with relevant references. Finally, three case studies were examined in order to understand some of the LIS/GIS implementation issues and problems.

The design, implementation, and maintenance of land and geographic systems (LIS/GIS) represent opportunities for the surveying profession. In a society where information is becoming a primary resource and the importance of reliable, consistent spatial data in decision making is increasingly being realized, greater involvement in LIS/GIS is essential for surveyors. Land rights are increasingly being defined and influenced by wetlands, floodplains, and other polygons that do not coincide with property parcels. If surveyors are to retain their expertise in the spatial definition of land rights, then a broader vision of these rights must be adopted. This can be achieved by becoming actively involved in the area of LIS/GIS and providing leadership on issues relating to the geodetic reference framework and cadastral overlay.

NOTES

1. J. McLaughlin. 1975. The nature, function and design concepts of a multipurpose cadastre. Ph.D. Dissertation, University of Wisconsin, Madison, WI.

2. National Research Council. 1980. *Need for a Multipurpose Cadastre*. Washington DC: National Academy of Sciences.

3. _____ . 1983. *Procedures and Standards for a Multipurpose Cadastre*. Washington, DC: National Academy of Sciences.

4.

5.

6. NRC. 1980. p. XX.

7. NRC. 1983. p. 16.

8. N. Chrisman, and B. Niemann. 1985. Alternative routes to a multipurpose cadastre: Merging institutional and technical reasoning. AutoCarto 7 Conference Proceedings. Washington, DC, pp. 84–94.

9. B. Niemann, J. Sullivan, S. Ventura, N. Chrisman, A. Vonderohe, D. Mezera, and D. Moyer. 1987. Results of the Dane County land records project. *Photogrammetric Engineering and Remote Sensing, 53* (10), 1371–1378.

10.

11.

12. B. Niemann. 1987. Better information for better decisions: No question about it. Proceedings of URPIS 15 Conference. Hobart, Australia, pp. 1371–1378.

13. M. E. Sedunary. 1984. LOTS: A South Australian approach to a total, fully integrated LIS. Proceedings of FIG Symposium. Edmonton, Alberta, 322–387.

14. G. D. Ralph, Ed. 1987. *Landform*. Newsletter of the South Australian LIS, 5, p. 1.

15. NRC. 1983. p. 20.

16. P. Dale, and J. McLaughlin. 1988. *Land Information Management*. Oxford, UK: Clarendon Press, p. 87.

17. P. Angus-Lepan. 1970. Land systems reform in Canada and their possible application in Australia. *The Australian Surveyor, 21.* p. 6.

18. E. Epstein, and T. Duchesneau. 1984. The use and value of a geodetic reference framework. FGCC Report, Washington, DC.

19. NRC. 1983. p. 24.

20. K. W. Bauer. 1984. Public planning and engineering: The role of maps and the geodetic base. Seminar on the Multipurpose Cadastre: Modernizing Land Records in North America IES Report 123. University of Wisconsin, Madison, WI, pp. 130–139.

21. NR3. 1983. p. 17.

22. R. E. Davis, F. S. Foote, J. M. Anderson, and E. M. Mikhail. 1981. *Surveying Theory and Practice*, 6th Ed. New York: McGraw-Hill.

23. Dale and McLaughlin. p. 103.

24. W. Huxhold. 1991. *An Introduction to Urban Geographic Information Systems*. New York: Oxford University Press, pp. 243–244.

25. J. Leininger. 1992. Can a GIS handle property surveys? *Professional Surveyor, 12* (1), Jan./Feb., pp. 36–38.

26. _____ . 1992. Boundary surveys: The debate continues. *Professional Surveyor, 12* (3), May–June, pp. 42–43.

27. G. Conradi. 1992. Boundary surveys are possible with GIS. *Professional Surveyor, 12* (2), March–April, pp. 38–43.

28. N. von Meyer. 1991. Cadastral evidence data in a land information system. *URISA Journal, 3* (1), Spring, 14–19.

29. D. Kjerne. 1986. Modeling location for cadastral maps using an object-oriented computer language. URISA Conference Proceedings, Vol. 1. pp. 174–189.

30. D. Kjerne, and K. Dueker. 1990. Modeling cadastral spatial relationships using smalltalk. *URISA Journal, 2* (1), 26–35.

31. R. J. Hintz, W. J. Blackman, B. M. Dana, and J. M. Kang. 1988. Least squares analysis in temporal coordinate and measurement management. *Surveying and Mapping, 48* (3). 173–183.

32. R. J. Hintz and H. J. Onsrud. 1990. Upgrading real property boundary information in a GIS. *URISA Journal, 2* (1), 2–10.

33. T. B. Buyong, W. Kuhn, and A. U. Frank. 1991. A conceptual model of measurement-based multipurpose cadastral systems. *URISA Journal, 3* (2), 35–49.

34. D. A. Waterman. 1986. *A Guide to Expert Systems*. Reading, MA: Addison-Wesley Publishing Co.

35. N. S. Tamim. 1992. Upgrading boundary information obtained from digitized tax maps for the purpose of creating a digital cadastral overlay. Ph.D. Dissertation, The Ohio State University, Columbus, OH.

36. J. McLaughlin, and S. Nichols. 1989. Resource management: The land administration and cadastral systems component. *Surveying and Mapping, 49* (2), 129.

37. S. R. Simpson. 1976. *Land Law and Registration*. Cambridge, UK: Cambridge University Press.

38. E. Dowson, and V. L. O. Sheppard. 1952. *Land registration*. London: HMSO.

39. W. Milligan, and A. Bowman. 1984. *Real Estate Law*. Englewood Cliffs, NJ: Prentice Hall.

40. C. M. Brown. 1969. *Boundary Control and Legal Principles*. New York: John Wiley & Sons.

41. Source unknown.

42. D. Wall. 1990. Chicago Title Insurance Co. v. Iowa. *Surveying and Land Information Systems, 50* (1), 37–42.

43. Simpson. p. 105.

44.

45. Torrens, Sir Robert 1885. An Essay on the Transfer of Land by Registration, HM Land Registry, Adelaide, Australia.

46. G. Barnes. 1985. Torrens—a name for all systems? Annual Meeting of the Wisconsin Society of Land Surveyors. Stevens Point, WI. Whole paper—not published.

47. M. Feindt. 1984. Setting the record straight for land surveyors on Torrens. *Title News*, Feb. Volume 10, p. 17.

48. B. C. Shick, and I. H. Plotkin. 1978. *Torrens in the U.S.: A Legal and Economic History and Analysis of American Land Registration System*. Lexington, MA: Lexington Books.

49. M. G. Williams. 1983. The Torrens registration system. *Surveying and Mapping, 43* (3), 331–335.

50. H. K. Larsen, 1972. On the economics of land and property information systems. *The Canadian Surveyor, 26* (5), 587–590.

51. G. Jeffress. 1991. Land ownership information use in real property market transactions. Ph.D. Dissertation, University of Maine, Orono, ME.

52. T. W. Mapp. 1978. *Torrens' Elusive Title: Basic Legal Principles of an Efficient Torrens System*. Edmonton: University of Alberta Press.

53. Simpson. p. 99.

54. Williams. pp. 333–334.

55. D. Marble. 1984. GIS and LIS: Differences and similarities. Proceedings of FIG. Alberta, pp. 35–43.

56. P. Burrough. 1986. *Principles of Geographical Information Systems for Land Resource Assessment*. Oxford, UK: Clarendon Press.

57. K. C. Clark. 1986. Recent trends in GIS research. *Geo-Processing, 3*, 1–15.

58. J. Antenucci, K. Brown, P. L. Crosswell, M. J. Kevany, and H. Archer. 1991. *Geographic Information Systems: A Guide to the Technology*. New York: Van Nostrand Reinhold.

59. T. Lillesand, and R. Kiefer. 1987. *Remote Sensing and Image Interpretation*. New York: John Wiley & Sons.

60. American Society of Photogrammetry. 1983. *Manual of Remote Sensing*. Alexandria, VA.

61. J. B. Campbell. 1987. *Introduction to Remote Sensing*. New York: The Guilford Press.

62. N. von Meyer. 1989. A conceptual model for spatial cadastral data in a land information system. Ph.D. Dissertation, University of Wisconsin, Madison, WI.

63. Huxhold.

64. Huxhold. p. 211.

65. Huxhold. p. 153.

66. Huxhold. p. 154.

67. R. Marx. 1990. The TIGER system: Automating the geographic structure of the United States Census. *Introductory Readings in Geo-*

graphic Information Systems (D. Peuquet and D. Marble), Bristol, PA: Eds., Taylor and Francis.

68. U.S. Bureau of Census. 1990. Technical description of the DIME system. *Introductory Readings in Geographic Information Systems* (D. Peuquet, and D. Marble, Eds.), Bristol, PA: Taylor and Francis, pp. 100–111.

69. J. Sobel. 1990. Principal components of the Census Bureau's TIGER file. *Introductory Readings in Geographic Information Systems* (D. Peuquet, and D. Marble, Eds.), Bristol, PA: Taylor and Francis, pp. 112–119.

70. M. Rhodes, and E. Crane. 1984. LANDS—a multipurpose land information system. Proceedings of FIG Symposium. Edmonton, Alberta.

71. D. W. Levings. 1976. Wyandotte hopefuls in tax trap. *The Kansas City Star*, July 30, 96(317).

72. Rhodes and Crane.

73. M. Rhodes. 1989. How geodetic control changed my life. ASPRS-ACSM Conference. Baltimore, MD, pp. 1–2.

74. Rhodes. 1989.

75. E. Crane. 1987. A multipurpose land information system for Wyandotte County, Kansas 53rd Annual Conference on Assessment Administration. New Orleans, LA, p. 1.

76. Rhodes and Crane.

77. Crane.

78. Rhodes and Crane. p. 8.

79. M. Rhodes. 1990. A parcel level multipurpose land information system in Wyandotte County, Kansas. *Surveying and Land Information Systems*, 50 (2), 109.

80. Franklin County Auditor's Office. 1987. Project overview of computerized appraisal data and mapping system. Franklin County, p. 9.

81. S. Davis. 1992. Personal communication.

82. Franklin County Auditor's Office. 1990. Franklin County appraisal geographic information system. Franklin County, KS (unpublished document).

83. J. Riskind. 1990. Mapping program's costs are debated. *Columbus Dispatch*, 9/16/90.

84. C. McNoldy. 1989. Why local jurisdictions need a parcel specific geographic information system. GIS/LIS '89 Conference Proceedings. Orlando, FL, pp. 250–259.

85. C. Musselman. 1989. Managing transition within a county GIS: The Prince William County Virginia story. GIS/LIS '89 Conference Proceedings. Orlando, FL, p. 90.

86. J. Phillips. 1987. The quest for a functional LIS: A case study of Prince William County. International GIS Conference Proceedings. Arlington, VA, pp. 201–214.

87. M. Marshall. 1988. A local government's approach to developing an integrated land information system: Prince William County's experience. URISA Conference Proceedings. Los Angeles, CA, pp. 195–206.

88. Musselman. 1989. p. 91.

89. Marshall. p. 200.

90. C. Musselman. 1992. Personal communication.

91. McNoldy. p. 252.

Appendix 1: State Boards of Registration for Land Surveying

Each state licenses land surveyors. Boards are established to test prospective land surveyors and ensure compliance with state laws applicable to land surveying. Rosters of licensees are also usually maintained. There is no national licensing of land surveyors. A list of boards for each state follows.

Alabama	State Board of Registration for Professional Engineers and Land Surveyors 301 Interstate Park Dr. Montgomery, AL 36109
Alaska	State Board of Registration for Architects, Engineers and Land Surveyors P.O.Box 110806 Juneau, AK 99811
Arkansas	State Board of Registration for Professional Engineers and Land Surveyors Twin City Bank Bldg., #660 North Little Rock, AR 72114
Arizona	State Board of Technical Registration 1951 W. Camelback Phoenix, AZ 85015
California	The Board of Registration for Professional Engineers and Land Surveyors 2355 Capitol Oaks Dr., Suite 300 Sacramento, CA 95831
Colorado	State Board of Registration for Professional Engineers and Professional Land Surveyors 1560 Broadway, #1370 Denver, CO 80202
Connecticut	State Board of Registration for Professional Engineers and Land Surveyors State Office Building, Room G-3A165 Capitol Ave. Hartford, CT 06106
Delaware	Delaware Association of Professional Engineers 2005 Concord Pike Wilmington, DE 19803
District of Columbia	Board of Registration for Professional Engineers 614 H St. N.W., Room 910 Washington DC, 20001
Florida	Board of Professional Engineers 130 N. Monroe St. Tallahassee, FL 32301
Georgia	State Board of Registration for Professional Engineers and Land Surveyors 166 Pryor St. S.W. Atlanta, GA 30303

Guam	Territorial Board of Registration for Professional Engineers, Architects, and Land Surveyors Department of Public Works Government of Guam P.O. Box 2950 Agana, GU 96910	**Maine**	State Board of Registration for Professional Engineers State House Station #92 Augusta, ME 04333
Hawaii	State Board of Registration for Professional Engineers, Architects, Land Surveyors, and Landscape Architects 1010 Richards St. Honolulu, HI 96801	**Maryland**	State Board of Registration for Professional Engineers 501 St. Paul Place, Suite 902 Baltimore, MD 21202
Idaho	Board of Professional Engineers and Land Surveyors 600 S. Orchard, #A Boise, ID 83705	**Massachusetts**	State Board of Registration for Professional Engineers and Land Surveyors Leverett Saltonstall Building 100 Cambridge St., Room 1512 Boston, MA 02202
Illinois	Department of Registration and Education Professional Engineer's Examining Committee 320 W. Washington, 3rd Floor Springfield, IL 62786	**Michigan**	State Board of Registration for Professional Engineers and Land Surveyors P.O. Box 30245 Lansing, MI 48909
Indiana	State Board of Registration for Professional Engineers and Land Surveyors 1021 State Office Bldg. 100 N. Senate Ave. Indianapolis, IN 46024	**Minnesota**	State Board of Registration 133 7th St. E. St. Paul, MN 55101
Iowa	State Board of Engineering Examiners 1918 S.E. Hilsizer Ankeny, IA 50021	**Mississippi**	State Board of Registration for Professional Engineers and Land Surveyors 239 N. Lamour, Suite 501 Jackson, MS 39205
Kansas	State Board of Technical Professions 900 Jackson, Room 507 Topeka, KS 66612	**Missouri**	Board of Architects, Professional Engineers and Land Surveyors P.O. Box 184, 3523 N. Ten Mile Dr. Jefferson City, MO 65102
Kentucky	State Board of Registration for Professional Engineers and Land Surveyors 160 Democrat Dr. Frankfort, KY 40601	**Montana**	State Board of Professional Engineers and Land Surveyors P.O. Box 200315, 111 N. Jackson Helena, MT 59620-0513
Lousiana	State Board of Registration for Professional Engineers and Land Surveyors 1055 St. Charles Ave., Suite 415 New Orleans, LA 70130	**Nebraska**	State Board of Examiners for Professional Engineers and Architects P.O. Box 94751, 301 Centennial Mall, S. Lincoln, NE 68509
		Nevada	State Board of Registered Professional Engineers and Land Surveyors 1755 E. Plum Lane, Suite 102 Reno, NV 89502

New Hampshire	State Board of Professional Engineers 67 Regional Dr. Concord, NH 03301	*Pennsylvania*	State Registration Board for Professional Engineers Transportation & Safety Building P.O. Box 2649 Commonwealth Ave. & Forester St., 6th Floor Harrisburg, PA 17105-2649
New Jersey	State Board of Professional Engineers and Land Surveyors 124 Halsey St. Newark, NJ 07102		
New Mexico	State Board of Registration for Professional Engineers and Land Surveyors 1040 Marquey Place Santa Fe, NM 87501	*Rhode Island*	State Board of Registration for Professional Engineers and Land Surveyors 10 Orms St., Suite 324 Providence, RI 02904
New York	State Board for Engineering and Land Surveying The State Education Department Cultural Education Center Madison Ave. Albany, NY 12230	*South Carolina*	State Board of Registration for Professional Engineers and Land Surveyors P.O. Box 50408, 2221 Devine St., Suite 404 Columbia, SC 29250
North Carolina	Board of Registration for Professional Engineers and Land Surveyors 3620 Six Forks Rd., Suite 300 Raleigh, NC 27609	*South Dakota*	State Commission of Engineering and Architectural Examiners 2040 W. Main St., Suite 304 Rapid City, SD 55702-2497
North Dakota	State Board of Registration for Professional Engineers and Surveyors P.O. Box 5503 Bismarck, ND 58502	*Tennessee*	State Board of Architectural and Engineering Examiners Volunteer Plaza 500 James Robertson Pkwy., 3rd Floor Nashville, TN 37243-1142
Northern Mariana Islands	Board of Professional Licensing P.O. Box 55 CHRB Siapan, MP 96950	*Texas*	Board of Land Surveying 7701 N. Lamar, Suite 400 Austin, TX 78752
Ohio	State Board of Registration for Professional Engineers and Surveyors 77 S. High, 16th Floor Columbus, OH 43266-0314	*Utah*	Representative Committee for Professional Engineers and Land Surveyors Division of Registration 8282 S. State, Unit 14 Midville, UT 84047
Oklahoma	State Board of Registration for Professional Engineers and Land Surveyors Oklahoma Engineering Center 201 N.E. 27th St., Room 120 Oklahoma City, OK 73105	*Vermont*	State Board of Registration for Professional Engineers Division of Licensing & Registration 109 State St. Montpelier, VT 05609-1106
Oregon	State Board of Engineering Examiners Department of Commerce 750 Front St. N.E., Suite 240 Salem, OR 97310	*Virginia*	State Board of Architects, Professional Engineers, Land Surveyors and Certified Landscape Architects Seaboard Building

3600 W. Broad St., 5th Floor
Richmond, VA 23230-4917

Virgin Islands   Board of Architects, Professional
Engineers and Land Surveyors
P.O. Box 476
St. Thomas, VI 00801

Washington   State Board of Registration for
Professional Engineers and Land
Surveyors
P.O. Box 9649, 2424 Bristol Ct.
Olympia, WA 98502

West Virginia   State Board of Registration for
Professional Engineers

608 Union Building
Charleston, WV 25301

Wisconsin   State Examining Board of
Architects, Professional
Engineers, Designers and
Land Surveyors
P.O. Box 8935
1400 E. Washington Ave.
Madison, WI 53708

Wyoming   State Board of Examining
Engineers
Hershel Building
122 W. 25th St., 4th Floor E.
Cheyenne, WY 82002

Appendix 2: Major Federal Surveying and Mapping Agencies

Some of the major federal agencies with extensive surveying and mapping programs are as follows.

NATIONAL GEODETIC SURVEY

NGS was the first civilian scientific agency, established as the Survey of the Coast by Thomas Jefferson in 1807. Its mission soon included surveys of the interior as the nation grew. From 1878 until 1970, the agency was known as the U.S. Coast and Geodetic Survey. Many references in the *Handbook* refer to this latter agency, often as USC and GS. The agency has always occupied a prominent place in scientific surveying in the United States and the world. Currently, it is the geodetic component of the U.S. Department of Commerce. The National Geodetic Survey (NGS) coordinates activities of the Federal Geodetic Control Committee, which develops standards and specifications for conducting federal geodetic surveys. It also assists state and regional agencies through cooperative programs.

Many NGS products are available on paper, magnetic tape, and floppy disks. To obtain more information on any of the products, programs, and services listed, contact:

National Geodetic Information Branch
1315 East-West Highway
Silver Spring, MD 20910-3282

U.S. GEOLOGICAL SURVEY

The U.S. Geological Survey was founded by Congress in 1879 to consolidate four earlier organizations that had been engaged in topographic and geologic mapping. A general plan was adopted in 1882 for the production of a series of topographic maps; each map covers an area bounded by meridians of longitude and parallels of latitude and is called a quadrangle map.

The plan of 1882 has been expanded and modified over the years to meet the needs of a growing nation and to satisfy numerous civil and military needs not contemplated in the 1879 Act. A principal objective of the Geological Survey's National Mapping Program is to provide multipurpose maps and related data of appropriate scale, content, and accuracy to satisfy modern requirements. A major element of this program is the series of topographic maps produced by the Geological Survey.

For information on all services, and for locations of the various offices of the agency, contact:

U.S. Geological Survey
582 National Center
Reston, VA 22092

EARTH SCIENCE INFORMATION OFFICE

The Earth Science Information Office, in the U.S. Geological Survey, informs the public of sources of maps, aerial photography, digital products, atlases, geographic data, and other cartographic or earth science products.

The office also maintains the Cartographic Catalog System to answer public inquiries regarding these sources. The cartographic catalog is a bibliographic database currently being converted to a CD-ROM format.

Earth Science Information Center
509 National Ctr.
Reston, VA 22092

BUREAU OF LAND MANAGEMENT, CADASTRAL SURVEY

The public-lands surveys and resurveys are carried out by this agency (otherwise referred to as the BLM). Substantial information about this program and its history is presented in the *Handbook's* chapter on public lands. Following

is a list of the offices that provide information about current and past public-land surveys:

Cadastral Survey
222 W. 7th Ave., No. 13
Anchorage, AK 99513-7599

Branch of Cadastral Survey
P.O. Box 16563, 3707 N. 7th St.
Phoenix, AZ 85011

Branch of Cadastral Survey
Federal Building
2800 Cottage Way, Room E-2841
Sacramento, CA 95825-1889

Branch of Cadastral Survey
2850 Youngfield St.
Lakewood, CO 80215

Branch of Cadastral Survey Development
P.O. Box 25047
Denver, CO 80225

Division of Cadastral Survey
1849 C St., N.W., MS L-302
Washington, DC 20240

Branch of Cadastral Survey
3380 Americana Terrace
Boise, ID 83706

Branch of Cadastral Survey
Granite Tower
222 N. 32nd St.
P.O. Box 36800,
Billings, MT 59107

Branch of Cadastral Survey
P.O. Box 12000, 850 Harvard Way,
Reno, NV 89520

Branch of Cadastral Survey
P.O. Box 27115, 1474 Rodeo Rd.,
Santa Fe, NM 87511-7115

Branch of Cadastral Survey
P.O. Box 2965, 825 N.E. Multnomah St.
Portland, OR 97208

Branch of Cadastral Survey
Coordinated Financial Center
324 S. State St., Suite 301
Salt Lake City, UT 84111-2303

Cadastral Survey
350 S. Pickett St.
Alexandria, VA 22304

Branch of Cadastral Survey
P.O. Box 1828, 2515 Warren Ave.
Cheyenne, WY 82003

Other federal agencies include:

Defense Mapping Agency
Hydrographic-Topographic Info.
6500 Brookes Lane
Washington, DC 20315

Defense Mapping Agency
Aerospace Center
2nd and Arsenal St.
St. Louis, MO 63118

Federal Highways Administration
Aerial Surveys Branch
400 7th St. SW
Washington, DC 20590

U.S. Forest Service
Division of Engineering
Independence Ave., 12th & 14th St. SW
Washington, DC 20250

Appendix 3: Sources of Information About Surveying

The first source to contact for information about surveying is:

American Congress of Surveying and Mapping (ACSM)
5410 Grosvenor Lane 100
Bathesda, MD 20814
Telephone: 301-493-0200
Fax: 301-493-8245.

ACSM is the national membership organization for all branches of surveying. It can provide information about survey education, addresses of state membership organizations, their own local affiliates, and survey organizations worldwide. A number of publications are available from them, as well as a journal and bulletins.

There are two magazines sent free to surveyors. They contain both topical articles and useful information. They are:

Professional Surveyor Magazine
2300 Ninth St., Suite 501
Arlington, VA 22204-2456

P.O.B. Magazine
5820 Lilley Rd., #5
Canton, MI 48187

The American Society of Civil Engineering has a Surveying and Mapping Division and a publication program. Information can be obtained by contacting the Society at:

345 East 47th St.
New York, NY 10017

Appendix 4: Standards and Specifications for Geodetic Control Networks

The document on the following pages was prepared by the Federal Geodetic Control Committee in September 1984, and reprinted in 1993. Although parts of it are used or cited in a number of chapters in this text, it is reproduced in its entirety to provide the context and continuity that can only be obtained from a complete document.

1. Introduction

The Government of the United States makes nation-wide surveys, maps, and charts of various kinds. These are necessary to support the conduct of public business at all levels of government, for planning and carrying out national and local projects, the development and utilization of natural resources, national defense, land management, and monitoring crustal motion. Requirements for geodetic control surveys are most critical where intense development is taking place, particularly offshore areas, where surveys are used in the exploration and development of natural resources, and in delineation of state and international boundaries.

State and local governments and industry regularly cooperate in various parts of the total surveying and mapping program. In surveying and mapping large areas, it is first necessary to establish frameworks of horizontal, vertical, and gravity control. These provide a common basis for all surveying and mapping operations to ensure a coherent product. A reference system, or datum, is the set of numerical quantities that serves as a common basis. Three National Geodetic Control Networks have been created by the Government to provide the datums. It is the responsibility of the National Geodetic Survey (NGS) to actively maintain the National Geodetic Control Networks (appendix A).

These control networks consist of stable, identifiable points tied together by extremely accurate observations. From these observations, datum values (coordinates or gravity) are computed and published. These datum values provide the common basis that is so important to surveying and mapping activities.

As stated, the United States maintains three control networks. A horizontal network provides geodetic latitudes and longitudes in the North American Datum reference system; a vertical network furnishes elevations in the National Geodetic Vertical Datum reference system; and a gravity network supplies gravity values in the U.S. absolute gravity reference system. A given station may be a control point in one, two, or all three control networks.

It is not feasible for all points in the control networks to be of the highest possible accuracy. Different levels of accuracy are referred to as the "order" of a point. Orders are often subdivided further by a "class" designation. Datum values for a station are assigned an order (and class) based upon the appropriate classification standard for each of the three control networks. Horizontal and vertical standards are defined in reasonable conformance with past practice. The recent development of highly accurate absolute gravity instrumentation now allows a gravity reference standard. In the section on "Standards," the classification standards for each of the control networks are described, sample computations performed, and monumentation requirements given.

Control networks can be produced only by making very accurate measurements which are referred to identifiable control points. The combination of survey design, instrumentation, calibration procedures, observational techniques, and data reduction methods is known as a measurement system. The section on "Specifications" describes important components and states permissible tolerances for a variety of measurement systems.

Clearly, the control networks would be of little use if the datum values were not published. The section entitled "Information" describes the various products and formats of available geodetic data.

Upon request, the National Geodetic Survey will accept data submitted in the correct formats with the proper supporting documentation (appendix C) for incorporation into the national networks. When a survey is submitted for inclusion into the national networks, the survey measurements are processed in a quality control procedure that leads to their classification of accuracy and storage in the National Geodetic Survey data base. To fully explain the process we shall trace a survey from the planning stage to admission into the data base. This example will provide an overview of the standards and specifications, and how they work together.

The user should first compare the distribution and accuracy of current geodetic control with both immediate and long-term needs. From this basis, requirements for the extent and accuracy of the planned survey are determined. The classification standards of the control networks will help in this formulation. Hereafter, the requirements for the accuracy of the planned survey will be referred to as the "intended accuracy" of the survey. A measurement system is then chosen, based on various factors such as: distribution and accuracy of present control; region of the country; extent, distribution, and accuracy of the desired control; terrain and accessibility of control; and economic factors.

Upon selection of the measurement system, a survey design can be started. The design will be strongly dependent upon the "Network Geometry" specifications for that measurement system. Of particular importance is the requirement to connect to previously established control points. If this is not done, then the survey cannot be placed on the national datum. An adequate number of existing control point connections are often required in the specifications in order to ensure strong network geometry for other users of the control, and to provide several closure checks to help measure accuracy. NGS can certify the results of a survey only if it is connected to the national network.

Situations will arise where one cannot, or prefers not to, conform to the specifications. NGS may downgrade the classification of a survey based upon failure to adhere to the measurement system specifications if the departure degrades the precision, accuracy, or utility of the survey. On the other hand, if specification requirements for the desired level of accuracy are exceeded, it may be possible to upgrade a survey to a higher classification.

Depending upon circumstances, one may wish to go into the field to recover old control and perform reconnaissance and site inspection for the new survey. Monumentation may be performed at this stage. Instruments should be checked to conform to the "Instrumentation" specifications, and to meet the "Calibration Procedures" specifications. Frequent calibration is an excellent method to help ensure accurate surveys.

In the field, the "Field Procedures" specifications are used to guide the methods for taking survey measurements. It must be stressed that the "Field Procedures" section is not an exhaustive account of how to perform observations. Reference should be made also to the appropriate manuals of observation methods and instruments.

Computational checks can be found in the "Field Procedures" as well as in the "Office Procedures" specifications, since one will probably want to perform some of the computations in the field to detect blunders. It is not necessary for the user to do the computations described in the "Office Procedures" specifications, since they will be done by NGS. However, it is certainly in the interest of the user to compute these checks before leaving the field, in case reobservations are necessary. With the tremendous increase in programmable calculator and small computer technology, any of the computations in the "Office Procedures" specifications could be done with ease in the field.

At this point the survey measurements have been collected, together with the new description and recovery notes of the stations in the new survey. They are then placed into the formats specified in the Federal Geodetic Control Committee (FGCC) publications *Input Formats and Specifications of the National Geodetic Survey Data Base*. Further details of this process can be found in appendix C, "Procedures for Submitting Data to the National Geodetic Survey."

The data and supporting documentation, after being received at NGS, are processed through a quality control procedure to make sure that all users may place confidence in the new survey points. First, the data and documentation are examined for compliance with the measurement system specifications for the intended accuracy of the new survey. Then office computations are performed, including a minimally constrained least squares adjustment. (See appendix B for details.) From this adjustment, accuracy measures can be computed by error propagation. The accuracy classification thus computed is called the "provisional accuracy" of the survey.

The provisional accuracy is compared to the intended accuracy. The difference indicates the departure of the accuracy of the survey from the specifications. If the difference is small, the intended accuracy has precedence because a possible shift in classification is not warranted. However, if the difference is substantial, the provisional accuracy will supersede the intended accuracy, either as a downgrade or an upgrade.

As the final step in the quality control procedure, the variance factor ratio computation using established control, as explained in the section on "Standards," is determined for the new survey. If this result meets the criteria stated there, then the survey is classified in accordance with the provisional accuracy (or intended accuracy, whichever has precedence).

Cases arise where the variance factor ratio is significantly larger than expected. Then the control network is at fault, or the new survey is subject to some unmodeled error source which degrades its accuracy. Both the established control measurements and the new survey measurements will be scrutinized by NGS to determine the source of the problem. In difficult cases, NGS may make diagnostic measurements in the field.

Upon completion of the quality control check, the survey measurements and datum values are placed into the data base. They become immediately available for electronic retrieval, and will be distributed in the next publication cycle by the National Geodetic Information Branch of NGS.

A final remark bears on the relationship between the classification standards and measurement system specifications. Specifications are combinations of rules of thumb and studies of error propagation, based upon experience, of how to best achieve a desired level of quality. Unfortunately, there is no guarantee that a particular standard will be met if the associated specifications are followed. However, the situation is ameliorated by a safety factor of two incorporated in the standards and specifications. Because of this safety factor, it is possible that one may fail to meet the specifications and still satisfy the desired standard. This is why the geodetic control is not automatically downgraded when one does not adhere to the specifications. Slight departures from the specifications can be accommodated. In practice, one should always strive to meet the measurement system specifications when extending a National Geodetic Control Network.

2. Standards

The classification standards of the National Geodetic Control Networks are based on accuracy. This means that when control points in a particular survey are classified, they are certified as having datum values consistent with all other points in the network, not merely those within that particular survey. It is not observation closures within a survey which are used to classify control points, but the ability of that survey to duplicate already established control values. This comparison takes into account models of crustal motion, refraction, and any other systematic effects known to influence the survey measurements.

The NGS procedure leading to classification covers four steps:

1. The survey measurements, field records, sketches, and other documentation are examined to verify compliance with the specifications for the intended accuracy of the survey. This examination may lead to a modification of the intended accuracy.
2. Results of a minimally constrained least squares adjustment of the survey measurements are examined to ensure correct weighting of the observations and freedom from blunders.
3. Accuracy measures computed by random error propagation determine the provisional accuracy. If the provisional accuracy is substantially different from the intended accuracy of the survey, then the provisional accuracy supersedes the intended accuracy.
4. A variance factor ratio for the new survey combined with network data is computed by the Iterated Almost Unbiased Estimator (IAUE) method (appendix B). If the variance factor ratio is reasonably close to 1.0 (typically less than 1.5), then the survey is considered to check with the network, and the survey is classified with the provisional (or intended) accuracy. If the variance factor ratio is much greater than 1.0 (typically 1.5 or greater), then the survey is considered to not check with the network, and both the survey and network measurements will be scrutinized for the source of the problem.

2.1 Horizontal Control Network Standards

When a horizontal control point is classified with a particular order and class, NGS certifies that the geodetic latitude and longitude of that control point bear a relation of specific accuracy to the coordinates of all other points in the horizontal control network. This relation is expressed as a distance accuracy, 1:a. A distance accuracy is the ratio of the relative positional error of a pair of control points to the horizontal separation of those points.

Table 2.1—Distance accuracy standards

Classification	Minimum distance accuracy
First-order	1:100,000
Second-order, class I	1: 50,000
Second-order, class II	1: 20,000
Third-order, class I	1: 10,000
Third-order, class II	1: 5,000

A distance accuracy, 1:a, is computed from a minimally constrained, correctly weighted, least squares adjustment by:

$$a = d/s$$

where
a = distance accuracy denominator
s = propagated standard deviation of distance between survey points obtained from the least squares adjustment
d = distance between survey points

The distance accuracy pertains to all pairs of points (but in practice is computed for a sampling of pairs of points). The worst distance accuracy (smallest denominator) is taken as the provisional accuracy. If this is substantially larger or smaller than the intended accuracy, then the provisional accuracy takes precedence.

As a test for systematic errors, the variance factor ratio of the new survey is computed by the Iterated Almost Unbiased Estimator (IAUE) method described in appendix B. This computation combines the new survey measurements with existing network data, which are assumed to be correctly weighted and free of systematic error. If the variance factor ratio is substantially greater than unity then the survey does not check with the network, and both the survey and the network data will be examined by NGS.

Computer simulations performed by NGS have shown that a variance factor ratio greater than 1.5 typically indicates systematic errors between the survey and the network. Setting a cutoff value higher than this could allow undetected systematic error to propagate into the national network. On the other hand, a higher cutoff value might be considered if the survey has only a small number of connections to the network, because this circumstance would tend to increase the variance factor ratio.

In some situations, a survey has been designed in which different sections provide different orders of control. For these multi-order surveys, the computed distance accuracy denominators should be grouped into sets appropriate to the different parts of the survey. Then, the smallest value of a in each set is used to classify the control points of that portion, as discussed above. If there are sufficient connections to the network, several variance factor ratios, one for each section of the survey, should be computed.

Horizontal Example

Suppose a survey with an intended accuracy of first-order (1:100,000) has been performed. A series of propagated distance accuracies from a minimally constrained adjustment is now computed.

Line	s (m)	d (m)	1:a
1-2	0.141	17,107	1:121,326
1-3	0.170	20,123	1:118,371
2-3	0.164	15,505	1: 94,543
......	.	.	.
......	.	.	.
......	.	.	.

Suppose that the worst distance accuracy is 1:94,543. This is not substantially different from the intended accuracy of 1:100,000, which would therefore have precedence for classification. It is not feasible to precisely quantify "substantially different." Judgment and experience are determining factors.

Now assume that a solution combining survey and network data has been obtained (as per appendix B), and that a variance factor ratio of 1.2 was computed for the survey. This would be reasonably close to unity, and would indicate that the survey checks with the network. The survey would then be classified as first-order using the intended accuracy of 1:100,000.

However, if a variance factor of, say, 1.9 was computed, the survey would not check with the network. Both the survey and network measurements then would have to be scrutinized to find the problem.

Monumentation

Control points should be part of the National Geodetic Horizontal Network only if they possess permanence, horizontal stability with respect to the Earth's crust, and a horizontal location which can be defined as a point. A 30-centimeter-long wooden stake driven into the ground, for example, would lack both permanence and horizontal stability. A mountain peak is difficult to define as a point. Typically, corrosion resistant metal disks set in a large concrete mass have the necessary qualities. First-order and second-order, class I, control points should have an underground mark, at least two monumented reference marks at right angles to one another, and at least one monumented azimuth mark no less than 400 m from the control point. Replacement of a temporary mark by a more permanent mark is not acceptable unless the two marks are connected in timely fashion by survey observations of sufficient accuracy. Detailed information may be found in C&GS *Special Publication* 247, "Manual of geodetic triangulation."

2.2 Vertical Control Network Standards

When a vertical control point is classified with a particular order and class, NGS certifies that the orthometric elevation at that point bears a relation of specific accuracy to the elevations of all other points in the vertical control network. That relation is expressed as an elevation difference accuracy, b. An elevation difference accuracy is the relative elevation error between a pair of control points that is scaled by the square root of their horizontal separation traced along existing level routes.

Table 2.2—Elevation accuracy standards

Classification	Maximum elevation difference accuracy
First-order, class I	0.5
First-order, class II	0.7
Second-order, class I	1.0
Second-order, class II	1.3
Third-order	2.0

An elevation difference accuracy, b, is computed from a minimally constrained, correctly weighted, least squares adjustment by

$$b = S/\sqrt{d}$$

where
d = approximate horizontal distance in kilometers between control point positions traced along existing level routes.
S = propagated standard deviation of elevation difference in millimeters between survey control points obtained from the least squares adjustment. Note that the units of b are $(mm)/\sqrt{(km)}$.

The elevation difference accuracy pertains to all pairs of points (but in practice is computed for a sample). The worst elevation difference accuracy (largest value) is taken as the provisional accuracy. If this is substantially larger or smaller than the intended accuracy, then the provisional accuracy takes precedence.

As a test for systematic errors, the variance factor ratio of the new survey is computed by the Iterated Almost Unbiased Estimator (IAUE) method described in appendix B. This computation combines the new survey measurements with existing network data, which are assumed to be correctly weighted and free of systematic error. If the variance factor ratio is substantially greater than unity, then the survey does not check with the network, and both the survey and the network data will be examined by NGS.

Computer simulations performed by NGS have shown that a variance factor ratio greater than 1.5 typically indicates systematic errors between the survey and the network. Setting a cutoff value higher than this could allow undetected systematic error to propagate into the national network. On the other hand, a higher cutoff value might be considered if the survey has only a small number of connections to the network, because this circumstance would tend to increase the variance factor ratio.

In some situations, a survey has been designed in which different sections provide different orders of control. For these multi-order surveys, the computed elevation difference accuracies should be grouped into sets appropriate to the different parts of the survey. Then, the largest value of b in each set is used to classify the control points of that portion, as discussed above. If there are sufficient connections to the network, several variance factor ratios, one for each section of the survey, should be computed.

Vertical Example

Suppose a survey with an intended accuracy of second-order, class II has been performed. A series of propagated elevation difference accuracies from a minimally constrained adjustment is now computed.

Line	S (mm)	d (km)	b (mm)/$\sqrt{(km)}$
1-2	1.574	1.718	1.20
1-3	1.743	2.321	1.14
2-3	2.647	4.039	1.32
....	.	.	.
....	.	.	.
....	.	.	.

Suppose that the worst elevation difference accuracy is 1.32. This is not substantially different from the intended accuracy of 1.3 which would therefore have precedence for classification. It is not feasible to precisely quantify "substantially different." Judgment and experience are determining factors.

Now assume that a solution combining survey and network data has been obtained (as per appendix B), and

that a variance factor ratio of 1.2 was computed for the survey. This would be reasonably close to unity and would indicate that the survey checks with the network. The survey would then be classified as second-order, class II, using the intended accuracy of 1.3.

However, if a survey variance factor ratio of, say, 1.9 was computed, the survey would not check with the network. Both the survey and network measurements then would have to be scrutinized to find the problem.

Monumentation

Control points should be part of the National Geodetic Vertical Network only if they possess permanence, vertical stability with respect to the Earth's crust, and a vertical location that can be defined as a point. A 30-centimeter-long wooden stake driven into the ground, for example, would lack both permanence and vertical stability. A rooftop lacks stability and is difficult to define as a point. Typically, corrosion resistant metal disks set in large rock outcrops or long metal rods driven deep into the ground have the necessary qualities. Replacement of a temporary mark by a more permanent mark is not acceptable unless the two marks are connected in timely fashion by survey observations of sufficient accuracy. Detailed information may be found in *NOAA Manual NOS NGS* 1, "Geodetic bench marks."

2.3 Gravity Control Network Standards

When a gravity control point is classified with a particular order and class, NGS certifies that the gravity value at that control point possesses a specific accuracy.

Gravity is commonly expressed in units of milligals (mGal) or microgals (μGal) equal, respectively, to (10^{-5}) meters/sec^2, and (10^{-8}) meters/sec^2. Classification order refers to measurement accuracies and class to site stability.

Table 2.3—Gravity accuracy standards

Classification	Gravity accuracy (μGal)
First-order, class I	20 (subject to stability verification)
First-order, class II	20
Second-order	50
Third-order	100

When a survey establishes only new points, and where only absolute measurements are observed, then each survey point is classified independently. The standard deviation from the mean of measurements observed at that point is corrected by the error budget for noise sources in accordance with the following formula:

$$c^2 = \sum_{i+1}^{n} \frac{(x_i - x_m)^2}{n - 1} + e^2$$

where
c = gravity accuracy
x_i = gravity measurement
n = number of measurements

$$x_m = (\sum_{i=1}^{n} x_i)/n$$

e = external random error

The value obtained for c is then compared directly against the gravity accuracy standards table.

When a survey establishes points at which both absolute and relative measurements are made, the absolute determination ordinarily takes precedence and the point is classified accordingly. (However, see Example D below for an exception.)

When a survey establishes points where only relative measurements are observed, and where the survey is tied to the National Geodetic Gravity Network, then the gravity accuracy is identified with the propagated gravity standard deviation from a minimally constrained, correctly weighted, least squares adjustment.

The worst gravity accuracy of all the points in the survey is taken as the provisional accuracy. If the provisional accuracy exceeds the gravity accuracy limit set for the intended survey classification, then the survey is classified using the provisional accuracy.

As a test for systematic errors, the variance factor ratio of the new survey is computed by the Iterated Almost Unbiased Estimator (IAUE) method described in appendix B. This computation combines the new survey measurements with existing network data which are assumed to be correctly weighted and free of systematic error. If the variance factor ratio is substantially greater than unity, then the survey does not check with the network, and both the survey and the network data will be examined by NGS.

Computer simulations performed by NGS have shown that a variance factor ratio greater than 1.5 typically indicates systematic errors between the survey and the network. Setting a cutoff value higher than this could allow undetected systematic error to propagate into the national network. On the other hand, a higher cutoff value might be considered if the survey has only a minimal number of connections to the network, because this circumstance would tend to increase the variance factor ratio.

In some situations, a survey has been designed in which different sections provide different orders of control. For these multi-order surveys, the computed gravity accuracies should be grouped into sets appropriate to the different parts of the survey. Then, the largest value of c in each set is used to classify the control points of that portion, as discussed above. If there are sufficient connections to the network, several variance factor ratios, one for each part of the survey, should be computed.

Gravity Examples

Example A. Suppose a gravity survey using absolute measurement techniques has been performed. These points are then unrelated. Consider one of these survey points.

Assume n = 750

$$\sum_{i=1}^{750} (x_i - x_m)^2 = .169 \text{ mGal}^2$$

$$e = 5 \text{ } \mu\text{Gal}$$

$$c^2 = \frac{0.169}{750-1} + (.005)^2$$

$$c = 16 \text{ } \mu\text{Gal}$$

The point is then classified as first-order, class II.

Example B. Suppose a relative gravity survey with an intended accuracy of second-order (50 µGal) has been performed. A series of propagated gravity accuracies from a minimally constrained adjustment is now computed.

Station	Gravity standard deviation (µGal)
1	38
2	44
3	55
.	.
.	.
.	.

Suppose that the worst gravity accuracy was 55 µGal. This is worse than the intended accuracy of 50 µGal. Therefore, the provisional accuracy of 55 µGal would have precedence for classification, which would be set to third-order.

Now assume that a solution combining survey and network data has been obtained (as per appendix B) and that a variance factor of 1.2 was computed for the survey. This would be reasonably close to unity, and would indicate that the survey checks with the network. The survey would then be classified as third-order using the provisional accuracy of 55 µGal.

However, if a variance factor of, say, 1.9 was computed, the survey would not check with the network. Both the survey and network measurements then would have to be scrutinized to find the problem.

Example C. Suppose a survey consisting of both absolute and relative measurements has been made at the same points. Assume the absolute observation at one of the points yielded a classification of first-order, class II, whereas the relative measurements produced a value to second-order standards. The point in question would be classified as first-order, class II, in accordance with the absolute observation.

Example D. Suppose we have a survey similar to Case C, where the absolute measurements at a particular point yielded a third-order classification due to an unusually noisy observation session, but the relative measurements still satisfied the second-order standard. The point in question would be classified as second-order, in accordance with the relative measurements.

Monumentation

Control points should be part of the National Geodetic Gravity Network only if they possess permanence, horizontal and vertical stability with respect to the Earth's crust, and a horizontal and vertical location which can be defined as a point. For all orders of accuracy, the mark should be imbedded in a stable platform such as flat, horizontal concrete. For first-order, class I stations, the platform should be imbedded in stable, hard rock, and

checked at least twice for the first year to ensure stability. For first-order, class II stations, the platform should be located in an extremely stable environment, such as the concrete floor of a mature structure. For second and third-order stations, standard bench mark monumentation is adequate. Replacement of a temporary mark by a more permanent mark is not acceptable unless the two marks are connected in timely fashion by survey observations of sufficient accuracy. Detailed information is given in *NOAA Manual NOS NGS* 1, "Geodetic bench marks." Monuments should not be near sources of electromagnetic interference.

It is recommended, but not necessary, to monument third-order stations. However, the location associated with the gravity value should be recoverable, based upon the station description.

3. Specifications

3.1 Introduction

All measurement systems regardless of their nature have certain common qualities. Because of this, the measurement system specifications follow a prescribed structure as outlined below. These specifications describe the important components and state permissible tolerances used in a general context of accurate surveying methods. The user is cautioned that these specifications are not substitutes for manuals that detail recommended field operations and procedures.

The observations will have spatial or temporal relationships with one another as given in the "Network Geometry" section. In addition, this section specifies the frequency of incorporation of old control into the survey. Computer simulations could be performed instead of following the "Network Geometry" and "Field Procedures" specifications. However, the user should consult the National Geodetic Survey before undertaking such a departure from the specifications.

The "Instrumentation" section describes the types and characteristics of the instruments used to make observations. An instrument must be able to attain the precision requirements given in "Field Procedures."

The section "Calibration Procedures" specifies the nature and frequency of instrument calibration. An instrument must be calibrated whenever it has been damaged or repaired.

The "Field Procedures" section specifies particular rules and limits to be met while following an appropriate method of observation. For a detailed account of how to perform observations, the user should consult the appropriate manuals.

Since NGS will perform the computations described under "Office Procedures," it is not necessary for the user to do them. However, these computations provide valuable checks on the survey measurements that could indicate the need for some reobservations. This section specifies commonly applied corrections to observations, and computations which monitor the precision and accuracy of the survey. It also discusses the correctly weighted, minimally constrained least squares adjustment used to ensure that the survey work is free from blunders and able to achieve the intended accuracy. Results of the least squares adjustment are used in the quality control and accuracy classification procedures. The adjustment

performed by NGS will use models of error sources, such as crustal motion, when they are judged to be significant to the level of accuracy of the survey.

3.2 Triangulation

Triangulation is a measurement system comprised of joined or overlapping triangles of angular observations supported by occasional distance and astronomic observations. Triangulation is used to extend horizontal control.

Network Geometry

Order Class	First	Second I	Second II	Third I	Third II
Station spacing not less than (km)	15	10	5	0.5	0.5
Average minimum distance angle† of figures not less than	40°	35°	30°	30°	25°
Minimum distance angle† of all figures not less than	30°	25°	25°	20°	20°
Base line spacing not more than (triangles)	5	10	12	15	15
Astronomic azimuth spacing not more than (triangles)	8	10	10	12	15

† Distance angle is angle opposite the side through which distance is propagated.

The new survey is required to tie to at least four network control points spaced well apart. These network points must have datum values equivalent to or better than the intended order (and class) of the new survey. For example, in an arc of triangulation, at least two network control points should be occupied at each end of the arc. Whenever the distance between two new unconnected survey points is less than 20 percent of the distance between those points traced along existing or new connections, then a direct connection should be made between those two survey points. In addition, the survey should tie into any sufficiently accurate network control points within the station spacing distance of the survey. These network stations should be occupied and sufficient observations taken to make these stations integral parts of the survey. Nonredundant geodetic connections to the network stations are not considered sufficient ties. Nonredundantly determined stations are not allowed. Control stations should not be determined by intersection or resection methods. Simultaneous reciprocal vertical angles or geodetic leveling are observed along base lines. A base line need not be observed if other base lines of sufficient accuracy were observed within the base line spacing specification in the network, and similarly for astronomic azimuths.

Instrumentation

Only properly maintained theodolites are adequate for observing directions and azimuths for triangulation. Only precisely marked targets, mounted stably on tripods or supported towers, should be employed. The target should have a clearly defined center, resolvable at the minimum control spacing. Optical plummets or collimators are required to ensure that the theodolites and targets are centered over the marks. Microwave-type electronic distance measurement (EDM) equipment is not sufficiently accurate for measuring higher-order base lines.

Order Class	First I	Second I	Second II	Third I	Third II
Theodolite, least count	0.2″	0.2″	1.0″	1.0″	1.0″

Calibration Procedures

Each year and whenever the difference between direct and reverse readings of the theodolite depart from 180° by more than 30″, the instrument should be adjusted for collimation error. Readjustment of the cross hairs and the level bubble should be done whenever their misadjustments affect the instrument reading by the amount of the least count.

All EDM devices and retroreflectors should be serviced regularly and checked frequently over lines of known distances. The National Geodetic Survey has established specific calibration base lines for this purpose. EDM instruments should be calibrated annually, and frequency checks made semiannually.

Field Procedures

Theodolite observations for first-order and second-order, class I surveys may only be made at night. Reciprocal vertical angles should be observed at times of best atmospheric conditions (between noon and late afternoon) for all orders of accuracy. Electronic distance measurements need a record at both ends of the line of wet and dry bulb temperatures to ±1°C, and barometric pressure to ±5 mm of mercury. The theodolite and targets should be centered to within 1 mm over the survey mark or eccentric point.

Order Class	First I	Second I	Second II	Third I	Third II
Directions					
Number of positions	16	16	8 or 12†	4	2

Order Class	First I	Second I	Second II	Third I	Third II
Standard deviation of mean not to exceed	0.4″	0.5″	0.8″	1.2″	2.0″
Rejection limit from the mean	4″	4″	5″	5″	5″
Reciprocal Vertical Angles (along distance sight path)					
Number of independent observations direct/reverse	3	3	2	2	2
Maximum spread	10″	10″	10″	10″	20″
Maximum time interval between reciprocal angles (hr)	1	1	1	1	1
Astronomic Azimuths					
Observations per night........	16	16	16	8	4
Number of nights	2	2	1	1	1
Standard deviation of mean not to exceed	0.45″	0.45″	0.6″	1.0″	1.7″
Rejection limit from the mean	5″	5″	5″	6″	6″
Electro-Optical Distances					
Minimum number of days ..	2*	2*	1	1	1
Minimum number of measurements/day	2§	2§	2§	1	1
Minimum number of concentric observations/ measurement	2	2	1	1	1
Minimum number of offset observations/ measurement	2	2	2	1	1
Maximum difference from mean of observations (mm)	40	40	50	60	60
Minimum number of readings/observation (or equivalent)	10	10	10	10	10
Maximum difference from mean of readings (mm) ..	‡	‡	‡	‡	‡
Infrared Distances					
Minimum number of days ..	—	2*	1	1	1
Minimum number of measurements	—	2§	2§	1	1
Minimum number of concentric observations/ measurement	—	1	1	1	1
Minimum number of offset observations/ measurement	—	2	1	1	1
Maximum difference from mean of observations (mm)	—	5	5	10	10
Minimum number of readings/observation (or equivalent)	—	10	10	10	10
Maximum difference from mean of readings (mm) ..	—	‡	‡	‡	‡
Microwave Distances					
Minimum number of measurements	—	—	—	2	1
Minimum time span between measurements (hr)	—	—	—	8	—

Order	First	Second	Second	Third	Third
Class		I	II	I	II
Maximum difference between measurements (mm)	—	—	—	100	—
Minimum number of concentric observations/ measurement	—	—	—	2**	1**
Maximum difference from mean of observations (mm)	—	—	—	100	150
Minimum number of readings/observation (or equivalent)	—	—	—	20 .	20
Maximum difference from mean of readings (mm) ..	—	—	—	‡	‡

† 8 if 0.2″, 12 if 1.0″ resolution.
* two or more instruments.
§ one measurement at each end of the line.
‡ as specified by manufacturer.
** carried out at both ends of the line.

Measurements of astronomic latitude and longitude are not required in the United States, except perhaps for first-order work, because sufficient information for determining deflections of the vertical exists. Detailed procedures can be found in Hoskinson and Duerksen (1952).

Office Procedures

Order	First	Second	Second	Third	Third
Class		I	II	I	II
Triangle Closure					
Average not to exceed	1.0″	1.2″	2.0″	3.0″	5.0″
Maximum not to exceed	3″	3″	5″	5″	10″
Side Checks					
Mean absolute correction by side equation not to exceed	0.3″	0.4″	0.6″	0.8″	2.0″

A minimally constrained least squares adjustment will be checked for blunders by examining the normalized residuals. The observation weights will be checked by inspecting the postadjustment estimate of the variance of unit weight. Distance standard errors computed by error propagation in this correctly weighted least squares adjustment will indicate the provisional accuracy classification. A survey variance factor ratio will be computed to check for systematic error. The least squares adjustment will use models which account for the following:

semimajor axis of the ellipsoid ....(a = 6378137 m)
reciprocal flattening of the ellipsoid ....(1/f = 298.257222)
mark elevation above mean sea level......................(known to ± 1 m)
geoid heights ....(known to ± 6 m)
deflections of the vertical ....(known to ± 3″)
geodesic correction
skew normal correction
height of instrument
height of target
sea level correction

arc correction
geoid height correction
second velocity correction
crustal motion

3.3 Traverse

Traverse is a measurement system comprised of joined distance and theodolite observations supported by occasional astronomic observations. Traverse is used to densify horizontal control.

Network Geometry

Order	First	Second	Second	Third	Third
Class		I	II	I	II
Station spacing not less than (km)	10	4	2	0.5	0.5
Maximum deviation of main traverse from straight line	20°	20°	25°	30°	40°
Minimum number of bench mark ties	2	2	2	2	2
Bench mark tie spacing not more than (segments)	6	8	10	15	20
Astronomic azimuth spacing not more than (segments)	6	12	20	25	40
Minimum number of network control points	4	3	2	2	2

The new survey is required to tie to a minimum number of network control points spaced well apart. These network points must have datum values equivalent to or better than the intended order (and class) of the new survey. Whenever the distance between two new unconnected survey points is less than 20 percent of the distance between those points traced along existing or new connections, then a direct connection must be made between those two survey points. In addition, the survey should tie into any sufficiently accurate network control points within the station spacing distance of the survey. These ties must include EDM or taped distances. Nonredundant geodetic connections to the network stations are not considered sufficient ties. Nonredundantly determined stations are not allowed. Reciprocal vertical angles or geodetic leveling are observed along all traverse lines.

Instrumentation

Only properly maintained theodolites are adequate for observing directions and azimuths for traverse. Only precisely marked targets, mounted stably on tripods or supported towers, should be employed. The target should have a clearly defined center, resolvable at the minimum control spacing. Optical plummets or collimators are required to ensure that the theodolites and targets are centered over the marks. Microwave-type electronic distance measurement equipment is not sufficiently accurate for measuring first-order traverses.

Order / Class	First	Second I	Second II	Third I	Third II
Theodolite, least count	0.2"	1.0"	1.0"	1.0"	1.0"

Calibration Procedures

Each year and whenever the difference between direct and reverse readings of the theodolite depart from 180° by more than 30", the instrument should be adjusted for collimation error. Readjustment of the cross hairs and the level bubble should be done whenever their misadjustments affect the instrument reading by the amount of the least count.

All electronic distance measuring devices and retroreflectors should be serviced regularly and checked frequently over lines of known distances. The National Geodetic Survey has established specific calibration base lines for this purpose. EDM instruments should be calibrated annually, and frequency checks made semiannually.

Field Procedures

Theodolite observations for first-order and second-order, class I surveys may be made only at night. Electronic distance measurements need a record at both ends of the line of wet and dry bulb temperatures to ±1°C and barometric pressure to ±5 mm of mercury. The theodolite, EDM, and targets should be centered to within 1 mm over the survey mark or eccentric point.

Order / Class	First	Second I	Second II	Third I	Third II
Directions					
Number of positions	16	8 or 12†	6 or 8*	4	2
Standard deviation of mean not to exceed	0.4"	0.5"	0.8"	1.2"	2.0"
Rejection limit from the mean	4"	5"	5"	5"	5"
Reciprocal Vertical Angles (along distance sight path)					
Number of independent observations direct/reverse	3	3	2	2	2
Maximum spread	10"	10"	10"	10"	20"
Maximum time interval between reciprocal angles (hr)	1	1	1	1	1
Astronomic Azimuths					
Observations per night	16	16	12	8	4
Number of nights	2	2	1	1	1
Standard deviation of mean not to exceed	0.45"	0.45"	0.6"	1.0"	1.7"
Rejection limit from the mean	5"	5"	5"	6"	6"
Electro-Optical Distances					
Minimum number of measurements	1	1	1	1	1
Minimum number of concentric observations/measurement	1	1	1	1	1
Minimum number of offset observations/measurement	1	1	—	—	—
Maximum difference from mean of observations (mm)	60	60	—	—	—

Order / Class	First	Second I	Second II	Third I	Third II
Minimum number of readings/observation (or equivalent)	10	10	10	10	10
Maximum difference from mean of readings (mm)	§	§	§	§	§
Infrared Distances					
Minimum number of measurements	1	1	1	1	1
Minimum number of concentric observations/measurement	1	1	1	1	1
Minimum number of offset observations/measurement	1	1	1‡	—	—
Maximum difference from mean of observations (mm)	10	10	10‡	—	—
Minimum number of readings/observation	10	10	10	10	10
Maximum difference from mean of readings (mm)	§	§	§	§	§
Microwave Distances					
Minimum number of measurements	—	1	1	1	1
Minimum number of concentric observations/measurement	—	2**	1**	1**	1**
Maximum difference from mean of observations (mm)	—	150	150	200	200
Minimum number of readings/observation	—	20	20	10	10
Maximum difference from mean of readings (mm)	—	§	§	§	§

† 8 if 0.2", 12 if 1.0" resolution.
* 6 if 0.2", 8 if 1.0" resolution.
§ as specified by manufacturer.
‡ only if decimal reading near 0 or high 9's.
** carried out at both ends of the line.

Measurements of astronomic latitude and longitude are not required in the United States, except perhaps for first-order work, because sufficient information for determining deflections of the vertical exists. Detailed procedures can be found in Hoskinson and Duerksen (1952).

Office Procedures

Order / Class	First	Second I	Second II	Third I	Third II
Azimuth closure at azimuth check point (seconds of arc)	1.7√N	3.0√N	4.5√N	10.0√N	12.0√N
Position closure after azimuth adjustment†	0.04√K or 1:100,000	0.08√K or 1:50,000	0.20√K or 1:20,000	0.40√K or 1:10,000	0.80√K or 1:5,000

(N is number of segments, K is route distance in km)
† The expression containing the square root is designed for longer lines where higher proportional accuracy is required. Use the formula that gives the smallest permissible closure. The closure (e.g., 1:100,000) is obtained by computing the difference between the computed and fixed values, and dividing this difference by K. Note: Do not confuse closure with distance accuracy of the survey.

A minimally constrained least squares adjustment will be checked for blunders by examining the normalized residuals. The observation weights will be checked by inspecting the postadjustment estimate of the variance of unit weight. Distance standard errors computed by error propagation in a correctly weighted least squares adjustment will indicate the provisional accuracy classification. A survey variance factor ratio will be computed to check for systematic error. The least squares adjustment will use models which account for the following:

semimajor axis of the ellipsoid	$(a = 6378137$ m$)$
reciprocal flattening of the ellipsoid	$(1/f = 298.257222)$
mark elevation above mean sea level	(known to ± 1 m)
geoid heights	(known to ± 6 m)
deflections of the vertical	(known to $\pm 3''$)
geodesic correction	
skew normal correction	
height of instrument	
height of target	
sea level correction	
arc correction	
geoid height correction	
second velocity correction	
crustal motion	

3.4 Inertial Surveying

Inertial surveying is a measurement system comprised of lines, or a grid, of Inertial Surveying System (ISS) observations. These specifications cover use of inertial systems only for horizontal control.

Network Geometry

Order	Second	Second	Third	Third
Class	I	II	I	II
Station spacing not less than (km)	10	4	2	1
Maximum deviation from straight line connecting endpoints	20°	25°	30°	35°

Each inertial survey line is required to tie into a minimum of four horizontal network control points spaced well apart and should begin and end at network control points. These network control points must have horizontal datum values better than the intended order (and class) of the new survey. Whenever the shortest distance between two new unconnected survey points is less than 20 percent of the distance between those points traced along existing or new connections, then a direct connection should be made between those two survey points. In addition, the survey should connect to any sufficiently accurate network control points within the distance specified by the station spacing. The connections may be measured by EDM or tape traverse, or by another ISS line. If an ISS line is used, then these lines should follow the same specifications as all other ISS lines in the survey.

For extended area surveys by ISS, a grid of intersecting lines that satisfies the 20 percent rule stated above can be designed. There must be a mark at each intersection of the lines. This mark need not be a permanent monument; it may be a stake driven into the ground. For a position to

receive an accuracy classification, it must be permanently monumented.

A grid of intersecting lines should contain a minimum of eight network points, and should have a network control point at each corner. The remaining network control points may be distributed about the interior or the periphery of the grid. However, there should be at least one network control point at an intersection of the grid lines near the center of the grid. If the required network points are not available, then they should be established by some other measurement system. Again, the horizontal datum values of these network control points must have an order (and class) better than the intended order (and class) of the new survey.

Instrumentation

ISS equipment falls into two types: analytic (or strapdown) and semianalytic. Analytic inertial units are not considered to possess geodetic accuracy. Semianalytic units are either "space stable" or "local level." Space stable systems maintain the orientation of the platform with respect to inertial space. Local level systems continuously torque the accelerometers to account for Earth rotation and movement of the inertial unit, and also torque the platform to coincide with the local level. This may be done on command at a coordinate update, or whenever the unit achieves zero velocity (Zero velocity UPdaTe, or "ZUPT"). Independently of the measurement technique, the recorded data may be filtered by an onboard computer. Because of the variable quality of individual ISS instruments, the user should test an instrument with existing geodetic control beforehand.

An offset measurement device accurate to within 5 mm should be affixed to the inertial unit or the vehicle.

Calibration Procedures

A static calibration should be performed yearly and immediately after repairs affecting the platform, gyroscopes, or accelerometers.

A dynamic or field calibration should be performed prior to each project or subsequent to a static calibration. The dynamic calibration should be performed only between horizontal control points of first-order accuracy and in each cardinal direction. The accelerometer scale factors from this calibration should be recorded and, if possible, stored in the onboard computer of the inertial unit.

Before each project or after repairs affecting the offset measurement device or the inertial unit, the relation between the center of the inertial unit and the zero point of the offset measurement device should be established.

Field Procedures

When surveying in a helicopter, the helicopter must come to rest on the ground for all ZUPT's and all measurements.

Order Class	Second I	Second II	Third I	Third II
Minimum number of complete runs per line	2	1	1	1
Maximum deviation from a uniform rate of travel (including ZUPT)	15%	20%	25%	30%
Maximum ZUPT interval (ZUPT to ZUPT) (sec)	200	240	300	300

A complete ISS measurement consists of measurement of the line while traveling in one direction, followed by measurement of the same line while traveling in the reverse direction (double-run). A coordinate update should not be performed at the far point or at midpoints of a line, even though those coordinates may be known.

The mark offset should be measured to the nearest 5 mm.

Office Procedures

Order Class	Second I	Second II	Third I	Third II
Maximum difference of smoothed coordinates between forward and reverse run (cm)	60	60	70	80

A minimally constrained least squares adjustment of the raw or filtered survey data will be checked for blunders by examining the normalized residuals. The observation weights will be checked by inspecting the postadjustment estimate of the variance of unit weight. Distance standard errors computed by error propagation in this correctly weighted least squares adjustment will indicate the provisional accuracy classification. A survey variance factor ratio will be computed to check for systematic error. The least squares adjustment will use the best available model for the particular inertial system. Weighted averages of individually smoothed lines are not considered substitutes for a combined least squares adjustment to achieve geodetic accuracy.

3.5 Geodetic Leveling

Geodetic leveling is a measurement system comprised of elevation differences observed between nearby rods. Leveling is used to extend vertical control.

Network Geometry

Order Class	First I	First II	Second I	Second II	Third
Bench mark spacing not more than (km)	3	3	3	3	3
Average bench mark spacing not more than (km)	1.6	1.6	1.6	3.0	3.0

Order Class	First I	First II	Second I	Second II	Third
Line length between network control points not more than (km)	300	100	50	50 (double-run) 25 (single-run)	25 10

New surveys are required to tie to existing network bench marks at the beginning and end of the leveling line. These network bench marks must have an order (and class) equivalent to or better than the intended order (and class) of the new survey. First-order surveys are required to perform check connections to a minimum of six bench marks, three at each end. All other surveys require a minimum of four check connections, two at each end. "Check connection" means that the observed elevation difference agrees with the adjusted elevation difference within the tolerance limit of the new survey. Checking the elevation difference between two bench marks located on the same structure, or so close together that both may have been affected by the same localized disturbance, is not considered a proper check. In addition, the survey is required to connect to any network control points within 3 km of its path. However, if the survey is run parallel to existing control, then the following table specifies the maximum spacing of extra connections between the survey and the control. At least one extra connection should always be made.

Distance, survey to network	Maximum spacing of extra connections (km)
0.5 km or less	5
0.5 km to 2.0 km	10
2.0 km to 3.0 km	20

Instrumentation

Order Class	First I	First II	Second I	Second II	Third
Leveling instrument Minimum repeatability of line of sight	0.25"	0.25"	0.50"	0.50"	1.00"
Leveling rod construction	IDS	IDS	IDS† or ISS	ISS	Wood or Metal
Instrument and rod resolution **(combined)** Least count (mm)	0.1	0.1	0.5-1.0*	1.0	1.0

(IDS—Invar, double scale)
(ISS—Invar, single scale)
† if optional micrometer is used.
* 1.0 mm if 3-wire method, 0.5 mm if optical micrometer.

Only a compensator or tilting leveling instrument with an optical micrometer should be used for first-order leveling. Leveling rods should be one piece. Wooden or metal rods may be employed only for third-order work. A turning point consisting of a steel turning pin with a driving cap should be utilized. If a steel pin cannot be driven, then a turning plate ("turtle") weighing at least 7 kg should be substituted. In situations allowing neither turning pins nor turning plates (sandy or marshy soils), a long wooden stake with a double-headed nail should be driven to a firm depth.

Calibration Procedures

Order Class	First I	First II	Second I	Second II	Third
Leveling instrument					
Maximum collimation error, single line of sight (mm/m)	0.05	0.05	0.05	0.05	0.10
Maximum collimation error, reversible compensator type instruments, mean of two lines of sight (mm/m)	0.02	0.02	0.02	0.02	0.04
Time interval between collimation error determinations not longer than (days)					
Reversible compensator	7	7	7	7	7
Other types	1	1	1	1	7
Maximum angular difference between two lines of sight, reversible compensator	40"	40"	40"	40"	60"
Leveling rod					
Minimum scale calibration standard	N	N	N	M	M
Time interval between scale calibrations (yr)	1	1	—	—	—
Leveling rod bubble verticality maintained to within	10'	10'	10'	10'	10'

(N—National standard)
(M—Manufacturer's standard)

Compensator-type instruments should be checked for proper operation at least every 2 weeks of use. Rod calibration should be repeated whenever the rod is dropped or damaged in any way. Rod levels should be checked for proper alignment once a week. The manufacturer's calibration standard should, as a minimum, describe scale behavior with respect to temperature.

Field Procedures

Order Class	First I	First II	Second I	Second II	Third
Minimal observation method	micrometer	micrometer	micrometer or 3-wire	3-wire	center wire
Section running	SRDS or DR or SP	SRDS or DR or SP	SRDS or DR† or SP	SRDS or DR*	SRDS or DR§

Field Procedures—Continued

Order Class	First I	First II	Second I	Second II	Third
Difference of forward and backward sight lengths never to exceed					
per setup (m)	2	5	5	10	10
per section (m)	4	10	10	10	10
Maximum sight length (m) ..	50	60	60	70	90
Minimum ground clearance of line of sight (m)	0.5	0.5	0.5	0.5	0.5
Even number of setups when not using leveling rods with detailed calibration	yes	yes	yes	yes	—
Determine temperature gradient for the vertical range of the line of sight at each setup	yes	yes	yes	—	—
Maximum section misclosure (mm)	$3\sqrt{D}$	$4\sqrt{D}$	$6\sqrt{D}$	$8\sqrt{D}$	$12\sqrt{D}$
Maximum loop misclosure (mm)	$4\sqrt{E}$	$5\sqrt{E}$	$6\sqrt{E}$	$8\sqrt{E}$	$12\sqrt{E}$
Single-run methods					
Reverse direction of single runs every half day	yes	yes	yes	—	—
Nonreversible compensator leveling instruments					
Off-level/relevel instrument between observing the high and low rod scales...........	yes	yes	yes	—	—
3-wire method					
Reading check (difference between top and bottom intervals) for one setup not to exceed (tenths of rod units)........................	—	—	2	2	3
Read rod 1 first in alternate setup method ...	—	—	yes	yes	yes
Double scale rods					
Low-high scale elevation difference for one setup not to exceed (mm)					
With reversible compensator	0.40	1.00	1.00	2.00	2.00
Other instrument types:					
Half-centimeter rods	0.25	0.30	0.60	0.70	1.30
Full-centimeter rods ...	0.30	0.30	0.60	0.70	1.30

(SRDS—Single-Run, Double Simultaneous procedure)
(DR—Double-Run)
(SP—SPur, less than 25 km, double-run)
D—shortest length of section (one-way) in km
E—perimeter of loop in km
† Must double-run when using 3-wire method.
* May single-run if line length between network control points is less than 25 km.
§ May single-run if line length between network control points is less than 10 km.

Double-run leveling may always be used, but single-run leveling done with the double simultaneous procedure may be used only where it can be evaluated by loop closures. Rods should be leap-frogged between setups (alternate setup method). The date, beginning and ending times, cloud coverage, air temperature (to the nearest degree), temperature scale, and average wind speed should be recorded for each section plus any changes in the date, instrumentation, observer or time zone. The instrument need not be off-leveled/releveled between observing the high and low scales when using an instrument with a reversible compensator. The low-high scale difference tolerance for a reversible compensator is used only for the control of blunders.

With double scale rods, the following observing sequence should be used:

    backsight, low-scale
    backsight, stadia
    foresight, low-scale
    foresight, stadia
    off-level/relevel or reverse compensator
    foresight, high-scale
    backsight, high-scale

Office Procedures

Order Class	First I	First II	Second I	Second II	Third
Section misclosures **(backward and forward)** Algebraic sum of all corrected section misclosures of a leveling line not to exceed (mm)	$3\sqrt{D}$	$4\sqrt{D}$	$6\sqrt{D}$	$8\sqrt{D}$	$12\sqrt{D}$
Section misclosure not to exceed (mm)	$3\sqrt{E}$	$4\sqrt{E}$	$6\sqrt{E}$	$8\sqrt{E}$	$12\sqrt{E}$
Loop misclosures Algebraic sum of all corrected misclosures not to exceed (mm)	$4\sqrt{F}$	$5\sqrt{F}$	$6\sqrt{F}$	$8\sqrt{F}$	$12\sqrt{F}$
Loop misclosure not to exceed (mm)	$4\sqrt{F}$	$5\sqrt{F}$	$6\sqrt{F}$	$8\sqrt{F}$	$12\sqrt{F}$

(D—shortest length of leveling line (one-way) in km)
(E—shortest one-way length of section in km)
(F—length of loop in km)

The normalized residuals from a minimally constrained least squares adjustment will be checked for blunders. The observation weights will be checked by inspecting the postadjustment estimate of the variance of unit weight. Elevation difference standard errors computed by error propagation in a correctly weighted least squares adjustment will indicate the provisional accuracy classification. A survey variance factor ratio will be computed to check for systematic error. The least squares adjustment will use models that account for:

    gravity effect or orthometric correction
    rod scale errors
    rod (Invar) temperature
    refraction—need latitude and longitude to 6″ or vertical temperature difference observations between 0.5 and 2.5 m above the ground
    earth tides and magnetic field
    collimation error
    crustal motion

3.6 Photogrammetry

Photogrammetry is a measurement system comprised of photographs taken by a precise metric camera and measured by a comparator. Photogrammetry is used for densification of horizontal control. The following specifications apply only to analytic methods.

Network Geometry

Order Class	Second I	Second II	Third I	Third II
Forward overlap not less than	66%	66%	60%	60%
Side overlap not less than	66%	66%	20%	20%
Intersecting rays per point not less than (design criteria)	9	8	3	3

The photogrammetric survey should be areal: single strips of photography are not acceptable. The survey should encompass, ideally, a minimum of eight horizontal control points and four vertical points spaced about the perimeter of the survey. In addition, the horizontal control points should be spaced no farther apart than seven air bases. The horizontal control points should have an order (and class) better than the intended order (and class) of the survey. The vertical points need not meet geodetic control standards. If the required control points are not available, then they must be established by some other measurement system.

Instrumentation

Order Class	Second I	Second II	Third I	Third II
Metric Camera Maximum warp of platen not more than (μm)	10	10	10	10
Dimensional control not less than	reseau with maximum spacing of 2 cm	8 fiducials	8 fiducials	8 fiducials
Comparator Least count (μm)	1	1	1	1

The camera should be of at least the quality of those employed for large-scale mapping. A platen should be included onto which the film must be satisfactorily flattened during exposure. Note that a reseau should be used for second-order, class I surveys.

Calibration Procedures

Order Class	Second I	Second II	Third I	Third II
Metric camera Root mean square of calibrated radial distortion not more than (μm)	1	3	3	5

Calibration Procedures—Continued

Order Class	Second I	Second II	Third I	Third II
Root mean square of calibrated decentering distortion not more than (μm)	1	5†	5†	5†
Root mean square of reseau coordinates not more than (μm)	1	1	3	3
Root mean square of fiducial coordinates not more than (μm)	—	1	3	3

† not usually treated separately in camera calibration facilities; manufacturer's certification is satisfactory.

The metric camera should be calibrated every 2 years, and the comparator should be calibrated every 6 months. These instruments should also be calibrated after repair or modifications.

Characteristics of the camera's internal geometry (radial symmetric distortion, decentered lens distortion, principal point and point of symmetry coordinates, and reseau coordinates) should be determined using recognized calibration techniques, like those described in the current edition of the *Manual of Photogrammetry*. These characteristics will be applied as corrections to the measured image coordinates.

Field Procedures

Photogrammetry involves hybrid measurements: a metric camera photographs targets and features in the field, and a comparator measures these photographs in an office environment. Although this section is entitled "Field Procedures," it deals with the actual measurement process and thus includes comparator specifications.

Order Class	Second I	Second II	Third I	Third II
Targets				
Control points targeted	yes	yes	yes	yes
Pass points targeted	yes	yes	optional	optional
Comparator				
Pointings per target not less than	4	3	2	2
Pointings per reseau (or fiducial) not less than	4	3	2	2
Number of different reseau intersections per target not less than	4	—	—	—
Rejection limit from mean of pointings per target (μm)	3	3	3	3

Office Procedures

Order Class	Second I	Second II	Third I	Third II
Root mean square of adjusted photocoordinates not more than (μm)	4	6	8	12

A least squares adjustment of the photocoordinates, constrained by the coordinates of the horizontal and vertical control points, will be checked for blunders by examining the normalized residuals. The observation weights will be checked by inspecting the postadjustment estimate of the variance of unit weight. Distance standard errors computed by error propagation in this correctly weighted least squares adjustment will indicate the provisional accuracy classification. A survey variance factor ratio will be computed to check for systematic error. The least squares adjustment will use models that incorporate the quantities determined by calibration.

3.7 Satellite Doppler Positioning

Satellite Doppler positioning is a three-dimensional measurement system based on the radio signals of the U.S. Navy Navigational Satellite System (NNSS), commonly referred to as the TRANSIT system. Satellite Doppler positioning is used primarily to establish horizontal control.

The Doppler observations are processed to determine station positions in Cartesian coordinates, which can be transformed to geodetic coordinates (geodetic latitude and longitude and height above reference ellipsoid). There are two methods by which station positions can be derived: point positioning and relative positioning.

Point positioning, for geodetic applications, requires that the processing of the Doppler data be performed with the precise ephemerides that are supplied by the Defense Mapping Agency. In this method, data from a single station is processed to yield the station coordinates.

Relative positioning is possible when two or more receivers are operated together in the survey area. The processing of the Doppler data can be performed in four modes: simultaneous point positioning, translocation, semishort arc, and short arc. The specifications for relative positioning are valid only for data reduced by the semishort or short arc methods. The semishort arc mode allows up to 5 degrees of freedom in the ephemerides; the short arc mode allows 6 or more degrees of freedom. These modes allow the use of the broadcast ephemerides in place of the precise ephemerides.

The specifications quoted in the following sections are based on the experience gained from the analysis of Doppler surveys performed by agencies of the Federal government. Since the data are primarily from surveys performed within the continental United States, the precisions and related specifications may not be appropriate for other areas of the world.

Network Geometry

The order of a Doppler survey is determined by: the spacing between primary Doppler stations, the order of the base network stations from which the primaries are established, and the method of data reduction that is used. The order and class of a survey cannot exceed the lowest order (and class) of the base stations used to establish the survey.

The primary stations should be spaced at regular intervals which meet or exceed the spacing required for the desired accuracy of the survey. The primary stations will carry the same order as the survey.

Supplemental stations may be established in the same survey as the primary stations. The lowest order (and class) of a supplemental station is determined either by its spacing with, or by the order of, the nearest Doppler or other horizontal control station. The processing mode determines the allowable station spacing.

In carrying out a Doppler survey, one should occupy, using the same Doppler equipment and procedures, at least two existing horizontal network (base) stations of order (and class) equivalent to, or better than, the intended order (and class) of the Doppler survey. If the Doppler survey is to be first-order, at least three base stations must be occupied. If relative positioning is to be used, all base station base lines must be directly observed during the survey. Base stations should be selected near the perimeter of the survey, so as to encompass the entire survey.

Stations which have a precise elevation referenced by geodetic leveling to the National Geodetic Vertical Datum (NGVD) are preferred. This will allow geoidal heights to be determined. As many base stations as possible should be tied to the NGVD. If a selection is to be made, those stations should be chosen which span the largest portion of the survey.

If none of the selected base stations is tied to the NGVD, at least two, preferably more, bench marks of the NGVD should be occupied. An attempt should be made to span the entire survey area.

Datum shifts for transformation of point position solutions should be derived from the observations made on the base stations.

The minimum spacing, D, of the Doppler stations may be computed by a formula determined by the processing mode to be employed. This spacing is also used in conjunction with established control, and other Doppler control, to determine the order and class of the supplemental stations.

By using the appropriate formula, tables can be constructed showing station spacing as a function of point or relative one-sigma position precision (s_p or s_r) and desired survey (or station) order.

Point Positioning

$$D = 2\sqrt{2}\, s_p a$$

where

a = denominator of distance accuracy classification standard (e.g., a = 100,000 for first-order standard).

Order Class	First	Second I	Second II	Third I	Third II
s_p (cm)		D (km)			
200	566	242	114	56	28
100	283	141	57	28	14
70	200	100	40	20	10
50	141	71	26	14	7

Relative Positioning

$$D = 2\, s_r a$$

where

a = denominator of distance accuracy classification standard (e.g., a = 100,000 for first-order standard).

Order Class	First	Second I	Second II	Third I	Third II
s_r (cm)		D (km)			
50	100	50	20	10	5
35	70	35	14	7	4
20	40	20	8	4	2

However, the spacing for relative positioning should not exceed 500 km.

Instrumentation

The receivers should receive the two carrier frequencies transmitted by the NNSS. The receivers should record the Doppler count of the satellite, the receiver clock times, and the signal strength. The integration interval should be approximately 4.6 sec. Typically six or seven of these intervals are accumulated to form a 30-second Doppler count observation. The reference frequency should be stable to within $5.0(10^{-11})$ per 100 sec. The maximum difference from the average receiver delay should not exceed 50 μsec. The best estimate of the mean electrical center of the antenna should be marked. This mark will be the reference point for all height-of-antenna measurements.

Calibration Procedures

Receivers should be calibrated at least once a year, or whenever a modification to the equipment is made. It is desirable to perform a calibration before every project to verify that the equipment is operational. The two-receiver method explained next is preferred and should be used whenever possible.

Two-Receiver Method

The observations are made on a vector base line, of internal accuracy sufficient to serve as a comparison standard, 10 to 50 m in length. The base line should be located in an area free of radio interference in the 150 and 400 MHz frequencies. The procedures found in the table on relative positioning in "Field Procedures" under the 20 cm column heading will be used. The data are reduced by either shortarc or semishort arc methods. The receivers will be considered operational if the differences between the Doppler and the terrestrial base line components do not exceed 40 cm (along any coordinate axis).

Single-Receiver Method

Observations are made on a first-order station using the procedures found in the table on relative positioning in "Field Procedures" under the 50 cm column heading. The data are reduced with the precise ephemerides. The resultant position must agree within 1 m of the network position.

Field Procedures

The following tables of field procedures are valid only for measurements made with the Navy Navigational Satellite System (TRANSIT).

Point Positioning

s_p (precise ephemerides)	50 cm	70 cm	100 cm	200 cm
Max. standard deviation of mean of counts/pass (cm), broadcast ephemerides	25	25	25	25
Period of observation not less than (hr)	48	36	24	12
Number of observed passes not less than†	40	30	15	8
Number of acceptable passes (evaluated by on-site point processing) not less than	30	20	9	4
Minimum number of acceptable passes within each quadrant*	6	4	2	1
Frequency standard warm-up time (hr)				
crystal	48	48	24	24
atomic	1.5	1.5	1.0	1.0
Maximum interval between meteorological observations (hr)	6	§	§	§

† Number of passes refers to those for which the precise ephemerides are available for reduction.
* There should be a nearly equal number of northward and southward passes.
§ each setup, visit and takedown.

Relative positioning

s_r	20 cm	35 cm	50 cm
Maximum standard deviation of mean of counts/pass (cm), broadcast ephemerides	25	25	25
Period of observation not less than (hr)	48	36	24
Number of observed passes not less than†	40	30	15
Number of acceptable passes (evaluated by on-site point position processing) not less than	30	20	9
Minimum number of acceptable passes within each quadrant*	6	4	2
Frequency standard warm-up time (hr)			
crystal	48	48	48
atomic	1.5	1.5	1.5
Maximum interval between meteorological observations (hr)	6	6	§

† Number of observed passes refers to all satellites available for tracking and reduction with the broadcast or precise ephemerides.
* Number of northward and southward passes should be nearly equal.
§ Each setup, visit and takedown.

The antenna should be located where radio interference is minimal for the 150 and 400 MHz frequencies. Medium frequency radar, high voltage power lines, transformers, excessive noise from automotive ignition systems, and high power radio and television transmission antennas should be avoided. The horizon should not be obstructed above 7.5°.

The antenna should not be located near metal structures, or, when on the roof of a building, less than 2 m from the edge. The antenna must be stably located within 1 mm over the station mark for the duration of the observations. The height difference between the mark and the reference point for the antenna phase center should be measured to the nearest millimeter. If an antenna is moved while a pass is in progress, that pass is not acceptable. If moved, the antenna should be relocated within 5 mm of the original antenna height; otherwise the data may have to be processed as if two separate stations were established. In the case of a reoccupation of an existing Doppler station, the antenna should be relocated within 5 mm of the original observing height.

Long-term reference frequency drift should be monitored to ensure it does not exceed the manufacturer's specifications.

Observations of temperature and relative humidity should be collected, if possible, at or near the height of the phase center of the antenna. Observations of wet-bulb and dry-bulb temperature readings should be recorded to the nearest 0.5°C. Barometric readings at the station site should be recorded to the nearest millibar and corrected for difference in height between the antenna and barometer.

Office Procedures

The processing constants and criteria for determining the quality of point and relative positioning results are as follows:

1. For all passes for a given station occupation, the average number of Doppler counts per pass should be at least 20 (before processing).
2. The cutoff angle for both data points and passes should be 7.5°.
3. For a given pass, the maximum allowable rejection of counts, 3 sigma postprocessing, will be 10.
4. Counts rejected (excluding cutoff angle) for a solution should be less than 10 percent.
5. Depending on number of passes and quality of data, the standard deviation of the range residuals for all passes of a solution should range between:
   Point positioning—10 to 20 cm
   Relative positioning—5 to 20 cm

A minimally constrained least squares adjustment will be checked for blunders by examining the normalized residuals. The observation weights will be checked by inspecting the postadjustment estimate of the variance of unit weight. Distance standard errors computed by error propagation between points in this correctly weighted least squares adjustment will indicate the maximum achievable accuracy classification. The formula presented in "Standards" will be used to arrive at the actual classification. The least squares adjustment will use models which account for:

tropospheric scale bias, 10 percent uncertainty
receiver time delay
satellite/receiver frequency offset
precise ephemeris
tropospheric refraction
ionospheric refraction
long-term ephemeris variations
crustal motion

3.8 Absolute Gravimetry

Absolute gravimetry is a measurement system which determines the magnitude of gravity at a station at a specific time. Absolute gravity measurements are used to establish and extend gravity control. Within the context of a geodetic gravity network, as discussed in "Standards," a series of absolute measurements at a control point is in itself sufficient to establish an absolute gravity value for that location.

The value of gravity at a point is time dependent, being subject to dynamic effects in the Earth. The extent of gravimetric stability can be determined only by repeated observations over many years.

Network Geometry

Network geometry cannot by systematized since absolute observations at a specific location are discrete and uncorrelated with other points. In absolute gravimetry, a network may consist of a single point.

A first-order, class I station must possess gravimetric stability, which only repeated measurements can determine. This gravimetric stability should not be confused with the accuracy determined at a specific time. It is possible for a value to be determined very precisely at two different dates and for the values at each of these respective dates to differ. Although the ultimate stability of a point cannot be determined by a single observation session, an attempt should be made to select sites which are believed to be tectonically stable, and sufficiently distant from large bodies of water to minimize ocean tide coastal loading.

The classification of first-order, class I is reserved for network points which have demonstrated long-term stability. To ensure this stability, the point should be reobserved at least twice during the year of establishment and thereafter at sufficient intervals to ensure the continuing stability of the point. The long-term drift should indicate that the value will not change by more than 20 μGal for at least 5 years. A point intended as first-order, class I will initially be classified as first-order, class II until stability during the first year is demonstrated.

Instrumentation

The system currently being used is a ballistic-laser device and is the only one at the current state of technolo-

gy considered sufficiently accurate for absolute gravity measurements. An absolute instrument measures gravity at a specific elevation above the surface, usually about 1 m. For this reason, the gravity value is referenced to that level. A measurement of the vertical gravity gradient, using a relative gravity meter and a tripod, must be made to transfer the gravity value to ground level. The accuracy of the relative gravimeter must satisfy the gravity gradient specifications found in "Field Procedures."

Calibration Procedures

Ballistic-laser instruments are extremely delicate and each one represents a unique entity with its own characteristics. It is impossible to identify common systematic errors for all instruments. Therefore, the manufacturer's recommendations for individual instrument calibration should be followed rigorously.

To identify any possible bias associated with a particular instrument, comparisons with other absolute devices are strongly recommended whenever possible. Comparisons with previously established first-order, class I network points, as well as first-order, class II network points tied to the class I points, are also useful.

Field Procedures

The following specifications were determined from results of a prototype device built by J. Faller and M. Zumberge (Zumberge, M., "A Portable Apparatus for Absolute Measurements of the Earth's Gravity," Department of Physics, University of Colorado, 1981) and are given merely as a guideline. It is possible that some of these values may be inappropriate for other instruments or models. Therefore, exceptions to these specifications are allowed on a case-by-case basis upon the recommendation of the manufacturer. Deviations from the specifications should be noted upon submission of data for classification.

Order	First	First	Second	Third
Class	I	II		
Absolute measurement				
Standard deviation of each accepted measurement set not to exceed (μGal)	20	20	50	100
Minimum number of sets/observation	5	5	5	5
Maximum difference of a measurement set from mean of all measurements (μGal)	12	12	37	48
Barometric pressure standard error (mbar)	4	4	—	—
Gradient measurement				
Standard deviation of measurement of vertical gravity gradient at time of observation (μGal/m)	5	5	5	5
Standard deviation of height of instrument above point (mm)	1	1	5	10

Office Procedures

The manufacturer of an absolute gravity instrument usually provides a reduction process which identifies and accounts for error sources and identifiable parameters. This procedure may be sufficient, making further office adjustments unnecessary.

A least squares adjustment will be checked for blunders by examining the normalized residuals. The observation weights will be checked by inspecting the postadjustment estimate of the variance of unit weight. Gravity value standard deviations computed by error propagation in a correctly weighted, least squares adjustment will indicate the provisional accuracy classification. The least squares adjustment, as well as digital filtering techniques and/or sampling, should use models which account for:

atmospheric mass attraction
microseismic activity
instrumental characteristics
lunisolar attraction
elastic and plastic response of the Earth (tidal loading)

3.9 Relative Gravimetry

Relative gravimetry is a measurement system which determines the difference in magnitude of gravity between two stations. Relative gravity measurements are used to extend and densify gravity control.

Network Geometry

A first-order, class I station must possess gravimetric stability, which only repeated measurements can determine. This gravimetric stability should not be confused with the accuracy determined at a specific time. It is possible for a value to be determined very precisely at two different dates, and for the values at each of these respective dates to differ. Although the ultimate stability of a point cannot be determined by a single observation session, an attempt should be made to select sites which are believed to be tectonically stable.

The classification of first-order, class I is reserved for network points that have demonstrated long-term stability. To ensure this stability, the point should be reobserved at least twice during the year of establishment and thereafter at sufficient intervals. The long-term drift should indicate that the value will not change by more than the 20 μGal for at least 5 years. A point intended as first-order, class I will initially be classified as first-order, class II until stability during the first year is demonstrated.

The new survey is required to tie at least two network points, which should have an order (and class) equivalent to or better than the intended order (and class) of the new survey. This is required to check the validity of existing network points as well as to ensure instrument calibration. Users are encouraged to exceed this minimal requirement. However, if one of the network stations is a first-order, class I mark, then that station alone can satisfy the minimum connecting requirement if the intended order of the new survey is less than first-order.

Instrumentation

Regardless of the type of a relative gravimeter, the internal error is of primary concern.

Order Class	First I	First II	Second	Third
Minimum instrument internal error (one-sigma), (μGal)	10	10	20	30

The instrument's internal accuracy may be determined by performing a relative survey over a calibration line (see below) and examining the standard deviation of a single reading. This determination should be performed after the instrument is calibrated using the latest calibration information. Thus the internal error is the measure of instrument uncertainty after all possible systematic error sources have been eliminated by calibration.

Calibration Procedures

An instrument should be properly calibrated before a geodetic survey is performed. The most important calibration item is the determination of the mathematical model that relates dial units, voltage, or some other observable to milligals. This may consist only of a scale factor. In other cases the model may demonstrate nonlinearity or periodicity. Most manufacturers provide tables or scale factors with each instrument. Care must be taken to ensure the validity of these data over time.

When performing first-order work, this calibration model should be determined by a combination of bench tests and field measurements. The bench tests are specified by the manufacturer. A field calibration should be performed over existing control points of first-order, class I or II. The entire usable gravimeter range interval should be sampled to ensure an uncertainty of less than 5 μGal. FGCC member agencies have established calibration lines for this specific purpose.

The response of an instrument to air pressure and temperature should be determined. The meter should be adjusted or calibrated for various pressures and temperatures so that the allowable uncertainty from these sources does not exceed the values in the table below.

The manufacturer's recommendations should be followed to ensure that all internal criteria, such as galvanometer sensitivity, long and cross level or tilt sensitivity, and reading line, are within the manufacturer's allowable tolerances.

The response of an instrument due to local orientation should also be determined. Systematic differences may be due to an instrument's sensitivity to local magnetic variations. Manufacturers attempt to limit or negate such a response. However, if a meter displays a variation with respect to orientation, then one must either have the instrument repaired by the manufacturer, or minimize the effect by fixing the orientation of the instrument throughout a survey.

Order Class	First I	First II	Second	Third
Necessary for user to determine calibration model	Yes	Yes	Yes	No
Allowable uncertainty of calibration model (μGal)	5	5	10	15
Allowable uncertainty due to external air temperature changes (μGal).............................	1	1	3	—
Maximum uncertainty due to external air pressure changes (μGal).............................	1	1	2	—
Allowable uncertainty due to other factors (μGal)	3	3	5	—

Field Procedures

A relative gravity survey is performed using a sequence of measurements known as a loop sequence. There are three common types: ladder, modified ladder, and line.

The ladder sequence begins and ends at the same network point, with the survey points being observed twice during the sequence: once in forward running and once in backward running. Of course, more than one network point may be present in a ladder sequence.

Order Class	First I	First II	Second	Third
Minimum number of instruments used in survey	2	2	2	1
Recommended number of instruments used in survey	3	3	2	1
Allowable loop sequence	a	a	a,b	a,b,c
Minimum number of readings at each observation/instrument	5	5	2†	1
Standard deviation of consecutive readings (unclamped) from mean* not to exceed (μGal)	2	2	5	—
Monitor external temperature and air pressure	Yes	Yes	No	No
Standard deviation of temperature measurements (˚C)	0.1	0.1	—	—
Standard deviation of air pressure measurement (mbar)	1	1	—	—
Standard deviation of height of instrument above point (mm)........	1	1	5	10

(a—ladder)　　(b—modified ladder) .　　　(c—line)
† Although two readings are required, only one reading need be recorded.
* corrected for lunisolar attraction.

The modified ladder sequence also begins and ends at the same network point. However, not all the survey points are observed twice during the sequence. Again, more than one network point may be observed in the sequence.

The line sequence begins at a network point and ends at a different network point. A survey point in a line sequence is usually observed only once.

One should always monitor the internal temperature of the instrument to ensure it does not fluctuate beyond the manufacturer's recommended limits. The time of each reading should be recorded to the nearest minute.

Office Procedures

Order	First I	First II	Second	Third
Rejection Limits				
Maximum standard error of a gravity value (μGal)	20	20	50	100
Total allowable instrument uncertainty (μGal)	10	10	20	30
Model Uncertainties				
Uncertainty of atmospheric mass model (μGal)	0.5	0.5	—	—
Uncertainty of lunisolar attraction (μGal)	1	1	5	5
Uncertainty of Earth elastic and plastic response to tidal loading (μGal)..............................	2	2	5	—

A least squares adjustment, constrained by the network configuration and precision of established gravity control, will be checked for blunders by examining the normalized residuals. The observation weights will be checked by inspecting the postadjustment estimate of the variance of unit weight. Gravity standard errors computed by error propagation in a correctly weighted least squares adjustment will indicate the provisional accuracy classification. A survey variance factor ratio will be computed to check for systematic error. The least squares adjustment will use models which account for:

instrument calibrations
　　1) conversion factors　　　　　(linear and higher order)
　　2) thermal response　　　　　　(if necessary)
　　3) atmospheric pressure response　(if necessary)

instrument drift
　　1) static
　　2) dynamic

atmospheric mass attraction　　　　(if necessary)

Earth tides
　　1) lunisolar attraction
　　2) Earth elastic and plastic response　(if necessary)

4. Information

Geodetic control data and cartographic information that pertain to the National Geodetic Control Networks are widely distributed by a component of the National Geodetic Survey, the National Geodetic Information Branch (NGIB). Users of this information include Federal, State, and local agencies, universities, private companies, and individuals. Data are furnished in response to individual orders, or by an automatic mailing service (the mechanism whereby users who maintain active geodetic files automatically receive newly published data for specified areas). Electronic retrieval of data can be carried out directly from the NGS data base by a user.

Geodetic control data for the national networks are primarily published as standard quadrangles of 30' in latitude by 30' in longitude. However, in congested areas, the standard quadrangles are 15' in latitude by 15' in longitude. In most areas of Alaska, because of the sparseness of control, quadrangle units are 1° in latitude by 1° in longitude. Data are now available in these formats for all horizontal control and approximately 65 percent of the vertical control. The remaining 35 percent are presented in the old formats; i.e., State leveling lines and description booklets. Until the old format data have been converted to the standard quadrangle formats, the vertical control data in the unconverted areas will be available only by complete county coverage. Field data and recently adjusted projects with data in manuscript form are available from NGS upon special request. The National Geodetic Control Networks are cartographically depicted on approximately 850 different control diagrams. NGS provides other related geodetic information: e.g., geoid heights, deflections of the vertical, calibration base lines, gravity values, astronomic positions, horizontal and vertical data for crustal movement studies, satellite-derived positions, UTM coordinates, computer programs, geodetic calculator programs, and reference materials from the NGS data bases.

The NGIB receives data from all NOAA geodetic field operations and mark-recovery programs. In addition, other Federal, State, and local governments, and private organizations contribute survey data from their field operations. These are incorporated into the NGS data base. NOAA has entered into formal agreements with several Federal and State Government agencies whereby NGIB publishes, maintains, and distributes geodetic data received from these organizations. Guidelines and formats have been established to standardize the data for processing and inclusion into the NGS data base. These formats are available to organizations interested in participating in the transfer of their files to NOAA (appendix C).

Upon completion of the geodetic data base management system, information generated from the data base will be automatically revised. A new data output format is being designed for both horizontal and vertical published control information. These formats, which were necessitated by the requirements of the new adjustments of the horizontal and vertical geodetic networks, will be more comprehensive than the present versions.

New micropublishing techniques are being introduced in the form of computer-generated microforms. Some geodetic data are available on magnetic tape, microfilm, and microfiche. These services will be expanded as the automation system is fully implemented. Charges for digital data are determined on the basis of the individual requests, and reflect processing time, materials, and postage. The booklets *Publications of the National Geodetic Survey* and *Products and Services of the National Geodetic Survey* are available from NGIB.

For additional information, write:

Chief, National Geodetic Information
  Branch, N/CG17
National Oceanic and Atmospheric Administration
Rockville, MD 20852

To order by telephone:

data:.. 301-443-8631
publications:..301-443-8316
computer programs or digital data:......... 301-443-8623

5. References

(Special reference lists also follow appendixes A and B)

Basic Geodetic Information

Bomford, G., 1980: *Geodesy* (4th ed.). Clarendon Press, Oxford, England, 855 pp.

Defense Mapping Agency, 1981: *Glossary of Mapping, Charting, and Geodetic Terms* (4th edition), Defense Mapping Agency Hydrographic/Topographic Center, Washington, D.C., 203 pp.

Mitchell, H., 1948: Definitions of terms used in geodetic and other surveys, *Special Publication 242*, U.S. Coast and Geodetic Survey, Washington, D.C., 87 pp.

Torge, W., 1980: *Geodesy*. Walter de Gruyter & Co., New York, N.Y., 254 pp.

Vanicek, P., and Krakiwsky, E., 1982: *Geodesy: The Concepts*. North-Holland Publishing Co., New York, N.Y., 691 pp.

Standards and Specifications

Director of National Mapping, 1981: *Standard Specifications and Recommended Practices for Horizontal and Vertical Control Surveys* (3rd edition), Director of National Mapping, Canberra, Australia, 51 pp.

Federal Geodetic Control Committee, 1974: *Classification, Standards of Accuracy, and General Specifications of Geodetic Control Surveys*, National Oceanic and Atmospheric Administration, Rockville, Md., 12 pp.

Federal Geodetic Control Committee, 1975, rev. 1980: *Specifications to Support Classification, Standards of Accuracy, and General Specifications of Geodetic Control Surveys*, National Oceanic and Atmospheric Administration, Rockville, Md., 46 pp.

Surveys and Mapping Branch, 1978 : *Specifications and Recommendations for Control Surveys and Survey Markers*, Surveys and Mapping Branch, Ottawa, Canada.

Manuals on Field Procedures

Baker, L., 1968: Specifications for horizontal control marks, *ESSA Tech. Memo. Coast and Geodetic Survey Publication 4*, U.S. Coast and Geodetic Survey, Rockville, Md., 14 pp. (revision of *Special Publication 247*, by Gossett, F., 1959, pp. 84-94).

Defense Mapping Agency, 1975: Field Operations Manual—Doppler Point Positioning. *Defense Mapping Agency Tech. Manual* TM-T-2-52220, Department of Defense, 76 pp.

Dewhurst, W., 1983: *Input Formats and Specifications of the National Geodetic Survey Data Base*, vol. III: Gravity control data, Federal Geodetic Control Committee, Rockville, Md., 163 pp.

Floyd, R., 1978: Geodetic bench marks, *NOAA Manual NOS NGS 1*, National Oceanic and Atmospheric Administration, Rockville, Md., 50 pp.

Gossett, F., 1950, rev. 1959: Manual of geodetic triangulation, *Special Publication 247*, U.S. Coast and Geodetic Survey, Washington, D.C., 205 pp.

Hoskinson, A., and Duerksen, J., 1952: Manual of geodetic astronomy: determination of longitude, latitude, and azimuth, *Special Publication 237*, U.S. Coast and Geodetic Survey, Washington, D.C., 205 pp.

Mussetter, W., 1941, rev. 1959: Manual of reconnaissance for triangulation, *Special Publication 225*, U.S. Coast and Geodetic Survey, Washington, D.C. 100 pp.

Pfeifer, L., 1980: *Input Formats and Specifications of the National Geodetic Survey Data Base*, vol. I: Horizontal control data, Federal Geodetic Control Committee, Rockville, Md., 205 pp.

Pfeifer, L., and Morrison, N., 1980: *Input Formats and Specifications of the National Geodetic Survey Data Base*, vol. II: Vertical control data, Federal Geodetic Control Committee, Rockville, Md. 136 pp.

Schomaker, M., and Berry, R., 1981: Geodetic leveling, *NOAA Manual NOS NGS 3*. National Oceanic and Atmospheric Administration, Rockville, Md. 209 pp.

Slama, C. (editor), 1980: *Manual of Photogrammetry* (4th edition), American Society of Photogrammetry, Falls Church, Va., 1056 pp.

APPENDIX A
Governmental Authority

A.1 Authority

The U.S. Department of Commerce's National Oceanic and Atmospheric Administration (NOAA) is responsible for establishing and maintaining the basic national horizontal, vertical, and gravity geodetic control networks to meet the needs of the Nation. Within NOAA this task is assigned to the National Geodetic Survey, a Division of the Office of Charting and Geodetic Services within the National Ocean Service. This responsibility has evolved from legislation dating back to the Act of February 10, 1807 (2 Stat. 413, which created the first scientific Federal agency, known as the "Survey of the Coast." Current authority is contained in United States Code, Title 33, USC 883a, as amended, and specifically defined by Executive Directive, Bureau of the Budget (now the Office of Management and Budget) Circular No. A-16, Revised (Bureau of the Budget 1967).

To coordinate national mapping, charting, and surveying activities, the Board of Surveys and Maps of the Federal Government was formed December 30, 1919, by Executive Order No. 3206. "Specifications for Horizontal and Vertical Control" were agreed upon by Federal surveying and mapping agencies and approved by the Board on May 9, 1933. When the Board was abolished March 10, 1942, its functions were transferred to the Bureau of the Budget, now the Office of Management and Budget, by Executive Order No. 9094. The basic survey specifications continued in effect. Bureau of the Budget Circular No. A-16, published January 16, 1953, and revised May 6, 1967 (Bureau of the Budget 1967), provides for the coordination of Federal surveying and mapping activities. "Classification and Standards of Accuracy of Geodetic Control Surveys," published March 1, 1957, replaced the 1933 specifications. Exhibit C to Circular A-16, dated October 10, 1958 (Bureau of the Budget 1958), established procedures for the required coordination of Federal geodetic and control surveys performed in accordance with the Bureau of the Budget classifications and standards.

The Federal Geodetic Control Committee (FGCC) was chartered December 11, 1968, and a Federal Coordinator for Geodetic Control and Related Surveys was appointed April 4, 1969. The FGCC Circular No. 1, "Exchange of Information," dated October 16, 1972, prescribes reporting procedures for the committee (vice Exhibit C of Circular A-16) (Federal Geodetic Control Committee 1972).

The Federal Coordinator for Geodetic Control and Related Surveys, Department of Commerce, is responsible for coordinating, planning, and executing national geodetic control surveys and related survey activities of Federal agencies, financed in whole or in part by Federal funds. The Executive Directive (Bureau of the Budget 1967: p. 2) states:

(1) The geodetic control needs of Government agencies and the public at large are met in the most expeditious and economical manner possible with available resources; and

(2) all surveying activities financed in whole or in part by Federal funds contribute to the National Networks of Geodetic Control when it is practicable and economical to do so.

The Federal Geodetic Control Committee assists and advises the Federal Coordinator for Geodetic Control and Related Surveys.

A.2 References

Bureau of the Budget, 1967: Coordination of surveying and mapping activities. *Circular* No. A-16, Revised, May 6, 3 pp. Executive Office of the President, Bureau of the Budget (now Office of Management and Budget), Washington, D.C. 20503.

Bureau of the Budget, 1958: Programing and coordination of geodetic control surveys. *Transmittal Memorandum* No. 2, 1 p., and Exhibit C of *Circular* No. A-16, 4 pp. Executive Office of the President, Bureau of the Budget (now Office of Management and Budget), Washington, D.C. 20503.

Federal Geodetic Control Committee, 1972: Exchange of Information. *Circular* No. 1, Federal Geodetic Control Committee, October 16, 6 pp.

APPENDIX B
Variance Factor Estimation

B.1 Introduction

The classification accuracies for the National Geodetic Control Networks measure how well a survey can provide position, elevation, and gravity. (More specifically, a distance accuracy is used for horizontal networks, and an elevation difference accuracy is used for vertical networks.) The interpretation of what is meant by "how well" contains two parts. A survey must be precise, i.e., fairly free of random error; it must also be accurate, i.e., relatively free of systematic error. This leads to a natural question of how to test for random and systematic error.

Testing for random error is an extremely broad subject, and is not examined here. It is assumed that the standard deviation of distance, elevation difference, or gravity provides an adequate basis to describe the amount of random error in a survey. Further, it is assumed that the selection of the worst instance of the classification accuracy computed at all points (or between all pairs of points) provides a satisfactory means of classifying a new survey. This procedure may seem harsh, but it allows the user of geodetic control to rely better upon a minimum quality of survey work. The nominal quality of a survey could be much higher.

Consider the method of observation equations (see Mikhail (1976) for a general discussion):

$$L_a = F(X_a)$$

where
L_a is a vector of computed values for the observations of dimension n,
X_a is a vector of coordinate and model parameters of dimension u, and
F is a vector of functions that describes the observations in terms of the parameters.

The design matrix, A, is defined as

$$A = \frac{\partial F}{\partial X_a}\bigg|\, X_a = X_o$$

where A is a matrix of differential changes in the observation model F with respect to the parameters, X_a, evaluated at a particular set of parameter values, X_o. A vector of observation misclosures is

$$L = L_b - L_a$$

where L_b is the vector of actual observations and L_a is the vector described above.

Associated with the observation vector L_b is a symmetric variance-covariance matrix Σ_{L_b}, which contains information on observation precision and correlation.

The observation equation may now be written in linearized form

$$AX = L + V$$

where V is a vector of residual errors and X is a vector of corrections to the parameter vector X_a. The least squares estimate of X is

$$X = (A^t(\Sigma_{L_b})^{-1}A)^{-1} A^t(\Sigma_{L_b})^{-1}L$$

where the superscripts t and $^{-1}$ denote transpose and inverse (of a matrix) respectively.

The estimate provides a new set of values for the parameters by

$$X_a + X \rightarrow X_a$$

If the observation model $F(X_a)$ is nonlinear (that is, A is not constant for any set of X_a), then the entire process, starting with the first equation, must be iterated until the vector X reaches a stationary point.

Once convergence is achieved, L_a, computed from the first equation, is the vector of adjusted observations. The vector of observation residual errors, V, is

$$V = L_a - L_b$$

Estimates of parameter precision and correlations are given by the adjusted parameter variance-covariance matrix, Σ_{X_a} computed by

$$\Sigma_{X_a} = (A^t(\Sigma_{L_b})^{-1}A)^{-1}.$$

The precision of any other quantity that can be derived from the parameters may also be computed. Suppose one wishes to compute a vector of quantities, S,

$$S = S(X_a)$$

from the adjusted parameters, X_a. A geometry matrix, G, is defined as

$$G = \frac{\partial S}{\partial X_a}\bigg|\; X_a = X_o$$

where G is a matrix of differential changes in the functions, S, with respect to the parameters, X_a, evaluated at a particular set of parameter values, X_o. By the principle of linear error propagation,

$$\Sigma_S = G\,\Sigma_{X_a}\,G^t$$

or

$$\Sigma_S = G(A^t(\Sigma_{L_b})^{-1}A)^{-1}\,G^t$$

where Σ_S is the variance-covariance matrix of the computed quantities.

This last equation is important since its terms are variances and covariances such as those for distance or height difference. Use of this equation assumes that the model is not too nonlinear, that the parameter vector X_a has been adequately estimated by the method of least squares, that the design matrix A, the geometry matrix G, and the variance-covariance matrix of the observations Σ_{L_b} are known. This last assumption is the focal point for the remainder of this appendix.

We must somehow estimate the $n\,(n+1)/2$ elements of Σ_L. Usually, we know Σ_L subject to some global variance factor, f. We would then assume that

$$\Sigma_L = f\,\Sigma_L^0$$

where
Σ_L = the "true" variance-covariance matrix of the observations
Σ_L^0 = initial estimate of variance-covariance matrix of the observations

Our assumption about the the structure of Σ_L^0 relative to a single factor usually suffices. But this assumption can be improved if we generalize the idea. Consider a partition of the observations into k homogeneous groups. We now estimate k different local variance factors

$$\Sigma_L = \begin{pmatrix} f_1\Sigma_{L_1}^0 & & & 0 \\ & f_2\Sigma_{L_2}^0 & & \\ & & \ddots & \\ 0 & & & f_k\Sigma_{L_k}^0 \end{pmatrix}$$

As will be discussed later, we may also detect systematic error if one of the variance components is based on certified network observations.

B.2 Global Variance Factor Estimation ($k = 1$)

The global variance factor, f, is simply the a posteriori variance of unit weight, $\hat{\sigma}_0^2$, when given an a priori variance of unit weight, σ_0^2, equal to 1.

It can be shown that

$$E(V^t(\Sigma_L)^{-1}V) = n - u. \quad \text{(Mikhail 1976: p. 287)}$$

For a single variance factor

$$\Sigma_L = f\,\Sigma_L^0$$

so that

$$\frac{1}{f}\,\Sigma(V^t(\Sigma_L^0)^{-1}V) = n - u$$

or for f to be unbiased (Hamilton 1964, p. 130)

$$f = \frac{E(V^t(\Sigma_L^0)^{-1}V)}{n - u} = \frac{V^t(\Sigma_L^0)^{-1}V}{n - u} \; .$$

This is identical to the form $\hat{\sigma}_0^2 = \dfrac{V^t P V}{n - u}$, where P is defined as $\sigma_0^2(\Sigma_L^0)^{-1}$

Since we are given that $\sigma_0^2 = 1$, then $P = (\Sigma_L^0)^{-1}$. Then $f = \hat{\sigma}_0^2$, as we wished to prove.

The derivation assumes that there is no bias in the residuals (Mikhail 1976), i.e.,

$$E(V) = 0.$$

However, outliers, as well as systematic errors, can produce a biased global variance factor. We must be satisfied that the observations contain no blunders, and that our mathematical model is satisfactory in order to use the global variance factor.

Particular types of systematic errors—global scale or orientation errors—are not detectable in a survey adjustment. They will not bias the residuals and will not influence the global variance factor. For example, to detect a global scale error, it must be transformed into a local scale error by addition of more data or measurements that can discriminate between global and local.

B.3 Local Variance Factor Estimation ($k = 2,3,...$)

Let us separate our observations into k homogeneous groups, and assume that we know the variance-covariance matrices of all k groups, $\Sigma_{L_i}^0$, subject to k local variance factors, f_i. Then

$$\Sigma_L = \begin{pmatrix} f_1\Sigma_{L_1}^0 & & & 0 \\ & f_2\Sigma_{L_2}^0 & & \\ & & \ddots & \\ 0 & & & f_k\Sigma_{L_k}^0 \end{pmatrix}$$

A variety of methods has been proposed that can be used to estimate local variance factors. Among them are MInimum Norm Quadratic Unbiased Estimation (MINQUE) (Rao 1971), Iterated MInimum Norm Quadratic Estimation (IMINQE) (Rao 1972), Almost Unbiased Estimation (AUE) (Horn et al. 1975), and Iterated Almost Unbiased Estimation (IAUE) (Lucas 1984). Underlying these methods is the assumption that there is no bias in any group of residuals; that is

$$E(V_k) = 0.$$

This assumption can be turned to our advantage in the detection of local systematic error.

Consider the partition of observations into a network group, subscript N, and a survey group, subscript S (k = 2). Then

$$\Sigma_L = \begin{pmatrix} f_N \Sigma_N^0 & 0 \\ 0 & f_S \Sigma_S^0 \end{pmatrix}$$

For an adjustment of the network only, we may estimate

$$\Sigma_N' = f_N' \Sigma_N^0$$

and for an adjustment of the survey only, we may estimate

$$\Sigma_S' = f_S' \Sigma_S^0$$

where f_S' is the global variance factor of the survey observations computed by a least squares adjustment free of outliers and known systematic errors.

With perfect information and an unbiased model we compute $f_N = f_N'$ and $f_S = f_S'$. On the other hand, if our model is biased, this may not be the case. In other words, we have a linkage between systematic error and consistent estimation of local variance factors.

Now assume that our network observations are certified as having no systematic error, and that we have perfect knowledge of their weights. Then $f_N' = 1$ and $\Sigma_N = \Sigma_N^0$. In the absence of residual bias in the survey, we should compute $f_N = 1$ and $f_S = f_S'$. In fact, we could impose a constraint on the computation, $f_N = 1$, to ensure this result. A survey systematic error could then manifest itself as an increase in f_S over f_S'.

There is no guarantee that systematic error in a survey will increase f_S over f_S'. For example, a survey may be connected to the network at only one control point. A scale error local to the survey would remain undetectable with combined variance factor estimation. With a second connection to the network, the survey scale error will begin to be detectable. As the survey is more closely connected to the network, the capability to detect a survey scale error becomes much better. We see that systematic error in a survey that is well-connected to a certified geodetic network can be discovered by local variance factor estimation. Of course a systematic error, such as a scale factor influencing both the network and the survey, would continue to remain hidden.

B.4 Iterated Almost Unbiased Estimation (IAUE)

The IAUE method (Lucas 1984) can be used to estimate covariance elements as well as the variance elements of Σ_L. However, in testing for systematic error we are concerned only with the survey and the network variance factors (k = 2).

As suggested by the title, the method is iterative. We start with the initial values

$$f_S^0 \text{ and } \Sigma_S^0, \text{ with } f_N^0 \text{ set to } 1 \ .$$

Let

$$\Sigma_L^0 = \begin{pmatrix} f_N^0 \Sigma_N^0 & 0 \\ 0 & f_S^0 \Sigma_S^0 \end{pmatrix}$$

$$P_L^0 = (\Sigma_L^0)^{-1} = \begin{pmatrix} P_N^0 & 0 \\ 0 & P_S^0 \end{pmatrix}$$

We now iterate from i = 0 to convergence

1) Perform least squares adjustment for

$$\hat{X} = (A'P_L^i A)^{-1} A'P_L^i L \ .$$

2) $\Sigma_{V_S}^i = (P_S^i)^{-1} - A_S(A'P_L^i A)^{-1} A_S^t \ .$

3) $f_S^{i+1} = \dfrac{(V_S^i)^t P_S^i V_S^i}{\text{tr}(\Sigma_{V_S}^i P_S^i)}$

where tr is the trace function.

4) $\Sigma_S^{i+1} = f_S^{i+1} \Sigma_S^i \ .$

We test for convergence by

$$\frac{f_S^{i+1} - f_S^i}{f_S^i} < \epsilon$$

where ϵ is a preset quantity > 0. The local survey variance factor is

$$f_S = \prod_{i=0}^{m} f_S^i$$

where m is the number of iterations to convergence. We can then compute a survey variance factor ratio,

$$f_S/f'_S$$

Computer simulations have shown that when the survey variance factor ratio exceeds 1.5, then the survey contains systematic error. This rule becomes less reliable when a survey is minimally connected to a network.

We note that for $k = 1$, the third step of the method yields

$$f^{i+1} = \frac{(V^t P V)^i}{n - u} .$$

It is immediately recognized as the a posteriori estimate of the variance of unit weight. In this special case, IAUE convergence is correct, immediate, and unbiased.

The IAUE method is particularly attractive from a computational point of view. If Σ_L is diagonal, or nearly so, then the requisite elements of Σ_L may be computed from elements of Σ_X that lie completely within the profile of the normal equations. Thus, the usual apparatus of sparse least squares adjustments can be retained.

B.5 References

Hamilton, Walter Clark, 1964: *Statistics in Physical Science,* The Ronald Press Company, New York.

Horn, S.D., Horn, R.A., and Duncan, D.B., 1975: Estimating heteroscedastic variances in linear models, *Journal of the American Statistical Association,* 70, 380-385.

Lucas, James R., 1984: A variance component estimation method for sparse matrix applications, unpublished manuscript, NGS, NOAA, Rockville, Md.

Mikhail, Edward M., 1976: *Observations and Least Squares,* IEP-A Dun-Donnelley publisher, New York.

Rao, C.R., 1972: Estimation of variance and covariance components in linear models, *Journal of the American Statistical Association,* 67, 112-115.

Rao, C.R., 1971: Estimation of variance and covariance components—MINQUE theory, *Journal of Multivariate Analysis,* 1, 257-275.

APPENDIX C
Procedures for Submitting Data to the National Geodetic Survey

The National Geodetic Survey (NGS) has determined that the value to the national network of geodetic observations performed by other Federal, State, and local organizations compensates for the costs of analyzing, adjusting, and publishing the associated data. Consequently, a procedure has been established for data from horizontal, vertical, and gravity control surveys to be submitted to NGS. Persons submitting data must adhere· to the requirements stated herein, but in any event, the final decision of acceptance on data will be the responsibility of the Chief, NGS.

The survey data must be submitted in the format specified in the Federal Geodetic Control Committee (FGCC) publication, *Input Formats and Specifications of the National Geodetic Survey Data Base,* which describes the procedures for submission of data for adjustment and assimilation into the National Geodetic Survey data base. Volume I (Horizontal control data), volume II (Vertical control data) or volume III (Gravity control data) may be purchased from:

National Geodetic Information Branch (N/CG17x2)
National Oceanic and Atmospheric Administration
Rockville, MD 20852

Horizontal control surveys must be accomplished to at least third-order, class I standards and tied to the National Geodetic Horizontal Network. Vertical control surveys must be accomplished in accordance with third-order or higher standards and tied to the National Geodetic Verti-

cal Network. Gravity control surveys must be accomplished to at least second-order standards and tied to the National Geodetic Gravity Network. Third-order gravity surveys ("detail" surveys) will be accepted by NGS for inclusion into the NGS Gravity Working Files only in accordance with the above mentioned FGCC publication. A clear and accurate station description should be provided for all control points.

The original field records (or acceptable copies), including sketches, record books, and project reports, are required. NGS will retain these records in the National Archives. This is necessary if questions arise concerning the surveys on which the adjusted data are based. In lieu of the original notes, high quality photo copies and microfilm are acceptable. The material in the original field books or sheets are needed, not the abstracts or intermediate computations.

Reconnaissance reports should be submitted before beginning the field measurements, describing proposed connections to the national network, the instrumentation, and the field procedures to be used. This will enable NGS to comment on the proposed survey, drawing on the information available in the NGS data base concerning the accuracy and condition of these points, and to determine if the proposed survey can meet its anticipated accuracy. This project review saves the submitting agency the expense of placing data that would fail to meet accuracy criteria into computer-readable form.

Index

The surveying handbook

ABBREVIATIONS

CONSTRUCTION SURVEYS

b.b.	batterboards
B.L.	building line
C.B.	catch basin
C.G.	center line of grade
℄	center line
const.	construction
C	cut
esmt.	easement
F	fill
F.G.	finish grade, Fin. Gr.
F.H.	fire hydrant
Ⅎ	fence line
F.L.	flow line (invert)
F.S.	finished surface
G.C.	grade change
G.P.	grade point
G.R.	grade rod (s.s. notes)
L	left (x-sect. notes)
M.H.	manhole
Ⅼ	property line
P.P.	power pole
pvmt.	pavement
R	right (x-sect. notes)
R/W	right-of-way
S.D.	storm drain
S.G.	subgrade
S.L.	spring line
spec.	specifications
Sq.	square
s.s.	slope stake, side slope
Std.	standard
Str. Gr.	straight grade
X-sect.	cross-section

PROPERTY SURVEYS

A	area
C.F.	curb face
ch "X"	chiseled X cross
C.I.	cast iron
diam.	diameter
Dr.	drive
ER	end of return
Ex.	existing

H & T	hub and tack
H.C.	house connection sewer
I.B.	iron bolt (bar)
I.P.	iron pipe; iron pin (confusing)
L & T	lead and tack
max.	maximum
min.	minimum
M.H.W.	mean high water
M.L.L.W.	mean low low water
M.L.W.	mean low water
Mon.	monument
No.	number
P	pipe; pin (confusing)
Rec.	record
St.	street
Std. Surv. Mon.	standard survey monument
Std. Trav. Mon.	standard traverse monument
2″ × 2″	two- × two-inch stake
X	cross cut in stone
yd	yard

PUBLIC LANDS SURVEYS

ac	acres
AMC	auxiliary meander corner
bdy., bdys.	boundary, boundaries
BT	bearing tree
CC	closing corner
ch, chs	chain, chains
cor., cors.	corner, corners
corr.	correction
decl.	declination
dist.	distance
frac.	fractional (sec., etc.)
Gr.	Greenwich
G.M.	guide meridian
lk, lks	link, links
meas.	measurement
mer.	meridian
mkd.	marked
Mi. Cor.	mile corner
MC	meander corner
M.S.	mineral survey